A PEARSON AUSTRALIA CUSTOM BOOK

Fundamentals of Engineering Mechanics for ENGG102 and ENGG100

This custom book is compiled from:

INTRODUCTION TO ENGINEERING ANALYSIS
4TH EDITION
HAGEN

MECHANICS FOR ENGINEERS
STATICS
13TH SI EDITION
HIBBELER & YAP

MECHANICS FOR ENGINEERS
DYNAMICS
13TH SI EDITION
HIBBELER & YAP

UNIVERSITY OF WOLLONGONG

Pearson Australia
Unit 4, Level 3
14 Aquatic Drive
Frenchs Forest, Sydney NSW 2086
Ph: 02 9454 2200
www.pearson.com.au

Project Management Team Leader:	Jill Gillies
Custom Specialist:	Nicki Barlow
Project Manager:	Linda Chryssavgis
Production Controller:	Emma Roberts

ISBN: 978 1 4886 1043 1

TABLE OF CONTENTS

ABOUT THIS CUSTOM BOOK

Welcome to **Fundamentals of Engineering Mechanics for ENGG102 and ENGG100.**

The material included in this custom book has been specifically chosen to meet your course requirements. Please be aware that chapter, section and page numbers from the original source texts still appear in this book.

The Table of Contents refers to the page numbers of this book, not the source text. These page numbers also appear in the **Navigation Bar** at the top of each page.

Navigation Bar – Use this information to navigate through the book. The page numbers here run continuously from beginning to end.

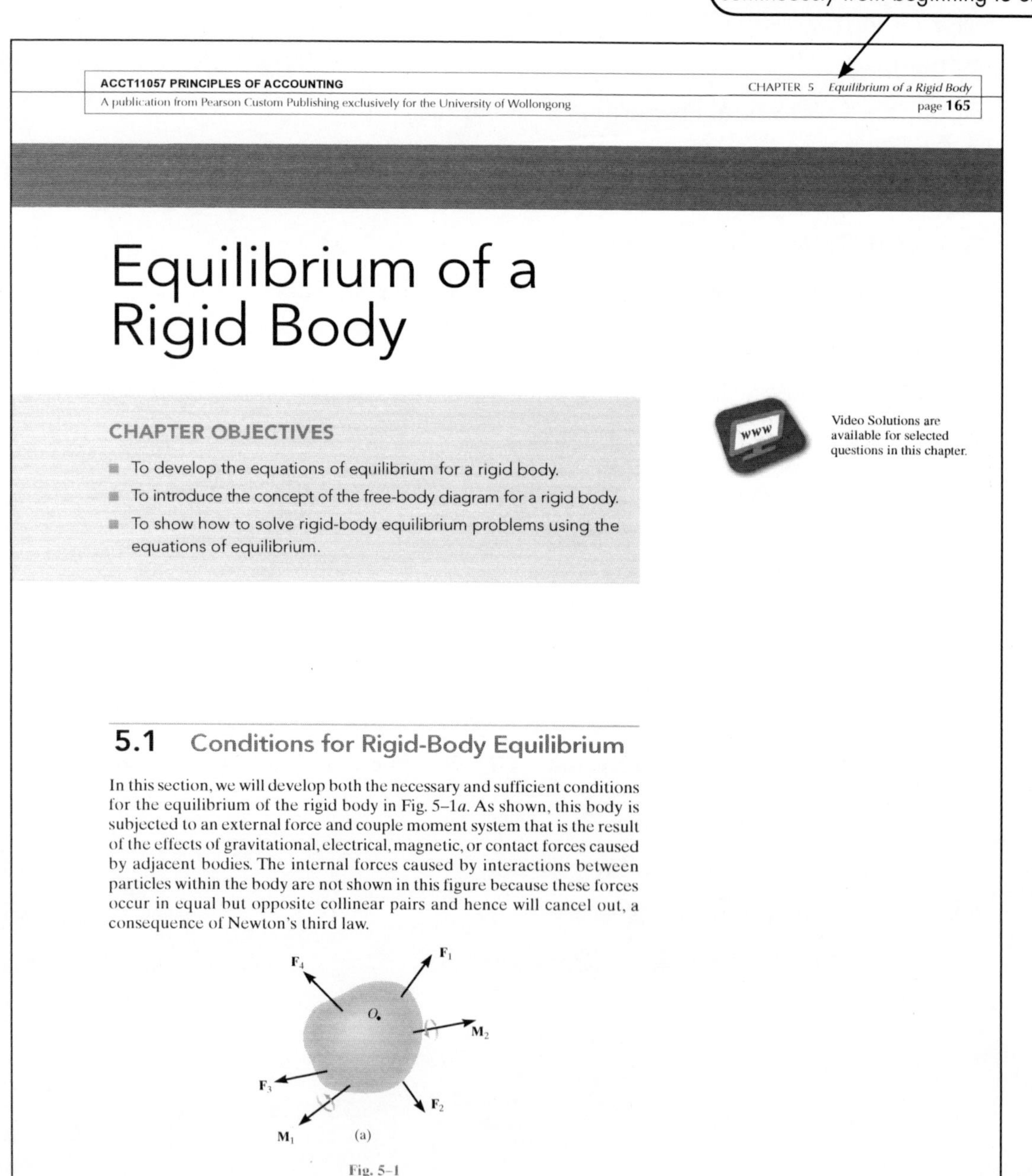

ACCT11057 PRINCIPLES OF ACCOUNTING — CHAPTER 5 *Equilibrium of a Rigid Body*

A publication from Pearson Custom Publishing exclusively for the University of Wollongong — page 165

Equilibrium of a Rigid Body

CHAPTER OBJECTIVES

- To develop the equations of equilibrium for a rigid body.
- To introduce the concept of the free-body diagram for a rigid body.
- To show how to solve rigid-body equilibrium problems using the equations of equilibrium.

Video Solutions are available for selected questions in this chapter.

5.1 Conditions for Rigid-Body Equilibrium

In this section, we will develop both the necessary and sufficient conditions for the equilibrium of the rigid body in Fig. 5–1*a*. As shown, this body is subjected to an external force and couple moment system that is the result of the effects of gravitational, electrical, magnetic, or contact forces caused by adjacent bodies. The internal forces caused by interactions between particles within the body are not shown in this figure because these forces occur in equal but opposite collinear pairs and hence will cancel out, a consequence of Newton's third law.

(a)

Fig. 5–1

Referencing

The correct referencing for the source texts is given on the relevant section openers throughout this book.

Section 1
Introduction to Engineering Analysis

The content in this section is sourced from Hagen's Introduction to Engineering Analysis, 4th edition

Hagen, K.D. (2014). *Introduction to engineering analysis* (4th ed.). Upper Saddle River, NJ: Pearson Education Inc.

CHAPTER 1

The Role of Analysis in Engineering

Objectives

After reading this chapter, you will have learned

- What engineering analysis is
- That analysis is a major component of the engineering curriculum
- How analysis is used in engineering design
- How analysis helps engineers prevent and diagnose failures

1.1 INTRODUCTION

What is ***analysis***? A dictionary definition of analysis might read something like this:

> the separation of a whole into its component parts, or an examination of a complex system, its elements, and their relationships.

Based on this general definition, analysis may refer to everything from the study of a person's mental state (psychoanalysis) to the determination of the amount of certain elements in an unknown metal alloy (elemental analysis). ***Engineering analysis***, however, has a specific meaning. A concise working definition is:

> analytical solution of an engineering problem, using mathematics and principles of science.

Engineering analysis relies heavily on ***basic mathematics*** such as algebra, geometry, trigonometry, calculus, and statistics. ***Higher level mathematics*** such as linear algebra, differential equations, and complex variables may also be used. Principles and laws from the ***physical sciences***, particularly physics and chemistry, are key ingredients of engineering analysis.

Engineering analysis involves more than searching for an equation that fits a problem, plugging numbers into the equation, and "turning the crank" to generate an answer. It is not a simple "plug and chug" procedure. Engineering analysis requires logical and systematic thinking about the engineering problem. The engineer must first be able to state the problem clearly, logically, and concisely. The engineer must understand the physical behavior of the system being analyzed and know which scientific principles to apply. He or she must recognize which mathematical tools to use and how to implement them by hand or on a computer. The engineer must be able to generate a solution that is consistent with the stated problem and any simplifying assumptions. The engineer must then ascertain that the solution is reasonable and contains no errors.

Engineering analysis may be regarded as a type of ***modeling*** or *simulation*. For example, suppose that a civil engineer wants to know the tensile stress in a cable of a suspension bridge that is being designed. The bridge exists only on paper, so a direct stress measurement cannot be made. A scale model of the bridge could be constructed, and a stress measurement taken on the model, but models are expensive and very time-consuming to develop. A better approach is to create an analytical model of the bridge or a portion of the bridge containing the cable. From this model, the tensile stress can be calculated.

Engineering courses that focus on analysis, such as statics, dynamics, mechanics of materials, thermodynamics, and electrical circuits, are considered *core* courses in the engineering curriculum. Because you will be taking many of these courses, it is vital that you gain a fundamental understanding of what analysis is and, more importantly, how to do analysis properly. As the bridge example illustrates, analysis is an integral part of engineering design. Analysis is also a key part of the study of engineering failures.

Engineers who perform engineering analyses on a regular basis are referred to as *engineering analysts* or *analytic engineers*. These functional titles are used to differentiate analysis from the other engineering functions such as research and development (R&D), design, testing, production, sales, and marketing. In some engineering companies, clear distinctions are made between the various engineering functions and the people who work in them. Depending on the organizational structure and the type of products involved, large companies may dedicate a separate department or group of engineers to be analysts. Engineers whose work is dedicated to analysis are considered specialists. In this capacity, the engineering analyst usually works in a support role for design engineering. It is not uncommon, however, for design and analysis functions to be combined in a single department because design and analysis are so closely related. In small firms that employ only a few engineers, the engineers often bear the responsibility of many technical functions, including analysis.

PROFESSIONAL SUCCESS—CHOOSING AN ENGINEERING MAJOR

Perhaps the biggest question facing the new engineering student (besides "How much money will I make after I graduate?") is "In which field of engineering should I major?" Engineering is a broad area, so the beginning student has numerous options. The new engineering student should be aware of a few facts. First, all engineering majors have the potential for preparing the student for a satisfying and rewarding engineering career. As a profession, engineering has historically enjoyed a fairly stable and well-paid market. There have been fluctuations in the engineering market in recent decades, but the demand for engineers in all the major disciplines is high, and the future looks bright for engineers. Second, all engineering majors are academically challenging, but some engineering majors may be more challenging than others. Study the differences between the various engineering programs. Compare the course requirements of each program by examining the course listings in your college or university catalog. Ask department chairs or advisors to discuss the similarities and differences between their engineering programs and the programs in other departments. (Just keep in mind that professors may be eager to tell you that *their* engineering discipline is the best.) Talk with people who are practicing engineers in the various disciplines and ask them about their educational experiences. Learn all you can from as many sources as you can about the various engineering disciplines. Third, and this is the most important point, try to answer the following question: "What kind of engineering will be the most gratifying for me?" It makes little sense to devote four or more years of intense study of X engineering just because it happens to be the highest paid discipline, because your uncle Vinny is an X engineer, because X engineering is the easiest program at your school, or because someone tells you that they are an X engineer, so you should be one too.

Engineering disciplines may be broadly categorized as either mainstream or narrowly focused. Mainstream disciplines are the broad-based, traditional disciplines that have been in existence for decades (or even centuries) and in which degrees are offered by most of the larger colleges and universities. Many colleges and universities do not offer engineering degrees in some of the narrowly focused disciplines. Chemical, civil, computer, electrical, and mechanical engineering are considered the core mainstream disciplines. These mainstream disciplines are broad in subject content and represent the majority of practicing engineers. Narrowly focused disciplines concentrate on a particular engineering subject by combining specific components from the mainstream disciplines. For example, biomedical engineering may combine portions of electrical and mechanical engineering plus components from biology. Construction engineering may combine elements from civil engineering and business or construction trades. Other narrowly focused disciplines include materials, aeronautical and aerospace, environmental, nuclear, ceramic, geological, manufacturing, automotive, metallurgical, corrosion, ocean, and cost and safety engineering.

Should you major in a mainstream area or a narrowly focused area? The safest thing to do, especially if you are uncertain about which discipline to study, is to major in one of the mainstream disciplines. By majoring in a mainstream area, you will graduate with a general engineering education that will make you marketable in a broad engineering industry. On the other hand, majoring in a narrowly focused discipline may lead you into an extremely satisfying career, particularly if your area of expertise, narrow as it may be, is in high demand. Perhaps your decision will be largely governed by geographical issues. The narrowly focused majors may not be offered at the school you wish to attend. These are important issues to consider when selecting an engineering major.

1.2 ANALYSIS AND ENGINEERING DESIGN

Design is the heart of engineering. In ancient times, people recognized a need for protection against the natural elements, for collecting and utilizing water, for finding and growing food, for transportation, and for defending themselves against other people with unfriendly intentions. Today, even though our world is much more advanced and complex than that of our ancestors, our basic needs are essentially the same. Throughout history, engineers have designed various devices and systems that met the changing needs of society. The following is a concise definition of ***engineering design***:

> a process of devising a component, system, or operation that meets a specific need.

The key word in this definition is *process*. The design process is like a road map that guides the designer from need recognition to problem solution. Design engineers make decisions based on a thorough understanding of engineering fundamentals, design constraints, cost, reliability, manufacturability, and human factors. A knowledge of design *principles* can be learned in school from professors and books, but in order to become a good design engineer, you must *practice* design. Design engineers are like artists and architects who harness their creative powers and skills to produce sculptures and buildings. The end products made by design engineers may be more functional than artistic, but their creation still requires knowledge, imagination, and creativity.

Engineering design is a process by which engineers meet the needs of society. This process may be described in a variety of ways, but it typically consists of the systematic sequence of steps shown in Figure 1.1.

Design has always been a key element of engineering programs in colleges and universities. Traditionally, engineering students take a "senior design" or a "capstone design project" course in their senior year. Recognizing that design is indeed the heart of engineering and that students need an earlier introduction to the subject, many schools integrate design experiences earlier in the curriculum, perhaps as early as the introductory course. By introducing design at the level that introductory mathematics and science courses are taught, engineering programs provide students a meaningful context within which mathematics and science are applied.

What is the relationship between engineering analysis and engineering design? As we defined it earlier, engineering analysis is the *analytical solution of an engineering problem, using mathematics and principles of science*. The false notion that engineering is merely mathematics and applied science is widely held by many beginning engineering students. This may lead a student to believe that engineering design is the equivalent of a "story problem" found in high school algebra books. However, unlike math problems, design problems are "open ended." This means, among other things, that such problems do not have a single "correct" solution. Design problems have many possible solutions, depending on the *decisions* made by the design engineer. The main goal of engineering design is to obtain the *best* or *optimum* solution within the specifications and constraints of the problem.

So, how does analysis fit in? One of the steps in the design process is to obtain a preliminary concept of the *design*. (Note that the word *design* here refers to the actual component, system, or operation that is being created.) At this point, the engineer begins to investigate design alternatives. Alternatives are different approaches, or options, that the design engineer considers to be viable at the conceptual stage

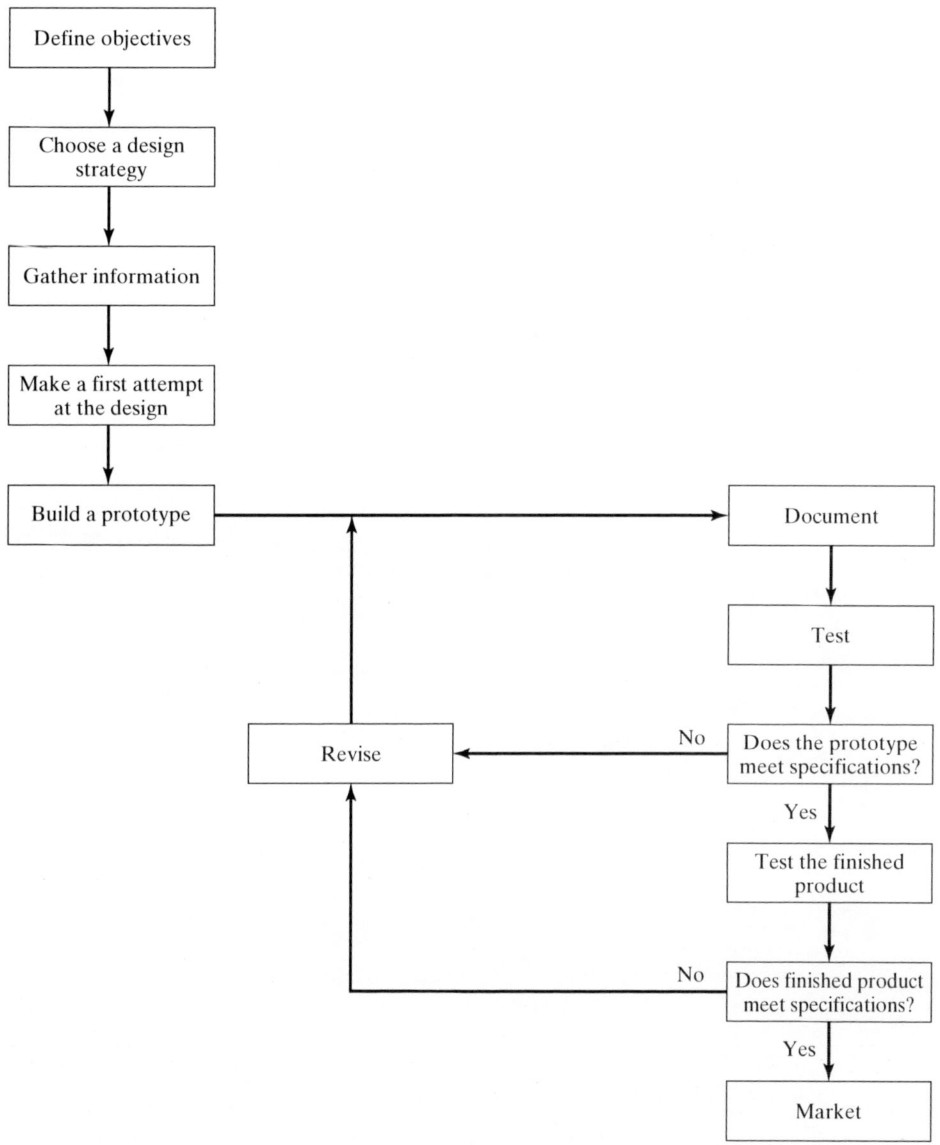

Figure 1.1
The engineering design process.

of the design. For example, some of these concepts may be used to design a better mousetrap:

- use a mechanical or an electronic sensor;
- insert cheese or peanut butter as bait;
- construct a wood, plastic, or metal cage;
- install an audible or a visible alarm;
- kill or catch and release the mouse.

Analysis is a *decision-making tool* for evaluating a set of design alternatives. By performing analysis, the design engineer zeroes in on the alternatives that yield the optimum solution, while eliminating alternatives that either violate design constraints or yield inferior solutions. In the mousetrap design, a dynamics analysis may show that a mechanical sensor is too slow, resulting in delaying the closing of a trap door and therefore freeing the mouse. Thus, an electronic sensor is chosen because it yields a superior solution.

The application that follows illustrates how analysis is used to design a machine component.

APPLICATION

DESIGNING A MACHINE COMPONENT

One of the major roles for mechanical engineers is the design of machines. Machines can be very complex systems consisting of numerous moving components. In order for a machine to work properly, each component must be designed so that it performs a specific function in unison with the other components. The components must be designed to withstand specified forces, vibrations, temperatures, corrosion, and other mechanical and environmental factors. An important aspect of machine design is determining the *dimensions* of the mechanical components.

Consider a machine component consisting of a 20-cm-long circular rod, as shown in Figure 1.2. As the machine operates, the rod is subjected to a 100-kN tensile force. (The unit "kN" stands for "kilo newton," which denotes 1000 newton. A newton is a unit of force). One of the design constraints is that the axial deformation (change in length) of the rod cannot exceed 0.5 mm if the rod is to interface properly with a mating component. Taking the rod length and the applied tensile force as given, what is the minimum diameter required for the rod?

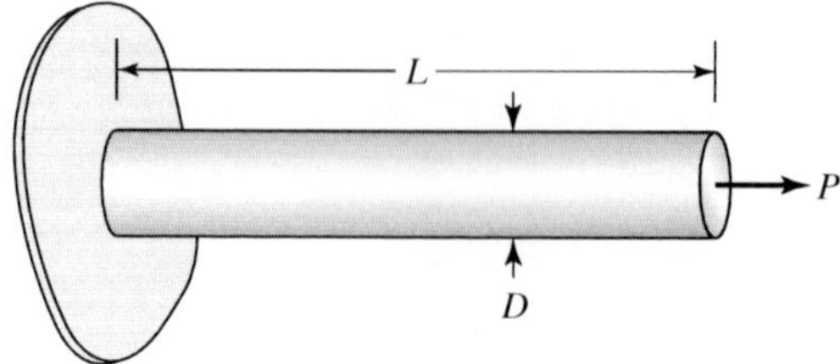

Figure 1.2 A machine component.

To solve this problem, we use an equation from mechanics of materials,

$$\delta = \frac{PL}{AE}$$

where

δ = axial deformation (m)
P = axial tensile force (N)
L = original length of rod (m)
$A = \pi D^2/4$ = cross-sectional area of rod (m^2)
E = modulus of elasticity (N/m^2).

The use of this equation assumes that the material behaves elastically (i.e., it does not undergo permanent deformation when subjected to a force). Upon substituting the formula for the rod's cross-sectional area into the equation and solving for the rod diameter D, we obtain

$$D = \sqrt{\frac{4\,PL}{\pi \delta E}}.$$

We know the tensile force P, the original rod length L, and the maximum axial deformation δ. But to find the diameter D, we must also know the modulus of elasticity E. The modulus of elasticity is a material property, a constant defined by the ratio of stress to strain. Suppose we choose 7075-T6 aluminum for the rod. This material has a modulus of elasticity of $E = 72$ GPa. (Note: a unit of stress, which is force divided by area, is the pascal (Pa). 1 Pa = 1 N/m^2 and 1 GPa = 10^9 Pa.) Substituting values into the equation gives the following diameter:

$$D = \sqrt{\frac{4(100 \times 10^3 \text{ N})(0.20 \text{ m})}{\pi(0.0005 \text{ m})(72 \times 10^9 \text{ N/m}^2)}}$$

$$= 0.0266 \text{ m} = 26.6 \text{ mm}.$$

As part of the design process, we wish to consider other materials for the rod. Let's find the diameter for a rod made of structural steel ($E = 200$ GPa). For structural steel, the rod diameter is

$$D = \sqrt{\frac{4(100 \times 10^3 \text{ N})(0.20 \text{ m})}{\pi(0.0005 \text{ m})(200 \times 10^9 \text{ N/m}^2)}}$$

$$= 0.0160 \text{ m} = 16.0 \text{ mm}.$$

Our analysis shows that the minimum diameter for the rod depends on the material we choose. Either 7075-T6 aluminum or structural steel will work as far as the axial deformation is concerned, but other design issues such as weight, strength, wear, corrosion, and cost should be considered. The important point to be learned here is that analysis is a fundamental step in machine design.

As the application example illustrates, analysis is used to ascertain what design features are required to make the component or system *functional*. Analysis is used to size the cable of a suspension bridge, to select a cooling fan for a computer, to size the heating elements for curing a plastic part in a manufacturing plant, and to design the solar panels that convert solar energy to electrical energy for a spacecraft. Analysis is a crucial part of virtually every design task because it guides the design engineer through a sequence of decisions that ultimately lead to the optimum design. It is important to point out that in design work, it is not enough to produce a *drawing* or CAD (computer-aided design) model of the component or system. A drawing by itself, while revealing the visual and dimensional characteristics of the design, may say little, or nothing, about the functionality of the design. Analysis must be included in the design process if the engineer is to know whether the design will actually work when it is placed into service. Also, once a working prototype of the design is constructed, testing is performed to validate analysis and to aid in the refinement of the design.

1.3 ANALYSIS AND ENGINEERING FAILURE

With the possible exception of farmers, engineers are probably the most taken-for-granted people in the world. Virtually all the man-made products and devices that people use in their personal and professional lives were designed by engineers. Think for a moment. What is the first thing you did when you arose from bed this morning? Did you hit the snooze button on your alarm clock? Your alarm clock was designed by engineers. What did you do next, go into the bathroom, perhaps? The bathroom fixtures — the sink, bathtub, shower, and toilet — were designed by engineers. Did you use an electrical appliance to fix breakfast? Your toaster, waffle maker, microwave oven, refrigerator, and other kitchen appliances were designed by engineers. Even if you ate cold cereal for breakfast, you still took advantage of engineering because engineers designed the processes by which the cereal and milk were produced, and they even designed the machinery for making the cereal box and milk container! What did you do after breakfast? If you brushed your teeth, you can thank engineers for designing the toothpaste tube and toothbrush and even formulating the toothpaste. Before leaving for school, you got dressed; engineers designed the machines that manufactured your clothes. Did you drive a car to school or ride a bicycle? In either case, engineers designed both transportation devices. What did you do when you arrived at school? You sat down in your favorite chair in a classroom, removed a pen or pencil and a note pad from your backpack, and began another day of learning. The chair you sat in, the writing instrument you used to take notes, the notepad you wrote on, and the bulging backpack you use to carry books, binders, paper, pens, and pencils, plus numerous other devices were designed by engineers.

We take engineers for granted, but we expect a lot from them. We expect everything they design, including alarm clocks, plumbing, toasters, automobiles, chairs, and pencils, to work and to work all the time. Unfortunately, they don't. We experience a relatively minor inconvenience when the heating coil in our toaster burns out, but when a bridge collapses, a commercial airliner crashes, or a space shuttle explodes, and people are injured or die, the story makes headline news, and engineers are suddenly thrust into the spotlight of public scrutiny. Are engineers to blame for every failure that occurs? Some failures occur because people misuse the products. For example, if you persist in using a screwdriver to pry lids off cans, to dig weeds from the garden, and to chisel masonry, it may soon stop functioning as a screwdriver. Although engineers try to design products that are "people proof," the types of failures that engineers take primary responsibility for are those caused by various types of errors during the design phase. After all, engineering is a human enterprise, and humans make mistakes.

Whether we like it or not, ***failure*** is part of engineering. It is part of the design process. When engineers design a new product, it seldom works exactly as expected the first time. Mechanical components may not fit properly, electrical components may be connected incorrectly, software glitches may occur, or materials may be incompatible. The list of potential causes of failure is long, and the cause of a specific failure in a design is probably unexpected because otherwise the design engineer would have accounted for it. Failure will always be part of engineering, because engineers cannot anticipate every mechanism by which failures *can* occur. Engineers should make a concerted effort to design systems that do not fail. If failures do arise, ideally they are revealed during the design phase and can be corrected before the product goes into service. One of the hallmarks of a good design engineer is one who turns failure into success.

The role of analysis in engineering failure is twofold. First, as discussed earlier, analysis is a crucial part of engineering design. It is one of the main decision-making tools the design engineer uses to explore alternatives. Analysis helps establish the functionality of the design. Analysis may therefore be regarded as a *failure prevention* tool. People expect kitchen appliances, automobiles, airplanes, televisions, and other systems to work as they are supposed to work, so engineers make every reasonable attempt to design products that are reliable. As part of the design phase, engineers use analysis to ascertain what the physical characteristics of the system must be in order to prevent system failure within a specified period of time. Do engineers ever design products to fail on purpose? Surprisingly, the answer is yes. Some devices rely on failure for their proper operation. For example, a fuse "fails" when the electrical current flowing through it exceeds a specified amperage. When this amperage is exceeded, a metallic element in the fuse melts, breaking the circuit, thereby protecting personnel or a piece of electrical equipment. Shear pins in transmission systems protect shafts, gears, and other components when the shear force exceeds a certain value. Some utility poles and highway signs are designed to safely break away when struck by an automobile.

The second role of failure analysis in engineering pertains to situations where design flaws escaped detection during the design phase, only to reveal themselves after the product was placed into service. In this role, analysis is utilized to address the questions "Why did the failure occur?" and "How can it be avoided in the future?" This type of detective work in engineering is sometimes referred to as *forensic engineering.* In failure investigations, analysis is used as a diagnostic tool of reevaluation and reconstruction. Following the explosion of the Space Shuttle *Challenger* in 1986, engineers at Thiokol used analysis (and testing) to reevaluate the joint design of the solid rocket boosters. Their analyses and tests showed that, under the unusually cold conditions on the day of launch, the rubber O-rings responsible for maintaining a seal between the segments of one of the solid rocket boosters lost resiliency and therefore the ability to contain the high-pressure gases inside the booster. Hot gases leaking past the O-rings developed into an impinging jet directed against the external (liquid hydrogen) tank and a lower strut attaching the booster to the external tank. Within seconds, the entire aft dome of the tank fell away, releasing massive amounts of liquid hydrogen. *Challenger* was immediately enveloped in the explosive burn, destroying the vehicle and killing all seven astronauts. In the aftermath of the *Challenger* disaster, engineers used analysis extensively to redesign the solid rocket booster joint.

APPLICATION

FAILURE OF THE TACOMA NARROWS BRIDGE

The collapse of the Tacoma Narrows Bridge was one of the most sensational failures in the history of engineering. This suspension bridge was the first of its kind spanning the Puget Sound, connecting Washington State with the Olympic Peninsula. Compared with existing suspension bridges, the Tacoma Narrows Bridge had an unconventional design. It had a narrow two-lane deck, and the stiffened-girder road structure was not very deep. This unusual design gave the bridge a slender, graceful appearance. Although the bridge was visually appealing, it had a problem: it oscillated in the wind. During the four months following its opening to traffic on July 1, 1940, the bridge earned the nickname "Galloping Gertie" from motorists who felt as though they were riding a giant roller coaster as they crossed the 2800-ft center span. (See Figure 1.3.) The design engineers failed to recognize that their bridge might behave more like the wing of an airplane subjected to severe turbulence than an earth-bound structure subjected to a steady load. The engineers' failure to consider the aerodynamic aspects of the design led to the destruction of the bridge on November 7, 1940, during a 42-mile-per-hour wind storm. (See Figure 1.4.) Fortunately, no people were injured or killed. A newspaper editor, who lost control of his car between the towers due to the violent undulations, managed to stumble and crawl his way to safety, only to look back to see the road rip away from the suspension cables and plunge, along with his car and presumably his dog, which he could not save, into the Narrows below.

Figure 1.3
The Tacoma Narrows Bridge twisting in the wind. (AP Images)

Figure 1.4
The center span of the Tacoma Narrows Bridge plunges into Puget Sound. (AP Images)

Even as the bridge was being torn apart by the windstorm, engineers were testing a scale model of the bridge at the University of Washington in an attempt to understand the problem. Within a few days following the bridge's demise, Theodore von Karman, a world renowned fluid dynamicist, who worked at the California Institute of Technology, submitted a letter to *Engineering News-Record* outlining an aerodynamic analysis of the bridge. In the analysis, he used a differential equation for an idealized bridge deck twisting like an airplane wing as the lift forces of the wind tend to twist the deck one way, while the steel in the bridge tends to twist it in another way. His analysis showed that the Tacoma Narrows Bridge should indeed have exhibited an aerodynamic instability more pronounced than any existing suspension bridge. Remarkably, von Karman's "back of the envelope" calculations predicted dangerous levels of vibration for a wind speed less than 10 miles per hour over the wind speed measured on the morning of November 7, 1940. The dramatic failure of Galloping Gertie forever established the importance of aerodynamic analysis in the design of suspension bridges.

The bridge was eventually redesigned with a deeper and stiffer open-truss structure that allowed the wind to pass through. The new and safer Tacoma Narrows Bridge was opened on October 14, 1950.

PROFESSIONAL SUCCESS—LEARN FROM FAILURE

The Tacoma Narrows Bridge and countless other engineering failures teach engineers a valuable lesson:

Learn from your own failures and the failures of other engineers.

Unfortunately, the designers of the Tacoma Narrows Bridge did not learn from the failures of others. Had they studied the history of suspension bridges dating back to the early nineteenth century, they would have discovered that 10 suspension bridges suffered severe damage or destruction by winds.

NASA and Thiokol learned that the pressure-seal design in the solid rocket-booster joint of the Space Shuttle *Challenger* was overly sensitive to a variety of factors such as temperature, physical dimensions, reusability, and joint loading. Not only did they learn some hard-core technical lessons, they also learned some lessons in engineering judgment. They learned that the decision-making process culminating in the launch of *Challenger* was flawed. To correct both types of errors, during the two-year period following the *Challenger* catastrophe, the joint was redesigned, additional safety-related measures were implemented, and the decision-making process leading to shuttle launches was improved.

In another catastrophic failure, NASA determined that fragments of insulation that broke away from the external fuel tank during the launch of the Space Shuttle *Columbia* impacted the left wing of the vehicle, severely damaging the wing's leading edge. The damage caused a breach in the wing's surface which, upon reentry of *Columbia*, precipitated a gradual burn-through of the wing, resulting in a loss of vehicle control. *Columbia* broke apart over the southwestern part of the United States, killing all seven astronauts aboard.

If we are to learn from engineering failures, the *history* of engineering becomes as relevant to our education as design, analysis, science, mathematics, and the liberal arts. Lessons learned not only from our own experiences, but also from those who have gone before us, contribute enormously to the improvement of our technology and the advancement of engineering as a profession. Errors in judgment made by Roman and Egyptian engineers are still relevent in modern times, notwithstanding a greatly improved chest of scientific and mathematical tools. Engineers have and will continue to make mistakes. We should learn from these mistakes.

KEY TERMS

analysis	engineering design	modeling
basic mathematics	failure	physical sciences
engineering analysis	higher level mathematics	

REFERENCES

Adams, J.L., *Flying Buttresses, Entropy and O-Rings: The World of an Engineer*, Cambridge, MA: Harvard University Press, 1991.

Horenstein, M.N., *Design Concepts for Engineers*, 4th ed., Upper Saddle River, NJ: Prentice Hall, 2010.

Howell, S.K., *Engineering Design and Problem Solving*, 2d ed., Upper Saddle River, NJ: Prentice Hall, 2002.

Hyman, B., *Fundamentals of Engineering Design*, 2d ed., Upper Saddle River, NJ: Prentice Hall, 2002.

Petroski, H., *Design Paradigms: Case Histories of Error and Judgment in Engineering*, Cambridge, UK: Cambridge University Press, 1994.

Petroski, H., *To Engineer is Human: The Role of Failure in Successful Design*, New York: Vintage Books, 1992.

Petroski, H., *Success Through Failure: The Paradox of Design*, Princeton, NJ: Princeton University Press, 2006.

Petroski, H., *The Evolution of Useful Things*, New York: Knopf Publishing Group, 1994.

Petroski, H., *Pushing the Limits: New Adventures in Engineering*, New York: Knopf Publishing Group, 2004.

Petroski, H., *Small Things Considered: Why There is No Perfect Design*, New York: Random House Publishing Group, 2004.

Petroski, H., *To Forgive Design: Understanding Failure*, Boston, MA: Harvard University Press, 2012.

PROBLEMS

Analysis and engineering design

1.1. The following basic devices are commonly found in a typical home or office. Discuss how analysis might be used to design these items.

a. clock
b. scissors
c. fork
d. mechanical pencil
e. door hinge
f. paper clip

g. toilet
h. incandescent light bulb
i. microwave oven
j. coat hanger
k. three-ring binder
l. light switch
m. doorknob
n. stapler
o. can opener
p. water faucet
q. kitchen sink
r. electrical outlet
s. window
t. toaster
u. dinner plate
v. chair
w. table
x. plastic CD case
y. drawer slide
z. soap dispenser.

1.2. A 1-m-long cantilevered beam of rectangular cross section carries a uniform load of $w = 15$ kN/m. The design specification calls for a 5-mm maximum deflection of the end of the beam. The beam is to be constructed of fir ($E = 13$ GPa). By analysis, determine at least five combinations of beam height h and beam width b that meet the specification. Use the equation

$$y_{max} = \frac{wL^4}{8EI}$$

where

y_{max} = deflection of end of beam (m)
w = uniform loading (N/m)
L = beam length (m)
E = modulus of elasticity of beam (Pa)
$I = bh^3/12$ = moment of inertia of beam cross section (m^4).

Note: 1 Pa = 1 N/m^2, 1 kN = 10^3 N, and 1 GPa = 10^9 Pa.
What design conclusions can you draw about the influence of beam height and width on the maximum deflection? Is the deflection more sensitive to h or b? If the beam were constructed of a different material, how would the deflection change? See Figure P1.2 for an illustration of the beam.

Figure P1.2

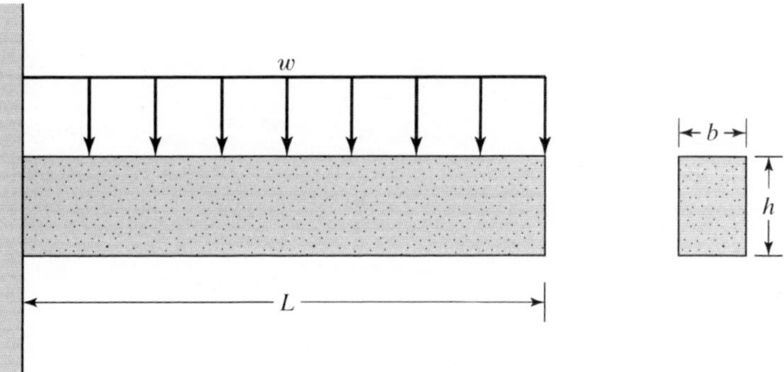

Analysis and engineering failure

1.3. Identify a device from your own experience that has failed. Discuss how it failed and how analysis might be used to redesign it.

1.4. Research the following notable engineering failures. Discuss how analysis was used or could have been used to investigate the failure.

a. Dee bridge, England, 1847
b. Boiler explosions, North America, 1870–1910

c. *Titanic*, North Atlantic, 1912
d. *Hindenburg* airship, New Jersey, 1937
e. *Apollo I* capsule fire, Cape Canaveral, Florida, 1967
f. *Apollo 13*, 1970
g. Ford Pinto gas tanks, 1970s
h. Teton dam, Idaho, 1976
i. Hartford Civic Center, Connecticut, 1978
j. Skylab, 1979
k. Three Mile Island nuclear power plant, Pennsylvania, 1979
l. American Airlines DC-10, Chicago, 1979
m. Hyatt Hotel, Kansas City, 1981
n. Union Carbide plant, India, 1984
o. Space shuttle *Challenger*, 1986
p. Chernobyl nuclear power plant, Soviet Union, 1986
q. Highway I-880, Loma Prieta, California earthquake, 1989
r. Green Bank radio telescope, West Virginia, 1989
s. Hubble space telescope, 1990
t. ValuJet Airlines DC-9, Miami, 1996
u. Mars Climate Orbiter, 1999
v. Space shuttle *Columbia*, 2003
w. Levees, New Orleands, Louisiana, 2005
x. BP oil spill, Gulf of Mexico, 2010
y. Nuclear power plant, Okuma, Fukushima, Japan, 2011

CHAPTER

2 Dimensions and Units

Objectives

After reading this chapter, you will have learned

- How to check equations for dimensional consistency
- The physical standards on which units are based
- Rules for proper usage of SI units
- Rules for proper usage of English units
- The difference between mass and weight
- How to do unit conversions between the SI and English unit systems

2.1 INTRODUCTION

Suppose for a moment that someone asks you to hurry to the grocery store to buy a few items for tonight's dinner. You get in your car, turn the ignition on, and drive down the road. Immediately you notice something strange. There are no numbers or divisions on your speedometer! As you accelerate and decelerate, the speedometer indicator changes position, but you do not know your speed because there are no markings to read. Bewildered, you notice that the speed limit and other road signs between your house and the store also lack numerical information. Realizing that you were instructed to arrive home with the groceries by 6 pm, you glance at your digital watch only to discover that the display is blank. Upon arriving at the store, you check your list: 1 pound of lean ground beef, 4 ounces of fresh mushrooms, and a 12-ounce can of tomato paste. You go to the meat counter first but the label on each package does not indicate the weight of the product. You grab what appears to be a 1-pound package and proceed to the produce section. Scooping up a bunch of mushrooms, you place them on the scale to weigh them, but the scale looks like your speedometer—it has no markings either! Once again, you estimate. One item is left: the tomato paste. The canned goods aisle contains many cans, but the labels on the cans have no numerical information—no weight, no volume, nothing to let you know the amount of tomato paste in the can. You make your purchase, drive home, and deliver the items mystified and shaken by the whole experience.

The preceding *Twilight Zone*-like story is, of course, fictitious, but it dramatically illustrates how strange our world would be without measures of physical quantities. Speed is a physical quantity that is measured by the speedometers in our automobiles and the radar gun of a traffic officer. Time is a physical quantity that is measured by the watch on our wrist and the clock on the wall. Weight is a physical quantity that is measured by the scale in the grocery store or at the health spa. The need for measurement was recognized by the ancients, who based standards of length on the breadth of the hand or palm, the length of the foot, or the distance from the elbow to the tip of the middle finger (referred to as a cubit). Such measurement standards were both changeable and perishable because they were based on human dimensions. In modern times, definite and unchanging standards of measurement have been adopted to help us quantify the physical world. These measurement standards are used by engineers and scientists to analyze physical phenomena by applying the laws of nature such as conservation of energy, the laws of thermodynamics, and the law of universal gravitation. As engineers design new products and processes by utilizing these laws, they use dimensions and units to describe the physical quantities involved. For instance, the design of a bridge primarily involves the dimensions of length and force. The units used to express the magnitudes of these quantities are usually either the meter and newton or the foot and pound. The thermal design of a boiler primarily involves the dimensions of pressure, temperature, and heat transfer, which are expressed in units of pascal, degrees Celsius, and watt, respectively. Dimensions and units are as important to engineers as the physical laws they describe. It is vitally important that engineering students learn how to work with dimensions and units. Without dimensions and units, analyses of engineering systems have little meaning.

2.2 DIMENSIONS

To most people, the term *dimension* denotes a measurement of length. Certainly, length is one type of dimension, but the term *dimension* has a broader meaning. A ***dimension*** is a *physical variable that is used to describe or specify the nature of a measurable quantity*. For example, the mass of a gear in a machine is a dimension of the gear. Obviously, the diameter is also a dimension of the gear. The compressive force in a concrete column holding up a bridge is a structural dimension of the column. The pressure and temperature of a liquid in a hydraulic cylinder are thermodynamic dimensions of the liquid. The velocity of a space probe orbiting a distant planet is also a dimension. Many other examples could be given. Any variable that engineers use to specify a physical quantity is, in the general sense, a dimension of the physical quantity. Hence, there are as many dimensions as there are physical quantities. Engineers always use dimensions in their analytical and experimental work. In order to specify a dimension fully, two characteristics must be given. First, the *numerical value* of the dimension is required. Second, the appropriate *unit* must be assigned. A dimension missing either of these two elements is incomplete and therefore cannot be fully used by the engineer. If the diameter of a gear is given as 3.85, we would ask the question, "3.85 what? Inches? Meters?" Similarly, if the compressive force in a concrete column is given as 150,000, we would ask, "150,000 what? Newtons? Pounds?"

Dimensions are categorized as either *base* or *derived.* A ***base dimension***, sometimes referred to as a *fundamental* dimension, is a dimension that has been internationally accepted as the most basic dimension of a physical quantity. There are seven base dimensions that have been formally defined for use in science and engineering:

1. length L
2. mass M
3. time t
4. temperature T
5. electric current I
6. amount of substance n
7. luminous intensity i.

A ***derived dimension*** is obtained by any combination of the base dimensions. For example, volume is length cubed, density is mass divided by length cubed, and velocity is length divided by time. Obviously, there are numerous derived dimensions. Table 2.1 lists some of the most commonly used derived dimensions in engineering, expressed in terms of base dimensions.

The single letters in Table 2.1 are symbols that designate each base dimension. These symbols are useful for checking the dimensional consistency of equations. Every mathematical relation used in science and engineering must be ***dimensionally consistent***, or *dimensionally homogeneous.* This means that the dimension on the left side of the equal sign must be the same as the dimension on the right side of the equal sign. The equality in any equation denotes not only a numerical equivalency

Table 2.1 Derived Dimensions Expressed in Terms of Base Dimensions

Quantity	Variable Name	Base Dimensions
Area	A	L^2
Volume	V	L^3
Velocity	v	Lt^{-1}
Acceleration	a	Lt^{-2}
Density	ρ	ML^{-3}
Force	F	MLt^{-2}
Pressure	P	$ML^{-1}t^{-2}$
Stress	σ	$ML^{-1}t^{-2}$
Energy	E	ML^2t^{-2}
Work	W	ML^2t^{-2}
Power	P	ML^2t^{-3}
Mass flow rate	$\dot{m}$	Mt^{-1}
Specific heat	c	$L^2t^{-2}T^{-1}$
Dynamic viscosity	μ	$ML^{-1}t^{-1}$
Molar mass	M	Mn^{-1}
Voltage	V	$ML^2t^{-3}I^{-1}$
Resistance	R	$ML^2t^{-3}I^{-2}$

but also a dimensional equivalency. To use a simple analogy, you cannot say that five apples equals four apples, nor can you say that five apples equals five oranges. You can only say that five apples equals five apples.

The following examples illustrate the concept of dimensional consistency.

EXAMPLE 2.1

Dynamics is a branch of engineering mechanics that deals with the motion of particles and rigid bodies. The straight-line motion of a particle, under the influence of gravity, may be analyzed by using the equation

$$y = y_0 + v_0 t - \frac{1}{2} g t^2.$$

where

y = height of particle at time t
y_0 = initial height of particle (at $t = 0$)
v_0 = initial velocity of particle (at $t = 0$)
t = time
g = gravitational acceleration.

Verify that this equation is dimensionally consistent.

Solution

We check the dimensional consistency of the equation by determining the dimensions on both sides of the equal sign. The heights, y_0 and y, are one-dimensional coordinates of the particle, so these quantities have a dimension of length L. The initial velocity v_0 is a derived dimension consisting of a length L divided by a time t. Gravitational acceleration g is also a derived dimension consisting of a length L, divided by time squared t^2. Of course, time t is a base dimension. Writing the equation in its dimensional form, we have

$$L = L + Lt^{-1}t - Lt^{-2}t^2$$

Note that the factor, $^1/_2$, in front of the gt^2 term is a pure number, and therefore has no dimension. In the second term on the right side of the equal sign, the dimension t cancels, leaving length L. Similarly, in the third term on the right side of the equal sign, the dimension t^2 cancels, leaving length L. This equation is dimensionally consistent because all terms have the dimension of length L.

EXAMPLE 2.2

Aerodynamics is the study of forces acting on bodies moving through air. An aerodynamics analysis could be used to determine the lift force on an airplane wing or the drag force on an automobile. A commonly used equation in aerodynamics relates the total drag force acting on a body to the velocity of the air approaching it. This equation is

$$F_D = \frac{1}{2} C_D A \rho U^2$$

where

F_D = drag force
C_D = drag coefficient
A = frontal area of body
ρ = air density
U = upstream air velocity.

Determine the dimensions of the drag coefficient, C_D.

Solution

The dimension of the drag coefficient C_D may be found by writing the equation in dimensional form and simplifying the equation by combining like dimensions. Using the information in Table 2.1, we write the dimensional equation as

$$\begin{aligned} \mathrm{MLt^{-2}} &= C_D \mathrm{L^2 ML^{-3} L^2 t^{-2}} \\ &= C_D\, \mathrm{MLt^{-2}} \end{aligned}$$

Compare the combination of base dimensions on the left and right sides of the equal sign. They are identical. This can only mean that the drag coefficient C_D has no dimension. If it did, the equation would not be dimensionally consistent. Thus, we say that C_D is *dimensionless*. In other words, the drag coefficient C_D has a numerical value, but no dimensional value. This is not as strange as it may sound. In engineering, there are many instances, particularly, in the disciplines of fluid mechanics and heat transfer, where a physical quantity is dimensionless. Dimensionless quantities enable engineers to form special ratios that reveal certain physical insights into properties and processes. In this instance, the drag coefficient is physically interpreted as a "shear stress" at the surface of the body, which means that there is an aerodynamic force acting on the body parallel to its surface that tends to retard the body's motion through the air. If you take a course in fluid mechanics, you will learn more about this important concept.

EXAMPLE 2.3

For the following dimensional equation, find the dimensions of the quantity k:

$$\mathrm{MLt^{-2}} = k\, \mathrm{Lt}.$$

Solution

To find the dimensions of k, we multiply both sides of the equation by $\mathrm{L^{-1}t^{-1}}$ to eliminate the dimensions on the right side of the equation, leaving k by itself. Thus, we obtain

$$\mathrm{MLt^{-2}L^{-1}t^{-1}} = k$$

which, after applying a law of exponents, reduces to

$$\mathrm{Mt}^{-3} = k$$

A closer examination of the given dimensional equation reveals that it is Newton's second law of motion:

$$F = ma.$$

Here F is force, m is mass, and a is acceleration. Referring to Table 2.1, force has dimensions of MLt^{-2}, which is a mass M multiplied by acceleration Lt^{-2}.

PRACTICE!

1. For the following dimensional equation, find the base dimensions of the parameter k:

$$\mathrm{ML}^2 = k\mathrm{LtM}^2.$$

Answer: $\mathrm{LM}^{-1}\mathrm{t}^{-1}$.

2. For the following dimensional equation, find the base dimensions of the parameter g:

$$\mathrm{T}^{-1}\mathrm{tL} = g\mathrm{L}^{-2}.$$

Answer: $\mathrm{L}^3\mathrm{tT}^{-1}$.

3. For the following dimensional equation, find the base dimensions of the parameter h:

$$\mathrm{It}^{-1}h = \mathrm{N}.$$

Answer: $\mathrm{NI}^{-1}\mathrm{t}$.

4. For the following dimensional equation, find the base dimensions of the parameter f:

$$\mathrm{MM}^{-3} = a\cos(f\mathrm{L}).$$

Answer: L^{-1}.

5. For the following dimensional equation, find the base dimensions of the parameter p:

$$\mathrm{T} = \mathrm{T}\log(\mathrm{T}^{-2}\mathrm{t}\,p).$$

Answer: $\mathrm{T}^2\mathrm{t}^{-1}$.

2.3 UNITS

A ***unit*** is *a standard measure of the magnitude of a dimension*. For example, the dimension length L may be expressed in units of meter (m), feet (ft), mile (mi), millimeter (mm), and many others. The dimension temperature T is expressed in units of degrees Celsius (°C), degrees Fahrenheit (°F), degrees Rankine (°R), or kelvin (K).

(By convention, the degree symbol (°) is not used for the Kelvin temperature scale.) In the United States, there are two unit systems commonly in use. The first unit system, and the one that is internationally accepted as the standard, is the ***SI*** (System International d'Unites) ***unit system***, commonly referred to as the *metric* system. The second unit system is the ***English*** (or **British**) ***unit system***, sometimes referred to as the *United States Customary System (USCS)*. With the exception of the United States, most of the industrialized nations of the world use the SI system exclusively. The SI system is preferred over the English system, because it is an internationally accepted standard and is based on simple powers of 10. To a limited extent, a transition to the SI system has been federally mandated in the United States. Unfortunately, this transition to total SI usage has been a slow one, but many American companies are using the SI system to remain internationally competitive. Until the United States makes a complete adaptation to the SI system, U.S. engineering students need to be conversant in both unit systems and know how to make unit conversions.

The seven base dimensions are expressed in terms of SI units that are based on ***physical standards***. These standards are defined such that, the corresponding SI units, except the mass unit, can be reproduced in a laboratory anywhere in the world. The reproducibility of these standards is important, because everyone with a suitably equipped laboratory has access to the same standards. Hence, all physical quantities, regardless of where in the world they are measured, are based on identical standards. This universality of physical standards eliminates the ancient problem of basing dimensions on the changing physical attributes of kings, rulers, and magistrates who reigned for a finite time. Modern standards are based on constants of nature and physical attributes of matter and energy.

The seven base dimensions and their associated SI units are summarized in Table 2.2. Note the symbol for each unit. These symbols are the accepted conventions for science and engineering. The discussion that follows outlines the physical standards by which the base units are defined.

Length

The unit of length in the SI system is the *meter* (m). As illustrated in Figure 2.1, the meter is defined as the distance traveled by light in a vacuum, during a time interval of 1/299,792,458 s. The definition is based on a physical standard, the speed of light in a vacuum. The speed of light in a vacuum is 299,792,458 m/s. Thus, light travels one meter during a time interval of the reciprocal of this number. Of course, the unit of time, the *second* (s), is itself a base unit.

Table 2.2 Base Dimensions and Their SI Units

Quantity	Unit	Symbol
Length	meter	m
Mass	kilogram	kg
Time	second	s
Temperature	kelvin	K
Electric current	ampere	A
Amount of substance	mole	mol
Luminous intensity	candela	cd

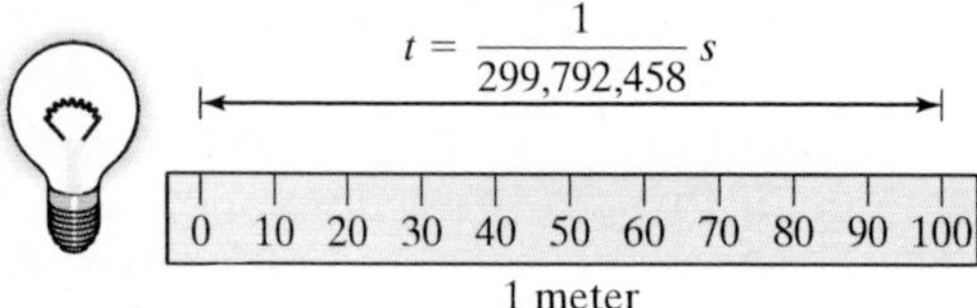

Figure 2.1
The physical standard for the meter is based on the speed of light in a vacuum.

Mass

The unit of mass in the SI system is the *kilogram* (kg). Unlike the other units, the kilogram is not based on a reproducible physical standard. The standard for the kilogram is a cylinder of platinum-iridium alloy, which is maintained by the International Bureau of Weights and Measures in Paris, France. A duplicate of this cylinder is kept in the United States by the National Institute of Standards and Technology (NIST). (See Figure 2.2.)

Mass is the only base dimension that is defined by an artifact. An artifact is a man-made object, not as easily reproduced as the other laboratory-based standards.

Time

The unit of time in the SI system is the *second* (s). The second is defined as the duration of 9,192,631,770 cycles of radiation of the cesium atom. An atomic clock incorporating this standard is maintained by NIST. (See Figure 2.3.)

Temperature

The unit of temperature in the SI system is the *kelvin* (K). The kelvin is defined as the fraction 1/273.16 of the temperature of the triple point of water. The triple point of water is the combination of pressure and temperature at which water exists as a solid, liquid, and gas at the same time. (See Figure 2.4.) This temperature is 273.16 K, 0.01°C, or 32.002°F. Absolute zero is the temperature at which all molecular activity ceases and has a value of 0 K.

Figure 2.2
A duplicate of the kilogram standard is a platinum-iridium cylinder maintained by NIST. (© Copyright Robert Rathe. Courtesy of the National Institute of Standards and Technology, Gaithersburg, MD.)

Figure 2.3
A cesium fountain atomic clock maintained by NIST keeps time to an accuracy of one second in 100 million years. (Source: National Institute of Standards and Technology, Boulder, CO.)

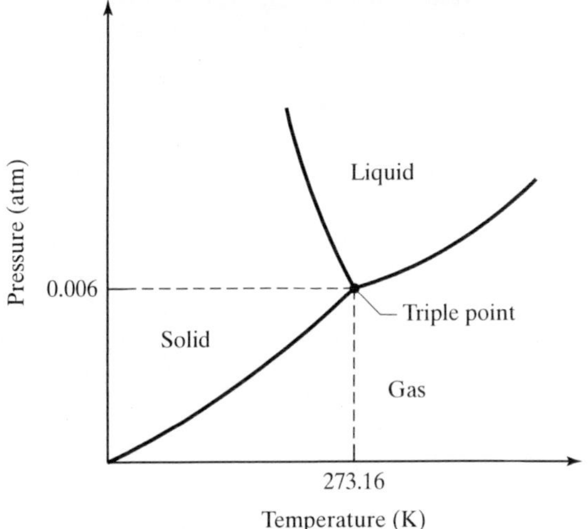

Figure 2.4
A phase diagram for water shows the triple point on which the kelvin temperature standard is based.

Electric Current

The unit of electric current in the SI system is the *ampere* (A). As shown in Figure 2.5, the ampere is defined as the steady current, which, if maintained in two straight parallel wires of infinite length and negligible circular cross section and placed one meter apart in a vacuum, produces a force of 2×10^{-7} newton per meter of wire length. Using Ohm's law, $I = V/R$, one ampere may also be denoted as the current that flows when one volt is applied across a 1-ohm resistor.

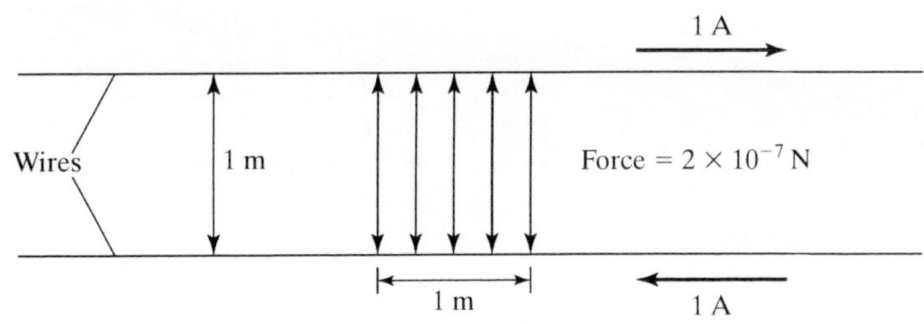

Figure 2.5
The standard for the ampere is based on the electrical force produced between two parallel wires, each carrying 1 A, located 1 m apart.

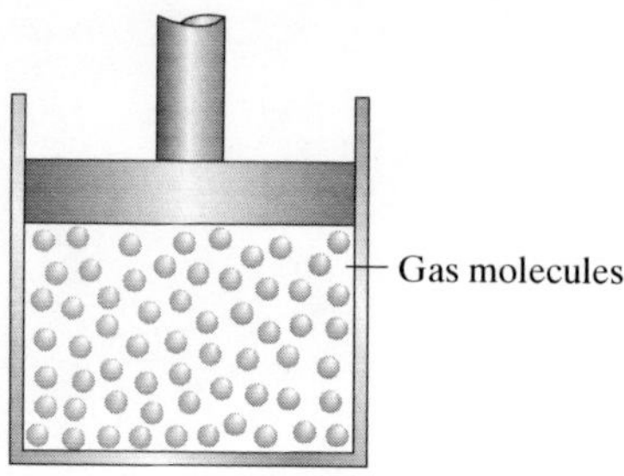

Figure 2.6
A mole of gas molecules in a piston-cylinder device contains 6.022×10^{23} molecules.

Amount of Substance

The unit used to denote the amount of substance is the *mole* (mol). One mole contains the same number of elements as there are atoms in 0.012 kg of carbon-12. This number is called Avogadro's number and has a value of approximately 6.022×10^{23}. (See Figure 2.6.)

Luminous Intensity

The unit for luminous intensity is the *candela* (cd). As illustrated in Figure 2.7, one candela is the luminous intensity of a source emitting light radiation at a frequency of 540×10^{12} Hz, that provides a power of 1/683 watt (W) per steradian. A steradian is a solid angle, which, having its vertex in the center of a sphere, subtends (cuts off) an area of the sphere equal to that of a square with sides of length equal to the radius of the sphere.

The unit for luminous intensity, the candela, utilizes the steradian, a dimension that may be unfamiliar to most students. The *radian* and *steradian* are called *supplementary* dimensions. These quantities, summarized in Table 2.3, refer

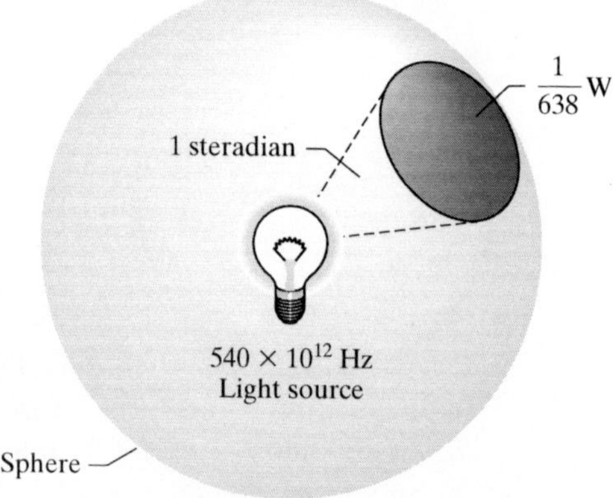

Figure 2.7
The candela standard for luminous intensity.

Table 2.3 Supplementary Dimensions

Quantity	Unit	Symbol
Plane angle	radian	rad
Solid angle	steradian	sr

to plane and solid angles, respectively. The radian is frequently used in engineering, and it is defined as the plane angle between two radii of a circle that subtends on the circumference of an arc equal in length to the radius. From trigonometry, you may recall that there are 2π radians in a circle (i.e., 2π radians equals 360°). Thus, one radian equals approximately 57.3°. The steradian, defined earlier, is used primarily for expressing radiation quantities such as light intensity and other electromagnetic parameters. These units appear dimensionless in measurements.

2.4 SI UNITS

Throughout the civilized world there are thousands of engineering companies that design and manufacture products for the benefit of society. The international buying and selling of these products is an integral part of a global network of industrialized countries, and the economic health of these countries, including the United States, depends to a large extent on international trade. Industries such as the automotive and electronics industries are heavily involved in international trade, so these industries have readily embraced the SI unit system in order to be economically competitive. The general adoption of the SI unit system by U.S. companies has been slow, but global economic imperatives are driving them to fall into step with the other industrialized nations of the world. SI units are now commonplace on food and beverage containers, gasoline pumps, and automobile speedometers. The SI unit system is the internationally accepted standard. In the United States, however, the English unit system is still widely used. We anticipate that it is only a matter of time before all U.S. companies use SI units exclusively. Until that time, the burden is upon you, the engineering student, to learn both unit systems. You will gladly discover, however, that most engineering textbooks emphasize SI units, but provide a list of unit conversions between the SI and English systems.

Table 2.2 summarizes the seven base dimensions and their SI units, and Table 2.3 summarizes the supplementary dimensions and their units. Derived dimensions consist of a combination of base and supplementary dimensions. Sometimes, the units of a derived dimension are given a specific name. For example, the derived dimension *force* consists of the SI base units $kg \cdot m \cdot s^{-2}$. This combination of SI base units is called a *newton* and is abbreviated N. Note that the unit name, in honor of Isaac Newton, is not capitalized when spelled out as a unit name. The same rule applies to other units named after people such as hertz (Hz), kelvin (K), and pascal (Pa). Another example is the *joule*, the SI unit for energy, work, and heat. The joule unit is abbreviated J and consists of the SI base units $kg \cdot m^2 \cdot s^{-2}$. A summary of the most commonly used SI derived dimensions and the corresponding SI unit names is given in Table 2.4.

Most derived dimensions do not have specific SI unit names, but their units may contain specific SI unit names. For example, the dimension *mass flow rate* is the mass of a fluid that flows past a point in a given time. The SI units for mass flow rate are $kg \cdot s^{-1}$, which we state as "kilograms per second." Note that units that are located in the denominator, that is, those that have a negative sign on their exponent, may also be written using a divisor line. Thus, the units for mass flow rate may be written as kg/s. Caution must be exercised, however, when utilizing this type of notation for some units. For example, the SI units for thermal conductivity, a quantity used in heat transfer, are $W \cdot m^{-1} \cdot K^{-1}$. How do we write these units with a

Table 2.4 Derived Dimensions and SI Units with Specific Names

Quantity	SI Unit	Unit Name	Base Units
Frequency	Hz	hertz	s^{-1}
Force	N	newton	$kg \cdot m \cdot s^{-2}$
Pressure	Pa	pascal	$kg \cdot m^{-1} \cdot s^{-2}$
Stress	Pa	pascal	$kg \cdot m^{-1} \cdot s^{-2}$
Energy	J	joule	$kg \cdot m^2 \cdot s^{-2}$
Work	J	joule	$kg \cdot m^2 \cdot s^{-2}$
Heat	J	joule	$kg \cdot m^2 \cdot s^{-2}$
Power	W	watt	$kg \cdot m^2 \cdot s^{-3}$
Electric charge	C	coulomb	$A \cdot s$
Electric potential (voltage)	V	volt	$kg \cdot m^2 \cdot s^{-3} \cdot A^{-1}$
Electric resistance	Ω	ohm	$kg \cdot m^2 \cdot s^{-3} \cdot A^{-2}$
Magnetic flux	Wb	weber	$kg^{-1} \cdot m \cdot s^{-2} \cdot A^{-1}$
Luminous flux	lm	lumen	$cd \cdot sr$

divisor line? Do we write these units as W/m/K? How about W/m · K? Either choice can cause some confusion. Does a "watt per meter per kelvin" mean that the kelvin unit is inverted twice and therefore goes above the divisor line? One glance at the units written as $W \cdot m^{-1} \cdot K^{-1}$ tells us that the temperature unit belongs "downstairs" because K has a negative exponent. If the kelvin unit were placed above the divisor line, and the thermal conductivity were used in an equation, a dimensional inconsistency would result. The second choice requires agreeing that multiplication takes precedence over division. Because the meter and kelvin units are located to the right of the divisor line and they are separated by a dot, both units are interpreted as being in the denominator. But to avoid all ambiguity, parentheses are used to group units above or below the divisor line. Units for thermal conductivity would then be written as $W/(m \cdot K)$. In any case, a dot or a dash should always be placed between adjacent units to separate them regardless of whether the units are above or below the divisor line. Some derived dimensions and their SI units are given in Table 2.5.

When a physical quantity has a numerical value that is very large or very small, it is cumbersome to write the number in standard decimal form. The general practice in engineering is to express numerical values between 0.1 and 1000 in standard decimal form. If a value cannot be expressed within this range, a *prefix* should be used. Because the SI unit system is based on powers of 10, it is more convenient to express such numbers by using prefixes. A prefix is a letter in front of a number that denotes multiples of powers of 10. For example, if the internal force in an I-beam is three million seven hundred and fifty thousand newtons, it would be awkward to write this number as 3,750,000 N. It is preferred to write the force as 3.75 MN, which is stated as "3.75 mega newtons." The prefix "M" denotes a multiple of a million. Hence, 3.75 MN equals 3.75×10^6 N. Electrical current is a good example of a quantity represented by a small number. Suppose the current flowing in a wire is 0.0082 A. This quantity would be expressed as 8.2 mA,

Table 2.5 Derived Dimensions and SI Units

Quantity	SI Units
Acceleration	$m \cdot s^{-2}$
Angular acceleration	$rad \cdot s^{-2}$
Angular velocity	$rad \cdot s^{-1}$
Area	m^2
Concentration	$mol \cdot m^{-3}$
Density	$kg \cdot m^{-3}$
Electric field strength	$V \cdot m^{-1}$
Energy	$N \cdot m$
Entropy	$J \cdot K^{-1}$
Heat	J
Heat transfer	W
Magnetic field strength	$A \cdot m^{-1}$
Mass flow rate	$kg \cdot s^{-1}$
Moment of force	$N \cdot m$
Radiant intensity	$W \cdot sr^{-1}$
Specific energy	$J \cdot kg^{-1}$
Surface tension	$N \cdot m^{-1}$
Thermal conductivity	$W \cdot m^{-1} \cdot K^{-1}$
Velocity	$m \cdot s^{-1}$
Viscosity, dynamic	$Pa \cdot s$
Viscosity, kinematic	$m^2 \cdot s^{-1}$
Volume	m^3
Volume flow rate	$m^3 \cdot s^{-1}$
Wavelength	m
Weight	N

which is stated as "8.2 milliamperes." The prefix "m" denotes a multiple of one-thousandth, or 1×10^{-3}.

A term we often hear in connection with computers is the storage capacity of hard disks. When personal computers first appeared in the early 1980s, most hard disks could hold around 10 or 20 MB (megabytes) of information. Nowadays, the typical storage capacity of a personal computer's hard disk is on the order of TB (terabytes). The standard prefixes for SI units are given in Table 2.6.

As indicated in Table 2.6, the most widely used SI prefixes for science and engineering quantities come in multiples of one thousand. For example, stress and pressure, which are generally large quantities for most structures and pressure vessels, are normally expressed in units of kPa, MPa, or GPa. Frequencies of electromagnetic waves such as radio, television, and telecommunications are also

Table 2.6 Standard Prefixes for SI Units

Multiple	Exponential Form	Prefix	Prefix Symbol
1,000,000,000,000,000	10^{15}	peta	P
1,000,000,000,000	10^{12}	tera	T
1,000,000,000	10^{9}	giga	G
1,000,000	10^{6}	mega	M
1000	10^{3}	kilo	k
0.01	10^{-2}	centi	c
0.001	10^{-3}	milli	m
0.000 001	10^{-6}	micro	μ
0.000 000 001	10^{-9}	nano	n
0.000 000 000 001	10^{-12}	pico	p
0.000 000 000 000 001	10^{-15}	femto	f

large numbers. Hence, they are generally expressed in units of kHz, MHz, or GHz. Electrical currents, on the other hand, are often small quantities, so they are usually expressed in units of μA or mA. Because frequencies of most electromagnetic waves are large quantities, the wavelengths of these waves are small. For example, the wavelength range of the visible light region of the electromagnetic spectrum is approximately 0.4 μm to 0.75 μm. It should be noted that the SI mass unit kilogram (kg) is the only base unit that has a prefix.

Here are some rules on how to use SI units properly that every beginning engineering student should know:

1. A unit symbol is never written as a plural with an "s." If a unit is pluralized, the "s" may be confused with the unit second (s).
2. A period is never used after a unit symbol, unless the symbol is at the end of a sentence.
3. Do not use invented unit symbols. For example, the unit symbol for "second" is (s), not (sec), and the unit symbol for "ampere" is (A), not (amp).
4. A unit symbol is always written by using lowercase letters, with two exceptions. The first exception applies to units named after people, such as the newton (N), joule (J), and watt (W). The second exception applies to units with the prefixes M, G, and T. (See Table 2.6.)
5. A quantity consisting of several units must be separated by dots or dashes to avoid confusion with prefixes. For example, if a dot is not used to express the units of "meter-second" $(\text{m} \cdot \text{s})$, the units could be interpreted as "millisecond" (ms).
6. An exponential power for a unit with a prefix refers to both the prefix and the unit; for example, $\text{ms}^2 = (\text{ms})^2 = \text{ms} \cdot \text{ms}$.
7. Do not use compound prefixes. For example, a "kilo MegaPascal" (kMPa) should be written as GPa, because the product of "kilo" (10^3) and "mega" (10^6) equals "giga" (10^9).
8. Put a space between the numerical value and the unit symbol.
9. Do not put a space between a prefix and a unit symbol.

Table 2.7 Correct and Incorrect Ways of Using SI Units

Correct	Incorrect	Rules
12.6 kg	12.6 kgs	1
450 N	450 Ns	1
36 kPa	36 kPa.	2
1.75 A	1.75 amps	1, 3
10.2 s	10.2 sec	3
20 kg	20 Kg	4
150 W	150 w.	2, 4
4.50 kg/m · s	4.50 kg/ms	5
750 GN	750 MkN	7
6 ms	6 kμs	7
650 Pa · s	650Pa · s	8
4.2 MΩ	4.2 M Ω	9
6 MN/m	6 N/μm	10

10. Do not use prefixes in the denominator of composite units. For example, the units N/mm should be written as kN/m.
Table 2.7 provides some additional examples of these rules.

APPLICATION

DERIVING FORMULAS FROM UNIT CONSIDERATIONS

To the beginning engineering student, it can seem as if there is an infinite number of formulas to learn. Formulas contain physical quantities that have numerical values plus units. Because formulas are written as equalities, formulas must be numerically and dimensionally equivalent across the equal sign. Can this feature be used to help us derive formulas that we do not know or have forgotten? Suppose that we want to know the mass of gasoline in an automobile's gas tank. The tank has a volume of 70 L, and a handbook of fluid properties states that the density of gasoline is 736 kg/m^3. (Note: 1 L $= 10^{-3}$ m^3). Thus, we write

$$\rho = 736\ \text{kg/m}^3, \quad V = 70\ \text{L} = 0.070\ \text{m}^3.$$

If the tank is completely filled with gasoline, what is the mass of the gasoline? Suppose that we have forgotten that density is defined as mass per volume, $\rho = m/V$. Because our answer will be a mass, the unit of our answer must be kilogram (kg). Looking at the units of the input quantities, we see that if we multiply density ρ by volume V, the volume unit (m^3) divides out, leaving mass (kg). Hence, the formula for mass in terms of ρ and V is

$$m = \rho V$$

so the mass of gasoline is

$$m = (736\ \text{kg/m}^3)(0.070\ \text{m}^3) = 51.5\ \text{kg}.$$

PROFESSIONAL SUCCESS—USING SI UNITS IN EVERYDAY LIFE

The SI unit system is used commercially to a limited extent in the United States, so the average person does not know the highway speed limit in kilometers per hour, his or her weight in newtons, atmospheric pressure in kilopascals, or the outdoor air temperature in kelvin or degrees Celsius. It is ironic that the leading industrialized nation on earth has yet to embrace this international standard. Admittedly, American beverage containers routinely show the volume of the liquid product in liters (L) or milliliters (mL), gasoline pumps often show liters of gasoline delivered, speedometers may indicate speed in kilometers per hour (km/h), and automobile tires indicate the proper inflation pressure in kilopascals (kPa) on the sidewall. On each of these products, and many others like them, a corresponding English unit is written along side the SI unit. The beverage container shows pints or quarts, the gasoline pump shows gallons, speedometers show miles per hour, and tires show pounds per square inch. Dual labeling of SI and English units on U.S. products are supposed to help people learn the SI system, "weaning" them from the antiquated English system in anticipation of the time when a full conversion to SI units occurs. This transition is analogous to the process of incrementally quitting smoking. Rather than quitting "cold turkey," we employ nicotine patches, gums, and other substitutes until our habit is broken. So, you may ask, "Why don't we make the total conversion now? Is it as painful as quitting smoking suddenly?" It probably is. As you might guess, the problem is largely an economic one. A complete conversion to SI units may not occur until we are willing to pay the price in actual dollars. People could learn the SI unit system fairly quickly if the conversion were done suddenly, but an enormous financial commitment would have to be made.

As long as dual product labeling of units is employed in the United States, most people will tend to ignore the SI unit and look only at the English unit, the unit with which they are most familiar. In U.S. engineering schools, SI units are emphasized. Therefore, the engineering student is not the average person on the street who does not know, or know how to calculate, his or her weight in newtons. So, what can engineering students in the United States do to accelerate the conversion process? A good place to start is with yourself. Start using SI units in your everyday life. When you make a purchase at the grocery store, look only at the SI unit on the label. Learn by inspection how many milliliters of liquid product are packaged in your favorite sized container. Abandon the use of inches, feet, yards, and miles as much as possible. How many kilometers lie between your home and school? What is 65 miles per hour in kilometers per hour? What is the mass of your automobile in kilograms? Determine your height in meters, your mass in kilograms, and your weight in newtons. How long is your arm in centimeters? What is your waist size in centimeters? What is the current outdoor air temperature in degrees Celsius? Most fast-food restaurants offer a "quarter pounder" on their menu. It turns out that $1 \text{ N} = 0.2248 \text{ lb}$, almost a quarter pound. On the next visit to your favorite fast-food place, order a "newton burger" and fries. (See Figure 2.8.)

Figure 2.8
An engineering student orders lunch (art by Kathryn Colton).

PRACTICE!

1. A structural engineer states that an I-beam in a truss has a design stress of "five million, six hundred thousand pascals." Write this stress, using the appropriate SI unit prefix.
   Answer. 5.6 MPa.
2. The power cord on an electric string trimmer carries a current of 5.2 A. How many milliamperes is this? How many microamperes?
   Answer. 5.2×10^3 mA, 5.2×10^6 μA.
3. Write the pressure 7.2 GPa in scientific notation.
   Answer. 7.2×10^9 Pa.
4. Write the voltage 0.000875 V, using the appropriate SI unit prefix.
   Answer. 0.875 mV or 875 μV.
5. In the following list, various quantities are written using SI units incorrectly. Write the quantities, using the correct form of SI units.
   a. 4.5 mw
   b. 8.75 M pa
   c. 200 Joules/sec

d. 20 W/m^2 K
e. 3 Amps.

Answer:
a. 4.5 mW
b. 8.75 MPa
c. 200 J/s
d. 20 W/m$^2 \cdot$ K
e. 3 A.

2.5 ENGLISH UNITS

The English unit system is known by various names. Sometimes it is referred to as the United States Customary System (USCS), the British System or the Foot-Pound-Second (FPS) system. The English unit system is still used extensively in the United States even though the rest of the industrialized world, including Great Britain, has adopted the SI unit system. English units have a long and colorful history. In ancient times, measures of length were based on human dimensions. The foot started out as the actual length of a man's foot. Because not all men were the same size, the foot varied in length by as much as three or four inches. Once the ancients started using feet and arms for measuring distance, it was only a matter of time before they began using hands and fingers. The unit of length that we refer to today as the inch was originally the width of a man's thumb. The inch was also once defined as the distance between the tip to the first joint of the forefinger. Twelve times that distance made one foot. Three times the length of a foot was the distance from the tip of a man's nose to the end of his outstretched arm. This distance closely approximates what we refer to today as the yard. Two yards equaled a fathom, which was defined as the distance across a man's outstretched arms. Half a yard was the 18-inch cubit, which was called a span. Half a span was referred to as a hand.

The pound, which uses the symbol lb, is named after the ancient Roman unit of weight called the libra. The British Empire retained this symbol into modern times. Today, there are actually two kinds of pound units, one for mass and one for weight and force. The first unit is called pound-mass $(\mathrm{lb_m})$, and the second is called pound-force $(\mathrm{lb_f})$. Because mass and weight are not the same quantity, the units $\mathrm{lb_f}$ and $\mathrm{lb_m}$ are different.

As discussed previously, the seven base dimensions are length, mass, time, temperature, electric current, amount of substance, and luminous intensity. These base dimensions, along with their corresponding English units, are given in Table 2.8. As with SI units, English units are not capitalized. The slug, which has no abbreviated symbol, is the mass unit in the English system, but the pound-mass $(\mathrm{lb_m})$ is frequently used. Electric current is based on SI units of meter and newton, and luminous intensity is based on SI units of watt. Hence, these two base dimensions do not have English units per se, and these quantities are rarely used in combination with other English units.

Recall that derived dimensions consist of a combination of base and supplementary dimensions. Table 2.9 summarizes some common derived dimensions expressed in English units. Note that Table 2.9 is the English counterpart of the SI version given by Table 2.5. The most notable English unit with a special name

Table 2.8 Base Dimensions and Their English Units

Quantity	Unit	Symbol
Length	foot	ft
Mass	slug(1)	slug
Time	second	s
Temperature	rankine	°R
Electric current	ampere(2)	A
Amount of substance	mole	mol
Luminous intensity	candela(2)	cd

(1) The unit poind-mass (lb_m) is also used. 1 slug = 32.174 lb_m.

(2) There are no English units for electrical current and luminous intensity. The SI units are given here for completeness only.

Table 2.9 Derived Dimensions and English Units

Quantity	English Units
Acceleration	$ft \cdot s^{-2}$
Angular acceleration	$rad \cdot s^{-2}$
Angular velocity	$rad \cdot s^{-1}$
Area	ft^2
Concentration	$mol \cdot ft^{-3}$
Density	$slug \cdot ft^{-3}$
Electric field strength	$V \cdot ft^{-1}$
Energy	Btu
Entropy	$Btu \cdot slug^{-1} \cdot {}^\circ R^{-1}$
Force	lb_f
Heat	Btu
Heat transfer	$Btu \cdot s^{-1}$
Magnetic field strength	$A \cdot ft^{-1}$
Mass flow rate	$slug \cdot s^{-1}$
Moment of force	$lb_f \cdot ft$
Radiant intensity	$Btu \cdot s^{-1} \cdot sr^{-1}$
Specific energy	$Btu \cdot slug^{-1}$
Surface tension	$lb_f \cdot ft^{-1}$
Thermal conductivity	$Btu \cdot s^{-1} \cdot ft^{-1} \cdot {}^\circ R$
Velocity	$ft \cdot s^{-1}$
Viscosity, dynamic	$slug \cdot ft^{-1} \cdot s^{-1}$
Viscosity, kinematic	$ft^2 \cdot s^{-1}$
Volume	ft^3
Volume flow rate	$ft^3 \cdot s^{-1}$
Wavelength	ft

is the British thermal unit (Btu), a unit of energy. One Btu is defined as the energy required to change the temperature of 1 lb_m of water at a temperature of 68°F by 1°F. One Btu is approximately the energy released by the complete burning of a single kitchen match. The magnitudes of the kilojoule and Btu are almost equal (1 Btu = 1.055 kJ). Unlike the kelvin (K), the temperature unit in the SI system, the rankine (°R) employs a degree symbol as do the Celsius (°C) and Fahrenheit (°F) units. The same rules for writing SI units apply for English units with one major exception: *prefixes are generally not used with English units.* Thus, units such as kft (kilo foot), Mslug (megaslug), and GBtu (gigaBtu) should not be used. Prefixes are reserved for SI units. Two exceptions are the units ksi, which refers to a stress of 1000 psi (pounds per square inch), and kip, which is a special name for a force of 1000 lb_f (pound-force).

There are some non-SI units that are routinely used in the United States and elsewhere. Table 2.10 summarizes some of these units and provides an equivalent value in the SI system. The inch is a common length unit, being found on virtually every student's ruler and carpenter's tape measure in the United States. There are exactly 2.54 centimeters per inch. Inches are still used as the primary length unit in many engineering companies. The yard is commonly used for measuring cloth, carpets, and loads of concrete (cubic yards), as well as ball advancement on the American football field. The ton is used in numerous industries, including shipping, construction, and transportation. Time subdivisions on clocks are measured in hours, minutes, and seconds. Radians and degrees are the most commonly used units for plane angles, whereas minutes and seconds are primarily used in navigational applications when referring to latitude and longitude on the earth's surface. The liter has made a lot of headway into the American culture, being found on beverage and food containers and many gasoline pumps. Virtually every American has seen the liter unit on a product, and many know that there are about four liters in a gallon (actually, 1 gal = 3.7854 L), but fewer people know that 1000 L = 1 m^3.

Table 2.10 Non-SI Units Commonly Used in the United States

Quantity	Unit Name	Symbol	SI Equivalent
Length	inch	in	0.0254 m[(1)]
	yard	yd	0.9144 m (36 in)
Mass	metric ton	t	1000 kg
	short ton	t	907.18 kg (2000 lb_m)
Time	minute	min	60 s
	hour	h	3600 s
	day	d	86,400 s
Plane angle	degree	°	$\pi/180$ rad
	minute	′	$\pi/10{,}800$ rad
	second	″	$\pi/648{,}000$ rad
Volume	liter	L	10^{-3} m^3
Land area	hectare	ha	10^4 m^2
Energy	electron-volt	eV	1.602177×10^{-19} J

(1) Exact conversion.

2.6 MASS AND WEIGHT

The concepts of *mass* and *weight* are fundamental to the proper use of dimensions and units in engineering analysis. Mass is one of the seven base dimensions used in science and engineering. Mass is a base dimension because it cannot be broken down into more fundamental dimensions. ***Mass*** is defined as a *quantity of matter*. This simple definition of mass may be expanded by exploring its basic properties. All matter possesses mass. The magnitude of a given mass is a measure of its resistance to a change in velocity. This property of matter is called *inertia*. A large mass offers more resistance to a change in velocity than a small mass, so a large mass has a greater inertia than a small mass. Mass may be considered in another way. Because all matter has mass, all matter exerts a gravitational attraction on other matter. Shortly after formulating his three laws of motion, Sir Isaac Newton postulated a law governing the gravitational attraction between two masses. Newton's law of universal gravitation is stated mathematically as

$$F = G\frac{m_1 m_2}{r^2} \tag{2.1}$$

where

$F =$ gravitational force between masses (N)
$G =$ universal gravitational constant $= 6.673 \times 10^{-11} \text{m}^3/\text{kg} \cdot \text{s}^2$
$m_1 =$ mass of body 1 (kg)
$m_2 =$ mass of body 2 (kg)
$r =$ distance between the centers of the two masses (m).

According to Equation (2.1), between any two masses there exists an attractive gravitational force whose magnitude varies inversely as the square of the distance between the masses. Because Newton's law of universal gravitation applies to *any* two masses, let's apply Equation (2.1) to a body resting on the surface of the earth. Accordingly, we let $m_1 = m_e$, the mass of the earth, and $m_2 = m$, the mass of the body. The distance, r, between the body and the earth may be taken as the mean radius of the earth, r_e. The quantities m_e and r_e have the approximate values

$$m_e = 5.979 \times 10^{24} \text{ kg} \qquad r_e = 6.378 \times 10^6 \text{ m}.$$

Thus, we have

$$\begin{aligned} F &= G\frac{m_e m}{r_e^2} \\ &= \frac{(6.673 \times 10^{-11} \text{ m}^3/\text{kg} \cdot \text{s}^2)(5.979 \times 10^{24} \text{ kg})}{(6.378 \times 10^6 \text{ m})^2} m \\ &= (9.808 \text{ m/s}^2)\ m. \end{aligned}$$

We can see that upon substituting values, the term Gm_e/r_e^2 yields approximately 9.81 m/s^2, the standard acceleration of gravity on the earth's surface. Redefining this term as g, and letting $F = W$, we express the law of universal gravitation in a special form as

$$W = mg \tag{2.2}$$

where

W = weight of body (N)

m = mass of body (kg)

g = standard gravitational acceleration = 9.81 m/s^2.

This derivation clearly shows the difference between mass and weight. We may therefore state the definition of ***weight*** as *a gravitational force exerted on a body by the earth.* Because mass is defined as a quantity of matter, the mass of a body is independent of its location in the universe. A body has the same mass whether it is located on the earth, the moon, Mars, or in outer space. The weight of the body, however, depends on its location. The mass of an 80 kg astronaut is the same whether or not he is on earth or in orbit above the earth. The astronaut weighs approximately 785 N on the earth, but while in orbit he is "weightless." His weight is zero while he orbits the earth, because he is continually "falling" toward earth. A similar weightless or "zero-g" condition is experienced by a skydiver as he begins falling.

The greatest source of confusion about mass and weight to the beginning engineering student is not the physical concept, but the units used to express each quantity. To see how units of mass and weight relate to each other, we employ a well-known scientific principle, ***Newton's second law*** of motion. Newton's second law of motion states that *a body of mass, m, acted upon by an unbalanced force, F, experiences an acceleration, a, that has the same direction of the force and a magnitude that is directly proportional to the force.* Stated mathematically, this law is

$$F = ma \tag{2.3}$$

where

F = force (N)

m = mass (kg)

a = acceleration (m/s^2)

Note that this relation resembles Equation (2.2). Weight is a particular type of force, and acceleration due to gravity is a particular type of acceleration, so Equation (2.2) is a special case of Newton's second law, given by Equation (2.3). In the SI unit system, the newton (N) is *defined* as the force that will accelerate a 1-kg mass at a rate of 1 m/s^2. Hence, we may write Newton's second law dimensionally as

$$1\text{ N} = 1\text{ kg}\cdot\text{m/s}^2$$

In the English unit system, the pound-force (lb$_f$) is *defined* as the force that will accelerate a 1-slug mass at a rate of 1 ft/s^2. Hence, we may write Newton's second law dimensionally as

$$1\text{ lb}_f = 1\text{ slug}\cdot\text{ft/s}^2$$

See Figure 2.9 for an illustration of Newton's second law. Confusion arises from the careless interchange of the English mass unit, pound-mass (lb$_m$), with the English force unit, pound-force (lb$_f$). These units are not the same thing! In accordance with our definitions of mass and weight, pound-mass refers to a quantity of matter, whereas pound-force refers to a force or weight. In order to write Newton's second law in terms of pound-mass instead of slug, we rewrite Equation (2.3) as

$$F = \frac{ma}{g_c} \tag{2.4}$$

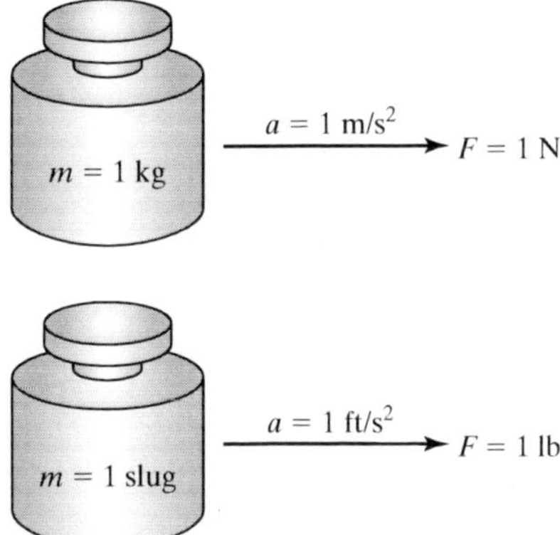

Figure 2.9
Definitions of the force units newton (N) and pound-force (lb_f).

where g_c is a constant that is required to make Newton's second law dimensionally consistent when mass, m, is expressed in lb_m, rather than slug. As stated previously, the English unit for force is lb_f, the English unit for acceleration is ft/s^2, and, as indicated in Table 2.8, 1 slug = 32.174 lb_m. Thus, the constant g_c is

$$g_c = \frac{ma}{F}$$

$$= \frac{(32.174\ lb_m)(ft/s^2)}{lb_f}$$

$$= 32.174 \frac{lb_m \cdot ft}{lb_f \cdot s^2}$$

This value is usually rounded to

$$g_c = 32.2 \frac{lb_m \cdot ft}{lb_f \cdot s^2}$$

Note that g_c has the same numerical value as g, the standard acceleration of gravity on the earth's surface. Newton's second law as expressed by Equation (2.4) is dimensionally consistent when the English unit of mass, lb_m, is used.

To verify that Equation (2.4) works, we recall that the pound-force is defined as the force that will accelerate a 1-slug mass at a rate of 1 ft/s^2. Recognizing that 1 slug = 32.2 lb_m, we have

$$F = \frac{ma}{g_c}$$

$$= \frac{(32.2\ lb_m)(1\ ft/s^2)}{32.2 \dfrac{lb_m \cdot ft}{lb_f \cdot s^2}} = 1\ lb_f$$

Note that in this expression, all the units, except lb_f, cancel. Hence, the pound-force (lb_f) is *defined* as the force that will accelerate a 32.2-lb_m mass at a rate of 1 ft/s^2. Therefore, we may write Newton's second law dimensionally as

$$1\ lb_f = 32.2\ lb_m \cdot ft/s^2$$

To have dimensional consistency when English units are involved, Equation (2.4) *must* be used when mass, m, is expressed in lb_m. When mass is expressed in slug,

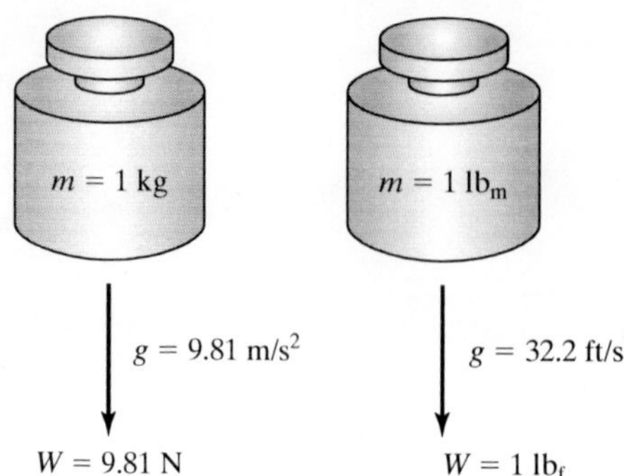

Figure 2.10 Definitions of weight for the standard value of gravitational acceleration.

however, the use of g_c in Newton's second law is not required for dimensional consistency because 1 lb_f is already defined as the force that will accelerate a 1-slug mass at a rate of 1 ft/s^2. Furthermore, because 1 N is already defined as the force that will accelerate a 1-kg mass at a rate of 1 m/s^2, the use of g_c is not required for dimensional consistency in the SI unit system. *Thus, Equation (2.3) suffices for all calculations, except for those in which mass is expressed in lb_m; in that case, Equation (2.4) must be used.* However, Equation (2.4) may be universally used when recognizing that the numerical value and units for g_c can be defined such that any consistent unit system will work. For example, substituting $F = 1$ N, $m = 1$ kg, and $a = 1$ m/s^2 into Equation (2.4) and solving for g_c, we obtain

$$g_c = \frac{1 \text{ kg} \cdot \text{m}}{\text{N} \cdot \text{s}^2}$$

Since the numerical value of g_c is 1, we can successfully use Equation (2.3) as long as we recognize that 1 N is the force that will accelerate a 1-kg mass at a rate of 1 m/s^2.

Sometimes, the units pound-mass (lb_m) and pound-force (lb_f) are casually interchanged because a body with a mass of 1 lb_m has a weight of 1 lb_f (i.e., the mass and weight are *numerically equivalent*). Let's see how this works: By definition, a body with a mass of 32.2 lb_m (1 slug) when accelerated at a rate of 1 ft/s^2 has a weight of 1 lb_f. Therefore, using Newton's second law in the form, $W = mg$, we can also state that a body with a mass of 1 lb_m, when accelerated at a rate of 32.2 ft/s^2 (the standard value of g), has a weight of 1 lb_f. Our rationale for making such a statement is that we maintained the same numerical value on the right side of Newton's second law by assigning the mass, m, a value of 1 lb_m and the gravitational acceleration, g, the standard value of 32.2 ft/s^2. The numerical values of the mass and weight are equal even though a pound-mass and a pound-force are conceptually different quantities. It must be emphasized, however, that mass in pound-mass and weight in pound-force are numerically equivalent only when the standard value, $g = 32.2$ ft/s^2, is used. See Figure 2.10 for an illustration. The next example illustrates the use of g_c.

EXAMPLE 2.4

Find the weight of some objects with the following masses:

a. 50 slug
b. 50 lb_m
c. 75 kg.

Solution

To find weight, we use Newton's second law, where the acceleration a is the standard acceleration of gravity, $g = 9.81\ \text{m/s}^2 = 32.2\ \text{ft/s}^2$.

a. The mass unit slug is the standard unit for mass in the English unit system. The weight is

$$\begin{aligned} W &= mg \\ &= (50\ \text{slug})(32.2\ \text{ft/s}^2) = 1{,}610\ \text{lb}_\text{f} \end{aligned}$$

b. When mass is expressed in terms of lb_m, we must use Equation (2.4):

$$W = \frac{mg}{g_c} = \frac{(50\ \text{lb}_\text{m})(32.2\ \text{ft/s}^2))}{32.2\dfrac{\text{lb}_\text{m}\cdot\text{ft}}{\text{lb}_\text{f}\cdot\text{s}^2}} = 50\ \text{lb}_\text{f}$$

Note that the mass and weight are numerically equivalent. This is true only in cases where the standard value of g is used, which means that an object with a mass of x lb_m will always have a weight of x lb_f on the earth's surface.

c. The mass unit kg is the standard unit for mass in the SI unit system. The weight is

$$\begin{aligned} W &= mg \\ &= (75\ \text{kg})(9.81\ \text{m/s}^2) = 736\ \text{N}. \end{aligned}$$

Alternatively, we can find weight by using Equation (2.4):

$$W = \frac{mg}{g_c} = \frac{(75\ \text{kg})(9.81\ \text{m/s}^2)}{1\dfrac{\text{kg}\cdot\text{m}}{\text{N}\cdot\text{s}^2}} = 736\ \text{N}$$

Now that we understand the difference between mass and weight and know how to use mass and weight units in the SI and English systems, let's revisit the astronaut we discussed earlier. (See Figure 2.11.) The mass of the astronaut is 80 kg, which equals about 5.48 slug. His mass does not change, regardless of where he

SI	English	SI	English
$m = 80$ kg	$m = 5.48$ slug	$m = 80$ kg	$m = 5.48$ slug
$W = mg = 785$ N	$W = mg = 176\ \text{lb}_\text{f}$	$W = mg = 130$ N	$W = mg = 29.1\ \text{lb}_\text{f}$

Earth — $g = 9.81\ \text{m/s}^2$, $g = 32.2\ \text{ft/s}^2$

Moon — $g = 1.62\ \text{m/s}^2$, $g = 5.31\ \text{ft/s}^2$

Figure 2.11 An astronaut's mass and weight on the earth and moon.

ventures. Prior to departing on a trip to the moon, he weighs in at 785 N (176 lb_f). What is the mass of the astronaut in pound-mass? Three days later, his vehicle lands on the moon, and he begins constructing a permanent base for future planetary missions. The value of the gravitational acceleration on the moon is only 1.62 m/s^2 (5.31 ft/s^2). The astronaut's mass is still 80 kg, but his weight is only 130 N (29.1 lb_f) due to the smaller value of g. Is the mass and weight of the astronaut in pound-mass and pound-force numerically equivalent? No, because the standard value of g is not used.

EXAMPLE 2.5

Special hoists are used in automotive repair shops to lift engines. As illustrated in Figure 2.12, a 200 kg engine is suspended in a fixed position by a chain attached to the cross member of an engine hoist. Neglecting the weight of the chain itself, what is the tension in portion AD of the chain?

Solution

This example is a simple problem in engineering statics. Statics is the branch of engineering mechanics that deals with forces acting on bodies at rest. The engine is held by the chain in a fixed position, so clearly the engine is at rest; that is, it is not in motion. This problem can be solved by recognizing that the entire weight of the engine is supported by portion AD of the chain. (The tension in portions AB and AC could also be calculated, but a thorough equilibrium analysis would be required.) Hence, the tension, which is a force that tends to elongate the chain, is equivalent to the weight of the engine. Using Equation (2.2), we have

$$\begin{aligned} F &= mg \\ &= (200\ \text{kg})(9.81\ \text{m/s}^2) = 1962\ \text{N}. \end{aligned}$$

Therefore, the tension in portion AD of the chain is 1962 N, the weight of the engine.

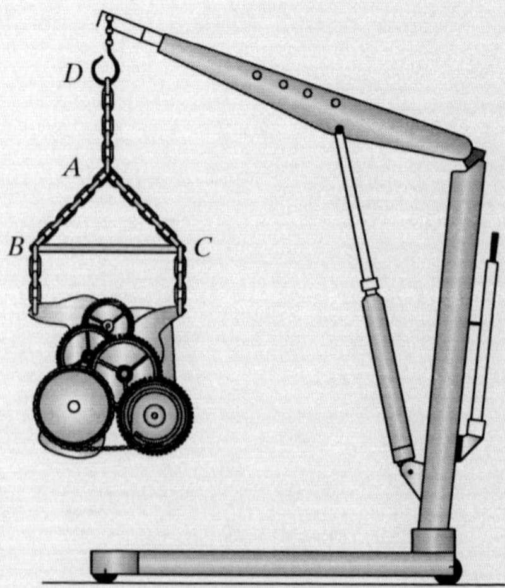

Figure 2.12
Engine hoist for Example 2.5.

PRACTICE!

1. It has been said that you do not fully understand a basic technical concept, unless you can explain it in terms simple enough that a second grader can understand it. Write an explanation of the difference between mass and weight for a second grader.
2. Which is larger, a slug or a pound-mass?
   Answer: slug.
3. Consider a professional linebacker who weighs 310 lb_f. What is his mass in slugs?
   Answer: 9.63 slug.
4. A rock $(\rho = 2300 \text{ kg/m}^3)$ is suspended by a single rope. Assuming the rock to be spherical, with a radius of 20 cm, what is the tension in the rope?
   Answer: 756 N.

2.7 UNIT CONVERSIONS

Although the SI unit system is the international standard, English units are in widespread use in the United States. Americans as a whole are much more familiar with English units than SI units. Students of science and engineering in U.S. schools primarily use SI units in their course work because most textbooks and the professors who teach out of them stress SI units. Unfortunately, when students of these disciplines go about their day-to-day activities outside of the academic environment, they tend to slip back into the English unit mode along with everyone else. It seems as if students have a "unit switch" in their brains. When they are in the classroom or laboratory, the switch is turned to the "SI position." When they are at home, in the grocery store, or driving their car, the switch is turned to the "English position." Ideally, there should be no unit switch at all, but as long as science and engineering programs at colleges and universities stress SI units and American culture stresses English units, our cerebral unit switch toggles. In this section, a systematic method for converting units between the SI and English systems is given.

A ***unit conversion*** enables us to convert from one unit system to the other by using ***conversion factors***. A conversion factor is an equivalency ratio that has a *unit value* of 1. Stated another way, a conversion factor simply relates the same physical quantity in two different unit systems. For example, 0.0254 m and 1 in are equivalent length quantities because 0.0254 m = 1 in. The ratio of these two quantities has a unit value of 1 because they are physically the same quantity. Obviously, the numerical value of the ratio is not 1, but depends on the numerical value of each individual quantity. Thus, when we multiply a given quantity by one or more conversion factors, we alter only the numerical value of the result and not its dimension. Table 2.11 summarizes some common conversion factors used in engineering analysis. A more extensive listing of unit conversions is given in Appendix B.

A systematic procedure for converting a quantity from one unit system to the other is as follows:

2.7.1 Unit Conversion Procedure

1. Write the given quantity in terms of its numerical value and units. Use a horizontal line to divide units in the numerator (upstairs) from those in the denominator (downstairs).
2. Determine the units *to* which you want to make the conversion.

Table 2.11 Some Common SI-to-English Unit Conversions

Quantity	Unit Conversion
Acceleration	1 m/s^2 = 3.2808 ft/s^2
Area	1 m^2 = 10.7636 ft^2 = 1550 in^2
Density	1 kg/m^3 = 0.06243 lb$_m$/ft^3
Energy, work, heat	1055.06 J = 1 Btu = 252 cal
Force	1 N = 0.22481 lb$_f$
Length	1 m = 3.2808 ft = 39.370 in 0.0254 m = 1 in[(1)]
Mass	1 kg = 2.20462 lb$_m$ = 0.06852 slug
Power	1 W = 3.4121 Btu/h 745.7 W = 1 hp
Pressure	1 kPa = 20.8855 lb$_f$/ft^2 = 0.14504 lb$_f$/in^2
Specific heat	1 kJ/kg·°C = 0.2388 Btu/lb$_m$·°F
Temperature	T(K) = T(°C) + 273.16 = T(°R)/1.8 = [T(°F) + 459.67]/1.8
Velocity	1 m/s = 2.2369 mi/h

(1) Exact conversion.

3. Multiply the given quantity by one or more conversion factors that, upon cancellation of units, leads to the desired units. Use a horizontal line to divide the units in the numerator and denominator of each conversion factor.
4. Draw a line through all canceled units.
5. Perform the numerical computations on a calculator, retaining maximum decimal place accuracy until the end of the computations.
6. Write the numerical value of the converted quantity by using the desired number of significant figures (three significant figures is standard practice for engineering) with the desired units.

Examples 2.6, 2.7, and 2.8 illustrate the unit conversion procedure.

EXAMPLE 2.6

An engineering student is late for an early morning class, so she runs across campus at a speed of 9 mi/h. Determine her speed in units of m/s.

Solution

The given quantity, expressed in English units, is 9 mi/h, but we want our answer to be in SI units of m/s. Thus, we need a conversion factor between mi and m and a conversion factor between h and s. To better illustrate the unit conversion procedure, we will use two length conversion factors rather than one. Following the procedure outlined, we have

$$\underbrace{9\frac{\cancel{\text{mi}}}{\cancel{\text{h}}}}_{\text{given quantity}} \times \underbrace{\frac{5280\ \cancel{\text{ft}}}{1\ \cancel{\text{mi}}} \times \frac{1\ \text{m}}{3.2808\ \cancel{\text{ft}}} \times \frac{1\ \cancel{\text{h}}}{3600\ \text{s}}}_{\text{conversion factors}} = \underbrace{4.02\frac{\text{m}}{\text{s}}}_{\text{answer}}$$

The key aspect of the unit conversion process is that the conversion factors must be written such that the appropriate units in the conversion factors cancel those in the given quantity. If we had inverted the conversion factor between ft and mi, writing it instead as 1 mi/5280 ft, the mi unit would not cancel and our unit conversion exercise would not work, because we would end up with units of mi^2 in the numerator. Similarly, the conversion factor between m and ft was written such that the ft unit canceled the ft unit in the first conversion factor. Also, the conversion factor between h and s was written such that the h unit canceled with the h unit in the given quantity. Writing conversion factors with the units in the proper locations, "upstairs" or "downstairs," requires some practice, but after doing several conversion problems, the correct placement of units will become second nature to you. Note that our answer is expressed in three significant figures.

EXAMPLE 2.7

Lead has one of the highest densities of all the pure metals. The density of lead is 11,340 kg/m^3. What is the density of lead in units of lb_m/in^3?

Solution

A direct conversion factor from kg/m^3 to lb_m/in^3 may be available, but to illustrate an important aspect of converting units with exponents, we will use a series of conversion factors for each length and mass unit. Thus, we write our unit conversion as

$$11{,}340\frac{\cancel{\text{kg}}}{\cancel{\text{m}^3}} \times \left(\frac{1\ \cancel{\text{m}}}{3.2808\ \cancel{\text{ft}}}\right)^3 \times \left(\frac{1\ \cancel{\text{ft}}}{12\ \text{in}}\right)^3 \times \frac{2.20462\ \text{lb}_\text{m}}{1\ \cancel{\text{kg}}} = 0.410\ \text{lb}_\text{m}/\text{in}^3.$$

We used two length conversion factors, one factor between m and ft and the other between ft and in. But the given quantity is a density that has a volume unit. When performing unit conversions involving exponents, *both* the numerical value and the unit must be raised to the exponent. A common error that students make is to raise the unit to the exponent, which properly cancels units, but to forget to raise the numerical value also. Failure to raise the numerical value to the exponent will lead to the wrong numerical answer even though the units in the answer will be correct. Using the direct conversion factor obtained from Appendix B, we obtain the same result:

$$11{,}340\ \cancel{\text{kg}}/\cancel{\text{m}^3} \times \frac{3.6127 \times 10^{-5}\ \text{lb}_\text{m}/\text{in}^3}{1\ \cancel{\text{kg}}/\cancel{\text{m}^3}} = 0.410\ \text{lb}_\text{m}/\text{in}^3.$$

EXAMPLE 2.8

Specific heat is defined as the energy required to raise the temperature of a unit mass of a substance by one degree. Pure aluminum has a specific heat of approximately 900 J/kg · °C. Convert this value to units of $Btu/lb_m \cdot °F$.

Solution

By following the unit conversion procedure, we write the given quantity and then multiply it by the appropriate conversion factors, which can be found in Appendix B:

$$\frac{900\,\cancel{J}}{\cancel{kg}\cdot\cancel{^\circ C}} \times \frac{1\text{ Btu}}{1055.06\,\cancel{J}} \times \frac{1\,\cancel{kg}}{2.20462\text{ lb}_\text{m}} \times \frac{1\cancel{^\circ C}}{1.8^\circ\text{F}} = 0.215\text{ Btu/lb}_\text{m}\cdot{}^\circ\text{F}.$$

The temperature unit °C in the original quantity has a unique interpretation. Because specific heat is the energy required to raise a unit mass of a substance by one degree, the temperature unit in this quantity denotes a temperature *change*, not an absolute temperature value. A temperature change of 1°C is equivalent to a temperature change of 1.8°F. Other thermal properties, such as thermal conductivity, involve the same temperature change interpretation.

This example can also be done by applying a single conversion factor $1\text{ kJ/kg}\cdot{}^\circ\text{C} = 0.2388\text{ Btu/lb}_\text{m}\cdot{}^\circ\text{F}$, which yields the same result.

PROFESSIONAL SUCCESS—UNIT CONVERSIONS AND CALCULATORS

Scientific pocket calculators have evolved from simple electronic versions of adding machines to complex portable computers. Today's high-end scientific calculators have numerous capabilities, including programming, graphing, numerical methods, and symbolic mathematics. Most scientific calculators also have an extensive compilation of conversion factors. Why, then, should students learn to do unit conversions by hand when calculators will do the work? This question lies at the root of a more fundamental question: why should students learn to do *any* computational task by hand when calculators or computers will do the work? Is it because "in the old days" engineers did not have the luxury of highly sophisticated computational tools, so professors, who perhaps lived in the "old days," forced their students to do things the old fashioned way? Not really.

Students will always need to learn engineering by *thinking* and *reasoning* their way through a problem, regardless of whether that problem is a unit conversion or a stress calculation in a machine component. Computers, and the software that runs on them, do not replace the thinking process. The calculator, like the computer, should never become a "black box" to the student. A black box is a mysterious device whose inner workings are largely unknown, but that, nonetheless, provides output for every input supplied. By the time you graduate with an engineering degree, or certainly by the time you have a few years of professional engineering practice, you will come to realize that a calculator program or computer software package exists for solving many types of engineering problems. This does not mean that you need to learn every one of these programs and software packages. It means that you should become proficient in the use of those computational tools that pertain to your particular engineering field *after* learning the underlying basis for each. By all means, use a calculator to perform unit conversions, but *first* know how to do them by hand, so you gain confidence in your own computational skills and have a way to verify the results of your calculator.

PRACTICE!

1. A microswitch is an electrical switch that requires only a small force to operate it. If a microswitch is activated by a 0.25-oz force, what is the force in units of N that will activate it?
   Answer: 0.0695 N.
2. At room temperature, water has a density of about $62.4\ \text{lb}_\text{m}/\text{ft}^3$. Convert this value to units of slug/in^3 and kg/m^3.
   Answer: $1.12 \times 10^{-3}\ \text{slug/in}^3$, $999.5\ \text{kg/m}^3$.
3. At launch, the Saturn V rocket that carried astronauts to the moon developed five million pounds of thrust. What is the thrust in units of MN?
   Answer: 22.2 MN.
4. Standard incandescent light bulbs produce more heat than light. Assuming that a typical house has twenty 60-W bulbs that are continuously on, how much heat in units of Btu/h is supplied to the house from light bulbs if 90 percent of the energy produced by the bulbs is in the form of heat?
   Answer: 3685 Btu/h.
5. Certain properties of animal (including human) tissue can be approximated by using those of water. Using the density of water at room temperature, $\rho = 62.4\ \text{lb}_\text{m}/\text{ft}^3$, calculate the weight of a human male by approximating him as a cylinder with a length and diameter of 6 ft and 10 in, respectively.
   Answer: $204\ \text{lb}_\text{f}$
6. The standard frequency for electrical power in the United States is 60 Hz. For an electrical device that operates on this power, how many times does the current alternate during a year?
   Answer: 1.89×10^9.

KEY TERMS

base dimension	English unit system	unit
conversion factors	mass	unit conversion
derived dimension	Newton's second law	weight
dimension	physical standards	
dimensionally consistent	SI unit system	

REFERENCES

Cardarelli, F., *Encyclopaedia of Scientific Units, Weights and Measures: Their SI Equivalences and Origins,* 3d ed., NY: Springer-Verlag, 2004.

Lewis, R., *Engineering Quantities and Systems of Units,* NY: Halsted Press, 1972.

Haynes, W.M., Editor, *CRC Handbook of Chemistry and Physics,* 92d ed., Boca Raton, FL: CRC Press, 2011.

PROBLEMS

Dimensions

2.1 For the following dimensional equations, find the base dimensions of the parameter *k*:

a. $MLt^{-2} = kML^{-1}t^{-2}$
b. $MLt^{-2}L^{-1} = k\,Lt^{-3}$
c. $L^2t^{-2} = k\,M^4T^2$
d. $ML^2t^{-3} = k\,LT$
e. $nLL^3k = T^2M^{-2}L$
f. $MI^2k = nTM^{-3}L^{-1}$
g. $IL^2t = k^2M^4t^2$
h. $k^3T^6M^3L^{-5} = T^{-3}t^{-6}L$
i. $T^{-1/2}L^{-1}I^2 = k^{-1/2}t^4T^{-5/2}L^{-3}$
j. $MLt^{-2} = MLt^{-2}\sin(kL^{-2}M^{-1})$
k. $T^2n = T^2n\ln(knT^{-1})$

2.2 Is the following dimensional equation dimensionally consistent? Explain.

$$ML = ML\cos(Lt).$$

2.3 Is the following dimensional equation dimensionally consistent? Explain.

$$t^2LT = tLT\log(tt^{-1}).$$

2.4 Is the following dimensional equation dimensionally consistent? Explain.

$$TnT = TnT\exp(MM^{-1}).$$

Units

2.5 In the following list, various quantities are written using SI units incorrectly: Write the quantities, using the correct form of SI units.

a. 10.6 secs
b. 4.75 amp
c. 120 M hz
d. 2.5 kw
e. 0.00846 kg/μs
f. 90 W/m^2 K
g. 650 mGPa
h. 25 MN.
i. 950 Joules
j. 1.5 m/s/s.

2.6 The dimension *moment*, sometimes referred to as *torque*, is defined as a force multiplied by a distance and is expressed in SI units of newton-meter (N·m). In addition to moment, what other physical quantities are expressed in SI units of N·m? What is the special name given to this combination of units?

2.7 Consider a 60-W light bulb. A watt (W) is defined as a joule per second (J/s). Write the quantity 60 W in terms of the units newton (N), meter (m), and second (s).

2.8 A commonly used formula in electrical circuit analysis is $P = IV$, power (W) equals current (A) multiplied by voltage (V). Using Ohm's law, write a formula for power in terms of current I and resistance R.

2.9 A particle undergoes an average acceleration of 8 m/s^2 as it travels between two points during a time interval of 2 s. Using unit considerations, derive a formula for the average velocity of a particle in terms of average acceleration and time interval. Calculate the average velocity of the particle for the numerical values given.

2.10 A crane hoists a large pallet of materials from the ground to the top of a building. In hoisting this load, the crane does 250 kJ of work during a time interval of 5 s. Using unit considerations, derive the formula for power in terms of work and time interval. Calculate the power expended by the crane in lifting the load.

Mass and weight

2.11 A spherical tank with a radius of 0.32 m is filled with water ($\rho = 1000$ kg/m^3). Calculate the mass and the weight of the water in SI units.

2.12 A large indoor sports arena is cylindrical in shape. The height and diameter of the cylinder are 120 m and 180 m, respectively. Calculate the mass and weight of air contained in the sports arena in SI units if the density of air is $\rho = 1.20$ kg/m^3.

2.13 A 90-kg astronaut biologist searches for microbial life on Mars where the gravitational acceleration is $g = 3.71$ m/s^2. What is the weight of the astronaut in units of N and lb$_f$?

2.14 A 90-kg astronaut biologist places a 4-lb$_m$ rock sample on two types of scales on Mars in order to measure the rock's weight. The first scale is a beam balance, which operates by comparing masses. The second scale operates by the compression of a spring. Calculate the weight of the rock sample in units of lb$_f$ using (a) the beam balance and (b) the spring scale.

2.15 A copper plate measuring 1.2 m $\times$ 0.8 m $\times$ 3 mm has a density of $\rho = 8940$ kg/m^3. Find the mass and weight of the plate in SI units.

2.16 A circular tube of stainless steel ($\rho = 7840$ kg/m^3) has an inside radius of 1.85 cm and an outside radius of 2.20 cm. If the tube is 35 cm long, what is the mass and weight of the tube in SI units?

2.17 The density of porcelain is $\rho = 144$ lb$_m$/ft^3. Approximating a porcelain dinner plate as a flat disk with a diameter and thickness of 9 in and 0.2 in, respectively, find the mass of the plate in units of slug and lb$_m$. What is the weight of the plate in units of lb$_f$?

2.18 In an effort to reduce the mass of an aluminum bulkhead for a spacecraft, a machinist drills an array of holes in the bulkhead. The bulkhead is a triangular-shaped plate with a base and height of 2.5 m and 1.6 m, respectively, and a thickness of 7 mm. How many 5-cm diameter holes must be drilled clear through the bulkhead to reduce its mass by 8 kg? For the density of aluminum, use $\rho = 2800$ kg/m^3.

Unit conversions

2.19 A world-class sprinter can run 100 m in a time of 9.80 s, an average speed of 10.2 m/s. Convert this speed to mi/h.

2.20 A world-class mile runner can run 1 mi in a time of 4 min. What is the runner's average speed in units of mi/h and m/s?

2.21 The typical home is heated by a forced-air furnace that burns natural gas or fuel oil. If the heat output of the furnace is 175,000 Btu/h, what is the heat output in units of kW?

2.22 Calculate the temperature at which the Celsius (°C) and Fahrenheit (°F) scales are numerically equal.

2.23 A large shipping container of ball bearings is suspended by a cable in a manufacturing plant. The combined mass of the container and ball bearings is 3250 lb_m. Find the tension in the cable in units of N.

2.24 A typical human adult loses about 65 Btu/h · ft^2 of heat while engaged in brisk walking. Approximating the human adult body as a cylinder with a height and diameter of 5.8 ft and 10 in, respectively, find the total amount of heat lost in units of J if the brisk walking is maintained for a period of 1 h. Include the two ends of the cylinder in the surface area calculation.

2.25 A symmetric I-beam of structural steel $(\rho = 7860 \text{ kg/m}^3)$ has the cross section shown in Figure P2.25. Calculate the weight per unit length of the I-beam in units of N/m and lb_f/ft.

Figure P2.25

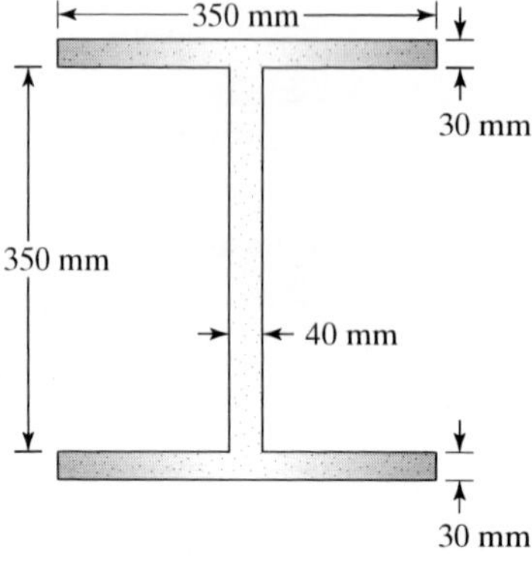

2.26 A sewer pipe carries waste away from a commercial building at a mass flow rate of 6 kg/s. What is this flow rate in units of lb_m/s and slug/h?

2.27 The rate at which solar radiation is intercepted by a unit area is called solar heat flux. Just outside the earth's atmosphere, the solar heat flux is approximately 1350 W/m^2. Determine the value of this solar heat flux in units of Btu/h · ft^2.

2.28 During a typical summer day in the arid southwest regions of the United States, the outdoor air temperature may range from 115°F during the late afternoon to 50°F several hours after sundown. What is this temperature range in units of °C, K, and °R?

2.29 An old saying is "an ounce of prevention is worth a pound of cure." Restate this maxim in terms of the SI unit newton.

2.30 How many seconds are there in the month of July?

2.31 What is your approximate age in hours?

2.32 A highway sign is supported by two posts as shown in Figure P2.32. The sign is constructed of a high-density pressboard material $(\rho = 900 \text{ kg/m}^3)$ and its thickness is 2 cm. Assuming that each post carries half the weight of the sign, calculate the compressive force in the posts in units of N and lb_f.

Figure P2.32

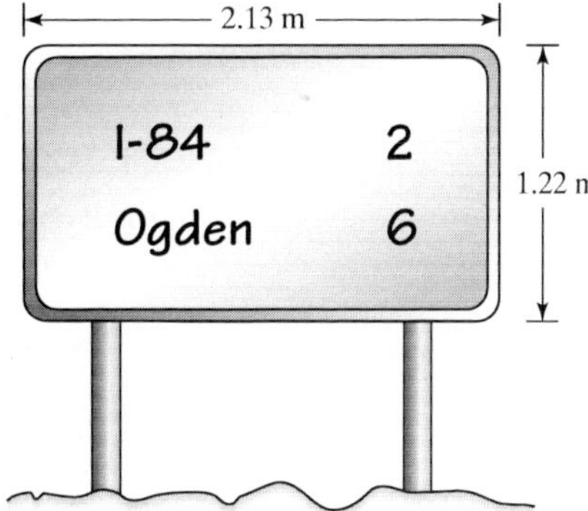

2.33 The steam exiting a turbine has a temperature and pressure of 400°C and 8 MPa, respectively. What is the temperature and pressure of the steam in units of K and psi, respectively?

2.34 A pressure gauge designed to measure small pressure differences in air ducts has an operating range of 0 to 16 inch H_2O. What is this pressure range in units of Pa and psi?

2.35 Resistors are electrical devices that retard the flow of current. These devices are rated by the maximum power they are capable of dissipating as heat to the surrounding area. How much heat does a 10-W resistor dissipate in units of Btu/h if the resistor operates at maximum capacity? Using the formula $P = I^2R$, what is the current flow I in the resistor if it has a resistance R of 100 Ω?

2.36 Chemical reactions can generate heat. This type of heat generation is often referred to as volume heat generation because the heat is produced internally by every small parcel of chemical. Consider a chemical reaction that generates heat at the rate of 125 MW/m^3. Convert this volume heat generation to units of $Btu/h \cdot ft^3$.

2.37 A sport-utility vehicle has an engine that delivers 290 hp. How much power does the engine produce in units of kW and Btu/h?

2.38 A copper tube carries hot water to a dishwasher at a volume flow rate of 3 gal/min. Convert this flow rate to units of m^3/s and ft^3/h.

2.39 Thermal conductivity is a property that describes the ability of a material to conduct heat. A material with a high thermal conductivity readily transports heat, whereas a material with a low thermal conductivity tends to retard heat flow. Fiberglass insulation and silver have thermal conductivities of 0.046 $W/m \cdot °C$ and 429 $W/m \cdot °C$, respectively. Convert these values to units of $Btu/h \cdot ft \cdot °F$.

2.40 A standard incandescent 60-W light bulb has an average life of 1000 h. What is the total amount of energy that this light bulb produces during its lifetime? Express the answer in units of J, Btu, and cal.

2.41 A steam power plant produces 750 MW of power. How much energy does the power plant produce in a year? Express your answer in units of J and Btu.

2.42 It is estimated that about 60 million Americans go on a new diet each year. If each of these people cuts 300 cal from their diets each day, how many 100-W light bulbs could be powered by this energy?

2.43 The standard acceleration of gravity at the earth's surface is $g = 9.81\ \text{m/s}^2$. Convert this acceleration to units of ft/h^2 and mi/s^2.

2.44 At room temperature, air has a specific heat of $1.007\ \text{kJ/kg}\cdot{}^\circ\text{C}$. Convert this value to units of $\text{J/kg}\cdot\text{K}$ and $\text{Btu/lb}_\text{m}\cdot{}^\circ\text{F}$.

2.45 The yield stress for structural steel is approximately 250 MPa. Convert this value to units of psi.

2.46 A 20-gage tungsten wire carries a current of 6.8 A. The electrical resistance of this wire is 106 Ω per kilometer of length. Find the power dissipated from this wire per meter of length in units of W. (Hint: Use the formula $P = I^2R$, where P = power, I = current, R = resistance). How much energy does this wire dissipate in one hour? One year?

2.47 An open pit copper mine yields 7×10^4 kg of copper per day. From the same ore, the mine also yields 2×10^3 kg of silver and 30 kg of gold per day. In units of lb_m, what is the annual production of these metals from the mine assuming year round operation?

CHAPTER

3 Analysis Methodology

Objectives

After reading this chapter, you will have learned

- How to make order-of-magnitude calculations
- The proper use of significant figures
- How to perform an analysis systematically
- The proper method of analysis presentation
- Advantages and disadvantages of using computers for analysis

3.1 INTRODUCTION

One of the most important things an engineering student learns during his or her program of study is how to approach an engineering problem in a systematic and logical fashion. In this respect, the study of engineering is somewhat similar to the study of science in that a science student learns how to think like a scientist by employing the scientific method. The scientific method is a process by which hypotheses about the physical world are stated, theories formulated, data collected and evaluated, and mathematical models constructed.

The ***engineering method*** may be thought of as a problem-solving process by which the needs of society are met through design and manufacturing of devices and systems. Engineering analysis is a major part of this problem-solving process. Admittedly, engineering and science are not the same, because they each play a different role in our technical society. Science seeks to explain how nature works through fundamental investigations of matter and energy. The objective of engineering is more pragmatic. Engineering, using science and mathematics as tools, seeks to design and build products and processes that enhance our standard of living. Generally, the scientific principles underlying the function of any engineering device were derived and established *before* the device was designed. For example, Newton's laws of motion and Kepler's orbital laws were well established scientific principles long before spacecraft orbited the earth or the other planets. Despite their contrasting objectives, both engineering and science employ tried-and-true methodologies that enable people working in each field to solve a variety of problems. To do science, the scientist must know how to employ the scientific method. To do engineering, the engineer must know how to employ the "engineering method."

Engineering analysis is the solution of an engineering problem by using mathematics and principles of science. Because of the close association between analysis and design, analysis is one of the key steps in the design process. Analysis also plays a major role in the study of engineering failures. The engineering method for conducting an analysis is a logical, systematic procedure characterized by a well-defined format. This procedure, when consistently and correctly applied, leads to the successful solution of any analytical engineering problem. Practicing engineers have been using this analysis procedure successfully for decades, and engineering graduates are expected to know how to apply it upon entering the technical workforce. Therefore, it behooves the engineering student to learn the analysis methodology as thoroughly as possible. The best way to do so is to practice solving analytical problems. As you advance in your engineering course work, you will have ample opportunities to apply the analysis methodology outlined in this chapter.

Courses such as statics, dynamics, mechanics of materials, thermodynamics, fluid mechanics, heat and mass transfer, electrical circuits, and engineering economics are analysis intensive. These courses, and others like them, focus almost exclusively on solving engineering problems that are analytical in nature. That is the character of these engineering subjects. The analysis methodology presented here is a *general* procedure that can be used to solve problems in any analytical subject. Clearly, engineering analysis heavily involves the use of numerical calculations.

3.2 NUMERICAL CALCULATIONS

As a college student, you are well aware of the rich diversity of academic programs and courses offered at institutions of higher learning. Because you are an engineering major, you are perhaps more familiar with the genre of engineering, science, and mathematics courses than with liberal arts courses such as sociology, philosophy, psychology, music, and languages. The tenor of liberal arts is vastly different from that of engineering. Suppose for a moment that you are enrolled in a literature class, studying Herman Melville's great book *Moby Dick.* While discussing the relationship between the whale and Captain Ahab, your literature professor asks the class, "What is your impression of Captain Ahab's attitude toward the whale?" As an engineering major, you are struck by the apparent looseness of this question. You are accustomed to answering questions that require a quantitative answer, not an "impression." What would engineering be like if our answers were "impressions"? Imagine an engineering professor asking a thermodynamics class, "What is your impression of the superheated steam temperature at the inlet of the turbine?" A more appropriate question would be, "What *is* the superheated steam temperature at the inlet of the turbine?"

Obviously, literature and the other liberal arts disciplines operate in a completely different mode from engineering. By its very nature, engineering is based on specific quantitative information. An answer of "hot" to the second thermodynamics question would be quantitative, but not specific and therefore insufficient. The temperature of the superheated steam at the inlet of the turbine could be calculated by conducting a thermodynamic analysis of the turbine, thereby providing a *specific* value for the temperature, 400°C for example. The analysis by which the temperature was obtained may consist of several numerical calculations involving different thermodynamic quantities. Numerical calculations are mathematical operations on numbers that represent physical quantities such as temperature, stress, voltage, mass, and flow rate. In this section, you will learn the proper numerical calculation techniques for engineering analysis.

3.2.1 Approximations

It is often useful, particularly during the early stages of design, to calculate an approximate answer to a given problem when the given information is uncertain or when little information is available. An approximation can be used to establish the cursory aspects of a design and to determine whether a more precise calculation is required. Approximations are usually based on assumptions, which must be modified or eliminated during the later stages of the design. Engineering approximations are sometimes referred to as "guesstimates," "ballpark calculations," or "back-of-the-envelope calculations." A more appropriate name for them is ***order-of-magnitude*** calculations. The term *order of magnitude* means a *power of* 10. Thus, an order-of-magnitude calculation refers to a calculation involving quantities whose numerical values are estimated to within a factor of 10. For example, if the estimate of a stress in a structure changes from about 1 kPa to about 1 MPa, we say that the stress has changed by three orders of magnitude, because 1 MPa is one thousand (10^3) times 1 kPa.

Engineers frequently conduct order-of-magnitude calculations to ascertain whether their initial design concepts are feasible. Order-of-magnitude calculations are therefore a useful decision-making tool in the design process. Order-of-magnitude calculations do not require the use of a calculator because all the quantities have simple power-of-10 values, so the arithmetic operations can be done by hand with pencil and paper or even in your head. The example that follows illustrates an order-of-magnitude calculation.

EXAMPLE 3.1

A warehouse with the approximate dimensions 200 ft × 150 ft × 20 ft is ventilated with 12 large industrial blowers. In order to maintain acceptable air quality in the warehouse, the blowers must provide two air changes per hour, meaning that the entire volume of air within the warehouse must be replenished with fresh outdoor air two times per hour. Using an order-of-magnitude analysis, find the required volume flow rate that each blower must deliver, assuming the blowers equally share the total flow rate.

Solution

To begin, we estimate the volume of the warehouse. The length, width, and height of the warehouse is 200 ft, 150 ft, and 20 ft, respectively. These lengths have order-of-magnitude values of 10^2, 10^2, and 10^1, respectively. Two air changes per hour are required. Thus, the total volume flow rate of air for the warehouse, including the factor of two air changes per hour, is:

$$Q_t \approx (10^2 \text{ ft})(10^2 \text{ ft})(10^1 \text{ ft})(2 \text{ air changes/h}) = 2 \times 10^5 \text{ ft}^3/\text{h}.$$

(Note that "air changes" is not a unit, so it does not appear in the answer.) The number of blowers (12) has an order-of-magnitude value of 10^1. Based on the assumption that each blower delivers the same flow rate, the flow rate per blower is the total volume flow rate divided by the number of blowers:

$$Q = Q_t/N = (2 \times 10^5 \text{ ft}^3/\text{h})(10^1 \text{ blowers})$$
$$= 2 \times 10^4 \text{ ft}^3/\text{h} \cdot \text{blower} \approx 10^4 \text{ ft}^3/\text{h} \cdot \text{blower}.$$

Our order-of-magnitude calculation shows that each blower must supply 10^4 ft^3/h of outdoor air to the warehouse.

How does our order-of-magnitude answer compare with the exact answer? The exact answer is:

$$Q = (200 \text{ ft})(150 \text{ ft})(20 \text{ ft})(2 \text{ air changes/h})/(12 \text{ blowers}) = 1 \times 10^5 \text{ ft}^3/\text{h} \cdot \text{blower}.$$

By dividing the exact answer by the approximate answer, we see that the approximate answer differs from the exact answer by a factor of 10, which shows that an order-of-magnitude analysis provides an acceptable approximate answer.

3.2.2 Significant Figures

After order-of-magnitude calculations have been made, engineers conduct more precise calculations to refine their design or to more fully characterize a particular failure mode. Accurate calculations demand more of the engineer than simply keeping track of powers of 10. Final design parameters must be determined with as much precision as possible to achieve the optimum design. Engineers must determine how many digits in their calculations are significant.

A ***significant figure*** or *significant digit* in a number is defined as *a digit that is considered reliable as a result of a measurement or calculation.* The number of significant figures in the answer of a calculation indicates the number of digits that can be used with confidence, thereby providing a way of telling the engineer how precise the answer is. No physical quantity can be specified with infinite precision because no physical quantity is *known* with infinite precision. Even the constants of nature such as the speed of light in a vacuum c, and the gravitational constant G, are known only to the precision with which they can be measured in a laboratory. Similarly, engineering material properties such as density, modulus of elasticity, and specific heat are known only to the precision with which these properties can be measured. A common mistake is to use more significant figures in an answer than are justified, giving the impression that the answer is more accurate than it really is. No answer can be more precise than the numbers used to generate that answer.

How do we determine how many significant figures (colloquially referred to as "sig figs") a number has? A set of rules has been established for counting the number of significant figures in a number. (All significant figures are underlined in the examples given for each rule.)

Rules for Significant Figures

1. All digits *other than zero* are significant. Examples: $\underline{8}.\underline{936}$, $\underline{456}$, $0.\underline{257}$.
2. All zeroes *between* significant figures are significant. Examples: $\underline{14}.\underline{06}$, $\underline{5}.\underline{0072}$.

3. For nondecimal numbers greater than one, all zeroes placed *after* the significant figures are *not* significant. Examples: 2500, 8,640,000. These numbers can be written in scientific notation as 2.5×10^3 and 8.64×10^6, respectively.
4. If a decimal point is used *after* a nondecimal number larger than one, the zeroes are significant. The decimal point establishes the precision of the number. Examples: 3200., 550,000.
5. Zeroes placed *after* a decimal point that are *not necessary* to set the decimal point are significant. The additional zeroes establish the precision of the number. Examples: 359.00, 1000.00.
6. For numbers smaller than one, all zeroes placed *before* the significant figures are *not* significant. These zeroes only serve to establish the location of the decimal point. Examples: 0.0254, 0.000609.

Do not confuse the number of significant figures with the number of decimal places in a number. The number of significant figures in a quantity is established by the precision with which a measurement of that quantity can be made. The primary exception to this are numbers such as π and the Naperian base e, which are derived from mathematical relations. These numbers are precise to an infinite number of significant figures but can be quite adequately approximated by 10-digit decimals.

Let's see how the rules for significant figures are used in calculations.

EXAMPLE 3.2

We wish to calculate the weight of a 25-kg object. Using Newton's second law $W = mg$, find the weight of the object in units of N. Express the answer, by using the appropriate number of significant figures.

Solution

We have $m = 25$ kg and $g = 9.81$ m/s^2. Suppose that our calculator is set to display six places to the right of the decimal point. We then multiply the numbers 25 and 9.81. In the display of the calculator, we see the number 245.250000. How many digits in this answer are we justified in writing? The number in the calculator's display implies that the answer is precise to six decimal places (i.e., to within one-millionth of a newton). Obviously, this kind of precision is not justified. The rule for significant figures for *multiplication* and *division* is that *the product or quotient should contain the number of significant figures that are contained in the number with the fewest significant figures.*

Another way to state this rule is to say that the quantity with the fewest number of significant figures *governs* the number of significant figures in the answer. The mass m contains two significant figures, and the acceleration of gravity g contains three. Therefore, we are only justified in writing the weight by using two significant figures, which is the fewest number of significant figures in our given values. Our answer can be written in two ways. First, we can write the weight as 250 N. According to rule 3, the zero is not significant, so our answer contains two significant figures, the "2" and the "5." Second, we can write the weight by using scientific notation as 2.5×10^2 N. In this form, we can immediately see that two significant figures are used without referring to the rules. Note that in both cases we *rounded* the answer *up* to the nearest tens place, because the value of the first digit dropped is 5 or greater. If our answer had been lower than 245 N, we would have rounded *down* to 240 N. If our answer had been precisely 245 N, the rules of rounding suggest rounding up, so our answer would again be 250 N.

The preceding example shows how significant figures are used for multiplication or division, but how are significant figures used for *addition* and *subtraction*?

EXAMPLE 3.3

Two collinear forces (forces that act in the same direction) of 875.4 N and 9.386 N act on a body. Add these two forces, expressing the result in the appropriate number of significant figures.

Solution

The best way to show how significant figures are used in addition or subtraction is to do the problem by hand. We have:

$$\begin{array}{r} 875.4\ \ \ \ \text{N} \\ +\ \ 9.386\ \text{N} \\ \hline 884.786\ \text{N} \end{array}$$

Both forces have four significant figures, but the first force reports one place past the decimal point, whereas the second force reports three places past the decimal point. The answer is written with six significant figures. Are six significant figures justified? Because addition and subtraction are arithmetic operations that require decimal point alignment, the rule for significant figures for *addition* and *subtraction* is different than for multiplication and division. For addition and subtraction, the answer should show *significant figures only as far to the right as is seen in the least precise number in the calculation.* The least precise number in the calculation is the 875.4-N force, because it reports accuracy to the first decimal place, whereas the second force 9.386 N reports precision to the third decimal place. We are not justified in writing the answer as 884.786 N. We may only write the answer by using the same number of places past the decimal point as seen in the least precise force. Hence, our answer, reported to the appropriate number of significant figures, is 884.8 N. Once again, we rounded the answer up because the value of the first digit dropped is 5 or greater.

In *combined* operations where multiplication and division are performed in the same operation as addition and subtraction, the multiplications and divisions should be performed first, establishing the proper number of significant figures in the intermediate answers, perform the additions and subtractions, and then round the answer to the proper number of significant figures. This procedure, while applicable to operations performed by hand, should not be used in calculator or computer applications, because intermediate rounding is cumbersome and may lead to a serious error in the answer. Perform the entire calculation, letting the calculator or computer software manage the numerical precision, and then express the final answer in the desired number of significant figures:

> *It is standard engineering practice to express final answers in three (or sometimes four) significant figures, because the given input values for geometry, loads, material properties, and other quantities are typically reported with this precision.*

Calculators and computer software such as spreadsheets and equation solvers keep track of and can display a large number of digits. How many digits will your calculator display? The number of digits displayed by a scientific calculator can be set by fixing the decimal point or specifying the numerical format. For example, by fixing the number of decimal places to one, the number 28.739 is displayed as 28.7. Similarly, the number 1.164 is displayed as 1.2. Because the first digit dropped is greater than 5, the calculator automatically rounds the answer up. Small and large numbers should be expressed in scientific notation. For example, the number 68,400 should be expressed as 6.84×10^4, and the number 0.0000359 should be expressed as 3.59×10^{-5}. Scientific calculators also have an *engineering notation* display setting because SI unit prefixes are primarily defined by multiples of one thousand (10^3). In engineering notation, the number 68,400 may be displayed as 68.4×10^3, and the number 0.0000359 may be displayed as 35.9×10^{-6}. Regardless of how numbers are displayed by calculators or computers, the engineering student who uses these computational tools must understand that significant figures have a physical meaning based on our ability to measure engineering and scientific quantities. The casual or sloppy handling of significant figures in engineering analysis may lead to solutions that are imprecise at best and completely wrong at worst.

APPLICATION

CALCULATING VISCOSITY BY USING THE FALLING-SPHERE METHOD

You know by experience that some fluids are thicker or more "gooey" than others. For example, pancake syrup and motor oil are thicker than water and alcohol. The technical term we use to describe the magnitude of a fluid's thickness is *viscosity*. Viscosity is a fluid property that characterizes the fluid's resistance to flow. Water and alcohol flow more readily than pancake syrup and motor oil under the same conditions. Hence, pancake syrup and motor oil are more viscous than water and alcohol. Gases have viscosities too, but their viscosities are much smaller than those of liquids.

One of the classical techniques for measuring viscosities of liquids is called the *falling-sphere method*. In the falling-sphere method, the viscosity of a liquid is calculated by measuring the time it takes for a small sphere to fall a prescribed distance in a large container of the liquid, as illustrated in Figure 3.1. As the sphere falls in the liquid under the influence of gravity, it accelerates until the downward force (the sphere's weight) is exactly balanced by the buoyancy force and drag force that act upward. From this time forward, the sphere falls with a constant velocity, referred to as terminal velocity. The buoyancy force, which is equal to the weight of the liquid allowed for displaced by the sphere, is usually small compared with the drag force, which is caused directly by viscosity. The terminal velocity of the sphere is inversely proportional to viscosity, since the sphere takes longer to fall a given

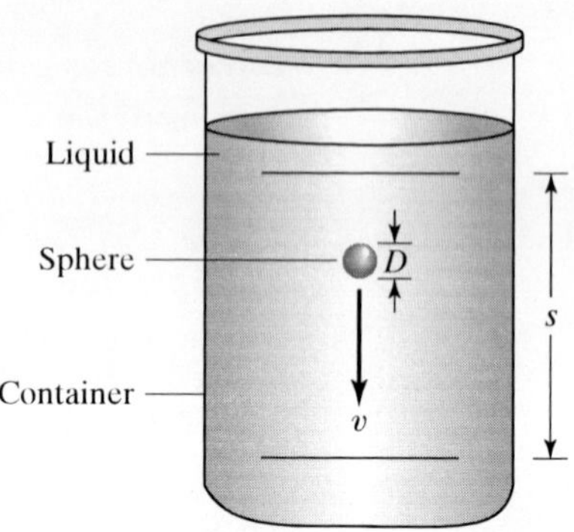

Figure 3.1
Experimental setup of the falling-sphere method for measuring viscosity.

distance in a very viscous liquid, such as motor oil, than in a less viscous liquid, such as water. By employing a force balance on the sphere and invoking some simple relations from fluid mechanics, we obtain the formula

$$\mu = \frac{(\gamma_s - \gamma_f)D^2}{18v}$$

where

μ = dynamic viscosity of liquid (Pa · s)
γ_s = specific weight of sphere (N/m^3)
γ_f = specific weight of liquid (N/m^3)
D = sphere diameter (m)
v = terminal velocity of sphere (m/s).

Note that the quantity *specific weight* is similar to *density*, except that it is a weight per volume, rather than a mass per volume. The word *dynamic* is used to avoid confusion with another measure of viscosity known as *kinematic* viscosity.

Using the falling-sphere method, let's calculate the viscosity of glycerine, a very viscous liquid used to make a variety of chemicals. We set up a large glass cylinder and place two marks, spaced $s = 200$ mm apart, on the outside surface. The marks are placed low enough on the cylinder to assure that the sphere will achieve terminal velocity before reaching the top mark. For the sphere, we use a steel ($\gamma_s = 76,800\ N/m^3$) ball bearing with a diameter of 2.381 mm (measured with a micrometer). From a previous measurement, the specific weight of the glycerin is $\gamma_f = 12,400\ N/m^3$. Now, we hold the steel sphere above the surface of the glycerin at the center of the cylinder and release the sphere. As accurately as we can determine with our eye, we start a handheld stopwatch when a part of the sphere reaches the top mark. Similarly, we stop the watch when the same part of sphere reaches the bottom mark. Our stopwatch is capable of displaying hundredths of a second, and it reads 11.32 s. Even though the stopwatch is capable of measuring time to the second decimal place, our crude visual timing method does not justify using a time interval with this precision. Sources of uncertainty such as human reaction time and thumb response do not justify the second decimal place. Thus, our time interval is reported as 11.3 s, which has three significant figures. We know that terminal velocity is distance divided by time:

$$v = \frac{s}{t} = \frac{0.200\ \text{m}}{11.3\ \text{s}} = 0.0177\ \text{m/s}$$

The distance was measured to the nearest millimeter, so the quantity s has three significant figures. Thus, terminal velocity may be written to three significant figures. (Remember that the zero, according to rule 6, is not significant.) Values of the given quantities for our calculation are summarized as follows:

$$\gamma_s = 76,800 \text{ N/m}^3 = 7.68 \times 10^4 \text{ N/m}^3$$
$$\gamma_f = 12,400 \text{ N/m}^3 = 1.24 \times 10^4 \text{ N/m}^3$$
$$v = 0.0177 \text{ m/s} = 1.77 \times 10^{-2} \text{ m/s}$$
$$D = 2.381 \text{ mm} = 2.381 \times 10^{-3} \text{ m}.$$

Each quantity, with the exception of D, which has four significant figures, has three significant figures. Upon substituting values into the equation for dynamic viscosity, we obtain:

$$\begin{aligned}\mu &= \frac{(\gamma_s - \gamma_f)D^2}{18v}\\ &= \frac{(76{,}800 - 12{,}400)\text{ N/m}^3\,(2.381 \times 10^{-3}\text{ m})^2}{18(0.0177\text{ m/s})}\\ &= 1.1459 \text{ Pa}\cdot\text{s}.\end{aligned}$$

(Where did the pressure unit Pa come from?) According to the rules of significant figures for multiplication and division, our answer should contain the same number of significant figures as the number with the fewest significant figures. Our answer should therefore have three significant figures, so the dynamic viscosity of glycerin, expressed in the proper number of significant figures is reported as:

$$\mu = 1.15 \text{ Pa}\cdot\text{s}.$$

Note that because the value of the first digit dropped is 5, we rounded our answer up.

PROFESSIONAL SUCCESS—LEARN HOW TO USE YOUR CALCULATOR

As an engineering student, you need a scientific calculator. If you do not yet own a quality scientific calculator, purchase one as soon as you can and begin learning how to use it. You cannot succeed in school without one. You will probably only need one calculator for your entire academic career, so purchase one that offers the greatest number of functions and features. Professors and fellow students may offer advice on which calculator to buy. Your particular engineering department or college may even require that you use a particular calculator because they have heavily integrated calculator usage in the curriculum, and it would be too cumbersome to accommodate several types of calculators. Your college bookstore or local office supply store may carry two or three name brands that have served engineering students and professionals for many years. Today's scientific calculators are remarkable engineering tools. A high-end scientific calculator has hundreds of built-in functions, large storage capacity graphics capabilities, and communication links to other calculators or personal computers.

Regardless of which scientific calculator you own or plan to purchase, *learn how to use it.* Begin with the basic arithmetic operations and the standard mathematical and statistical functions. Learn how to set the number of decimal places in the display and how to display numbers in scientific and engineering notation. After you are confident with performing unit conversions by hand, learn how to do them with your calculator. Learn how to write simple programs on your calculator. This skill will come in handy numerous times throughout your course work. Learn how to use the equation-solving functions, matrix operations and calculus routines. By the time you learn most of the calculator's operations, you will probably have devoted many hours. The time spent mastering your calculator is perhaps as valuable as the time spent attending lectures, conducting experiments in a laboratory, doing homework problems, or studying for exams. Knowing your calculator thoroughly will help you succeed in your engineering program. Your engineering courses will be challenging enough. Do not make them an even bigger challenge by failing to adequately learn how to use your principal computational asset, your calculator.

PRACTICE!

1. Using an order-of-magnitude analysis, estimate the surface area of your state or country in units of cm^2.
2. Using an order-of-magnitude analysis, estimate the number of hairs on your head.
3. Use an order-of-magnitude analysis to estimate the number of cell phones in use in the world.
4. Use an order-of-magnitude analysis to estimate the electrical energy in kWh used by your city in one month.
5. Underline the significant figures in the following numbers (the first number is done for you):
   a. 0.000369
   b. 42.07
   c. 9001
   d. 403.50
   e. 0.0330
   f. 700
   Answer: b. 42.07 c. 9001 d. 403.50 e. 0.0330 f. 700
6. Perform the following calculations, reporting the answers with the correct number of significant figures:
   a. $5.64/1.9$
   b. $500./0.0025$
   c. $(45.8 - 8.1)/1.922$
   d. $2\pi/2.50$
   e. $(5.25 \times 10^4)/(100 + 10.5)$
   f. $0.0008/(1.2 \times 10^{-5})$.
   Answer: a. 3.0 b. 2.0×10^5 c. 19.6 d. 2.51 e. 500 f. 7×10^1.

7. A ball bearing is reported to have a radius of 3.256 mm. Using the correct number of significant figures, what is the weight of this bearing in units of N if its density is $\rho = 1675\ \text{kg/m}^3$? Use $g = 9.81\ \text{m/s}^2$.
Answer: 2.38×10^{-3} N.
8. The cylinder of an internal combustion engine is reported to have a diameter of 4.000 in. If the stroke (length) of the cylinder is 6.25 in, what is the volume of the cylinder in units of in^3? Write the answer using the correct number of significant figures.
Answer: $78.5\ \text{in}^3$

3.3 GENERAL ANALYSIS PROCEDURE

Engineers are problem solvers. In order to solve an engineering analysis problem thoroughly and accurately, engineers employ a solution method that is systematic, logical, and orderly. This method, when consistently and correctly applied, leads the engineer to a successful solution of the analytical problem at hand. The problem-solving method is an integral part of a good engineer's thought process. To the engineer, the procedure is second nature. When challenged by a new analysis, a good engineer knows precisely how to approach the problem. The problem may be fairly short and simple or extremely long and complex. Regardless of the size or complexity of the problem, the same solution method applies. Because of the *general* nature of the procedure, it applies to analytical problems associated with *any* engineering discipline: chemical, civil, electrical, mechanical, or other.

Practicing engineers in all disciplines have been using the ***general analysis procedure*** in one form or another for a long time, and the history of engineering achievements is a testament to its success. While you are a student, it is vitally important that you learn the steps of the general analysis procedure. After you have learned the steps in the procedure and feel confident that you can use the procedure to solve problems, apply it in your analytical course work. Practice the procedure over and over again until it becomes a habit. Establishing good habits while still in school will make it that much easier for you to make a successful transition into professional engineering practice.

General Analysis Procedure

The general analysis procedure consists of the following seven steps:

1. **Problem Statement** The problem statement is a written description of the analytical problem to be solved. It should be written clearly, concisely, and logically. The problem statement summarizes the given information, including all input data provided to solve the problem. The problem statement also states what is to be determined by performing the analysis.
2. **Diagram** The diagram is a sketch, drawing, or schematic of the system being analyzed. Typically, it is a simplified pictorial representation of the actual system, showing only those aspects of the system that are necessary to perform the analysis. The diagram should show all given information contained in the problem statement such as geometry, applied forces, energy flows, mass flows, electrical currents, temperatures, or other physical quantities as required.
3. **Assumptions** Engineering analysis almost always involves some assumptions. Assumptions are special assertions about the physical characteristics of the

problem that simplify or refine the analysis. A very complex analytical problem would be difficult or even impossible to solve without making some assumptions.

4. **Governing Equations** All physical systems may be described by mathematical relations. Governing equations are those mathematical relations that specifically pertain to the physical system being analyzed. These equations may represent physical laws, such as Newton's laws of motion, conservation of mass, conservation of energy, and Ohm's law; or they may represent fundamental engineering definitions such as velocity, stress, moment of force, and heat flux. The equations may also be basic mathematical or geometrical formulas involving angles, lines, areas, and volumes.
5. **Calculations** In this step, the solution is generated. First, the solution is developed algebraically as far as possible. Then numerical values of known physical quantities are substituted for the corresponding algebraic variables. All necessary calculations are performed, using a calculator or computer to produce a numerical result with the correct units and the proper number of significant figures.
6. **Solution Check** This step is crucial. Immediately after obtaining the result, examine it carefully. Using established knowledge of similar analytical solutions and common sense, try to ascertain whether the result is reasonable. However, whether the result seems reasonable or not, double-check every step of the analysis. Flush out defective diagrams, bad assumptions, erroneously applied equations, incorrect numerical manipulations, and improper use of units.
7. **Discussion** After the solution has been thoroughly checked and corrected, discuss the result. The discussion may include an assessment of the assumptions, a summary of the main conclusions, a proposal on how the result may be verified experimentally in a laboratory, or a parametric study demonstrating the sensitivity of the result to a range of input parameters.

Now that the seven-step procedure has been summarized, further discussion of each step is warranted.

1. Problem statement In your engineering textbook, the problem statement will generally be supplied to you in the form of a problem or question at the end of each chapter. These problem statements are written by the textbook authors, professors or practicing engineers, who have expertise in the subject area. The great majority of end-of-chapter problems in engineering texts are well organized and well written, so you do not have to fret too much about the problem statement. Alternatively, your engineering professor may give you problem statements from sources outside your textbook or from his or her own engineering experience. In either case, the problem statement should be well posed, contain all the necessary input information, and clearly state what is to be determined by the analysis. What is known and what is unknown in the problem should be clearly identified. If the problem statement is flawed in any way, a meaningful analysis is difficult or even impossible.

2. Diagram The old saying, "One picture is worth a thousand words," is certainly applicable to engineering analysis. A complete diagram of the system being analyzed is critical. A good diagram helps the engineer visualize the physical processes or characteristics of the system. It also helps the engineer identify reasonable assumptions and the appropriate governing equations. A diagram might even reveal flaws in the problem statement or alternative methods of solution. Engineers use a variety of diagrams in their analytical work.

One of the most widely used diagrams in engineering is the *free-body diagram*. Free-body diagrams are used to solve engineering mechanics (statics, dynamics, mechanics of materials) problems. These diagrams are called "free-body" diagrams because they represent a specific body, isolated from all other bodies that are in physical contact with, or that may be in the vicinity of, the body in question. The influences of nearby bodies are represented as external forces acting on the body being analyzed. Hence, a free-body diagram is a sketch of the body in question, showing all external forces applied to the body. A free-body diagram is a pictorial representation of a "force balance" on the body. Diagrams are also used in the analysis of thermal systems.

Unlike a free-body diagram, which shows forces applied to the body, a diagram of a thermal system shows all the various forms of energy entering and leaving the system. This type of diagram is a pictorial representation of an "energy balance" on the system. Another type of diagram represents a system that transports mass at known rates. Common examples include pipe and duct systems, conveyors, and storage systems. A diagram for these systems shows all the mass entering and leaving the system. This type of diagram is a pictorial representation of a "mass balance" on the system. Still another type of diagram is an electrical circuit schematic. Electrical schematics show how components are connected and the currents, voltages, and other electrical quantities in the circuit. Some examples of diagrams used in analysis are given in Figure 3.2.

3. Assumptions An "atmospheric scientist" who studied various processes that occur in the upper atmosphere once gave a lecture and recounted an accomplishment that seemed truly remarkable. After convincing the audience that atmospheric processes are some of the most complex phenomena in physics, he boasted that he had developed, over the space of a few months, an analytical model of the upper atmosphere that contained *no* assumptions. There was only one problem: his model

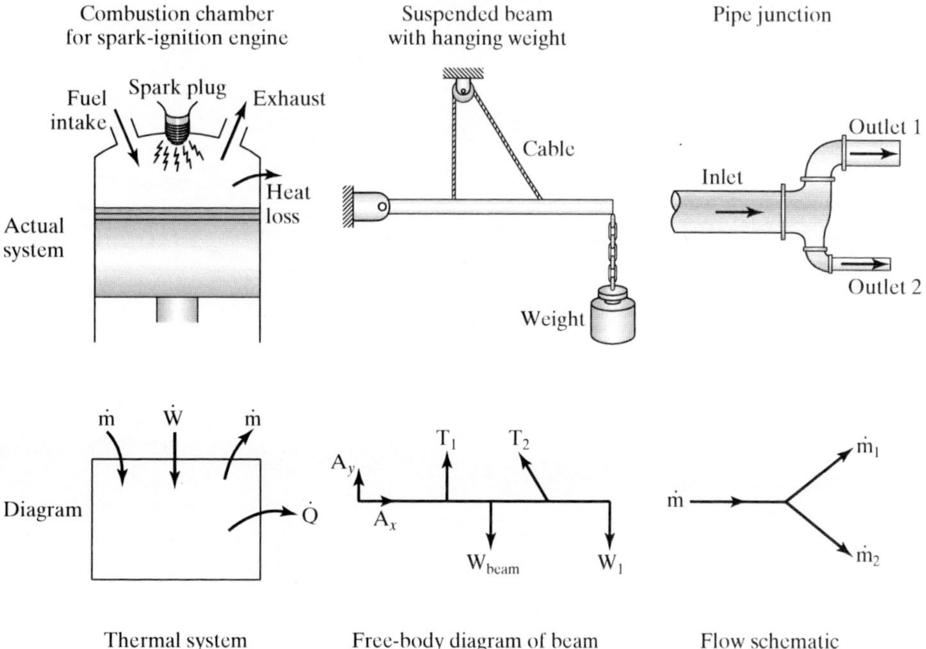

Figure 3.2 Examples of typical diagrams used in engineering analysis.

had no solution either. By including every physical mechanism to the minutest detail in the model, his analysis was so mathematically convoluted that it could not generate a solution. Had he made some simplifying assumptions, his atmospheric model could have worked even though the results would have been approximate.

Engineers and scientists routinely employ assumptions to simplify a problem. As this story illustrates, an approximate answer is better than no answer at all. Failure to invoke one or more simplifying assumptions in the analysis, particularly a complex one, can increase the complexity of the problem by an order of magnitude, leading the engineer down a very long road, only to reach a dead end. How do we determine which assumptions to use and whether our assumptions are good or bad? To a large extent, the application of good assumptions is an acquired skill, a skill that comes with engineering experience. However, you can begin to learn this skill in school through repeated application of the general analysis procedure in your engineering courses. As you apply the procedure to a variety of engineering problems, you will gain a basic understanding of how assumptions are used in engineering analysis. Then, after you graduate and accept a position with an engineering firm, you can refine this skill as you apply the analysis procedure to solve problems that are specific to the company. Sometimes, a problem can be overly constrained by assumptions such that the problem is simplified to the point where it becomes grossly inaccurate or even meaningless. The engineer must therefore be able to apply the proper *number* as well as the proper *type* of assumptions in a given analysis. A common assumption made in the stress analysis of a column is shown in Figure 3.3.

4. Governing equations The governing equations are the "workhorses" of the analysis and describe the physical problem at hand. If the wrong governing equations are used, the analysis may lead to a result that does not reflect the true physical nature of the problem, or the analysis will not be possible at all because the governing equations are not in harmony with the problem statement or assumptions. When using a governing equation to a solve a problem, the engineer must ascertain that the equation being used *actually* applies to the specific problem at hand. As an extreme (and probably absurd) example, imagine an engineer attempting to use

Figure 3.3
A common assumption made in the stress analysis of a column.

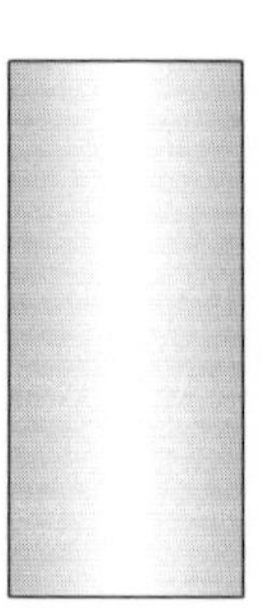

A column

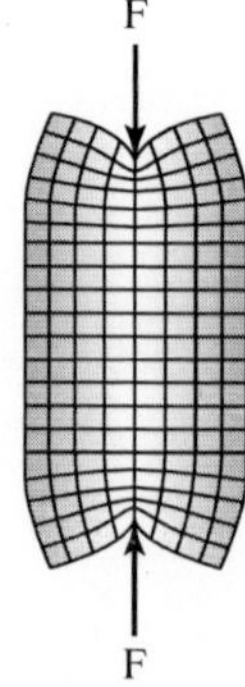

When a concentrated force is applied to the column, stresses are concentrated near the points of application, but the stresses far away from the ends are nearly uniform.

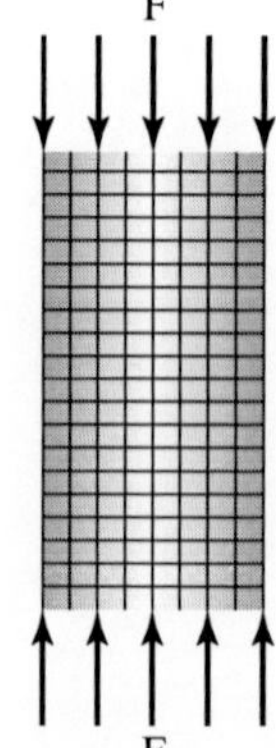

To simplify the stress analysis, the concentrated force is assumed to be uniformly distributed, thereby producing a uniform stress in all regions of the column.

Newton's second law $F = ma$ to calculate the heat loss from a boiler. How about trying to apply Ohm's law $V = IR$ to find the stress in a concrete column that supports a bridge deck?

The problem of matching governing equations to the problem at hand is usually more subtle than these absurd examples. In thermodynamics, for example, the engineer must determine whether the thermal system is "closed" or "open" (i.e., whether the system allows mass to cross the system boundary). After the type of thermal system has been identified, the thermodynamic equations which apply to that type of system are chosen, and the analysis proceeds. Governing equations must also be consistent with the assumptions. It is counterproductive to invoke simplifying assumptions if the governing equations do not make allowances for them. Some governing equations, particularly those that are experimentally derived, have built-in restrictions that limit the use of the equations to specific numerical values of key variables. A common mistake made in the application of a governing equation in this situation is failing to recognize the restrictions by forcing the equation to accept numerical values that lie outside the equation's range of applicability.

5. Calculations A common practice, particularly among beginning students, is to substitute numerical values of quantities into equations *too early* in the calculations. It seems that some students are more comfortable working with *numbers* than *algebraic variables*, so their first impulse is to substitute numerical values for all parameters at the beginning of the calculation. Avoid this impulse. To the extent that it is practical, develop the solution *analytically* prior to assigning physical quantities to their numerical values. Before rushing to "plug" numbers into equations, carefully examine the equations to see if they can be mathematically manipulated to yield simpler expressions. A variable from one equation can often be substituted into another equation to reduce the total number of variables. Perhaps an expression can be simplified by factoring. By developing the solution analytically first, you might uncover certain physical characteristics about the system or even make the problem easier to solve. The analytical skills you learned in your algebra, trigonometry, and calculus courses are meant to be used for performing mathematical operations on *symbolic* quantities, not numbers.

The calculations step demands more of an engineer than the ability to simply "crunch numbers" on a calculator or computer. The numbers have to be meaningful, and the equations containing the numbers must be fully understood and properly used. All mathematical relations must be dimensionally consistent, and all physical quantities must have a numerical value plus the correct units. Here is a tip concerning units that will save you time and help you avoid mistakes: *if the quantities given in the problem statement are not expressed in terms of a consistent set of units, convert all quantities to a consistent set of units before performing any calculations.* If some of the input parameters are expressed as a mixture of SI units and English units, convert all parameters to either SI units or English units, and then perform the calculations. Students tend to make more mistakes when they attempt to perform unit conversions *within* the governing equations. If all unit conversions are done prior to substituting numerical values into the equations, unit consistency is assured throughout the remainder of the calculations, because a consistent set of units is established at the onset. Dimensional consistency should still be verified, however, by substituting all quantities along with their units into the governing equations.

6. Solution check This step is perhaps the easiest one to overlook. Even good engineers sometimes neglect to thoroughly check their solution. The solution may "look" good at first glance, but a mere glance is not good enough. Much effort has gone into formulating the problem statement, constructing diagrams of the system, determining the appropriate number and type of assumptions, invoking governing equations, and performing a sequence of calculations. All this work may be for naught if the solution is not carefully checked. Checking the solution of an engineering analysis is analogous to checking the operation of an automobile immediately following a major repair. It's always a good idea if the mechanic checks to verify that it works before returning the vehicle to its owner.

There are two main aspects of the solution check. First, the result itself should be checked. Ask the question, "Is this result reasonable?" There are several ways to answer this question. The result must be consistent with the information given in the problem statement. For example, suppose you wish to calculate the temperature of a microprocessor chip in a computer. In the problem statement, the ambient air temperature is given as 25°C, but your analysis indicates that the chip temperature is only 20°C. This result is not consistent with the given information because it is physically impossible for a heat-producing component, a microprocessor chip in this case, to have a lower temperature than the surrounding environment. If the answer had been 60°C, it is at least consistent with the problem statement, but it may still be incorrect. Another way to check the result is to compare it with that of similar analysis performed by you or other engineers. If the result of a similar analysis is not available, an alternative analysis that utilizes a different solution approach may have to be conducted. In some cases, a laboratory test may be needed to verify the solution experimentally. Testing is a normal part of engineering design anyway, so a test to verify an analytical result may be customary.

The second aspect of the solution check is a thorough inspection and review of each step of the analysis. Returning to our microprocessor example, if no mathematical or numerical errors are committed, the answer of 60°C may be considered correct insofar as the calculations are concerned, but the answer could still be in error due to bad assumptions. For example, suppose that the microprocessor chip is air cooled by a small fan, so we assert that forced convection is the dominant mechanism by which heat is transferred from the chip. Accordingly, we assume that conduction and radiation heat transfer are negligible, so we do not include these mechanisms in the analysis. A temperature of 60°C seems a little high, so we revise our assumptions. A second analysis that includes conduction and radiation reveals that the microprocessor chip is much cooler, about 42°C. Knowing whether assumptions are good or bad comes through increased knowledge of physical processes and practical engineering experience.

7. Discussion This step is valuable from the standpoint of communicating to others what the results of the analysis mean. By discussing the analysis, you are in effect writing a "mini-technical report." This report summarizes the major conclusions of the analysis. In the microprocessor example given earlier, the main conclusion may be that 42°C is below the recommended operating temperature for the chip and therefore, the chip will operate reliably in the computer for a minimum of 10,000 hours before failing. If the chip temperature was actually measured at 45°C shortly after performing the analysis, the discussion might include an examination of why the predicted and measured temperatures differ and particularly why the predicted temperature is lower than the measured temperature. A brief parametric study may be included that shows how the chip temperature varies as a function of ambient air temperature. The discussion may even include an entirely separate analysis that predicts the chip temperature in the event of a fan failure. In the discussion step, the engineer is given one last opportunity to gain additional insights into the problem.

PROFESSIONAL SUCCESS—REAL-WORLD PROBLEM STATEMENTS

Engineering programs strive to give students a sense of what it is like to actually practice engineering in the "real world." But *studying* engineering in school and *practicing* engineering in the real world are not the same thing. One difference is amply illustrated by considering the origins of problem statements for analysis. In school, problem statements are typically found at the end of each chapter of your engineering texts. (The answers to many of these problems are even provided at the back of the book.) Sometimes your professors obtain problem statements from other texts or invent new ones (especially for exams). In any case, problem statements are supplied to you in a nice, neat little package all ready for you to tackle the problem.

If textbooks and professors supply problem statements to students in school, who or what supplies problem statements to practicing engineers in industry? Real-world engineering problems are not typically found in textbooks (answers are never found in the back of the book either), and your engineering professors are not going to follow you around after you graduate. So, where do the real-world problem statements come from? They are *formulated* by the engineer who is going to perform the analysis. As stated before, analysis is an integral part of engineering design. As a design matures, quantitative parameters that characterize the design begin to emerge. When an analysis is called for, these parameters are woven into a problem statement from which an analysis may be conducted. The engineer must be able to formulate a coherent, logical problem statement from the design information available. Because engineering design is an intuitive process, the values of some or all of the input parameters may be uncertain. The engineer must therefore be able to write the problem statement in such a way as to allow for these uncertainties. The analysis will have to be repeated several times until the parameters are no longer in a state of flux, at which time the design is complete.

The seven-step procedure for performing an engineering analysis is a time-tested method. In order to effectively communicate an analysis to others, the analysis must be presented in a format that can be readily understood and followed. Engineers are known for their ability to present analyses and other technical information with clarity in a thorough, neat, and careful manner. As an engineering student, you can begin to develop this ability by consistently applying the analysis procedure outlined in this section. Your engineering professors will insist that you follow the procedure, or a procedure similar to it, in your engineering courses. You will probably be graded not only on how well you perform the analysis itself, but how well you *present* the analysis on paper. This grading practice is meant to convince students of the importance of presentation standards in engineering and to assist them in developing good presentation skills. An engineering analysis is of little value to anyone unless it can be read and understood. A good analysis is one that can be easily read and understood by others. Apply the presentation guidelines given in this section to the point where they become second nature. Then, after you graduate and begin practicing engineering, you can hone your presentation skills as you gain industrial experience.

The 10 guidelines that follow will help you present an engineering analysis in a clear and complete manner. These guidelines are applicable to analysis work in school as well as industrial engineering practice. It should be noted that the guidelines apply specifically to analyses performed by hand with the use of pencil and paper, as opposed to computer-generated analyses.

Analysis Presentation Guidelines

1. A standard practice of engineers who do analysis is to use a special type of *paper.* This paper is usually referred to as "engineer calculation pad" or "engineer's computation paper." The paper is light green in color, and should be available in your college or university bookstore. The back side of the paper is ruled horizontally and vertically with five squares per inch, with only heading and margin rulings on the front side. The rulings on the back side are faintly visible through the paper to help the engineer maintain the proper position and orientation for lettering, diagrams, and graphs. (See Figure 3.4.) All work is to be done on the *front* side of the paper. The back side is not used. The paper usually comes pre-punched with a standard three-hole pattern at the left edge for placement in a three-ring binder.
2. No more than *one* problem should be placed on a page. This practice helps maintain clarity by keeping different problems separate. Even if a problem occupies a small fraction of a page, the next problem should be started on a separate page.
3. The *heading* area at the top of the page should indicate your name, date, course number, and assignment number. The upper right corner of the heading area is usually reserved for page numbers. To alert the reader to the total number of pages present, page numbers are often reported, for example, as

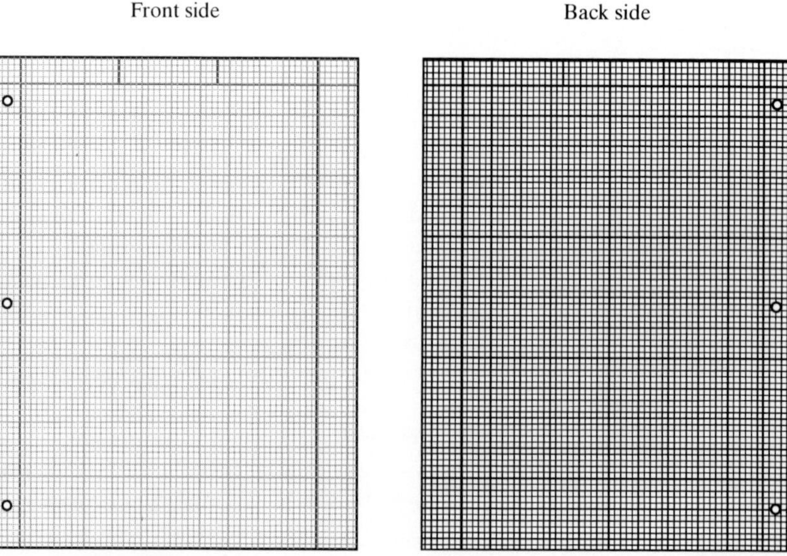

Figure 3.4
Engineer's computation paper is standard issue for analysis work.

"1/3," which is read as "page 1 of 3." Page 1 is the current page, and there are a total of three pages. When multiple pages are used, they should be stapled in the upper left corner. Each page should nonetheless be identified with your name, in the unlikely event the pages become separated.

4. The *problem statement* should be written out completely, not summarized or condensed. All figures that accompany the problem statement should be shown. If the problem statement originates from a textbook, it should be written *verbatim* so the reader does not have to refer back to the textbook for the full version. One way to do this is to photocopy the problem statement, along with any figures given, and then cut and attach it by using rubber cement or transparent tape directly beneath the heading area on the engineer's computation paper. The problem statement could also be electronically scanned and printed directly onto the paper.
5. Work should be done in *pencil*, not ink. Everyone makes mistakes. If the analysis is written in pencil, mistakes can be easily erased and corrected. If the analysis is written in ink, mistakes will have to be crossed out, and the presentation will not have a neat appearance. To avoid smudges, use a pencil lead with the appropriate hardness. All markings should be dark enough to reproduce a legible copy if photocopies are needed.
6. Lettering should be *printed*. The lettering style should be consistent throughout.
7. Correct *spelling* and *grammar* must be used. Even if the technical aspects of the presentation are flawless, the engineer will lose some credibility if the writing is poor.
8. There are seven steps in the general analysis procedure. These steps should be sufficiently *spaced* so that the reader can easily follow the analysis from problem statement to discussion. A horizontal line drawn across the page is one way of providing this separation.
9. Good *diagrams* are a must. A straight edge, drawing templates, and other manual drafting tools should be used. All pertinent quantitative information such as geometry, forces, energy flows, mass flows, electrical currents, and pressures, should be shown on the diagrams.
10. Answers should be *double underlined* or *boxed* for ready identification. To enhance the effect, colored pencils may be used.

These 10 guidelines for analysis presentation are recommended to the engineering student. You may find that your particular engineering department or professors may advocate guidelines that are slightly different. By all means, follow the guidelines given to you. Your professors may have special reasons for teaching their students certain methods of analysis presentation. Methods may vary somewhat from course to course and professor to professor, but should still reflect the major points contained in the guidelines given in this section.

The next four examples illustrate the general analysis procedure and the recommended guidelines for analysis presentation. Each example represents a basic analysis taken from the subject areas of statics, electrical circuits, thermodynamics, and fluid mechanics. You probably have not yet taken courses in these subjects, so do not be overly concerned if you do not understand all the technical aspects of the examples. Therefore, do not focus on the theoretical and mathematical details. Focus instead on how the general analysis procedure is used to solve problems from different engineering areas and the systematic manner in which the analyses are presented.

EXAMPLE 3.4

OCT. 12, 2012	EXAMPLE 3.4	BERT DILLON	1/1

Problem Statement

A 200-kg crate is suspended by ropes as shown. Rope AC is horizontal. Find the tension in ropes AB and AC.

Diagram (Free-Body Diagram)

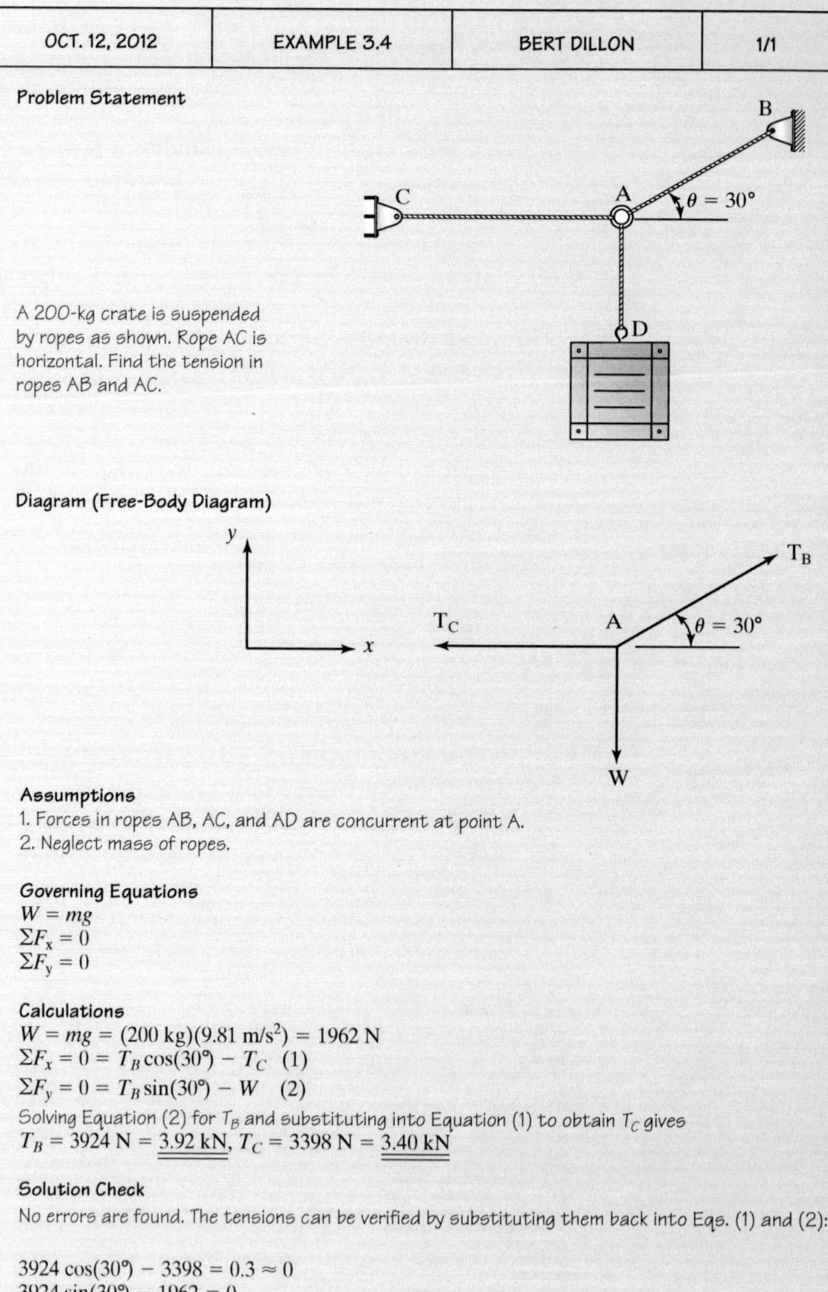

Assumptions

1. Forces in ropes AB, AC, and AD are concurrent at point A.
2. Neglect mass of ropes.

Governing Equations

$W = mg$
$\Sigma F_x = 0$
$\Sigma F_y = 0$

Calculations

$W = mg = (200\text{ kg})(9.81\text{ m/s}^2) = 1962\text{ N}$
$\Sigma F_x = 0 = T_B\cos(30°) - T_C$ (1)
$\Sigma F_y = 0 = T_B\sin(30°) - W$ (2)

Solving Equation (2) for T_B and substituting into Equation (1) to obtain T_C gives

$T_B = 3924\text{ N} = \underline{\underline{3.92\text{ kN}}}, T_C = 3398\text{ N} = \underline{\underline{3.40\text{ kN}}}$

Solution Check

No errors are found. The tensions can be verified by substituting them back into Eqs. (1) and (2):

$3924\cos(30°) - 3398 = 0.3 \approx 0$
$3924\sin(30°) - 1962 = 0$

The negligible nonzero result in Equation (1) is due to roundoff.

Discussion

As θ increases, T_B and T_C decrease. When $\theta = 90°$, $T_C = 0$ (rope AC is slack) and $T_B = W = 1962$ N.

EXAMPLE 3.5

JAN. 03, 2013	EXAMPLE 3.5	LUCY SMITH	1/2

Problem Statement
Two resistors with resistances of 10 Ω and 270 Ω are connected in parallel across a 10 V battery. Find the current in each resistor.

Diagram (Electrical Schematic)

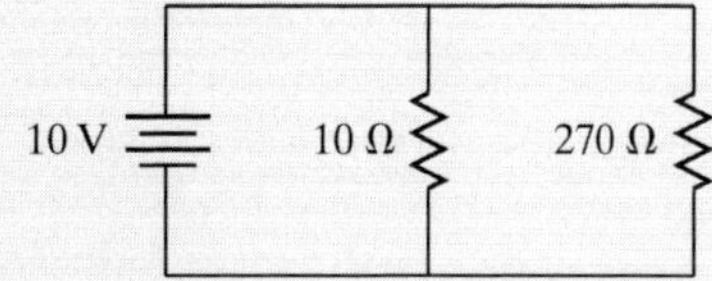

Assumptions
1. Neglect resistance of wires.
2. Battery voltage is a constant 10 V.

Governing Equations (Ohm's law)

$V = IR$ $\quad V$ = Voltage (V)
I = Current (A)
R = Resistance (Ω)

Calculations

Rearranging Ohm's law: $I = \frac{V}{R}$.

Define: $R_1 = 10\ \Omega, R_2 = 270\ \Omega$
Because resistors are connected in parallel with battery,
$V = V_1 = V_2 = 10\ V$.

$$\therefore I_1 = \frac{V_1}{R_1} = \frac{10\ V}{10\ \Omega} = \underline{\underline{1.0\ \text{A}}}, I_2 = \frac{V_2}{R_2} = \frac{10\ V}{270\ \Omega} = \underline{\underline{0.037\ \text{A}}}$$

Solution Check
The assumptions are reasonable, and there are no errors in the calculations.

JAN. 03, 2013	EXAMPLE 3.5	LUCY SMITH	2/2

Discussion
Current flow in a resistor is inversely proportional to the resistance.
Total current is split according to the ratio of resistances:

$$\frac{I_1}{I_2} = \frac{R_2}{R_1} = \frac{1.0\ \text{A}}{0.037\ \text{A}} = \frac{270\ \Omega}{10\ \Omega} \approx 27.0$$

Total current:

$$I_T = I_1 + I_2$$
$$= 1.0\ \text{A} + 0.037\ \text{A} = 1.037\ \text{A}$$

Total current may also be found by finding total resistance and then using Ohm's law.

Resistors in parallel as follows:

$$R_T = \frac{1}{\frac{1}{R_1} + \frac{1}{R_2}} = \frac{1}{\frac{1}{10} + \frac{1}{270}}$$

$$R_T = 9.643\ \Omega$$

$$I_T = \frac{V}{R_T} = \frac{10\ V}{9.643\ \Omega} = 1.037\ \text{A}$$

EXAMPLE 3.6

MAR. 24, 2013	EXAMPLE 3.6	CY BRAYTON	1/2

Problem Statement

A classroom occupied by 50 students is to be air-conditioned with window-mounted air-conditioning units with a 4 kW rating. There are 20 fluorescent lights in the room, each rated at 60 W. While sitting at their desks, each student dissipates 100 W. If the heat transfer to the classroom through the roof, walls, and windows is 5 kW, how many air-conditioning units are required to maintain the classroom at a constant temperature of 22°C?

Diagram (Thermodynamic System)

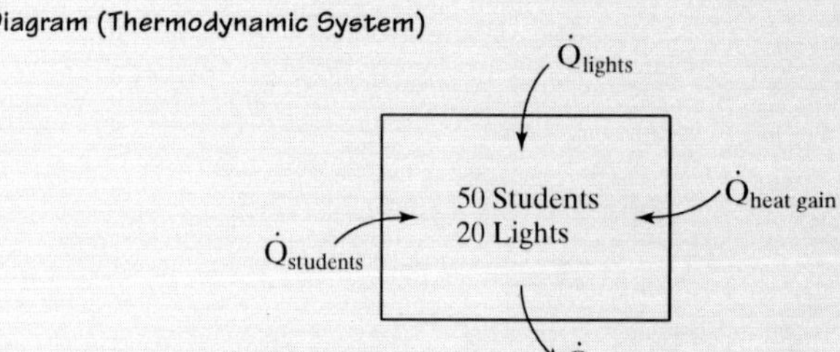

Assumptions

1. Classroom is a closed system (i.e., no mass flows).
2. All heat flows are steady.
3. No other heat occurs in classroom such as from computers and TVs.

Governing Equations (Conservation of Energy)

$$\dot{E}_{in} - \dot{E}_{out} = \Delta E_{system}$$

Calculations

$$\dot{E}_{in} = \dot{Q}_{students} + \dot{Q}_{lights} + \dot{Q}_{heat\ gain}$$
$$= (50)(100\ \text{W}) + (20)(60\ \text{W}) + 5000\ \text{W} = 11{,}200\ \text{W} = 11.2\ \text{kW}$$

$\Delta E_{system} = 0$ (Classroom is maintained at constant temperature)

$$\dot{E}_{out} = \dot{Q}_{cool}$$

Thus,

$$\dot{E}_{in} = \dot{Q}_{cool}$$

$$\text{Number of A.C. units required} = \frac{\dot{Q}_{cool}}{4\ \text{kW}} = \frac{11.2\ \text{kW}}{4\ \text{kW}} = 2.8$$

Fractions of A.C. units are impossible, so round up answer to next integer.

MAR. 24, 2013	EXAMPLE 3.6	CY BRAYTON	2/2

Number of A.C. units required = $\underline{\underline{3}}$.

Solution check

In order to maintain a constant temperature, the net heat transfer into the classroom must be equivalent to the heat removed by the air-conditioner. Reviewing the assumptions, equations, and calculations, no errors are found.

Discussion

The classroom temperature of 22°C was not used in the calculation because this temperature, as well as the outdoor air temperature, are inferred in the given heat gain by a prior heat transfer analysis.

Suppose that the classroom was a computer lab containing 30 computers each dissipating 250 W. We eliminate assumption 3 by including heat input by the computers

$$\dot{Q}_{\text{cool}} = \dot{Q}_{\text{students}} + \dot{Q}_{\text{lights}} + \dot{Q}_{\text{heat gain}} + \dot{Q}_{\text{computers}}$$
$$= 11{,}200 \text{ W} + 30(250 \text{ W}) = 18{,}700 \text{ W} = 18.7 \text{ kW}$$

$$\text{Number of A.C. units required} = \frac{\dot{Q}_{\text{cool}}}{4 \text{ kW}} = \frac{18.7 \text{ kW}}{4 \text{ kW}} = 4.7$$

Number of A.C. units required = 5.

This example illustrates the effect that computers have on air-conditioning requirements.

EXAMPLE 3.7

JULY 12, 2013	EXAMPLE 3.7	MAX POWER	1/2

Problem Statement

Water enters a pipe junction at a mass flow rate of 3.6 kg/s. If the mass flow rate in the small branch is 1.4 kg/s, what is the mass flow rate in the large pipe branch? If the inside diameter of the large pipe branch is 5 cm, what is the velocity in the large pipe branch?

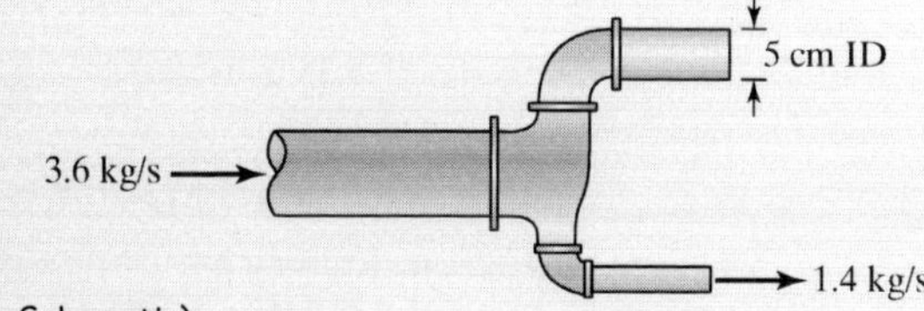

Diagram (Flow Schematic)

$\dot{m} = 3.6$ kg/s → $\dot{m}_2$, $\dot{m}_1 = 1.4$ kg/s

Assumptions

1. Steady, incompressible flow
2. Density of water: $\rho = 1000 \text{ kg/m}^3$

Governing Equations

Conservation of mass:	$\dot{m}_{\text{in}} = \dot{m}_{\text{out}}$	$\dot{m}$ = mass flow rate (kg/s)
Mass flow rate:	$\dot{m} = \rho A v$	ρ = fluid density (kg/m^3)
		A = flow cross-sectional area (m^2)
		v = velocity (m/s)

Calculations

$$\dot{m} = \dot{m}_1 + \dot{m}_2$$
$$\dot{m}_2 = \dot{m} - \dot{m}_1 = 3.6 \text{ kg/s} - 1.4 \text{ kg/s}$$
$$= 2.2 \text{ kg/s}$$

JULY 12, 2013	EXAMPLE 3.7	MAX POWER	2/2

$$\dot{m}_2 = \rho A_2 v_2 = \rho \frac{\pi D_2^2}{4} v_2$$

$$v_2 = \frac{4\,\dot{m}_2}{\pi \rho D_2^2} = \frac{4\,(2.2\text{ kg/s})}{\pi(1000\text{ kg/m}^3)(0.05\text{ m})^2}$$

$$= \underline{\underline{1.12\text{ m/s}}}$$

Solution check

The calculated value for the velocity in the large branch seems reasonable for a typical plumbing system. No errors are found in the calculations.

Discussion

The velocity is an average value because there is a velocity profile across the pipe. The velocity profile is caused by viscosity. If the flow condition is laminar, the velocity profile is parabolic, as shown in the following sketch.

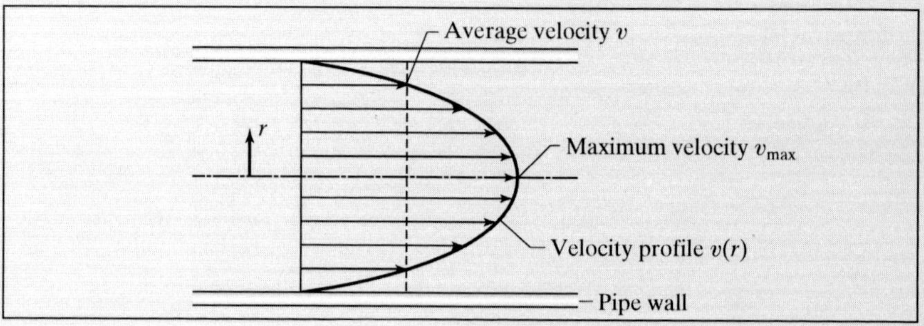

In contrast to Examples 3.4 through 3.7 that illustrated hand calculations, the final example incorporates a computer analysis tool, TK Solver. This software is an equation solver (see Section 3.4.2) that is useful for performing the calculations step of the general analysis procedure. The other steps of the procedure are done in the usual manner.

EXAMPLE 3.8

JANUARY 2, 2013	EXAMPLE 3.8	FRANK GRIMES	1/3

Problem Statement

A 6-kg block falls 20 cm into a spring with a spring constant of 1750 N/m. When the block comes into contact with the spring, it sticks to the spring. If the block falls from rest, what is the deformation of the spring when the block momentarily comes to rest?

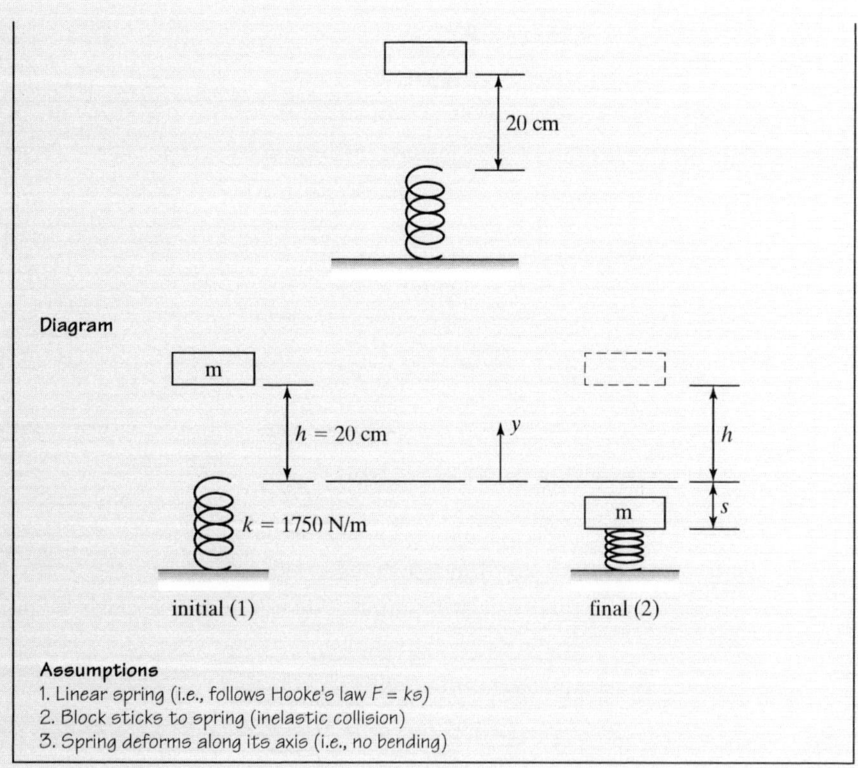

Governing Equations

Conservation of energy: $V_1 + T_1 = V_2 + T_2$
Gravitational potential energy: $V = mgh$
Spring potential energy: $V = \frac{1}{2}ks^2$

Calculations

The *rule sheet* shows the governing equations, and the *variable sheet* shows the inputs and outputs for all the physical quantities. TK Solver does not require the user to perform any algebraic manipulations; the software is capable of solving the governing equations in their original form. Due to the spring potential energy term, the conservation of energy equation becomes a quadratic equation, which has two roots. To generate these roots, a "guess" value for the spring deformation s is entered in the input column of the variable sheet. Then, a G (for guess) is entered in the *status* column beside the output variable. The iterative solver is initiated, generating *one* of the two roots. The root that is calculated depends on how close the guess value is to that root.

Rule sheet

Status	Rule
Satisfied	V1 + T1 = V2 + T2
Satisfied	V1 = m*g*h
Satisfied	V2 = −m*g*s + 0.5*k*s^2

Variable Sheet

Status	Input	Name	Output	Unit	Comment
		V1	11.772	J	initial potential energy
	0	T1		J	initial kinetic energy
		V2	11.772	J	final potential energy
	0	T2		J	final kinetic energy
	6	m		kg	mass of block
	9.81	g		m/s^2	gravitational acceleration
	.2	h		m	initial height of block
	1750	k		N/m	spring constant
G		s	.15440257	m	spring deformation

As shown in the variable sheet, the spring deformation is:

$$s = 0.1544 \text{ m} \approx \underline{\underline{15.4 \text{ cm}}}$$

Solution check

The solution may be checked by substituting values into the conservation of energy equation. Rearranging the equation, we have:

$$V_1 + T_1 - (V_2 + T_2) = 0$$

$$mgh + 0 - (-mgs + {}^1\!/_2\, ks^2 + 0) = 0$$

$$(6)(9.81)(0.20) + 0 - [-(6)(9.81)(0.1544) + {}^1\!/_2\,(1750)(0.1544)^2 + 0] = 0$$

$$11.7720 - [-9.0880 + 20.8594] = 0$$

$$11.7720 - 11.7714 = 5.84 \times 10^{-4} \approx 0$$

The small nonzero answer is due to round-off, so our answer is verified.

Discussion

The second root is $s = -0.0871$, which is obtained by using a guess value of zero or less. Because the spring deformation s is defined as a positive quantity in the spring potential energy equation, the second root is nonphysical, that is, has no physical significance.

It can be easily shown that the location of the origin is arbitrary.

PROFESSIONAL SUCCESS—AVOIDING A "COOKBOOK" LEARNING APPROACH TO ENGINEERING ANALYSIS

A good engineer is a person who solves an engineering analysis problem by reasoning through it, rather than simply following a prepared "recipe" consisting of step-by-step instructions written by someone else. Similarly, a good engineering student is a person who learns engineering analysis by thinking conceptually about each problem, rather than simply memorizing a collection of disjointed solution sequences and mathematical formulas. This "cookbook" learning approach promotes fragmented rather than integrative learning. A student who embraces this type of learning method will soon discover that it will be difficult and take a long time to solve new engineering problems, unless identical or very similar problems have been previously solved by using an established recipe. An analogy may be drawn from the familiar maxim "Give a man a fish, and you have fed him for a day. Teach a man to fish, and you have fed him for a lifetime." A recipe enables a student to solve only one specific type of problem, whereas a more general conceptual-based learning approach enables a student to solve many engineering problems.

PRACTICE!

Use the general analysis procedure to solve the following problems (present the analysis by using the guidelines for analysis presentation covered in this section):

1. Radioactive waste is to be permanently encased in concrete and buried in the ground. The vessel containing the waste measures 30 cm × 30 cm × 80 cm. Federal regulations dictate that there must be a minimum concrete thickness of 50 cm surrounding the vessel on all sides. What is the minimum volume of concrete required to safely encase the radioactive waste?
   Answer: 2.97 m^3
2. An elevator in an office building has an operating capacity of 15 passengers with a maximum weight of 180 lb_f each. The elevator is suspended by a special pulley system with four cables, two of which support 20 percent of the total load and two of which support 80 percent of the total load. Find the maximum tension in each elevator cable.
   Answer: 270 lb_f, 1080 lb_f
3. A technician measures a voltage drop of 25 V across a 47-Ω resistor by using a digital voltmeter. Ohm's law states that $V = IR$. What is the current flow through the resistor? How much power is consumed by the resistor? (*Hint*: $P = I^2R$.)
   Answer: 532 mA, 13.3 W
4. Air flows through a main duct at a mass flow rate of 4 kg/s. The main duct enters a junction that splits into two branch ducts, one with a cross section of 20 cm × 30 cm and one with a cross section of 40 cm × 60 cm. If the mass flow rate in the large branch is 2.8 kg/s, what is the mass flow rate in the small branch? If the density of air is $\rho = 1.16$ kg/m^3, what is the velocity in each branch?
   Answer: 1.2 kg/s, 10.1 m/s, 17.2 m/s

3.4 THE COMPUTER AS AN ANALYSIS TOOL

Computers are an integral part of the civilized world. They affect virtually every aspect of our everyday lives, including communications, transportation, financial transactions, information processing, food production, and health care. The world is a much different place today than it was prior to the advent of computers. Like everyone else, engineers use computers in their personal lives in the same ways just mentioned, but they also depend heavily on computers in their professional work. To the engineer, the computer is an indispensable tool. Without the computer, engineers would not be able to do their work as accurately or efficiently. Engineers use computers for computer-aided design (CAD), word processing, communications, information access, graphing, process control, simulation, data acquisition, and of course, analysis.

The computer is one of the most powerful analysis tools available to the engineer, but the computer does not replace the engineer's thinking. When faced with a new analysis, the engineer must reason through the problem by using sound scientific principles, applied mathematics, and engineering judgement. A computer can only carry out the instructions supplied to it, but it does so with remarkable speed and efficiency. A computer yields wrong answers just as quickly as it yields right ones. The burden is upon the engineer to supply the computer with correct input. An often-used engineering acronym is *GIGO* (*G*arbage *I*n, *G*arbage *O*ut), which refers to a situation in which erroneous input data is supplied to a computer, thereby producing erroneous output. When GIGO is at work, the calculations are numerically correct, but the results of those calculations are meaningless, because the engineer supplied the computer with bad input. The computer is capable of accurately performing enormous numbers of computations in a very short time, but it is incapable of composing a problem statement, constructing a diagram of the engineering system, formulating assumptions, selecting the appropriate governing equations, checking the reasonableness of the solution, or discussing and evaluating the results of the analysis. Thus, the only step in the analysis procedure for which a computer is perfectly suited is step 5: calculations. This is not to say that a computer cannot be used to write problem statements, assumptions, and equations as well as draw diagrams. These steps may also be performed by using the computer, but as directed by the engineer, whereas calculations are performed automatically once the equations and numerical inputs are supplied.

Engineers use analysis primarily as a design tool and as a means of predicting or investigating failures. Specifically, how does an engineer use the computer to perform an analysis? Steps 1 through 4 and steps 6 and 7 of the analysis procedure are largely unchanged, whether a computer is employed or not. So, exactly how are the calculations in step 5 carried out on a computer? There are basically five categories of computer tools for doing engineering analysis work:

1. Spreadsheets
2. Equation solvers and mathematics software
3. Programming languages
4. Specialty software
5. Finite element software.

3.4.1 Spreadsheets

The term ***spreadsheet*** originally referred to a special tabulation of rows and columns for doing financial calculations. The computer-based spreadsheet is a modern electronic version of the paper spreadsheet and was initially used for business and accounting applications. By virtue of their general structure, spreadsheets are useful not only for doing financial calculations, but can also be used for performing a variety of scientific and engineering calculations. Like the original paper version, the computer-based spreadsheet consists of any array of rows and columns. The intersection of a row with a column is called a *cell.* Cells serve as locations for input and output data such as text, numbers, or formulas. For example, a cell may contain an equation representing Newton's second law of motion, $F = ma$. A nearby cell would contain a number for the mass m, while another cell would contain a number for the acceleration a. Immediately after entering these two input values in their respective cells, the spreadsheet automatically evaluates the formula, inserting the numerical value of the force F in the cell containing the formula for Newton's second law. If the values of the mass or acceleration are changed, the spreadsheet automatically updates the value of the force.

The forgoing example is very simple, but spreadsheets are capable of doing calculations that involve hundreds or even thousands of variables. Suppose that our analysis involves 100 variables and that we want to know how changing only *one* of those variables affects the solution. We simply change the variable of interest and the entire spreadsheet automatically updates all calculations to reflect the change. The spreadsheet is an excellent analysis tool for rapidly answering "what if" questions. Numerous design alternatives can be efficiently investigated by performing the analysis on a spreadsheet. In addition to numerical functions, spreadsheets also have graphics capabilities. Excel[1] and Quattro Pro[2] are popular spreadsheet products. Figure 3.5 shows a simple example of calculating force using Newton's second law by using Excel.

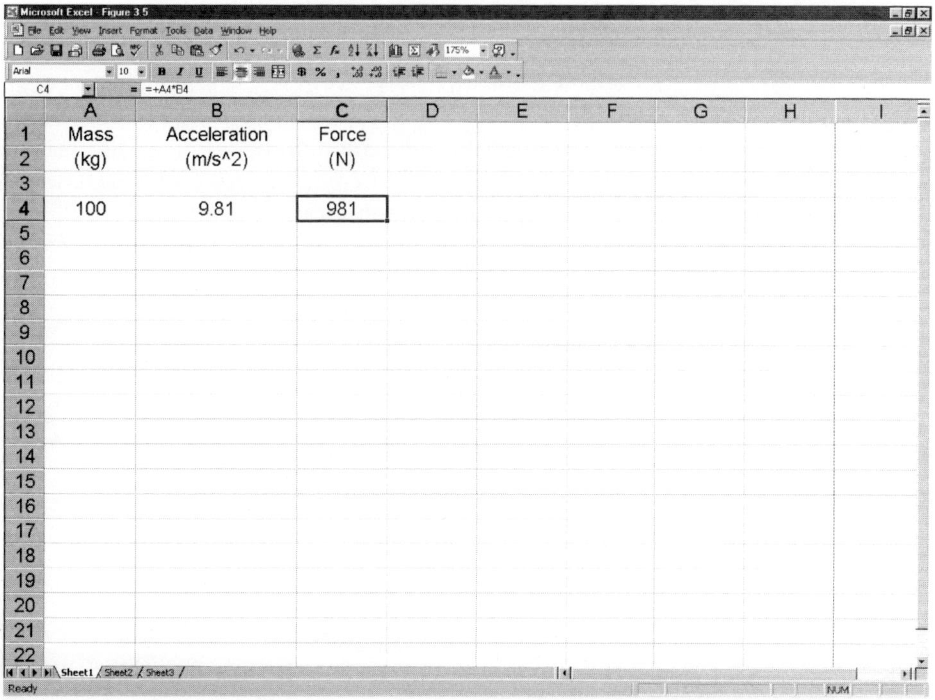

Figure 3.5
A calculation of Newton's second law using Excel. Note the formula for the force, +A4*B4, entered in cell C4.

Microsoft Excel - Figure 3.5

C4 = =+A4*B4

	A	B	C
1	Mass	Acceleration	Force
2	(kg)	(m/s^2)	(N)
3			
4	100	9.81	981

[1] Excel is a registered trademark of Microsoft® Corporation.
[2] Quattro® Pro is a registered trademark of Corel® Corporation.

3.4.2 Equation Solvers and Mathematics Software

Equation solvers and ***mathematics software*** packages are general-purpose scientific and engineering tools for solving equations and performing symbolic mathematical operations. Equation solvers are primarily designed for solving problems that involve *numerical* inputs and outputs, whereas mathematics packages are primarily suited for performing *symbolic* mathematical operations much as you would do in a mathematics course. Equation solvers accept a set of equations that represent the mathematical model of the analytical problem. The equations can be linear or nonlinear. The equations may be written in their usual form without prior mathematical manipulation to isolate the unknown quantities on one side of the equals sign. For example, Newton's second law would be written in its usual form as $F = ma$ even if the unknown quantity was the acceleration a. Solving this problem by hand, however, we would have to write the equation as $a = F/m$ because we are solving for the acceleration. This is not necessary when we use equation solvers. After we supply the numerical values for the known quantities, equation solvers solve for the remaining unknown values. Equation solvers have a large built-in library of functions for use in trigonometry, linear algebra, statistics, and calculus. Equation solvers can perform a variety of mathematical operations, including differentiation, integration, and matrix operations. In addition to these mathematical features, equation solvers also do unit conversions. Equation solvers also have the capability of displaying results in graphical form. Programming can also be done within equation solvers. Although all equation solvers have some symbolic capabilities, some have the capacity for data acquisition, image analysis, and signal processing. Popular equation solvers are TK Solver,[3] Mathcad,[4] and MATLAB.[5]

The strength of mathematics packages is their ability to perform symbolic mathematical operations. A symbolic mathematical operation is one that involves the manipulation of symbols (variables), using mathematical operators such as the vector product, differentiation, integration, and transforms. These packages are capable of performing very complex and sophisticated mathematical procedures. They also have extensive graphical capabilities. Even though mathematics packages are primarily designed for symbolic operations, they can also perform numerical computations. Mathematica[6] and Maple[7] are popular mathematics software products.

3.4.3 Programming Languages

Spreadsheets, equation solvers, and mathematics software packages may not always meet the computational demands of every engineering analysis. In such cases, engineers may choose to write their own computer programs with the use of a programming language. ***Programming languages*** refer to sequential instructions supplied to a computer for carrying out specific calculations. Computer languages are generally categorized according to their level. *Machine language* is a low-level language, based on a binary system of "zeroes" and "ones." Machine language is the most primitive language, because computers are digital devices whose rudimentary logic functions are carried out by using solid state switches in the "on" or "off" positions. *Assembly language* is also a low-level language, but its instructions are written in English-like

[3] TK Solver is a registered trademark of Universal Technical Systems, Incorporated.
[4] Mathcad® is a registered trademark of Mathsoft™, Incorporated.
[5] MATLAB® is a registered trademark of The MathWorks, Incorporated.
[6] MATHEMATICA® is a registered trademark of Wolfram Research, Incorporated.
[7] Maple™ is a registered trademark of Maplesoft™, a division of Waterloo Maple, Incorporated.

statements rather than binary. Assembly language does not have many commands, and it must be written specifically for the computer hardware. Computer programs written in low-level languages run very fast because these languages are tied closely to the hardware, but writing the programs is very tedious.

Due to the tediousness of writing programs in low-level languages, engineers usually write programs in high-level languages that consist of straightforward, English-like commands. The most commonly used high-level languages by engineers are Fortran, C, C++, Pascal, Ada, and BASIC. Fortran is the patriarch of all scientific programming languages. The first version of Fortran (FORmula TRANslation) was developed by IBM between 1954 and 1957. Since its inception, Fortran has been the workhorse of scientific and engineering programming languages. It has undergone several updates and improvements and is still in widespread use today. The C language evolved from two languages, BCPL and B, which were developed during the late 1960s. In 1972, the first C program was compiled. The C++ language grew out of C and was developed during the early 1980s. Both C and C++ are popular programming languages for engineering applications because they use powerful commands and data structures. Pascal was developed during the early 1970s and is a popular programming language for beginning computer science students who are learning programming for the first time. The U.S. Department of Defense prompted the development of Ada during the 1970s in order to have a high-level language suitable for embedded computer systems. BASIC (Beginner's All-purpose Symbolic Instruction Code) was developed during the mid-1960s as a simple learning tool for secondary school students as well as college students.

Writing programs in high-level languages is easier than writing programs in low-level languages, but the high-level languages utilize a larger number of commands. Furthermore, high-level languages must be written with specific grammatical rules, referred to as *syntax*. Rules of syntax govern how punctuation, arithmetic operators, parentheses, and other characters are used in writing commands. To illustrate the syntactical differences between programming languages, equation solvers, and mathematics packages, Table 3.1 shows how a simple equation is written. Note the similarities and differences in the equals sign, the constant π, and the operator for exponentiation.

Table 3.1 Comparison of Computer Statements for the Equation, $V = 4/3\pi R^3$, the Volume of a Sphere

Computer Tool	Statement
Mathcad	$V{:} = 4/3*\pi*R^3$
TK Solver	$V = 4/3*\text{pi}(\,)*R^3$
MATLAB	$V = 4/3*\text{pi}*R^3;$
Mathematica	$V = 4/3*\text{Pi}*R^3$
Maple	$V{:} = 4/3*\text{pi}*R^3;$
Fortran	V = 4/3*3.141593*R**3
C, C++	V = 4/3*3.141593*pow(R, 3);
Pascal	V: = 4/3*3.141593*R*R*R;
Ada	V: = 4/3*3.141593*R**3;
BASIC	$V = 4/3*3.141593*R^3$

3.4.4 Specialty Software

Considered as a whole, engineering is a broad field that covers a variety of disciplines. Some of the main engineering disciplines are chemical engineering, civil engineering, electrical and computer engineering, environmental engineering, and mechanical engineering. Given the variety of specific problems that engineers who work in these fields encounter, it comes as no surprise that numerous *specialty software* packages are available to help the engineer analyze specific problems relating to a particular engineering system. For example, specialty software packages are available to electrical engineers for analyzing and simulating electrical circuits. Mechanical and chemical engineers can take advantage of software packages designed specifically for calculating flow parameters in pipe networks. Special software is available to civil and structural engineers for calculating forces and stresses in trusses and other structures. Other specialty software packages are available for performing analysis of heat exchangers, machinery, pressure vessels, propulsion systems, turbines, pneumatic and hydraulic systems, manufacturing processes, mechanical fasteners, and many others too numerous to list. After you graduate and begin working for a company that produces a specific product or process, you will probably become familiar with one or more of these specialty software packages.

3.4.5 Finite Element Software

Some engineering analysis problems are far too complex to solve using any of the aforementioned computer tools. *Finite element* software packages enable the engineer to analyze systems that have irregular configurations, variable material properties, complex conditions at the boundaries, and nonlinear behavior. The finite element method originated in the aerospace industry during the early 1950s when it was used for stress analysis of aircraft. Later, as the method matured, it found application in other analysis areas such as fluid flow, heat transfer, vibrations, impacts, acoustics, and electromagnetics. The basic concept behind the finite-element method is to subdivide a continuous region (i.e., the system to be analyzed is divided into a set of simple geometric shapes called "finite elements"). The elements are interconnected at common points called "nodes." Material properties, conditions at the system boundaries, and other pertinent inputs are supplied. With the use of an advanced mathematical procedure, the finite element software calculates the value of parameters such as stress, temperature, flow rate, or vibration frequency at each node in the region. Hence, the engineer is provided with a set of output parameters at discrete points that approximates a continuous distribution of those parameters for the entire region. The finite element method is an advanced analysis method and is normally introduced in colleges and universities at the senior level or the first-year graduate level.

PROFESSIONAL SUCCESS—PITFALLS OF USING COMPUTERS

The vital role that computers play in engineering analysis cannot be overstated. Given the tremendous advantages of using computers for engineering analysis, however, it may be difficult to accept the fact that there are also pitfalls. A common hazard that entangles some engineers is the tendency to treat the computer as a "black box," a wondrous electronic device whose inner workings are largely unknown, but that nonetheless provides output for every input supplied. Engineers who treat the computer as a black box are not effectively employing the general analysis procedure and in so doing are in danger of losing their ability to systematically reason their way through a problem.

The computer is a remarkable computational machine, but it does not replace the engineer's thinking, reasoning, and judgment. Computers, and the software that runs on them, produce output that *precisely* reflects the input supplied to them. If the input is good, the output will be good. If the input is bad, the output will be bad. Computers are not smart enough to compensate for an engineer's inability to make good assumptions or employ the correct governing equations. Engineers must have a thorough understanding of the physical aspects of the problem at hand and the underlying mathematical principles *before* implementing the solution on the computer. A good engineer understands *what* the computer does when it "crunches the numbers" in the analysis. A good engineer is confident that the input data will result in reasonable output because a lot of sound thinking and reasoning has gone into the formulation of that input.

Can the computer be used too much? In a sense, it can. The tendency of some engineers is to use the computer to analyze problems that may not require a computer at all. Upon beginning a new problem, their first impulse is to set up the problem on the computer without even checking to see whether the problem can be solved by hand. For example, a problem in engineering statics may be represented by the quadratic equation, $x^2 + 4x - 12 = 0$.

This problem can be solved analytically by factoring $(x + 6)(x - 2) = 0$, which yields the two roots $x = -6$ and $x = 2$. To use the computer in a situation like this is to rely on the computer as a "crutch" to compensate for weak analytical skills. Continued reliance on the computer to solve problems that do not require a computer will gradually dull your ability to solve problems with pencil and paper. Do not permit this to happen. Examine the equations carefully to see whether a computer solution is justified. If it is, use one of the computer tools discussed earlier. If not, solve the problem by hand. Then, if you have time and wish to check your solution with the use of the computer, by all means do so.

APPLICATION

COMPUTERS FOR NUMERICAL ANALYSIS

Most of the equations that you will encounter in school can be solved analytically; that is, they can be solved by employing standard algebraic operations to isolate the desired variable on one side of the equation. Some equations, however, cannot be solved analytically with standard algebraic operations. These equations are referred to as *transcendental* equations because they contain one or more transcendental functions such as a logarithm or trigonometric function. Transcendental equations occur often in engineering analysis work, and techniques for solving them are known as *numerical methods.* For example, consider the transcendental equation:

$$e^x - 3x = 0.$$

This equation looks straightforward enough, but try solving it by hand. If we add $3x$ to both sides and take the natural logarithm of both sides to undo the exponential function, we obtain:

$$x = \ln(3x) \qquad \text{(a)}$$

which, unfortunately, does not isolate the variable x because we still have the term $\ln(3x)$ on the right side of the equation. If we add $3x$ to both sides and then divide both sides by 3, we obtain:

$$\frac{e^x}{3} = x. \tag{b}$$

The variable x is still not isolated without leaving a transcendental function in the equation. Clearly, this equation cannot be solved analytically, so it must be solved numerically. To solve it numerically, we utilize a method called *iteration*, a process by which we repeat the calculation until an answer is obtained.

Before solving this problem by using the computer, we will work it manually to illustrate how iteration works. To begin, we rewrite Equation (a) in the iterative form:

$$x_{i+1} = \ln(3x_i).$$

The "i" and the "$i + 1$" subscripts refer to "old" and "new" values of x, respectively. The iteration process requires that we begin the calculation by immediately substituting a number into the iteration formula. This first number constitutes an estimate for the root (or roots) of the variable x, that satisfy the formula. To keep track of the iterations, we use an iteration table, illustrated in Table 3.2. To start the iterations, we estimate a value of x by letting $x_1 = 1$. We now substitute this number into the right side of the formula, yielding a new value of $x_2 = 1.098612$. We again substitute into the right side of the formula, yielding the next new value, $x_3 = 1.192660$. This process is repeated until the value of x stops changing significantly. At this point, we say that the calculation has *converged* to an answer. Table 3.2 shows the first five iterations and indicates that 41 iterations are required for the calculation to converge to an answer that is accurate to the sixth decimal place. Upon substituting $x = 1.512135$ into the original equation, we see that the equation is satisfied. As this example illustrates, numerous iterations may be required to obtain an accurate solution. The accuracy of the answer depends on how many iterations are taken. Some equations converge to a precise answer in a few iterations but others, like this one, require several iterations. It is important to note that 1.512135 is not the only root of this equation. The equation has a second root at $x = 0.619061$. If we attempt to find this root by using Equation (a), we discover that our calculation either converges again to 1.512135 or does not converge at all by leading us to an illegal operation,

Table 3.2 Iteration Table for Finding One Root of the Equation $e^x - 3x = 0$

Iteration	x_i	x_{i+1}
1	1	1.098612
2	1.098612	1.192660
3	1.192660	1.274798
4	1.274798	1.341400
5	1.341400	1.392326
.		
.		
.		1.512134
41	1.512134	1.512135

Figure 3.6
BASIC computer program for finding one root of the equation $e^x - 3x = 0$.

```
INPUT "ESTIMATE = ", XOLD
DO
    XNEW = LOG (3*XOLD)
    DIFF = ABS (XNEW - XOLD)
    XOLD = XNEW
LOOP WHILE DIFF > 0.0000001
PRINT XNEW
END
```

that is, taking the logarithm of a negative number. To find the second root, we iterate on Equation (b), writing it in the iterative form:

$$x_{i+1} = \frac{e^{x_i}}{3}.$$

With numerical methods, there are often no guarantees that a certain iteration formula will converge rapidly or even converge at all. The success of the iteration formula may also depend on the initial estimate chosen to start the iterations. If our initial estimate for Equation (a) is less than $\frac{1}{3}$, the new value of x immediately goes negative, leading to an illegal operation. If our initial estimate for Equation (b) is too large, the new value of x grows large very rapidly, leading to an exponential overflow. These and other kinds of numerical difficulties can occur whether the iterations are performed by hand or by using a computer.

As Table 3.2 suggests, performing iterations by hand can be a long and tedious task. The computer is tailor-made for performing repetitive calculations. The roots of our transcendental equation can readily be found by using one of the computer tools discussed earlier. Figure 3.6 shows a computer program, written in the BASIC language for finding the first root $x = 1.512135$. In the first line the user inputs an initial estimate, which is assigned the variable name XOLD. The program then executes what is referred to as a DO loop that performs the iterations. Each time through the loop, a new value of x is calculated from the old value and an absolute value of the difference between the old and new values is calculated. This value is called DIFF. While DIFF is larger than a preselected convergence tolerance of 0.0000001, the new value of x, XNEW, is reset to the old value XOLD and looping continues. When DIFF is less than or equal to the convergence tolerance, convergence has been achieved, and looping is halted. The root is then printed. The same program, with the third line replaced with XNEW = EXP(XOLD)/3, could be used to find the second root. There are more sophisticated numerical methods for finding roots than the simple iteration technique illustrated here, and you will study them in your engineering or mathematics courses.

PRACTICE!

Using one of the computer tools discussed in this section, work the following problems:

(Note: These problems are identical to those in Section 3.3.)

1. Radioactive waste is to be permanently encased in concrete and buried in the ground. The vessel containing the waste measures 30 cm × 30 cm × 80 cm. Federal regulations dictate that there must

be a minimum concrete thickness of 50 cm surrounding the vessel on all sides. What is the minimum volume of concrete required to safely encase the radioactive waste?
Answer: 2.97 m^3

2. An elevator in an office building has an operating capacity of 15 passengers with a maximum weight of 180 lb_f each. The elevator is suspended by a special pulley system with four cables, two of which support 20 percent of the total load and two of which support 80 percent of the total load. Find the maximum tension in each elevator cable.
Answer: 270 lb_f, 1080 lb_f

3. A technician measures a voltage drop of 25 V across a 47-Ω resistor by using a digital voltmeter. Using Ohm's law, we find that $V = IR$. What is the current flow through the resistor? How much power is consumed by the resistor? (*Hint*: $P = I^2R$.)
Answer: 532 mA, 13.3 W

4. Air flows through a main duct at a mass flow rate of 4 kg/s. The main duct enters a junction that splits into two branch ducts, one with a cross section of 20 cm × 30 cm and one with a cross section of 40 cm × 60 cm. If the mass flow rate in the large branch is 2.8 kg/s, what is the mass flow rate in the small branch? If the density of air is $\rho = 1.16$ kg/m^3, what is the velocity in each branch?
Answer: 1.2 kg/s, 10.1 m/s, 17.2 m/s

KEY TERMS

engineering method
equation solver
general analysis procedure
mathematics software
order of magnitude
programming language
significant figure
spreadsheet

REFERENCES

Borwein, J.M. and M.P. Skerritt, *An Introduction to Modern Mathematical Computing: With Maple*, New York, NY: Springer, 2011.

Chivers, I. and J. Sleightholme, *Introduction to Programming with FORTRAN: With Coverage of FORTRAN 90, 95, 2003, 2008 and 77*, 2nd ed., New York, NY: Springer, 2012.

Etter, D.M., *Engineering Problem Solving with C++*, 3rd ed., Upper Saddle River, NJ: Prentice Hall, 2011.

Larsen, R.W., *Introduction to Mathcad 15*, 3rd ed., Upper Saddle River, NJ: Prentice Hall, 2010.

Larsen, R.W., *Engineering with Excel*, 4th ed., Upper Saddle River, NJ: Prentice Hall, 2012.

Moore, H., *MATLAB for Engineers*, 3rd ed., Upper Saddle River, NJ: Prentice Hall, 2011.

Torrence, B.F. and E.A. Torrence, *The Student's Introduction to MATHEMATICA: A Handbook for Precalculus, Calculus and Linear Algebra*, 2nd ed., Boston, MA: Cambridge University Press, 2009.

PROBLEMS

Order-of-magnitude analysis

3.1 Using an order-of-magnitude analysis, estimate the number of gallons of gasoline used by all automobiles in the United States each year.

3.2 Using an order-of-magnitude analysis, estimate the number of 4 ft × 8 ft plywood sheets required for the floor, roof, and exterior sheathing of a 6000-ft^2 house.

3.3 Using an order-of-magnitude analysis, estimate the number of Post-it® notes that would just cover the land surface of the earth.

3.4 Using an order-of-magnitude analysis, estimate the number of spam e-mail messages received by residents of the United States each year.

3.5 Using an order-of-magnitude analysis, estimate the number of breaths you will take during your lifetime.

3.6 Use an order-of-magnitude analysis to estimate the number of tons of human waste produced worldwide each year.

3.7 The earth has a mean radius of about 6.37×10^6 m. Assuming the earth is made of granite ($\rho = 2770\ kg/m^3$), estimate the mass of the earth using an order-of-magnitude analysis.

3.8 The solar radiation flux just outside the earth's atmosphere is about 1350 W/m^2. Using an order-of-magnitude analysis, estimate the amount of solar energy that is intercepted by the Pacific Ocean each year.

3.9 Using an order-of-magnitude analysis, estimate the total textbook expenditure incurred by all engineering majors at all U.S. colleges and universities per year.

Significant figures

3.10 Underline the significant figures in the following numbers (the first number is done for you):

a. $\underline{3450}$
b. 9.807
c. 0.00216
d. 9000
e. 7000.
f. 12.00
g. 1066
h. 106.07
i. 0.02880
j. 163.07
k. 1.207×10^{-3}

3.11 Perform the following calculations, reporting the answers with the correct number of significant figures:

a. (8.14)(260)
b. 456/4.9
c. (6.74)(41.07)/4.13
d. (10.78 − 4.5)/300
e. (10.78 − 4.50)/300.0

f. $(65.2 - 13.9)/240.0$
g. $(1.2 \times 10^6)/(4.52 \times 10^3 + 769)$
h. $(1.764 - 0.0391)/(8.455 \times 10^4)$
i. $1000/(1.003 \times 10^9)$
j. $(8.4 \times 10^{-3})/5000$
k. $(8.40 \times 10^3)/5000.0$
l. 8π
m. $(2\pi - 5)/10$.

3.12 A 650.0-kg mass hangs by a cable from the ceiling. Using the standard value of gravitational acceleration $g = 9.81\ \text{m/s}^2$, what is the tension in the cable? Express your answer with the correct number of significant figures.

3.13 A 9-slug mass hangs by a rope from the ceiling. Using the standard value of gravitational acceleration $g = 32.2\ \text{ft/s}^2$, what is the tension in the rope? Express your answer with the correct number of significant figures. Redo the problem, using a mass of 9.00 slug. Is the answer different? Why?

3.14 A 72.9 mA current flows through a 360-Ω resistor. Using Ohm's law $V = IR$, what is the voltage across the resistor? Express your answer with the correct number of significant figures.

3.15 A rectangular building lot is reported to have the dimensions 200 ft × 300 ft. Using the correct number of significant figures, what is the area of this lot in units of acre?

General analysis procedure

For problems 16 through 31, use the general analysis procedure of (1) problem statement, (2) diagram, (3) assumptions, (4) governing equations, (5) calculations, (6) solution check, and (7) discussion.

3.16 An excavation crew digs a hole in the ground measuring 60 yd × 50 yd × 8 yd to facilitate a basement for an office building. Five dump trucks, each with a capacity of 20 yd^3, are used to haul the material away. How many trips must each truck make to remove all the material?

3.17 Find the current in each resistor and the total current for the circuit shown in Figure P3.17.

Figure P3.17

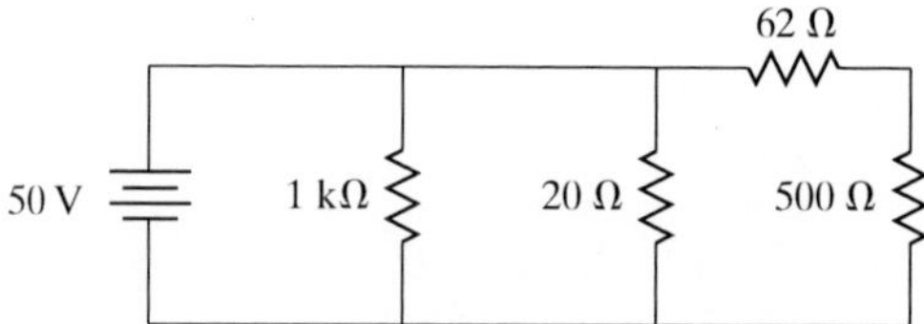

3.18 For easy handling, long sheets of steel for manufacturing automobile body panels are tightly rolled up into a cylinder-shaped package. Consider a roll of steel with an inside and outside diameter of 45 cm and 1.6 m, respectively, that is suspended by a single cable. If the length of the roll is 2.25 m and the density of steel is $\rho = 7850\ \text{kg/m}^3$, what is the tension in the cable?

3.19 In a chemical processing plant, glycerin flows toward a pipe junction at a mass flow rate of 30 kg/s as shown in Figure P3.19. If the mass flow rate in the small pipe branch is 8 kg/s, find the velocity in both branches. The density of glycerin is $\rho = 1260 \text{ kg/m}^3$.

Figure P3.19

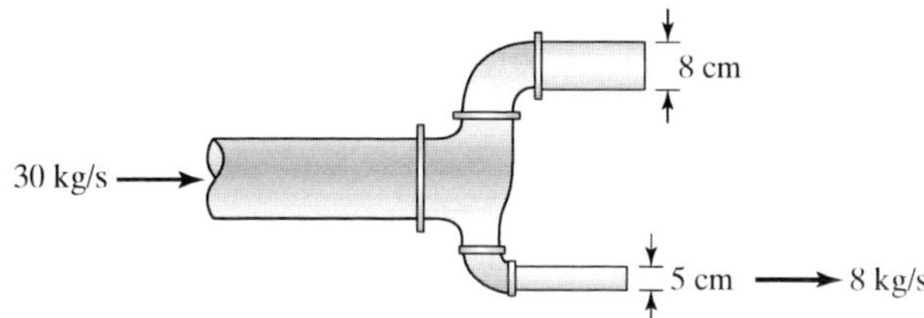

3.20 Water begins filling an empty 50-gallon bathtub at a volume flow rate of 0.045 ft^3/s. If the drain, which is partially open, allows water to drain out of the bathtub at a volume flow rate of 7.5 gal/min, how much time will it take for the bathtub to over flow?

3.21 A man pushes on a barrel with a force of $P = 30 \text{ N}_f$ as shown. Assuming that the barrel does not move, what is the friction force between the barrel and the floor? (*Hint:* The friction force acts parallel to the floor toward the man. See Figure P3.21.)

Figure P3.21

3.22 The total resistance for resistors connected in series is the arithmetic sum of the resistances. Find the total resistance for the series circuit shown in Figure P3.22. Because the resistors are connected in series, the current is the same in each resistor. Using Ohm's law, find this current. Also, find the voltage drop across each resistor.

Figure P3.22

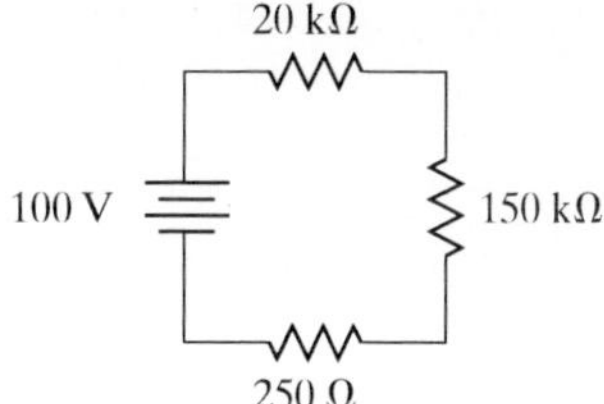

3.23 The pressure exerted by a static liquid on a vertical submerged surface is calculated from the relation:

$$P = \rho g h$$

where
P = pressure
ρ = density of the liquid
g = gravitational acceleration = $9.81\ \text{m/s}^2$
h = height of vertical surface that is submerged.

Consider the dam shown in Figure P3.23. What is the pressure exerted on the dam's surface at depths of 1 m, 5 m, and 25 m? For the density of water, use $\rho = 1000\ \text{kg/m}^3$.

Figure P3.23

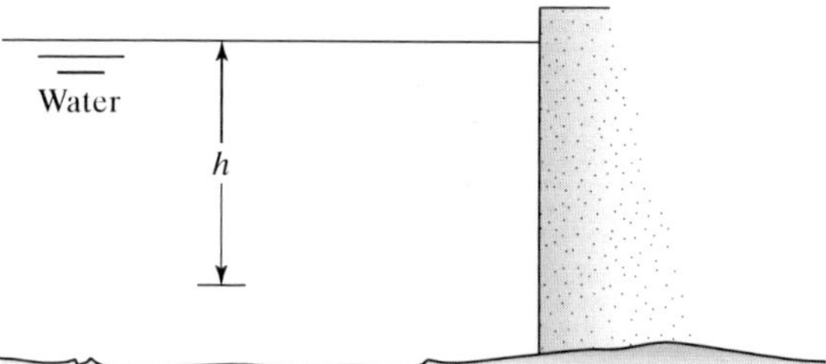

3.24 Work Problem 3.16, using one of the computer tools discussed in this chapter.

3.25 Work Problem 3.17, using one of the computer tools discussed in this chapter.

3.26 Work Problem 3.18, using one of the computer tools discussed in this chapter.

3.27 Work Problem 3.19, using one of the computer tools discussed in this chapter.

3.28 Work Problem 3.20, using one of the computer tools discussed in this chapter.

3.29 Work Problem 3.21, using one of the computer tools discussed in this chapter.

3.30 Work Problem 3.22, using one of the computer tools discussed in this chapter.

3.31 Work Problem 3.23, using one of the computer tools discussed in this chapter.

CHAPTER 4 Mechanics

Objectives

After reading this chapter, you will have learned

- The importance of mechanics in engineering
- The difference between a scalar and a vector
- How to perform basic vector operations
- How to add forces vectorially
- How to construct free-body diagrams
- How to use equilibrium principles to find unknown forces on a particle
- How to calculate normal stress, strain, and deformation
- How to apply a factor of safety to stress

4.1 INTRODUCTION

Mechanics is one of the most important fields of study in engineering. Mechanics was the first analytical science, and its historical roots can be traced to such great mathematicians and scientists as Archimedes (287–212 B.C.), Galileo Galilei (1564–1642), and Isaac Newton (1642–1727). ***Mechanics*** is the *study of the state of rest or motion of bodies that are subjected to forces.* As a discipline, mechanics is divided into three general areas: *rigid-body mechanics, deformable-body mechanics,* and *fluid mechanics.* As the term implies, rigid-body mechanics deals with the mechanical characteristics of bodies that are rigid (i.e., bodies that do not deform under the influence of forces). Rigid-body mechanics is subdivided into two main areas: *statics* and *dynamics.* Statics deals with rigid bodies in equilibrium. Equilibrium is a state in which a body is at rest with respect to its surroundings. When a body is in equilibrium, the forces that act on it are balanced, resulting in no motion. A state of equilibrium also exists when a body moves with a constant velocity, but this type of equilibrium is a dynamic equilibrium, not a static equilibrium. Dynamics deals with rigid bodies that are in motion with respect to its surroundings or to other rigid bodies. The body may have a constant velocity, in which case the acceleration is zero, but, generally, the body undergoes an acceleration due to the application of an unbalanced force.

Deformable-body mechanics, often referred to as *mechanics of materials* or *strength of materials*, deals with solid bodies that deform under the application of external forces. In this branch of mechanics, the relationships between externally applied forces and the resulting internal forces and deformations are studied. Deformable-body mechanics is often subdivided into two specific areas: *elasticity* and *plasticity*. Elasticity deals with the behavior of solid materials that return to their original size and shape after a force is removed, whereas plasticity deals with the behavior of solid materials that experience a permanent deformation after a force is removed.

Fluid mechanics deals with the behavior of liquids and gases at rest and in motion. The study of fluids at rest is called *fluid statics*; the study of fluids in motion is called *fluid dynamics*. Even though fluids are, strictly speaking, deformable materials, deformable-body mechanics is set apart from fluid mechanics because deformable-body mechanics deals exclusively with *solid* materials that have the ability, unlike fluids, to sustain shear forces. The topical structure of engineering mechanics is shown schematically in Figure 4.1.

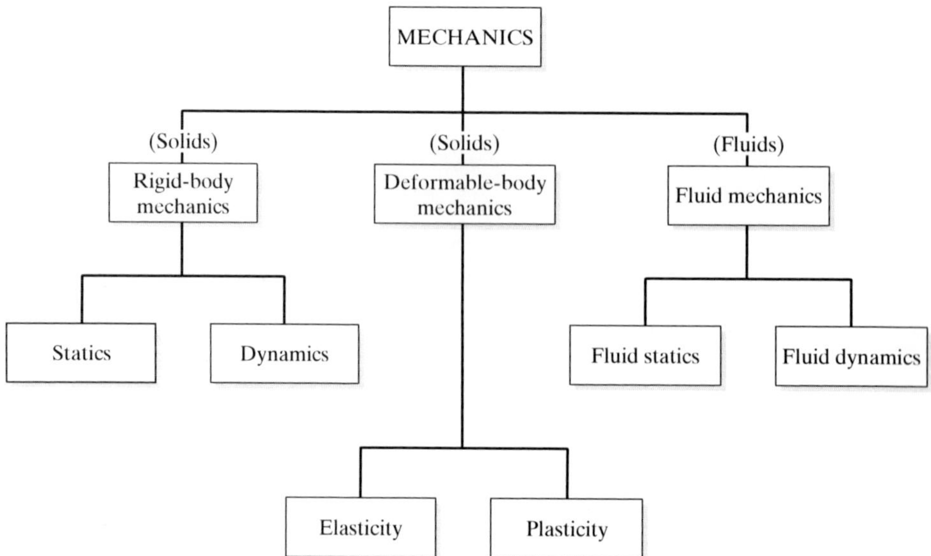

Figure 4.1
Topical structure of engineering mechanics.

In most colleges and universities, the branches of mechanics just outlined are generally taught as separate and distinct engineering courses. Hence, a typical engineering program consists of individual courses in statics, dynamics, mechanics of materials, and fluid mechanics. Other analytically oriented courses such as electrical circuits and thermodynamics are also offered. Mechanics is so essential to engineering education that students majoring in “nonmechanical” fields such as electrical engineering, environmental engineering, and chemical engineering gain a deeper understanding of energy, power, potential, equilibrium, and stability by first studying these principles in their mechanical contexts. However, depending on the specific curricular policies of your school or department, students in all engineering majors may or may not be required to take all of the aforementioned mechanics courses. In any case, the main purpose of this chapter is to introduce the beginning engineering student to the most fundamental principles of mechanics and to show how the general analysis procedure is applied to mechanics problems. In order to focus on the basics and to assist the student in the transition to more advanced material, our treatment of mechanics in this chapter is limited to a few fundamental principles of statics and mechanics of materials. However, dynamics is not covered in this book.

Engineers use principles of mechanics to analyze and design a wide variety of devices and systems. Look around you. Are you reading this book in a building? The structural members in the floor, roof, and walls were designed by structural or civil engineers to withstand forces exerted on them by the contents of the building, winds, earthquakes, snow, and other structural members. Bridges, dams, canals, underground pipelines, and other large, earthbound structures are designed with the use of mechanics. Do you see any mechanical devices nearby? The automobile is an excellent example of a single engineering system that embodies virtually every branch of engineering mechanics, as well as other engineering disciplines. The chassis, bumpers, suspension system, power train, brakes, steering system, engine, air bag, doors, trunk, and even the windshield wipers were designed with the use of mechanics. Even the design of simple mechanisms such as staple removers, paper punches, door locks, and pencil sharpeners involves principles of mechanics. Principles of mechanics are used to analyze and design virtually every type of engineering system that can be devised. Figures 4.2, 4.3, and 4.4 show some familiar engineering systems that involved the use of engineering mechanics in their design.

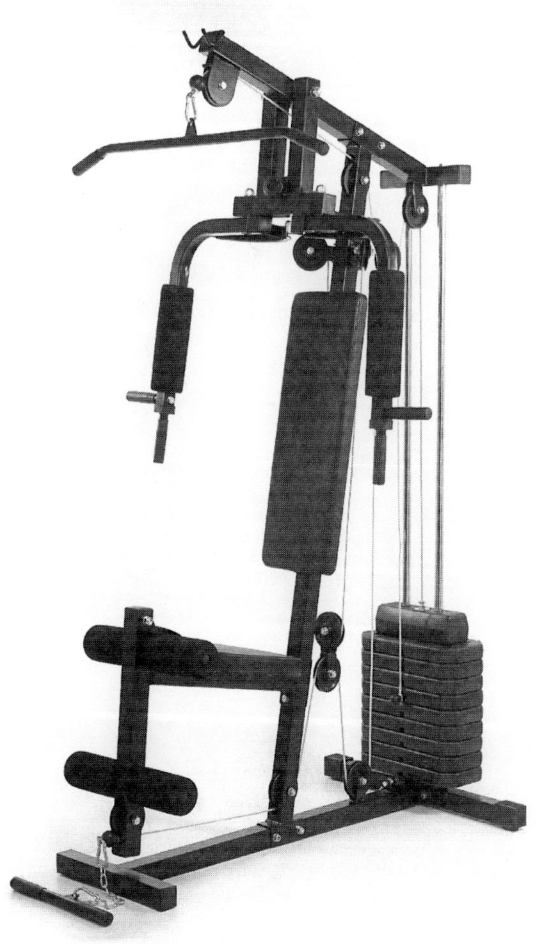

Figure 4.2
Engineers use principles of engineering mechanics to design fitness equipment. (Julian Rovagnati/ Shutterstock).

Figure 4.3
Principles of engineering mechanics are used to design heavy construction equipment.

Figure 4.4
Engineers used principles of engineering mechanics to design the Beijing National Stadium in Beijing, China. This stadium was used for the 2008 Summer Olympics and Paralympics.
(© James Doro/Alamy)

4.2 SCALARS AND VECTORS

Every physical quantity used in mechanics, and in all of engineering and science, is classified as either a ***scalar*** or a ***vector***. A scalar is a *quantity having magnitude, but no direction.* Having magnitude only, a scalar may be positive or negative, but has no directional characteristics. Common scalar quantities are length, mass, temperature, energy, volume, and density. A vector is a *quantity having both magnitude and direction.* A vector may be positive or negative and has a specified direction in space. Common vector quantities are displacement, force, velocity, acceleration, stress, and momentum. A scalar quantity can be fully defined by a single parameter, its magnitude, whereas a vector requires that both its magnitude and direction be specified. For example, speed is a scalar but velocity is a vector. A typical speedometer of an automobile indicates how fast the vehicle is traveling, but does not reveal the direction of travel. The temperature of water boiling in an open container at sea level can be completely defined by a single number, 100°C. The force exerted on a beam used as a floor joist, however, must be defined by specifying a magnitude, 2 kN for example, and a direction down. The effect of the force on the beam (i.e., the stress and deformation) cannot be determined unless the direction of the force is specified. A force directed along the axis of the beam, for example, would produce a completely different stress and deformation than a normal downward force. Table 4.1 is a summary of some scalar and vector quantities.

When writing scalars and vectors, standard nomenclature should be followed. Scalars are often printed in italic font such as m for mass, T for temperature, ρ for density. To differentiate vectors from scalars, vectors are written in a special way. For handwritten work, a vector is usually written as a letter with a bar $^{-}$, arrow $^{\rightarrow}$, or caret ^ over it such as $\overline{\mathrm{A}}$, $\vec{\mathrm{A}}$, and $\hat{\mathrm{A}}$. In books and other printed matter, vectors are typically written in boldface type. For example, **A** is used to denote a vector "A." The magnitude of a vector, which is always a positive quantity, is normally written by hand, using "absolute value" notation. Thus, the magnitude of **A** is written as $|\mathbf{A}|$. In books and other printed matter, the magnitude of **A** is usually written in italic type as A.

As shown in Figure 4.5, a vector is represented graphically by a straight arrow with a specified *magnitude and direction*. The magnitude is the length of the arrow and the direction is defined by the angles between the arrow and reference axes. The *line of action* of the vector is a line that is collinear with the vector, locating its direction in space; note this is an additional attribute and a vector need not have a specified location. The vector **A** in Figure 4.5 has a magnitude of 5 units and a direction of 30° with respect to the x-axis, upward and to the right. Point O is called the *tail* of the vector and point P is called the *head* of the vector. The units of the vector depend on what physical quantity the vector represents. For example, if the vector is a force, the units would be N or $\mathrm{lb_f}$.

Table 4.1 Scalar and Vector Quantities

Scalar	Vector
Length	Force
Mass	Pressure
Time	Stress
Temperature	Moment of force
Speed	Velocity
Density	Acceleration
Volume	Momentum
Energy	Impulse
Work	Electric field
Resistance	Magnetic field

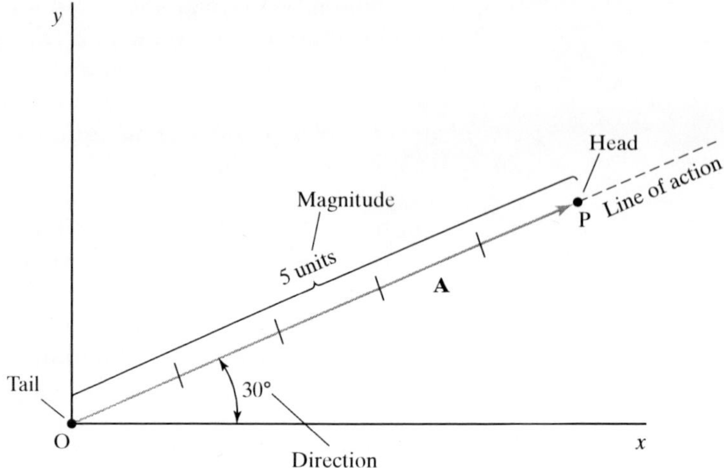

Figure 4.5
A vector has magnitude and direction.

4.2.1 Vector Operations

In order to utilize principles of mechanics to carry out an analysis, engineers must be able to mathematically manipulate vector quantities. Due to the directional character of vectors, the rules for performing algebraic operations with vectors are different from those of scalars. The product of a positive scalar k and a vector **A**, denoted by k**A**, has the effect of changing the length of the vector **A**, but its direction is unaffected. For example, the product 3**A** increases the magnitude of the vector **A** by a factor of three, but the direction of **A** is the same. The product −2**A** increases the magnitude of **A** by a factor of two, but reverses the direction of **A** because the scalar is negative. Graphical examples of the product of scalars and a vector are illustrated in Figure 4.6. Two vectors **A** and **B** are *equal* if they have the same magnitude and direction regardless of the location of their tails and heads. As shown in Figure 4.6, **A** = **B**.

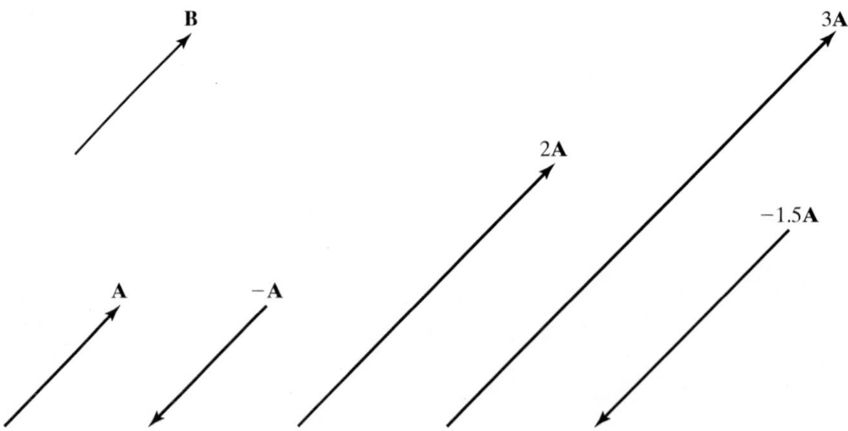

Figure 4.6
Scalar multiplication and vector equality, **A** = **B**.

The addition of two scalars results in a simple algebraic sum, such as $c = a + b$. The addition of two vectors, however, cannot be obtained by simply adding the magnitudes of each vector. Vectors must be added such that their directions as well as their magnitudes are accounted for. Consider the vectors **A** and **B** in Figure 4.7(a). Vectors **A** and **B** may be added by using the *parallelogram law.* To form this sum, **A** and **B** are joined at their tails. Parallel lines are drawn from the

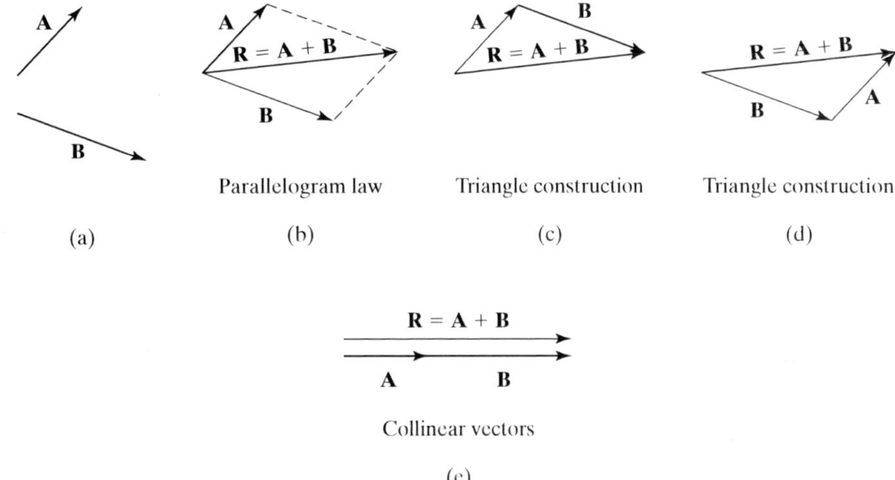

Figure 4.7
Vector addition.

head of each vector, intersecting at a common point, forming adjacent sides of a parallelogram. The vector sum of **A** and **B**, referred to as the *resultant vector* or simply ***resultant***, is the diagonal of the parallelogram that extends from the vector tails to the intersection point, as illustrated in Figure 4.7(b). Hence, we may write the vector sum as $\mathbf{R} = \mathbf{A} + \mathbf{B}$, where **R** is the resultant. The vector sum may also be obtained by constructing a triangle, which is actually half of a parallelogram. In this technique, the tail of **B** is connected to the head of **A**. The resultant $\mathbf{R} = \mathbf{A} + \mathbf{B}$ extends from the tail of **A** to the head of **B**, as shown in Figure 4.7(c).

Alternatively, the triangle may be constructed such that the tail of **A** is connected to the head of **B**, in which case we have $\mathbf{R} = \mathbf{B} + \mathbf{A}$, as shown in Figure 4.7(d). In both triangles, the same resultant is obtained, so we conclude that vector addition is *commutative* (i.e., the vectors can be added in either order). Hence, $\mathbf{R} = \mathbf{A} + \mathbf{B} = \mathbf{B} + \mathbf{A}$. A special case of the parallelogram law is when the two vectors are parallel (such as when they have the same line of action). In that case, the parallelogram is degenerate, and the vector sum reduces to a scalar sum $R = A + B$, as indicated in Figure 4.7(e).

4.2.2 Vector Components

A powerful method for finding the resultant of two vectors is to first find the *rectangular components* of each vector and then add the corresponding components to obtain the resultant. To see how this method works, we draw the two vectors **A** and **B** in Figure 4.7 on a set of (x, y) coordinate axes, shown in Figure 4.8. For convenience, both vectors are drawn with their tails at the origin, and the directions of **A** and **B** with respect to the positive x-axis are defined by the angles α and β, respectively. For the moment, let's consider each vector separately. Using a modified form of the parallelogram law, we draw lines parallel to the x- and y-axes such that the vector **A** becomes the diagonal of a rectangle, which is a special type of parallelogram. The sides of the rectangle that lie along the x- and y-axes are called the rectangular components of vector **A**, and are denoted $\mathbf{A}_x$ and $\mathbf{A}_y$ respectively. Because vector **A** is the diagonal of the rectangle, **A** becomes the resultant of vectors $\mathbf{A}_x$ and $\mathbf{A}_y$. Thus, we may write the vector as $\mathbf{A} = \mathbf{A}_x + \mathbf{A}_y$. Similarly, lines parallel to the x- and y-axes are drawn such that the vector **B** becomes the diagonal of a rectangle. The sides of the rectangle that lie along the x- and y-axes are the rectangular components of

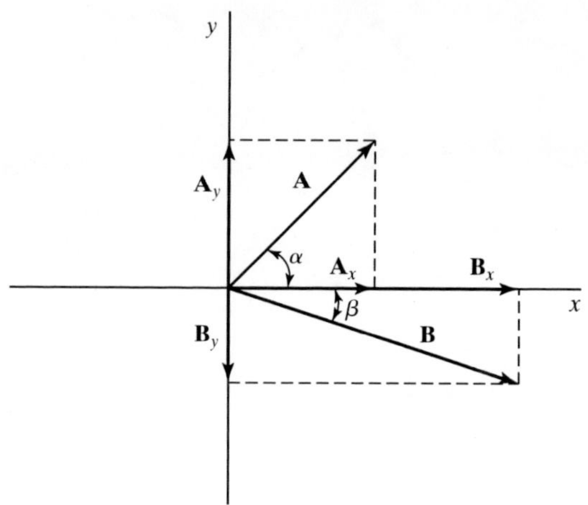

Figure 4.8
Vector components.

vector **B** and are denoted $\mathbf{B}_x$ and $\mathbf{B}_y$, respectively. Hence, we may write the vector as $\mathbf{B} = \mathbf{B}_x + \mathbf{B}_y$. The resultant of **A** and **B** may now be written as:

$$\mathbf{R} = \mathbf{A} + \mathbf{B} = (\mathbf{A}_x + \mathbf{B}_x) + (\mathbf{A}_y + \mathbf{B}_y). \tag{4.1}$$

The magnitude of the components of **A** and **B** may be written in terms of the angles that define the vectors' directions. From the definitions of the trigonometric functions for cosine and sine, the x and y components of **A** are:

$$A_x = A \cos \alpha \tag{4.2}$$

and

$$A_y = A \sin \alpha \tag{4.3}$$

where A is the magnitude of **A**. Similarly, the x and y components of **B** are:

$$B_x = B \cos \beta \tag{4.4}$$

and

$$B_y = B \sin \beta \tag{4.5}$$

where B is the magnitude of **B**. Alternatively, we can also see from trigonometry that:

$$A_y = A_x \tan \alpha \tag{4.6}$$

and

$$B_y = B_x \tan \beta. \tag{4.7}$$

The magnitudes of **A** and **B** form the hypotenuse of their respective right triangles, so from the theorem of Pythagoras, we may write:

$$A = \sqrt{A_x^2 + A_y^2} \tag{4.8}$$

and

$$B = \sqrt{B_x^2 + B_y^2}. \tag{4.9}$$

4.2.3 Unit Vectors

The justification for grouping the x components of each vector and the y components of each vector in Equation (4.1) is based on the concept of *unit vectors*. A unit vector is a *dimensionless vector of unit length used to specify a given direction.* Unit vectors have no other physical meaning. The most common unit vectors are the *rectangular* or ***Cartesian unit vectors***, denoted **i**, **j**, and **k**. The unit vectors **i**, **j**, and **k** coincide with the positive x-, y-, and z-axes respectively, as shown in Figure 4.9. The rectangular unit vectors form a set of mutually perpendicular vectors and are used to specify the direction of a vector in three-dimensional space.

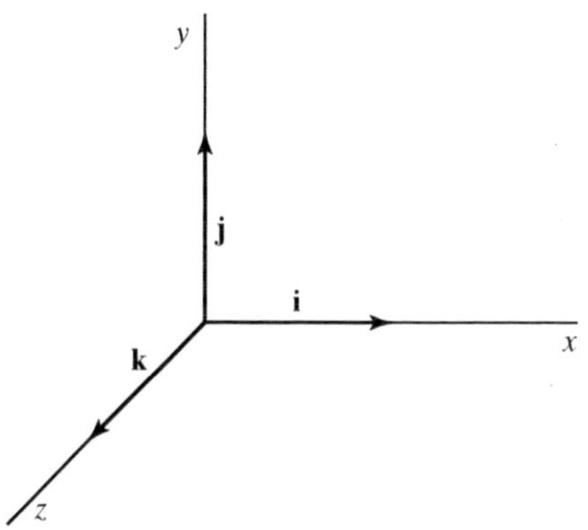

Figure 4.9
Rectangular unit vectors.

If the quantity of interest can be described by a two-dimensional vector, only the **i** and **j** unit vectors are required. The vectors **A** and **B**, shown in Figure 4.8, lie in the x–y plane, so they can be represented by the **i** and **j** unit vectors. The x component of **A** has a magnitude of A_x, and the y component of **A** has a magnitude of A_y. Note that the quantities A_x and A_y are not vectors but scalars, because they represent magnitudes only.

The vector components $\mathbf{A}_x$ and $\mathbf{A}_y$ can be written as products of a scalar and a unit vector as $\mathbf{A}_x = A_x\mathbf{i}$ and $\mathbf{A}_y = A_y\mathbf{j}$. Thus, vector **A** is expressed as:

$$\mathbf{A} = A_x\mathbf{i} + A_y\mathbf{j} \tag{4.10}$$

and vector **B** is expressed as:

$$\mathbf{B} = B_x\mathbf{i} + B_y\mathbf{j}. \tag{4.11}$$

Rewriting Equation (4.1) in terms of the x and y component groups, the resultant of **A** and **B** is:

$$\mathbf{R} = \mathbf{A} + \mathbf{B} = (A_x + B_x)\mathbf{i} + (A_y + B_y)\mathbf{j}. \tag{4.12}$$

The rectangular components of the resultant vector **R** are given by:

$$R_x = A_x + B_x \tag{4.13}$$

and

$$R_y = A_y + B_y. \tag{4.14}$$

Hence, Equation (4.12) can be written:

$$\mathbf{R} = R_x\mathbf{i} + R_y\mathbf{j} \tag{4.15}$$

where R_x and R_y are the x and y components of **R**, as shown in Figure 4.10. By trigonometry, we may write:

$$R_x = R\cos\theta \tag{4.16}$$

$$R_y = R\sin\theta \tag{4.17}$$

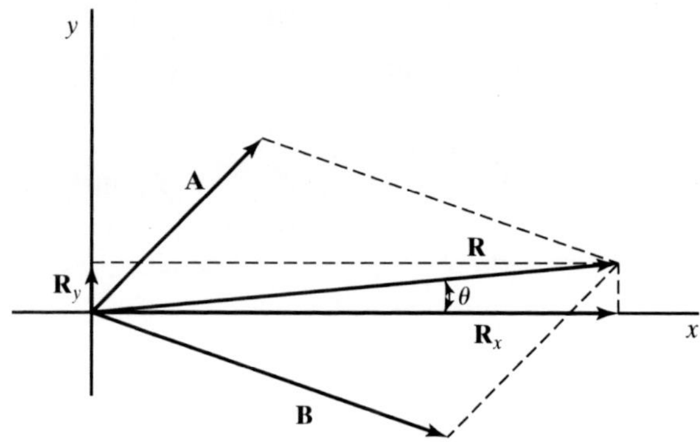

Figure 4.10
Resultant vector.

and

$$R_y = R_x \tan \theta. \tag{4.18}$$

The magnitude of **R** forms the hypotenuse of a right triangle, hence, from the theorem of Pythagoras, we have:

$$R = \sqrt{R_x^2 + R_y^2}. \tag{4.19}$$

EXAMPLE 4.1

Two vectors have magnitudes of $A = 8$ and $B = 6$ and directions as shown in Figure 4.11(a). Find the resultant vector, using (a) the parallelogram law and (b) by resolving the vectors into their x and y components.

Solution

(a) Parallelogram law

The parallelogram for vectors **A** and **B** is shown in Figure 4.11(b). In order to find the magnitude and direction of the resultant vector **R**, some angles must be determined. By subtraction, the acute angle between the vectors is 45°. The sum of the interior angles of a quadrilateral is 360°, so the adjacent angle is found to be 135°. The magnitude of **R** may be found by using the law of cosines:

$$R = \sqrt{6^2 + 8^2 - 2(6)(8) \cos 135°}$$

$$R = \sqrt{36 + 64 - 96(-0.7071)}$$

$$= 12.96.$$

The direction of **R** is found by calculating the angle θ. Using the law of sines, we have:

$$\frac{\sin \theta}{6} = \frac{\sin 135°}{12.96}$$

$$\sin \theta = 0.3274$$

$$\theta = \sin^{-1}(0.3274) = 19.1°.$$

Thus, the angle of **R** with respect to the positive x-axis is:

$$\phi = 19.1° + 15° = 34.1°.$$

The resultant vector **R** has now been completely defined, because both its direction and magnitude have been determined.

Figure 4.11
Example 4.1.

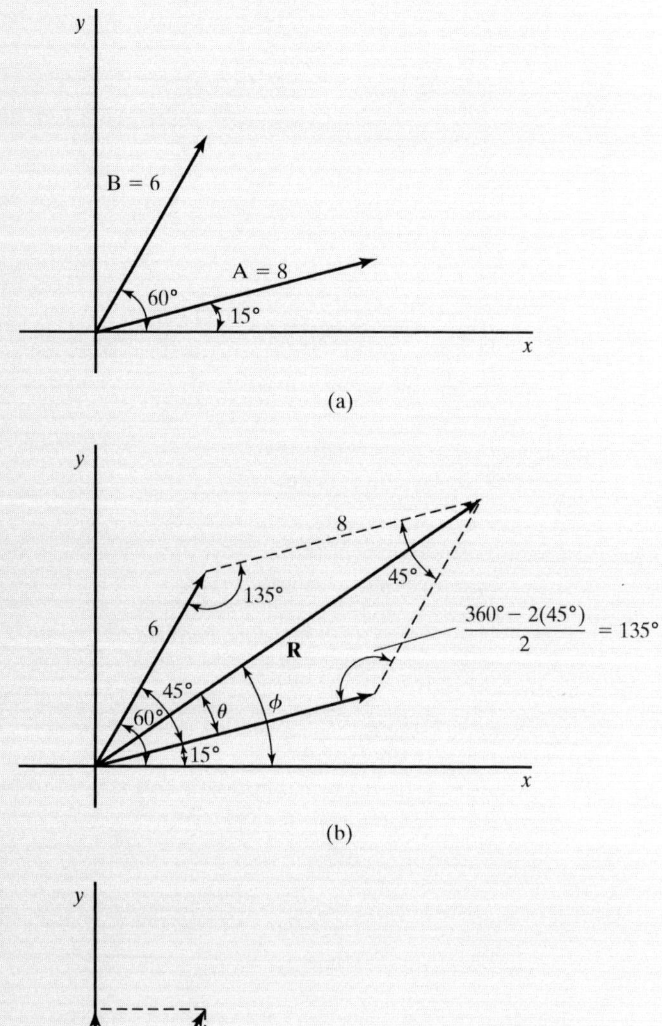

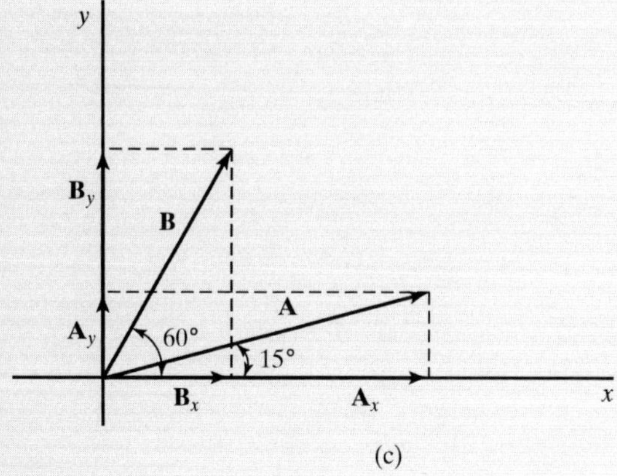

(b) Vector components

In Figure 4.11(c), vectors **A** and **B** are resolved into their x and y components. The magnitudes of these components are:

$$
\begin{aligned}
A_x &= A\cos 15^\circ = 8\cos 15^\circ = 7.7274 \\
A_y &= A\sin 15^\circ = 8\sin 15^\circ = 2.0706 \\
B_x &= B\cos 60^\circ = 6\cos 60^\circ = 3 \\
B_y &= B\sin 60^\circ = 6\sin 60^\circ = 5.1962.
\end{aligned}
$$

Vectors **A** and **B** may now be written in terms of the unit vectors **i** and **j**:

$$\mathbf{A} = A_x\mathbf{i} + A_y\mathbf{j} = 7.7274\mathbf{i} + 2.0706\mathbf{j}$$
$$\mathbf{B} = B_x\mathbf{i} + B_y\mathbf{j} = 3\mathbf{i} + 5.1962\mathbf{j}.$$

The resultant vector **R** is:

$$\mathbf{R} = \mathbf{A} + \mathbf{B} = R_x\mathbf{i} + R_y\mathbf{j} = (7.7274 + 3)\mathbf{i} + (2.0706 + 5.1962)\mathbf{j}$$
$$= 10.7274\mathbf{i} + 7.2668\mathbf{j}.$$

This is the answer, but to compare it with the answer obtained by the parallelogram law, we must find the magnitude of **R** and its direction with respect to the positive x-axis. Using the theorem of Pythagoras, we find that the magnitude of **R** is:

$$R = \sqrt{10.7274^2 + 7.2668^2} = 12.96.$$

The direction is given by:

$$R_y = R_x \tan \phi.$$

Solving for the angle ϕ, we get:

$$\phi = \tan^{-1}(R_y/R_x) = \tan^{-1}(7.2668/10.7274) = 34.1°.$$

We have obtained the same result with two different methods of vector addition. The second method may appear to involve more work. However, many mechanics problems involve more than two vectors, and the problem may be three dimensional. In these cases, resolving the vectors into their rectangular components is the preferred approach, since the parallelogram law is too cumbersome. In the example, to ensure that both methods yielded the same answers to three significant figures, four decimal places were used.

EXAMPLE 4.2

For the vectors $\mathbf{A} = 3\mathbf{i} - 6\mathbf{j} + \mathbf{k}$, $\mathbf{B} = 5\mathbf{i} + \mathbf{j} - 2\mathbf{k}$, and $\mathbf{C} = -2\mathbf{i} + 4\mathbf{j} + 3\mathbf{k}$, find the resultant vector and its magnitude.

Solution

These vectors, unlike those in the previous example, are three-dimensional. They are already expressed in terms of the Cartesian unit vectors **i**, **j**, and **k**, so it is a straightforward matter to add them vectorially. Recall that the **i**, **j**, and **k** unit vectors correspond to the positive x, y, and z directions, respectively. To find the resultant, we simply add the x components, the y components, and the z components of each vector. To help us avoid errors as we perform the addition, it is useful to write the vectors with their components aligned in columns:

$$\mathbf{A} = 3\mathbf{i} - 6\mathbf{j} + 1\mathbf{k}$$
$$\mathbf{B} = 5\mathbf{i} + 1\mathbf{j} - 2\mathbf{k}$$
$$\mathbf{C} = -2\mathbf{i} + 4\mathbf{j} + 3\mathbf{k}.$$

Performing the additions, the resultant vector is:

$$\mathbf{R} = (3 + 5 - 2)\mathbf{i} + (-6 + 1 + 4)\mathbf{j} + (1 - 2 + 3)\mathbf{k}$$
$$= 6\mathbf{i} - \mathbf{j} + 2\mathbf{k}.$$

The magnitude of the resultant vector is found by extending the theorem of Pythagoras to three dimensions:

$$R = \sqrt{R_x^2 + R_y^2 + R_z^2}$$
$$= \sqrt{6^2 + (-1)^2 + 2^2} = 6.40.$$

4.3 FORCES

From our early childhood experiences, we all have a basic understanding of the concept of force. We commonly use terms such as *push*, *pull*, and *lift* to describe forces that we encounter in our daily lives. *Mechanics* is the study of the state of rest or motion of bodies that are subjected to forces. To the engineer, ***force*** is defined as *an influence that causes a body to deform or accelerate*. For example, when you push or pull on a lump of clay, the clay deforms into a different shape. When you pull on a rubber band, the rubber band increases in length. The forces required to deform clay and rubber bands are much smaller than those required to deform engineering structures such as buildings, bridges, dams, and machines, but these objects deform nevertheless.

What happens when you push on the wall with your hand? In accordance with Newton's third law, as you push on the wall, the wall pushes back on your hand with the same force. When you push on a book in an attempt to slide it across the table, the book will not move unless the frictional force between the table and book is exceeded by the horizontal pushing force. These types of situations are encountered in virtually all engineering systems that are in static equilibrium. Forces are present, but motion does not occur because the forces cause the body to be in a state of balance. When the forces acting on a body are unbalanced, the body undergoes an acceleration. For example, the propulsive force delivered to the wheels of an automobile can exceed the frictional forces that tend to retard the automobile's motion, so the automobile accelerates. Similarly, the thrust and lift forces acting on an aircraft can exceed the weight and drag forces, thereby allowing the aircraft to accelerate vertically and horizontally.

Forces commonly encountered in the majority of engineering systems may be generally categorized as a *contact force, gravitational force, cable force, pressure force*, or *fluid dynamic force.* These five types of forces are depicted in Figure 4.12. A contact force is a force produced by two or more bodies in direct contact. The force produced by pushing on a wall is a contact force because the hand is in direct contact with the wall. When two billiard balls collide, a contact force is produced at the region where the balls touch each other. Friction is a type of contact force. A gravitational force, referred to as *weight*, is exerted on an object on or near the earth's surface. Gravitational forces are directed downward, toward the center of the earth, and act through a point in the body called the *center of gravity*. For a body that is uniform in density, the center of gravity lies at the geometric center of the body. This point is referred to as the *centroid*. The force in a cable is actually a special type of contact force, since the cable is in contact with a body, but it occurs so frequently that it deserves a separate definition. Cables, ropes, and cords are used in pulley systems, suspension bridges, and other engineering structures. A cable, due to its limp and flexible nature, can support tension forces only. Forces in cables are always

Figure 4.12
Types of forces commonly encountered in engineering applications.

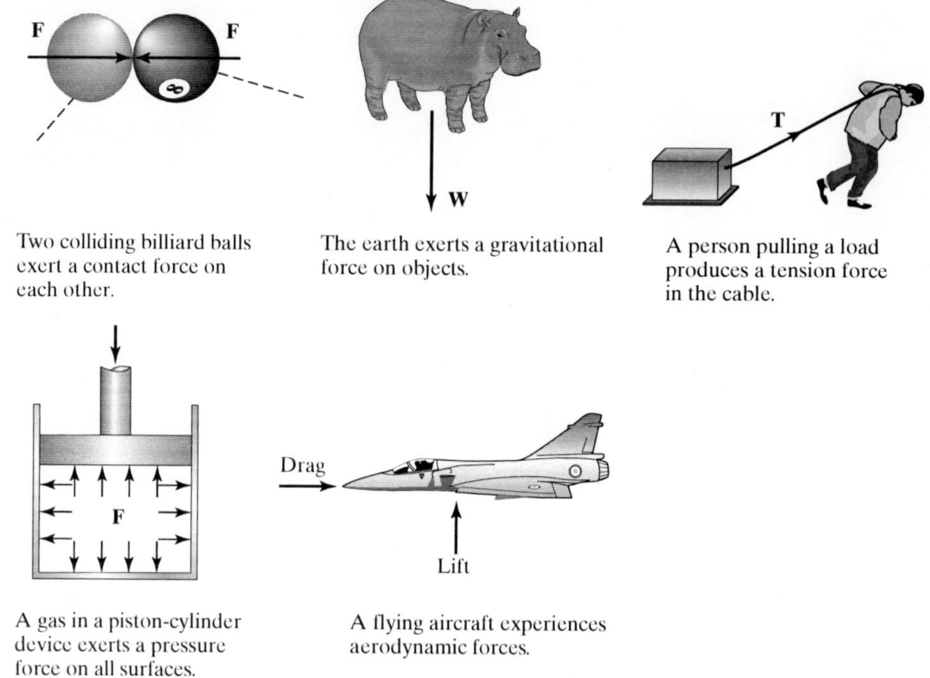

directed along the axis of the cable, regardless of whether the cable is straight or not. Pressure forces are normally associated with static fluids. A gas in a cylinder exerts a pressure force on all surfaces of the cylinder. A static liquid, such as the water behind a dam, exerts a pressure force on the dam. Pressure forces always act in a direction normal to the surface. A fluid dynamic force is produced when a fluid flows around a body or through a pipe or conduit. When a fluid flows such as air around a body (or when a body moves through a fluid) aerodynamic forces act on the body. There are basically two types of aerodynamic forces: pressure forces and viscous forces. Pressure forces are caused by pressure distributions around the body and are produced by certain fluid-related mechanisms and body geometry. Viscous forces, sometimes called friction or shear forces, are caused by fluid viscosity. Any object (for example, airplane, missile, ship, submarine, automobile, baseball) that moves through a fluid experiences aerodynamic forces. When a fluid flows through a pipe, a friction force is produced between the fluid and the inside surface of the pipe. This friction force, which is caused by fluid viscosity, has the effect of retarding the flow. The five types of forces just mentioned are the most common, but there are other kinds of forces that engineers sometimes encounter. These include electric, magnetic, nuclear, and surface tension forces.

Forces are vectors, so all the mathematical operations and expressions that apply to vectors apply to forces. Because a force is a vector, a force has magnitude and direction. For example, the weight of a 170-pound person is a vector with a magnitude of 170 lb_f and a direction downward. A situation in which more than one force acts on a body is referred to as a ***force system***. A system of forces is *coplanar* or *two dimensional* if the lines of action of the forces lie in the same plane. Otherwise, the system of forces is *three dimensional*. Forces are *concurrent* if their lines of action pass through the same point and *parallel* if their lines of action are parallel. *Collinear* forces have the same line of action. These force concepts are illustrated in Figure 4.13.

Figure 4.13
Force systems.

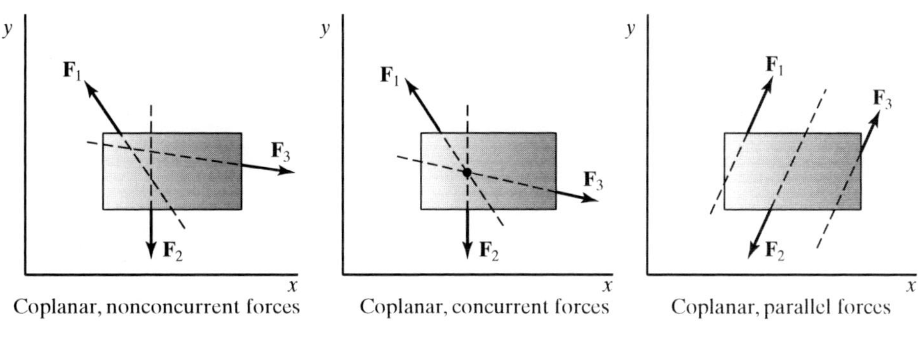

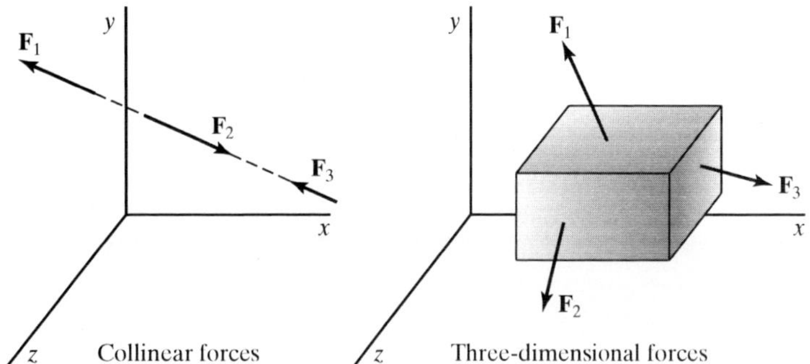

EXAMPLE 4.3

Three coplanar forces act as shown in Figure 4.14. Find the resultant force, its magnitude and its direction with respect to the positive x-axis.

Solution

We have three coplanar forces that act concurrently at the origin. Note that force $\mathbf{F}_1$ lies along the x-axis. First, we resolve the forces into their x and y components:

$$F_{1x} = F_1 \cos 0° = 10 \cos 0° = 10 \text{ kN}$$
$$F_{1y} = F_1 \sin 0° = 10 \sin 0° = 0 \text{ kN}$$

Figure 4.14
Concurrent forces for Example 4.3.

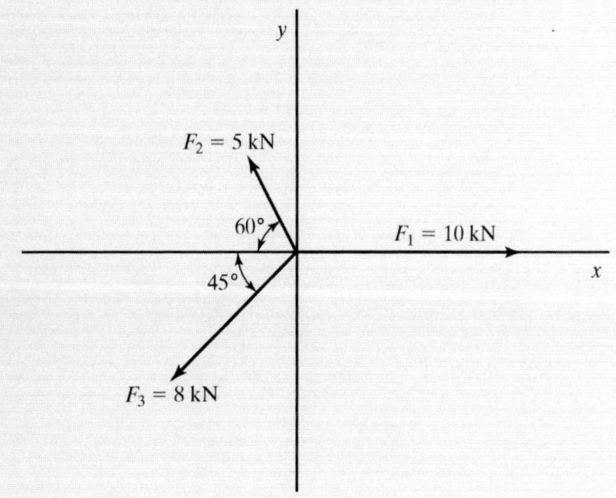

$$F_{2x} = -F_2 \cos 60° = -5 \cos 60° = -2.5 \text{ kN}$$
$$F_{2y} = F_2 \sin 60° = 5 \sin 60° = 4.330 \text{ kN}$$
$$F_{3x} = -F_3 \cos 45° = -8 \cos 45° = -5.657 \text{ kN}$$
$$F_{3y} = -F_3 \sin 45° = -8 \cos 45° = -5.657 \text{ kN}.$$

Notice that F_{2x}, F_{3x}, and F_{3y} are *negative* quantities to reflect the proper directions of the vectors with respect to the positive *x*- and *y*-axes. The forces may now be written in terms of the unit vectors **i** and **j**:

$$F_1 = F_{1x}\mathbf{i} + F_{1y}\mathbf{j} = 10\mathbf{i} + 0\mathbf{j} = 10\mathbf{i} \text{ kN}$$
$$F_2 = F_{2x}\mathbf{i} + F_{2y}\mathbf{j} = -2.5\mathbf{i} + 4.330\mathbf{j} \text{ kN}$$
$$F_3 = F_{3x}\mathbf{i} + F_{3y}\mathbf{j} = -5.657\mathbf{i} - 5.657\mathbf{j} \text{ kN}.$$

Earlier in this chapter we learned that a resultant is the sum of two or more vectors. Here we define a **resultant force** as the sum of two or more forces. Therefore, the resultant force $\mathbf{F}_R$ is the vector sum of the three forces. Adding corresponding components, we obtain:

$$\mathbf{F}_R = (10 - 2.5 - 5.657)\mathbf{i} + (0 + 4.330 - 5.657)\mathbf{j}$$
$$= 1.843\mathbf{i} - 1.327\mathbf{j} \text{ kN}.$$

The signs on the *x* and *y* components of $\mathbf{F}_R$ are significant. A positive sign on the *x* component and a negative sign on the *y* component means that the resultant force lies in the fourth quadrant. The magnitude of $\mathbf{F}_R$ is:

$$F_R = \sqrt{1.843^2 + (-1.327)^2}$$
$$= 2.271 \text{ kN}.$$

The direction of $\mathbf{F}_R$ with respect to the positive *x*-axis is:

$$\phi = \tan^{-1}(-1.327/1.843) = -35.8°$$

where the minus sign on the angle is consistent with the fact that $\mathbf{F}_R$ lies in the fourth quadrant, as shown in Figure 4.15.

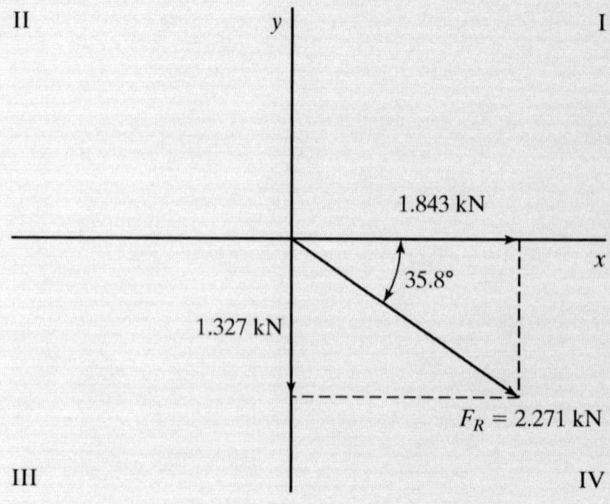

Figure 4.15
Resultant force for Example 4.4.

PRACTICE!

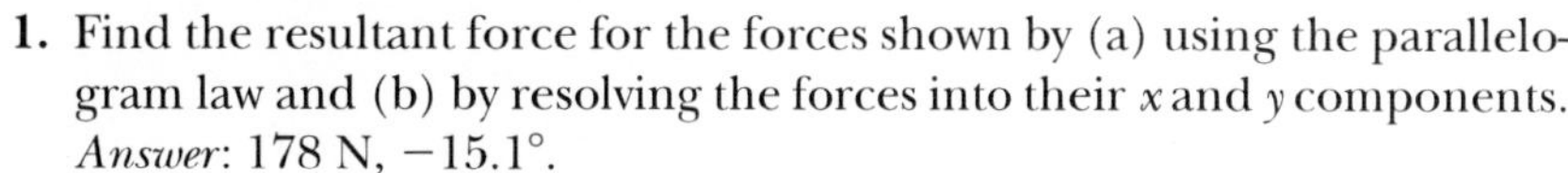

1. Find the resultant force for the forces shown by (a) using the parallelogram law and (b) by resolving the forces into their x and y components. *Answer*: 178 N, −15.1°.

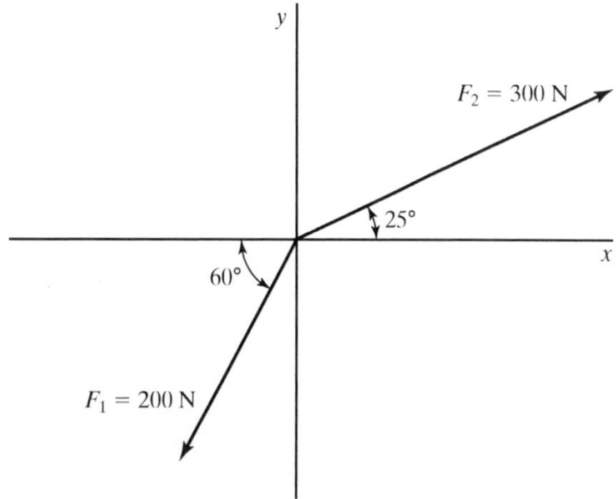

2. Find the resultant force for the forces shown by (a) using the parallelogram law and (b) by resolving the forces into their x and y components. *Answer*: 166 N, 5.5°.

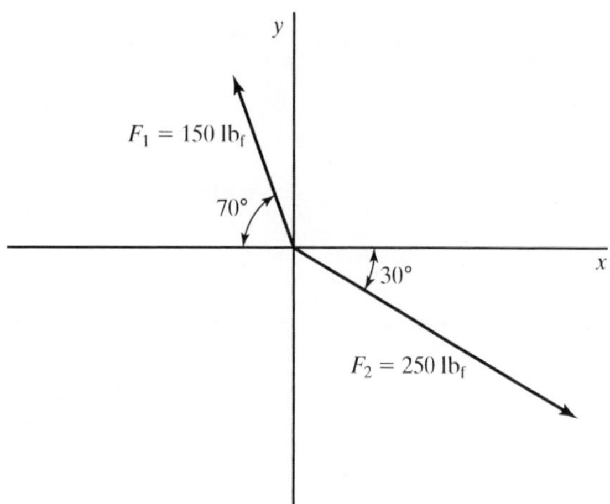

3. Find the resultant force for the forces shown by (a) using the parallelogram law and (b) by resolving the forces into their x and y components. *Answer*: 26.0 kN, 75.0°.

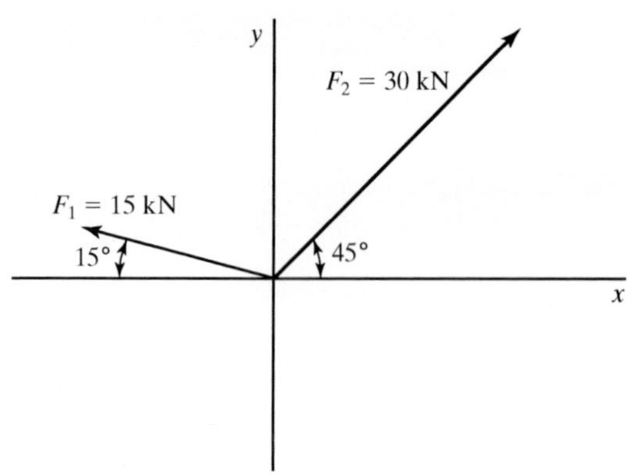

4. Consider the three forces $\mathbf{F}_1 = 5\mathbf{i} + 2\mathbf{j}$ kN, $\mathbf{F}_2 = -3\mathbf{i} - 8\mathbf{j}$ kN, and $\mathbf{F}_3 = -4\mathbf{i} + 7\mathbf{j}$ kN. Find the resultant force, its magnitude, and its direction with respect to the positive x-axis.
Answer: $-2\mathbf{i} + \mathbf{j}$ kN, 2.24 kN, 153°.
5. Consider the three forces $\mathbf{F}_1 = 10\mathbf{i} + 9\mathbf{j}$ kN, $\mathbf{F}_2 = -8\mathbf{i} - 4\mathbf{j}$ kN, and $\mathbf{F}_3 = 4\mathbf{i} + 5\mathbf{j}$ kN. Find the resultant force, its magnitute, and its direction with respect to the positive x-axis.
Answer: $6\mathbf{i} + 10\mathbf{j}$ kN, 11.7 kN, 59.0°.
6. Consider the three forces $\mathbf{F}_1 = 2\mathbf{i} + 8\mathbf{j} + 7\mathbf{k}$ lb_f, $\mathbf{F}_2 = 3\mathbf{i} - \mathbf{j} - \mathbf{k}$ lb_f, and $\mathbf{F}_3 = 3\mathbf{i} + \mathbf{j} + 2\mathbf{k}$ lb_f. Find the resultant force and its magnitude.
Answer: $8(\mathbf{i} + \mathbf{j} + \mathbf{k})$ lb_f, 13.9 lb_f

APPLICATION

STABILIZING A COMMUNICATIONS TOWER WITH CABLES

Tall slender structures often incorporate cables to stabilize them. The cables, which are connected at various points around the structure and along its length, are connected to concrete anchors buried deep in the ground. Shown in Figure 4.16(a) is a typical communications tower that is stabilized with several cables. On this particular tower each ground anchor facilitates two cables that are connected at a common point, as shown in Figure 4.16(b). The upper and lower cables exert forces of 15 kN and 25 kN, respectively, and their directions are 45° and 32°, respectively, as measured from the ground (Figure 4.16(c)). What is the resultant force exerted by the cables on the ground anchor?

Any two forces in three-dimensional space lie in a single plane, so we may arbitrarily locate our two cable forces in the x–y plane. Thus, we have two coplanar forces that act concurrently at the origin. We let $F_1 = 15$ kN and $F_2 = 25$ kN. We resolve the forces into their x and y components:

$$\begin{aligned} F_{1x} &= F_1 \cos 45° = 15 \cos 45° = 10.607 \text{ kN} \\ F_{1y} &= F_1 \sin 45° = 15 \sin 45° = 10.607 \text{ kN} \\ F_{2x} &= F_2 \cos 32° = 25 \cos 32° = 21.201 \text{ kN} \\ F_{2y} &= F_2 \sin 32° = 25 \sin 32° = 13.248 \text{ kN}. \end{aligned}$$

The forces may now be written in terms of the unit vectors **i** and **j**:

$$\begin{aligned} \mathbf{F}_1 &= F_{1x}\mathbf{i} + F_{1y}\mathbf{j} = 10.607\mathbf{i} + 10.607\mathbf{j} \text{ kN} \\ \mathbf{F}_2 &= F_{2x}\mathbf{i} + F_{2y}\mathbf{j} = 21.201\mathbf{i} + 13.248\mathbf{j} \text{ kN}. \end{aligned}$$

(a) (b)

(c)

Figure 4.16
A communications tower stabilized with cables.

The resultant force $\mathbf{F}_R$ is the vector sum of the two forces. Adding corresponding components, we obtain:

$$\mathbf{F}_R = (10.607 + 21.201)\mathbf{i} + (10.607 + 13.248)\mathbf{j}$$
$$= 31.808\mathbf{i} + 23.855\mathbf{j}\ \text{kN}.$$

The magnitude of $\mathbf{F}_R$ is:

$$F_R = \sqrt{F_{Rx}^2 + F_{Ry}^2}$$
$$= \sqrt{31.808^2 + 23.855^2}$$
$$= 39.76\ \text{kN}$$

and the direction of $\mathbf{F}_R$ with respect to the ground is:

$$\phi = \tan^{-1}(23.855/31.808)$$
$$= 36.9^\circ.$$

What does our answer mean, and how would it be used? The resultant force would be used by an engineer (probably a civil engineer) to design the concrete anchor. A force of nearly 40 kN directed at an angle of about 37° with respect to the ground would have a tendency to pull the anchor out of the ground. If not designed properly, the anchor could become loose or break under the load, thereby causing an unbalanced force on the tower. Look carefully at Figure 4.16(b). Notice that the two cables connect via turnbuckles at a ring assembly connected to a single rod that goes into the concrete anchor, which is not shown. The resultant force would also be used to ascertain the structural integrity of the ring assembly and rod.

4.4 FREE-BODY DIAGRAMS

One of the most important steps in the general analysis procedure is to construct a diagram of the system being analyzed. In engineering mechanics, this diagram is referred to as a free-body diagram. A ***free-body diagram*** is a *diagram that shows all external forces acting on the body*. As the term implies, a free-body diagram shows only the body in question, being isolated or "free" from all other bodies. The body is conceptually removed from all supports, connections, and regions of contact with other bodies. All forces produced by these external influences are schematically represented on the free-body diagram. In a free-body diagram, only the *external* forces acting on the body in question are considered in the analysis. There may be ***internal forces*** (i.e., forces originating from inside the body that act on other parts of the body), but it can be shown that these forces cancel one another and therefore do not contribute to the overall mechanical state of the body. The free-body diagram is one of the most critical parts of a mechanical analysis. It focuses the engineer's attention on the body being analyzed and helps to identify all the external forces acting on the body. The free-body diagram also helps the engineer to write the correct governing equations.

Free-body diagrams are used in statics, dynamics, and mechanics of materials, but their application to statics and mechanics of materials will be emphasized here. ***Statics*** is the branch of engineering mechanics that deals with bodies in static equilibrium. If a body is in static equilibrium, the external forces cause the body to be in a state of balance. Even though the body does not move, it experiences stresses and deformations that must be determined if its performance as a structural member is to be evaluated. In order to determine the forces that act on the body, a free-body diagram must be properly constructed.

Procedure for Constructing Free-Body Diagrams

The following procedure should be followed when constructing free-body diagrams:

1. *Identify* the body you wish to isolate and make a *simple drawing* of it.
2. Draw the appropriate *force vectors* at all locations of supports, connections and contacts with other bodies.
3. Draw a force vector for the *weight* of the body, unless the gravitational force is to be neglected in the analysis.
4. *Label* all forces that are known with a numerical value and those that are unknown with a letter.
5. Draw a *coordinate system* on, or near, the free-body to establish directions of the forces.
6. Add *geometric data* such as lengths and angles as required.

Free-body diagrams for some of the most common force configurations are illustrated in Figure 4.17.

Figure 4.17
Free-body diagrams for some common force configurations.

Configuration	Free-body diagram	Comments
Gravitational force 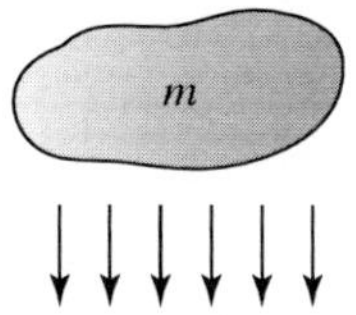 m	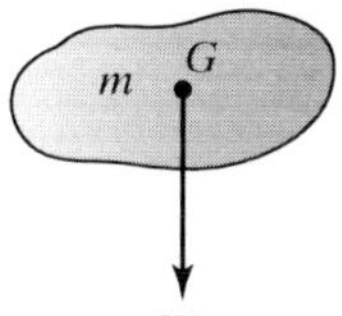 m, G $W = mg$	The gravitational force acts through the center of gravity G.
Cable force 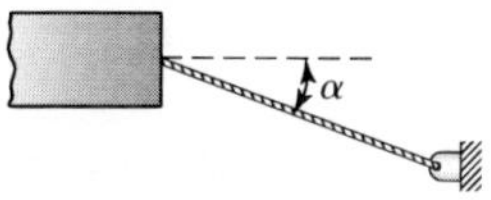 α Weight of cable neglected 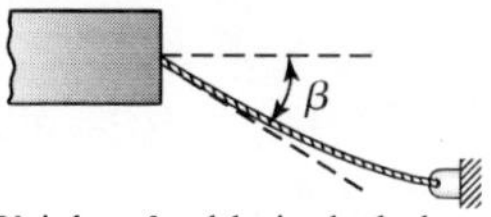 β Weight of cable included	α, T β, T	The tension force T in a cable is always directed along the axis of the cable.
Contact force 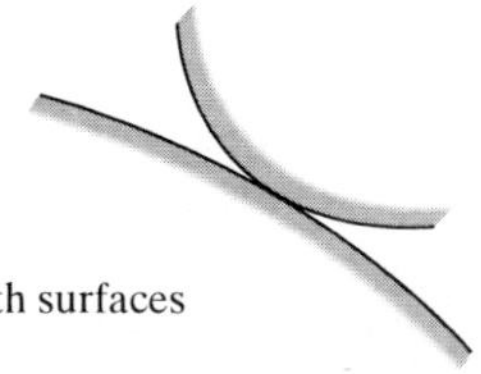 Smooth surfaces 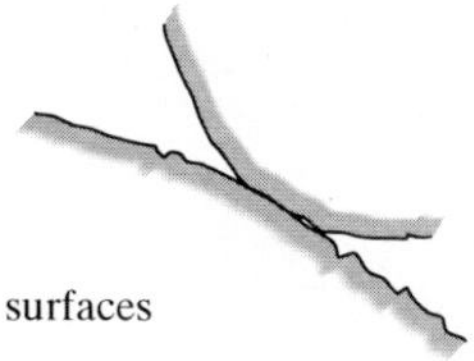	N	For smooth surfaces, the contact force N is toward the body, normal to the tangent drawn through the point of contact.
Rough surfaces	F, N	For rough surfaces, there are two forces, a normal force N and a friction force, F. These two forces are perpendicular to each other. The friction force F acts in the direction opposing the impeding motion.
Roller support 	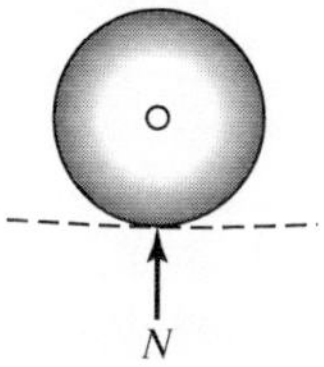 N	A roller supports a normal force but no friction force because a friction force would cause the roller to rotate.
Pin connection 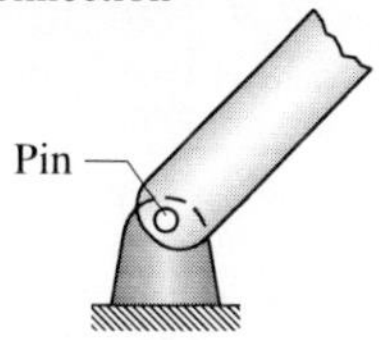 Pin	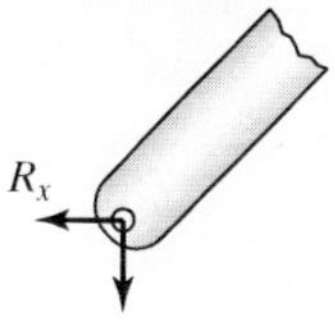 R_x, R_y	A pin connection can support a reaction force in any direction in the plane normal to the pin's axis. This force may be resolved into its x and y components, R_x and R_y.

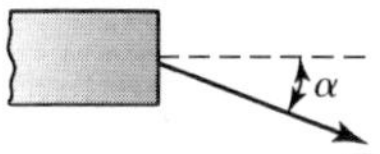

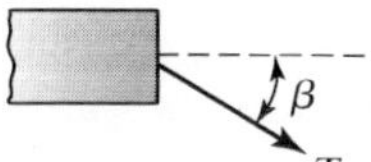

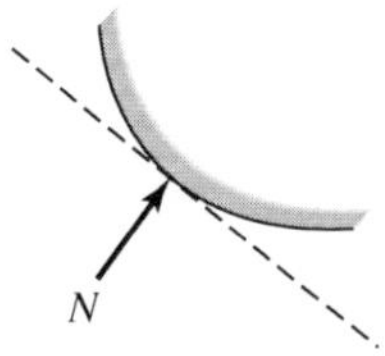

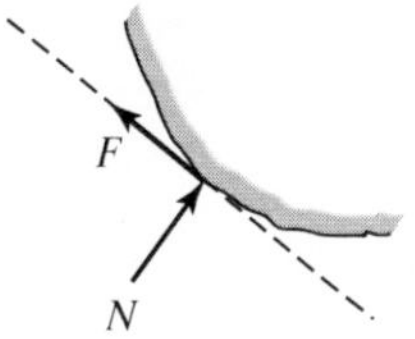

PROFESSIONAL SUCCESS—DON'T BEGIN IN THE MIDDLE OF A PROBLEM

It's human nature to want to finish a job in the least amount of time. Sometimes, we take shortcuts without taking enough time to assure that the job is done thoroughly. Like everyone else, engineers are only human and may sometimes take shortcuts in the solution of a problem. Engineers may take shortcuts for a variety of reasons. Perhaps the engineer is simply overloaded with work, and the only way to meet deadlines is to spend less time on each problem. Perhaps the engineer's manager has unrealistic expectations and does not budget enough time for each project. Time and budget-related reasons, while serious enough to warrant corrective action, are not usually the reasons that engineers take shortcuts in their analytical work. They take shortcuts because they either have become lax in their problem-solving practices or have forgotten how to perform a thorough analysis of a problem. Perhaps they have forgotten some of the steps in the general analysis procedure, or even worse, never learned them at all.

Regardless of the underlying reasons, the practice of taking problem-solving shortcuts may provoke an engineer to begin an analysis "in the middle of the problem." How does this happen? In an attempt to solve the problem more efficiently, the engineer may want to get right to the equations and calculations. By going directly to the *governing equations* and *calculations* steps of the analysis procedure, three crucial steps are omitted: problem statement, diagram, and assumptions. How can an engineer solve a problem if he or she does not even state what the problem is? The engineer may defensively exclaim, "But I know what the problem statement is. It's in my head." A problem statement not written is not a problem statement! Others who will review the analysis cannot read minds. A good engineer documents everything in writing, including problem statements. The engineer may say, "Everyone knows exactly what the component looks like, and the forces acting on it are straightforward. A free-body diagram is unnecessary." Everyone may be intimately familiar with the component's configuration and loading today, but 18 months from now, when the analysis is re-evaluated because the component failed in its first year of service, everyone, including the engineer who did the analysis, may not remember all the details. Once again, written documentation is essential. The formulation of good assumptions is as much an art as it is a science. A hurried engineer may declare, "The assumptions are obvious. It's no big deal," The assumptions may or may not be obvious, but they are critical to the outcome of the problem. Assumptions must be explicitly stated, and the governing equations and calculations must be consistent with those assumptions. If the component failed in its first year of service, it is perhaps because the engineer *thought* the assumptions were obvious, and they were not, resulting in a flawed analysis and a failed component.

While you are in school, develop the habit of conscientiously applying the general analysis procedure to all your analytical problem-solving work. Then, as you make the transition from student to engineering professional, you will not experience the pitfalls of beginning "in the middle of a problem."

PRACTICE!

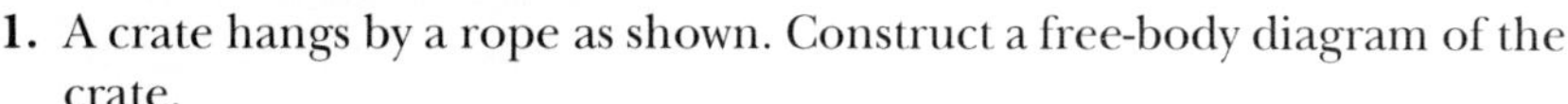

1. A crate hangs by a rope as shown. Construct a free-body diagram of the crate.

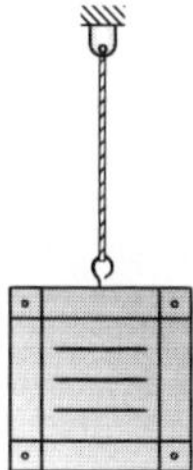

Answer:

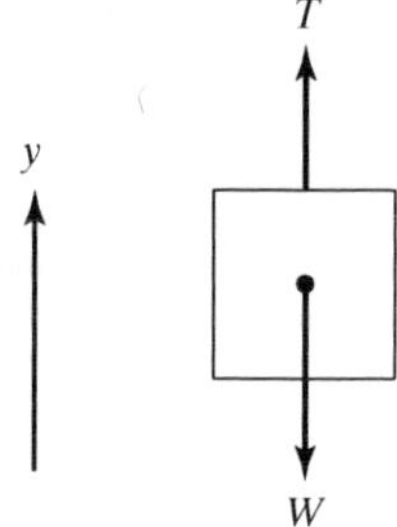

2. Two crates hang by ropes from a ceiling as shown. Construct a free-body diagram of (a) crate A and (b) crate B.

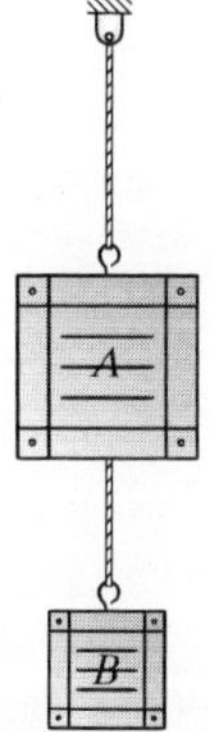

Answer:

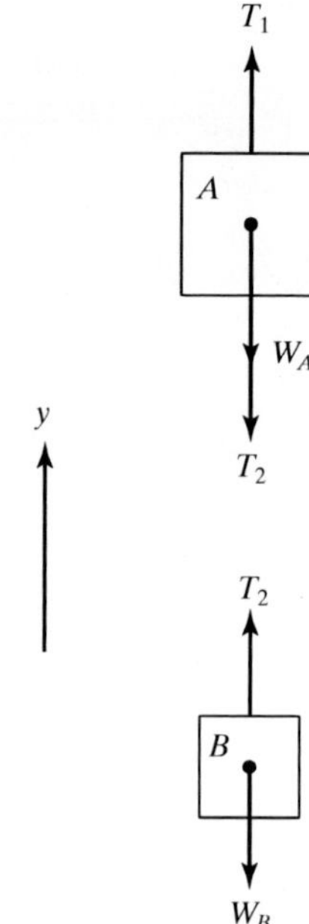

3. A wooden block rests on a rough inclined plane as shown. Construct a free-body diagram of the block.

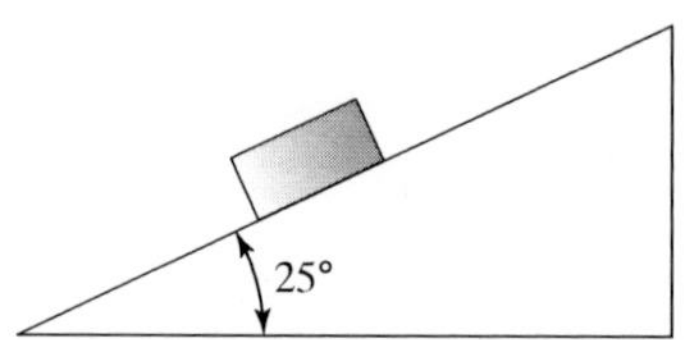

Answer:

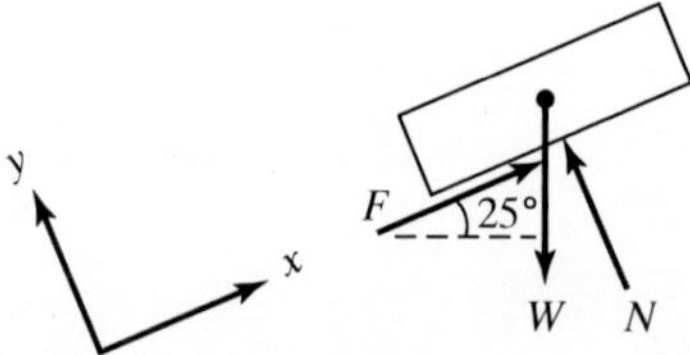

4. An obliquely loaded I-beam is supported by a roller at A and a pin at B as shown. Construct a free-body diagram of the beam. Include the weight of the beam.

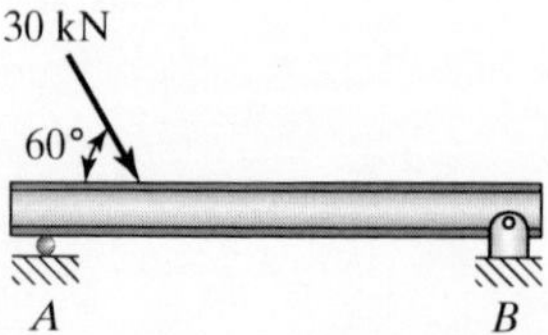

Answer:

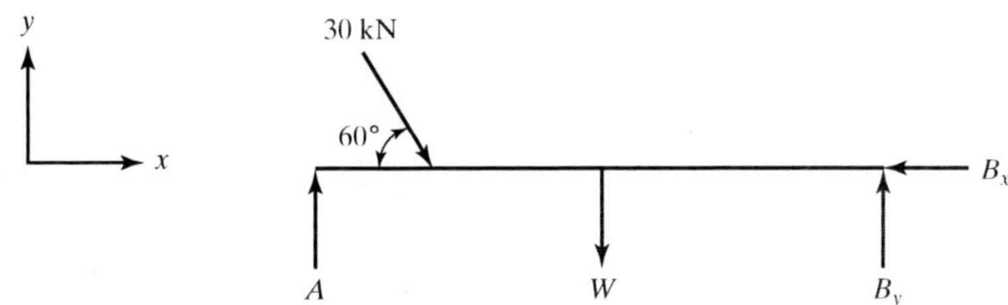

5. A horizontal pulling force P acts on block A as shown. Block B, which rests on block A, is tied to a rigid wall by a cable. The force P is not sufficient to cause block A to move. If all surfaces are rough, construct a free-body diagram of each block.

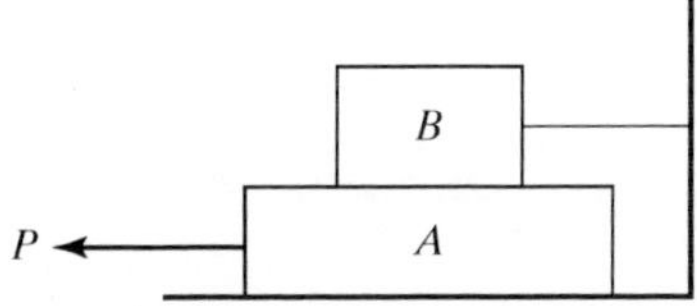

Answer:

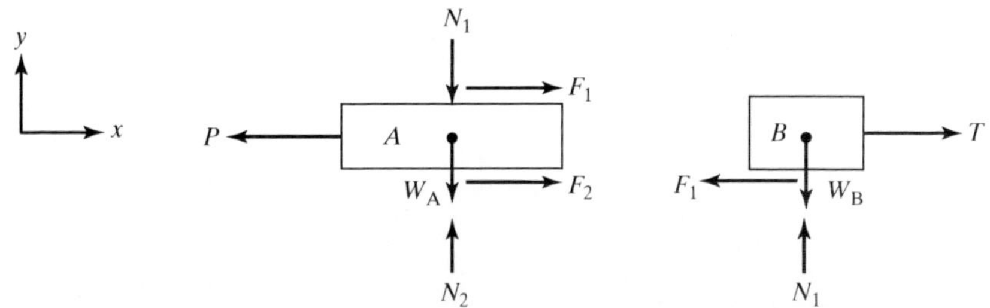

6. A box is held in position on the bed of a truck by a cable as shown. The surface of the truck bed is smooth. Construct a free-body diagram of the box.

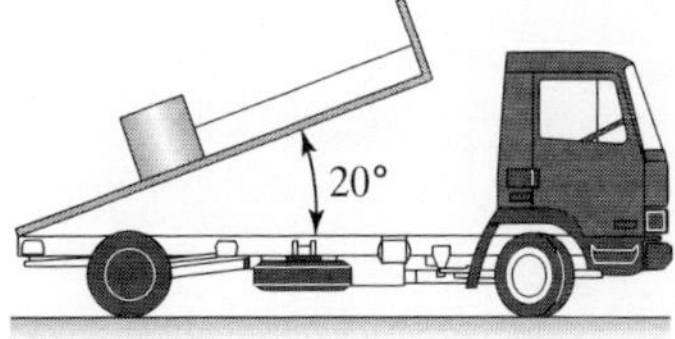

Answer:

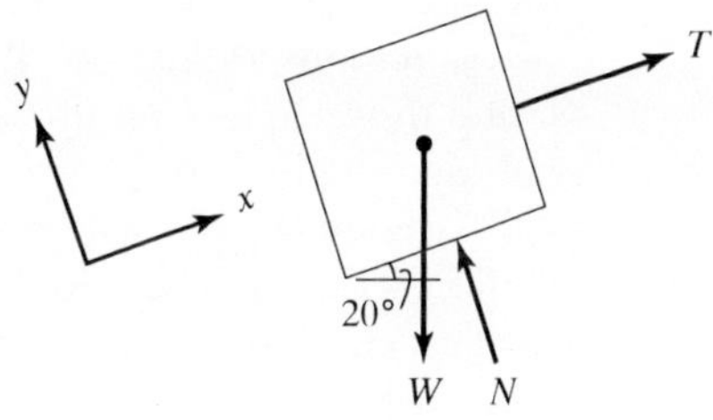

4.5 EQUILIBRIUM

Equilibrium is a state of balance between or among opposing forces, and it is one of the most important concepts in engineering mechanics. There are two types of equilibrium in engineering mechanics: static and dynamic. If a body is in static equilibrium, the body does not move, whereas if a body is in dynamic equilibrium, the body moves with a constant velocity. In this book, we will restrict our discussion to static equilibrium. Furthermore, we will confine our treatment of static equilibrium to *concurrent* force systems. In a concurrent force system, the lines of action of all forces pass through a single point, so the forces do not have a tendency to rotate the body. Therefore, there are no *moments of force* to deal with, only the forces themselves. Because the forces act concurrently, the body effectively becomes a *particle* (i.e., a dimensionless point in space through which the forces act). The actual body may or may not be a particle, but is modeled as such for purposes of the analysis. This concept will be demonstrated in some examples later.

A body is in static or dynamic equilibrium if the vector sum of all external forces is zero. Consistent with this definition, the condition of equilibrium may be stated mathematically as:

$$\Sigma \mathbf{F} = \mathbf{0} \tag{4.20}$$

where the summation symbol Σ denotes a sum of all external forces. Note that the zero is written as a vector to preserve the vector character of the equation across the equal sign. Equation (4.20) is a necessary and sufficient condition for equilibrium according to Newton's second law, which can be written as $\Sigma\mathbf{F} = m\mathbf{a}$. If the sum of the forces is zero, then $m\mathbf{a} = \mathbf{0}$. The quantity m is a scalar that can be divided out, leaving $\mathbf{a} = \mathbf{0}$. Thus, the acceleration is zero, so the body either moves with a constant velocity or remains at rest. Equation (4.20) is a vector equation that may be broken into its scalar components. Writing the equation in terms of the unit vectors **i**, **j**, and **k**, we obtain:

$$\Sigma F_x\mathbf{i} + \Sigma F_y\mathbf{j} + \Sigma F_z\mathbf{k} = \mathbf{0} \tag{4.21}$$

where the three terms on the left side are the total *scalar* forces in the x, y, and z directions, respectively. Equation (4.21) can only be satisfied if the sum of the scalar forces in each coordinate direction is zero. Hence, we have three scalar equations:

$$\Sigma F_x = 0,\ \Sigma F_y = 0,\ \Sigma F_z = 0. \tag{4.22}$$

These relations are referred to as the *equations of equilibrium for a particle.* Each of these three scalar equations must be satisfied for the particle to be in equilibrium. If *any one* of these scalar equations is not satisfied, the particle is not in equilibrium. For example, if $\Sigma F_x = 0$ and $\Sigma F_y = 0$, but $\Sigma F_z \neq 0$, the particle will be in equilibrium in the x and y directions, but will accelerate in the z direction. Similarly, if $\Sigma F_x = 0$, but $\Sigma F_y \neq 0$ and $\Sigma F_z \neq 0$, the particle will be in equilibrium in the x direction, but will have components of acceleration in the y and z directions.

Equations (4.22) are the governing equations for a particle in equilibrium. Using those equations and a free-body diagram of the particle, the unknown external forces can be determined. Consider the particle in Figure 4.18(a). A force of 2 kN acts on the particle in the positive x direction. An unknown force F, whose direction is *assumed* to act in the positive x direction, also acts on the particle. Applying the first equation of equilibrium, we have:

$$\Sigma F_x = 0 = +F + 2.$$

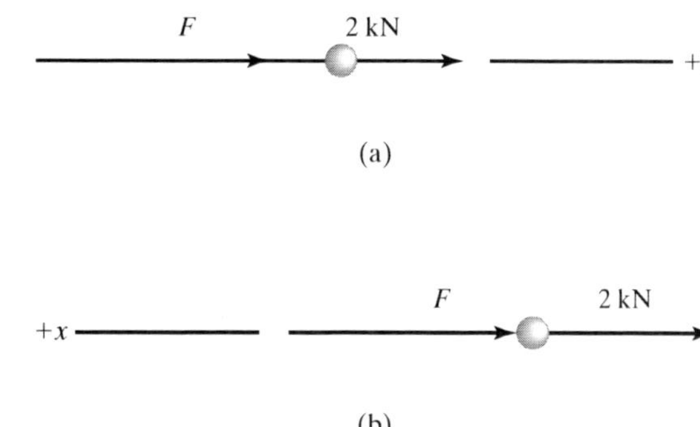

Figure 4.18
A force of $F = -2$ kN is required to maintain equilibrium, regardless of the orientation of the coordinate system.

Both forces are positive because they act in the positive *x* direction. Solving for the unknown force *F*, we obtain:

$$F = -2 \text{ kN}.$$

Thus, in order for the particle to be in equilibrium, a 2-kN force acting to the *left* must be applied. The negative sign on the answer is consistent with the direction of the positive *x*-axis. In mechanics, the orientation of the coordinate system is arbitrary (i.e., does not affect the solution), as long as it is used consistently. Let's rework the example by reversing the direction of the *x*-axis but keep positive *F* pointing right as before. As shown in Figure 4.18(b), the positive *x*-axis is now directed to the *left*, but the forces remain unchanged. Writing the equation of equilibrium, we have:

$$\Sigma F_x = 0 = -F - 2.$$

Solving yields:

$$F = -2 \text{ kN}$$

and we obtain the same answer as before. The direction of the *x*-axis has no influence on the answer. In both cases, the negative sign indicates that the direction of *F* required to maintain the particle in equilibrium is *opposite* to the assumed direction.

The examples that follow demonstrate how to find forces acting on a particle. Each example is worked in detail using the general analysis procedure of (1) problem statement, (2) diagram, (3) assumptions, (4) governing equations, (5) calculations, (6) solution check, and (7) discussion. For the sake of simplicity, the examples are limited to coplanar force systems.

EXAMPLE 4.4

Problem statement

Two blocks hang from cords as shown in Figure 4.19. Find the tension in each cord.

Diagram

In order to find the tension in each cord, a separate free-body diagram is constructed for each block. The most critical part of a free-body diagram is the inclusion of every external force acting on the body in question. Two forces act on block *A*,

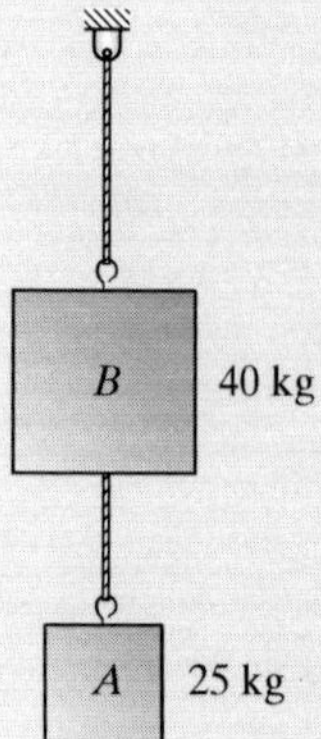

Figure 4.19
Suspended blocks for Example 4.4.

its weight and the tension force in the lower cord. Three forces act on block *B*, its weight, the tension force in the lower cord, and the tension force in the upper cord. All forces are concurrent, so we treat the boxes as particles. The free-body diagrams for the blocks are shown in Figure 4.20(a).

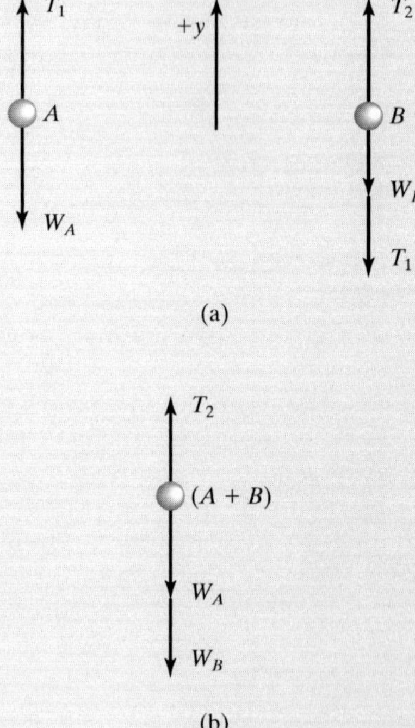

Figure 4.20
Free-body diagrams for Example 4.4.

Assumptions

1. All forces are concurrent.
2. The weights of the cords are negligible.
3. The cords are sufficiently flexible to hang straight down.

Governing equations

Because forces act in one direction only, there is only one governing equation: the equation of equilibrium for the vertical direction. Thus, for both blocks, we have:

$$\sum F_y = 0.$$

Calculations

To solve the problem, the equation of equilibrium must be written for both blocks. Noting the direction of the positive *y*-axis and using the free-body diagrams in Figure 4.20(a), we have:

Block *A*:

$$\begin{aligned}\sum F_y = 0 &= T_1 - W_A \\ &= T_1 - (25\text{ kg})(9.81\text{ m/s}^2).\end{aligned}$$

Block *B*:

$$\begin{aligned}\sum F_y = 0 &= T_2 - T_1 - W_B \\ &= T_2 - T_1 - (40\text{ kg})(9.81\text{ m/s}^2).\end{aligned}$$

Solving the first equation for T_1, we obtain:

$$T_1 = 245.25\text{ N}.$$

Substituting this value of T_1 into the second equation and solving for T_2, we obtain:

$$T_2 = 637.65\text{ N}.$$

We commonly express engineering answers in three significant figures, so our answers are reported as:

$$T_1 = \underline{\underline{245\text{ N}}},\ T_2 = \underline{\underline{638\text{ N}}}.$$

Solution Check

No mathematical or calculation-related errors are detected. Do the answers seem reasonable? The lower cord supports block *A* only, so tension T_1 is simply the weight of block *A*. Because the upper cord supports both blocks, the tension T_2 should be the sum of the weights:

$$\begin{aligned}W_A + W_B &= (m_A + m_B)\,g \\ &= (25\text{ kg} + 40\text{ kg})(9.81\text{ m/s}^2) \\ &= 637.65\text{ N}.\end{aligned}$$

Our solution checks out.

Discussion

An alternative method for finding the tension in the upper cord T_2 is to construct a free-body diagram of both blocks as a *single* particle. The interesting aspect of this approach is that the tension forces produced by the lower cord on both blocks are ignored because they are *internal* forces, not external forces. The internal forces exerted on each block by the lower cord are equal in magnitude, but opposite in direction; hence, they cancel, thereby having no overall mechanical effect on the system. There are three external forces acting on the combined blocks, the weights of each block, and the tension T_2. Using the free-body diagram in Figure 4.20(b), we have:

$$\Sigma F_y = 0 = T_2 - W_A - W_B$$
$$= T_2 - (25 \text{ kg} + 40 \text{ kg})(9.81 \text{ m/s}^2)$$

which yields:

$$T_2 = 637.65 \text{ N}.$$

EXAMPLE 4.5

Problem statement

A 200 kg engine block hangs from a system of cables as shown in Figure 4.21. Find the tension in cables AB and AC. Cable AB is horizontal.

Diagram

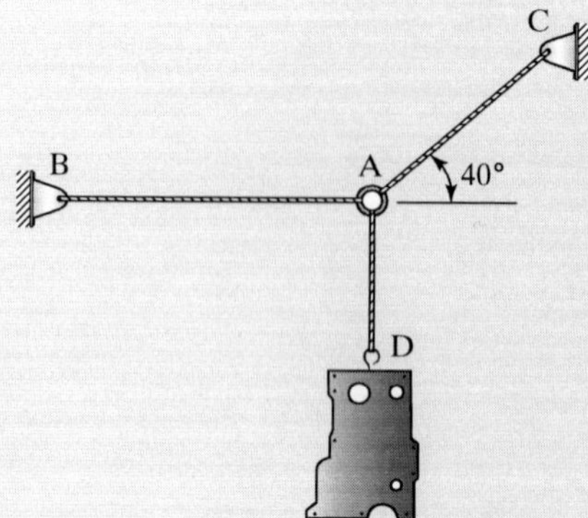

Figure 4.21
Suspended engine block for Example 4.5.

We have a coplanar force system in which the force in each cable acts concurrently at A, so we construct a free-body diagram for a "particle" at A. (See Figure 4.22.) The tension force in cable AB acts to the left along the x-axis, and the tension force in cable AC acts along a line 40° with respect to the x-axis. The tension force in cable AD, which is equivalent to the engine block's weight, acts straight down.

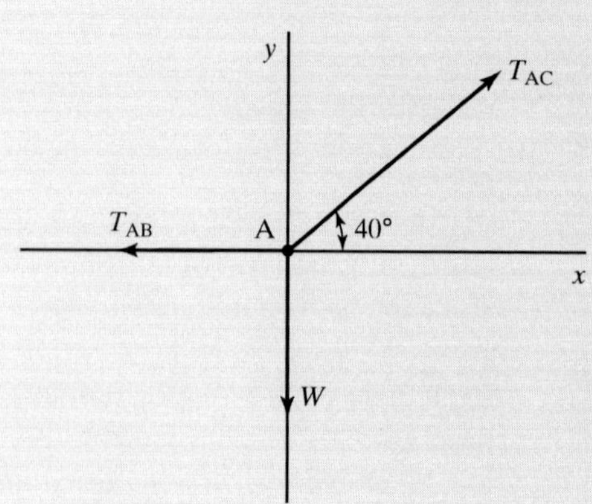

Figure 4.22
Free-body diagram for Example 4.5.

Assumptions

1. All forces are concurrent at A.
2. The weights of the cables are negligible.
3. All cables are taut.

Governing equations

The governing equations are Newton's second law and the equations of equilibrium in the x and y directions:

$$W = mg$$
$$\Sigma F_x = 0$$
$$\Sigma F_y = 0.$$

Calculations

Using the free-body diagram in Figure 4.22, the equations of equilibrium are:

$$\sum F_x = 0 = -T_{AB} + T_{AC} \cos 40°$$
$$\sum F_y = 0 = T_{AC} \sin 40° - W$$

where $W = mg = (200 \text{ kg})(9.81 \text{ m/s}^2) = 1962 \text{ N}$. The second equation can be immediately solved for T_{AC}:

$$T_{AC} = \underline{\underline{3052 \text{ N.}}}$$

Substituting this value of T_{AC} into the first equation and solving for T_{AB}, we get:

$$T_{AB} = \underline{\underline{2338 \text{ N.}}}$$

Solution check

To verify that our answers are correct, we substitute them back into the equilibrium equations. If they satisfy the equations, they are correct.

$$\Sigma F_x = -2338 \text{ N} + (3052 \text{ N}) \cos 40° = -0.032 \approx 0$$
$$\Sigma F_y = (3052 \text{ N}) \sin 40° - (200 \text{ kg})(9.81 \text{ m/s}^2) = -0.212 \approx 0.$$

Within the numerical precision of the calculations, the sum of the forces in the x and y directions is zero. Our answers are therefore correct.

Discussion

Now that we know the tension forces in the cables, what do we do with them? Knowing the forces per se does not tell us how the cables perform structurally. The next step in the analysis would be to determine the stress in each cable. If the calculated stresses are less than an allowable or design stress, the cables will support the engine block without experiencing failure. In this situation, failure most likely means cable breakage, but may also mean permanent cable strain. Stress and strain would have to be calculated in order to make a full structural assessment of the cables.

PRACTICE!

For the following practice problems, use the general analysis procedure of (1) problem statement, (2) diagram, (3) assumptions, (4) governing equations, (5) calculations, (6) solution check, and (7) discussion.

1. A 30-cm diameter solid steel sphere hangs from cables as shown. Find the tension in cables AB and AC. For the density of steel, use $\rho = 7270\ \text{kg/m}^3$.
   Answer: $T_{AB} = T_{AC} = 712.9$ N.

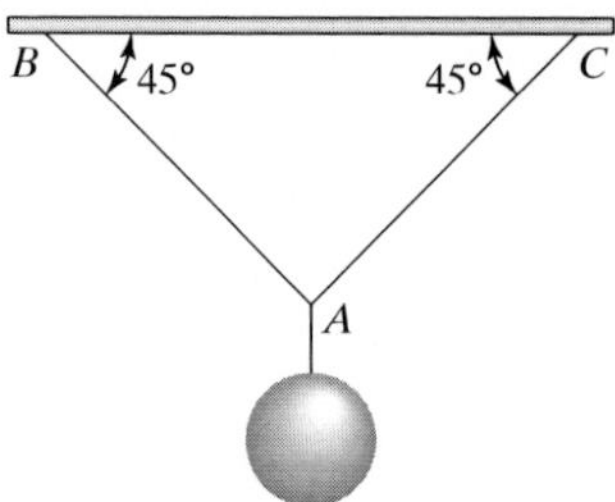

2. A 250-kg cylinder rests in a long channel as shown. Find the forces acting on the cylinder by the sides of the channel.
   Answer: 1999 N, 893 N.

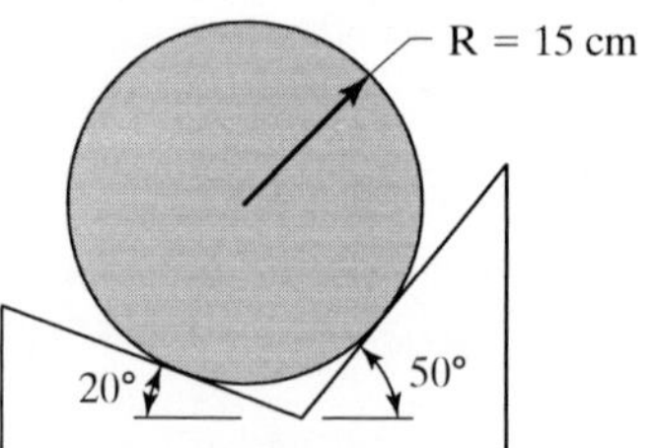

3. Three coplanar forces are applied to a box in an attempt to slide it across the floor, as shown. If the box remains at rest, what is the friction force between the box and the floor?
   Answer: 116.5 N.

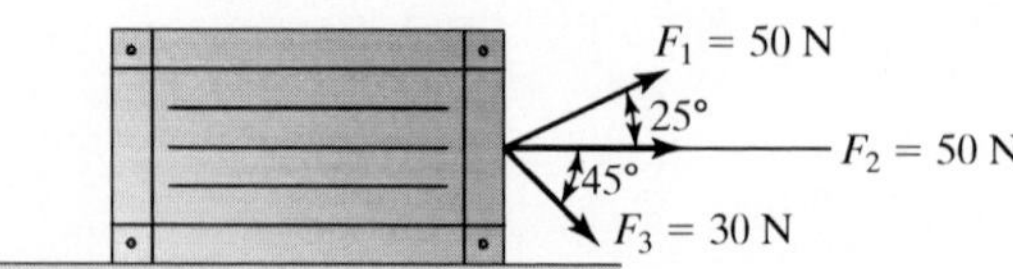

4. A 15-kg flowerpot hangs from wires as shown. Find the tension in wires AB and AC.
   Answer: $T_{AB} = 88.3$ N, $T_{AC} = 117.7$ N.

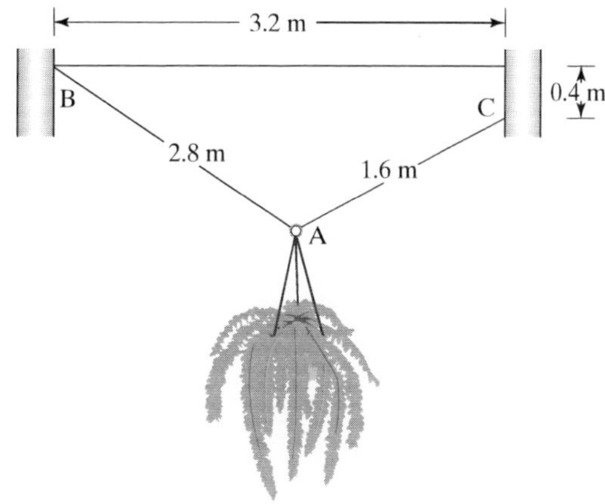

4.6 STRESS AND STRAIN

If the vector sum of the external forces acting on a body is zero, the body is in a state of equilibrium. There are also internal forces acting on the body. Internal forces are *caused* by external forces, but internal forces do not affect the equilibrium of the body. So, if internal forces do not affect equilibrium, how are they important? To illustrate the importance of internal forces, let's use a familiar example from the sport of weight lifting. (See Figure 4.23.) As a weight lifter holds a heavy set of weights, his body and the weights are momentarily in a state of static equilibrium. The gravitational force of the weights is balanced by the force exerted on the bar

Figure 4.23
A weight lifter is in equilibrium, but his body is in a state of stress.
(Art by Kathryn Colton).

by the weight lifter's hands or shoulders, and the gravitational forces of the weights plus his body are balanced by the force exerted by the floor on his feet. Are there internal forces acting on the weight lifter? Most definitely, yes. If the weight he is holding is great, he is painfully aware of those internal forces. The external forces of the weights and the reaction at the floor cause internal forces in his arms, torso, and legs. The magnitude of these internal forces usually limits the time the weight lifter can sustain his position to only a few seconds. Like the weight lifter, engineering structures such as buildings, bridges, and machines experience internal forces when external forces are applied to them. Engineering structures, however, must usually sustain internal forces for long periods of time, perhaps years. Principles of statics alone, which yield the external forces acting on a body, are insufficient to define the mechanical state of the body. In order for an engineer to make a complete assessment of the structural integrity of any body, internal forces must be considered. From the internal forces and the deformations resulting from them, stress and strain can be determined.

4.6.1 Stress

The concept of stress is of prime importance in mechanics of materials. ***Stress*** is the primary physical quantity that engineers use to ascertain whether a structure can withstand the external forces applied to it. By finding stresses, engineers have a standard method of comparing the abilities of given materials to withstand external forces. There are two types of stress: *normal stress* and *shear stress*. In this book, we will confine our attention to normal stress. Normal stress is the *stress that acts normal (perpendicular) to a selected plane or along an axis within a body*. Normal stress is often associated with the stress in the axial direction in long slender members such as rods, beams, and columns. Consider the slender bar shown in Figure 4.24. An axial force F acts on each end of the bar, maintaining the bar in equilibrium, as indicated in Figure 4.24(a). Now, suppose that we pass an imaginary plane through the bar perpendicular to its axis, as shown in Figure 4.24(b). Conceptually, we then remove the bottom portion of the bar that was "cut away" by the imaginary plane. In removing the bottom portion of the bar, we also removed the force applied at the bottom end of the bar that

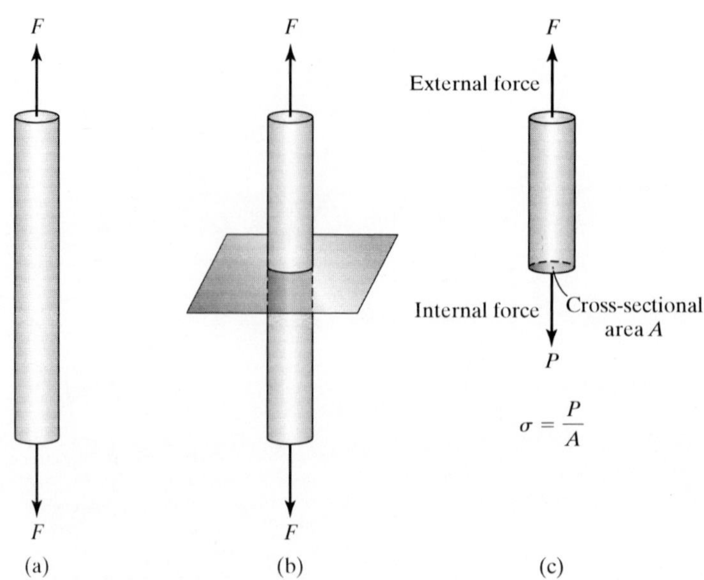

Figure 4.24
Normal stress in a rod.

balanced the force applied at the top end. To restore equilibrium, we must apply an equivalent force P at the "cut" end. This force, unlike the external force applied at the top of the bar, is an *internal* force because it acts *within* the bar. The internal force P acts perpendicular to the cross-sectional area created by passing the imaginary plane through the bar, as indicated in Figure 4.24(c). The normal stress σ in the bar is defined as the internal force P divided by the cross-sectional area A:

$$\sigma = \frac{P}{A}. \tag{4.23}$$

This mathematical definition of normal stress is actually an *average* normal stress, because there may be a variation of stress across the cross section of the bar. Stress variations are normally present only near points where the external forces are applied, however, so Equation (4.23) may be used in the majority of stress calculations without regard to stress variations. The cross section of the bar in Figure 4.24 is circular, but the quantity A represents the cross-sectional area of a member of any shape (e.g., circular, rectangular, triangular). Note that the definition of stress is very similar to that of pressure. Both quantities are defined as a force divided by an area. Accordingly, stress has the same units as pressure. Typical units for stress are kPa or MPa in the SI system and psi or ksi in the English system.

In Figure 4.24, the force vectors are directed away from each other, indicating that the bar is stretched. The normal stress associated with this force configuration is referred to as *tensile stress* because the forces place the body in tension. Conversely, if the force vectors are directed toward each other, the bar is compressed. The normal stress associated with this force configuration is referred to as *compressive stress* because the forces place the body in compression. These two force configurations are illustrated in Figure 4.25. One may think that the type of normal stress, tensile or compressive, does not matter, since Equation (4.23) says nothing about direction. However, some materials can withstand one type of stress more readily than the other. For example, concrete is stronger in compression than in tension. Consequently, concrete is typically used in applications where the stresses are compressive, such as columns that support bridge decks and highway overpasses. When concrete members are designed for applications that involve tensile stresses, reinforcing bars are used.

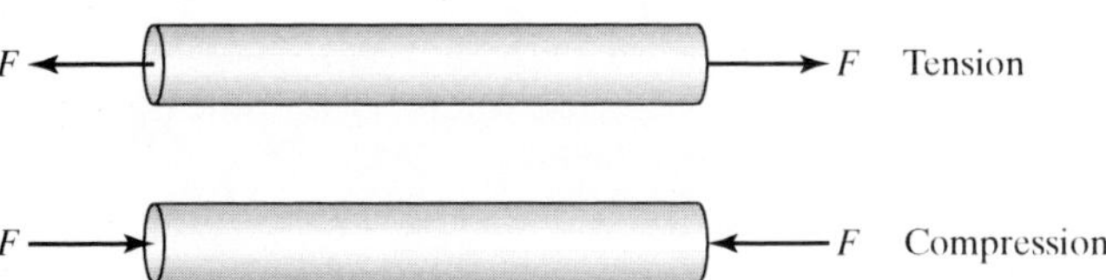

Figure 4.25
External forces for tension and compression.

4.6.2 Strain

External forces are responsible for producing stress, and they are also responsible for producing deformation. Deformation may also be caused by temperature changes. ***Deformation*** is defined as a *change in the size or shape of a body*. No material is perfectly rigid; hence, when external forces are applied to a body, the body changes its size or shape according to the magnitude and direction of the external forces applied to it. We have all stretched a rubber band and noticed that its length changes appreciably under a small tensile force. All materials—steel, concrete, wood, and other structural materials—deform to some extent under applied forces, but the deformations are usually too small to detect visually, so special measuring instruments are employed. Consider the bar shown in Figure 4.26. Prior to applying an external force, the bar has a length L. Now, the bar is placed in tension, applying an external force F at each end. The tensile force causes the bar to increase in length by an amount δ. The quantity δ is called the *normal deformation* or *axial deformation,* since the change in length is normal to the direction of the force, which is along the axis of the rod. Depending on the bar's material and the magnitude of the applied force, the normal deformation may be small, perhaps only a few thousandths of an inch. In order to normalize the change in size or shape of a body with respect to the body's original geometry, engineers use a quantity called ***strain***. There are two types of strain, *normal strain* and *shear strain*. In this book, we confine our attention to normal strain. Normal strain ε is defined as the *normal deformation δ divided by the original length L*:

$$\varepsilon = \frac{\delta}{L}. \tag{4.24}$$

Because strain is a ratio of two lengths, it is a dimensionless quantity. It is customary however, to express strain as a ratio of two length units. In the SI unit system, strain is usually expressed in units of μm/m because, as mentioned before, deformations are typically small. In the English unit system, strain is usually expressed in units of in/in. Since strain is a dimensionless quantity, it is sometimes expressed as a percentage. The normal strain illustrated in Figure 4.26 is for a body in tension, but the definition given by Equation (4.24) also applies to bodies in compression.

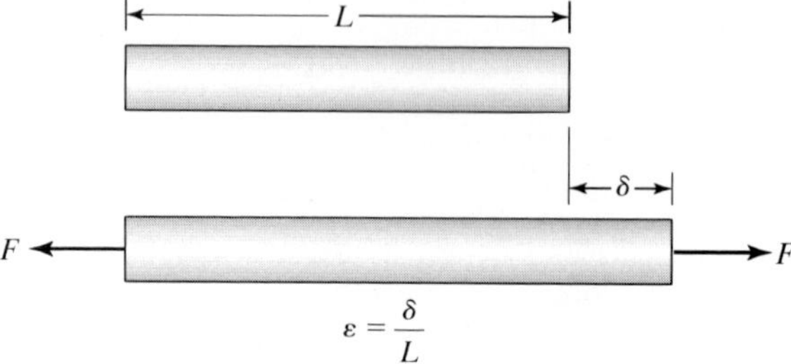

Figure 4.26
Normal strain in a rod.

4.6.3 Hooke's Law

About three centuries ago, the English mathematician Robert Hooke (1635–1703) discovered that the force required to stretch or compress a spring is proportional to the displacement of a point on the spring. The law describing this phenomenon, known as Hooke's law, is expressed mathematically as:

$$F = kx \tag{4.25}$$

where F is force, x is displacement, and k is a constant of proportionality called the spring constant. Equation (4.25) applies only if the spring is not deformed beyond its ability to resume its original length after the force is removed. A more useful form of Hooke's law for engineering materials has the same mathematical form as Equation (4.25), but is expressed in terms of stress and strain:

$$\sigma = E\varepsilon. \tag{4.26}$$

Hooke's law, given by Equation (4.26), states that the stress σ in a material is proportional to the strain ε. The constant of proportionality E is called the ***modulus of elasticity*** or *Young's modulus,* after the English mathematician Thomas Young

(1773–1829). Like the spring equation, the engineering version of Hooke's law applies only if the material is not deformed beyond its ability to resume its original size after the force is removed. A material that obeys Hooke's law is said to be *elastic* because it returns to its original size after the removal of the force. As strain ε is a dimensionless quantity, the modulus of elasticity E has the same units as stress.

Equation (4.26) describes a straight line with E as the slope. The modulus of elasticity is an experimentally derived quantity. A sample of the material in question is subjected to tensile stresses in a special apparatus that facilitates a sequence of stress and strain measurements in the elastic range of the material. The ***elastic range*** is the distance or extent a material can be deformed and still be capable of returning to its original shape. Stress–strain data points are plotted on a linear scale, and a best-fit straight line is drawn through the points. The slope of this line is the modulus of elasticity E.

A useful relationship may be obtained by combining Equations (4.23), (4.24), and (4.26). The axial deformation δ may be expressed directly in terms of the internal force P and the geometrical and material properties of the member. This is done by substituting the definition of strain ε given by Equation (4.24) into Equation (4.26), Hooke's law, and noting that normal stress is the internal force divided by the cross-sectional area given in Equation (4.23). Thus, the resulting expression is:

$$\delta = \frac{PL}{AE}. \tag{4.27}$$

Equation (4.27) is useful because the strain does not have to be calculated first to find the deformation of the member. However, this equation is valid only over the linear region of the stress-strain curve.

4.6.4 Stress–Strain Diagram

A ***stress–strain diagram*** *is a graph of stress as a function of strain in a given material.* The shape of this graph varies somewhat with material, but stress–strain diagrams have some common features. A typical stress–strain diagram is illustrated in Figure 4.27.

Figure 4.27
A typical stress–strain diagram.

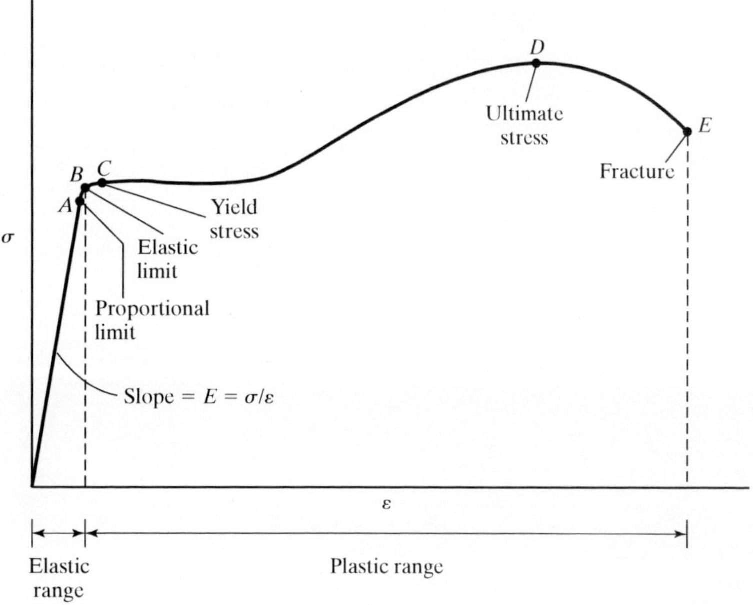

The upper stress limit of the linear relationship described by Hooke's law is called the *proportional limit*, labeled point *A*. At any stress between point *A* and the *elastic limit*, labeled point *B*, stress is not proportional to strain, but the material will still return to its original size after the force is removed. For many materials, the proportional and elastic limits are very close together. Point *C* is called the ***yield stress*** or *yield strength*. Any stress above the yield stress will result in *plastic* deformation of the material (i.e., the material will not return to its original size, but will deform permanently). As the stress increases beyond the yield stress, the material experiences a large increase in strain for a small increase in stress. At about point *D*, called the ***ultimate stress*** or *ultimate strength*, the cross-sectional area of the material begins to decrease rapidly until the material experiences *fracture* at point *E*.

In the next example, we use the general analysis procedure of (1) problem statement, (2) diagram, (3) assumptions, (4) governing equations, (5) calculations, (6) solution check, and (7) discussion.

EXAMPLE 4.6

Problem statement

A 200-kg engine block hangs from a system of cables as shown in Figure 4.28. Find the normal stress and axial deformation in cables AB and AC. The cables are 0.7 m long and have a diameter of 4 mm. The cables are steel with a modulus of elasticity of $E = 200$ GPa.

Diagram

We will presume that the statics portion of the problem has been solved, so a free-body diagram of the entire system is unnecessary. Diagrams showing a cross section of the cables and the corresponding internal forces are sufficient. (See Figure 4.29.)

Assumptions

1. Cables are circular in cross section.
2. Cables have the same modulus of elasticity.
3. Stress is uniform in the cables.

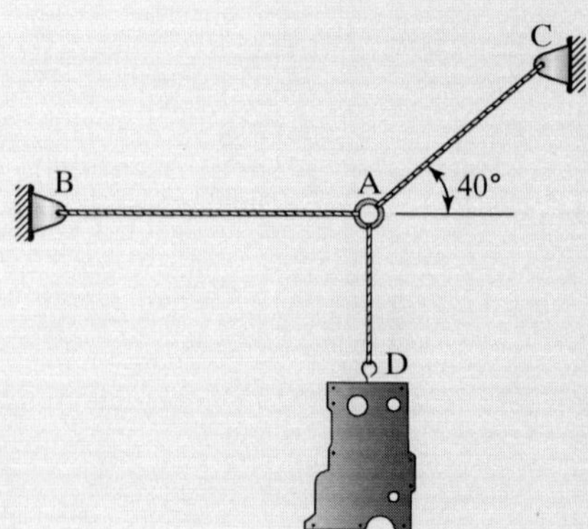

Figure 4.28
Suspended engine block for Example 4.6.

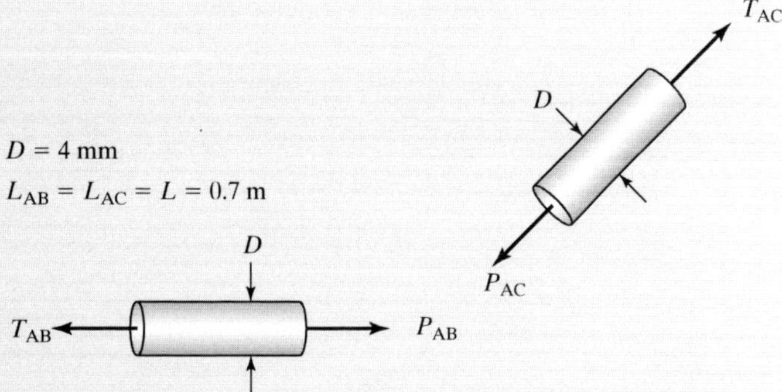

Figure 4.29 Cables for Example 4.6.

Governing equations

Cross-sectional area:

$$A = \frac{\pi D^2}{4}$$

Normal stress:

$$\sigma = \frac{P}{A}$$

Axial deformation:

$$\delta = \frac{PL}{AE}.$$

Calculations

The cross-sectional area of the cables is:

$$A = \frac{\pi D^2}{4} = \frac{\pi(0.004\text{ m})^2}{4} = 1.2566 \times 10^{-5}\text{ m}^2.$$

From a prior statics analysis, the tensions in cables AB and AC are 2338 N and 3052 N, respectively. Hence, the normal stress in each cable is:

$$\sigma_{AB} = \frac{P_{AB}}{A} = \frac{2338\text{ N}}{1.2566 \times 10^{-5}\text{ m}^2} = 186.1 \times 10^6\text{ N/m}^2 = \underline{\underline{186.1\text{ MPa}}}$$

$$\sigma_{AC} = \frac{P_{AC}}{A} = \frac{3052\text{ N}}{1.2566 \times 10^{-5}\text{ m}^2} = 242.9 \times 10^6\text{ N/m}^2 = \underline{\underline{242.9\text{ MPa}}}.$$

The deformation in each cable is

$$\delta_{AB} = \frac{P_{AB}L}{AE}$$

$$= \frac{(2338\ \text{N})(0.7\ \text{m})}{(1.2566 \times 10^{-5}\ \text{m}^2)(200 \times 10^9\ \text{N/m}^2)} = 6.51 \times 10^{-4}\ \text{m} = \underline{\underline{0.651\ \text{mm}}}$$

$$\delta_{AC} = \frac{P_{AC}L}{AE}$$

$$= \frac{(3052\ \text{N})(0.7\ \text{m})}{(1.2566 \times 10^{-5}\ \text{m}^2)(200 \times 10^9\ \text{N/m}^2)} = 8.50 \times 10^{-4}\ \text{m} = \underline{\underline{0.850\ \text{mm}}}.$$

Solution check

One way to check the validity of the results is to compare the relative magnitudes of the stress and deformation in each cable. The internal force in cable AC is greater than the normal stress in cable AB. Consequently, the normal stress and axial deformation in cable AC must also be greater because the cables are geometrically and materially identical. Our calculations show that this is indeed the case.

Discussion

The deformations are small, less than a millimeter in both cables. These deformations would probably not be significant in an engine hoist application and would not be perceptible by the naked eye. Are the stresses excessive? Will they plastically deform the cables? To answer these questions, we would have to know something about the yield stress of the cable material and the stresses for which the cables were designed.

PRACTICE!

In the following practice problems, use the general analysis procedure of (1) problem statement, (2) diagram, (3) assumptions, (4) governing equations, (5) calculations, (6) solution check, and (7) discussion.

1. A solid rod of stainless steel ($E = 190$ GPa) is 50 cm in length and has a 4 mm × 4 mm cross section. The rod is subjected to an axial tensile force of 8 kN. Find the normal stress, strain, and axial deformation.
   Answer: 500 MPa, 0.00263, 1.32 mm.
2. A 25-cm-long 10-gauge wire of yellow brass ($E = 105$ GPa) is subjected to an axial tensile force of 1.75 kN. Find the normal stress and deformation in the wire. A 10-gauge wire has a diameter of 2.588 mm.
   Answer: 333 MPa, 0.792 mm.
3. An 8-m-high granite column sustains an axial compressive load of 500 kN. If the column shortens 0.12 mm under the load, what is the diameter of the column? For granite, $E = 70$ GPa.
   Answer: 0.779 m.

4. A solid rod with a length and diameter of 1 m and 5 mm, respectively, is subjected to an axial tensile force of 20 kN. If the axial deformation is measured as $\delta = 1$ cm, what is the modulus of elasticity of the material?
   Answer: 1018 GPa.
5. A plastic ($E = 3$ GPa) tube with an outside and inside diameter of 6 cm and 5.4 cm, respectively, is subjected to an axial compressive force of 12 kN. If the tube is 25 cm long, how much does the tube shorten under the load?
   Answer: 1.86 mm.

4.7 DESIGN STRESS

Most engineering structures are not designed to deform permanently or fracture. Every member in a structure must maintain a certain degree of dimensional control to assure that it does not plastically deform, thereby losing its size or shape, interfering with surrounding structures or other members in the same structure. Obviously, the members must not fracture either, because this would lead to a catastrophic failure that would result in material and financial loss and perhaps the loss of human life. Therefore, members in most structures are designed to sustain a maximum stress that is *below* the yield stress on the stress–strain diagram for the particular material used to construct that member. This maximum stress is called the *design stress* or ***allowable stress.*** When a properly designed member is subjected to a load, the stress in the member will not exceed the design stress. Because the design stress is within the elastic range of the material, the member will return to its original dimensions after the load is removed. A bridge, for example, sustains stresses in its members while traffic passes over it. When there is no traffic, the members in the bridge return to their original dimensions. Similarly, while a boiler is operating, the pressure vessel sustains stresses that deform it, but when the pressure is reduced to atmospheric pressure, the vessel returns to its original dimensions.

If a structural member is designed to carry stresses below the yield stress, how does an engineer choose what the allowable stress should be? And why choose a stress below the yield stress in the first place? Why not design the member by using the yield stress itself, since that would allow the member to carry the maximum possible load? Engineering design is not an exact science. If it was, structures could be designed with ultimate precision by using the yield stress, or any other stress for that matter, as the design stress, and the design stress would never be exceeded while the structure was in service. Because design is not an exact science, engineers incorporate an allowance in their designs that takes into account the following uncertainties:

1. *Loadings* The design engineer may not anticipate every type of loading or the number of loadings that may occur. Vibration, impact, or accidental loadings may occur that were not accounted for in the design of the structure.
2. *Failure modes* Materials can fail by one or more of several different mechanisms. The design engineer may not have anticipated every failure mode by which the structure can possibly fail.
3. *Material properties* Physical properties of materials are subject to variations during manufacture, and there are experimental uncertainties in their numerical

values. Properties may also be altered by heating or by deformation during manufacture, handling, and storage.

4. *Deterioration* Exposure to the elements, poor maintenance, or unexpected natural phenomena may cause a material to deteriorate, thereby compromising its structural integrity. Various types of corrosion are the most common forms of material deterioration.
5. *Analysis* Engineering analysis is a critical part of design, and analysis involves making simplifying assumptions. Thus, analytical results are not precise, but are approximations.

To account for the uncertainties listed, engineers use a design or allowable stress based on a parameter called the ***factor of safety***. The factor of safety (FS) is defined as the *ratio of the failure stress to the allowable stress*:

$$\text{FS} = \frac{\sigma_{\text{fail}}}{\sigma_{\text{allow}}}. \tag{4.28}$$

Because the yield stress is the stress above which a material plastically deforms, the yield stress σ_y is commonly used as the failure stress σ_{fail}. The ultimate stress σ_u may also be used. The failure stress is always greater than the allowable stress, so FS > 1. The value chosen for the factor of safety depends on the type of engineering structure, the relative importance of the member compared with other members in the structure, the risk to property and life, and the severity of the design uncertainties previously listed. For example, to minimize weight, the factor of safety for some aircraft and spacecraft structures may be in the range of 1.05 to 1.2. However, the factor of safety for ground-based structures such as dams, bridges and buildings may be higher, perhaps 1.5 or 2. High-risk structures that pose a safety hazard to people in the event of failure, such as certain nuclear power plant components, may have a factor of safety as high as 3. Factors of safety for structural members in specific engineering systems have been standardized through many years of testing and industrial evaluation. Factors of safety are often defined by building codes or engineering standards established by city, state, or federal agencies and professional engineering societies.

APPLICATION

DESIGNING A TURNBUCKLE

Turnbuckles are special mechanical fasteners that facilitate connections between cables, chains, or cords. A basic turnbuckle consists of a slender, cylindrical shaped body threaded on each end to accept an eyebolt, a hook, or other type of tying component. The tension in the cables that are tied to a turnbuckle is adjusted by rotating the body of the turnbuckle. Turnbuckles are designed such that tightening or loosening may be accomplished without twisting the cables. Like the cables that are connected to them, turnbuckles must sustain the tensile stresses to which they are subjected. Consider a turnbuckle used to adjust the tension in a cable that stabilizes a communications tower. From a prior analysis, the tension in the cable is determined to be 25 kN. The loaded turnbuckle is shown in Figure 4.30(a). Let us suppose that, as a new engineer, your first job is to select a turnbuckle for this application. Turnbuckles are available in a variety of sizes and materials from several suppliers. Hardware suppliers specify the maximum recommended load that a particular turnbuckle can sustain without failing. It is, therefore, a simple matter for you, the end user, to select a turnbuckle with a recommended maximum load that

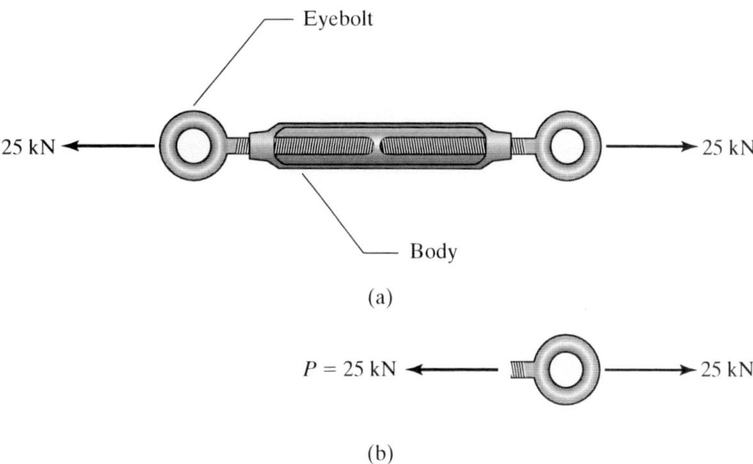

Figure 4.30
A loaded turnbuckle.

is somewhat greater than the actual load of 25 kN. But how did the engineers who designed the turnbuckle obtain this load value? The example that follows shows how fundamental concepts of stress and factor of safety may be used to design the eyebolt portion of a turnbuckle. The general analysis procedure of (1) problem statement, (2) diagram, (3) assumptions, (4) governing equations, (5) calculations, (6) solution check, and (7) discussion is used.

Problem statement

Determine the minimum-diameter eyebolt in a turnbuckle used to stabilize a communications tower. The tensile force in the cable is 25 kN. The eyebolt is to be made of AISI 4130 steel, a high-strength forging steel. (AISI is an abbreviation for American Iron and Steel Institute.) To account for potential high wind loads and other uncertainties, use a factor of safety of 2.0.

Diagram

The internal and external forces acting on the eyebolt are shown in Figure 4.30(b).

Assumptions

1. Stress is uniform in the eyebolt.
2. Stress in the eyebolt is purely axial.
3. Consider stress in the main body of the eyebolt only, not the threads.

Governing equations

The governing equations for this problem are the cross-sectional area for a circular bolt, the definition of normal stress, and the factor of safety.

$$A = \frac{\pi D^2}{4} \tag{a}$$

$$\sigma_{\text{allow}} = \frac{P}{A} \tag{b}$$

$$\text{FS} = \frac{\sigma_{\text{fail}}}{\sigma_{\text{allow}}} = \frac{\sigma_y}{\sigma_{\text{allow}}}. \tag{c}$$

Calculations

In the third governing equation, we have used the yield stress σ_y as the failure stress. The yield stress of AISI 4130 steel is 760 MPa. The objective of the analysis is to find the diameter D of the eyebolt required to sustain the applied load. There are three unknown quantities: σ_{allow}, A, and D. Because the three governing equations are not dependent, we may combine them algebraically to obtain the diameter, D. Upon substituting Equation (a) into Equation (b) and then Equation (b) into Equation (c) we obtain:

$$D = \left(\frac{4\,P\,\mathrm{FS}}{\pi\sigma_y}\right)^{1/2}$$

$$= \left(\frac{4(25 \times 10^3\ \mathrm{N})(2.0)}{\pi(760 \times 10^6\ \mathrm{Pa})}\right)^{1/2}$$

$$= 9.15 \times 10^{-3}\ \mathrm{m} = \underline{\underline{9.15\ \mathrm{mm}}}.$$

Solution check

No errors were found. The answer seems reasonable based on our knowledge of turnbuckles and other mechanical fasteners.

Discussion

The minimum eyebolt diameter that will sustain the applied load with a factor of safety of 2.0 is 9.15 mm. In English units, this diameter is:

$$D = 9.15\ \cancel{\mathrm{mm}} \times \frac{1\ \mathrm{in}}{25.4\ \cancel{\mathrm{mm}}} = 0.360\ \mathrm{in}.$$

Bolts come in standard diameters, and 0.360 is not a standard size. Bolts are typically available in standard sizes such as $\frac{1}{4}$ in, $\frac{5}{16}$ in, and $\frac{3}{8}$ in. A $\frac{5}{16}$ in (0.3125 in) bolt is too small, so the $\frac{3}{8}$ in (0.375 in) should be chosen, even though it is slightly larger than required. It should be emphasized that this analysis reflects only a part of the analysis that would be required in the total design of a turnbuckle. Stresses in the threads of the eyebolt and the turnbuckle body, as well as the main body of the turnbuckle itself, would also have to be calculated.

PRACTICE!

In the practice problems, use the general analysis procedure of (1) problem statement, (2) diagram, (3) assumptions, (4) governing equations, (5) calculations, (6) solution check, and (7) discussion.

1. A rod of aluminum 6061-T6 has a square cross section measuring 0.25 in × 0.25 in. Using the yield stress as the failure stress, find the maximum tensile load that the rod can sustain for a factor of safety of 1.5. The yield stress of aluminum 6061-T6 is 240 MPa.
   Answer: 6.45 kN.
2. A concrete column with a diameter of 60 cm supports a portion of a highway overpass. Using the ultimate stress as the failure stress, what is

the maximum compressive load that the column can carry for a factor of safety of 1.25? For the ultimate stress of concrete, use $\sigma_u = 40$ MPa.
Answer: 9.05 MN.

3. A column of rectangular cross section constructed from fir timber is subjected to a compressive load of 3.5 MN. If the width of the column is 25 cm, find the depth required to sustain the load with a factor of safety of 1.6. The ultimate stress of fir is $\sigma_u = 50$ MPa.
Answer: 44.8 cm.

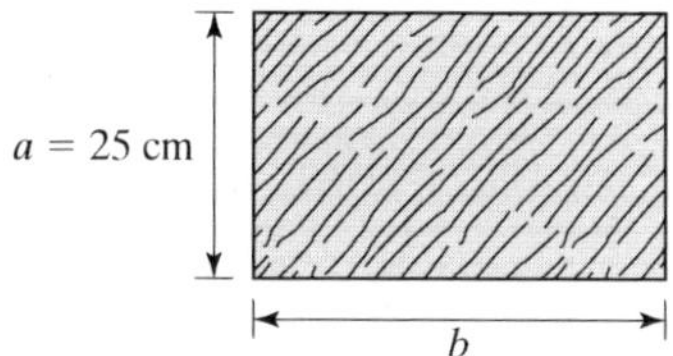

KEY TERMS

allowable stress
Cartesian unit vector
deformation
elastic range
equilibrium
factor of safety
force
force system
free-body diagram
internal force
mechanics
modulus of elasticity
resultant
resultant force
scalar
statics
strain
stress
stress–strain diagram
ultimate stress
vector
yield stress

REFERENCES

Bedford, A. and W. Fowler, *Engineering Mechanics: Statics*, 5th ed., Upper Saddle River, NJ: Prentice Hall, 2008.
Beer, F.P., E.R. Johnston, E.R. Eisenberg, and D. Mazurek, *Vector Mechanics for Engineers: Statics*, 9th ed. NY: McGraw-Hill, 2009.
Beer, F., E.R. Johnston, J. DeWolf, and D. Mazurek, *Mechanics of Materials*, 6th ed., New York, NY: McGraw-Hill, 2011.
Hibbeler, R.C., *Engineering Mechanics: Statics*, 13th ed., Upper Saddle River, NJ: Prentice Hall, 2012.
Hibbeler, R.C., *Mechanics of Materials*, 8th ed., Upper Saddle River, NJ: Prentice Hall, 2010.

PROBLEMS

Forces

4.1 Find the resultant force for the forces shown in Figure P4.1 (a) by using the parallelogram law and (b) by resolving the forces into their x and y components.

Figure P4.1

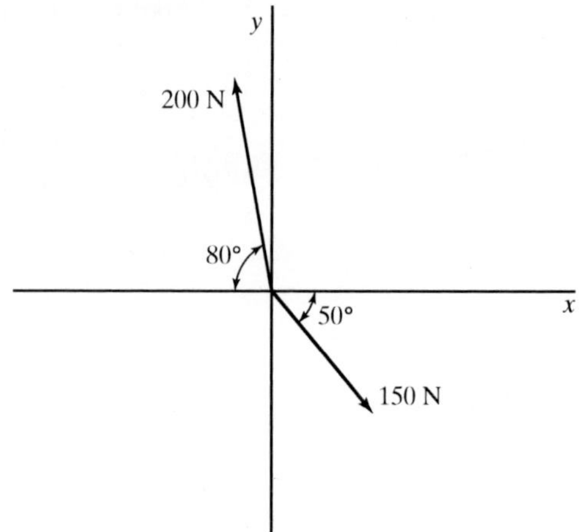

4.2 Find the resultant force for the forces shown in Figure P4.2 (a) by using the parallelogram law and (b) by resolving the forces into their x and y components.

Figure P4.2

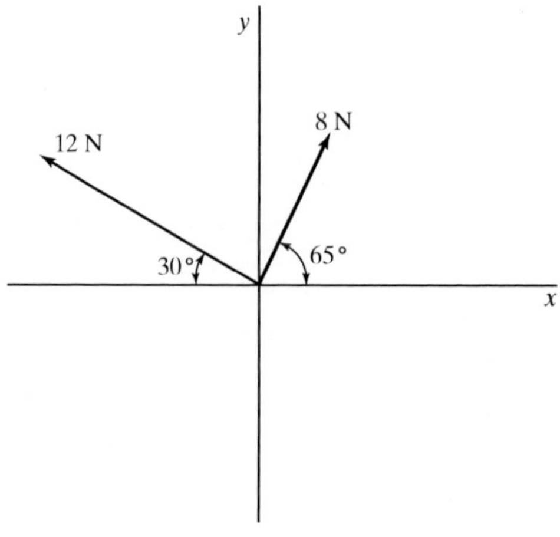

4.3 For the three forces shown in Figure P4.3, find the resultant force, its magnitude, and direction with respect to the x-axis.

Figure P4.3

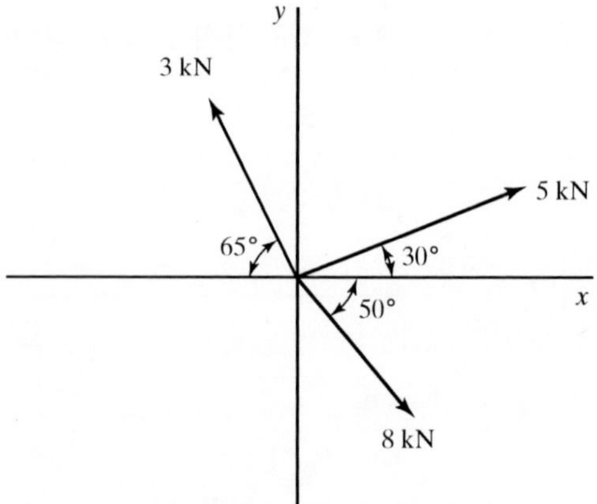

4.4 For the three forces shown in Figure P4.4, find the resultant force, its magnitude, and direction with respect to the *x*-axis.

Figure P4.4

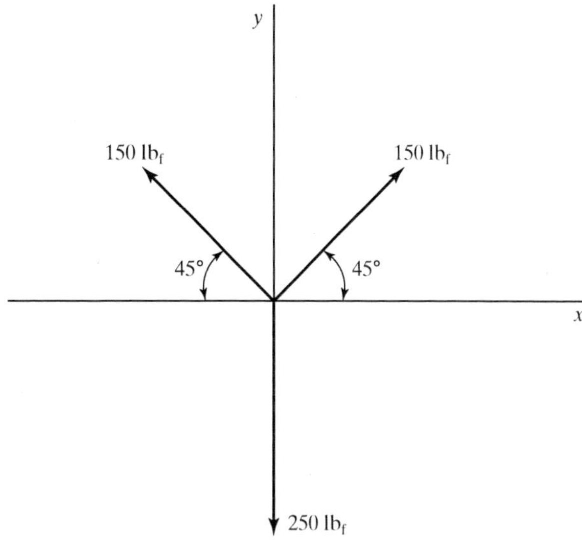

4.5 Consider the three forces $\mathbf{F}_1 = 8\mathbf{i} + 2\mathbf{j} - 10\mathbf{k}$ N, $\mathbf{F}_2 = -4\mathbf{i} - 7\mathbf{j} + 6\mathbf{k}$ N, and $\mathbf{F}_3 = \mathbf{i} - \mathbf{j} - \mathbf{k}$ N. Find the resultant force and its magnitude.

Free-body diagrams

4.6 Cylinders *A* and *B* are suspended by a system of cords as shown in Figure P4.6. Points *C* and *F* are fixed connections, and *E* is a pulley of negligible size and mass. Draw a free-body diagram of (a) cylinder *A*, (b) cylinder *B*, (c) connection point *D*, and (d) pulley *E*.

Figure P4.6

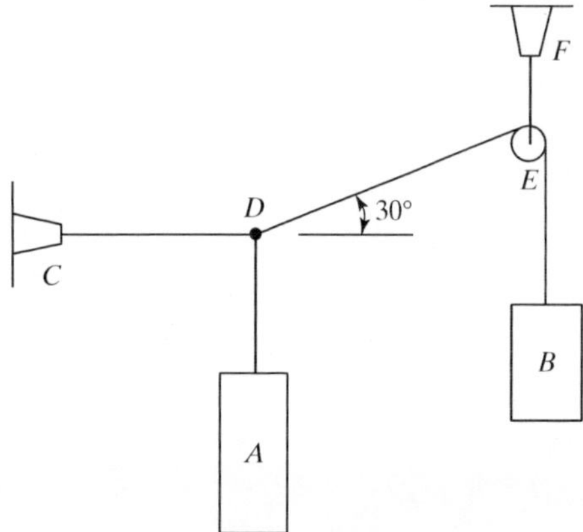

4.7 The beam shown in Figure P4.7 is connected to a pin at *A* and rests on a roller at *B*. Neglecting the weight of the beam, draw a free-body diagram of the beam.

Figure P4.7

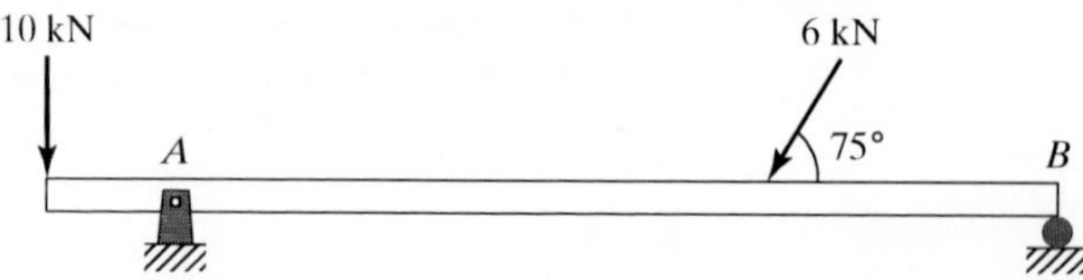

4.8 A file cabinet is dragged across a rough floor at a constant velocity, as shown in Figure P4.8. The center of mass of the file cabinet is located at *G*. Draw a free body diagram of the file cabinet.

Figure P4.8

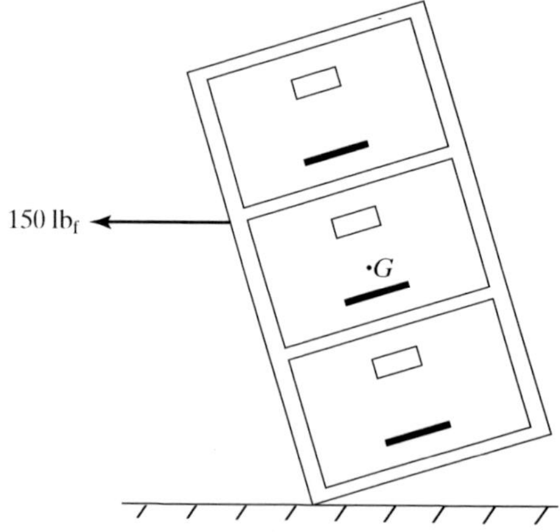

Equilibrium

4.9 A particle is subjected to three forces: $\mathbf{F}_1 = 3\mathbf{i} + 5\mathbf{j} - 8\mathbf{k}$ kN, $\mathbf{F}_2 = -2\mathbf{i} - 3\mathbf{j} + 4\mathbf{k}$ kN, and $\mathbf{F}_3 = -\mathbf{i} - 2\mathbf{j} + 4\mathbf{k}$ kN. Is this particle in equilibrium? Explain.

4.10 A particle is subjected to three forces: $\mathbf{F}_1 = 3\mathbf{i} - a\mathbf{j} - 7\mathbf{k}$ N, $\mathbf{F}_2 = -4\mathbf{i} - 2\mathbf{j} + b\mathbf{k}$ N, and $\mathbf{F}_3 = c\mathbf{i} - 6\mathbf{j} + 4\mathbf{k}$ N. Find the values of the scalars a, b, and c such that the particle is in equilibrium.

4.11 Find the magnitude of the force **F** and its direction θ in Figure P4.11 so that the particle P is in equilibrium.

Figure P4.11

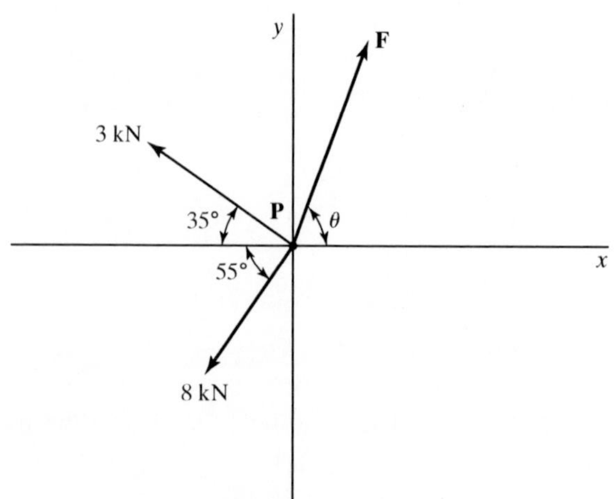

4.12 A gusset plate is subjected to the forces shown in Figure P4.12. Find the magnitude and direction θ of the force in member B so that the plate is in equilibrium.

Figure P4.12

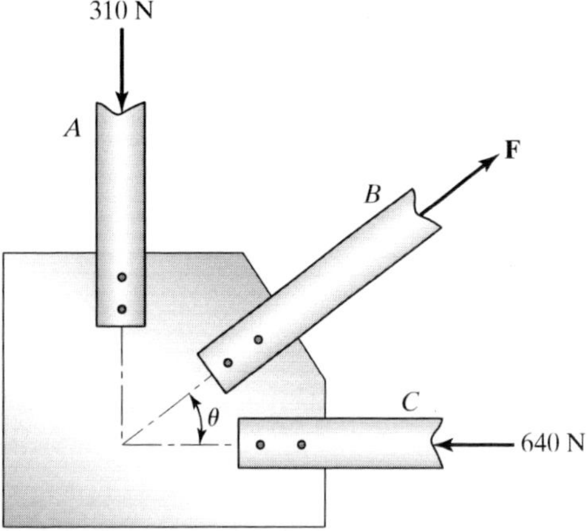

For problems 13 through 29, use the general analysis procedure of (1) problem statement, (2) diagram, (3) assumptions, (4) governing equations, (5) calculations, (6) solution check, and (7) discussion.

4.13 A 160-kg crate hangs from ropes as shown in Figure P4.13. Find the tension in ropes AB and AC.

Figure P4.13

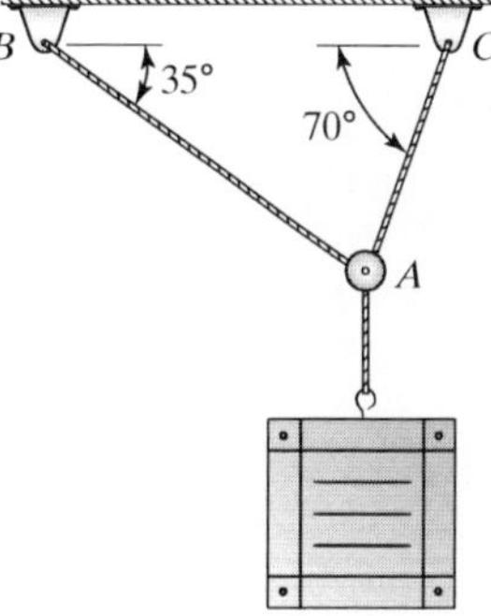

4.14 A 250-lb_m box is held in place by a cord with a spring scale on an inclined plane as shown in Figure P4.14. If all surfaces are smooth, what is the force reading on the scale?

Figure P4.14

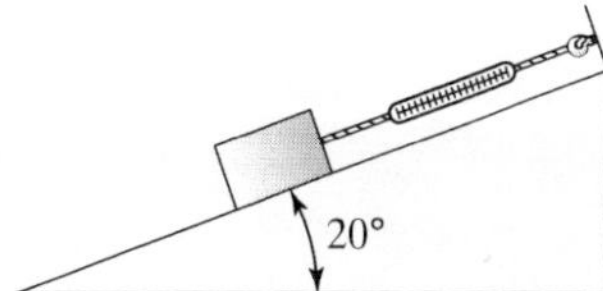

4.15 A concrete pipe with an inside and outside diameter of 60 cm and 70 cm, respectively, hangs from cables as shown in Figure P4.15. The pipe is supported at two locations, and a spreader bar maintains cable segments AB

and AC at 45°. Each support carries half the total weight of the pipe. If the density of concrete is $\rho = 2320 \text{ kg/m}^3$, find the tension in cable segments AB and AC.

Figure P4.15

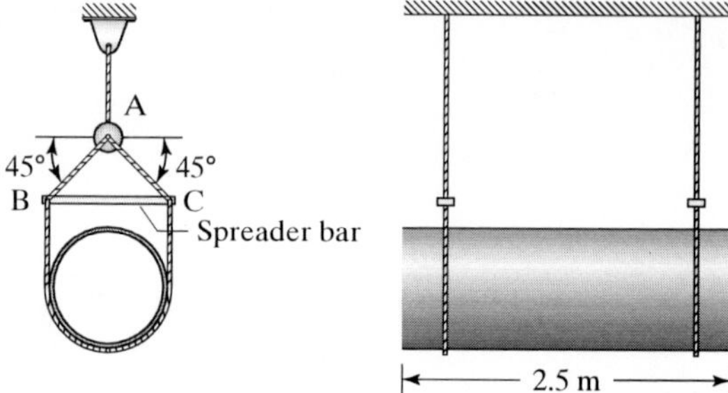

4.16 A construction worker holds a 500-kg crate in the position shown in Figure P4.16. What force must the worker exert on the cable?

Figure P4.16

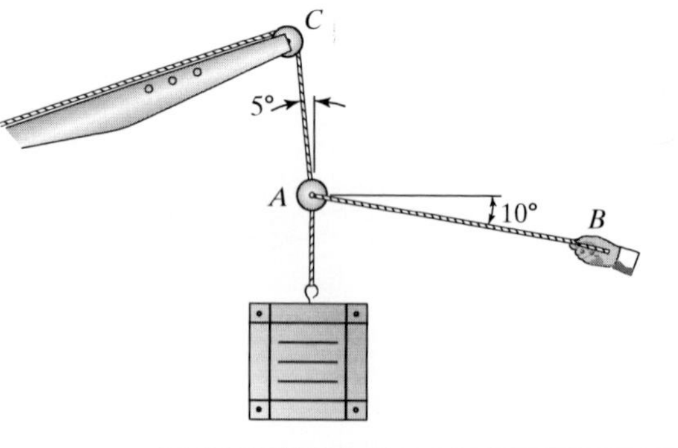

4.17 A 25-slug cylinder is suspended by a cord and frictionless pulley system as shown in Figure P4.17. A person standing on the floor pulls on the free end of the cord to hold the cylinder in a stationary position. What is the person's minimum weight for this to be possible?

Figure P4.17

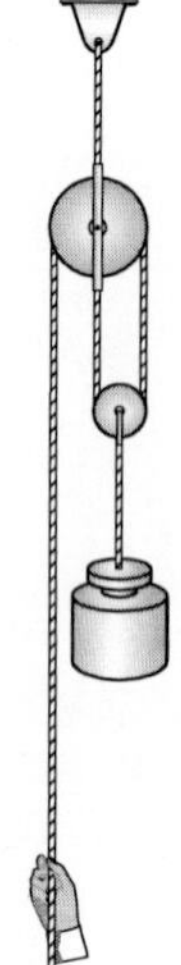

4.18 A 15.3 kg traffic signal is suspended from a symmetrical system of cables as shown in Figure P4.18. Find the tension in all cables. Cable BC is horizontal.

Figure P4.18

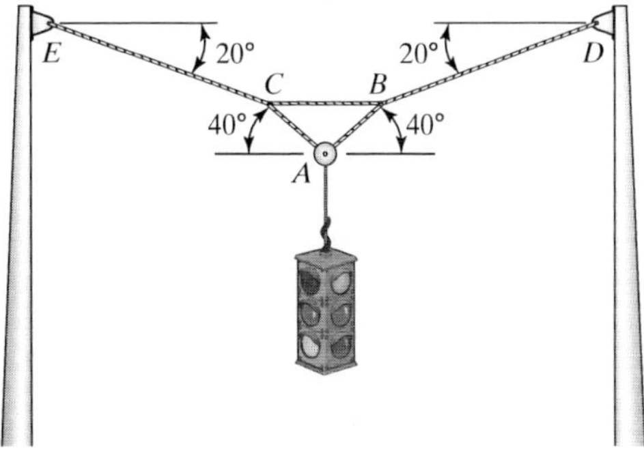

Stress and strain

4.19 A 90-cm-diameter wrecking ball hangs motionless from a 1.75-cm diameter cable. The wrecking ball is solid and is constructed of steel ($\rho = 7800 \text{ kg/m}^3$). If the cable is 16 m long, how much does the cable stretch? For the cable, use $E = 175$ GPa.

4.20 A 30-cm-long cylindrical rod of red brass ($E = 120$ GPa) is loaded axially causing the rod to elongate 0.580 mm. If the diameter of the rod is 1.75 cm, what is the load?

4.21 The column is subjected to a 15-kN force as shown in Figure P4.21. Find the average normal stress in the column.

Figure P4.21

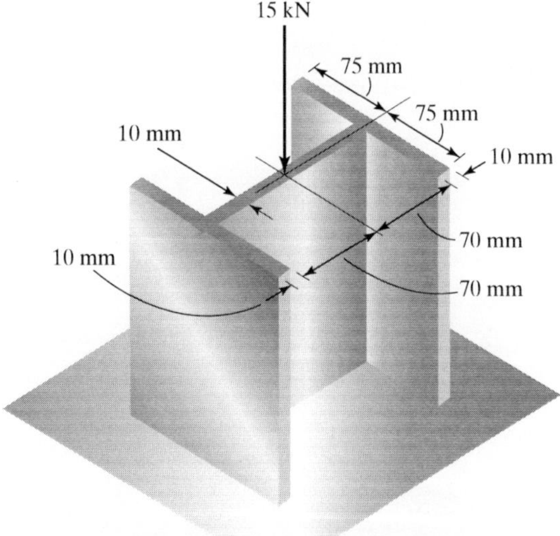

4.22 A solid composite shaft is subjected to a 2-MN force as shown in Figure P4.22. Section AB is red brass ($E = 120$ GPa), and section BC is AISI 1010 steel ($E = 200$ GPa). Find the normal stress in each section and the total axial deformation of the shaft.

Figure P4.22

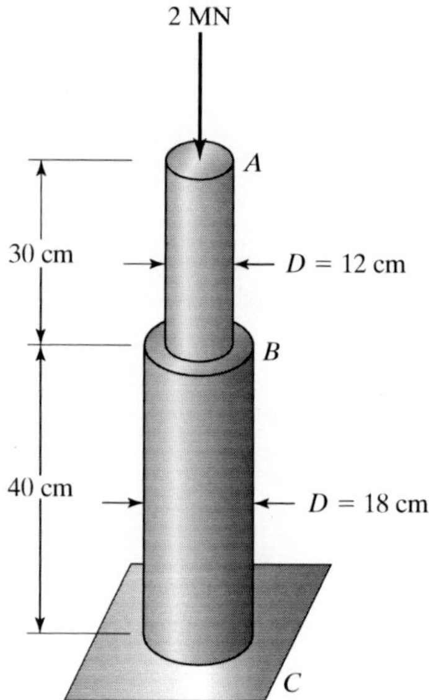

4.23 A tapered column of concrete ($E = 30$ GPa) is subjected to a 200-kN force as shown in Figure P4.23. Find the axial deformation of the column. *Hint*: express the cross-sectional area A as a function of x and perform the integration:

$$\delta = \int_0^L \frac{P\,dx}{A(x)E}.$$

Figure P4.23

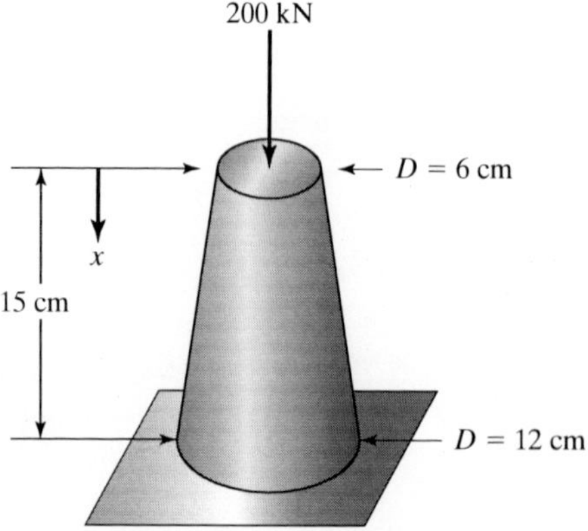

4.24 A 12 × 12 cm square plate of titanium ($E = 115$ GPa) is subjected to normal tensile forces of 15 kN and 20 kN on the top and right edges as shown in Figure P4.24. The thickness of the plate is 5 mm, and the left and bottom edges of the plate are fixed. Find the normal strain and deformation of the plate in the horizontal and vertical directions.

Figure P4.24

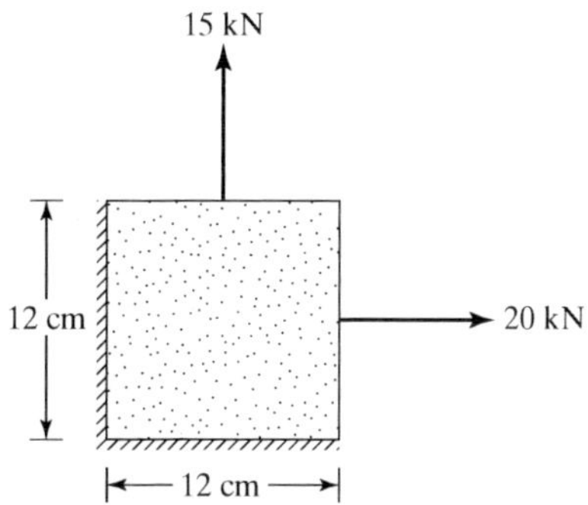

4.25 A tensile test is conducted on a steel specimen with a diameter of 8.0 mm and a test length of 6.0 cm. The data is shown in the table. Plot the stress–strain diagram, and find the approximate value of the modulus of elasticity for the steel.

Load (kN)	Deformation (mm)
2.0	0.0119
5.0	0.0303
10.0	0.0585
15.0	0.0895
20.0	0.122
25.0	0.145

Design stress

4.26 A 1-cm-diameter, 0.4-m-long steel rod is to be used in an application where it will be subjected to an axial tensile force of 15 kN. The factor of safety based on yield stress must be at least 1.5, and the axial deformation must not exceed 2 cm. Is the steel whose stress–strain diagram is shown in Figure P4.26 suitable for this application? Explain.

Figure P4.26

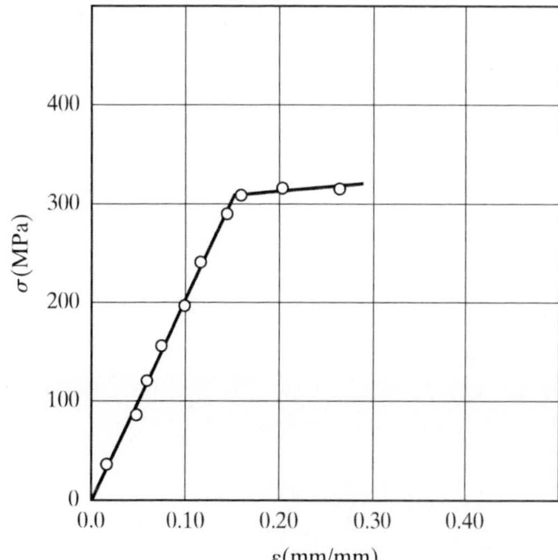

4.27 A 4-cm-diameter rod of aluminum 7075-T6 is to be subjected to an axial tensile force such that the factor of safety based on the yield stress is 1.25. Find the maximum allowable tensile force. The yield stress of aluminum 7075-T6 is $\sigma_y = 500$ MPa.

4.28 An 18-in-diameter sandstone column is subjected to an axial compressive force of 2×10^6 lb_f. Find the factor of safety based on the ultimate compressive stress. For the ultimate compressive stress of sandstone, use $\sigma_u = 85$ MPa.

4.29 A slender cylindrical member in a child's toy made of polystyrene plastic is to be subjected to an axial tensile force of 300N. Find the minimum diameter of this member for a factor of safety of 1.75 based on yield stress. The yield stress for polystyrene is $\sigma_y = 55$ MPa.

APPENDIX

A Mathematical Formulas

A.1 ALGEBRA

A.1.1 Quadratic equation

The quadratic equation:

$$ax^2 + bx + c = 0 \quad (a \neq 0)$$

has the solution:

$$x = \frac{-b \pm \sqrt{b^2 - 4ac}}{2a}$$

The two roots of the quadratic equation are either (a) both real or (b) complex conjugates.

A.1.2 Laws of exponents

$$x^m x^n = x^{m+n}$$

$$(x^m)^n = x^{mn}$$

$$(xy)^n = x^n y^n$$

$$(x/y)^n = x^n/y^n \qquad (y \neq 0)$$

$$(x^m/x^n) = x^{m-n} \qquad (x \neq 0)$$

$$x^m/x^n = 1 \qquad \text{if } m = n$$

$$x^0 = 1$$

$$x^{-n} = 1/x^n \qquad (x \neq 0)$$

A.1.3 Logarithms

In the relations that follow, the parameter a is called the *base.* These relations hold when $a > 0$ and $a \neq 1$. Typically, $a = 10$ (the common logarithm), or $a = e$ (the natural logarithm). The common logarithm is usually written as log (x), and the natural logarithm is usually written as $\ln(x)$:

$$
\begin{aligned}
v &= \log_a u && \textit{if } a^v = u \\
x &= \log_a a^x \\
\log_a(xy) &= \log_a x + \log_a y \\
\log_a(x/y) &= \log_a x - \log_a y \\
\log_a 1 &= 0 \\
\log_a(x^c) &= c \log_a x \\
\log_a a &= 1
\end{aligned}
$$

A.1.4 Exponential Function

The same relations that apply to the laws of exponents apply to the exponential function $\exp(x) = e^x$:

$$
\begin{aligned}
e^m e^n &= e^{m+n} \\
(e^m)^n &= e^{mn} \\
(e^m/e^n) &= e^{m-n} \\
e^m/e^n &= 1 && \text{if } m = n \\
e^0 &= 1 \\
e^{-n} &= 1/e^n
\end{aligned}
$$

The exponential function $\exp(x)$ and the natural logarithm $\ln(x)$ are inverse functions. Thus,

$$
\begin{aligned}
\ln(\exp(x)) &= x \\
\exp(\ln(x)) &= x
\end{aligned}
$$

A.2 GEOMETRY

A.2.1 Areas

Rectangle $A = ab$

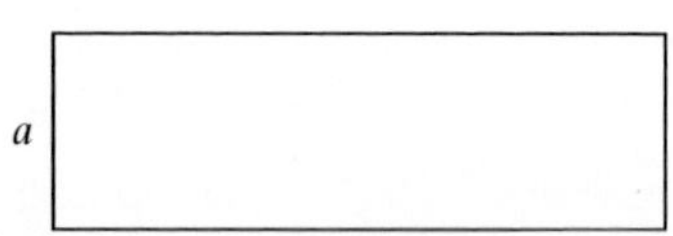

Parallelogram $A = bh$

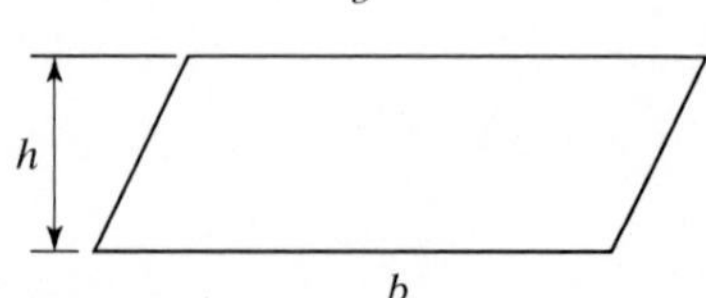

Trapezoid $A = \frac{1}{2}\, h(a + b)$

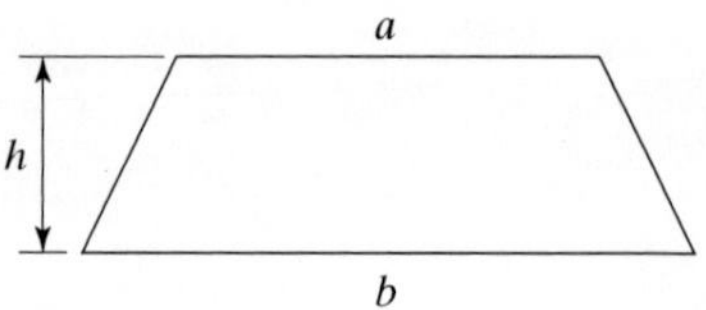

Triangle $A = \frac{1}{2} bh$

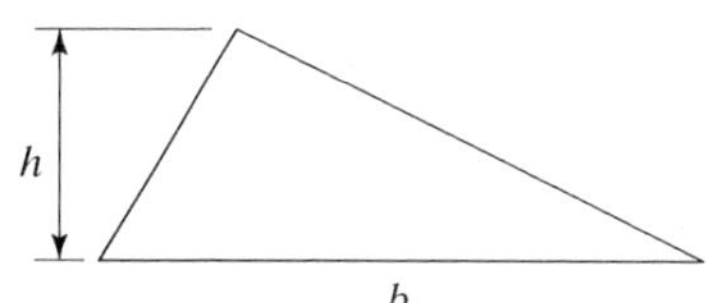

Circle $A = \pi R^2 = \frac{1}{4} \pi D^2$

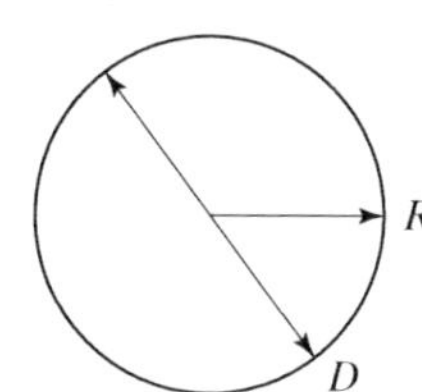

Circular sector $A = \frac{1}{2} R^2\theta$ (θ in radians)

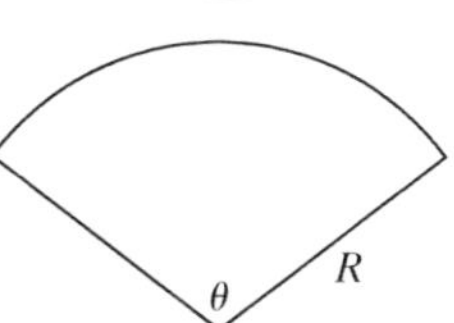

Circular segment $A = \frac{1}{2} R^2(\theta - \sin\theta)$
(θ in radians)

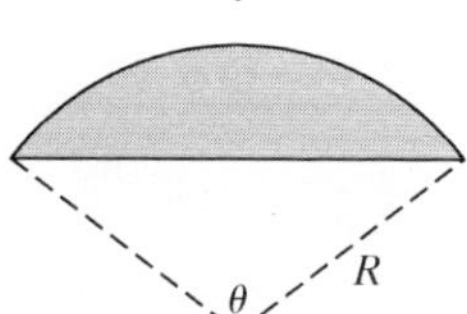

Regular polygon $A = nr^2 \tan(180°/n)$
$= \frac{1}{2} nR^2 \sin(360°/n)$
$n =$ number of sides

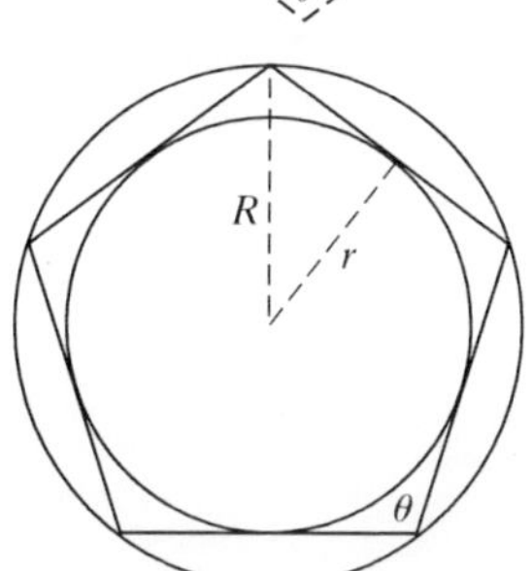

A.2.2 Solids

$A =$ Surface area
$V =$ Volume

Parallelpiped $A = 2(ab + ac + bc)$
$V = abc$

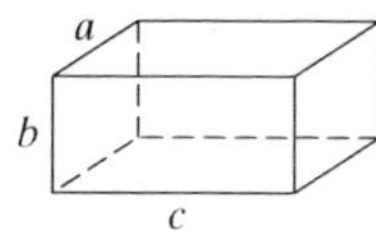

Cylinder $A = 2\pi RL = \pi DL$ (ends excluded)
$A = 2\pi R(L + R)$ (total)
$V = \pi R^2 L$

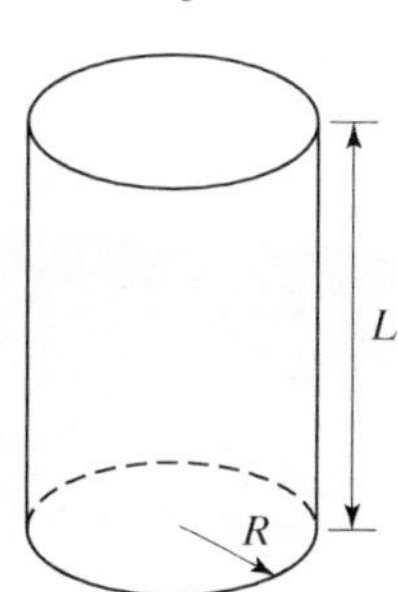

Sphere

$$A = 4\pi R^2 = \pi D^2$$
$$V = \frac{4\pi R^3}{3} = \frac{\pi D^3}{6}$$

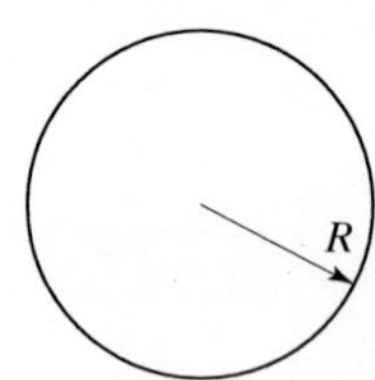

Cone

$$A = \pi R(R^2 + h^2)^{1/2} \text{ (base excluded)}$$
$$A = \pi R[R + (R^2 + h^2)^{1/2}] \text{ (total)}$$
$$V = \frac{\pi R^2 h}{3}$$

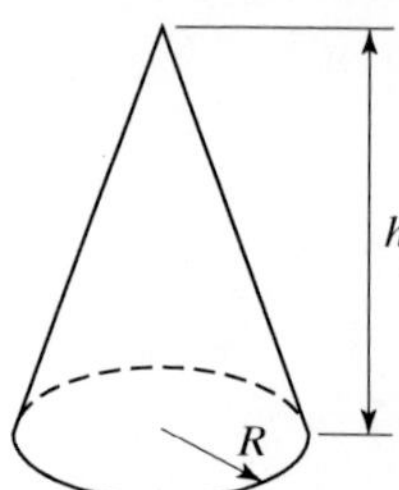

Torus

$$A = 4\pi^2 Rr$$
$$V = 2\pi^2 Rr^2$$

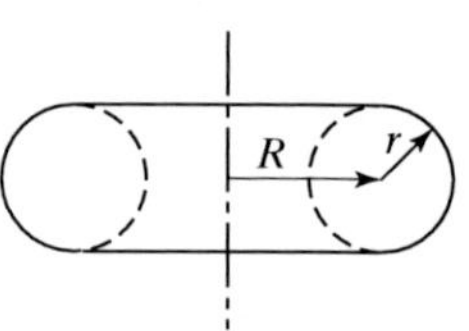

A.3 TRIGONOMETRY

A.3.1 Trigonometric Functions

$$\sin\theta = \frac{\text{opposite side}}{\text{hypotenuse}} = \frac{b}{c}$$
$$\cos\theta = \frac{\text{adjacent side}}{\text{hypotenuse}} = \frac{a}{c}$$
$$\tan\theta = \frac{\sin\theta}{\cos\theta} = \frac{\text{opposite side}}{\text{adjacent side}} = \frac{b}{a}$$
$$\cot\theta = \frac{1}{\tan\theta} = \frac{a}{b}$$
$$\sec\theta = \frac{1}{\cos\theta} = \frac{c}{a}$$
$$\csc\theta = \frac{1}{\sin\theta} = \frac{c}{b}$$

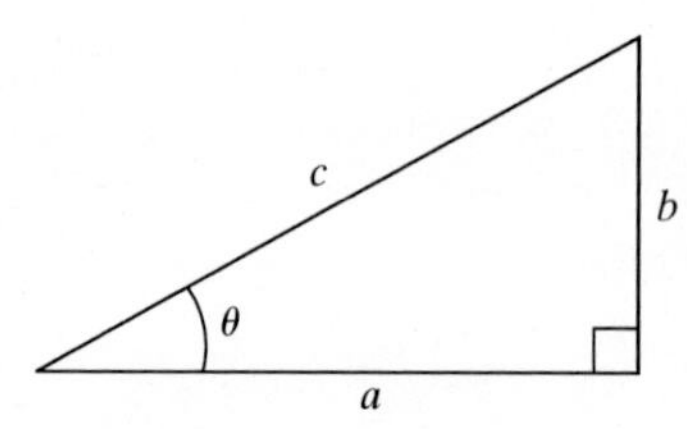

A.3.2 Identities and Relationships

$$\sin(-\theta) = -\sin(\theta)$$
$$\cos(-\theta) = \cos(\theta)$$
$$\tan(-\theta) = -\tan(\theta)$$
$$\sin^2\theta + \cos^2\theta = 1$$
$$1 + \tan^2\theta = \sec^2\theta$$

$$1 + \cot^2\theta = \csc^2\theta$$
$$\sin\theta = \cos(90° - \theta) = \sin(180° - \theta)$$
$$\cos\theta = \sin(90° - \theta) = -\cos(180° - \theta)$$
$$\tan\theta = \cot(90° - \theta) = -\tan(180° - \theta)$$
$$\sin(\theta \pm \alpha) = \sin\theta\cos\alpha \pm \cos\theta\sin\alpha$$
$$\cos(\theta \pm \alpha) = \cos\theta\cos\alpha \mp \sin\theta\sin\alpha$$
$$\tan(\theta \pm \alpha) = \frac{\tan\theta \pm \tan\alpha}{1 \mp \tan\theta\tan\alpha}$$
$$\sin 2\theta = 2\sin\theta\cos\theta$$
$$\cos 2\theta = \cos^2\theta - \sin^2\theta = 2\cos^2\theta - 1 = 1 - 2\sin^2\theta$$
$$\tan 2\theta = \frac{2\tan\theta}{1 - \tan^2\theta}$$
$$\sin(\theta/2) = \pm\sqrt{\frac{1 - \cos\theta}{2}}$$
$$\cos(\theta/2) = \pm\sqrt{\frac{1 + \cos\theta}{2}}$$
$$\tan(\theta/2) = \frac{\sin\theta}{1 + \cos\theta} = \frac{1 - \cos\theta}{\sin\theta}$$

A.3.3 Laws of Sines and Cosines

Law of sines $\dfrac{\sin A}{a} = \dfrac{\sin B}{b} = \dfrac{\sin C}{c}$

Law of cosines
$$a^2 = b^2 + c^2 - 2bc\cos A$$
$$b^2 = a^2 + c^2 - 2ac\cos B$$
$$c^2 = a^2 + b^2 - 2ab\cos C$$

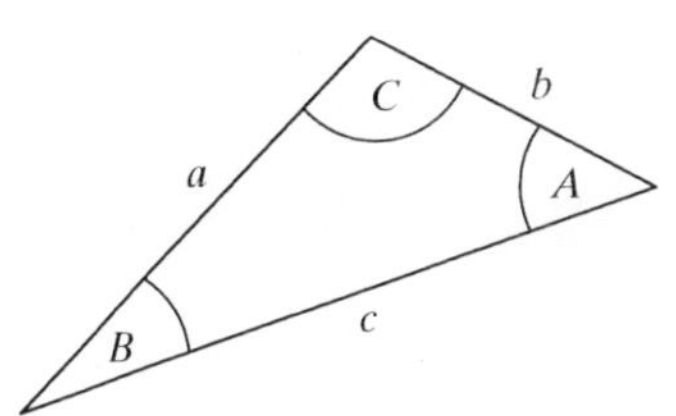

A.4 CALCULUS

In the following formulas, u and v represent functions of x, while a and n represent constants:

A.4.1 Derivatives

$$\frac{d(a)}{dx} = 0$$
$$\frac{d(x)}{dx} = 1$$
$$\frac{d(au)}{dx} = a\frac{du}{dx}$$
$$\frac{d(uv)}{dx} = u\frac{dv}{dx} + v\frac{du}{dx}$$

$$\frac{d(u^n)}{dx} = nu^{n-1}\frac{du}{dx}$$

$$\frac{d(\ln u)}{dx} = \frac{1}{u}\frac{du}{dx}$$

$$\frac{d(e^u)}{dx} = e^u\frac{du}{dx}$$

$$\frac{d(\sin u)}{dx} = \frac{du}{dx}(\cos u)$$

$$\frac{d(\cos u)}{dx} = -\frac{du}{dx}(\sin u)$$

A.4.2 Integrals

$$\int a\,dx = ax$$

$$\int a f(x)\,dx = a\int f(x)\,dx$$

$$\int x^n\,dx = \frac{x^{n+1}}{n+1} \quad (n \neq -1)$$

$$\int e^x\,dx = e^x$$

$$\int e^{ax}\,dx = \frac{e^{ax}}{a}$$

$$\int \ln(x)\,dx = x\ln(x) - x$$

APPENDIX B Unit Conversions

Acceleration	1 m/s^2	$= 3.2808\ ft/s^2$
		$= 39.370\ in/s^2$
		$= 4.252 \times 10^7\ ft/h^2$
		$= 8053\ mi/h^2$
Area	1 m^2	$= 10^4\ cm^2 = 10^6\ mm^2$
		$= 10.7636\ ft^2$
		$= 1550\ in^2$
	1 acre	$= 43{,}560\ ft^2$
Density	1 kg/m^3	$= 1000\ g/m^3 = 0.001\ g/cm^3$
		$= 0.06243\ lb_m/ft^3$
		$= 3.6127 \times 10^{-5}\ lb_m/in^3$
		$= 0.001940\ slug/ft^3$
Energy, work, heat	1055.06 J	= 1 Btu
	1.35582 J	$= 1\ ft \cdot b_f$
	4.1868 J	= 1 cal
	252 cal	= 1 Btu
	1 kWh	= 3412 Btu = 3600 kJ

Force	1 N	$= 10^5$ dyne
		$= 0.22481\ lb_f$
	1 lb_f	$= 32.174\ lb_m \cdot ft/s^2$
Heat transfer, power	1 W	= 1 J/s
		= 3.6 kJ/h
		= 3.4121 Btu/h
	745.7 W	= 1 hp
		$= 550\ lb_f \cdot ft/s$
		= 2544.4 Btu/h
	1.3558 W	$= 1\ lb_f \cdot ft/s$
Length	1 m	= 100 cm = 1000 mm
		= 3.2808 ft
		= 39.370 in
		= 1.0936 yd
	2.54 cm	= 1 in
	1 ft	= 12 in
	5280 ft	= 1 mi
	1 km	= 0.6214 mi
		= 0.5400 nautical mi
Mass	1 kg	= 1000 g
		$= 2.20462\ lb_m$
		= 0.06852 slug
	1 slug	$= 32.174\ lb_m$
	1 short ton	$= 2000\ lb_m$
	1 long ton	$= 2240\ lb_m$
Mass flow rate	1 kg/s	$= 2.20462\ lb_m/s$
		$= 7937\ lb_m/h$
		= 0.06852 slug/s
		= 246.68 slug/h
Pressure	1 kN/m^2	= 1 kPa
		$= 20.8855\ lb_f/ft^2$
		$= 0.14504\ lb_f/in^2 = 0.14504$ psi
		= 0.2953 in Hg
		$= 4.0146$ in H_2O
	101.325 kPa	= 1 atm
		$= 14.6959\ lb_f/in^2 = 14.6959$ psi
		= 760 mm Hg at 0°C
	1 bar	$= 10^5$ Pa

Specific heat	1 kJ/kg · °C	= 1 kJ/kg · K = 1 J/g · °C
		= 0.2388 Btu/lb_m · °F
		= 0.2388 Btu/lb_m · °R
Stress, modulus	1 kN/m^2	= 1 kPa
		= 0.14504 lb_f/in^2 = 0.14504 psi
	1 MN/m^2	= 1 MPa
		= 1000 kPa
		= 145.04 lb_f/in^2 = 145.04 psi
	1 GN/m^2	= 1 GPa
		= 1000 MPa
		= 1.4504 × 10^5 lb_f/in^2
		= 1.4504 × 10^5 psi
		= 145 ksi
Temperature	T(K)	= T(°C) + 273.15
		= T(°R)/1.8
		= [T(°F) + 459.67]/1.8
	T(°F)	= 1.8 T(°C) + 32
Temperature difference	ΔT(K)	= ΔT(°C)
		= ΔT(°F)/1.8
		= ΔT(°R)/1.8
Velocity	1 m/s	= 3.2808 ft/s
		= 11,811 ft/h
		= 2.2369 mi/h
		= 3.6000 km/h
		= 0.5400 knot
Viscosity (dynamic)	1 kg/m · s	= 1 Pa · s = 10 poise
		= 0.6720 lb_m/ft · s
		= 2419 lb_m/ft · h
Viscosity (kinematic)	1 m^2/s	= 10,000 stoke
		= 10.7639 ft^2/s
		= 38,750 ft^2/h
Volume	1 m^3	= 1000 L
		= 35.3134 ft^3
		= 61,022 in^3
		= 264.17 gal

C Physical Properties of Materials

Table C.1 Physical Properties of Solids at 20°C

Property Definitions:

ρ = density

c_p = specific heat at constant pressure

E = modulus of elasticity

σ_y = yield stress, tension

σ_u = ultimate stress, tension[a]

Material	ρ (kg/m³)	c_p (J/kg · °C)	E (GPa)	σ_y (MPa)	σ_u (MPa)
Metals					
Aluminum (99.6% Al)	2710	921	70	100	110
Aluminum 2014-T6	2800	875	75	400	455
Aluminum 6061-T6	2710	963	70	240	260
Aluminum 7075-T6	2800	963	72	500	570
Copper					
Oxygen-free (99.9% Cu)	8940	385	120	70	220
Red brass, cold rolled	8710	385	120	435	585
Yellow brass, cold rolled	8470	377	105	410	510

Material	ρ (kg/m³)	c_p (J/kg·°C)	E (GPa)	σ_y (MPa)	σ_u (MPa)
Iron alloys					
Structural steel	7860	420	200	250	400
Cast iron, gray	7270	420	69		655[a]
AISI 1010 steel, cold rolled	7270	434	200	300	365
AISI 4130 steel, cold rolled	7840	460	200	760	850
AISI 302 stainless, cold rolled	8055	480	190	520	860
Magnesium (AZ31)	1770	1026	45	200	255
Monel 400 (67% Ni, 32% Cu) cold worked	8830	419	180	585	675
Titanium (6% Al, 4% V)	4420	610	115	830	900
Nonmetals					
Concrete, high strength	2320	900	30		40[a]
Glass	2190	750	65		50[a]
Plastic					
Acrylic	1180	1466	2.8	52	
Nylon 6/6	1140	1680	2.8	45	75
Polypropylene	905	1880	1.3	34	
Polystyrene	1030	1360	3.1	55	90
Polyvinylchloride (PVC)	1440	1170	3.1	45	70
Rock					
Granite	2770	775	70		240[a]
Sandstone	2300	745	40		85[a]
Wood					
Fir	470		13		50[a]
Oak	660		12		47[a]
Pine	415		9		36[a]

[a]Ultimate stress in compression.

Table C.2 Physical Properties of Fluids at 20°C

Property Definitions:

ρ = density

c_p = specific heat at constant pressure

μ = dynamic viscosity

v = kinematic viscosity

(Continued)

Fluid	ρ **(kg/m³)**	C_p **(J/kg · °C)**	μ **(kg/m · s)**	v **(m²/s)**
Liquids				
Ammonia	600	4825	1.31×10^{-4}	2.18×10^{-7}
Engine oil (SAE10W-30)	878	1800	0.191	2.17×10^{-4}
Ethyl alcohol	802	2457	1.05×10^{-3}	1.31×10^{-6}
Gasoline	751	2060	5.29×10^{-4}	7.04×10^{-7}
Glycerin	1260	2350	1.48	1.18×10^{-3}
Mercury	13,550	140	1.56×10^{-3}	1.15×10^{-7}
Water	998	4182	1.00×10^{-3}	1.00×10^{-6}
Water/ethylene glycol (50/50 mixture)	1073	3281	3.94×10^{-3}	3.67×10^{-6}
Gases (at 1 atm pressure)				
Air	1.194	1006	1.81×10^{-5}	1.52×10^{-5}
Carbon dioxide (CO_2)	1.818	844	1.46×10^{-5}	8.03×10^{-6}
Carbon monoxide (CO)	1.152	1043	1.72×10^{-5}	1.49×10^{-5}
Helium (He)	0.165	5193	1.95×10^{-5}	1.18×10^{-4}
Hydrogen (H)	0.0830	14,275	8.81×10^{-6}	1.06×10^{-4}
Nitrogen (N_2)	1.155	1041	1.75×10^{-5}	1.52×10^{-5}
Oxygen (O_2)	1.320	919	2.03×10^{-5}	1.54×10^{-5}
Water vapor, saturated (H_2O)	0.0173	1874	8.85×10^{-6}	5.12×10^{-4}

APPENDIX

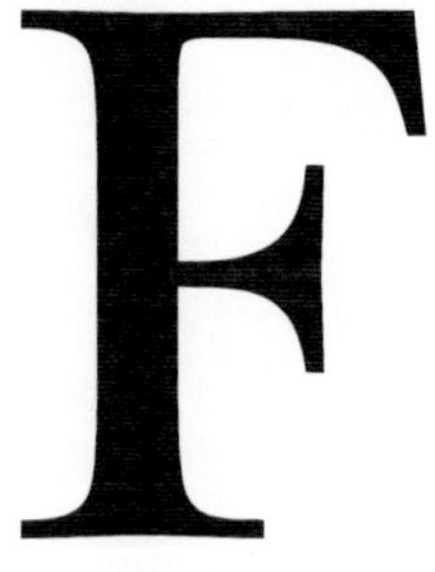

F Answers to Selected Problems

CHAPTER 1

1.2

y_{max} (mm)	*h* (mm)	*b* (mm)
2.16	200	100
2.70	200	80
4.04	175	80
1.11	250	100
3.94	125	225
1.85	175	175
3.42	150	150

CHAPTER 2

2.2 The argument of any mathematical function must be dimensionless. Because the argument [L][t] is not a dimensionless quantity, the equation is not dimensionally consistent.

2.4 Yes, because the argument MM^{-1} of the cosine function is dimensionless and TNT appears on both sides of the equation.

2.6 $1\ N \cdot m = 1\ J$(joule); work, energy, heat.

2.8 $P = I^2R$.

2.10 50 kW

2.12 3.66×10^6 kg, 36.0 MN

2.14 a. 4 lb_f, b. 1.51 lb_f

2.16 1.22 kg, 12.0 N

2.18 208

2.20 15.0 mi/h, 6.71 m/s

2.22 −40°

2.24 1.12 MJ

2.26 13.2 lb_m/s, 1480 slug/h

2.28 36.1°C, 36.1 K, 65°R

2.30 2.678×10^6 s

2.32 229 N, 51.6 lb_f

2.34 3986 Pa, 0.578 psi

2.36 1.21×10^7 Btu/h · ft^3

2.38 1.89×10^{-4} m^3/s, 24.1 ft^3/h

2.40 216 MJ, 2.05×10^5 Btu, 5.16×10^7 cal

2.42 8.72×10^6

2.44 1007 J/kg · K, 0.241 Btu/lb_m · °F

2.46 490 W, 17.6 kJ, 155 MJ

CHAPTER 3

3.10 b. $\underline{9.807}$, c. $0.00\underline{216}$, d. $\underline{9}000$ e. $\underline{7000}$. f. $\underline{12.00}$ g. $\underline{1066}$ h. $\underline{106.07}$ i. $0.0\underline{2880}$ j. $\underline{163}.\underline{07}$ k. $\underline{1.207} \times 10^{-3}$

3.12 6.38 kN

3.14 2.6×10^4 V

3.16 240

3.18 0.32 MN

3.20 3.94 min

3.22 170 kΩ, 0.587 mA
20 kΩ: $V = 11.7$ V; 150 kΩ: $V = 88.1$ V; 250 Ω: $V = 0.147$ V

CHAPTER 4

4.2 $-7.01\,\mathbf{i} + 13.3\,\mathbf{j}$ N

4.4 $\mathbf{F}_R = 0\,\mathbf{i} - 37.9\,\mathbf{j}$ lb_f, $F_R = 37.9$ lb_f, $\theta = -90°$

4.10 a = −8, b = 3, c = 1

4.12 F = 711 N, $\theta = 25.8°$

4.14 85.5 lb_f

4.16 99.5 lb_f

4.18 $T_{AB} = T_{AC} = T_{BC} = 117$ N, $T_{BD} = T_{CE} = 219$ N

4.20 55.8 kN

4.22 $\sigma_{AB} = 177$ MPa, $\sigma_{BC} = 78.6$ MPa, $\delta = 0.599$ mm

4.24 horizontal: $\varepsilon = 2.90 \times 10^{-4}$, $\delta = 0.0348$ mm
vertical: $\varepsilon = 2.18 \times 10^{-4}$, $\delta = 0.0261$ mm

4.26 No. The factor of safety is acceptable at FS = 1.57, but the deformation is $\delta = 3.82$ cm, which exceeds the allowable deformation.

4.28 1.57

Section 2
Mechanics for Engineers – Statics

Hibbeler and Yap's Mechanics for Engineers – Statics, 13th SI edition

Hibbeler, R.C. & Yap, K.B. (2013). *Mechanics for Engineers SI – Statics* (13th ed.). Jurong, Singapore: Pearson Education South Asia Pte Ltd.

Chapter 5

It is important to be able to determine the forces in the cables used to support this submarine to insure that they do not fail. In this chapter we will study how to apply equilibrium methods to determine the forces acting on the supports of a rigid body such as this.

Equilibrium of a Rigid Body

CHAPTER OBJECTIVES

- To develop the equations of equilibrium for a rigid body.
- To introduce the concept of the free-body diagram for a rigid body.
- To show how to solve rigid-body equilibrium problems using the equations of equilibrium.

Video Solutions are available for selected questions in this chapter.

5.1 Conditions for Rigid-Body Equilibrium

In this section, we will develop both the necessary and sufficient conditions for the equilibrium of the rigid body in Fig. 5–1*a*. As shown, this body is subjected to an external force and couple moment system that is the result of the effects of gravitational, electrical, magnetic, or contact forces caused by adjacent bodies. The internal forces caused by interactions between particles within the body are not shown in this figure because these forces occur in equal but opposite collinear pairs and hence will cancel out, a consequence of Newton's third law.

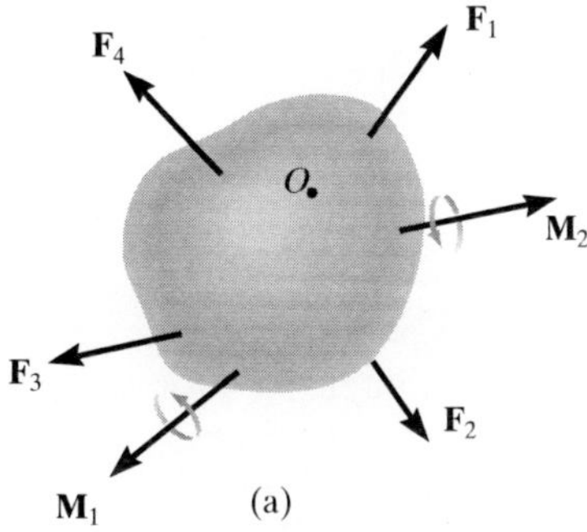

(a)

Fig. 5–1

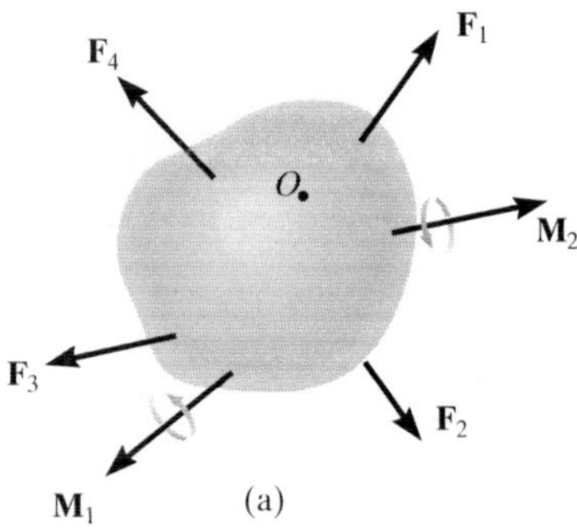

(a)

Using the methods of the previous chapter, the force and couple moment system acting on a body can be reduced to an equivalent resultant force and resultant couple moment at any arbitrary point O on or off the body, Fig. 5–1b. If this resultant force and couple moment are both equal to zero, then the body is said to be in *equilibrium*. Mathematically, the equilibrium of a body is expressed as

$$\mathbf{F}_R = \Sigma \mathbf{F} = \mathbf{0}$$
$$(\mathbf{M}_R)_O = \Sigma \mathbf{M}_O = \mathbf{0} \qquad (5\text{–}1)$$

The first of these equations states that the sum of the forces acting on the body is equal to *zero*. The second equation states that the sum of the moments of all the forces in the system about point O, added to all the couple moments, is equal to *zero*. These two equations are not only necessary for equilibrium, they are also sufficient. To show this, consider summing moments about some other point, such as point A in Fig. 5–1c. We require

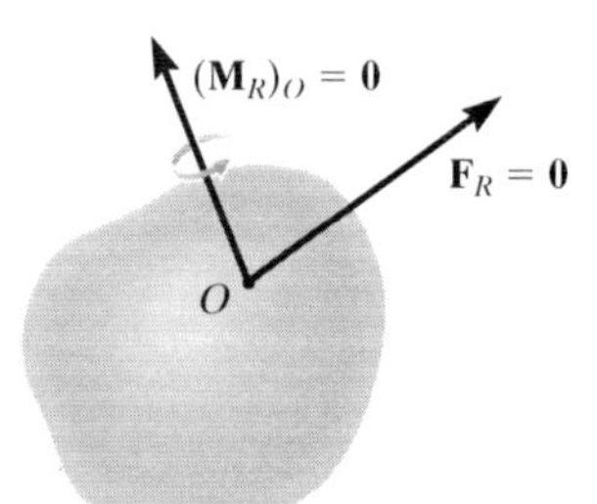

(b)

$$\Sigma \mathbf{M}_A = \mathbf{r} \times \mathbf{F}_R + (\mathbf{M}_R)_O = \mathbf{0}$$

Since $\mathbf{r} \neq \mathbf{0}$, this equation is satisfied if Eqs. 5–1 are satisfied, namely $\mathbf{F}_R = \mathbf{0}$ and $(\mathbf{M}_R)_O = \mathbf{0}$.

When applying the equations of equilibrium, we will assume that the body remains rigid. In reality, however, all bodies deform when subjected to loads. Although this is the case, most engineering materials such as steel and concrete are very rigid and so their deformation is usually very small. Therefore, when applying the equations of equilibrium, we can generally assume that the body will remain *rigid* and *not deform* under the applied load without introducing any significant error. This way the direction of the applied forces and their moment arms with respect to a fixed reference remain the same both before and after the body is loaded.

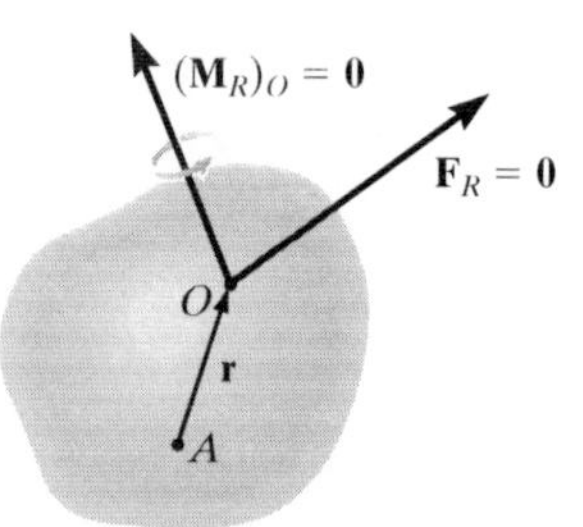

(c)

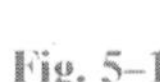
Fig. 5–1

EQUILIBRIUM IN TWO DIMENSIONS

In the first part of the chapter, we will consider the case where the force system acting on a rigid body lies in or may be projected onto a *single* plane and, furthermore, any couple moments acting on the body are directed perpendicular to this plane. This type of force and couple system is often referred to as a two-dimensional or *coplanar* force system. For example, the airplane in Fig. 5–2 has a plane of symmetry through its center axis, and so the loads acting on the airplane are symmetrical with respect to this plane. Thus, each of the two wing tires will support the same load $\mathbf{T}$, which is represented on the side (two-dimensional) view of the plane as $2\mathbf{T}$.

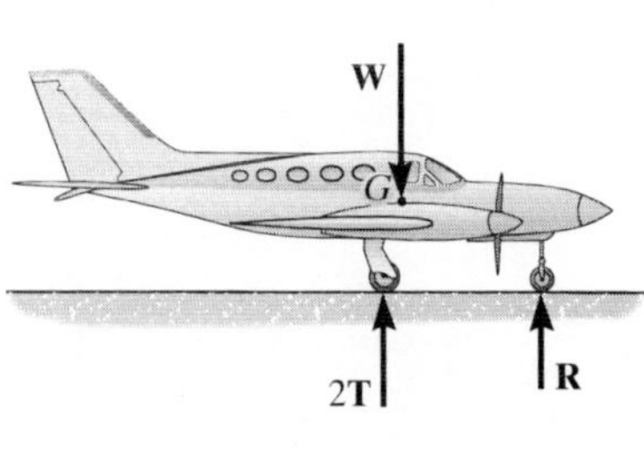

Fig. 5–2

5.2 Free-Body Diagrams

Successful application of the equations of equilibrium requires a complete specification of *all* the known and unknown external forces that act *on* the body. The best way to account for these forces is to draw a free-body diagram. This diagram is a sketch of the outlined shape of the body, which represents it as being *isolated* or "free" from its surroundings, i.e., a "free body." On this sketch it is necessary to show *all* the forces and couple moments that the surroundings exert *on the body* so that these effects can be accounted for when the equations of equilibrium are applied. *A thorough understanding of how to draw a free-body diagram is of primary importance for solving problems in mechanics.*

Support Reactions. Before presenting a formal procedure as to how to draw a free-body diagram, we will first consider the various types of reactions that occur at supports and points of contact between bodies subjected to coplanar force systems. As a general rule,

- If a support prevents the translation of a body in a given direction, then a force is developed on the body in that direction.
- If rotation is prevented, a couple moment is exerted on the body.

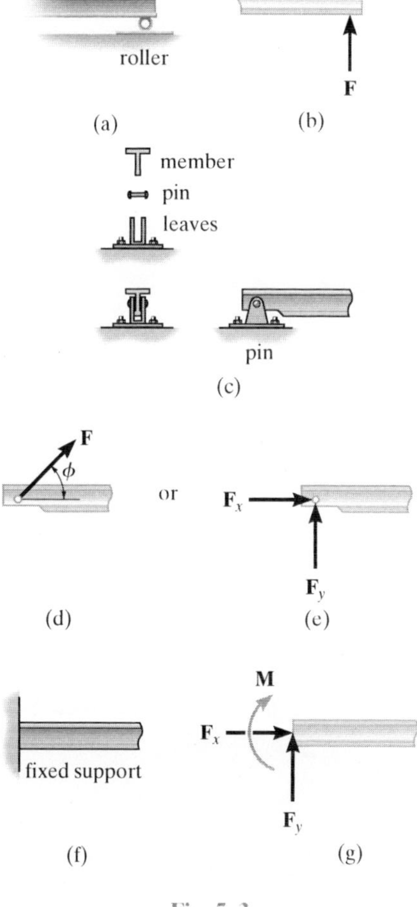

Fig. 5–3

For example, let us consider three ways in which a horizontal member, such as a beam, is supported at its end. One method consists of a *roller* or cylinder, Fig. 5–3*a*. Since this support only prevents the beam from *translating* in the vertical direction, the roller will only exert a *force* on the beam in this direction, Fig. 5–3*b*.

The beam can be supported in a more restrictive manner by using a *pin,* Fig. 5–3*c*. The pin passes through a hole in the beam and two leaves which are fixed to the ground. Here the pin can prevent *translation* of the beam in *any direction* ϕ, Fig. 5–3*d*, and so the pin must exert a *force* **F** on the beam in this direction. For purposes of analysis, it is generally easier to represent this resultant force **F** by its two rectangular components $\mathbf{F}_x$ and $\mathbf{F}_y$, Fig. 5–3*e*. If F_x and F_y are known, then F and ϕ can be calculated.

The most restrictive way to support the beam would be to use a *fixed support* as shown in Fig. 5–3*f*. This support will prevent both *translation and rotation* of the beam. To do this a *force and couple moment* must be developed on the beam at its point of connection, Fig. 5–3*g*. As in the case of the pin, the force is usually represented by its rectangular components $\mathbf{F}_x$ and $\mathbf{F}_y$.

Table 5–1 lists other common types of supports for bodies subjected to coplanar force systems. (In all cases the angle θ is assumed to be known.) Carefully study each of the symbols used to represent these supports and the types of reactions they exert on their contacting members.

5

TABLE 5–1 Supports for Rigid Bodies Subjected to Two-Dimensional Force Systems

Types of Connection	Reaction	Number of Unknowns
(1)  θ cable	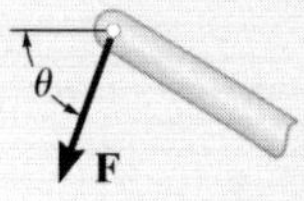 θ F	One unknown. The reaction is a tension force which acts away from the member in the direction of the cable.
(2) 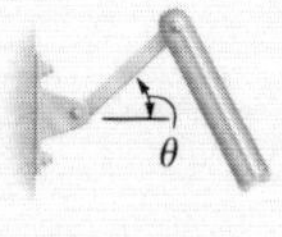 θ weightless link	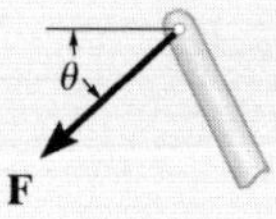 θ F or θ F 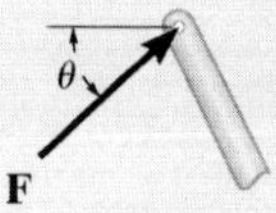	One unknown. The reaction is a force which acts along the axis of the link.
(3) 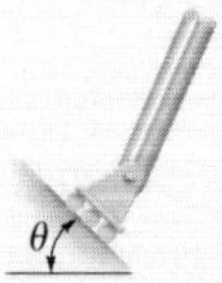 θ roller	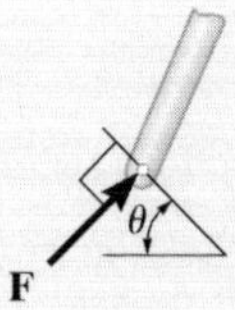 θ F	One unknown. The reaction is a force which acts perpendicular to the surface at the point of contact.
(4) 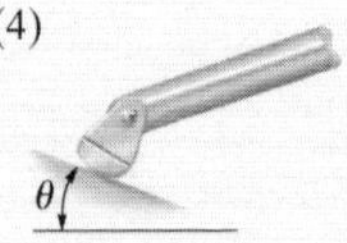 θ rocker	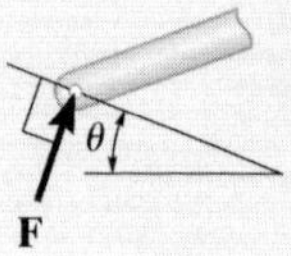 θ F	One unknown. The reaction is a force which acts perpendicular to the surface at the point of contact.
(5) 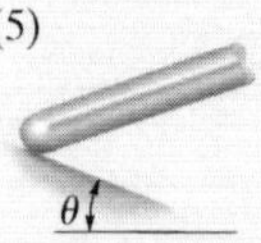 θ smooth contacting surface	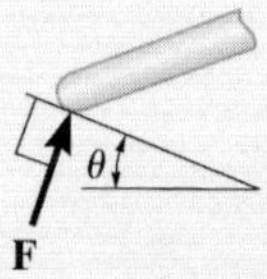 θ F	One unknown. The reaction is a force which acts perpendicular to the surface at the point of contact.
(6) 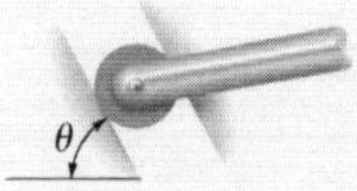 θ roller or pin in confined smooth slot	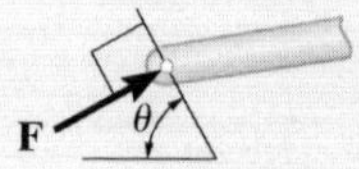 F θ or F θ 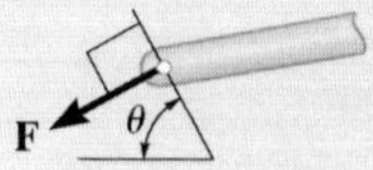	One unknown. The reaction is a force which acts perpendicular to the slot.
(7) 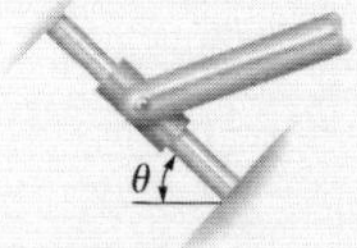 θ member pin connected to collar on smooth rod	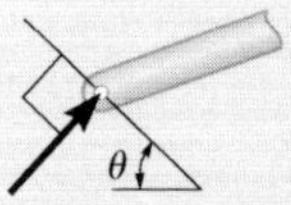 θ or θ F 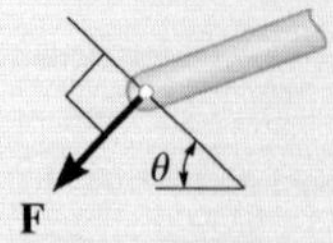	One unknown. The reaction is a force which acts perpendicular to the rod.

continued

TABLE 5–1 Continued

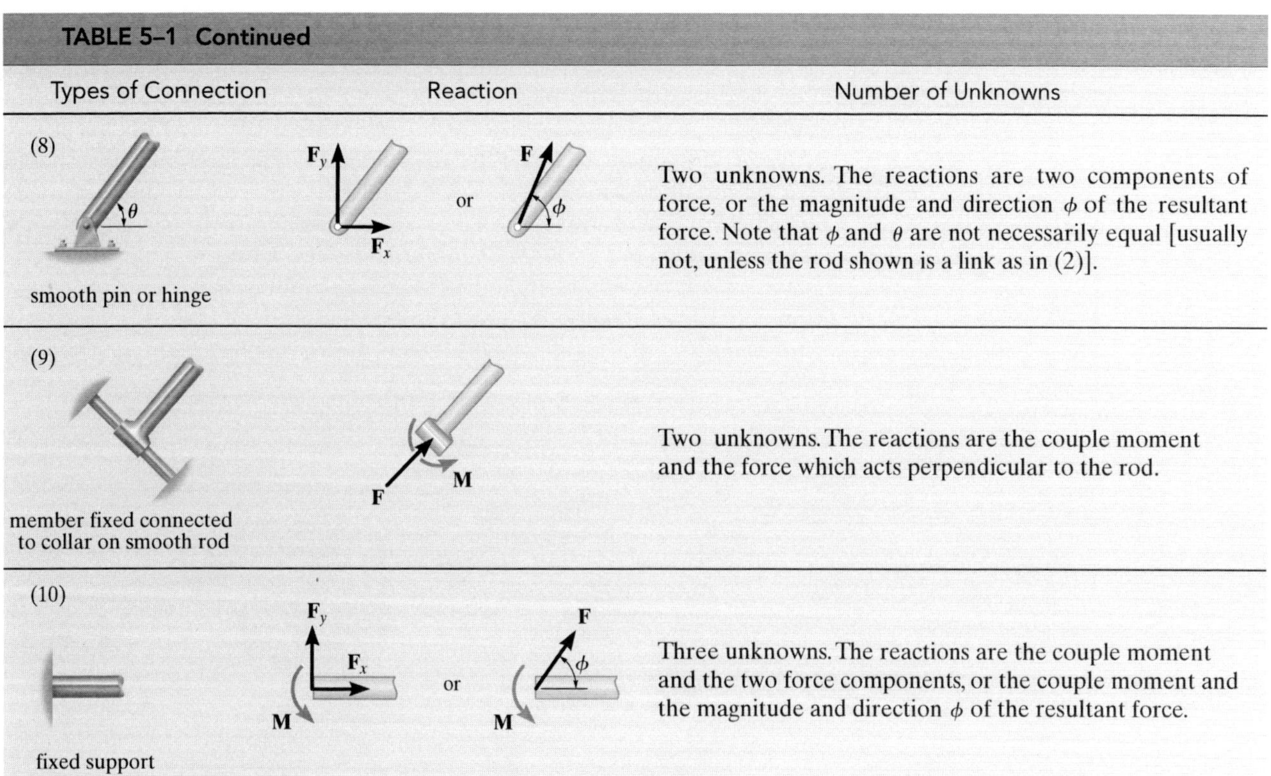

Types of Connection	Reaction	Number of Unknowns
(8) smooth pin or hinge	F_y, F_x or F, ϕ	Two unknowns. The reactions are two components of force, or the magnitude and direction ϕ of the resultant force. Note that ϕ and θ are not necessarily equal [usually not, unless the rod shown is a link as in (2)].
(9) member fixed connected to collar on smooth rod	F, M	Two unknowns. The reactions are the couple moment and the force which acts perpendicular to the rod.
(10) fixed support	F_y, F_x, M or F, ϕ, M	Three unknowns. The reactions are the couple moment and the two force components, or the couple moment and the magnitude and direction ϕ of the resultant force.

Typical examples of actual supports are shown in the following sequence of photos. The numbers refer to the connection types in Table 5–1.

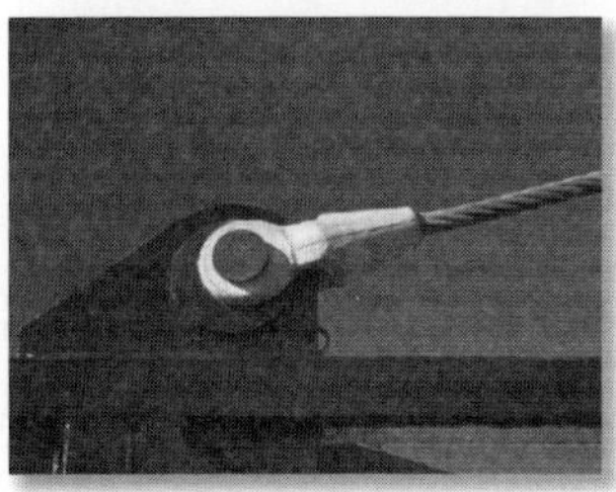

The cable exerts a force on the bracket in the direction of the cable. (1)

The rocker support for this bridge girder allows horizontal movement so the bridge is free to expand and contract due to a change in temperature. (4)

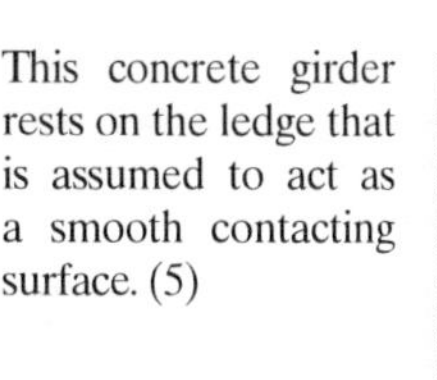

This concrete girder rests on the ledge that is assumed to act as a smooth contacting surface. (5)

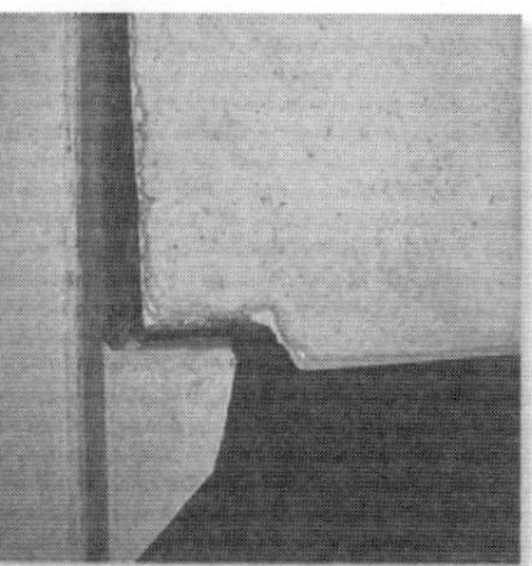

This utility building is pin supported at the top of the column. (8)

The floor beams of this building are welded together and thus form fixed connections. (10)

Internal Forces. As stated in Sec. 5.1, the internal forces that act between adjacent particles in a body always occur in collinear pairs such that they have the same magnitude and act in opposite directions (Newton's third law). Since these forces cancel each other, they will not create an *external effect* on the body. It is for this reason that the internal forces should not be included on the free-body diagram if the entire body is to be considered. For example, the engine shown in Fig. 5–4*a* has a free-body diagram shown in Fig. 5–4*b*. The internal forces between all its connected parts, such as the screws and bolts, will cancel out because they form equal and opposite collinear pairs. Only the external forces $\mathbf{T}_1$ and $\mathbf{T}_2$, exerted by the chains and the engine weight $\mathbf{W}$, are shown on the free-body diagram.

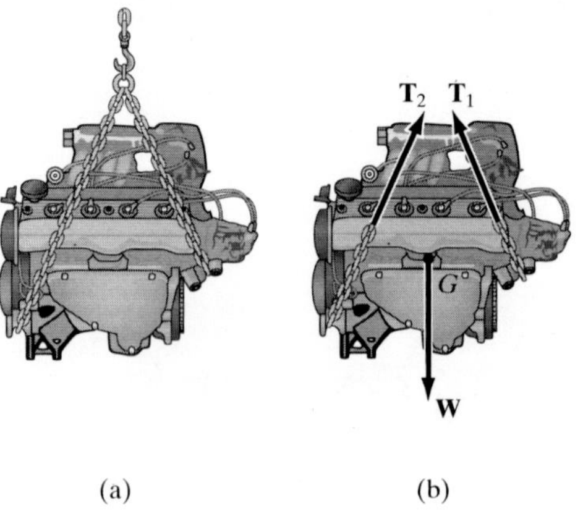

Fig. 5–4

Weight and the Center of Gravity. When a body is within a gravitational field, then each of its particles has a specified weight. It was shown in Sec. 4.8 that such a system of forces can be reduced to a single resultant force acting through a specified point. We refer to this force resultant as the *weight* $\mathbf{W}$ of the body and to the location of its point of application as the *center of gravity*. The methods used for its determination will be developed in Chapter 9.

In the examples and problems that follow, if the weight of the body is important for the analysis, this force will be reported in the problem statement. Also, when the body is *uniform* or made from the same material, the center of gravity will be located at the body's *geometric center* or *centroid*; however, if the body consists of a nonuniform distribution of material, or has an unusual shape, then the location of its center of gravity G will be given.

Idealized Models. When an engineer performs a force analysis of any object, he or she considers a corresponding analytical or idealized model that gives results that approximate as closely as possible the actual situation. To do this, careful choices have to be made so that selection of the type of supports, the material behavior, and the object's dimensions can be justified. This way one can feel confident that any design or analysis will yield results which can be trusted. In complex cases this process may require developing several different models of the object that must be analyzed. In any case, this selection process requires both skill and experience.

The following two cases illustrate what is required to develop a proper model. In Fig. 5–5*a*, the steel beam is to be used to support the three roof joists of a building. For a force analysis it is reasonable to assume the material (steel) is rigid since only very small deflections will occur when the beam is loaded. A bolted connection at *A* will allow for any slight rotation that occurs here when the load is applied, and so a *pin* can be considered for this support. At *B* a *roller* can be considered since this support offers no resistance to horizontal movement. Building code is used to specify the roof loading *A* so that the joist loads **F** can be calculated. These forces will be larger than any actual loading on the beam since they account for extreme loading cases and for dynamic or vibrational effects. Finally, the weight of the beam is generally neglected when it is small compared to the load the beam supports. The idealized model of the beam is therefore shown with average dimensions *a, b, c,* and *d* in Fig. 5–5*b*.

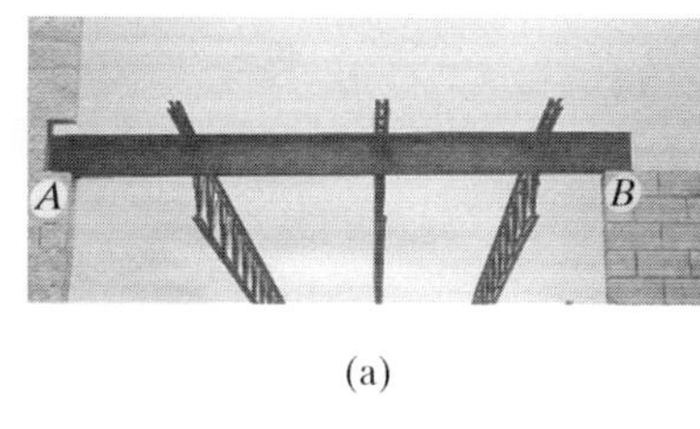

(a)

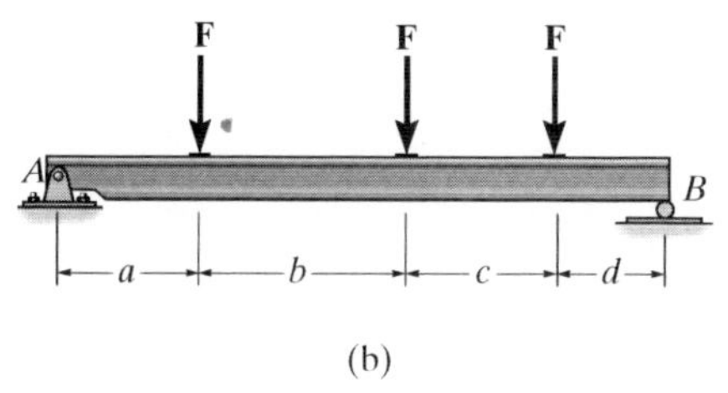

(b)

Fig. 5–5

As a second case, consider the lift boom in Fig. 5–6*a*. By inspection, it is supported by a pin at *A* and by the hydraulic cylinder *BC*, which can be approximated as a weightless link. The material can be assumed rigid, and with its density known, the weight of the boom and the location of its center of gravity *G* are determined. When a design loading **P** is specified, the idealized model shown in Fig. 5–6*b* can be used for a force analysis. Average dimensions (not shown) are used to specify the location of the loads and the supports.

Idealized models of specific objects will be given in some of the examples throughout the text. It should be realized, however, that each case represents the reduction of a practical situation using simplifying assumptions like the ones illustrated here.

(a)

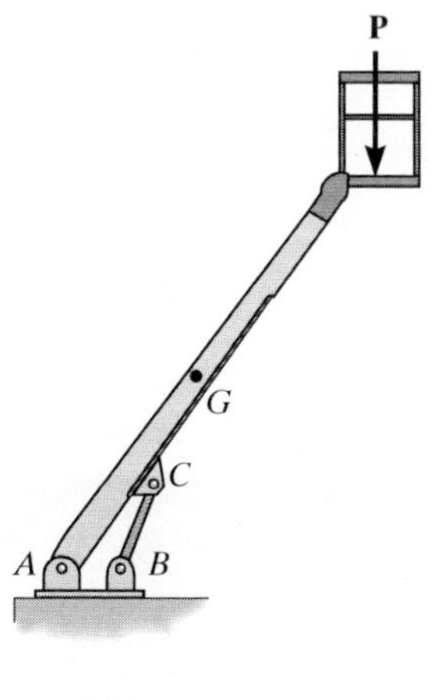

(b)

Fig. 5–6

Procedure for Analysis

To construct a free-body diagram for a rigid body or any group of bodies considered as a single system, the following steps should be performed:

Draw Outlined Shape.

Imagine the body to be *isolated* or cut "free" from its constraints and connections and draw (sketch) its outlined shape.

Show All Forces and Couple Moments.

Identify all the known and unknown *external forces* and couple moments that *act on the body*. Those generally encountered are due to (1) applied loadings, (2) reactions occurring at the supports or at points of contact with other bodies (see Table 5–1), and (3) the weight of the body. To account for all these effects, it may help to trace over the boundary, carefully noting each force or couple moment acting on it.

Identify Each Loading and Give Dimensions.

The forces and couple moments that are known should be labeled with their proper magnitudes and directions. Letters are used to represent the magnitudes and direction angles of forces and couple moments that are unknown. Establish an x, y coordinate system so that these unknowns, A_x, A_y, etc., can be identified. Finally, indicate the dimensions of the body necessary for calculating the moments of forces.

Important Points

- No equilibrium problem should be solved without *first drawing the free-body diagram*, so as to account for all the forces and couple moments that act on the body.
- If a support *prevents translation* of a body in a particular direction, then the support exerts a *force* on the body in that direction.
- If *rotation is prevented*, then the support exerts a *couple moment* on the body.
- Study Table 5–1.
- Internal forces are never shown on the free-body diagram since they occur in equal but opposite collinear pairs and therefore cancel out.
- The weight of a body is an external force, and its effect is represented by a single resultant force acting through the body's center of gravity G.
- *Couple moments* can be placed anywhere on the free-body diagram since they are *free vectors*. *Forces* can act at any point along their lines of action since they are *sliding vectors*.

EXAMPLE 5.1

Draw the free-body diagram of the uniform beam shown in Fig. 5–7*a*. The beam has a mass of 100 kg.

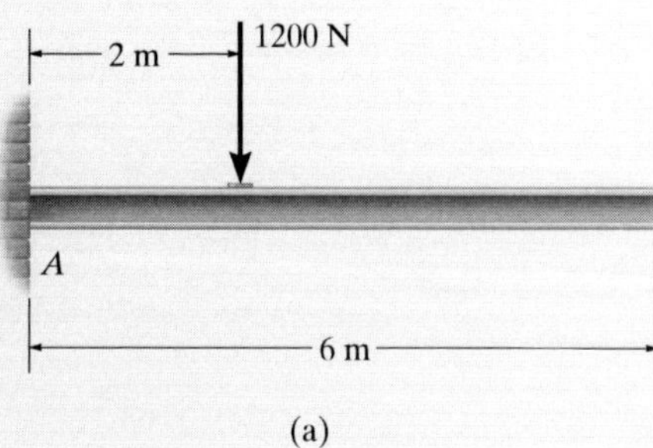

(a)

SOLUTION

The free-body diagram of the beam is shown in Fig. 5–7*b*. Since the support at A is fixed, the wall exerts three reactions *on the beam*, denoted as $\mathbf{A}_x$, $\mathbf{A}_y$, and $\mathbf{M}_A$. The magnitudes of these reactions are *unknown*, and their sense has been *assumed*. The weight of the beam, $W = 100(9.81)\text{ N} = 981\text{ N}$, acts through the beam's center of gravity G, which is 3 m from A since the beam is uniform.

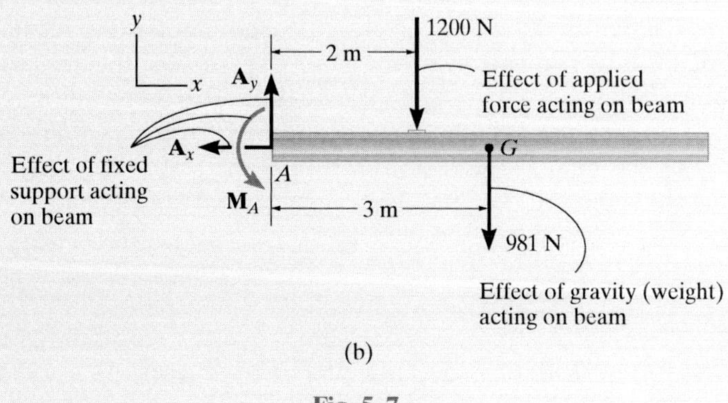

(b)

Fig. 5–7

EXAMPLE 5.2

Draw the free-body diagram of the foot lever shown in Fig. 5–8*a*. The operator applies a vertical force to the pedal so that the spring is stretched 36 mm and the force on the link at *B* is 100 N.

(a)

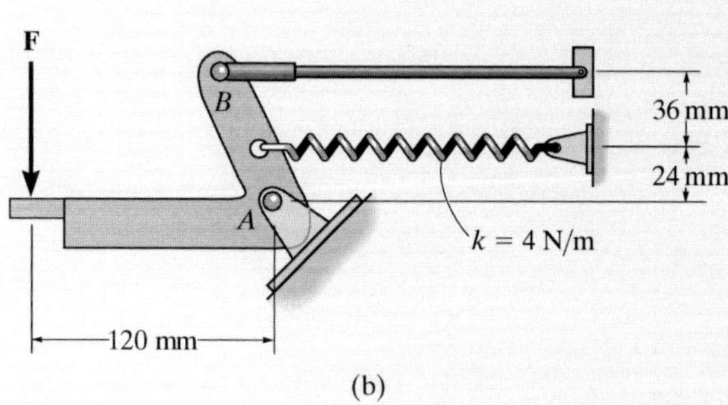

(b)

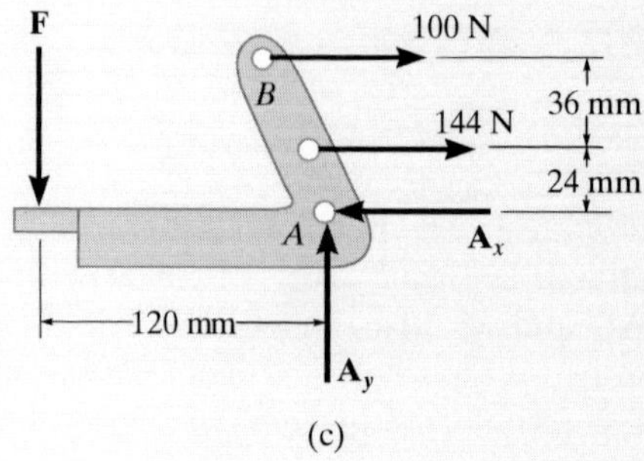

(c)

Fig. 5–8

SOLUTION

By inspection of the photo the lever is loosely bolted to the frame at *A* and so this bolt acts as a pin. (See (8) in Table 5–1.) Although not shown here the link at *B* is pinned at both ends and so it is like (2) in Table 5-1. After making the proper measurements, the idealized model of the lever is shown in Fig. 5–8*b*. From this, the free-body diagram is shown in Fig. 5–8*c*. The pin at *A* exerts force components $\mathbf{A}_x$ and $\mathbf{A}_y$ on the lever. The link exerts a force of 100 N, acting in the direction of the link. In addition the spring also exerts a horizontal force on the lever. If the stiffness is measured and found to be $k = 4\text{ N/mm}$, then since the stretch $s = 36\text{ mm}$, using Eq. 3–2, $F_s = ks = 4\text{ N/mm}\,(36\text{ mm}) = 144\text{ N}$. Finally, the operator's shoe applies a vertical force of **F** on the pedal. The dimensions of the lever are also shown on the free-body diagram, since this information will be useful when calculating the moments of the forces. As usual, the senses of the unknown forces at *A* have been assumed. The correct senses will become apparent after solving the equilibrium equations.

EXAMPLE 5.3

Two smooth pipes, each having a mass of 300 kg, are supported by the forked tines of the tractor in Fig. 5–9*a*. Draw the free-body diagrams for each pipe and both pipes together.

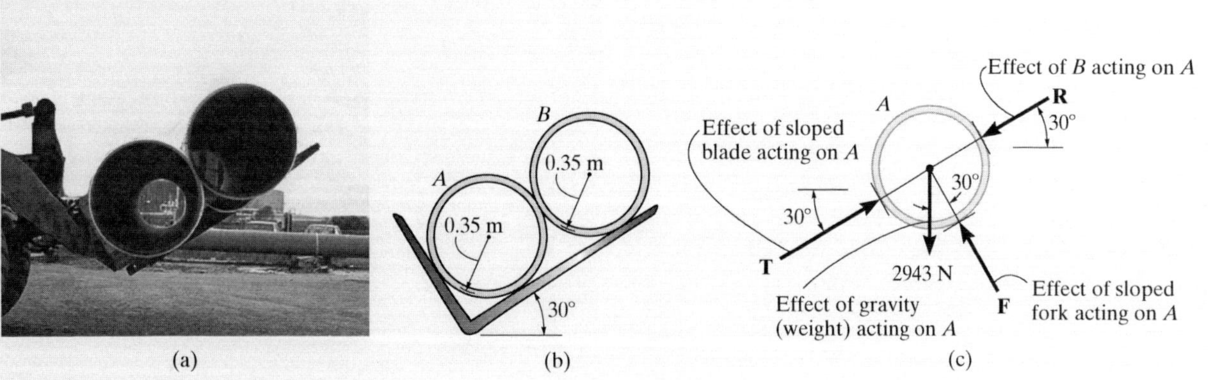

(a) (b) (c)

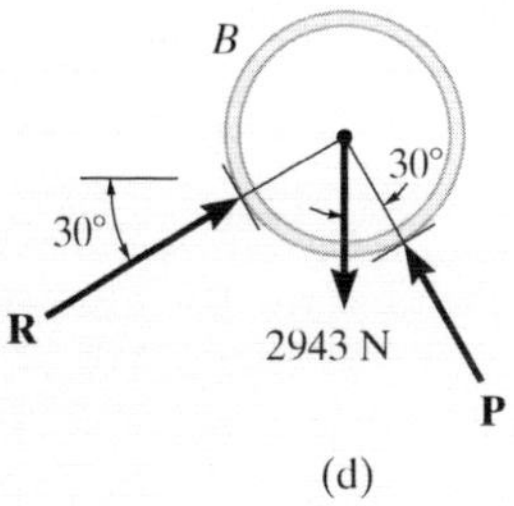

(d)

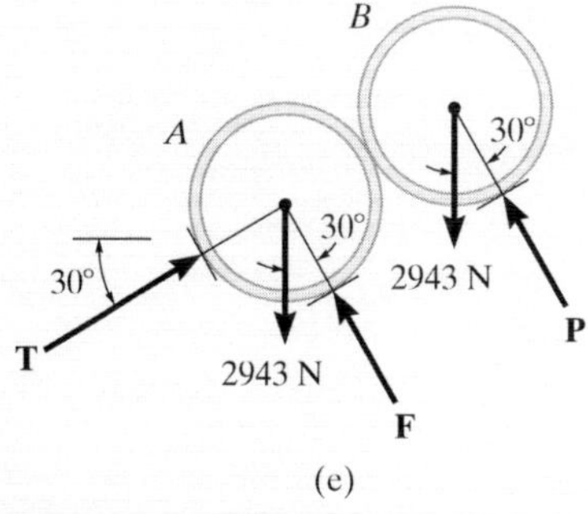

(e)

Fig. 5–9

SOLUTION

The idealized model from which we must draw the free-body diagrams is shown in Fig. 5–9*b*. Here the pipes are identified, the dimensions have been added, and the physical situation reduced to its simplest form.

The free-body diagram for pipe *A* is shown in Fig. 5–9*c*. Its weight is $W = 300(9.81)\text{ N} = 2943\text{ N}$. Assuming all contacting surfaces are *smooth*, the reactive forces **T**, **F**, **R** act in a direction *normal* to the tangent at their surfaces of contact.

The free-body diagram of pipe *B* is shown in Fig. 5–9*d*. Can you identify each of the three forces acting *on this pipe*? In particular, note that **R**, representing the force of *A* on *B*, Fig. 5–9*d*, is equal and opposite to **R** representing the force of *B* on *A*, Fig. 5–9*c*. This is a consequence of Newton's third law of motion.

The free-body diagram of both pipes combined ("system") is shown in Fig. 5–9*e*. Here the contact force **R**, which acts between *A* and *B*, is considered as an *internal* force and hence is not shown on the free-body diagram. That is, it represents a pair of equal but opposite collinear forces which cancel each other.

EXAMPLE 5.4

Draw the free-body diagram of the unloaded platform that is suspended off the edge of the oil rig shown in Fig. 5–10*a*. The platform has a mass of 200 kg.

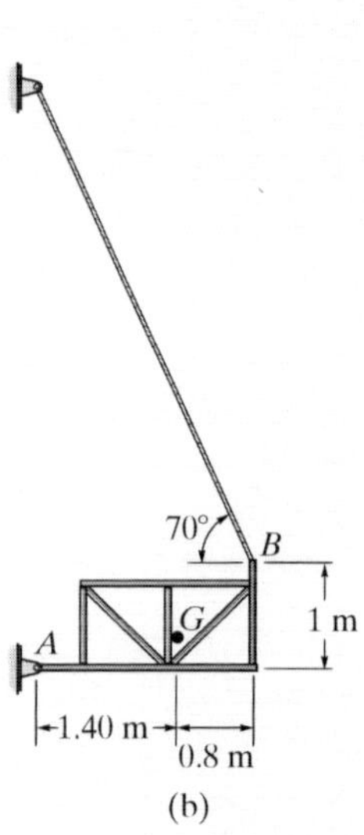

(b)

(a)

Fig. 5–10

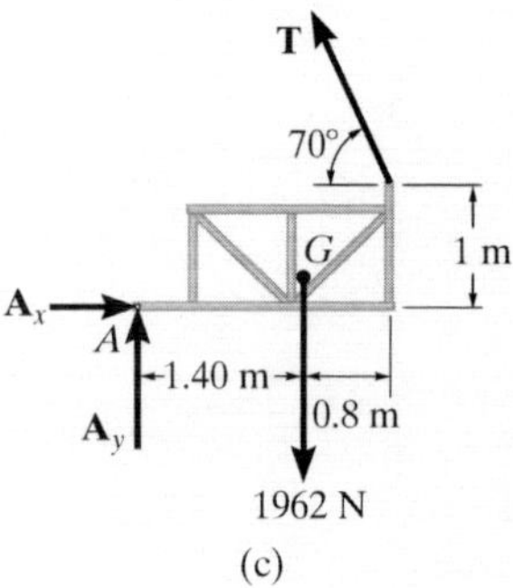

(c)

SOLUTION

The idealized model of the platform will be considered in two dimensions because by observation the loading and the dimensions are all symmetrical about a vertical plane passing through its center, Fig. 5–10*b*. The connection at A is considered to be a pin, and the cable supports the platform at B. The direction of the cable and average dimensions of the platform are listed, and the center of gravity G has been determined. It is from this model that we have drawn the free-body diagram shown in Fig. 5–10*c*. The platform's weight is $200(9.81) = 1962$ N. The force components $\mathbf{A}_x$ and $\mathbf{A}_y$ along with the cable force $\mathbf{T}$ represent the reactions that *both* pins and *both* cables exert on the platform, Fig. 5–10*a*. As a result, half their magnitudes are developed on each side of the platform.

PROBLEMS

5–1. Draw the free-body diagram of the dumpster D of the truck, which has a mass of 2.5 Mg and a center of gravity at G. It is supported by a pin at A and a pin-connected hydraulic cylinder BC (short link). Explain the significance of each force on the diagram. (See Fig. 5–7*b*.)

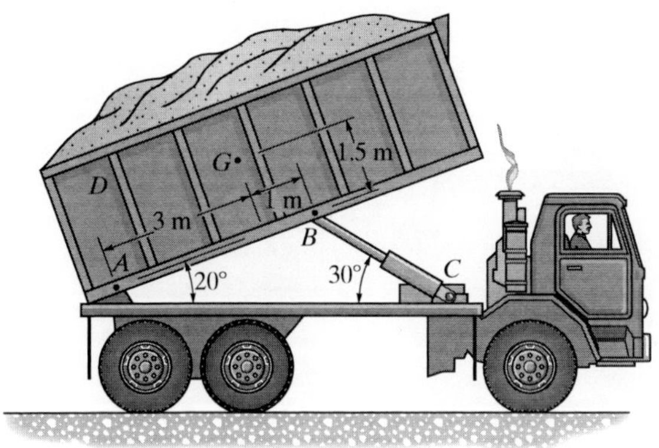

Prob. 5–1

5–2. Draw the free-body diagram of member ABC which is supported by a smooth collar at A, rocker at B, and short link CD. Explain the significance of each force acting on the diagram. (See Fig. 5–7*b*.)

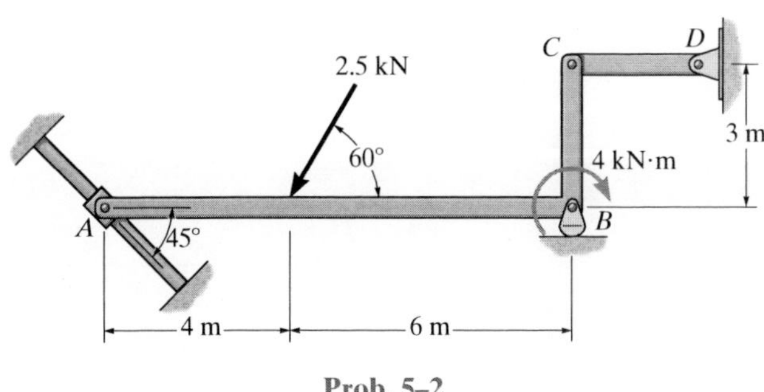

Prob. 5–2

5–3. Draw the free-body diagram of the beam which supports the 80-kg load and is supported by the pin at A and a cable which wraps around the pulley at D. Explain the significance of each force on the diagram. (See Fig. 5–7*b*.)

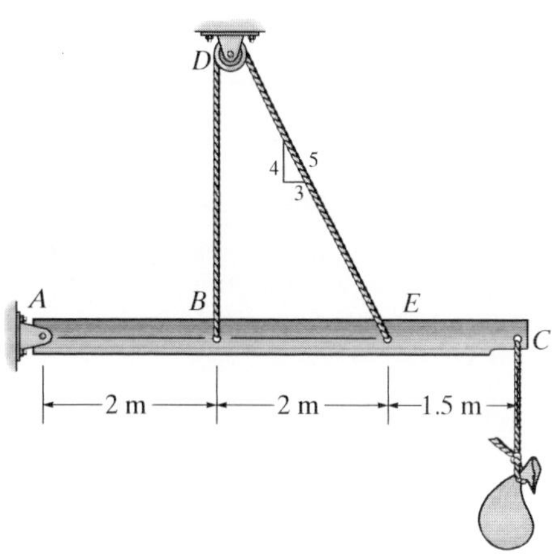

Prob. 5–3

***5–4.** Draw the free-body diagram of the hand punch, which is pinned at A and bears down on the smooth surface at B.

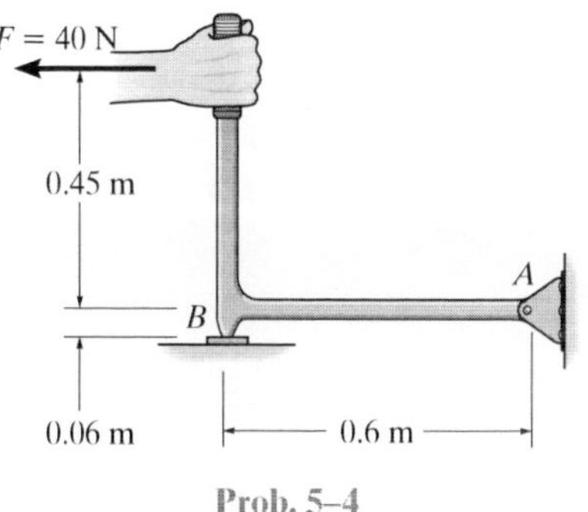

Prob. 5–4

5–5. Draw the free-body diagram of the uniform bar, which has a mass of 100 kg and a center of mass at G. The supports A, B, and C are smooth.

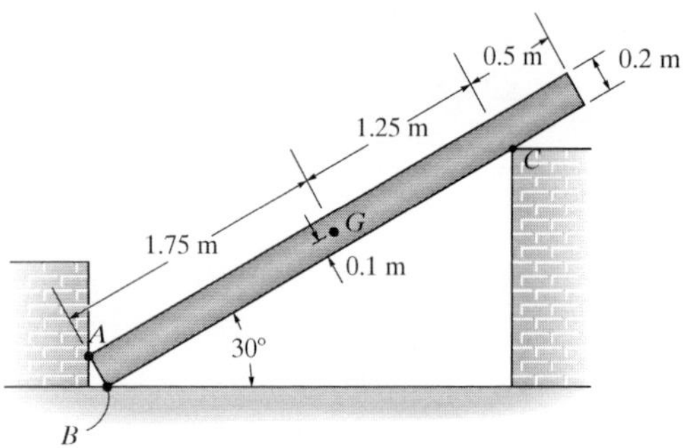

Prob. 5–5

5–6. Draw the free-body diagram of the jib crane AB, which is pin-connected at A and supported by member (link) BC.

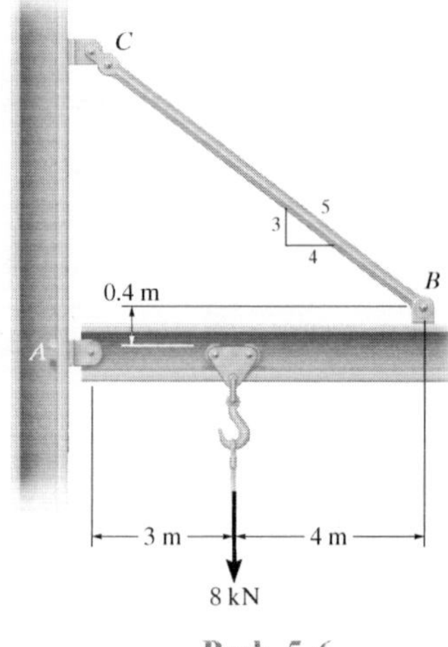

Prob. 5–6

5–7. Draw the free-body diagram of the beam, which is pin connected at A and rocker-supported at B.

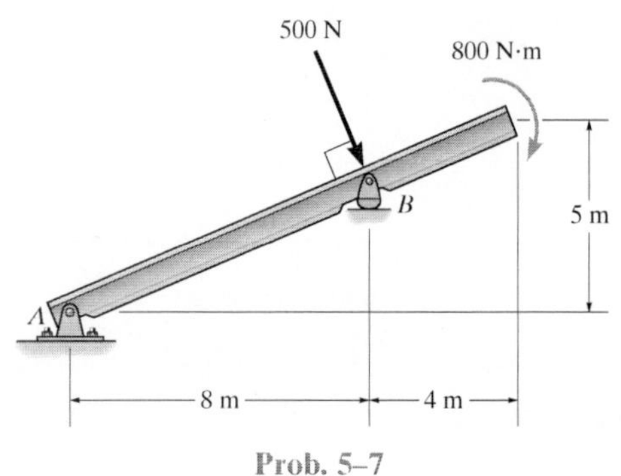

Prob. 5–7

***5–8.** Draw the free-body diagram of the truss that is supported by the cable AB and pin C. Explain the significance of each force acting on the diagram. (See Fig. 5–7b.)

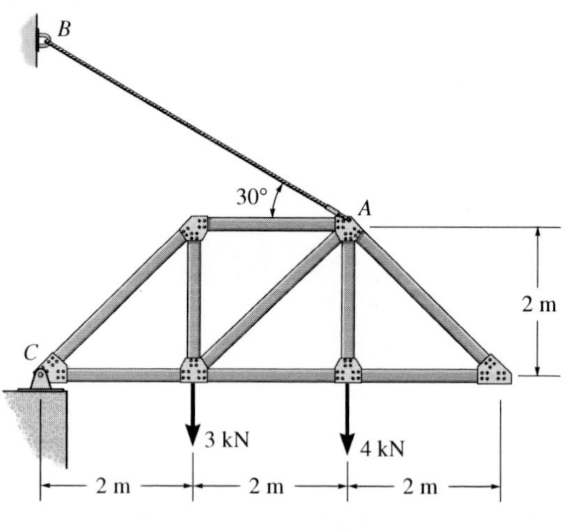

Prob. 5–8

5–9. Draw the free-body diagram of the jib crane AB, which is pin connected at A and supported by member (link) BC.

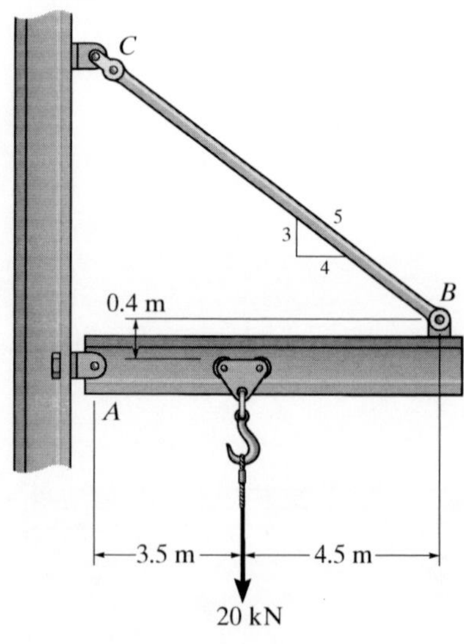

Prob. 5–9

CONCEPTUAL PROBLEMS

P5–1. Draw the free-body diagram of the uniform trash bucket which has a significant weight. It is pinned at A and rests against the smooth horizontal member at B. Show your result in side view. Label any necessary dimensions.

P5–1

P5–2. Draw the free-body diagram of the outrigger ABC used to support a backhoe. The pin B is connected to the hydraulic cylinder, which can be considered a short link (two-force member), the bearing shoe at A is smooth, and the outrigger is pinned to the frame at C.

P5–2

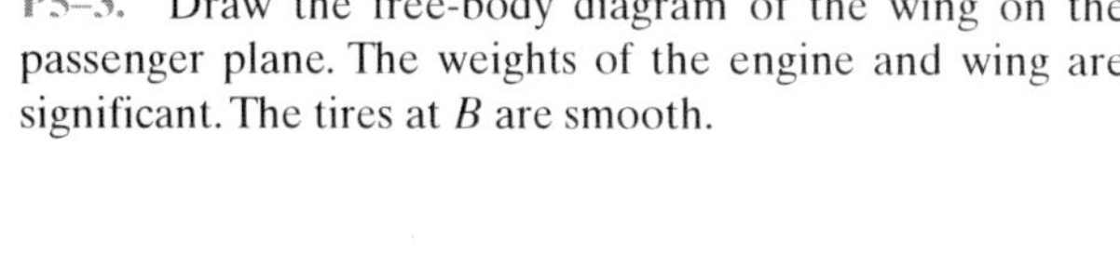

P5–3. Draw the free-body diagram of the wing on the passenger plane. The weights of the engine and wing are significant. The tires at B are smooth.

P5–3

P5–4. Draw the free-body diagrams of the wheel and member ABC used as part of the landing gear on a jet plane. The hydraulic cylinder AD acts as a two-force member, and there is a pin connection at B.

P5–4

Please refer to the Companion Website for the animation: *Equilibrium of a Free Body*

5.3 Equations of Equilibrium

In Sec. 5.1 we developed the two equations which are both necessary and sufficient for the equilibrium of a rigid body, namely, $\Sigma\mathbf{F} = \mathbf{0}$ and $\Sigma\mathbf{M}_O = \mathbf{0}$. When the body is subjected to a system of forces, which all lie in the x–y plane, then the forces can be resolved into their x and y components. Consequently, the conditions for equilibrium in two dimensions are

$$\Sigma F_x = 0 \qquad \Sigma F_y = 0 \qquad \Sigma M_O = 0 \tag{5–2}$$

Here ΣF_x and ΣF_y represent, respectively, the algebraic sums of the x and y components of all the forces acting on the body, and ΣM_O represents the algebraic sum of the couple moments and the moments of all the force components about the z axis, which is perpendicular to the x–y plane and passes through the arbitrary point O.

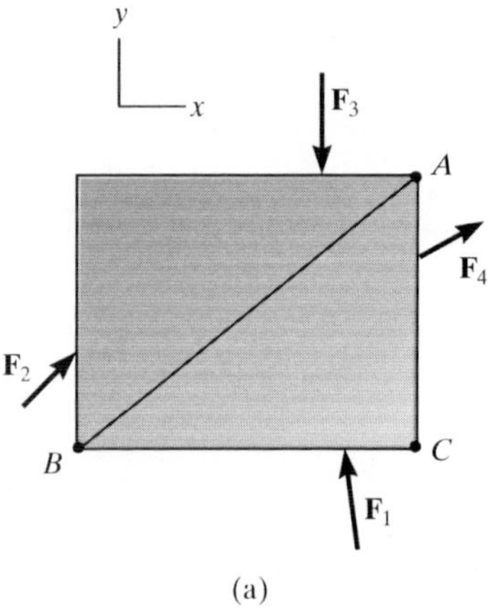

(a)

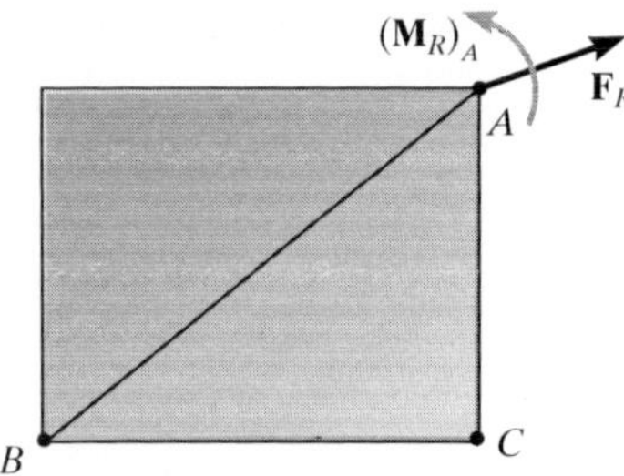

(b)

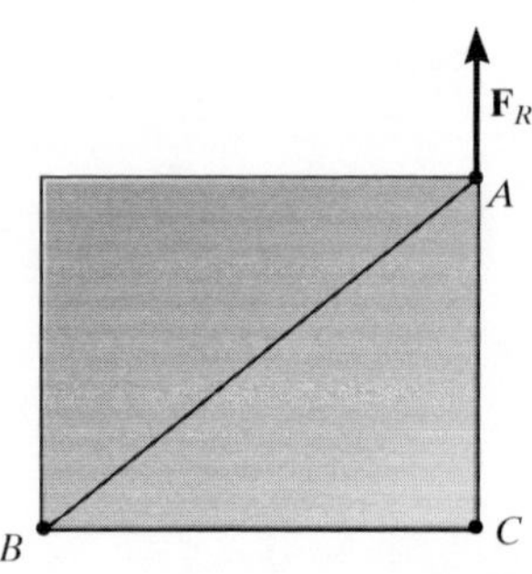

(c)

Fig. 5–11

Alternative Sets of Equilibrium Equations. Although Eqs. 5–2 are *most often* used for solving coplanar equilibrium problems, two *alternative* sets of three independent equilibrium equations may also be used. One such set is

$$\Sigma F_x = 0 \qquad \Sigma M_A = 0 \qquad \Sigma M_B = 0 \tag{5–3}$$

When using these equations it is required that a line passing through points A and B is *not parallel* to the y axis. To prove that Eqs. 5–3 provide the *conditions* for equilibrium, consider the free-body diagram of the plate shown in Fig. 5–11*a*. Using the methods of Sec. 4.7, all the forces on the free-body diagram may be replaced by an equivalent resultant force $\mathbf{F}_R = \Sigma\mathbf{F}$, acting at point A, and a resultant couple moment $(\mathbf{M}_R)_A = \Sigma\mathbf{M}_A$, Fig. 5–11*b*. If $\Sigma M_A = 0$ is satisfied, it is necessary that $(\mathbf{M}_R)_A = \mathbf{0}$. Furthermore, in order that $\mathbf{F}_R$ satisfy $\Sigma F_x = 0$, it must have *no component* along the x axis, and therefore $\mathbf{F}_R$ must be parallel to the y axis, Fig. 5–11*c*. Finally, if it is required that $\Sigma M_B = 0$, where B does not lie on the line of action of $\mathbf{F}_R$, then $\mathbf{F}_R = \mathbf{0}$. Since Eqs. 5–3 show that both of these resultants are zero, indeed the body in Fig. 5–11*a* must be in equilibrium.

A second alternative set of equilibrium equations is

$$\begin{aligned} \Sigma M_A &= 0 \\ \Sigma M_B &= 0 \\ \Sigma M_C &= 0 \end{aligned} \qquad (5\text{–}4)$$

Here it is necessary that points A, B, and C do not lie on the same line. To prove that these equations, when satisfied, ensure equilibrium, consider again the free-body diagram in Fig. 5–11*b*. If $\Sigma M_A = 0$ is to be satisfied, then $(\mathbf{M}_R)_A = \mathbf{0}$. $\Sigma M_C = 0$ is satisfied if the line of action of $\mathbf{F}_R$ passes through point C as shown in Fig. 5–11*c*. Finally, if we require $\Sigma M_B = 0$, it is necessary that $\mathbf{F}_R = \mathbf{0}$, and so the plate in Fig. 5–11*a* must then be in equilibrium.

Procedure for Analysis

Coplanar force equilibrium problems for a rigid body can be solved using the following procedure.

Free-Body Diagram.

- Establish the x, y coordinate axes in any suitable orientation.
- Draw an outlined shape of the body.
- Show all the forces and couple moments acting on the body.
- Label all the loadings and specify their directions relative to the x or y axis. The sense of a force or couple moment having an *unknown* magnitude but known line of action can be *assumed*.
- Indicate the dimensions of the body necessary for computing the moments of forces.

Equations of Equilibrium.

- Apply the moment equation of equilibrium, $\Sigma M_O = 0$, about a point (O) that lies at the intersection of the lines of action of two unknown forces. In this way, the moments of these unknowns are zero about O, and a *direct solution* for the third unknown can be determined.
- When applying the force equilibrium equations, $\Sigma F_x = 0$ and $\Sigma F_y = 0$, orient the x and y axes along lines that will provide the simplest resolution of the forces into their x and y components.
- If the solution of the equilibrium equations yields a negative scalar for a force or couple moment magnitude, this indicates that the sense is opposite to that which was assumed on the free-body diagram.

EXAMPLE 5.5

Determine the horizontal and vertical components of reaction on the beam caused by the pin at B and the rocker at A as shown in Fig. 5–12a. Neglect the weight of the beam.

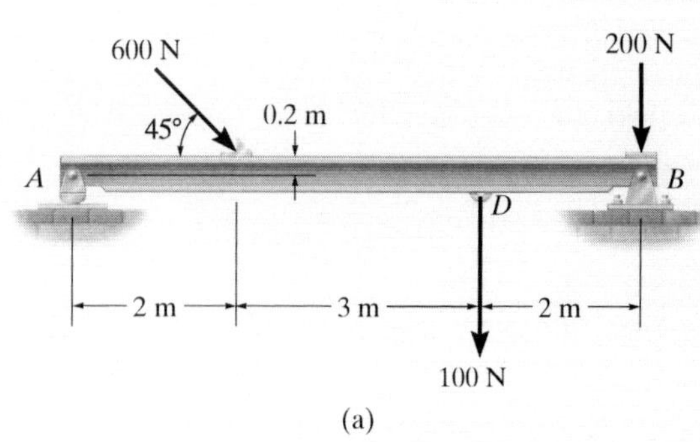

(a)

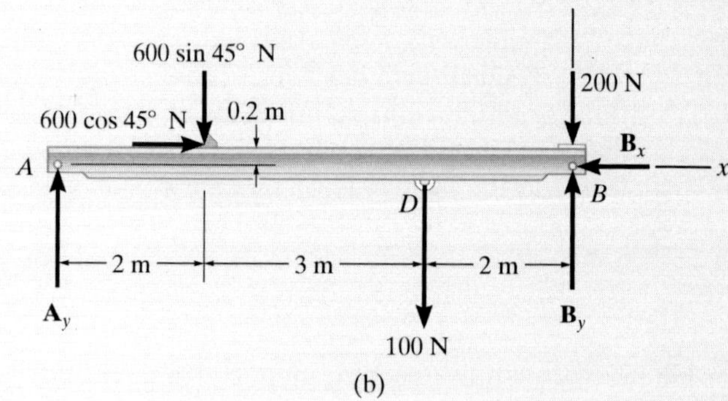

(b)

Fig. 5–12

5

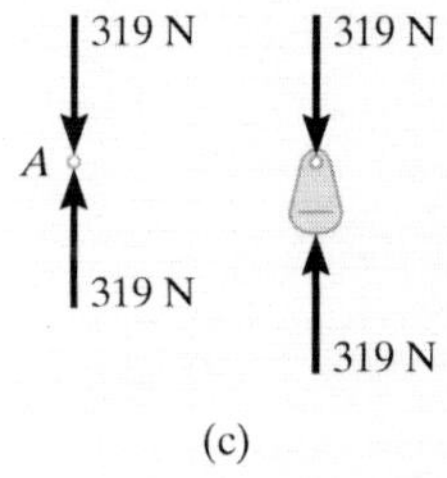

(c)

SOLUTION

Free-Body Diagram. Identify each of the forces shown on the free-body diagram of the beam, Fig. 5–12b. (See Example 5.1.) For simplicity, the 600-N force is represented by its x and y components as shown in Fig. 5–12b.

Equations of Equilibrium. Summing forces in the x direction yields

$$\overset{+}{\rightarrow}\Sigma F_x = 0; \qquad 600 \cos 45^\circ \text{ N} - B_x = 0$$

$$B_x = 424 \text{ N} \qquad \textit{Ans.}$$

A direct solution for $\mathbf{A}_y$ can be obtained by applying the moment equation $\Sigma M_B = 0$ about point B.

$$\circlearrowleft + \Sigma M_B = 0; \qquad 100 \text{ N}(2 \text{ m}) + (600 \sin 45^\circ \text{ N})(5 \text{ m})$$

$$- (600 \cos 45^\circ \text{ N})(0.2 \text{ m}) - A_y(7 \text{ m}) = 0$$

$$A_y = 319 \text{ N} \qquad \textit{Ans.}$$

Summing forces in the y direction, using this result, gives

$$+\uparrow \Sigma F_y = 0; \qquad 319 \text{ N} - 600 \sin 45^\circ \text{ N} - 100 \text{ N} - 200 \text{ N} + B_y = 0$$

$$B_y = 405 \text{ N} \qquad \textit{Ans.}$$

NOTE: Remember, the support forces in Fig. 5–12b are the result of pins that *act on the beam*. The opposite forces act on the pins. For example, Fig. 5–12c shows the equilibrium of the pin at A and the rocker.

EXAMPLE 5.6

The cord shown in Fig. 5–13*a* supports a force of 500 N and wraps over the frictionless pulley. Determine the tension in the cord at *C* and the horizontal and vertical components of reaction at pin *A*.

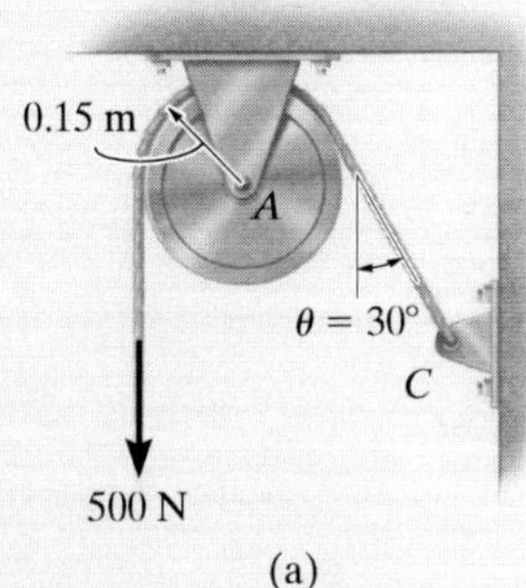

(a)

Fig. 5–13

SOLUTION

Free-Body Diagrams. The free-body diagrams of the cord and pulley are shown in Fig. 5–13*b*. Note that the principle of action, equal but opposite reaction must be carefully observed when drawing each of these diagrams: the cord exerts an unknown load distribution *p* on the pulley at the contact surface, whereas the pulley exerts an equal but opposite effect on the cord. For the solution, however, it is simpler to *combine* the free-body diagrams of the pulley and this portion of the cord, so that the distributed load becomes *internal* to this "system" and is therefore eliminated from the analysis, Fig. 5–13*c*.

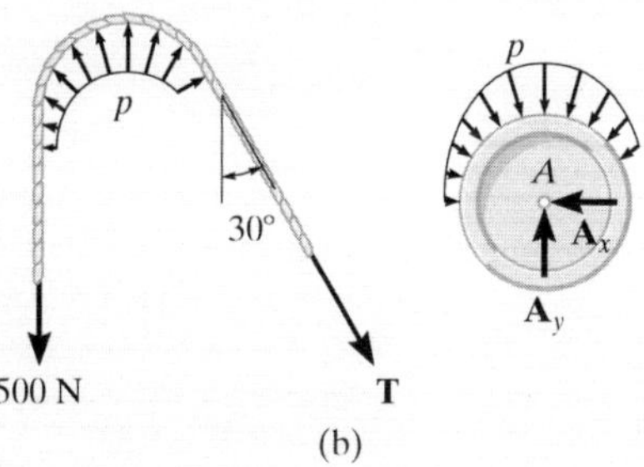

(b)

Equations of Equilibrium. Summing moments about point *A* to eliminate $\mathbf{A}_x$ and $\mathbf{A}_y$, Fig. 5–13*c*, we have

$$\circlearrowleft+\Sigma M_A = 0; \quad 500\text{ N}(0.15\text{ m}) - T(0.15\text{ m}) = 0$$

$$T = 500\text{ N} \qquad \textit{Ans.}$$

Using this result,

$$\xrightarrow{+}\Sigma F_x = 0; \quad -A_x + 500\sin 30^\circ\text{ N} = 0$$

$$A_x = 250\text{ N} \qquad \textit{Ans.}$$

$$+\uparrow\Sigma F_y = 0; \quad A_y - 500\text{ N} - 500\cos 30^\circ\text{ N} = 0$$

$$A_y = 933\text{ N} \qquad \textit{Ans.}$$

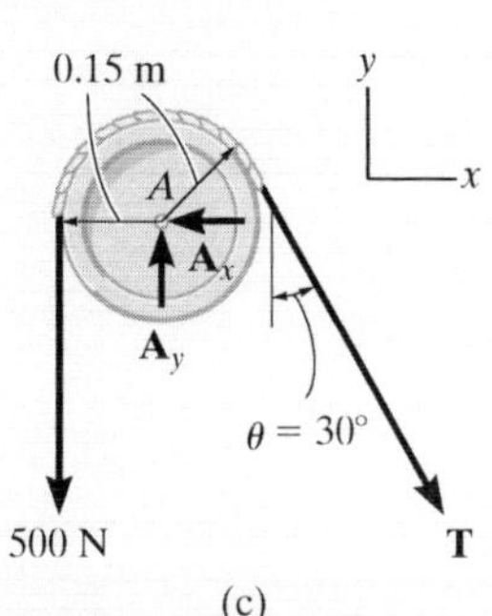

(c)

NOTE: From the moment equation, it is seen that the tension remains *constant* as the cord passes over the pulley. (This of course is true for *any angle θ* at which the cord is directed and for *any radius r* of the pulley.)

5

5

EXAMPLE 5.7

The member shown in Fig. 5–14*a* is pin connected at *A* and rests against a smooth support at *B*. Determine the horizontal and vertical components of reaction at the pin *A*.

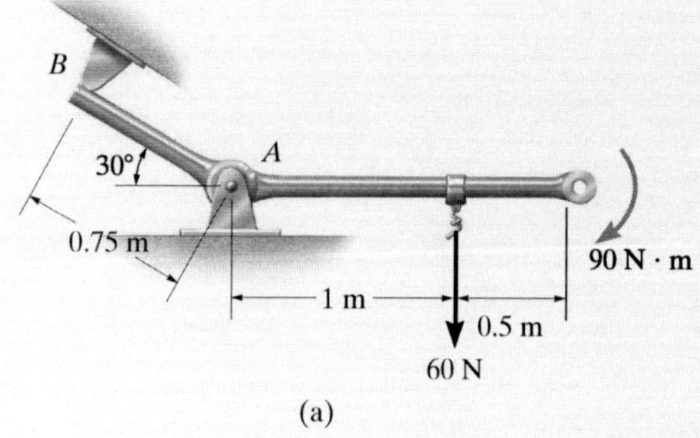

(a)

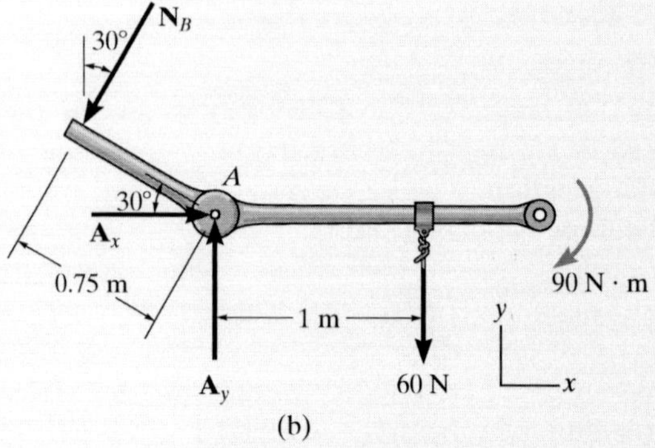

(b)

Fig. 5–14

SOLUTION

Free-Body Diagram. As shown in Fig. 5–14*b*, the reaction $\mathbf{N}_B$ is perpendicular to the member at *B*. Also, horizontal and vertical components of reaction are represented at *A*.

Equations of Equilibrium. Summing moments about *A*, we obtain a direct solution for N_B,

$$\zeta+\Sigma M_A = 0; \quad -90\text{ N}\cdot\text{m} - 60\text{ N}(1\text{ m}) + N_B(0.75\text{ m}) = 0$$

$$N_B = 200\text{ N}$$

Using this result,

$$\overset{+}{\rightarrow}\Sigma F_x = 0; \quad A_x - 200\sin 30^\circ\text{ N} = 0$$

$$A_x = 100\text{ N} \qquad \textit{Ans.}$$

$$+\uparrow\Sigma F_y = 0; \quad A_y - 200\cos 30^\circ\text{ N} - 60\text{ N} = 0$$

$$A_y = 233\text{ N} \qquad \textit{Ans.}$$

EXAMPLE 5.8

The box wrench in Fig. 5–15*a* is used to tighten the bolt at *A*. If the wrench does not turn when the load is applied to the handle, determine the torque or moment applied to the bolt and the force of the wrench on the bolt.

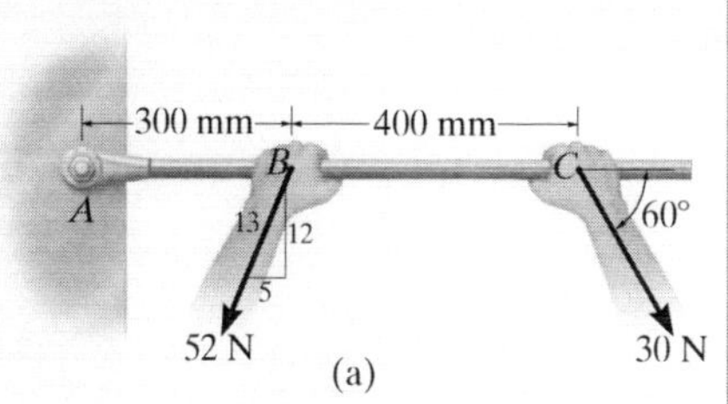

(a)

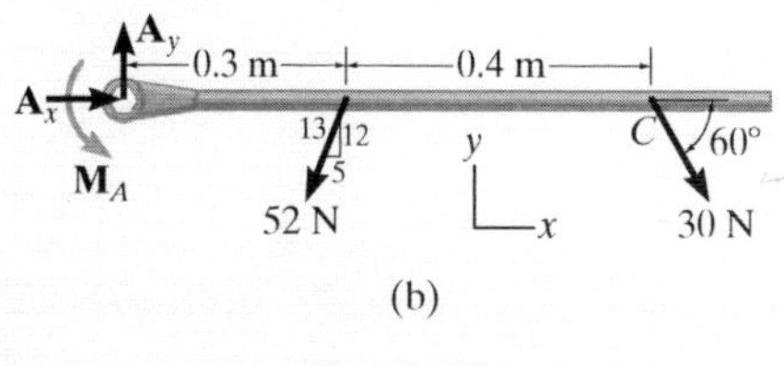

(b)

Fig. 5–15

SOLUTION

Free-Body Diagram. The free-body diagram for the wrench is shown in Fig. 5–15*b*. Since the bolt acts as a "fixed support," it exerts force components $\mathbf{A}_x$ and $\mathbf{A}_y$ and a moment $\mathbf{M}_A$ on the wrench at *A*.

Equations of Equilibrium.

$$\overset{+}{\rightarrow}\Sigma F_x = 0; \qquad A_x - 52\left(\tfrac{5}{13}\right)\text{ N} + 30\cos 60^\circ\text{ N} = 0$$

$$A_x = 5.00\text{ N} \qquad \textit{Ans.}$$

$$+\uparrow\Sigma F_y = 0; \qquad A_y - 52\left(\tfrac{12}{13}\right)\text{ N} - 30\sin 60^\circ\text{ N} = 0$$

$$A_y = 74.0\text{ N} \qquad \textit{Ans.}$$

$$\circlearrowleft +\Sigma M_A = 0; \quad M_A - \left[52\left(\tfrac{12}{13}\right)\text{N}\right](0.3\text{ m}) - (30\sin 60^\circ\text{ N})(0.7\text{ m}) = 0$$

$$M_A = 32.6\text{ N}\cdot\text{m} \qquad \textit{Ans.}$$

Note that $\mathbf{M}_A$ must be *included* in this moment summation. This couple moment is a free vector and represents the twisting resistance of the bolt on the wrench. By Newton's third law, the wrench exerts an equal but opposite moment or torque on the bolt. Furthermore, the resultant force on the wrench is

$$F_A = \sqrt{(5.00)^2 + (74.0)^2} = 74.1\text{ N} \qquad \textit{Ans.}$$

NOTE: Although only *three* independent equilibrium equations can be written for a rigid body, it is a good practice to *check* the calculations using a fourth equilibrium equation. For example, the above computations may be verified in part by summing moments about point *C*:

$$\circlearrowleft +\Sigma M_C = 0; \quad \left[52\left(\tfrac{12}{13}\right)\text{N}\right](0.4\text{ m}) + 32.6\text{ N}\cdot\text{m} - 74.0\text{ N}(0.7\text{ m}) = 0$$

$$19.2\text{ N}\cdot\text{m} + 32.6\text{ N}\cdot\text{m} - 51.8\text{ N}\cdot\text{m} = 0$$

5

EXAMPLE 5.9

Please refer to the Companion Website for the animation: *Free-Body Diagram for a Beam on Slanting Support*

Determine the horizontal and vertical components of reaction on the member at the pin A, and the normal reaction at the roller B in Fig. 5–16a.

SOLUTION

Free-Body Diagram. The free-body diagram is shown in Fig. 5–16b. The pin at A exerts two components of reaction on the member, $\mathbf{A}_x$ and $\mathbf{A}_y$.

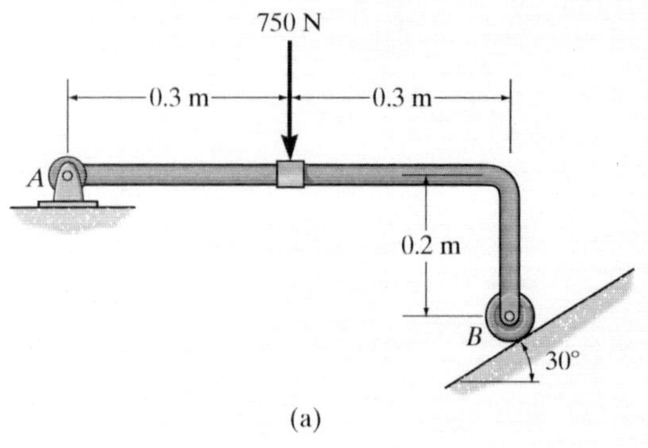

(a)

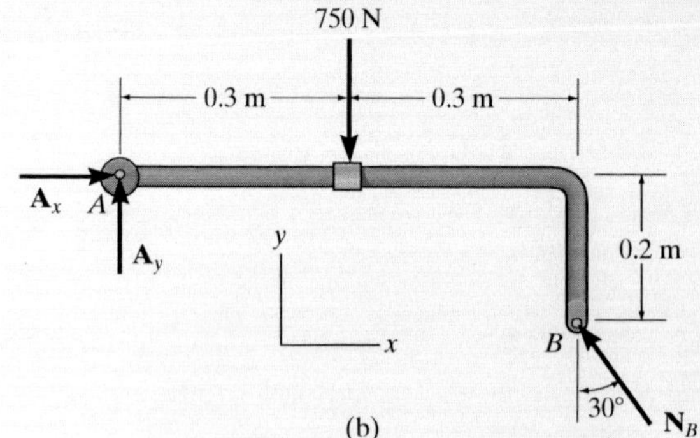

(b)

Fig. 5–16

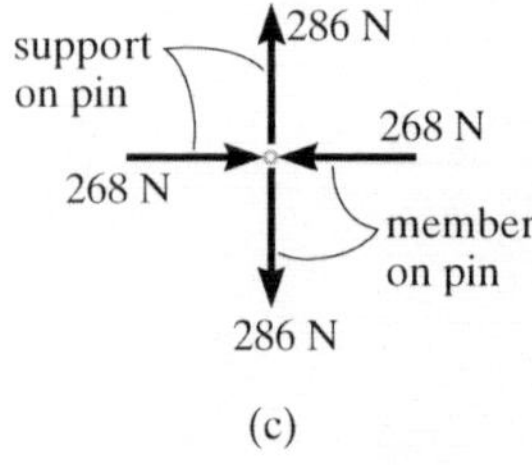

(c)

Equations of Equilibrium. The reaction N_B can be obtained *directly* by summing moments about point A, since $\mathbf{A}_x$ and $\mathbf{A}_y$ produce no moment about A.

$\circlearrowleft + \Sigma M_A = 0;$

$$[N_B \cos 30°](0.6 \text{ m}) - [N_B \sin 30°](0.2 \text{ m}) - 750 \text{ N}(0.3 \text{ m}) = 0$$

$$N_B = 536.2 \text{ N} = 536 \text{ N} \qquad \textit{Ans.}$$

Using this result,

$\overset{+}{\rightarrow} \Sigma F_x = 0; \quad A_x - (536.2 \text{ N}) \sin 30° = 0$

$$A_x = 268 \text{ N} \qquad \textit{Ans.}$$

$+\uparrow \Sigma F_y = 0; \quad A_y + (536.2 \text{ N}) \cos 30° - 750 \text{ N} = 0$

$$A_y = 286 \text{ N} \qquad \textit{Ans.}$$

Details of the equilibrium of the pin at A are shown in Fig. 5–16c.

EXAMPLE 5.10

The uniform smooth rod shown in Fig. 5–17*a* is subjected to a force and couple moment. If the rod is supported at *A* by a smooth wall and at *B* and *C* either at the top or bottom by rollers, determine the reactions at these supports. Neglect the weight of the rod.

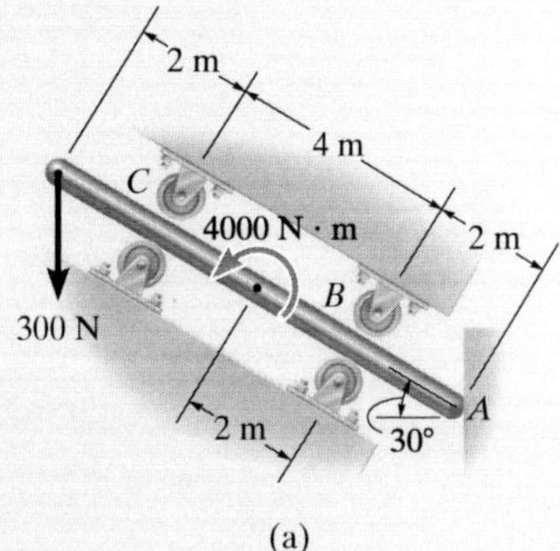

(a)

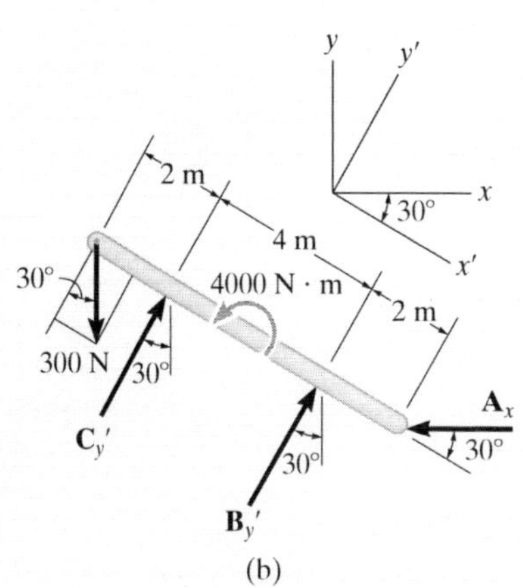

(b)

Fig. 5–17

SOLUTION

Free-Body Diagram. As shown in Fig. 5–17*b*, all the support reactions act normal to the surfaces of contact since these surfaces are smooth. The reactions at *B* and *C* are shown acting in the positive y' direction. This assumes that only the rollers located on the bottom of the rod are used for support.

Equations of Equilibrium. Using the *x, y* coordinate system in Fig. 5–17*b*, we have

$$\overset{+}{\rightarrow}\Sigma F_x = 0; \qquad C_{y'} \sin 30^\circ + B_{y'} \sin 30^\circ - A_x = 0 \qquad (1)$$

$$+\uparrow \Sigma F_y = 0; \qquad -300\text{ N} + C_{y'} \cos 30^\circ + B_{y'} \cos 30^\circ = 0 \qquad (2)$$

$$\circlearrowleft + \Sigma M_A = 0; \qquad -B_{y'}(2\text{ m}) + 4000\text{ N}\cdot\text{m} - C_{y'}(6\text{ m})$$
$$+ (300 \cos 30^\circ\text{ N})(8\text{ m}) = 0 \qquad (3)$$

When writing the moment equation, it should be noted that the line of action of the force component 300 sin 30° N passes through point *A*, and therefore this force is not included in the moment equation.

Solving Eqs. 2 and 3 simultaneously, we obtain

$$B_{y'} = -1000.0\text{ N} = -1\text{ kN} \qquad \textit{Ans.}$$
$$C_{y'} = 1346.4\text{ N} = 1.35\text{ kN} \qquad \textit{Ans.}$$

Since $B_{y'}$ is a negative scalar, the sense of $\mathbf{B}_{y'}$ is opposite to that shown on the free-body diagram in Fig. 5–17*b*. Therefore, the top roller at *B* serves as the support rather than the bottom one. Retaining the negative sign for $B_{y'}$ (Why?) and substituting the results into Eq. 1, we obtain

$$1346.4 \sin 30^\circ\text{ N} + (-1000.0 \sin 30^\circ\text{ N}) - A_x = 0$$
$$A_x = 173\text{ N} \qquad \textit{Ans.}$$

5

EXAMPLE 5.11

(a)

The uniform truck ramp shown in Fig. 5–18*a* has a weight of 2000 N and is pinned to the body of the truck at each side and held in the position shown by the two side cables. Determine the tension in the cables.

SOLUTION

The idealized model of the ramp, which indicates all necessary dimensions and supports, is shown in Fig. 5–18*b*. Here the center of gravity is located at the midpoint since the ramp is considered to be uniform.

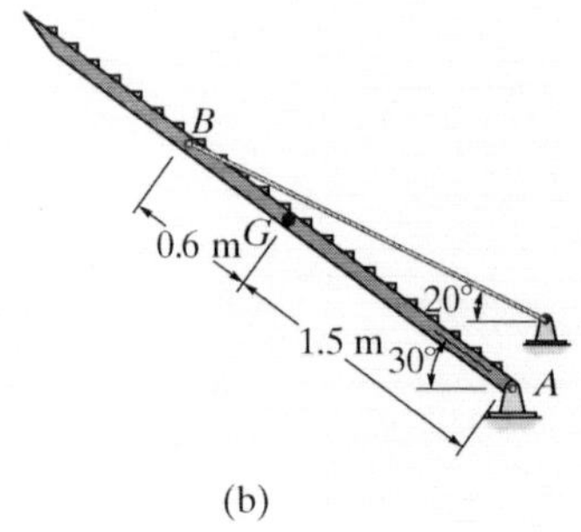

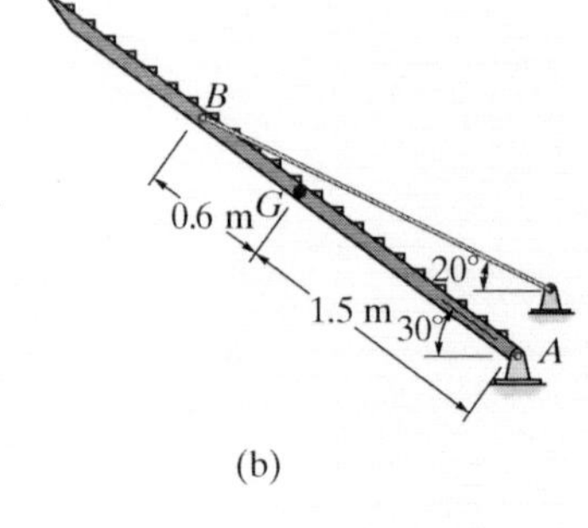

(b)

Free-Body Diagram. Working from the idealized model, the ramp's free-body diagram is shown in Fig. 5–18*c*.

Equations of Equilibrium. Summing moments about point A will yield a direct solution for the cable tension. Using the principle of moments, there are several ways of determining the moment of $\mathbf{T}$ about A. If we use x and y components, with $\mathbf{T}$ applied at B, we have

$$\circlearrowleft + \Sigma M_A = 0; \qquad -T\cos 20^\circ(2.1 \sin 30^\circ \text{ m}) + T\sin 20^\circ(2.1\cos 30^\circ \text{ m})$$

$$+ 2000 \text{ N } (1.5 \cos 30^\circ \text{ m}) = 0$$

$$T = 7124.6 \text{ N}$$

We can also determine the moment of $\mathbf{T}$ about A by resolving it into components along and perpendicular to the ramp at B. Then the moment of the component along the ramp will be zero about A, so that

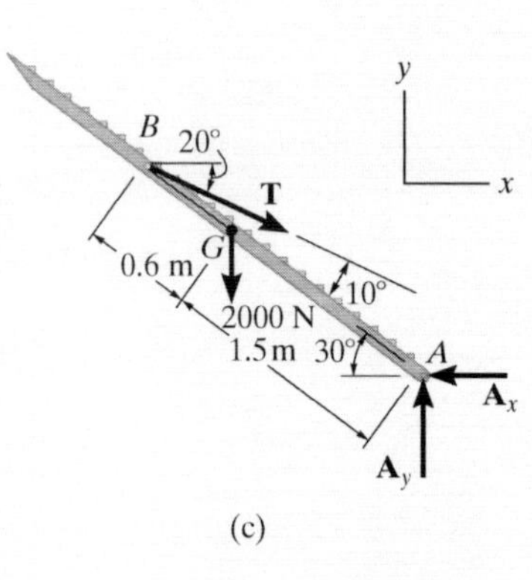

(c)

Fig. 5–18

$$\circlearrowleft + \Sigma M_A = 0; \qquad -T\sin 10^\circ(2.1 \text{ m}) + 2000 \text{ N } (1.5\cos 30^\circ \text{ m}) = 0$$

$$T = 7124.6 \text{ N}$$

Since there are two cables supporting the ramp,

$$T' = \frac{T}{2} = 3562.3 \text{ N} = 3.56 \text{ kN} \qquad \textit{Ans.}$$

NOTE: As an exercise, show that $A_x = 6695$ N and $A_y = 4437$ N.

EXAMPLE 5.12

Determine the support reactions on the member in Fig. 5–19*a*. The collar at *A* is fixed to the member and can slide vertically along the vertical shaft.

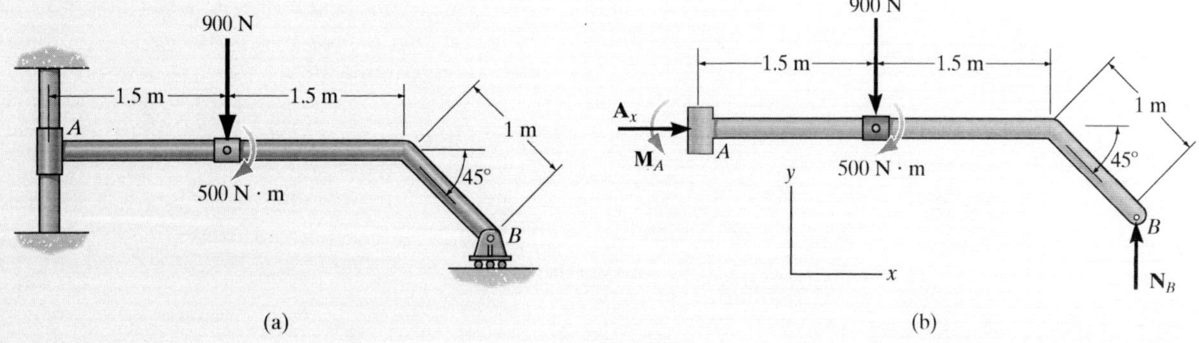

Fig. 5–19

SOLUTION

Free-Body Diagram. The free-body diagram of the member is shown in Fig. 5–19*b*. The collar exerts a horizontal force $\mathbf{A}_x$ and moment $\mathbf{M}_A$ on the member. The reaction $\mathbf{N}_B$ of the roller on the member is vertical.

Equations of Equilibrium. The forces A_x and N_B can be determined directly from the force equations of equilibrium.

$\overset{+}{\rightarrow} \Sigma F_x = 0;$ $\quad A_x = 0$ *Ans.*

$+\uparrow \Sigma F_y = 0;$ $\quad N_B - 900\text{ N} = 0$

$N_B = 900\text{ N}$ *Ans.*

The moment M_A can be determined by summing moments either about point *A* or point *B*.

$\circlearrowleft + \Sigma M_A = 0;$

$$M_A - 900\text{ N}(1.5\text{ m}) - 500\text{ N}\cdot\text{m} + 900\text{ N}\,[3\text{ m} + (1\text{ m})\cos 45^\circ] = 0$$

$$M_A = -1486\text{ N}\cdot\text{m} = 1.49\text{ kN}\cdot\text{m} \circlearrowright$$ *Ans.*

or

$$\circlearrowleft + \Sigma M_B = 0; \quad M_A + 900\text{ N}\,[1.5\text{ m} + (1\text{ m})\cos 45^\circ] - 500\text{ N}\cdot\text{m} = 0$$

$$M_A = -1486\text{ N}\cdot\text{m} = 1.49\text{ kN}\cdot\text{m} \circlearrowright$$ *Ans.*

The negative sign indicates that $\mathbf{M}_A$ has the opposite sense of rotation to that shown on the free-body diagram.

The hydraulic cylinder AB is a typical example of a two-force member since it is pin connected at its ends and, provided its weight is neglected, only the pin forces act on this member.

The link used for this railroad car brake is a three-force member. Since the force $\mathbf{F}_B$ in the tie rod at B and $\mathbf{F}_C$ from the link at C are parallel, then for equilibrium the resultant force $\mathbf{F}_A$ at the pin A must also be parallel with these two forces.

The boom and bucket on this lift is a three-force member, provided its weight is neglected. Here the lines of action of the weight of the worker, $\mathbf{W}$, and the force of the two-force member (hydraulic cylinder) at B, $\mathbf{F}_B$, intersect at O. For moment equilibrium, the resultant force at the pin A, $\mathbf{F}_A$, must also be directed towards O.

5.4 Two- and Three-Force Members

The solutions to some equilibrium problems can be simplified by recognizing members that are subjected to only two or three forces.

Two-Force Members. As the name implies, a *two-force member* has forces applied at only two points on the member. An example of a two-force member is shown in Fig. 5–20*a*. To satisfy force equilibrium, $\mathbf{F}_A$ and $\mathbf{F}_B$ must be equal in magnitude, $F_A = F_B = F$, but opposite in direction ($\Sigma\mathbf{F} = \mathbf{0}$), Fig. 5–20*b*. Furthermore, moment equilibrium requires that $\mathbf{F}_A$ and $\mathbf{F}_B$ share the same line of action, which can only happen if they are directed along the line joining points A and B ($\Sigma\mathbf{M}_A = \mathbf{0}$ or $\Sigma\mathbf{M}_B = \mathbf{0}$), Fig. 5–20*c*. Therefore, for any two-force member to be in equilibrium, the two forces acting on the member *must have the same magnitude, act in opposite directions, and have the same line of action, directed along the line joining the two points where these forces act.*

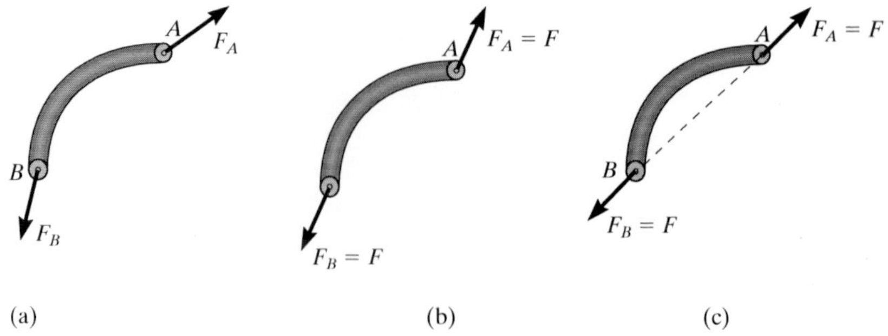

Two-force member
Fig. 5–20

Three-Force Members. If a member is subjected to only *three forces*, it is called a *three-force member*. Moment equilibrium can be satisfied only if the three forces form a *concurrent* or *parallel* force system. To illustrate, consider the member subjected to the three forces $\mathbf{F}_1$, $\mathbf{F}_2$, and $\mathbf{F}_3$, shown in Fig. 5–21*a*. If the lines of action of $\mathbf{F}_1$ and $\mathbf{F}_2$ intersect at point O, then the line of action of $\mathbf{F}_3$ must *also* pass through point O so that the forces satisfy $\Sigma\mathbf{M}_O = \mathbf{0}$. As a special case, if the three forces are all parallel, Fig. 5–21*b*, the location of the point of intersection, O, will approach infinity.

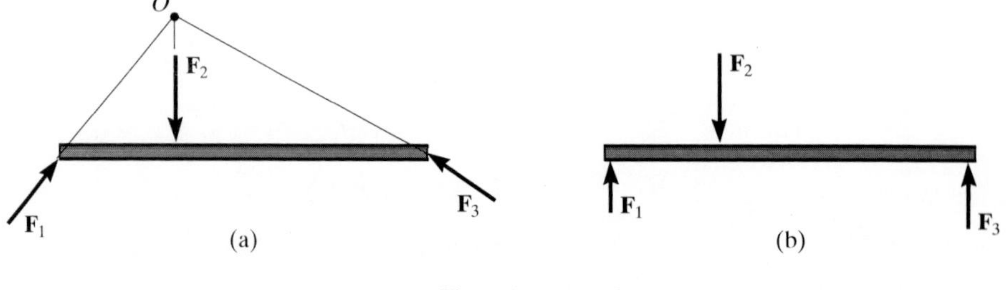

Three-force member
Fig. 5–21

EXAMPLE 5.13

The lever ABC is pin supported at A and connected to a short link BD as shown in Fig. 5–22a. If the weight of the members is negligible, determine the force of the pin on the lever at A.

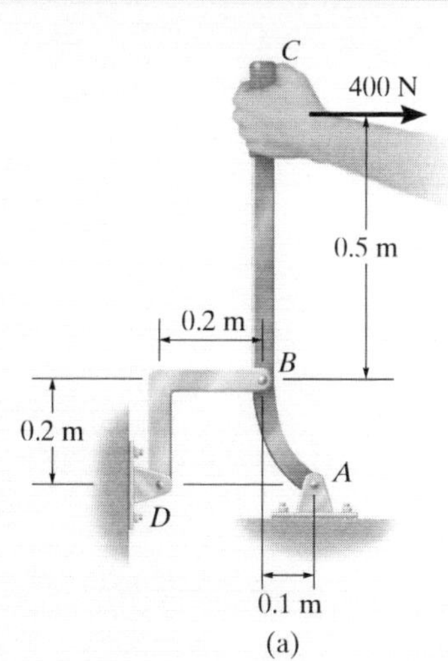

(a)

SOLUTION

Free-Body Diagrams. As shown in Fig. 5–22b, the short link BD is a *two-force member*, so the *resultant forces* at pins D and B must be equal, opposite, and collinear. Although the magnitude of the force is unknown, the line of action is known since it passes through B and D.

Lever ABC is a *three-force member*, and therefore, in order to satisfy moment equilibrium, the three nonparallel forces acting on it must be concurrent at O, Fig. 5–22c. In particular, note that the force **F** on the lever at B is equal but opposite to the force **F** acting at B on the link. Why? The distance CO must be 0.5 m since the lines of action of **F** and the 400-N force are known.

Equations of Equilibrium. By requiring the force system to be concurrent at O, since $\Sigma M_O = 0$, the angle θ which defines the line of action of $\mathbf{F}_A$ can be determined from trigonometry,

$$\theta = \tan^{-1}\left(\frac{0.7}{0.4}\right) = 60.3^\circ$$

Using the x, y axes and applying the force equilibrium equations,

$$\overset{+}{\rightarrow}\Sigma F_x = 0; \qquad F_A \cos 60.3^\circ - F \cos 45^\circ + 400 \text{ N} = 0$$

$$+\uparrow \Sigma F_y = 0; \qquad F_A \sin 60.3^\circ - F \sin 45^\circ = 0$$

Solving, we get

$$F_A = 1.07 \text{ kN} \qquad \textit{Ans.}$$

$$F = 1.32 \text{ kN}$$

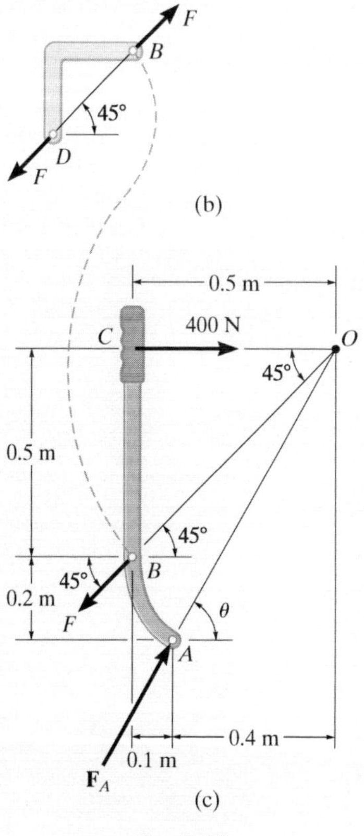

(b)

(c)

Fig. 5–22

NOTE: We can also solve this problem by representing the force at A by its two components $\mathbf{A}_x$ and $\mathbf{A}_y$ and applying $\Sigma M_A = 0$, $\Sigma F_x = 0$, $\Sigma F_y = 0$ to the lever. Once A_x and A_y are determined, we can get F_A and θ.

FUNDAMENTAL PROBLEMS

All problem solutions must include an FBD.

F5–1. Determine the horizontal and vertical components of reaction at the supports. Neglect the thickness of the beam.

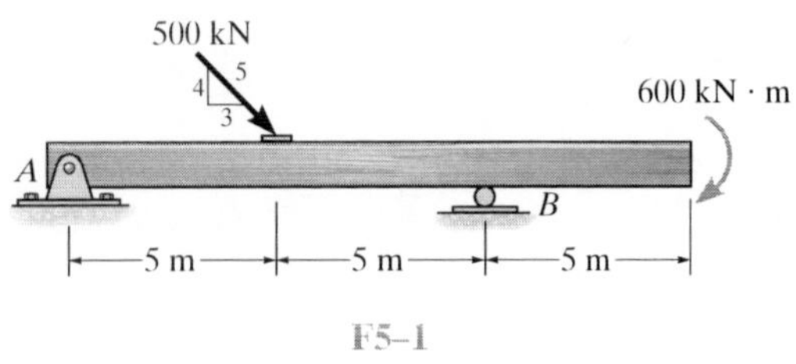

F5–1

F5–2. Determine the horizontal and vertical components of reaction at the pin A and the reaction on the beam at C.

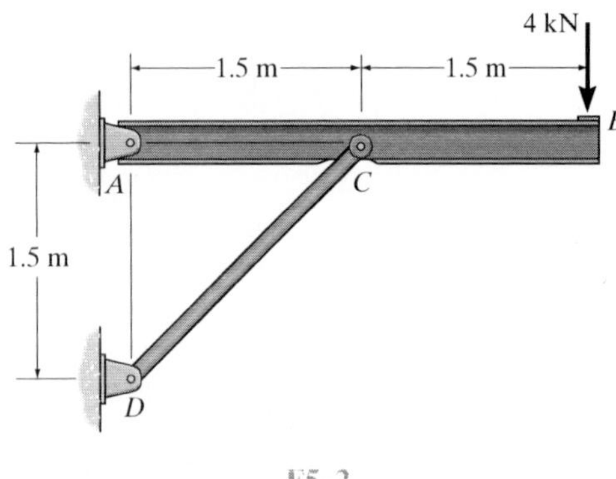

F5–2

F5–3. The truss is supported by a pin at A and a roller at B. Determine the support reactions.

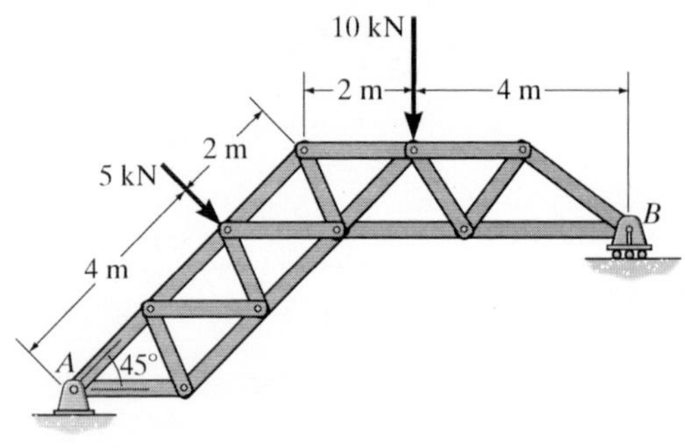

F5–3

F5–4. Determine the components of reaction at the fixed support A. Neglect the thickness of the beam.

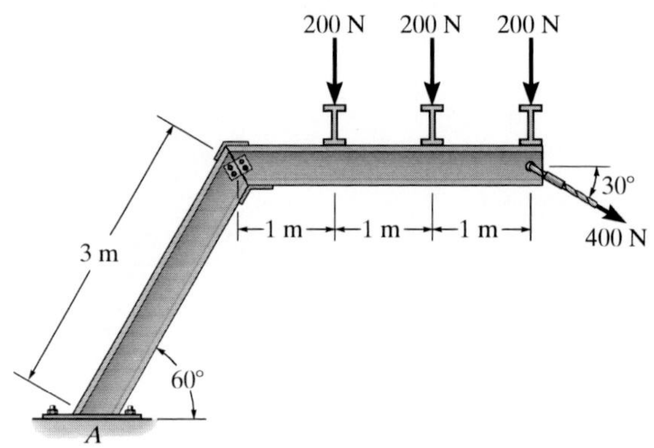

F5–4

F5–5. The 25-kg bar has a center of mass at G. If it is supported by a smooth peg at C, a roller at A, and cord AB, determine the reactions at these supports.

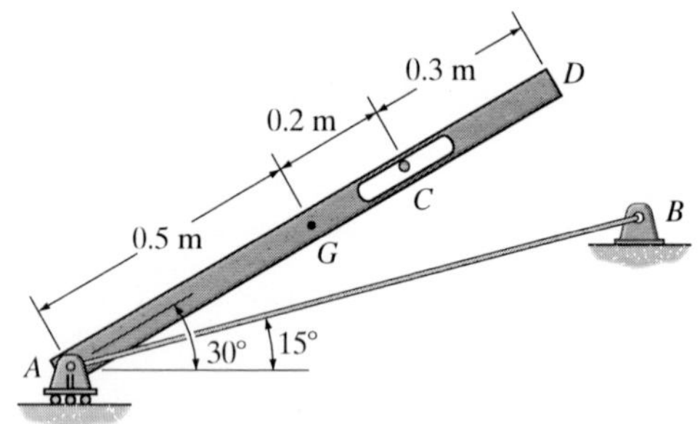

F5–5

F5–6. Determine the reactions at the smooth contact points A, B, and C on the bar.

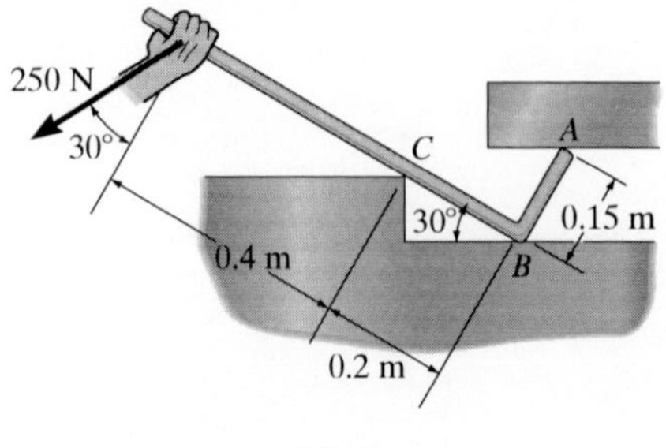

F5–6

5

PROBLEMS

All problem solutions must include an FBD.

5–10. Determine the horizontal and vertical components of reaction at the pin A and the reaction of the rocker B on the beam.

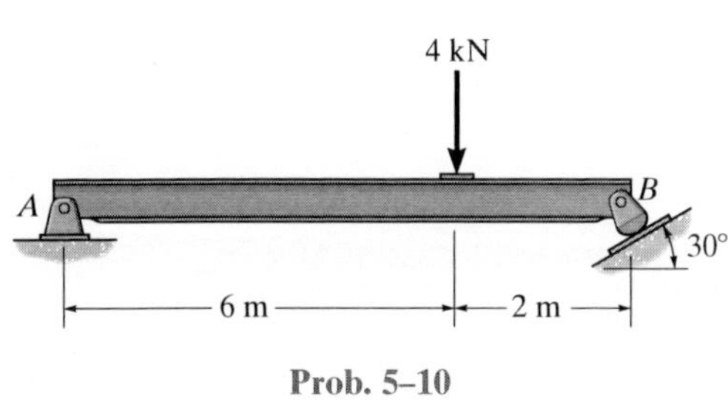

Prob. 5–10

5–11. Determine the magnitude of the reactions on the beam at A and B. Neglect the thickness of the beam.

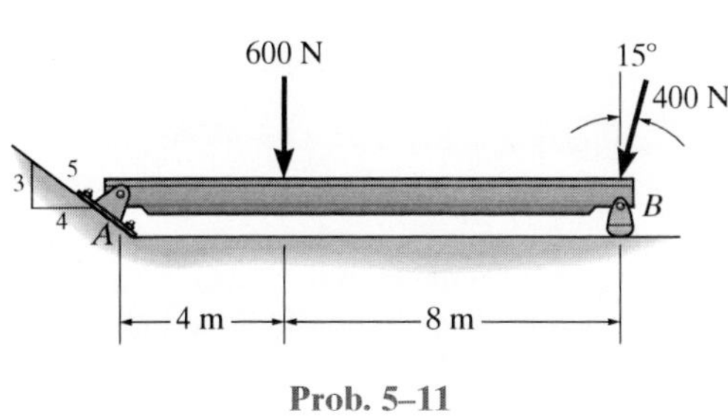

Prob. 5–11

***5–12.** Determine the components of the support reactions at the fixed support A on the cantilevered beam.

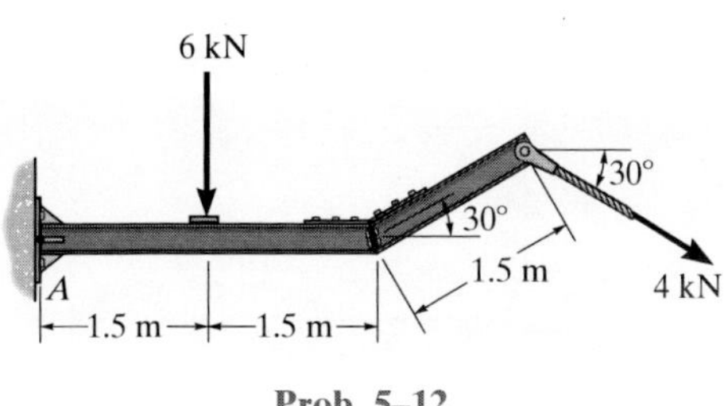

Prob. 5–12

5–13. The 75-kg gate has a center of mass located at G. If A supports only a horizontal force and B can be assumed as a pin, determine the components of reaction at these supports.

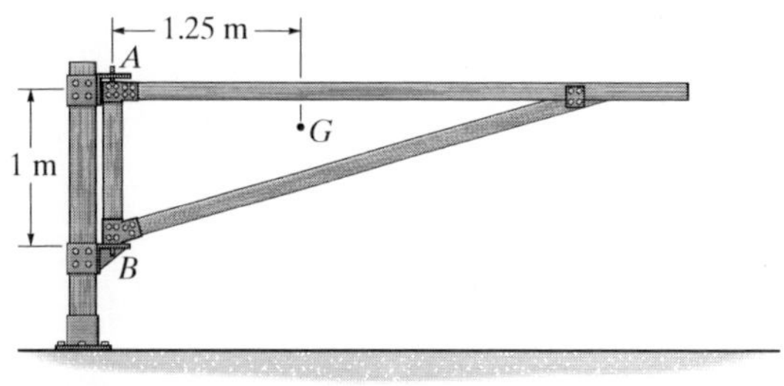

Prob. 5–13

5–14. The overhanging beam is supported by a pin at A and the two-force strut BC. Determine the horizontal and vertical components of reaction at A and the reaction at B on the beam.

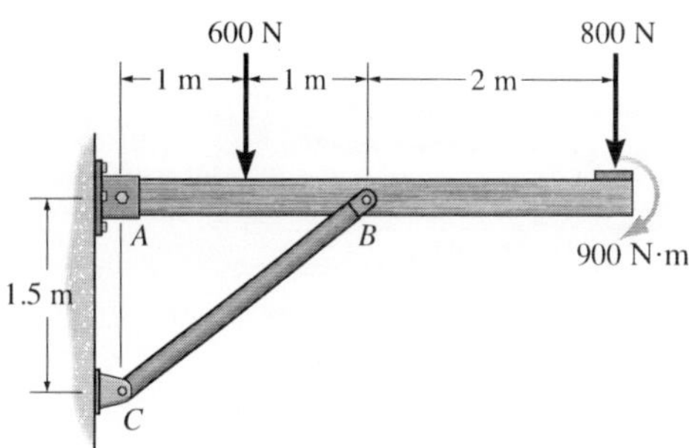

Prob. 5–14

5–15. Determine the reactions at the pins A and B. The spring has an unstretched length of 80 mm.

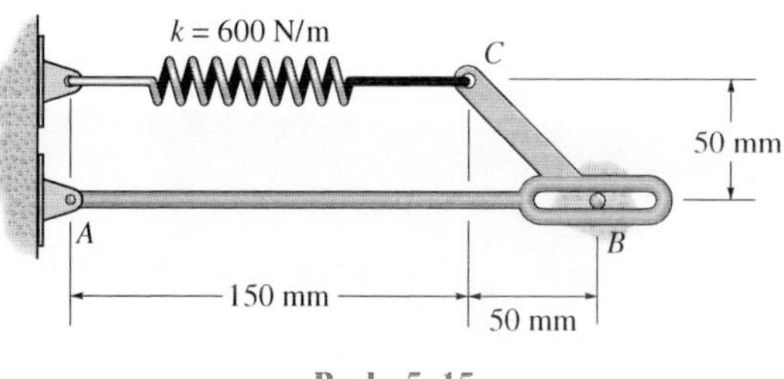

Prob. 5–15

*5–16. Determine the components of reaction at the supports A and B on the rod.

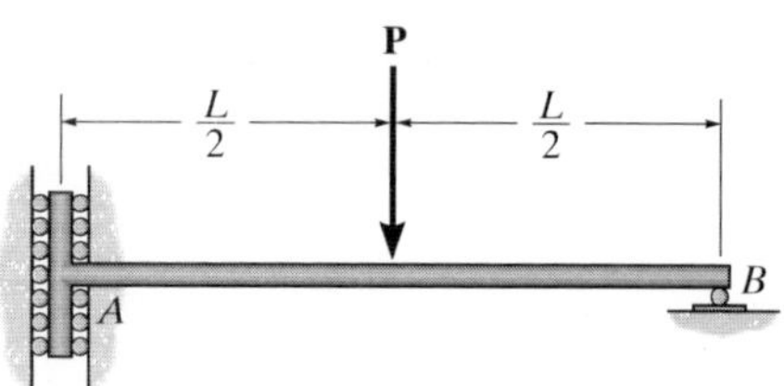

Prob. 5–16

5–17. If the wheelbarrow and its contents have a mass of 60 kg and center of mass at G, determine the magnitude of the resultant force which the man must exert on *each* of the two handles in order to hold the wheelbarrow in equilibrium.

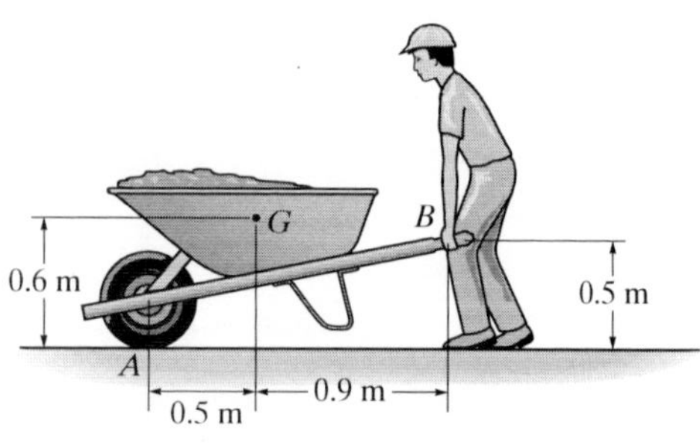

Prob. 5–17

5–18. Determine the reactions at the roller A and pin B.

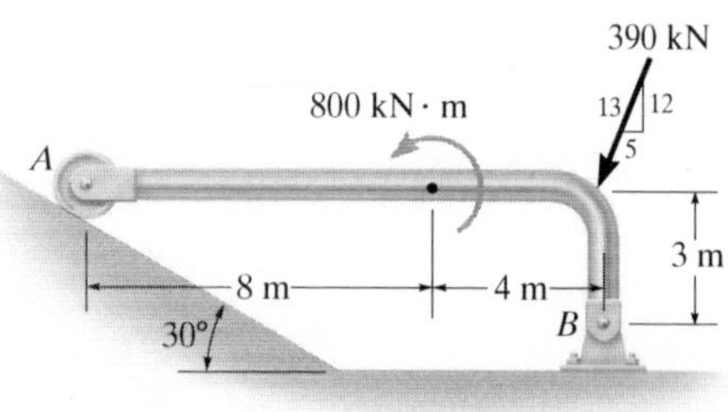

Prob. 5–18

5–19. The shelf supports the electric motor which has a mass of 15 kg and mass center at G_m. The platform upon which it rests has a mass of 4 kg and mass center at G_p. Assuming that a single bolt B holds the shelf up and the bracket bears against the smooth wall at A, determine this normal force at A and the horizontal and vertical components of reaction of the bolt on the bracket.

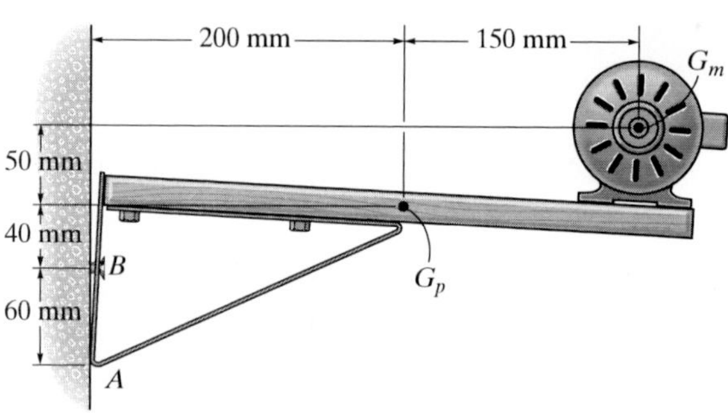

Prob. 5–19

*5–20. The bulk head AD is subjected to both water and soil-backfill pressures. Assuming AD is "pinned" to the ground at A, determine the horizontal and vertical reactions there and also the required tension in the ground anchor BC necessary for equilibrium. The bulk head has a mass of 800 kg.

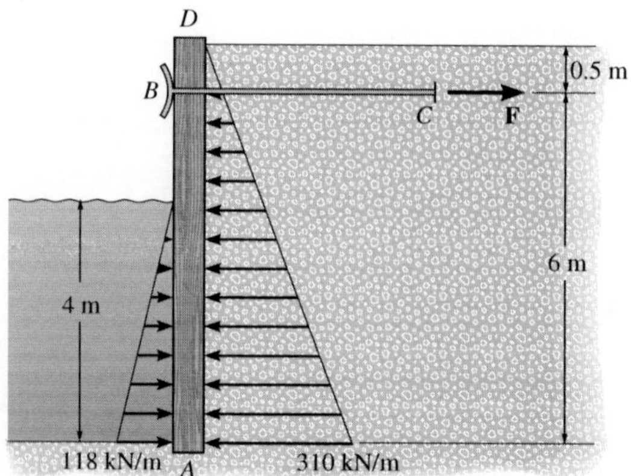

Prob. 5–20

5–21. A cantilever beam, having an extended length of 3 m, is subjected to a vertical force of 500 N. Assuming that the wall resists this load with linearly varying distributed loads over the 0.15-m length of the beam portion inside the wall, detemine the intensities w_1 and w_2 for equilibrium.

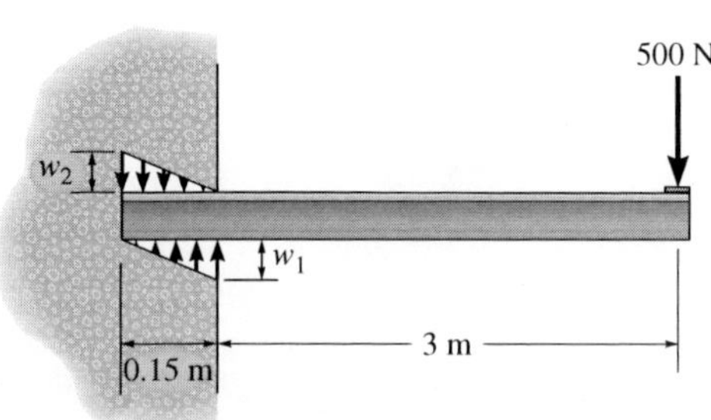

Prob. 5–21

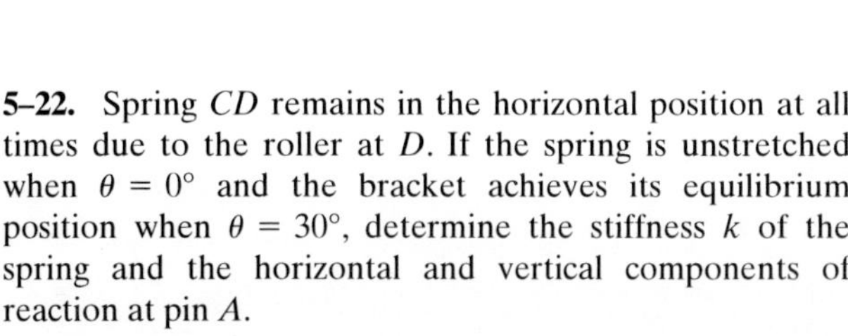

5–22. Spring CD remains in the horizontal position at all times due to the roller at D. If the spring is unstretched when $\theta = 0°$ and the bracket achieves its equilibrium position when $\theta = 30°$, determine the stiffness k of the spring and the horizontal and vertical components of reaction at pin A.

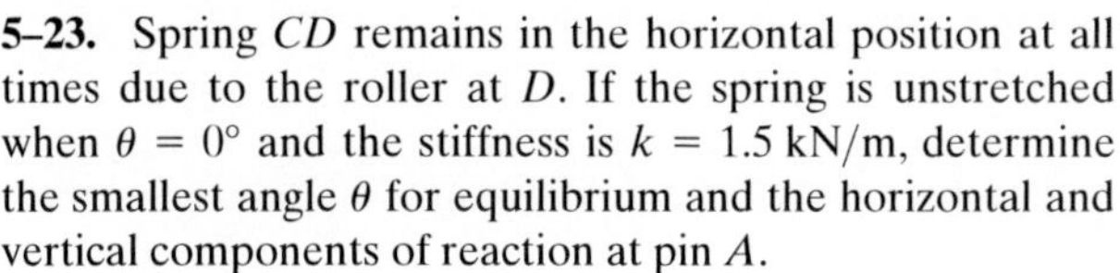

5–23. Spring CD remains in the horizontal position at all times due to the roller at D. If the spring is unstretched when $\theta = 0°$ and the stiffness is $k = 1.5$ kN/m, determine the smallest angle θ for equilibrium and the horizontal and vertical components of reaction at pin A.

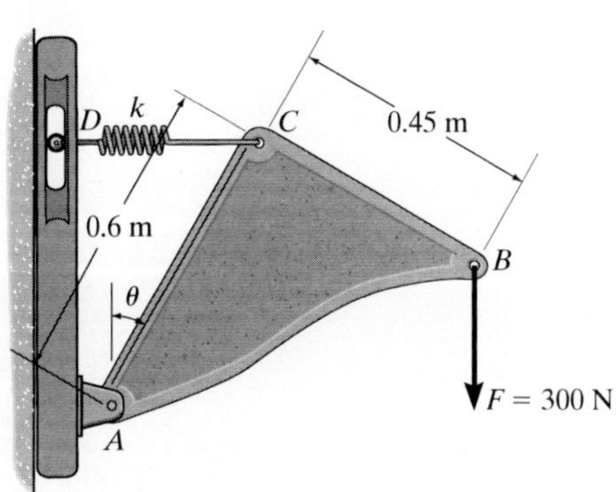

Probs. 5–22/23

*__5–24.__ The platform assembly has a weight of 1000 N ($\approx$ 100 kg) and center of gravity at G_1. If it is intended to support a maximum load of 1600 N placed, at point G_2 determine the smallest counterweight W that should be placed at B in order to prevent the platform from tipping over.

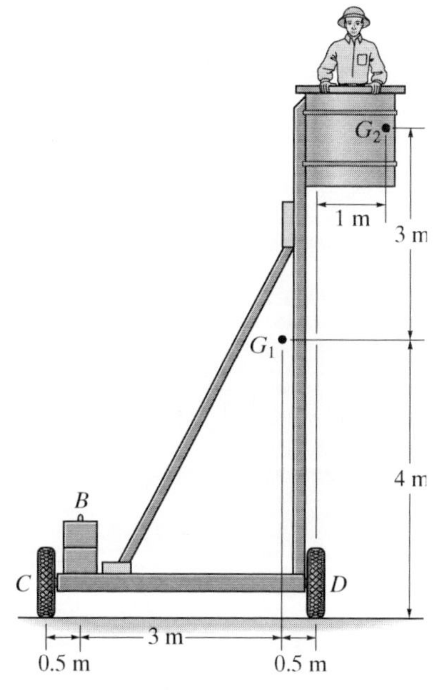

Prob. 5–24

5–25. Determine the reactions on the bent rod which is supported by a smooth surface at B and by a collar at A, which is fixed to the rod and is free to slide over the fixed inclined rod.

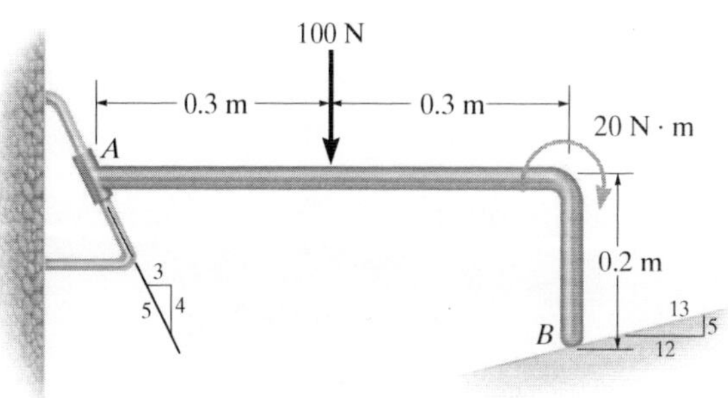

Prob. 5–25

5–26. The toggle switch consists of a cocking lever that is pinned to a fixed frame at A and held in place by the spring which has an unstretched length of 200 mm. Determine the magnitude of the resultant force at A and the normal force on the peg at B when the lever is in the position shown.

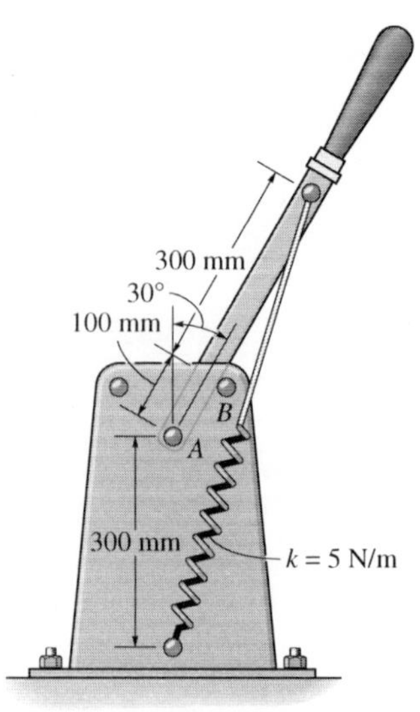

Prob. 5–26

5–27. The sports car has a mass of 1.5 Mg and mass center at G. If the front two springs each have a stiffness of $k_A = 58$ kN/m and the rear two springs each have a stiffness of $k_B = 65$ kN/m, determine their compression when the car is parked on the 30° incline. Also, what friction force $\mathbf{F}_B$ must be applied to each of the rear wheels to hold the car in equilibrium? *Hint:* First determine the normal force at A and B, then determine the compression in the springs.

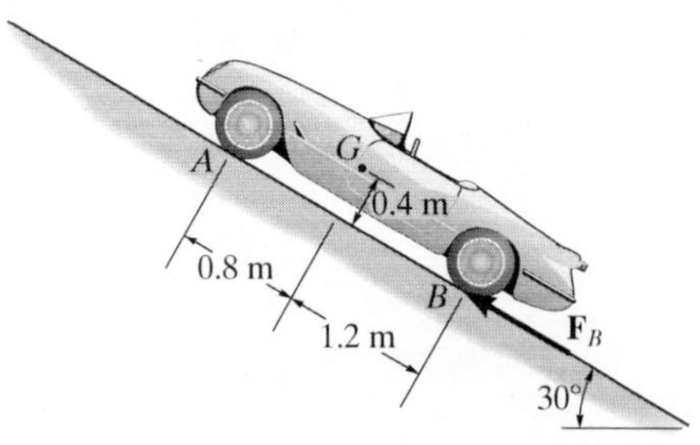

Prob. 5–27

*__5–28.__ Determine the magnitude and direction θ of the minimum force P needed to pull the 50-kg roller over the smooth step.

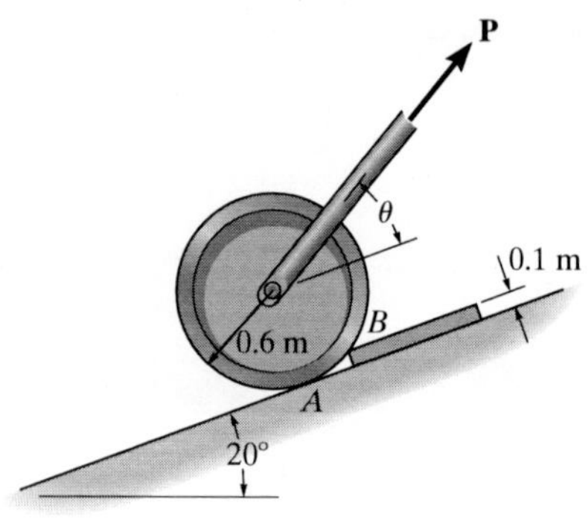

Prob. 5–28

5–29. The horizontal beam is supported by springs at its ends. Each spring has a stiffness of $k = 5$ kN/m and is originally unstretched when the beam is in the horizontal position. Determine the angle of tilt of the beam if a load of 800 N is applied at point C as shown.

5–30. The horizontal beam is supported by springs at its ends. If the stiffness of the spring at A is $k_A = 5$ kN/m, determine the required stiffness of the spring at B so that if the beam is loaded with the 800 N it remains in the horizontal position. The springs are originally constructed so that the beam is in the horizontal position when it is unloaded.

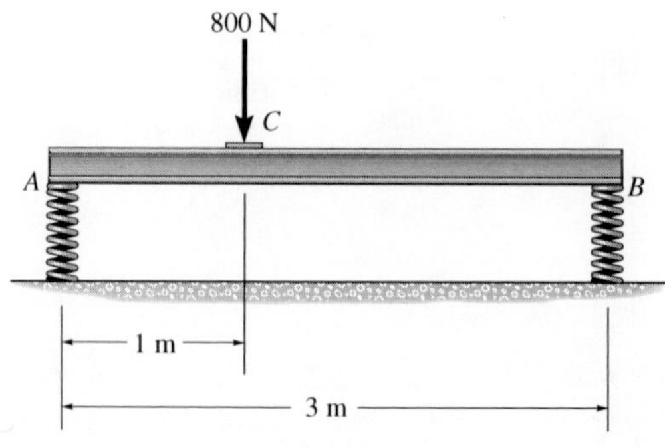

Probs. 5–29/30

5–31. The jib crane is supported by a pin at C and rod AB. If the load has a mass of 2 Mg with its center of mass located at G, determine the horizontal and vertical components of reaction at the pin C and the force developed in rod AB on the crane when $x = 5$ m.

***5–32.** The jib crane is supported by a pin at C and rod AB. The rod can withstand a maximum tension of 40 kN. If the load has a mass of 2 Mg, with its center of mass located at G, determine its maximum allowable distance x and the corresponding horizontal and vertical components of reaction at C.

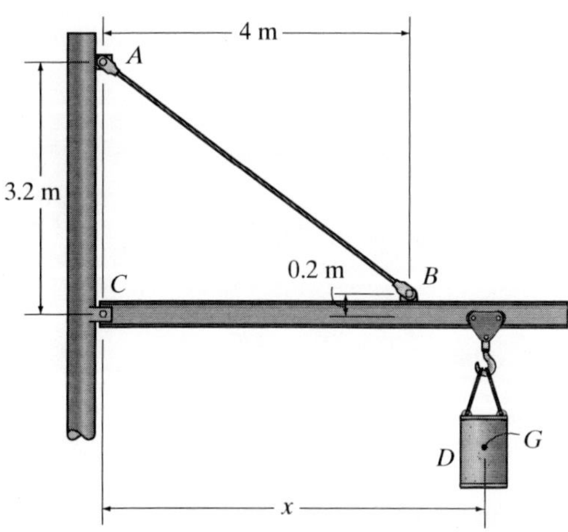

Probs. 5–31/32

5–33. The woman exercises on the rowing machine. If she exerts a holding force of $F = 200$ N on handle ABC, determine the horizontal and vertical components of reaction at pin C and the force developed along the hydraulic cylinder BD on the handle.

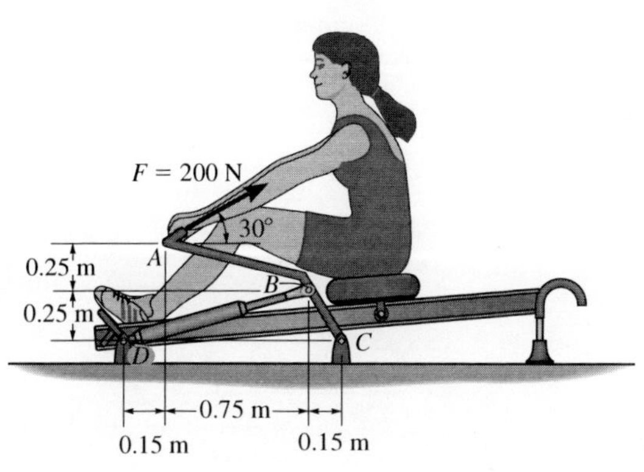

Prob. 5–33

5–34. A uniform glass rod having a length L is placed in the smooth hemispherical bowl having a radius r. Determine the angle of inclination θ for equilibrium.

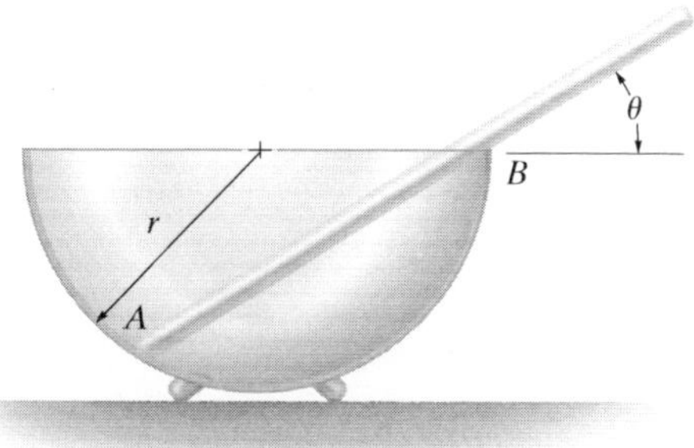

Prob. 5–34

5–35. The toggle switch consists of a cocking lever that is pinned to a fixed frame at A and held in place by the spring which has an unstretched length of 250 mm. Determine the magnitude of the resultant force at A and the normal force on the peg at B when the lever is in the position shown.

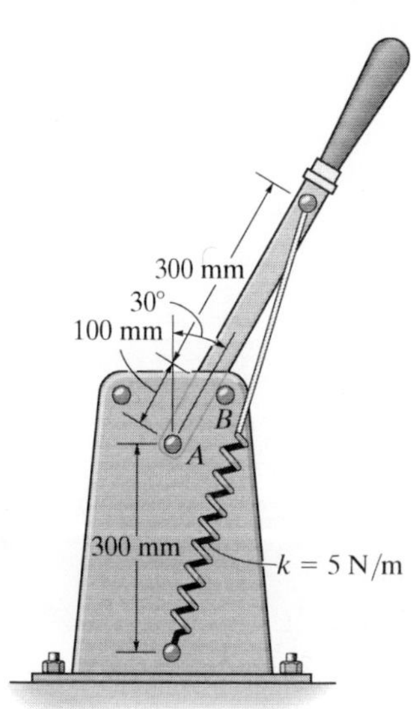

Prob. 5–35

***5–36.** The 1.4-Mg drainpipe is held in the tines of the fork lift. Determine the normal forces at A and B as functions of the blade angle θ and plot the results of force (ordinate) versus θ (abscissa) for $0 \leq \theta \leq 90°$.

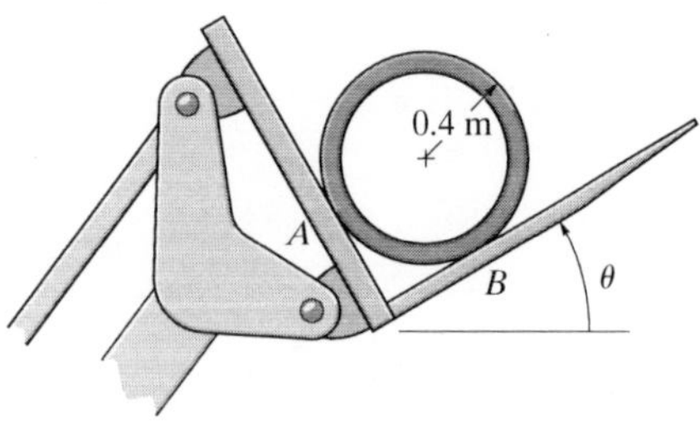

Prob. 5–36

5–37. The boom supports the two vertical loads. Neglect the size of the collars at D and B and the thickness of the boom, and compute the horizontal and vertical components of force at the pin A and the force in cable CB. Set $F_1 = 800$ N and $F_2 = 350$ N.

5–38. The boom is intended to support two vertical loads, $\mathbf{F}_1$ and $\mathbf{F}_2$. If the cable CB can sustain a maximum load of 1500 N before it fails, determine the critical loads if $F_1 = 2F_2$. Also, what is the magnitude of the maximum reaction at pin A?

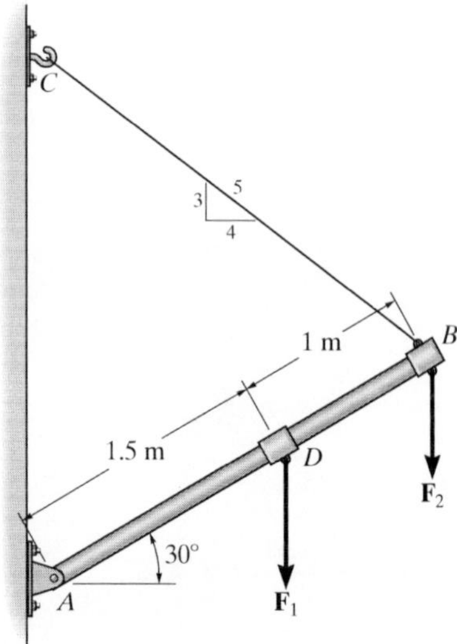

Probs. 5–37/38

5–39. The airstroke actuator at D is used to apply a force of $F = 200$ N on the member at B. Determine the horizontal and vertical components of reaction at the pin A and the force of the smooth shaft at C on the member.

***5–40.** The airstroke actuator at D is used to apply a force of $\mathbf{F}$ on the member at B. The normal reaction of the smooth shaft at C on the member is 300 N. Determine the magnitude of $\mathbf{F}$ and the horizontal and vertical components of reaction at pin A.

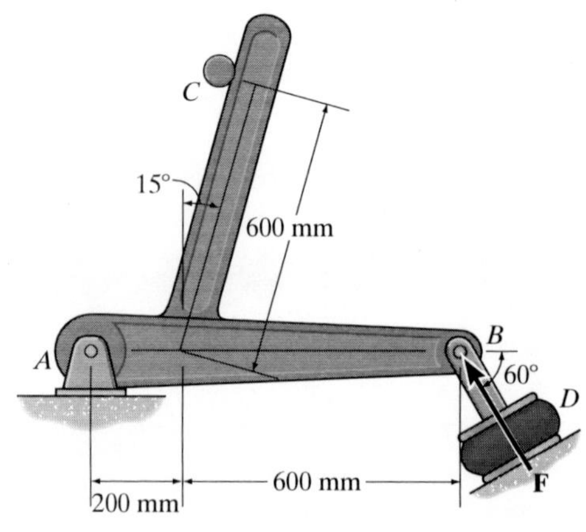

Probs. 5–39/40

5–41. The uniform beam has a weight W and length $2l$ and is supported by a pin at A and a cable BC. Determine the horizontal and vertical components of reaction at A and the tension in the cable necessary to hold the beam in the position shown.

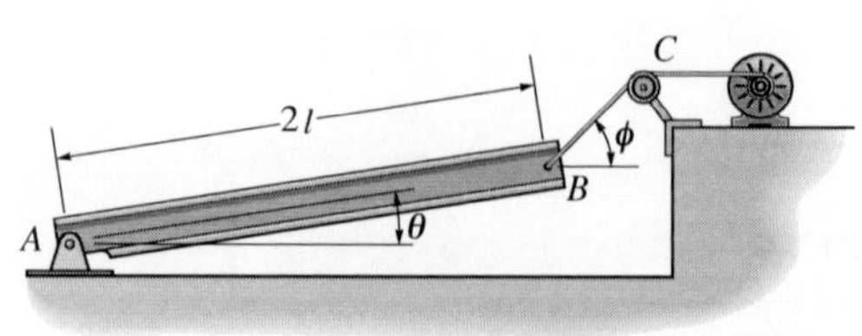

Prob. 5–41

5–42. The rigid metal strip of negligible weight is used as part of an electromagnetic switch. If the stiffness of the springs at A and B is $k = 5$ N/m, and the strip is originally horizontal when the springs are unstretched, determine the smallest force needed to close the contact gap at C.

5–43. The rigid metal strip of negligible weight is used as part of an electromagnetic switch. Determine the maximum stiffness k of the springs at A and B so that the contact at C closes when the vertical force developed there is 0.5 N. Originally the strip is horizontal as shown.

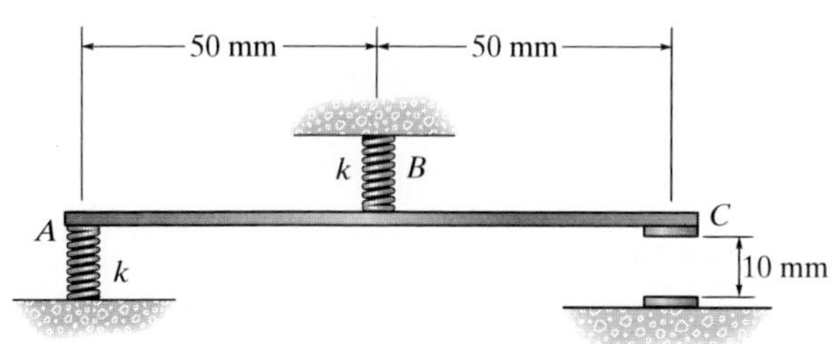

Probs. 5–42/43

***5–44.** The upper portion of the crane boom consists of the jib AB, which is supported by the pin at A, the guy line BC, and the backstay CD, each cable being separately attached to the mast at C. If the 5-kN load is supported by the hoist line, which passes over the pulley at B, determine the magnitude of the resultant force the pin exerts on the jib at A for equilibrium, the tension in the guy line BC, and the tension T in the hoist line. Neglect the weight of the jib. The pulley at B has a radius of 0.1 m.

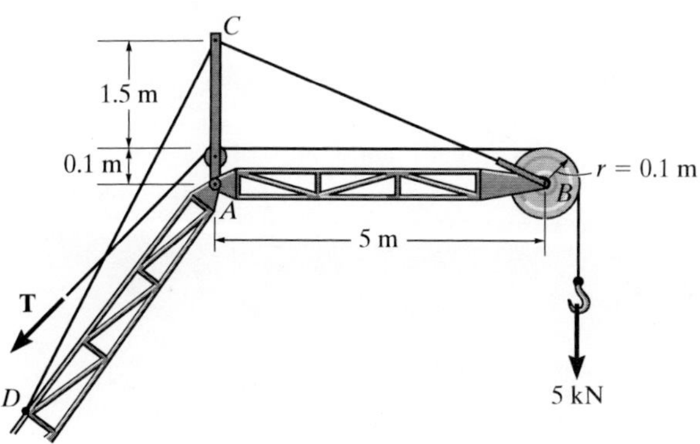

Prob. 5–44

5–45. The device is used to hold an elevator door open. If the spring has a stiffness of $k = 40$ N/m and it is compressed 0.2 m, determine the horizontal and vertical components of reaction at the pin A and the resultant force at the wheel bearing B.

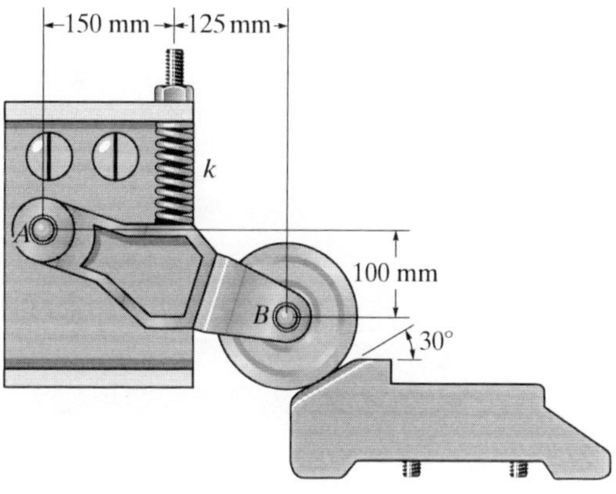

Prob. 5–45

5–46. Three uniform books, each having a weight W and length a, are stacked as shown. Determine the maximum distance d that the top book can extend out from the bottom one so the stack does not topple over.

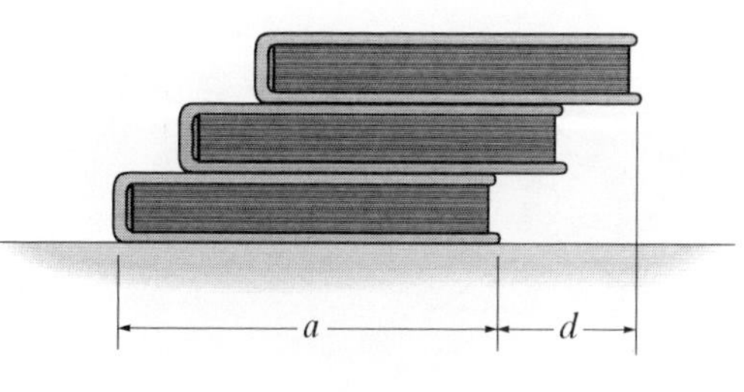

Prob. 5–46

5–47. The horizontal beam is supported by springs at its ends. Each spring has a stiffness of $k = 5\text{ kN/m}$ and is originally unstretched when the beam is in the horizontal position. Determine the angle of tilt of the beam if a load of 800 N is applied at point C as shown.

***5–48.** The horizontal beam is supported by springs at its ends. If the stiffness of the spring at A is $k_A = 5\text{ kN/m}$, determine the required stiffness of the spring at B so that if the beam is loaded with the 800-N force it remains in the horizontal position. The springs are originally constructed so that the beam is in the horizontal position when it is unloaded.

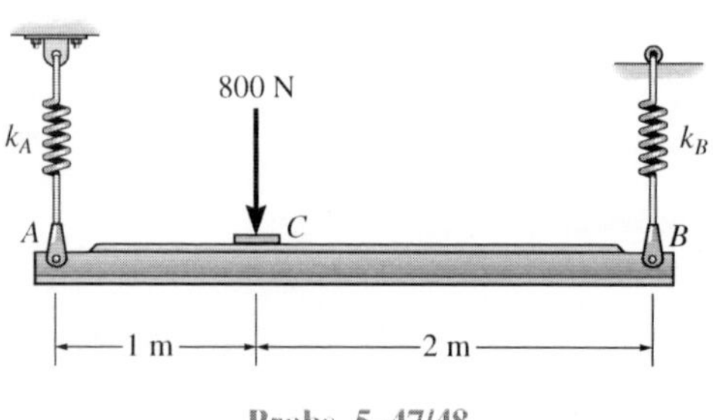

Probs. 5–47/48

5–49. The wheelbarrow and its contents have a mass of $m = 60$ kg with a center of mass at G. Determine the normal reaction on the tire and the vertical force on each hand to hold it at $\theta = 30°$. Take $a = 0.3$ m, $b = 0.45$ m, $c = 0.75$ m and $d = 0.1$ m.

5–50. The wheelbarrow and its contents have a mass m and center of mass at G. Determine the greatest angle of tilt θ without causing the wheelbarrow to tip over.

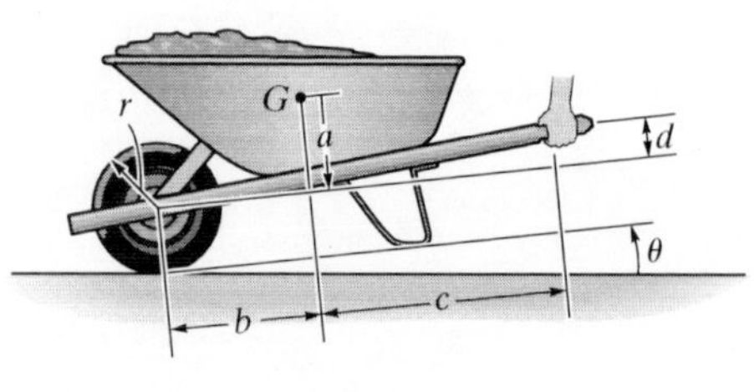

Probs. 5–49/50

5–51. The rigid beam of negligible weight is supported horizontally by two springs and a pin. If the springs are uncompressed when the load is removed, determine the force in each spring when the load **P** is applied. Also, compute the vertical deflection of end C. Assume the spring stiffness k is large enough so that only small deflections occur. *Hint:* The beam rotates about A so the deflections in the springs can be related.

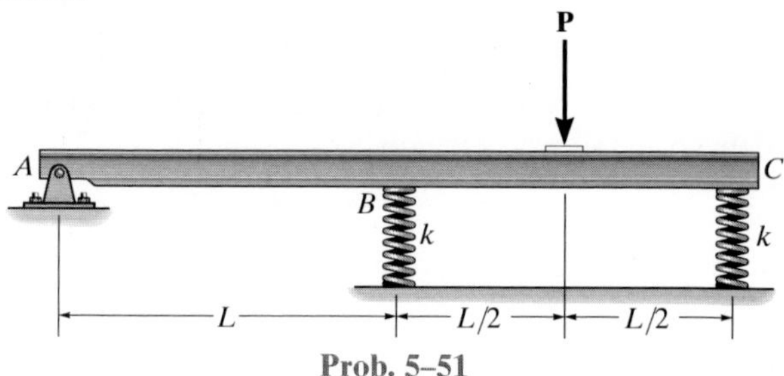

Prob. 5–51

***5–52.** A boy stands out at the end of the diving board, which is supported by two springs A and B, each having a stiffness of $k = 15\text{ kN/m}$. In the position shown the board is horizontal. If the boy has a mass of 40 kg, determine the angle of tilt which the board makes with the horizontal after he jumps off. Neglect the weight of the board and assume it is rigid.

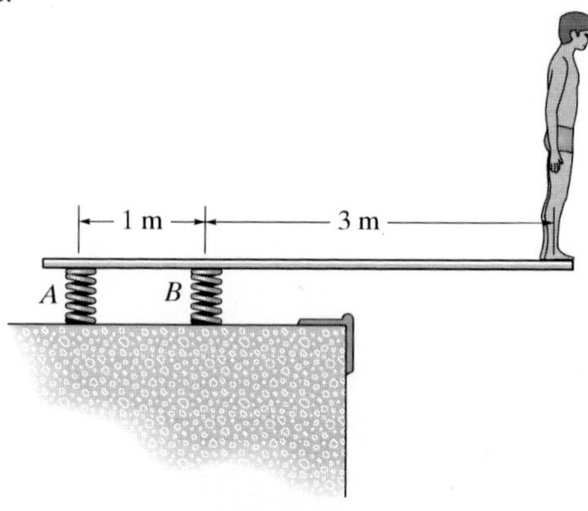

Prob. 5–52

5–53. The uniform beam has a weight W and length l and is supported by a pin at A and a cable BC. Determine the horizontal and vertical components of reaction at A and the tension in the cable necessary to hold the beam in the position shown.

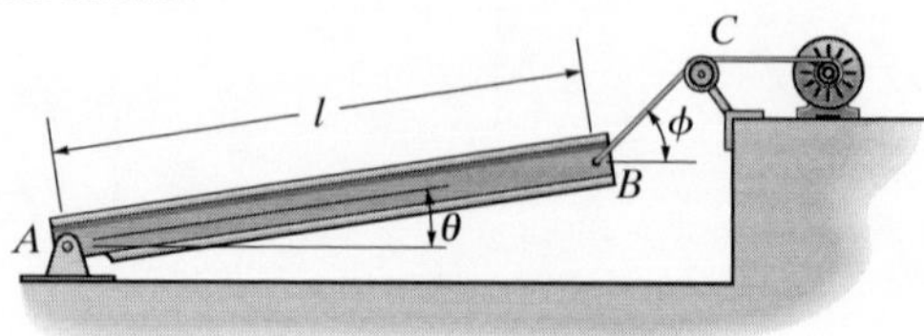

Prob. 5–53

5–54. Determine the distance d for placement of the load **P** for equilibrium of the smooth bar in the position θ as shown. Neglect the weight of the bar.

5–55. If $d = 1$ m, and $\theta = 30°$, determine the normal reaction at the smooth supports and the required distance a for the placement of the roller if $P = 600$ N. Neglect the weight of the bar.

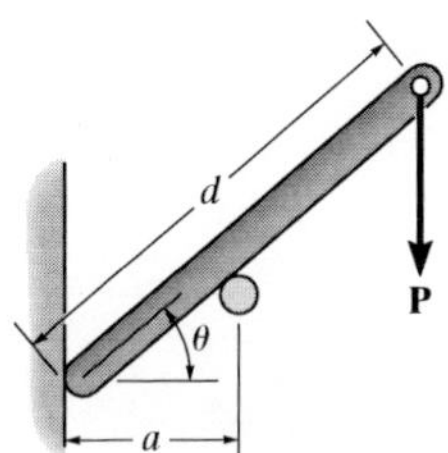

Probs. 5–54/55

*__5–56.__ The disk B has a mass of 20 kg and is supported on the smooth cylindrical surface by a spring having a stiffness of $k = 400$ N/m and unstretched length of $l_0 = 1$ m. The spring remains in the horizontal position since its end A is attached to the small roller guide which has negligible weight. Determine the angle θ for equilibrium of the roller.

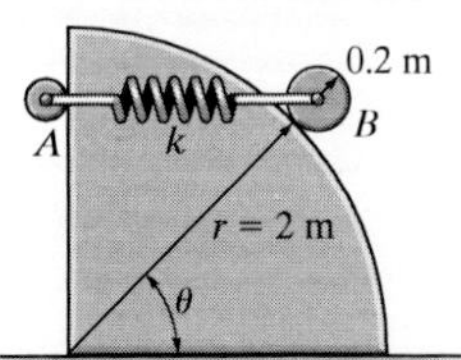

Prob. 5–56

5–57. The winch cable on a tow truck is subjected to a force of $T = 6$ kN when the cable is directed at $\theta = 60°$. Determine the magnitudes of the total brake frictional force **F** for the rear set of wheels B and the total normal forces at *both* front wheels A and both rear wheels B for equilibrium. The truck has a total mass of 4 Mg and mass center at G.

5–58. Determine the minimum cable force T and critical angle θ which will cause the tow truck to start tipping, i.e., for the normal reaction at A to be zero. Assume that the truck is braked and will not slip at B. The truck has a total mass of 4 Mg and mass center at$G$.x

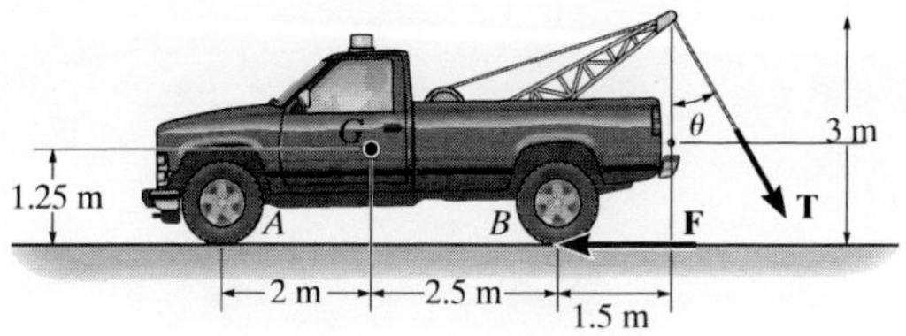

Probs. 5–57/58

5–59. The thin rod of length l is supported by the smooth tube. Determine the distance a needed for equilibrium if the applied load is **P**.

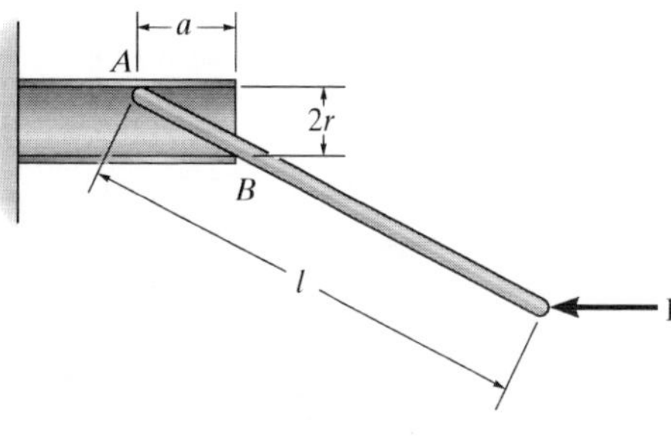

Prob. 5–59

5

*__5–60.__ The 30-N uniform rod has a length of $l = 1$ m. If $s = 1.5$ m, determine the distance h of placement at the end A along the smooth wall for equilibrium.

5–61. The uniform rod has a length l and weight W. It is supported at one end A by a smooth wall and the other end by a cord of length s which is attached to the wall as shown. Determine the placement h for equilibrium.

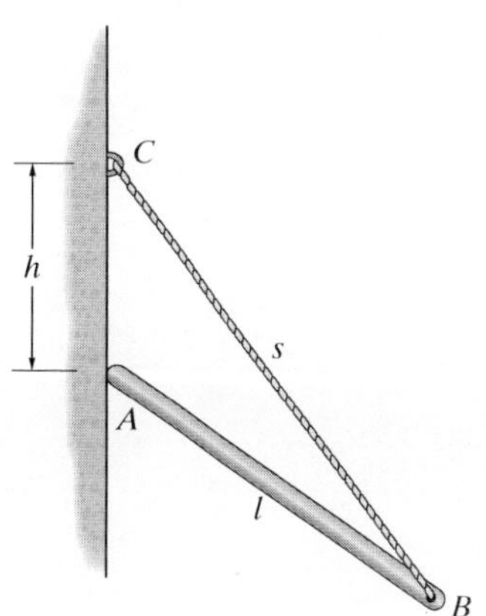

Probs. 5–60/61

CONCEPTUAL PROBLEMS

P5–5. The tie rod is used to support this overhang at the entrance of a building. If it is pin connected to the building wall at A and to the center of the overhang B, determine if the force in the rod will increase, decrease, or remain the same if (a) the support at A is moved to a lower position D, and (b) the support at B is moved to the outer position C. Explain your answer with an equilibrium analysis, using dimensions and loads. Assume the overhang is pin supported from the building wall.

P5–5

P5–6. The man attempts to pull the four wheeler up the incline and onto the trailer. From the position shown, is it more effective to pull on the rope at A, or would it be better to pull on the rope at B? Draw a free-body diagram for each case, and do an equilibrium analysis to explain your answer. Use appropriate numerical values to do your calculations.

P5–6

P5–7. Like all aircraft, this jet plane rests on three wheels. Why not use an additional wheel at the tail for better support? (Can you think of any other reason for not including this wheel?) If there was a fourth tail wheel, draw a free-body diagram of the plane from a side (2 D) view, and show why one would not be able to determine all the wheel reactions using the equations of equilibrium.

P5–7

P5–8. Where is the best place to arrange most of the logs in the wheelbarrow so that it minimizes the amount of force on the backbone of the person transporting the load? Do an equilibrium analysis to explain your answer.

P5–8

EQUILIBRIUM IN THREE DIMENSIONS

5.5 Free-Body Diagrams

The first step in solving three-dimensional equilibrium problems, as in the case of two dimensions, is to draw a free-body diagram. Before we can do this, however, it is first necessary to discuss the types of reactions that can occur at the supports.

Support Reactions. The reactive forces and couple moments acting at various types of supports and connections, when the members are viewed in three dimensions, are listed in Table 5–2. It is important to recognize the symbols used to represent each of these supports and to understand clearly how the forces and couple moments are developed. As in the two-dimensional case:

- A force is developed by a support that restricts the translation of its attached member.
- A couple moment is developed when rotation of the attached member is prevented.

For example, in Table 5–2, item (4), the ball-and-socket joint prevents any translation of the connecting member; therefore, a force must act on the member at the point of connection. This force has three components having unknown magnitudes, F_x, F_y, F_z. Provided these components are known, one can obtain the magnitude of force, $F = \sqrt{F_x^2 + F_y^2 + F_z^2}$, and the force's orientation defined by its coordinate direction angles α, β, γ, Eqs. 2–5.* Since the connecting member is allowed to rotate freely about *any* axis, no couple moment is resisted by a ball-and-socket joint.

It should be noted that the *single* bearing supports in items (5) and (7), the *single* pin (8), and the *single* hinge (9) are shown to resist both force and couple-moment components. If, however, these supports are used in conjunction with *other* bearings, pins, or hinges to hold a rigid body in equilibrium and the supports are *properly aligned* when connected to the body, then the *force reactions* at these supports *alone* are adequate for supporting the body. In other words, the couple moments become redundant and are not shown on the free-body diagram. The reason for this should become clear after studying the examples which follow.

* The three unknowns may also be represented as an unknown force magnitude F and two unknown coordinate direction angles. The third direction angle is obtained using the identity $\cos^2\alpha + \cos^2\beta + \cos^2\gamma = 1$, Eq. 2–8.

5

TABLE 5–2 Supports for Rigid Bodies Subjected to Three-Dimensional Force Systems

Types of Connection	Reaction	Number of Unknowns
(1)  cable	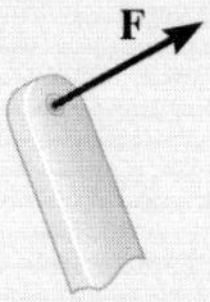 **F**	One unknown. The reaction is a force which acts away from the member in the known direction of the cable.
(2) 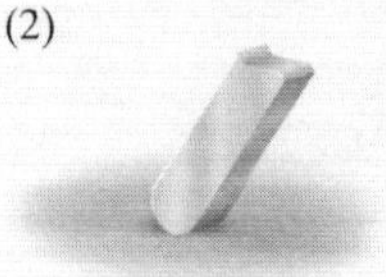 smooth surface support	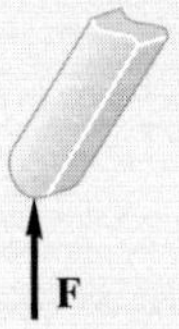 **F**	One unknown. The reaction is a force which acts perpendicular to the surface at the point of contact.
(3) 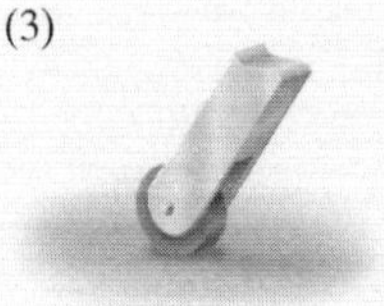 roller	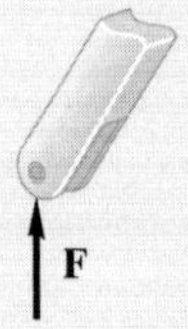 **F**	One unknown. The reaction is a force which acts perpendicular to the surface at the point of contact.
(4) 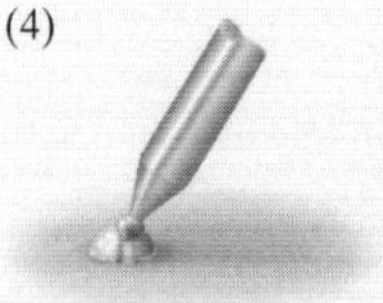 ball and socket	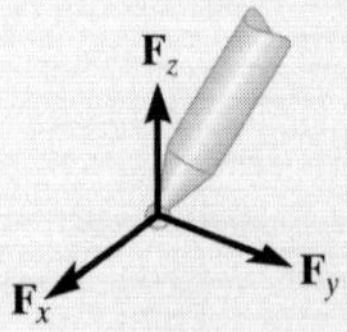 $\mathbf{F}_z$, $\mathbf{F}_x$, $\mathbf{F}_y$	Three unknowns. The reactions are three rectangular force components.
(5) 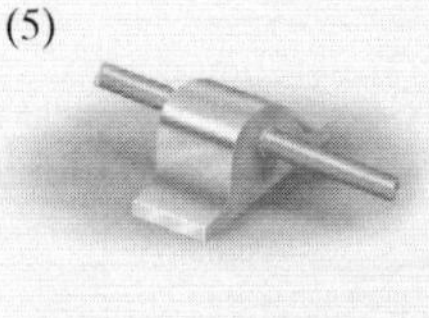 single journal bearing	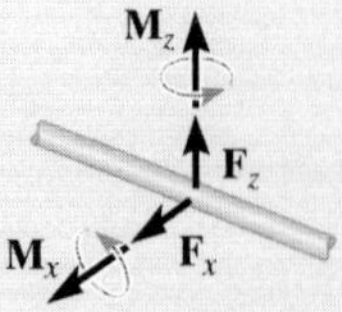 $\mathbf{M}_z$, $\mathbf{F}_z$, $\mathbf{M}_x$, $\mathbf{F}_x$	Four unknowns. The reactions are two force and two couple-moment components which act perpendicular to the shaft. Note: The couple moments are *generally not applied* if the body is supported elsewhere. See the examples.

5

continued

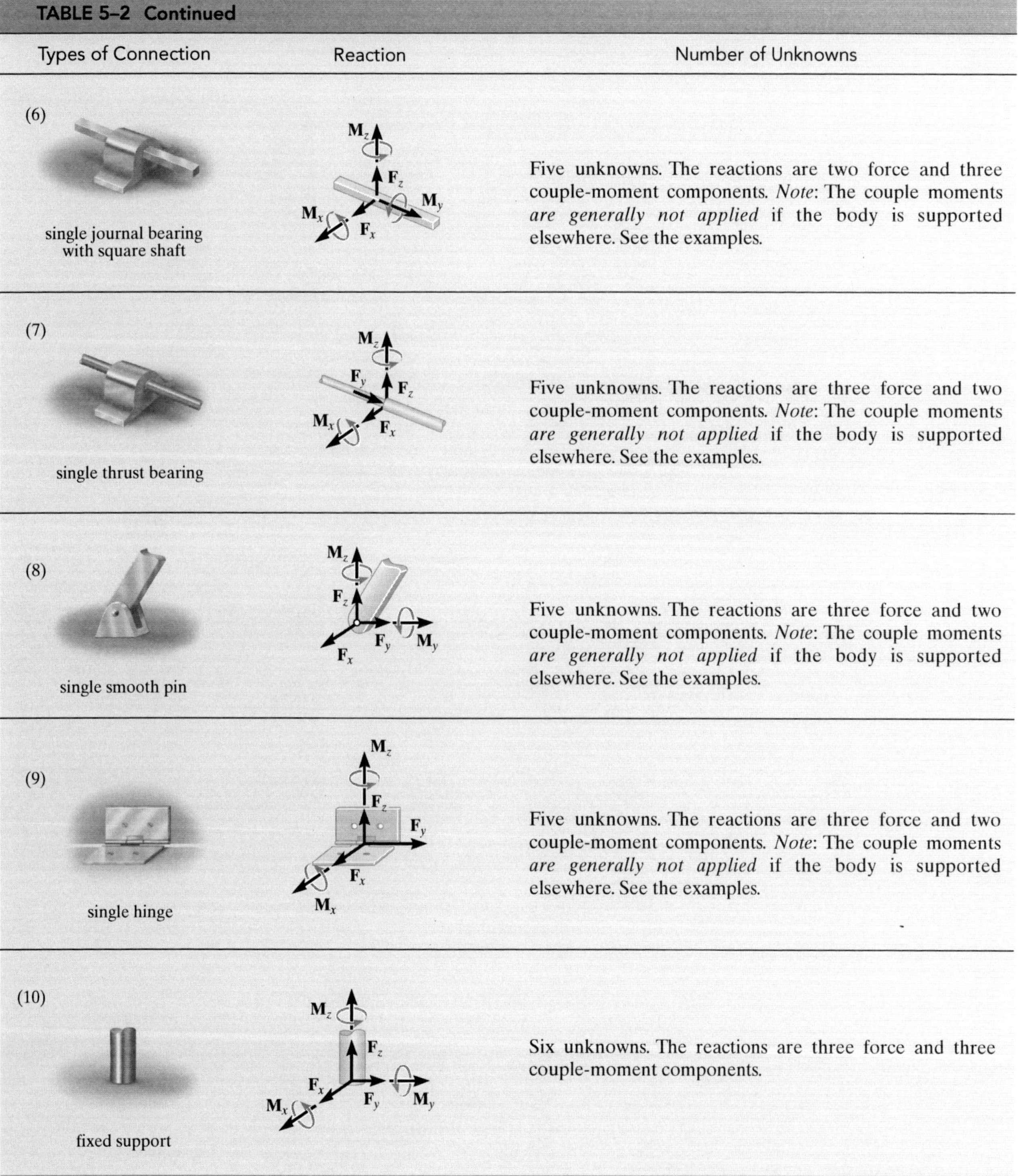

TABLE 5–2 Continued

Types of Connection	Reaction	Number of Unknowns
(6) single journal bearing with square shaft	$\mathbf{M}_z$, $\mathbf{F}_z$, $\mathbf{M}_y$, $\mathbf{M}_x$, $\mathbf{F}_x$	Five unknowns. The reactions are two force and three couple-moment components. *Note*: The couple moments *are generally not applied* if the body is supported elsewhere. See the examples.
(7) single thrust bearing	$\mathbf{M}_z$, $\mathbf{F}_y$, $\mathbf{F}_z$, $\mathbf{M}_x$, $\mathbf{F}_x$	Five unknowns. The reactions are three force and two couple-moment components. *Note*: The couple moments *are generally not applied* if the body is supported elsewhere. See the examples.
(8) single smooth pin	$\mathbf{M}_z$, $\mathbf{F}_z$, $\mathbf{F}_x$, $\mathbf{F}_y$, $\mathbf{M}_y$	Five unknowns. The reactions are three force and two couple-moment components. *Note*: The couple moments *are generally not applied* if the body is supported elsewhere. See the examples.
(9) single hinge	$\mathbf{M}_z$, $\mathbf{F}_z$, $\mathbf{F}_y$, $\mathbf{F}_x$, $\mathbf{M}_x$	Five unknowns. The reactions are three force and two couple-moment components. *Note*: The couple moments *are generally not applied* if the body is supported elsewhere. See the examples.
(10) fixed support	$\mathbf{M}_z$, $\mathbf{F}_z$, $\mathbf{F}_x$, $\mathbf{M}_x$, $\mathbf{F}_y$, $\mathbf{M}_y$	Six unknowns. The reactions are three force and three couple-moment components.

Typical examples of actual supports that are referenced to Table 5–2 are shown in the following sequence of photos.

This ball-and-socket joint provides a connection for the housing of an earth grader to its frame. (4)

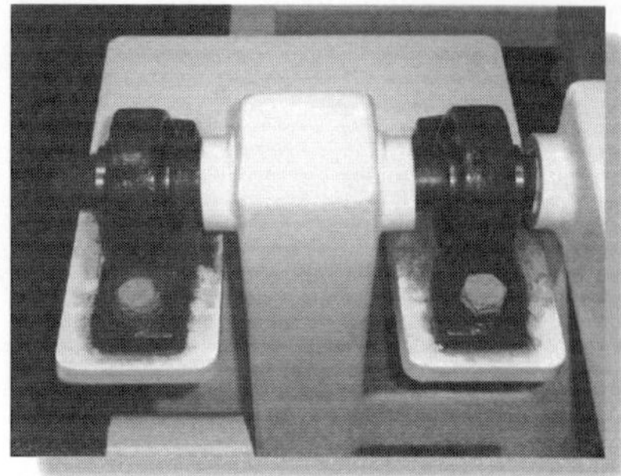

The journal bearings support the ends of the shaft. (5)

This thrust bearing is used to support the drive shaft on a machine. (7)

This pin is used to support the end of the strut used on a tractor. (8)

Free-Body Diagrams. The general procedure for establishing the free-body diagram of a rigid body has been outlined in Sec. 5.2. Essentially it requires first "isolating" the body by drawing its outlined shape. This is followed by a careful *labeling* of *all* the forces and couple moments with reference to an established x, y, z coordinate system. As a general rule, it is suggested to show the unknown components of reaction as acting on the free-body diagram in the *positive sense*. In this way, if any negative values are obtained, they will indicate that the components act in the negative coordinate directions.

EXAMPLE 5.14

Consider the two rods and plate, along with their associated free-body diagrams, shown in Fig. 5–23. The *x, y, z* axes are established on the diagram and the unknown reaction components are indicated in the *positive sense*. The weight is neglected.

SOLUTION

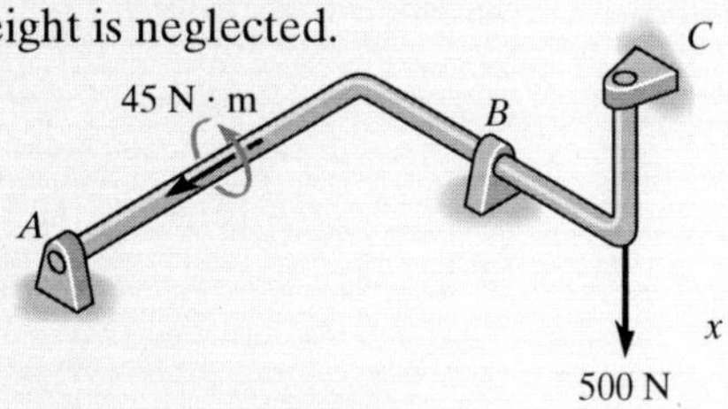

Properly aligned journal bearings at *A*, *B*, *C*.

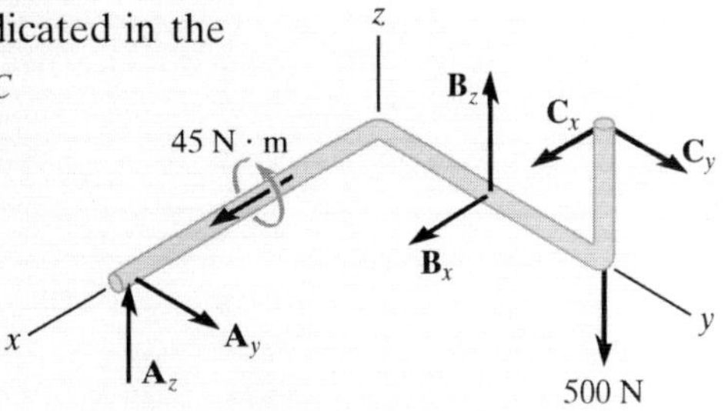

The force reactions developed by the bearings are *sufficient* for equilibrium since they prevent the shaft from rotating about each of the coordinate axes. No couple moments at each bearing are developed.

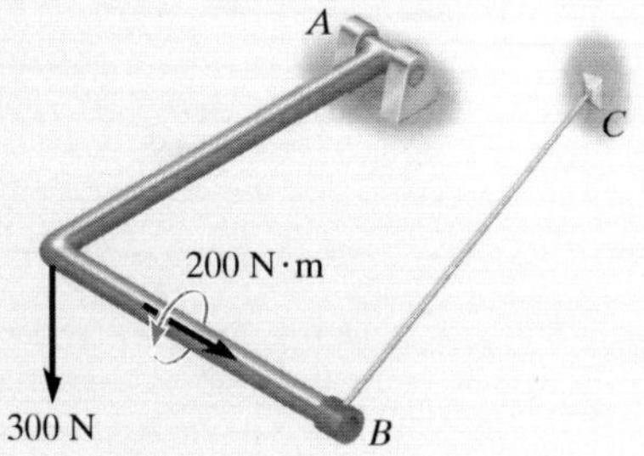

Pin at *A* and cable *BC*.

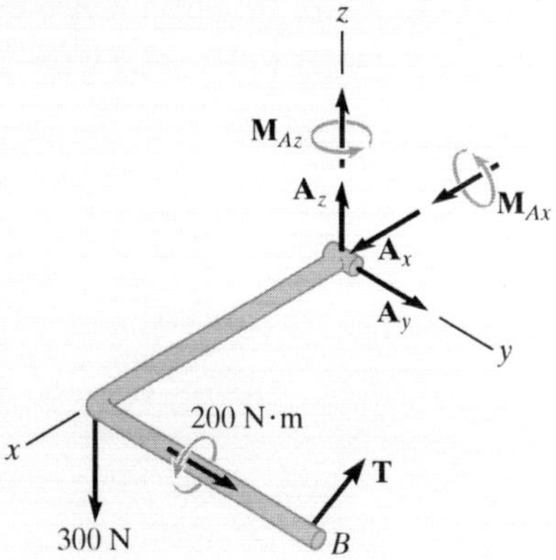

Moment components are developed by the pin on the rod to prevent rotation about the *x* and *z* axes.

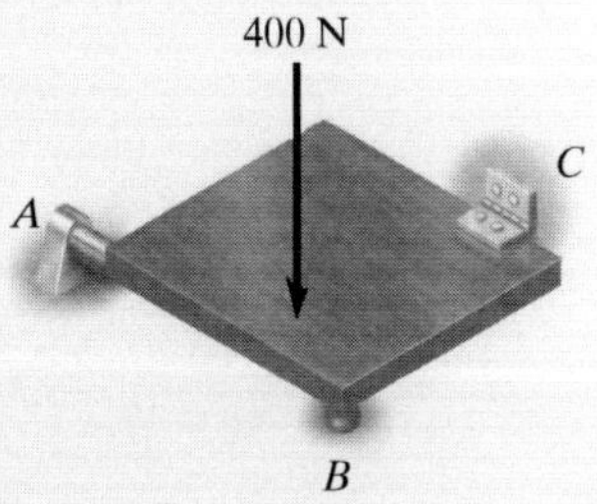

Properly aligned journal bearing at *A* and hinge at *C*. Roller at *B*.

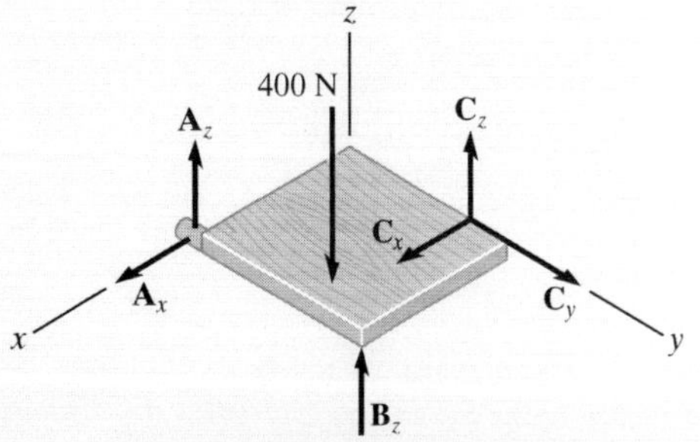

Only force reactions are developed by the bearing and hinge on the plate to prevent rotation about each coordinate axis. No moments are developed at the hinge.

Fig. 5–23

5.6 Equations of Equilibrium

As stated in Sec. 5.1, the conditions for equilibrium of a rigid body subjected to a three-dimensional force system require that both the *resultant* force and *resultant* couple moment acting on the body be equal to *zero*.

Vector Equations of Equilibrium. The two conditions for equilibrium of a rigid body may be expressed mathematically in vector form as

$$\Sigma \mathbf{F} = \mathbf{0} \quad \Sigma \mathbf{M}_O = \mathbf{0} \tag{5–5}$$

where $\Sigma\mathbf{F}$ is the vector sum of all the external forces acting on the body and $\Sigma\mathbf{M}_O$ is the sum of the couple moments and the moments of all the forces about any point O located either on or off the body.

5

Scalar Equations of Equilibrium. If all the external forces and couple moments are expressed in Cartesian vector form and substituted into Eqs. 5–5, we have

$$\Sigma \mathbf{F} = \Sigma F_x \mathbf{i} + \Sigma F_y \mathbf{j} + \Sigma F_z \mathbf{k} = \mathbf{0}$$

$$\Sigma \mathbf{M}_O = \Sigma M_x \mathbf{i} + \Sigma M_y \mathbf{j} + \Sigma M_z \mathbf{k} = \mathbf{0}$$

Since the **i**, **j**, and **k** components are independent from one another, the above equations are satisfied provided

$$\Sigma F_x = 0 \quad \Sigma F_y = 0 \quad \Sigma F_z = 0 \tag{5–6a}$$

and

$$\Sigma M_x = 0 \quad \Sigma M_y = 0 \quad \Sigma M_z = 0 \tag{5–6b}$$

These *six scalar equilibrium equations* may be used to solve for at most six unknowns shown on the free-body diagram. Equations 5–6*a* require the sum of the external force components acting in the x, y, and z directions to be zero, and Eqs. 5–6*b* require the sum of the moment components about the x, y, and z axes to be zero.

5.7 Constraints and Statical Determinacy

To ensure the equilibrium of a rigid body, it is not only necessary to satisfy the equations of equilibrium, but the body must also be properly held or constrained by its supports. Some bodies may have more supports than are necessary for equilibrium, whereas others may not have enough or the supports may be arranged in a particular manner that could cause the body to move. Each of these cases will now be discussed.

Redundant Constraints. When a body has redundant supports, that is, more supports than are necessary to hold it in equilibrium, it becomes statically indeterminate. *Statically indeterminate* means that there will be more unknown loadings on the body than equations of equilibrium available for their solution. For example, the beam in Fig. 5–24*a* and the pipe assembly in Fig. 5–24*b*, shown together with their free-body diagrams, are both statically indeterminate because of additional (or redundant) support reactions. For the beam there are five unknowns, M_A, A_x, A_y, B_y, and C_y, for which only three equilibrium equations can be written ($\Sigma F_x = 0$, $\Sigma F_y = 0$, and $\Sigma M_O = 0$, Eq. 5–2). The pipe assembly has eight unknowns, for which only six equilibrium equations can be written, Eqs. 5–6.

The additional equations needed to solve statically indeterminate problems of the type shown in Fig. 5–24 are generally obtained from the deformation conditions at the points of support. These equations involve the physical properties of the body which are studied in subjects dealing with the mechanics of deformation, such as "mechanics of materials."*

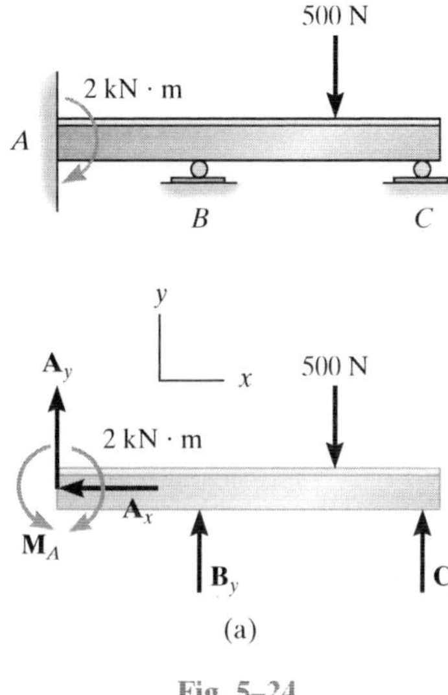

(a)

Fig. 5–24

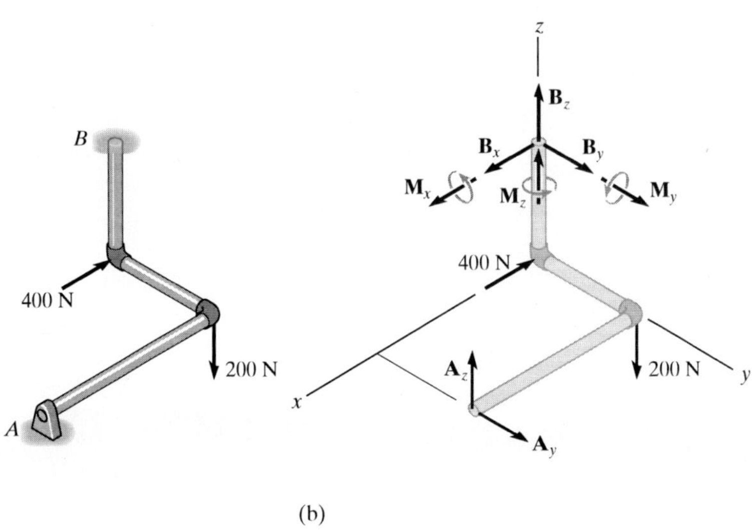

(b)

* See R. C. Hibbeler, *Mechanics of Materials*, 8th edition, Pearson Education/Prentice Hall, Inc.

5

Improper Constraints. Having the same number of unknown reactive forces as available equations of equilibrium does not always guarantee that a body will be stable when subjected to a particular loading. For example, the pin support at A and the roller support at B for the beam in Fig. 5–25*a* are placed in such a way that the lines of action of the reactive forces are *concurrent* at point A. Consequently, the applied loading **P** will cause the beam to rotate slightly about A, and so the beam is improperly constrained, $\Sigma M_A \neq 0$.

In three dimensions, a body will be improperly constrained if the lines of action of all the reactive forces intersect a common axis. For example, the reactive forces at the ball-and-socket supports at A and B in Fig. 5–25*b* all intersect the axis passing through A and B. Since the moments of these forces about A and B are all zero, then the loading **P** will rotate the member about the AB axis, $\Sigma M_{AB} \neq 0$.

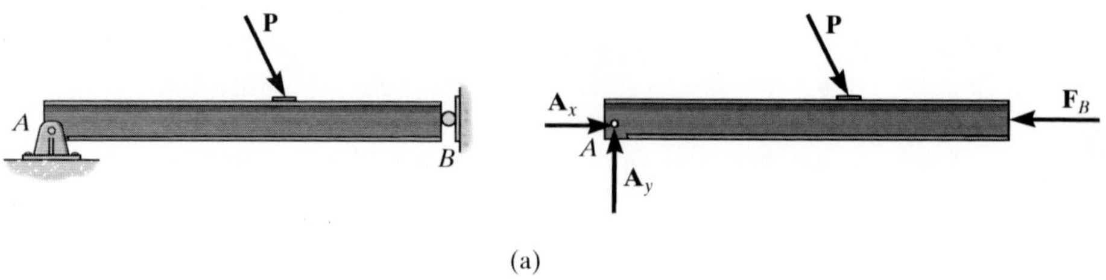

(a)

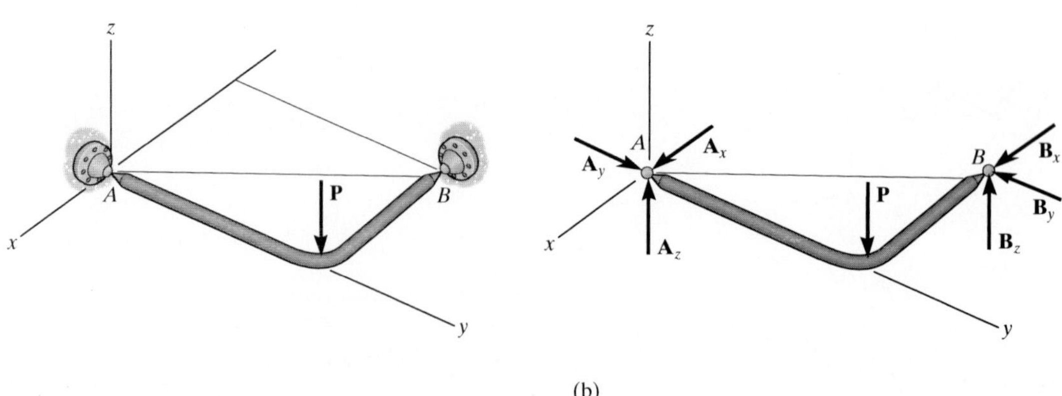

(b)

Fig. 5–25

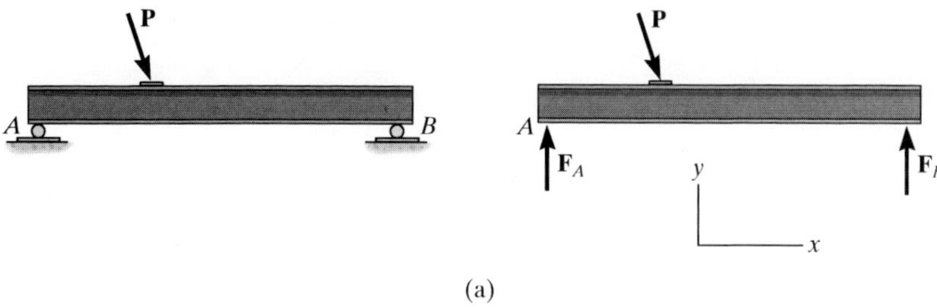

(a)

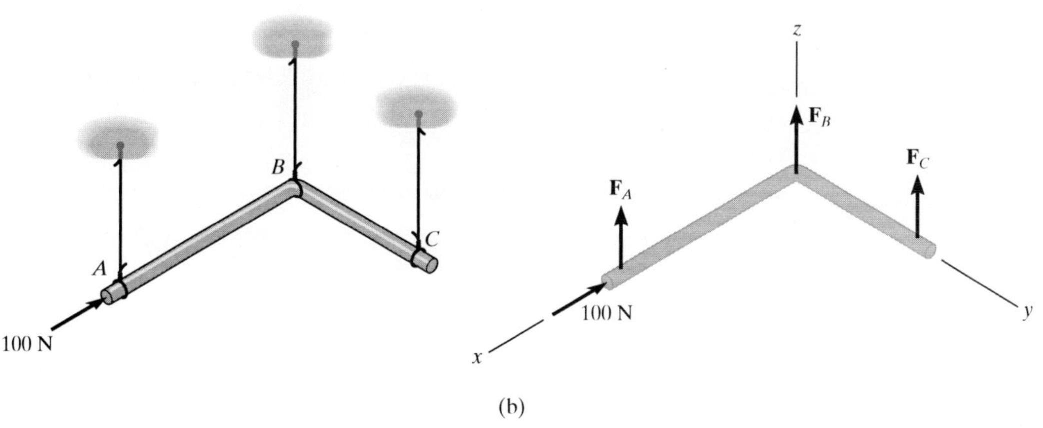

(b)

Fig. 5–26

Another way in which improper constraining leads to instability occurs when the *reactive forces* are all *parallel*. Two- and three-dimensional examples of this are shown in Fig. 5–26. In both cases, the summation of forces along the x axis will not equal zero.

In some cases, a body may have *fewer* reactive forces than equations of equilibrium that must be satisfied. The body then becomes only *partially constrained*. For example, consider member AB in Fig. 5–27*a* with its corresponding free-body diagram in Fig. 5–27*b*. Here $\Sigma F_y = 0$ will not be satisfied for the loading conditions and therefore equilibrium will not be maintained.

To summarize these points, a body is considered *improperly constrained* if all the reactive forces intersect at a common point or pass through a common axis, or if all the reactive forces are parallel. In engineering practice, these situations should be avoided at all times since they will cause an unstable condition.

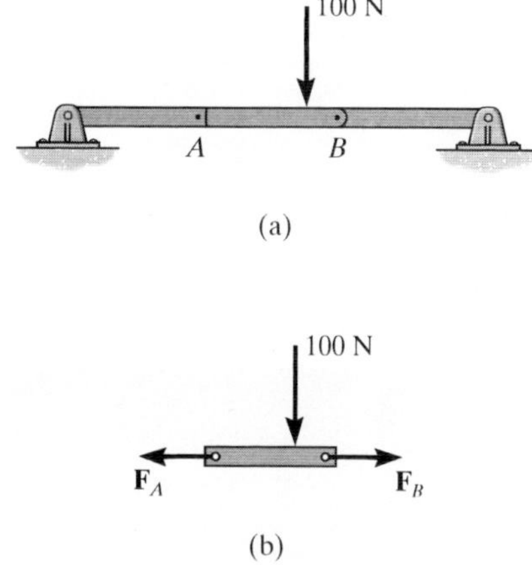

Fig. 5–27

Important Points

- Always draw the free-body diagram first when solving any equilibrium problem.
- If a support *prevents translation* of a body in a specific direction, then the support exerts a *force* on the body in that direction.
- If a support *prevents rotation about an axis*, then the support exerts a *couple moment* on the body about the axis.
- If a body is subjected to more unknown reactions than available equations of equilibrium, then the problem is *statically indeterminate*.
- A stable body requires that the lines of action of the reactive forces do not intersect a common axis and are not parallel to one another.

Procedure for Analysis

Three-dimensional equilibrium problems for a rigid body can be solved using the following procedure.

Free-Body Diagram.

- Draw an outlined shape of the body.
- Show all the forces and couple moments acting on the body.
- Establish the origin of the x, y, z axes at a convenient point and orient the axes so that they are parallel to as many of the external forces and moments as possible.
- Label all the loadings and specify their directions. In general, show all the *unknown* components having a *positive sense* along the x, y, z axes.
- Indicate the dimensions of the body necessary for computing the moments of forces.

Equations of Equilibrium.

- If the x, y, z force and moment components seem easy to determine, then apply the six scalar equations of equilibrium; otherwise use the vector equations.
- It is not necessary that the set of axes chosen for force summation coincide with the set of axes chosen for moment summation. Actually, an axis in any arbitrary direction may be chosen for summing forces and moments.
- Choose the direction of an axis for moment summation such that it intersects the lines of action of as many unknown forces as possible. Realize that the moments of forces passing through points on this axis and the moments of forces which are parallel to the axis will then be zero.
- If the solution of the equilibrium equations yields a negative scalar for a force or couple moment magnitude, it indicates that the sense is opposite to that assumed on the free-body diagram.

EXAMPLE 5.15

The homogeneous plate shown in Fig. 5–28*a* has a mass of 100 kg and is subjected to a force and couple moment along its edges. If it is supported in the horizontal plane by a roller at *A*, a ball-and-socket joint at *B*, and a cord at *C*, determine the components of reaction at these supports.

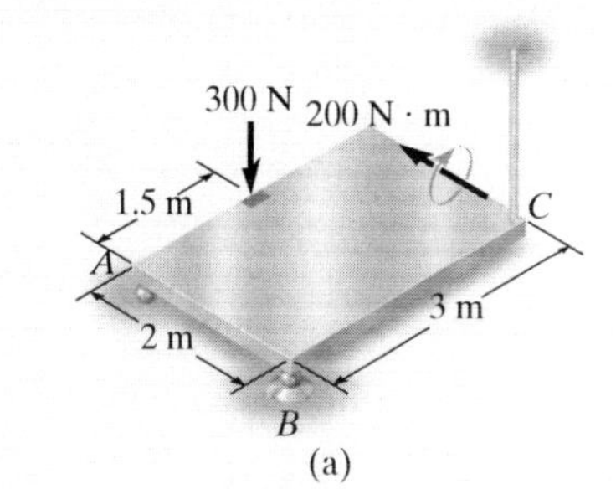

(a)

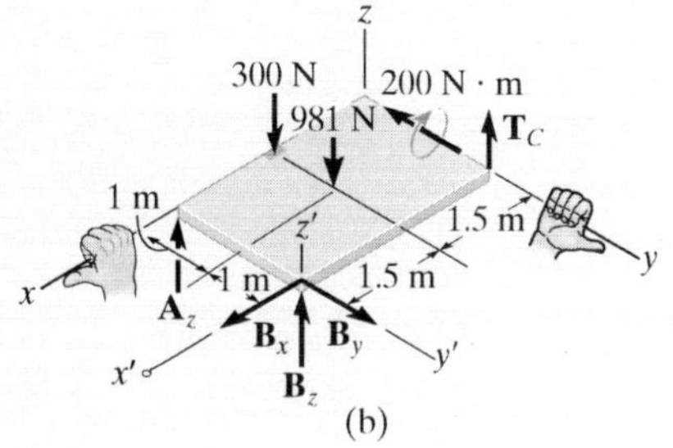

(b)

Fig. 5–28

SOLUTION (*SCALAR ANALYSIS*)

Free-Body Diagram. There are five unknown reactions acting on the plate, as shown in Fig. 5–28*b*. Each of these reactions is assumed to act in a positive coordinate direction.

Equations of Equilibrium. Since the three-dimensional geometry is rather simple, a *scalar analysis* provides a *direct solution* to this problem. A force summation along each axis yields

$\Sigma F_x = 0; \quad B_x = 0$ *Ans.*

$\Sigma F_y = 0; \quad B_y = 0$ *Ans.*

$\Sigma F_z = 0; \quad A_z + B_z + T_C - 300\text{ N} - 981\text{ N} = 0 \qquad (1)$

Recall that the moment of a force about an axis is equal to the product of the force magnitude and the perpendicular distance (moment arm) from the line of action of the force to the axis. Also, forces that are parallel to an axis or pass through it create no moment about the axis. Hence, summing moments about the positive *x* and *y* axes, we have

$\Sigma M_x = 0; \quad T_C(2\text{ m}) - 981\text{ N}(1\text{ m}) + B_z(2\text{ m}) = 0 \qquad (2)$

$\Sigma M_y = 0; \quad 300\text{ N}(1.5\text{ m}) + 981\text{ N}(1.5\text{ m}) - B_z(3\text{ m}) - A_z(3\text{ m}) - 200\text{ N}\cdot\text{m} = 0 \qquad (3)$

The components of the force at *B* can be eliminated if moments are summed about the x' and y' axes. We obtain

$\Sigma M_{x'} = 0; \quad 981\text{ N}(1\text{ m}) + 300\text{ N}(2\text{ m}) - A_z(2\text{ m}) = 0 \qquad (4)$

$\Sigma M_{y'} = 0; \quad -300\text{ N}(1.5\text{ m}) - 981\text{ N}(1.5\text{ m}) - 200\text{ N}\cdot\text{m} + T_C(3\text{ m}) = 0 \qquad (5)$

Solving Eqs. 1 through 3 or the more convenient Eqs. 1, 4, and 5 yields

$A_z = 790\text{ N} \quad B_z = -217\text{ N} \quad T_C = 707\text{ N}$ *Ans.*

The negative sign indicates that $\mathbf{B}_z$ acts downward.

NOTE: The solution of this problem does not require a summation of moments about the *z* axis. The plate is partially constrained since the supports cannot prevent it from turning about the *z* axis if a force is applied to it in the *x*–*y* plane.

EXAMPLE 5.16

Determine the components of reaction that the ball-and-socket joint at A, the smooth journal bearing at B, and the roller support at C exert on the rod assembly in Fig. 5–29*a*.

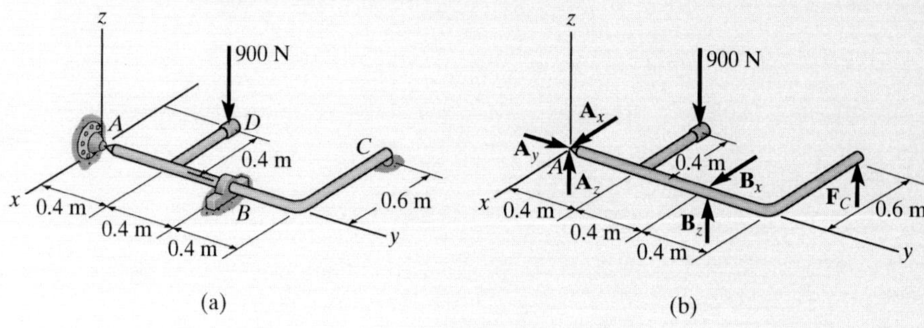

Fig. 5–29

SOLUTION

Free-Body Diagram. As shown on the free-body diagram, Fig. 5–29*b*, the reactive forces of the supports will prevent the assembly from rotating about each coordinate axis, and so the journal bearing at B only exerts reactive forces on the member. No couple moments are required.

Equations of Equilibrium. A direct solution for A_y can be obtained by summing forces along the y axis.

$\Sigma F_y = 0; \qquad A_y = 0$ *Ans.*

The force F_C can be determined directly by summing moments about the y axis.

$\Sigma M_y = 0; \qquad F_C(0.6 \text{ m}) - 900 \text{ N}(0.4 \text{ m}) = 0$

$F_C = 600 \text{ N}$ *Ans.*

Using this result, B_z can be determined by summing moments about the x axis.

$\Sigma M_x = 0; \qquad B_z(0.8 \text{ m}) + 600 \text{ N}(1.2 \text{ m}) - 900 \text{ N}(0.4 \text{ m}) = 0$

$B_z = -450 \text{ N}$ *Ans.*

The negative sign indicates that $\mathbf{B}_z$ acts downward. The force B_x can be found by summing moments about the z axis.

$\Sigma M_z = 0; \qquad -B_x(0.8 \text{ m}) = 0 \quad B_x = 0$ *Ans.*

Thus,

$\Sigma F_x = 0; \qquad A_x + 0 = 0 \quad A_x = 0$ *Ans.*

Finally, using the results of B_z and F_C.

$\Sigma F_z = 0; \qquad A_z + (-450 \text{ N}) + 600 \text{ N} - 900 \text{ N} = 0$

$A_z = 750 \text{ N}$ *Ans.*

EXAMPLE 5.17

The boom is used to support the 375-N (≈37.5-kg) flowerpot in Fig. 5–30*a*. Determine the tension developed in wires *AB* and *AC*.

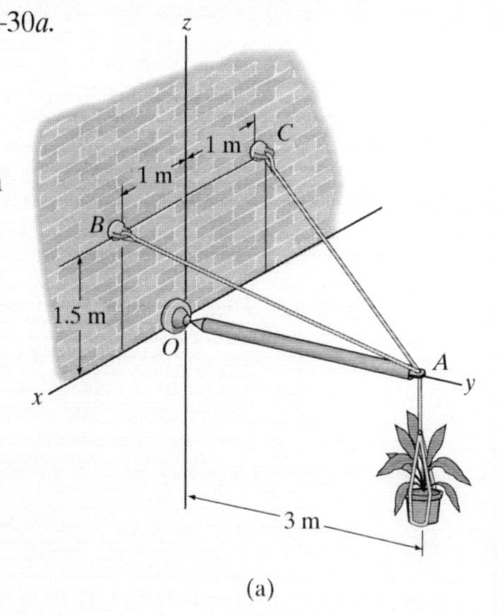

(a)

Fig. 5–30

SOLUTION

Free-Body Diagram. The free-body diagram of the boom is shown in Fig. 5–30*b*.

Equations of Equilibrium. We will use a vector analysis.

$$\mathbf{F}_{AB} = F_{AB}\left(\frac{\mathbf{r}_{AB}}{r_{AB}}\right) = F_{AB}\left(\frac{\{1\mathbf{i} - 3\mathbf{j} + 1.5\mathbf{k}\}\text{ m}}{\sqrt{(1\text{ m})^2 + (-3\text{ m})^2 + (1.5\text{ m})^2}}\right)$$

$$= \tfrac{2}{7}F_{AB}\mathbf{i} - \tfrac{6}{7}F_{AB}\mathbf{j} + \tfrac{3}{7}F_{AB}\mathbf{k}$$

$$\mathbf{F}_{AC} = F_{AC}\left(\frac{\mathbf{r}_{AC}}{r_{AC}}\right) = F_{AC}\left(\frac{\{-1\mathbf{i} - 3\mathbf{j} + 1.5\mathbf{k}\}\text{ m}}{\sqrt{(-1\text{ m})^2 + (-3\text{ m})^2 + (1.5\text{ m})^2}}\right)$$

$$= -\tfrac{2}{7}F_{AC}\mathbf{i} - \tfrac{6}{7}F_{AC}\mathbf{j} + \tfrac{3}{7}F_{AC}\mathbf{k}$$

We can eliminate the force reaction at *O* by writing the moment equation of equilibrium about point *O*.

$$\Sigma\mathbf{M}_O = \mathbf{0}; \qquad \mathbf{r}_A \times (\mathbf{F}_{AB} + \mathbf{F}_{AC} + \mathbf{W}) = \mathbf{0}$$

$$(6\mathbf{j}) \times \left[\left(\tfrac{2}{7}F_{AB}\mathbf{i} - \tfrac{6}{7}F_{AB}\mathbf{j} + \tfrac{3}{7}F_{AB}\mathbf{k}\right) + \left(-\tfrac{2}{7}F_{AC}\mathbf{i} - \tfrac{6}{7}F_{AC}\mathbf{j} + \tfrac{3}{7}F_{AC}\mathbf{k}\right) + (-375\mathbf{k})\right] = \mathbf{0}$$

$$\left(\tfrac{18}{7}F_{AB} + \tfrac{18}{7}F_{AC} - 2250\right)\mathbf{i} + \left(-\tfrac{12}{7}F_{AB} + \tfrac{12}{7}F_{AC}\right)\mathbf{k} = \mathbf{0}$$

$$\Sigma M_x = 0; \qquad \tfrac{18}{7}F_{AB} + \tfrac{18}{7}F_{AC} - 2250 = 0 \qquad (1)$$

$$\Sigma M_y = 0; \qquad 0 = 0$$

$$\Sigma M_z = 0; \qquad -\tfrac{12}{7}F_{AB} + \tfrac{12}{7}F_{AC} = 0 \qquad (2)$$

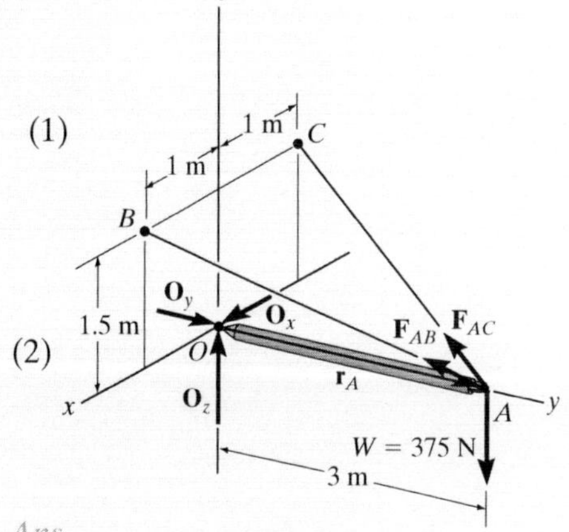

(b)

Solving Eqs. (1) and (2) simultaneously,

$$F_{AB} = F_{AC} = 437.5\text{ N} \qquad \textit{Ans.}$$

EXAMPLE 5.18

Rod AB shown in Fig. 5–31a is subjected to the 200-N force. Determine the reactions at the ball-and-socket joint A and the tension in the cables BD and BE. The collar at C is fixed to the rod.

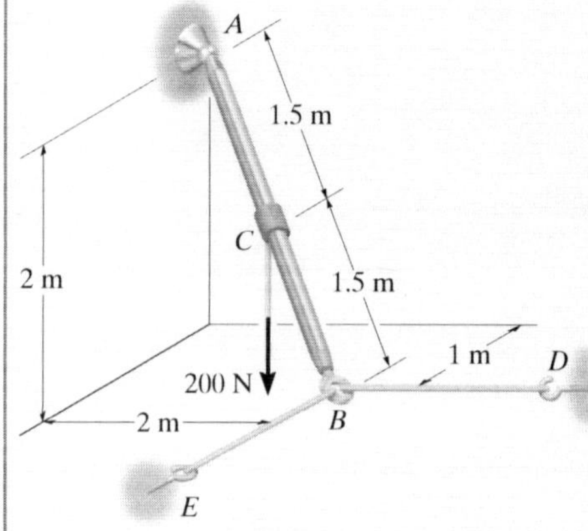

(a)

(b)

Fig. 5–31

SOLUTION (*VECTOR ANALYSIS*)

Free-Body Diagram. Fig. 5–31b.

Equations of Equilibrium. Representing each force on the free-body diagram in Cartesian vector form, we have

$$\mathbf{F}_A = A_x\mathbf{i} + A_y\mathbf{j} + A_z\mathbf{k}$$
$$\mathbf{T}_E = T_E\mathbf{i}$$
$$\mathbf{T}_D = T_D\mathbf{j}$$
$$\mathbf{F} = \{-200\mathbf{k}\}\text{ N}$$

Applying the force equation of equilibrium.

$$\Sigma\mathbf{F} = \mathbf{0}; \qquad \mathbf{F}_A + \mathbf{T}_E + \mathbf{T}_D + \mathbf{F} = \mathbf{0}$$

$$(A_x + T_E)\mathbf{i} + (A_y + T_D)\mathbf{j} + (A_z - 200)\mathbf{k} = \mathbf{0}$$

$$\Sigma F_x = 0; \qquad A_x + T_E = 0 \qquad (1)$$
$$\Sigma F_y = 0; \qquad A_y + T_D = 0 \qquad (2)$$
$$\Sigma F_z = 0; \qquad A_z - 200 = 0 \qquad (3)$$

Summing moments about point A yields

$$\Sigma\mathbf{M}_A = \mathbf{0}; \qquad \mathbf{r}_C \times \mathbf{F} + \mathbf{r}_B \times (\mathbf{T}_E + \mathbf{T}_D) = \mathbf{0}$$

Since $\mathbf{r}_C = \frac{1}{2}\mathbf{r}_B$, then

$$(0.5\mathbf{i} + 1\mathbf{j} - 1\mathbf{k}) \times (-200\mathbf{k}) + (1\mathbf{i} + 2\mathbf{j} - 2\mathbf{k}) \times (T_E\mathbf{i} + T_D\mathbf{j}) = \mathbf{0}$$

Expanding and rearranging terms gives

$$(2T_D - 200)\mathbf{i} + (-2T_E + 100)\mathbf{j} + (T_D - 2T_E)\mathbf{k} = \mathbf{0}$$

$$\Sigma M_x = 0; \qquad 2T_D - 200 = 0 \qquad (4)$$
$$\Sigma M_y = 0; \qquad -2T_E + 100 = 0 \qquad (5)$$
$$\Sigma M_z = 0; \qquad T_D - 2T_E = 0 \qquad (6)$$

Solving Eqs. 1 through 5, we get

$$T_D = 100\text{ N} \qquad \textit{Ans.}$$
$$T_E = 50\text{ N} \qquad \textit{Ans.}$$
$$A_x = -50\text{ N} \qquad \textit{Ans.}$$
$$A_y = -100\text{ N} \qquad \textit{Ans.}$$
$$A_z = 200\text{ N} \qquad \textit{Ans.}$$

NOTE: The negative sign indicates that $\mathbf{A}_x$ and $\mathbf{A}_y$ have a sense which is opposite to that shown on the free-body diagram, Fig. 5–31b.

EXAMPLE 5.19

The bent rod in Fig. 5–32*a* is supported at *A* by a journal bearing, at *D* by a ball-and-socket joint, and at *B* by means of cable *BC*. Using only *one equilibrium equation*, obtain a direct solution for the tension in cable *BC*. The bearing at *A* is capable of exerting force components only in the *z* and *y* directions since it is properly aligned on the shaft. In other words, no couple moments are required at this support.

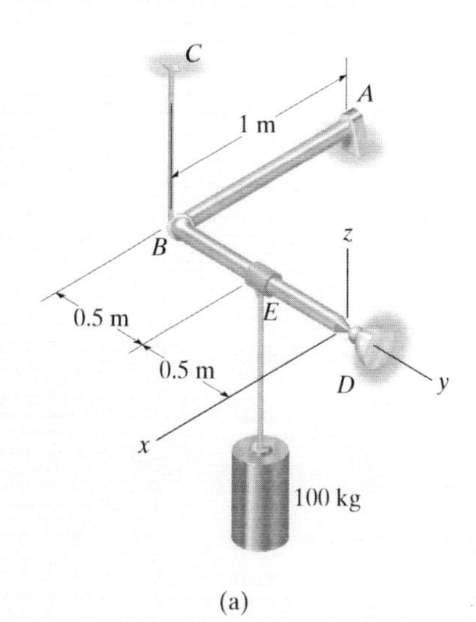

(a)

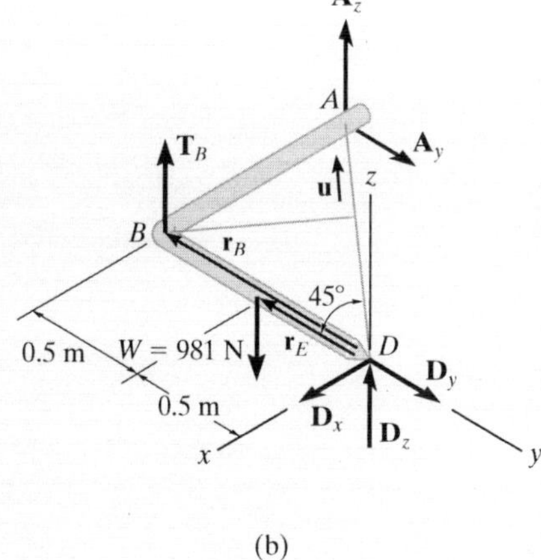

(b)

Fig. 5–32

SOLUTION (*VECTOR ANALYSIS*)

Free-Body Diagram. As shown in Fig. 5–32*b*, there are six unknowns.

Equations of Equilibrium. The cable tension $\mathbf{T}_B$ may be obtained *directly* by summing moments about an axis that passes through points *D* and *A*. Why? The direction of this axis is defined by the unit vector **u**, where

$$\mathbf{u} = \frac{\mathbf{r}_{DA}}{r_{DA}} = -\frac{1}{\sqrt{2}}\mathbf{i} - \frac{1}{\sqrt{2}}\mathbf{j}$$
$$= -0.7071\mathbf{i} - 0.7071\mathbf{j}$$

Hence, the sum of the moments about this axis is zero provided

$$\Sigma M_{DA} = \mathbf{u} \cdot \Sigma(\mathbf{r} \times \mathbf{F}) = 0$$

Here **r** represents a position vector drawn from *any point* on the axis *DA* to any point on the line of action of force **F** (see Eq. 4–11). With reference to Fig. 5–32*b*, we can therefore write

$$\mathbf{u} \cdot (\mathbf{r}_B \times \mathbf{T}_B + \mathbf{r}_E \times \mathbf{W}) = \mathbf{0}$$
$$(-0.7071\mathbf{i} - 0.7071\mathbf{j}) \cdot \big[(-1\mathbf{j}) \times (T_B\mathbf{k}) + (-0.5\mathbf{j}) \times (-981\mathbf{k})\big] = \mathbf{0}$$
$$(-0.7071\mathbf{i} - 0.7071\mathbf{j}) \cdot [(-T_B + 490.5)\mathbf{i}] = \mathbf{0}$$
$$-0.7071(-T_B + 490.5) + 0 + 0 = 0$$
$$T_B = 490.5 \text{ N} \qquad \textit{Ans.}$$

Since the moment arms from the axis to $\mathbf{T}_B$ and **W** are easy to obtain, we can also determine this result using a scalar analysis. As shown in Fig. 5–32*b*,

$$\Sigma M_{DA} = 0; \quad T_B(1 \text{ m} \sin 45^\circ) - 981 \text{ N}(0.5 \text{ m} \sin 45^\circ) = 0$$
$$T_B = 490.5 \text{ N} \qquad \textit{Ans.}$$

FUNDAMENTAL PROBLEMS

All problem solutions must include an FBD.

F5–7. The uniform plate has a weight of 50 kN. Determine the tension in each of the supporting cables.

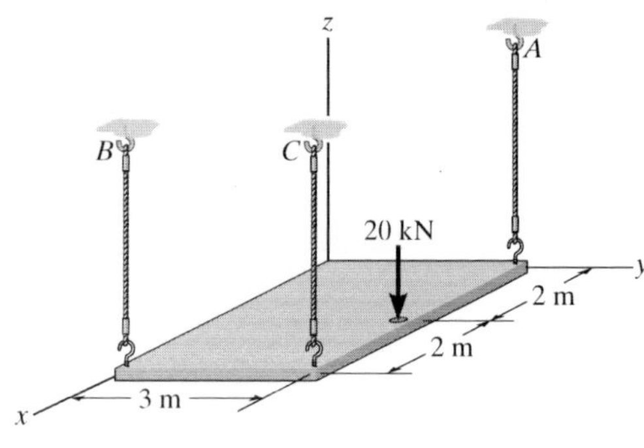

F5–7

5

F5–8. Determine the reactions at the roller support A, the ball-and-socket joint D, and the tension in cable BC for the plate.

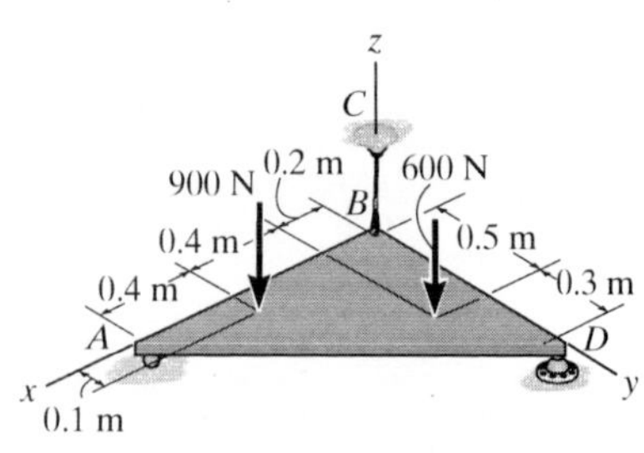

F5–8

F5–9. The rod is supported by smooth journal bearings at A, B, and C and is subjected to the two forces. Determine the reactions at these supports.

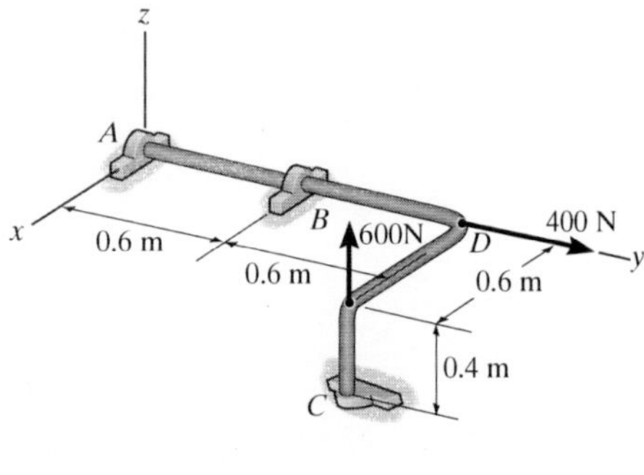

F5–9

F5–10. Determine the support reactions at the smooth journal bearings A, B, and C of the pipe assembly.

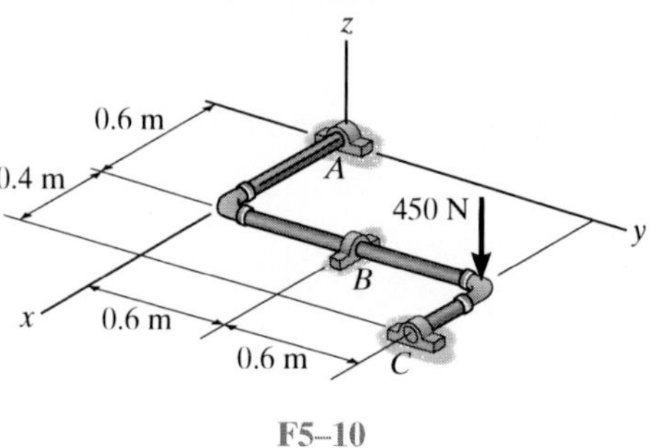

F5–10

F5–11. Determine the force developed in the short link BD, and the tension in the cords CE and CF, and the reactions of the ball-and-socket joint A on the block.

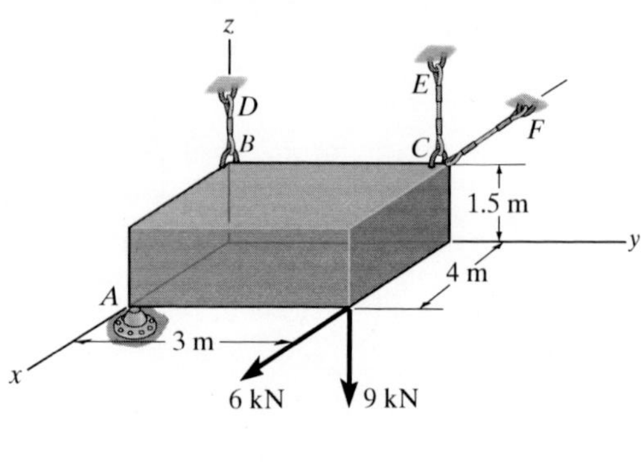

F5–11

F5–12. Determine the components of reaction that the thrust bearing A and cable BC exert on the bar.

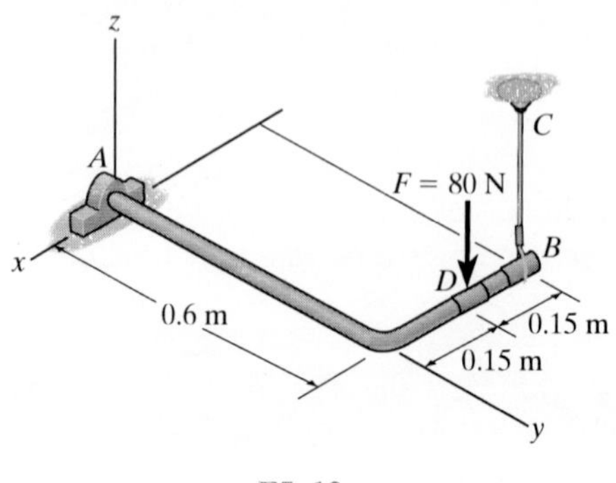

F5–12

PROBLEMS

All problem solutions must include an FBD.

5–62. The uniform load has a mass of 600 kg and is lifted using a uniform 30-kg strongback beam BAC and four ropes as shown. Determine the tension in each rope and the force that must be applied at A.

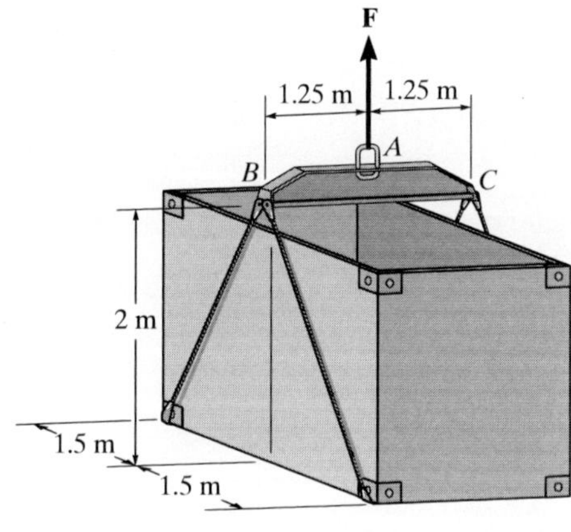

Prob. 5–62

***5–64.** The wing of the jet aircraft is subjected to a thrust of $T = 8$ kN from its engine and the resultant lift force $L = 45$ kN. If the mass of the wing is 2.1 Mg and the mass center is at G, determine the x, y, z components of reaction where the wing is fixed to the fuselage A.

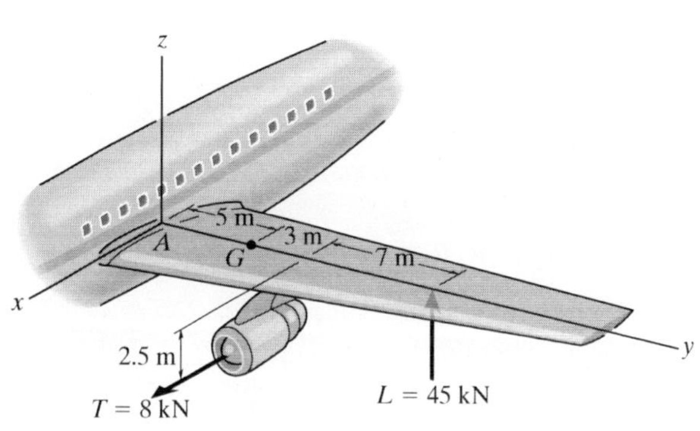

Prob. 5–64

5–63. Due to an unequal distribution of fuel in the wing tanks, the centers of gravity for the airplane fuselage A and wings B and C are located as shown. If these components have weights $W_A = 225$ kN, $W_B = 40$ kN, and $W_C = 30$ kN, determine the normal reactions of the wheels D, E, and F on the ground.

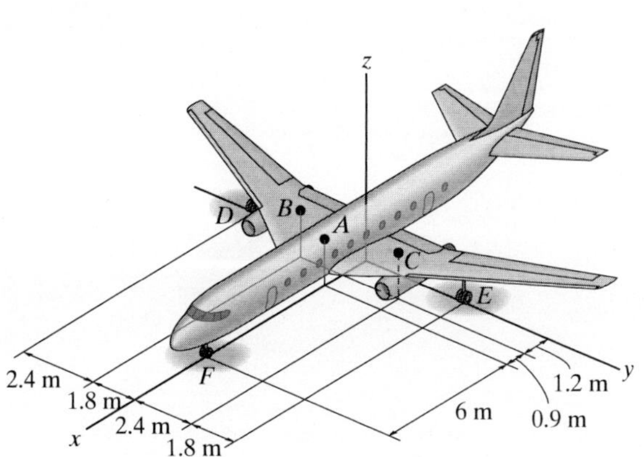

Prob. 5–63

5–65. The cart supports the uniform crate having a mass of 85 kg. Determine the vertical reactions on the three casters at A, B, and C. The caster at B is not shown. Neglect the mass of the cart and assume the casters at A and B are at the rear corner of the cart.

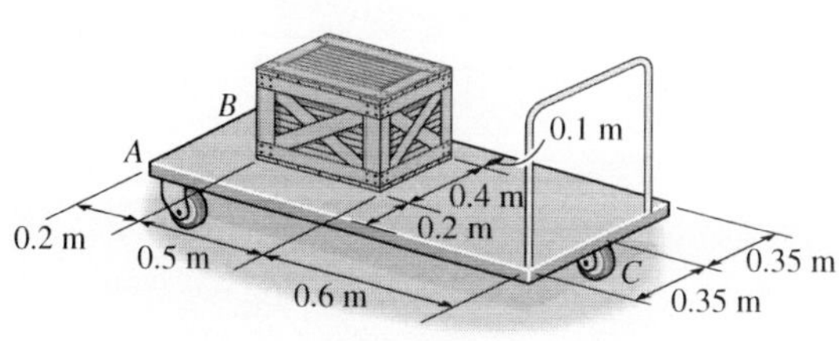

Prob. 5–65

5–66. The pipe assembly supports the vertical loads shown. Determine the components of reaction at the ball-and-socket joint A and the tension in the supporting cables BC and BD.

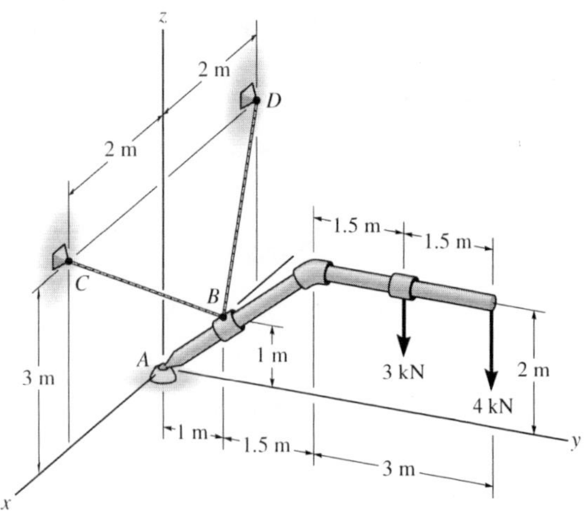

Prob. 5–66

5–67. The boom is supported by a ball-and-socket joint at A and a guy wire at B. If the 5-kN loads lie in a plane which is parallel to the x–y plane, determine the x, y, z components of reaction at A and the tension in the cable at B.

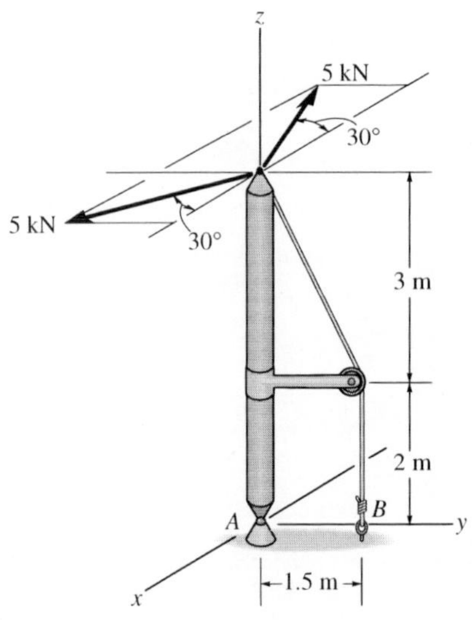

Prob. 5–67

***5–68.** Determine the force components acting on the ball-and-socket at A, the reaction at the roller B and the tension on the cord CD needed for equilibrium of the quarter circular plate.

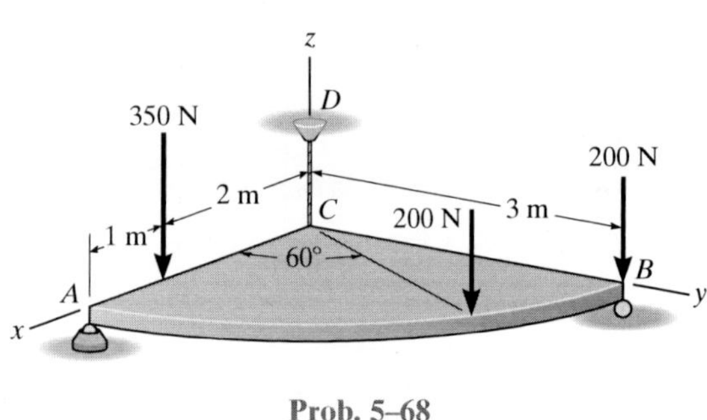

Prob. 5–68

5–69. Determine the components of reaction acting at the smooth journal bearings A, B, and C.

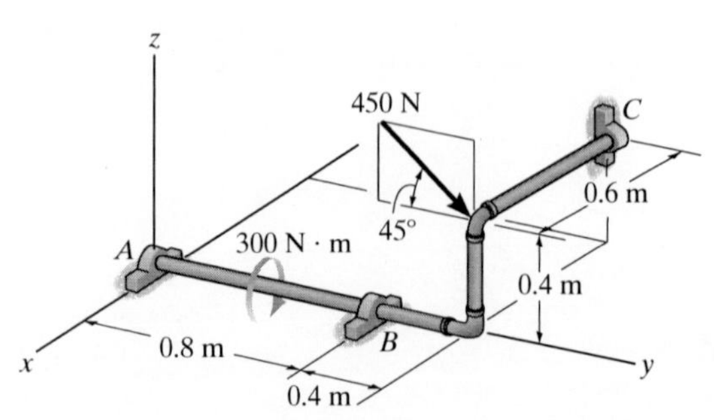

Prob. 5–69

5–70. The circular plate has a weight W and center of gravity at its center. If it is supported by three vertical cords tied to its edge, determine the largest distance d from the center to where any vertical force $\mathbf{P}$ can be applied so as not to cause the force in any one of the cables to become zero.

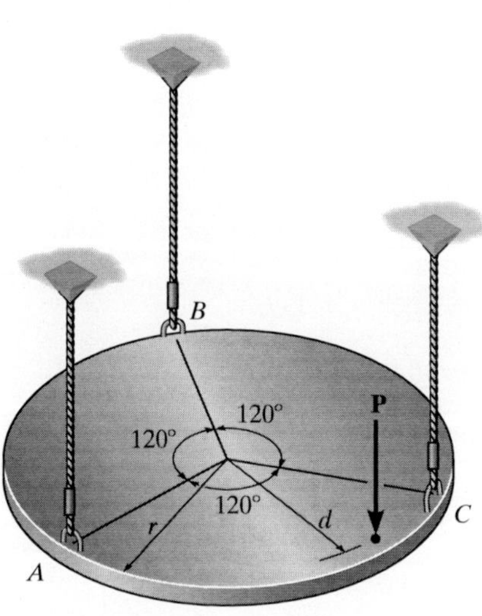

Prob. 5–70

***5–72.** The pole is subjected to the two forces shown. Determine the components of reaction of A assuming it to be a ball-and-socket joint. Also, compute the tension in each of the guy wires, BC and ED.

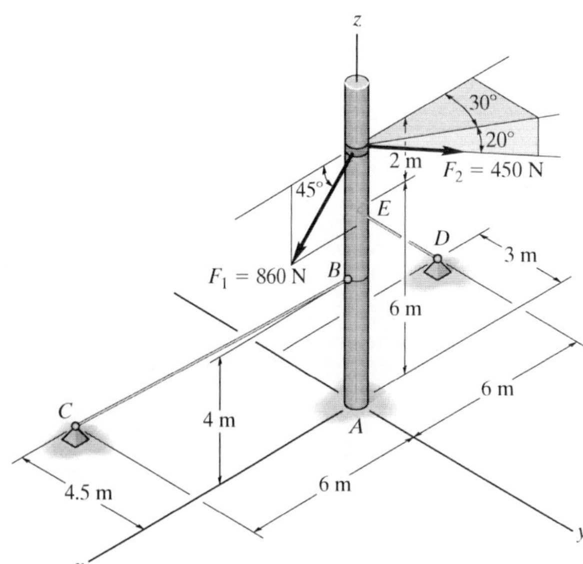

Prob. 5–72

5–73. If $P = 6$ kN, $x = 0.75$ m, and $y = 1$ m, determine the tension developed in cables AB, CD, and EF. Neglect the weight of the plate.

5–74. Determine the location x and y of the point of application of force $\mathbf{P}$ so that the tension developed in cables AB, CD, and EF is the same. Neglect the weight of the plate.

5–71. Determine the support reactions at the smooth collar A and the normal reaction at the roller support B.

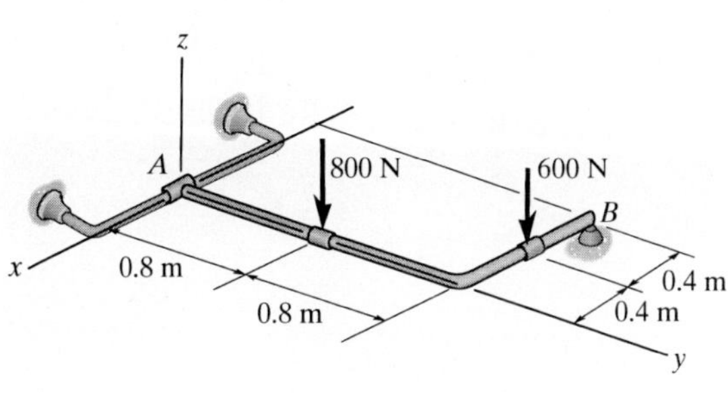

Prob. 5–71

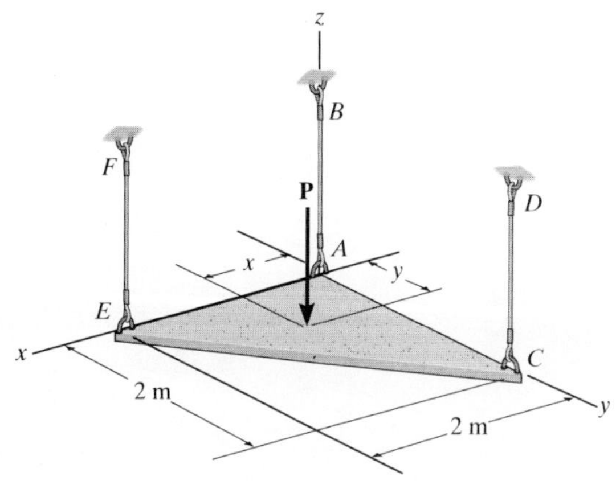

Probs. 5–73/74

5–75. If the pulleys are fixed to the shaft, determine the magnitude of tension **T** and the x, y, z components of reaction at the smooth thrust bearing A and smooth journal bearing B.

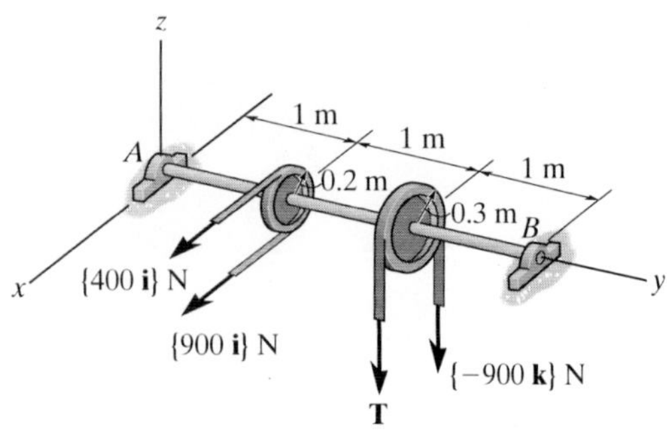

Prob. 5–75

5–77. A uniform square table having a weight W and sides a is supported by three vertical legs. Determine the smallest vertical force **P** that can be applied to its top that will cause it to tip over.

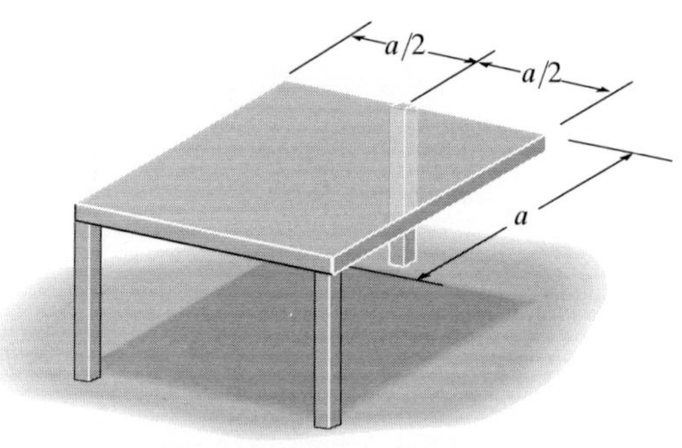

Prob. 5–77

***5–76.** The 2-kg ball rests between the 45° grooves A and B of the 10° incline and against a vertical wall at C. If all three surfaces of contact are smooth, determine the reactions of the surfaces on the ball. *Hint:* Use the x, y, z axes, with origin at the center of the ball, and the z axis inclined as shown.

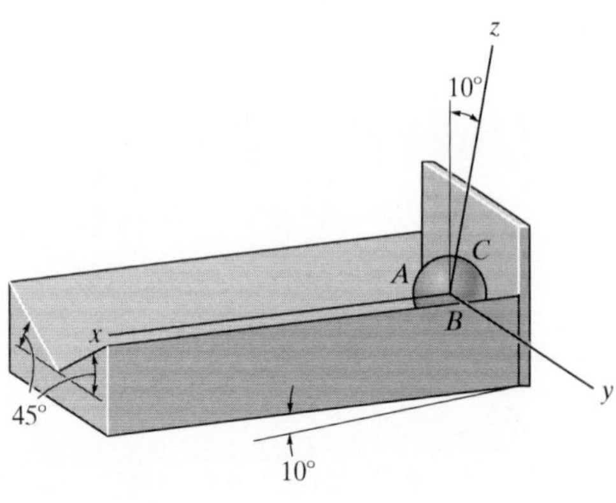

Prob. 5–76

5–78. The shaft is supported by three smooth journal bearings at A, B, and C. Determine the components of reaction at these bearings.

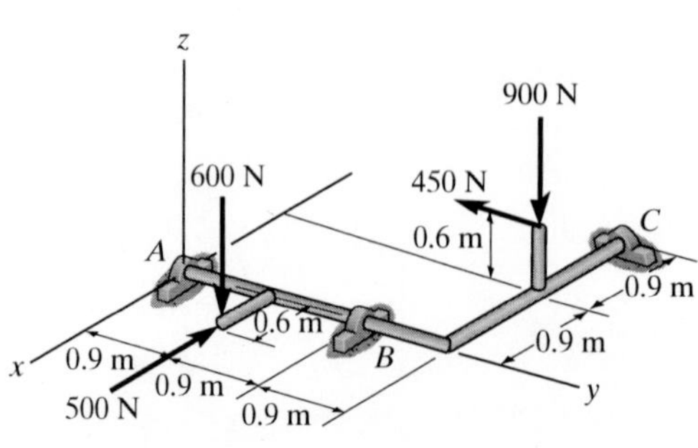

Prob. 5–78

5–79. The platform has a mass of 3 Mg and center of mass located at G. If it is lifted with constant velocity using the three cables, determine the force in each of the cables.

***5–80.** The platform has a mass of 2 Mg and center of mass located at G. If it is lifted using the three cables, determine the force in each of the cables. Solve for each force by using a single moment equation of equilibrium.

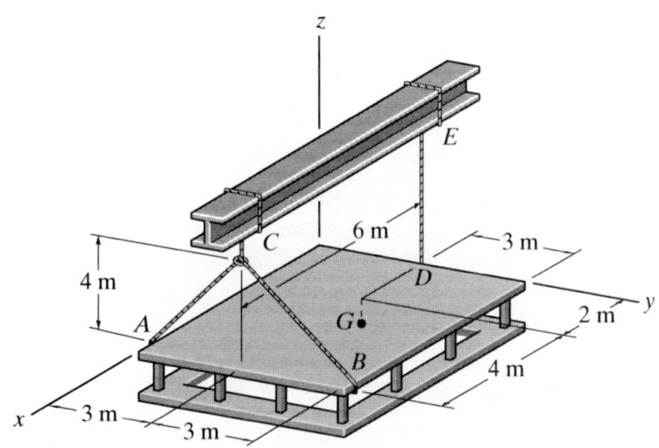

Probs. 5–79/80

5–81. The sign has a mass of 100 kg with center of mass at G. Determine the x, y, z components of reaction at the ball-and-socket joint A and the tension in wires BC and BD.

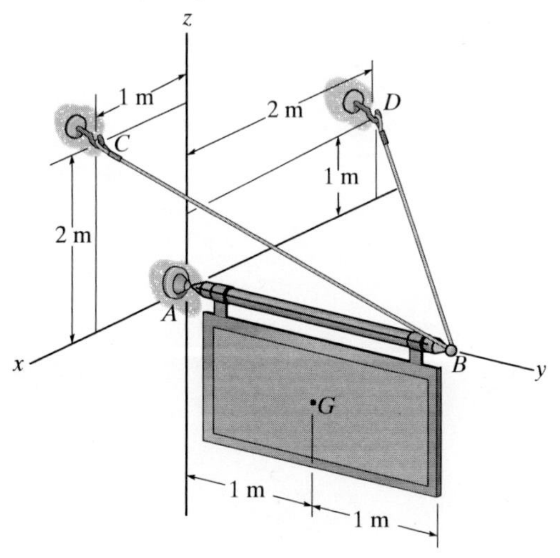

Prob. 5–81

5–82. Determine the tension in cables BD and CD and the x, y, z components of reaction at the ball-and-socket joint at A.

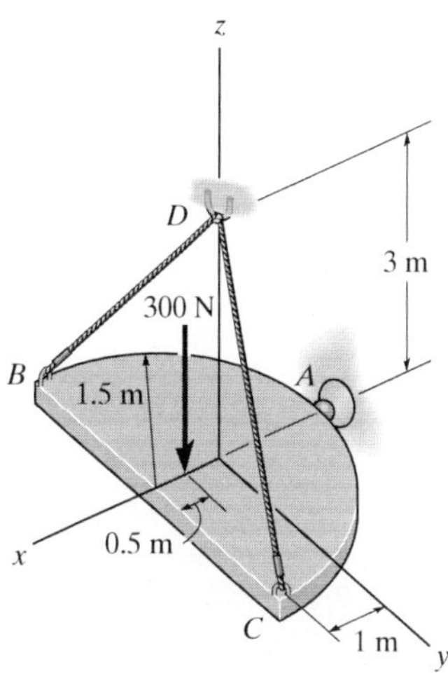

Prob. 5–82

5–83. Both pulleys are fixed to the shaft and as the shaft turns with constant angular velocity, the power of pulley A is transmitted to pulley B. Determine the horizontal tension $\mathbf{T}$ in the belt on pulley B and the x, y, z components of reaction at the journal bearing C and thrust bearing D if $\theta = 0°$. The bearings are in proper alignment and exert only force reactions on the shaft.

***5–84.** Both pulleys are fixed to the shaft and as the shaft turns with constant angular velocity, the power of pulley A is transmitted to pulley B. Determine the horizontal tension $\mathbf{T}$ in the belt on pulley B and the x, y, z components of reaction at the journal bearing C and thrust bearing D if $\theta = 45°$. The bearings are in proper alignment and exert only force reactions on the shaft.

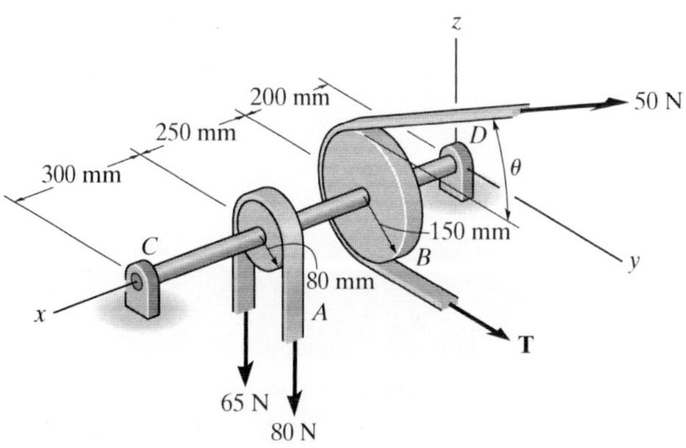

Probs. 5–83/84

5

CHAPTER REVIEW

Equilibrium

A body in equilibrium is at rest or can translate with constant velocity.

$$\Sigma \mathbf{F} = \mathbf{0}$$
$$\Sigma \mathbf{M} = \mathbf{0}$$

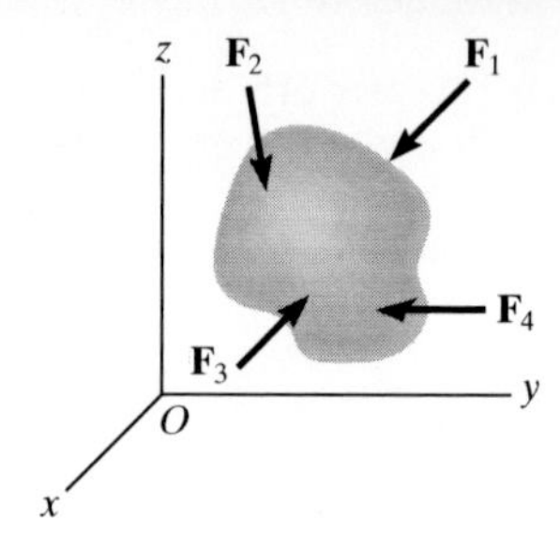

Two Dimensions

Before analyzing the equilibrium of a body, it is first necessary to draw its free-body diagram. This is an outlined shape of the body, which shows all the forces and couple moments that act on it.

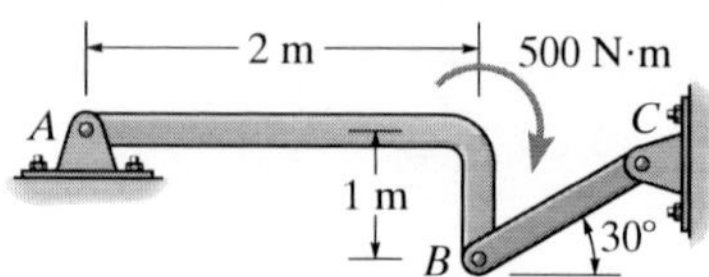

Couple moments can be placed anywhere on a free-body diagram since they are free vectors. Forces can act at any point along their line of action since they are sliding vectors.

Angles used to resolve forces, and dimensions used to take moments of the forces, should also be shown on the free-body diagram.

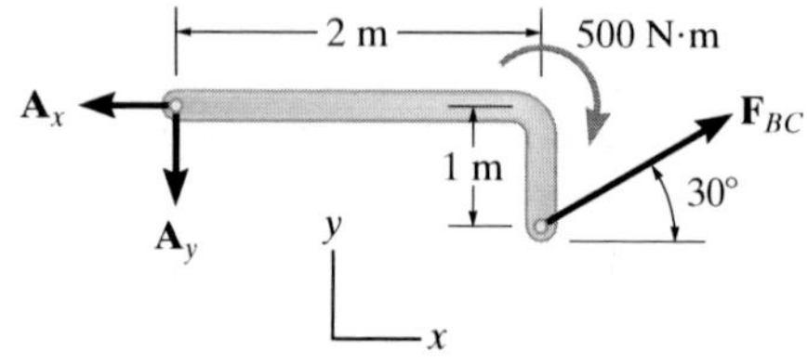

Some common types of supports and their reactions are shown below in two dimensions.

Remember that a support will exert a force on the body in a particular direction if it prevents translation of the body in that direction, and it will exert a couple moment on the body if it prevents rotation.

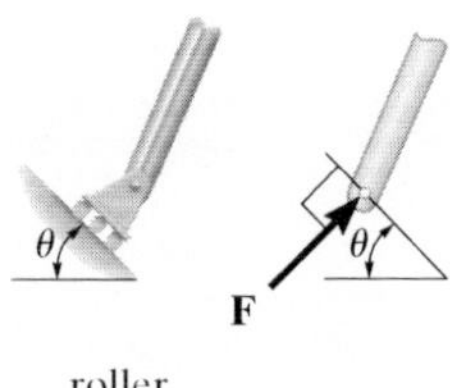

roller

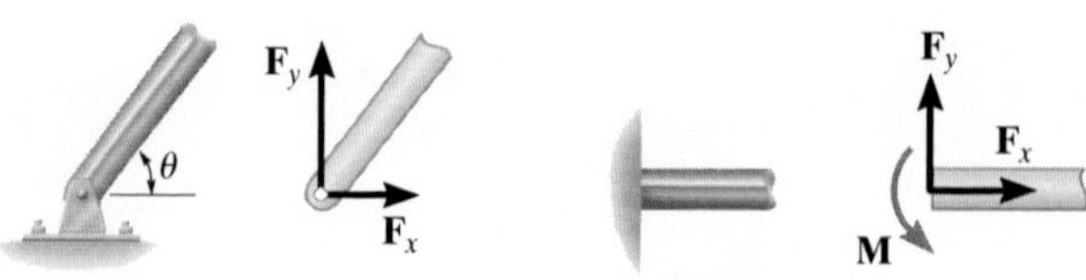

smooth pin or hinge

fixed support

The three scalar equations of equilibrium can be applied when solving problems in two dimensions, since the geometry is easy to visualize.

$$\Sigma F_x = 0$$
$$\Sigma F_y = 0$$
$$\Sigma M_O = 0$$

For the most direct solution, try to sum forces along an axis that will eliminate as many unknown forces as possible. Sum moments about a point A that passes through the line of action of as many unknown forces as possible.

$\Sigma F_x = 0;$

$$A_x - P_2 = 0 \quad A_x = P_2$$

$\Sigma M_A = 0;$

$$P_2 d_2 + B_y d_B - P_1 d_1 = 0$$

$$B_y = \frac{P_1 d_1 - P_2 d_2}{d_B}$$

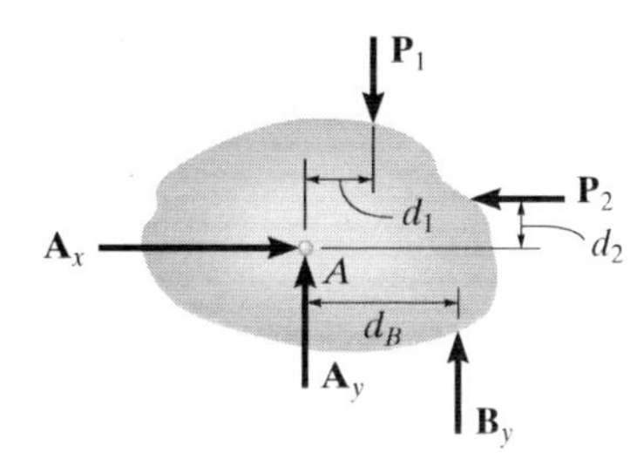

Three Dimensions

Some common types of supports and their reactions are shown here in three dimensions.

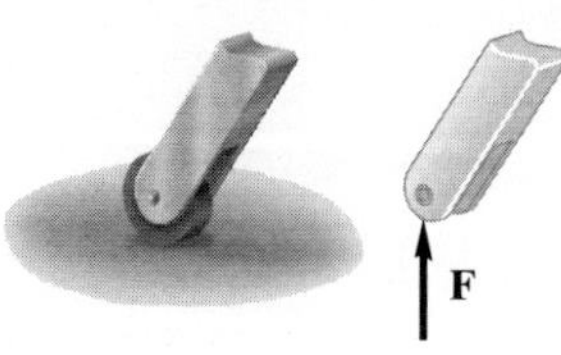

roller

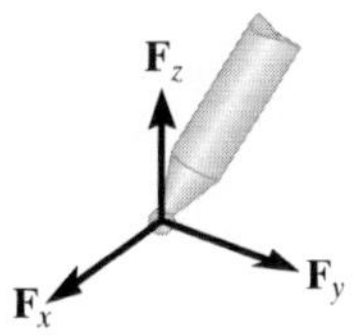

ball and socket

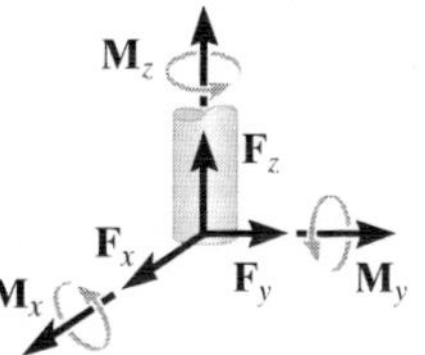

fixed support

In three dimensions, it is often advantageous to use a Cartesian vector analysis when applying the equations of equilibrium. To do this, first express each known and unknown force and couple moment shown on the free-body diagram as a Cartesian vector. Then set the force summation equal to zero. Take moments about a point O that lies on the line of action of as many unknown force components as possible. From point O direct position vectors to each force, and then use the cross product to determine the moment of each force.

$$\Sigma \mathbf{F} = \mathbf{0}$$
$$\Sigma \mathbf{M}_O = \mathbf{0}$$

The six scalar equations of equilibrium are established by setting the respective **i**, **j**, and **k** components of these force and moment summations equal to zero.

$$\Sigma F_x = 0 \qquad \Sigma M_x = 0$$
$$\Sigma F_y = 0 \qquad \Sigma M_y = 0$$
$$\Sigma F_z = 0 \qquad \Sigma M_z = 0$$

Determinacy and Stability

If a body is supported by a minimum number of constraints to ensure equilibrium, then it is statically determinate. If it has more constraints than required, then it is statically indeterminate.

To properly constrain the body, the reactions must not all be parallel to one another or concurrent.

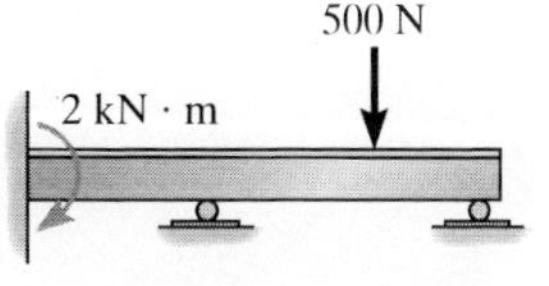

Statically indeterminate, five reactions, three equilibrium equations

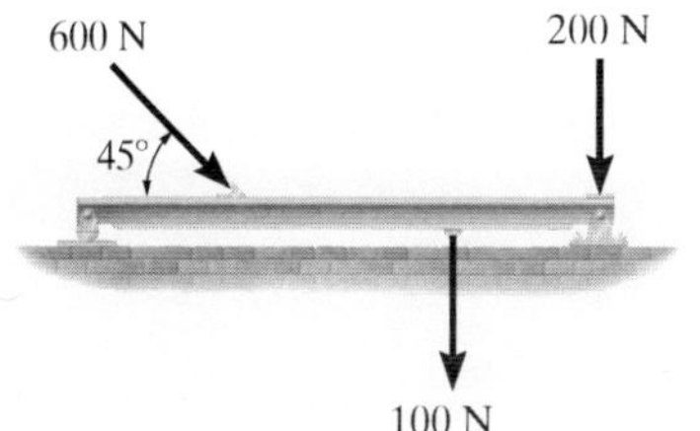

Proper constraint, statically determinate

REVIEW PROBLEMS

5–85. If the roller at B can sustain a maximum load of 3 kN, determine the largest magnitude of each of the three forces $\mathbf{F}$ that can be supported by the truss.

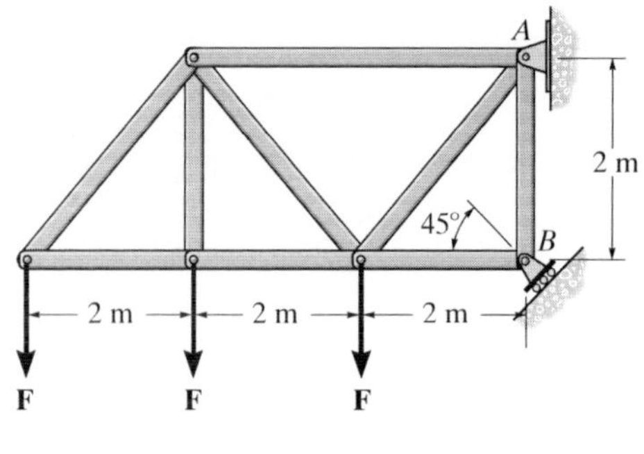

Prob. 5–85

5–86. Determine the normal reaction at the roller A and horizontal and vertical components at pin B for equilibrium of the member.

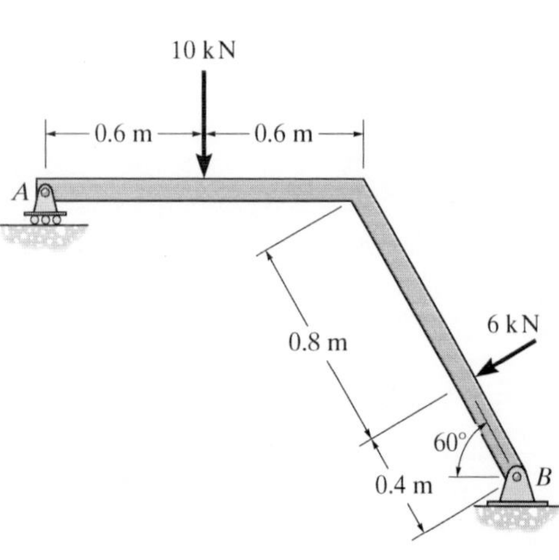

Prob. 5–86

5–87. The symmetrical shelf is subjected to a uniform load of 4 kPa. Support is provided by a bolt (or pin) located at each end A and A' and by the symmetrical brace arms, which bear against the smooth wall on both sides at B and B'. Determine the force resisted by each bolt at the wall and the normal force at B for equilibrium.

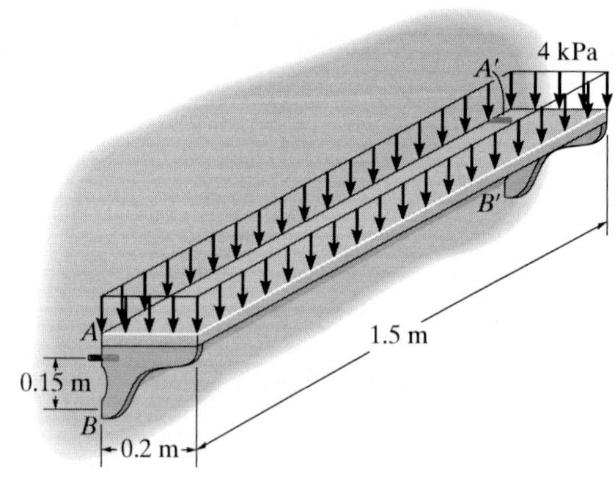

Prob. 5–87

***5–88.** Determine the x and z components of reaction at the journal bearing A and the tension in cords BC and BD necessary for equilibrium of the rod.

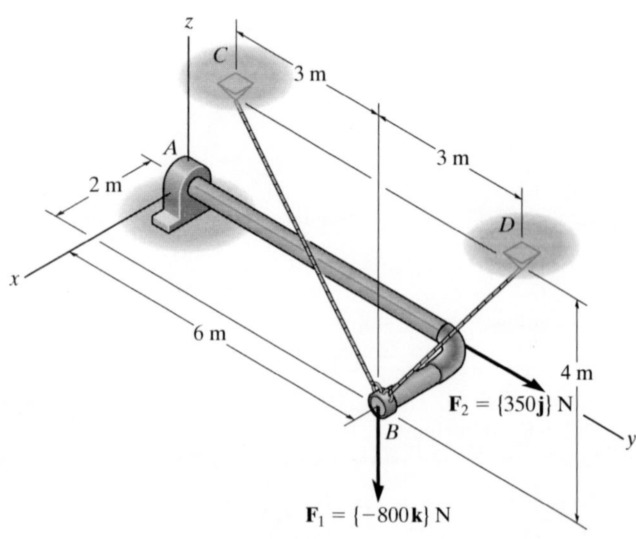

Prob. 5–88

5–89. The uniform rod of length L and weight W is supported on the smooth planes. Determine its position θ for equilibrium. Neglect the thickness of the rod.

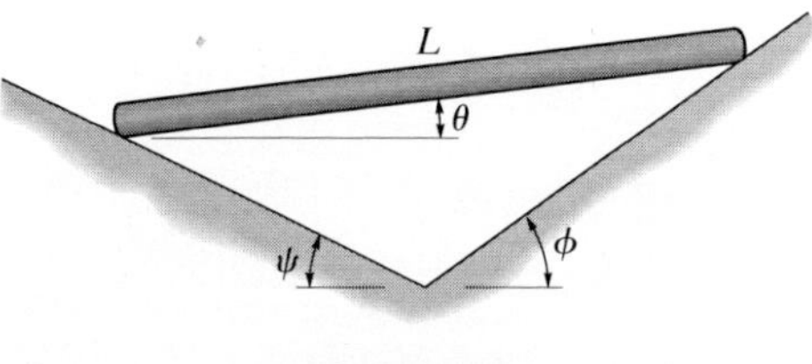

Prob. 5–89

5–90. Determine the horizontal and vertical components of force at the pin A and the reaction at the rocker B of the curved beam.

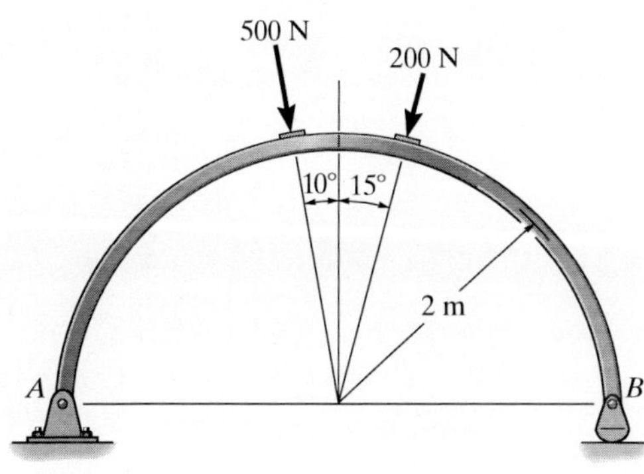

Prob. 5–90

5–91. Determine the x, y, z components of reaction at the fixed wall A. The 150-N force is parallel to the z axis and the 200-N force is parallel to the y axis.

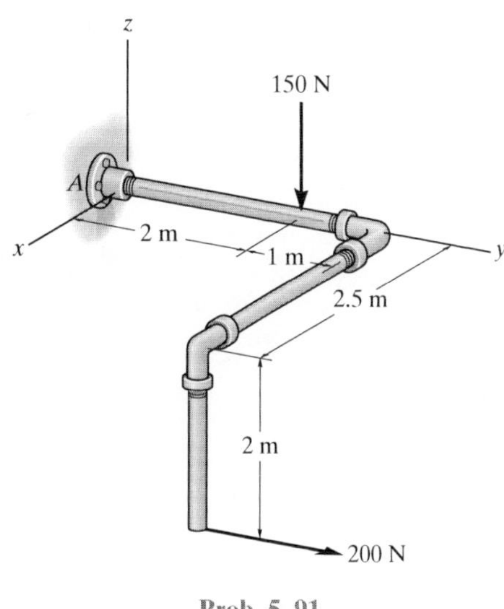

Prob. 5–91

***5–92.** Determine the reactions at the supports A and B for equilibrium of the beam.

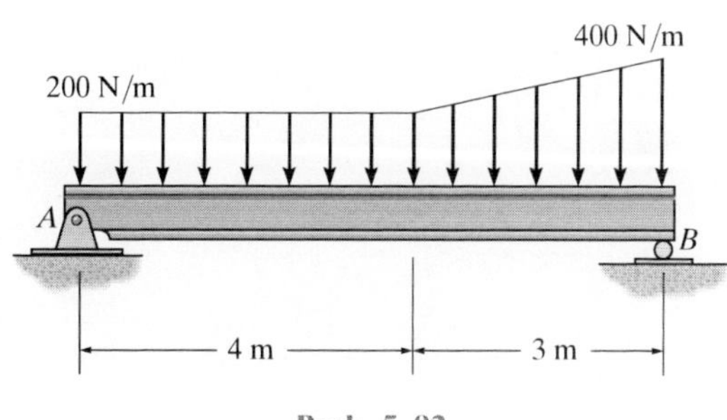

Prob. 5–92

5

Chapter 6

In order to design the many parts of this boom assembly it is required that we know the forces that they must support. In this chapter we will show how to analyze such structures using the equations of equilibrium.

Structural Analysis

CHAPTER OBJECTIVES

- To show how to determine the forces in the members of a truss using the method of joints and the method of sections.
- To analyze the forces acting on the members of frames and machines composed of pin-connected members.

Video Solutions are available for selected questions in this chapter.

6.1 Simple Trusses

A *truss* is a structure composed of slender members joined together at their end points. The members commonly used in construction consist of wooden struts or metal bars. In particular, *planar* trusses lie in a single plane and are often used to support roofs and bridges. The truss shown in Fig. 6–1*a* is an example of a typical roof-supporting truss. In this figure, the roof load is transmitted to the truss *at the joints* by means of a series of *purlins*. Since this loading acts in the same plane as the truss, Fig. 6–1*b*, the analysis of the forces developed in the truss members will be two-dimensional.

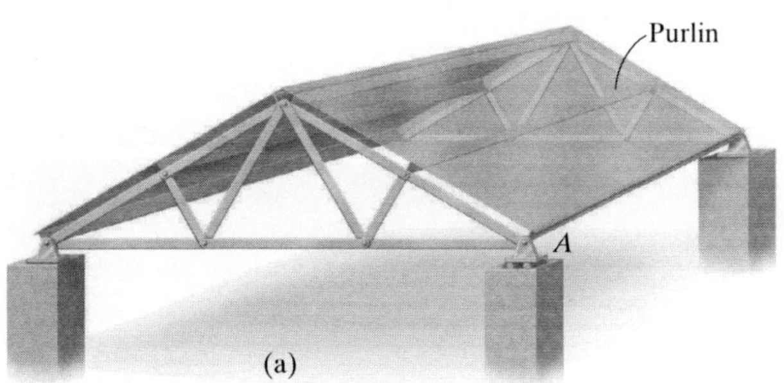

(a)

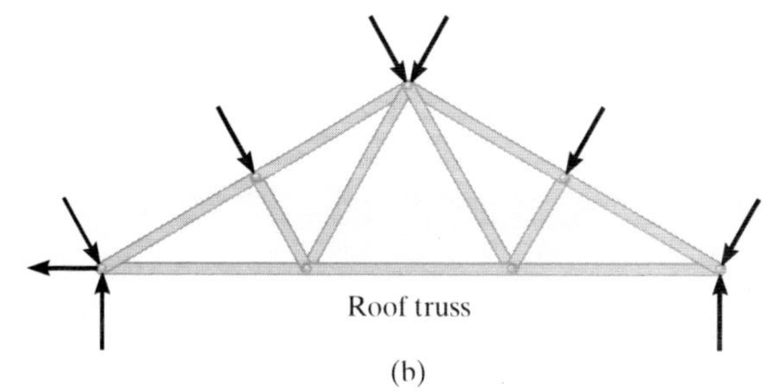

(b)

Fig. 6–1

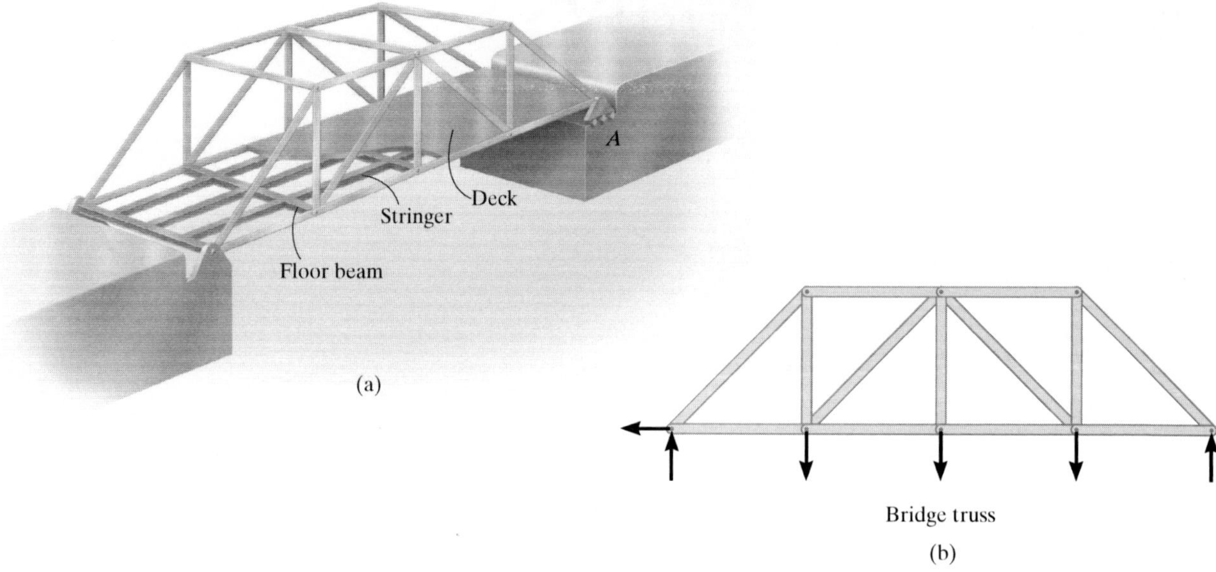

Fig. 6–2

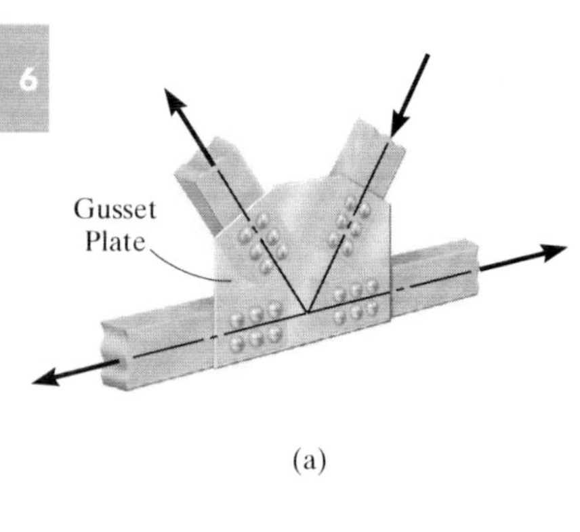

(a)

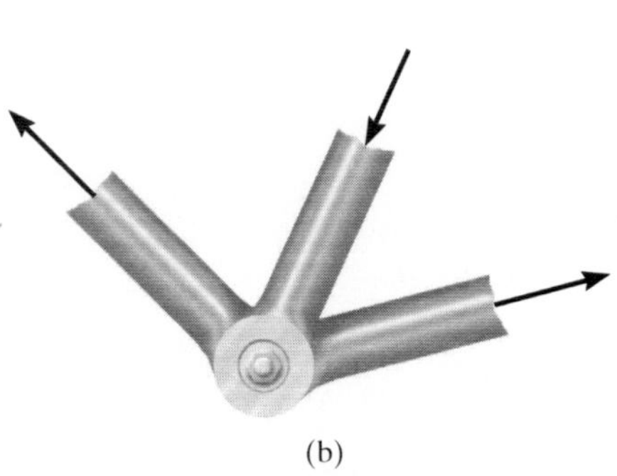

(b)

Fig. 6–3

In the case of a bridge, such as shown in Fig. 6–2*a*, the load on the deck is first transmitted to *stringers*, then to *floor beams*, and finally to the *joints* of the two supporting side trusses. Like the roof truss, the bridge truss loading is also coplanar, Fig. 6–2*b*.

When bridge or roof trusses extend over large distances, a rocker or roller is commonly used for supporting one end, for example, joint A in Figs. 6–1*a* and 6–2*a*. This type of support allows freedom for expansion or contraction of the members due to a change in temperature or application of loads.

Assumptions for Design. To design both the members and the connections of a truss, it is necessary first to determine the *force* developed in each member when the truss is subjected to a given loading. To do this we will make two important assumptions:

- ***All loadings are applied at the joints.*** In most situations, such as for bridge and roof trusses, this assumption is true. Frequently the weight of the members is neglected because the force supported by each member is usually much larger than its weight. However, if the weight is to be included in the analysis, it is generally satisfactory to apply it as a vertical force, with half of its magnitude applied at each end of the member.
- ***The members are joined together by smooth pins.*** The joint connections are usually formed by bolting or welding the ends of the members to a common plate, called a *gusset plate*, as shown in Fig. 6–3*a*, or by simply passing a large bolt or pin through each of the members, Fig. 6–3*b*. We can assume these connections act as pins provided the center lines of the joining members are *concurrent*, as in Fig. 6–3.

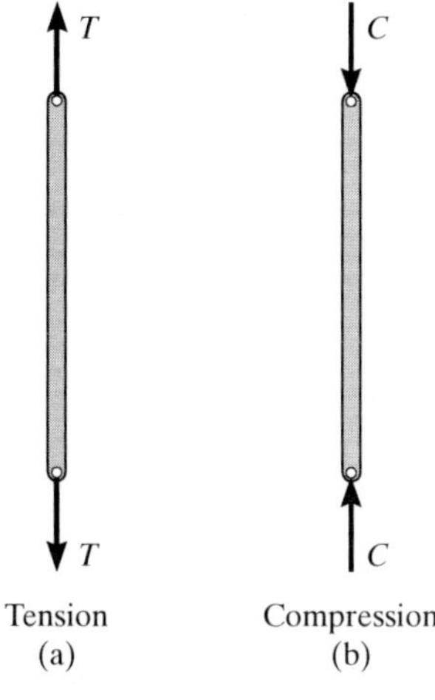

Fig. 6–4

Because of these two assumptions, *each truss member will act as a two-force member*, and therefore the force acting at each end of the member will be directed along the axis of the member. If the force tends to *elongate* the member, it is a *tensile force* (T), Fig. 6–4*a*; whereas if it tends to *shorten* the member, it is a *compressive force* (C), Fig. 6–4*b*. In the actual design of a truss it is important to state whether the nature of the force is tensile or compressive. Often, compression members must be made *thicker* than tension members because of the buckling or column effect that occurs when a member is in compression.

The use of metal gusset plates in the construction of these Warren trusses is clearly evident.

Simple Truss. If three members are pin connected at their ends, they form a *triangular truss* that will be *rigid*, Fig. 6–5. Attaching two more members and connecting these members to a new joint D forms a larger truss, Fig. 6–6. This procedure can be repeated as many times as desired to form an even larger truss. If a truss can be constructed by expanding the basic triangular truss in this way, it is called a *simple truss*.

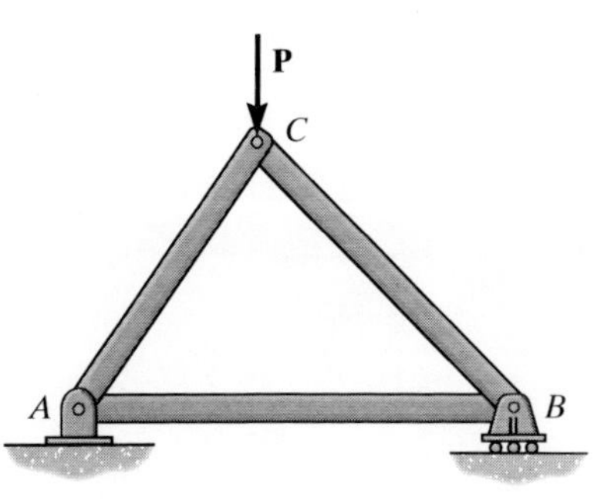

Fig. 6–5

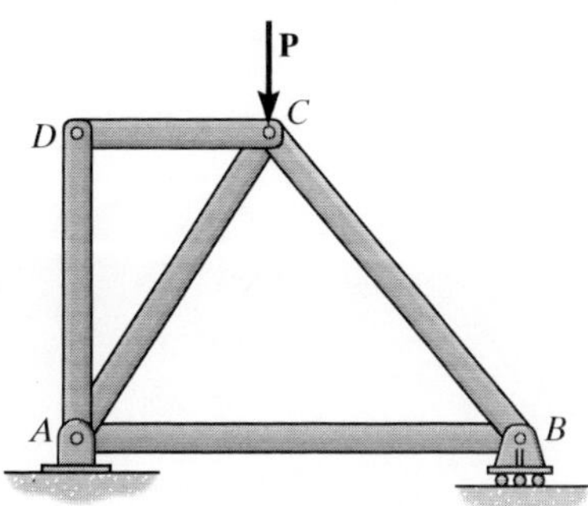

Fig. 6–6

6.2 The Method of Joints

Please refer to the Companion Website for the animation: *Free-Body Diagram for an A-Frame*

In order to analyze or design a truss, it is necessary to determine the force in each of its members. One way to do this is to use the method of joints. This method is based on the fact that if the entire truss is in equilibrium, then each of its joints is also in equilibrium. Therefore, if the free-body diagram of each joint is drawn, the force equilibrium equations can then be used to obtain the member forces acting on each joint. Since the members of a *plane truss* are straight two-force members lying in a single plane, each joint is subjected to a force system that is *coplanar and concurrent*. As a result, only $\Sigma F_x = 0$ and $\Sigma F_y = 0$ need to be satisfied for equilibrium.

For example, consider the pin at joint B of the truss in Fig. 6–7*a*. Three forces act on the pin, namely, the 500-N force and the forces exerted by members BA and BC. The free-body diagram of the pin is shown in Fig. 6–7*b*. Here, $\mathbf{F}_{BA}$ is "pulling" on the pin, which means that member BA is in *tension;* whereas $\mathbf{F}_{BC}$ is "pushing" on the pin, and consequently member BC is in *compression.* These effects are clearly demonstrated by isolating the joint with small segments of the member connected to the pin, Fig. 6–7*c*. The pushing or pulling on these small segments indicates the effect of the member being either in compression or tension.

When using the method of joints, always start at a joint having at least one known force and at most two unknown forces, as in Fig. 6–7*b*. In this way, application of $\Sigma F_x = 0$ and $\Sigma F_y = 0$ yields two algebraic equations which can be solved for the two unknowns. When applying these equations, the correct sense of an unknown member force can be determined using one of two possible methods.

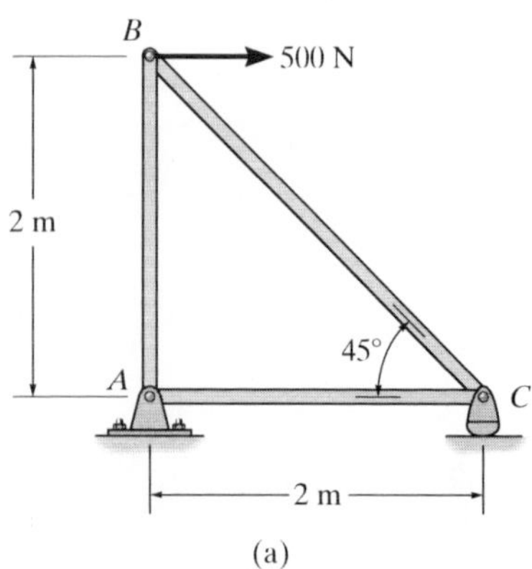

(a)

B
500 N
45°
$\mathbf{F}_{BA}$(tension)
$\mathbf{F}_{BC}$ (compression)

(b)

B
500 N
45°
$\mathbf{F}_{BA}$(tension)
$\mathbf{F}_{BC}$ (compression)

(c)

Fig. 6–7

- The *correct* sense of direction of an unknown member force can, in many cases, be determined "by inspection." For example, $\mathbf{F}_{BC}$ in Fig. 6–7*b* must push on the pin (compression) since its horizontal component, $F_{BC} \sin 45°$, must balance the 500-N force ($\Sigma F_x = 0$). Likewise, $\mathbf{F}_{BA}$ is a tensile force since it balances the vertical component, $F_{BC} \cos 45°$ ($\Sigma F_y = 0$). In more complicated cases, the sense of an unknown member force can be *assumed*; then, after applying the equilibrium equations, the assumed sense can be verified from the numerical results. A *positive* answer indicates that the sense is *correct*, whereas a *negative* answer indicates that the sense shown on the free-body diagram must be *reversed*.
- *Always assume* the *unknown member forces* acting on the joint's free-body diagram to be in *tension*; i.e., the forces "pull" on the pin. If this is done, then numerical solution of the equilibrium equations will yield *positive scalars for members in tension and negative scalars for members in compression*. Once an unknown member force is found, use its *correct* magnitude and sense (T or C) on subsequent joint free-body diagrams.

The forces in the members of this simple roof truss can be determined using the method of joints.

Procedure for Analysis

The following procedure provides a means for analyzing a truss using the method of joints.

- Draw the free-body diagram of a joint having at least one known force and at most two unknown forces. (If this joint is at one of the supports, then it may be necessary first to calculate the external reactions at the support.)
- Use one of the two methods described above for establishing the sense of an unknown force.
- Orient the x and y axes such that the forces on the free-body diagram can be easily resolved into their x and y components and then apply the two force equilibrium equations $\Sigma F_x = 0$ and $\Sigma F_y = 0$. Solve for the two unknown member forces and verify their correct sense.
- Using the calculated results, continue to analyze each of the other joints. Remember that a member in *compression* "pushes" on the joint and a member in *tension* "pulls" on the joint. Also, be sure to choose a joint having at most two unknowns and at least one known force.

6

EXAMPLE 6.1

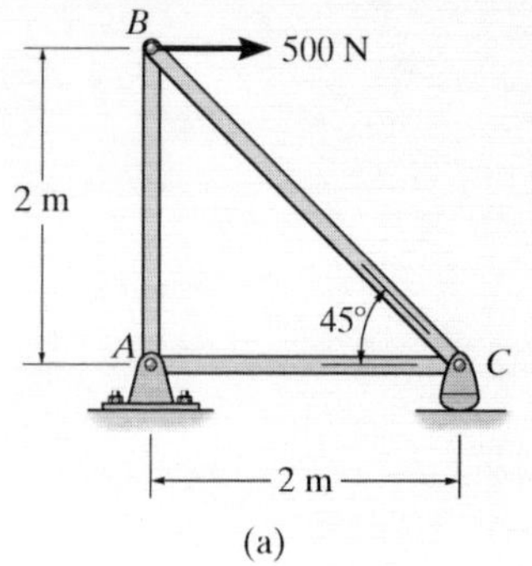

(a)

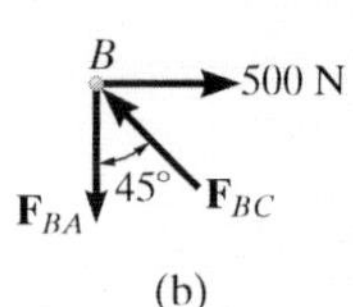

(b)

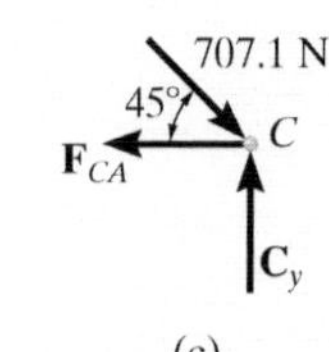

(c)

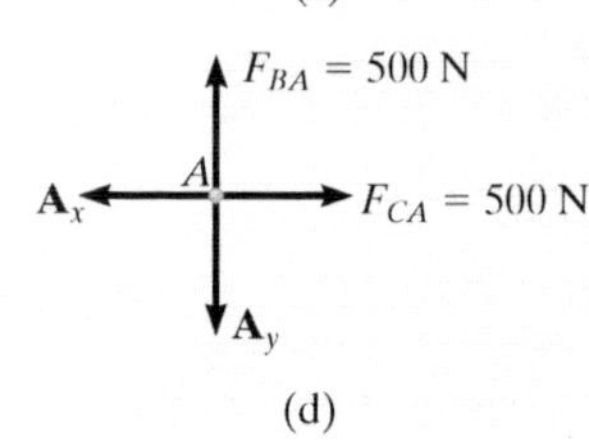

(d)

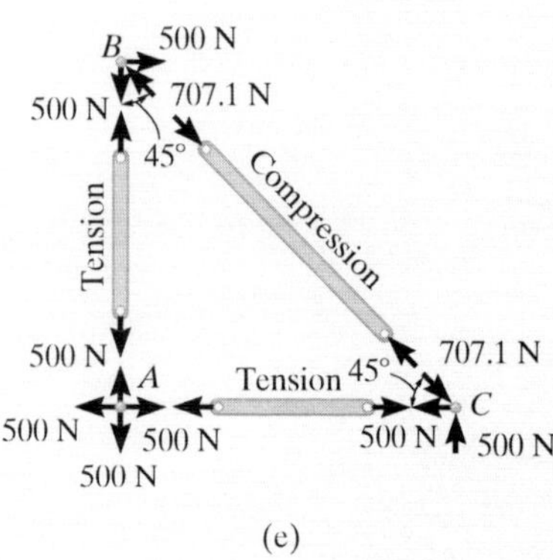

(e)

Fig. 6–8

Determine the force in each member of the truss shown in Fig. 6–8*a* and indicate whether the members are in tension or compression.

SOLUTION

Since we should have no more than two unknown forces at the joint and at least one known force acting there, we will begin our analysis at joint *B*.

Joint *B*. The free-body diagram of the joint at *B* is shown in Fig. 6–8*b*. Applying the equations of equilibrium, we have

$$\overset{+}{\rightarrow}\Sigma F_x = 0; \quad 500\text{ N} - F_{BC}\sin 45^\circ = 0 \quad F_{BC} = 707.1\text{ N (C)} \quad \textit{Ans.}$$

$$+\uparrow\Sigma F_y = 0; \quad F_{BC}\cos 45^\circ - F_{BA} = 0 \quad F_{BA} = 500\text{ N (T)} \quad \textit{Ans.}$$

Since the force in member *BC* has been calculated, we can proceed to analyze joint *C* to determine the force in member *CA* and the support reaction at the rocker.

Joint *C*. From the free-body diagram of joint *C*, Fig. 6–8*c*, we have

$$\overset{+}{\rightarrow}\Sigma F_x = 0; \quad -F_{CA} + 707.1\cos 45^\circ\text{ N} = 0 \quad F_{CA} = 500\text{ N (T)} \quad \textit{Ans.}$$

$$+\uparrow\Sigma F_y = 0; \quad C_y - 707.1\sin 45^\circ\text{ N} = 0 \quad C_y = 500\text{ N} \quad \textit{Ans.}$$

Joint *A*. Although it is not necessary, we can determine the components of the support reactions at joint *A* using the results of F_{CA} and F_{BA}. From the free-body diagram, Fig. 6–8*d*, we have

$$\overset{+}{\rightarrow}\Sigma F_x = 0; \quad 500\text{ N} - A_x = 0 \quad A_x = 500\text{ N}$$

$$+\uparrow\Sigma F_y = 0; \quad 500\text{ N} - A_y = 0 \quad A_y = 500\text{ N}$$

NOTE: The results of the analysis are summarized in Fig. 6–8*e*. Note that the free-body diagram of each joint (or pin) shows the effects of all the connected members and external forces applied to the joint, whereas the free-body diagram of each member shows only the effects of the end joints on the member.

EXAMPLE 6.2

Determine the forces acting in all the members of the truss shown in Fig. 6–9*a*.

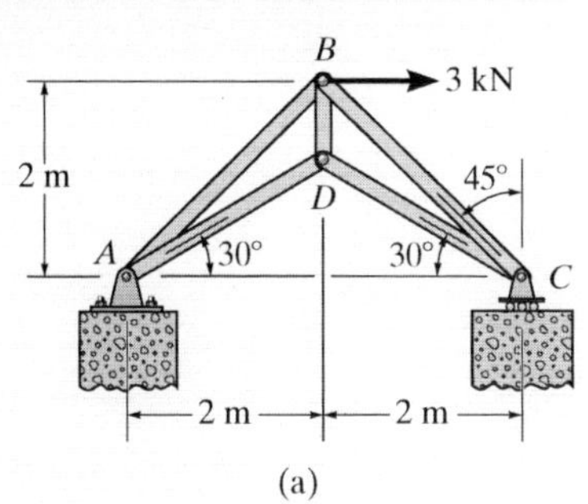

(a)

SOLUTION

By inspection, there are more than two unknowns at each joint. Consequently, the support reactions on the truss must first be determined. Show that they have been correctly calculated on the free-body diagram in Fig. 6–9*b*. We can now begin the analysis at joint *C*. Why?

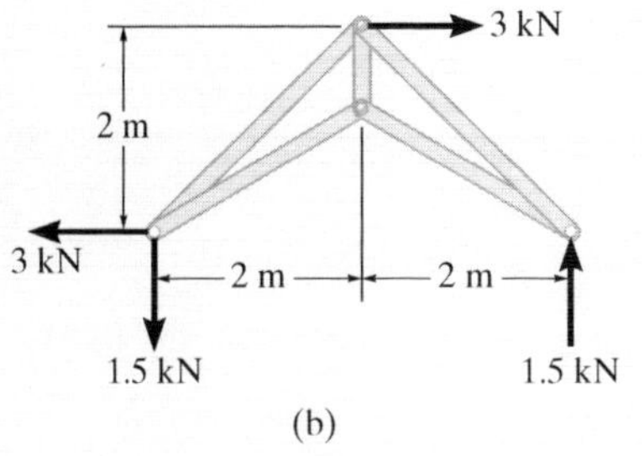

(b)

Joint *C*. From the free-body diagram, Fig. 6–9*c*,

$$\overset{+}{\rightarrow}\Sigma F_x = 0; \qquad -F_{CD}\cos 30° + F_{CB}\sin 45° = 0$$

$$+\uparrow\Sigma F_y = 0; \qquad 1.5\text{ kN} + F_{CD}\sin 30° - F_{CB}\cos 45° = 0$$

These two equations must be solved *simultaneously* for each of the two unknowns. Note, however, that a *direct solution* for one of the unknown forces may be obtained by applying a force summation along an axis that is *perpendicular* to the direction of the other unknown force. For example, summing forces along the y' axis, which is perpendicular to the direction of $\mathbf{F}_{CD}$, Fig. 6–9*d*, yields a *direct solution* for F_{CB}.

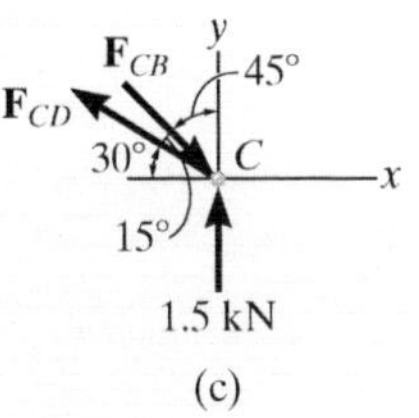

(c)

$$+\nearrow\Sigma F_{y'} = 0; \qquad 1.5\cos 30°\text{ kN} - F_{CB}\sin 15° = 0$$

$$F_{CB} = 5.019\text{ kN} = 5.02\text{ kN (C)} \qquad \textit{Ans.}$$

Then,

$$+\searrow\Sigma F_{x'} = 0;$$

$$-F_{CD} + 5.019\cos 15° - 1.5\sin 30° = 0;\ F_{CD} = 4.10\text{ kN (T)} \qquad \textit{Ans.}$$

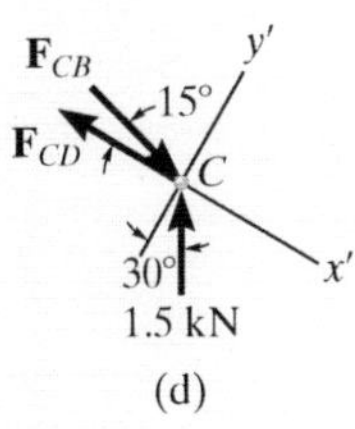

(d)

Joint *D*. We can now proceed to analyze joint *D*. The free-body diagram is shown in Fig. 6–9*e*.

$$\overset{+}{\rightarrow}\Sigma F_x = 0; \qquad -F_{DA}\cos 30° + 4.10\cos 30°\text{ kN} = 0$$

$$F_{DA} = 4.10\text{ kN (T)} \qquad \textit{Ans.}$$

$$+\uparrow\Sigma F_y = 0; \qquad F_{DB} - 2(4.10\sin 30°\text{ kN}) = 0$$

$$F_{DB} = 4.10\text{ kN (T)} \qquad \textit{Ans.}$$

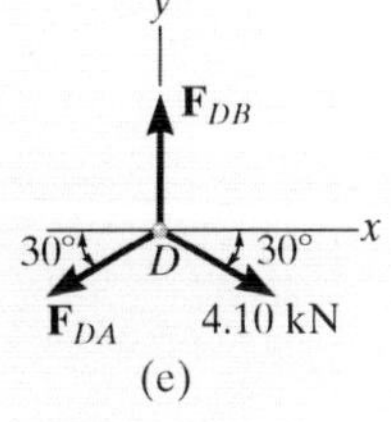

(e)

Fig. 6–9

NOTE: The force in the last member, *BA*, can be obtained from joint *B* or joint *A*. As an exercise, draw the free-body diagram of joint *B*, sum the forces in the horizontal direction, and show that $F_{BA} = 0.776$ kN (C).

6

EXAMPLE 6.3

Determine the force in each member of the truss shown in Fig. 6–10*a*. Indicate whether the members are in tension or compression.

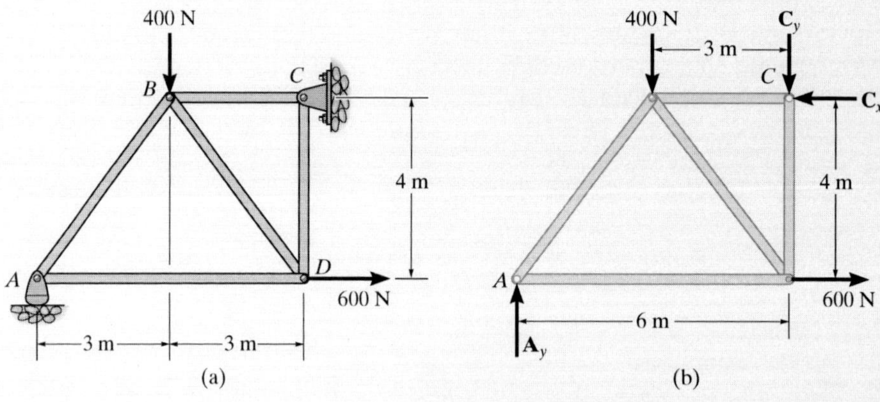

Fig. 6–10

SOLUTION

Support Reactions. No joint can be analyzed until the support reactions are determined, because each joint has at least three unknown forces acting on it. A free-body diagram of the entire truss is given in Fig. 6–10*b*. Applying the equations of equilibrium, we have

$$\overset{+}{\rightarrow}\Sigma F_x = 0; \qquad 600\text{ N} - C_x = 0 \qquad C_x = 600\text{ N}$$

$$\circlearrowleft + \Sigma M_C = 0; \qquad -A_y(6\text{ m}) + 400\text{ N}(3\text{ m}) + 600\text{ N}(4\text{ m}) = 0$$

$$A_y = 600\text{ N}$$

$$+\uparrow \Sigma F_y = 0; \qquad 600\text{ N} - 400\text{ N} - C_y = 0 \qquad C_y = 200\text{ N}$$

The analysis can now start at either joint A or C. The choice is arbitrary since there are one known and two unknown member forces acting on the pin at each of these joints.

Joint A. (Fig. 6–10*c*). As shown on the free-body diagram, $\mathbf{F}_{AB}$ is assumed to be compressive and $\mathbf{F}_{AD}$ is tensile. Applying the equations of equilibrium, we have

(c)

$$+\uparrow \Sigma F_y = 0; \qquad 600\text{ N} - \tfrac{4}{5}F_{AB} = 0 \qquad F_{AB} = 750\text{ N} \quad (\text{C}) \qquad \textit{Ans.}$$

$$\overset{+}{\rightarrow}\Sigma F_x = 0; \qquad F_{AD} - \tfrac{3}{5}(750\text{ N}) = 0 \qquad F_{AD} = 450\text{ N} \quad (\text{T}) \qquad \textit{Ans.}$$

Joint D. (Fig. 6–10*d*). Using the result for F_{AD} and summing forces in the horizontal direction, Fig. 6–10*d*, we have

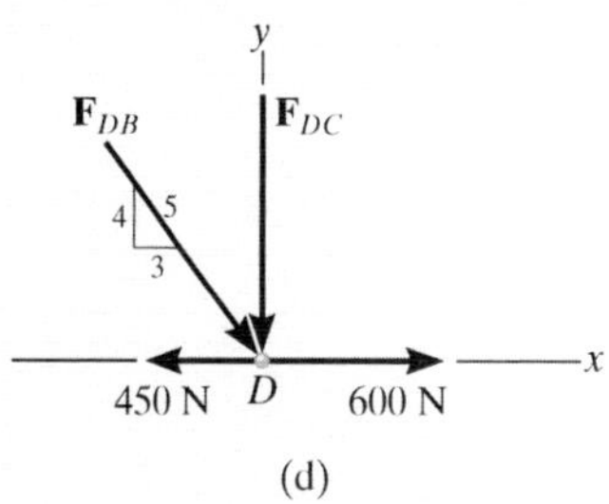

(d)

$$\xrightarrow{+} \Sigma F_x = 0; \qquad -450\text{ N} + \tfrac{3}{5}F_{DB} + 600\text{ N} = 0 \qquad F_{DB} = -250\text{ N}$$

The negative sign indicates that $\mathbf{F}_{DB}$ acts in the *opposite sense* to that shown in Fig. 6–10*d*.* Hence,

$$F_{DB} = 250\text{ N (T)} \qquad \textit{Ans.}$$

To determine $\mathbf{F}_{DC}$, we can either correct the sense of $\mathbf{F}_{DB}$ on the free-body diagram, and then apply $\Sigma F_y = 0$, or apply this equation and retain the negative sign for F_{DB}, i.e.,

$$+\uparrow \Sigma F_y = 0; \qquad -F_{DC} - \tfrac{4}{5}(-250\text{ N}) = 0 \qquad F_{DC} = 200\text{ N (C)} \qquad \textit{Ans.}$$

Joint C. (Fig. 6–10*e*).

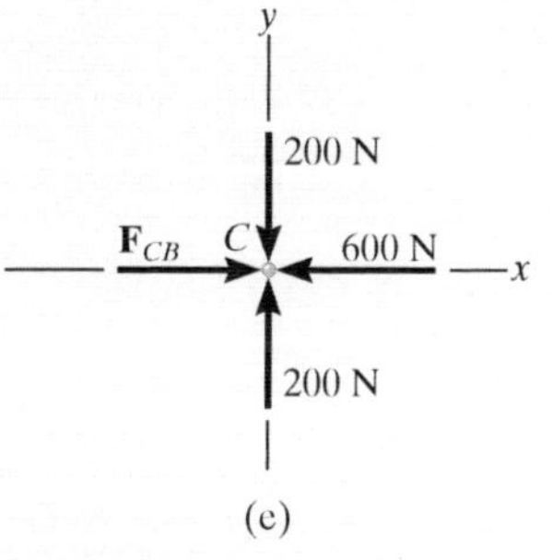

(e)

$$\xrightarrow{+} \Sigma F_x = 0; \qquad F_{CB} - 600\text{ N} = 0 \qquad F_{CB} = 600\text{ N (C)} \qquad \textit{Ans.}$$

$$+\uparrow \Sigma F_y = 0; \qquad 200\text{ N} - 200\text{ N} \equiv 0 \quad \text{(check)}$$

NOTE: The analysis is summarized in Fig. 6–10*f*, which shows the free-body diagram for each joint and member.

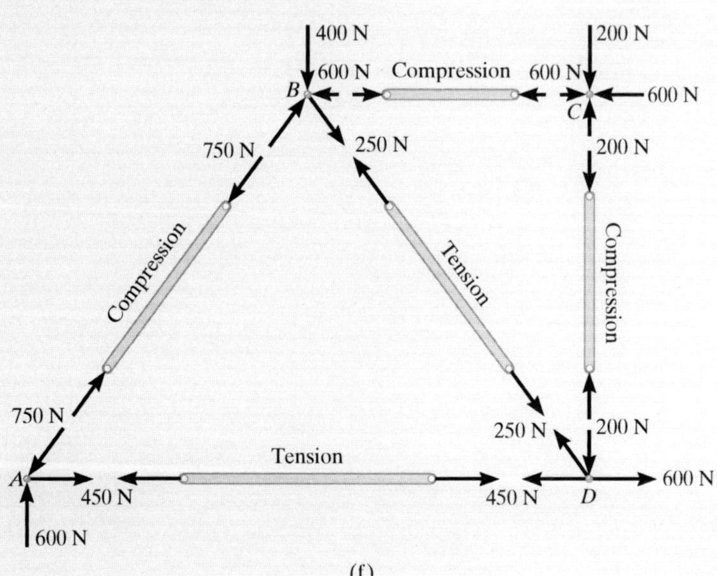

(f)

Fig. 6–10 (cont.)

*The proper sense could have been determined by inspection, prior to applying $\Sigma F_x = 0$.

6.3 Zero-Force Members

Truss analysis using the method of joints is greatly simplified if we can first identify those members which support *no loading*. These *zero-force members* are used to increase the stability of the truss during construction and to provide added support if the loading is changed.

The zero-force members of a truss can generally be found *by inspection* of each of the joints. For example, consider the truss shown in Fig. 6–11*a*. If a free-body diagram of the pin at joint A is drawn, Fig. 6–11*b*, it is seen that members AB and AF are zero-force members. (We could not have come to this conclusion if we had considered the free-body diagrams of joints F or B simply because there are five unknowns at each of these joints.) In a similar manner, consider the free-body diagram of joint D, Fig. 6–11*c*. Here again it is seen that DC and DE are zero-force members. From these observations, we can conclude that *if only two non-collinear members form a truss joint and no external load or support reaction is applied to the joint, the two members must be zero-force members.* The load on the truss in Fig. 6–11*a* is therefore supported by only five members as shown in Fig. 6–11*d*.

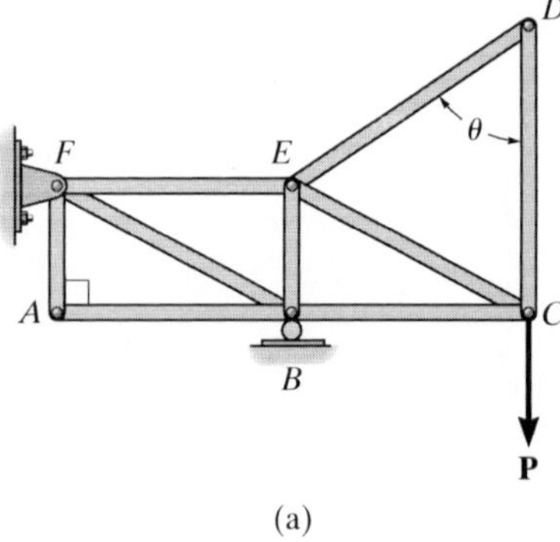

(a)

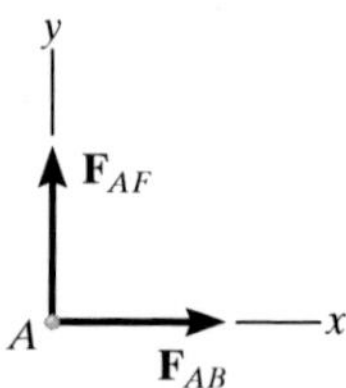

$\overset{+}{\rightarrow} \Sigma F_x = 0;\ F_{AB} = 0$
$+\uparrow \Sigma F_y = 0;\ F_{AF} = 0$

(b)

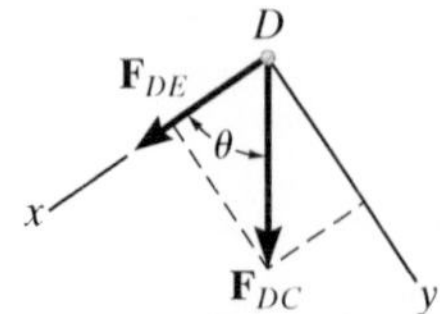

$+\searrow \Sigma F_y = 0;\ F_{DC} \sin\theta = 0;\quad F_{DC} = 0 \text{ since } \sin\theta \neq 0$
$+\swarrow \Sigma F_x = 0;\ F_{DE} + 0 = 0;\quad F_{DE} = 0$

(c)

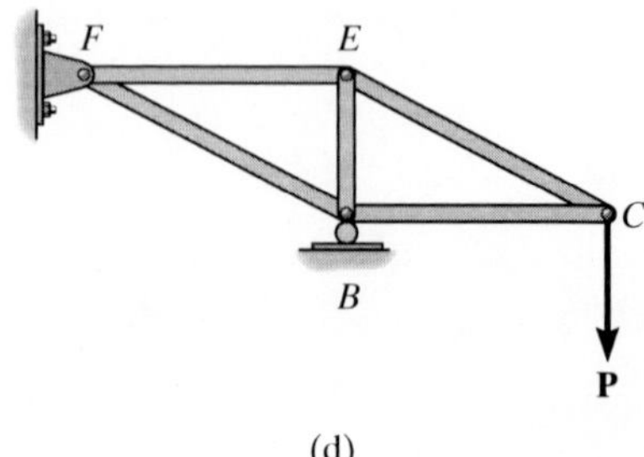

(d)

Fig. 6–11

Now consider the truss shown in Fig. 6–12*a*. The free-body diagram of the pin at joint D is shown in Fig. 6–12*b*. By orienting the y axis along members DC and DE and the x axis along member DA, it is seen that DA is a zero-force member. Note that this is also the case for member CA, Fig. 6–12*c*. In general then, *if three members form a truss joint for which two of the members are collinear, the third member is a zero-force member provided no external force or support reaction is applied to the joint.* The truss shown in Fig. 6–12*d* is therefore suitable for supporting the load **P**.

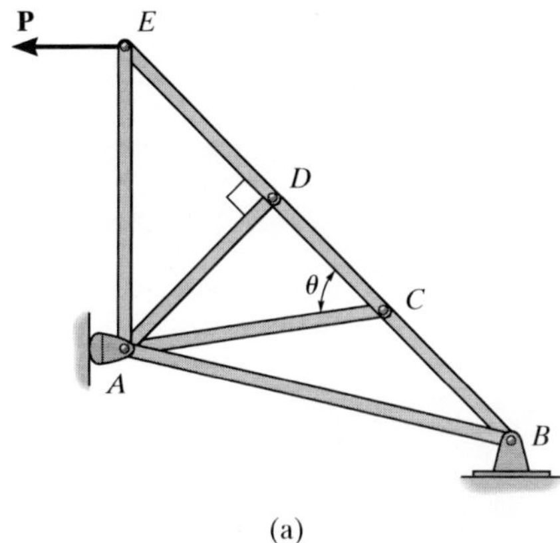

(a)

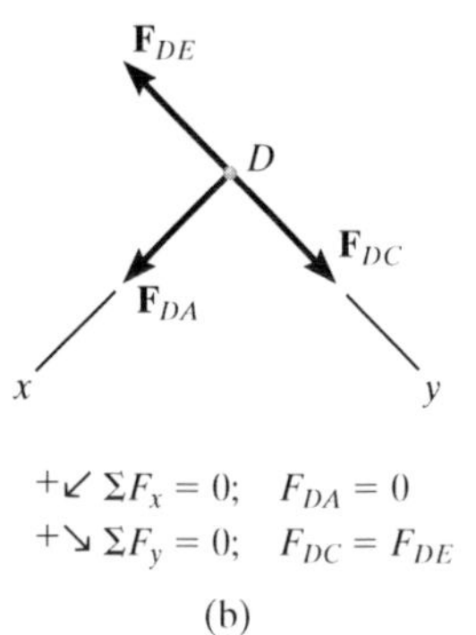

$+\swarrow \Sigma F_x = 0;\quad F_{DA} = 0$
$+\searrow \Sigma F_y = 0;\quad F_{DC} = F_{DE}$

(b)

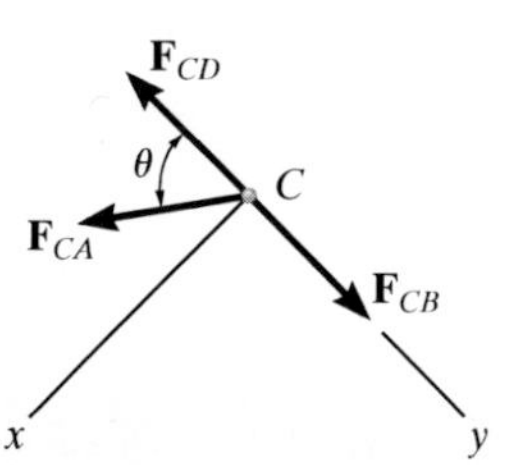

$+\swarrow \Sigma F_x = 0;\quad F_{CA} \sin \theta = 0;\quad F_{CA} = 0 \text{ since } \sin \theta \neq 0;$
$+\searrow \Sigma F_y = 0;\quad F_{CB} = F_{CD}$

(c)

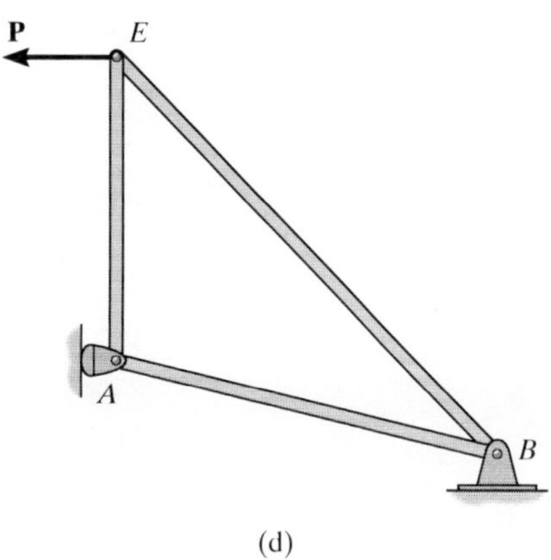

(d)

Fig. 6–12

6

EXAMPLE 6.4

Using the method of joints, determine all the zero-force members of the *Fink roof truss* shown in Fig. 6–13*a*. Assume all joints are pin connected.

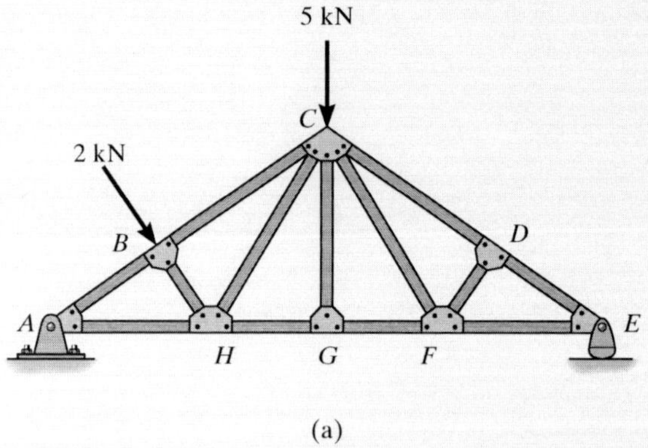

(a)

Fig. 6–13

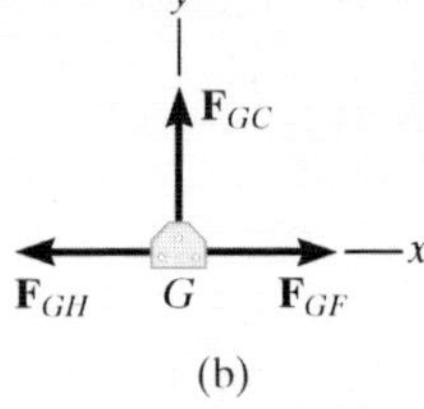

(b)

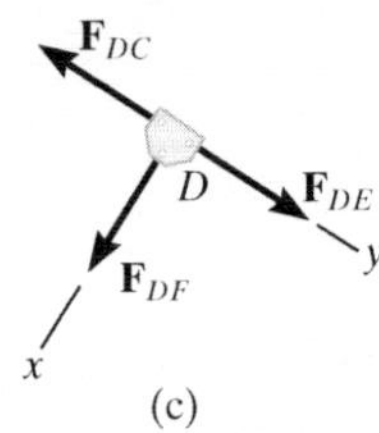

(c)

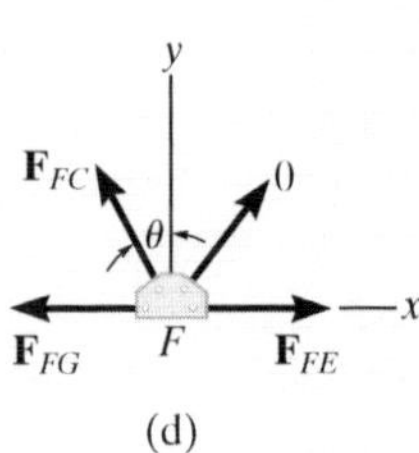

(d)

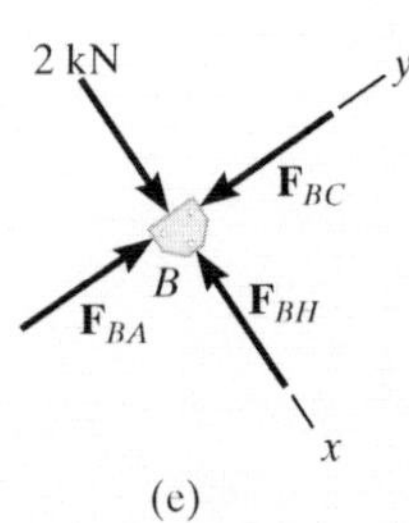

(e)

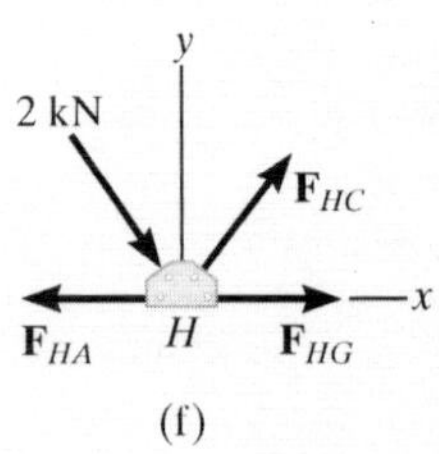

(f)

SOLUTION

Look for joint geometries that have three members for which two are collinear. We have

Joint G. (Fig. 6–13*b*).

$$+\uparrow \Sigma F_y = 0; \qquad F_{GC} = 0 \qquad \textit{Ans.}$$

Realize that we could not conclude that *GC* is a zero-force member by considering joint *C*, where there are five unknowns. The fact that *GC* is a zero-force member means that the 5-kN load at *C* must be supported by members *CB*, *CH*, *CF*, and *CD*.

Joint D. (Fig. 6–13*c*).

$$+\swarrow \Sigma F_x = 0; \qquad F_{DF} = 0 \qquad \textit{Ans.}$$

Joint F. (Fig. 6–13*d*).

$$+\uparrow \Sigma F_y = 0; \qquad F_{FC} \cos\theta = 0 \quad \text{Since } \theta \neq 90^\circ, \quad F_{FC} = 0 \qquad \textit{Ans.}$$

NOTE: If joint *B* is analyzed, Fig. 6–13*e*,

$$+\searrow \Sigma F_x = 0; \qquad 2\text{ kN} - F_{BH} = 0 \quad F_{BH} = 2\text{ kN} \quad \text{(C)}$$

Also, F_{HC} must satisfy $\Sigma F_y = 0$, Fig. 6–13*f*, and therefore *HC* is *not* a zero-force member.

FUNDAMENTAL PROBLEMS

All problem solutions must include FBDs.

F6–1. Determine the force in each member of the truss. State if the members are in tension or compression.

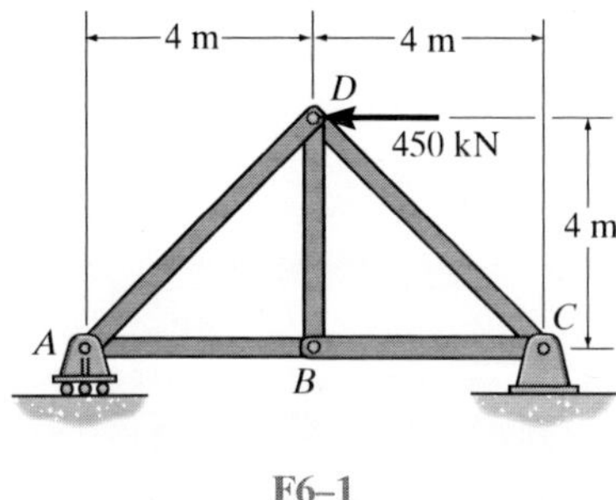

F6–1

F6–2. Determine the force in each member of the truss. State if the members are in tension or compression.

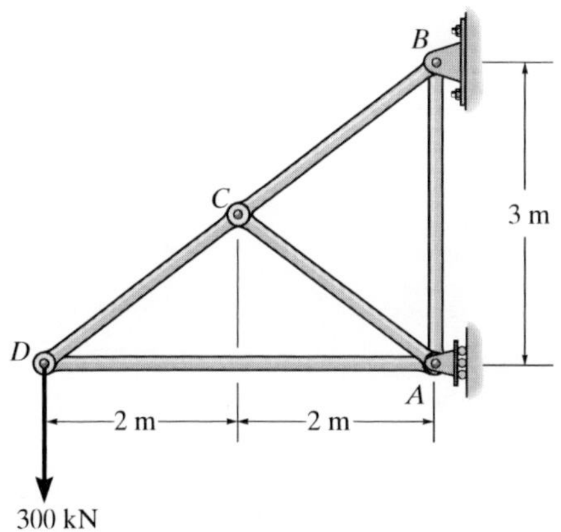

F6–2

F6–3. Determine the force in members AE and DC. State if the members are in tension or compression.

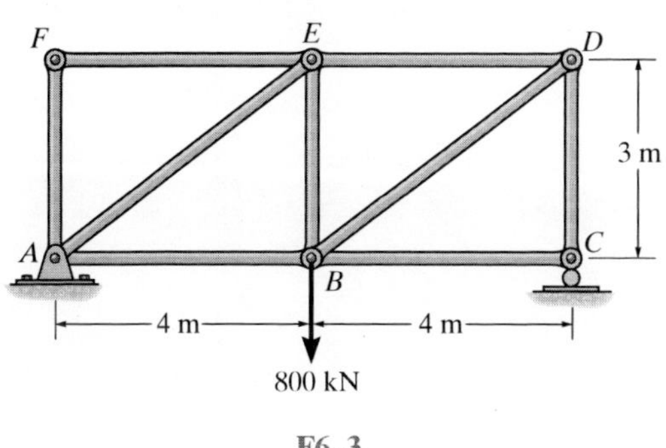

F6–3

F6–4. Determine the greatest load P that can be applied to the truss so that none of the members are subjected to a force exceeding either 2 kN in tension or 1.5 kN in compression.

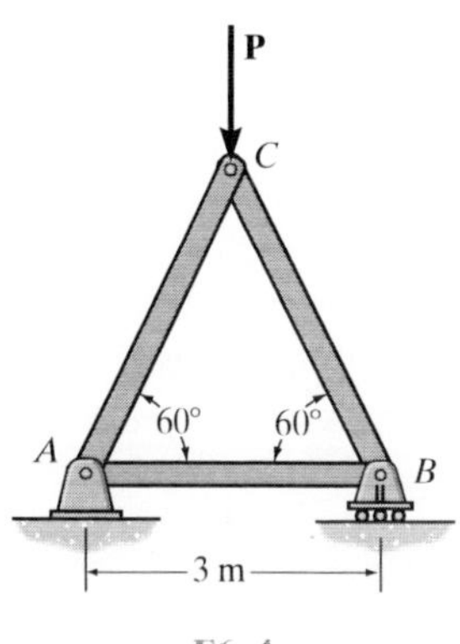

F6–4

F6–5. Identify the zero-force members in the truss.

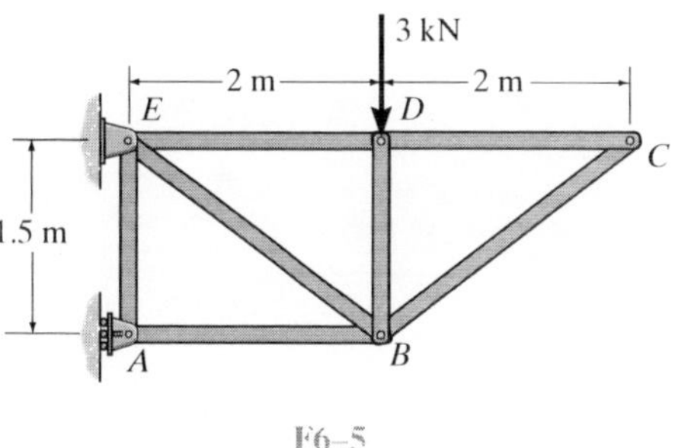

F6–5

F6–6. Determine the force in each member of the truss. State if the members are in tension or compression.

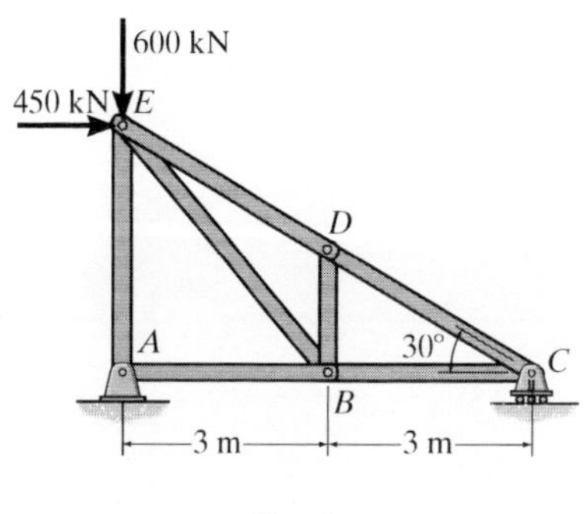

F6–6

PROBLEMS

All problem solutions must include FBDs.

6–1. The truss, used to support a balcony, is subjected to the loading shown. Approximate each joint as a pin and determine the force in each member. State whether the members are in tension or compression. Set $P_1 = 60$ kN, $P_2 = 40$ kN.

6–2. The truss, used to support a balcony, is subjected to the loading shown. Approximate each joint as a pin and determine the force in each member. State whether the members are in tension or compression. Set $P_1 = 80$ kN, $P_2 = 0$.

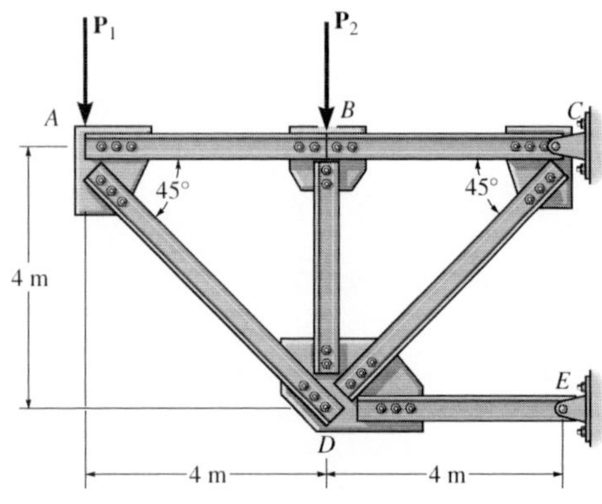

Probs. 6–1/2

6–3. Determine the force in each member of the truss, and state if the members are in tension or compression. Set $\theta = 0°$.

***6–4.** Determine the force in each member of the truss, and state if the members are in tension or compression. Set $\theta = 30°$.

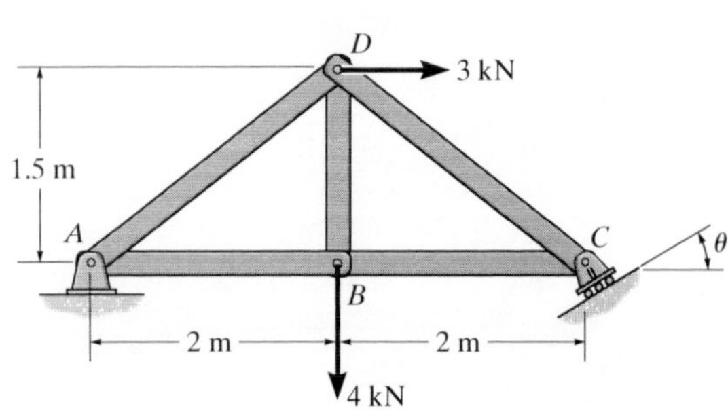

Probs. 6–3/4

6–5. Determine the force in each member of the truss, and state if the members are in tension or compression.

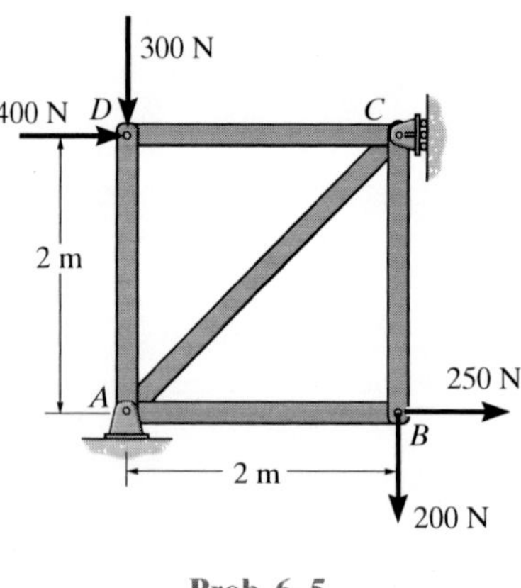

Prob. 6–5

6–6. Determine the force in each member of the truss, and state if the members are in tension or compression.

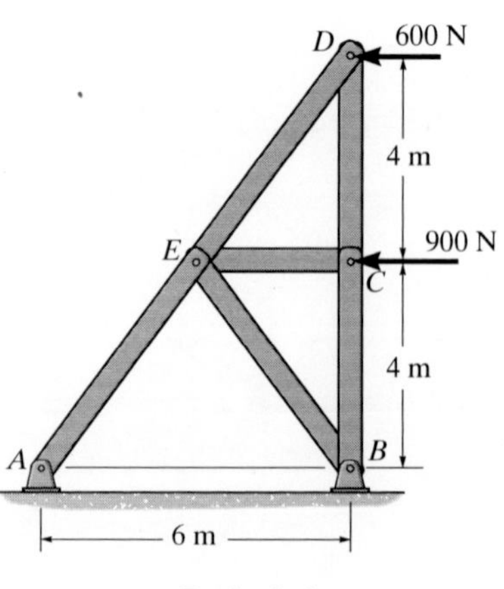

Prob. 6–6

6–7. Determine the force in each member of the *Pratt truss*, and state if the members are in tension or compression.

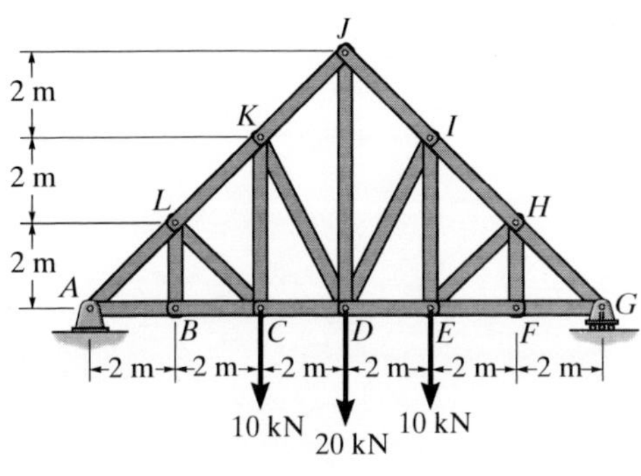

Prob. 6–7

***6–8.** Determine the force in each member of the truss and state if the members are in tension or compression. *Hint:* The horizontal force component at A must be zero. Why?

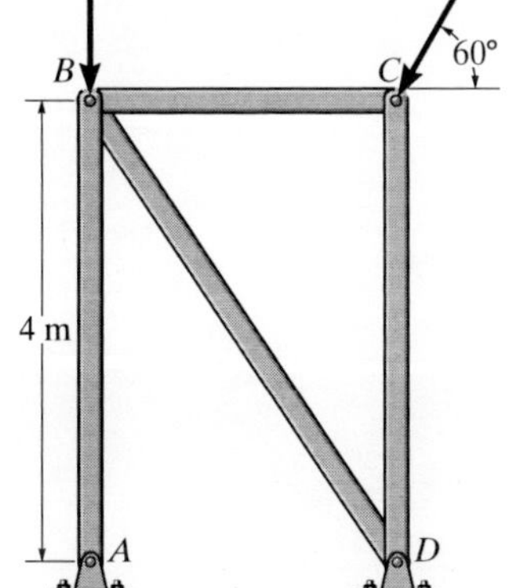

Prob. 6–8

6–9. Determine the force in each member of the truss and state if the members are in tension or compression. *Hint:* The vertical component of force at C must equal zero. Why?

6–10. Each member of the truss is uniform and has a mass of 8 kg/m. Remove the external loads of 6 kN and 8 kN and determine the approximate force in each member due to the weight of the truss. State if the members are in tension or compression. Solve the problem by *assuming* the weight of each member can be represented as a vertical force, half of which is applied at each end of the member.

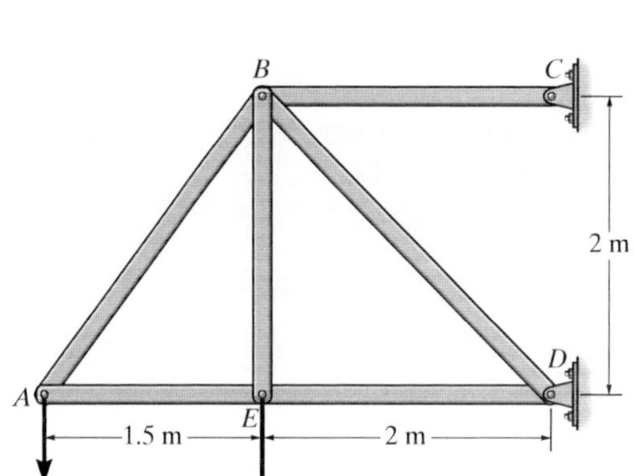

Probs. 6–9/10

6–11. Determine the force in each member of the truss and state if the members are in tension or compression.

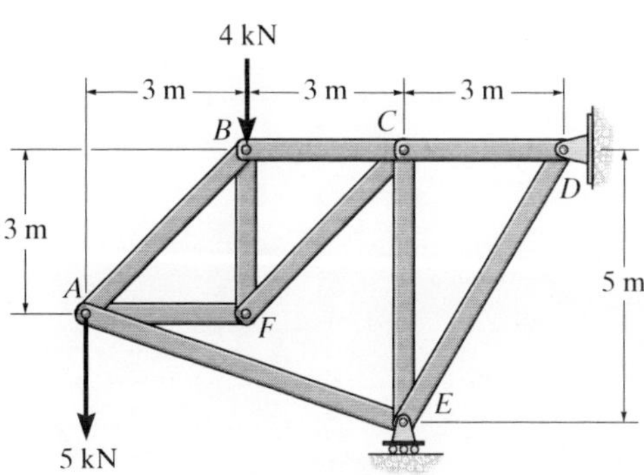

Prob. 6–11

*__6–12.__ Determine the force in each member of the truss and state if the members are in tension or compression. Set $P_1 = 10$ kN, $P_2 = 15$ kN.

6–13. Determine the force in each member of the truss and state if the members are in tension or compression. Set $P_1 = 0$, $P_2 = 20$ kN.

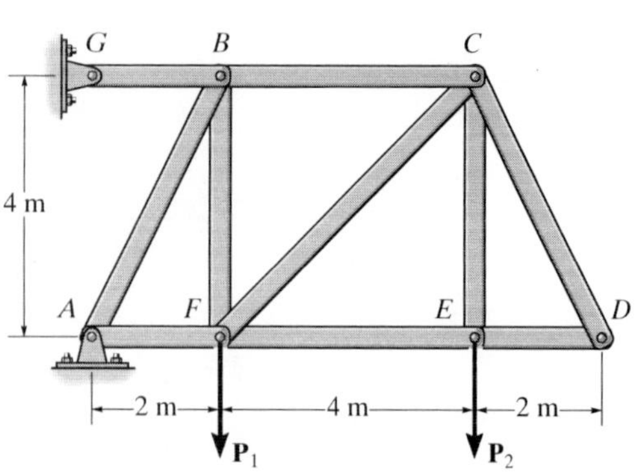

Probs. 6–12/13

6–14. Determine the force in each member of the truss and state if the members are in tension or compression. Set $P_1 = 2$ kN and $P_2 = 1.5$ kN.

6–15. Determine the force in each member of the truss and state if the members are in tension or compression. Set $P_1 = P_2 = 4$ kN.

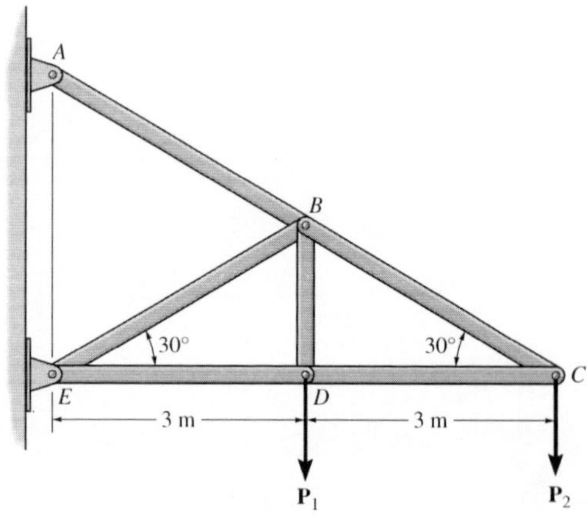

Probs. 6–14/15

*__6–16.__ Determine the force in each member of the truss. State whether the members are in tension or compression. Set $P = 8$ kN.

6–17. If the maximum force that any member can support is 8 kN in tension and 6 kN in compression, determine the maximum force P that can be supported at joint D.

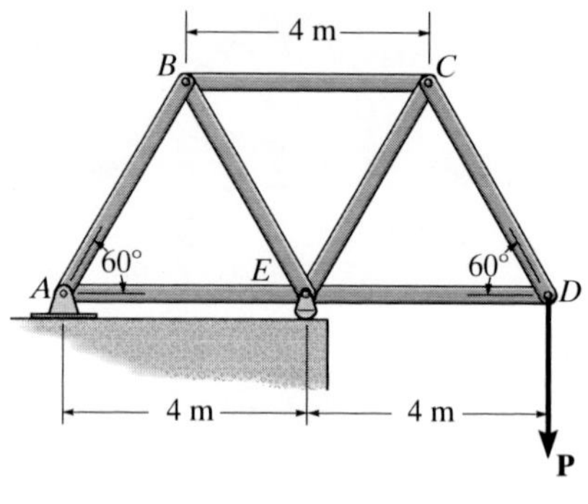

Probs. 6–16/17

6–18. Determine the force in each member of the truss and state if the members are in tension or compression. *Hint:* The resultant force at the pin E acts along member ED. Why?

6–19. Each member of the truss is uniform and has a mass of 8 kg/m. Remove the external loads of 3 kN and 2 kN and determine the approximate force in each member due to the weight of the truss. State if the members are in tension or compression. Solve the problem by *assuming* the weight of each member can be represented as a vertical force, half of which is applied at each end of the member.

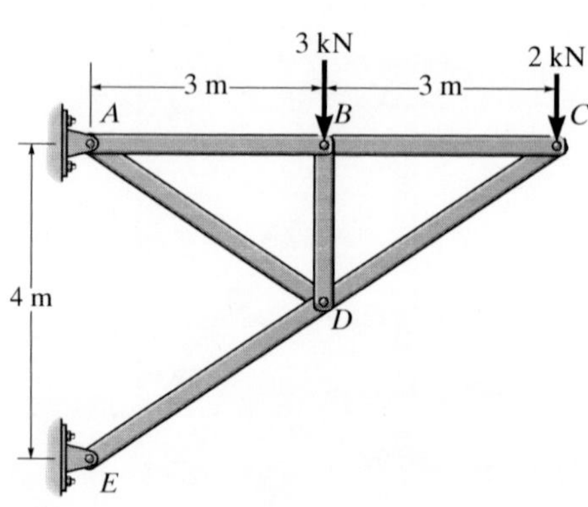

Probs. 6–18/19

*6–20. Determine the force in each member of the truss in terms of the load P, and indicate whether the members are in tension or compression.

6–21. If the maximum force that any member can support is 4 kN in tension and 3 kN in compression, determine the maximum force P that can be supported at point B. Take $d = 1$ m.

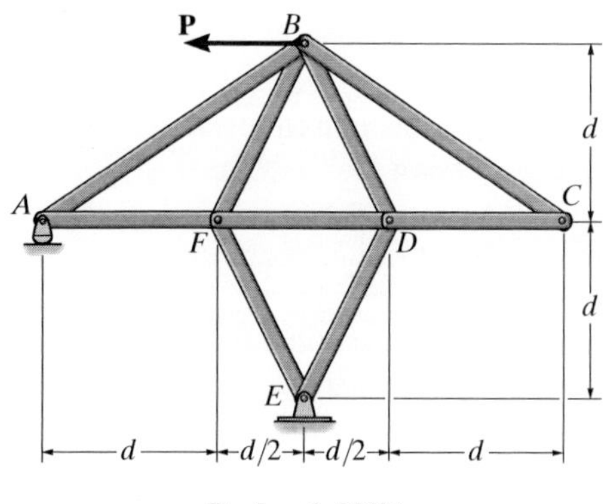

Probs. 6–20/21

6–23. Determine the force in each member of the truss in terms of the load P and state if the members are in tension or compression.

*6–24. Each member of the truss is uniform and has a weight W. Remove the external forces **P** and determine the approximate force in each member due to the weight of the truss. State if the members are in tension or compression. Solve the problem by *assuming* the weight of each member can be represented as a vertical force, half of which is applied at each end of the member.

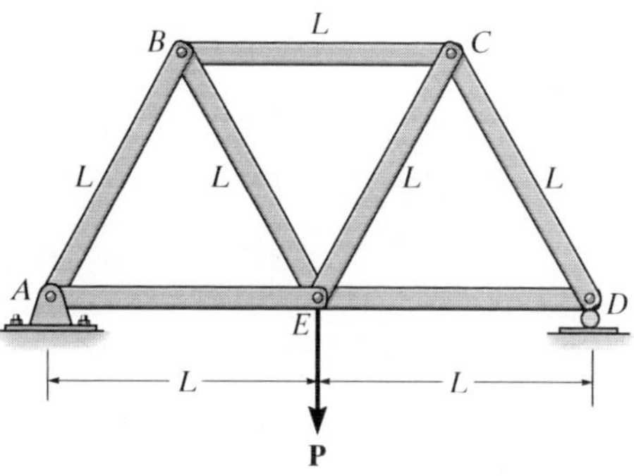

Probs. 6–23/24

6–22. Determine the force in each member of the double scissors truss in terms of the load P and state if the members are in tension or compression.

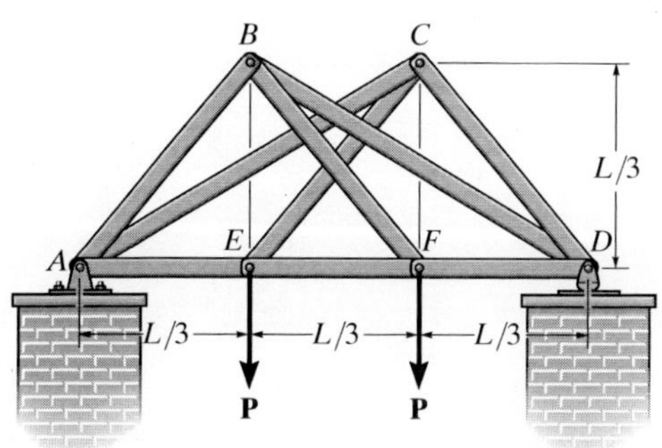

Prob. 6–22

6–25. Determine the force in each member of the truss in terms of the external loading and state if the members are in tension or compression.

6–26. The maximum allowable tensile force in the members of the truss is $(F_t)_{max} = 2$ kN, and the maximum allowable compressive force is $(F_c)_{max} = 1.2$ kN. Determine the maximum magnitude P of the two loads that can be applied to the truss. Take $L = 2$ m and $\theta = 30°$.

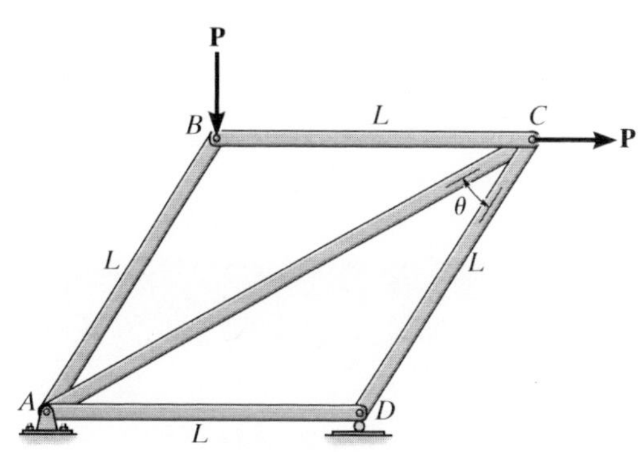

Probs. 6–25/26

6.4 The Method of Sections

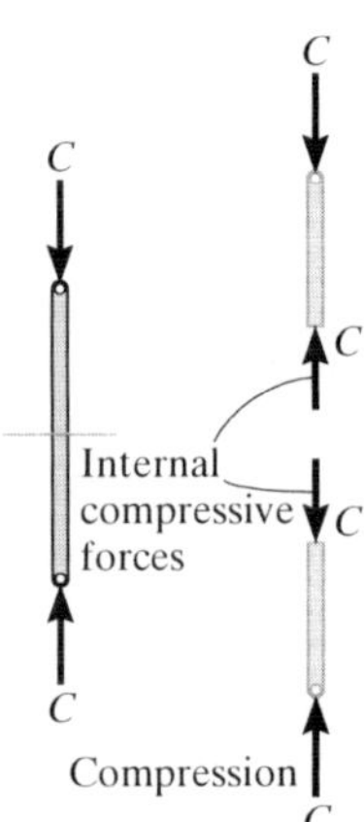

Fig. 6–14

When we need to find the force in only a few members of a truss, we can analyze the truss using the *method of sections*. It is based on the principle that if the truss is in equilibrium then any segment of the truss is also in equilibrium. For example, consider the two truss members shown on the left in Fig. 6–14. If the forces within the members are to be determined, then an imaginary section, indicated by the blue line, can be used to cut each member into two parts and thereby "expose" each internal force as "external" to the free-body diagrams shown on the right. Clearly, it can be seen that equilibrium requires that the member in tension (T) be subjected to a "pull," whereas the member in compression (C) is subjected to a "push."

The method of sections can also be used to "cut" or section the members of an entire truss. If the section passes through the truss and the free-body diagram of either of its two parts is drawn, we can then apply the equations of equilibrium to that part to determine the member forces at the "cut section." Since only *three* independent equilibrium equations ($\Sigma F_x = 0$, $\Sigma F_y = 0$, $\Sigma M_O = 0$) can be applied to the free-body diagram of any segment, then we should try to select a section that, in general, passes through not more than *three* members in which the forces are unknown. For example, consider the truss in Fig. 6–15*a*. If the forces in members *BC*, *GC*, and *GF* are to be determined, then section *aa* would be appropriate. The free-body diagrams of the two segments are shown in Figs. 6–15*b* and 6–15*c*. Note that the line of action of each member force is specified from the *geometry* of the truss, since the force in a member is along its axis. Also, the member forces acting on one part of the truss are equal but opposite to those acting on the other part—Newton's third law. Members *BC* and *GC* are assumed to be in *tension* since they are subjected to a "pull," whereas *GF* in *compression* since it is subjected to a "push."

The three unknown member forces $\mathbf{F}_{BC}$, $\mathbf{F}_{GC}$, and $\mathbf{F}_{GF}$ can be obtained by applying the three equilibrium equations to the free-body diagram in Fig. 6–15*b*. If, however, the free-body diagram in Fig. 6–15*c* is considered, the three support reactions $\mathbf{D}_x$, $\mathbf{D}_y$ and $\mathbf{E}_x$ will have to be known, because only three equations of equilibrium are available. (This, of course, is done in the usual manner by considering a free-body diagram of the *entire truss*.)

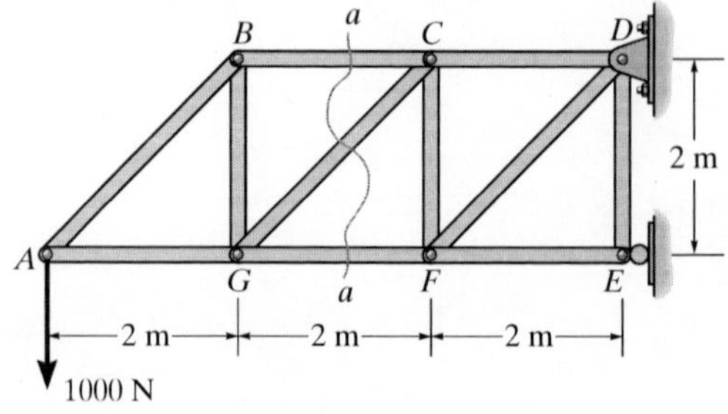

Fig. 6–15

When applying the equilibrium equations, we should carefully consider ways of writing the equations so as to yield a *direct solution* for each of the unknowns, rather than having to solve simultaneous equations. For example, using the truss segment in Fig. 6–15*b* and summing moments about *C* would yield a direct solution for $\mathbf{F}_{GF}$ since $\mathbf{F}_{BC}$ and $\mathbf{F}_{GC}$ create zero moment about *C*. Likewise, $\mathbf{F}_{BC}$ can be directly obtained by summing moments about *G*. Finally, $\mathbf{F}_{GC}$ can be found directly from a force summation in the vertical direction since $\mathbf{F}_{GF}$ and $\mathbf{F}_{BC}$ have no vertical components. This ability to *determine directly* the force in a particular truss member is one of the main advantages of using the method of sections.*

The forces in selected members of this Pratt truss can readily be determined using the method of sections.

As in the method of joints, there are two ways in which we can determine the correct sense of an unknown member force:

- The correct sense of an unknown member force can in many cases be determined "by inspection." For example, $\mathbf{F}_{BC}$ is a tensile force as represented in Fig. 6–15*b* since moment equilibrium about *G* requires that $\mathbf{F}_{BC}$ create a moment opposite to that of the 1000-N force. Also, $\mathbf{F}_{GC}$ is tensile since its vertical component must balance the 1000-N force which acts downward. In more complicated cases, the sense of an unknown member force may be *assumed.* If the solution yields a *negative* scalar, it indicates that the force's sense is *opposite* to that shown on the free-body diagram.
- *Always assume* that the unknown member forces at the cut section are *tensile* forces, i.e., "pulling" on the member. By doing this, the numerical solution of the equilibrium equations will yield *positive scalars for members in tension and negative scalars for members in compression.*

*Notice that if the method of joints were used to determine, say, the force in member *GC*, it would be necessary to analyze joints *A*, *B*, and *G* in sequence.

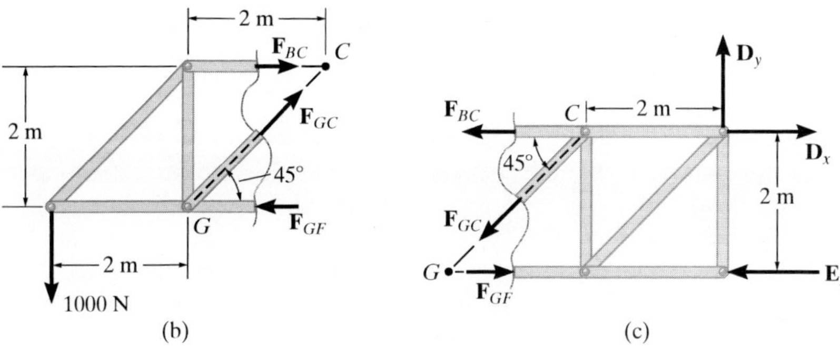

Fig. 6–15 (cont.)

6

Simple trusses are often used in the construction of large cranes in order to reduce the weight of the boom and tower.

6

Procedure for Analysis

The forces in the members of a truss may be determined by the method of sections using the following procedure.

Free-Body Diagram.

- Make a decision on how to "cut" or section the truss through the members where forces are to be determined.
- Before isolating the appropriate section, it may first be necessary to determine the truss's support reactions. If this is done then the three equilibrium equations will be available to solve for member forces at the section.
- Draw the free-body diagram of that segment of the sectioned truss which has the least number of forces acting on it.
- Use one of the two methods described above for establishing the sense of the unknown member forces.

Equations of Equilibrium.

- Moments should be summed about a point that lies at the intersection of the lines of action of two unknown forces, so that the third unknown force can be determined directly from the moment equation.
- If two of the unknown forces are *parallel*, forces may be summed *perpendicular* to the direction of these unknowns to determine *directly* the third unknown force.

EXAMPLE 6.5

Determine the force in members *GE*, *GC*, and *BC* of the truss shown in Fig. 6–16*a*. Indicate whether the members are in tension or compression.

Please refer to the Companion Website for the animation: *Free-Body Diagram for a Truss*

SOLUTION

Section *aa* in Fig. 6–16*a* has been chosen since it cuts through the *three* members whose forces are to be determined. In order to use the method of sections, however, it is *first* necessary to determine the external reactions at *A* or *D*. Why? A free-body diagram of the entire truss is shown in Fig. 6–16*b*. Applying the equations of equilibrium, we have

$$\overset{+}{\rightarrow}\Sigma F_x = 0; \qquad 400\text{ N} - A_x = 0 \qquad A_x = 400\text{ N}$$

$$\circlearrowleft+\Sigma M_A = 0; \qquad -1200\text{ N}(8\text{ m}) - 400\text{ N}(3\text{ m}) + D_y(12\text{ m}) = 0$$

$$D_y = 900\text{ N}$$

$$+\uparrow\Sigma F_y = 0; \qquad A_y - 1200\text{ N} + 900\text{ N} = 0 \qquad A_y = 300\text{ N}$$

Free-Body Diagram. For the analysis the free-body diagram of the left portion of the sectioned truss will be used, since it involves the least number of forces, Fig. 6–16*c*.

Equations of Equilibrium. Summing moments about point *G* eliminates $\mathbf{F}_{GE}$ and $\mathbf{F}_{GC}$ and yields a direct solution for F_{BC}.

$$\circlearrowleft+\Sigma M_G = 0; \quad -300\text{ N}(4\text{ m}) - 400\text{ N}(3\text{ m}) + F_{BC}(3\text{ m}) = 0$$

$$F_{BC} = 800\text{ N} \quad \text{(T)} \qquad \textit{Ans.}$$

In the same manner, by summing moments about point *C* we obtain a direct solution for F_{GE}.

$$\circlearrowleft+\Sigma M_C = 0; \quad -300\text{ N}(8\text{ m}) + F_{GE}(3\text{ m}) = 0$$

$$F_{GE} = 800\text{ N} \quad \text{(C)} \qquad \textit{Ans.}$$

Since $\mathbf{F}_{BC}$ and $\mathbf{F}_{GE}$ have no vertical components, summing forces in the *y* direction directly yields F_{GC}, i.e.,

$$+\uparrow\Sigma F_y = 0; \qquad 300\text{ N} - \tfrac{3}{5}F_{GC} = 0$$

$$F_{GC} = 500\text{ N} \quad \text{(T)} \qquad \textit{Ans.}$$

NOTE: Here it is possible to tell, by inspection, the proper direction for each unknown member force. For example, $\Sigma M_C = 0$ requires $\mathbf{F}_{GE}$ to be *compressive* because it must balance the moment of the 300-N force about *C*.

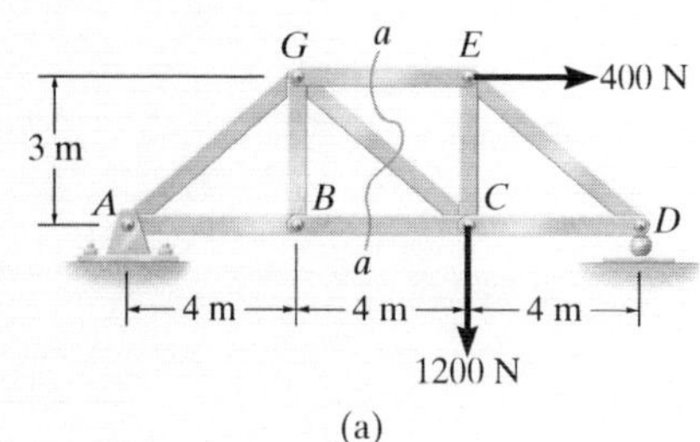

(a)

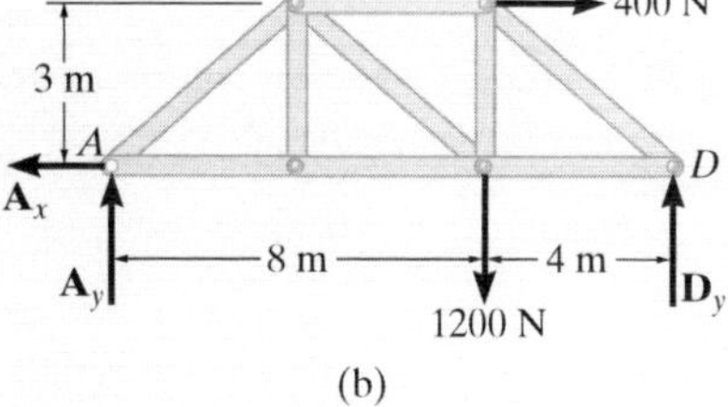

(b)

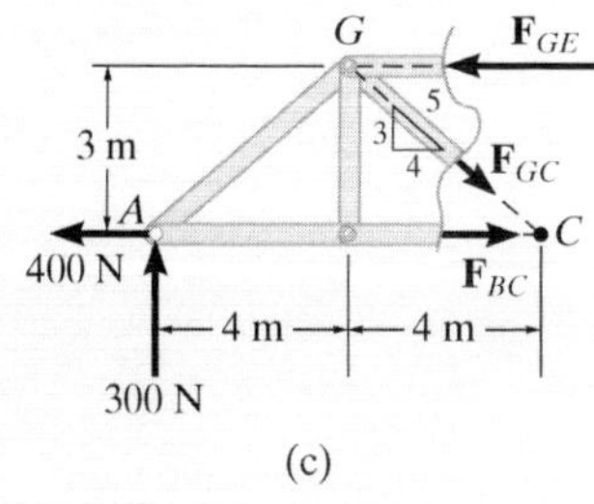

(c)

Fig. 6–16

EXAMPLE 6.6

Determine the force in member CF of the truss shown in Fig. 6–17a. Indicate whether the member is in tension or compression. Assume each member is pin connected.

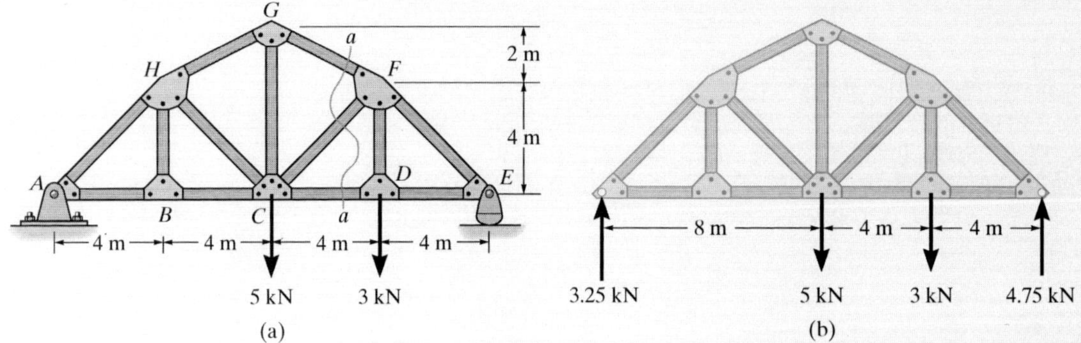

Fig. 6–17

SOLUTION

Free-Body Diagram. Section aa in Fig. 6–17a will be used since this section will "expose" the internal force in member CF as "external" on the free-body diagram of either the right or left portion of the truss. It is first necessary, however, to determine the support reactions on either the left or right side. Verify the results shown on the free-body diagram in Fig. 6–17b.

The free-body diagram of the right portion of the truss, which is the easiest to analyze, is shown in Fig. 6–17c. There are three unknowns, F_{FG}, F_{CF}, and F_{CD}.

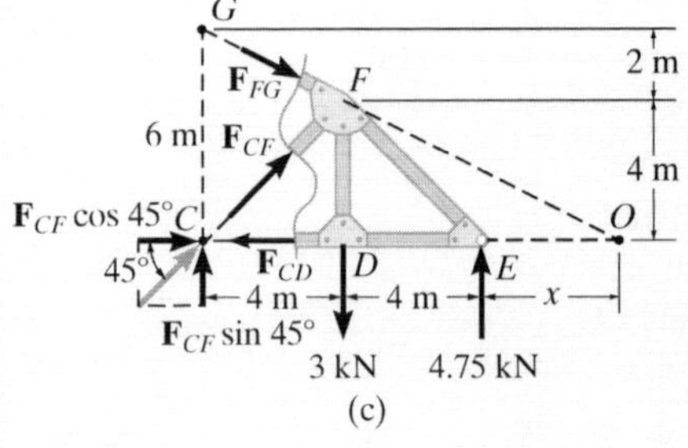

Equations of Equilibrium. We will apply the moment equation about point O in order to eliminate the two unknowns F_{FG} and F_{CD}. The location of point O measured from E can be determined from proportional triangles, i.e., $4/(4 + x) = 6/(8 + x)$, $x = 4$ m. Or, stated in another manner, the slope of member GF has a drop of 2 m to a horizontal distance of 4 m. Since FD is 4 m, Fig. 6–17c, then from D to O the distance must be 8 m.

An easy way to determine the moment of $\mathbf{F}_{CF}$ about point O is to use the principle of transmissibility and slide $\mathbf{F}_{CF}$ to point C, and then resolve $\mathbf{F}_{CF}$ into its two rectangular components. We have

$$\circlearrowleft +\Sigma M_O = 0;$$

$$-F_{CF}\sin 45°(12\text{ m}) + (3\text{ kN})(8\text{ m}) - (4.75\text{ kN})(4\text{ m}) = 0$$

$$F_{CF} = 0.589\text{ kN}\quad\text{(C)}\qquad\textit{Ans.}$$

EXAMPLE 6.7

Determine the force in member *EB* of the roof truss shown in Fig. 6–18*a*. Indicate whether the member is in tension or compression.

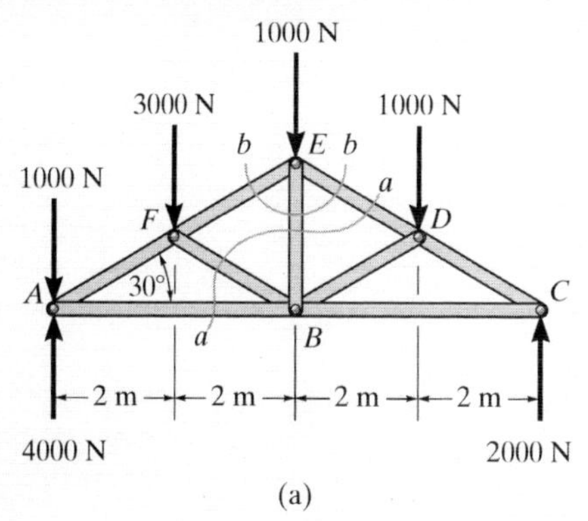

(a)

SOLUTION

Free-Body Diagrams. By the method of sections, any imaginary section that cuts through *EB*, Fig. 6–18*a*, will also have to cut through three other members for which the forces are unknown. For example, section *aa* cuts through *ED, EB, FB*, and *AB*. If a free-body diagram of the left side of this section is considered, Fig. 6–18*b*, it is possible to obtain $\mathbf{F}_{ED}$ by summing moments about *B* to eliminate the other three unknowns; however, $\mathbf{F}_{EB}$ cannot be determined from the remaining two equilibrium equations. One possible way of obtaining $\mathbf{F}_{EB}$ is first to determine $\mathbf{F}_{ED}$ from section *aa*, then use this result on section *bb*, Fig. 6–18*a*, which is shown in Fig. 6–18*c*. Here the force system is concurrent and our sectioned free-body diagram is the same as the free-body diagram for the joint at *E*.

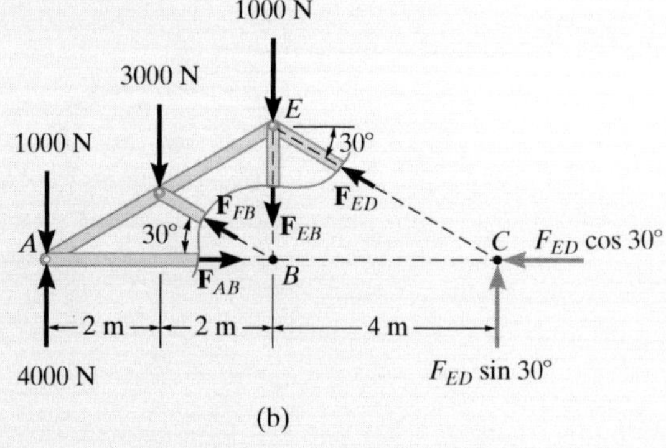

(b)

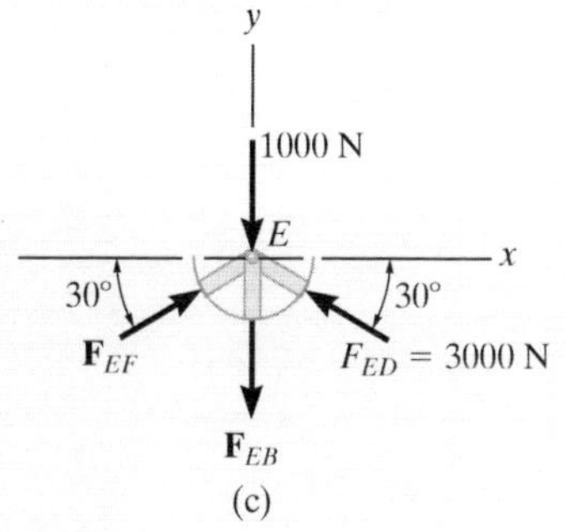

(c)

Fig. 6–18

Equations of Equilibrium. In order to determine the moment of $\mathbf{F}_{ED}$ about point *B*, Fig. 6–18*b*, we will use the principle of transmissibility and slide the force to point *C* and then resolve it into its rectangular components as shown. Therefore,

$$\curvearrowleft +\Sigma M_B = 0; \qquad 1000\text{ N}(4\text{ m}) + 3000\text{ N}(2\text{ m}) - 4000\text{ N}(4\text{ m}) + F_{ED}\sin 30°(4\text{ m}) = 0$$

$$F_{ED} = 3000\text{ N} \quad \text{(C)}$$

Considering now the free-body diagram of section *bb*, Fig. 6–18*c*, we have

$$\overset{+}{\rightarrow} \Sigma F_x = 0; \qquad F_{EF}\cos 30° - 3000\cos 30°\text{ N} = 0$$

$$F_{EF} = 3000\text{ N} \quad \text{(C)}$$

$$+\uparrow \Sigma F_y = 0; \qquad 2(3000\sin 30°\text{ N}) - 1000\text{ N} - F_{EB} = 0$$

$$F_{EB} = 2000\text{ N} \quad \text{(T)} \qquad \textit{Ans.}$$

FUNDAMENTAL PROBLEMS

All problem solutions must include FBDs.

F6–7. Determine the force in members *BC*, *CF*, and *FE*. State if the members are in tension or compression.

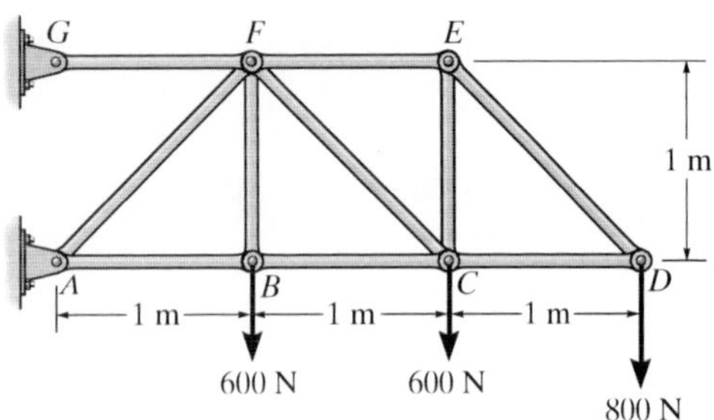

F6–7

F6–8. Determine the force in members *LK*, *KC*, and *CD* of the Pratt truss. State if the members are in tension or compression.

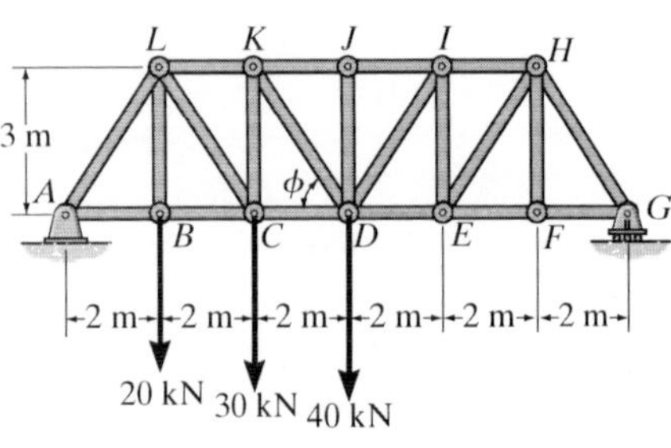

F6–8

F6–9. Determine the force in members *KJ*, *KD*, and *CD* of the Pratt truss. State if the members are in tension or compression.

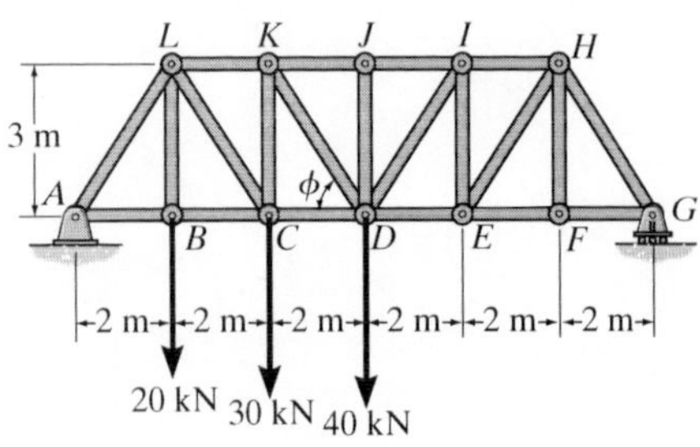

F6–9

F6–10. Determine the force in members *EF*, *CF*, and *BC* of the truss. State if the members are in tension or compression.

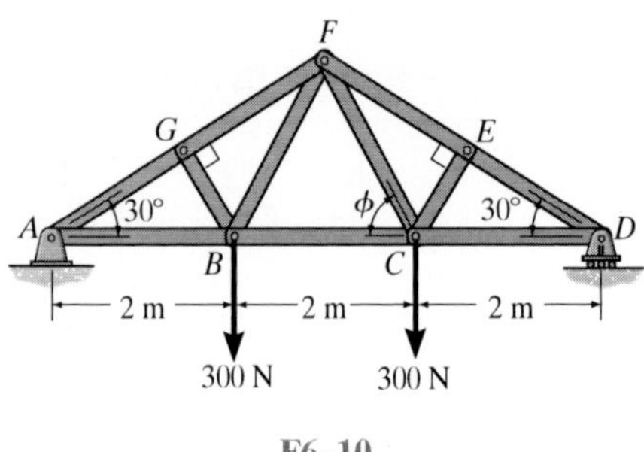

F6–10

F6–11. Determine the force in members *GF*, *GD*, and *CD* of the truss. State if the members are in tension or compression.

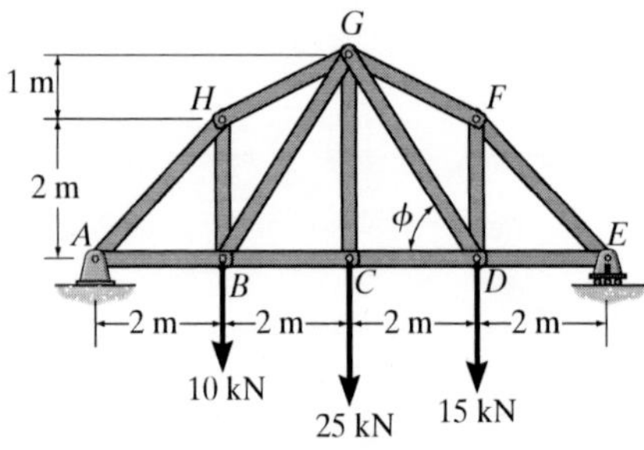

F6–11

F6–12. Determine the force in members *DC*, *HI*, and *JI* of the truss. State if the members are in tension or compression. *Suggestion:* Use the sections shown.

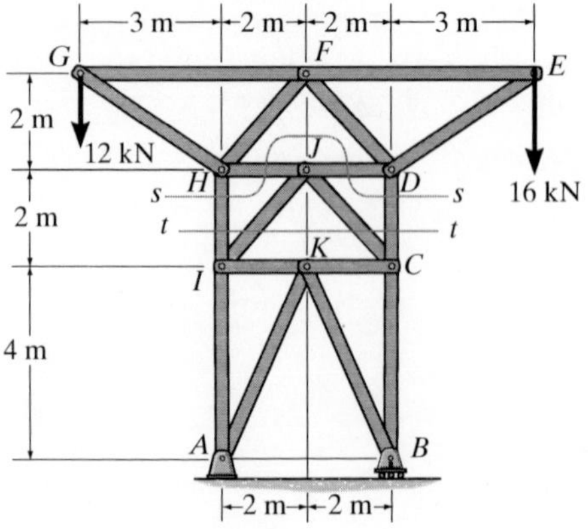

F6–12

PROBLEMS

All problem solutions must include FBDs.

6–27. Determine the force in members BC, CG, and GF of the *Warren truss*. Indicate if the members are in tension or compression.

***6–28.** Determine the force in members CD, CF, and FG of the *Warren truss*. Indicate if the members are in tension or compression.

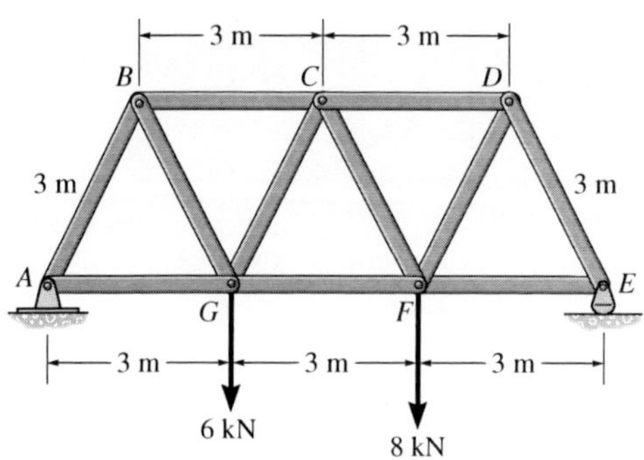

Probs. 6–27/28

6–29. The *Pratt bridge truss* is subjected to the loading shown. Determine the force in members LD, LK, CD, and KD, and state if the members are in tension or compression.

6–30. The *Pratt bridge truss* is subjected to the loading shown. Determine the force in members JI, JE, and DE, and state if the members are in tension or compression.

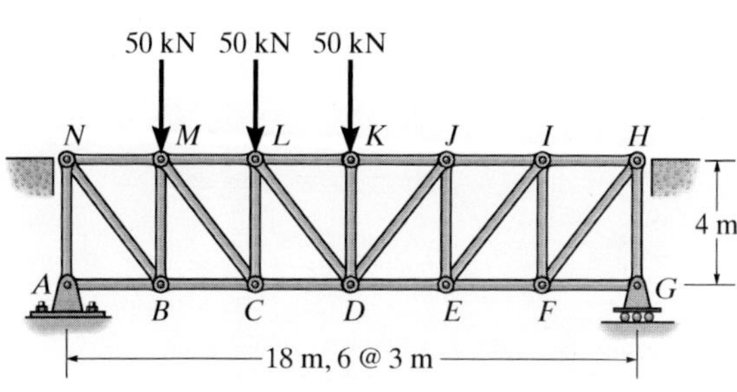

Probs. 6–29/30

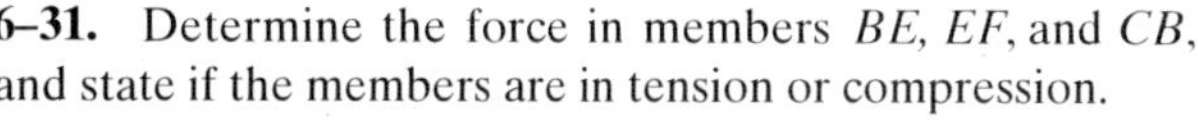

6–31. Determine the force in members BE, EF, and CB, and state if the members are in tension or compression.

***6–32.** Determine the force in members BF, BG, and AB, and state if the members are in tension or compression.

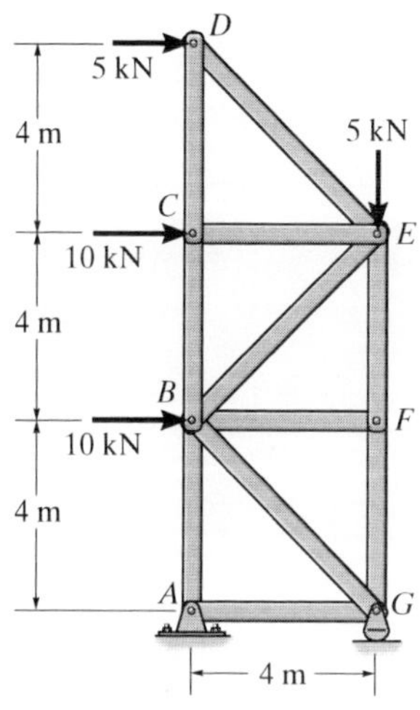

Probs. 6–31/32

6–33. Determine the force developed in members BC and CH of the roof truss and state if the members are in tension or compression.

6–34. Determine the force in members CD and GF of the truss and state if the members are in tension or compression. Also indicate all zero-force members.

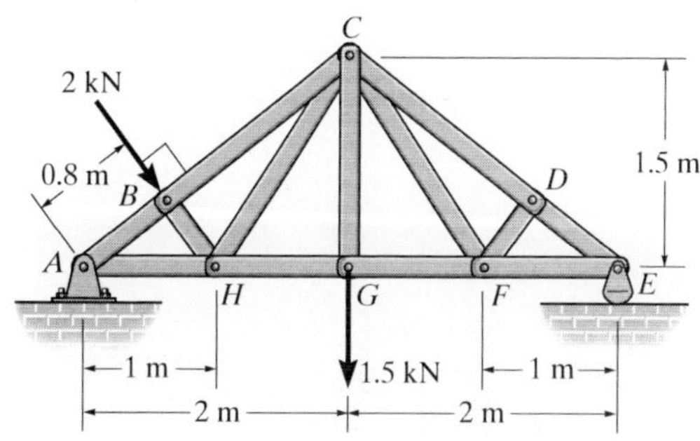

Probs. 6–33/34

6–35. Determine the force in members *BC*, *HC*, and *HG*. After the truss is sectioned use a single equation of equilibrium for the calculation of each force. State if these members are in tension or compression.

***6–36.** Determine the force in members *CD*, *CF*, and *CG* and state if these members are in tension or compression.

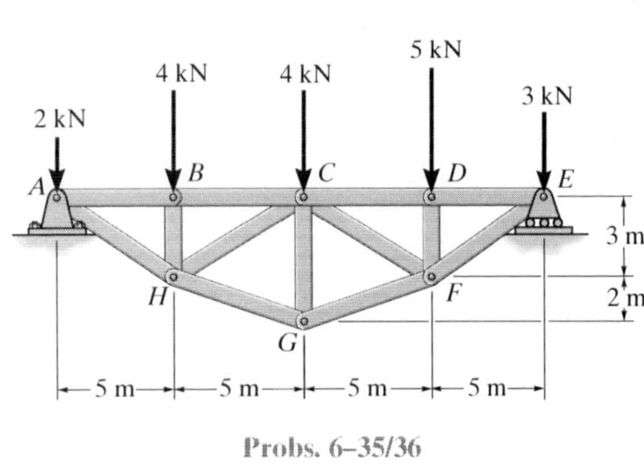

Probs. 6–35/36

6–39. Determine the force in members *IC* and *CG* of the truss and state if these members are in tension or compression. Also, indicate all zero-force members.

***6–40.** Determine the force in members *JE* and *GF* of the truss and state if these members are in tension or compression. Also, indicate all zero-force members.

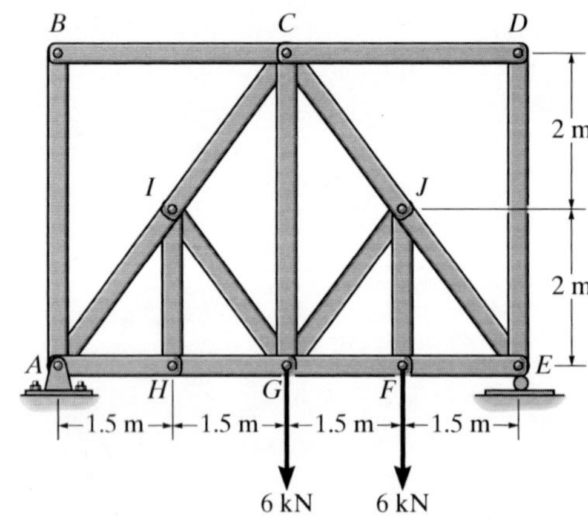

Probs. 6–39/40

6–37. Determine the force in members *KJ*, *NJ*, *ND*, and *CD* of the *K truss*. Indicate if the members are in tension or compression. *Hint:* Use sections *aa* and *bb*.

6–38. Determine the force in members *JI* and *DE* of the *K truss*. Indicate if the members are in tension or compression.

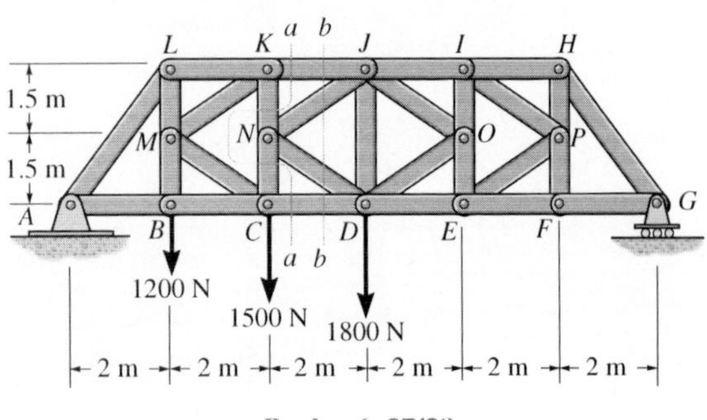

Probs. 6–37/38

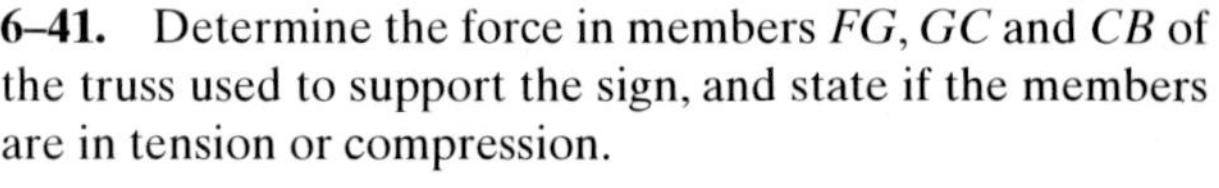

6–41. Determine the force in members *FG*, *GC* and *CB* of the truss used to support the sign, and state if the members are in tension or compression.

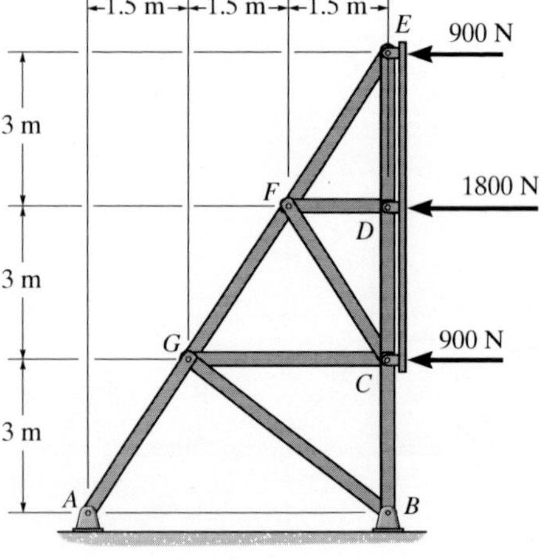

Prob. 6–41

6–42. Determine the force in members DE, DL, and ML of the roof truss and state if the members are in tension or compression.

6–43. Determine the force in members EF and EL of the roof truss and state if the members are in tension or compression.

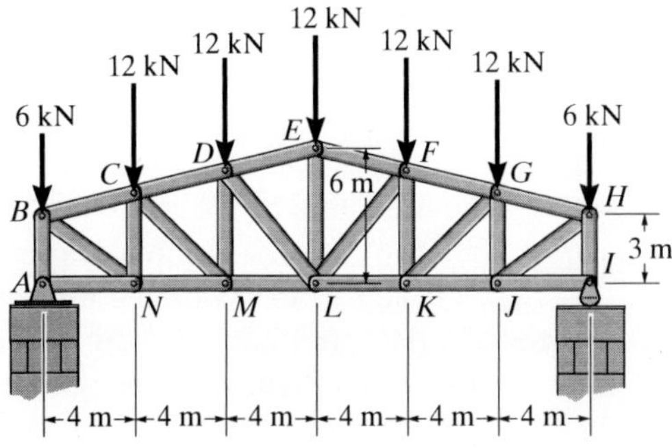

Probs. 6–42/43

***6–44.** The skewed truss carries the load shown. Determine the force in members CB, BE, and EF and state if these members are in tension or compression. Assume that all joints are pinned.

6–45. The skewed truss carries the load shown. Determine the force in members AB, BF, and EF and state if these members are in tension or compression. Assume that all joints are pinned.

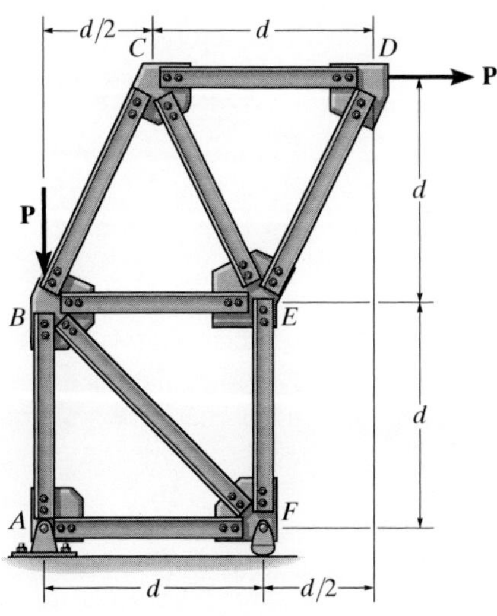

Probs. 6–44/45

6–46. Determine the force in members CD and CM of the *Baltimore bridge truss* and state if the members are in tension or compression. Also, indicate all zero-force members.

6–47. Determine the force in members EF, EP, and LK of the *Baltimore bridge truss* and state if the members are in tension or compression. Also, indicate all zero-force members.

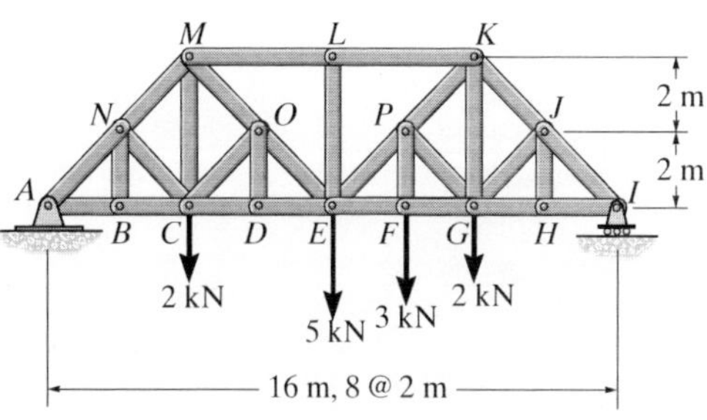

Probs. 6–46/47

***6–48.** The truss supports the vertical load of 600 N. If $L = 2$ m, determine the force on members HG and HB of the truss and state if the members are in tension or compression.

6–49. The truss supports the vertical load of 600 N. Determine the force in members BC, BG, and HG as the dimension L varies. Plot the results of F (ordinate with tension as positive) versus L (abscissa) for $0 \leq L \leq 3$ m.

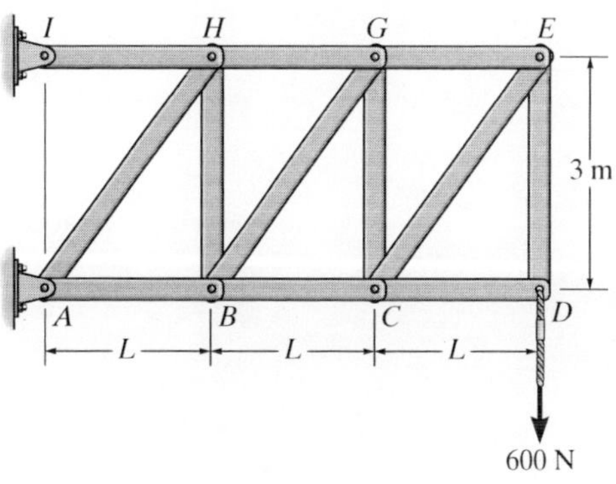

Probs. 6–48/49

6

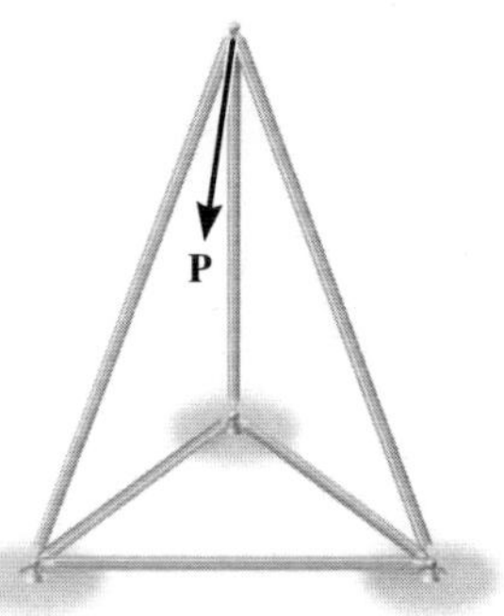

Fig. 6–19

Typical roof-supporting space truss. Notice the use of ball-and-socket joints for the connections.

For economic reasons, large electrical transmission towers are often constructed using space trusses.

*6.5 Space Trusses

A *space truss* consists of members joined together at their ends to form a stable three-dimensional structure. The simplest form of a space truss is a *tetrahedron*, formed by connecting six members together, as shown in Fig. 6–19. Any additional members added to this basic element would be redundant in supporting the force **P**. A *simple space truss* can be built from this basic tetrahedral element by adding three additional members and a joint, and continuing in this manner to form a system of multiconnected tetrahedrons.

Assumptions for Design. The members of a space truss may be treated as two-force members provided the external loading is applied at the joints and the joints consist of ball-and-socket connections. These assumptions are justified if the welded or bolted connections of the joined members intersect at a common point and the weight of the members can be neglected. In cases where the weight of a member is to be included in the analysis, it is generally satisfactory to apply it as a vertical force, half of its magnitude applied at each end of the member.

Procedure for Analysis

Either the method of joints or the method of sections can be used to determine the forces developed in the members of a simple space truss.

Method of Joints.

If the forces in *all* the members of the truss are to be determined, then the method of joints is most suitable for the analysis. Here it is necessary to apply the three equilibrium equations $\Sigma F_x = 0$, $\Sigma F_y = 0$, $\Sigma F_z = 0$ to the forces acting at each joint. Remember that the solution of many simultaneous equations can be avoided if the force analysis begins at a joint having at least one known force and at most three unknown forces. Also, if the three-dimensional geometry of the force system at the joint is hard to visualize, it is recommended that a Cartesian vector analysis be used for the solution.

Method of Sections.

If only a *few* member forces are to be determined, the method of sections can be used. When an imaginary section is passed through a truss and the truss is separated into two parts, the force system acting on one of the segments must satisfy the *six* equilibrium equations: $\Sigma F_x = 0$, $\Sigma F_y = 0$, $\Sigma F_z = 0$, $\Sigma M_x = 0$, $\Sigma M_y = 0$, $\Sigma M_z = 0$ (Eqs. 5–6). By proper choice of the section and axes for summing forces and moments, many of the unknown member forces in a space truss can be computed *directly*, using a single equilibrium equation.

EXAMPLE 6.8

Determine the forces acting in the members of the space truss shown in Fig. 6–20*a*. Indicate whether the members are in tension or compression.

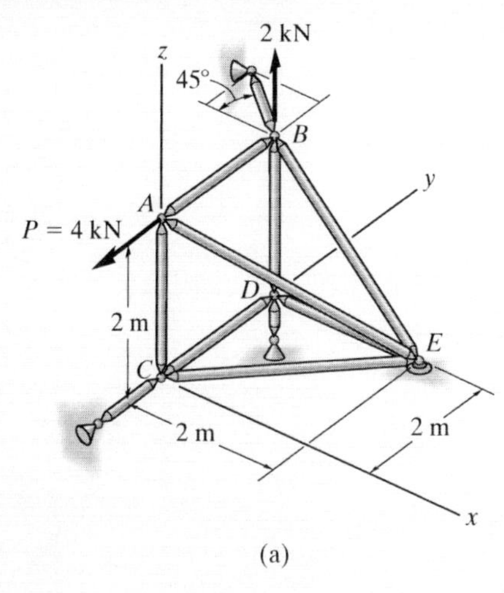

(a)

SOLUTION

Since there are one known force and three unknown forces acting at joint A, the force analysis of the truss will begin at this joint.

Joint A. (Fig. 6–20*b*). Expressing each force acting on the free-body diagram of joint A as a Cartesian vector, we have

$$\mathbf{P} = \{-4\mathbf{j}\} \text{ kN}, \qquad \mathbf{F}_{AB} = F_{AB}\mathbf{j}, \quad \mathbf{F}_{AC} = -F_{AC}\mathbf{k},$$

$$\mathbf{F}_{AE} = F_{AE}\left(\frac{\mathbf{r}_{AE}}{r_{AE}}\right) = F_{AE}(0.577\mathbf{i} + 0.577\mathbf{j} - 0.577\mathbf{k})$$

For equilibrium,

$$\Sigma\mathbf{F} = \mathbf{0}; \qquad \mathbf{P} + \mathbf{F}_{AB} + \mathbf{F}_{AC} + \mathbf{F}_{AE} = \mathbf{0}$$

$$-4\mathbf{j} + F_{AB}\mathbf{j} - F_{AC}\mathbf{k} + 0.577F_{AE}\mathbf{i} + 0.577F_{AE}\mathbf{j} - 0.577F_{AE}\mathbf{k} = \mathbf{0}$$

$$\Sigma F_x = 0; \qquad 0.577F_{AE} = 0$$

$$\Sigma F_y = 0; \qquad -4 + F_{AB} + 0.577F_{AE} = 0$$

$$\Sigma F_z = 0; \qquad -F_{AC} - 0.577F_{AE} = 0$$

$$F_{AC} = F_{AE} = 0 \qquad \textit{Ans.}$$

$$F_{AB} = 4 \text{ kN (T)} \qquad \textit{Ans.}$$

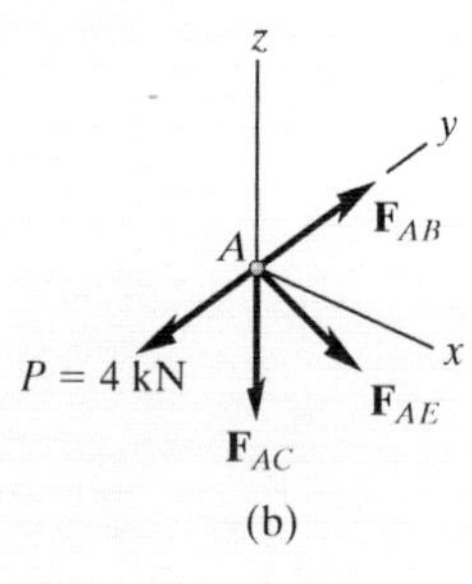

(b)

Since F_{AB} is known, joint B can be analyzed next.

Joint B. (Fig. 6–20*c*).

$$\Sigma F_x = 0; \qquad -R_B \cos 45° + 0.707F_{BE} = 0$$

$$\Sigma F_y = 0; \qquad -4 + R_B \sin 45° = 0$$

$$\Sigma F_z = 0; \qquad 2 + F_{BD} - 0.707F_{BE} = 0$$

$$R_B = F_{BE} = 5.66 \text{ kN (T)}, \qquad F_{BD} = 2 \text{ kN (C)} \qquad \textit{Ans.}$$

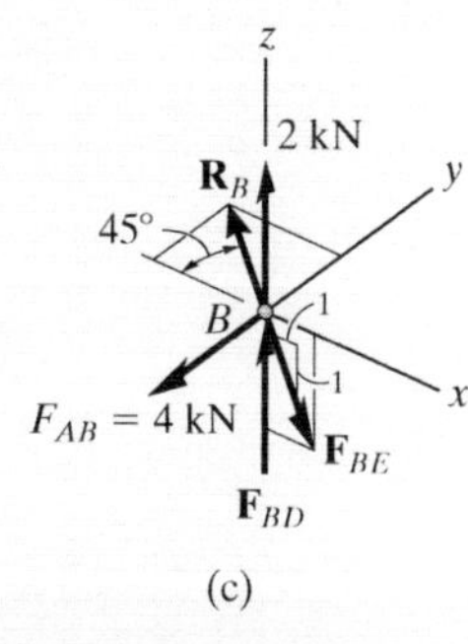

(c)

Fig. 6–20

The *scalar* equations of equilibrium may also be applied directly to the forces acting on the free-body diagrams of joints D and C since the force components are easily determined. Show that

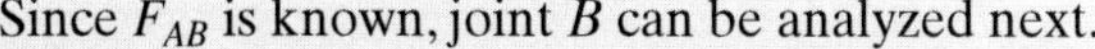

$$F_{DE} = F_{DC} = F_{CE} = 0 \qquad \textit{Ans.}$$

6

PROBLEMS

All problem solutions must include FBDs.

6–50. Two space trusses are used to equally support the uniform 50-kg sign. Determine the force developed in members AB, AC, and BC of truss $ABCD$ and state if the members are in tension or compression. Horizontal short links support the truss at joints B and D and there is a ball-and-socket joint at C.

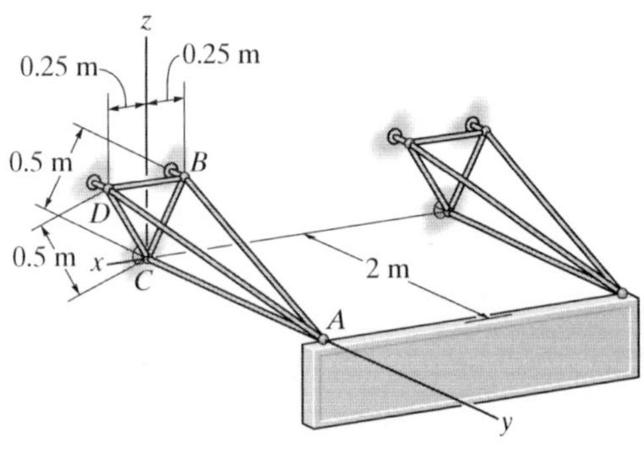

Prob. 6–50

6–51. Determine the force in each member of the space truss and state if the members are in tension or compression. *Hint:* The support reaction at E acts along member EB. Why?

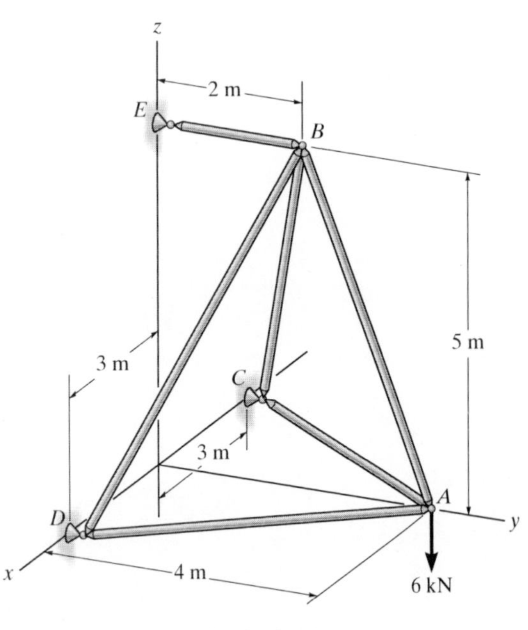

Prob. 6–51

***6–52.** Determine the force in each member of the space truss and state if the members are in tension or compression. The truss is supported by rollers at A, B, and C.

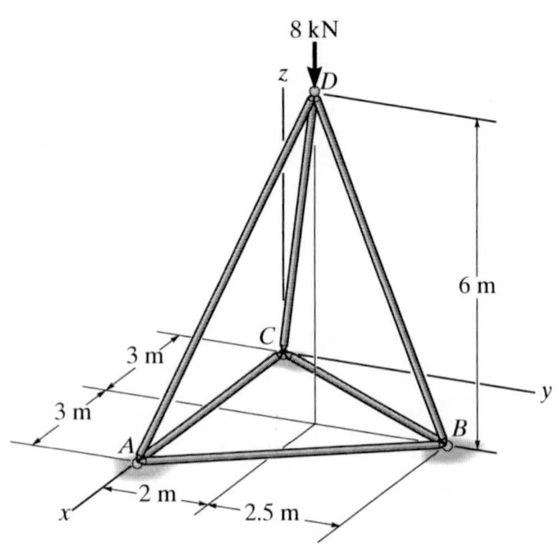

Prob. 6–52

6–53. The space truss supports a force $\mathbf{F} = [300\mathbf{i} + 400\mathbf{j} - 500\mathbf{k}]$ N. Determine the force in each member, and state if the members are in tension or compression.

6–54. The space truss supports a force $\mathbf{F} = [-400\mathbf{i} + 500\mathbf{j} + 600\mathbf{k}]$ N. Determine the force in each member, and state if the members are in tension or compression.

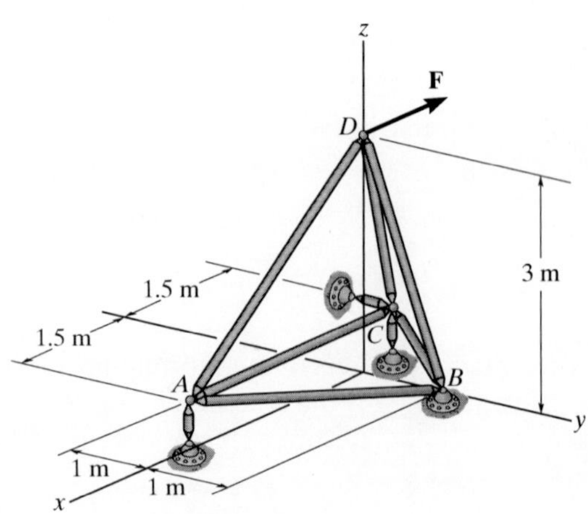

Probs. 6–53/54

6–55. Determine the force in each member of the space truss and state if the members are in tension or compression. The truss is supported by ball-and-socket joints at *C, D, E,* and *G*.

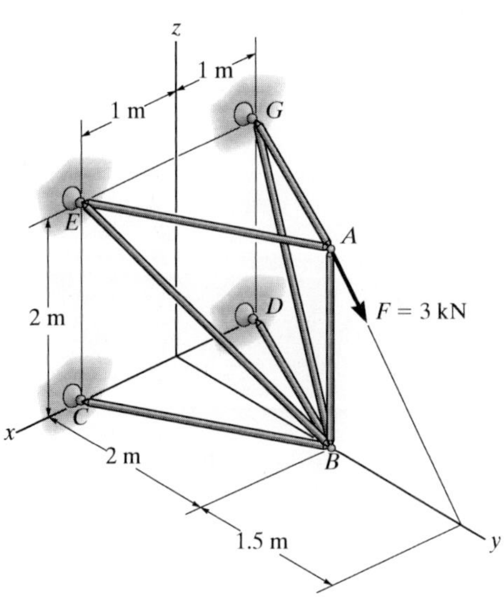

Prob. 6–55

***6–56.** The space truss is used to support vertical forces at joints *B, C,* and *D*. Determine the force in each member and state if the members are in tension or compression. There is a roller at *E,* and *A* and *F* are ball-and-socket joints.

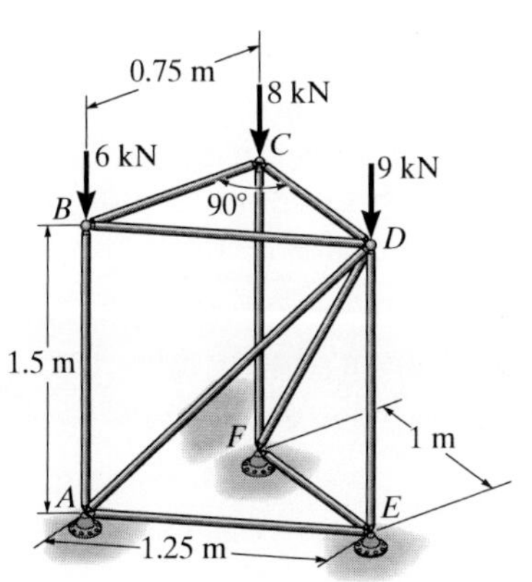

Prob. 6–56

6–57. Determine the force in members *BE, BC, BF,* and *CE* of the space truss, and state if the members are in tension or compression.

6–58. Determine the force in members *AF, AB, AD, ED, FD,* and *BD* of the space truss, and state if the members are in tension or compression.

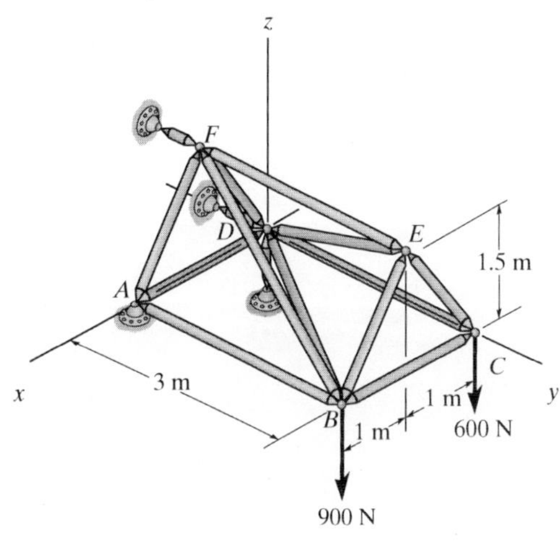

Probs. 6–57/58

6–59. If the truss supports a force of $F = 200$ N, determine the force in each member and state if the members are in tension or compression.

***6–60.** If each member of the space truss can support a maximum force of 600 N in compression and 800 N in tension, determine the greatest force *F* the truss can support.

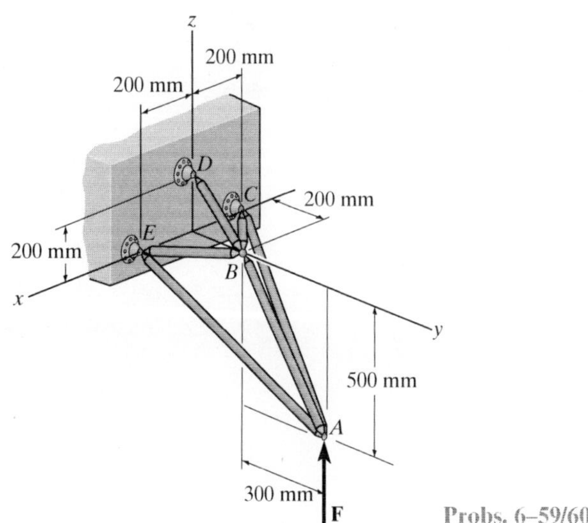

Probs. 6–59/60

6.6 Frames and Machines

This large crane is a typical example of a framework.

Frames and machines are two types of structures which are often composed of pin-connected *multiforce members*, i.e., members that are subjected to more than two forces. *Frames* are used to support loads, whereas *machines* contain moving parts and are designed to transmit and alter the effect of forces. Provided a frame or machine contains no more supports or members than are necessary to prevent its collapse, the forces acting at the joints and supports can be determined by applying the equations of equilibrium to each of its members. Once these forces are obtained, it is then possible to *design* the size of the members, connections, and supports using the theory of mechanics of materials and an appropriate engineering design code.

Free-Body Diagrams. In order to determine the forces acting at the joints and supports of a frame or machine, the structure must be disassembled and the free-body diagrams of its parts must be drawn. The following important points *must* be observed:

Common tools such as these pliers act as simple machines. Here the applied force on the handles creates a much larger force at the jaws.

- Isolate each part by drawing its *outlined shape*. Then show all the forces and/or couple moments that act on the part. Make sure to *label* or *identify* each known and unknown force and couple moment with reference to an established *x, y* coordinate system. Also, indicate any dimensions used for taking moments. Most often the equations of equilibrium are easier to apply if the forces are represented by their rectangular components. As usual, the sense of an unknown force or couple moment can be assumed.
- Identify all the two-force members in the structure and represent their free-body diagrams as having two equal but opposite collinear forces acting at their points of application. (See Sec. 5.4.) By recognizing the two-force members, we can avoid solving an unnecessary number of equilibrium equations.
- Forces common to *any* two *contacting* members act with equal magnitudes but opposite sense on the respective members. If the two members are treated as a *"system" of connected members*, then these forces are *"internal"* and are *not shown* on the *free-body diagram of the system*; however, if the free-body diagram of *each member* is drawn, the forces are *"external"* and *must* be shown as equal in magnitude and opposite in direction on each of the two free-body diagrams.

The following examples graphically illustrate how to draw the free-body diagrams of a dismembered frame or machine. In all cases, the weight of the members is neglected.

EXAMPLE 6.9

For the frame shown in Fig. 6–21*a*, draw the free-body diagram of (a) each member, (b) the pins at *B* and *A*, and (c) the two members connected together.

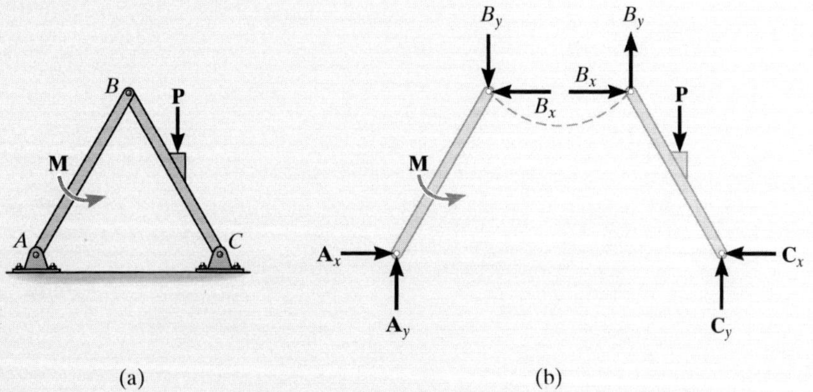

(a) (b)

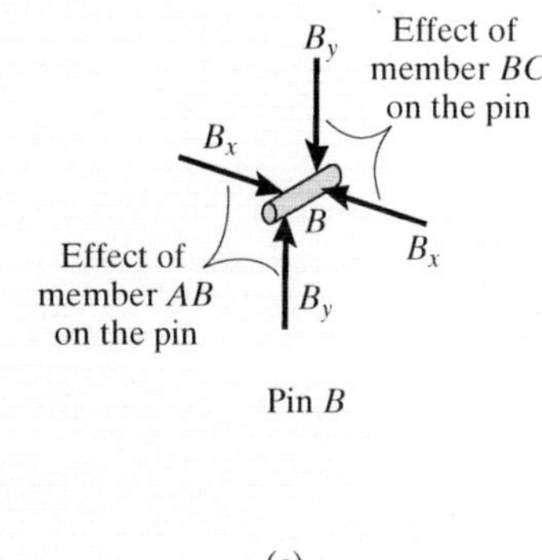

(c)

SOLUTION

Part (a). By inspection, members *BA* and *BC* are *not* two-force members. Instead, as shown on the free-body diagrams, Fig. 6–21*b*, *BC* is subjected to a force from each of the pins at *B* and *C* and the external force **P**. Likewise, *AB* is subjected to a force from each of the pins at *A* and *B* and the external couple moment **M**. The pin forces are represented by their *x* and *y* components.

Part (b). The pin at *B* is subjected to only *two forces*, i.e., the force of member *BC* and the force of member *AB*. For *equilibrium* these forces (or their respective components) must be equal but opposite, Fig. 6–21*c*. Realize that Newton's third law is applied between the pin and its connected members, i.e., the effect of the pin on the two members, Fig. 6–21*b*, and the equal but opposite effect of the two members on the pin, Fig. 6–21*c*. In the same manner, there are three forces on pin *A*, Fig. 6–21*d*, caused by the force components of member *AB* and each of the two pin leafs.

Part (c). The free-body diagram of both members connected together, yet removed from the supporting pins at *A* and *C*, is shown in Fig. 6–21*e*. The force components $\mathbf{B}_x$ and $\mathbf{B}_y$ are *not shown* on this diagram since they are *internal* forces (Fig. 6–21*b*) and therefore cancel out. Also, to be consistent when later applying the equilibrium equations, the unknown force components at *A* and *C* must act in the *same sense* as those shown in Fig. 6–21*b*.

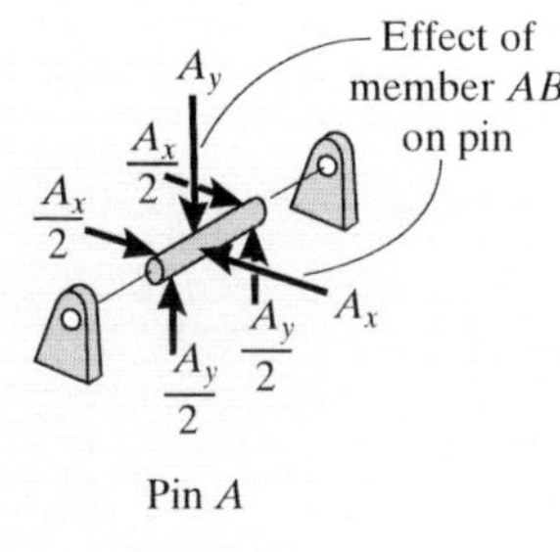

(d)

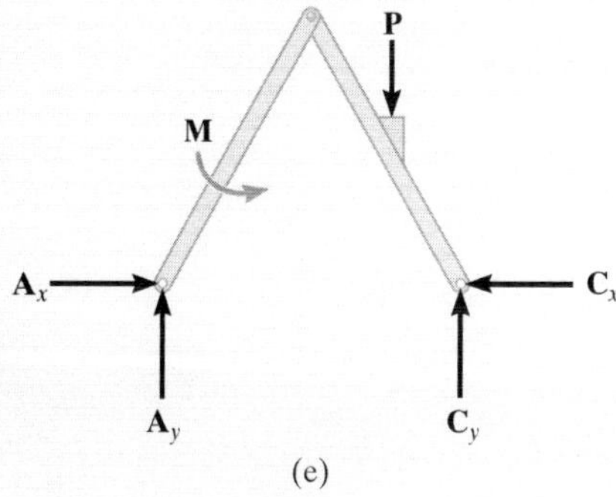

(e)

Fig. 6–21

EXAMPLE 6.10

A constant tension in the conveyor belt is maintained by using the device shown in Fig. 6–22*a*. Draw the free-body diagrams of the frame and the cylinder (or pulley) that the belt surrounds. The suspended block has a weight of *W*.

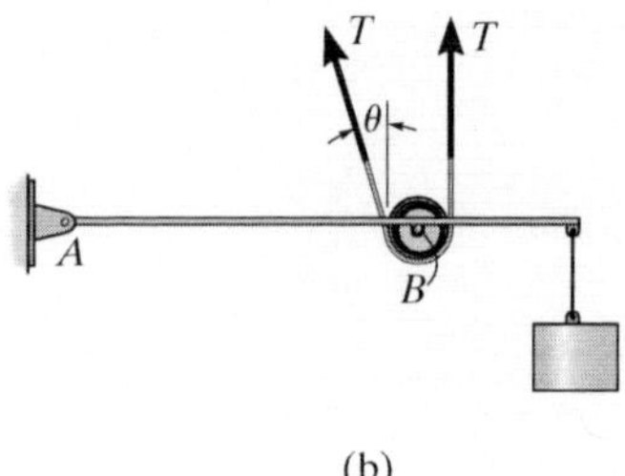

(b)

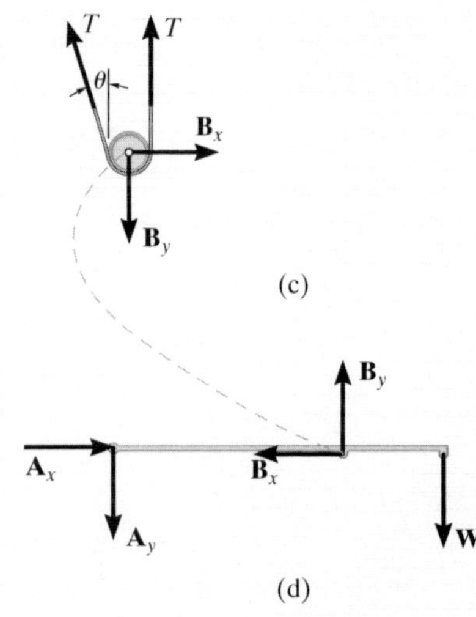

(d)

(a)

Fig. 6–22

SOLUTION

The idealized model of the device is shown in Fig. 6–22*b*. Here the angle θ is assumed to be known. From this model, the free-body diagrams of the pulley and frame are shown in Figs. 6–22*c* and 6–22*d*, respectively. Note that the force components $\mathbf{B}_x$ and $\mathbf{B}_y$ that the pin at *B* exerts on the pulley must be equal but opposite to the ones acting on the frame. See Fig. 6–21*c* of Example 6.9.

EXAMPLE 6.11

For the frame shown in Fig. 6–23*a*, draw the free-body diagrams of (a) the entire frame including the pulleys and cords, (b) the frame without the pulleys and cords, and (c) each of the pulleys.

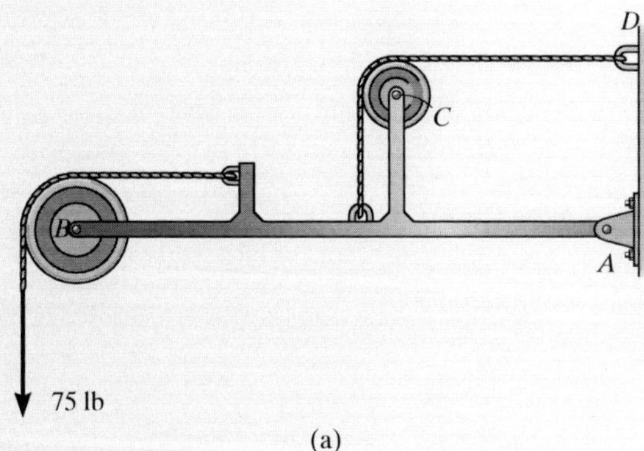

(a)

SOLUTION

Part (a). When the entire frame including the pulleys and cords is considered, the interactions at the points where the pulleys and cords are connected to the frame become pairs of *internal* forces which cancel each other and therefore are not shown on the free-body diagram, Fig. 6–23*b*.

Part (b). When the cords and pulleys are removed, their effect *on the frame* must be shown, Fig. 6–23*c*.

Part (c). The force components $\mathbf{B}_x$, $\mathbf{B}_y$, $\mathbf{C}_x$, $\mathbf{C}_y$ of the pins on the pulleys, Fig. 6–23*d*, are equal but opposite to the force components exerted by the pins on the frame, Fig. 6–23*c*. See Example 6.9.

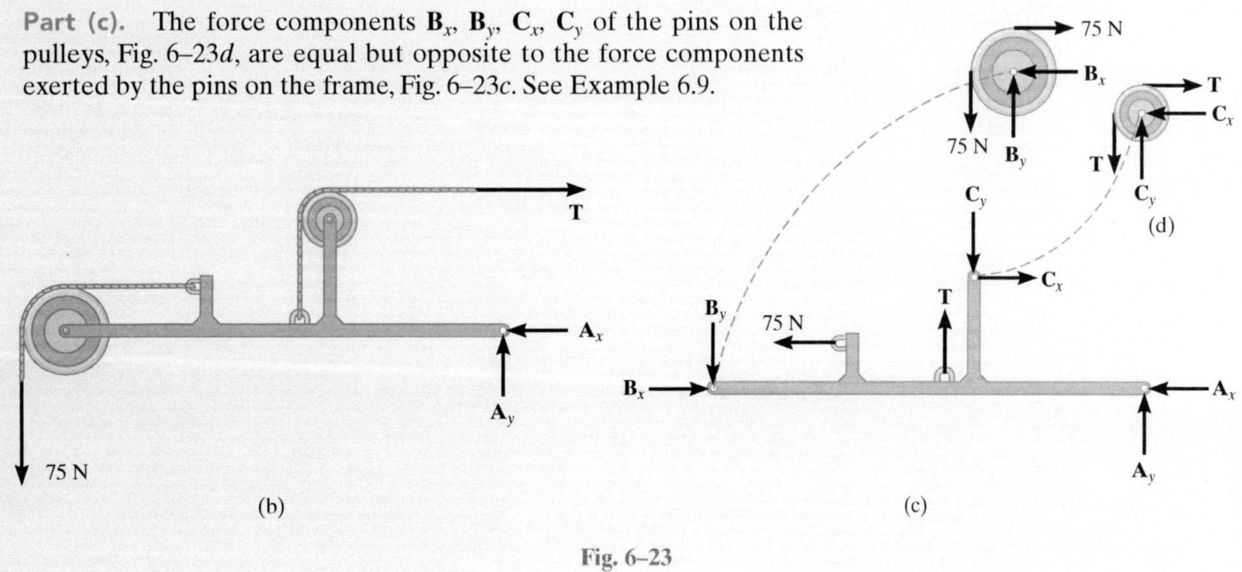

Fig. 6–23

6

EXAMPLE 6.12

(a)

Fig. 6–24

Draw the free-body diagrams of the members of the backhoe, shown in the photo, Fig. 6–24*a*. The bucket and its contents have a weight W.

SOLUTION

The idealized model of the assembly is shown in Fig. 6–24*b*. By inspection, members AB, BC, BE, and HI are all two-force members since they are pin connected at their end points and no other forces act on them. The free-body diagrams of the bucket and the stick are shown in Fig. 6–24*c*. Note that pin C is subjected to only two forces, whereas the pin at B is subjected to three forces, Fig. 6–24*d*. The free-body diagram of the entire assembly is shown in Fig. 6–24*e*.

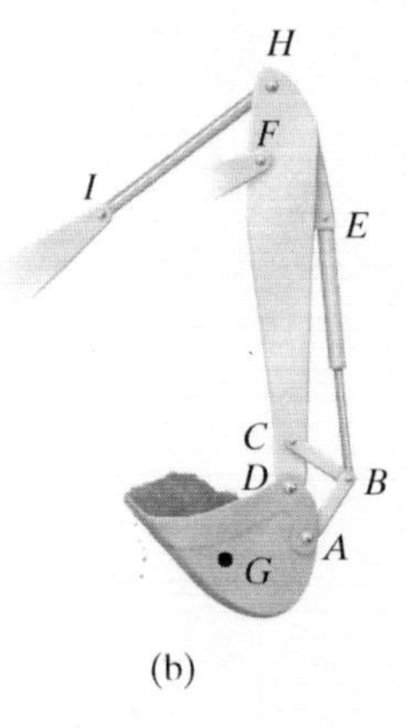

(b)

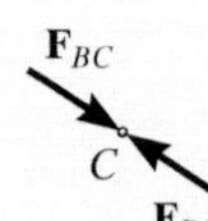

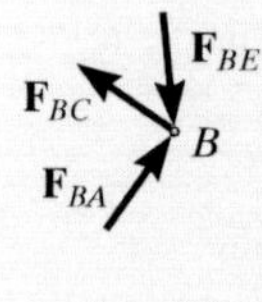

(d)

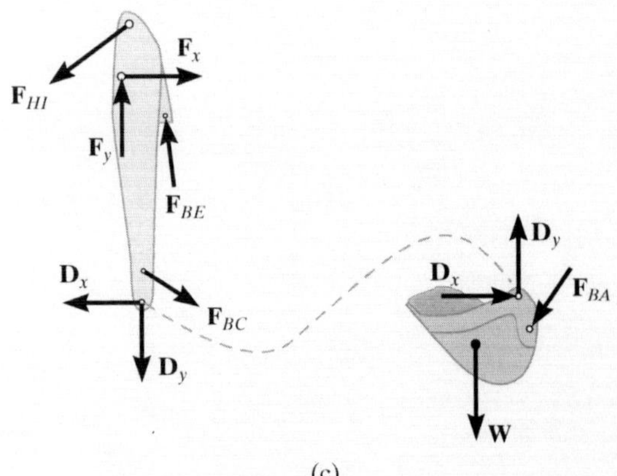

(c)

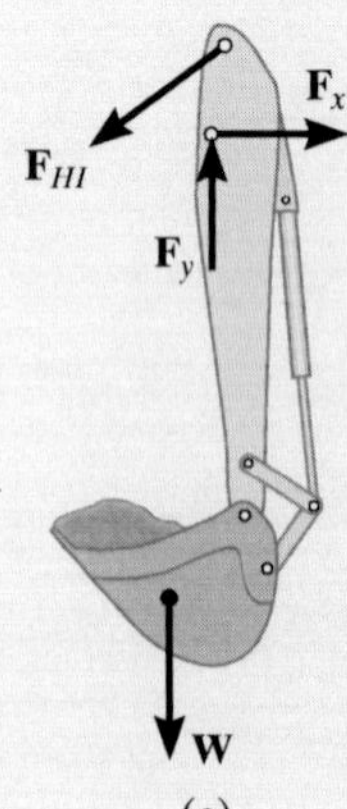

(e)

EXAMPLE 6.13

Draw the free-body diagram of each part of the smooth piston and link mechanism used to crush recycled cans, Fig. 6–25*a*.

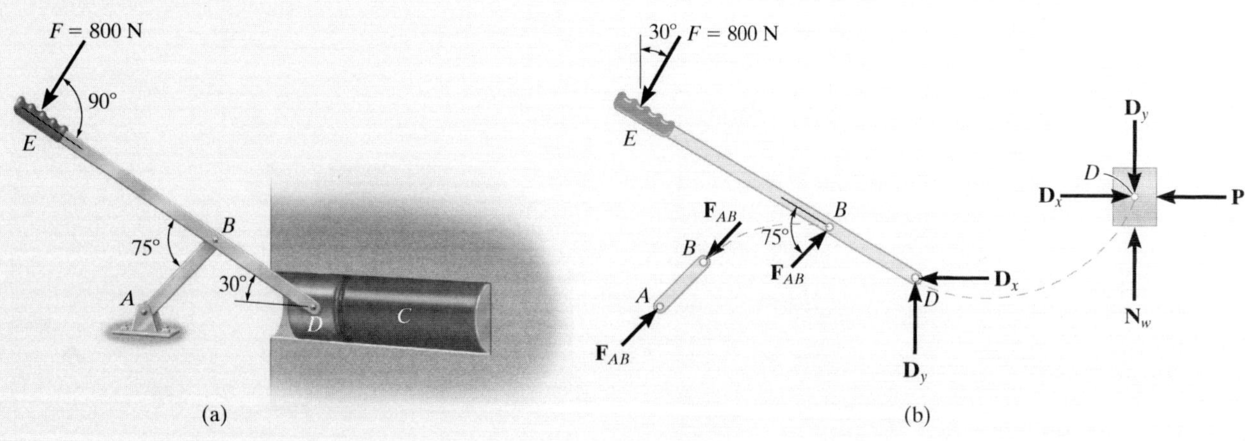

Fig. 6–25

SOLUTION

By inspection, member AB is a two-force member. The free-body diagrams of the three parts are shown in Fig. 6–25*b*. Since the pins at B and D *connect only two parts together*, the forces there are shown as equal but opposite on the separate free-body diagrams of their connected members. In particular, four components of force act on the piston: $\mathbf{D}_x$ and $\mathbf{D}_y$ represent the effect of the pin (or lever EBD), $\mathbf{N}_w$ is the *resultant force* of the wall support, and $\mathbf{P}$ is the resultant compressive force caused by the can C. The directional sense of each of the unknown forces is assumed, and the correct sense will be established after the equations of equilibrium are applied.

NOTE: A free-body diagram of the entire assembly is shown in Fig. 6–25*c*. Here the forces between the components are internal and are not shown on the free-body diagram.

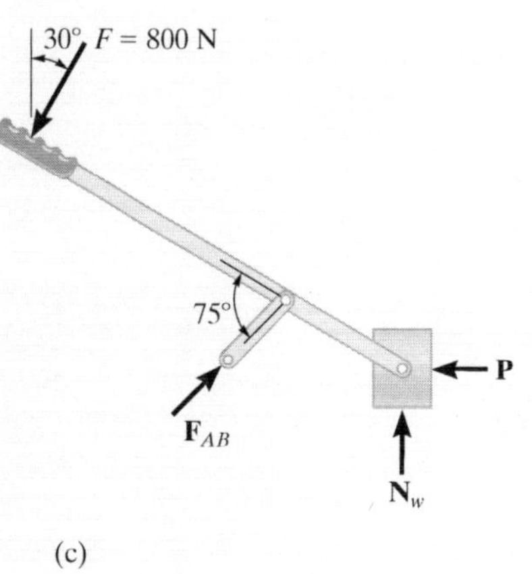

(c)

Before proceeding, it is highly recommended that you cover the solutions to the previous examples and attempt to draw the requested free-body diagrams. When doing so, make sure the work is neat and that all the forces and couple moments are properly labeled. When finished, challenge yourself and solve the following four problems.

CONCEPTUAL PROBLEMS

P6–1. Draw the free-body diagrams of each of the crane boom segments *AB*, *BC*, and *BD*. Only the weights of *AB* and *BC* are significant. Assume *A* and *B* are pins.

P6–1

P6–2. Draw the free-body diagrams of the boom *ABCD* and the stick *EDFGH* of the backhoe. The weights of these two members are significant. Neglect the weights of all the other members, and assume all indicated points of connection are pins.

P6–2

P6–3. Draw the free-body diagrams of the boom *ABCDF* and the stick *FGH* of the bucket lift. Neglect the weights of the members. The bucket weighs *W*. The two–force members are *BI*, *CE*, *DE* and *GE*. Assume all indicated points of connection are pins.

P6–3

P6–4. To operate the can crusher one pushes down on the lever arm *ABC* which rotates about the fixed pin at *B*. This moves the side links *CD* downward, which causes the guide plate *E* to also move downward and thereby crush the can. Draw the free-body diagrams of the lever, side link, and guide plate. Make up some reasonable numbers and do an equilibrium analysis to show how much an applied vertical force at the handle is magnified when it is transmitted to the can. Assume all points of connection are pins and the guides for the plate are smooth.

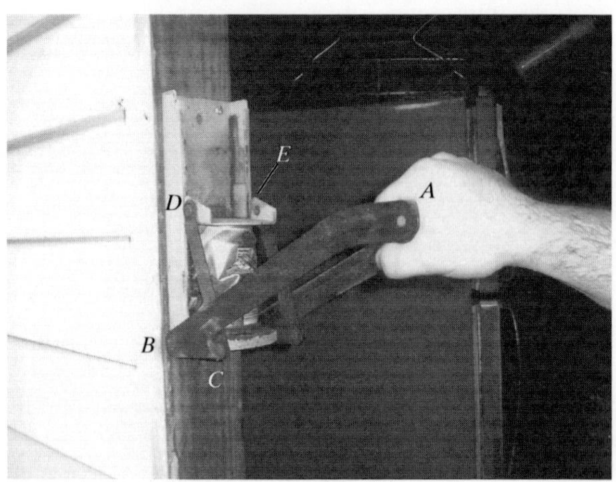

P6–4

Procedure for Analysis

The joint reactions on frames or machines (structures) composed of multiforce members can be determined using the following procedure.

Free-Body Diagram.

- Draw the free-body diagram of the entire frame or machine, a portion of it, or each of its members. The choice should be made so that it leads to the most direct solution of the problem.
- When the free-body diagram of a group of members of a frame or machine is drawn, the forces between the connected parts of this group are internal forces and are not shown on the free-body diagram of the group.
- Forces common to two members which are in contact act with equal magnitude but opposite sense on the respective free-body diagrams of the members.
- Two-force members, regardless of their shape, have equal but opposite collinear forces acting at the ends of the member.
- In many cases it is possible to tell by inspection the proper sense of the unknown forces acting on a member; however, if this seems difficult, the sense can be assumed.
- Remember that a couple moment is a free vector and can act at any point on the free-body diagram. Also, a force is a sliding vector and can act at any point along its line of action.

Equations of Equilibrium.

- Count the number of unknowns and compare it to the total number of equilibrium equations that are available. In two dimensions, there are three equilibrium equations that can be written for each member.
- Sum moments about a point that lies at the intersection of the lines of action of as many of the unknown forces as possible.
- If the solution of a force or couple moment magnitude is found to be negative, it means the sense of the force is the reverse of that shown on the free-body diagram.

EXAMPLE 6.14

Determine the tension in the cables and also the force **P** required to support the 600-N force using the frictionless pulley system shown in Fig. 6–26*a*.

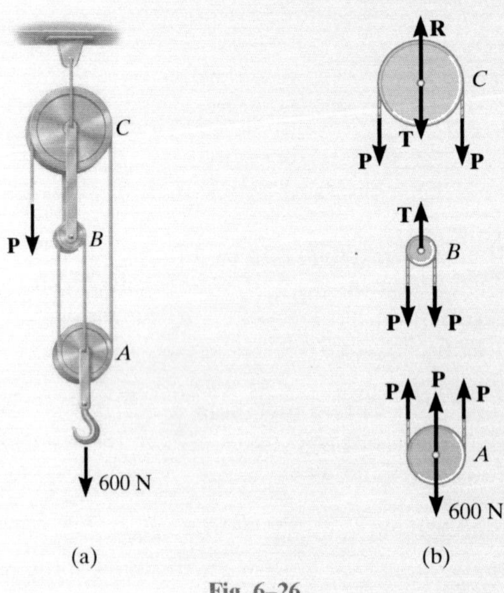

Fig. 6–26

6

SOLUTION

Free-Body Diagram. A free-body diagram of each pulley *including* its pin and a portion of the contacting cable is shown in Fig. 6–26*b*. Since the cable is *continuous*, it has a *constant tension* P acting throughout its length. The link connection between pulleys B and C is a two-force member, and therefore it has an unknown tension T acting on it. Notice that the *principle of action, equal but opposite reaction* must be carefully observed for forces **P** and **T** when the *separate* free-body diagrams are drawn.

Equations of Equilibrium. The three unknowns are obtained as follows:

Pulley A

$+\uparrow \Sigma F_y = 0;$ $\quad 3P - 600\text{ N} = 0 \quad P = 200\text{ N}$ *Ans.*

Pulley B

$+\uparrow \Sigma F_y = 0;$ $\quad T - 2P = 0 \quad T = 400\text{ N}$ *Ans.*

Pulley C

$+\uparrow \Sigma F_y = 0;$ $\quad R - 2P - T = 0 \quad R = 800\text{ N}$ *Ans.*

EXAMPLE 6.15

A 500-kg elevator car in Fig. 6–27*a* is being hoisted by motor *A* using the pulley system shown. If the car is traveling with a constant speed, determine the force developed in the two cables. Neglect the mass of the cable and pulleys.

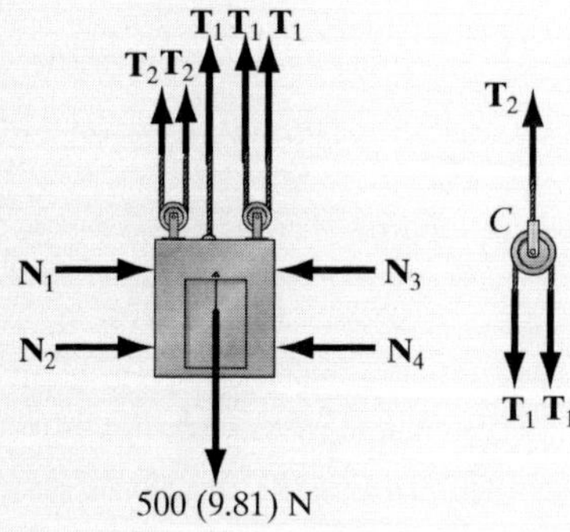

(b)

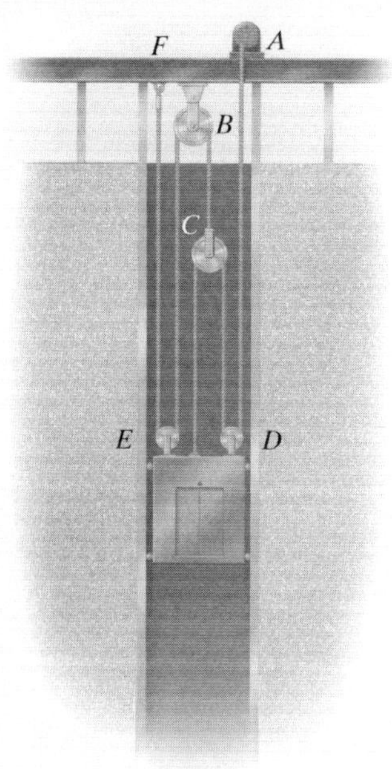

(a)

Fig. 6–27

SOLUTION

Free-Body Diagram. We can solve this problem using the free-body diagrams of the elevator car and pulley *C*, Fig. 6–27*b*. The tensile forces developed in the cables are denoted as T_1 and T_2.

Equations of Equilibrium. For pulley *C*,

$$+\uparrow \Sigma F_y = 0; \qquad T_2 - 2T_1 = 0 \qquad \text{or} \qquad T_2 = 2T_1 \qquad (1)$$

For the elevator car,

$$+\uparrow \Sigma F_y = 0; \qquad 3T_1 + 2T_2 - 500(9.81)\text{ N} = 0 \qquad (2)$$

Substituting Eq. (1) into Eq. (2) yields

$$3T_1 + 2(2T_1) - 500(9.81)\text{ N} = 0$$

$$T_1 = 700.71\text{ N} = 701\text{ N} \qquad \textit{Ans.}$$

Substituting this result into Eq. (1),

$$T_2 = 2(700.71)\text{ N} = 1401\text{ N} = 1.40\text{ kN} \qquad \textit{Ans.}$$

EXAMPLE 6.16

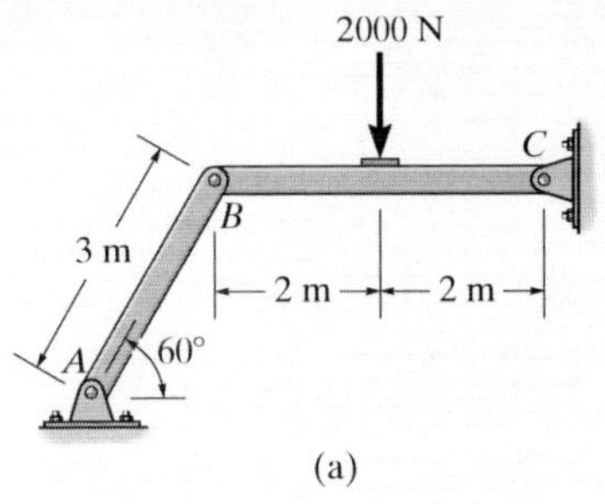

(a)

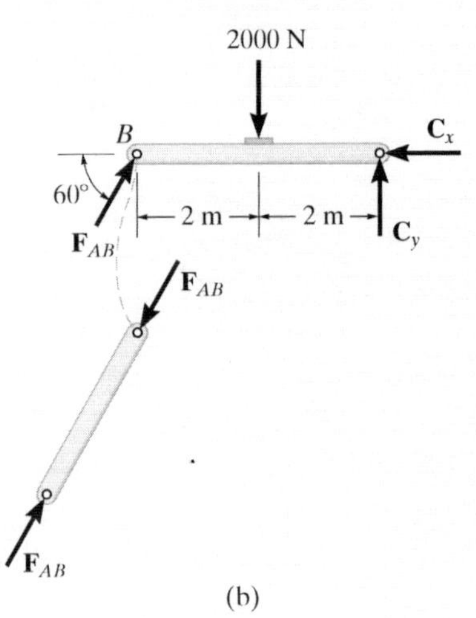

(b)

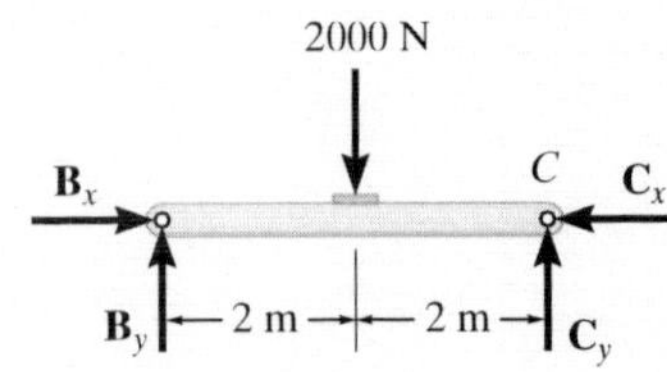

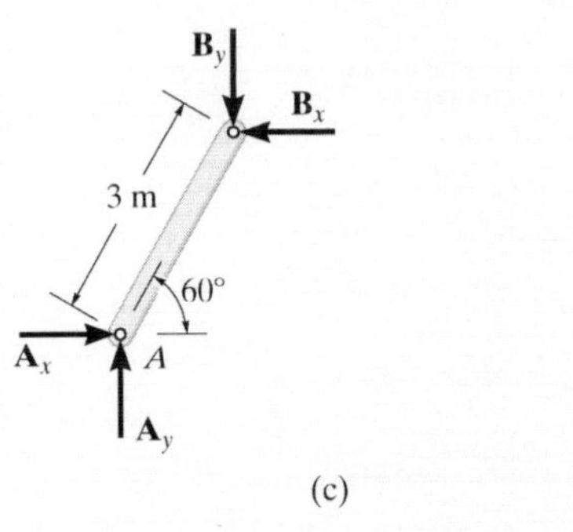

(c)

Fig. 6–28

Determine the horizontal and vertical components of force which the pin at C exerts on member BC of the frame in Fig. 6–28a.

SOLUTION I

Free-Body Diagrams. By inspection it can be seen that AB is a two-force member. The free-body diagrams are shown in Fig. 6–28b.

Equations of Equilibrium. The *three unknowns* can be determined by applying the three equations of equilibrium to member CB.

$$\circlearrowleft+\Sigma M_C = 0;\quad 2000\text{ N}(2\text{ m}) - (F_{AB}\sin 60^\circ)(4\text{ m}) = 0 \quad F_{AB} = 1154.7\text{ N}$$

$$\overset{+}{\rightarrow}\Sigma F_x = 0;\quad 1154.7\cos 60^\circ\text{ N} - C_x = 0 \quad C_x = 577\text{ N} \qquad \textit{Ans.}$$

$$+\uparrow\Sigma F_y = 0;\quad 1154.7\sin 60^\circ\text{ N} - 2000\text{ N} + C_y = 0$$

$$C_y = 1000\text{ N} \qquad \textit{Ans.}$$

SOLUTION II

Free-Body Diagrams. If one does not recognize that AB is a two-force member, then more work is involved in solving this problem. The free-body diagrams are shown in Fig. 6–28c.

Equations of Equilibrium. The *six unknowns* are determined by applying the three equations of equilibrium to each member.

Member AB

$$\circlearrowleft+\Sigma M_A = 0;\quad B_x(3\sin 60^\circ\text{ m}) - B_y(3\cos 60^\circ\text{ m}) = 0 \qquad (1)$$

$$\overset{+}{\rightarrow}\Sigma F_x = 0;\quad A_x - B_x = 0 \qquad (2)$$

$$+\uparrow\Sigma F_y = 0;\quad A_y - B_y = 0 \qquad (3)$$

Member BC

$$\circlearrowleft+\Sigma M_C = 0;\quad 2000\text{ N}(2\text{ m}) - B_y(4\text{ m}) = 0 \qquad (4)$$

$$\overset{+}{\rightarrow}\Sigma F_x = 0;\quad B_x - C_x = 0 \qquad (5)$$

$$+\uparrow\Sigma F_y = 0;\quad B_y - 2000\text{ N} + C_y = 0 \qquad (6)$$

The results for C_x and C_y can be determined by solving these equations in the following sequence: 4, 1, 5, then 6. The results are

$$B_y = 1000\text{ N}$$

$$B_x = 577\text{ N}$$

$$C_x = 577\text{ N} \qquad \textit{Ans.}$$

$$C_y = 1000\text{ N} \qquad \textit{Ans.}$$

By comparison, Solution I is simpler since the requirement that F_{AB} in Fig. 6–28b be equal, opposite, and collinear at the ends of member AB automatically satisfies Eqs. 1, 2, and 3 above and therefore eliminates the need to write these equations. ***As a result, save yourself some time and effort by always identifying the two-force members before starting the analysis!***

EXAMPLE 6.17

The compound beam shown in Fig. 6–29*a* is pin connected at *B*. Determine the components of reaction at its supports. Neglect its weight and thickness.

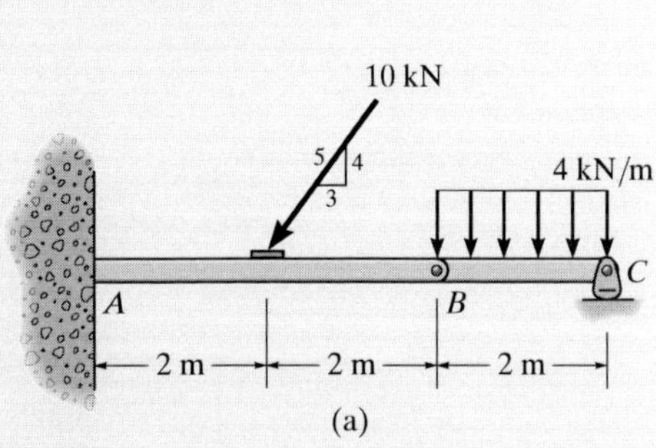

(a)

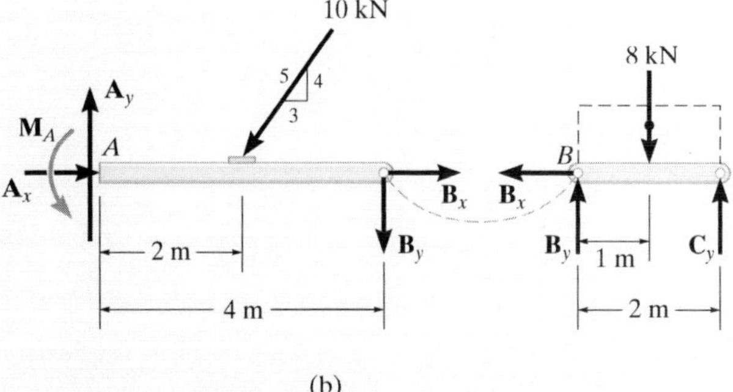

(b)

Fig. 6–29

SOLUTION

Free-Body Diagrams. By inspection, if we consider a free-body diagram of the *entire beam ABC*, there will be three unknown reactions at *A* and one at *C*. These four unknowns cannot all be obtained from the three available equations of equilibrium, and so for the solution it will become necessary to dismember the beam into its two segments, as shown in Fig. 6–29*b*.

Equations of Equilibrium. The six unknowns are determined as follows:

Segment BC

$\stackrel{+}{\leftarrow} \Sigma F_x = 0; \qquad B_x = 0$

$\zeta + \Sigma M_B = 0; \qquad -8\text{ kN}(1\text{ m}) + C_y(2\text{ m}) = 0$

$+\uparrow \Sigma F_y = 0; \qquad B_y - 8\text{ kN} + C_y = 0$

Segment AB

$\stackrel{+}{\rightarrow} \Sigma F_x = 0; \qquad A_x - (10\text{ kN})\left(\frac{3}{5}\right) + B_x = 0$

$\zeta + \Sigma M_A = 0; \qquad M_A - (10\text{ kN})\left(\frac{4}{5}\right)(2\text{ m}) - B_y(4\text{ m}) = 0$

$+\uparrow \Sigma F_y = 0; \qquad A_y - (10\text{ kN})\left(\frac{4}{5}\right) - B_y = 0$

Solving each of these equations successively, using previously calculated results, we obtain

$A_x = 6\text{ kN} \qquad A_y = 12\text{ kN} \qquad M_A = 32\text{ kN}\cdot\text{m}$ *Ans.*

$B_x = 0 \qquad B_y = 4\text{ kN}$

$C_y = 4\text{ kN}$ *Ans.*

6

EXAMPLE 6.18

The two planks in Fig. 6–30*a* are connected together by cable *BC* and a smooth spacer *DE*. Determine the reactions at the smooth supports *A* and *F*, and also find the force developed in the cable and spacer.

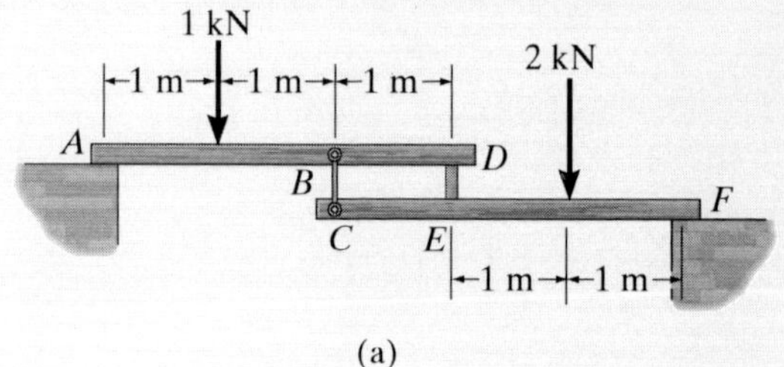

(a)

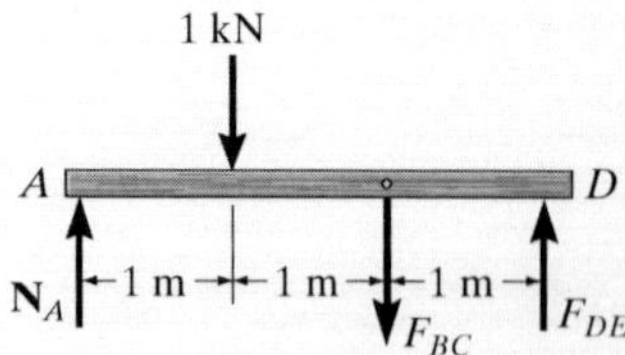

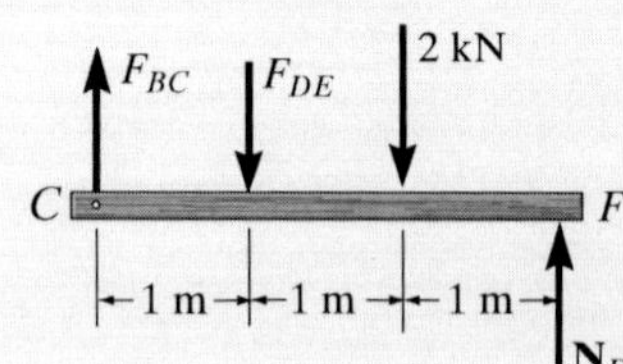

(b)

Fig. 6–30

SOLUTION

Free-Body Diagrams. The free-body diagram of each plank is shown in Fig. 6–30*b*. It is important to apply Newton's third law to the interaction forces F_{BC} and F_{DE} as shown.

Equations of Equilibrium. For plank *AD*,

$$\circlearrowleft +\Sigma M_A = 0; \qquad F_{DE}(3\text{ m}) - F_{BC}(2\text{ m}) - 1\text{ kN}(1\text{ m}) = 0$$

For plank *CF*,

$$\circlearrowleft +\Sigma M_F = 0; \qquad F_{DE}(2\text{ m}) - F_{BC}(3\text{ m}) + 2\text{ kN}(1\text{ m}) = 0$$

Solving simultaneously,

$$F_{DE} = 1.40\text{ kN} \quad F_{BC} = 1.60\text{ kN} \qquad \textit{Ans.}$$

Using these results, for plank *AD*,

$$+\uparrow\Sigma F_y = 0; \qquad N_A + 1.40\text{ kN} - 1.60\text{ kN} - 1\text{ kN} = 0$$

$$N_A = 1.20\text{ kN} \qquad \textit{Ans.}$$

And for plank *CF*,

$$+\uparrow\Sigma F_y = 0; \qquad N_F + 1.60\text{ kN} - 1.40\text{ kN} - 2\text{ kN} = 0$$

$$N_F = 1.80\text{ kN} \qquad \textit{Ans.}$$

NOTE: Draw the free-body diagram of the system of *both* planks and apply $\Sigma M_A = 0$ to determine N_F. Then use the free-body diagram of *CEF* to determine F_{DE} and F_{BC}.

EXAMPLE 6.19

The 75-kg man in Fig. 6–31*a* attempts to lift the 40-kg uniform beam off the roller support at *B*. Determine the tension developed in the cable attached to *B* and the normal reaction of the man on the beam when this is about to occur.

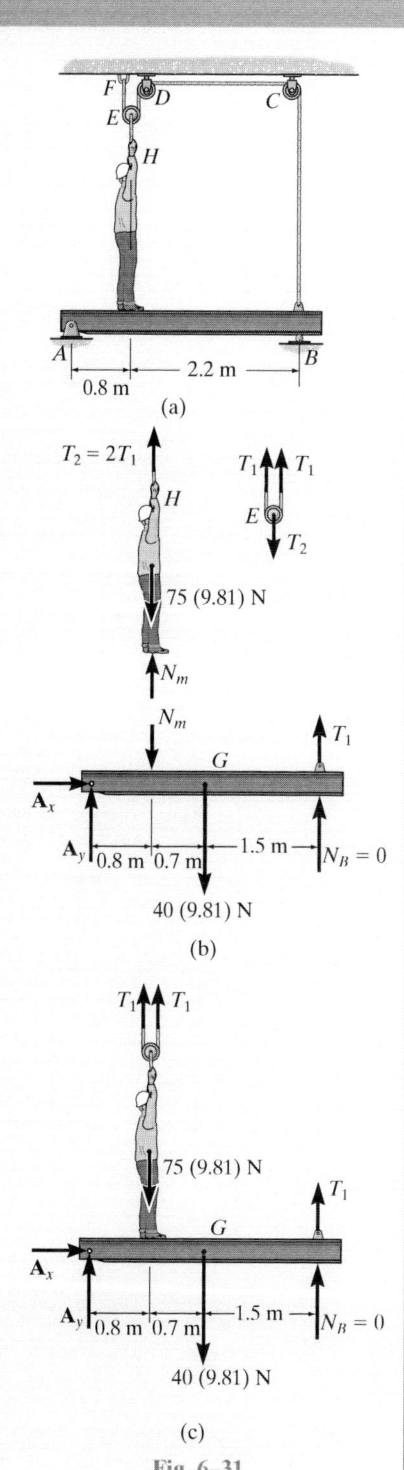

Fig. 6–31

SOLUTION

Free-Body Diagrams. The tensile force in the cable will be denoted as T_1. The free-body diagrams of the pulley *E*, the man, and the beam are shown in Fig. 6–31*b*. Since the man must lift the beam off the roller *B* then $N_B = 0$. When drawing each of these diagrams, it is very important to apply Newton's third law.

Equations of Equilibrium. Using the free-body diagram of pulley *E*,

$$+\uparrow\Sigma F_y = 0; \qquad 2T_1 - T_2 = 0 \quad \text{or} \quad T_2 = 2T_1 \qquad (1)$$

Referring to the free-body diagram of the man using this result,

$$+\uparrow\Sigma F_y = 0 \qquad N_m + 2T_1 - 75(9.81)\text{ N} = 0 \qquad (2)$$

Summing moments about point *A* on the beam,

$$\circlearrowleft+\Sigma M_A = 0; \; T_1(3\text{ m}) - N_m(0.8\text{ m}) - [40(9.81)\text{ N}](1.5\text{ m}) = 0 \qquad (3)$$

Solving Eqs. 2 and 3 simultaneously for T_1 and N_m, then using Eq. (1) for T_2, we obtain

$$T_1 = 256\text{ N} \qquad N_m = 224\text{ N} \qquad T_2 = 512\text{ N} \qquad \textit{Ans.}$$

SOLUTION II

A direct solution for T_1 can be obtained by considering the beam, the man, and pulley *E* as a *single system*. The free-body diagram is shown in Fig. 6–31*c*. Thus,

$$\circlearrowleft+\Sigma M_A = 0; \qquad 2T_1(0.8\text{ m}) - [75(9.81)\text{ N}](0.8\text{ m})$$
$$- [40(9.81)\text{ N}](1.5\text{ m}) + T_1(3\text{ m}) = 0$$
$$T_1 = 256\text{ N} \qquad \textit{Ans.}$$

With this result Eqs. 1 and 2 can then be used to find N_m and T_2.

EXAMPLE 6.20

The smooth disk shown in Fig. 6–32*a* is pinned at *D* and has a weight of 20 lb. Neglecting the weights of the other members, determine the horizontal and vertical components of reaction at pins *B* and *D*.

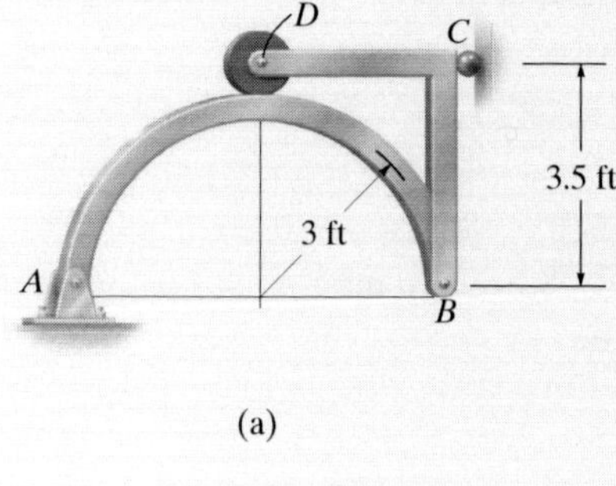

(a)

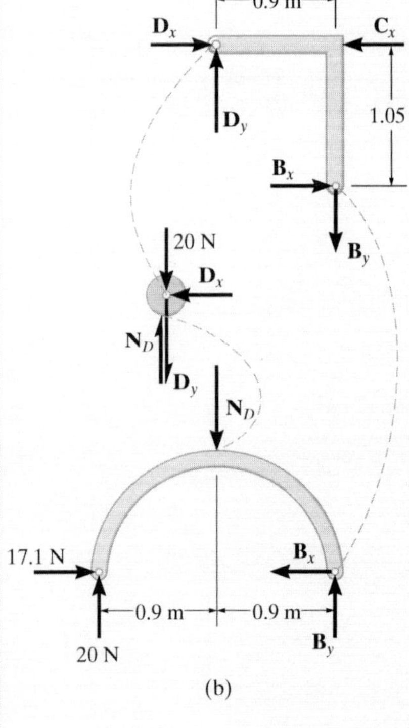

(b)

Fig. 6–32

SOLUTION

Free-Body Diagrams. The free-body diagrams of the entire frame and each of its members are shown in Fig. 6–32*b*.

Equations of Equilibrium. The eight unknowns can of course be obtained by applying the eight equilibrium equations to each member—three to member *AB*, three to member *BCD*, and two to the disk. (Moment equilibrium is automatically satisfied for the disk.) If this is done, however, all the results can be obtained only from a simultaneous solution of some of the equations. (Try it and find out.) To avoid this situation, it is best first to determine the three support reactions on the *entire* frame; then, using these results, the remaining five equilibrium equations can be applied to two other parts in order to solve successively for the other unknowns.

Entire Frame

$\circlearrowleft + \Sigma M_A = 0; \quad -20\text{ N}(0.9\text{ m}) + C_x(1.05\text{ m}) = 0 \quad C_x = 17.1\text{ N}$

$\overset{+}{\rightarrow} \Sigma F_x = 0; \quad A_x - 17.1\text{ N} = 0 \quad A_x = 17.1\text{ N}$

$+\uparrow \Sigma F_y = 0; \quad A_y - 20\text{ N} = 0 \quad A_y = 20\text{ N}$

Member AB

$\overset{+}{\rightarrow} \Sigma F_x = 0; \quad 17.1\text{ N} - B_x = 0 \quad B_x = 17.1\text{ N}$ *Ans.*

$\circlearrowleft + \Sigma M_B = 0; \quad -20\text{ N}(1.8\text{ m}) + N_D(0.9\text{ m}) = 0 \quad N_D = 40\text{ N}$

$+\uparrow \Sigma F_y = 0; \quad 20\text{ N} - 40\text{ N} + B_y = 0 \quad B_y = 20\text{ N}$ *Ans.*

Disk

$\overset{+}{\rightarrow} \Sigma F_x = 0; \quad D_x = 0$ *Ans.*

$+\uparrow \Sigma F_y = 0; \quad 40\text{ N} - 20\text{ N} - D_y = 0 \quad D_y = 20\text{ N}$ *Ans.*

EXAMPLE 6.21

The frame in Fig. 6–33*a* supports the 50-kg cylinder. Determine the horizontal and vertical components of reaction at *A* and the force at *C*.

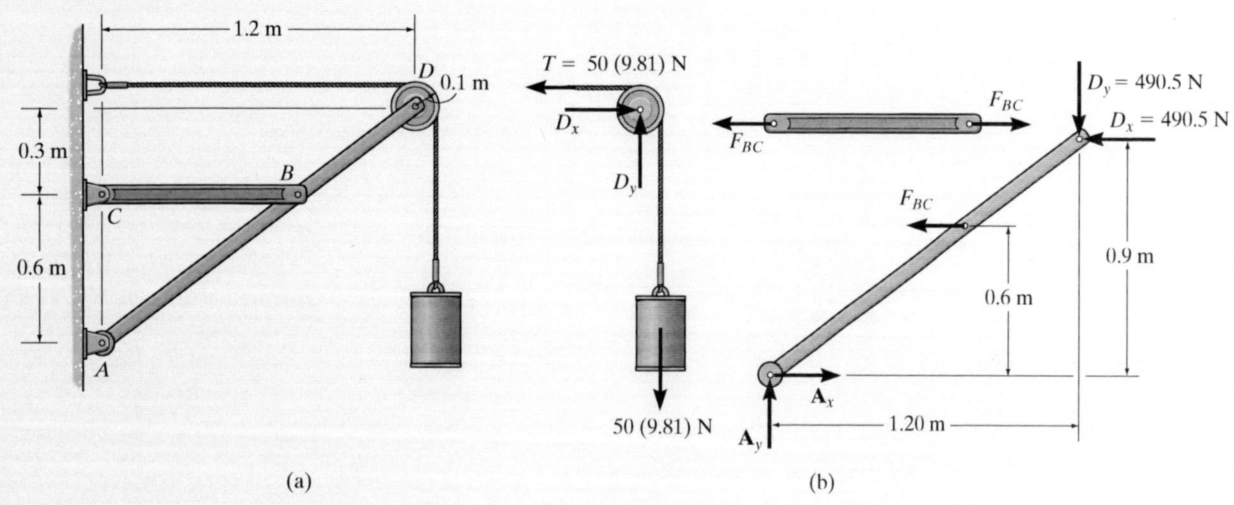

Fig. 6–33

6

SOLUTION

Free-Body Diagrams. The free-body diagram of pulley *D*, along with the cylinder and a portion of the cord (a system), is shown in Fig. 6–33*b*. Member *BC* is a two-force member as indicated by its free-body diagram. The free-body diagram of member *ABD* is also shown.

Equations of Equilibrium. We will begin by analyzing the equilibrium of the pulley. The moment equation of equilibrium is automatically satisfied with $T = 50(9.81)$ N, and so

$$\overset{+}{\rightarrow}\Sigma F_x = 0; \quad D_x - 50(9.81)\text{ N} = 0 \quad D_x = 490.5\text{ N}$$

$$+\uparrow\Sigma F_y = 0; \quad D_y - 50(9.81)\text{ N} = 0 \quad D_y = 490.5\text{ N} \qquad \textit{Ans.}$$

Using these results, F_{BC} can be determined by summing moments about point *A* on member *ABD*.

$$\circlearrowleft +\Sigma M_A = 0; \; F_{BC}(0.6\text{ m}) + 490.5\text{ N}(0.9\text{ m}) - 490.5\text{ N}(1.20\text{ m}) = 0$$

$$F_{BC} = 245.25\text{ N} \qquad \textit{Ans.}$$

Now A_x and A_y can be determined by summing forces.

$$\overset{+}{\rightarrow}\Sigma F_x = 0; \quad A_x - 245.25\text{ N} - 490.5\text{ N} = 0 \quad A_x = 736\text{ N} \qquad \textit{Ans.}$$

$$+\uparrow\Sigma F_y = 0; \quad A_y - 490.5\text{ N} = 0 \quad A_y = 490.5\text{ N} \qquad \textit{Ans.}$$

EXAMPLE 6.22

Determine the force the pins at A and B exert on the two-member frame shown in Fig. 6–34*a*.

SOLUTION I

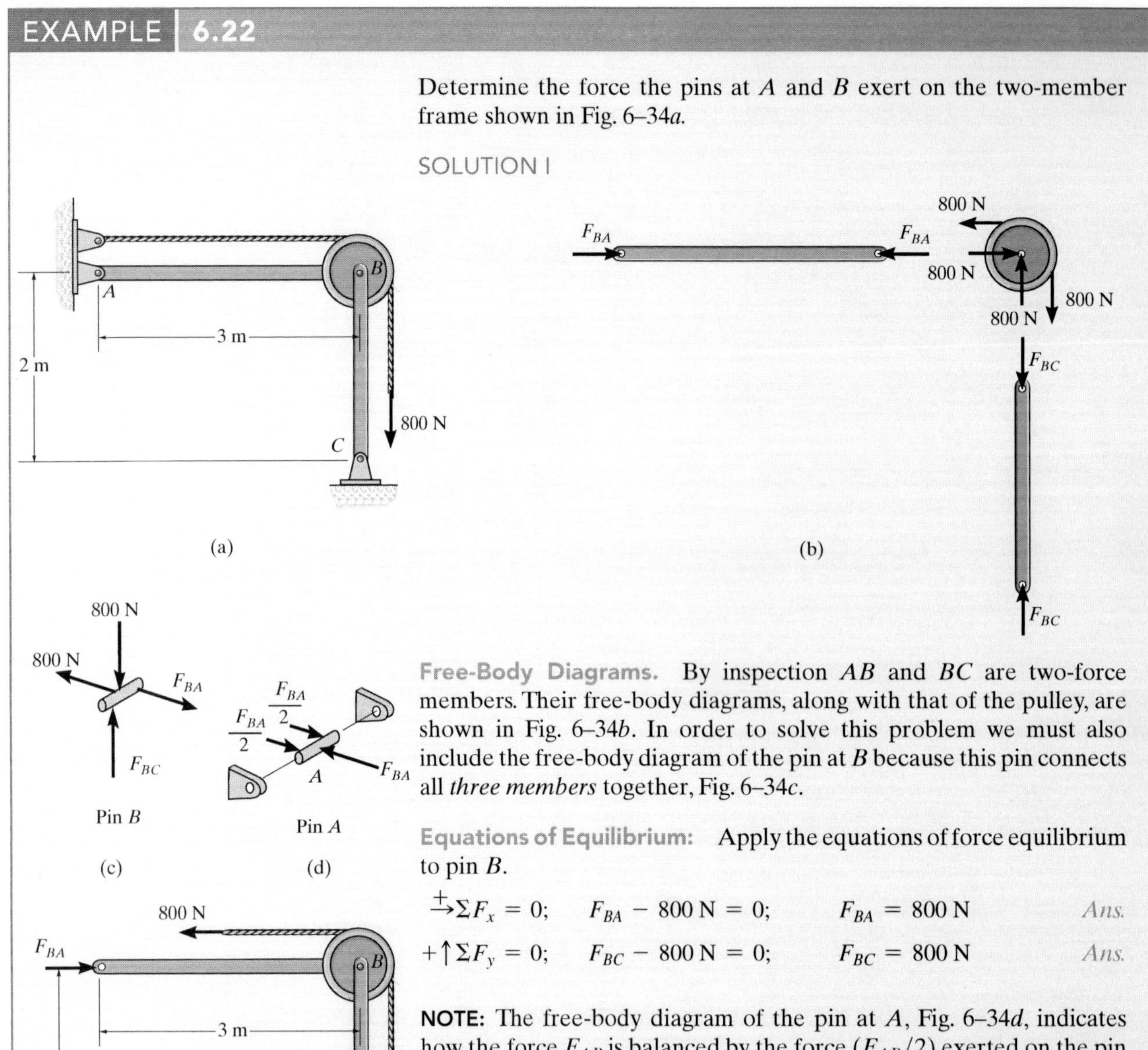

Free-Body Diagrams. By inspection AB and BC are two-force members. Their free-body diagrams, along with that of the pulley, are shown in Fig. 6–34*b*. In order to solve this problem we must also include the free-body diagram of the pin at B because this pin connects all *three members* together, Fig. 6–34*c*.

Equations of Equilibrium: Apply the equations of force equilibrium to pin B.

$$\overset{+}{\rightarrow}\Sigma F_x = 0; \quad F_{BA} - 800 \text{ N} = 0; \quad F_{BA} = 800 \text{ N} \quad \textit{Ans.}$$

$$+\uparrow\Sigma F_y = 0; \quad F_{BC} - 800 \text{ N} = 0; \quad F_{BC} = 800 \text{ N} \quad \textit{Ans.}$$

NOTE: The free-body diagram of the pin at A, Fig. 6–34*d*, indicates how the force F_{AB} is balanced by the force $(F_{AB}/2)$ exerted on the pin by each of the two pin leaves.

Fig. 6–34

SOLUTION II

Free-Body Diagram. If we realize that AB and BC are two-force members, then the free-body diagram of the *entire frame* produces an easier solution, Fig. 6–34*e*. The force equations of equilibrium are the same as those above. Note that moment equilibrium will be satisfied, regardless of the radius of the pulley.

FUNDAMENTAL PROBLEMS

All problem solutions must include FBDs.

F6–13. Determine the force P needed to hold the 60-N weight in equilibrium.

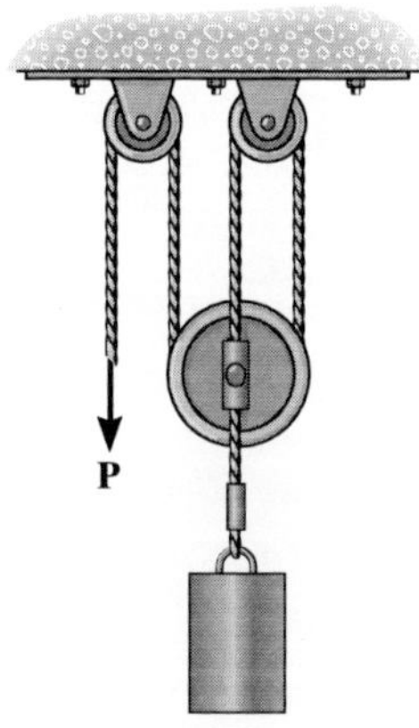

F6–13

F6–14. Determine the horizontal and vertical components of reaction at pin C.

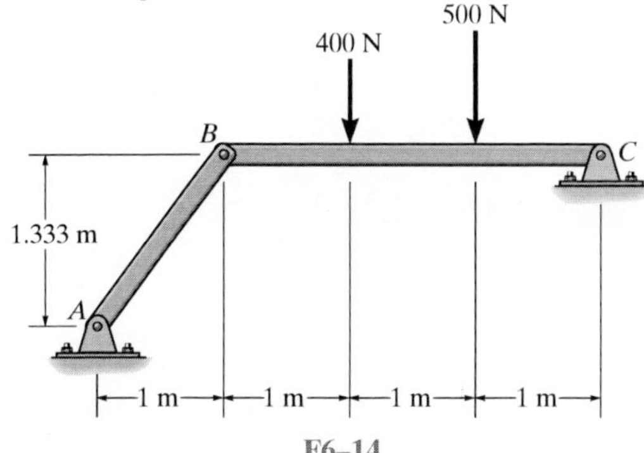

F6–14

F6–15. If a 100-N force is applied to the handles of the pliers, determine the clamping force exerted on the smooth pipe B and the magnitude of the resultant force that one of the members exerts on pin A.

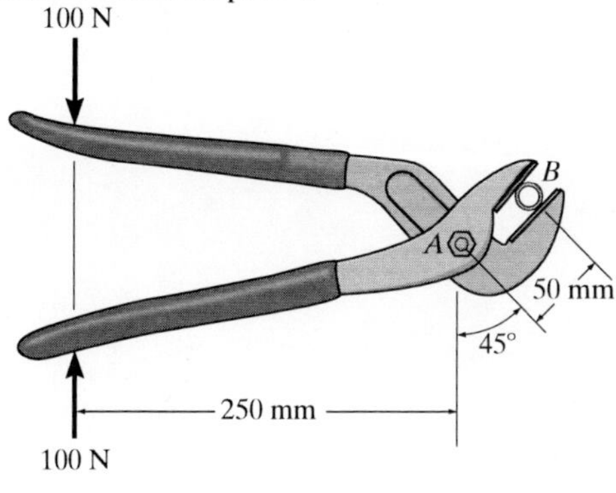

F6–15

F6–16. Determine the horizontal and vertical components of reaction at pin C.

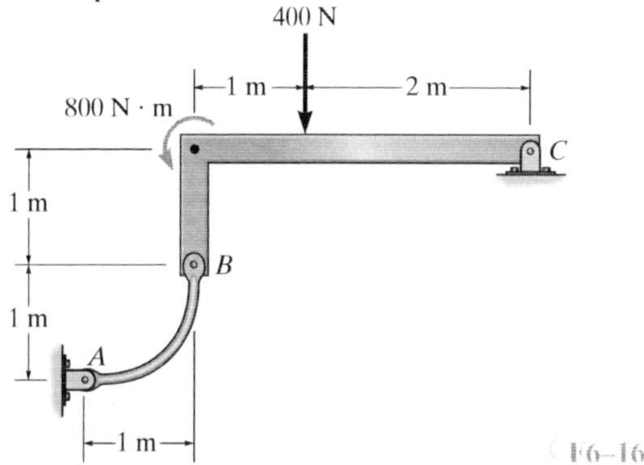

F6–16

F6–17. Determine the normal force that the 100-N plate A exerts on the 30-N plate B.

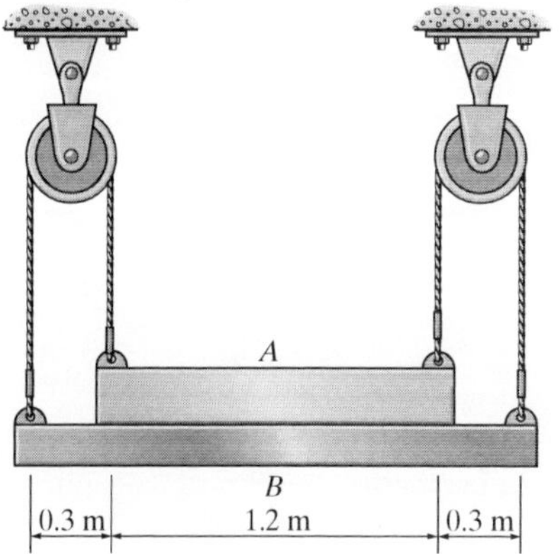

F6–17

F6–18. Determine the force P needed to lift the load. Also, determine the proper placement x of the hook for equilibrium. Neglect the weight of the beam.

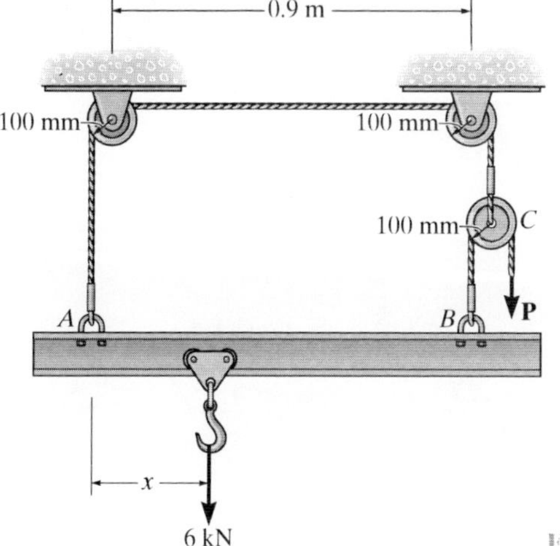

F6–18

F6–19. Determine the components of reaction at A and B.

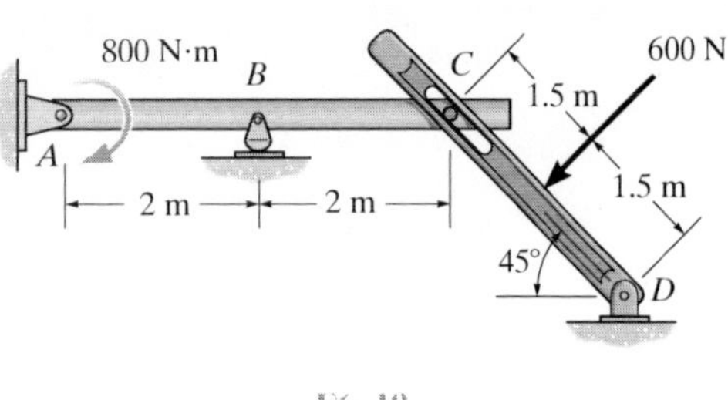

F6–19

F6–20. Determine the components of reaction at D.

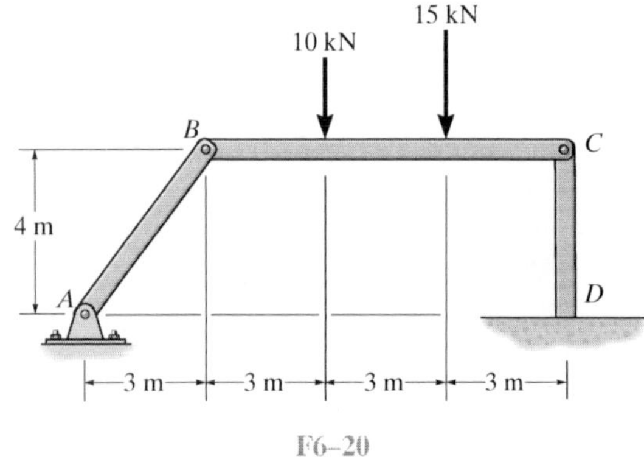

F6–20

F6–21. Determine the components of reaction at A and C.

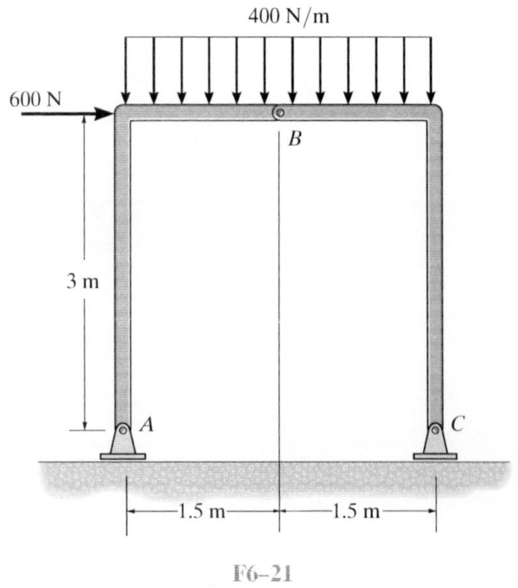

F6–21

F6–22. Determine the components of reaction at C.

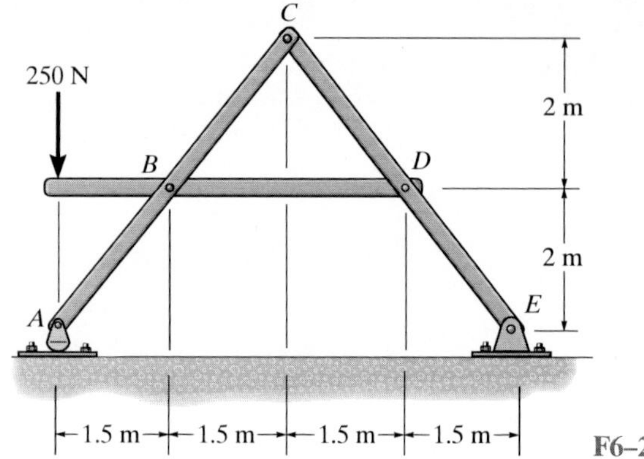

F6–22

F6–23. Determine the components of reaction at E.

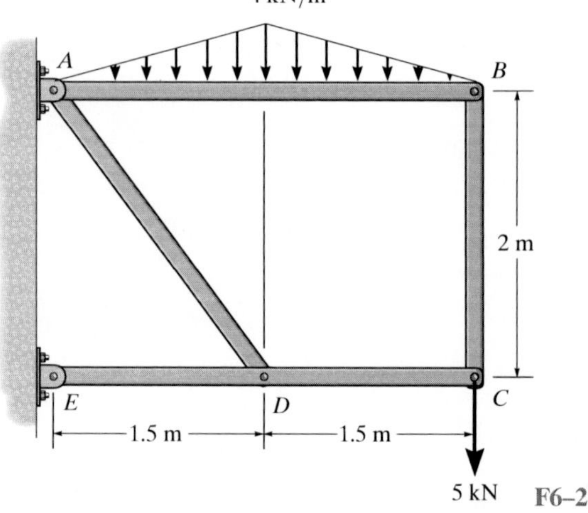

F6–23

F6–24. Determine the components of reaction at D and the components of reaction the pin at A exerts on member BA.

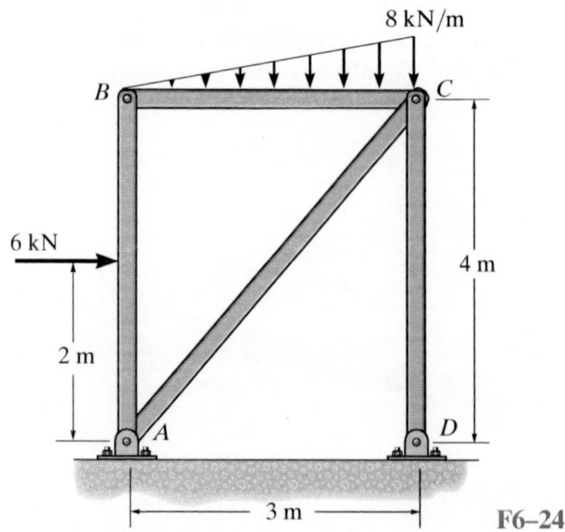

F6–24

6

PROBLEMS

All problem solutions must include FBDs.

6–61. Determine the force **P** required to hold the 50-kg mass in equilibrium.

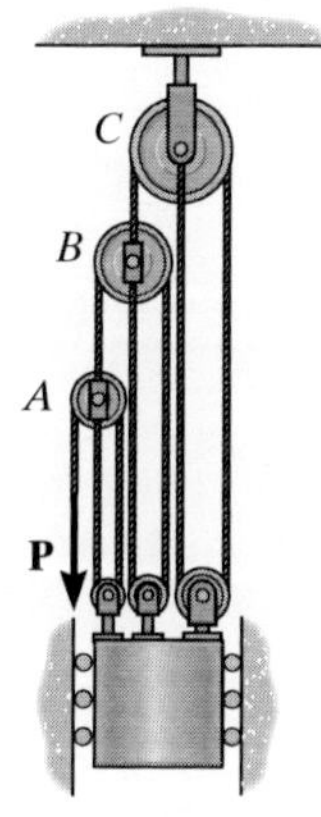

Prob. 6–61

6–62. Determine the force **P** required to hold the 150-kg crate in equilibrium. The two cables are connected to the bottom of the hanger.

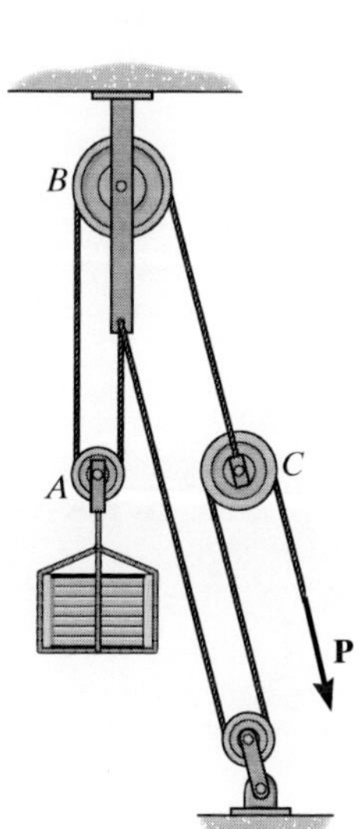

Prob. 6–62

6–63. The principles of a *differential chain block* are indicated schematically in the figure. Determine the magnitude of force **P** needed to support the 800-N force. Also, find the distance x where the cable must be attached to bar AB so the bar remains horizontal. All pulleys have a radius of 60 mm.

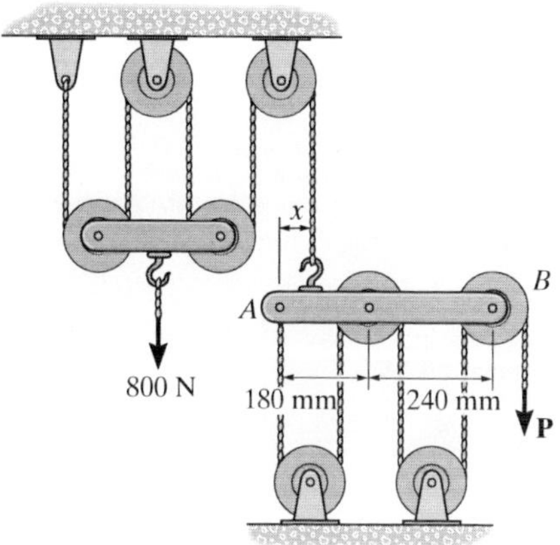

Prob. 6–63

*__6–64.__ Determine the force P needed to support the 20-kg mass using the *Spanish Burton rig*. Also, what are the reactions at the supporting hooks A, B, and C?

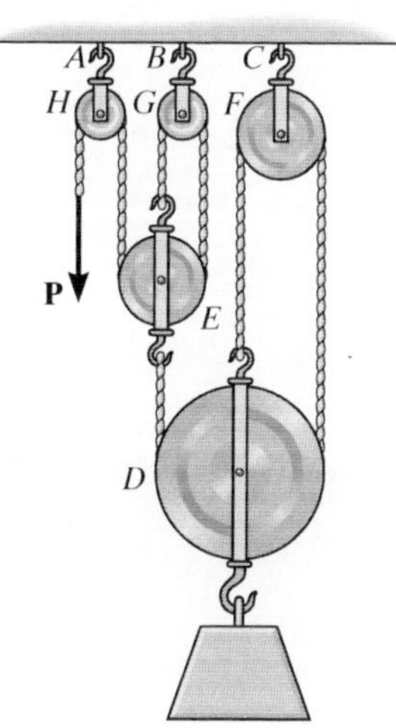

Prob. 6–64

6–65. Determine the greatest force P that can be applied to the frame if the largest force resultant acting at A can have a magnitude of 5 kN.

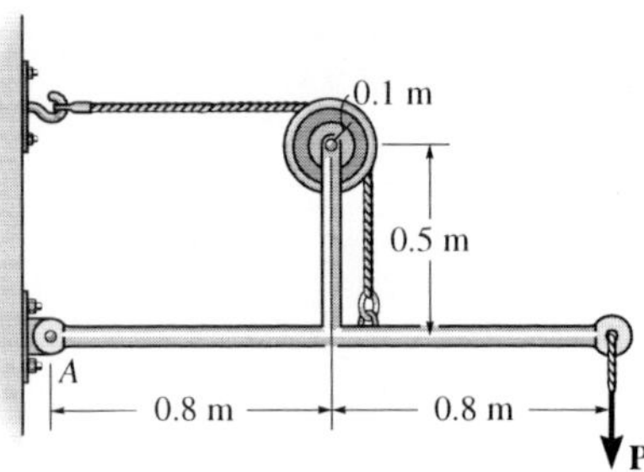

Prob. 6–65

6–66. Determine the horizontal and vertical components of force that the pins at A, B, and C exert on their connecting members.

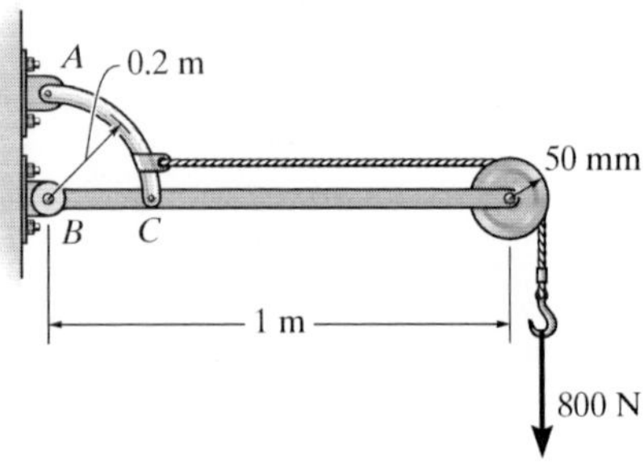

Prob. 6–66

6–67. The double link grip is used to lift the beam. If the beam weighs 4 kN, determine the horizontal and vertical components of force acting on the pin at A and the horizontal and vertical components of force that the flange of the beam exerts on the jaw at B.

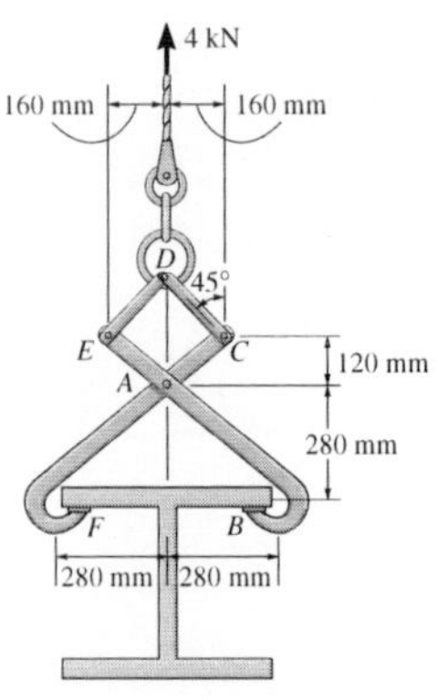

Prob. 6–67

***6–68.** Determine the greatest force P that can be applied to the frame if the largest force resultant acting at A can have a magnitude of 2 kN.

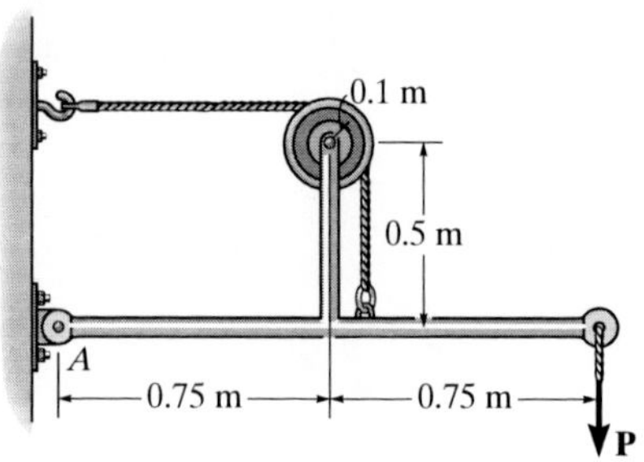

Prob. 6–68

6–69. Determine the horizontal and vertical components of force that pins A and C exert on the frame.

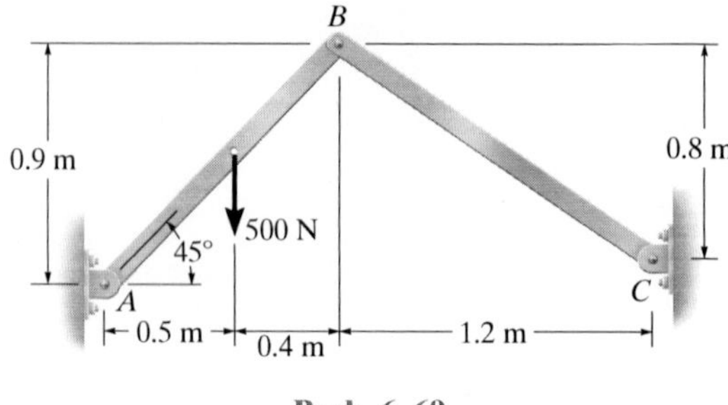

Prob. 6–69

6–70. The toggle clamp is subjected to a force **F** at the handle. Determine the vertical clamping force acting at E.

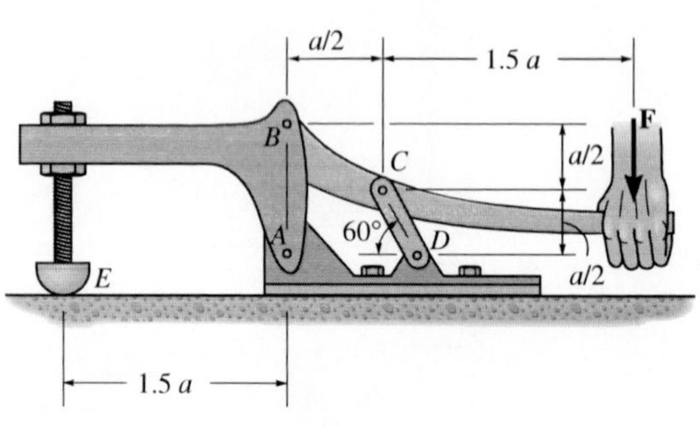

Prob. 6–70

6

6–71. Determine the support reactions at A, C, and E on the compound beam which is pin connected at B and D.

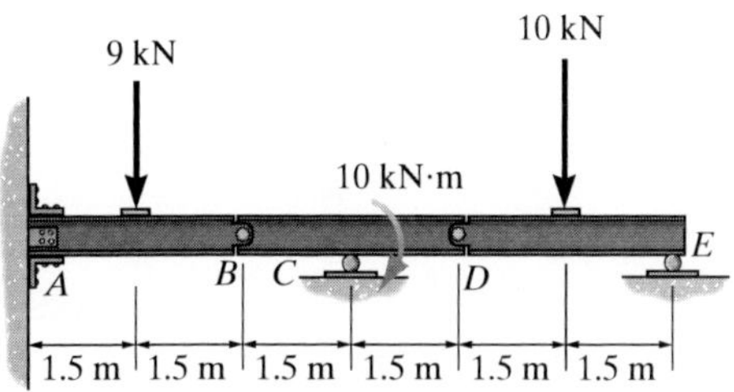

Prob. 6–71

***6–72.** Determine the horizontal and vertical components of force at pins A, B, and C, and the reactions at the fixed support D of the three-member frame.

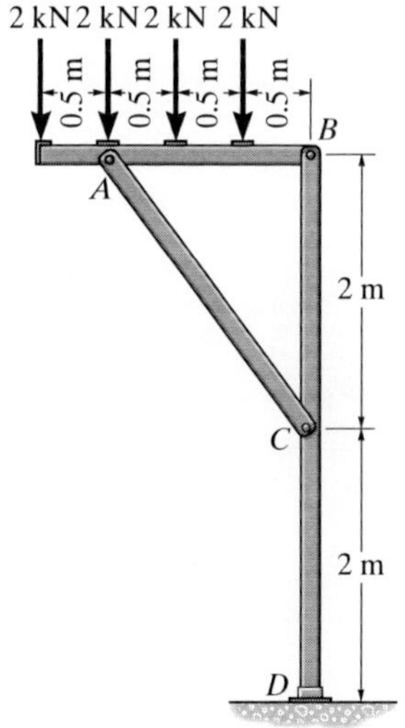

Prob. 6–72

6–73. The compound beam is fixed at A and supported by a rocker at B and C. There are hinges (pins) at D and E. Determine the reactions at the supports.

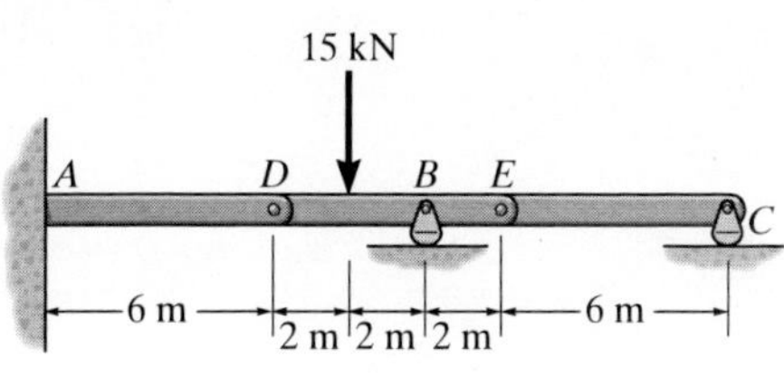

Prob. 6–73

6–74. The platform scale consists of a combination of third and first class levers so that the load on one lever becomes the effort that moves the next lever. Through this arrangement, a small weight can balance a massive object. If $x = 450$ mm, determine the required mass of the counterweight S required to balance a 90-kg load, L.

6–75. The platform scale consists of a combination of third and first class levers so that the load on one lever becomes the effort that moves the next lever. Through this arrangement, a small weight can balance a massive object. If $x = 450$ mm and, the mass of the counterweight S is 2 kg, determine the mass of the load L required to maintain the balance.

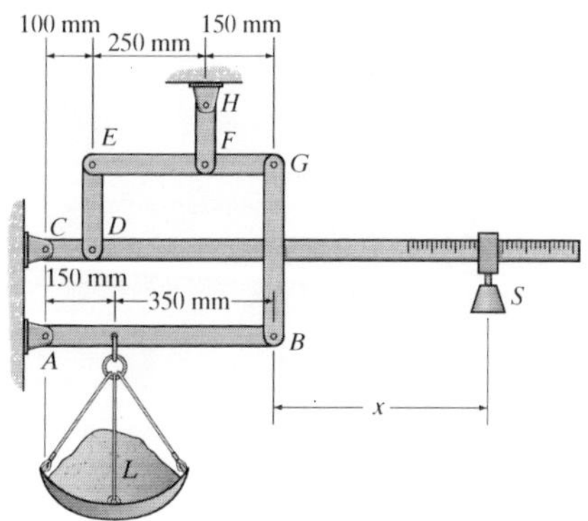

Probs. 6–74/75

***6–76.** Determine the horizontal and vertical components of force which the pins at A, B, and C exert on member ABC of the frame.

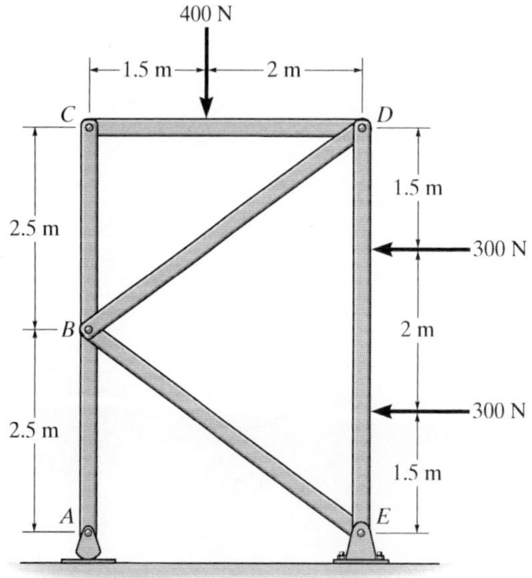

Prob. 6–76

6–77. Determine the required mass of the suspended cylinder if the tension in the chain wrapped around the freely turning gear is to be 2 kN. Also, what is the magnitude of the resultant force on pin A?

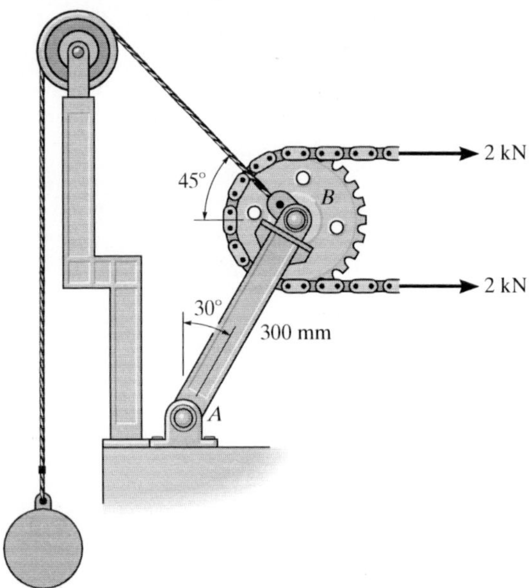

Prob. 6–77

6–78. Determine the reactions on the collar at A and the pin at C. The collar fits over a smooth rod, and rod AB is fixed connected to the collar.

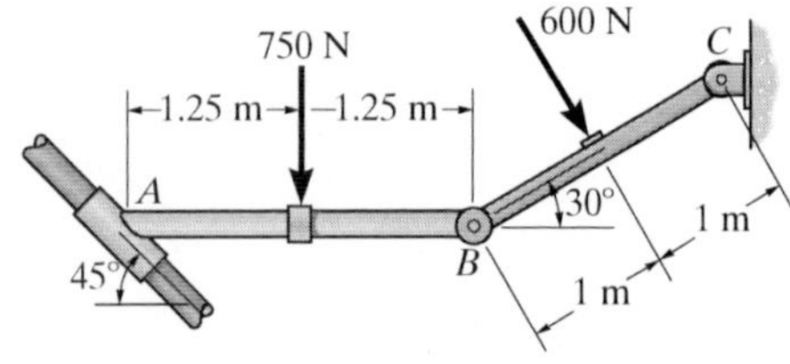

Prob. 6–78

6–79. The toggle clamp is subjected to a force **F** at the handle. Determine the vertical clamping force acting at E.

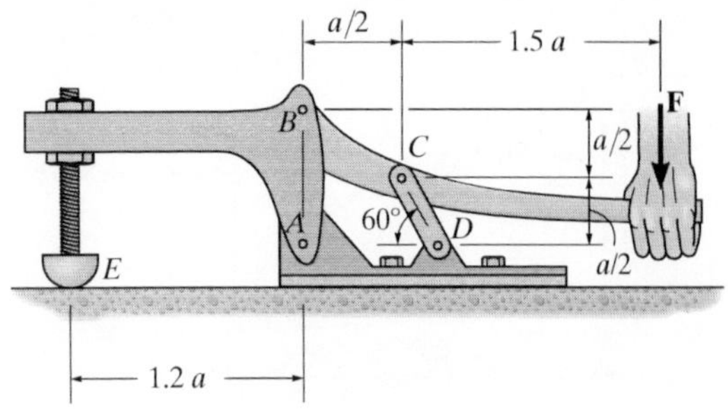

Prob. 6–79

***6–80.** Determine the force that the jaws J of the metal cutters exert on the smooth cable C if 100-N forces are applied to the handles. The jaws are pinned at E and A, and D and B. There is also a pin at F.

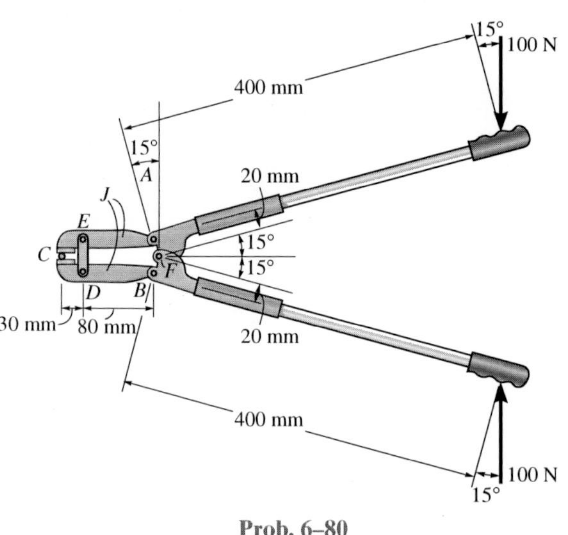

Prob. 6–80

6–81. The engine hoist is used to support the 200-kg engine. Determine the force acting in the hydraulic cylinder AB, the horizontal and vertical components of force at the pin C, and the reactions at the fixed connection D.

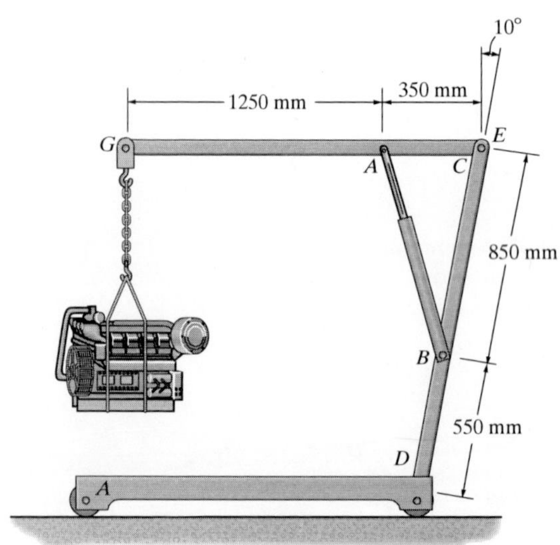

Prob. 6–81

6–82. Determine the horizontal and vertical components of force that the pins at A, B, and C exert on the frame. The cylinder has a mass of 80 kg. The pulley has a radius of 0.1 m.

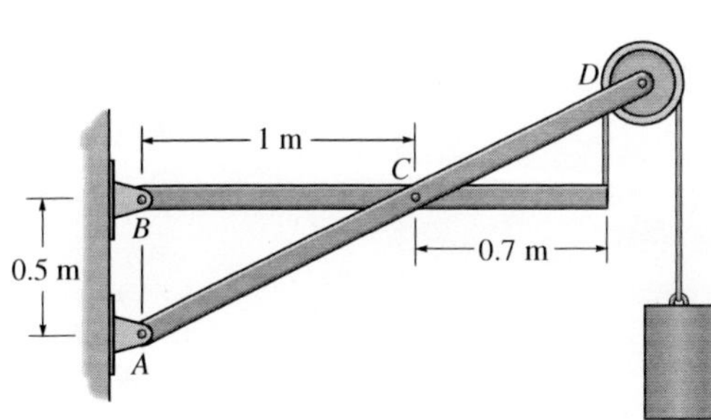

Prob. 6–82

***6–84.** The compound beam is fixed supported at A and supported by rockers at B and C. If there are hinges (pins) at D and E, determine the reactions at the supports A, B, and C.

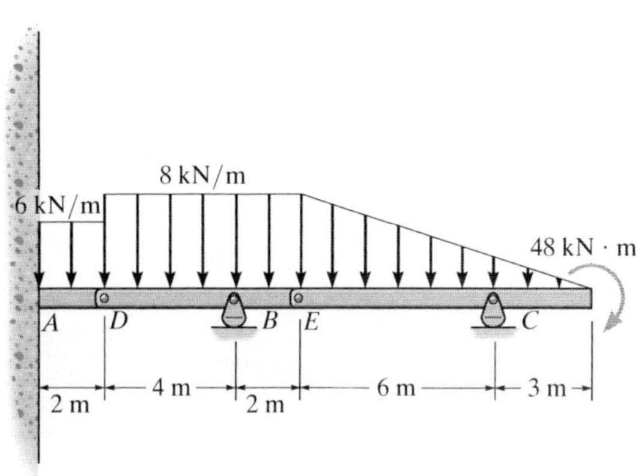

Prob. 6–84

6–83. Determine the horizontal and vertical components of force that pins A and C exert on the frame.

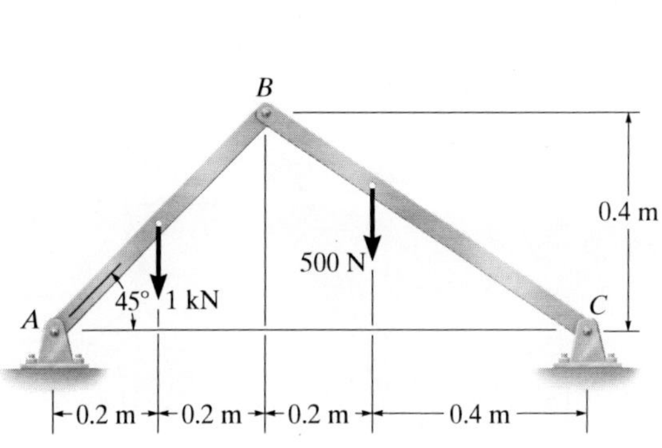

Prob. 6–83

6–85. The pruner multiplies blade-cutting power with the compound leverage mechanism. If a 20-N force is applied to the handles, determine the cutting force generated at A. Assume that the contact surface at A is smooth.

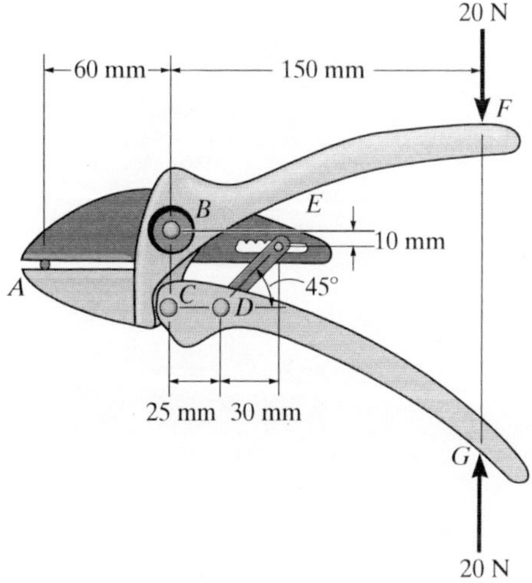

Prob. 6–85

6–86. The pipe cutter is clamped around the pipe P. If the wheel at A exerts a normal force of $F_A = 80$ N on the pipe, determine the normal forces of wheels B and C on the pipe. Also compute the pin reaction on the wheel at C. The three wheels each have a radius of 7 mm and the pipe has an outer radius of 10 mm.

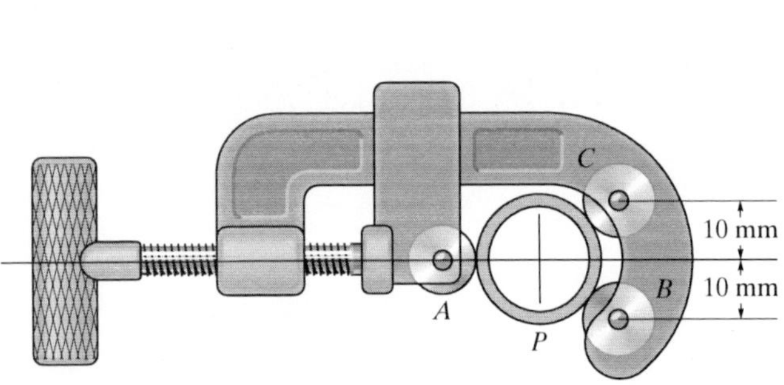

Prob. 6–86

***6–88.** Show that the weight W_1 of the counterweight at H required for equilibrium is $W_1 = (b/a)W$, and so it is independent of the placement of the load W on the platform.

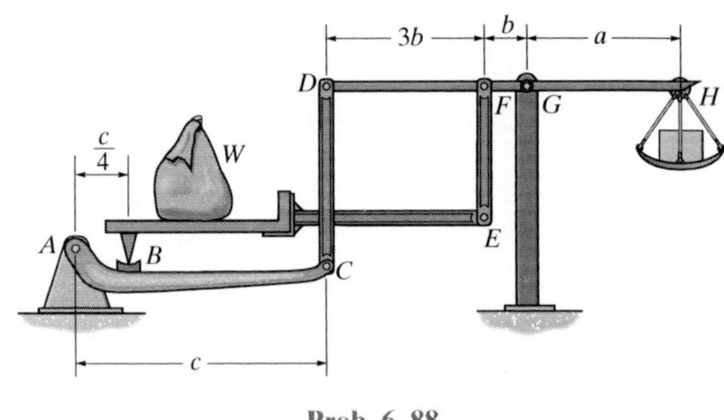

Prob. 6–88

6–87. The link is used to hold the rod in place. Determine the required axial force on the screw at E if the largest force to be exerted on the rod at B, C, or D is to be 100 N. Also, find the magnitude of the force reaction at pin A. Assume all surfaces of contact are smooth.

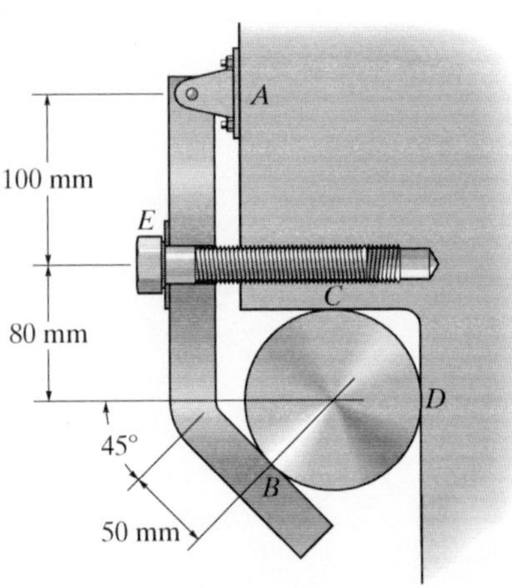

Prob. 6–87

6–89. The derrick is pin connected to the pivot at A. Determine the largest mass that can be supported by the derrick if the maximum force that can be sustained by the pin at A is 18 kN.

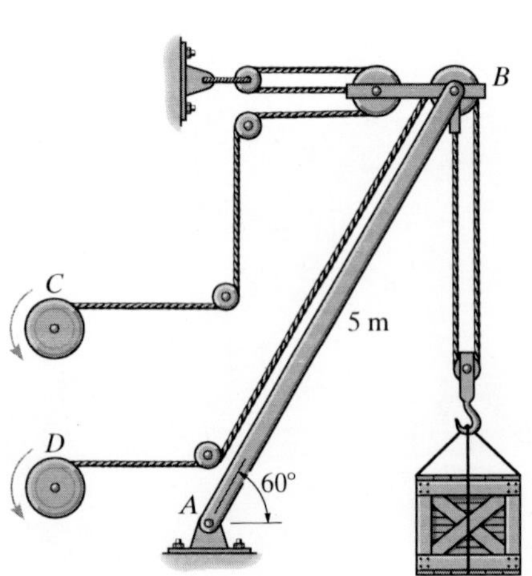

Prob. 6–89

6–90. Determine the force that the jaws J of the metal cutters exert on the smooth cable C if 200-N forces are applied to the handles. The jaws are pinned at E and A, and D and B. There is also a pin at F.

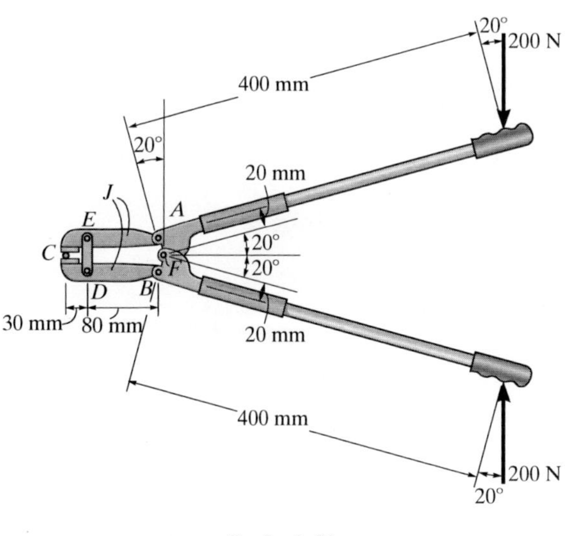

Prob. 6–90

***6–92.** The scissors lift consists of *two* sets of cross members and *two* hydraulic cylinders, DE, symmetrically located on *each side* of the platform. The platform has a uniform mass of 60 kg, with a center of gravity at G_1. The load of 85 kg, with center of gravity at G_2, is centrally located between each side of the platform. Determine the force in each of the hydraulic cylinders for equilibrium. Rollers are located at B and D.

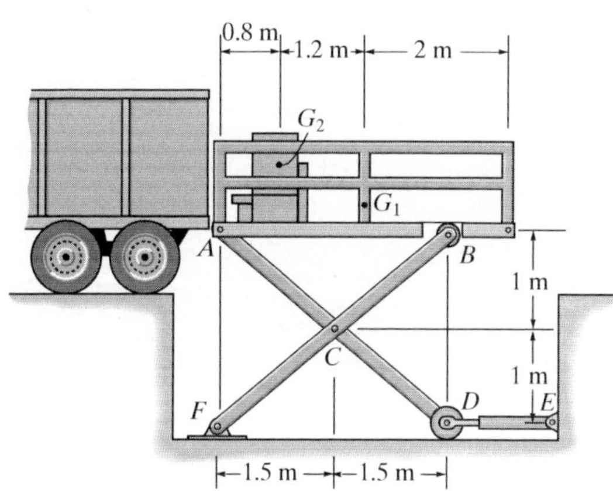

Prob. 6–92

6–91. The compound beam is pin supported at B and supported by rockers at A and C. There is a hinge (pin) at D. Determine the reactions at the supports.

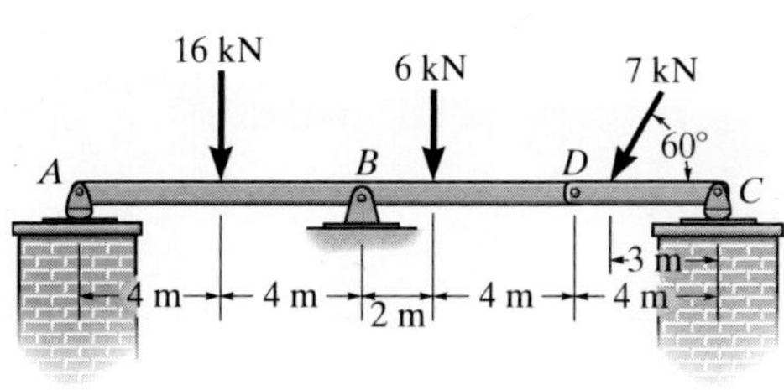

Prob. 6–91

6–93. The two disks each have a mass of 20 kg and are attached at their centers by an elastic cord that has a stiffness of $k = 2$ kN/m. Determine the stretch of the cord when the system is in equilibrium and the angle θ of the cord.

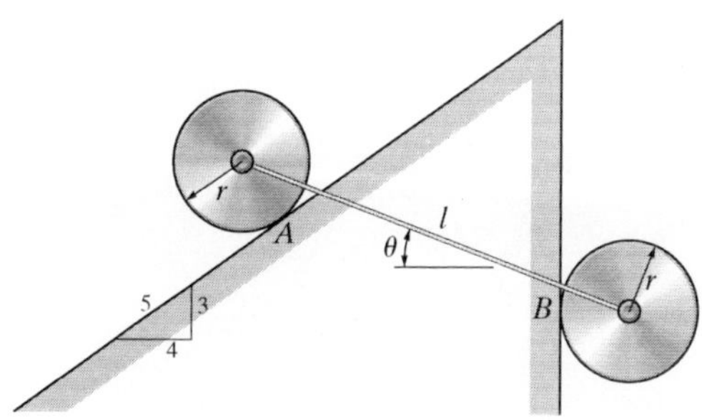

Prob. 6–93

6

6–94. A man having a weight of 875 N (≈ 87.5 kg) attempts to hold himself using one of the two methods shown. Determine the total force he must exert on bar AB in each case and the normal reaction he exerts on the platform at C. Neglect the weight of the platform.

6–95. A man having a weight of 875 N (≈ 87.5 kg) attempts to hold himself using one of the two methods shown. Determine the total force he must exert on bar AB in each case and the normal reaction he exerts on the platform at C. The platform has a weight of 150 N (≈ 15 kg).

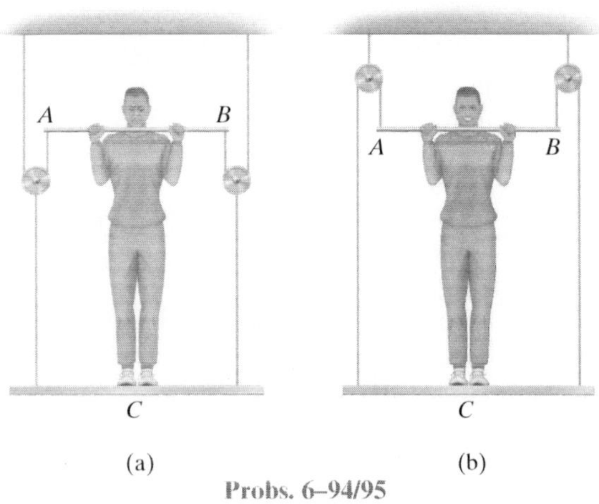

Probs. 6–94/95

*6–96. The double link grip is used to lift the beam. If the beam weighs 8 kN, determine the horizontal and vertical components of force acting on the pin at A and the horizontal and vertical components of force that the flange of the beam exerts on the jaw at B.

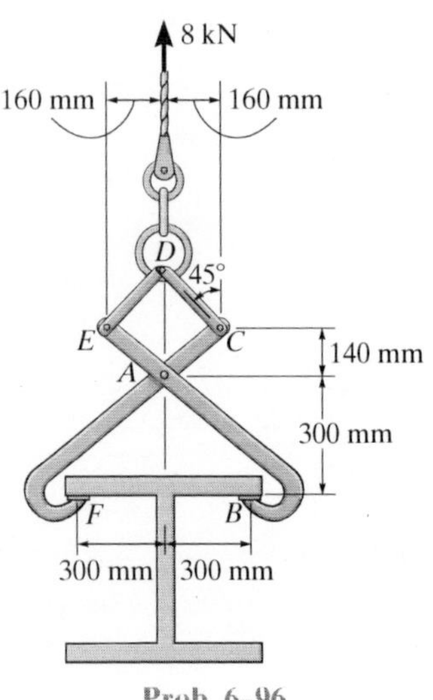

Prob. 6–96

6–97. Operation of exhaust and intake valves in an automobile engine consists of the cam C, push rod DE, rocker arm EFG which is pinned at F, and a spring and valve, V. If the compression in the spring is 20 mm when the valve is open as shown, determine the normal force acting on the cam lobe at C. Assume the cam and bearings at H, I, and J are smooth. The spring has a stiffness of 300 N/m.

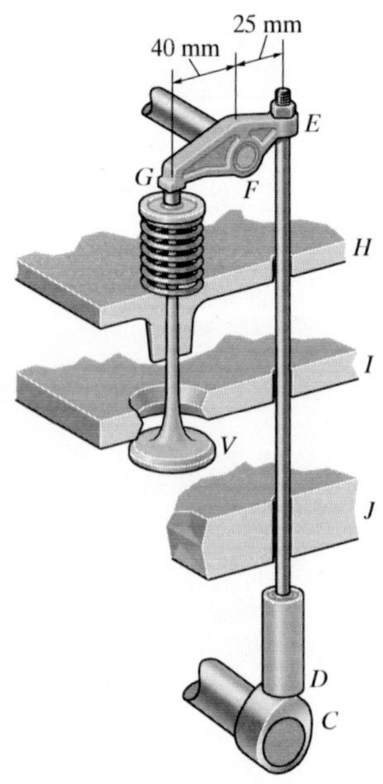

Prob. 6–97

6–98. Determine the horizontal and vertical components of force that the pins at A, B, and C exert on their connecting members.

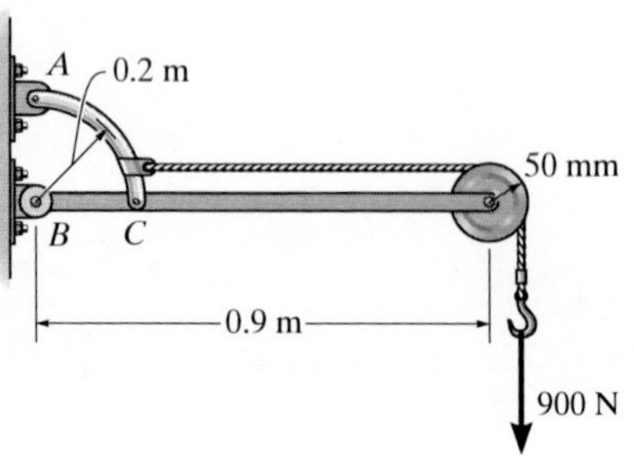

Prob. 6–98

6–99. If a clamping force of 300 N is required at A, determine the amount of force **F** that must be applied to the handle of the toggle clamp.

***6–100.** If a force of $F = 350$ N is applied to the handle of the toggle clamp, determine the resulting clamping force at A.

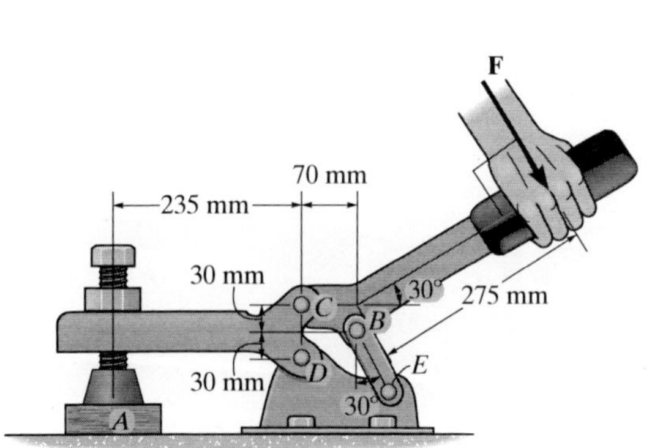

Probs. 6–99/100

6–101. The skid steer loader has a mass of 1.18 Mg, and in the position shown the center of mass is at G_1. If there is a 300-kg stone in the bucket, with center of mass at G_2, determine the reactions of each pair of wheels A and B on the ground and the force in the hydraulic cylinder CD and at the pin E. There is a similar linkage on each side of the loader.

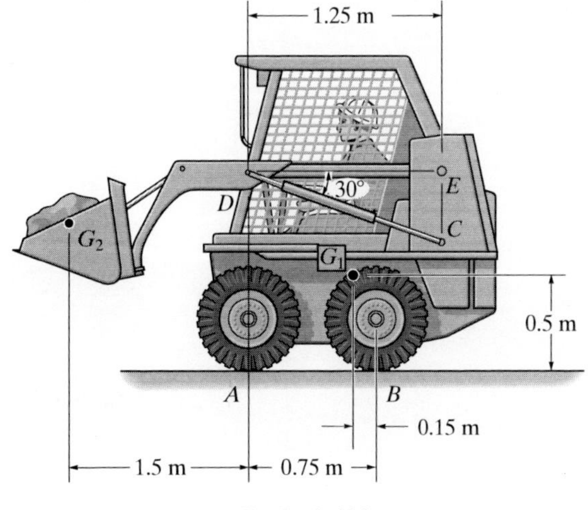

Prob. 6–101

6–102. The tractor boom supports the uniform mass of 600 kg in the bucket which has a center of mass at G. Determine the force in each hydraulic cylinder AB and CD and the resultant force at pins E and F. The load is supported equally on each side of the tractor by a similar mechanism.

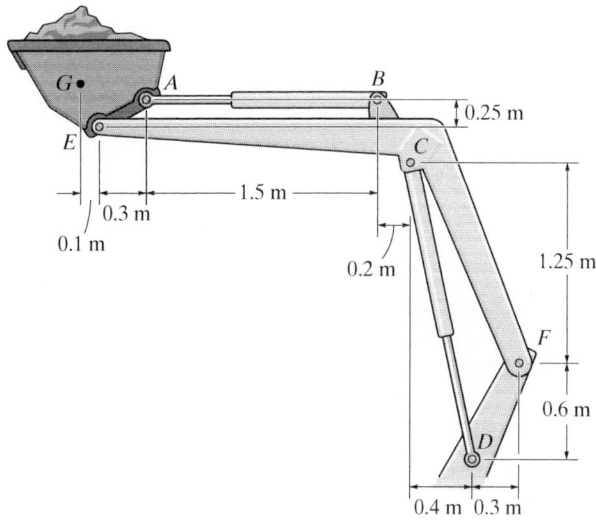

Prob. 6–102

6–103. The nail cutter consists of the handle and the two cutting blades. Assuming the blades are pin connected at B and the surface at D is smooth, determine the normal force on the fingernail when a force of 5 N is applied to the handles as shown. The pin AC slides through a smooth hole at A and is attached to the bottom member at C.

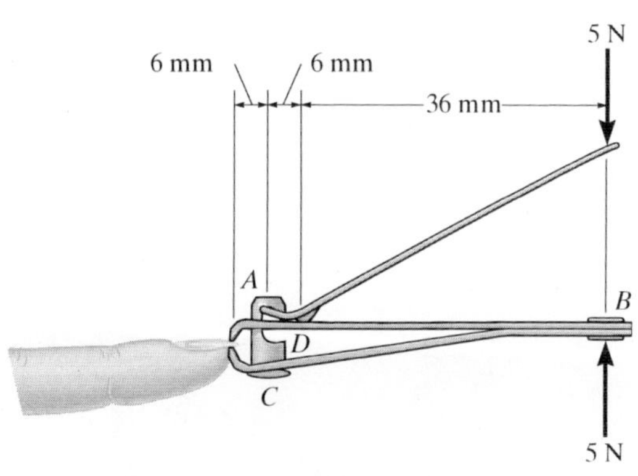

Prob. 6–103

***6–104.** Determine the force created in the hydraulic cylinders EF and AD in order to hold the shovel in equilibrium. The shovel load has a mass of 1.25 Mg and a center of gravity at G. All joints are pin connected.

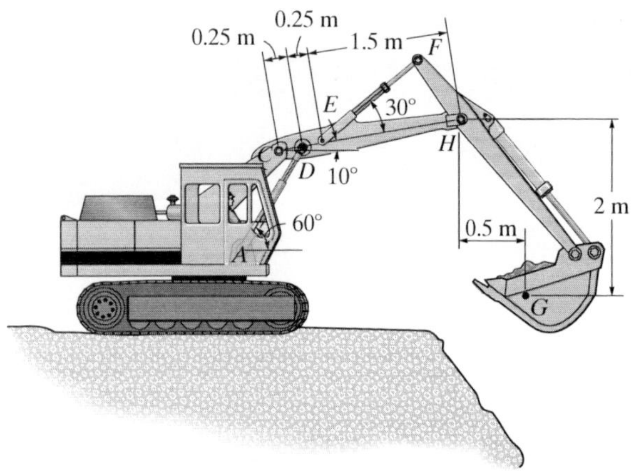

Prob. 6–104

6–106. If $P = 75$ N, determine the force F that the toggle clamp exerts on the wooden block.

6–107. If the wooden block exerts a force of $F = 600$ N on the toggle clamp, determine the force P applied to the handle.

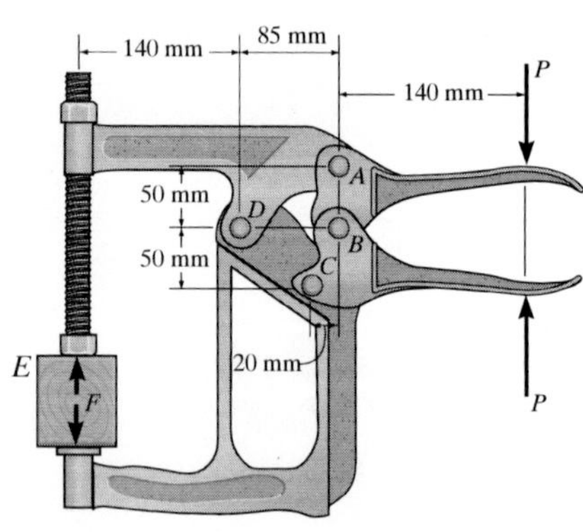

Probs. 6–106/107

6–105. The hoist supports the 125-kg engine. Determine the force in member DB and in the hydraulic cylinder H of member FB.

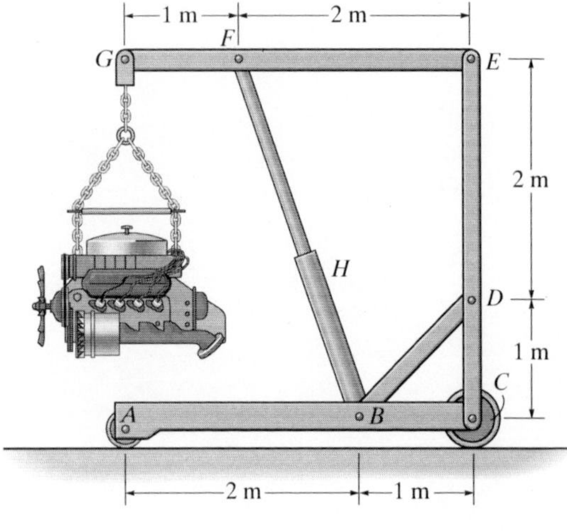

Prob. 6–105

***6–108.** The pillar crane is subjected to the load having a mass of 500 kg. Determine the force developed in the tie rod AB and the horizontal and vertical reactions at the pin support C when the boom is tied in the position shown.

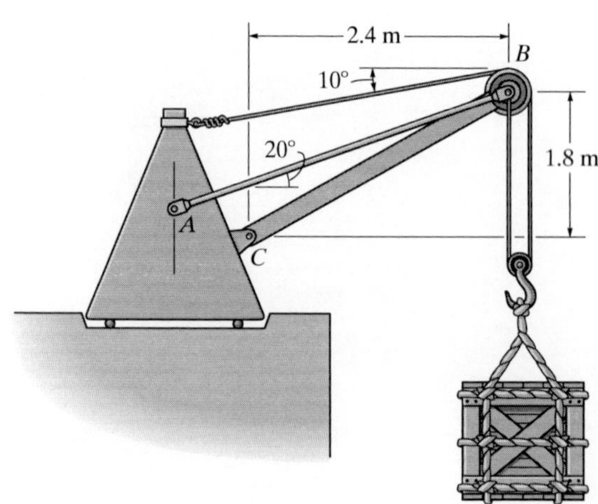

Prob. 6–108

6–109. The symmetric coil tong supports the coil which has a mass of 800 kg and center of mass at G. Determine the horizontal and vertical components of force the linkage exerts on plate $DEIJH$ at points D and E. The coil exerts only vertical reactions at K and L.

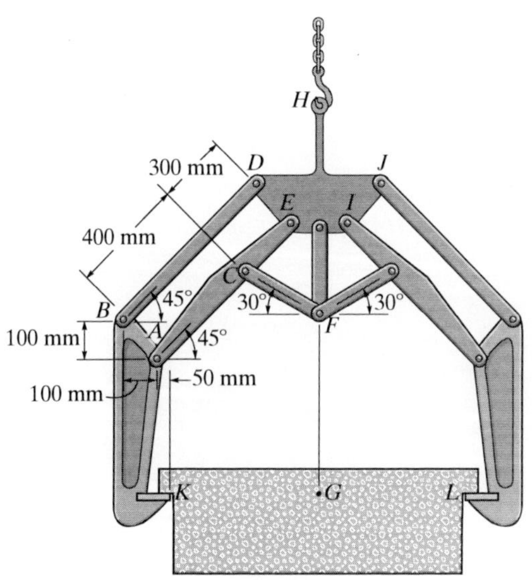

Prob. 6–109

6–110. A 300-kg counterweight, with center of mass at G, is mounted on the pitman crank AB of the oil-pumping unit. If a force of $F = 5$ kN is to be developed in the fixed cable attached to the end of the walking beam DEF, determine the torque M that must be supplied by the motor.

6–111. A 300-kg counterweight, with center of mass at G, is mounted on the pitman crank AB of the oil-pumping unit. If the motor supplies a torque of $M = 2500$ N · m, determine the force **F** developed in the fixed cable attached to the end of the walking beam DEF.

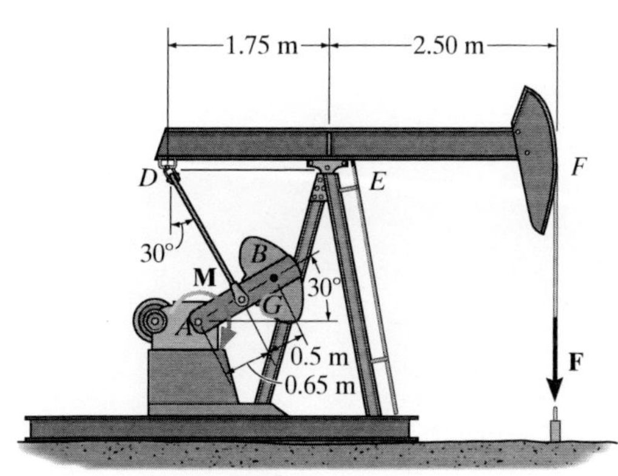

Probs. 6–110/111

*__6–112.__ The aircraft-hangar door opens and closes slowly by means of a motor which draws in the cable AB. If the door is made in two sections (bifold) and each section has a uniform weight W and length L, determine the force in the cable as a function of the door's position θ. The sections are pin-connected at C and D and the bottom is attached to a roller that travels along the vertical track.

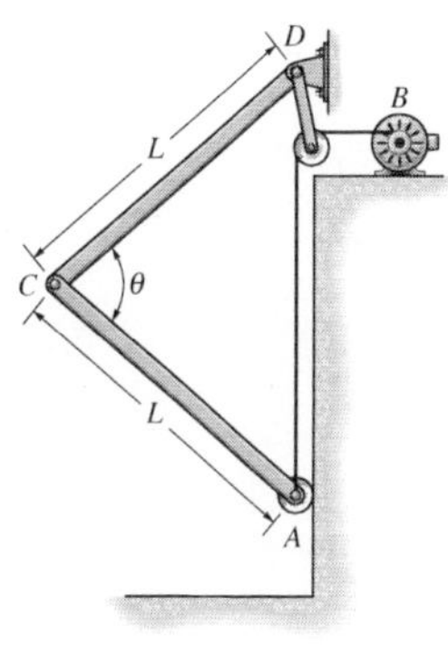

Prob. 6–112

6–113. Determine the horizontal and vertical components of reaction which the pins exert on member AB of the frame.

6–114. Determine the horizontal and vertical components of reaction which the pins exert on member EDC of the frame.

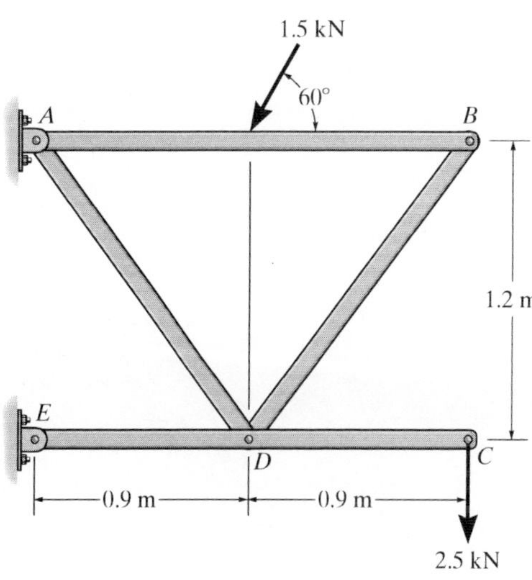

Probs. 6–113/114

6–115. The handle of the sector press is fixed to gear G, which in turn is in mesh with the sector gear C. Note that AB is pinned at its ends to gear C and the underside of the table EF, which is allowed to move vertically due to the smooth guides at E and F. If the gears exert tangential forces between them, determine the compressive force developed on the cylinder S when a vertical force of 40 N is applied to the handle of the press.

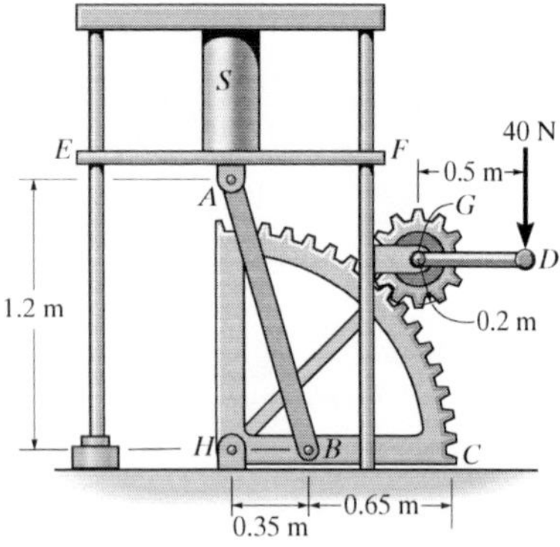

Prob. 6–115

6–117. The four-member "A" frame is supported at A and E by smooth collars and at G by a pin. All the other joints are ball-and-sockets. If the pin at G will fail when the resultant force there is 800 N, determine the largest vertical force P that can be supported by the frame. Also, what are the x, y, z force components which member BD exerts on members EDC and ABC? The collars at A and E and the pin at G only exert force components on the frame.

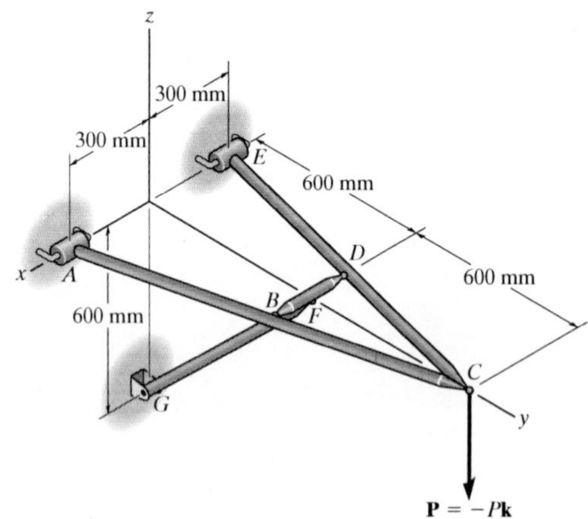

Prob. 6–117

***6–116.** The structure is subjected to the loading shown. Member AD is supported by a cable AB and roller at C and fits through a smooth circular hole at D. Member ED is supported by a roller at D and a pole that fits in a smooth snug circular hole at E. Determine the x, y, z components of reaction at E and the tension in cable AB.

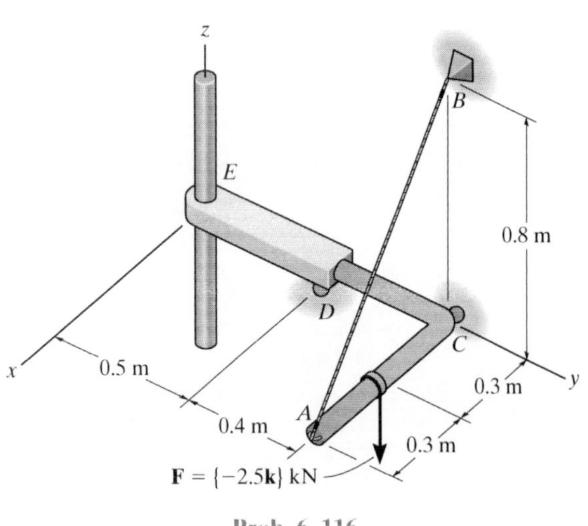

Prob. 6–116

6–118. The structure is subjected to the loadings shown. Member AB is supported by a ball-and-socket at A and smooth collar at B. Member CD is supported by a pin at C. Determine the x, y, z components of reaction at A and C.

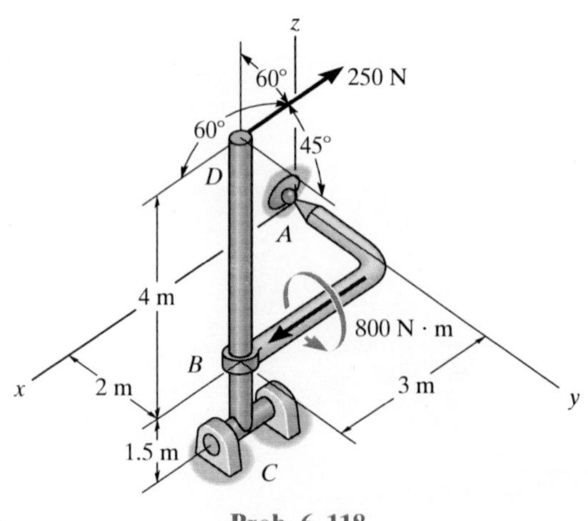

Prob. 6–118

CHAPTER REVIEW

Simple Truss

A simple truss consists of triangular elements connected together by pinned joints. The forces within its members can be determined by assuming the members are all two-force members, connected concurrently at each joint. The members are either in tension or compression, or carry no force.

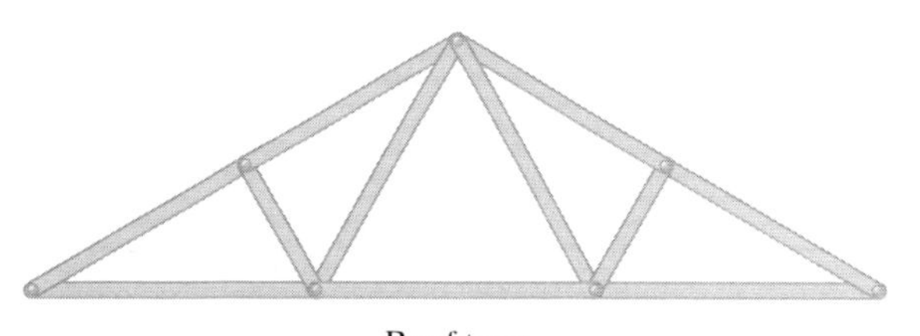

Roof truss

Method of Joints

The method of joints states that if a truss is in equilibrium, then each of its joints is also in equilibrium. For a plane truss, the concurrent force system at each joint must satisfy force equilibrium.

To obtain a numerical solution for the forces in the members, select a joint that has a free-body diagram with at most two unknown forces and one known force. (This may require first finding the reactions at the supports.)

Once a member force is determined, use its value and apply it to an adjacent joint.

Remember that forces that are found to *pull* on the joint are *tensile forces*, and those that *push* on the joint are *compressive forces*.

To avoid a simultaneous solution of two equations, set one of the coordinate axes along the line of action of one of the unknown forces and sum forces perpendicular to this axis. This will allow a direct solution for the other unknown.

The analysis can also be simplified by first identifying all the zero-force members.

$$\Sigma F_x = 0$$

$$\Sigma F_y = 0$$

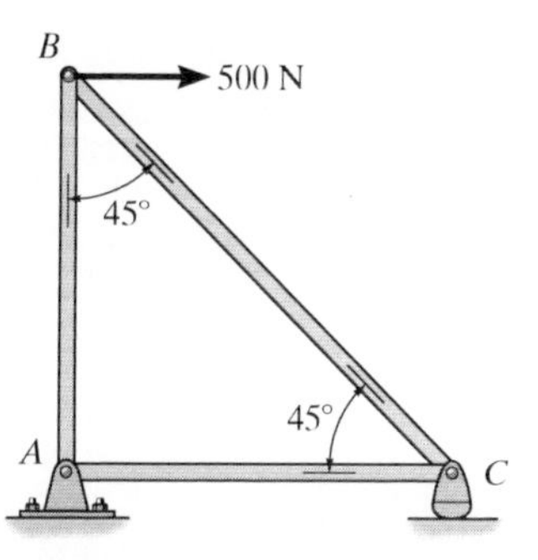

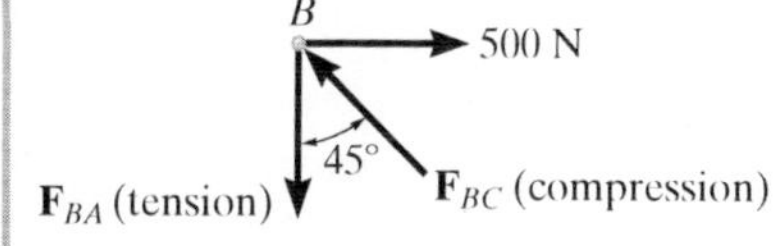

Method of Sections

The method of sections states that if a truss is in equilibrium, then each segment of the truss is also in equilibrium. Pass a section through the truss and the member whose force is to be determined. Then draw the free-body diagram of the sectioned part having the least number of forces on it.

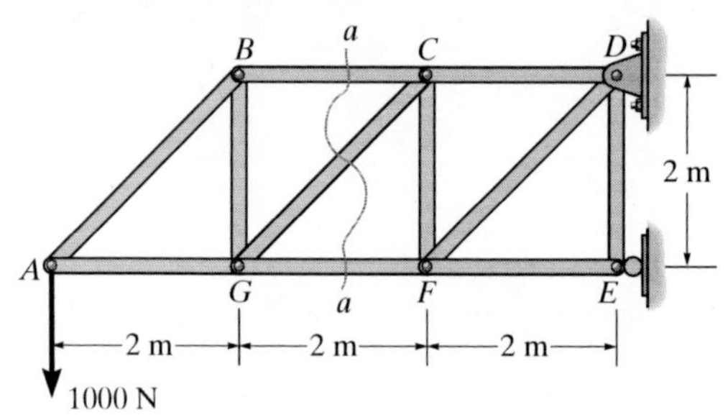

Sectioned members subjected to *pulling* are in *tension*, and those that are subjected to *pushing* are in *compression*.

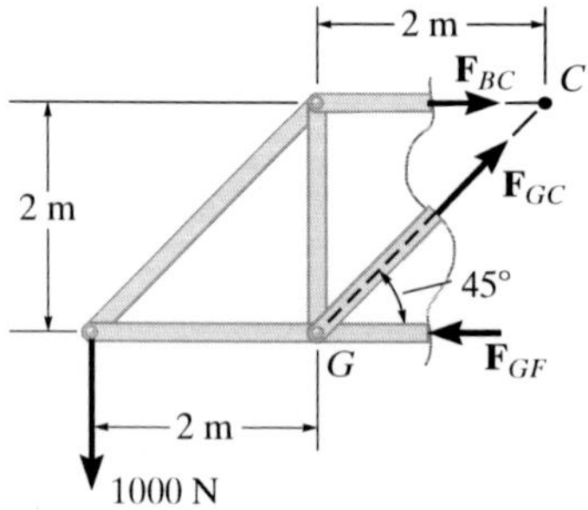

Three equations of equilibrium are available to determine the unknowns.

$$\Sigma F_x = 0$$
$$\Sigma F_y = 0$$
$$\Sigma M_O = 0$$

If possible, sum forces in a direction that is perpendicular to two of the three unknown forces. This will yield a direct solution for the third force.

$$+\uparrow \Sigma F_y = 0$$
$$-1000 \text{ N} + F_{GC} \sin 45^\circ = 0$$
$$F_{GC} = 1.41 \text{ kN (T)}$$

Sum moments about the point where the lines of action of two of the three unknown forces intersect, so that the third unknown force can be determined directly.

$$\circlearrowleft + \Sigma M_C = 0$$
$$1000 \text{ N}(4 \text{ m}) - F_{GF}(2 \text{ m}) = 0$$
$$F_{GF} = 2 \text{ kN (C)}$$

Space Truss

A space truss is a three-dimensional truss built from tetrahedral elements, and is analyzed using the same methods as for plane trusses. The joints are assumed to be ball-and-socket connections.

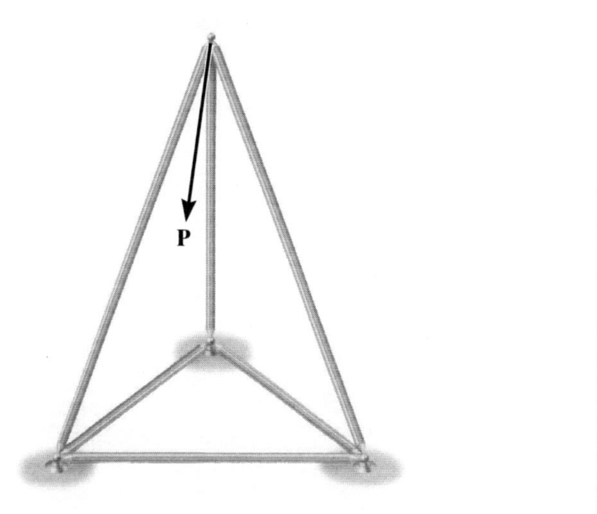

Frames and Machines

Frames and machines are structures that contain one or more multiforce members, that is, members with three or more forces or couples acting on them. Frames are designed to support loads, and machines transmit and alter the effect of forces.

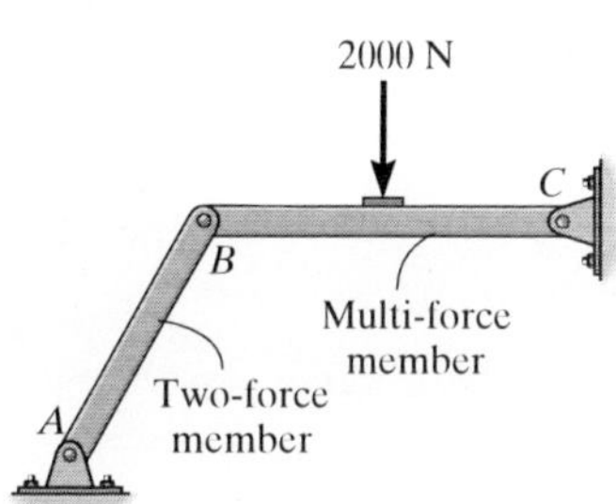

The forces acting at the joints of a frame or machine can be determined by drawing the free-body diagrams of each of its members or parts. The principle of action–reaction should be carefully observed when indicating these forces on the free-body diagram of each adjacent member or pin. For a coplanar force system, there are three equilibrium equations available for each member.

To simplify the analysis, be sure to recognize all two-force members. They have equal but opposite collinear forces at their ends.

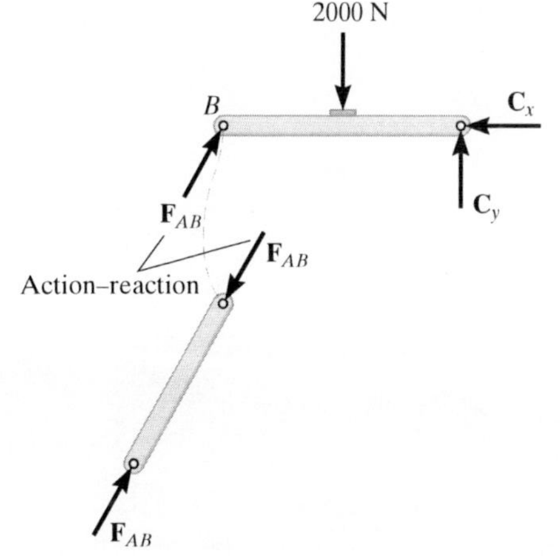

REVIEW PROBLEMS

6–119. The tractor boom supports the uniform mass of 500 kg in the bucket which has a center of mass at G. Determine the force in each hydraulic cylinder AB and CD and the resultant force at pins E and F. The load is supported equally on each side of the tractor by a similar mechanism.

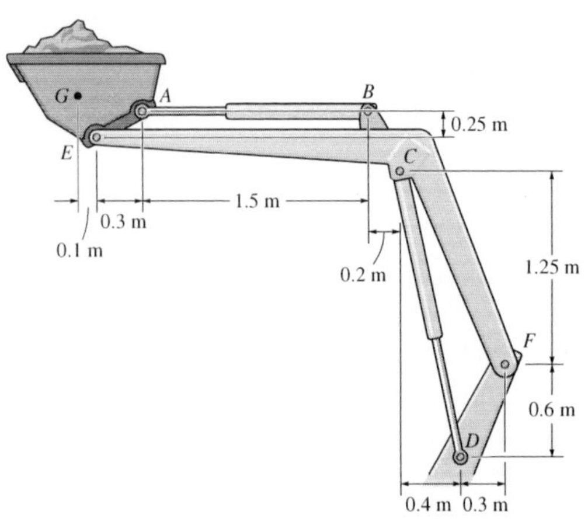

Prob. 6–119

*__6–120.__ Determine the force in each member of the truss and state if the members are in tension or compression.

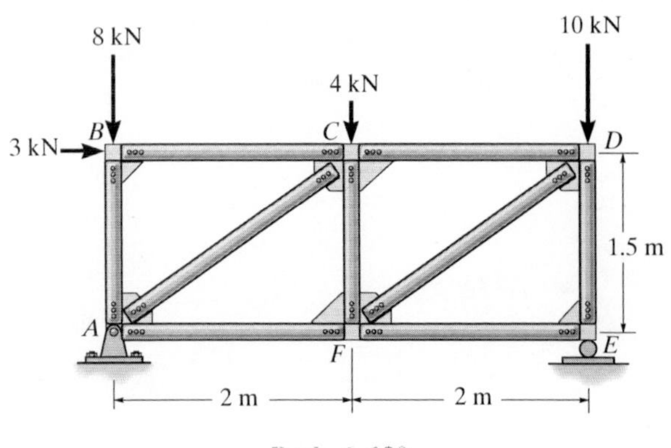

Prob. 6–120

6–121. Determine the horizontal and vertical components of force at pins A and C of the two-member frame.

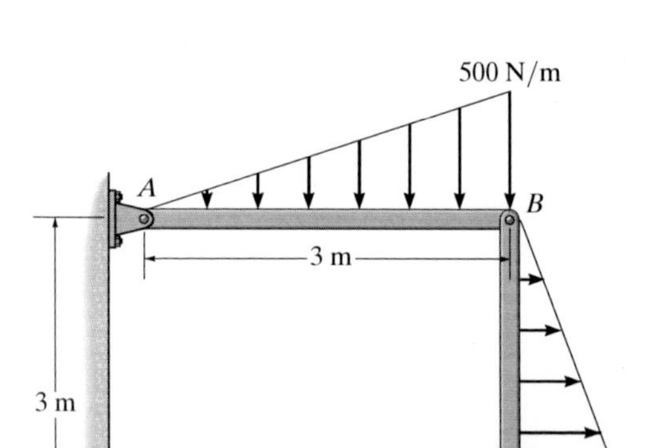

Prob. 6–121

6–122. The clamping hooks are used to lift the uniform smooth 500-kg plate. Determine the resultant compressive force that the hook exerts on the plate at A and B, and the pin reaction at C.

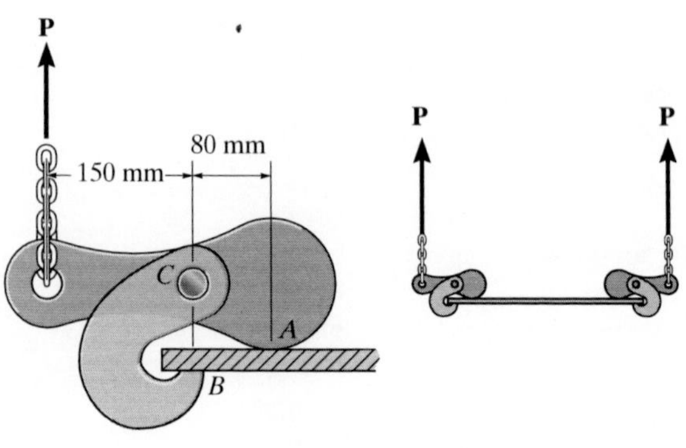

Prob. 6–122

6

6–123. The spring has an unstretched length of 0.3 m. Determine the mass m of each uniform link if the angle $\theta = 20^\circ$ for equilibrium.

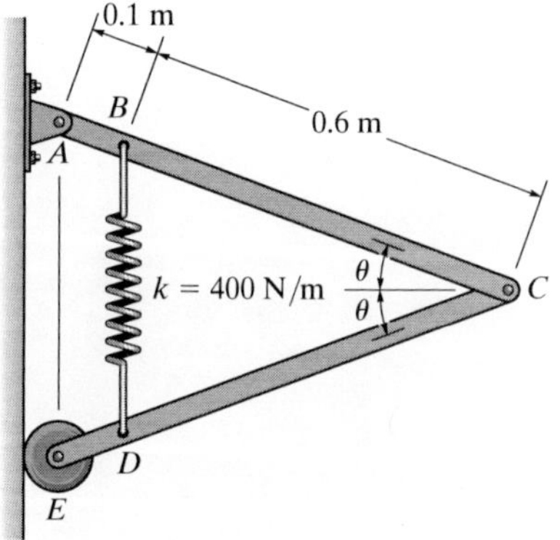

Prob. 6–123

***6–124.** Determine the horizontal and vertical components of force that the pins A and B exert on the two-member frame. Set $F = 0$.

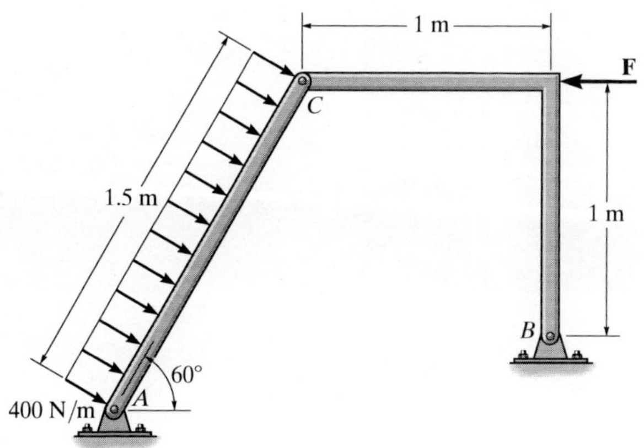

Prob. 6–124

6–125. Determine the horizontal and vertical components of force that pins A and B exert on the two-member frame. Set $F = 500$ N.

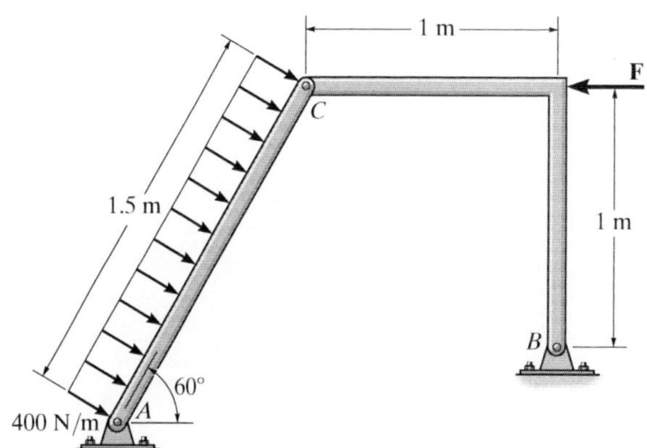

Prob. 6–125

6–126. Determine the force in each member of the truss and state if the members are in tension or compression.

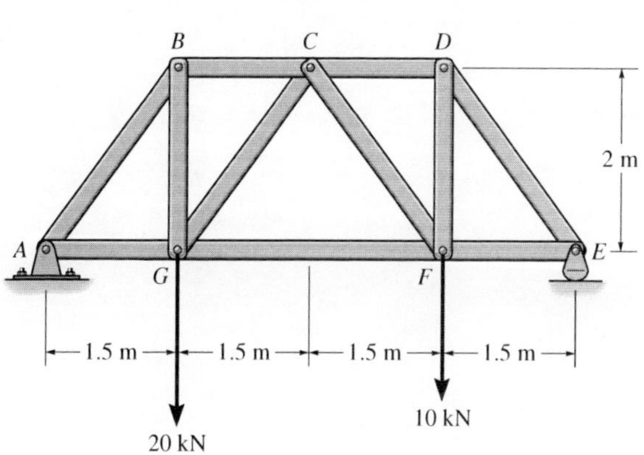

Prob. 6–126

Chapter 7

When external loads are placed upon these beams and columns, the loads within them must be determined if they are to be properly designed. In this chapter we will study how to determine these internal loadings.

Internal Forces

CHAPTER OBJECTIVES

- To show how to use the method of sections to determine the internal loadings in a member.
- To generalize this procedure by formulating equations that can be plotted so that they describe the internal shear and moment throughout a member.
- To analyze the forces and study the geometry of cables supporting a load.

Video Solutions are available for selected questions in this chapter.

7.1 Internal Loadings Developed in Structural Members

To design a structural or mechanical member it is necessary to know the loading acting within the member in order to be sure the material can resist this loading. Internal loadings can be determined by using the *method of sections*. To illustrate this method, consider the cantilever beam in Fig. 7–1*a*. If the internal loadings acting on the cross section at point B are to be determined, we must pass an imaginary section a–a perpendicular to the axis of the beam through point B and then separate the beam into two segments. The internal loadings acting at B will then be exposed and become *external* on the free-body diagram of each segment, Fig. 7–1*b*.

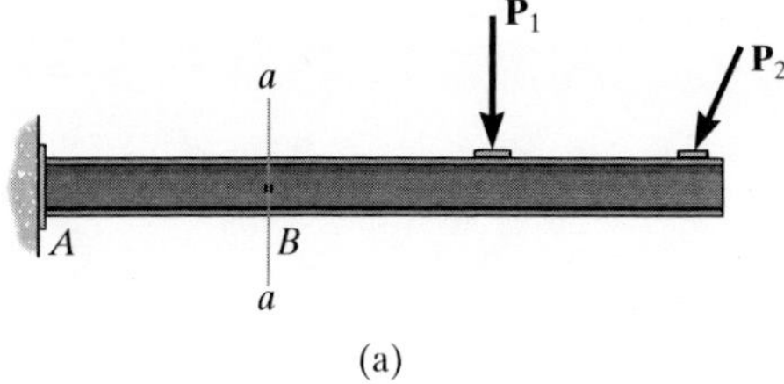

(a)

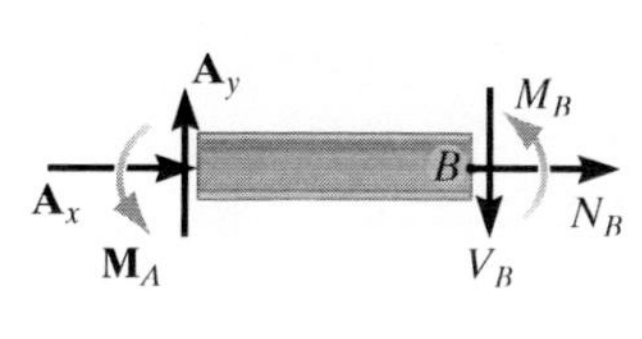

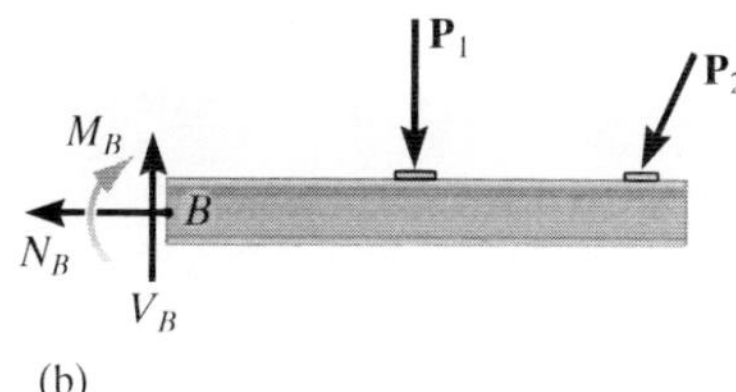

(b)

Fig. 7–1

In each case, the link on the backhoe is a two-force member. In the top photo it is subjected to both bending and an axial load at its center. By making the member straight, as in the bottom photo, then only an axial force acts within the member.

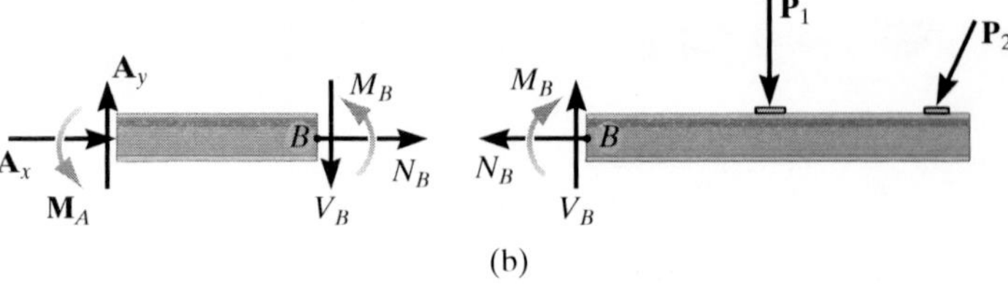

Fig. 7–1 (Repeated)

The force component $\mathbf{N}_B$ that acts *perpendicular* to the cross section is termed the *normal force*. The force component $\mathbf{V}_B$ that is tangent to the cross section is called the *shear force*, and the couple moment $\mathbf{M}_B$ is referred to as the *bending moment*. The force components prevent the relative translation between the two segments, and the couple moment prevents the relative rotation. According to Newton's third law, these loadings must act in opposite directions on each segment, as shown in Fig. 7–1*b*. They can be determined by applying the equations of equilibrium to the free-body diagram of either segment. In this case, however, the right segment is the better choice since it does not involve the unknown support reactions at A. A direct solution for $\mathbf{N}_B$ is obtained by applying $\Sigma F_x = 0$, $\mathbf{V}_B$ is obtained from $\Sigma F_y = 0$, and $\mathbf{M}_B$ can be obtained by applying $\Sigma M_B = 0$, since the moments of $\mathbf{N}_B$ and $\mathbf{V}_B$ about B are zero.

In two dimensions, we have shown that three internal loading resultants exist, Fig. 7–2*a*; however in three dimensions, a general internal force and couple moment resultant will act at the section. The *x, y, z* components of these loadings are shown in Fig. 7–2*b*. Here $\mathbf{N}_y$ is the *normal force*, and $\mathbf{V}_x$ and $\mathbf{V}_z$ are *shear force components*. $\mathbf{M}_y$ is a *torsional or twisting moment*, and $\mathbf{M}_x$ and $\mathbf{M}_z$ are *bending moment components*. For most applications, these *resultant loadings* will act at the geometric center or centroid (C) of the section's cross-sectional area. Although the magnitude for each loading generally will be different at various points along the axis of the member, the method of sections can always be used to determine their values.

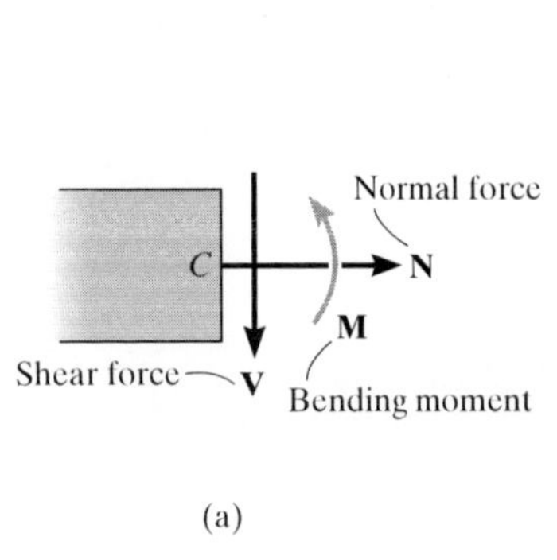

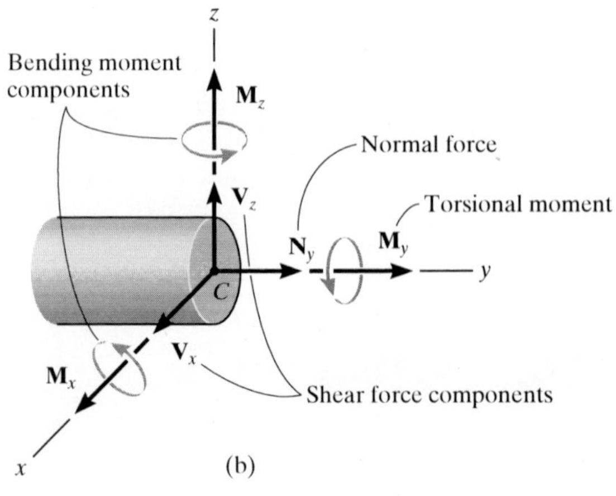

Fig. 7–2

Sign Convention. For problems in two dimensions engineers generally use a sign convention to report the three internal loadings **N**, **V**, and **M**. Although this sign convention can be arbitrarily assigned, the one that is widely accepted will be used here, Fig. 7–3. The normal force is said to be positive if it creates *tension*, a positive shear force will cause the beam segment on which it acts to rotate clockwise, and a positive bending moment will tend to bend the segment on which it acts in a concave upward manner. Loadings that are opposite to these are considered negative.

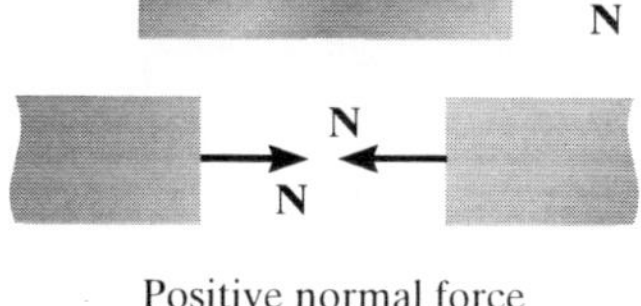

Positive normal force

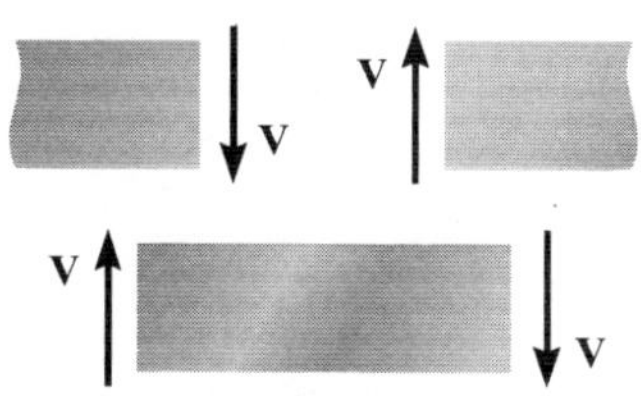

Positive shear

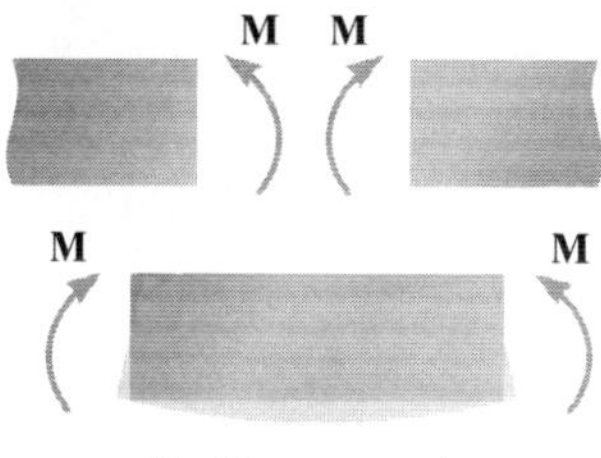

Positive moment

Fig. 7–3

Procedure for Analysis

The method of sections can be used to determine the internal loadings on the cross section of a member using the following procedure.

Support Reactions.

- Before the member is sectioned, it may first be necessary to determine its support reactions. Once obtained, the equilibrium equations can then be used to solve for the internal loadings after the member is sectioned.

Free-Body Diagram.

- It is important to *keep* all distributed loadings, couple moments, and forces acting on the member in their *exact locations*, *then* pass an imaginary section through the member, perpendicular to its axis at the point where the internal loadings are to be determined.
- After the section is made, draw a free-body diagram of the segment that has the least number of loads on it, and indicate the components of the internal force and couple moment resultants at the cross section acting in their positive directions in accordance with the established sign convention.

Equations of Equilibrium.

- Moments should be summed at the section. This way the normal and shear forces at the section are elminated, and we can obtain a direct solution for the moment.
- If the solution of the equilibrium equations yields a negative scalar, the sense of the quantity is opposite to that shown on the free-body diagram.

The designer of this shop crane realized the need for additional reinforcement around the joint in order to prevent severe internal bending of the joint when a large load is suspended from the chain hoist.

7

EXAMPLE 7.1

Please refer to the Companion Website for the animation: *Free-Body Diagram for a Simple Beam*

Determine the normal force, shear force, and bending moment acting just to the left, point *B*, and just to the right, point *C*, of the 6-kN force on the beam in Fig. 7–4*a*.

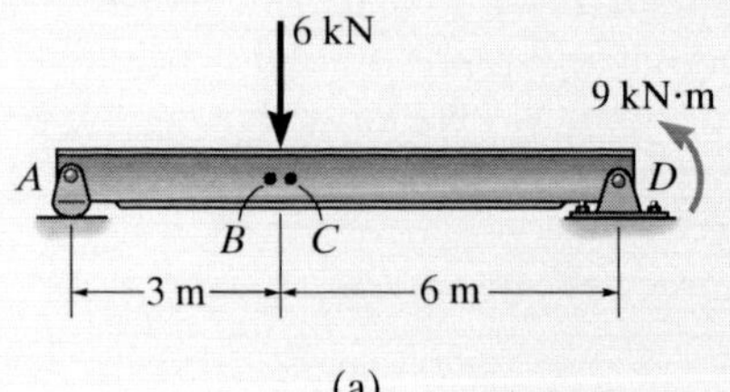

(a)

SOLUTION

Support Reactions. The free-body diagram of the beam is shown in Fig. 7–4*b*. When determining the *external reactions*, realize that the 9-kN · m couple moment is a free vector and therefore it can be placed *anywhere* on the free-body diagram of the entire beam. Here we will only determine $\mathbf{A}_y$, since the left segments will be used for the analysis.

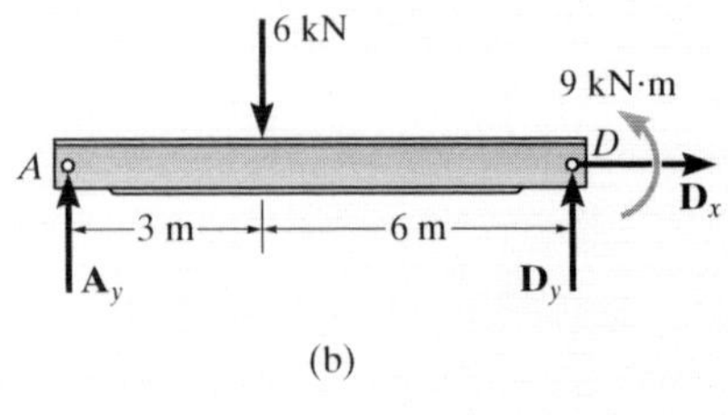

(b)

$$\zeta+\Sigma M_D = 0; \quad 9\text{ kN}\cdot\text{m} + (6\text{ kN})(6\text{ m}) - A_y(9\text{ m}) = 0$$

$$A_y = 5\text{ kN}$$

Free-Body Diagrams. The free-body diagrams of the left segments *AB* and *AC* of the beam are shown in Figs. 7–4*c* and 7–4*d*. In this case the 9-kN · m couple moment is *not included* on these diagrams since it must be kept in its *original position* until *after* the section is made and the appropriate segment is isolated.

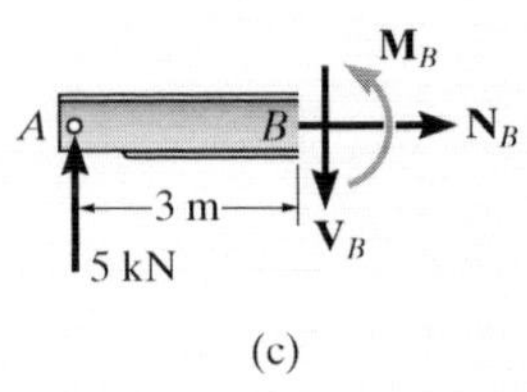

(c)

Equations of Equilibrium.

Segment AB

$$\overset{+}{\rightarrow}\Sigma F_x = 0; \quad N_B = 0 \qquad \textit{Ans.}$$

$$+\uparrow\Sigma F_y = 0; \quad 5\text{ kN} - V_B = 0 \quad V_B = 5\text{ kN} \qquad \textit{Ans.}$$

$$\zeta+\Sigma M_B = 0; \quad -(5\text{ kN})(3\text{ m}) + M_B = 0 \quad M_B = 15\text{ kN}\cdot\text{m} \qquad \textit{Ans.}$$

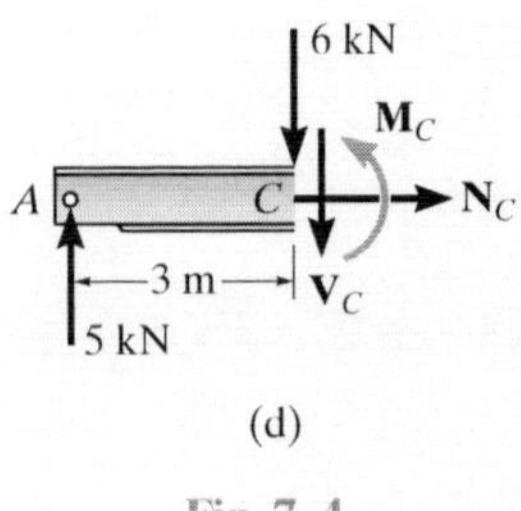

(d)

Fig. 7–4

Segment AC

$$\overset{+}{\rightarrow}\Sigma F_x = 0; \quad N_C = 0 \qquad \textit{Ans.}$$

$$+\uparrow\Sigma F_y = 0; \quad 5\text{ kN} - 6\text{ kN} - V_C = 0 \quad V_C = -1\text{ kN} \qquad \textit{Ans.}$$

$$\zeta+\Sigma M_C = 0; \quad -(5\text{ kN})(3\text{ m}) + M_C = 0 \quad M_C = 15\text{ kN}\cdot\text{m} \qquad \textit{Ans.}$$

NOTE: The negative sign indicates that $\mathbf{V}_C$ acts in the opposite sense to that shown on the free-body diagram. Also, the moment arm for the 5-kN force in both cases is approximately 3 m since *B* and *C* are "almost" coincident.

EXAMPLE 7.2

Determine the normal force, shear force, and bending moment at C of the beam in Fig. 7–5a.

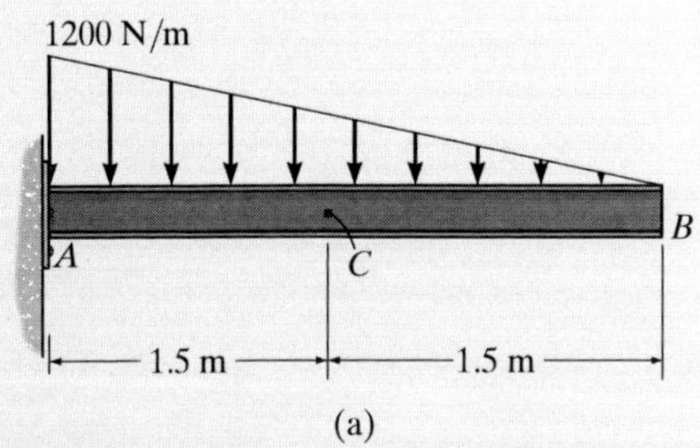

(a)

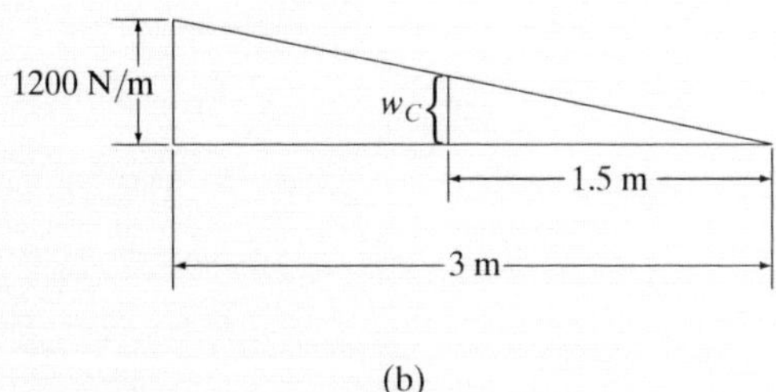

(b)

Fig. 7–5

SOLUTION

Free-Body Diagram. It is not necessary to find the support reactions at A since segment BC of the beam can be used to determine the internal loadings at C. The intensity of the triangular distributed load at C is determined using similar triangles from the geometry shown in Fig. 7–5b, i.e.,

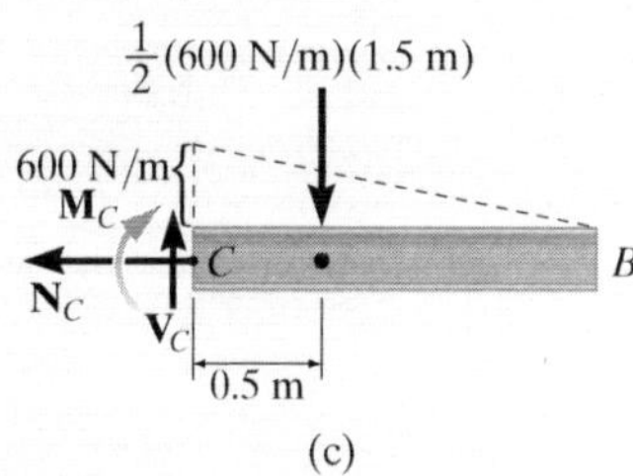

(c)

$$w_C = (1200\text{ N/m})\left(\frac{1.5\text{ m}}{3\text{ m}}\right) = 600\text{ N/m}$$

The distributed load acting on segment BC can now be replaced by its resultant force, and its location is indicated on the free-body diagram, Fig. 7–5c.

Equations of Equilibrium.

$$\xrightarrow{+} \Sigma F_x = 0; \qquad N_C = 0 \qquad \textit{Ans.}$$

$$+\uparrow \Sigma F_y = 0; \qquad V_C - \tfrac{1}{2}(600\text{ N/m})(1.5\text{ m}) = 0$$

$$V_C = 450\text{ N} \qquad \textit{Ans.}$$

$$\circlearrowleft + \Sigma M_C = 0; \quad -M_C - \tfrac{1}{2}(600\text{ N/m})(1.5\text{ m})(0.5\text{ m}) = 0$$

$$M_C = -225\text{ N} \qquad \textit{Ans.}$$

The negative sign indicates that $\mathbf{M}_C$ acts in the opposite sense to that shown on the free-body diagram.

EXAMPLE 7.3

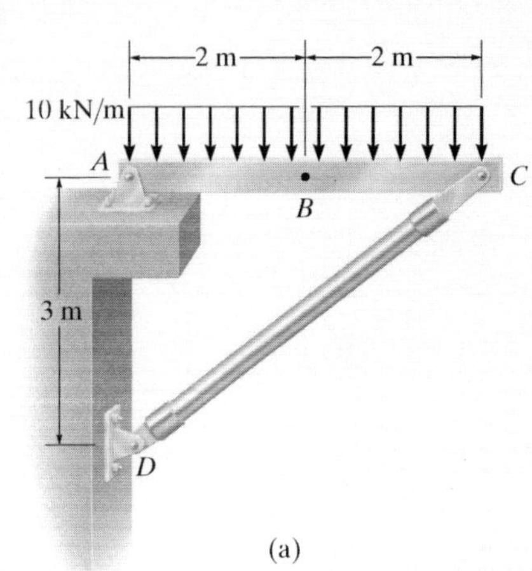

(a)

Determine the normal force, shear force, and bending moment acting at point B of the two-member frame shown in Fig. 7–6a.

SOLUTION

Support Reactions. A free-body diagram of each member is shown in Fig. 7–6b. Since CD is a two-force member, the equations of equilibrium need to be applied only to member AC.

$$\circlearrowleft +\Sigma M_A = 0; \quad -40 \text{ kN } (2 \text{ m}) + \left(\tfrac{3}{5}\right) F_{DC}(4 \text{ m}) = 0 \quad F_{DC} = 33.33 \text{ kN}$$

$$\xrightarrow{+} \Sigma F_x = 0; \quad -A_x + \left(\tfrac{4}{5}\right)(33.33 \text{ kN}) = 0 \quad A_x = 26.67 \text{ kN}$$

$$+\uparrow \Sigma F_y = 0; \quad A_y - 40 \text{ kN} + \left(\tfrac{3}{5}\right)(33.33 \text{ kN}) = 0 \quad A_y = 20 \text{ kN}$$

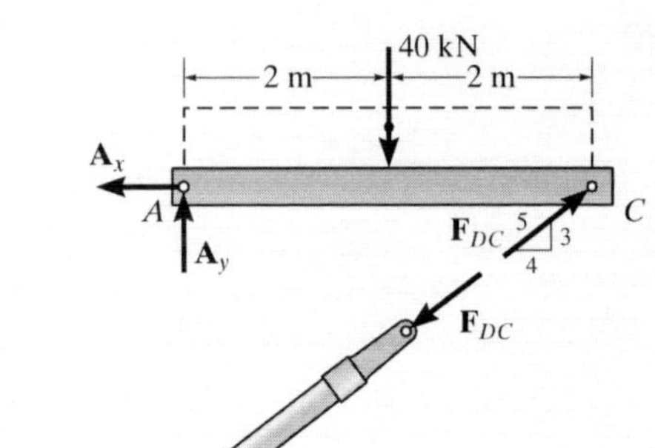

(b)

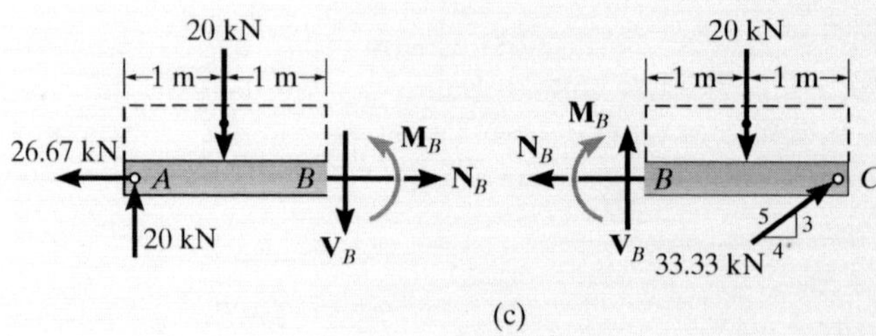

(c)

Fig. 7–6

Free-Body Diagrams. Passing an imaginary section perpendicular to the axis of member AC through point B yields the free-body diagrams of segments AB and BC shown in Fig. 7–6c. When constructing these diagrams it is important to keep the distributed loading where it is until *after the section is made.* Only then can it be replaced by a single resultant force.

Equations of Equilibrium. Applying the equations of equilibrium to segment AB, we have

$$\xrightarrow{+} \Sigma F_x = 0; \quad N_B - 26.67 \text{ kN} = 0 \quad N_B = 26.7 \text{ kN} \qquad \textit{Ans.}$$

$$+\uparrow \Sigma F_y = 0; \quad 20 \text{ kN} - 20 \text{ kN} - V_B = 0 \quad V_B = 0 \qquad \textit{Ans.}$$

$$\circlearrowleft +\Sigma M_B = 0; \quad M_B - 20 \text{ kN } (2 \text{ m}) + 20 \text{ kN } (1 \text{ m}) = 0$$

$$M_B = 20 \text{ kN} \cdot \text{m} \qquad \textit{Ans.}$$

NOTE: As an exercise, try to obtain these same results using segment BC.

EXAMPLE 7.4

Determine the normal force, shear force, and bending moment acting at point E of the frame loaded as shown in Fig. 7–7*a*.

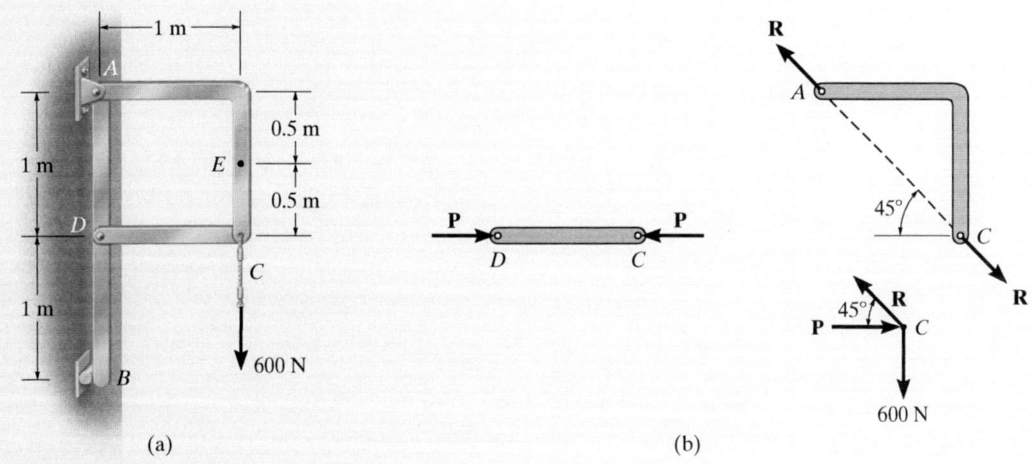

SOLUTION

Support Reactions. By inspection, members AC and CD are two-force members, Fig. 7–7*b*. In order to determine the internal loadings at E, we must first determine the force **R** acting at the end of member AC. To obtain it, we will analyze the equilibrium of the pin at C.

Summing forces in the vertical direction on the pin, Fig. 7–7*b*, we have

$$+\uparrow \Sigma F_y = 0; \qquad R \sin 45^\circ - 600 \text{ N} = 0 \qquad R = 848.5 \text{ N}$$

Free-Body Diagram. The free-body diagram of segment CE is shown in Fig. 7–7*c*.

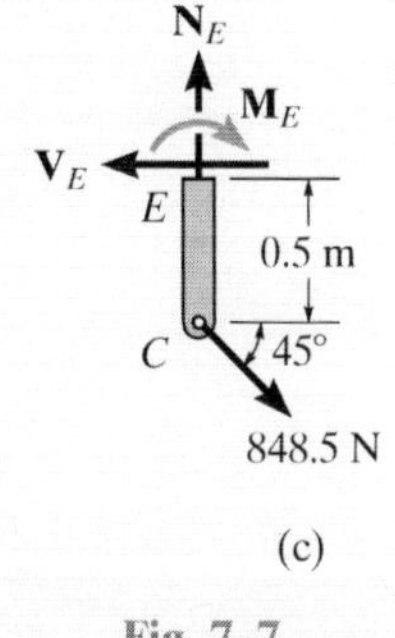

(c)

Fig. 7–7

Equations of Equilibrium.

$$\overset{+}{\rightarrow} \Sigma F_x = 0; \qquad 848.5 \cos 45^\circ \text{ N} - V_E = 0 \qquad V_E = 600 \text{ N} \qquad \textit{Ans.}$$

$$+\uparrow \Sigma F_y = 0; \qquad -848.5 \sin 45^\circ \text{ N} + N_E = 0 \qquad N_E = 600 \text{ N} \qquad \textit{Ans.}$$

$$\circlearrowleft + \Sigma M_E = 0; \qquad 848.5 \cos 45^\circ \text{ N}(0.5 \text{ m}) - M_E = 0$$

$$M_E = 300 \text{ N} \cdot \text{m} \qquad \textit{Ans.}$$

NOTE: These results indicate a poor design. Member AC should be *straight* (from A to C) so that bending within the member is *eliminated*. If AC were straight then the internal force would only create tension in the member.

EXAMPLE 7.5

(a)

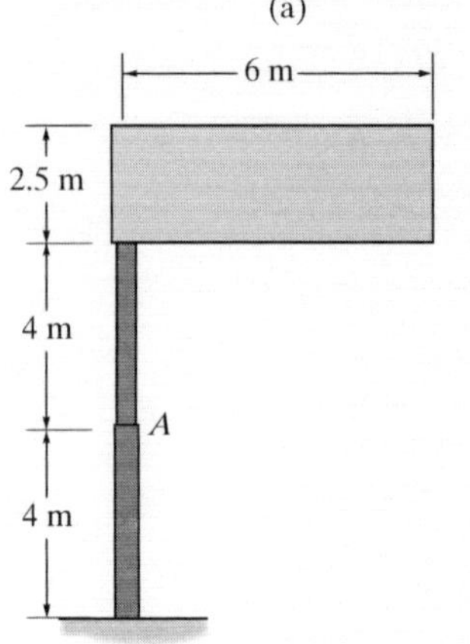

(b)

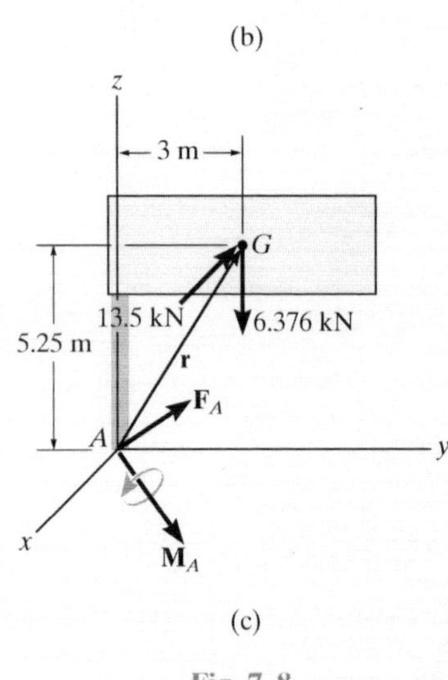

(c)

Fig. 7–8

The uniform sign shown in Fig. 7–8*a* has a mass of 650 kg and is supported on the fixed column. Design codes indicate that the expected maximum uniform wind loading that will occur in the area where it is located is 900 Pa. Determine the internal loadings at *A*.

SOLUTION

The idealized model for the sign is shown in Fig. 7–8*b*. Here the necessary dimensions are indicated. We can consider the free-body diagram of a section above point *A* since it does not involve the support reactions.

Free-Body Diagram. The sign has a weight of $W = 650(9.81)\text{ N} = 6.376\text{ kN}$, and the wind creates a resultant force of $F_w = 900\text{ N/m}^2(6\text{ m})(2.5\text{ m}) = 13.5\text{ kN}$, which acts perpendicular to the face of the sign. These loadings are shown on the free-body diagram, Fig. 7–8*c*.

Equations of Equilibrium. Since the problem is three dimensional, a vector analysis will be used.

$$\Sigma \mathbf{F} = \mathbf{0}; \qquad \mathbf{F}_A - 13.5\mathbf{i} - 6.376\mathbf{k} = \mathbf{0}$$

$$\mathbf{F}_A = \{13.5\mathbf{i} + 6.38\mathbf{k}\}\text{ kN} \qquad \textit{Ans.}$$

$$\Sigma \mathbf{M}_A = \mathbf{0}; \qquad \mathbf{M}_A + \mathbf{r} \times (\mathbf{F}_w + \mathbf{W}) = \mathbf{0}$$

$$\mathbf{M}_A + \begin{vmatrix} \mathbf{i} & \mathbf{j} & \mathbf{k} \\ 0 & 3 & 5.25 \\ -13.5 & 0 & -6.376 \end{vmatrix} = \mathbf{0}$$

$$\mathbf{M}_A = \{19.1\mathbf{i} + 70.9\mathbf{j} - 40.5\mathbf{k}\}\text{ kN}\cdot\text{m} \qquad \textit{Ans.}$$

NOTE: Here $\mathbf{F}_{A_z} = \{6.38\mathbf{k}\}$ kN represents the normal force, whereas $\mathbf{F}_{A_x} = \{13.5\mathbf{i}\}$ kN is the shear force. Also, the torsional moment is $\mathbf{M}_{A_z} = \{-40.5\mathbf{k}\}$ kN·m, and the bending moment is determined from its components $\mathbf{M}_{A_x} = \{19.1\mathbf{i}\}$ kN·m and $\mathbf{M}_{A_y} = \{70.9\mathbf{j}\}$ kN·m; i.e., $(M_b)_A = \sqrt{(M_A)_x^2 + (M_A)_y^2} = 73.4\text{ kN}\cdot\text{m}$.

FUNDAMENTAL PROBLEMS

All problem solutions must include FBDs.

F7–1. Determine the normal force, shear force, and moment at point C.

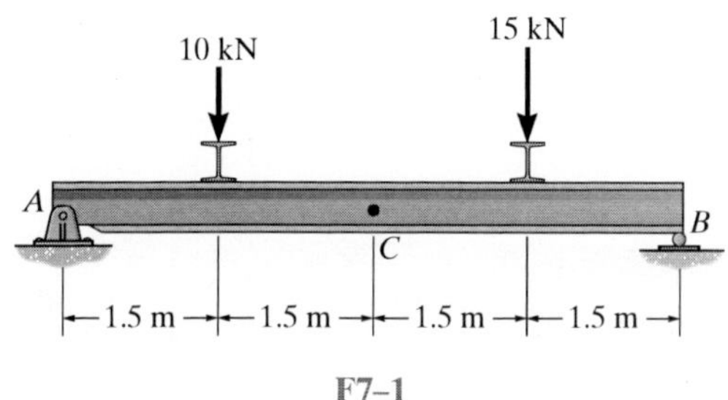

F7–1

F7–2. Determine the normal force, shear force, and moment at point C.

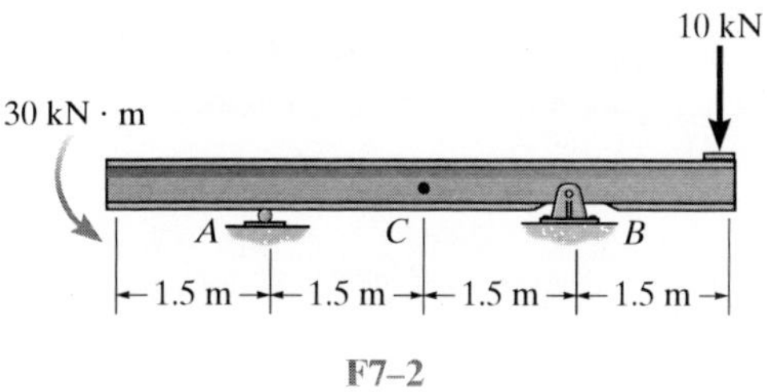

F7–2

F7–3. Determine the normal force, shear force, and moment at point C.

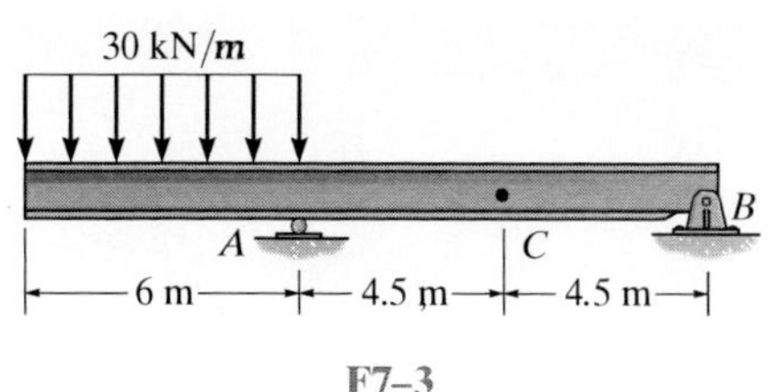

F7–3

F7–4. Determine the normal force, shear force, and moment at point C.

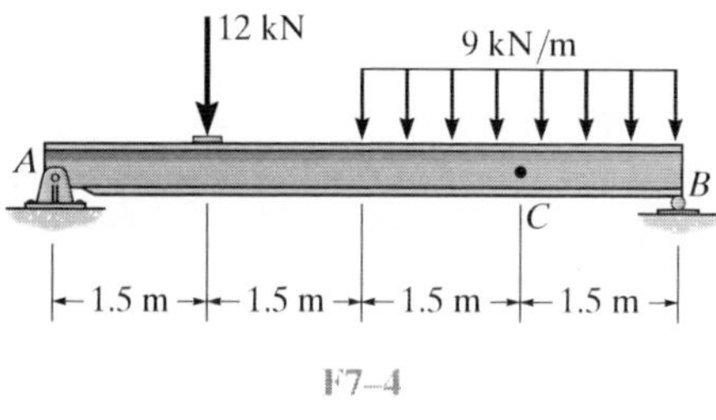

F7–4

F7–5. Determine the normal force, shear force, and moment at point C.

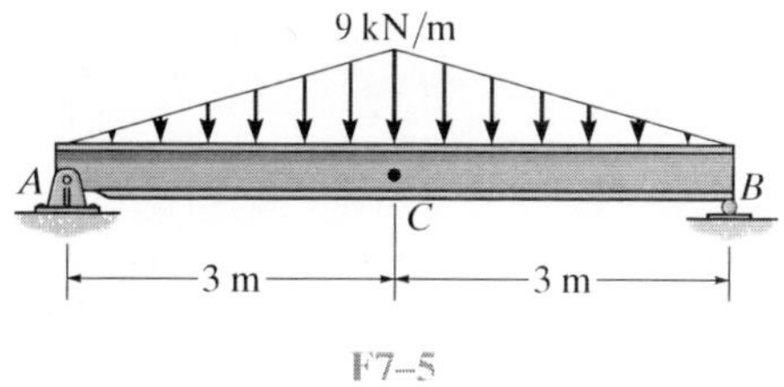

F7–5

F7–6. Determine the normal force, shear force, and moment at point C. Assume A is pinned and B is a roller.

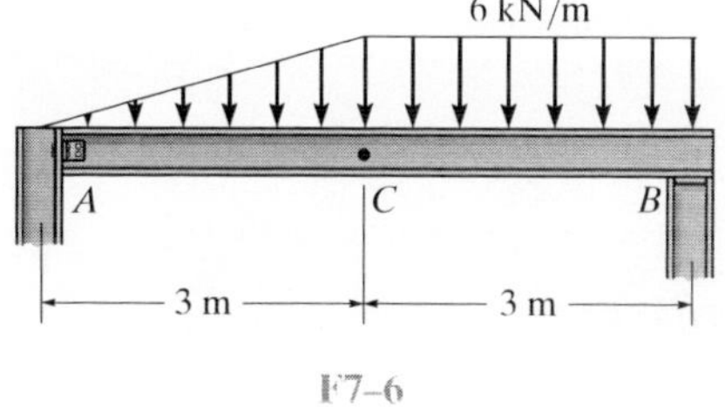

F7–6

7

PROBLEMS

All problem solutions must include FBDs.

7–1. The forces act on the shaft shown. Determine the internal normal force at points A, B, and C.

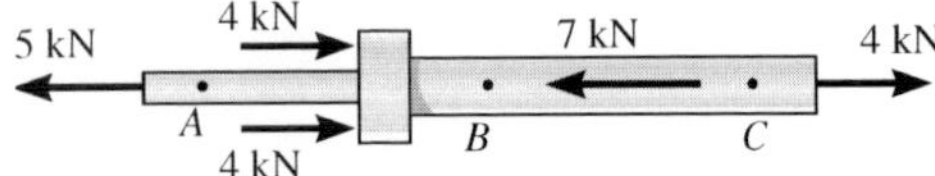

Prob. 7–1

7–2. Three torques act on the shaft. Determine the internal torque at points A, B, C, and D.

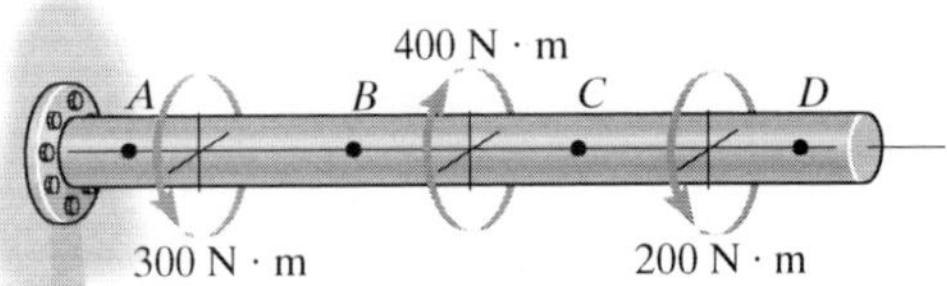

Prob. 7–2

7–3. The strongback or lifting beam is used for materials handling. If the suspended load has a weight of 2 kN and a center of gravity of G, determine the placement d of the padeyes on the top of the beam so that there is no moment developed within the length AB of the beam. The lifting bridle has two legs that are positioned at 45°, as shown.

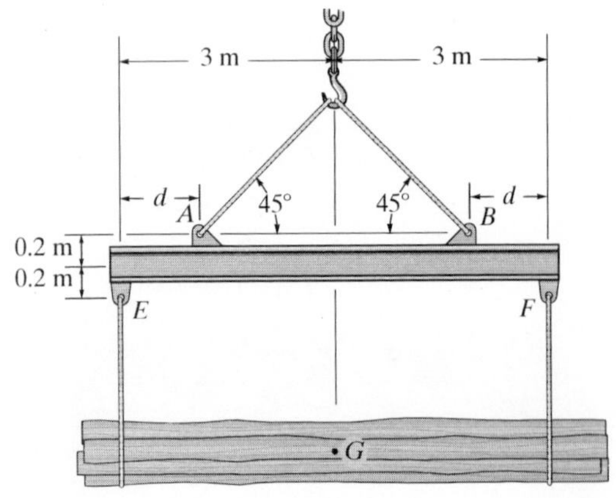

Prob. 7–3

***7–4.** Determine the normal force, shear force, and moment at a section passing through point D of the two-member frame.

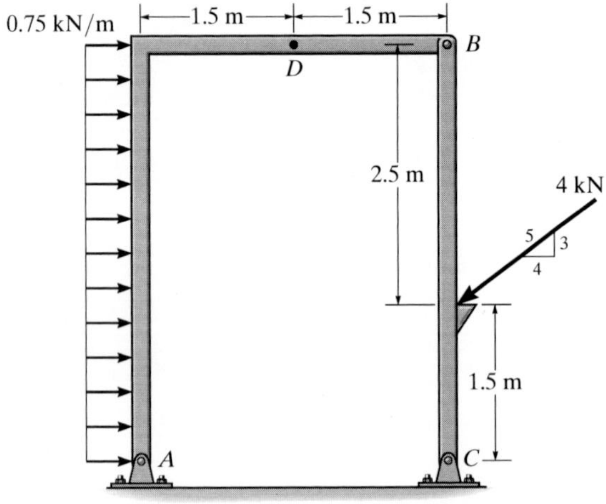

Prob. 7–4

7–5. Determine the internal normal force, shear force, and moment at points A and B in the column.

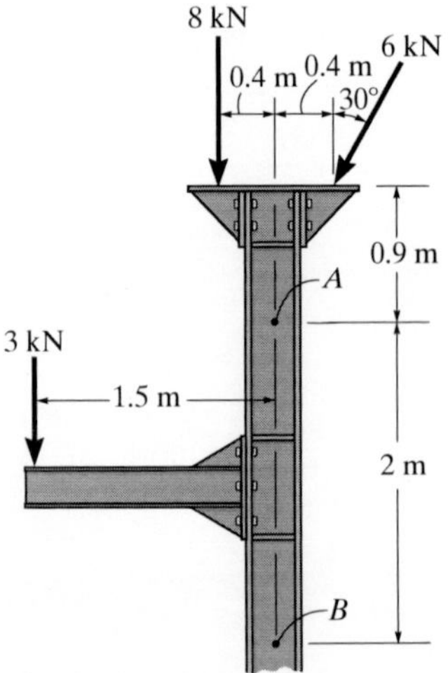

Prob. 7–5

7–6. Determine the distance a as a fraction of the beam's length L for locating the roller support so that the moment in the beam at B is zero.

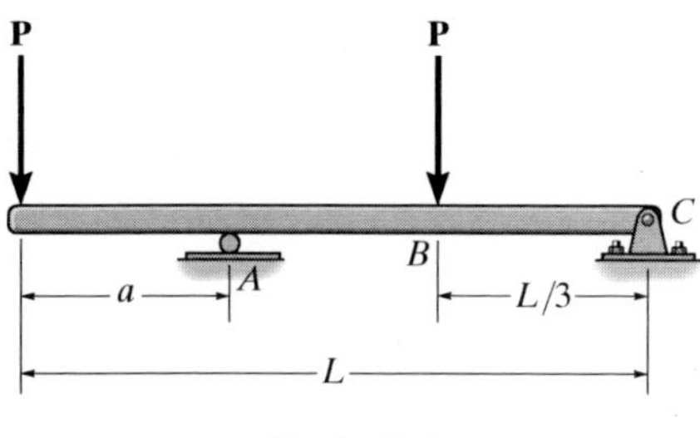

Prob. 7–6

7–7. Determine the internal normal force, shear force, and moment at points E and D of the compound beam.

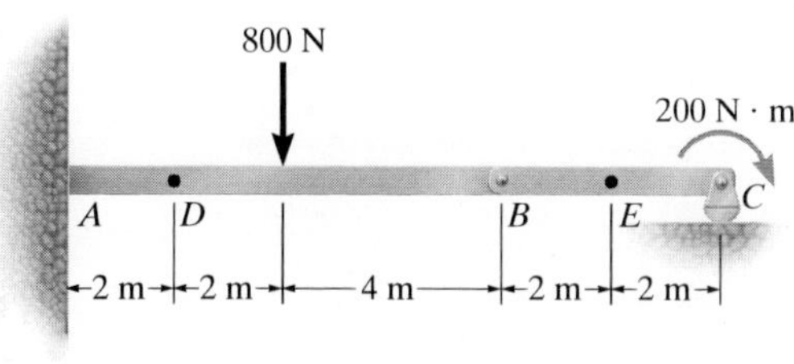

Prob. 7–7

***7–8.** Determine the internal normal force, shear force, and moment at point C.

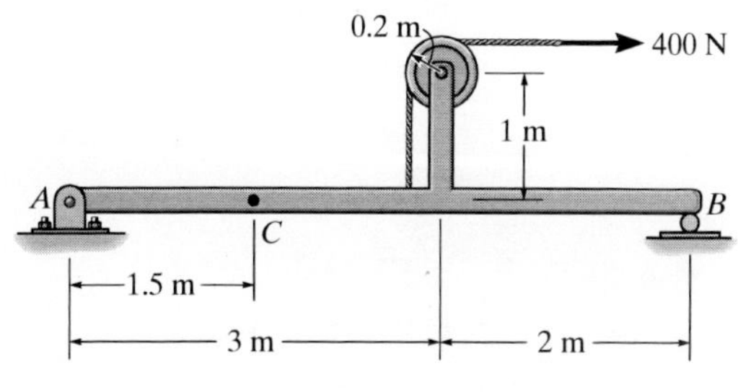

Prob. 7–8

7–9. Determine the normal force, shear force, and moment at a section passing through point C. Take $P = 8$ kN.

7–10. The cable will fail when subjected to a tension of 2 kN. Determine the largest vertical load P the frame will support and calculate the internal normal force, shear force, and moment at a section passing through point C for this loading.

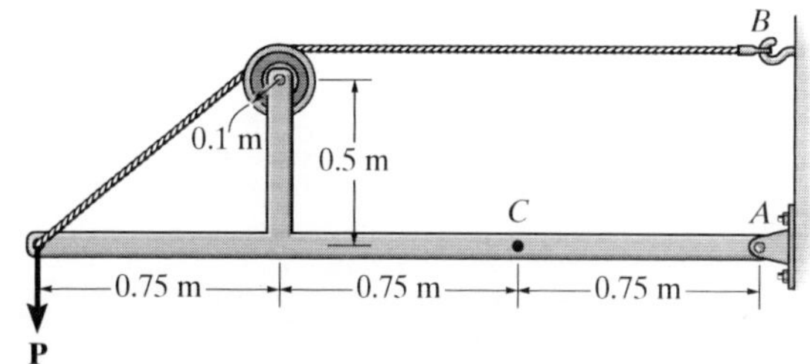

Probs. 7–9/10

7–11. Determine the internal normal force, shear force, and moment at points D and E in the compound beam. Point E is located just to the left of the 10-kN concentrated load. Assume the support at A is fixed and the connection at B is a pin.

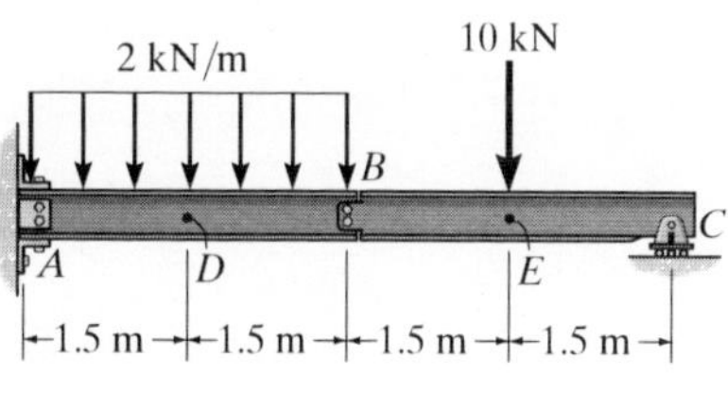

Prob. 7–11

***7–12.** Determine the internal normal force, shear force, and the moment at points C and D.

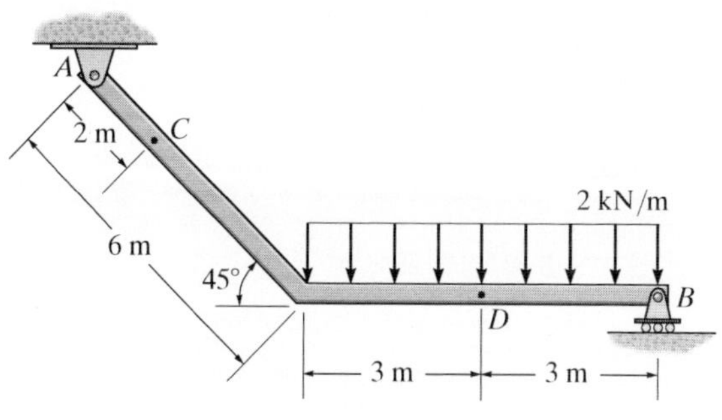

Prob. 7–12

7

7–13. Determine the internal normal force, shear force, and moment at point C in the simply supported beam.

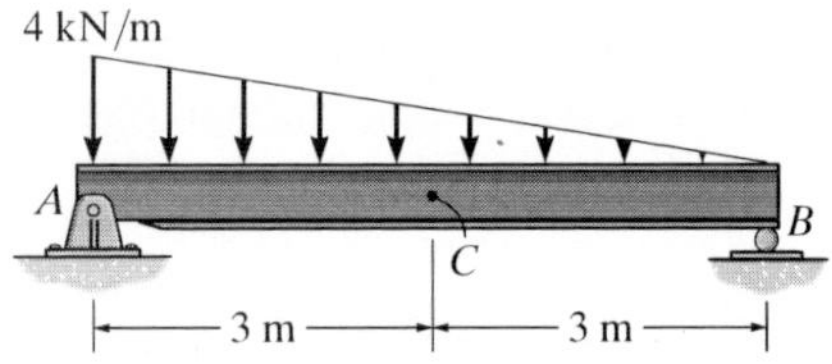

Prob. 7–13

7–14. Determine the normal force, shear force, and moment at a section passing through point D. Take $w = 150$ N/m.

7–15. The beam AB will fail if the maximum internal moment at D reaches 800 N·m or the normal force in member BC becomes 1500 N. Determine the largest load w it can support.

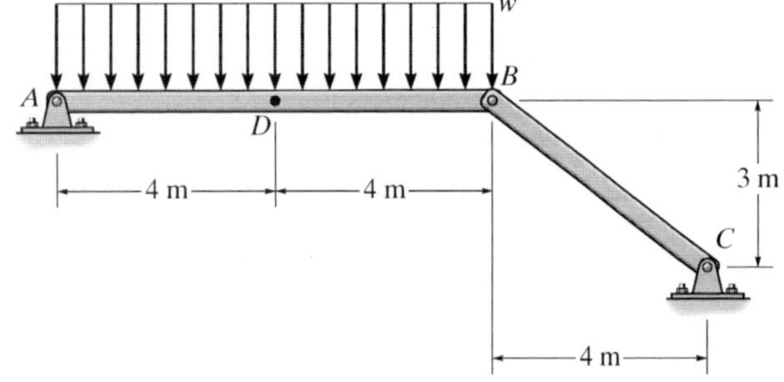

Probs. 7–14/15

*__7–16.__ Determine the internal normal force, shear force, and moment at point D in the beam.

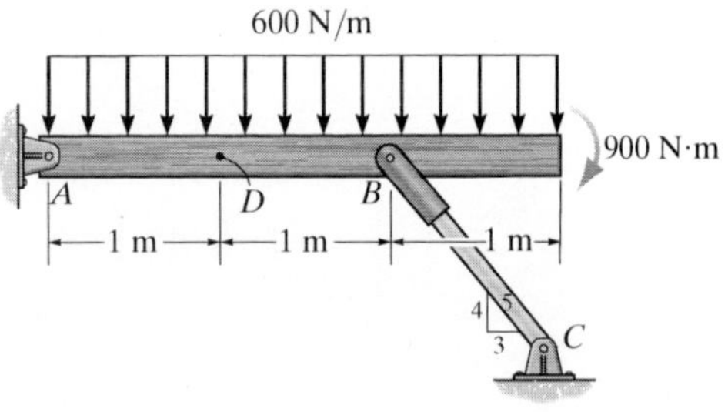

Prob. 7–16

7–17. Determine the normal force, shear force, and moment at a section passing through point E of the two-member frame.

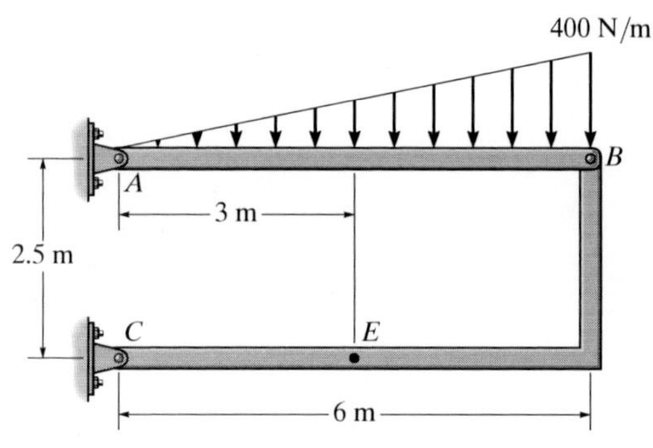

Prob. 7–17

7–18. Determine the internal normal force, shear force, and moment at point C in the cantilever beam.

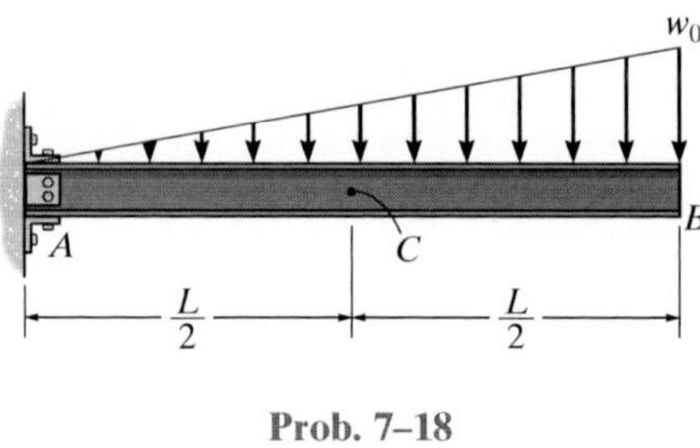

Prob. 7–18

7–19. Determine the internal normal force, shear force, and moment at points E and F in the beam.

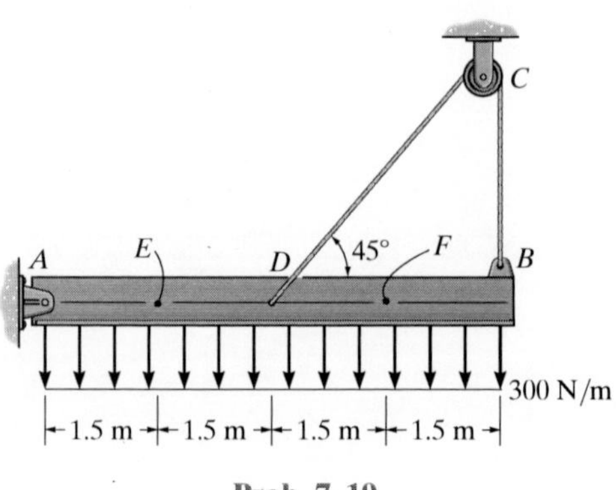

Prob. 7–19

7

***7–20.** Determine the internal normal force, shear force, and moment at points C and D in the simply supported beam. Point D is located just to the left of the 5-kN force.

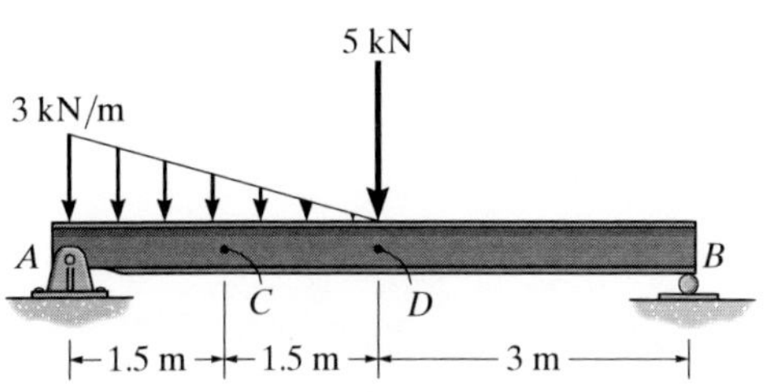

Prob. 7–20

7–21. Determine the internal normal force, shear force, and moment at points D and E in the overhang beam. Point D is located just to the left of the roller support at B, where the couple moment acts.

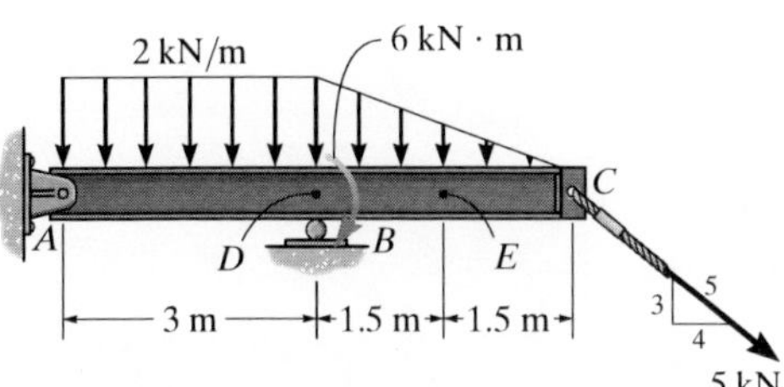

Prob. 7–21

7–22. Determine the internal normal force, shear force, and moment at points E and F in the compound beam. Point F is located just to the left of the 15-kN force and 25-kN · m couple moment.

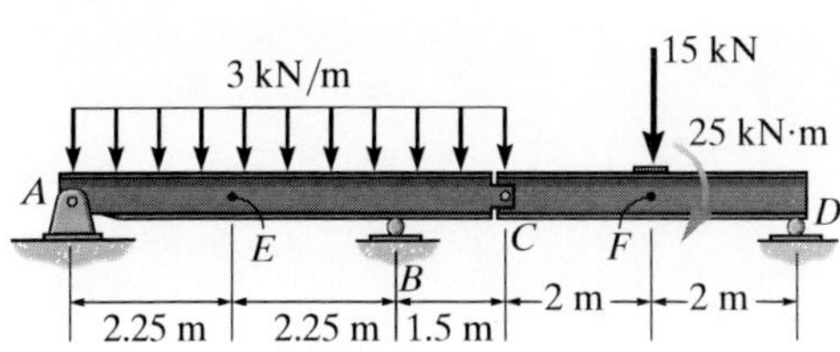

Prob. 7–22

7–23. Determine the internal normal force, shear force, and moment at points D and E in the frame. Point D is located just above the 400-N force.

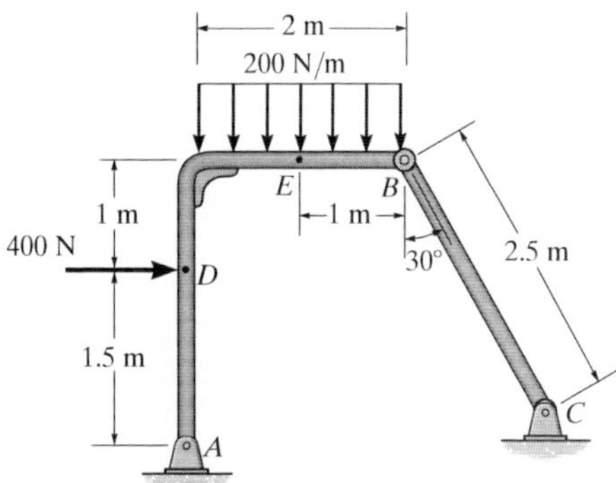

Prob. 7–23

***7–24.** Determine the internal normal force, shear force, and bending moment at point C.

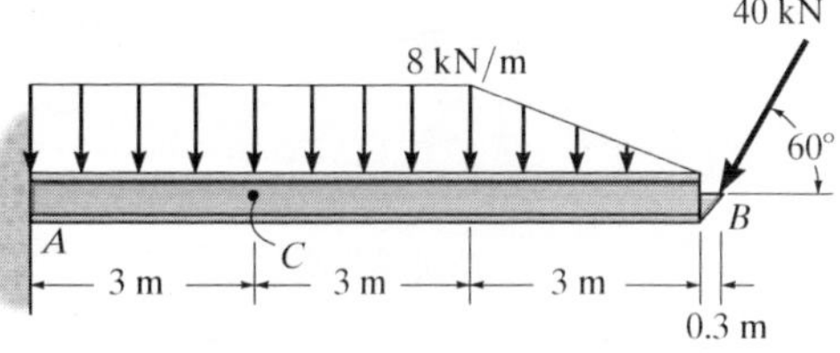

Prob. 7–24

7–25. Determine the internal normal force, shear force, and moment at point C in the double-overhang beam.

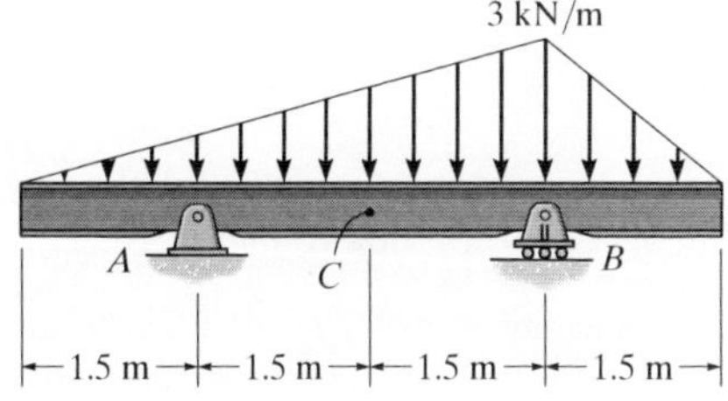

Prob. 7–25

7–26. Determine the ratio of a/b for which the shear force will be zero at the midpoint C of the beam.

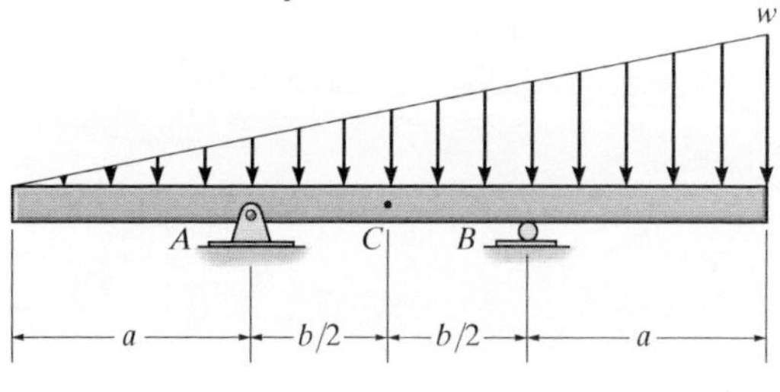

Prob. 7–26

7

7–27. Determine the normal force, shear force, and moment at a section passing through point D of the two-member frame.

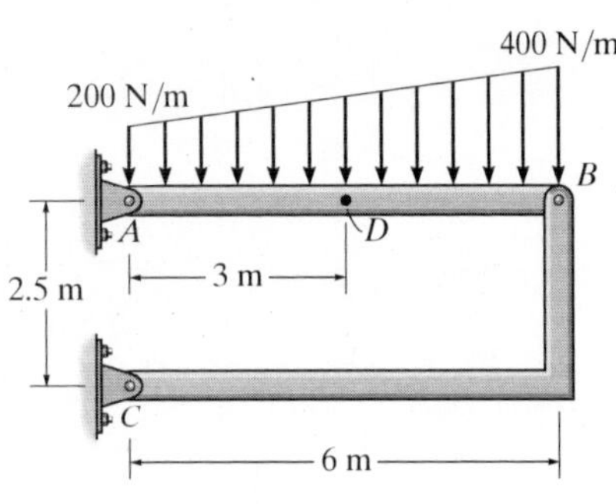

Prob. 7–27

***7–28.** Determine the internal normal force, shear force, and moment at point C in the simply supported beam. Point C is located just to the right of the 2.5-kN · m couple moment.

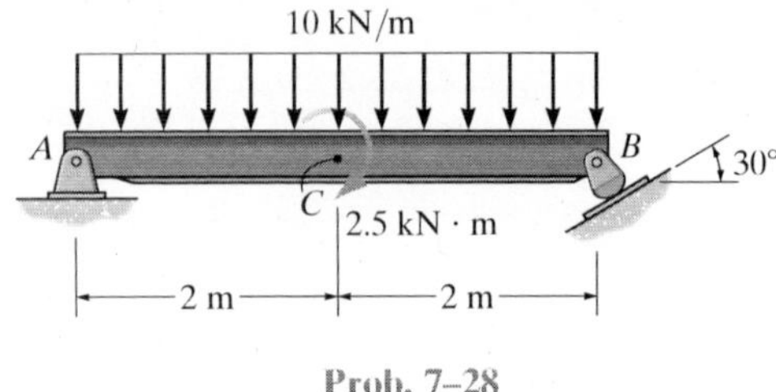

Prob. 7–28

7–29. Determine the internal normal force, shear force, and moment at point D of the two-member frame.

7–30. Determine the internal normal force, shear force, and moment at point E of the two-member frame.

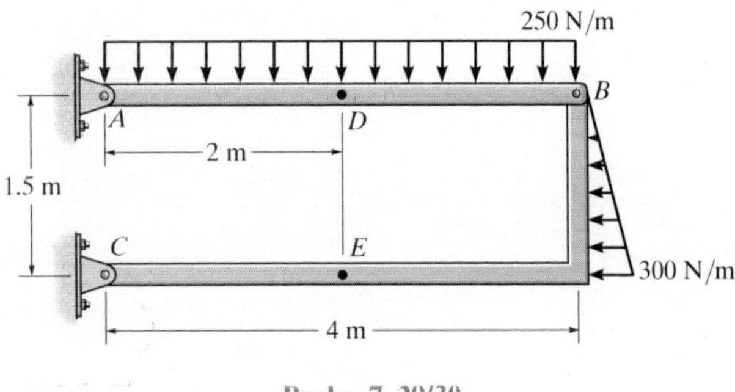

Probs. 7–29/30

7–31. The hook supports the 4-kN load. Determine the internal normal force, shear force, and moment at point A.

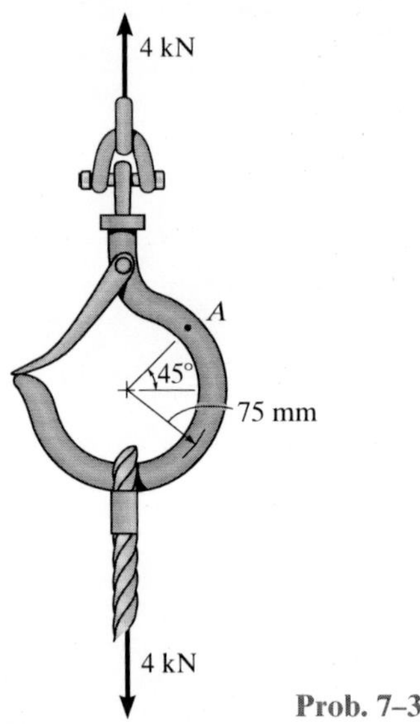

Prob. 7–31

***7–32.** Determine the internal normal force, shear force, and moment at points C and D in the simply supported beam. Point D is located just to the left of the 10-kN concentrated load.

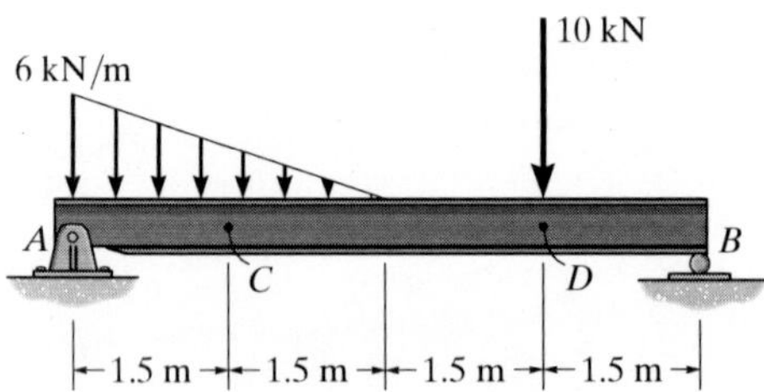

Prob. 7–32

7–33. Determine the distance a in terms of the beam's length L between the symmetrically placed supports A and B so that the internal moment at the center of the beam is zero.

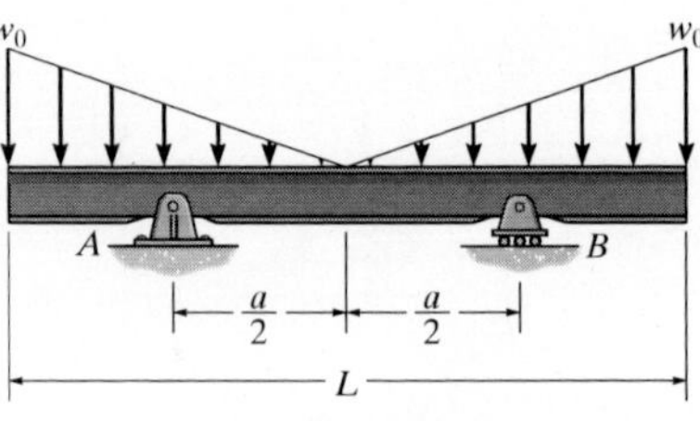

Prob. 7–33

7

7–34. The beam has a weight w per unit length. Determine the internal normal force, shear force, and moment at point C due to its weight.

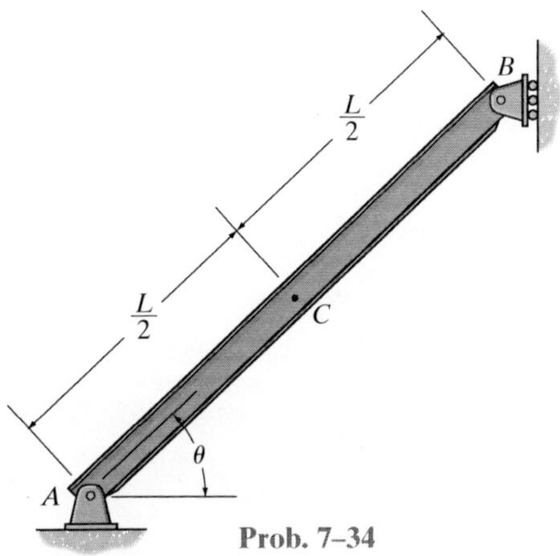

Prob. 7–34

7–35. Determine the internal normal force, shear force, and bending moment at points E and F of the frame.

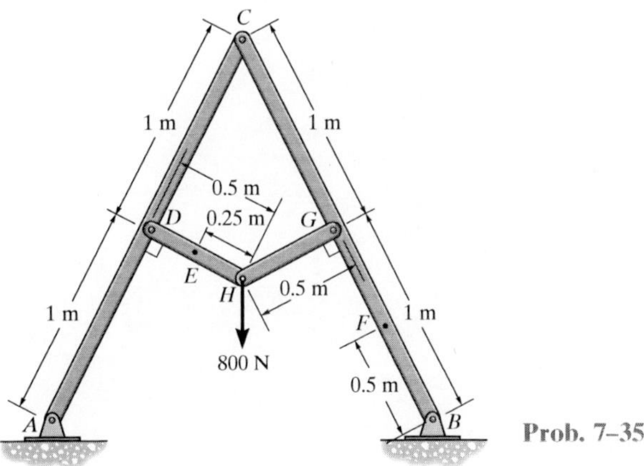

Prob. 7–35

***7–36.** Determine the distance a between the supports in terms of the shaft's length L so that the bending moment in the *symmetric* shaft is zero at the shaft's center. The intensity of the distributed load at the center of the shaft is w_0. The supports are journal bearings.

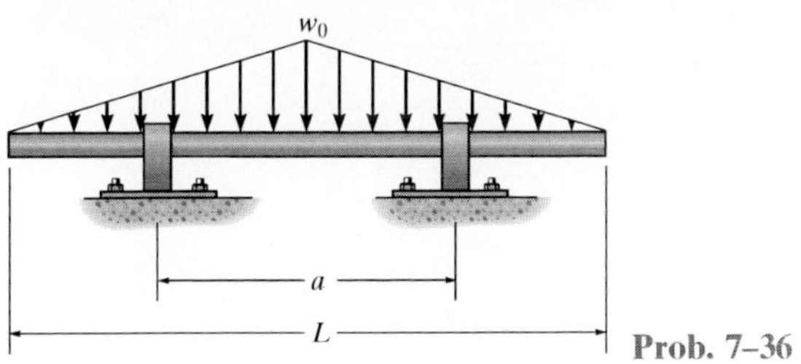

Prob. 7–36

7–37. Determine the internal normal force, shear force, and moment acting at point C. The cooling unit has a total mass of 225 kg with a center of mass at G.

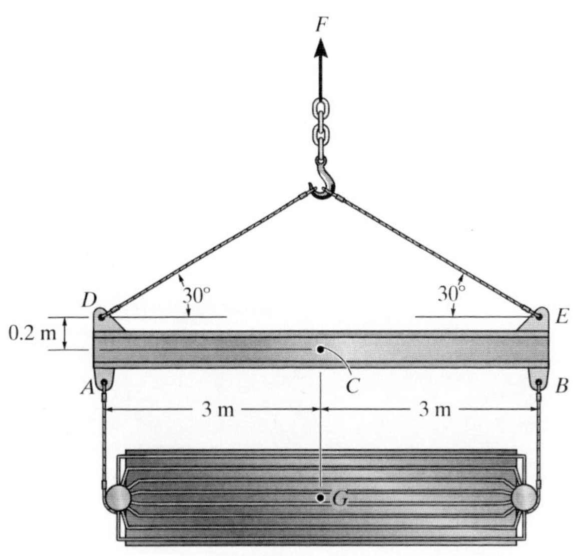

Prob. 7–37

7–38. Determine the internal normal force, shear force, and moment at points D and E of the frame which supports the 100-kg crate. Neglect the size of the smooth peg at C.

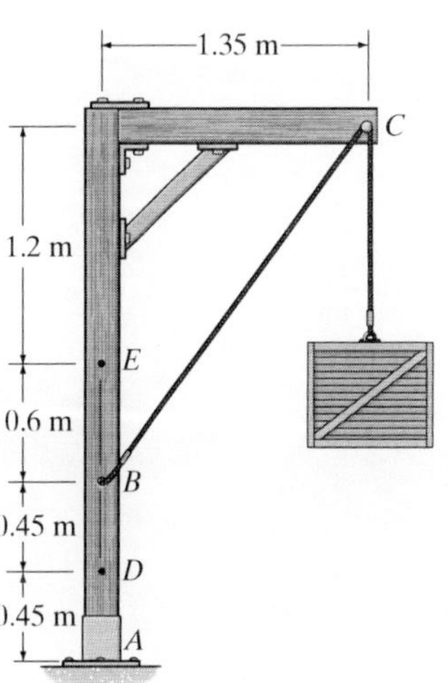

Prob. 7–38

7–39. The semicircular arch is subjected to a uniform distributed load along its axis of w_0 per unit length. Determine the internal normal force, shear force, and moment in the arch at $\theta = 45°$.

***7–40.** The semicircular arch is subjected to a uniform distributed load along its axis of w_0 per unit length. Determine the internal normal force, shear force, and moment in the arch at $\theta = 120°$.

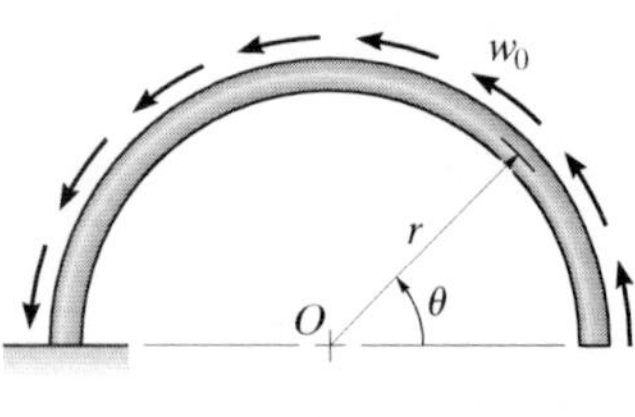

Probs. 7–39/40

7–41. Determine the x, y, z components of internal loading at a section passing through point C in the pipe assembly. Neglect the weight of the pipe. Take $\mathbf{F}_1 = \{350\mathbf{j} - 400\mathbf{k}\}$ kN and $\mathbf{F}_2 = \{150\mathbf{i} - 300\mathbf{k}\}$ kN.

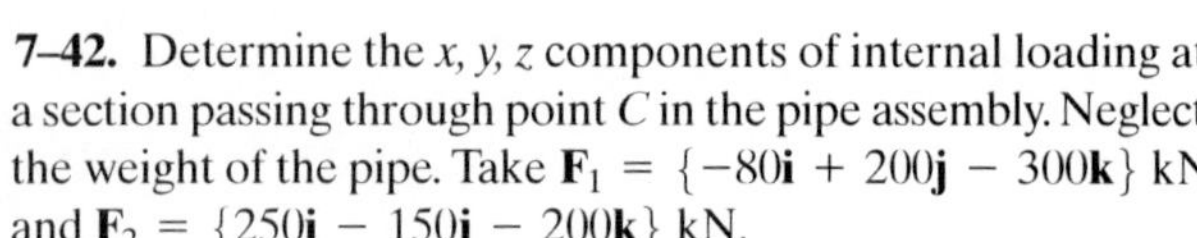

7–42. Determine the x, y, z components of internal loading at a section passing through point C in the pipe assembly. Neglect the weight of the pipe. Take $\mathbf{F}_1 = \{-80\mathbf{i} + 200\mathbf{j} - 300\mathbf{k}\}$ kN and $\mathbf{F}_2 = \{250\mathbf{i} - 150\mathbf{j} - 200\mathbf{k}\}$ kN.

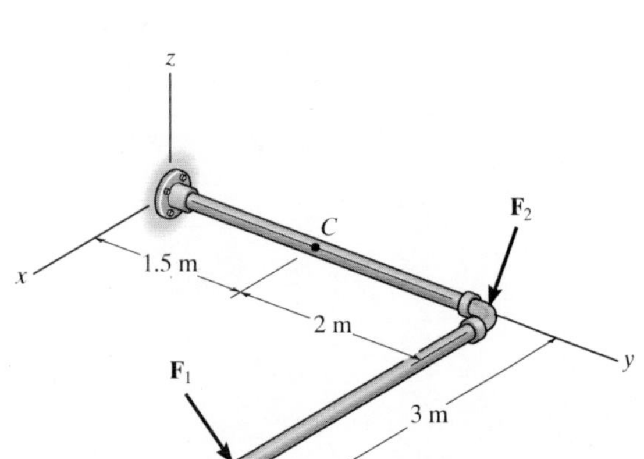

Probs. 7–41/42

7–43. Determine the x, y, z components of internal loading in the rod at point D. $\mathbf{F} = \{7\mathbf{i} - 12\mathbf{j} - 5\mathbf{k}\}$ kN.

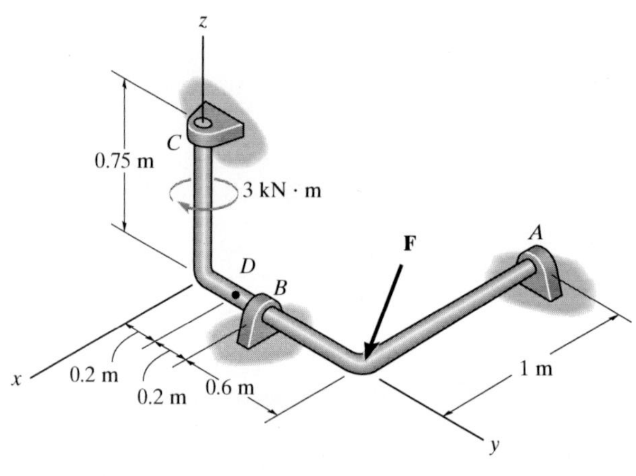

Prob. 7–43

***7–44.** Determine the x, y, z components of internal loading in the rod at point E. $\mathbf{F} = \{7\mathbf{i} - 12\mathbf{j} - 5\mathbf{k}\}$ kN.

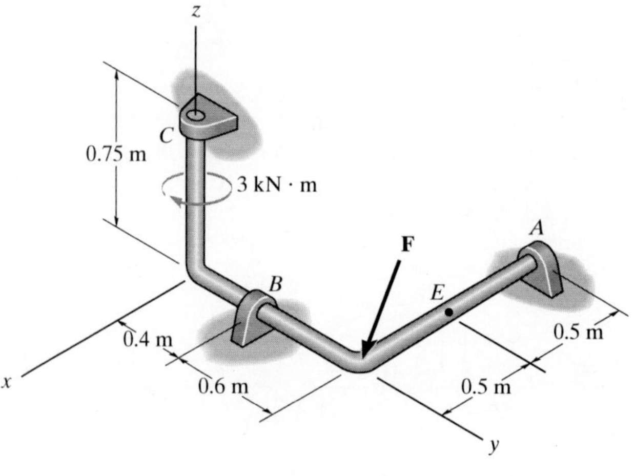

Prob. 7–44

*7.2 Shear and Moment Equations and Diagrams

Beams are structural members designed to support loadings applied perpendicular to their axes. In general, they are long and straight and have a constant cross-sectional area. They are often classified as to how they are supported. For example, a *simply supported beam* is pinned at one end and roller supported at the other, as in Fig. 7–9*a*, whereas a *cantilevered beam* is fixed at one end and free at the other. The actual design of a beam requires a detailed knowledge of the *variation* of the internal shear force V and bending moment M acting at *each point* along the axis of the beam.*

These *variations* of V and M along the beam's axis can be obtained by using the method of sections discussed in Sec. 7.1. In this case, however, it is necessary to section the beam at an arbitrary distance x from one end and then apply the equations of equilibrium to the segment having the length x. Doing this we can then obtain V and M as functions of x.

To save on material and thereby produce an efficient design, these beams, also called girders, have been tapered, since the internal moment in the beam will be larger at the supports, or piers, than at the center of the span.

In general, the internal shear and bending-moment functions will be discontinuous, or their slopes will be discontinuous, at points where a distributed load changes or where concentrated forces or couple moments are applied. Because of this, these functions must be determined for *each segment* of the beam located between any two discontinuities of loading. For example, segments having lengths x_1, x_2, and x_3 will have to be used to describe the variation of V and M along the length of the beam in Fig. 7–9*a*. These functions will be valid *only* within regions from 0 to a for x_1, from a to b for x_2, and from b to L for x_3. If the resulting functions of x are plotted, the graphs are termed the *shear diagram* and *bending-moment diagram*, Fig. 7–9*b* and Fig. 7–9*c*, respectively.

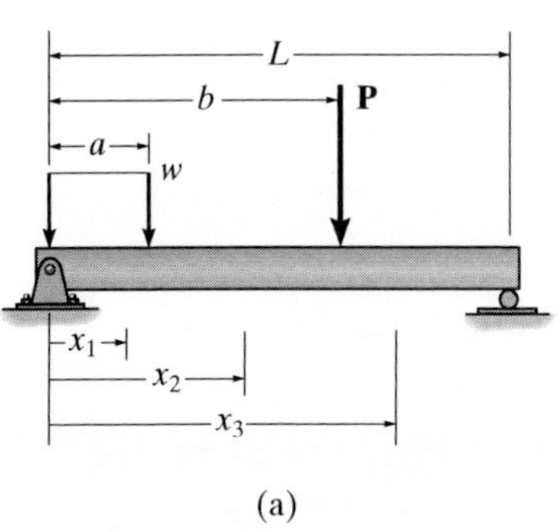

(a)

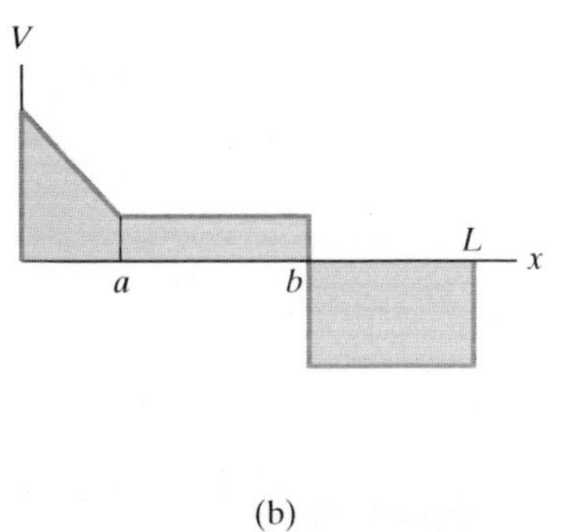

(b)

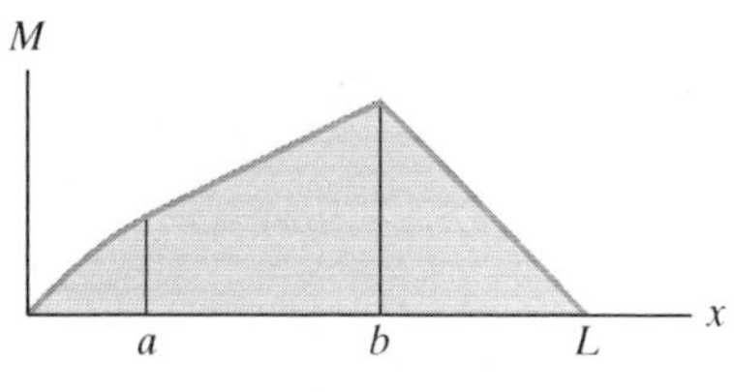

(c)

Fig. 7–9

*The internal normal force is not considered for two reasons. In most cases, the loads applied to a beam act perpendicular to the beam's axis and hence produce only an internal shear force and bending moment. And for design purposes, the beam's resistance to shear, and particularly to bending, is more important than its ability to resist a normal force.

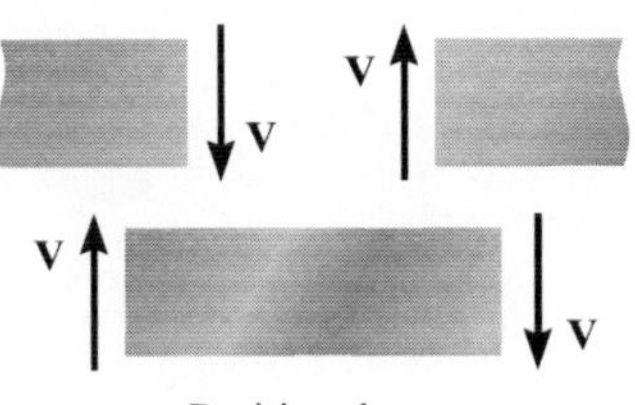

Positive shear

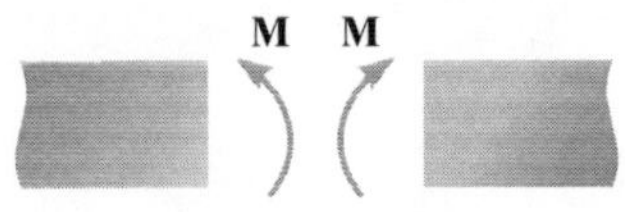

Positive moment

Beam sign convention

Fig. 7–10

This extended towing arm must resist both bending and shear loadings throughout its length due to the weight of the vehicle. The variation of these loadings must be known if the arm is to be properly designed.

Procedure for Analysis

The shear and bending-moment diagrams for a beam can be constructed using the following procedure.

Support Reactions.

- Determine all the reactive forces and couple moments acting on the beam and resolve all the forces into components acting perpendicular and parallel to the beam's axis.

Shear and Moment Functions.

- Specify separate coordinates x having an origin at the beam's left end and extending to regions of the beam *between* concentrated forces and/or couple moments, or where the distributed loading is continuous.
- Section the beam at each distance x and draw the free-body diagram of one of the segments. Be sure **V** and **M** are shown acting in their *positive sense*, in accordance with the sign convention given in Fig. 7–10.
- The shear V is obtained by summing forces perpendicular to the beam's axis.
- The moment M is obtained by summing moments about the sectioned end of the segment.

Shear and Moment Diagrams.

- Plot the shear diagram (V versus x) and the moment diagram (M versus x). If computed values of the functions describing V and M are *positive*, the values are plotted above the x axis, whereas *negative* values are plotted below the x axis.
- Generally, it is convenient to plot the shear and bending-moment diagrams directly below the free-body diagram of the beam.

7

EXAMPLE 7.6

Draw the shear and moment diagrams for the shaft shown in Fig. 7–11*a*. The support at *A* is a thrust bearing and the support at *C* is a journal bearing.

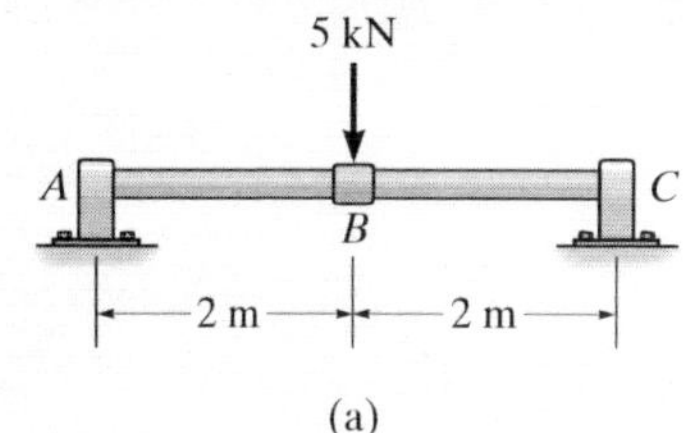

(a)

SOLUTION

Support Reactions. The support reactions are shown on the shaft's free-body diagram, Fig. 7–11*d*.

Shear and Moment Functions. The shaft is sectioned at an arbitrary distance *x* from point *A*, extending within the region *AB*, and the free-body diagram of the left segment is shown in Fig. 7–11*b*. The unknowns **V** and **M** are assumed to act in the *positive sense* on the right-hand face of the segment according to the established sign convention. Applying the equilibrium equations yields

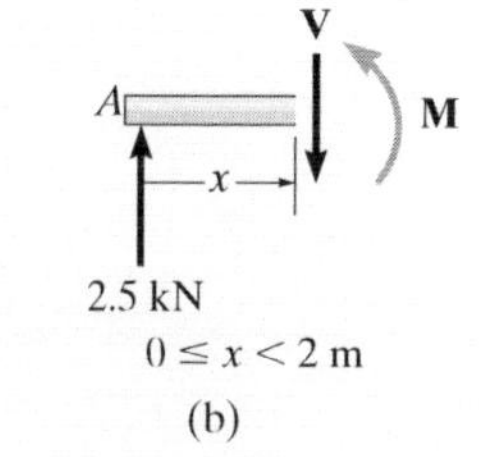

(b)

$$+\uparrow\Sigma F_y = 0; \qquad V = 2.5 \text{ kN} \qquad (1)$$

$$\circlearrowleft +\Sigma M = 0; \qquad M = 2.5x \text{ kN}\cdot\text{m} \qquad (2)$$

A free-body diagram for a left segment of the shaft extending from *A* a distance *x*, within the region *BC* is shown in Fig. 7–11*c*. As always, **V** and **M** are shown acting in the positive sense. Hence,

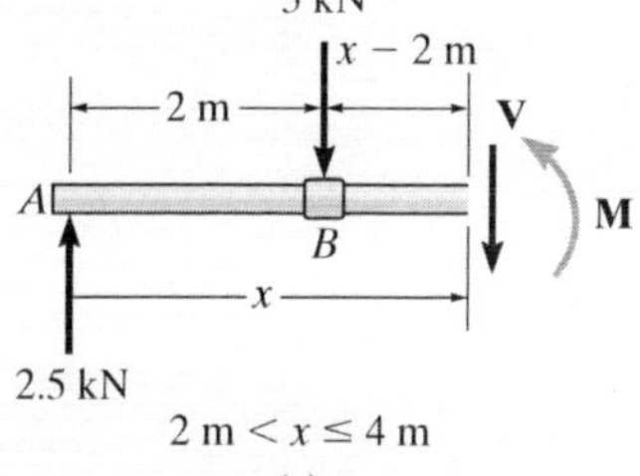

(c)

$$+\uparrow\Sigma F_y = 0; \qquad 2.5 \text{ kN} - 5 \text{ kN} - V = 0$$

$$V = -2.5 \text{ kN} \qquad (3)$$

$$\circlearrowleft +\Sigma M = 0; \qquad M + 5 \text{ kN}(x - 2 \text{ m}) - 2.5 \text{ kN}(x) = 0$$

$$M = (10 - 2.5x) \text{ kN}\cdot\text{m} \qquad (4)$$

Shear and Moment Diagrams. When Eqs. 1 through 4 are plotted within the regions in which they are valid, the shear and moment diagrams shown in Fig. 7–11*d* are obtained. The shear diagram indicates that the internal shear force is always 2.5 kN (positive) within segment *AB*. Just to the right of point *B*, the shear force changes sign and remains at a constant value of −2.5 kN for segment *BC*. The moment diagram starts at zero, increases linearly to point *B* at $x = 2$ m, where $M_{max} = 2.5$ kN(2 m) $= 5$ kN·m, and thereafter decreases back to zero.

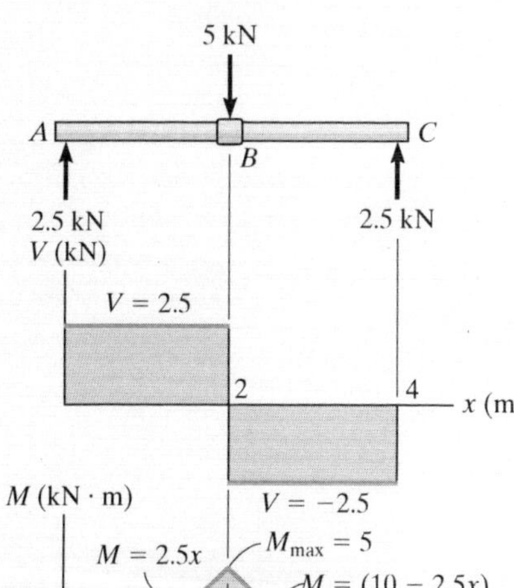

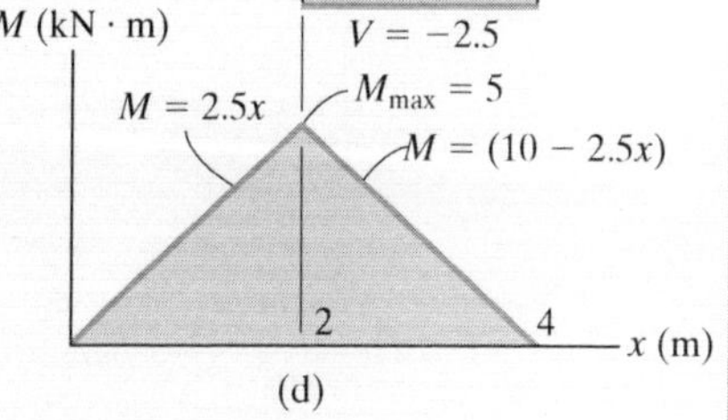

(d)

Fig. 7–11

NOTE: It is seen in Fig. 7–11*d* that the graphs of the shear and moment diagrams are discontinuous where the concentrated force acts, i.e., at points *A*, *B*, and *C*. For this reason, as stated earlier, it is necessary to express both the shear and moment functions separately for regions between concentrated loads. It should be realized, however, that all loading discontinuities are mathematical, arising from the *idealization of a concentrated force and couple moment*. Physically, loads are always applied over a finite area, and if the actual load variation could be accounted for, the shear and moment diagrams would then be continuous over the shaft's entire length.

7

EXAMPLE 7.7

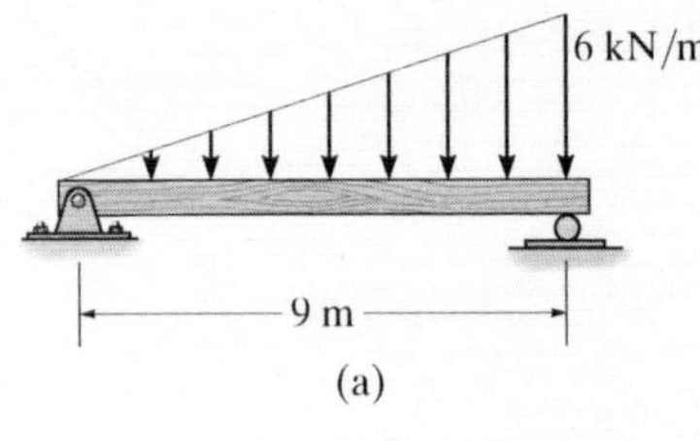

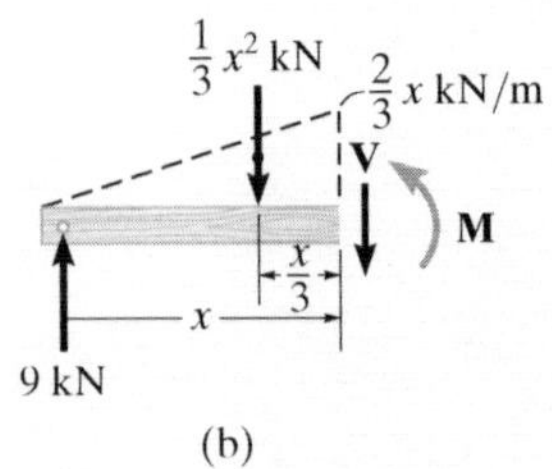

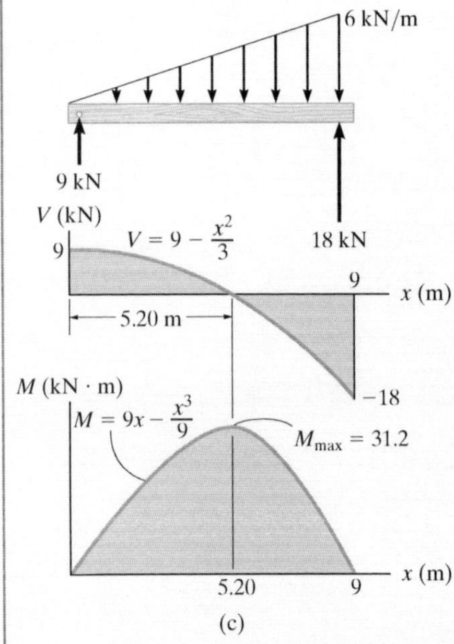

Fig. 7–12

Draw the shear and moment diagrams for the beam shown in Fig. 7–12*a*.

SOLUTION

Support Reactions. The support reactions are shown on the beam's free-body diagram, Fig. 7–12*c*.

Shear and Moment Functions. A free-body diagram for a left segment of the beam having a length x is shown in Fig. 7–12*b*. Due to proportional triangles, the distributed loading acting at the end of this segment has an intensity of $w/x = 6/9$ or $w = (2/3)x$. It is replaced by a resultant force *after* the segment is isolated as a free-body diagram. The *magnitude* of the resultant force is equal to $\frac{1}{2}(x)\left(\frac{2}{3}x\right) = \frac{1}{3}x^2$. This force *acts through the centroid* of the distributed loading area, a distance $\frac{1}{3}x$ from the right end. Applying the two equations of equilibrium yields

$$+\uparrow \Sigma F_y = 0; \qquad 9 - \frac{1}{3}x^2 - V = 0$$

$$V = \left(9 - \frac{x^2}{3}\right) \text{kN} \qquad (1)$$

$$\circlearrowleft + \Sigma M = 0; \qquad M + \frac{1}{3}x^2\left(\frac{x}{3}\right) - 9x = 0$$

$$M = \left(9x - \frac{x^3}{9}\right) \text{kN} \cdot \text{m} \qquad (2)$$

Shear and Moment Diagrams. The shear and moment diagrams shown in Fig. 7–12*c* are obtained by plotting Eqs. 1 and 2.

The point of *zero shear* can be found using Eq. 1:

$$V = 9 - \frac{x^2}{3} = 0$$

$$x = 5.20 \text{ m}$$

NOTE: It will be shown in Sec. 7.3 that this value of x happens to represent the point on the beam where the *maximum moment* occurs. Using Eq. 2, we have

$$M_{max} = \left(9(5.20) - \frac{(5.20)^3}{9}\right) \text{kN} \cdot \text{m}$$

$$= 31.2 \text{ kN} \cdot \text{m}$$

FUNDAMENTAL PROBLEMS

F7–7. Determine the shear and moment as a function of x, and then draw the shear and moment diagrams.

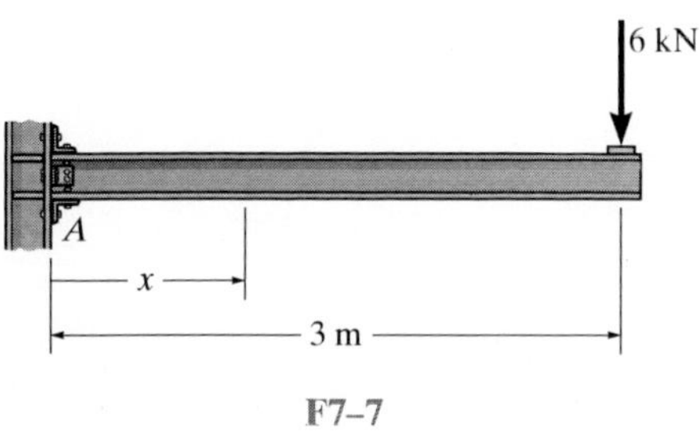

F7–7

F7–8. Determine the shear and moment as a function of x, and then draw the shear and moment diagrams.

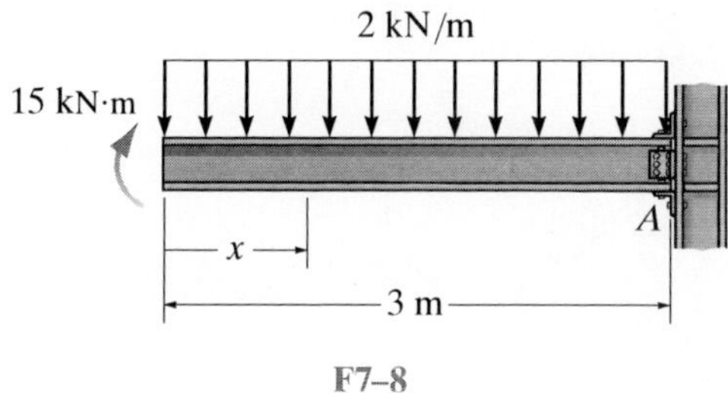

F7–8

F7–9. Determine the shear and moment as a function of x, and then draw the shear and moment diagrams.

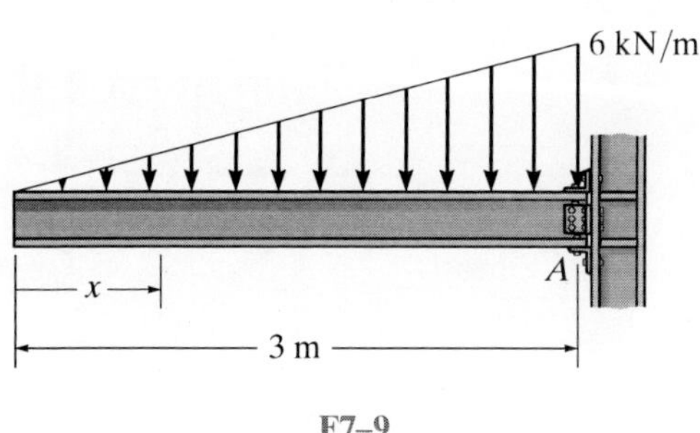

F7–9

F7–10. Determine the shear and moment as a function of x, and then draw the shear and moment diagrams.

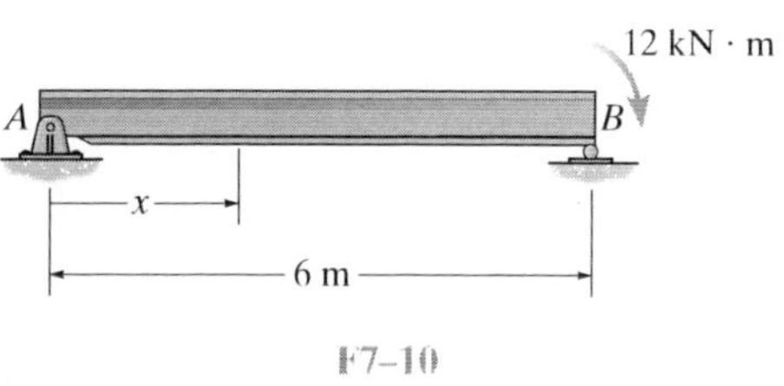

F7–10

F7–11. Determine the shear and moment as a function of x, where $0 \le x < 3$ m and $3 \text{ m} < x \le 6$ m, and then draw the shear and moment diagrams.

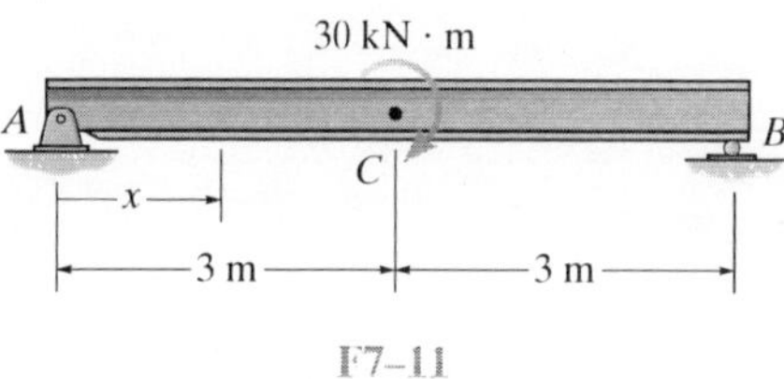

F7–11

F7–12. Determine the shear and moment as a function of x, where $0 \le x < 3$ m and $3 \text{ m} < x \le 6$ m, and then draw the shear and moment diagrams.

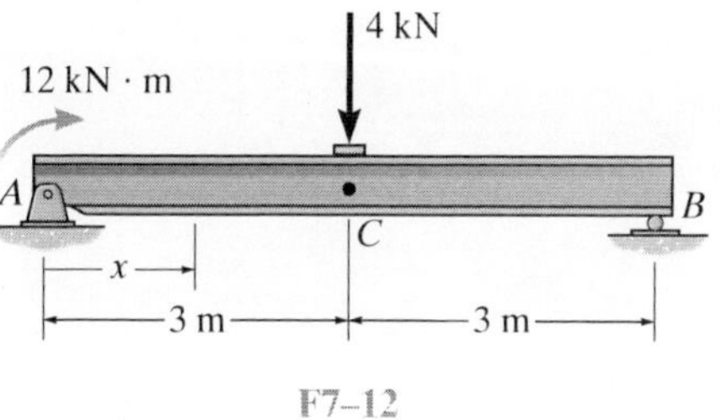

F7–12

7

PROBLEMS

For each of the following problems, establish the x axis with the origin at the left side of the beam, and obtain the internal shear and moment as a function of x. Use these results to plot the shear and moment diagrams.

7–45. Draw the shear and moment diagrams for the overhang beam.

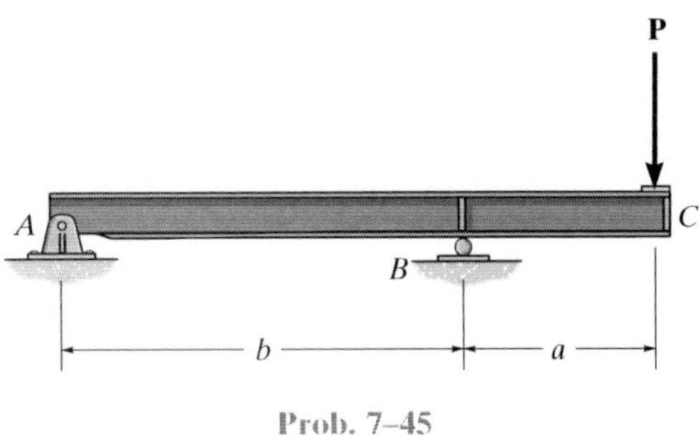

Prob. 7–45

7–46. Draw the shear and moment diagrams for the beam.

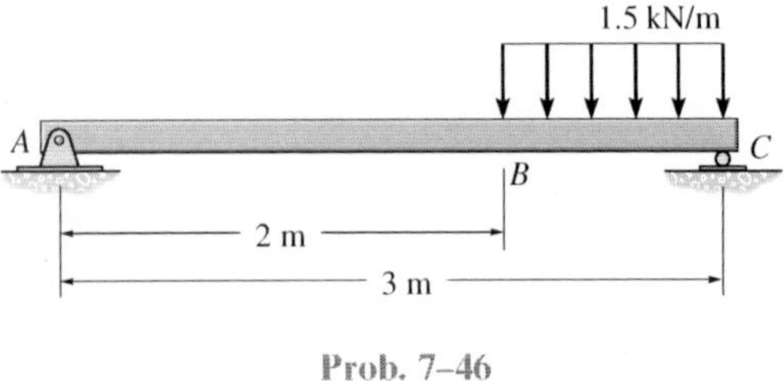

Prob. 7–46

7–47. Draw the shear and moment diagrams for the beam (a) in terms of the parameters shown; (b) set $P = 4$ kN, $a = 1.5$ m, $L = 3.6$ m.

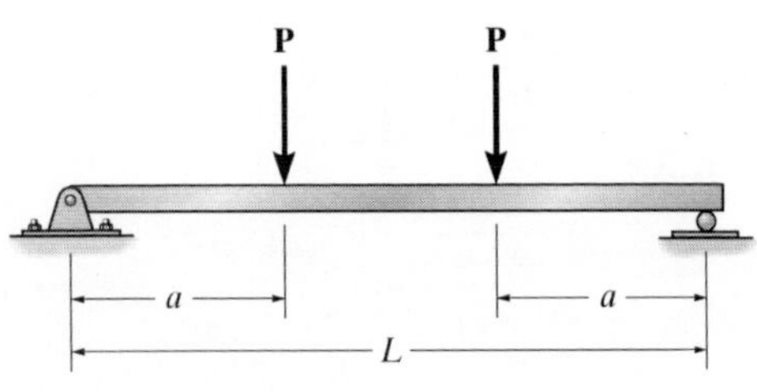

Prob. 7–47

***7–48.** Draw the shear and moment diagrams for the beam (a) in terms of the parameters shown; (b) set $M_0 = 500\ \text{N}\cdot\text{m}$, $L = 8$ m.

7–49. If $L = 9$ m, the beam will fail when the maximum shear force is $V_{max} = 5$ kN or the maximum bending moment is $M_{max} = 2\ \text{kN}\cdot\text{m}$. Determine the magnitude M_0 of the largest couple moments it will support.

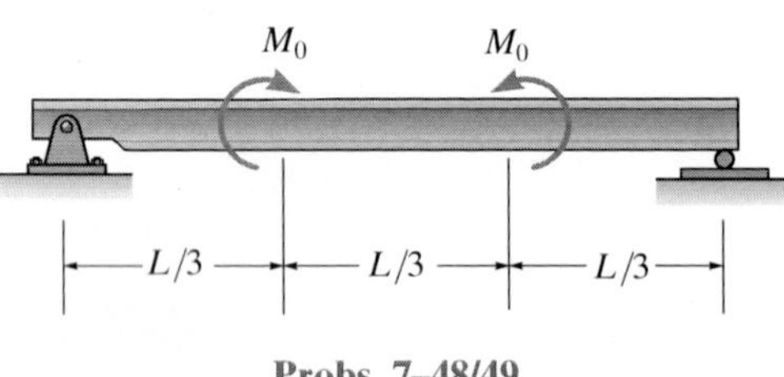

Probs. 7–48/49

7–50. Draw the shear and moment diagrams for the cantilever beam.

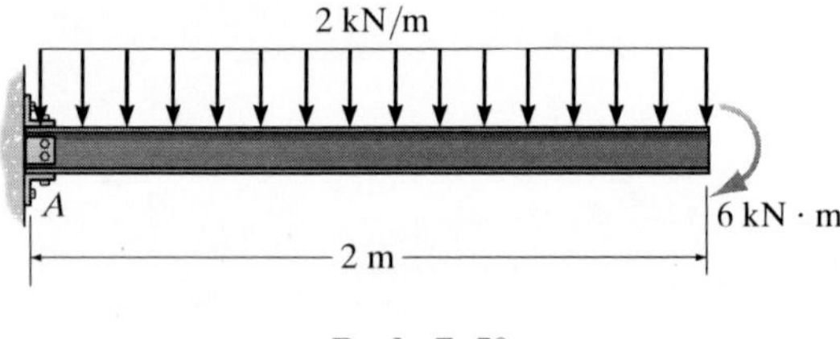

Prob. 7–50

7–51. Draw the shear and moment diagrams for the beam.

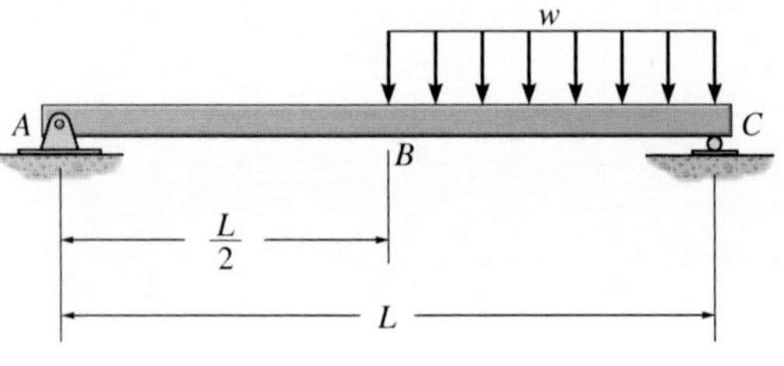

Prob. 7–51

*7–52. Draw the shear and moment diagrams for the beam.

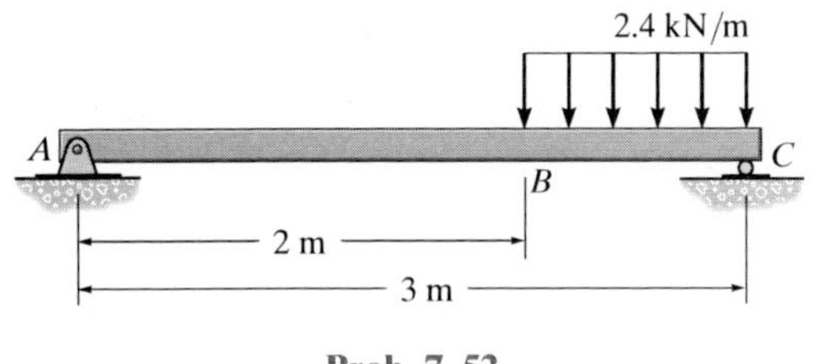

Prob. 7–52

7–53. Draw the shear and moment diagrams for the beam.

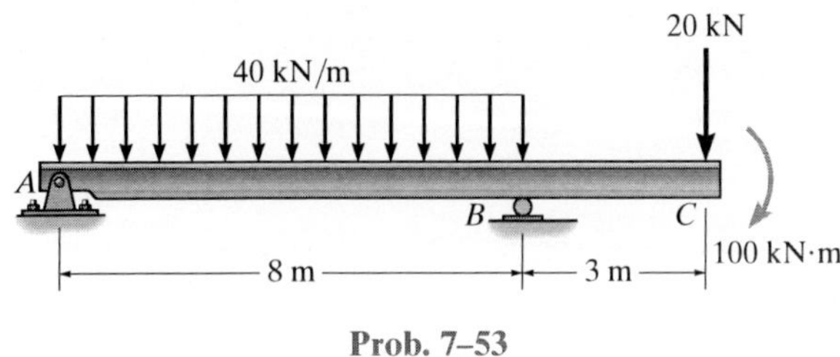

Prob. 7–53

7–54. Draw the shear and moment diagrams for the simply supported beam.

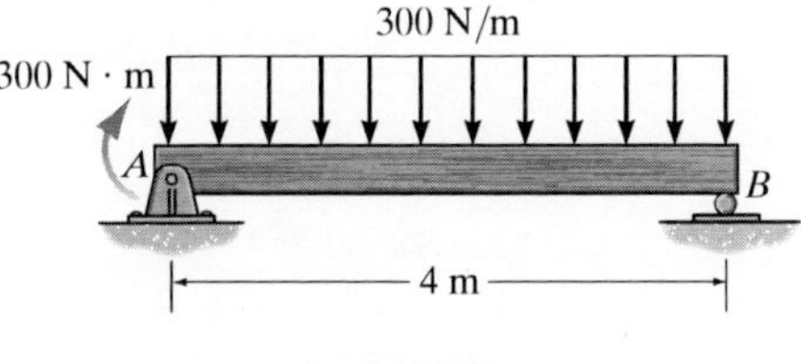

Prob. 7–54

7–55. Draw the shear and moment diagrams for the simply supported beam.

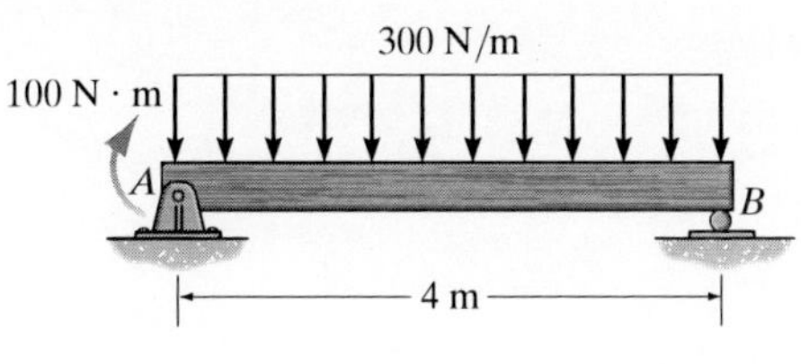

Prob. 7–55

*7–56. Draw the shear and bending-moment diagrams for beam ABC. Note that there is a pin at B.

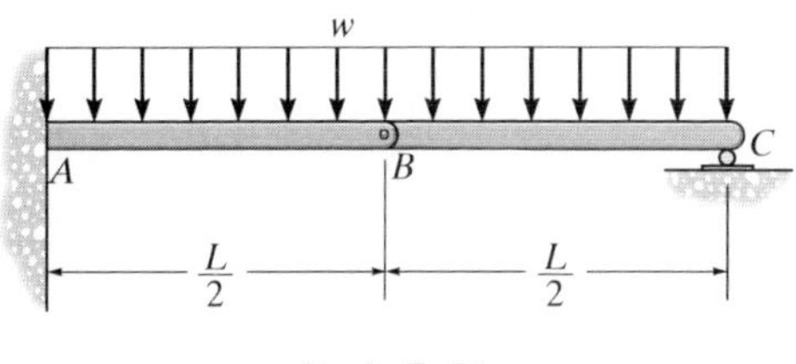

Prob. 7–56

7–57. Draw the shear and moment diagrams for the beam.

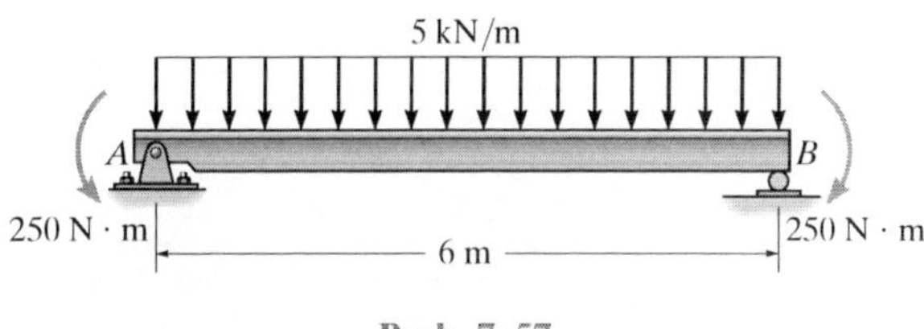

Prob. 7–57

7–58. Draw the shear and moment diagrams for the compound beam. The beam is pin-connected at E and F.

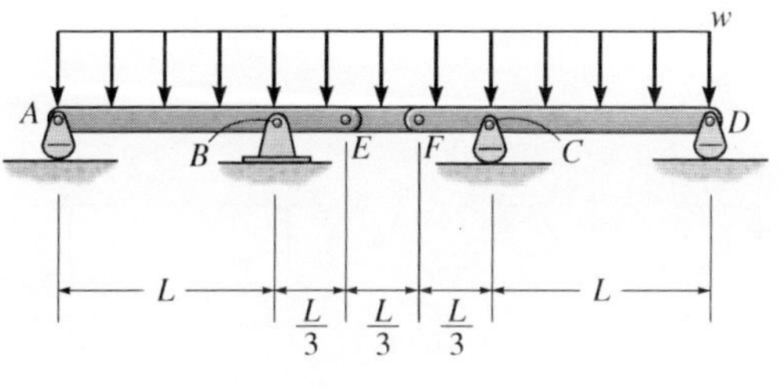

Prob. 7–58

7–59. Draw the shear and moment diagrams for the beam.

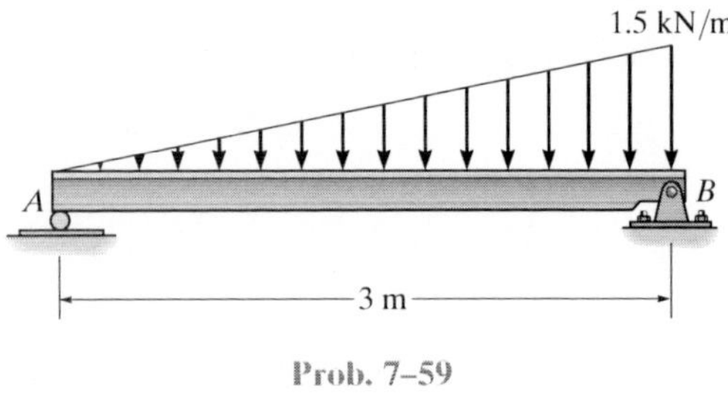

Prob. 7–59

***7–60.** The beam will fail when the maximum internal moment is M_{max}. Determine the position x of the concentrated force **P** and its smallest magnitude that will cause failure.

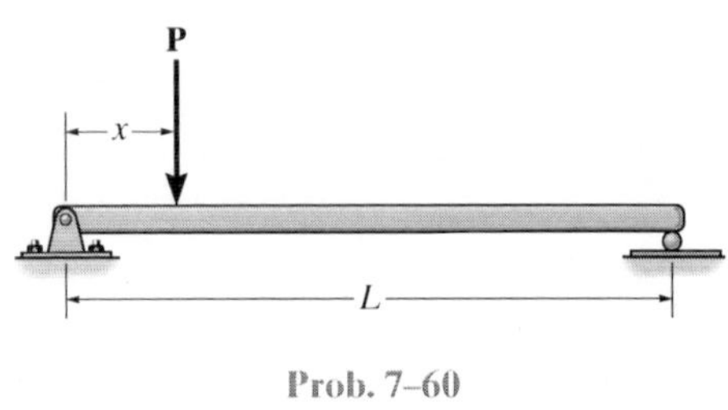

Prob. 7–60

7–61. Draw the shear and moment diagrams for the beam.

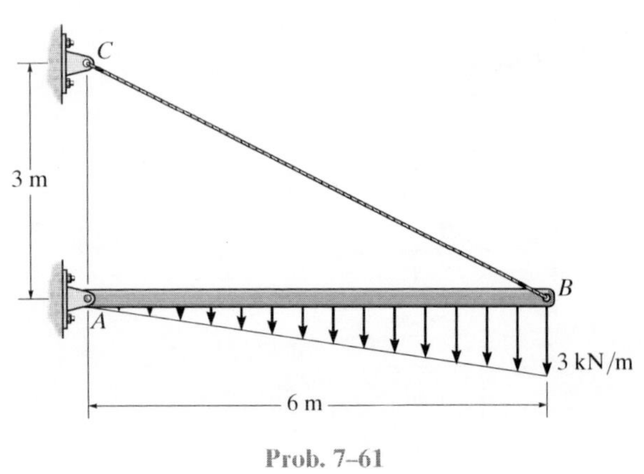

Prob. 7–61

7–62. The cantilevered beam is made of material having a specific weight γ. Determine the shear and moment in the beam as a function of x.

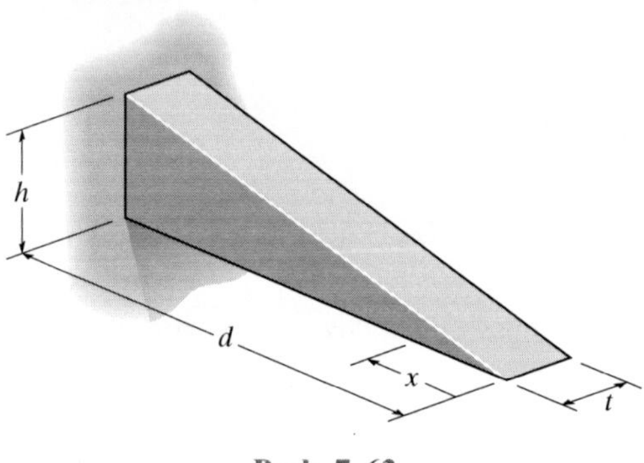

Prob. 7–62

7–63. Draw the shear and moment diagrams for the overhang beam.

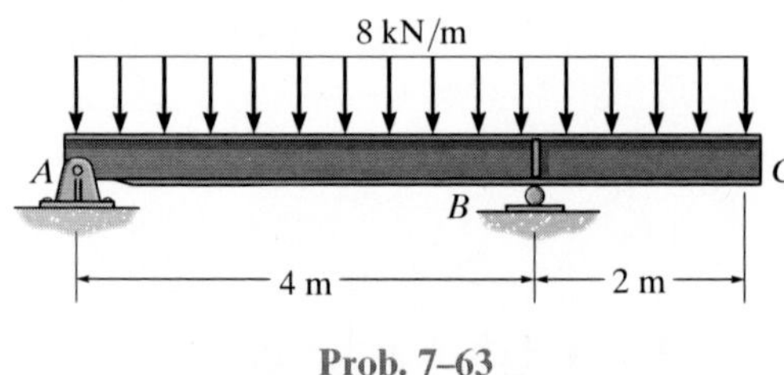

Prob. 7–63

***7–64.** Draw the shear and moment diagrams for the beam.

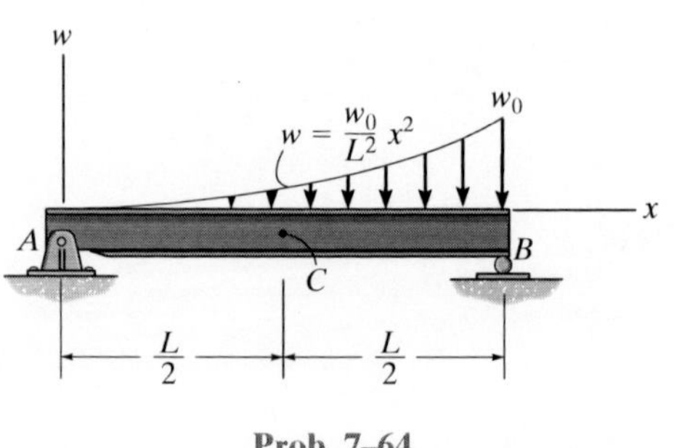

Prob. 7–64

7–65. Draw the shear and bending-moment diagrams for the beam.

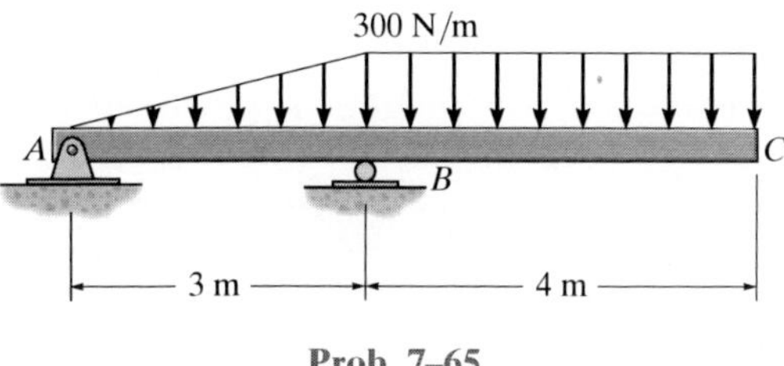

Prob. 7–65

7–66. Draw the shear and moment diagrams for the beam.

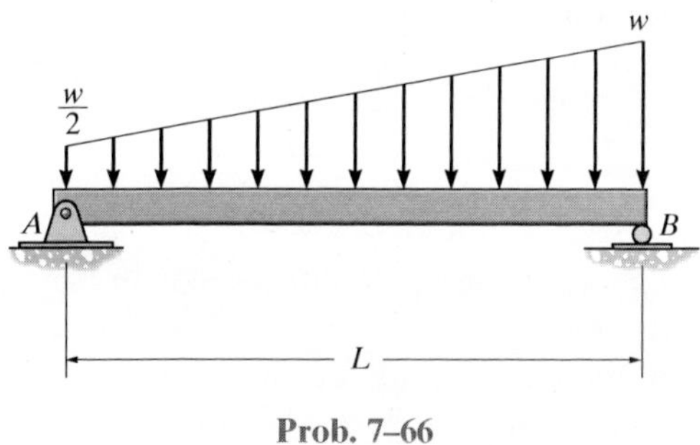

Prob. 7–66

7–67. Determine the internal normal force, shear force, and moment in the curved rod as a function of θ, where $0° \leq \theta \leq 90°$.

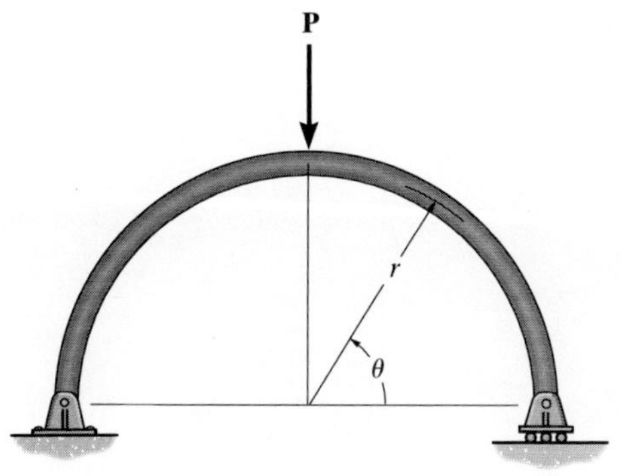

Prob. 7–67

***7–68.** Determine the normal force, shear force, and moment in the curved rod as a function of θ.

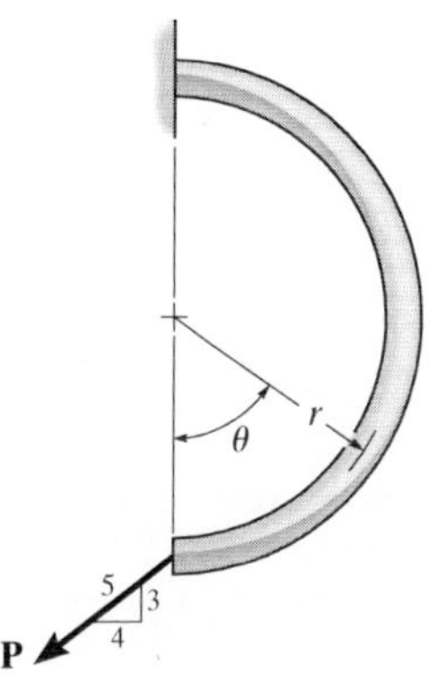

Prob. 7–68

7–69. The quarter circular rod lies in the horizontal plane and supports a vertical force **P** at its end. Determine the magnitudes of the components of the internal shear force, moment, and torque acting in the rod as a function of the angle θ.

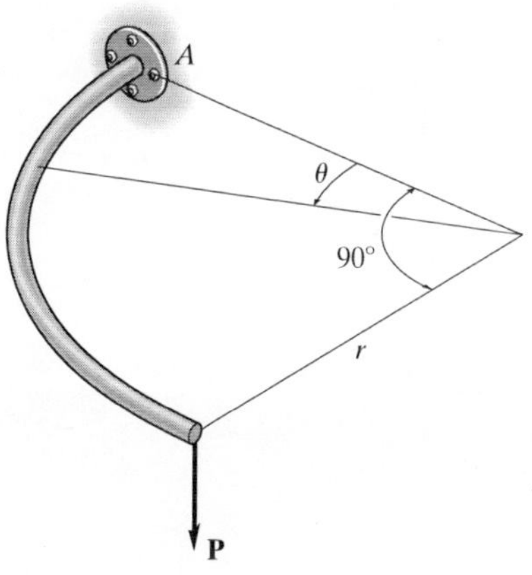

Prob. 7–69

*7.3 Relations between Distributed Load, Shear, and Moment

In order to design the beam used to support these power lines, it is important to first draw the shear and moment diagrams for the beam.

If a beam is subjected to several concentrated forces, couple moments, and distributed loads, the method of constructing the shear and bending-moment diagrams discussed in Sec. 7.2 may become quite tedious. In this section a simpler method for constructing these diagrams is discussed—a method based on differential relations that exist between the load, shear, and bending moment.

Distributed Load. Consider the beam AD shown in Fig. 7–13*a*, which is subjected to an arbitrary load $w = w(x)$ and a series of concentrated forces and couple moments. In the following discussion, the *distributed load* will be considered *positive* when the *loading acts upward* as shown. A free-body diagram for a small segment of the beam having a length Δx is chosen at a point x along the beam which is *not* subjected to a concentrated force or couple moment, Fig. 7–13*b*. Hence any results obtained will not apply at these points of concentrated loading. The internal shear force and bending moment shown on the free-body diagram are assumed to act in the *positive sense* according to the established sign convention. Note that both the shear force and moment acting on the right-hand face must be increased by a small, finite amount in order to keep the segment in equilibrium. The distributed loading has been replaced by a resultant force $\Delta F = w(x)\,\Delta x$ that acts at a fractional distance $k(\Delta x)$ from the right end, where $0 < k < 1$ [for example, if $w(x)$ is *uniform*, $k = \frac{1}{2}$].

(a)

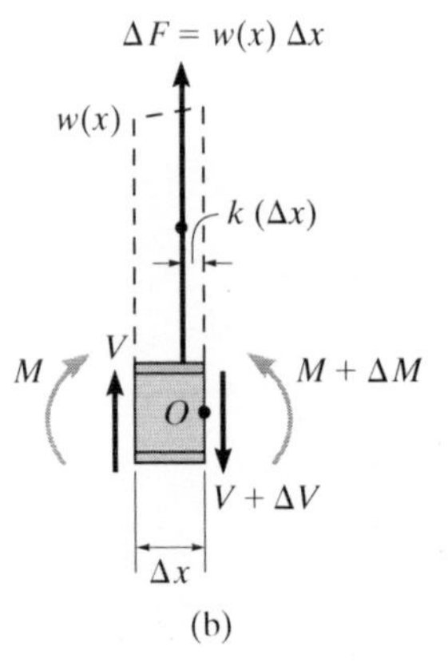

(b)

Fig. 7–13

Relation Between the Distributed Load and Shear. If we apply the force equation of equilibrium to the segment, then

$$+\uparrow \Sigma F_y = 0; \qquad V + w(x)\Delta x - (V + \Delta V) = 0$$
$$\Delta V = w(x)\Delta x$$

Dividing by Δx, and letting $\Delta x \to 0$, we get

$$\frac{dV}{dx} = w(x) \tag{7–1}$$

slope of shear diagram = distributed load intensity

If we rewrite the above equation in the form $dV = w(x)dx$ and perform an integration between any two points B and C on the beam, we see that

$$\Delta V = \int w(x)\,dx \qquad (7\text{–}2)$$

Change in shear = Area under loading curve

Relation Between the Shear and Moment. If we apply the moment equation of equilibrium about point O on the free-body diagram in Fig. 7–13b, we get

$$\circlearrowleft + \Sigma M_O = 0; \quad (M + \Delta M) - [w(x)\Delta x]\,k\Delta x - V\Delta x - M = 0$$
$$\Delta M = V\Delta x + k\,w(x)\Delta x^2$$

Dividing both sides of this equation by Δx, and letting $\Delta x \rightarrow 0$, yields

$$\frac{dM}{dx} = V \qquad (7\text{–}3)$$

Slope of moment diagram = Shear

In particular, notice that the maximum bending moment $|M|_{max}$ will occur at the point where the slope $dM/dx = 0$, since this is where the shear is equal to zero.

If Eq. 7–3 is rewritten in the form $dM = \int V\,dx$ and integrated between any two points B and C on the beam, we have

$$\Delta M = \int V\,dx \qquad (7\text{–}4)$$

Change in moment = Area under shear diagram

As stated previously, the above equations do not apply at points where a *concentrated* force or couple moment acts. These two special cases create *discontinuities* in the shear and moment diagrams, and as a result, each deserves separate treatment.

Force. A free-body diagram of a small segment of the beam in Fig. 7–13a, taken from under one of the forces, is shown in Fig. 7–14a. Here force equilibrium requires

$$+\uparrow \Sigma F_y = 0; \qquad \Delta V = F \qquad (7\text{–}5)$$

Since the *change in shear is positive*, the shear diagram will "jump" *upward when* **F** *acts upward* on the beam. Likewise, the jump in shear (ΔV) is downward when **F** acts downward.

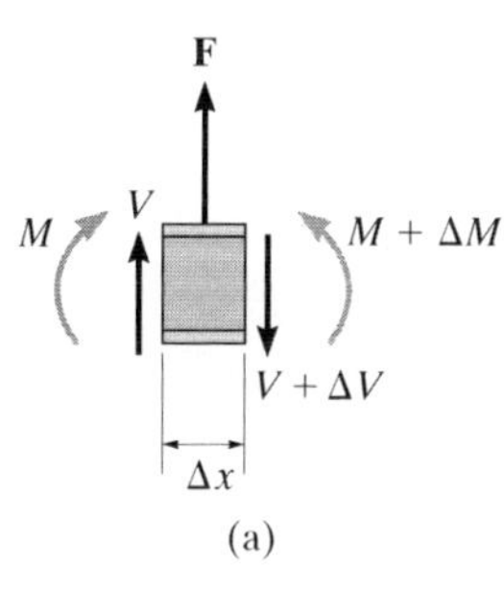

Fig. 7–14

7

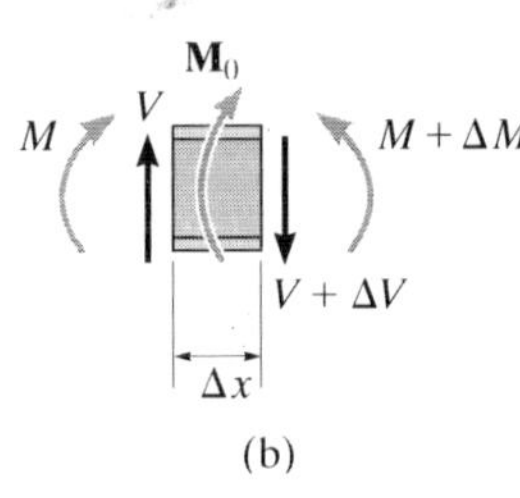

(b)

Fig. 7–14 (cont.)

Couple Moment. If we remove a segment of the beam in Fig. 7–13*a* that is located at the couple moment $\mathbf{M}_0$, the free-body diagram shown in Fig. 7–14*b* results. In this case letting $\Delta x \rightarrow 0$, moment equilibrium requires

$$\circlearrowleft + \Sigma M = 0; \qquad \Delta M = M_0 \qquad (7\text{–}6)$$

Thus, the *change in moment is positive*, or the moment diagram will "jump" *upward if* $\mathbf{M}_0$ *is clockwise*. Likewise, the jump ΔM is downward when $\mathbf{M}_0$ is counterclockwise.

The examples which follow illustrate application of the above equations when used to construct the shear and moment diagrams. After working through these examples, it is recommended that you also go back and solve Examples 7.6 and 7.7 using this method.

Important Points

- The slope of the shear diagram at a point is equal to the intensity of the distributed loading, where positive distributed loading is upward, i.e., $dV/dx = w(x)$.
- The change in the shear ΔV between two points is equal to *the area* under the distributed-loading curve between the points.
- If a concentrated force acts upward on the beam, the shear will jump upward by the same amount.
- The slope of the moment diagram at a point is equal to the shear, i.e., $dM/dx = V$.
- The change in the moment ΔM between two points is equal to the *area* under the shear diagram between the two points.
- If a *clock*wise couple moment acts on the beam, the shear will not be affected; however, the moment diagram will jump *upward* by the amount of the moment.
- Points of *zero shear* represent points of *maximum or minimum moment* since $dM/dx = 0$.
- Because two integrations of $w = w(x)$ are involved to first determine the change in shear, $\Delta V = \int w(x)\,dx$, then to determine the change in moment, $\Delta M = \int V\,dx$, then if the loading curve $w = w(x)$ is a polynomial of degree n, $V = V(x)$ will be a curve of degree $n + 1$, and $M = M(x)$ will be a curve of degree $n + 2$.

This concrete beam is used to support the deck. Its size and the placement of steel reinforcement within it can be determined once the shear and moment diagrams have been established.

7

EXAMPLE 7.8

Draw the shear and moment diagrams for the cantilever beam in Fig. 7–15*a*.

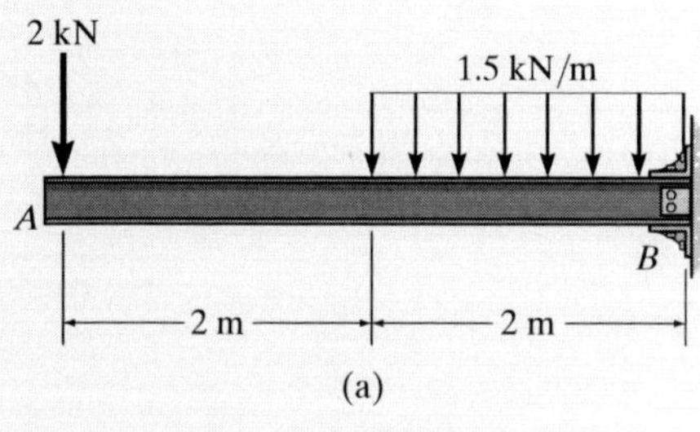

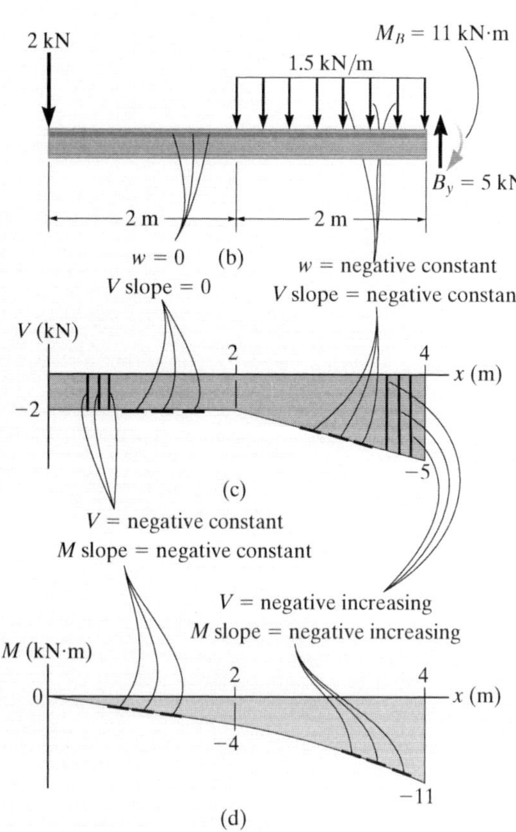

Fig. 7–15

SOLUTION

The support reactions at the fixed support B are shown in Fig. 7–15*b*.

Shear Diagram. The shear at end A is -2 kN. This value is plotted at $x = 0$, Fig. 7–15*c*. Notice how the shear diagram is constructed by following the slopes defined by the loading w. The shear at $x = 4$ m is -5 kN, the reaction on the beam. This value can be verified by finding the area under the distributed loading; i.e.,

$$V|_{x=4\text{ m}} = V|_{x=2\text{ m}} + \Delta V = -2\text{ kN} - (1.5\text{ kN/m})(2\text{ m}) = -5\text{ kN}$$

Moment Diagram. The moment of zero at $x = 0$ is plotted in Fig. 7–15*d*. Construction of the moment diagram is based on knowing that its slope is equal to the shear at each point. The change of moment from $x = 0$ to $x = 2$ m is determined from the area under the shear diagram. Hence, the moment at $x = 2$ m is

$$M|_{x=2\text{ m}} = M|_{x=0} + \Delta M = 0 + [-2\text{ kN}(2\text{ m})] = -4\text{ kN}\cdot\text{m}$$

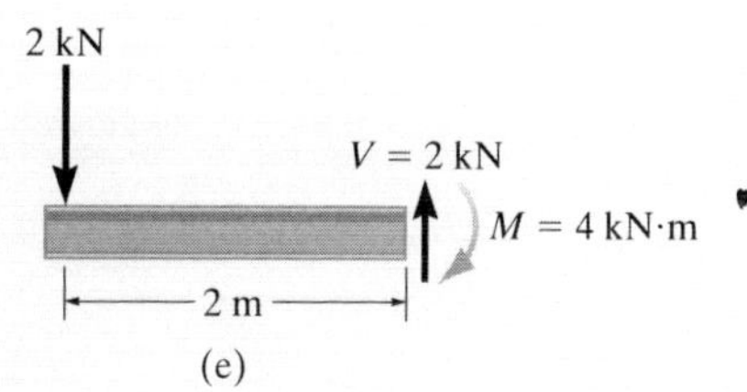

This same value can be determined from the method of sections, Fig. 7–15*e*.

7

EXAMPLE 7.9

Draw the shear and moment diagrams for the overhang beam in Fig. 7–16*a*.

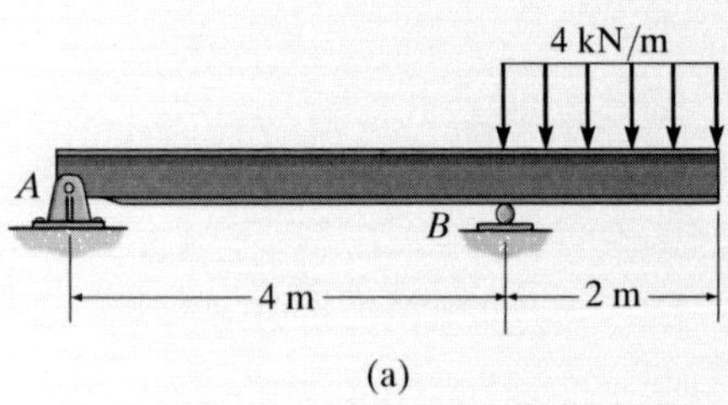

(a)

Fig. 7–16

SOLUTION

The support reactions are shown in Fig. 7–16*b*.

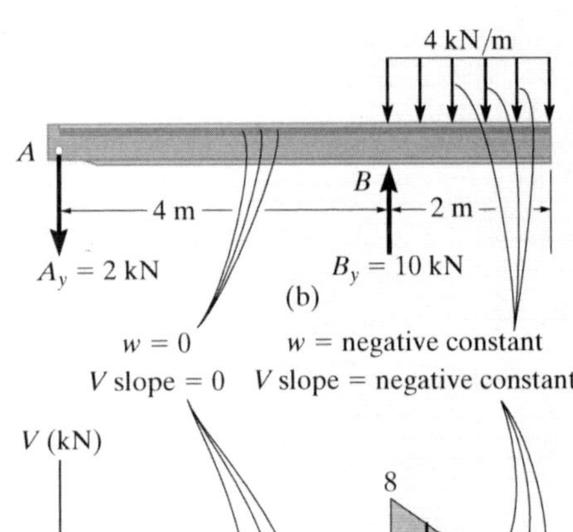

(b)

Shear Diagram. The shear of −2 kN at end A of the beam is plotted at $x = 0$, Fig. 7–16*c*. The slopes are determined from the loading and from this the shear diagram is constructed, as indicated in the figure. In particular, notice the positive jump of 10 kN at $x = 4$ m due to the force B_y, as indicated in the figure.

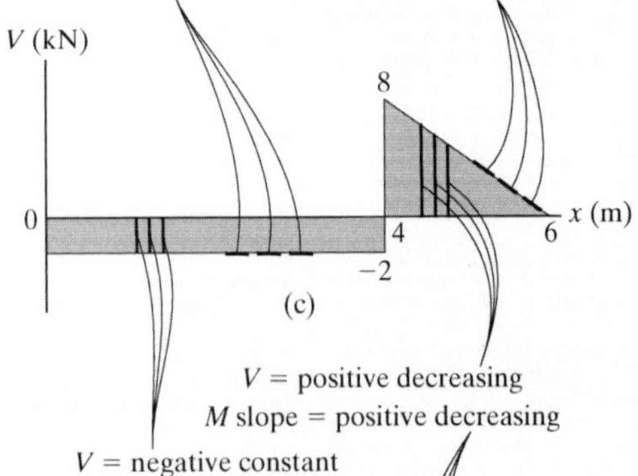

(c)

Moment Diagram. The moment of zero at $x = 0$ is plotted, Fig. 7–16*d*, then following the behavior of the slope found from the shear diagram, the moment diagram is constructed. The moment at $x = 4$ m is found from the area under the shear diagram.

$$M|_{x=4\text{ m}} = M|_{x=0} + \Delta M = 0 + [-2\text{ kN}(4\text{ m})] = -8\text{ kN}\cdot\text{m}$$

We can also obtain this value by using the method of sections, as shown in Fig. 7–16*e*.

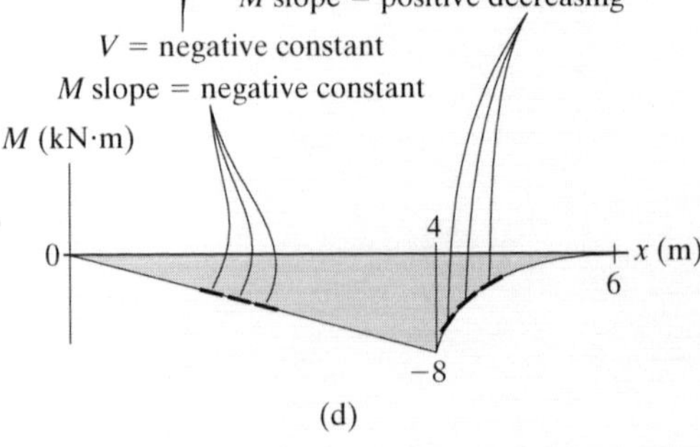

(d)

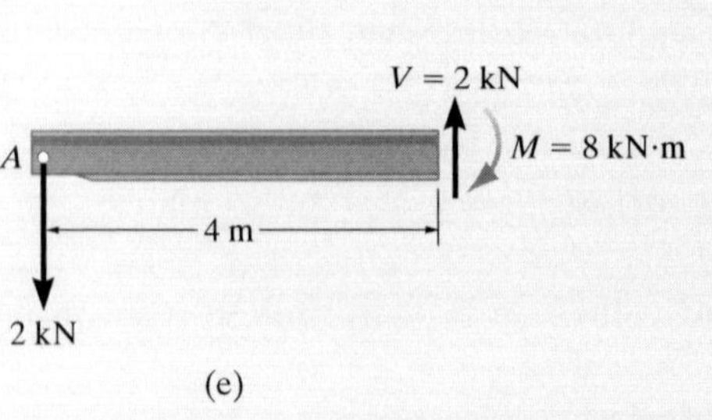

(e)

EXAMPLE 7.10

The shaft in Fig. 7–17*a* is supported by a thrust bearing at A and a journal bearing at B. Draw the shear and moment diagrams.

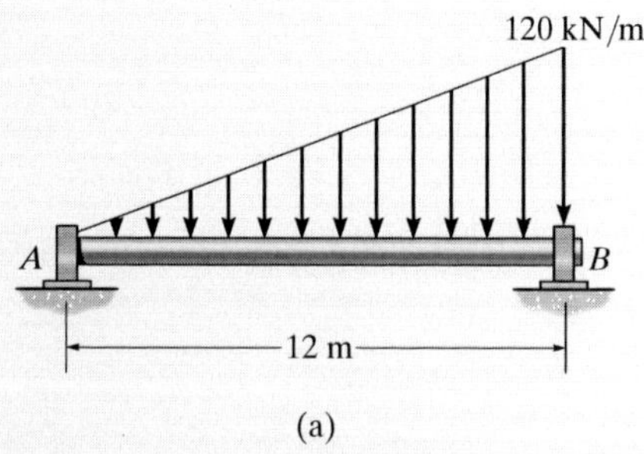

Fig. 7–17

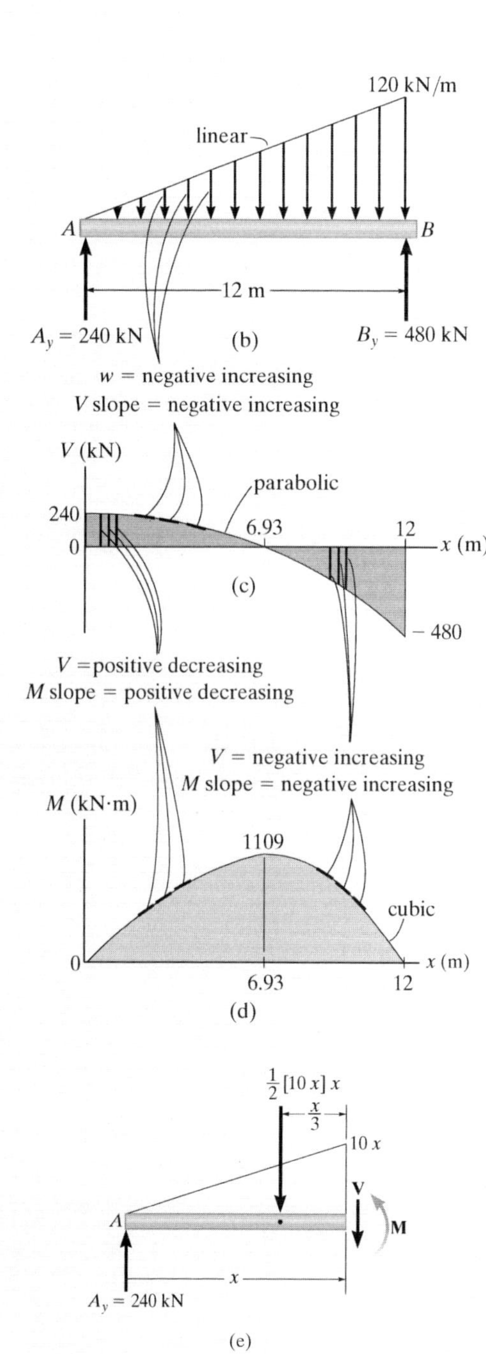

SOLUTION

The support reactions are shown in Fig. 7–17*b*.

Shear Diagram. As shown in Fig. 7–17*c*, the shear at $x = 0$ is +240. Following the slope defined by the loading, the shear diagram is constructed, where at B its value is −480 kN. Since the shear changes sign, the point where $V = 0$ must be located. To do this we will use the method of sections. The free-body diagram of the left segment of the shaft, sectioned at an arbitrary position x within the region $0 \le x < 12$ m, is shown in Fig. 7–17*e*. Notice that the intensity of the distributed load at x is $w = 10x$, which has been found by proportional triangles, i.e., $120/12 = w/x$.

Thus, for $V = 0$,

$$+\uparrow \Sigma F_y = 0; \qquad 240 \text{ kN} - \tfrac{1}{2}(10x)x = 0$$

$$x = 6.93 \text{ m}$$

Moment Diagram. The moment diagram starts at 0 since there is no moment at A, then it is constructed based on the slope as determined from the shear diagram. The maximum moment occurs at $x = 6.93$ m, where the shear is equal to zero, since $dM/dx = V = 0$, Fig. 7–17*e*,

$$\circlearrowleft + \Sigma M = 0;$$

$$M_{max} + \tfrac{1}{2}[(10)(6.93)]\, 6.93\left(\tfrac{1}{3}(6.93)\right) - 240(6.93) = 0$$

$$M_{max} = 1109 \text{ kN} \cdot \text{m}$$

Finally, notice how integration, first of the loading w which is linear, produces a shear diagram which is parabolic, and then a moment diagram which is cubic.

7

FUNDAMENTAL PROBLEMS

F7–13. Draw the shear and moment diagrams for the beam.

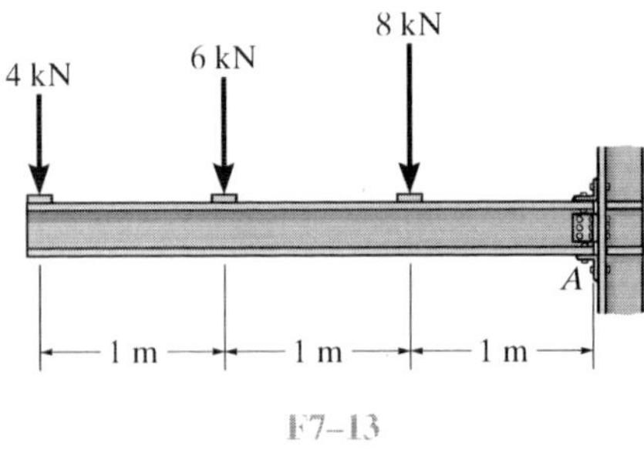

F7–13

F7–14. Draw the shear and moment diagrams for the beam.

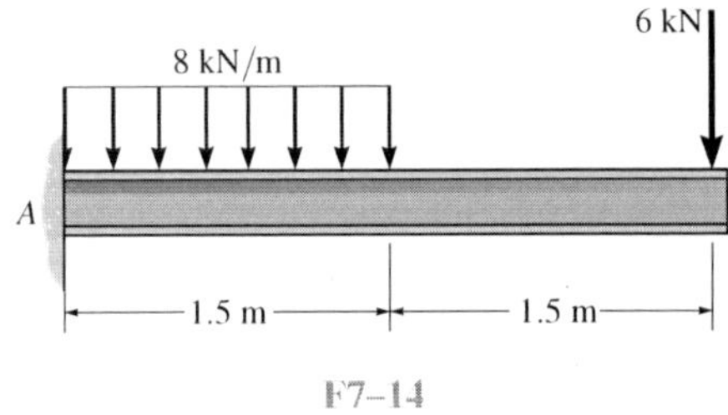

F7–14

F7–15. Draw the shear and moment diagrams for the beam.

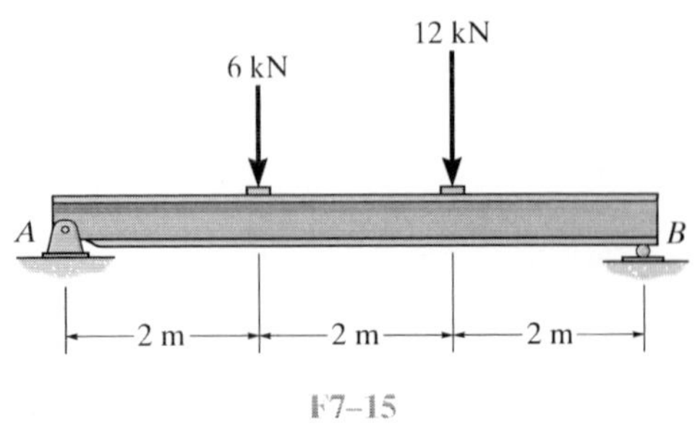

F7–15

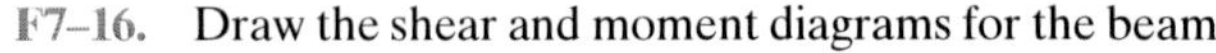

F7–16. Draw the shear and moment diagrams for the beam.

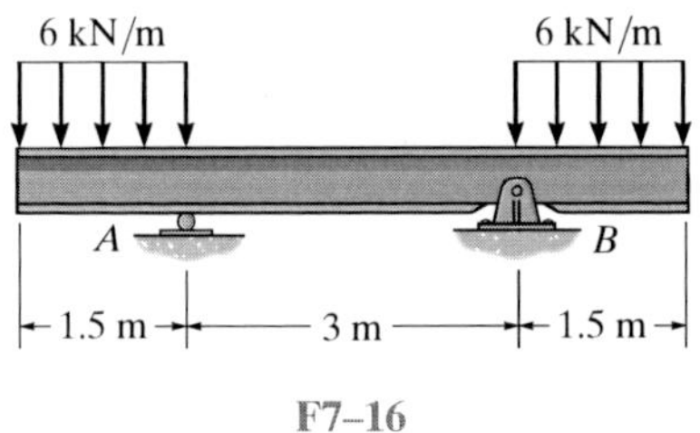

F7–16

F7–17. Draw the shear and moment diagrams for the beam.

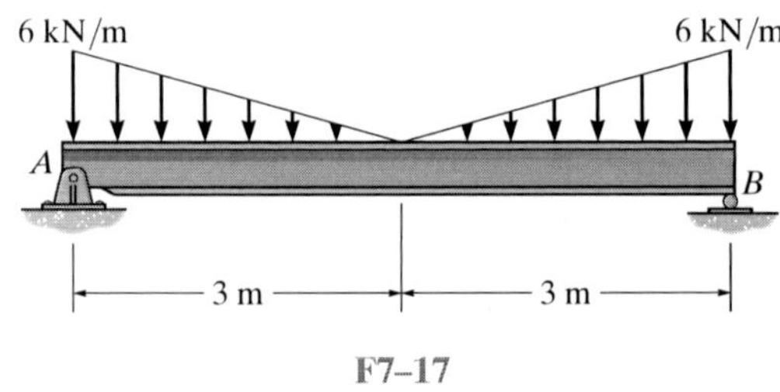

F7–17

F7–18. Draw the shear and moment diagrams for the beam.

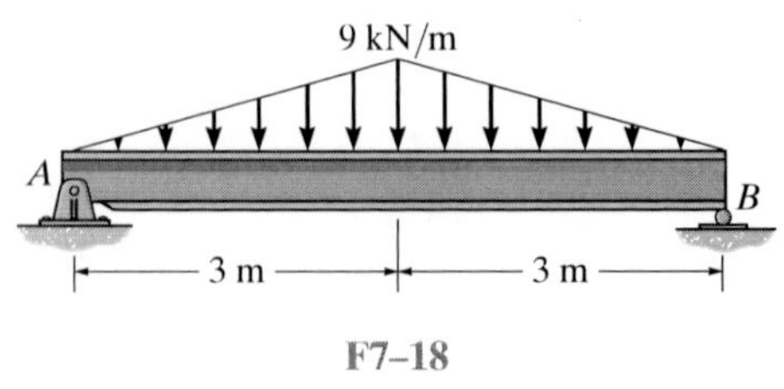

F7–18

PROBLEMS

7–70. Draw the shear and moment diagrams for the beam.

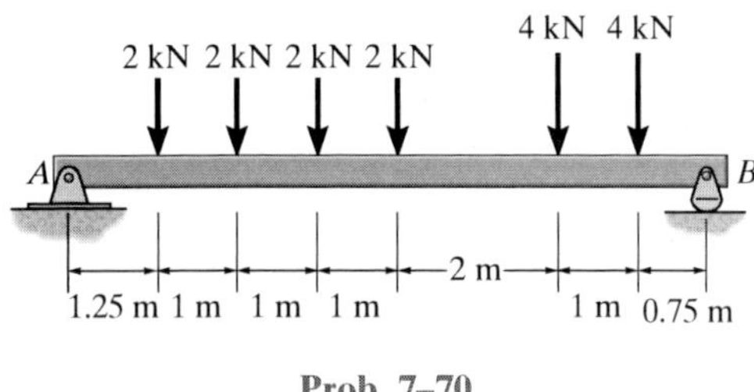

Prob. 7–70

7–71. Draw the shear and moment diagrams for the beam.

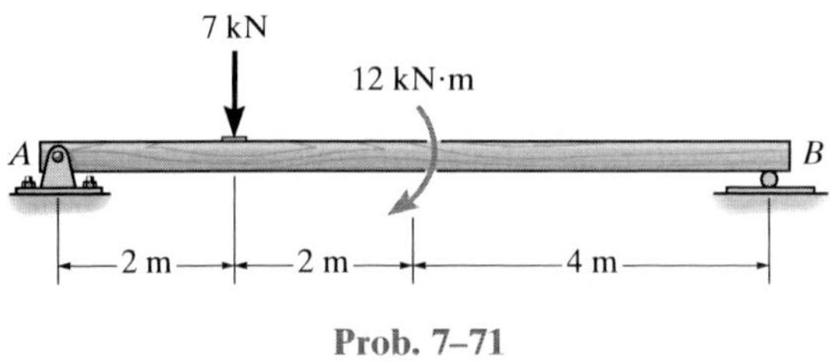

Prob. 7–71

***7–72.** Draw the shear and moment diagrams for the beam.

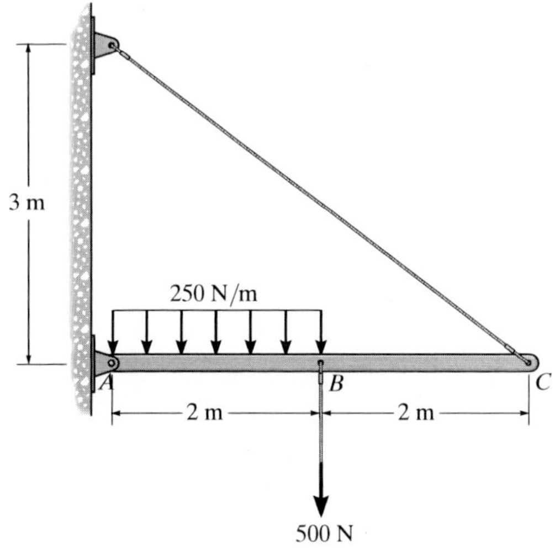

Prob. 7–72

7–72. Draw the shear and moment diagrams for the beam.

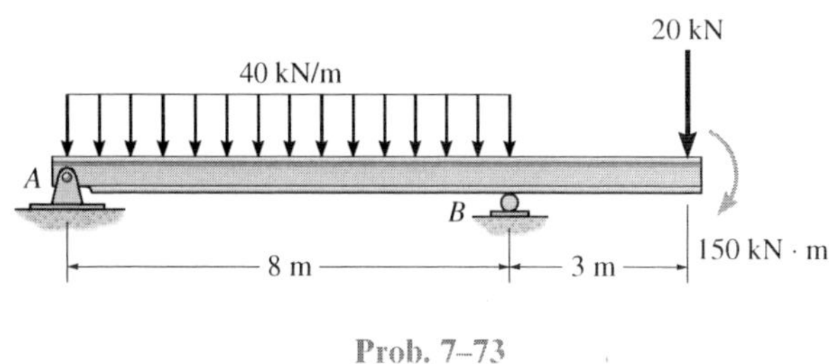

Prob. 7–73

7–74. Draw the shear and moment diagrams for the simply-supported beam.

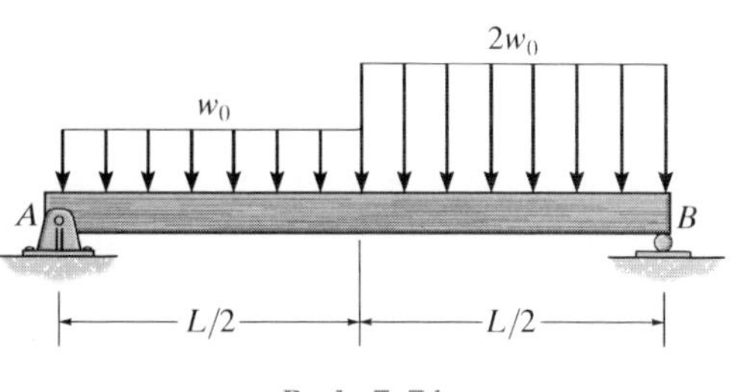

Prob. 7–74

7–75. Draw the shear and moment diagrams for the beam. The support at A offers no resistance to vertical load.

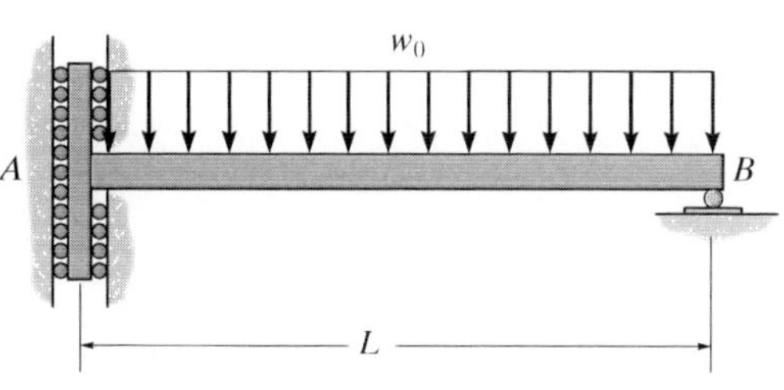

Prob. 7–75

***7–76.** Draw the shear and moment diagrams for the beam.

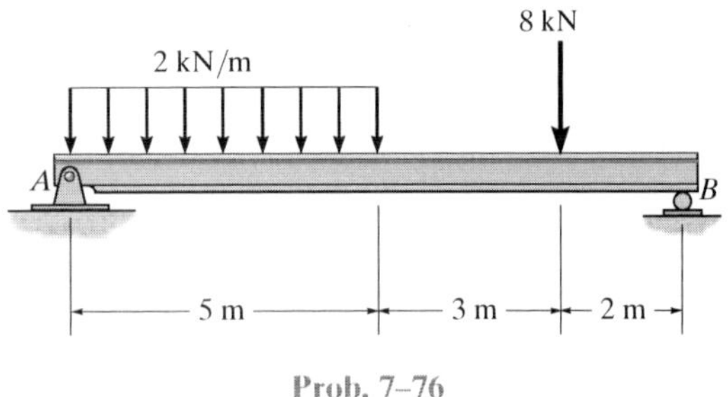

Prob. 7–76

7–77. The shaft is supported by a thrust bearing at A and a journal bearing at B. Draw the shear and moment diagrams for the shaft.

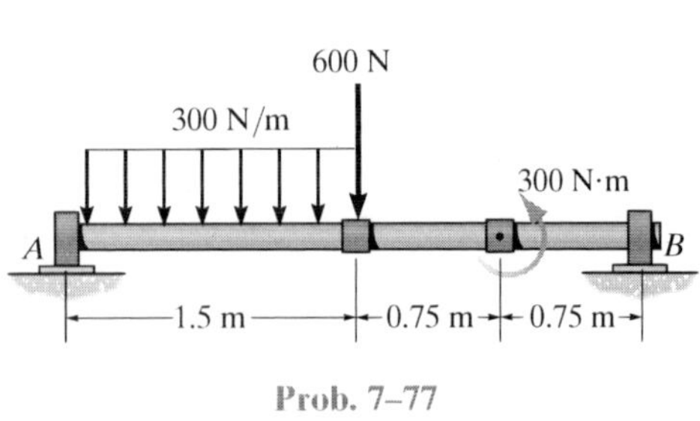

Prob. 7–77

7–78. Draw the shear and moment diagrams for the shaft. The support at A is a thrust bearing and at B it is a journal bearing.

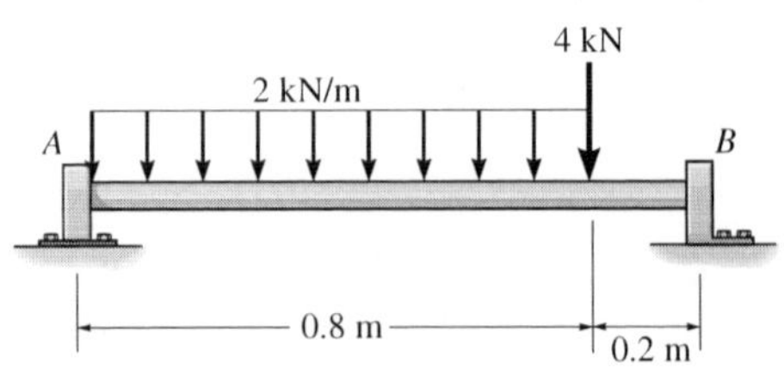

Prob. 7–78

7–79. Draw the shear and moment diagrams for the beam.

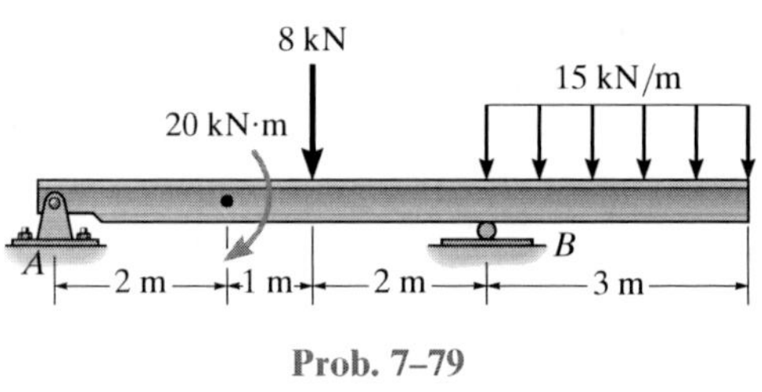

Prob. 7–79

***7–80.** Draw the shear and moment diagrams for the compound supported beam.

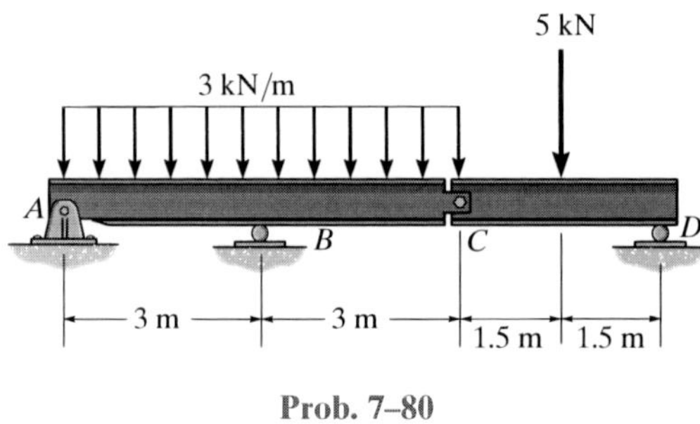

Prob. 7–80

7–81. Draw the shear and moment diagrams for the beam.

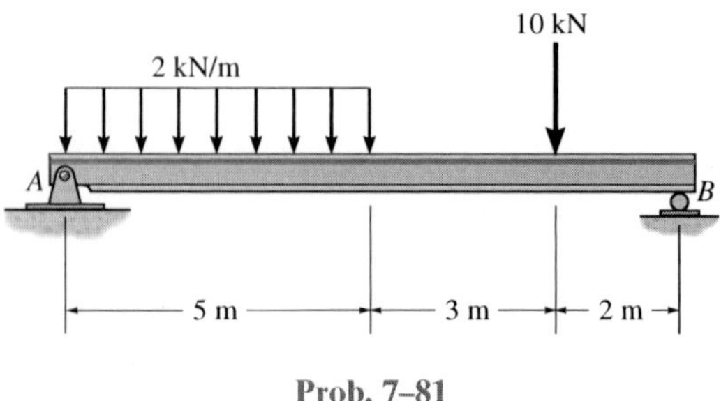

Prob. 7–81

7–82. Draw the shear and moment diagrams for the overhang beam.

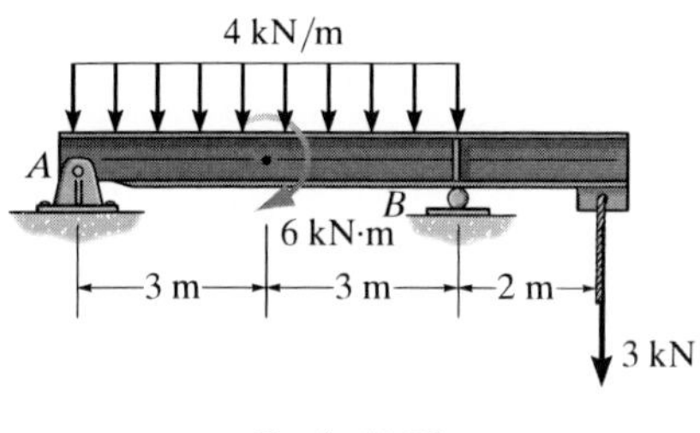

Prob. 7–82

7–83. Draw the shear and moment diagrams for the shaft. The support at A is a journal bearing and at B it is a thrust bearing.

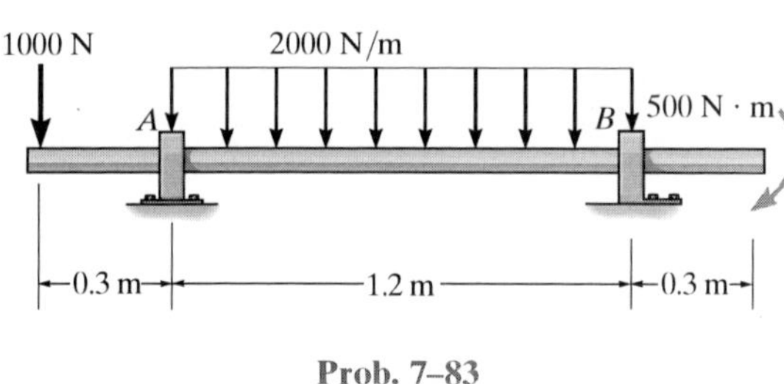

Prob. 7–83

***7–84.** Draw the shear and moment diagrams for the beam.

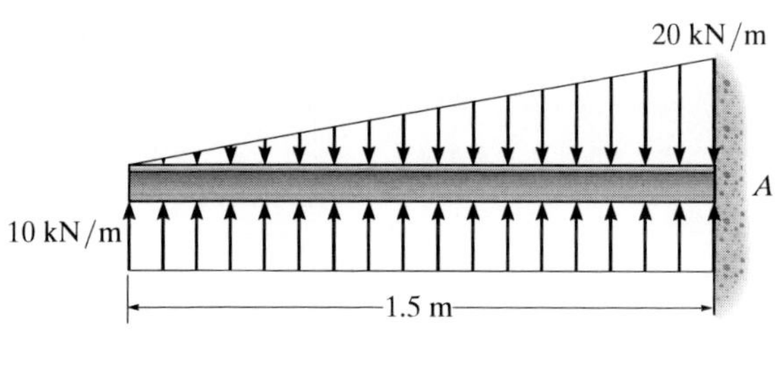

Prob. 7–84

7–85. Draw the shear and moment diagrams for the beam.

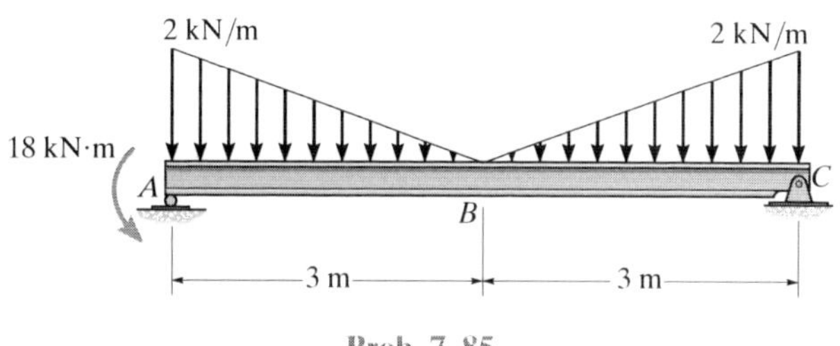

Prob. 7–85

7–86. Draw the shear and moment diagrams for the beam.

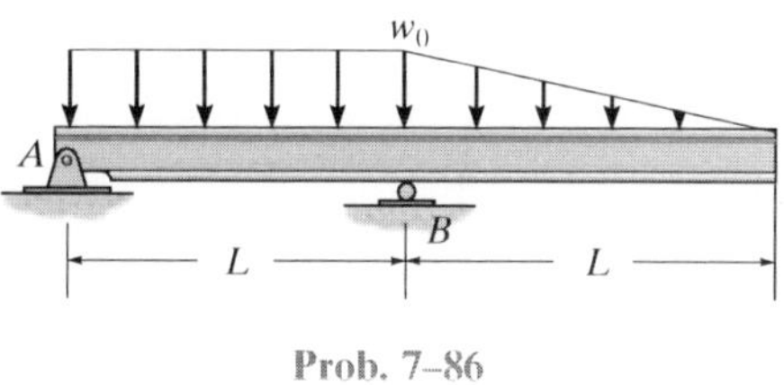

Prob. 7–86

7–87. Draw the shear and moment diagrams for the beam.

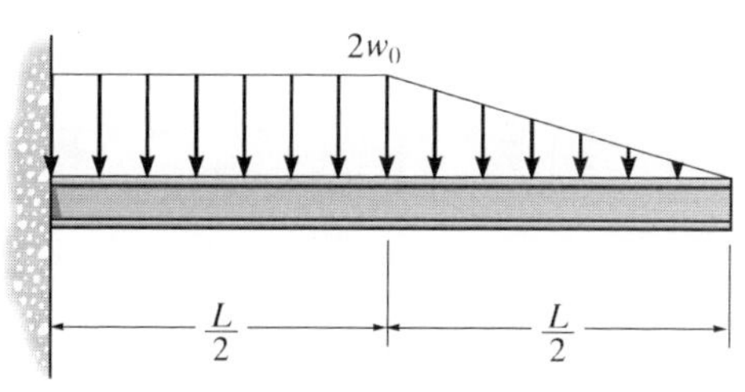

Prob. 7–87

7

*7–88. Draw the shear and moment diagrams for the beam.

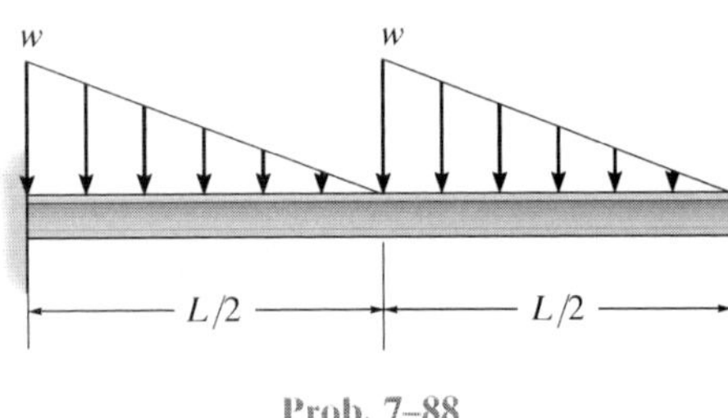

Prob. 7–88

7–89. The shaft is supported by a smooth thrust bearing at A and a smooth journal bearing at B. Draw the shear and moment diagrams for the shaft.

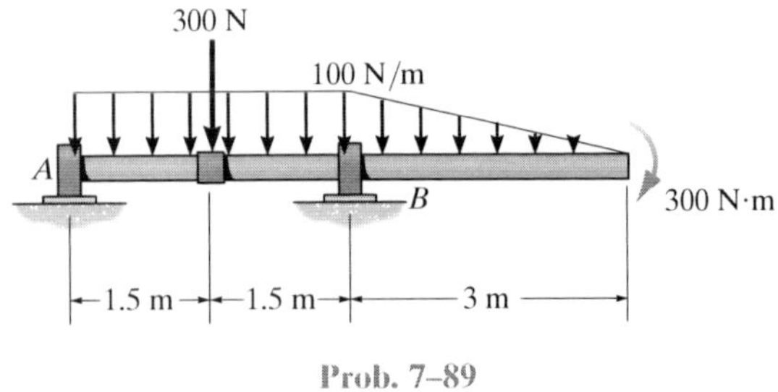

Prob. 7–89

7–90. Draw the shear and moment diagrams for the overhang beam.

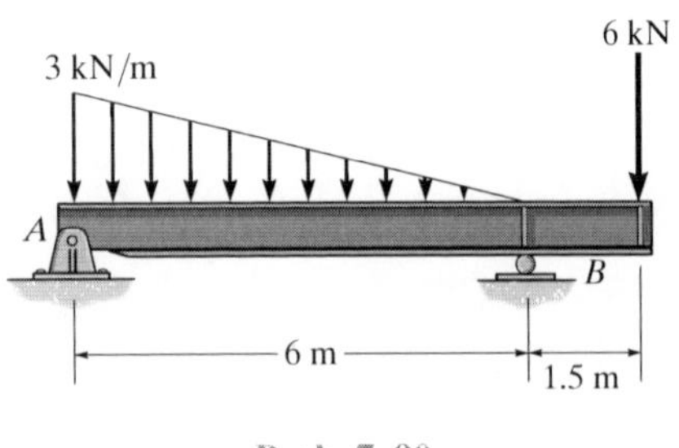

Prob. 7–90

7–91. Draw the shear and moment diagrams for the beam.

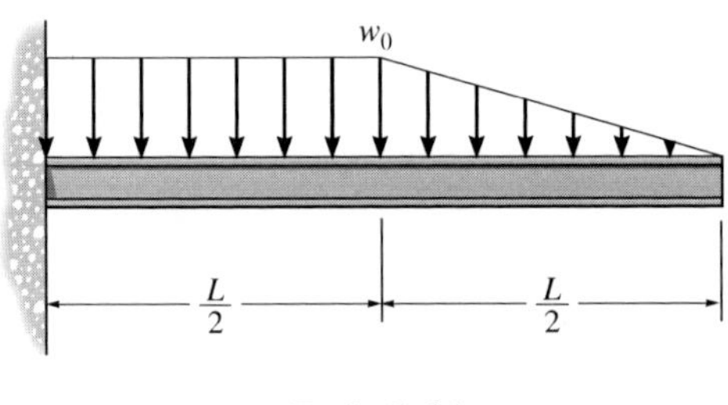

Prob. 7–91

*7–92. The beam consists of three segments pin connected at B and E. Draw the shear and moment diagrams for the beam.

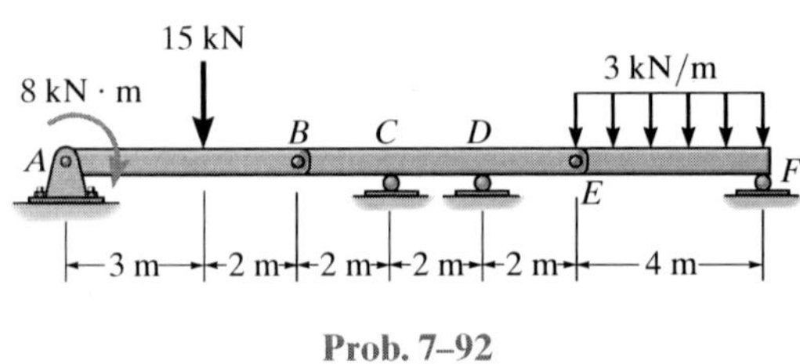

Prob. 7–92

7–93. Draw the shear and moment diagrams for the beam.

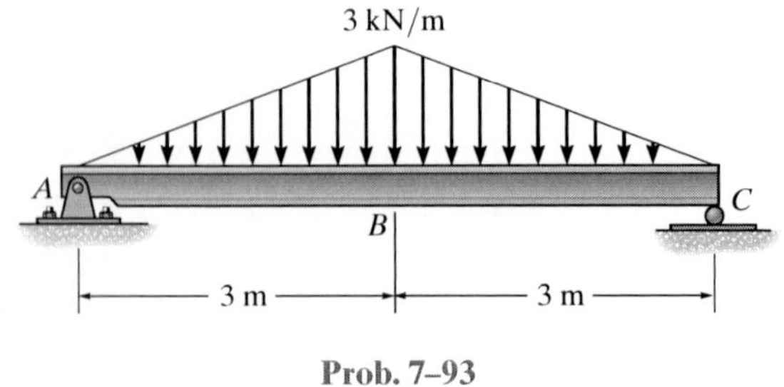

Prob. 7–93

*7.4 Cables

Flexible cables and chains combine strength with lightness and often are used in structures for support and to transmit loads from one member to another. When used to support suspension bridges and trolley wheels, cables form the main load-carrying element of the structure. In the force analysis of such systems, the weight of the cable itself may be neglected because it is often small compared to the load it carries. On the other hand, when cables are used as transmission lines and guys for radio antennas and derricks, the cable weight may become important and must be included in the structural analysis.

Each of the cable segments remains approximately straight as they support the weight of these traffic lights.

Three cases will be considered in the analysis that follows. In each case we will make the assumption that the cable is *perfectly flexible* and *inextensible*. Due to its flexibility, the cable offers no resistance to bending, and therefore, the tensile force acting in the cable is always tangent to the cable at points along its length. Being inextensible, the cable has a constant length both before and after the load is applied. As a result, once the load is applied, the geometry of the cable remains unchanged, and the cable or a segment of it can be treated as a rigid body.

Cable Subjected to Concentrated Loads. When a cable of negligible weight supports several concentrated loads, the cable takes the form of several straight-line segments, each of which is subjected to a constant tensile force. Consider, for example, the cable shown in Fig. 7–18, where the distances h, L_1, L_2, and L_3 and the loads $\mathbf{P}_1$ and $\mathbf{P}_2$ are known. The problem here is to determine the *nine unknowns* consisting of the tension in each of the *three* segments, the *four* components of reaction at A and B, and the *two* sags y_C and y_D at points C and D. For the solution we can write *two* equations of force equilibrium at each of points A, B, C, and D. This results in a total of *eight equations*.* To complete the solution, we need to know something about the geometry of the cable in order to obtain the necessary ninth equation. For example, if the cable's total *length* L is specified, then the Pythagorean theorem can be used to relate each of the three segmental lengths, written in terms of h, y_C, y_D, L_1, L_2, and L_3, to the total length L. Unfortunately, this type of problem cannot be solved easily by hand. Another possibility, however, is to specify one of the sags, either y_C or y_D, instead of the cable length. By doing this, the equilibrium equations are then sufficient for obtaining the unknown forces and the remaining sag. Once the sag at each point of loading is obtained, the length of the cable can then be determined by trigonometry. The following example illustrates a procedure for performing the equilibrium analysis for a problem of this type.

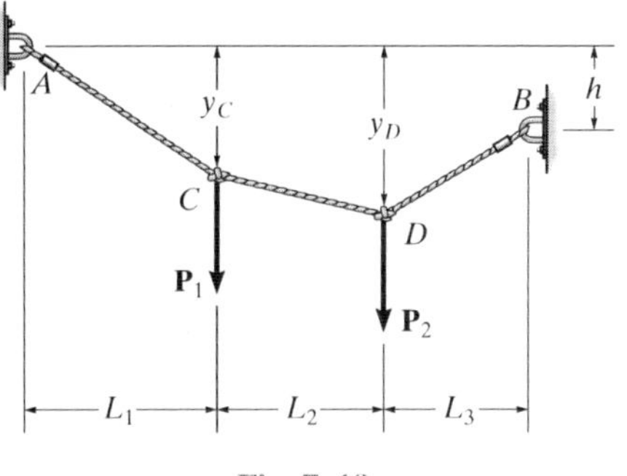

Fig. 7–18

7

*As will be shown in the following example, the eight equilibrium equations *also* can be written for the entire cable, or any part thereof. But *no more* than *eight* independent equations are available.

EXAMPLE 7.11

Determine the tension in each segment of the cable shown in Fig. 7–19*a*.

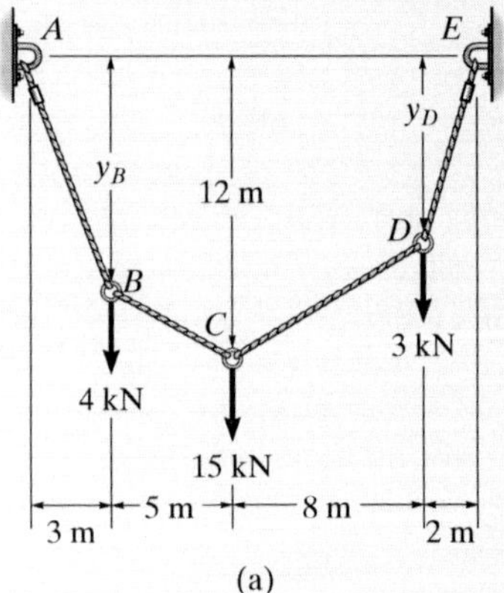

(a)

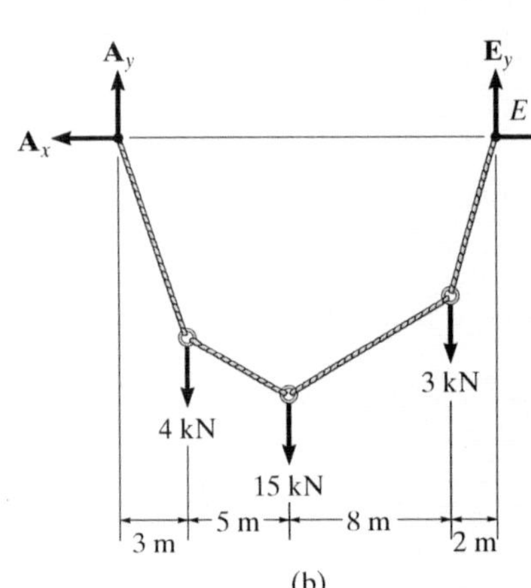

(b)

SOLUTION

By inspection, there are four unknown external reactions (A_x, A_y, E_x, and E_y) and four unknown cable tensions, one in each cable segment. These eight unknowns along with the two unknown sags y_B and y_D can be determined from *ten* available equilibrium equations. One method is to apply the force equations of equilibrium ($\Sigma F_x = 0$, $\Sigma F_y = 0$) to each of the five points A through E. Here, however, we will take a more direct approach.

Consider the free-body diagram for the entire cable, Fig. 7–19*b*. Thus,

$$\overset{+}{\rightarrow}\Sigma F_x = 0; \qquad -A_x + E_x = 0$$

$$\circlearrowleft+\Sigma M_E = 0;$$

$$-A_y(18\text{ m}) + 4\text{ kN}(15\text{ m}) + 15\text{ kN}(10\text{ m}) + 3\text{ kN}(2\text{ m}) = 0$$

$$A_y = 12\text{ kN}$$

$$+\uparrow\Sigma F_y = 0; \qquad 12\text{ kN} - 4\text{ kN} - 15\text{ kN} - 3\text{ kN} + E_y = 0$$

$$E_y = 10\text{ kN}$$

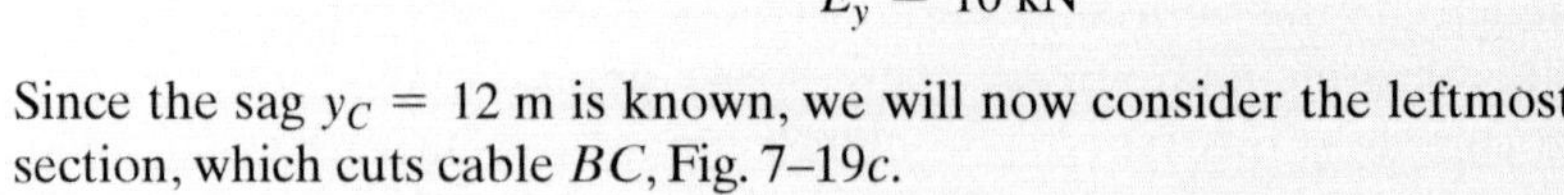

Since the sag $y_C = 12$ m is known, we will now consider the leftmost section, which cuts cable BC, Fig. 7–19*c*.

$$\circlearrowleft+\Sigma M_C = 0;\ A_x(12\text{ m}) - 12\text{ kN}(8\text{ m}) + 4\text{ kN}(5\text{ m}) = 0$$

$$A_x = E_x = 6.33\text{ kN}$$

$$\overset{+}{\rightarrow}\Sigma F_x = 0; \qquad T_{BC}\cos\theta_{BC} - 6.33\text{ kN} = 0$$

$$+\uparrow\Sigma F_y = 0; \quad 12\text{ kN} - 4\text{ kN} - T_{BC}\sin\theta_{BC} = 0$$

Thus,

$$\theta_{BC} = 51.6^\circ$$

$$T_{BC} = 10.2\text{ kN} \qquad \textit{Ans.}$$

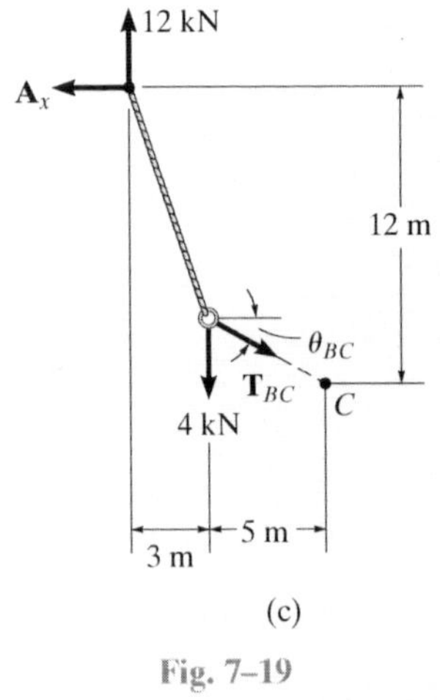

(c)

Fig. 7–19

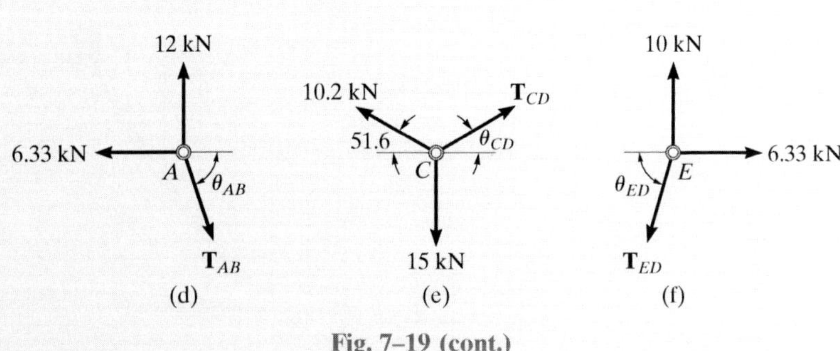

Fig. 7–19 (cont.)

Proceeding now to analyze the equilibrium of points A, C, and E in sequence, we have

Point A. (Fig. 7–19d).

$$\xrightarrow{+} \Sigma F_x = 0; \qquad T_{AB} \cos \theta_{AB} - 6.33 \text{ kN} = 0$$

$$+\uparrow \Sigma F_y = 0; \qquad -T_{AB} \sin \theta_{AB} + 12 \text{ kN} = 0$$

$$\theta_{AB} = 62.2^\circ$$

$$T_{AB} = 13.6 \text{ kN} \qquad \textit{Ans.}$$

Point C. (Fig. 7–19e).

$$\xrightarrow{+} \Sigma F_x = 0; \qquad T_{CD} \cos \theta_{CD} - 10.2 \cos 51.6^\circ \text{ kN} = 0$$

$$+\uparrow \Sigma F_y = 0; \qquad T_{CD} \sin \theta_{CD} + 10.2 \sin 51.6^\circ \text{ kN} - 15 \text{ kN} = 0$$

$$\theta_{CD} = 47.9^\circ$$

$$T_{CD} = 9.44 \text{ kN} \qquad \textit{Ans.}$$

Point E. (Fig. 7–19f).

$$\xrightarrow{+} \Sigma F_x = 0; \qquad 6.33 \text{ kN} - T_{ED} \cos \theta_{ED} = 0$$

$$+\uparrow \Sigma F_y = 0; \qquad 10 \text{ kN} - T_{ED} \sin \theta_{ED} = 0$$

$$\theta_{ED} = 57.7^\circ$$

$$T_{ED} = 11.8 \text{ kN} \qquad \textit{Ans.}$$

NOTE: By comparison, the maximum cable tension is in segment AB since this segment has the greatest slope (θ) and it is required that for any cable segment the horizontal component $T \cos \theta = A_x = E_x$ (a constant). Also, since the slope angles that the cable segments make with the horizontal have now been determined, it is possible to determine the sags y_B and y_D, Fig. 7–19a, using trigonometry.

7

The cable and suspenders are used to support the uniform load of a gas pipe which crosses the river.

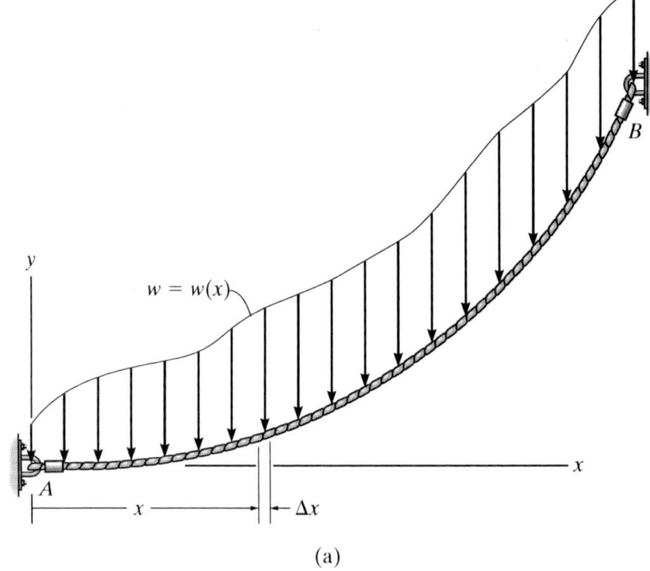

(a)

Fig. 7–20

7

Cable Subjected to a Distributed Load. Let us now consider the weightless cable shown in Fig. 7–20*a*, which is subjected to a distributed loading $w = w(x)$ that is *measured in the x direction*. The free-body diagram of a small segment of the cable having a length Δs is shown in Fig. 7–20*b*. Since the tensile force changes in both magnitude and direction along the cable's length, we will denote this change on the free-body diagram by ΔT. Finally, the distributed load is represented by its resultant force $w(x)(\Delta x)$, which acts at a fractional distance $k(\Delta x)$ from point O, where $0 < k < 1$. Applying the equations of equilibrium, we have

$$\overset{+}{\rightarrow}\Sigma F_x = 0; \qquad -T\cos\theta + (T + \Delta T)\cos(\theta + \Delta\theta) = 0$$

$$+\uparrow\Sigma F_y = 0; \qquad -T\sin\theta - w(x)(\Delta x) + (T + \Delta T)\sin(\theta + \Delta\theta) = 0$$

$$\circlearrowleft +\Sigma M_O = 0; \qquad w(x)(\Delta x)k(\Delta x) - T\cos\theta\,\Delta y + T\sin\theta\,\Delta x = 0$$

Dividing each of these equations by Δx and taking the limit as $\Delta x \rightarrow 0$, and therefore $\Delta y \rightarrow 0$, $\Delta\theta \rightarrow 0$, and $\Delta T \rightarrow 0$, we obtain

$$\frac{d(T\cos\theta)}{dx} = 0 \tag{7–7}$$

$$\frac{d(T\sin\theta)}{dx} - w(x) = 0 \tag{7–8}$$

$$\frac{dy}{dx} = \tan\theta \tag{7–9}$$

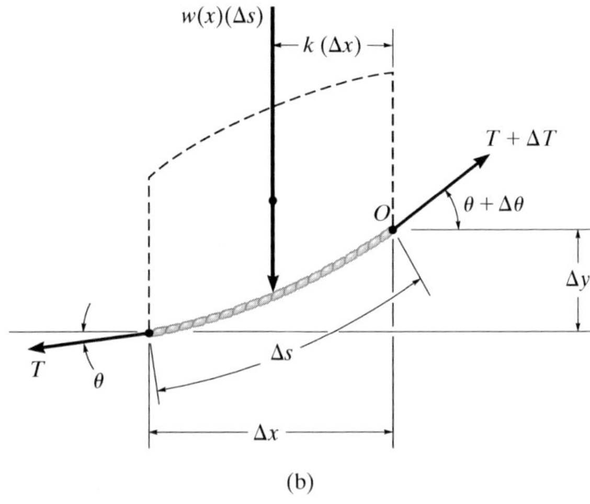

(b)

Fig. 7–20 (cont.)

Integrating Eq. 7–7, we have

$$T \cos \theta = \text{constant} = F_H \qquad (7\text{–}10)$$

where F_H represents the horizontal component of tensile force at *any point* along the cable.

Integrating Eq. 7–8 gives

$$T \sin \theta = \int w(x)\, dx \qquad (7\text{–}11)$$

Dividing Eq. 7–11 by Eq. 7–10 eliminates T. Then, using Eq. 7–9, we can obtain the slope of the cable.

$$\tan \theta = \frac{dy}{dx} = \frac{1}{F_H} \int w(x)\, dx$$

Performing a second integration yields

$$y = \frac{1}{F_H} \int \left(\int w(x)\, dx \right) dx \qquad (7\text{–}12)$$

This equation is used to determine the curve for the cable, $y = f(x)$. The horizontal force component F_H and the additional two constants, say C_1 and C_2, resulting from the integration are determined by applying the boundary conditions for the curve.

The cables of the suspension bridge exert very large forces on the tower and the foundation block which have to be accounted for in their design.

7

EXAMPLE 7.12

The cable of a suspension bridge supports half of the uniform road surface between the two towers at A and B, Fig. 7–21a. If this distributed loading is w_0, determine the maximum force developed in the cable and the cable's required length. The span length L and sag h are known.

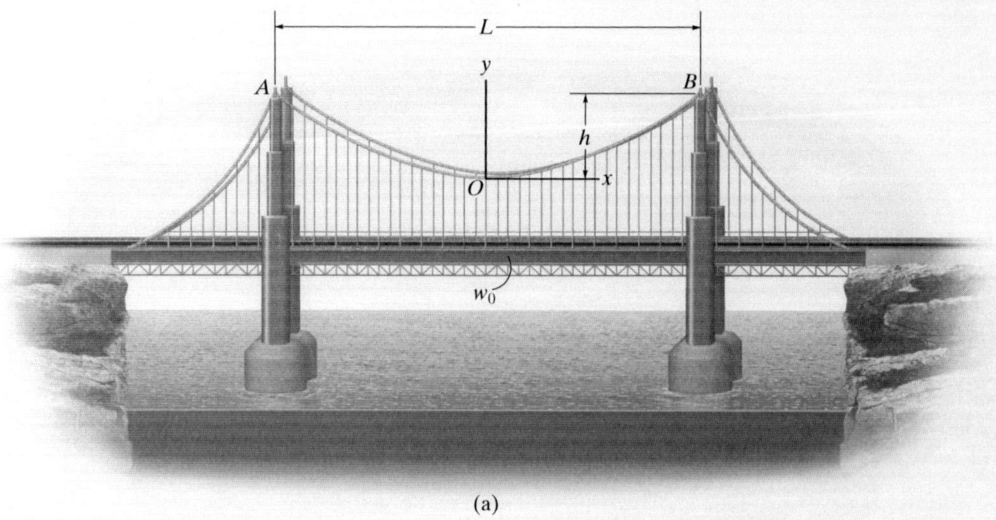

(a)

Fig. 7–21

SOLUTION

We can determine the unknowns in the problem by first finding the equation of the curve that defines the shape of the cable using Eq. 7–12. For reasons of symmetry, the origin of coordinates has been placed at the cable's center. Noting that $w(x) = w_0$, we have

$$y = \frac{1}{F_H}\int\left(\int w_0\,dx\right)dx$$

Performing the two integrations gives

$$y = \frac{1}{F_H}\left(\frac{w_0x^2}{2} + C_1x + C_2\right) \qquad (1)$$

The constants of integration may be determined using the boundary conditions $y = 0$ at $x = 0$ and $dy/dx = 0$ at $x = 0$. Substituting into Eq. 1 and its derivative yields $C_1 = C_2 = 0$. The equation of the curve then becomes

$$y = \frac{w_0}{2F_H}x^2 \qquad (2)$$

This is the equation of a *parabola*. The constant F_H may be obtained using the boundary condition $y = h$ at $x = L/2$. Thus,

$$F_H = \frac{w_0 L^2}{8h} \tag{3}$$

Therefore, Eq. 2 becomes

$$y = \frac{4h}{L^2}x^2 \tag{4}$$

Since F_H is known, the tension in the cable may now be determined using Eq. 7–10, written as $T = F_H/\cos\theta$. For $0 \leq \theta < \pi/2$, the maximum tension will occur when θ is *maximum*, i.e., at point B, Fig. 7–21*a*. From Eq. 2, the slope at this point is

$$\left.\frac{dy}{dx}\right|_{x=L/2} = \tan\theta_{max} = \left.\frac{w_0}{F_H}x\right|_{x=L/2}$$

or

$$\theta_{max} = \tan^{-1}\left(\frac{w_0 L}{2F_H}\right) \tag{5}$$

Therefore,

$$T_{max} = \frac{F_H}{\cos(\theta_{max})} \tag{6}$$

Using the triangular relationship shown in Fig. 7–21*b*, which is based on Eq. 5, Eq. 6 may be written as

$$T_{max} = \frac{\sqrt{4F_H^2 + w_0^2 L^2}}{2}$$

Substituting Eq. 3 into the above equation yields

$$T_{max} = \frac{w_0 L}{2}\sqrt{1 + \left(\frac{L}{4h}\right)^2} \qquad \textit{Ans.}$$

For a differential segment of cable length ds, we can write

$$ds = \sqrt{(dx)^2 + (dy)^2} = \sqrt{1 + \left(\frac{dy}{dx}\right)^2}\,dx$$

Hence, the total length of the cable can be determined by integration. Using Eq. 4, we have

$$\mathcal{L} = \int ds = 2\int_0^{L/2}\sqrt{1 + \left(\frac{8h}{L^2}x\right)^2}\,dx \tag{7}$$

Integrating yields

$$\mathcal{L} = \frac{L}{2}\left[\sqrt{1 + \left(\frac{4h}{L}\right)^2} + \frac{L}{4h}\sinh^{-1}\left(\frac{4h}{L}\right)\right] \qquad \textit{Ans.}$$

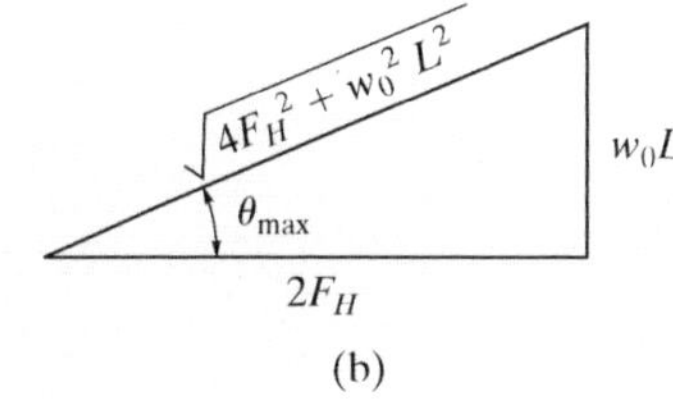

(b)

Fig. 7–21 (cont.)

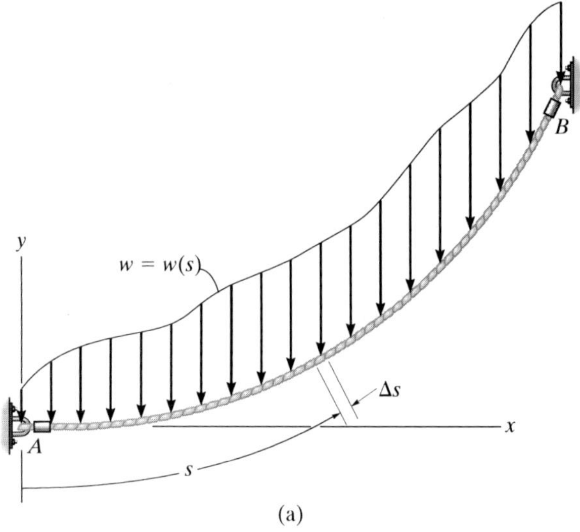

Fig. 7–22

Cable Subjected to Its Own Weight. When the weight of a cable becomes important in the force analysis, the loading function along the cable will be a function of the arc length s rather than the projected length x. To analyze this problem, we will consider a generalized loading function $w = w(s)$ acting along the cable as shown in Fig. 7–22a. The free-body diagram for a small segment Δs of the cable is shown in Fig. 7–22b. Applying the equilibrium equations to the force system on this diagram, one obtains relationships identical to those given by Eqs. 7–7 through 7–9, but with s replacing x in Eqs.7–7 and 7–8. Therefore, we can show that

$$T \cos \theta = F_H$$

$$T \sin \theta = \int w(s)\, ds \qquad (7\text{–}13)$$

$$\frac{dy}{dx} = \frac{1}{F_H} \int w(s)\, ds \qquad (7\text{–}14)$$

To perform a direct integration of Eq. 7–14, it is necessary to replace dy/dx by ds/dx. Since

$$ds = \sqrt{dx^2 + dy^2}$$

then

$$\frac{dy}{dx} = \sqrt{\left(\frac{ds}{dx}\right)^2 - 1}$$

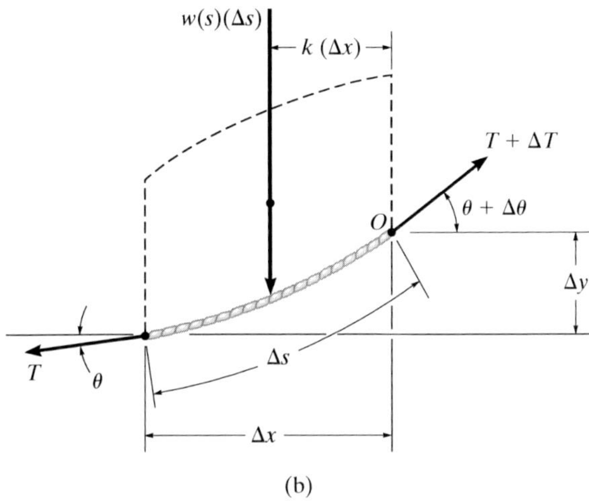

(b)

Fig. 7–22 (cont.)

Therefore,

$$\frac{ds}{dx} = \left[1 + \frac{1}{F_H^2}\left(\int w(s)\,ds\right)^2\right]^{1/2}$$

Separating the variables and integrating we obtain

$$x = \int \frac{ds}{\left[1 + \frac{1}{F_H^2}\left(\int w(s)\,ds\right)^2\right]^{1/2}} \qquad (7\text{–}15)$$

The two constants of integration, say C_1 and C_2, are found using the boundary conditions for the curve.

Electrical transmission towers must be designed to support the weights of the suspended power lines. The weight and length of the cables can be determined since they each form a catenary curve.

EXAMPLE 7.13

Determine the deflection curve, the length, and the maximum tension in the uniform cable shown in Fig. 7–23. The cable has a weight per unit length of $w_0 = 5 \text{ N/m}$.

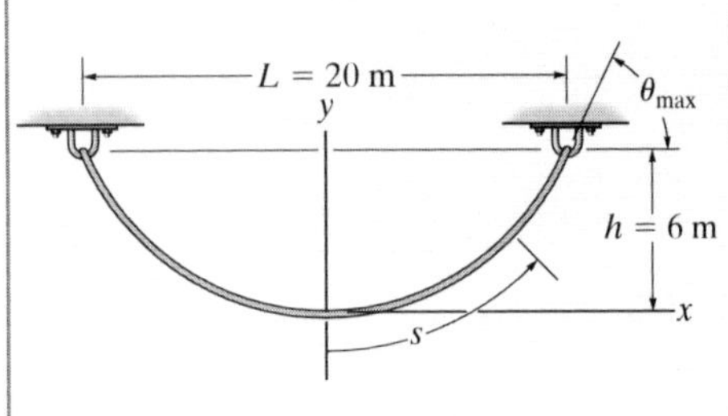

Fig. 7–23

SOLUTION

For reasons of symmetry, the origin of coordinates is located at the center of the cable. The deflection curve is expressed as $y = f(x)$. We can determine it by first applying Eq. 7–15, where $w(s) = w_0$.

$$x = \int \frac{ds}{\left[1 + (1/F_H^2)\left(\int w_0\, ds\right)^2\right]^{1/2}}$$

Integrating the term under the integral sign in the denominator, we have

$$x = \int \frac{ds}{[1 + (1/F_H^2)(w_0 s + C_1)^2]^{1/2}}$$

Substituting $u = (1/F_H)(w_0 s + C_1)$ so that $du = (w_0/F_H)\, ds$, a second integration yields

$$x = \frac{F_H}{w_0}(\sinh^{-1} u + C_2)$$

or

$$x = \frac{F_H}{w_0}\left\{\sinh^{-1}\left[\frac{1}{F_H}(w_0 s + C_1)\right] + C_2\right\} \tag{1}$$

To evaluate the constants note that, from Eq. 7–14,

$$\frac{dy}{dx} = \frac{1}{F_H}\int w_0\, ds \quad \text{or} \quad \frac{dy}{dx} = \frac{1}{F_H}(w_0 s + C_1)$$

Since $dy/dx = 0$ at $s = 0$, then $C_1 = 0$. Thus,

$$\frac{dy}{dx} = \frac{w_0 s}{F_H} \tag{2}$$

The constant C_2 may be evaluated by using the condition $s = 0$ at $x = 0$ in Eq. 1, in which case $C_2 = 0$. To obtain the deflection curve, solve for s in Eq. 1, which yields

$$s = \frac{F_H}{w_0}\sinh\left(\frac{w_0}{F_H}x\right) \tag{3}$$

Now substitute into Eq. 2, in which case

$$\frac{dy}{dx} = \sinh\left(\frac{w_0}{F_H}x\right)$$

Hence,

$$y = \frac{F_H}{w_0}\cosh\left(\frac{w_0}{F_H}x\right) + C_3$$

If the boundary condition $y = 0$ at $x = 0$ is applied, the constant $C_3 = -F_H/w_0$, and therefore the deflection curve becomes

$$y = \frac{F_H}{w_0}\left[\cosh\left(\frac{w_0}{F_H}x\right) - 1\right] \qquad (4)$$

This equation defines the shape of a *catenary curve*. The constant F_H is obtained by using the boundary condition that $y = h$ at $x = L/2$, in which case

$$h = \frac{F_H}{w_0}\left[\cosh\left(\frac{w_0 L}{2F_H}\right) - 1\right] \qquad (5)$$

Since $w_0 = 5\text{ N/m}$, $h = 6\text{ m}$, and $L = 20\text{ m}$, Eqs. 4 and 5 become

$$y = \frac{F_H}{5\text{ N/m}}\left[\cosh\left(\frac{5\text{ N/m}}{F_H}x\right) - 1\right] \qquad (6)$$

$$6\text{ m} = \frac{F_H}{5\text{ N/m}}\left[\cosh\left(\frac{50\text{ N}}{F_H}\right) - 1\right] \qquad (7)$$

Equation 7 can be solved for F_H by using a trial-and-error procedure. The result is

$$F_H = 45.9\text{ N}$$

and therefore the deflection curve, Eq. 6, becomes

$$y = 9.19[\cosh(0.109x) - 1]\text{ m} \qquad \textit{Ans.}$$

Using Eq. 3, with $x = 10\text{ m}$, the half-length of the cable is

$$\frac{\mathscr{L}}{2} = \frac{45.9\text{ N}}{5\text{ N/m}}\sinh\left[\frac{5\text{ N/m}}{45.9\text{ N}}(10\text{ m})\right] = 12.1\text{ m}$$

Hence,

$$\mathscr{L} = 24.2\text{ m} \qquad \textit{Ans.}$$

Since $T = F_H/\cos\theta$, the maximum tension occurs when θ is maximum, i.e., at $s = \mathscr{L}/2 = 12.1\text{ m}$. Using Eq. 2 yields

$$\left.\frac{dy}{dx}\right|_{s=12.1\text{ m}} = \tan\theta_{max} = \frac{5\text{ N/m}(12.1\text{ m})}{45.9\text{ N}} = 1.32$$

$$\theta_{max} = 52.8^\circ$$

And so,

$$T_{max} = \frac{F_H}{\cos\theta_{max}} = \frac{45.9\text{ N}}{\cos 52.8^\circ} = 75.9\text{ N} \qquad \textit{Ans.}$$

7

PROBLEMS

Neglect the weight of the cable in the following problems, unless specified.

7–94. Determine the tension in each segment of the cable and the cable's total length. Set $P = 400$ N.

7–95. If each cable segment can support a maximum tension of 375 N, determine the largest load P that can be applied.

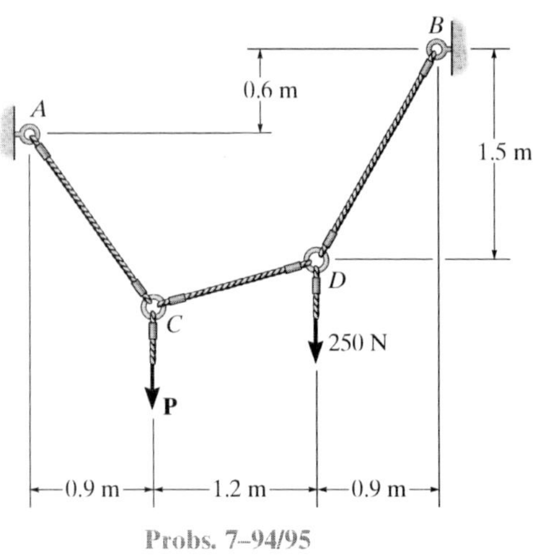

Probs. 7–94/95

7

***7–96.** Cable $ABCD$ supports the 10-kg lamp E and the 15-kg lamp F. Determine the maximum tension in the cable and the sag of point B.

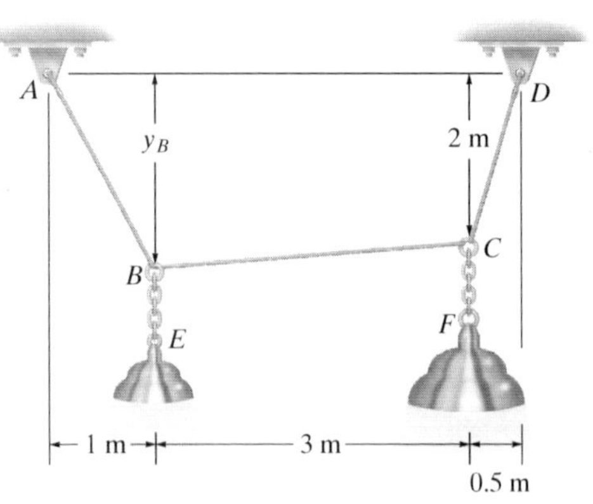

Prob. 7–96

7–97. The cable supports the three loads shown. Determine the sags y_B and y_D of points B and D. Take $P_1 = 2000$ N, $P_2 = 1250$ N.

7–98. The cable supports the three loads shown. Determine the magnitude of $\mathbf{P}_1$ if $P_2 = 1500$ N and $y_B = 2.4$ m. Also find the sag y_D.

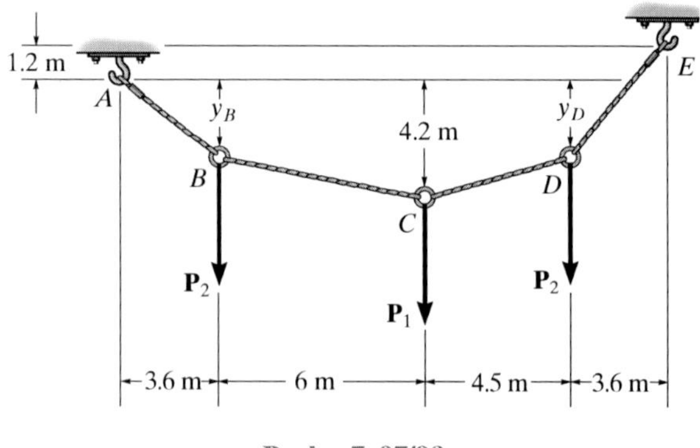

Probs. 7–97/98

7–99. If cylinders E and F have a mass of 20 kg and 40 kg, respectively, determine the tension developed in each cable and the sag y_C.

***7–100.** If cylinder E has a mass of 20 kg and each cable segment can sustain a maximum tension of 400 N, determine the largest mass of cylinder F that can be supported. Also, what is the sag y_C?

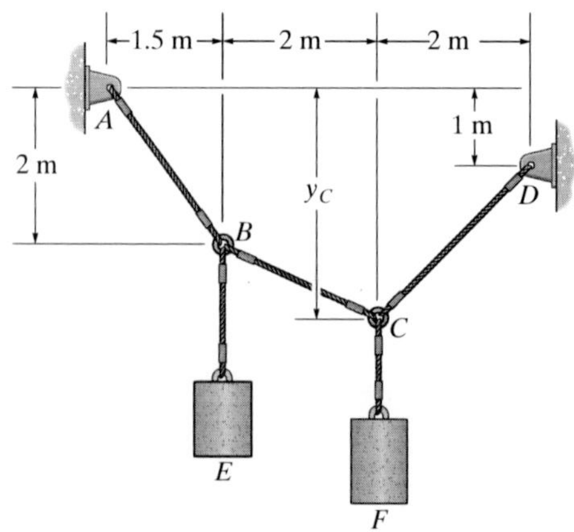

Probs. 7–99/100

7–101. If $h = 5$ m, determine the maximum tension developed in the chain and its length. The chain has a mass per unit length of 8 kg/m.

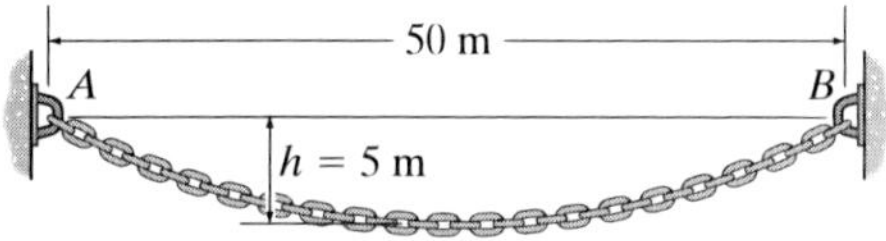

Prob. 7–101

7–102. The cable supports the uniform distributed load of $w_0 = 12$ kN/m. Determine the tension in the cable at each support A and B.

7–103. Determine the maximum uniform distributed load w_0 the cable can support if the maximum tension the cable can sustain is 20 kN.

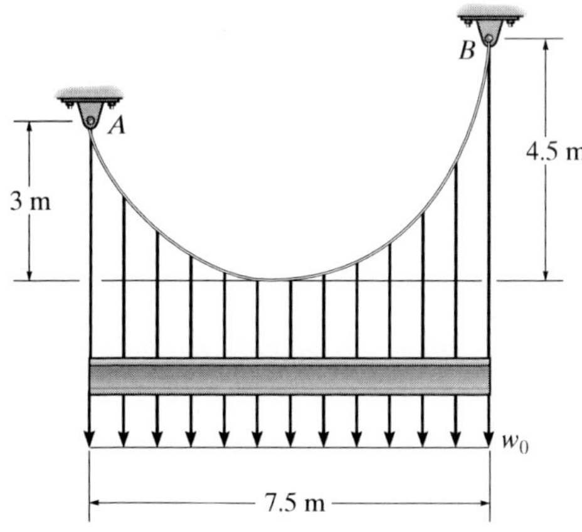

Probs. 7–102/103

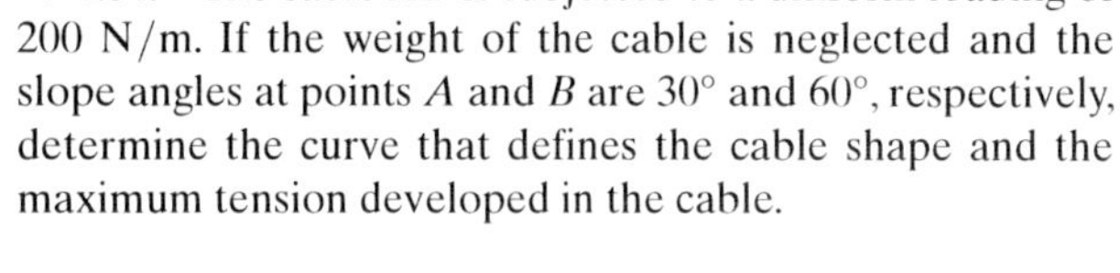

***7–104.** The cable AB is subjected to a uniform loading of 200 N/m. If the weight of the cable is neglected and the slope angles at points A and B are 30° and 60°, respectively, determine the curve that defines the cable shape and the maximum tension developed in the cable.

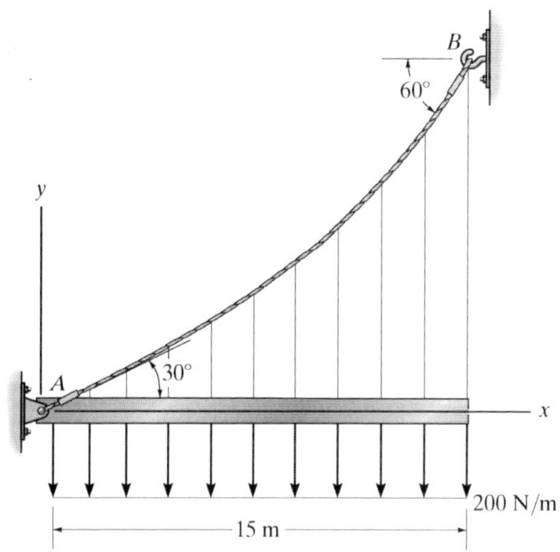

Prob. 7–104

7–105. The bridge deck has a weight per unit length of 80 kN/m. It is supported on each side by a cable. Determine the tension in each cable at the piers A and B.

7–106. If each of the two side cables that support the bridge deck can sustain a maximum tension of 50 MN, determine the allowable uniform distributed load w_0 caused by the weight of the bridge deck.

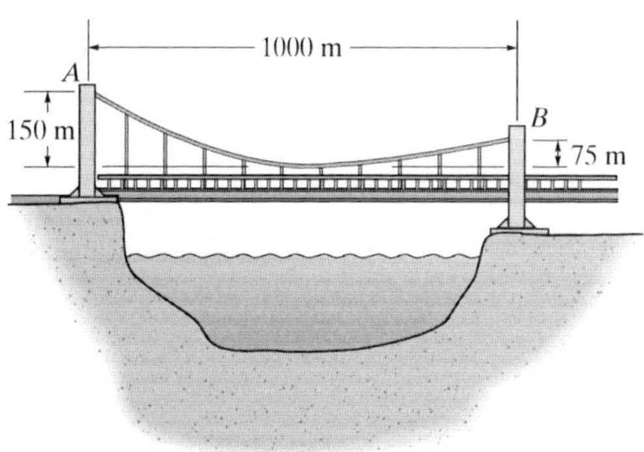

Probs. 7–105/106

7–107. Cylinders C and D are attached to the end of the cable. If D has a mass of 600 kg, determine the required mass of C, the maximum sag h of the cable, and the length of the cable between the pulleys A and B. The beam has a mass per unit length of 50 kg/m.

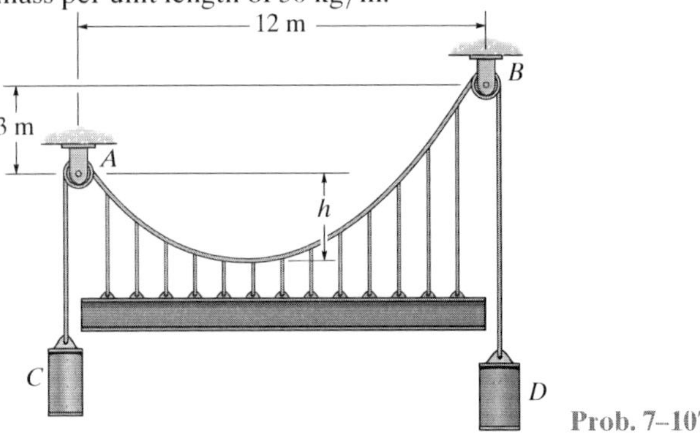

Prob. 7–107

***7–108.** The cable will break when the maximum tension reachs T_{max} = 10 kN. Determine the minimum sag h if it supports the uniform distributed load of w = 600 N/m.

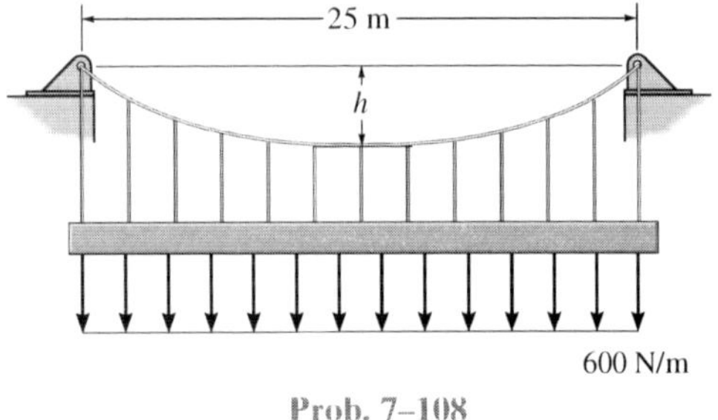

Prob. 7–108

7–109. If the pipe has a mass per unit length of 1500 kg/m, determine the maximum tension developed in the cable.

7–110. If the pipe has a mass per unit length of 1500 kg/m, determine the minimum tension developed in the cable.

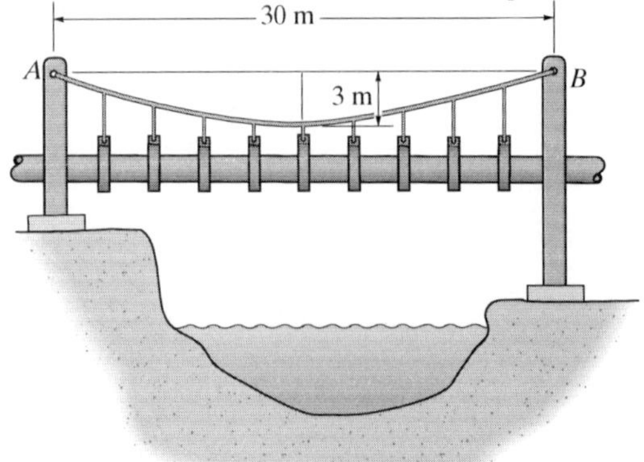

Probs. 7–109/110

7–111. If the slope of the cable at support A is zero, determine the deflection curve $y = f(x)$ of the cable and the maximum tension developed in the cable.

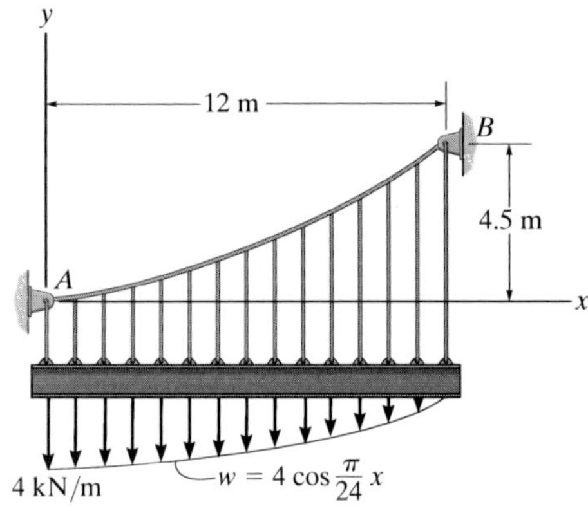

Prob. 7–111

***7–112.** Determine the maximum tension developed in the cable if it is subjected to a uniform load of 600 N/m.

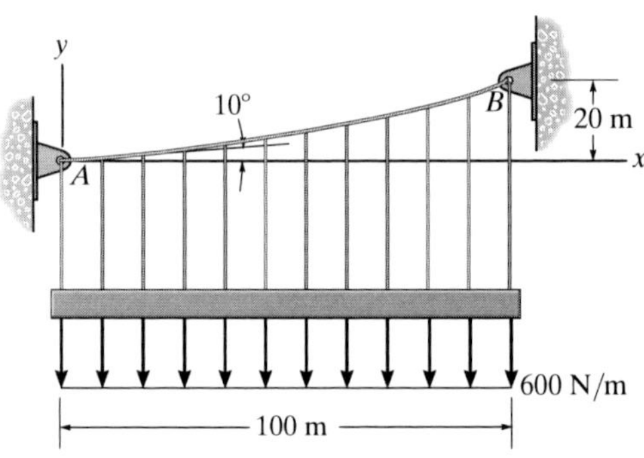

Prob. 7–112

7–113. Determine the maximum uniform distributed loading w_0 N/m that the cable can support if it is capable of sustaining a maximum tension of 60 kN.

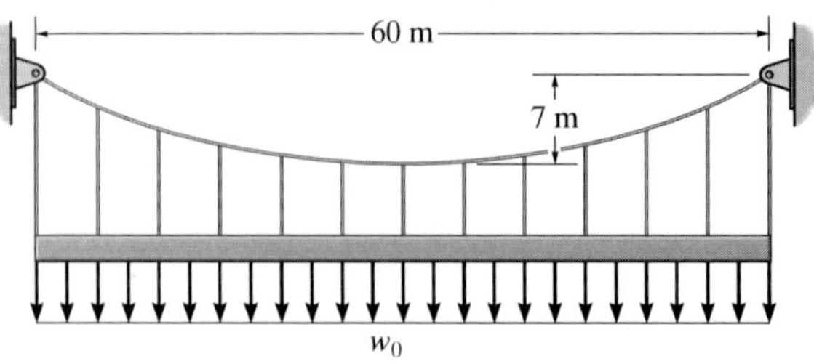

Prob. 7–113

7–114. A cable has a weight of 30 N/m ($\approx$ 3 kg/m) and is supported at points that are 25 m apart and at the same elevation. If it has a length of 26 m, determine the sag.

7–115. A wire has a weight of 2 N/m ($\approx$ 0.2 kg/m). If it can span 10 m and has a sag of 1.2 m, determine the length of the wire. The ends of the wire are supported from the same elevation.

***7–116.** The 10 kg/m cable is suspended between the supports A and B. If the cable can sustain a maximum tension of 1.5 kN and the maximum sag is 3 m, determine the maximum distance L between the supports

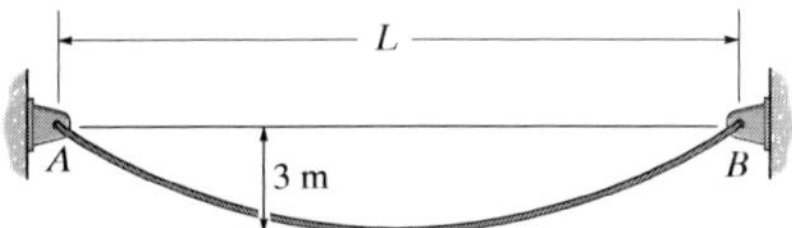

Prob. 7–116

7–117. Show that the deflection curve of the cable discussed in Example 7.13 reduces to Eq. 4 in Example 7.12 when the *hyperbolic cosine function* is expanded in terms of a series and only the first two terms are retained. (The answer indicates that the *catenary* may be replaced by a *parabola* in the analysis of problems in which the sag is small. In this case, the cable weight is assumed to be uniformly distributed along the horizontal.)

7–118. If the horizontal towing force is T = 20 kN and the chain has a mass per unit length of 15 kg/m, determine the maximum sag h. Neglect the buoyancy effect of the water on the chain. The boats are stationary.

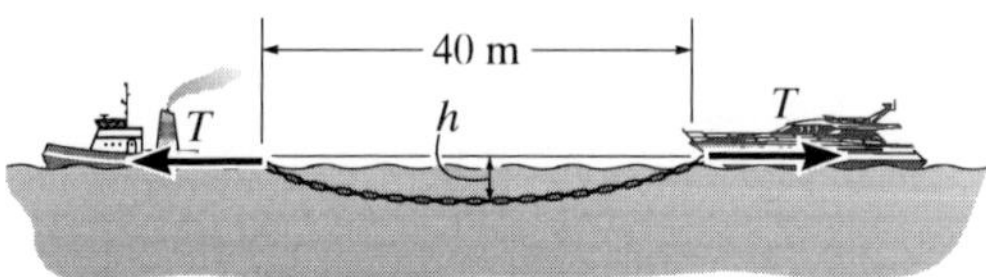

Prob. 7–118

7–119. The cable has a mass of 0.5 kg/m and is 25 m long. Determine the vertical and horizontal components of force it exerts on the top of the tower.

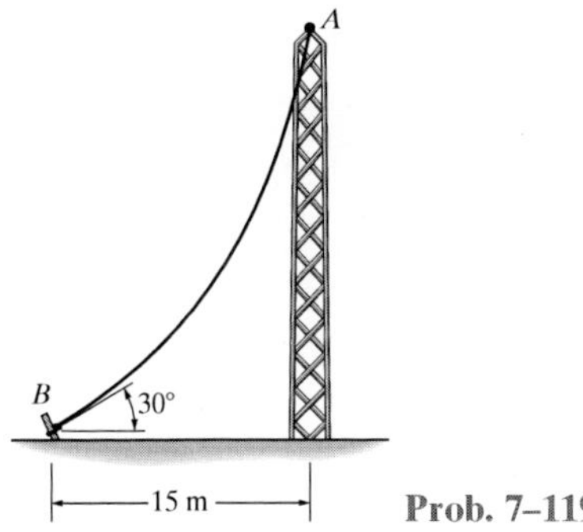

Prob. 7–119

***7–120.** A 50-m cable is suspended between two points a distance of 15 m apart and at the same elevation. If the minimum tension in the cable is 200 kN, determine the total weight of the cable and the maximum tension developed in the cable.

7–121. The 80-m-long chain is fixed at its ends and hoisted at its midpoint B using a crane. If the chain has a weight of 0.5 kN / m, determine the minimum height h of the hook in order to lift the chain *completely* off the ground. What is the horizontal force at pin A or C when the chain is in this position? *Hint:* When h is a minimum, the slope at A and C is zero.

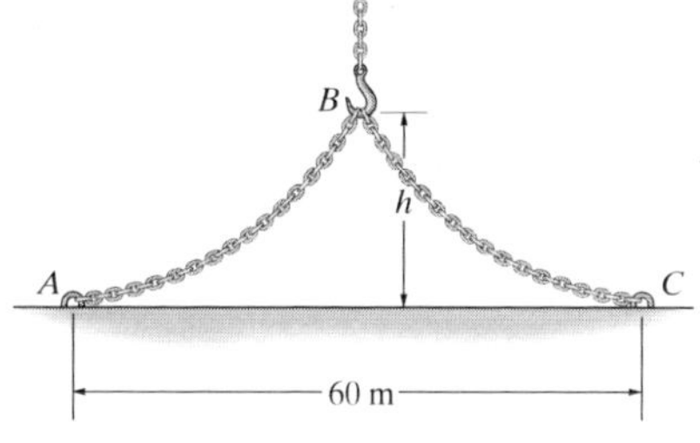

Prob. 7–121

7–122. A uniform cord is suspended between two points having the same elevation. Determine the sag-to-span ratio so that the maximum tension in the cord equals the cord's total weight.

7–123. A cable having a weight per unit length of 0.1 kN/m is suspended between supports A and B Determine the equation of the catenary curve of the cable and the cable's length.

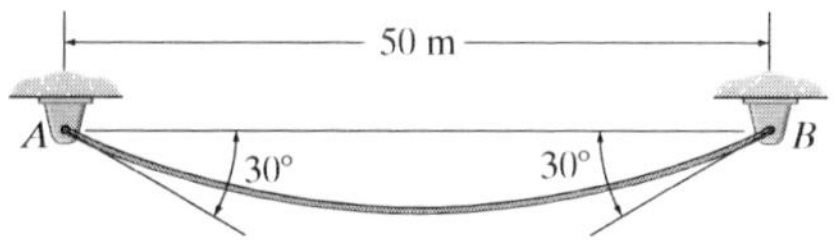

Prob. 7–123

***7–124.** A fiber optic cable is suspended over the poles so that the angle at the supports is $\theta = 22°$. Determine the minimum tension in the cable and the sag. The cable has a mass of 0.9 kg/m and the supports are at the same elevation.

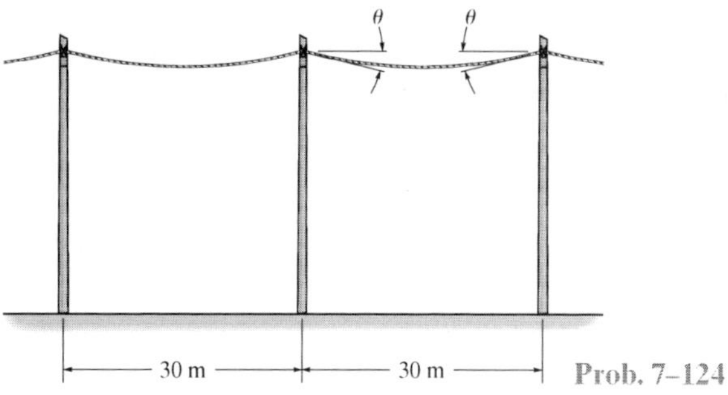

Prob. 7–124

7

CHAPTER REVIEW

Internal Loadings

If a coplanar force system acts on a member, then in general a resultant internal *normal force* **N**, *shear force* **V**, and *bending moment* **M** will act at any cross section along the member. For two-dimensional problems the positive directions of these loadings are shown in the figure.

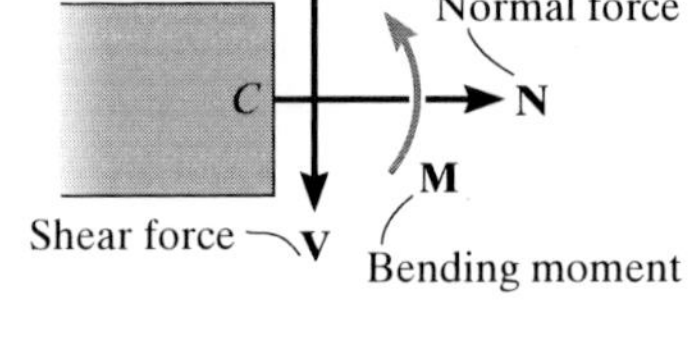

The resultant internal normal force, shear force, and bending moment are determined using the method of sections. To find them, the member is sectioned at the point C where the internal loadings are to be determined. A free-body diagram of one of the sectioned parts is then drawn and the internal loadings are shown in their positive directions.

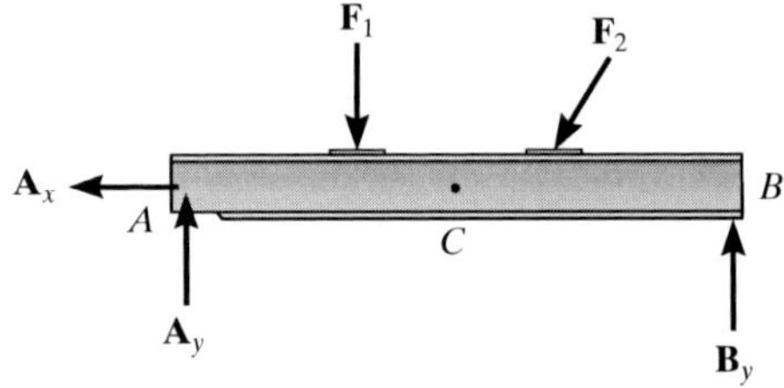

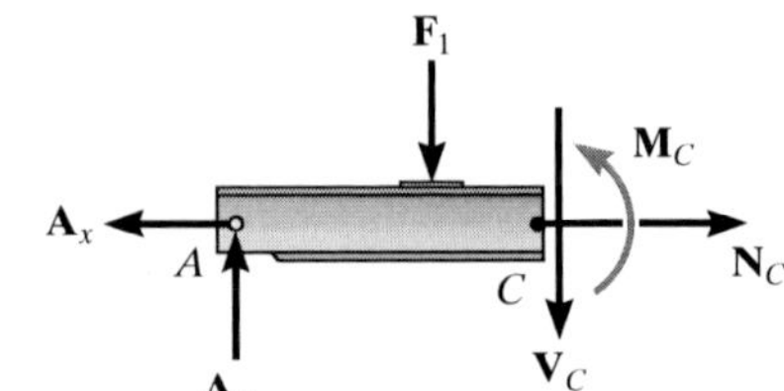

7

The resultant normal force is determined by summing forces normal to the cross section. The resultant shear force is found by summing forces tangent to the cross section, and the resultant bending moment is found by summing moments about the geometric center or centroid of the cross-sectional area.

$$\Sigma F_x = 0$$

$$\Sigma F_y = 0$$

$$\Sigma M_C = 0$$

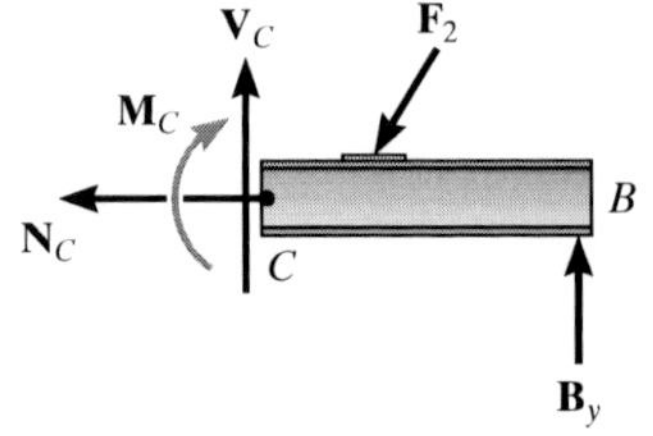

If the member is subjected to a three-dimensional loading, then, in general, a *torsional moment* will also act on the cross section. It can be determined by summing moments about an axis that is perpendicular to the cross section and passes through its centroid.

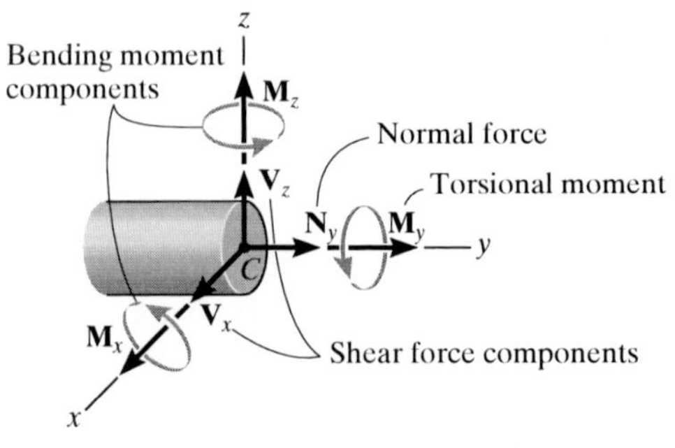

Shear and Moment Diagrams

To construct the shear and moment diagrams for a member, it is necessary to section the member at an arbitrary point, located a distance x from the left end.

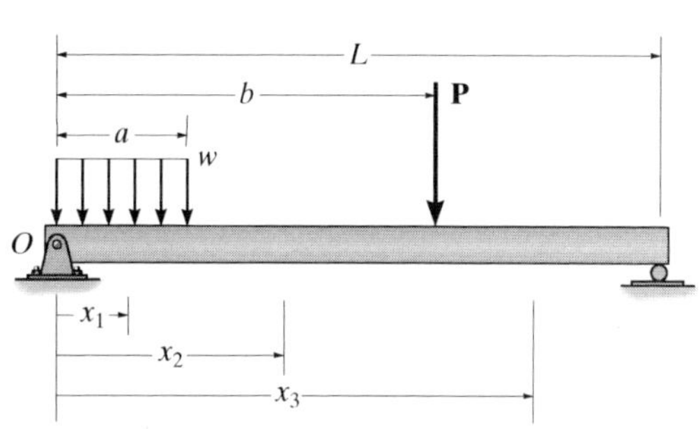

If the external loading consists of changes in the distributed load, or a series of concentrated forces and couple moments act on the member, then different expressions for V and M must be determined within regions between any load discontinuities.

The unknown shear and moment are indicated on the cross section in the positive direction according to the established sign convention, and then the internal shear and moment are determined as functions of x.

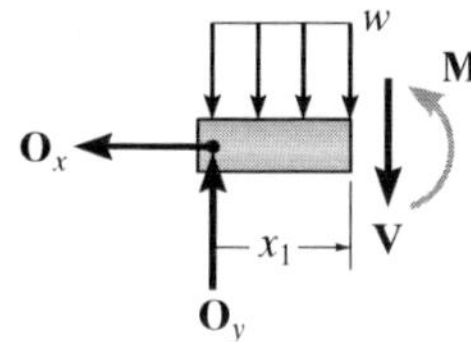

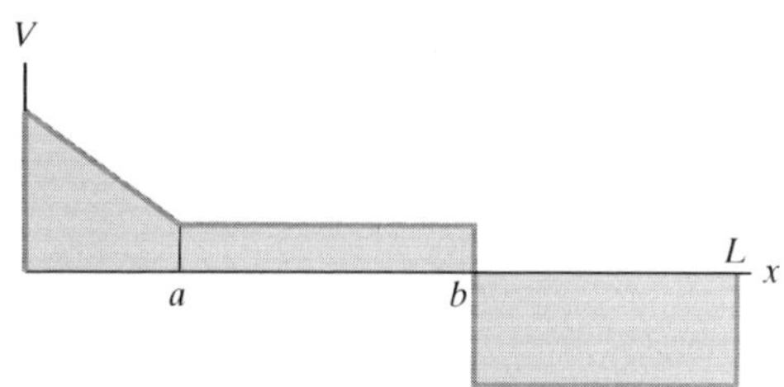

Each of the functions of the shear and moment is then plotted to create the shear and moment diagrams.

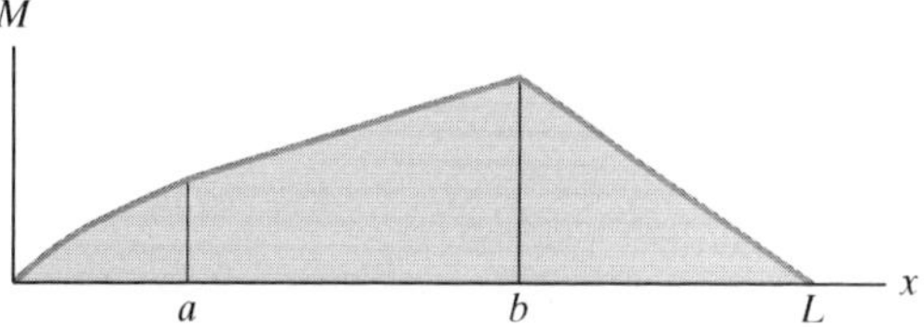

Relations between Shear and Moment

It is possible to plot the shear and moment diagrams quickly by using differential relationships that exist between the distributed loading w, V and M.

The slope of the shear diagram is equal to the distributed loading at any point. The slope is positive if the distributed load acts upward, and vice-versa.

$$\frac{dV}{dx} = w$$

The slope of the moment diagram is equal to the shear at any point. The slope is positive if the shear is positive, or vice-versa.

$$\frac{dM}{dx} = V$$

The change in shear between any two points is equal to the area under the distributed loading between the points.

$$\Delta V = \int w\,dx$$

The change in the moment is equal to the area under the shear diagram between the points.

$$\Delta M = \int V\,dx$$

Cables

When a flexible and inextensible cable is subjected to a series of concentrated forces, then the analysis of the cable can be performed by using the equations of equilibrium applied to free-body diagrams of either segments or points of application of the loading.

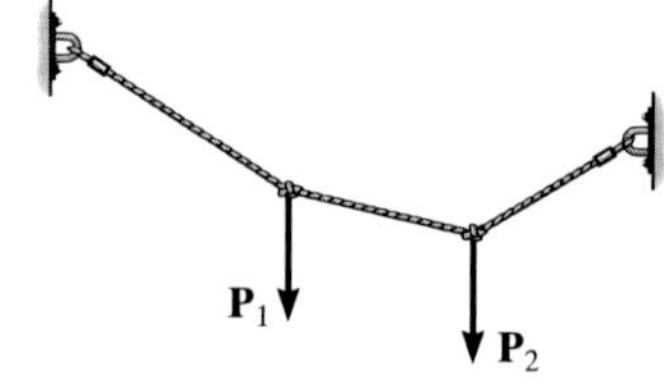

If external distributed loads or the weight of the cable are to be considered, then the shape of the cable must be determined by first analyzing the forces on a differential segment of the cable and then integrating this result. The two constants, say C_1 and C_2, resulting from the integration are determined by applying the boundary conditions for the cable.

$$y = \frac{1}{F_H}\int\left(\int w(x)\,dx\right)dx$$

Distributed load

$$x = \int \frac{ds}{\left[1 + \frac{1}{F_H^2}\left(\int w(s)\,ds\right)^2\right]^{1/2}}$$

Cable weight

REVIEW PROBLEMS

7–125. The beam is supported by a pin at C and a rod AB. Determine the internal normal force, shear force, and moment at point D.

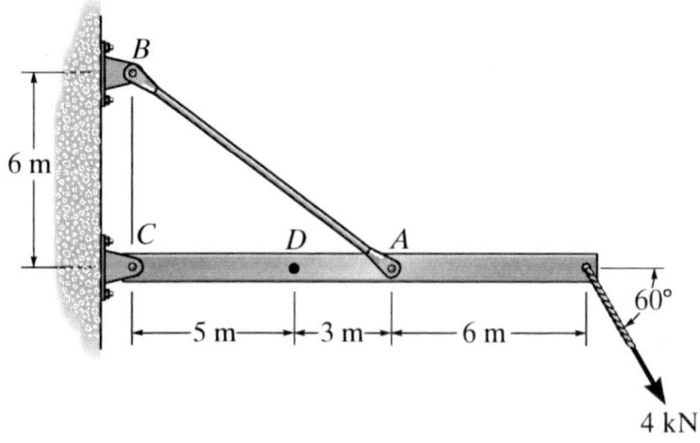

Prob. 7–125

7–126. Draw the shear and moment diagrams for the beam.

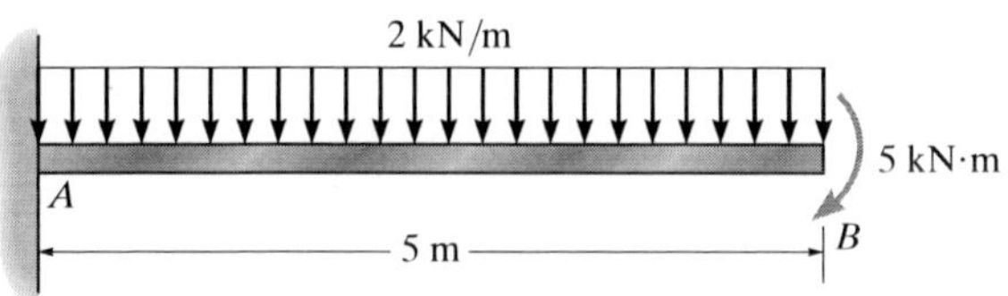

Prob. 7–126

7–127. Determine the distance a between the supports in terms of the beam's length L so that the moment in the *symmetric* beam is zero at the beam's center.

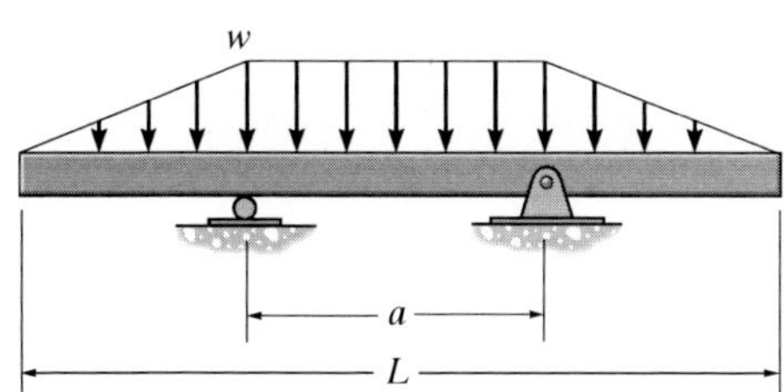

Prob. 7–127

***7–128.** The stacker crane supports a 1.5-Mg boat with the center of mass at G. Determine the internal normal force, shear force, and moment at point D in the girder. The trolley is free to roll along the girder rail and is located at the position shown. Only vertical reactions occur at A and B.

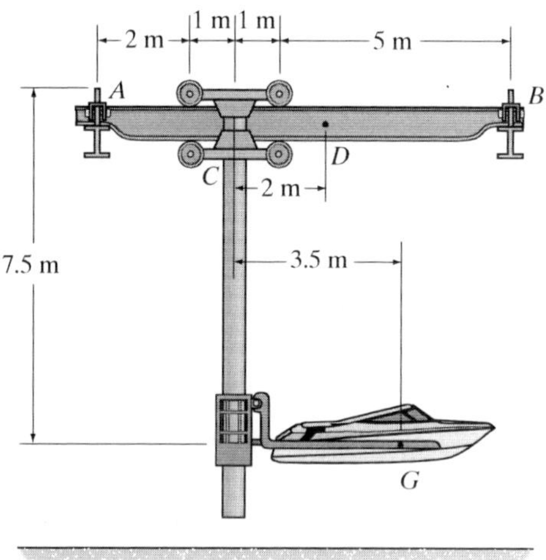

Prob. 7–128

7–129. The yacht is anchored with a chain that has a total length of 40 m and a mass per unit length of 18 kg/m, and the tension in the chain at A is 7 kN. Determine the length of chain l_d which is lying at the bottom of the sea. What is the distance d? Assume that buoyancy effects of the water on the chain are negligible. *Hint:* Establish the origin of the coordinate system at B as shown in order to find the chain length BA.

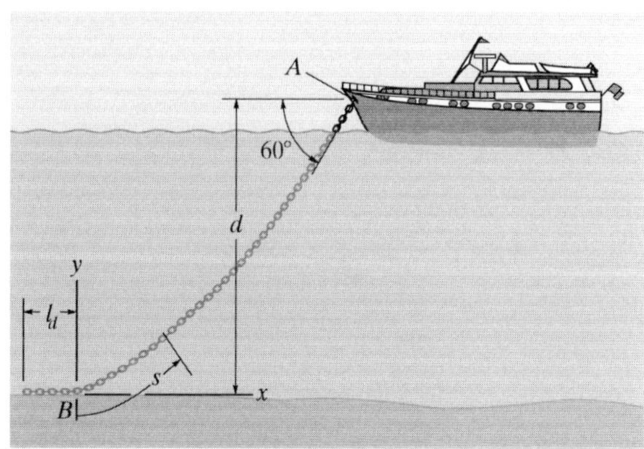

Prob. 7–129

7–130. Draw the shear and moment diagrams for the beam ABC.

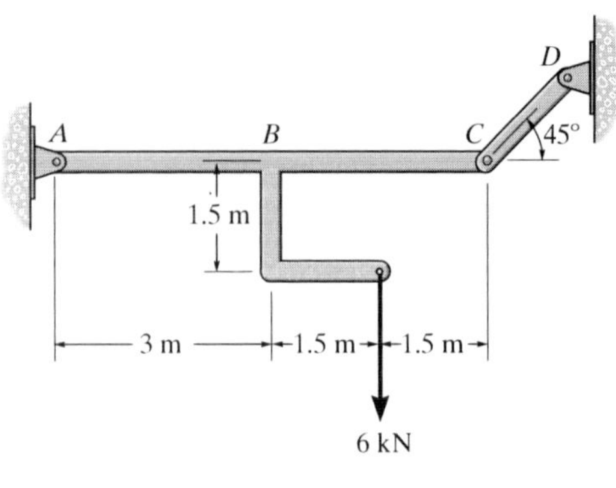

Prob. 7–130

***7–132.** A chain is suspended between points at the same elevation and spaced a distance of 60 m apart. If it has a weight per unit length of 0.5 kN/m and the sag is 3 m, determine the maximum tension in the chain.

7–133. Draw the shear and moment diagrams for the beam.

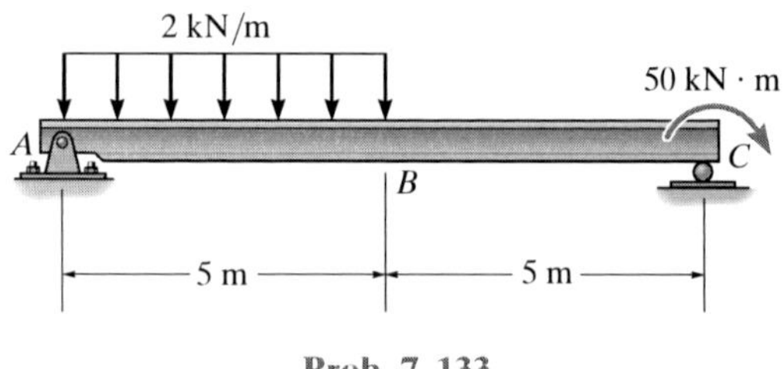

Prob. 7–133

7–131. The shaft is supported by a thrust bearing at A and a journal bearing at B. Determine the x, y, z components of internal loading at point C.

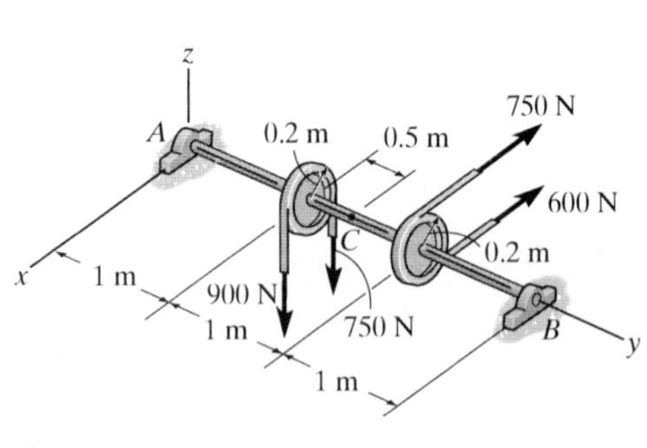

Prob. 7–131

7–134. Determine the normal force, shear force, and moment at points B and C of the beam.

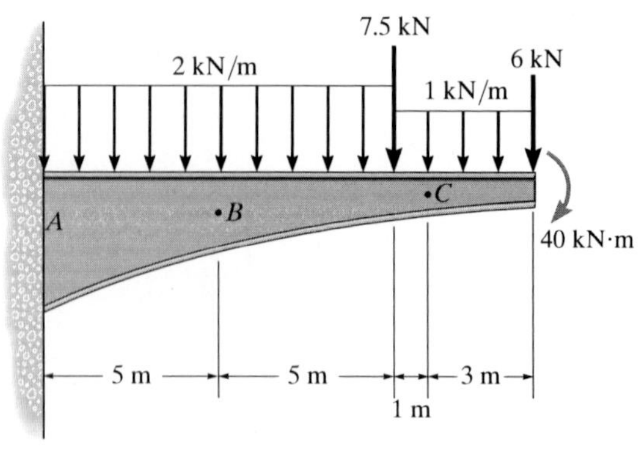

Prob. 7–134

7–135. Draw the shear and moment diagrams for the beam.

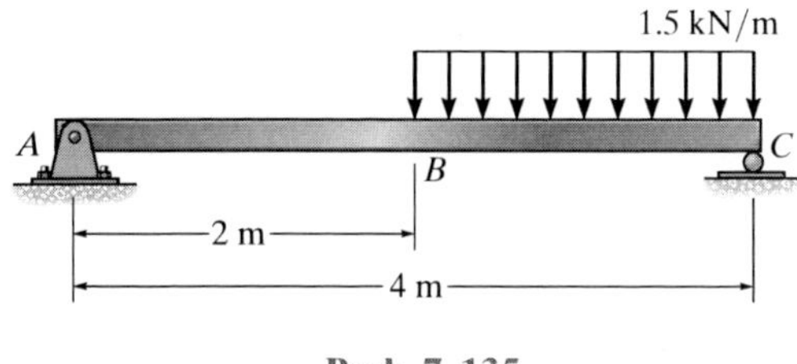

Prob. 7–135

***7–136.** If the 45-m-long cable has a mass per unit length of 5 kg/m, determine the equation of the catenary curve of the cable and the maximum tension developed in the cable.

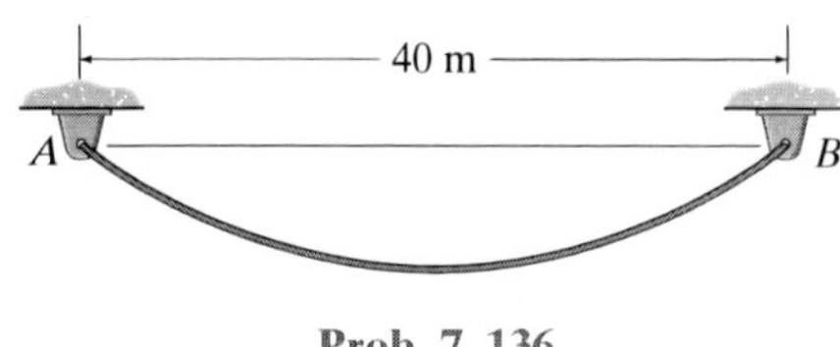

Prob. 7–136

7–137. The traveling crane consists of a 5-m-long beam having a uniform mass per unit length of 20 kg/m. The chain hoist and its supported load exert a force of 8 kN on the beam when $x = 2$ m. Draw the shear and moment diagrams for the beam. The guide wheels at the ends A and B exert only vertical reactions on the beam. Neglect the size of the trolley at C.

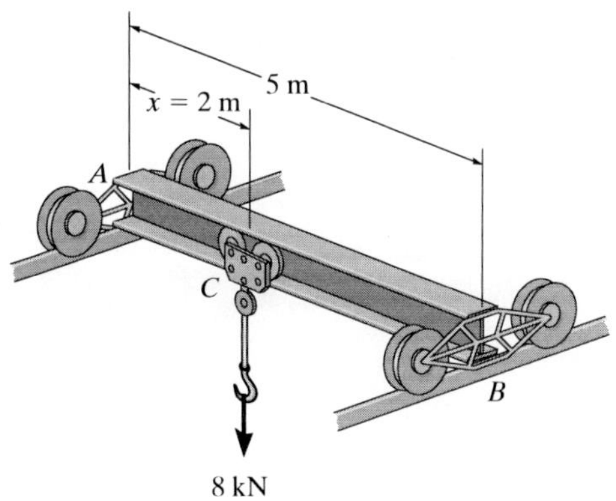

Prob. 7–137

7–138. Draw the shear and moment diagrams for the beam.

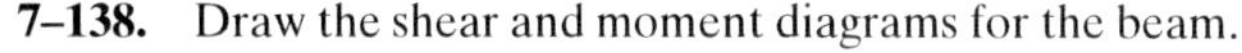

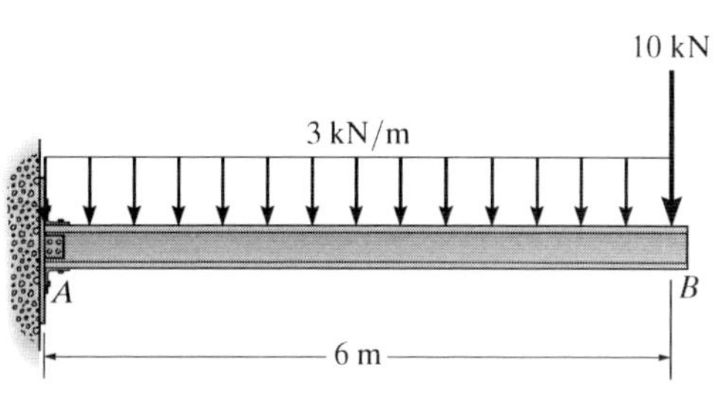

Prob. 7–138

7–139. The suspender bar supports the 300-kg engine. Draw the shear and moment diagrams for the bar.

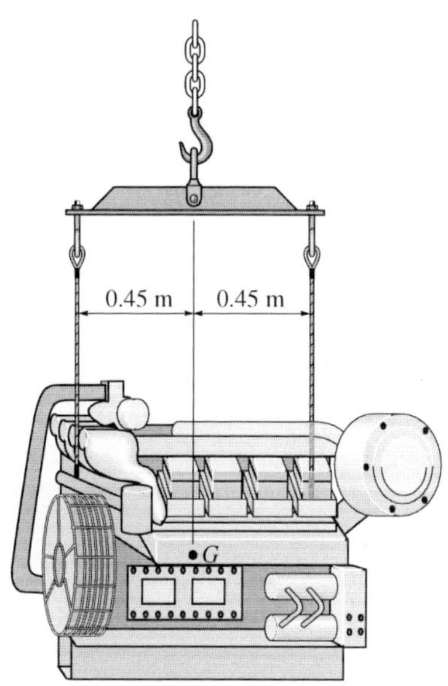

Prob. 7–139

7

Chapter 8

The effective design of each brake on this railroad wheel requires that it resist the frictional forces developed between it and the wheel. In this chapter we will study dry friction, and show how to analyze friction forces for various engineering applications.

Friction

CHAPTER OBJECTIVES

- To introduce the concept of dry friction and show how to analyze the equilibrium of rigid bodies subjected to this force.
- To present specific applications of frictional force analysis on wedges, screws, belts, and bearings.
- To investigate the concept of rolling resistance.

Video Solutions are available for selected questions in this chapter.

8.1 Characteristics of Dry Friction

Friction is a force that resists the movement of two contacting surfaces that slide relative to one another. This force always acts *tangent* to the surface at the points of contact and is directed so as to oppose the possible or existing motion between the surfaces.

In this chapter, we will study the effects of *dry friction*, which is sometimes called *Coulomb friction* since its characteristics were studied extensively by C. A. Coulomb in 1781. Dry friction occurs between the contacting surfaces of bodies when there is no lubricating fluid.*

The heat generated by the abrasive action of friction can be noticed when using this grinder to sharpen a metal blade.

*Another type of friction, called fluid friction, is studied in fluid mechanics.

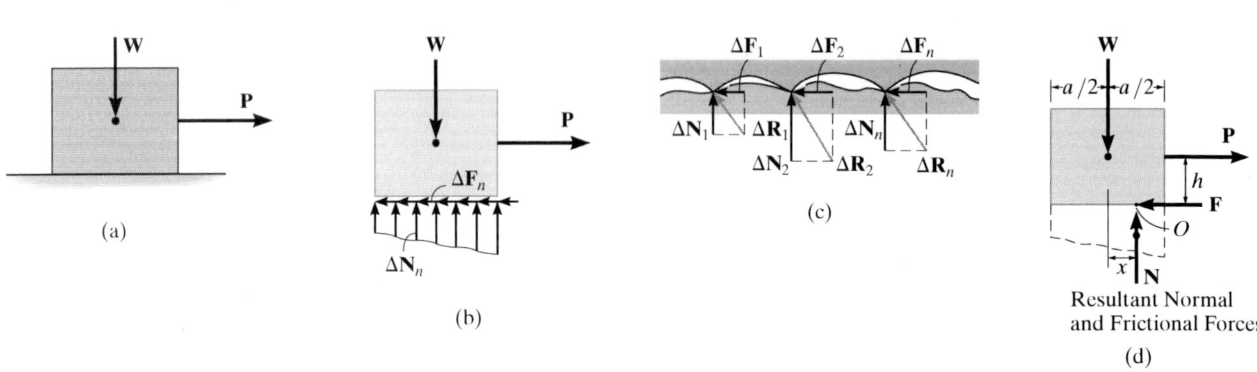

Fig. 8–1

Regardless of the weight of the rake or shovel that is suspended, the device has been designed so that the small roller holds the handle in equilibrium due to frictional forces that develop at the points of contact, A, B, C.

Theory of Dry Friction. The theory of dry friction can be explained by considering the effects caused by pulling horizontally on a block of uniform weight $\mathbf{W}$ which is resting on a rough horizontal surface that is *nonrigid or deformable,* Fig. 8–1*a*. The upper portion of the block, however, can be considered rigid. As shown on the free-body diagram of the block, Fig. 8–1*b*, the floor exerts an uneven *distribution* of both *normal force* $\Delta\mathbf{N}_n$ and *frictional force* $\Delta\mathbf{F}_n$ along the contacting surface. For equilibrium, the normal forces must act *upward* to balance the block's weight $\mathbf{W}$, and the frictional forces act to the left to prevent the applied force $\mathbf{P}$ from moving the block to the right. Close examination of the contacting surfaces between the floor and block reveals how these frictional and normal forces develop, Fig. 8–1*c*. It can be seen that many microscopic irregularities exist between the two surfaces and, as a result, reactive forces $\Delta\mathbf{R}_n$ are developed at each point of contact.* As shown, each reactive force contributes both a frictional component $\Delta\mathbf{F}_n$ and a normal component $\Delta\mathbf{N}_n$.

8

Equilibrium. The effect of the *distributed* normal and frictional loadings is indicated by their *resultants* $\mathbf{N}$ and $\mathbf{F}$ on the free-body diagram, Fig. 8–1*d*. Notice that $\mathbf{N}$ acts a distance x to the right of the line of action of $\mathbf{W}$, Fig. 8–1*d*. This location, which coincides with the centroid or geometric center of the normal force distribution in Fig. 8–1*b*, is necessary in order to balance the "tipping effect" caused by $\mathbf{P}$. For example, if $\mathbf{P}$ is applied at a height h from the surface, Fig. 8–1*d*, then moment equilibrium about point O is satisfied if $Wx = Ph$ or $x = Ph/W$.

*Besides mechanical interactions as explained here, which is referred to as a classical approach, a detailed treatment of the nature of frictional forces must also include the effects of temperature, density, cleanliness, and atomic or molecular attraction between the contacting surfaces. See J. Krim, *Scientific American*, October, 1996.

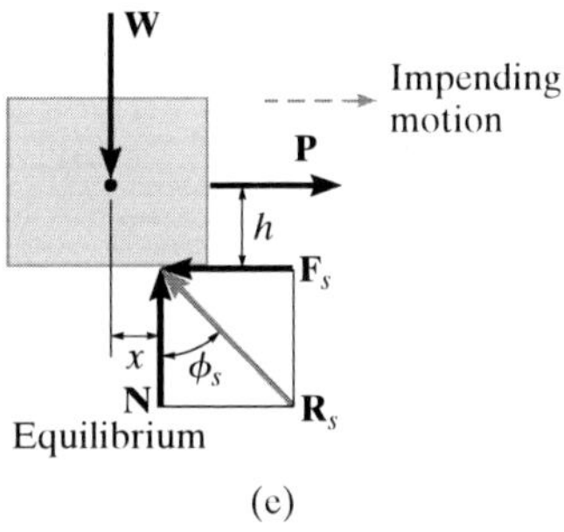

(e)

Fig. 8–1 (cont.)

Impending Motion. In cases where the surfaces of contact are rather "slippery," the frictional force **F** may *not* be great enough to balance **P**, and consequently the block will tend to slip. In other words, as P is slowly increased, F correspondingly increases until it attains a certain *maximum value* F_s, called the *limiting static frictional force*, Fig. 8–1*e*. When this value is reached, the block is in *unstable equilibrium* since any further increase in P will cause the block to move. Experimentally, it has been determined that this limiting static frictional force F_s is *directly proportional* to the resultant normal force N. Expressed mathematically,

$$F_s = \mu_s N \qquad (8\text{–}1)$$

where the constant of proportionality, μ_s (mu "sub" s), is called the *coefficient of static friction*.

Thus, when the block is on the *verge of sliding*, the normal force **N** and frictional force $\mathbf{F}_s$ combine to create a resultant $\mathbf{R}_s$, Fig. 8–1*e*. The angle ϕ_s (phi "sub" s) that $\mathbf{R}_s$ makes with **N** is called the *angle of static friction*. From the figure,

$$\phi_s = \tan^{-1}\left(\frac{F_s}{N}\right) = \tan^{-1}\left(\frac{\mu_s N}{N}\right) = \tan^{-1}\mu_s$$

Typical values for μ_s are given in Table 8–1. Note that these values can vary since experimental testing was done under variable conditions of roughness and cleanliness of the contacting surfaces. For applications, therefore, it is important that both caution and judgment be exercised when selecting a coefficient of friction for a given set of conditions. When a more accurate calculation of F_s is required, the coefficient of friction should be determined directly by an experiment that involves the two materials to be used.

8

Table 8–1 Typical Values for μ_s

Contact Materials	Coefficient of Static Friction (μ_s)
Metal on ice	0.03–0.05
Wood on wood	0.30–0.70
Leather on wood	0.20–0.50
Leather on metal	0.30–0.60
Aluminum on aluminum	1.10–1.70

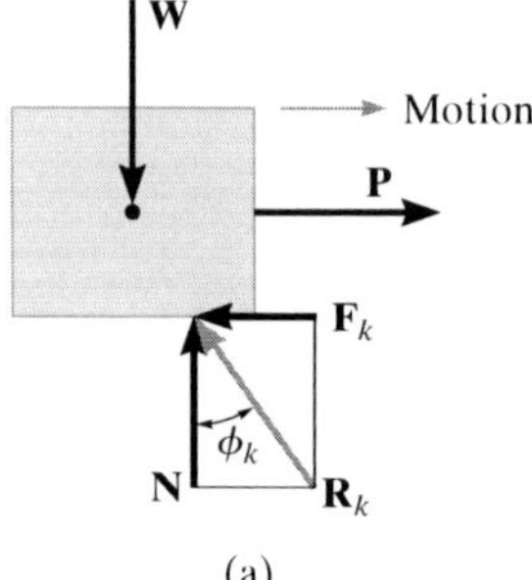

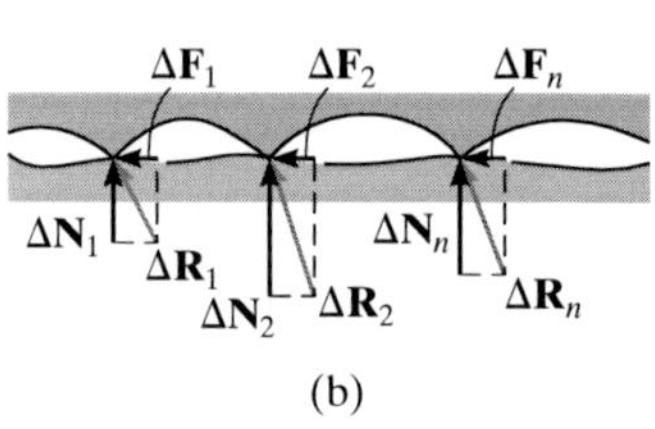

Fig. 8–2

Motion. If the magnitude of **P** acting on the block is increased so that it becomes slightly greater than F_s, the frictional force at the contacting surface will drop to a smaller value F_k, called the *kinetic frictional force*. The block will begin to slide with increasing speed, Fig. 8–2*a*. As this occurs, the block will "ride" on top of these peaks at the points of contact, as shown in Fig. 8–2*b*. The continued breakdown of the surface is the dominant mechanism creating kinetic friction.

Experiments with sliding blocks indicate that the magnitude of the kinetic friction force is directly proportional to the magnitude of the resultant normal force, expressed mathematically as

$$F_k = \mu_k N \qquad (8\text{–}2)$$

8

Here the constant of proportionality, μ_k, is called the *coefficient of kinetic friction*. Typical values for μ_k are approximately 25 percent *smaller* than those listed in Table 8–1 for μ_s.

As shown in Fig. 8–2*a*, in this case, the resultant force at the surface of contact, $\mathbf{R}_k$, has a line of action defined by ϕ_k. This angle is referred to as the *angle of kinetic friction*, where

$$\phi_k = \tan^{-1}\left(\frac{F_k}{N}\right) = \tan^{-1}\left(\frac{\mu_k N}{N}\right) = \tan^{-1}\mu_k$$

By comparison, $\phi_s \geq \phi_k$.

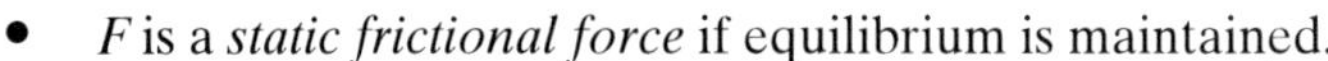

The above effects regarding friction can be summarized by referring to the graph in Fig. 8–3, which shows the variation of the frictional force F versus the applied load P. Here the frictional force is categorized in three different ways:

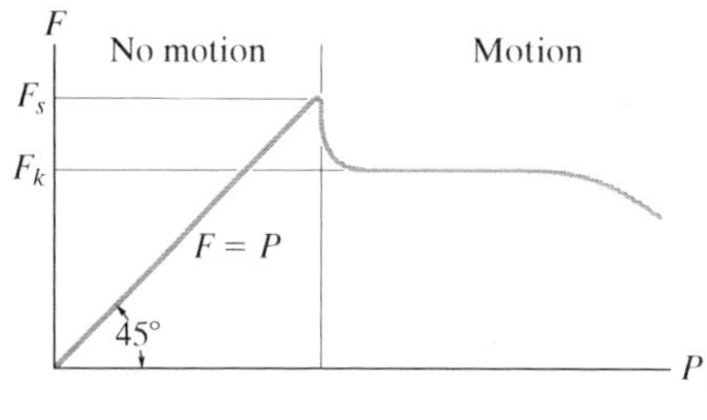

Fig. 8–3

- F is a *static frictional force* if equilibrium is maintained.
- F is a *limiting static frictional force* F_s when it reaches a maximum value needed to maintain equilibrium.
- F is a *kinetic frictional force* F_k when sliding occurs at the contacting surface.

Notice also from the graph that for very large values of P or for high speeds, aerodynamic effects will cause F_k and likewise μ_k to begin to decrease.

Characteristics of Dry Friction. As a result of *experiments* that pertain to the foregoing discussion, we can state the following rules which apply to bodies subjected to dry friction.

- The frictional force acts *tangent* to the contacting surfaces in a direction *opposed* to the *motion* or tendency for motion of one surface relative to another.
- The maximum static frictional force F_s that can be developed is independent of the area of contact, provided the normal pressure is not very low nor great enough to severely deform or crush the contacting surfaces of the bodies.
- The maximum static frictional force is generally greater than the kinetic frictional force for any two surfaces of contact. However, if one of the bodies is moving with a *very low velocity* over the surface of another, F_k becomes approximately equal to F_s, i.e., $\mu_s \approx \mu_k$.
- When *slipping* at the surface of contact is *about to occur*, the maximum static frictional force is proportional to the normal force, such that $F_s = \mu_s N$.
- When *slipping* at the surface of contact is *occurring*, the kinetic frictional force is proportional to the normal force, such that $F_k = \mu_k N$.

8

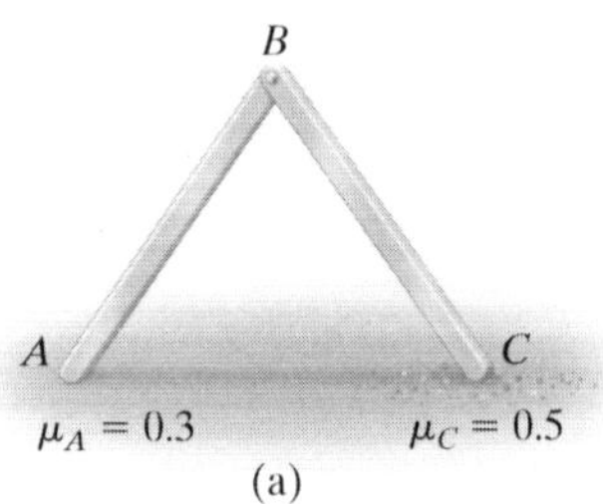

(a)

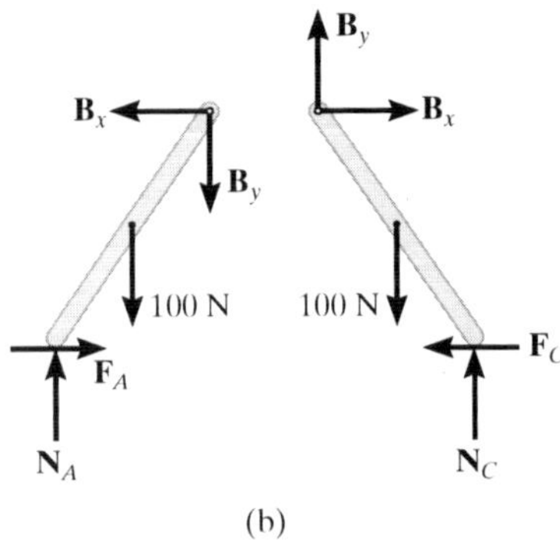

(b)

Fig. 8–4

8.2 Problems Involving Dry Friction

If a rigid body is in equilibrium when it is subjected to a system of forces that includes the effect of friction, the force system must satisfy not only the equations of equilibrium but *also* the laws that govern the frictional forces.

Types of Friction Problems. In general, there are three types of mechanics problems involving dry friction. They can easily be classified once free-body diagrams are drawn and the total number of unknowns are identified and compared with the total number of available equilibrium equations.

No Apparent Impending Motion. Problems in this category are strictly equilibrium problems, which require the number of unknowns to be *equal* to the number of available equilibrium equations. Once the frictional forces are determined from the solution, however, their numerical values must be checked to be sure they satisfy the inequality $F \leq \mu_s N$; otherwise, slipping will occur and the body will not remain in equilibrium. A problem of this type is shown in Fig. 8–4*a*. Here we must determine the frictional forces at A and C to check if the equilibrium position of the two-member frame can be maintained. If the bars are uniform and have known weights of 100 N each, then the free-body diagrams are as shown in Fig. 8–4*b*. There are six unknown force components which can be determined *strictly* from the six equilibrium equations (three for each member). Once F_A, N_A, F_C, and N_C are determined, then the bars will remain in equilibrium provided $F_A \leq 0.3N_A$ and $F_C \leq 0.5N_C$ are satisfied.

8

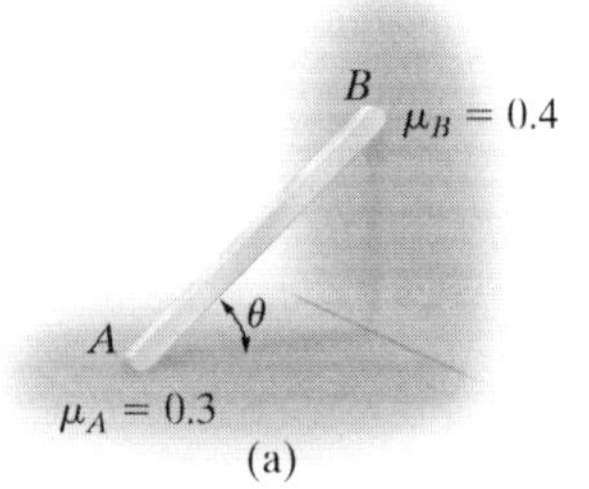

(a)

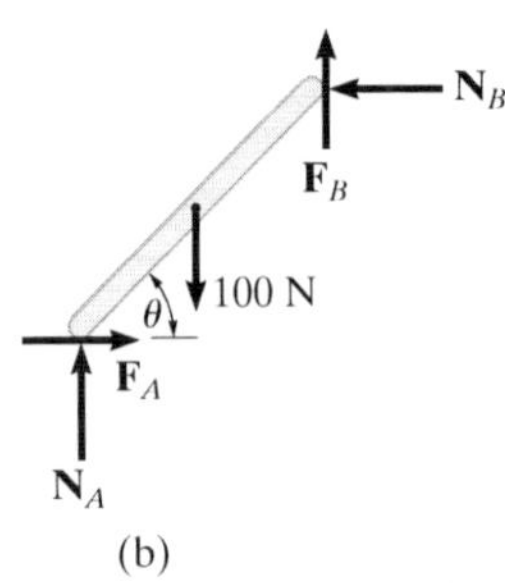

(b)

Fig. 8–5

Impending Motion at All Points of Contact. In this case the total number of unknowns will *equal* the total number of available equilibrium equations *plus* the total number of available frictional equations, $F = \mu N$. When *motion is impending* at the points of contact, then $F_s = \mu_s N$; whereas if the body is *slipping*, then $F_k = \mu_k N$. For example, consider the problem of finding the smallest angle θ at which the 100-N bar in Fig. 8–5*a* can be placed against the wall without slipping. The free-body diagram is shown in Fig. 8–5*b*. Here the *five* unknowns are determined from the *three* equilibrium equations and *two* static frictional equations which apply at *both* points of contact, so that $F_A = 0.3N_A$ and $F_B = 0.4N_B$.

Impending Motion at Some Points of Contact. Here the number of unknowns will be *less* than the number of available equilibrium equations plus the number of available frictional equations or conditional equations for tipping. As a result, several possibilities for motion or impending motion will exist and the problem will involve a determination of the kind of motion which actually occurs. For example, consider the two-member frame in Fig. 8–6*a*. In this problem we wish to determine the horizontal force P needed to cause movement. If each member has a weight of 100 N, then the free-body diagrams are as shown in Fig. 8–6*b*. There are *seven* unknowns. For a unique solution we must satisfy the *six* equilibrium equations (three for each member) and only *one* of two possible static frictional equations. This means that as P increases it will either cause slipping at A and no slipping at C, so that $F_A = 0.3N_A$ and $F_C \leq 0.5N_C$; or slipping occurs at C and no slipping at A, in which case $F_C = 0.5N_C$ and $F_A \leq 0.3N_A$. The actual situation can be determined by calculating P for each case and then choosing the case for which P is *smaller*. If in both cases the *same value* for P is calculated, which would be highly improbable, then slipping at both points occurs simultaneously; i.e., the *seven unknowns* would satisfy *eight equations*.

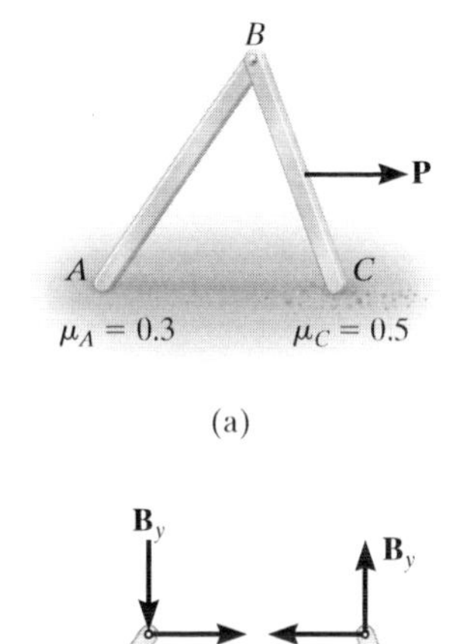

(a)

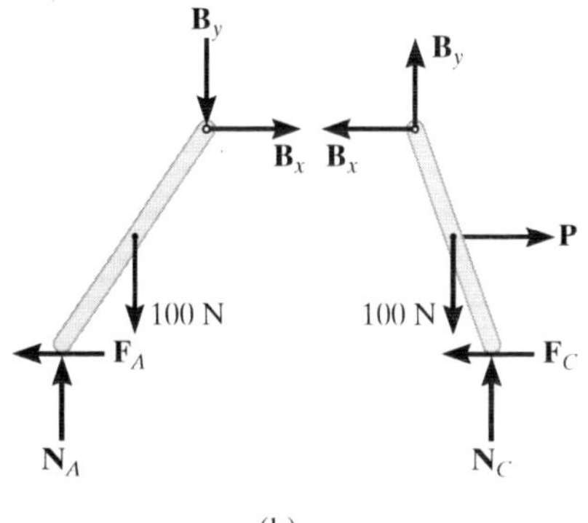

(b)

Fig. 8–6

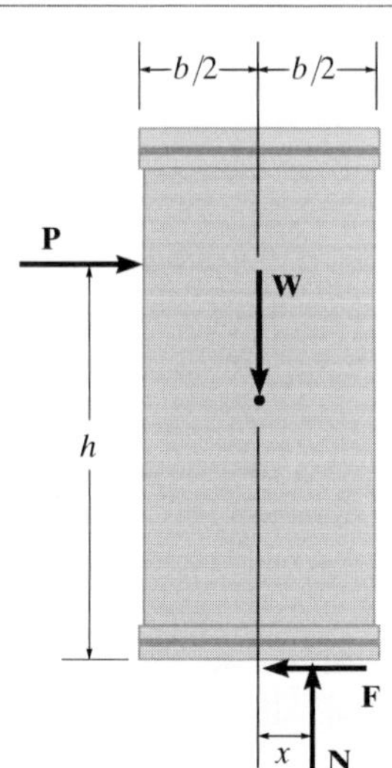

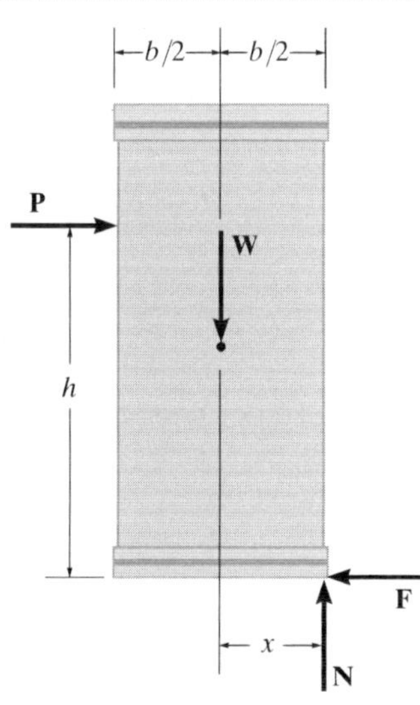

Consider pushing on the uniform crate that has a weight W and sits on the rough surface. As shown on the first free-body diagram, if the magnitude of **P** is small, the crate will remain in equilibrium. As P increases the crate will either be on the verge of slipping on the surface ($F = \mu_s N$), or if the surface is very rough (large μ_s) then the resultant normal force will shift to the corner, $x = b/2$, as shown on the second free-body diagram. At this point the crate will begin to tip over. The crate also has a greater chance of tipping if **P** is applied at a greater height h above the surface, or if its width b is smaller.

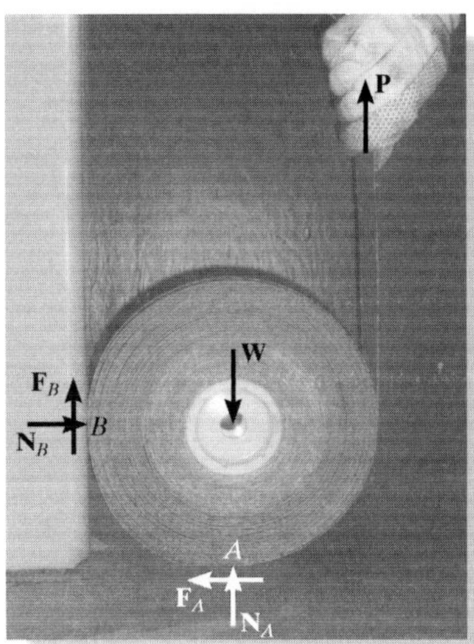

The applied vertical force **P** on this roll must be large enough to overcome the resistance of friction at the contacting surfaces *A* and *B* in order to cause rotation.

Equilibrium Versus Frictional Equations. Whenever we solve a problem such as the one in Fig. 8–4, where the friction force F is to be an "equilibrium force" and satisfies the inequality $F < \mu_s N$, then we can assume the sense of direction of F on the free-body diagram. The correct sense is made known *after* solving the equations of equilibrium for F. If F is a negative scalar the sense of **F** is the reverse of that which was assumed. This convenience of *assuming* the sense of **F** is possible because the equilibrium equations equate to zero the *components of vectors* acting in the *same direction*. However, in cases where the frictional equation $F = \mu N$ is used in the solution of a problem, as in the case shown in Fig. 8–5, then the convenience of *assuming* the sense of **F** is *lost*, since the frictional equation relates only the *magnitudes* of two *perpendicular* vectors. Consequently, **F** *must always* be shown acting with its *correct sense* on the free-body diagram, *whenever* the frictional equation is used for the solution of a problem.

Procedure for Analysis

Equilibrium problems involving dry friction can be solved using the following procedure.

Free-Body Diagrams.

- Draw the necessary free-body diagrams, and unless it is stated in the problem that impending motion or slipping occurs, *always* show the frictional forces as unknowns (i.e., *do not assume* $F = \mu N$).
- Determine the number of unknowns and compare this with the number of available equilibrium equations.
- If there are more unknowns than equations of equilibrium, it will be necessary to apply the frictional equation at some, if not all, points of contact to obtain the extra equations needed for a complete solution.
- If the equation $F = \mu N$ is to be used, it will be necessary to show **F** acting in the correct sense of direction on the free-body diagram.

Equations of Equilibrium and Friction.

- Apply the equations of equilibrium and the necessary frictional equations (or conditional equations if tipping is possible) and solve for the unknowns.
- If the problem involves a three-dimensional force system such that it becomes difficult to obtain the force components or the necessary moment arms, apply the equations of equilibrium using Cartesian vectors.

8

EXAMPLE 8.1

The uniform crate shown in Fig. 8–7*a* has a mass of 20 kg. If a force $P = 80$ N is applied to the crate, determine if it remains in equilibrium. The coefficient of static friction is $\mu_s = 0.3$.

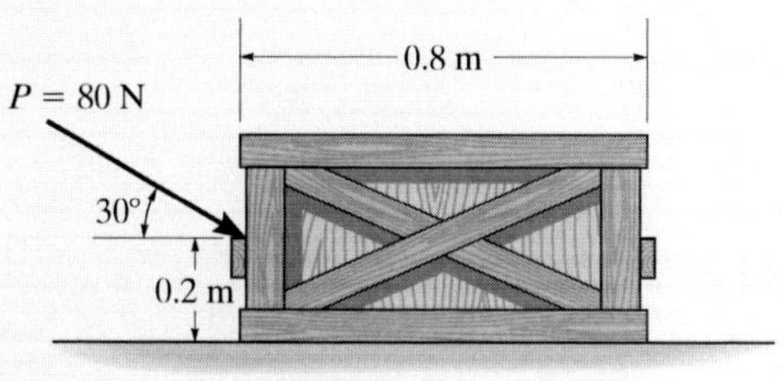

(a)

Fig. 8–7

SOLUTION

Free-Body Diagram. As shown in Fig. 8–7*b*, the *resultant* normal force $\mathbf{N}_C$ must act a distance x from the crate's center line in order to counteract the tipping effect caused by $\mathbf{P}$. There are *three unknowns*, F, N_C, and x, which can be determined strictly from the *three* equations of equilibrium.

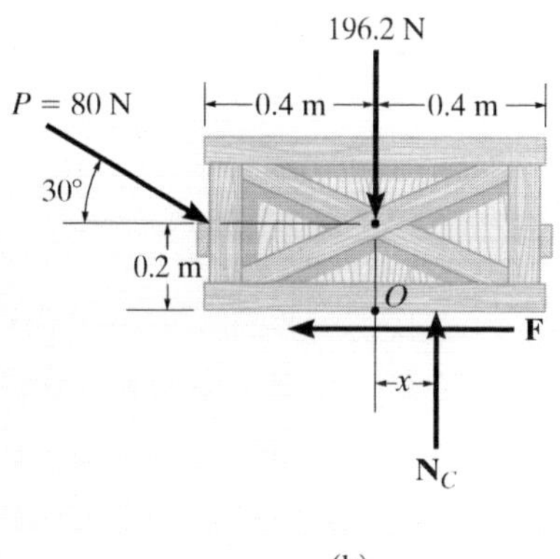

(b)

Equations of Equilibrium.

$$\overset{+}{\rightarrow}\Sigma F_x = 0; \qquad 80 \cos 30^\circ \text{ N} - F = 0$$

$$+\uparrow\Sigma F_y = 0; \qquad -80 \sin 30^\circ \text{ N} + N_C - 196.2 \text{ N} = 0$$

$$\circlearrowleft +\Sigma M_O = 0; \quad 80 \sin 30^\circ \text{ N}(0.4 \text{ m}) - 80 \cos 30^\circ \text{ N}(0.2 \text{ m}) + N_C(x) = 0$$

Solving,

$$F = 69.3 \text{ N}$$

$$N_C = 236.2 \text{ N}$$

$$x = -0.00908 \text{ m} = -9.08 \text{ mm}$$

Since x is negative it indicates the *resultant* normal force acts (slightly) to the *left* of the crate's center line. No tipping will occur since $x < 0.4$ m. Also, the *maximum* frictional force which can be developed at the surface of contact is $F_{max} = \mu_s N_C = 0.3(236.2 \text{ N}) = 70.9$ N. Since $F = 69.3 \text{ N} < 70.9$ N, the crate will *not slip*, although it is very close to doing so.

8

EXAMPLE 8.2

(a)

It is observed that when the bed of the dump truck is raised to an angle of $\theta = 25°$ the vending machines will begin to slide off the bed, Fig. 8–8*a*. Determine the static coefficient of friction between a vending machine and the surface of the truckbed.

SOLUTION

An idealized model of a vending machine resting on the truckbed is shown in Fig. 8–8*b*. The dimensions have been measured and the center of gravity has been located. We will assume that the vending machine weighs W.

(b)

Free-Body Diagram. As shown in Fig. 8–8*c*, the dimension x is used to locate the position of the resultant normal force **N**. There are four unknowns, N, F, μ_s, and x.

Equations of Equilibrium.

$$+\searrow \Sigma F_x = 0; \qquad W \sin 25° - F = 0 \qquad (1)$$

$$+\nearrow \Sigma F_y = 0; \qquad N - W \cos 25° = 0 \qquad (2)$$

$$\circlearrowleft + \Sigma M_O = 0; \quad -W \sin 25°(0.75 \text{ m}) + W \cos 25°(x) = 0 \qquad (3)$$

Since slipping impends at $\theta = 25°$, using Eqs. 1 and 2, we have

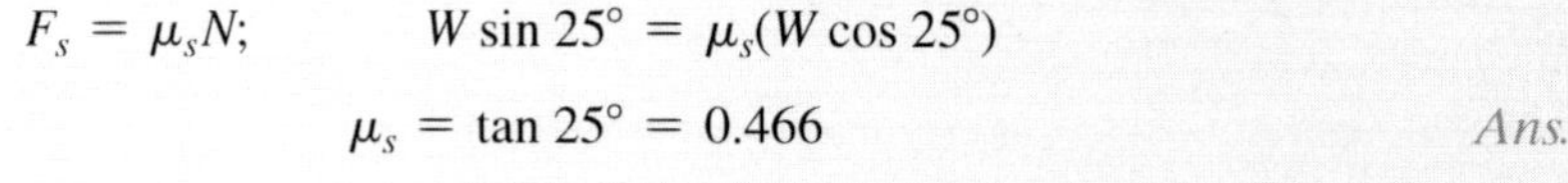

$$F_s = \mu_s N; \qquad W \sin 25° = \mu_s(W \cos 25°)$$

$$\mu_s = \tan 25° = 0.466 \qquad \textit{Ans.}$$

The angle of $\theta = 25°$ is referred to as the *angle of repose*, and by comparison, it is equal to the angle of static friction, $\theta = \phi_s$. Notice from the calculation that θ is independent of the weight of the vending machine, and so knowing θ provides a convenient method for determining the coefficient of static friction.

NOTE: From Eq. 3, we find $x = 0.350$ m. Since $0.350 \text{ m} < 0.45 \text{ m}$, indeed the vending machine will slip before it can tip as observed in Fig. 8–8*a*.

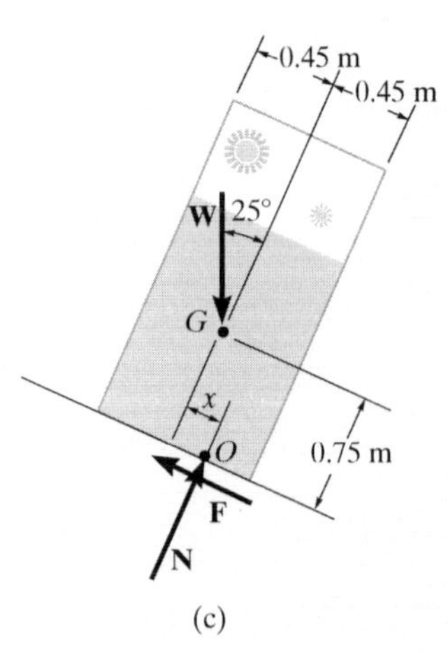

(c)

Fig. 8–8

8

EXAMPLE 8.3

The uniform 10-kg ladder in Fig. 8–9*a* rests against the smooth wall at *B*, and the end *A* rests on the rough horizontal plane for which the coefficient of static friction is $\mu_s = 0.3$. Determine the angle of inclination θ of the ladder and the normal reaction at *B* if the ladder is on the verge of slipping.

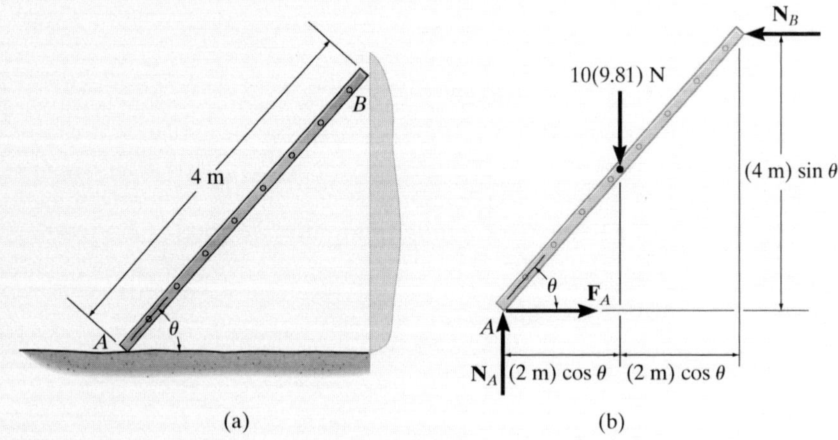

Fig. 8–9

SOLUTION

Free-Body Diagram. As shown on the free-body diagram, Fig. 8–9*b*, the frictional force $\mathbf{F}_A$ must act to the right since impending motion at *A* is to the left.

Equations of Equilibrium and Friction. Since the ladder is on the verge of slipping, then $F_A = \mu_s N_A = 0.3N_A$. By inspection, N_A can be obtained directly.

$+\uparrow \Sigma F_y = 0; \qquad N_A - 10(9.81)\text{ N} = 0 \qquad N_A = 98.1\text{ N}$

Using this result, $F_A = 0.3(98.1\text{ N}) = 29.43\text{ N}$. Now N_B can be found.

$\overset{+}{\rightarrow} \Sigma F_x = 0; \qquad 29.43\text{ N} - N_B = 0$

$$N_B = 29.43\text{ N} = 29.4\text{ N} \qquad \textit{Ans.}$$

Finally, the angle θ can be determined by summing moments about point *A*.

$\circlearrowleft + \Sigma M_A = 0; \qquad (29.43\text{ N})(4\text{ m})\sin\theta - [10(9.81)\text{ N}](2\text{ m})\cos\theta = 0$

$$\frac{\sin\theta}{\cos\theta} = \tan\theta = 1.6667$$

$$\theta = 59.04^\circ = 59.0^\circ \qquad \textit{Ans.}$$

8

EXAMPLE 8.4

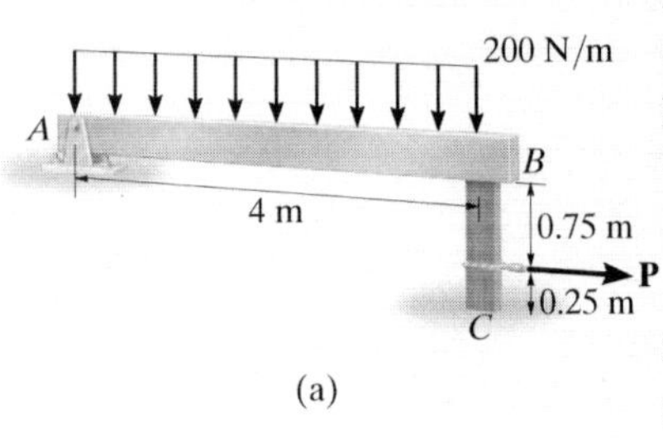

(a)

Beam AB is subjected to a uniform load of 200 N/m and is supported at B by post BC, Fig. 8–10*a*. If the coefficients of static friction at B and C are $\mu_B = 0.2$ and $\mu_C = 0.5$, determine the force **P** needed to pull the post out from under the beam. Neglect the weight of the members and the thickness of the beam.

SOLUTION

Free-Body Diagrams. The free-body diagram of the beam is shown in Fig. 8–10*b*. Applying $\Sigma M_A = 0$, we obtain $N_B = 400$ N. This result is shown on the free-body diagram of the post, Fig. 8–10*c*. Referring to this member, the *four* unknowns F_B, P, F_C, and N_C are determined from the *three* equations of equilibrium and *one* frictional equation applied either at B or C.

Equations of Equilibrium and Friction.

$$\overset{+}{\rightarrow}\Sigma F_x = 0; \qquad P - F_B - F_C = 0 \qquad (1)$$

$$+\uparrow\Sigma F_y = 0; \qquad N_C - 400\text{ N} = 0 \qquad (2)$$

$$\circlearrowleft +\Sigma M_C = 0; \qquad -P(0.25\text{ m}) + F_B(1\text{ m}) = 0 \qquad (3)$$

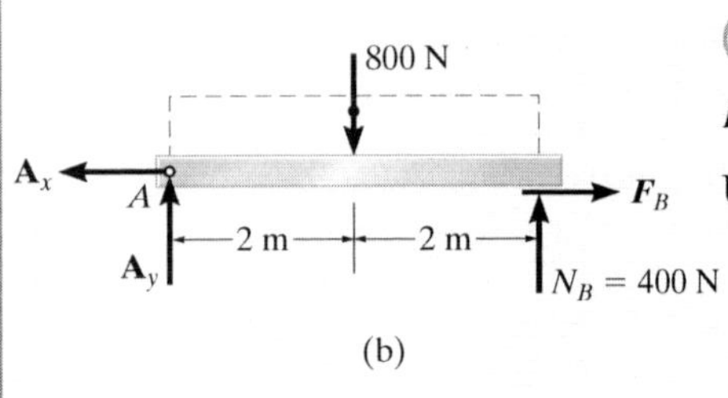

(b)

(Post Slips at *B* and Rotates about *C*.) This requires $F_C \leq \mu_C N_C$ and

$$F_B = \mu_B N_B; \qquad F_B = 0.2(400\text{ N}) = 80\text{ N}$$

Using this result and solving Eqs. 1 through 3, we obtain

$$P = 320\text{ N}$$
$$F_C = 240\text{ N}$$
$$N_C = 400\text{ N}$$

Since $F_C = 240\text{ N} > \mu_C N_C = 0.5(400\text{ N}) = 200\text{ N}$, slipping at C occurs. Thus the other case of movement must be investigated.

(Post Slips at *C* and Rotates about *B*.) Here $F_B \leq \mu_B N_B$ and

$$F_C = \mu_C N_C; \qquad F_C = 0.5N_C \qquad (4)$$

Solving Eqs. 1 through 4 yields

$$P = 267\text{ N} \qquad \textit{Ans.}$$
$$N_C = 400\text{ N}$$
$$F_C = 200\text{ N}$$
$$F_B = 66.7\text{ N}$$

(c)

Fig. 8–10

Obviously, this case occurs first since it requires a *smaller* value for P.

EXAMPLE 8.5

Blocks A and B have a mass of 3 kg and 9 kg, respectively, and are connected to the weightless links shown in Fig. 8–11a. Determine the largest vertical force **P** that can be applied at the pin C without causing any movement. The coefficient of static friction between the blocks and the contacting surfaces is $\mu_s = 0.3$.

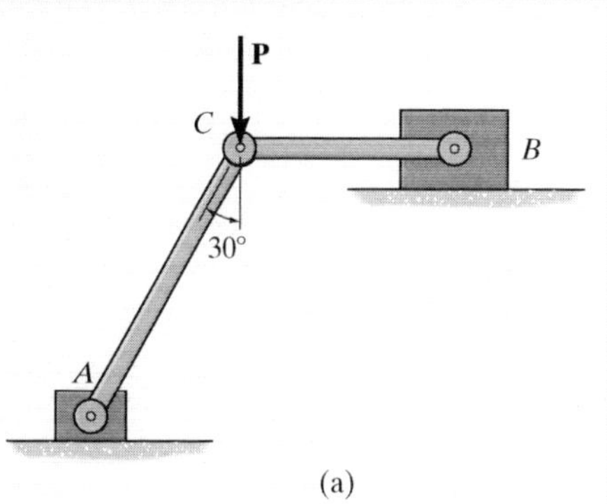

(a)

SOLUTION

Free-Body Diagram. The links are two-force members and so the free-body diagrams of pin C and blocks A and B are shown in Fig. 8–11b. Since the horizontal component of $\mathbf{F}_{AC}$ tends to move block A to the left, $\mathbf{F}_A$ must act to the right. Similarly, $\mathbf{F}_B$ must act to the left to oppose the tendency of motion of block B to the right, caused by $\mathbf{F}_{BC}$. There are seven unknowns and six available force equilibrium equations, two for the pin and two for each block, so that *only one* frictional equation is needed.

Equations of Equilibrium and Friction. The force in links AC and BC can be related to P by considering the equilibrium of pin C.

$$+\uparrow \Sigma F_y = 0; \qquad F_{AC}\cos 30^\circ - P = 0; \qquad F_{AC} = 1.155P$$

$$\xrightarrow{+} \Sigma F_x = 0; \qquad 1.155P\sin 30^\circ - F_{BC} = 0; \qquad F_{BC} = 0.5774P$$

Using the result for F_{AC}, for block A,

$$\xrightarrow{+} \Sigma F_x = 0; \qquad F_A - 1.155P\sin 30^\circ = 0; \qquad F_A = 0.5774P \qquad (1)$$

$$+\uparrow \Sigma F_y = 0; \qquad N_A - 1.155P\cos 30^\circ - 3(9.81\text{ N}) = 0;$$

$$N_A = P + 29.43\text{ N} \qquad (2)$$

Using the result for F_{BC}, for block B,

$$\xrightarrow{+} \Sigma F_x = 0; \qquad (0.5774P) - F_B = 0; \qquad F_B = 0.5774P \qquad (3)$$

$$+\uparrow \Sigma F_y = 0; \qquad N_B - 9(9.81)\text{ N} = 0; \qquad N_B = 88.29\text{ N}$$

Movement of the system may be caused by the initial slipping of *either* block A or block B. If we assume that block A slips first, then

$$F_A = \mu_s N_A = 0.3N_A \qquad (4)$$

Substituting Eqs. 1 and 2 into Eq. 4,

$$0.5774P = 0.3(P + 29.43)$$

$$P = 31.8\text{ N} \qquad \textit{Ans.}$$

Substituting this result into Eq. 3, we obtain $F_B = 18.4$ N. Since the maximum static frictional force at B is $(F_B)_{max} = \mu_s N_B = 0.3(88.29\text{ N}) = 26.5\text{ N} > F_B$, block B will not slip. Thus, the above assumption is correct. Notice that if the inequality were not satisfied, we would have to assume slipping of block B and then solve for P.

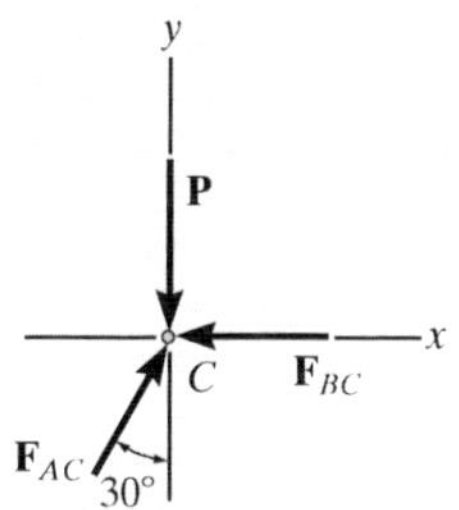

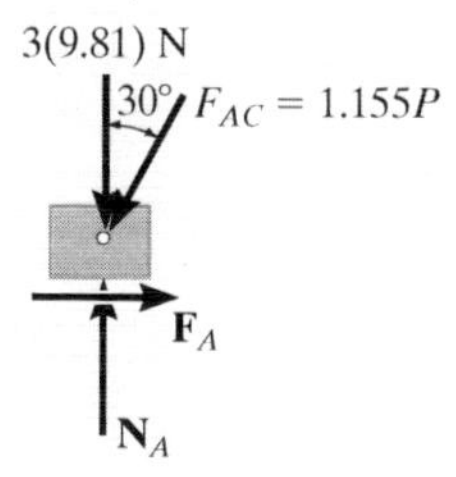

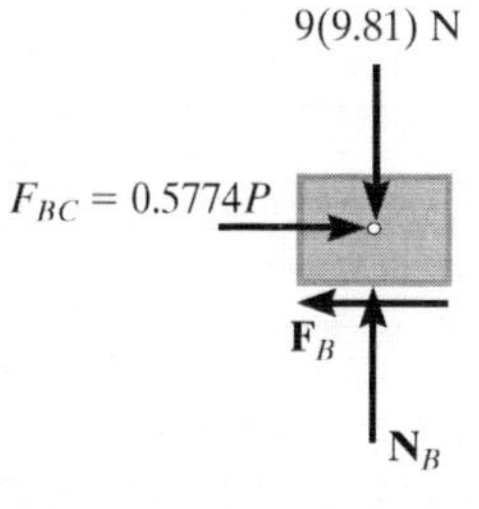

(b)

Fig. 8–11

FUNDAMENTAL PROBLEMS

All problem solutions must include FBDs.

F8–1. If $P = 200$ N, determine the friction developed between the 50-kg crate and the ground. The coefficient of static friction between the crate and the ground is $\mu_s = 0.3$.

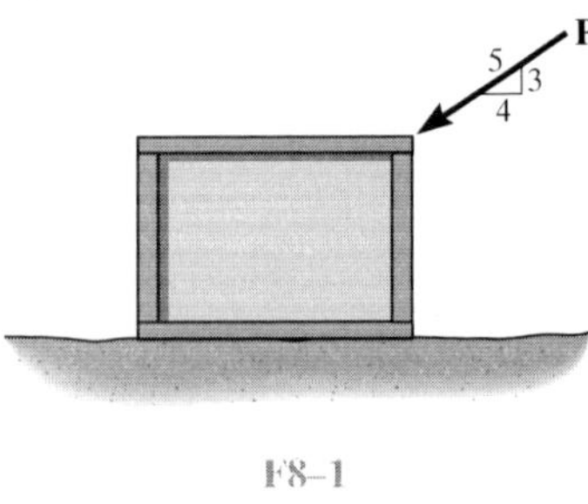

F8–1

F8–2. Determine the minimum force P to prevent the 30-kg rod AB from sliding. The contact surface at B is smooth, whereas the coefficient of static friction between the rod and the wall at A is $\mu_s = 0.2$.

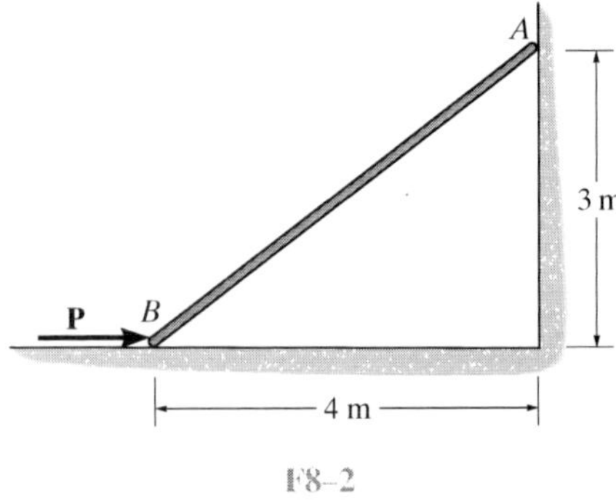

F8–2

F8–3. Determine the maximum force P that can be applied without causing the two 50-kg crates to move. The coefficient of static friction between each crate and the ground is $\mu_s = 0.25$.

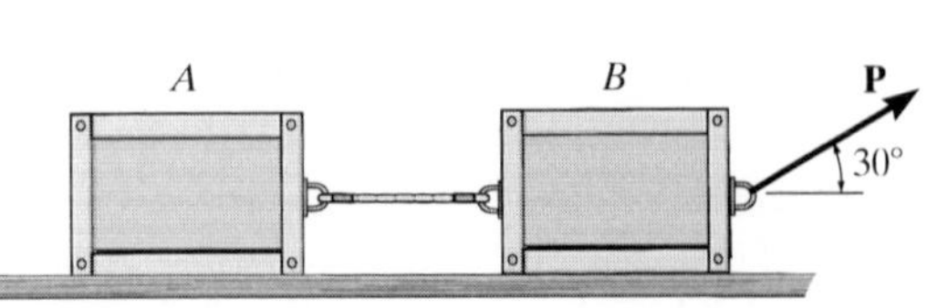

F8–3

F8–4. If the coefficient of static friction at contact points A and B is $\mu_s = 0.3$, determine the maximum force P that can be applied without causing the 100-kg spool to move.

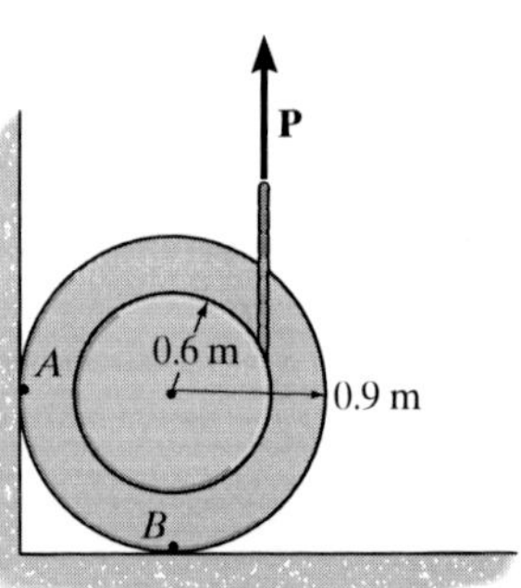

F8–4

F8–5. Determine the maximum force P that can be applied without causing movement of the 100-kg crate that has a center of gravity at G. The coefficient of static friction at the floor is $\mu_s = 0.4$.

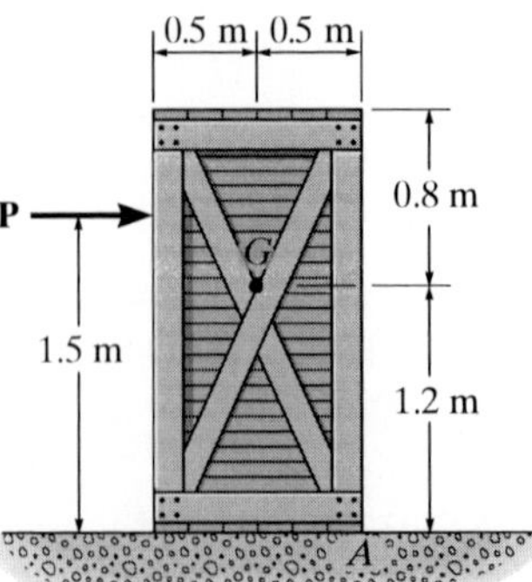

F8–5

8

F8–6. Determine the minimum coefficient of static friction between the uniform 50-kg spool and the wall so that the spool does not slip.

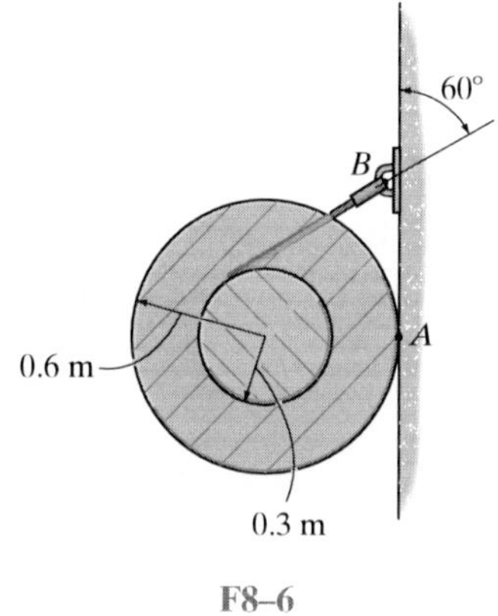

F8–6

F8–7. Blocks A, B, and C have weights of 50 N, 25 N, and 15 N, respectively. Determine the smallest horizontal force P that will cause impending motion. The coefficient of static friction between A and B is $\mu_s = 0.3$, between B and C, $\mu_s' = 0.4$, and between block C and the ground, $\mu_s'' = 0.35$.

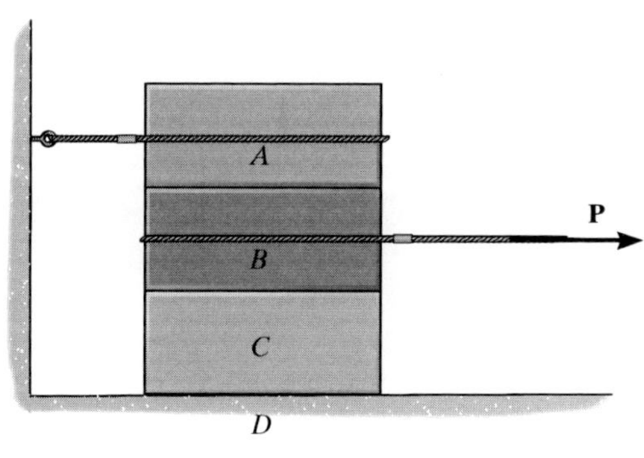

F8–7

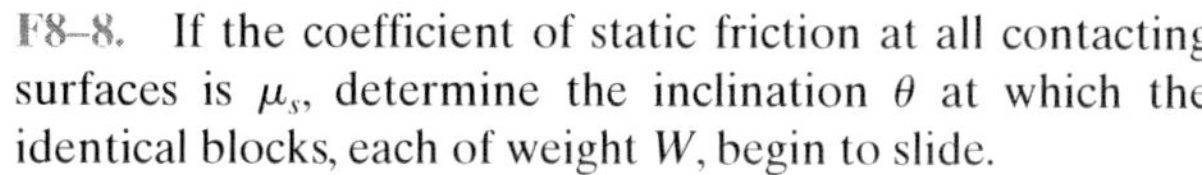

F8–8. If the coefficient of static friction at all contacting surfaces is μ_s, determine the inclination θ at which the identical blocks, each of weight W, begin to slide.

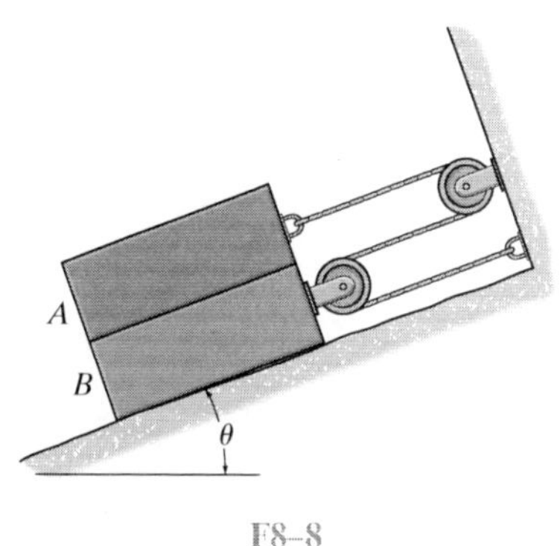

F8–8

F8–9. Blocks A and B have a mass of 7 kg and 10 kg, respectively. Using the coefficients of static friction indicated, determine the largest force P which can be applied to the cord without causing motion. There are pulleys at C and D.

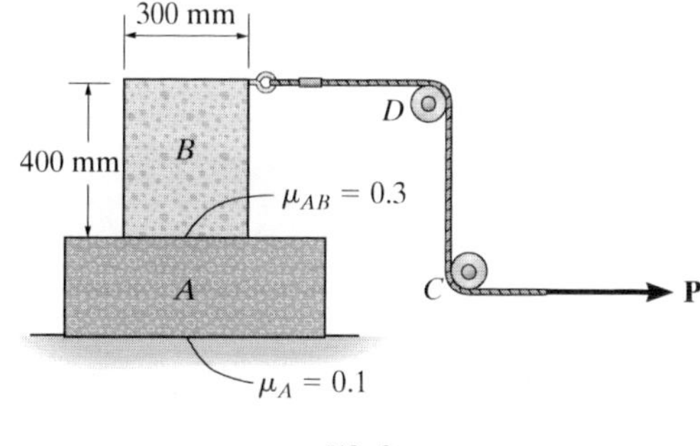

F8–9

PROBLEMS

All problem solutions must include FBDs.

8–1. The mine car and its contents have a total mass of 6 Mg and a center of gravity at G. If the coefficient of static friction between the wheels and the tracks is $\mu_s = 0.4$ when the wheels are locked, find the normal force acting on the front wheels at B and the rear wheels at A when the brakes at both A and B are locked. Does the car move?

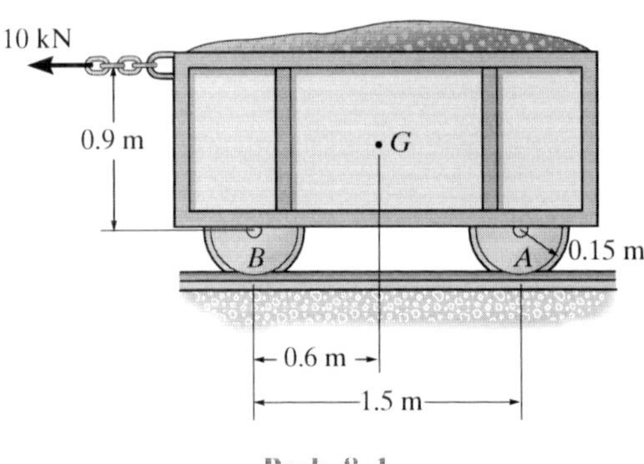

Prob. 8–1

8–2. Determine the maximum force P the connection can support so that no slipping occurs between the plates. There are four bolts used for the connection and each is tightened so that it is subjected to a tension of 4 kN. The coefficient of static friction between the plates is $\mu_s = 0.4$.

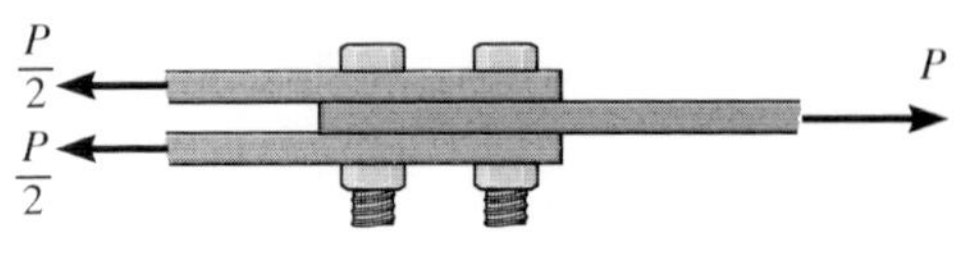

Prob. 8–2

8–3. If the coefficient of static friction at A is $\mu_s = 0.4$ and the collar at B is smooth so it only exerts a horizontal force on the pipe, determine the minimum distance x so that the bracket can support the cylinder of any mass without slipping. Neglect the mass of the bracket.

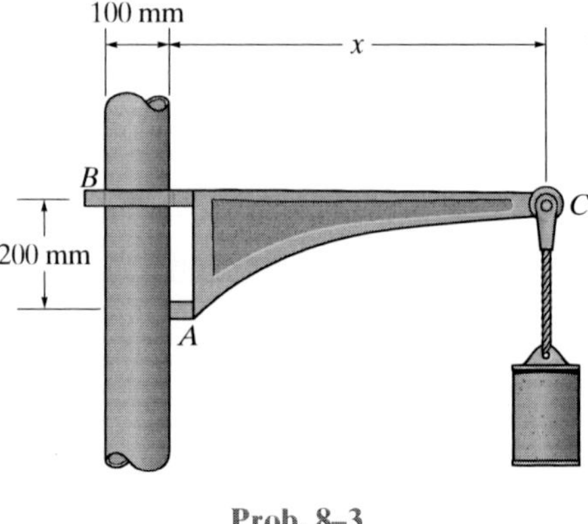

Prob. 8–3

***8–4.** If the coefficient of static friction at contacting surface between blocks A and B is μ_s, and that between block B and bottom is $2\mu_s$, determine the inclination θ at which the identical blocks, each of weight W, begin to slide.

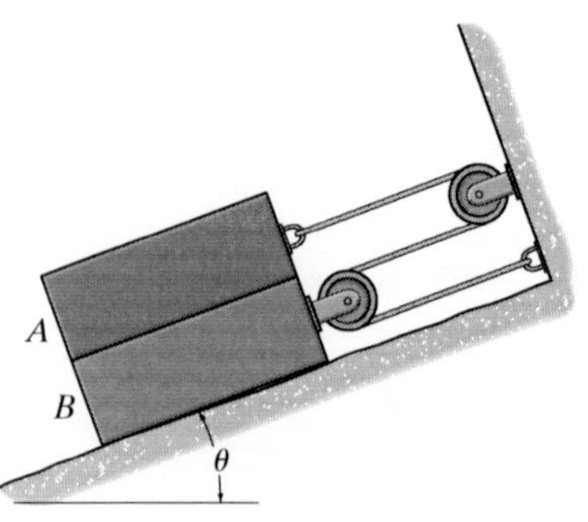

Prob. 8–4

8

8–5. The coefficients of static and kinetic friction between the drum and brake bar are $\mu_s = 0.4$ and $\mu_k = 0.3$, respectively. If $M = 50 \text{ N} \cdot \text{m}$ and $P = 85 \text{ N}$ determine the horizontal and vertical components of reaction at the pin O. Neglect the weight and thickness of the brake. The drum has a mass of 25 kg.

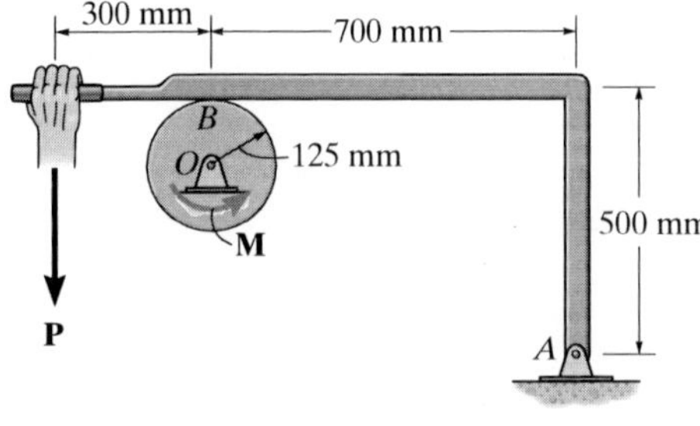

Prob. 8–5

8–6. The coefficient of static friction between the drum and brake bar is $\mu_s = 0.4$. If the moment $M = 35 \text{ N} \cdot \text{m}$, determine the smallest force P that needs to be applied to the brake bar in order to prevent the drum from rotating. Also determine the corresponding horizontal and vertical components of reaction at pin O. Neglect the weight and thickness of the brake bar. The drum has a mass of 25 kg.

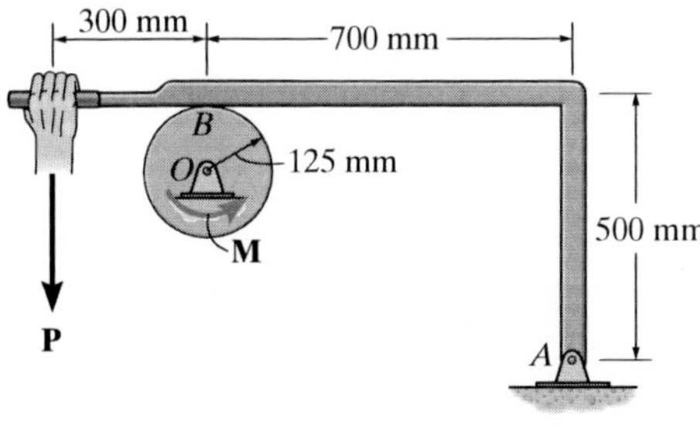

Prob. 8–6

8–7. The block brake consists of a pin-connected lever and friction block at B. The coefficient of static friction between the wheel and the lever is $\mu_s = 0.3$, and a torque of $5 \text{ N} \cdot \text{m}$ is applied to the wheel. Determine if the brake can hold the wheel stationary when the force applied to the lever is (a) $P = 30$ N, (b) $P = 70$ N.

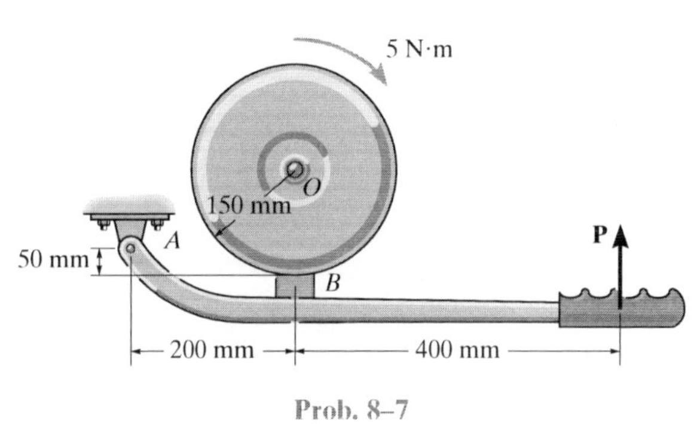

Prob. 8–7

***8–8.** The block brake consists of a pin-connected lever and friction block at B. The coefficient of static friction between the wheel and the lever is $\mu_s = 0.3$, and a torque of $5 \text{ N} \cdot \text{m}$ is applied to the wheel. Determine if the brake can hold the wheel stationary when the force applied to the lever is (a) $P = 30$ N, (b) $P = 70$ N.

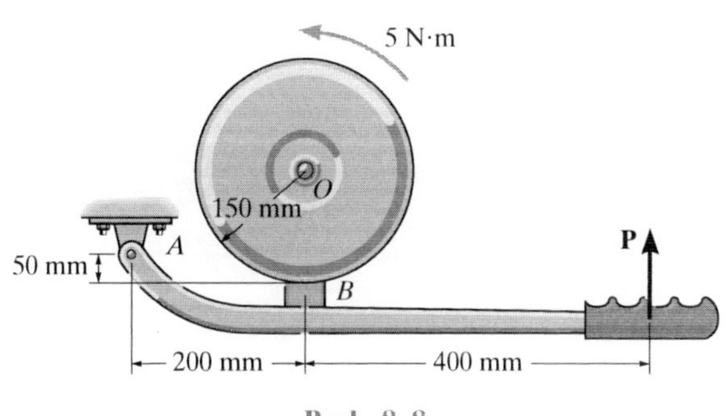

Prob. 8–8

8

8–9. The uniform hoop of weight W is suspended from the peg at A and a horizontal force **P** is slowly applied at B. If the hoop begins to slip at A when $\theta = 30^\circ$, determine the coefficient of static friction between the hoop and the peg.

8–10. The uniform hoop of weight W is suspended from the peg at A and a horizontal force **P** is slowly applied at B. If the coefficient of static friction between the hoop and peg is $\mu_s = 0.2$, determine if it is possible for the angle $\theta = 30^\circ$ before the hoop begins to slip.

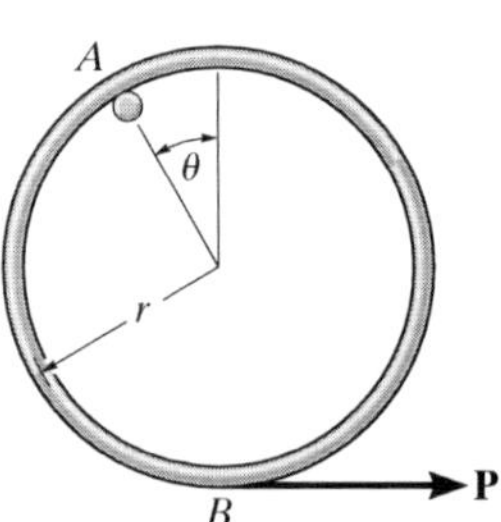

Probs. 8–9/10

***8–12.** If a torque of $M = 300\ \text{N}\cdot\text{m}$ is applied to the flywheel, determine the force that must be developed in the hydraulic cylinder CD to prevent the flywheel from rotating. The coefficient of static friction between the friction pad at B and the flywheel is $\mu_s = 0.4$.

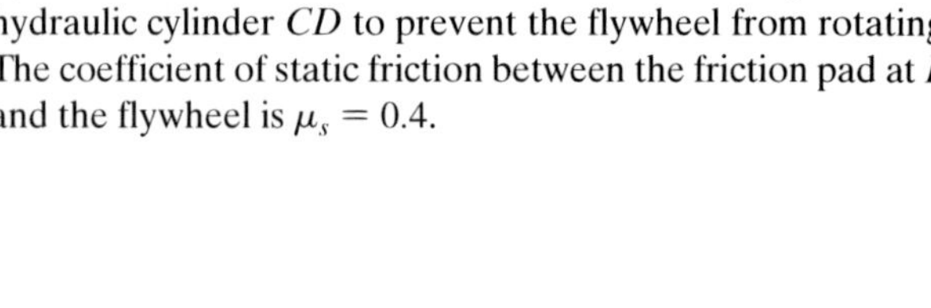

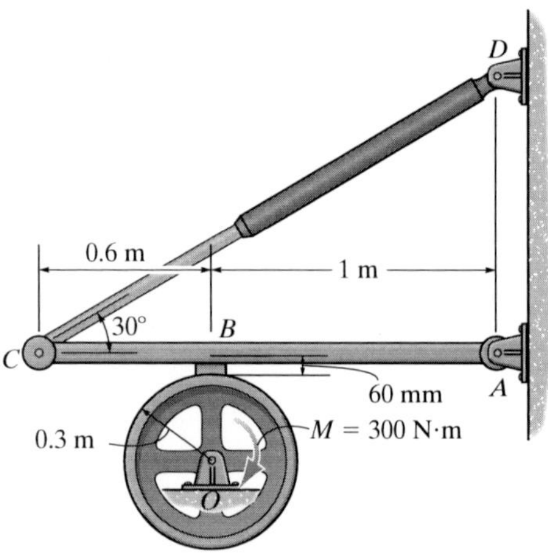

Prob. 8–12

8–11. The fork lift has a weight of 12 kN and a center of gravity at G. If the rear wheels are powered, whereas the front wheels are free to roll, determine the maximum number of 150-kg crates the fork lift can push forward. The coefficient of static friction between the wheels and the ground is $\mu_s = 0.4$, and between each crate and the ground $\mu_s' = 0.35$.

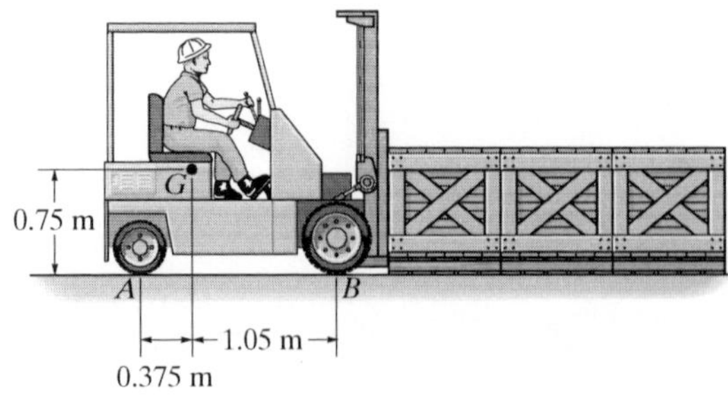

Prob. 8–11

8–13. The cam is subjected to a couple moment of $5\ \text{N}\cdot\text{m}$ Determine the minimum force P that should be applied to the follower in order to hold the cam in the position shown. the coefficient of static friction between the cam and the follower is $\mu = 0.4$. The guide at A is smooth.

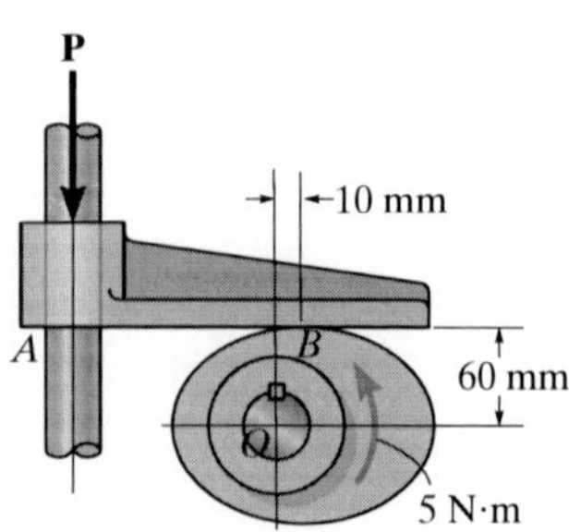

Prob. 8–13

8–14. A 35-kg disk rests on an inclined surface for which $\mu_s = 0.2$. Determine the maximum vertical force **P** that may be applied to link AB without causing the disk to slip at C.

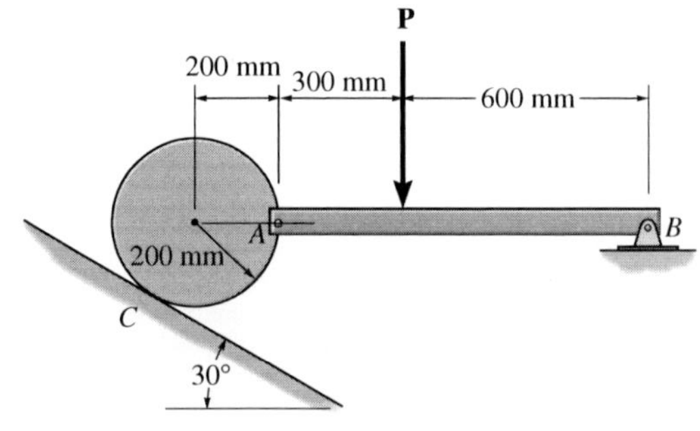

Prob. 8–14

8–15. The car has a mass of 1.6 Mg and center of mass at G. If the coefficient of static friction between the shoulder of the road and the tires is $\mu_s = 0.4$, determine the greatest slope θ the shoulder can have without causing the car to slip or tip over if the car travels along the shoulder at constant velocity.

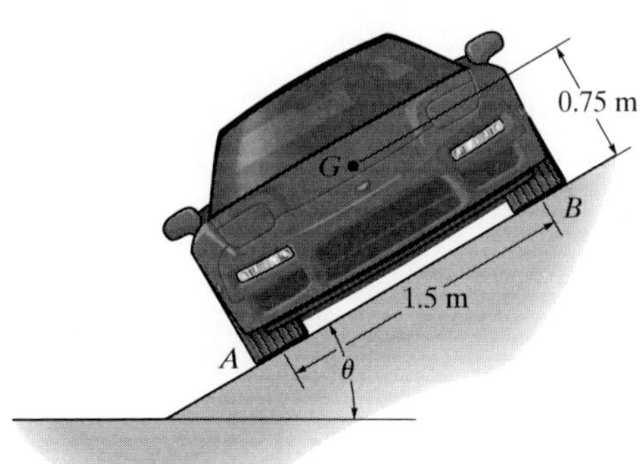

Prob. 8–15

***8–16.** If the coefficient of static friction between the collars A and B and the rod is $\mu_s = 0.6$, determine the maximum angle θ for the system to remain in equilibrium, regardless of the weight of cylinder D. Links AC and BC have negligible weight and are connected together at C by a pin.

8–17. If $\theta = 15°$, determine the minimum coefficient of static friction between the collars A and B and the rod required for the system to remain in equilibrium, regardless of the weight of cylinder D. Links AC and BC have negligible weight and are connected together at C by a pin.

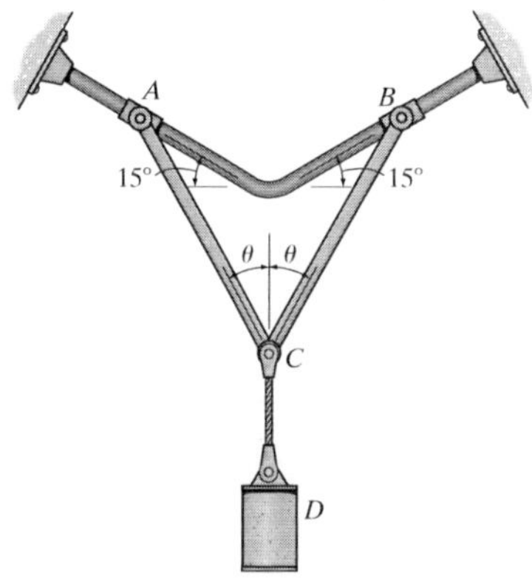

Probs. 8–16/17

8–18. The 5-kg cylinder is suspended from two equal-length cords. The end of each cord is attached to a ring of negligible mass that passes along a horizontal shaft. If the rings can be separated by the greatest distance $d = 400$ mm and still support the cylinder, determine the coefficient of static friction between each ring and the shaft.

8–19. The 5-kg cylinder is suspended from two equal-length cords. The end of each cord is attached to a ring of negligible mass that passes along a horizontal shaft. If the coefficient of static friction between each ring and the shaft is $\mu_s = 0.5$, determine the greatest distance d by which the rings can be separated and still support the cylinder.

8

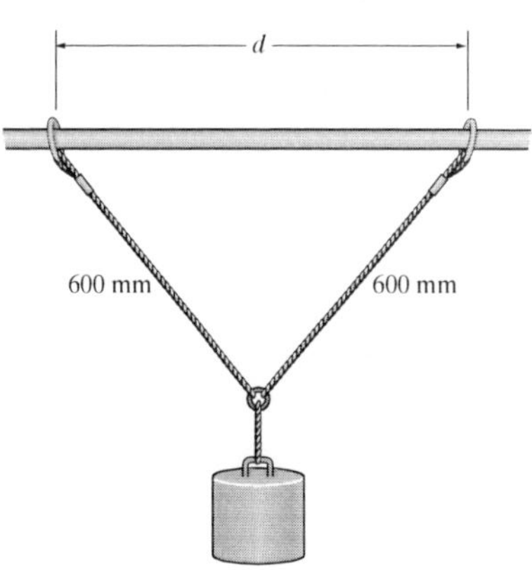

Probs. 8–18/19

***8–20.** The pipe is hoisted using the tongs. If the coefficient of static friction at A and B is μ_s, determine the smallest dimension b so that any pipe of inner diameter d can be lifted.

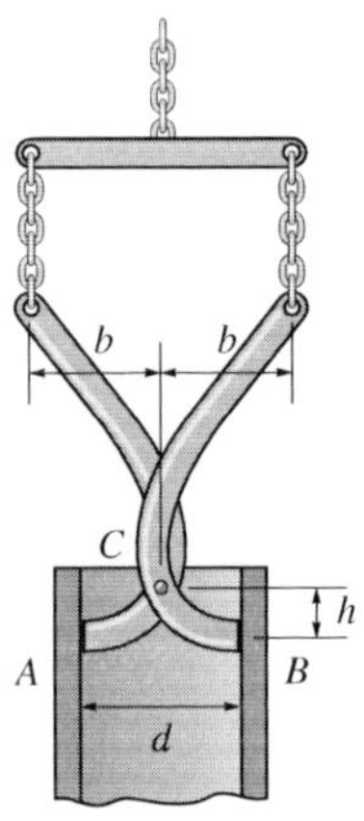

Prob. 8–20

8–21. The uniform pole has a weight W and length L. Its end B is tied to a supporting cord, and end A is placed against the wall, for which the coefficient of static friction is μ_s. Determine the largest angle θ at which the pole can be placed without slipping.

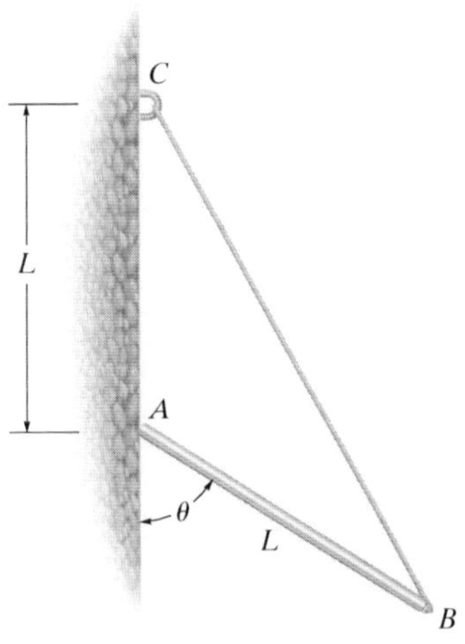

Prob. 8–21

8–22. If the clamping force is $F = 200$ N and each board has a mass of 2 kg, determine the maximum number of boards the clamp can support. The coefficient of static friction between the boards is $\mu_s = 0.3$, and the coefficient of static friction between the boards and the clamp is $\mu_s' = 0.45$.

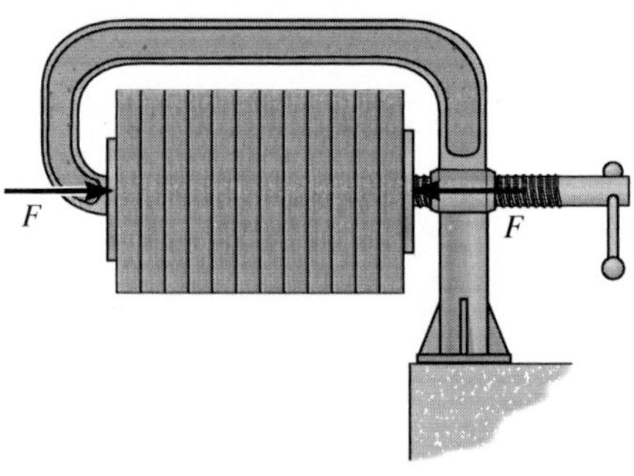

Prob. 8–22

8–23. A 35-kg disk rests on an inclined surface for which $\mu_s = 0.3$. Determine the maximum vertical force **P** that may be applied to bar AB without causing the disk to slip at C. Neglect the mass of the bar.

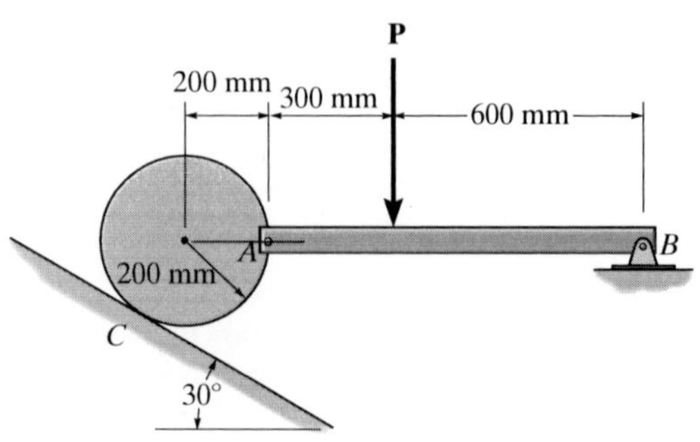

Prob. 8–23

8

***8–24.** The coefficient of static friction between the shoes at A and B of the tongs and the pallet is $\mu_s' = 0.5$, and between the pallet and the floor $\mu_s = 0.4$. If a horizontal towing force of $P = 300$ N is applied to the tongs, determine the largest mass that can be towed.

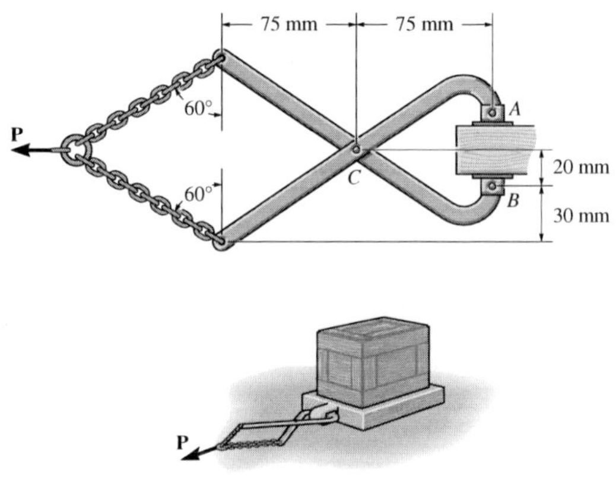

Prob. 8–24

8–25. The block brake is used to stop the wheel from rotating when the wheel is subjected to a couple moment $\mathbf{M}_0$. If the coefficient of static friction between the wheel and the block is μ_s, determine the smallest force P that should be applied.

8–26. Show that the brake in Prob. 8–25 is self locking, i.e., $P \le 0$, provided $b/c \le \mu_s$.

8–27. Solve Prob. 8–25 if the couple moment $\mathbf{M}_0$ is applied counterclockwise.

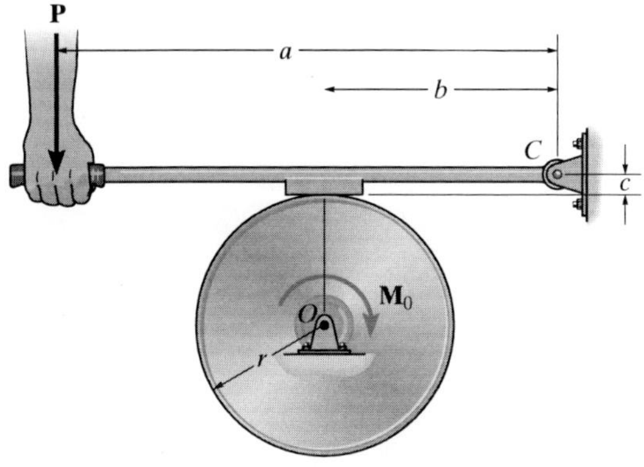

Probs. 8–25/26/27

***8–28.** The 10-kg cylinder is suspended from two equal-length cords. The end of each cord is attached to a ring of negligible mass, which passes along a horizontal shaft. If the coefficient of static friction between each ring and the shaft is $\mu_s = 0.5$, determine the greatest distance d by which the rings can be separated and still support the cylinder.

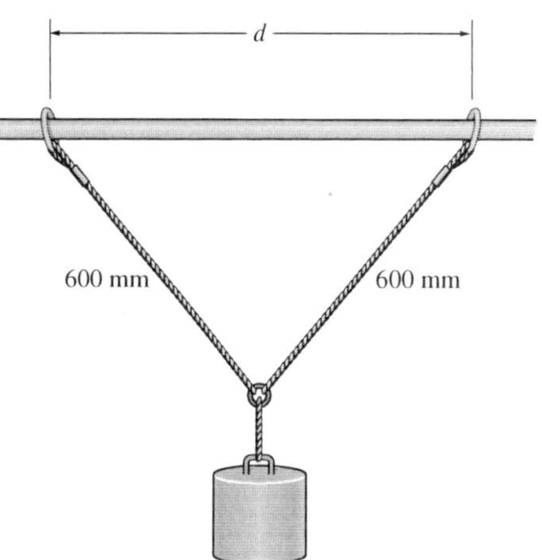

Prob. 8–28

8–29. The friction pawl is pinned at A and rests against the wheel at B. It allows freedom of movement when the wheel is rotating counterclockwise about C. Clockwise rotation is prevented due to friction of the pawl which tends to bind the wheel. If $(\mu_s)_B = 0.6$, determine the design angle θ which will prevent clockwise motion for any value of applied moment M. *Hint:* Neglect the weight of the pawl so that it becomes a two-force member.

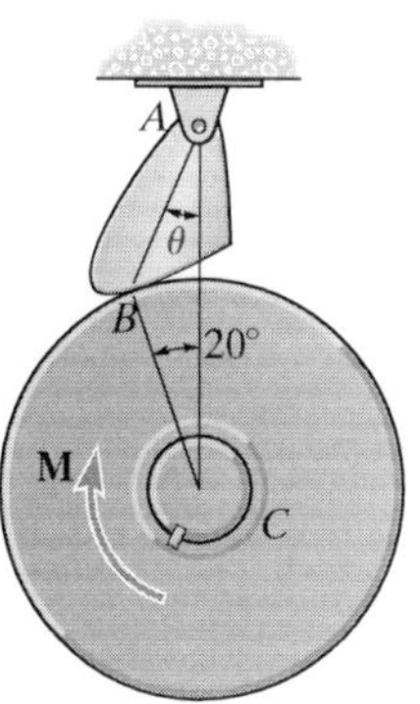

Prob. 8–29

8–30. If $\theta = 30°$, determine the minimum coefficient of static friction at A and B so that equilibrium of the supporting frame is maintained regardless of the mass of the cylinder C. Neglect the mass of the rods.

8–31. If the coefficient of static friction at A and B is $\mu_s = 0.6$, determine the maximum angle θ so that the frame remains in equilibrium, regardless of the mass of the cylinder. Neglect the mass of the rods.

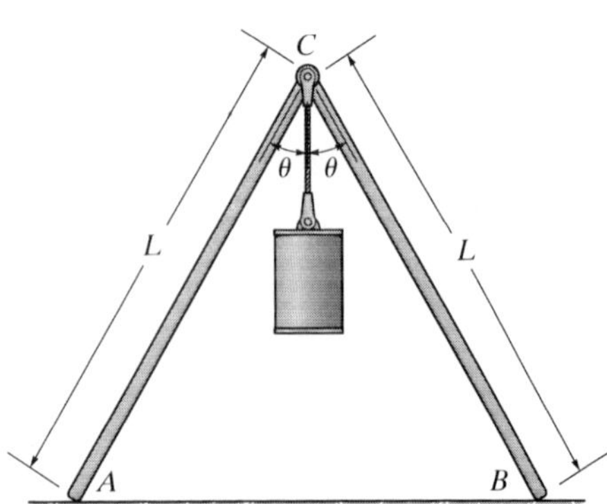

Probs. 8–30/31

8–34. The coefficient of static friction between the 150-kg crate and the ground is $\mu_s = 0.3$, while the coefficient of static friction between the 80-kg man's shoes and the ground is $\mu_s' = 0.4$. Determine if the man can move the crate.

8–35. If the coefficient of static friction between the crate and the ground in Prob. 8–34 is $\mu_s = 0.3$, determine the minimum coefficient of static friction between the man's shoes and the ground so that the man can move the crate.

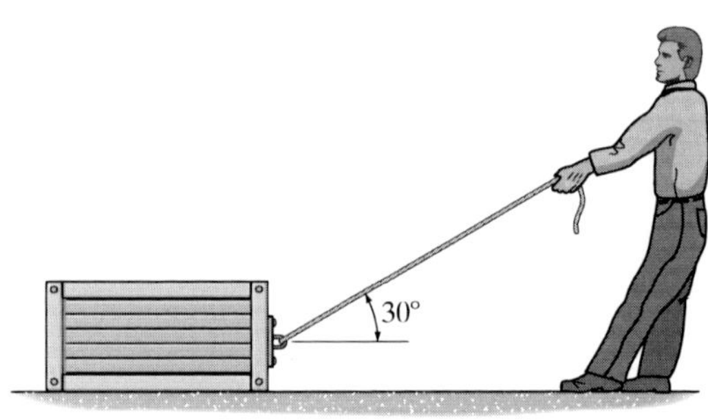

Probs. 8–34/35

*__8–32.__ The semicylinder of mass m and radius r lies on the rough inclined plane for which $\phi = 10°$ and the coefficient of static friction is $\mu_s = 0.3$. Determine if the semicylinder slides down the plane, and if not, find the angle of tip θ of its base AB.

8–33. The semicylinder of mass m and radius r lies on the rough inclined plane. If the inclination $\phi = 15°$, determine the smallest coefficient of static friction which will prevent the semicylinder from slipping.

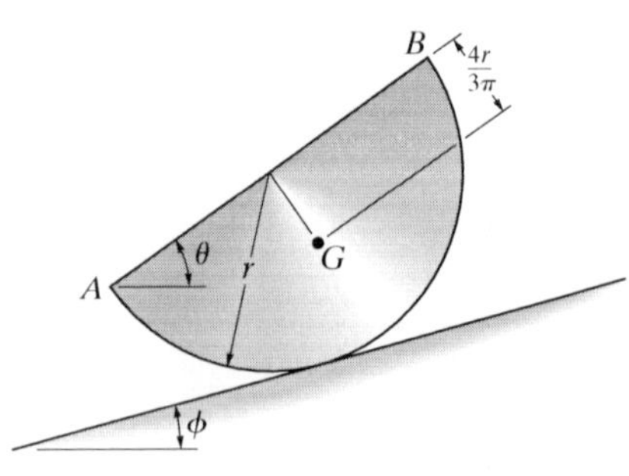

Probs. 8–32/33

*__8–36.__ The rod has a weight W and rests against the floor and wall for which the coefficients of static friction are μ_A and μ_B, respectively. Determine the smallest value of θ for which the rod will not move.

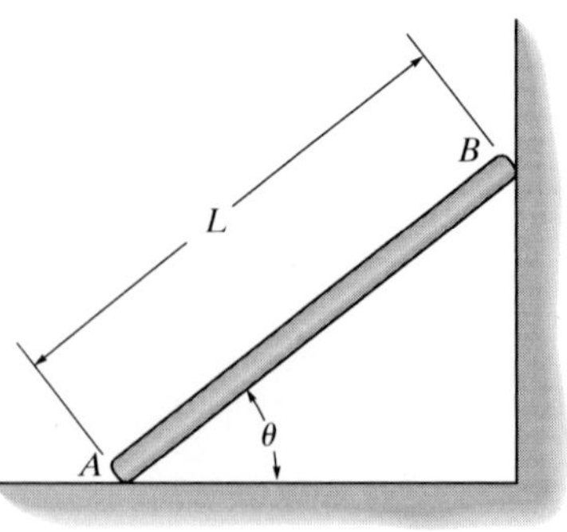

Prob. 8–36

8–37. Determine the magnitude of force **P** needed to start towing the 40-kg crate. Also determine the location of the resultant normal force acting on the crate, measured from point A. Take $\mu_s = 0.3$.

8–38. Determine the friction force on the 40-kg crate, and the resultant normal force and its position x, measured from point A, if the force $P = 300$ N. Take $\mu_s = 0.5$ and $\mu_k = 0.2$.

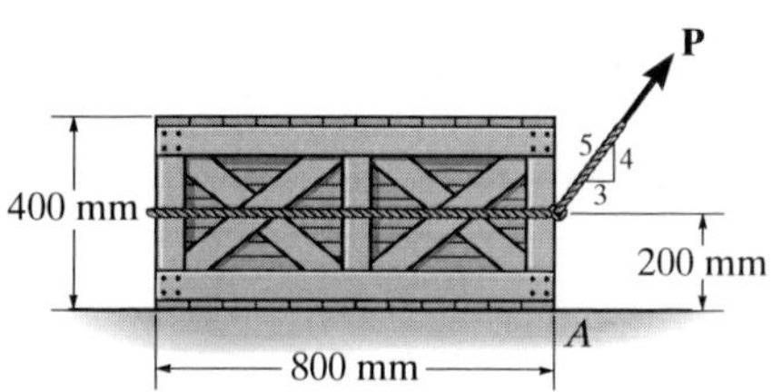

Probs. 8–37/38

***8–40.** The spool of wire having a mass of 150 kg rests on the ground at A and against the wall at B. Determine the force P required to begin pulling the wire horizontally off the spool. The coefficient of static friction between the spool and its points of contact is $\mu_s = 0.25$.

8–41. The spool of wire having a mass of 150 kg rests on the ground at A and against the wall at B. Determine the forces acting on the spool at A and B if $P = 800$ N. The coefficient of static friction between the spool and the ground at point A is $\mu_s = 0.35$. The wall at B is smooth.

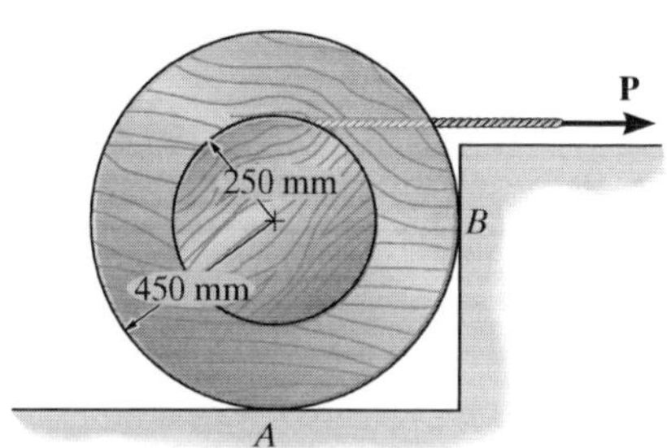

Probs. 8–40/41

8–39. Determine the smallest force the man must exert on the rope in order to move the 80-kg crate. Also, what is the angle θ at this moment? The coefficient of static friction between the crate and the floor is $\mu_s = 0.3$.

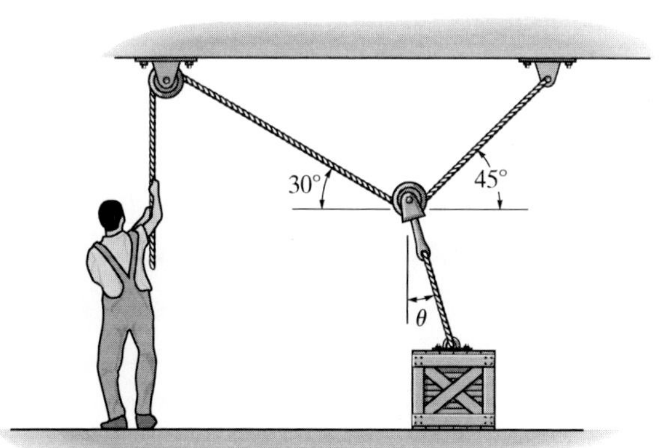

Prob. 8–39

8–42. The friction hook is made from a fixed frame and a cylinder of negligible weight. A piece of paper is placed between the wall and the cylinder. If $\theta = 20°$, determine the smallest coefficient of static friction μ at all points of contact so that any weight W of paper p can be held.

Prob. 8–42

8

8–43. The uniform rod has a mass of 10 kg and rests on the inside of the smooth ring at B and on the ground at A. If the rod is on the verge of slipping, determine the coefficient of static friction between the rod and the ground.

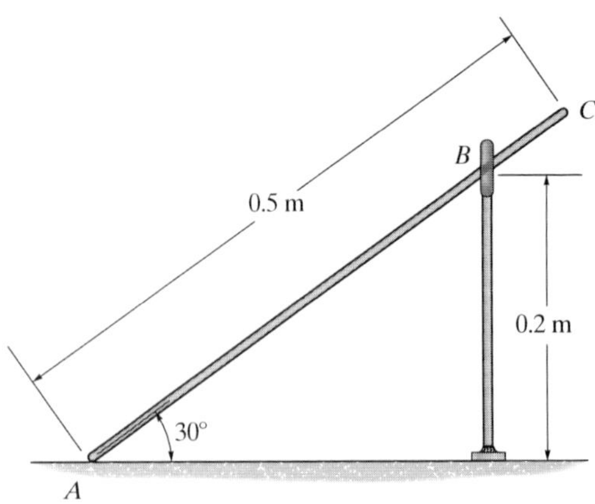

Prob. 8–43

*8–44. The rings A and C each weigh W and rest on the rod, which has a coefficient of static friction of μ_s. If the suspended ring at B has a weight of $2W$, determine the largest distance d between A and C so that no motion occurs. Neglect the weight of the wire. The wire is smooth and has a total length of l.

8

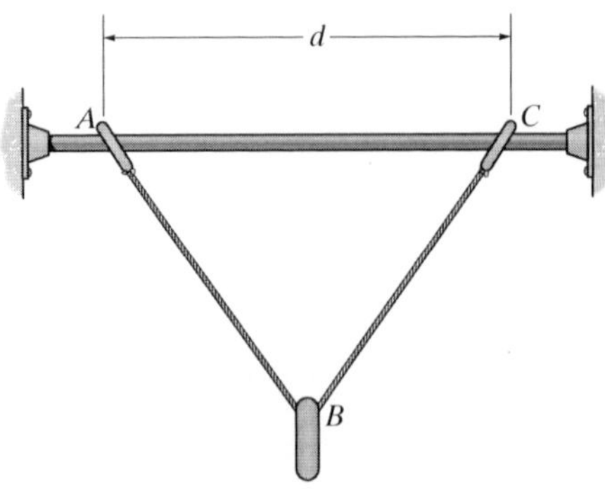

Prob. 8–44

8–45. Car A has a mass of 1.4 Mg and mass center at G. If car B exerts a horizontal force on A of 2 kN, determine if this force is great enough to move car A. The coefficients of static and kinetic friction between the tires and the road are $\mu_s = 0.5$ and $\mu_k = 0.35$. Assume B's bumper is smooth.

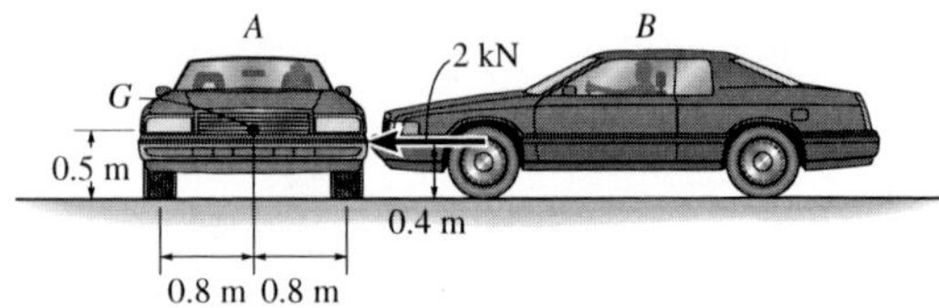

Prob. 8–45

8–46. The beam AB has a negligible mass and thickness and is subjected to a triangular distributed loading. It is supported at one end by a pin and at the other end by a post having a mass of 50 kg and negligible thickness. Determine the minimum force P needed to move the post. The coefficients of static friction at B and C are $\mu_B = 0.4$ and $\mu_C = 0.2$, respectively.

8–47. The beam AB has a negligible mass and thickness and is subjected to a triangular distributed loading. It is supported at one end by a pin and at the other end by a post having a mass of 50 kg and negligible thickness. Determine the two coefficients of static friction at B and at C so that when the magnitude of the applied force is increased to $P = 150$ N, the post slips at both B and C simultaneously.

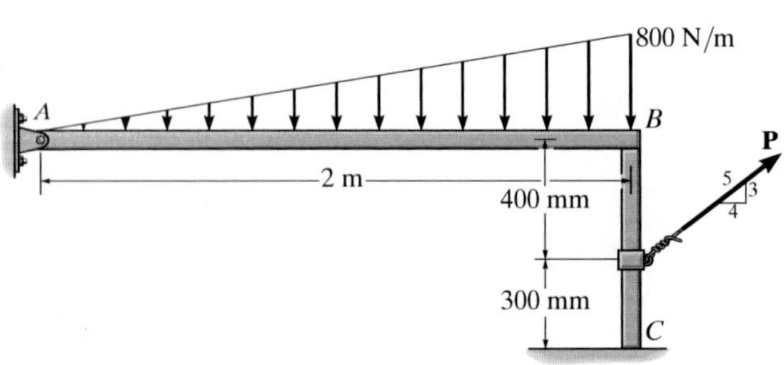

Probs. 8–46/47

***8–48.** The beam AB has a negligible mass and thickness and is subjected to a force of 200 N. It is supported at one end by a pin and at the other end by a spool having a mass of 40 kg. If a cable is wrapped around the inner core of the spool, determine the minimum cable force P needed to move the spool. The coefficients of static friction at B and D are $\mu_B = 0.4$ and $\mu_D = 0.2$, respectively.

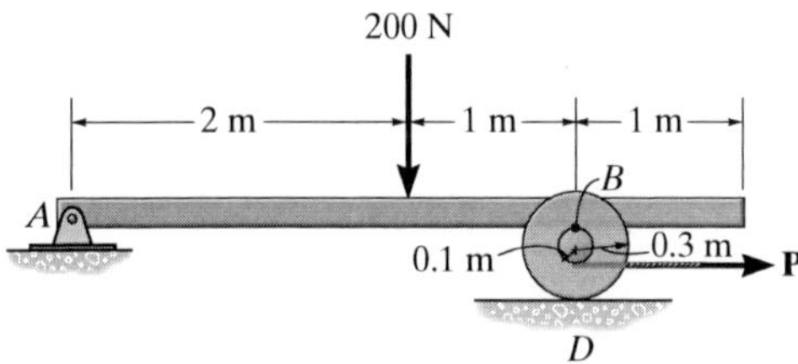

Prob. 8–48

8–49. The 45-kg disk rests on the surface for which the coefficient of static friction is $\mu_A = 0.2$. Determine the largest couple moment M that can be applied to the bar without causing motion.

8–50. The 45-kg disk rests on the surface for which the coefficient of static friction is $\mu_A = 0.15$. If $M = 50\ \text{N}\cdot\text{m}$, determine the friction force at A.

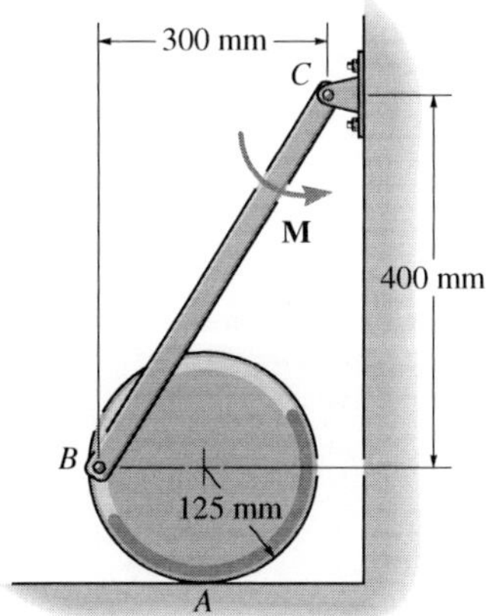

Probs. 8–49/50

8–51. The block of weight W is being pulled up the inclined plane of slope α using a force **P**. If **P** acts at the angle ϕ as shown, show that for slipping to occur, $P = W\sin(\alpha + \theta)/\cos(\phi - \theta)$, where θ is the angle of friction; $\theta = \tan^{-1}\mu$.

***8–52.** Determine the angle ϕ at which **P** should act on the block so that the magnitude of **P** is as small as possible to begin pulling the block up the incline. What is the corresponding value of P? The block weighs W and the slope α is known.

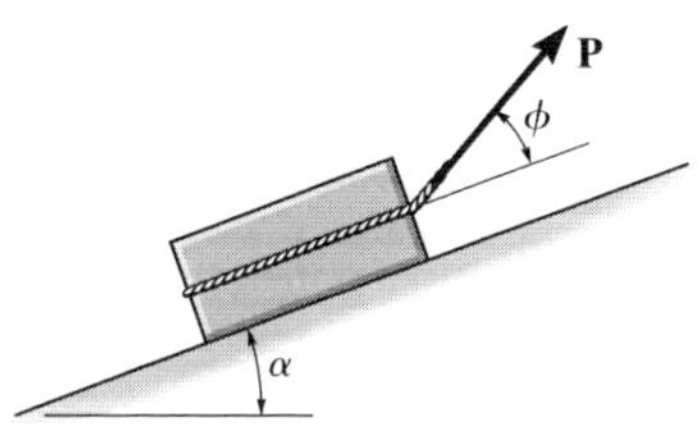

Probs. 8–51/52

8–53. The homogeneous semicylinder has a mass m and mass center at G. Determine the largest angle θ of the inclined plane upon which it rests so that it does not slip down the plane. The coefficient of static friction between the plane and the cylinder is $\mu_s = 0.3$. Also, what is the angle ϕ for this case?

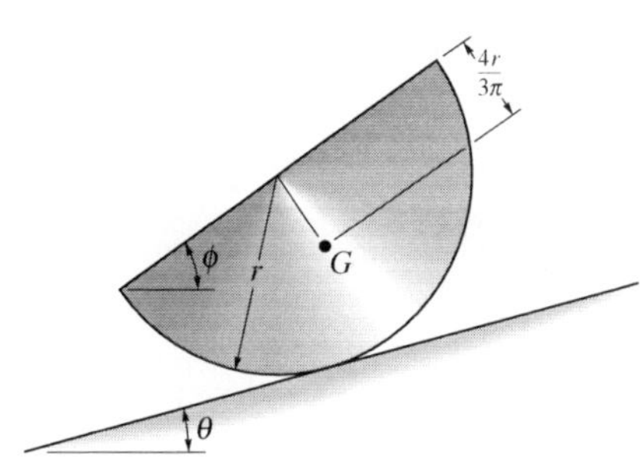

Prob. 8–53

8–54. The uniform beam has a weight W and length $4a$. It rests on the fixed rails at A and B. If the coefficient of static friction at the rails is μ_s, determine the horizontal force P, applied perpendicular to the face of the beam, which will cause the beam to move.

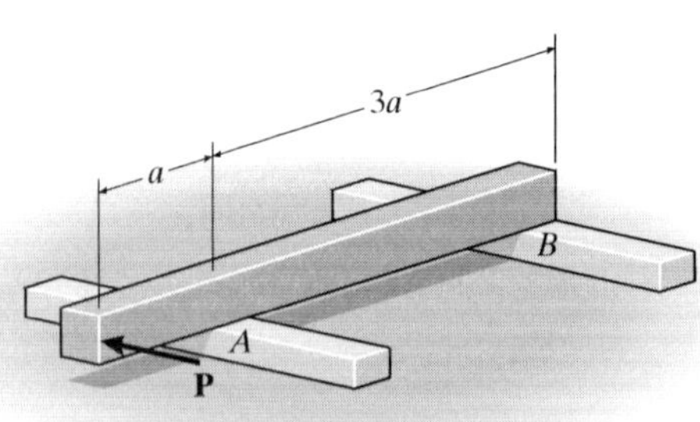
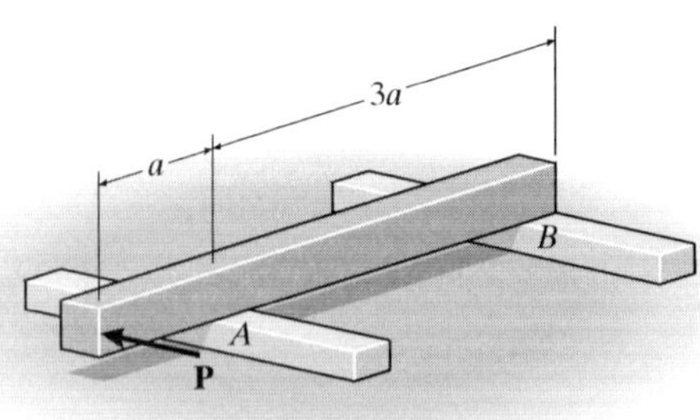

Prob. 8–54

***8–56.** The uniform 6-kg slender rod rests on the top center of the 3-kg block. If the coefficients of static friction at the points of contact are $\mu_A = 0.4$, $\mu_B = 0.6$, and $\mu_C = 0.3$, determine the largest couple moment M which can be applied to the rod without causing motion of the rod.

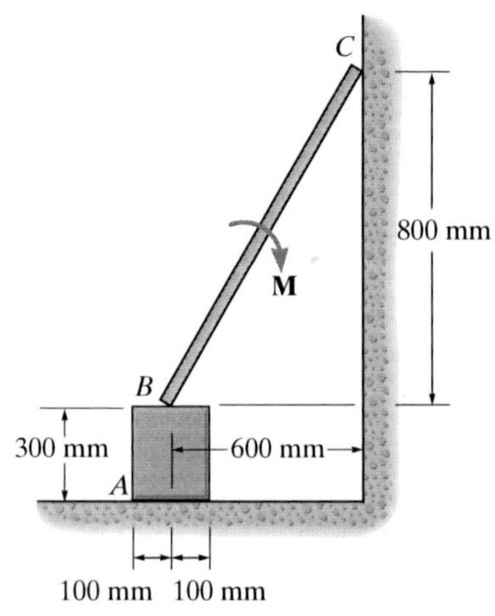

Prob. 8–56

8–55. Determine the greatest angle θ so that the ladder does not slip when it supports the 75-kg man in the position shown. The surface is rather slippery, where the coefficient of static friction at A and B is $\mu_s = 0.3$.

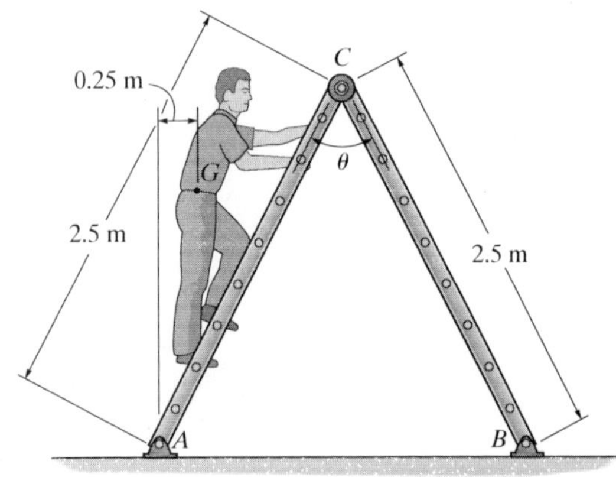

Prob. 8–55

8–57. The disk has a weight W and lies on a plane that has a coefficient of static friction μ. Determine the maximum height h to which the plane can be lifted without causing the disk to slip.

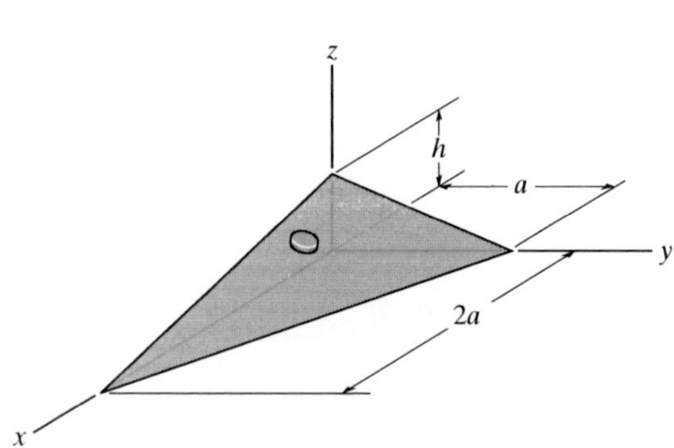

Prob. 8–57

CONCEPTUAL PROBLEMS

P8–1. Is it more effective to move the load forward at constant velocity with the boom fully extended as shown, or should the boom be fully retracted? Power is supplied to the rear wheels. The front wheels are free to roll. Do an equilibrium analysis to explain your answer.

P8–1

P8–2. The lug nut on the free-turning wheel is to be removed using the wrench. Which is the most effective way to apply force to the wrench? Also, why is it best to keep the car tire on the ground rather than first jacking it up? Explain your answers with an equilibrium analysis.

P8–2

P8–3. The rope is used to tow the refrigerator. Is it best to pull slightly up on the rope as shown, pull horizontally, or pull somewhat downwards? Also, is it best to attach the rope at a high position as shown, or at a lower position? Do an equilibrium analysis to explain your answer.

P8–4. The rope is used to tow the refrigerator. In order to prevent yourself from slipping while towing, is it best to pull up as shown, pull horizontally, or pull downwards on the rope? Do an equilibrium analysis to explain your answer.

P8–3/4

P8–5. Is it easier to tow the load by applying a force along the tow bar when it is in an almost horizontal position as shown, or is it better to pull on the bar when it has a steeper slope? Do an equilibrium analysis to explain your answer.

P8–5

8.3 Wedges

Please refer to the Companion Website for the animation: *Free-Body Diagram for a Load and a Wedge*

Wedges are often used to adjust the elevation of structural or mechanical parts. Also, they provide stability for objects such as this pipe.

A *wedge* is a simple machine that is often used to transform an applied force into much larger forces, directed at approximately right angles to the applied force. Wedges also can be used to slightly move or adjust heavy loads.

Consider, for example, the wedge shown in Fig. 8–12*a*, which is used to *lift* the block by applying a force to the wedge. Free-body diagrams of the block and wedge are shown in Fig. 8–12*b*. Here we have excluded the weight of the wedge since it is usually *small* compared to the weight **W** of the block. Also, note that the frictional forces $\mathbf{F}_1$ and $\mathbf{F}_2$ must oppose the motion of the wedge. Likewise, the frictional force $\mathbf{F}_3$ of the wall on the block must act downward so as to oppose the block's upward motion. The locations of the resultant normal forces are not important in the force analysis since neither the block nor wedge will "tip." Hence the moment equilibrium equations will not be considered. There are seven unknowns, consisting of the applied force **P**, needed to cause motion of the wedge, and six normal and frictional forces. The seven available equations consist of four force equilibrium equations, $\Sigma F_x = 0, \Sigma F_y = 0$ applied to the wedge and block, and three frictional equations, $F = \mu N$, applied at each surface of contact.

If the block is to be *lowered*, then the frictional forces will all act in a sense opposite to that shown in Fig. 8–12*b*. Provided the coefficient of friction is very *small* or the wedge angle θ is *large*, then the applied force **P** must act to the right to hold the block. Otherwise, **P** may have a reverse sense of direction in order to *pull* on the wedge to remove it. If **P** is *not applied* and friction forces hold the block in place, then the wedge is referred to as *self-locking*.

8

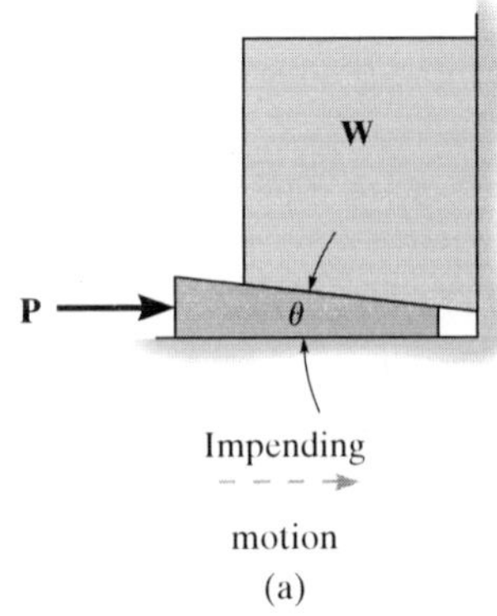

(a)

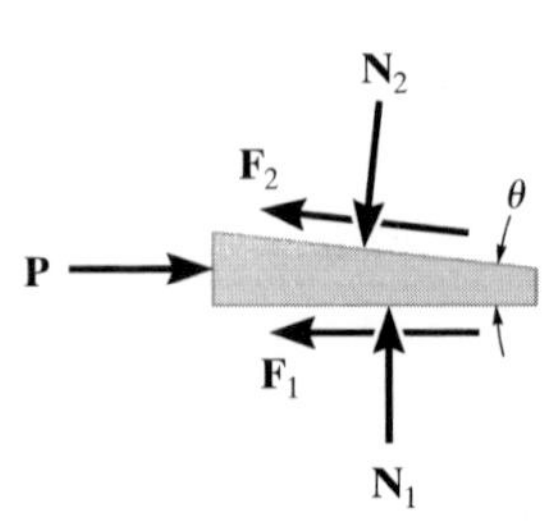

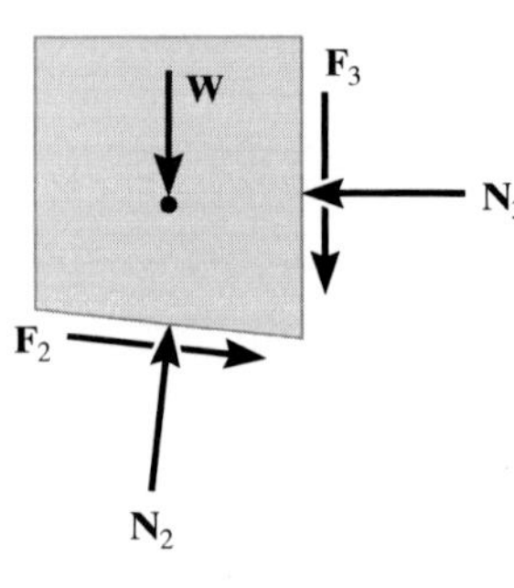

(b)

Fig. 8–12

EXAMPLE 8.6

The uniform stone in Fig. 8–13*a* has a mass of 500 kg and is held in the horizontal position using a wedge at *B*. If the coefficient of static friction is $\mu_s = 0.3$ at the surfaces of contact, determine the minimum force *P* needed to remove the wedge. Assume that the stone does not slip at *A*.

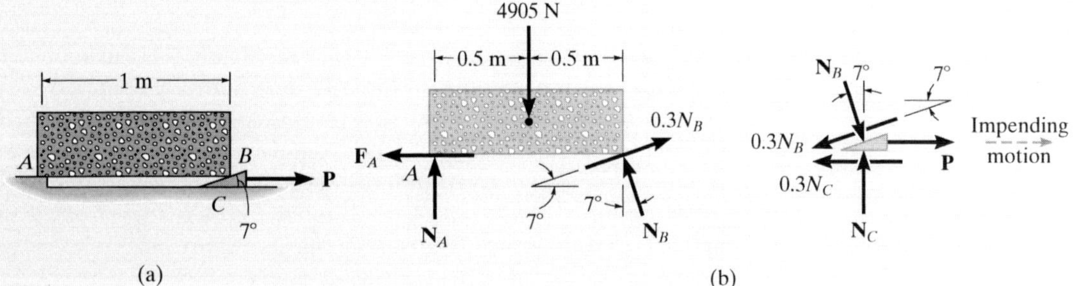

Fig. 8–13

SOLUTION

The minimum force *P* requires $F = \mu_s N$ at the surfaces of contact with the wedge. The free-body diagrams of the stone and wedge are shown in Fig. 8–13*b*. On the wedge the friction force opposes the impending motion, and on the stone at A, $F_A \leq \mu_s N_A$, since slipping does not occur there. There are five unknowns. Three equilibrium equations for the stone and two for the wedge are available for solution. From the free-body diagram of the stone,

$$\circlearrowleft +\Sigma M_A = 0; \quad -4905\text{ N}(0.5\text{ m}) + (N_B \cos 7^\circ\text{ N})(1\text{ m}) + (0.3N_B \sin 7^\circ\text{ N})(1\text{ m}) = 0$$

$$N_B = 2383.1\text{ N}$$

Using this result for the wedge, we have

$$+\uparrow \Sigma F_y = 0; \quad N_C - 2383.1 \cos 7^\circ\text{ N} - 0.3(2383.1 \sin 7^\circ\text{ N}) = 0$$

$$N_C = 2452.5\text{ N}$$

$$\overset{+}{\rightarrow} \Sigma F_x = 0; \quad 2383.1 \sin 7^\circ\text{ N} - 0.3(2383.1 \cos 7^\circ\text{ N}) + P - 0.3(2452.5\text{ N}) = 0$$

$$P = 1154.9\text{ N} = 1.15\text{ kN} \qquad \textit{Ans.}$$

NOTE: Since *P* is positive, indeed the wedge must be pulled out. If *P* were zero, the wedge would remain in place (self-locking) and the frictional forces developed at *B* and *C* would satisfy $F_B < \mu_s N_B$ and $F_C < \mu_s N_C$.

Square-threaded screws find applications on valves, jacks, and vises, where particularly large forces must be developed along the axis of the screw.

8.4 Frictional Forces on Screws

In most cases, screws are used as fasteners; however, in many types of machines they are incorporated to transmit power or motion from one part of the machine to another. A *square-threaded screw* is commonly used for the latter purpose, especially when large forces are applied along its axis. In this section, we will analyze the forces acting on square-threaded screws. The analysis of other types of screws, such as the V-thread, is based on these same principles.

For analysis, a square-threaded screw, as in Fig. 8–14, can be considered a cylinder having an inclined square ridge or *thread* wrapped around it. If we unwind the thread by one revolution, as shown in Fig. 8–14*b*, the slope or the *lead angle* θ is determined from $\theta = \tan^{-1}(l/2\pi r)$. Here l and $2\pi r$ are the vertical and horizontal distances between A and B, where r is the mean radius of the thread. The distance l is called the *lead* of the screw and it is equivalent to the distance the screw advances when it turns one revolution.

Upward Impending Motion. Let us now consider the case of the square-threaded screw jack in Fig. 8–15 that is subjected to upward impending motion caused by the applied torsional moment ***M**. A free-body diagram of the *entire unraveled thread h* in contact with the jack can be represented as a block as shown in Fig. 8–16*a*. The force **W** is the vertical force acting on the thread or the axial force applied to the shaft, Fig. 8–15, and M/r is the resultant horizontal force produced by the couple moment M about the axis of the shaft. The reaction **R** of the groove on the thread has both frictional and normal components, where $F = \mu_s N$. The angle of static friction is $\phi_s = \tan^{-1}(F/N) = \tan^{-1}\mu_s$. Applying the force equations of equilibrium along the horizontal and vertical axes, we have

$$\overset{+}{\rightarrow}\Sigma F_x = 0; \qquad M/r - R\sin(\theta + \phi_s) = 0$$

$$+\uparrow\Sigma F_y = 0; \qquad R\cos(\theta + \phi_s) - W = 0$$

Eliminating R from these equations, we obtain

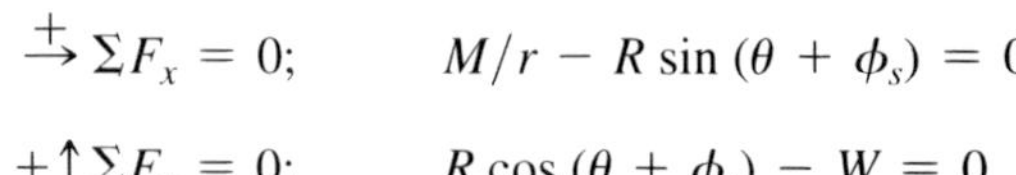

$$M = rW\tan(\theta + \phi_s) \qquad (8\text{–}3)$$

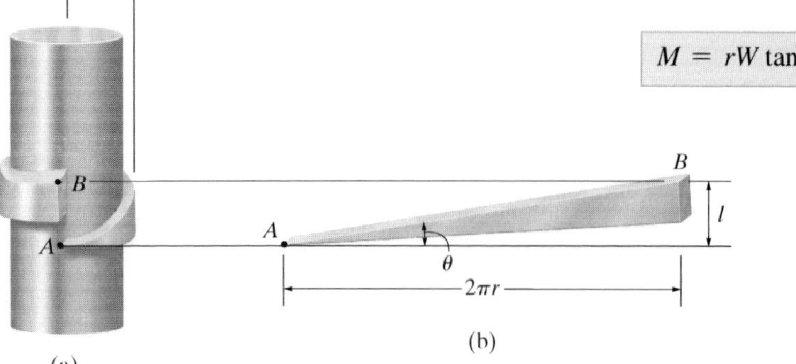

(a) (b)

Fig. 8–14

*For applications, **M** is developed by applying a horizontal force **P** at a right angle to the end of a lever that would be fixed to the screw.

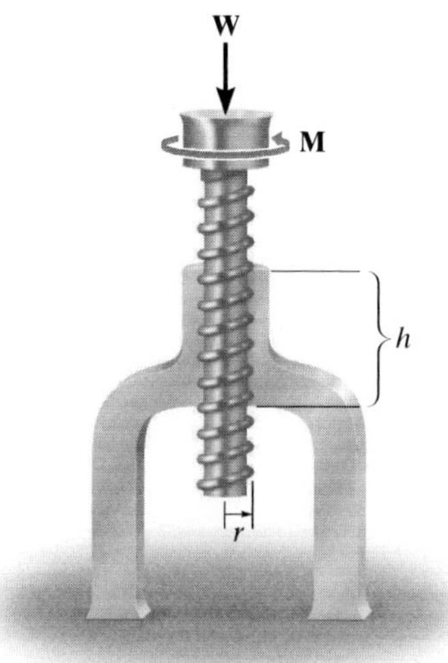

Fig. 8–15

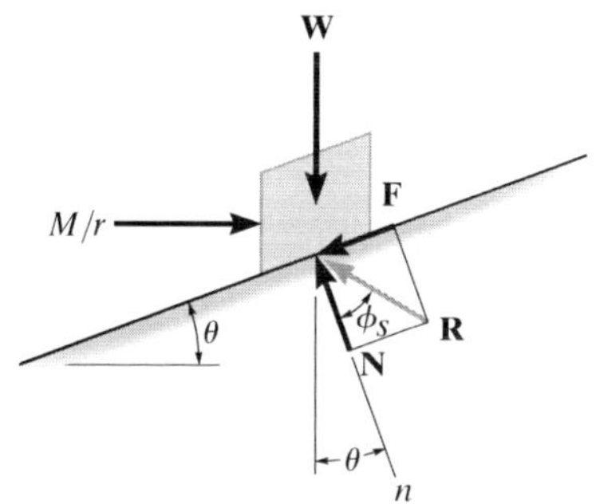

Upward screw motion

(a)

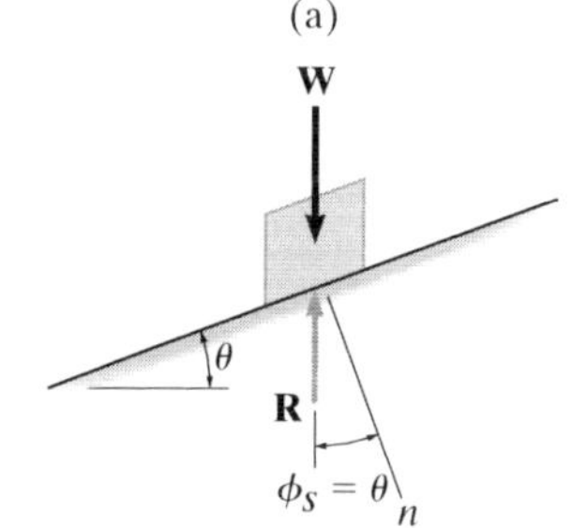

Self-locking screw ($\theta = \phi_s$)
(on the verge of rotating downward)

(b)

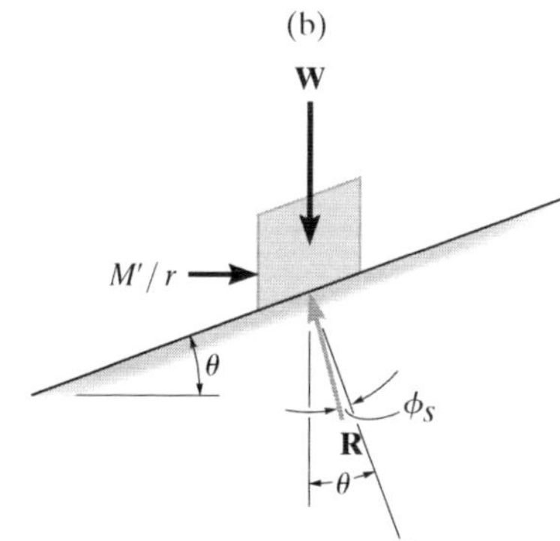

Downward screw motion ($\theta > \phi_s$)

(c)

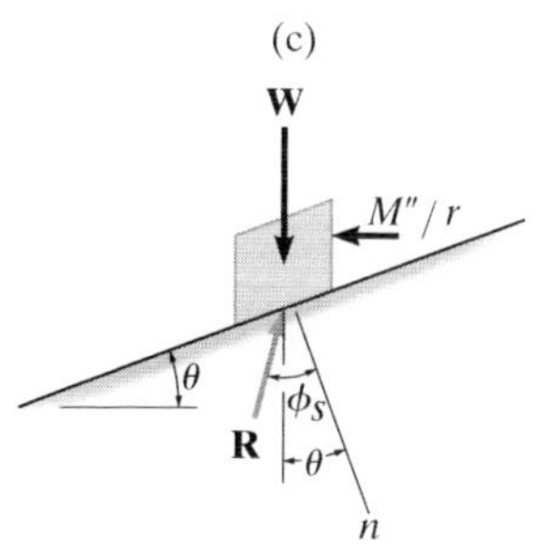

Downward screw motion ($\theta < \phi_s$)

(d)

Fig. 8–16

Self-Locking Screw. A screw is said to be *self-locking* if it remains in place under any axial load **W** when the moment **M** is removed. For this to occur, the direction of the frictional force must be reversed so that **R** acts on the other side of **N**. Here the angle of static friction ϕ_s becomes greater than or equal to θ, Fig. 8–16*d*. If $\phi_s = \theta$, Fig. 8–16*b*, then **R** will act vertically to balance **W**, and the screw will be on the verge of winding downward.

Downward Impending Motion, ($\theta > \phi_s$). If the screw is not self-locking, it is necessary to apply a moment **M**′ to prevent the screw from winding downward. Here, a horizontal force M'/r is required to push against the thread to prevent it from sliding down the plane, Fig. 8–16*c*. Using the same procedure as before, the magnitude of the moment **M**′ required to prevent this unwinding is

$$M' = rW\tan(\theta - \phi_s) \qquad (8\text{–}4)$$

Downward Impending Motion, ($\phi_s > \theta$). If a screw is self-locking, a couple moment **M**″ must be applied to the screw in the opposite direction to wind the screw downward ($\phi_s > \theta$). This causes a reverse horizontal force M''/r that pushes the thread down as indicated in Fig. 8–16*d*. In this case, we obtain

$$M'' = rW\tan(\phi_s - \theta) \qquad (8\text{–}5)$$

If *motion of the screw* occurs, Eqs. 8–3, 8–4, and 8–5 can be applied by simply replacing ϕ_s with ϕ_k.

8

EXAMPLE 8.7

The turnbuckle shown in Fig. 8–17 has a square thread with a mean radius of 5 mm and a lead of 2 mm. If the coefficient of static friction between the screw and the turnbuckle is $\mu_s = 0.25$, determine the moment **M** that must be applied to draw the end screws closer together.

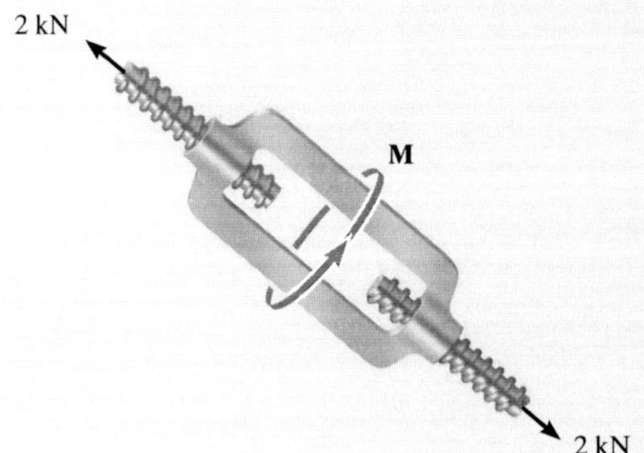

Fig. 8–17

8

SOLUTION

The moment can be obtained by applying Eq. 8–3. Since friction at *two screws* must be overcome, this requires

$$M = 2[rW\tan(\theta + \phi_s)] \qquad (1)$$

Here $W = 2000\text{ N}$, $\phi_s = \tan^{-1}\mu_s = \tan^{-1}(0.25) = 14.04°$, $r = 5\text{ mm}$, and $\theta = \tan^{-1}(l/2\pi r) = \tan^{-1}(2\text{ mm}/[2\pi(5\text{ mm})]) = 3.64°$. Substituting these values into Eq. 1 and solving gives

$$M = 2[(2000\text{ N})(5\text{ mm})\tan(14.04° + 3.64°)]$$

$$= 6374.7\text{ N}\cdot\text{mm} = 6.37\text{ N}\cdot\text{m} \qquad \textit{Ans.}$$

NOTE: When the moment is *removed*, the turnbuckle will be self-locking; i.e., it will not unscrew since $\phi_s > \theta$.

PROBLEMS

8–58. Determine the largest angle θ that will cause the wedge to be self-locking regardless of the magnitude of horizontal force P applied to the blocks. The coefficient of static friction between the wedge and the blocks is $\mu_s = 0.3$. Neglect the weight of the wedge.

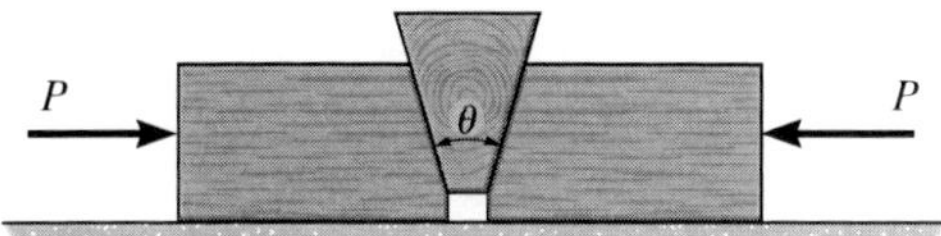

Prob. 8–58

8–59. If the beam AD is loaded as shown, determine the horizontal force P which must be applied to the wedge in order to remove it from under the beam. The coefficients of static friction at the wedge's top and bottom surfaces are $\mu_{CA} = 0.25$ and $\mu_{CB} = 0.35$, respectively. If $P = 0$, is the wedge self-locking? Neglect the weight and size of the wedge and the thickness of the beam.

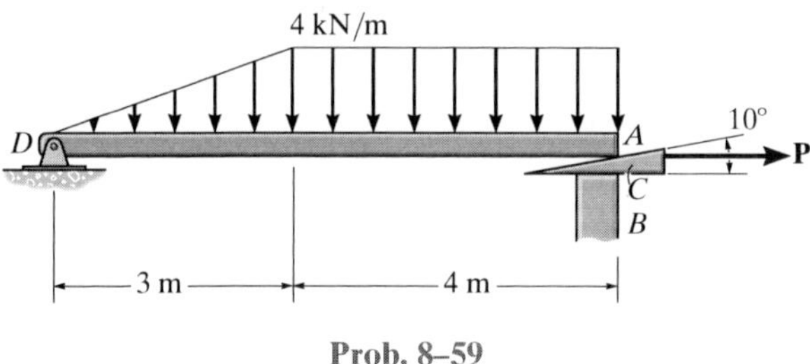

Prob. 8–59

***8–60.** The wedge has a negligible weight and a coefficient of static friction $\mu_s = 0.35$ with all contacting surfaces. Determine the largest angle θ so that it is "self-locking." This requires no slipping for any magnitude of the force **P** applied to the joint.

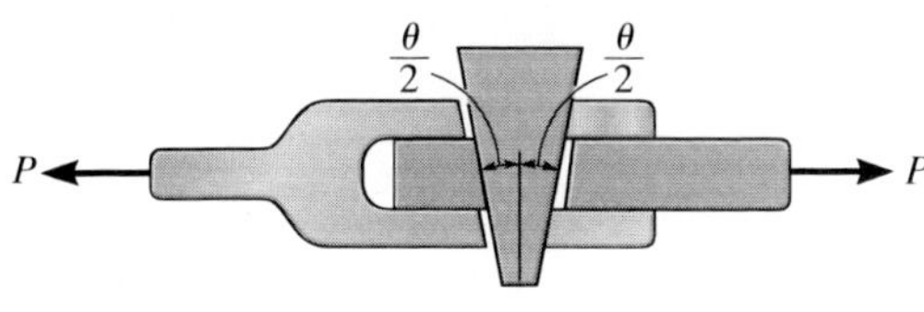

Prob. 8–60

8–61. If the spring is compressed 60 mm and the coefficient of static friction between the tapered stub S and the slider A is $\mu_{SA} = 0.5$, determine the horizontal force **P** needed to move the slider forward. The stub is free to move without friction within the fixed collar C. The coefficient of static friction between A and surface B is $\mu_{AB} = 0.4$. Neglect the weights of the slider and stub.

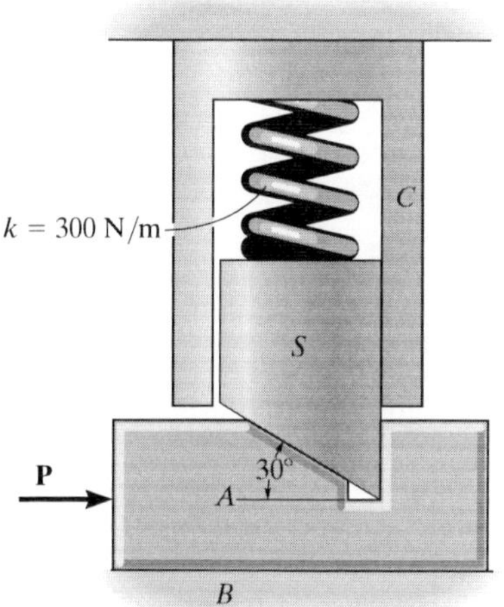

Prob. 8–61

8–62. If $P = 250$ N, determine the required minimum compression in the spring so that the wedge will not move to the right. Neglect the weight of A and B. The coefficient of static friction for all contacting surfaces is $\mu_s = 0.35$. Neglect friction at the rollers.

8–63. Determine the minimum applied force **P** required to move wedge A to the right. The spring is compressed a distance of 175 mm. Neglect the weight of A and B. The coefficient of static friction for all contacting surfaces is $\mu_s = 0.35$. Neglect friction at the rollers.

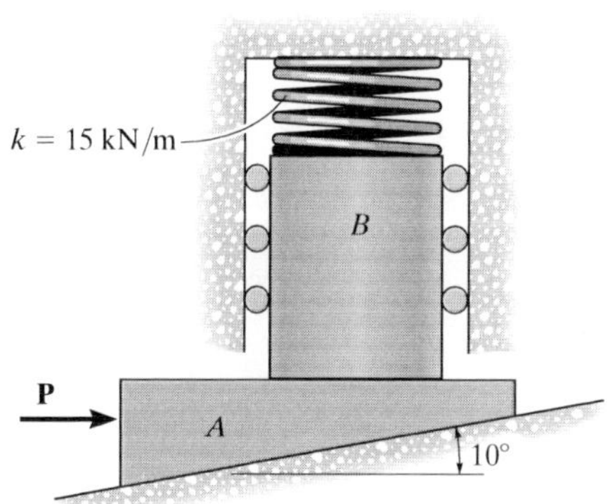

Probs. 8–62/63

*8–64. The wedge blocks are used to hold the specimen in a tension testing machine. Determine the design angle θ of the wedges so that the specimen will not slip regardless of the applied load. The coefficients of static friction are $\mu_A = 0.1$ at A and $\mu_B = 0.6$ at B. Neglect the weight of the blocks.

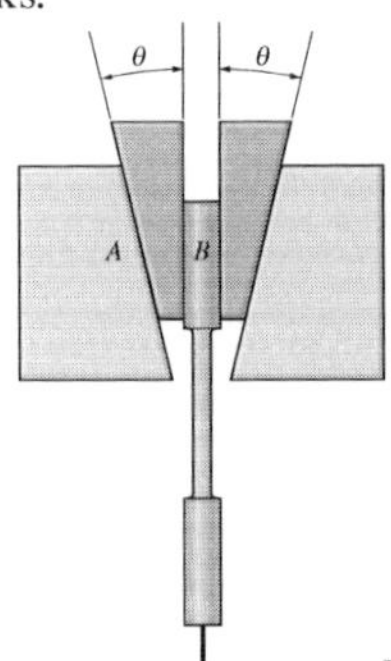

Prob. 8–64

8–65. The coefficient of static friction between wedges B and C is $\mu_s = 0.6$ and between the surfaces of contact B and A and C and D, $\mu_s' = 0.4$. If the spring is compressed 200 mm when in the position shown, determine the smallest force P needed to move wedge C to the left. Neglect the weight of the wedges.

8–66. The coefficient of static friction between the wedges B and C is $\mu_s = 0.6$ and between the surfaces of contact B and A and C and D, $\mu_s' = 0.4$. If $P = 50$ N, determine the smallest allowable compression of the spring without causing wedge C to move to the left. Neglect the weight of the wedges.

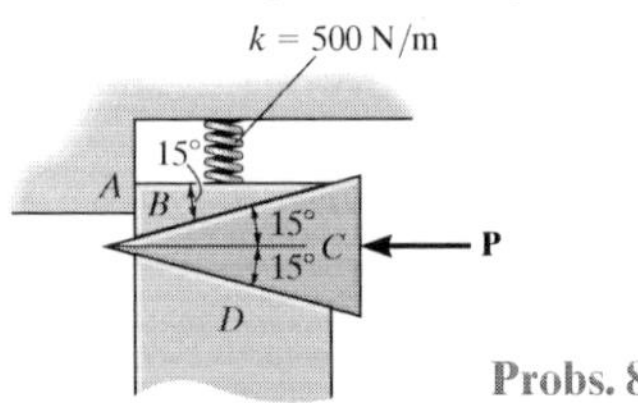

Probs. 8–65/66

8–67. The vise is used to grip the pipe. If a horizontal force of 100 N is applied perpendicular to the end of the 250 mm handle, determine the compressive force F developed in the pipe. The square threads have a mean diameter of 37.5 mm and a lead of 5 mm. How much force must be applied perpendicular to the handle to loosen the vise? Take $\mu_s = 0.3$.

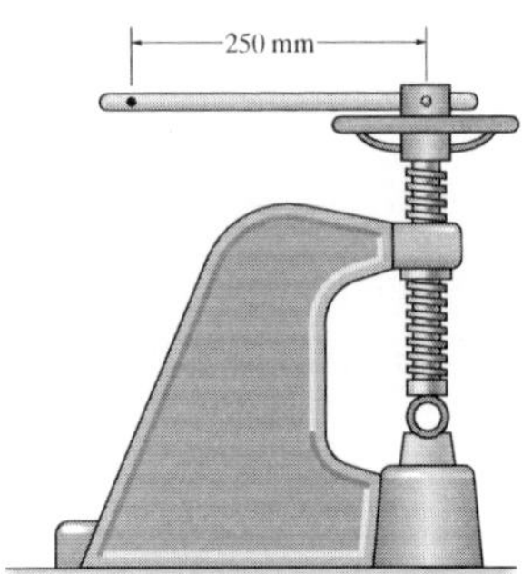

Prob. 8–67

*8–68. Determine the couple forces F that must be applied to the handle of the machinist's vise in order to create a compressive force of 400 N in the block. Neglect friction at the bearing A. The guide at B is smooth so that the axial force on the screw is 400 N. The single square-threaded screw has a mean radius of 6 mm and a lead of 8 mm, and the coefficient of static friction is $\mu_s = 0.27$.

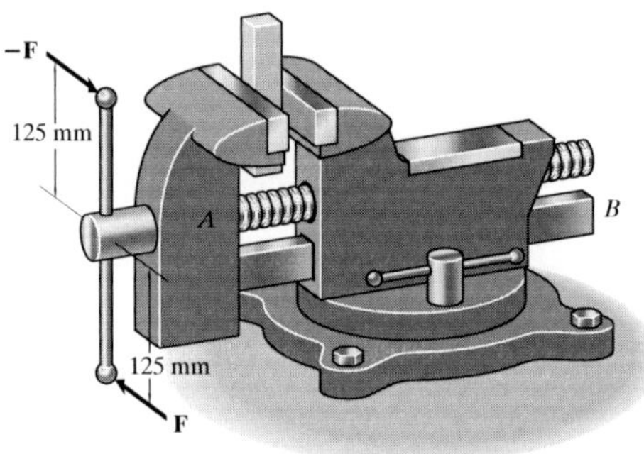

Prob. 8–68

8–69. The column is used to support the upper floor. If a force $F = 80$ N is applied perpendicular to the handle to tighten the screw, determine the compressive force in the column. The square-threaded screw on the jack has a coefficient of static friction of $\mu_s = 0.4$, mean diameter of 25 mm, and a lead of 3 mm.

8–70. If the force **F** is removed from the handle of the jack in Prob. 8–69, determine if the screw is self-locking.

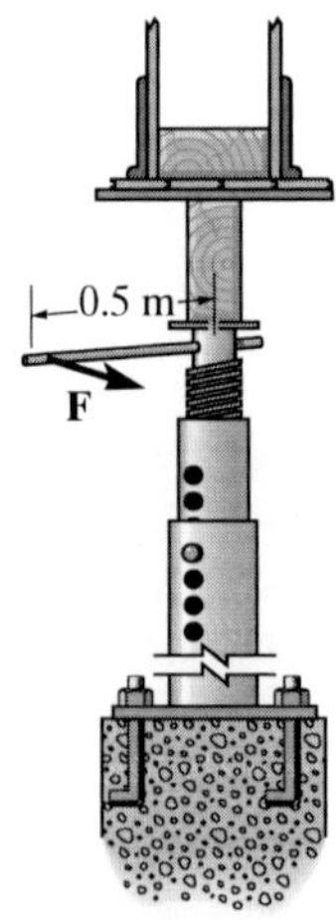

Probs. 8–69/70

8–71. If the clamping force at G is 900 N, determine the horizontal force **F** that must be applied perpendicular to the handle of the lever at E. The mean diameter and lead of both single square-threaded screws at C and D are 25 mm and 5 mm, respectively. The coefficient of static friction is $\mu_s = 0.3$.

8

***8–72.** If a horizontal force of $F = 50$ N is applied perpendicular to the handle of the lever at E, determine the clamping force developed at G. The mean diameter and lead of the single square-threaded screw at C and D are 25 mm and 5 mm, respectively. The coefficient of static friction is $\mu_s = 0.3$.

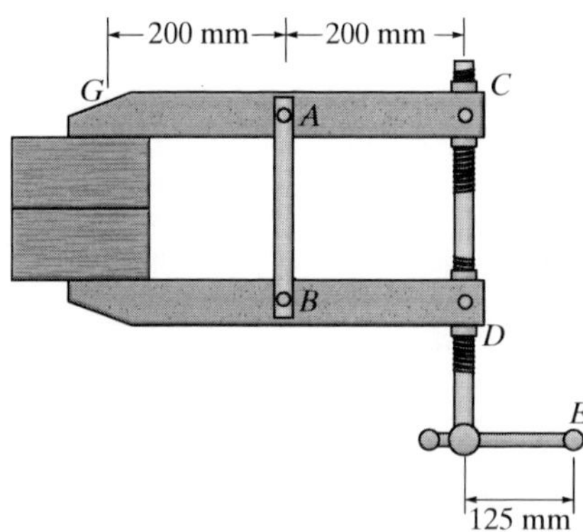

Probs. 8–71/72

8–73. A turnbuckle, similar to that shown in Fig. 8–17, is used to tension member AB of the truss. The coefficient of the static friction between the square threaded screws and the turnbuckle is $\mu_s = 0.5$. The screws have a mean radius of 6 mm and a lead of 3 mm. If a torque of $M = 10$ N · m is applied to the turnbuckle, to draw the screws closer together, determine the force in each member of the truss. No external forces act on the truss.

8–74. A turnbuckle, similar to that shown in Fig. 8–17, is used to tension member AB of the truss. The coefficient of the static friction between the square-threaded screws and the turnbuckle is $\mu_s = 0.5$. The screws have a mean radius of 6 mm and a lead of 3 mm. Determine the torque M which must be applied to the turnbuckle to draw the screws closer together, so that the compressive force of 500 N is developed in member BC.

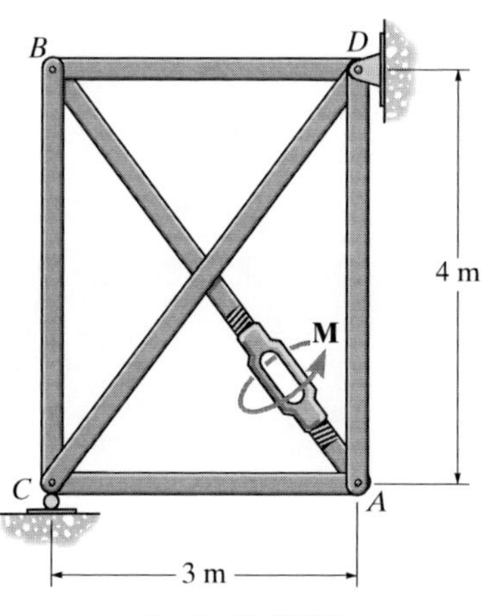

Probs. 8–73/74

8–75. The shaft has a square-threaded screw with a lead of 8 mm and a mean radius of 15 mm. If it is in contact with a plate gear having a mean radius of 30 mm, determine the resisting torque **M** on the plate gear which can be overcome if a torque of 7 N · m is applied to the shaft. The coefficient of static friction at the screw is $\mu_B = 0.2$. Neglect friction of the bearings located at A and B.

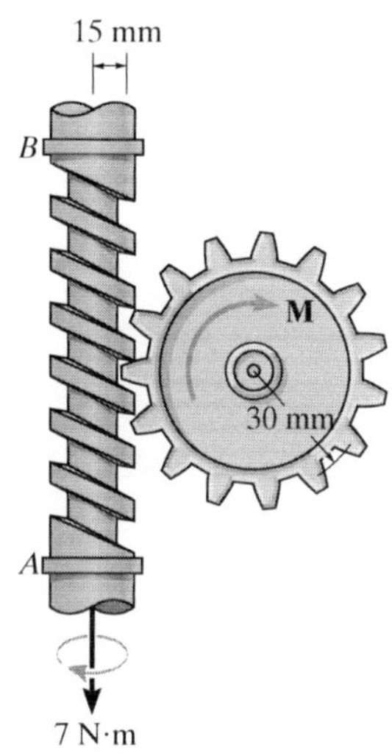

Prob. 8–75

***8–76.** The square threaded screw of the clamp has a mean diameter of 14 mm and a lead of 6 mm. If $\mu_s = 0.2$ for the threads, and the torque applied to the handle is 1.5 N · m, determine the compressive force **F** on the block.

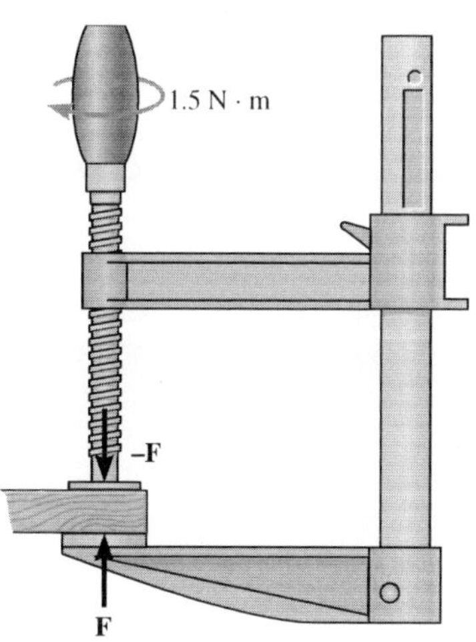

Prob. 8–76

8–77. The fixture clamp consist of a square-threaded screw having a coefficient of static friction of $\mu_s = 0.3$, mean diameter of 3 mm, and a lead of 1 mm. The five points indicated are pin connections. Determine the clamping force at the smooth blocks D and E when a torque of $M = 0.08$ N · m is applied to the handle of the screw.

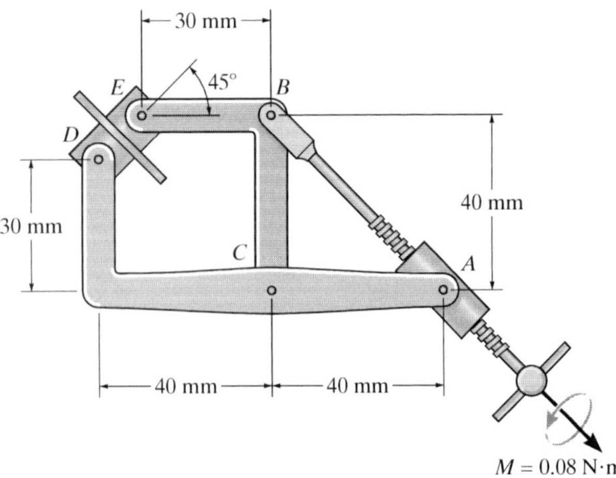

Prob. 8–77

8–78. The braking mechanism consists of two pinned arms and a square-threaded screw with left- and right-hand threads. Thus when turned, the screw draws the two arms together. If the lead of the screw is 4 mm, the mean diameter 12 mm, and the coefficient of static friction is $\mu_s = 0.35$, determine the tension in the screw when a torque of 5 N · m is applied to tighten the screw. If the coefficient of static friction between the brake pads A and B and the circular shaft is $\mu_s' = 0.5$, determine the maximum torque M the brake can resist.

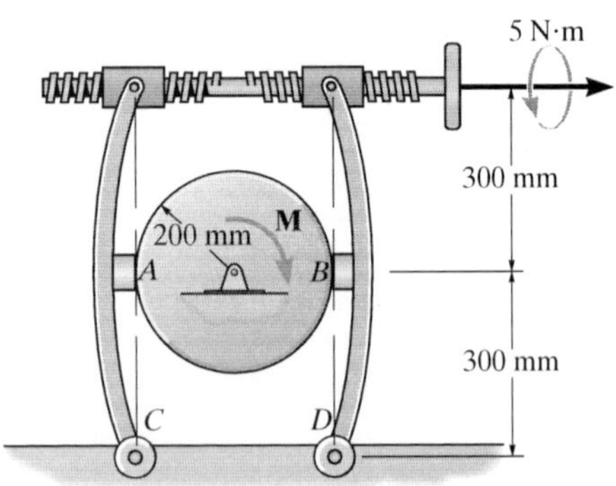

Prob. 8–78

8–79. If a horizontal force of $P = 100$ N is applied perpendicular to the handle of the lever at A, determine the compressive force **F** exerted on the material. Each single square-threaded screw has a mean diameter of 25 mm and a lead of 7.5 mm. The coefficient of static friction at all contacting surfaces of the wedges is $\mu_s = 0.2$, and the coefficient of static friction at the screw is $\mu_s' = 0.15$.

***8–80.** Determine the horizontal force **P** that must be applied perpendicular to the handle of the lever at A in order to develop a compressive force of 12 kN on the material. Each single square-threaded screw has a mean diameter of 25 mm and a lead of 7.5 mm. The coefficient of static friction at all contacting surfaces of the wedges is $\mu_s = 0.2$, and the coefficient of static friction at the screw is $\mu_s' = 0.15$.

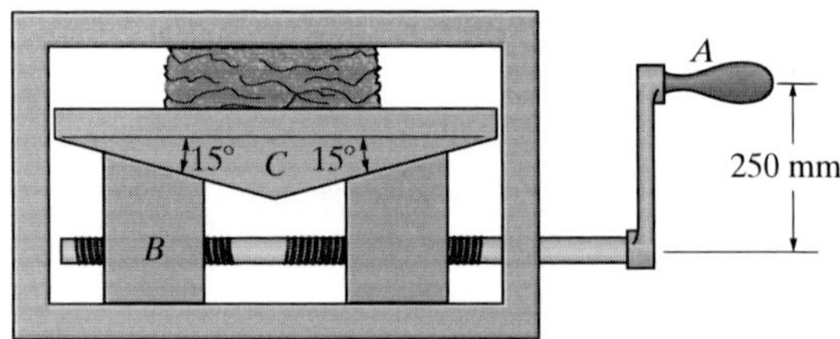

Probs. 8–79/80

8–81. Determine the clamping force on the board A if the screw of the "C" clamp is tightened with a twist of $M = 8$ N · m. The single square-threaded screw has a mean radius of 10 mm, a lead of 3 mm, and the coefficient of static friction is $\mu_s = 0.35$.

8–82. If the required clamping force at the board A is to be 50 N, determine the torque M that must be applied to the handle of the "C" clamp to tighten it down. The single square-threaded screw has a mean radius of 10 mm, a lead of 3 mm, and the coefficient of static friction is $\mu_s = 0.35$.

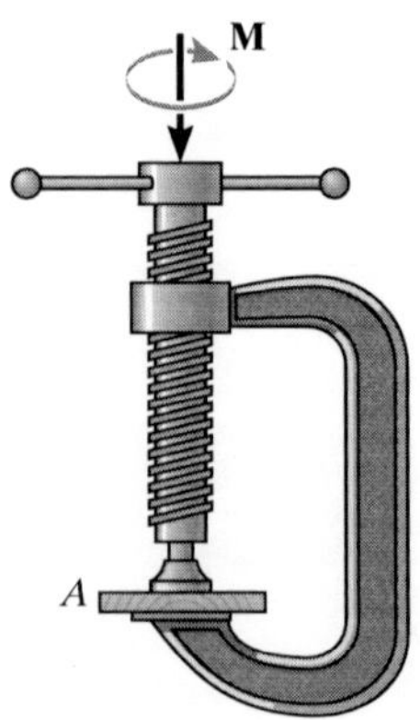

Probs. 8–81/82

8.5 Frictional Forces on Flat Belts

Whenever belt drives or band brakes are designed, it is necessary to determine the frictional forces developed between the belt and its contacting surface. In this section we will analyze the frictional forces acting on a flat belt, although the analysis of other types of belts, such as the V-belt, is based on similar principles.

Consider the flat belt shown in Fig. 8–18*a*, which passes over a fixed curved surface. The total angle of belt to surface contact in radians is β, and the coefficient of friction between the two surfaces is μ. We wish to determine the tension T_2 in the belt, which is needed to pull the belt counterclockwise over the surface, and thereby overcome both the frictional forces at the surface of contact and the tension T_1 in the other end of the belt. Obviously, $T_2 > T_1$.

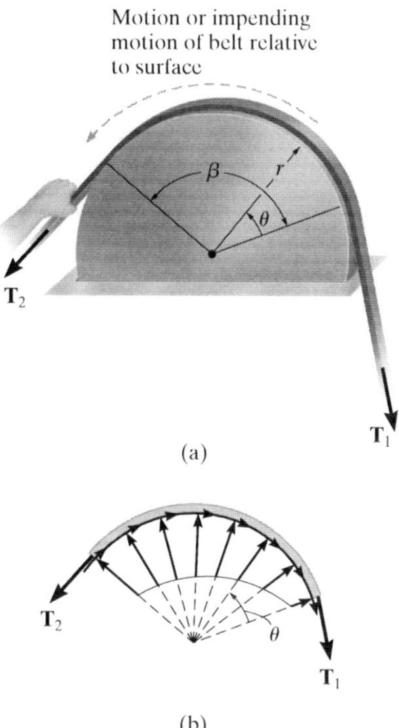

Frictional Analysis. A free-body diagram of the belt segment in contact with the surface is shown in Fig. 8–18*b*. As shown, the normal and frictional forces, acting at different points along the belt, will vary both in magnitude and direction. Due to this *unknown* distribution, the analysis of the problem will first require a study of the forces acting on a differential element of the belt.

A free-body diagram of an element having a length ds is shown in Fig. 8–18*c*. Assuming either impending motion or motion of the belt, the magnitude of the frictional force $dF = \mu\, dN$. This force opposes the sliding motion of the belt, and so it will increase the magnitude of the tensile force acting in the belt by dT. Applying the two force equations of equilibrium, we have

$$\searrow + \Sigma F_x = 0; \qquad T\cos\left(\frac{d\theta}{2}\right) + \mu\, dN - (T + dT)\cos\left(\frac{d\theta}{2}\right) = 0$$

$$+\nearrow \Sigma F_y = 0; \qquad dN - (T + dT)\sin\left(\frac{d\theta}{2}\right) - T\sin\left(\frac{d\theta}{2}\right) = 0$$

Since $d\theta$ is of *infinitesimal size*, $\sin(d\theta/2) = d\theta/2$ and $\cos(d\theta/2) = 1$. Also, the *product* of the two infinitesimals dT and $d\theta/2$ may be neglected when compared to infinitesimals of the first order. As a result, these two equations become

$$\mu\, dN = dT$$

and

$$dN = T\, d\theta$$

Eliminating dN yields

$$\frac{dT}{T} = \mu\, d\theta$$

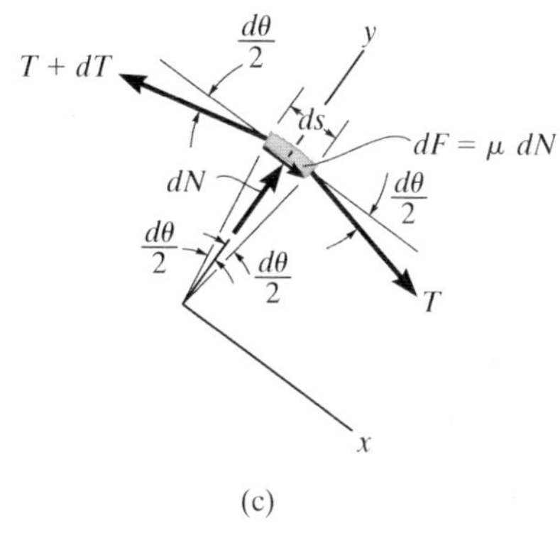

Fig. 8–18

8

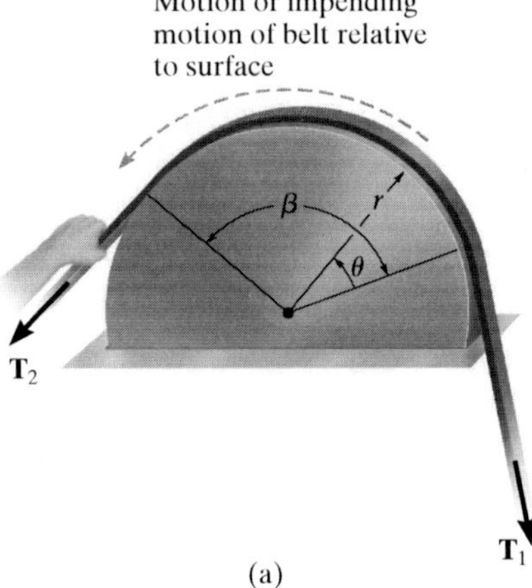

Fig. 8–18 (Repeated)

Flat or V-belts are often used to transmit the torque developed by a motor to a wheel attached to a pump, fan or blower.

Integrating this equation between all the points of contact that the belt makes with the drum, and noting that $T = T_1$ at $\theta = 0$ and $T = T_2$ at $\theta = \beta$, yields

$$\int_{T_1}^{T_2} \frac{dT}{T} = \mu \int_0^{\beta} d\theta$$

$$\ln \frac{T_2}{T_1} = \mu\beta$$

Solving for T_2, we obtain

$$T_2 = T_1 e^{\mu\beta} \tag{8–6}$$

where

T_2, T_1 = belt tensions; T_1 opposes the direction of motion (or impending motion) of the belt measured relative to the surface, while T_2 acts in the direction of the relative belt motion (or impending motion); because of friction, $T_2 > T_1$

μ = coefficient of static or kinetic friction between the belt and the surface of contact

β = angle of belt to surface contact, measured in radians

e = 2.718 . . . , base of the natural logarithm

Note that T_2 is *independent* of the *radius* of the drum, and instead it is a function of the angle of belt to surface contact, β. As a result, this equation is valid for flat belts passing over any curved contacting surface.

8

EXAMPLE 8.8

The maximum tension that can be developed in the cord shown in Fig. 8–19*a* is 500 N. If the pulley at *A* is free to rotate and the coefficient of static friction at the fixed drums *B* and *C* is $\mu_s = 0.25$, determine the largest mass of the cylinder that can be lifted by the cord.

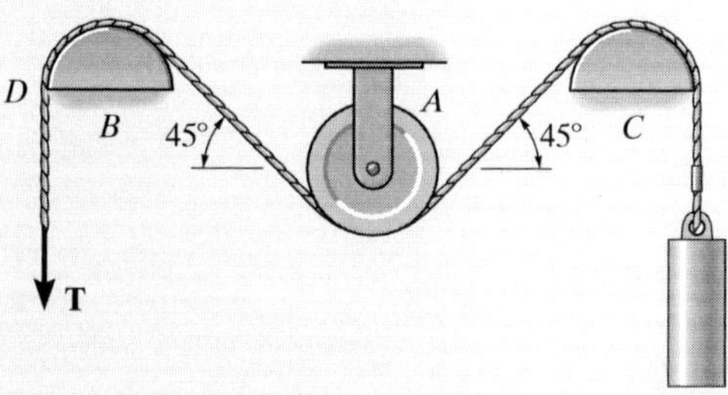

(a)

SOLUTION

Lifting the cylinder, which has a weight $W = mg$, causes the cord to move counterclockwise over the drums at *B* and *C*; hence, the maximum tension T_2 in the cord occurs at *D*. Thus, $F = T_2 = 500$ N. A section of the cord passing over the drum at *B* is shown in Fig. 8–19*b*. Since $180° = \pi$ rad the angle of contact between the drum and the cord is $\beta = (135°/180°)\pi = 3\pi/4$ rad. Using Eq. 8–6, we have

Impending motion
135°
B
500 N
$\mathbf{T}_1$

(b)

$$T_2 = T_1 e^{\mu_s \beta}; \qquad 500\text{ N} = T_1 e^{0.25[(3/4)\pi]}$$

Hence,

$$T_1 = \frac{500\text{ N}}{e^{0.25[(3/4)\pi]}} = \frac{500\text{ N}}{1.80} = 277.4\text{ N}$$

Since the pulley at *A* is free to rotate, equilibrium requires that the tension in the cord remains the *same* on both sides of the pulley.

The section of the cord passing over the drum at *C* is shown in Fig. 8–19*c*. The weight $W < 277.4$ N. Why? Applying Eq. 8–6, we obtain

$$T_2 = T_1 e^{\mu_s \beta}; \qquad 277.4\text{ N} = W e^{0.25[(3/4)\pi]}$$

$$W = 153.9\text{ N}$$

so that

$$m = \frac{W}{g} = \frac{153.9\text{ N}}{9.81\text{ m/s}^2}$$

$$= 15.7\text{ kg} \qquad \textit{Ans.}$$

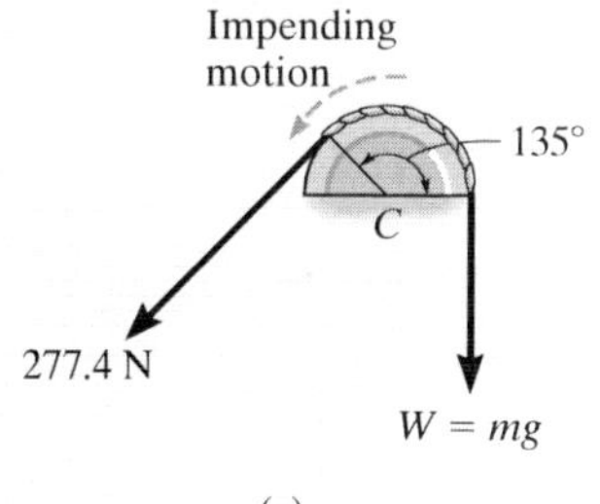

(c)

Fig. 8–19

PROBLEMS

8–83. A cylinder having a mass of 250 kg is to be supported by the cord that wraps over the pipe. Determine the smallest vertical force **F** needed to support the load if the cord passes (a) once over the pipe, $\beta = 180°$, and (b) two times over the pipe, $\beta = 540°$. Take $\mu_s = 0.2$.

*__8–84.__ A cylinder having a mass of 250 kg is to be supported by the cord that wraps over the pipe. Determine the largest vertical force **F** that can be applied to the cord without moving the cylinder. The cord passes (a) once over the pipe, $\beta = 180°$, and (b) two times over the pipe, $\beta = 540°$. Take $\mu_s = 0.2$.

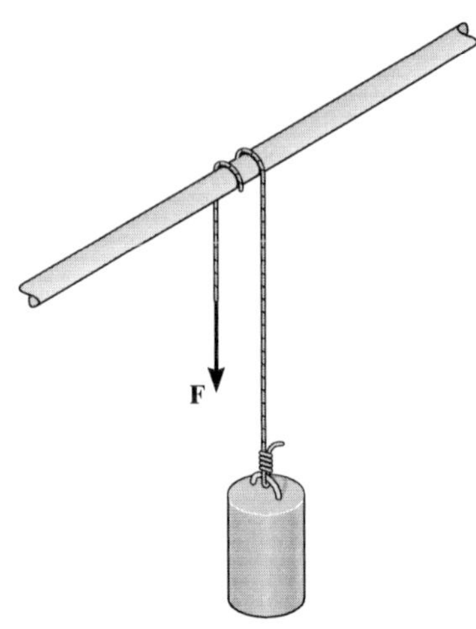

Probs. 8–83/84

8

8–85. The cord supporting the 6-kg cylinder passes around three pegs, *A*, *B*, *C*, where $\mu_s = 0.2$. Determine the range of values for the magnitude of the horizontal force **P** for which the cylinder will not move up or down.

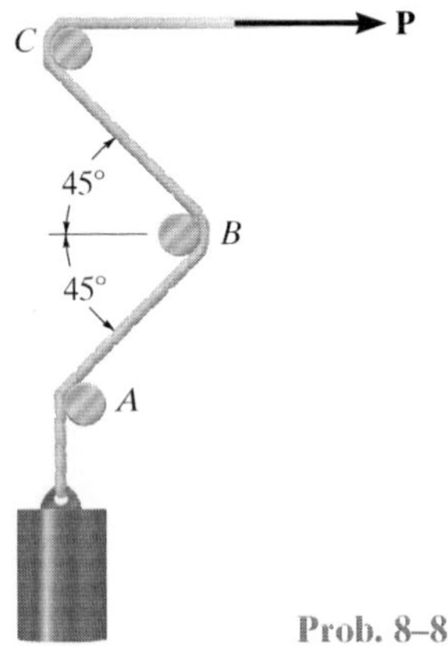

Prob. 8–85

8–86. A force of $P = 25$ N is just sufficient to prevent the 20-kg cylinder from descending. Determine the required force **P** to begin lifting the cylinder. The rope passes over a rough peg with two and half turns.

Prob. 8–86

8–87. The 20-kg cylinder *A* and 50-kg cylinder *B* are connected together using a rope that passes around a rough peg two and a half turns. If the cylinders are on the verge of moving, determine the coefficient of static friction between the rope and the peg.

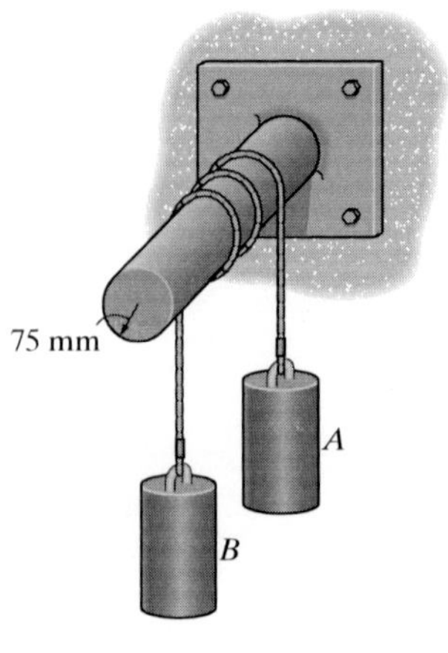

Prob. 8–87

***8–88.** The uniform bar AB is supported by a rope that passes over a frictionless pulley at C and a fixed peg at D. If the coefficient of static friction between the rope and the peg is $\mu_D = 0.3$, determine the smallest distance x from the end of the bar at which a 20-N force may be placed and not cause the bar to move.

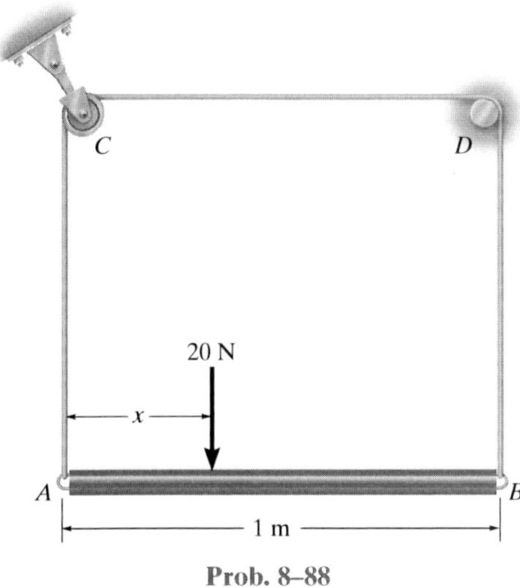

Prob. 8–88

8–89. The truck, which has a mass of 3.4 Mg, is to be lowered down the slope by a rope that is wrapped around a tree. If the wheels are free to roll and the man at A can resist a pull of 300 N, determine the minimum number of turns the rope should be wrapped around the tree to lower the truck at a constant speed. The coefficient of kinetic friction between the tree and rope is $\mu_k = 0.3$.

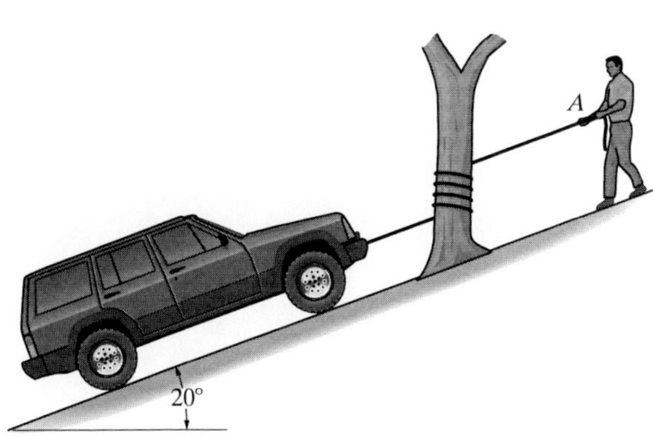

Prob. 8–89

8–90. The smooth beam is being hoisted using a rope that is wrapped around the beam and passes through a ring at A as shown. If the end of the rope is subjected to a tension **T** and the coefficient of static friction between the rope and ring is $\mu_s = 0.3$, determine the smallest angle of θ for equilibrium.

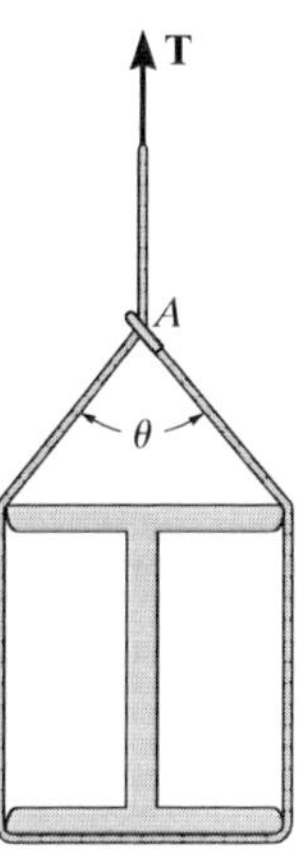

Prob. 8–90

8–91. A cable is attached to the 20-kg plate B, passes over a fixed peg at C, and is attached to the block at A. Using the coefficients of static friction shown, determine the smallest mass of block A so that it will prevent sliding motion of B down the plane.

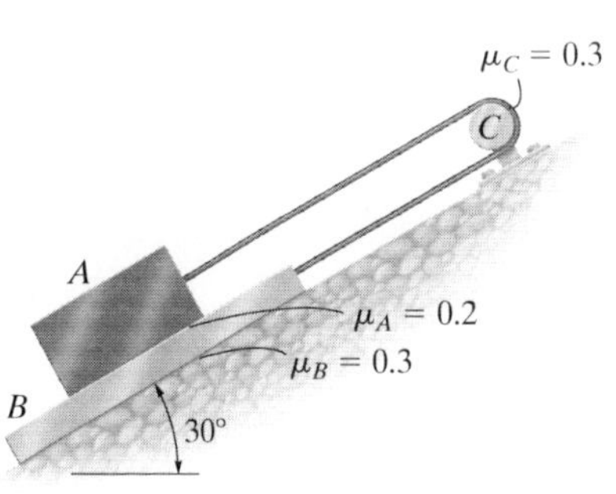

Prob. 8–91

8

***8–92.** Determine the smallest lever force P needed to prevent the wheel from rotating if it is subjected to a torque of $M = 250\ \text{N}\cdot\text{m}$. The coefficient of static friction between the belt and the wheel is $\mu_s = 0.3$. The wheel is pin-connected at its center, B.

8–93. Determine the torque M that can be resisted by the band brake if a force of $P = 30$ N is applied to the handle of the lever. The coefficient of static friction between the belt and the wheel is $\mu_s = 0.3$. The wheel is pin-connected at its center, B.

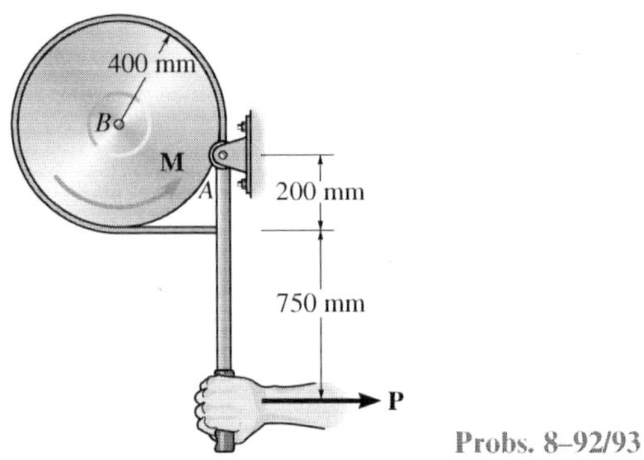

Probs. 8–92/93

***8–96.** The 20-kg motor has a center of gravity at G and is pin-connected at C to maintain a tension in the drive belt. Determine the smallest counterclockwise twist or torque **M** that must be supplied by the motor to turn the disk B if wheel A locks and causes the belt to slip over the disk. No slipping occurs at A. The coefficient of static friction between the belt and the disk is $\mu_s = 0.3$.

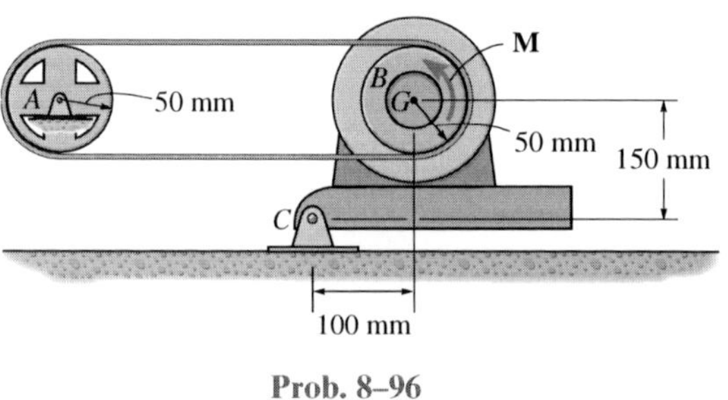

Prob. 8–96

8–94. Block A has a mass of 50 kg and rests on surface for which $\mu_s = 0.25$. If the coefficient of static friction between the cord and the fixed peg at C is $\mu_s' = 0.3$, determine the greatest mass of the suspended cylinder D without causing motion.

8–95. Block A rests on the surface for which $\mu_s = 0.25$. If the mass of the suspended cylinder D is 4 kg, determine the smallest mass of block A so that it does not slip or tip. The coefficient of static friction between the cord and the fixed peg at C is $\mu_s' = 0.3$.

8

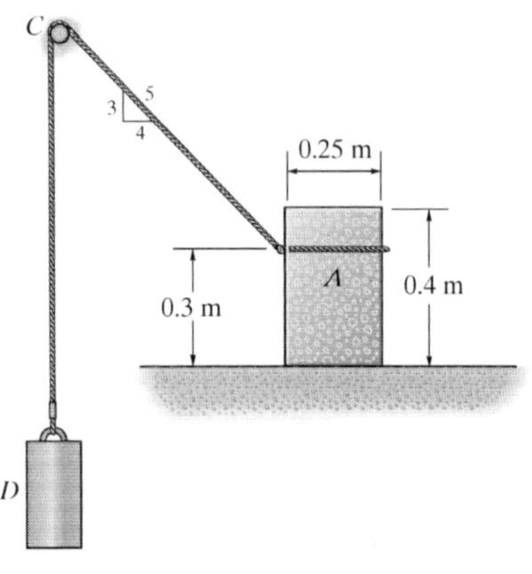

Probs. 8–94/95

8–97. Determine the smallest force **P** required to lift the 40-kg crate. The coefficient of static friction between the cable and each peg is $\mu_s = 0.1$.

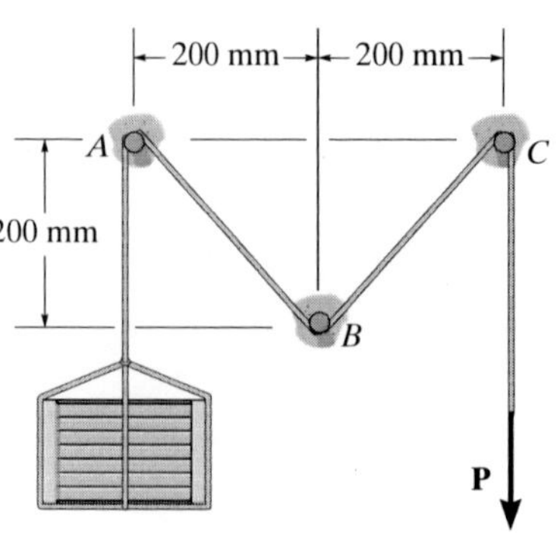

Prob. 8–97

8–98. Show that the frictional relationship between the belt tensions, the coefficient of friction μ, and the angular contacts α and β for the V-belt is $T_2 = T_1 e^{\mu\beta/\sin(\alpha/2)}$.

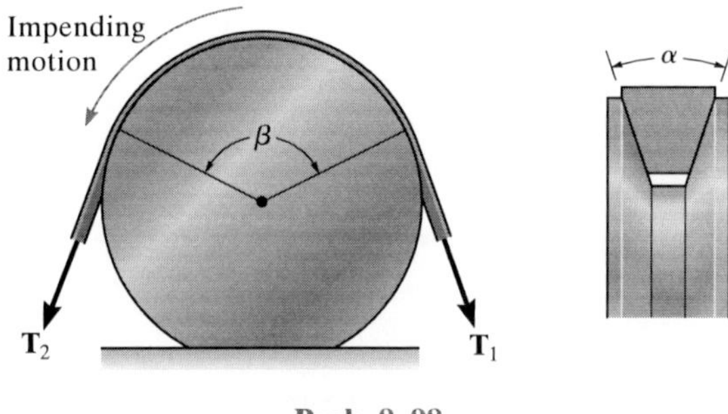

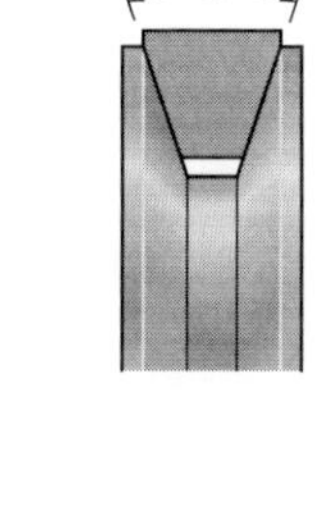

Prob. 8–98

*__8–100.__ A 10-kg cylinder D, which is attached to a small pulley B, is placed on the cord as shown. Determine the largest angle θ so that the cord does not slip over the peg at C. The cylinder at E also has a mass of 10 kg, and the coefficient of static friction between the cord and the peg is $\mu_s = 0.1$.

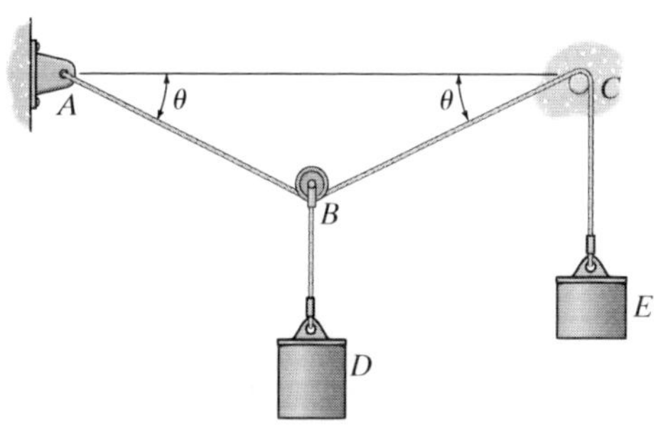

Prob. 8–100

8–99. If a force of $P = 200$ N is applied to the handle of the bell crank, determine the maximum torque **M** that can be resisted so that the flywheel does not rotate clockwise. The coefficient of static friction between the brake band and the rim of the wheel is $\mu_s = 0.3$.

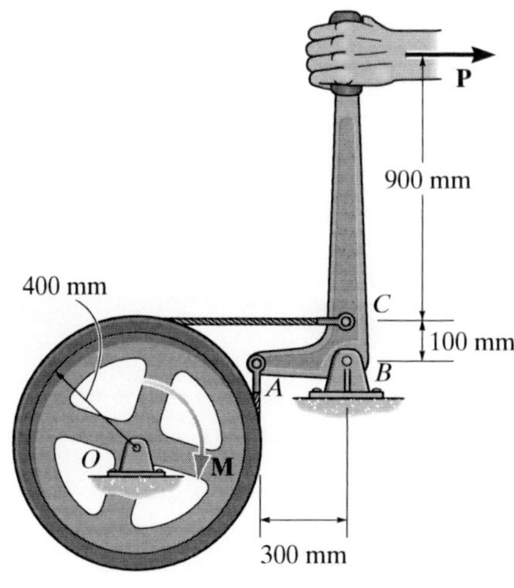

Prob. 8–99

8–101. A V-belt is used to connect the hub A of the motor to wheel B. If the belt can withstand a maximum tension of 1200 N, determine the largest mass of cylinder C that can be lifted and the corresponding torque **M** that must be supplied to A. The coefficient of static friction between the hub and the belt is $\mu_s = 0.3$, and between the wheel and the belt is $\mu_s' = 0.20$. *Hint*: See Prob. 8–98.

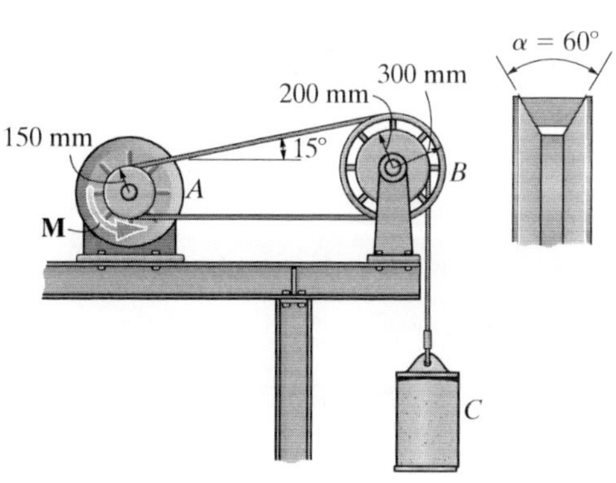

Prob. 8–101

8

8–102. The 20-kg motor has a center of gravity at G and is pin-connected at C to maintain a tension in the drive belt. Determine the smallest counterclockwise twist or torque **M** that must be supplied by the motor to turn the disk B if wheel A locks and causes the belt to slip over the disk. No slipping occurs at A. The coefficient of static friction between the belt and the disk is $\mu_s = 0.35$.

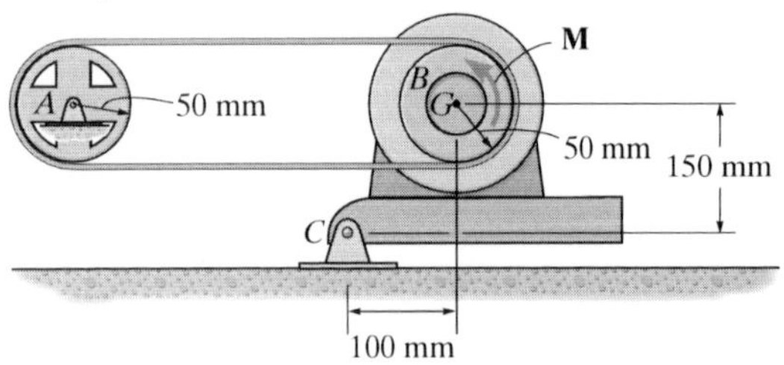

Prob. 8–102

8–103. Blocks A and B have a mass of 100 kg and 150 kg, respectively. If the coefficient of static friction between A and B and between B and C is $\mu_s = 0.25$ and between the ropes and the pegs D and E $\mu'_s = 0.5$ determine the smallest force F needed to cause motion of block B if $P = 30$ N.

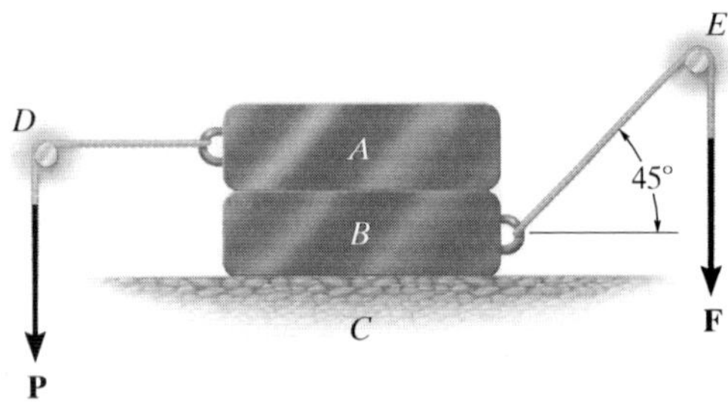

Prob. 8–103

***8–104.** Determine the minimum coefficient of static friction μ_s between the cable and the peg and the placement d of the 3-kN force for the uniform 100-kg beam to maintain equilibrium.

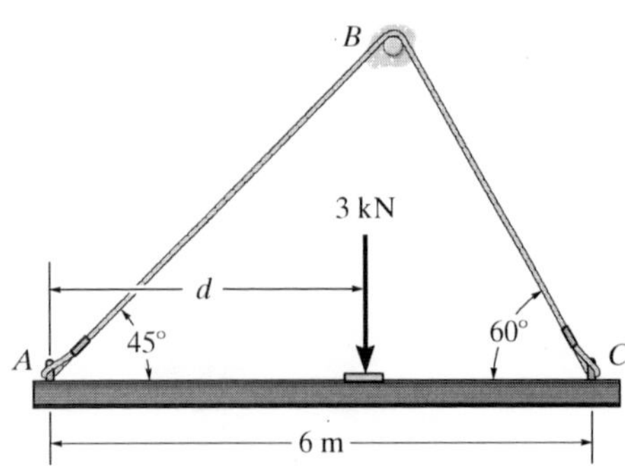

Prob. 8–104

8–105. A conveyer belt is used to transfer granular material and the frictional resistance on the top of the belt is $F = 500$ N. Determine the smallest stretch of the spring attached to the moveable axle of the idle pulley B so that the belt does not slip at the drive pulley A when the torque **M** is applied. What minimum torque **M** is required to keep the belt moving? The coefficient of static friction between the belt and the wheel at A is $\mu_s = 0.2$.

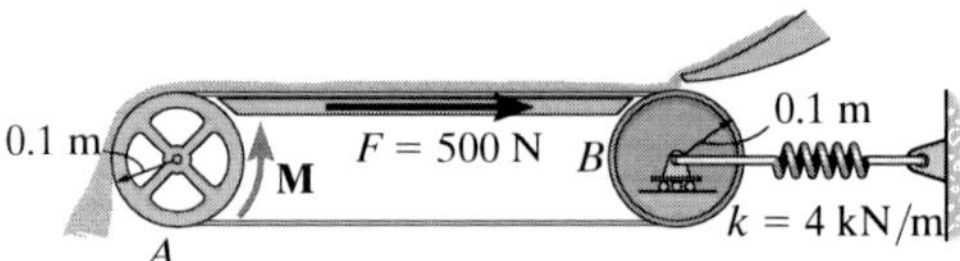

Prob. 8–105

8–106. The belt on the portable dryer wraps around the drum D, idler pulley A, and motor pulley B. If the motor can develop a maximum torque of $M = 0.80$ N · m, determine the smallest spring tension required to hold the belt from slipping. The coefficient of static friction between the belt and the drum and motor pulley is $\mu_s = 0.3$.

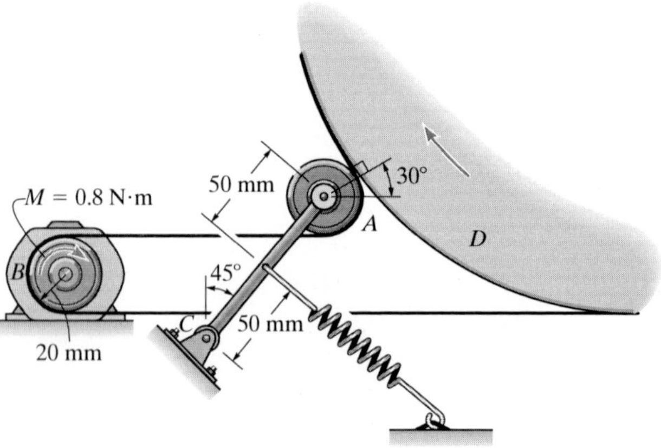

Prob. 8–106

*8.6 Frictional Forces on Collar Bearings, Pivot Bearings, and Disks

Pivot and *collar bearings* are commonly used in machines to support an *axial load* on a rotating shaft. Typical examples are shown in Fig. 8–20. Provided these bearings are not lubricated, or are only partially lubricated, the laws of dry friction may be applied to determine the moment needed to turn the shaft when it supports an axial force.

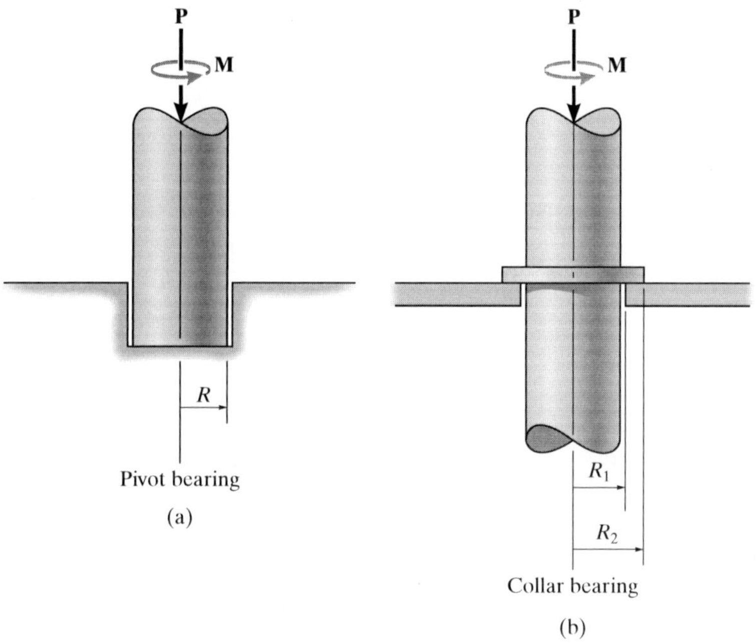

Fig. 8–20

Frictional Analysis. The collar bearing on the shaft shown in Fig. 8–21 is subjected to an axial force **P** and has a total bearing or contact area $\pi(R_2^2 - R_1^2)$. Provided the bearing is new and evenly supported, then the normal pressure p on the bearing will be *uniformly distributed* over this area. Since $\Sigma F_z = 0$, then p, measured as a force per unit area, is $p = P/\pi(R_2^2 - R_1^2)$.

The moment needed to cause impending rotation of the shaft can be determined from moment equilibrium about the z axis. A differential area element $dA = (r\,d\theta)(dr)$, shown in Fig. 8–21, is subjected to both a normal force $dN = p\,dA$ and an associated frictional force,

$$dF = \mu_s\,dN = \mu_s p\,dA = \frac{\mu_s P}{\pi(R_2^2 - R_1^2)}dA$$

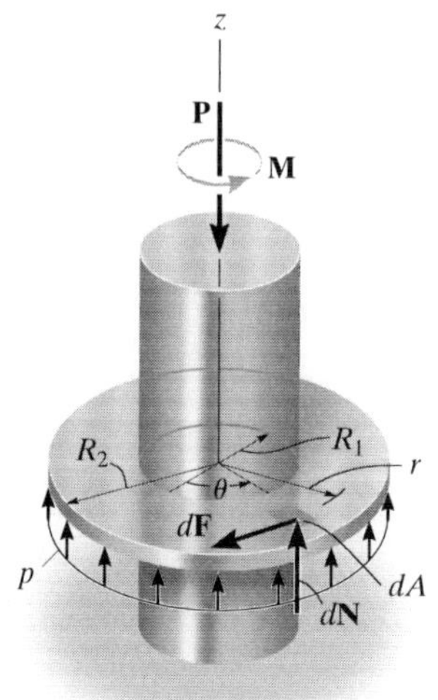

Fig. 8–21

8

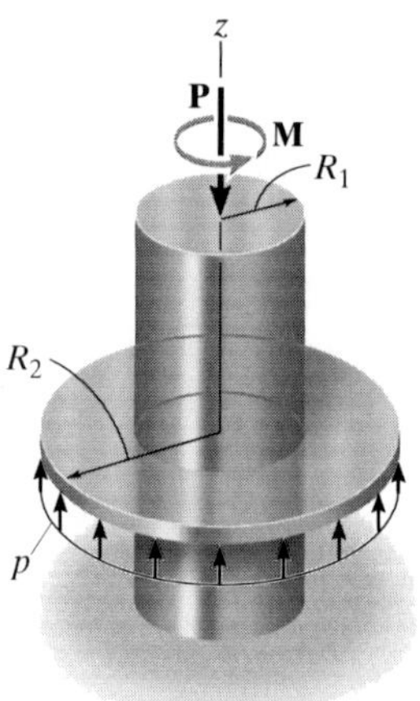

Fig. 8–21 (Repeated)

The normal force does not create a moment about the z axis of the shaft; however, the frictional force does; namely, $dM = r\,dF$. Integration is needed to compute the applied moment **M** needed to overcome all the frictional forces. Therefore, for impending rotational motion,

$$\Sigma M_z = 0; \qquad M - \int_A r\,dF = 0$$

Substituting for dF and dA and integrating over the entire bearing area yields

$$M = \int_{R_1}^{R_2}\int_0^{2\pi} r\left[\frac{\mu_s P}{\pi(R_2^2 - R_1^2)}\right](r\,d\theta\,dr) = \frac{\mu_s P}{\pi(R_2^2 - R_1^2)}\int_{R_1}^{R_2} r^2\,dr\int_0^{2\pi} d\theta$$

or

$$M = \frac{2}{3}\mu_s P\left(\frac{R_2^3 - R_1^3}{R_2^2 - R_1^2}\right) \qquad (8\text{–}7)$$

The moment developed at the end of the shaft, when it is *rotating* at constant speed, can be found by substituting μ_k for μ_s in Eq. 8–7.

In the case of a pivot bearing, Fig. 8–20*a*, then $R_2 = R$ and $R_1 = 0$, and Eq. 8–7 reduces to

$$M = \frac{2}{3}\mu_s PR \qquad (8\text{–}8)$$

Remember that Eqs. 8–7 and 8–8 apply only for bearing surfaces subjected to *constant pressure*. If the pressure is not uniform, a variation of the pressure as a function of the bearing area must be determined before integrating to obtain the moment. The following example illustrates this concept.

8

The motor that turns the disk of this sanding machine develops a torque that must overcome the frictional forces acting on the disk.

EXAMPLE 8.9

The uniform bar shown in Fig. 8–22*a* has a weight of 20 N(≈2 kg). If it is assumed that the normal pressure acting at the contacting surface varies linearly along the length of the bar as shown, determine the couple moment **M** required to rotate the bar. Assume that the bar's width is negligible in comparison to its length. The coefficient of static friction is equal to $\mu_s = 0.3$.

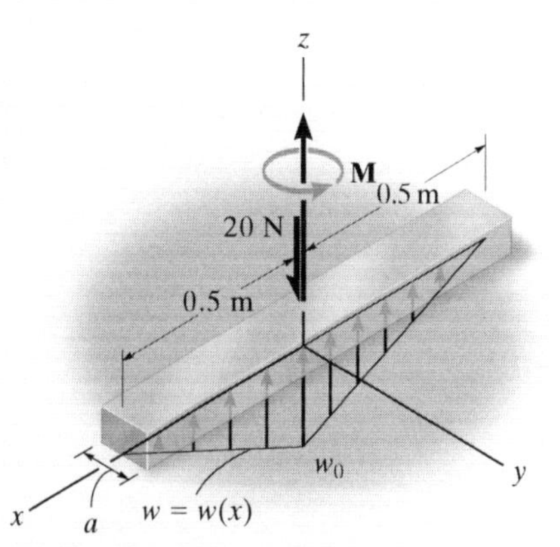

(a)

SOLUTION

A free-body diagram of the bar is shown in Fig. 8–22*b*. The intensity w_0 of the distributed load at the center ($x = 0$) is determined from vertical force equilibrium, Fig. 8–22*a*.

$$+\uparrow\Sigma F_z = 0; \quad -20\text{ N} + 2\left[\frac{1}{2}\left(0.5\text{ m}\right)w_0\right] = 0 \qquad w_0 = 40\text{ N/m}$$

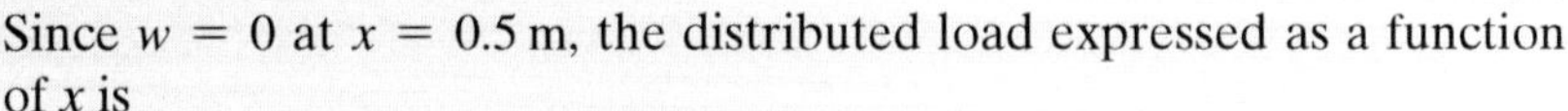

Since $w = 0$ at $x = 0.5$ m, the distributed load expressed as a function of x is

$$w = (40\text{ N/m})\left(1 - \frac{x}{0.5\text{ m}}\right) = 40 - 80x$$

The magnitude of the normal force acting on a differential segment of area having a length dx is therefore

$$dN = w\,dx = (40 - 80x)dx$$

The magnitude of the frictional force acting on the same element of area is

$$dF = \mu_s\,dN = 0.3(40 - 80x)dx$$

Hence, the moment created by this force about the z axis is

$$dM = x\,dF = 0.3(40 - 80x^2)dx$$

The summation of moments about the z axis of the bar is determined by integration, which yields

$$\Sigma M_z = 0; \qquad M - 2\int_0^{0.5\text{ m}} (0.3)(40x - 80x^2)\,dx = 0$$

$$M = 6\left(x^2 - \frac{4x^3}{3}\right)\Big|_0^{0.5}$$

$$M = 0.5\text{ N}\cdot\text{m}$$ *Ans.*

(b)

Fig. 8–22

8

8.7 Frictional Forces on Journal Bearings

Unwinding the cable from this spool requires overcoming friction from the supporting shaft.

When a shaft or axle is subjected to lateral loads, a *journal bearing* is commonly used for support. Provided the bearing is not lubricated, or is only partially lubricated, a reasonable analysis of the frictional resistance on the bearing can be based on the laws of dry friction.

Frictional Analysis. A typical journal-bearing support is shown in Fig. 8–23*a*. As the shaft rotates, the contact point moves up the wall of the bearing to some point A where slipping occurs. If the vertical load acting at the end of the shaft is $\mathbf{P}$, then the bearing reactive force $\mathbf{R}$ acting at A will be equal and opposite to $\mathbf{P}$, Fig. 8–23*b*. The moment needed to maintain constant rotation of the shaft can be found by summing moments about the z axis of the shaft; i.e.,

$$\Sigma M_z = 0; \qquad M - (R \sin \phi_k)r = 0$$

or

$$M = Rr \sin \phi_k \tag{8–9}$$

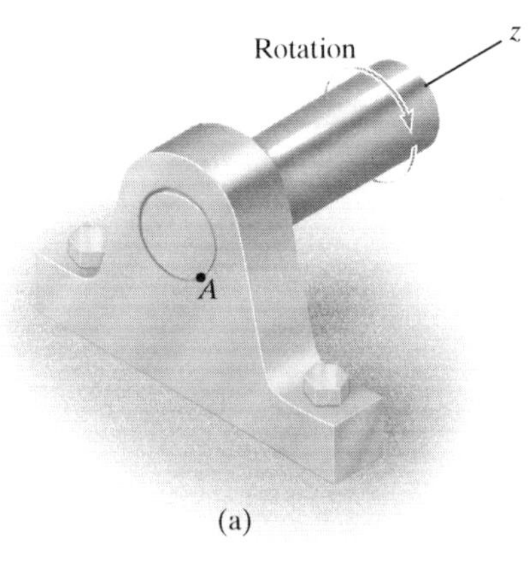

(a)

Fig. 8–23

where ϕ_k is the angle of kinetic friction defined by $\tan \phi_k = F/N = \mu_k N/N = \mu_k$. In Fig. 8–23*c*, it is seen that $r \sin \phi_k = r_f$. The dashed circle with radius r_f is called the *friction circle*, and as the shaft rotates, the reaction $\mathbf{R}$ will always be tangent to it. If the bearing is partially lubricated, μ_k is small, and therefore $\sin \phi_k \approx \tan \phi_k \approx \mu_k$. Under these conditions, a reasonable *approximation* to the moment needed to overcome the frictional resistance becomes

$$M \approx Rr\mu_k \tag{8–10}$$

Notice that to minimize friction the bearing radius r should be as small as possible. In practice, however, this type of journal bearing is not suitable for long service since friction between the shaft and bearing will eventually wear down the surfaces. Instead, designers will incorporate "ball bearings" or "rollers" in journal bearings to minimize frictional losses.

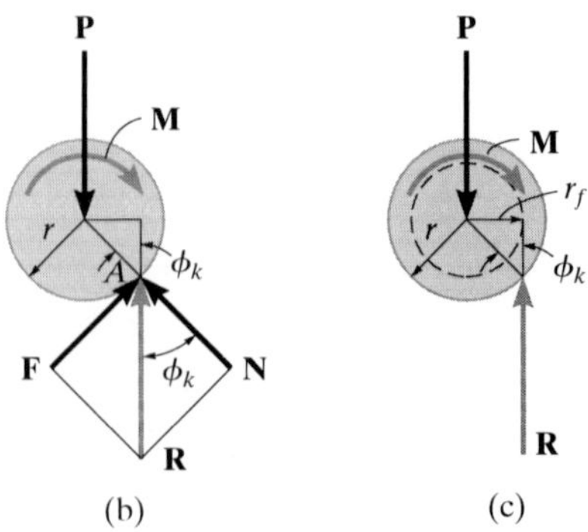

8

EXAMPLE 8.10

The 100-mm-diameter pulley shown in Fig. 8–24*a* fits loosely on a 10-mm-diameter shaft for which the coefficient of static friction is $\mu_s = 0.4$. Determine the minimum tension T in the belt needed to (a) raise the 100-kg block and (b) lower the block. Assume that no slipping occurs between the belt and pulley and neglect the weight of the pulley.

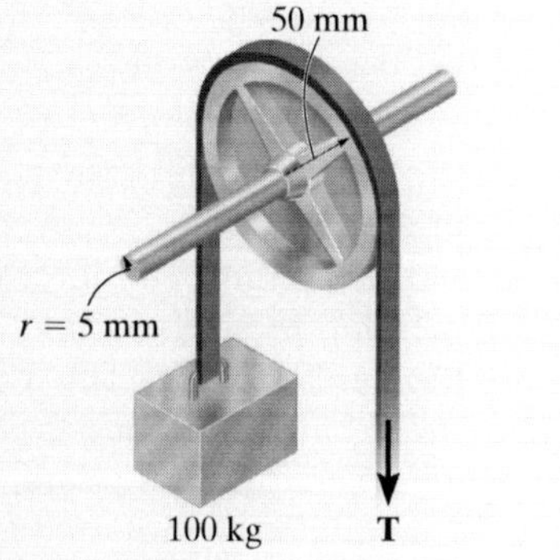

(a)

SOLUTION

Part (a). A free-body diagram of the pulley is shown in Fig. 8–24*b*. When the pulley is subjected to belt tensions of 981 N each, it makes contact with the shaft at point P_1. As the tension T is *increased*, the contact point will move around the shaft to point P_2 before motion impends. From the figure, the friction circle has a radius $r_f = r \sin \phi_s$. Using the simplification that $\sin \phi_s \approx \tan \phi_s \approx \mu_s$ then $r_f \approx r\mu_s = (5 \text{ mm})(0.4) = 2 \text{ mm}$, so that summing moments about P_2 gives

$$\circlearrowleft +\Sigma M_{P_2} = 0; \qquad 981 \text{ N}(52 \text{ mm}) - T(48 \text{ mm}) = 0$$
$$T = 1063 \text{ N} = 1.06 \text{ kN} \qquad \textit{Ans.}$$

If a more exact analysis is used, then $\phi_s = \tan^{-1} 0.4 = 21.8°$. Thus, the radius of the friction circle would be $r_f = r \sin \phi_s = 5 \sin 21.8° = 1.86 \text{ mm}$. Therefore,

$$\circlearrowleft +\Sigma M_{P_2} = 0;$$
$$981 \text{ N}(50 \text{ mm} + 1.86 \text{ mm}) - T(50 \text{ mm} - 1.86 \text{ mm}) = 0$$
$$T = 1057 \text{ N} = 1.06 \text{ kN} \qquad \textit{Ans.}$$

Part (b). When the block is lowered, the resultant force **R** acting on the shaft passes through point as shown in Fig. 8–24*c*. Summing moments about this point yields

$$\circlearrowleft +\Sigma M_{P_3} = 0; \qquad 981 \text{ N}(48 \text{ mm}) - T(52 \text{ mm}) = 0$$
$$T = 906 \text{ N} \qquad \textit{Ans.}$$

NOTE: Using the approximate analysis, the difference between raising and lowering the block is thus 157 N.

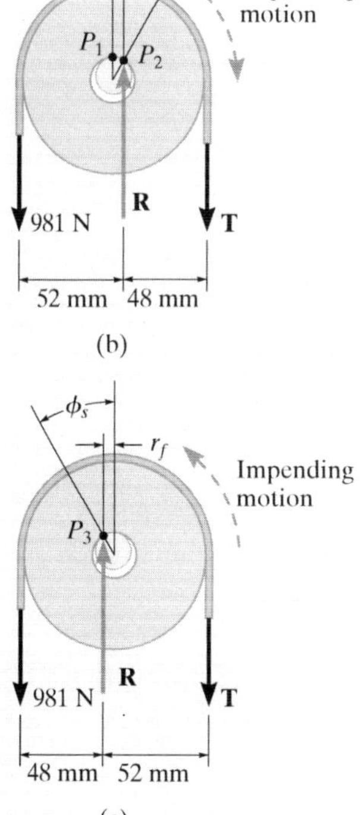

Fig. 8–24

8

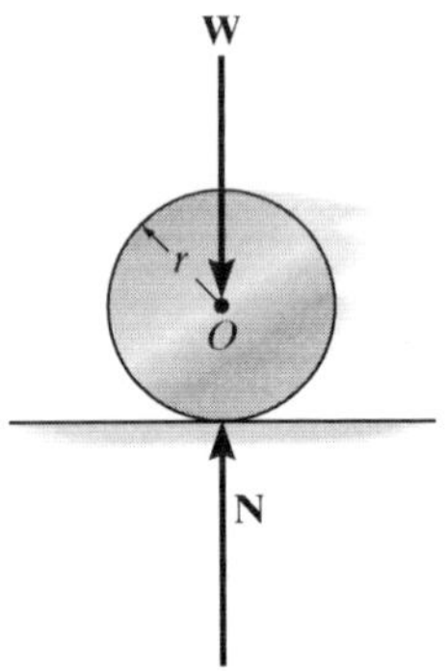

Rigid surface of contact

(a)

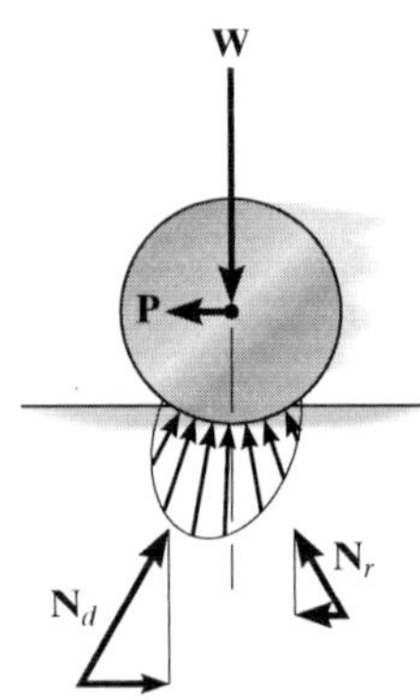

Soft surface of contact

(b)

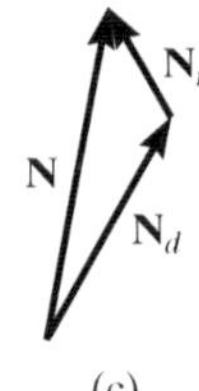

(c)

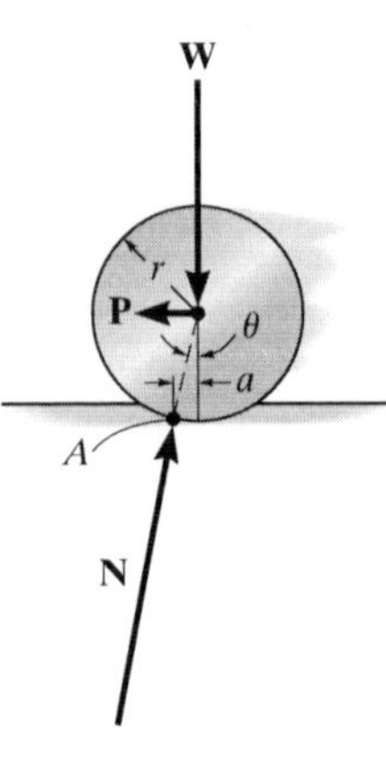

(d)

Fig. 8–25

*8.8 Rolling Resistance

When a *rigid* cylinder rolls at constant velocity along a *rigid* surface, the normal force exerted by the surface on the cylinder acts perpendicular to the tangent at the point of contact, as shown in Fig. 8–25*a*. Actually, however, no materials are perfectly rigid, and therefore the reaction of the surface on the cylinder consists of a distribution of normal pressure. For example, consider the cylinder to be made of a very hard material, and the surface on which it rolls to be relatively soft. Due to its weight, the cylinder compresses the surface underneath it, Fig. 8–25*b*. As the cylinder rolls, the surface material in front of the cylinder *retards* the motion since it is being *deformed*, whereas the material in the rear is *restored* from the deformed state and therefore tends to *push* the cylinder forward. The normal pressures acting on the cylinder in this manner are represented in Fig. 8–25*b* by their resultant forces $\mathbf{N}_d$ and $\mathbf{N}_r$. The magnitude of the force of *deformation*, $\mathbf{N}_d$, and its horizontal component is *always greater* than that of *restoration*, $\mathbf{N}_r$, and consequently a horizontal driving force **P** must be applied to the cylinder to maintain the motion. Fig. 8–25*b*.*

Rolling resistance is caused primarily by this effect, although it is also, to a lesser degree, the result of surface adhesion and relative micro-sliding between the surfaces of contact. Because the actual force **P** needed to overcome these effects is difficult to determine, a simplified method will be developed here to explain one way engineers have analyzed this phenomenon. To do this, we will consider the resultant of the *entire* normal pressure, $\mathbf{N} = \mathbf{N}_d + \mathbf{N}_r$, acting on the cylinder, Fig. 8–25*c*. As shown in Fig. 8–25*d*, this force acts at an angle θ with the vertical. To keep the cylinder in equilibrium, i.e., rolling at a constant rate, it is necessary that **N** be *concurrent* with the driving force **P** and the weight **W**. Summing moments about point A gives $Wa = P(r\cos\theta)$. Since the deformations are generally very small in relation to the cylinder's radius, $\cos\theta \approx 1$; hence,

$$Wa \approx Pr$$

or

$$P \approx \frac{Wa}{r} \qquad (8\text{–}11)$$

The distance a is termed the *coefficient of rolling resistance,* which has the dimension of length. For instance, $a \approx 0.5$ mm for a wheel rolling on a rail, both of which are made of mild steel. For hardened steel ball

*Actually, the deformation force $\mathbf{N}_d$ causes *energy* to be stored in the material as its magnitude is increased, whereas the restoration force $\mathbf{N}_r$, as its magnitude is decreased, allows some of this energy to be released. The remaining energy is *lost* since it is used to heat up the surface, and if the cylinder's weight is very large, it accounts for permanent deformation of the surface. Work must be done by the horizontal force **P** to make up for this loss.

bearings on steel, $a \approx 0.1$ mm. Experimentally, though, this factor is difficult to measure, since it depends on such parameters as the rate of rotation of the cylinder, the elastic properties of the contacting surfaces, and the surface finish. For this reason, little reliance is placed on the data for determining a. The analysis presented here does, however, indicate why a heavy load (W) offers greater resistance to motion (P) than a light load under the same conditions. Furthermore, since Wa/r is generally very small compared to $\mu_k W$, the force needed to *roll* a cylinder over the surface will be much less than that needed to *slide* it across the surface. It is for this reason that a roller or ball bearings are often used to minimize the frictional resistance between moving parts.

Rolling resistance of railroad wheels on the rails is small since steel is very stiff. By comparison, the rolling resistance of the wheels of a tractor in a wet field is very large.

EXAMPLE 8.11

A 10-kg steel wheel shown in Fig. 8–26*a* has a radius of 100 mm and rests on an inclined plane made of soft wood. If θ is increased so that the wheel begins to roll down the incline with constant velocity when $\theta = 1.2°$, determine the coefficient of rolling resistance.

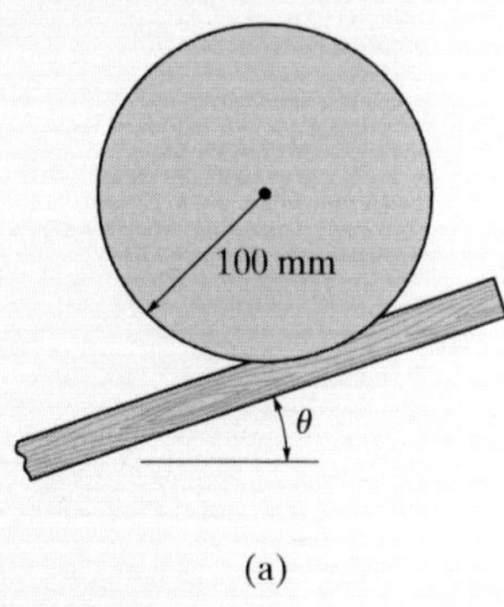

(a)

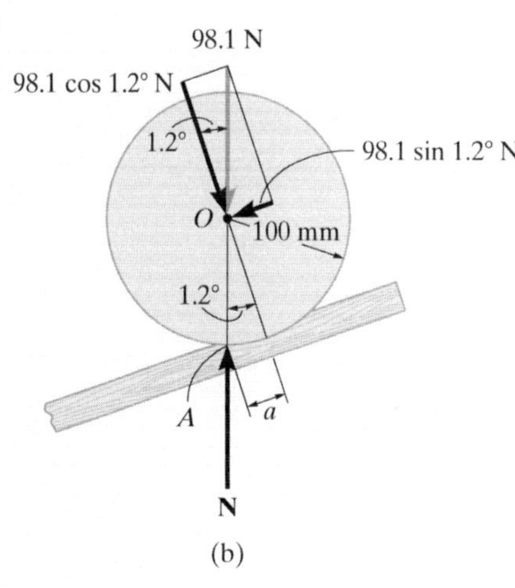

(b)

Fig. 8–26

SOLUTION

As shown on the free-body diagram, Fig. 8–26*b*, when the wheel has impending motion, the normal reaction **N** acts at point A defined by the dimension a. Resolving the weight into components parallel and perpendicular to the incline, and summing moments about point A, yields

$\circlearrowleft + \Sigma M_A = 0;$

$$-(98.1 \cos 1.2° \text{ N})(a) + (98.1 \sin 1.2° \text{ N})(100 \cos 1.2° \text{ mm}) = 0$$

Solving, we obtain

$$a = 2.09 \text{ mm} \qquad \textit{Ans.}$$

PROBLEMS

8–107. The collar bearing uniformly supports an axial force of $P = 2000$ N. If the coefficient of static friction is $\mu_s = 0.3$, determine the torque M required to overcome friction.

***8–108.** The collar bearing uniformly supports an axial force of $P = 2000$ N. If a torque of $M = 3.6$ N · m is applied to the shaft and causes it to rotate at constant velocity, determine the coefficient of kinetic friction at the surface of contact.

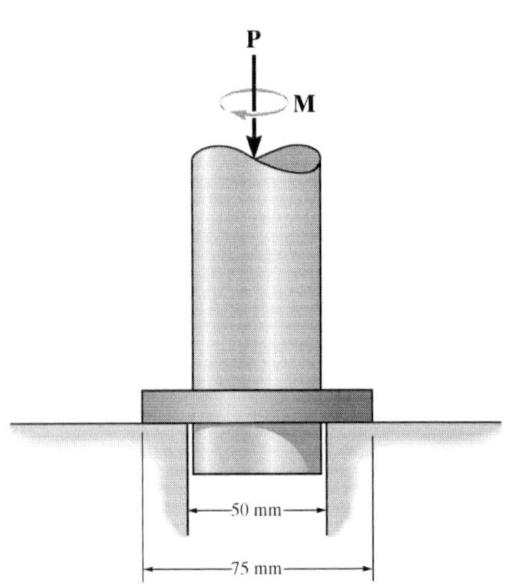

Probs. 8–107/108

8–109. The annular ring bearing is subjected to a thrust of 4000 N. If $\mu_s = 0.35$, determine the torque M that must be applied to overcome friction.

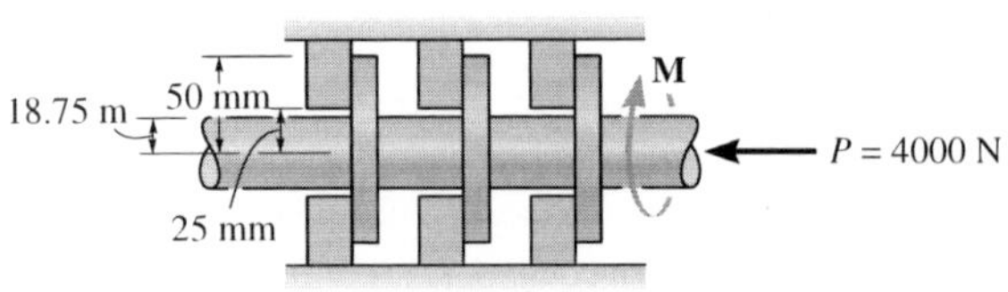

Prob. 8–109

8–110. The shaft is supported by a thrust bearing A and a journal bearing B. Determine the torque **M** required to rotate the shaft at constant angular velocity. The coefficient of kinetic friction at the thrust bearing is $\mu_k = 0.2$. Neglect friction at B.

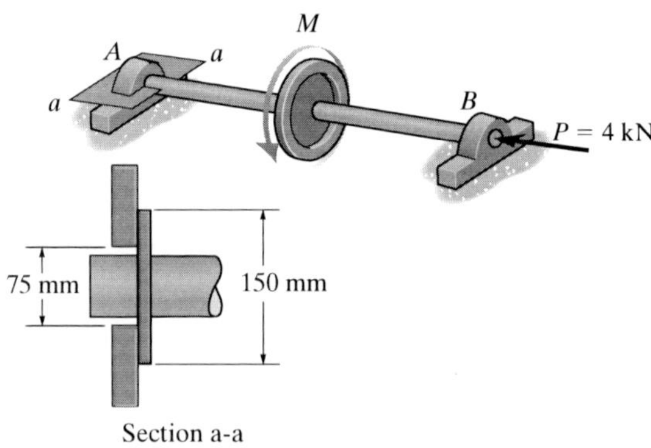

Prob. 8–110

8–111. The thrust bearing supports an axial load of $P = 6$ kN. If a torque of $M = 150$ N·m is required to rotate the shaft, determine the coefficient of static friction at the constant surface.

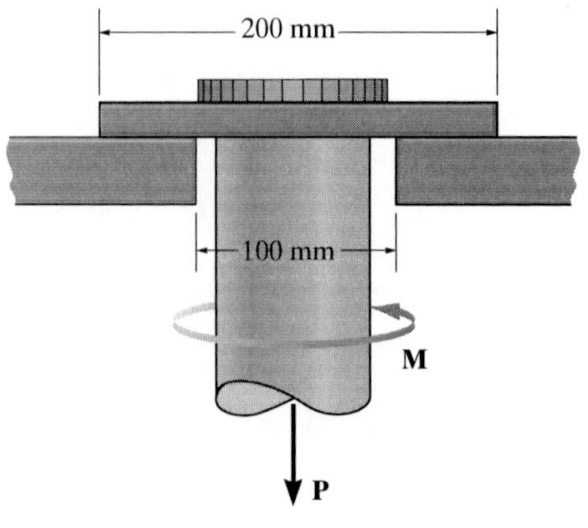

Prob. 8–111

***8–112.** Assuming that the variation of pressure at the bottom of the pivot bearing is defined as $p = p_0(R_2/r)$, determine the torque M needed to overcome friction if the shaft is subjected to an axial force **P**. The coefficient of static friction is μ_s. For the solution, it is necessary to determine p_0 in terms of P and the bearing dimensions R_1 and R_2.

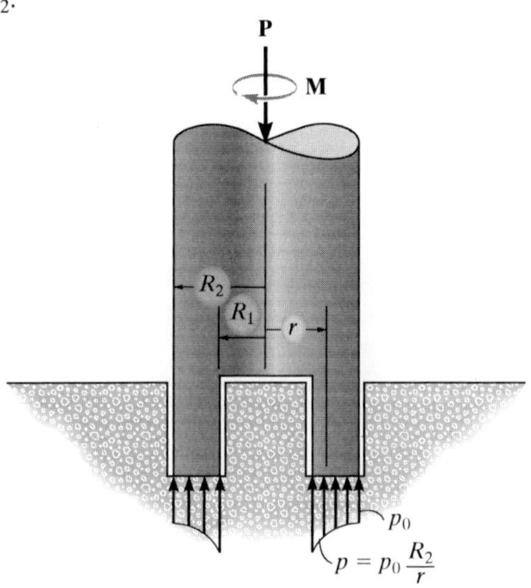

Prob. 8–112

8–113. The plate clutch consists of a flat plate A that slides over the rotating shaft S. The shaft is fixed to the driving plate gear B. If the gear C, which is in mesh with B, is subjected to a torque of $M = 0.8\ \text{N}\cdot\text{m}$, determine the smallest force P, that must be applied via the control arm, to stop the rotation. The coefficient of static friction between the plates A and D is $\mu_s = 0.4$. Assume the bearing pressure between A and D to be uniform.

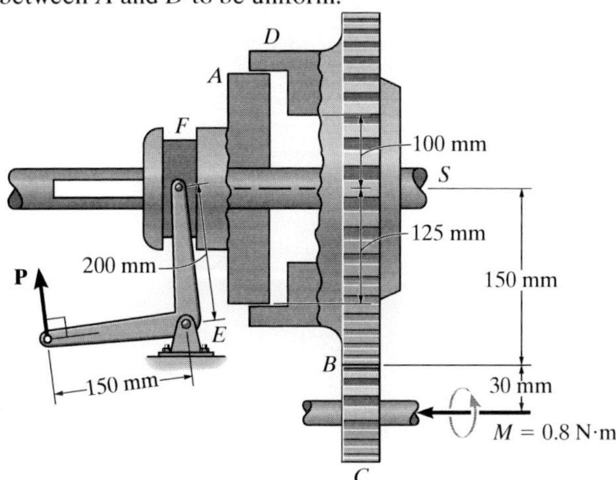

Prob. 8–113

8–114. The conical bearing is subjected to a constant pressure distribution at its surface of contact. If the coefficient of static friction is μ_s, determine the torque M required to overcome friction if the shaft supports an axial force **P**.

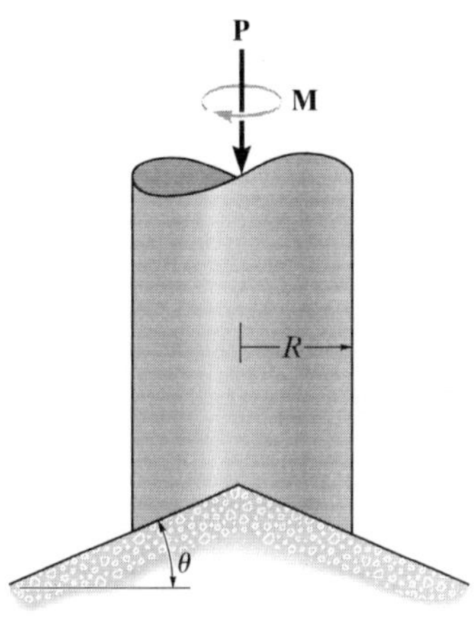

Prob. 8–114

8–115. The pivot bearing is subjected to a pressure distribution at its surface of contact which varies as shown. If the coefficient of static friction is μ, determine the torque M required to overcome friction if the shaft supports an axial force **P**.

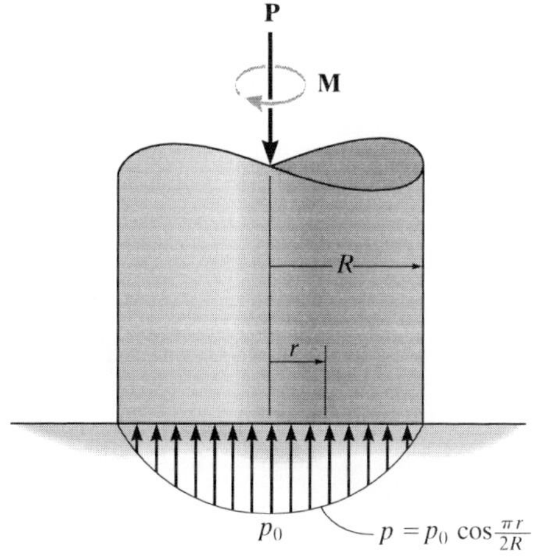

Prob. 8–115

8

***8–116.** A 200-mm-diameter post is driven 3 m into sand for which $\mu_s = 0.3$. If the normal pressure acting *completely around the post* varies linearly with depth as shown, determine the frictional torque **M** that must be overcome to rotate the post.

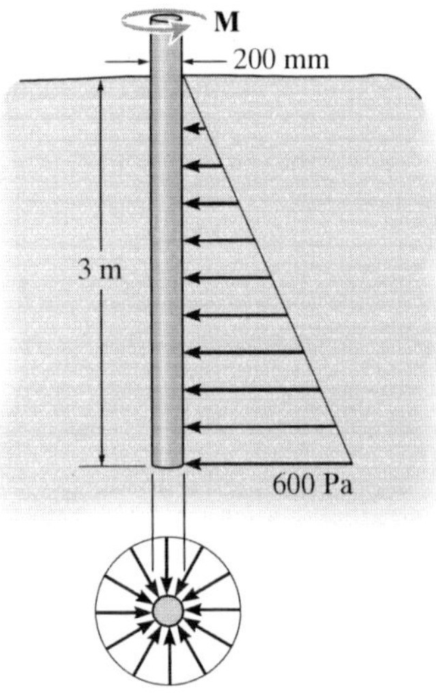

Prob. 8–116

8–117. A beam having a uniform weight W rests on the rough horizontal surface having a coefficient of static friction μ_s. If the horizontal force **P** is applied perpendicular to the beam's length, determine the location d of the point O about which the beam begins to rotate.

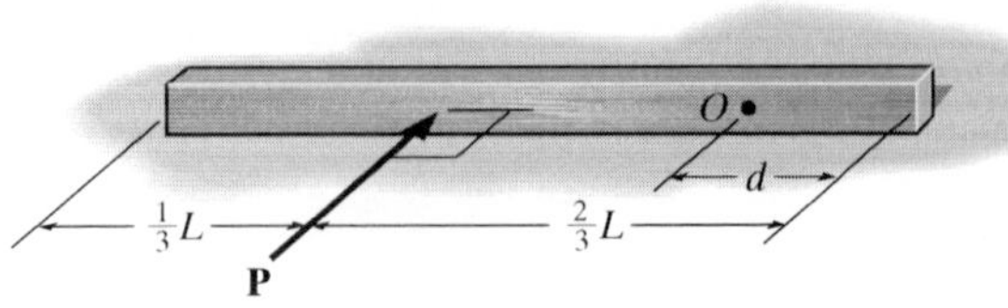

Prob. 8–117

8–118. The connecting rod is attached to the piston by a 20-mm-diameter pin at B and to the crank shaft by a 50-mm-diameter bearing A. If the piston is moving downwards, and the coefficient of static friction at these points is $\mu_s = 0.2$, determine the radius of the friction circle at each connection.

8–119. The connecting rod is attached to the piston by a 20-mm-diameter pin at B and to the crank shaft by a 50-mm-diameter bearing A. If the piston is moving upwards, and the coefficient of static friction at these points is $\mu_s = 0.3$, determine the radius of the friction circle at each connection.

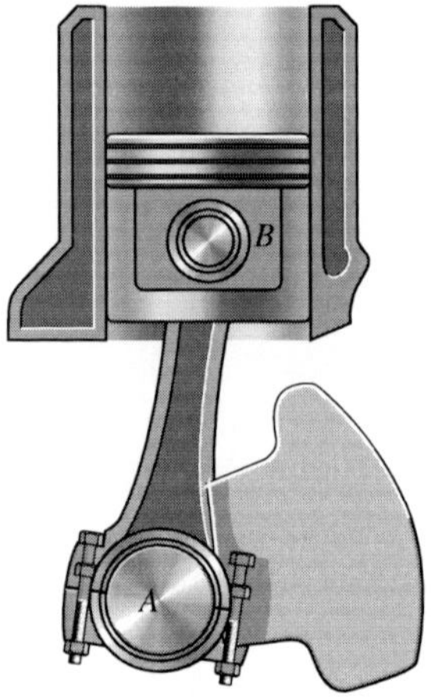

Probs. 8–118/119

***8–120.** The 5-kg pulley has a diameter of 240 mm and the axle has a diameter of 40 mm. If the coefficient of kinetic friction between the axle and the pulley is $\mu_k = 0.15$, determine the vertical force P on the rope required to lift the 80-kg block at constant velocity.

8–121. Solve Prob. 8–120 if the force **P** is applied horizontally to the right.

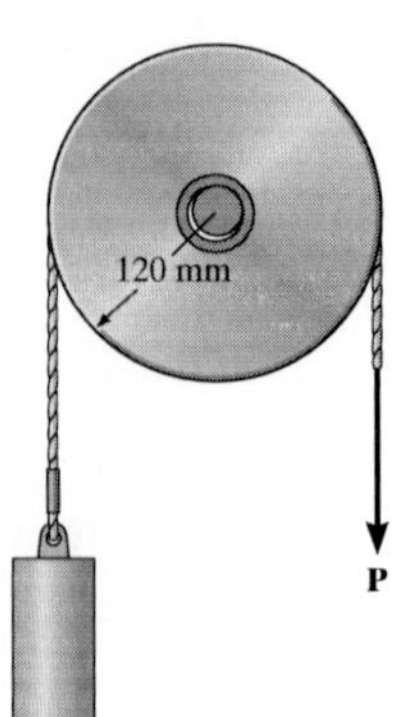

Probs. 8–120/121

8–122. The collar fits *loosely* around a fixed shaft that has a radius of 50 mm. If the coefficient of kinetic friction between the shaft and the collar is $\mu_k = 0.3$, determine the force P on the horizontal segment of the belt so that the collar rotates *counterclockwise* with a constant angular velocity. Assume that the belt does not slip on the collar; rather, the collar slips on the shaft. Neglect the weight and thickness of the belt and collar. The radius, measured from the center of the collar to the mean thickness of the belt, is 56.25 mm.

8–123. The collar fits *loosely* around a fixed shaft that has a radius of 50 mm. If the coefficient of kinetic friction between the shaft and the collar is $\mu_k = 0.3$, determine the force P on the horizontal segment of the belt so that the collar rotates *clockwise* with a constant angular velocity. Assume that the belt does not slip on the collar; rather, the collar slips on the shaft. Neglect the weight and thickness of the belt and collar. The radius, measured from the center of the collar to the mean thickness of the belt, is 56.25 mm.

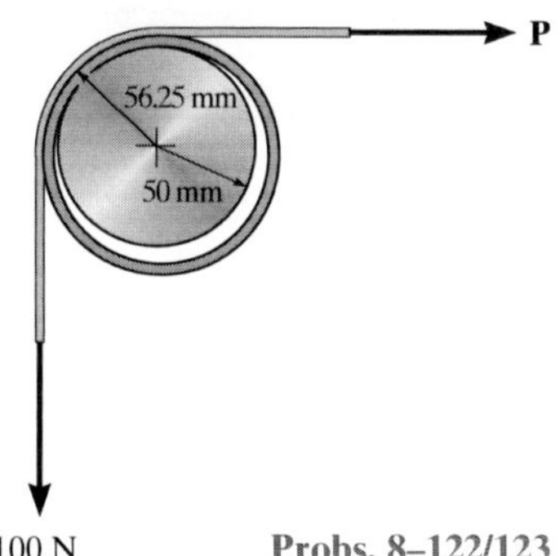

Probs. 8–122/123

***8–124.** A pulley having a diameter of 80 mm and mass of 1.25 kg is supported loosely on a shaft having a diameter of 20 mm. Determine the torque M that must be applied to the pulley to cause it to rotate with constant motion. The coefficient of kinetic friction between the shaft and pulley is $\mu_k = 0.4$. Also calculate the angle θ which the normal force at the point of contact makes with the horizontal. The shaft itself cannot rotate.

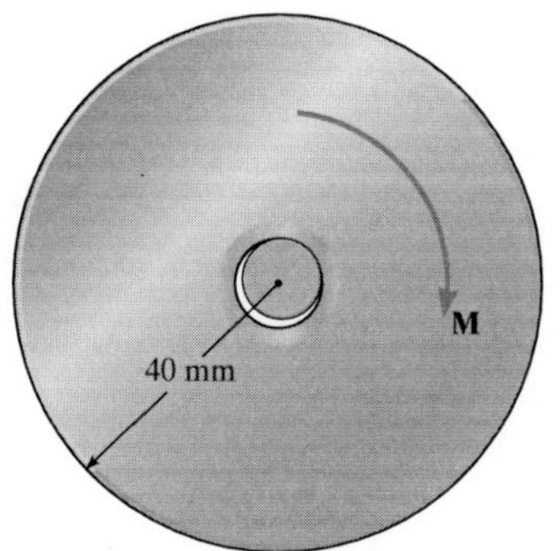

Prob. 8–124

8–125. The 5-kg skateboard rolls down the 5° slope at constant speed. If the coefficient of kinetic friction between the 12.5 mm diameter axles and the wheels is $\mu_k = 0.3$, determine the radius of the wheels. Neglect rolling resistance of the wheels on the surface. The center of mass for the skateboard is at G.

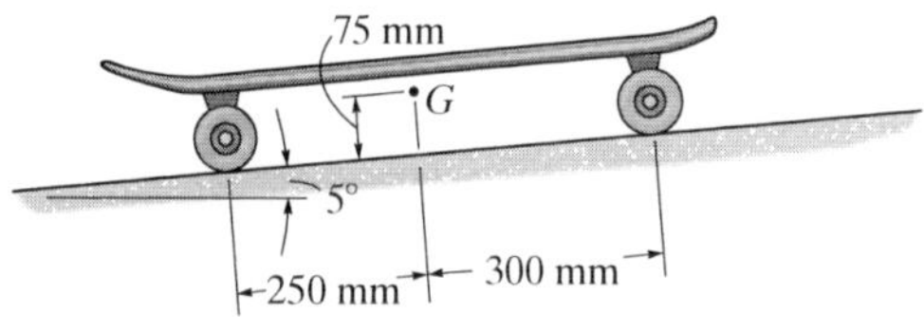

Prob. 8–125

8–126. Determine the force P required to overcome rolling resistance and pull the 50-kg roller up the inclined plane with constant velocity. The coefficient of rolling resistance is $a = 15$ mm.

8–127. Determine the force P required to overcome rolling resistance and support the 50-kg roller if it rolls down the inclined plane with constant velocity. The coefficient of rolling resistance is $a = 15$ mm.

8

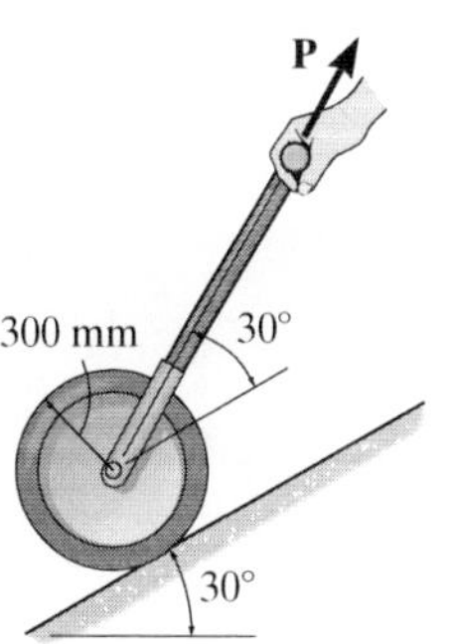

Probs. 8–126/127

*8–128. The lawn roller has a mass of 80 kg. If the arm BA is held at an angle of 30° from the horizontal and the coefficient of rolling resistance for the roller is 25 mm, determine the force P needed to push the roller at constant speed. Neglect friction developed at the axle, A, and assume that the resultant force **P** acting on the handle is applied along arm BA.

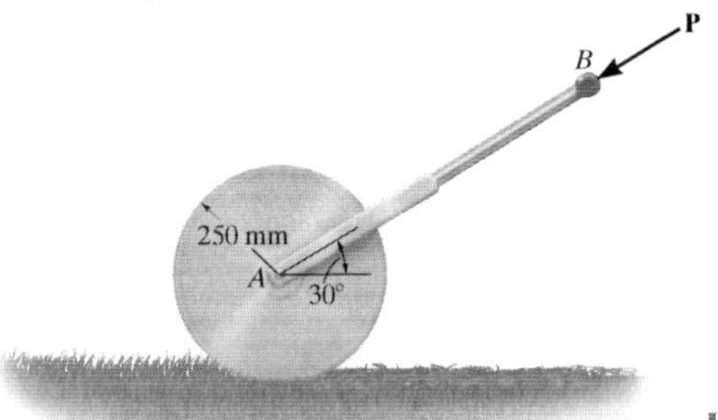

Prob. 8–128

8–129. The 1.4-Mg machine is to be moved over a level surface using a series of rollers for which the coefficient of rolling resistance is 0.5 mm at the ground and 0.2 mm at the bottom surface of the machine. Determine the appropriate diameter of the rollers so that the machine can be pushed forward with a horizontal force of $P = 250$ N. *Hint:* Use the result of Prob. 8–131.

Prob. 8–129

8 8–130. The hand cart has wheels with a diameter of 80 mm. If a crate having a mass of 500 kg is placed on the cart so that each wheel carries an equal load, determine the horizontal force P that must be applied to the handle to overcome the rolling resistance. The coefficient of rolling resistance is 2 mm. Neglect the mass of the cart.

Prob. 8–130

8–131. The cylinder is subjected to a load that has a weight W. If the coefficients of rolling resistance for the cylinder's top and bottom surfaces are a_A and a_B, respectively, show that a horizontal force having a magnitude of $P = [W(a_A + a_B)]/2r$ is required to move the load and thereby roll the cylinder forward. Neglect the weight of the cylinder.

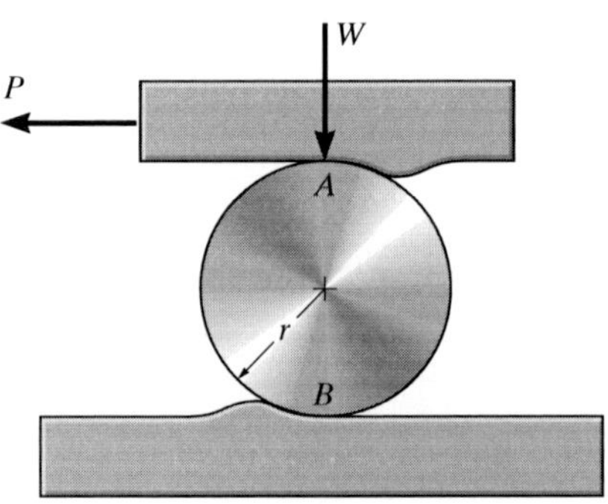

Prob. 8–131

*8–132. A large crate having a mass of 200 kg is moved along the floor using a series of 150-mm-diameter rollers for which the coefficient of rolling resistance is 3 mm at the ground and 7 mm at the bottom surface of the crate. Determine the horizontal force **P** needed to push the crate forward at a constant speed. *Hint:* Use the result of Prob. 8–131.

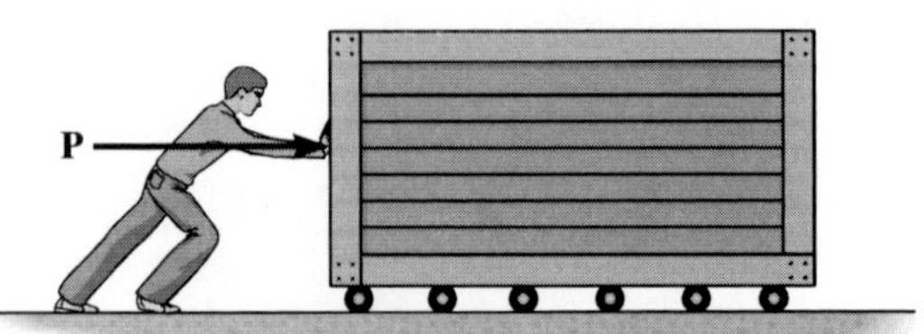

Prob. 8–132

CHAPTER REVIEW

Dry Friction

Frictional forces exist between two rough surfaces of contact. These forces act on a body so as to oppose its motion or tendency of motion.

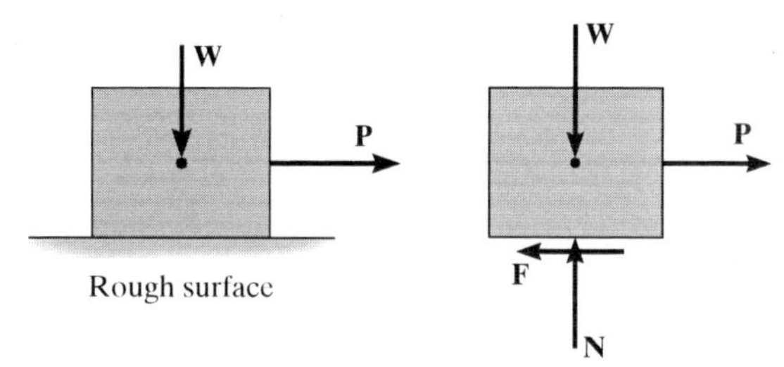

A static frictional force approaches a maximum value of $F_s = \mu_s N$, where μ_s is the *coefficient of static friction*. In this case, motion between the contacting surfaces is *impending*.

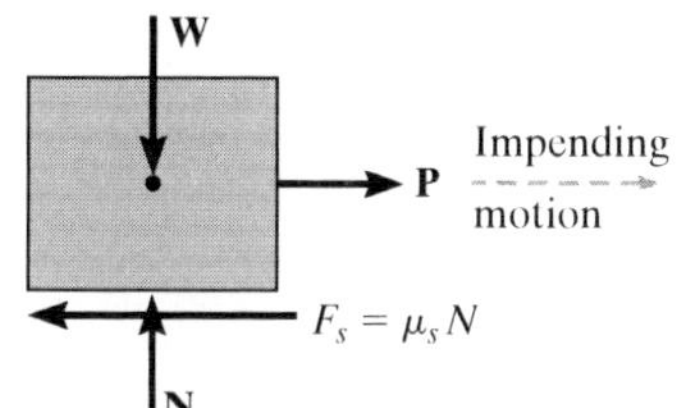

If slipping occurs, then the friction force remains essentially constant and equal to $F_k = \mu_k N$. Here μ_k is the *coefficient of kinetic friction*.

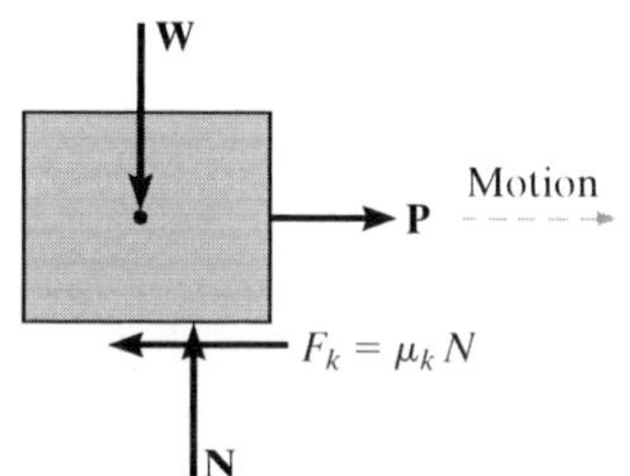

The solution of a problem involving friction requires first drawing the free-body diagram of the body. If the unknowns cannot be determined strictly from the equations of equilibrium, and the possibility of slipping occurs, then the friction equation should be applied at the appropriate points of contact in order to complete the solution.

It may also be possible for slender objects, like crates, to tip over, and this situation should also be investigated.

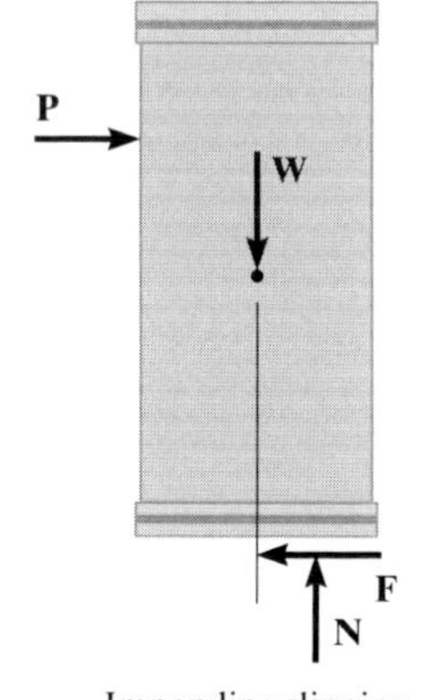

Impending slipping
$F = \mu_s N$

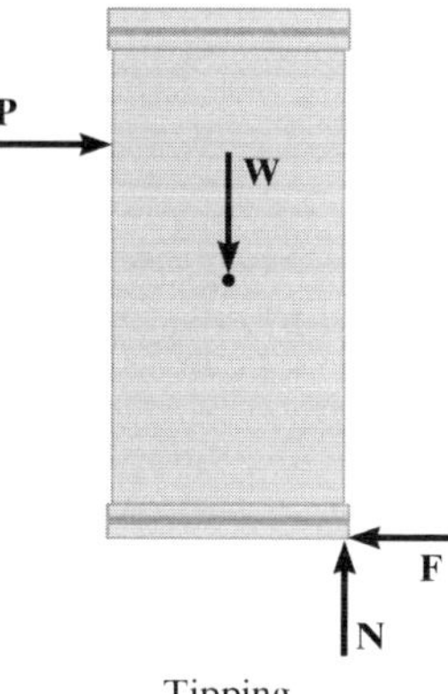

Tipping

8

Wedges

Wedges are inclined planes used to increase the application of a force. The two force equilibrium equations are used to relate the forces acting on the wedge.

An applied force **P** must push on the wedge to move it to the right.

If the coefficients of friction between the surfaces are large enough, then **P** can be removed, and the wedge will be self-locking and remain in place.

$$\Sigma F_x = 0$$
$$\Sigma F_y = 0$$

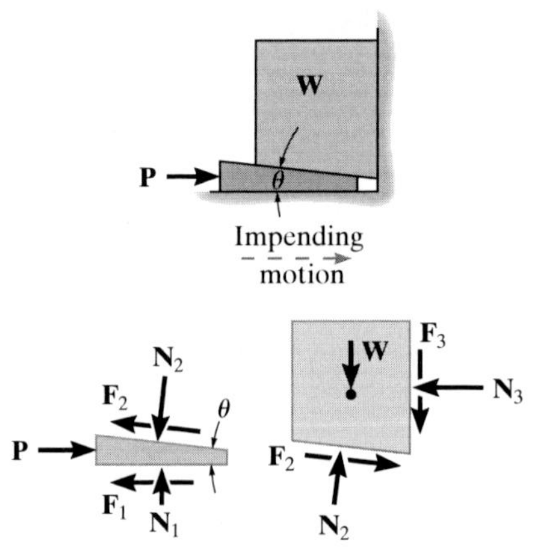

Screws

Square-threaded screws are used to move heavy loads. They represent an inclined plane, wrapped around a cylinder.

The moment needed to turn a screw depends upon the coefficient of friction and the screw's lead angle θ.

If the coefficient of friction between the surfaces is large enough, then the screw will support the load without tending to turn, i.e., it will be self-locking.

$$M = rW\tan(\theta + \phi_s)$$
Upward Impending Screw Motion

$$M' = rW\tan(\theta - \phi_s)$$
Downward Impending Screw Motion

$$\theta > \phi_s$$

$$M'' = rW\tan(\phi_s - \theta)$$
Downward Screw Motion

$$\phi_s > \theta$$

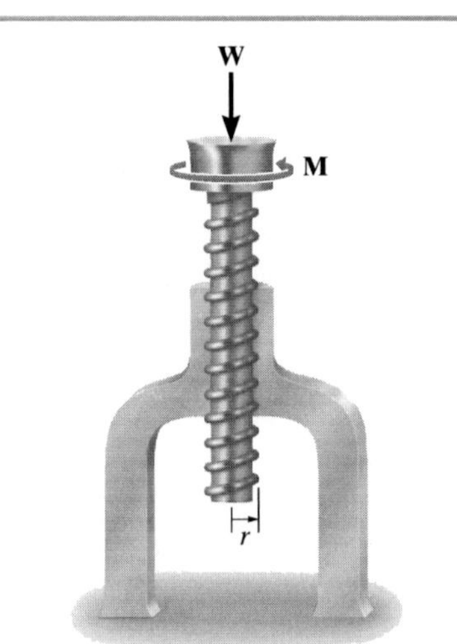

Flat Belts

The force needed to move a flat belt over a rough curved surface depends only on the angle of belt contact, β, and the coefficient of friction.

$$T_2 = T_1 e^{\mu\beta}$$
$$T_2 > T_1$$

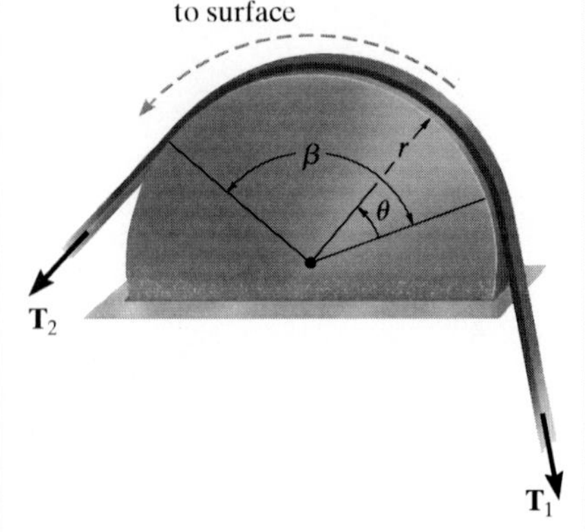

Collar Bearings and Disks The frictional analysis of a collar bearing or disk requires looking at a differential element of the contact area. The normal force acting on this element is determined from force equilibrium along the shaft, and the moment needed to turn the shaft at a constant rate is determined from moment equilibrium about the shaft's axis. If the pressure on the surface of a collar bearing is uniform, then integration gives the result shown.	$M = \frac{2}{3}\mu_s P\left(\frac{R_2^3 - R_1^3}{R_2^2 - R_1^2}\right)$	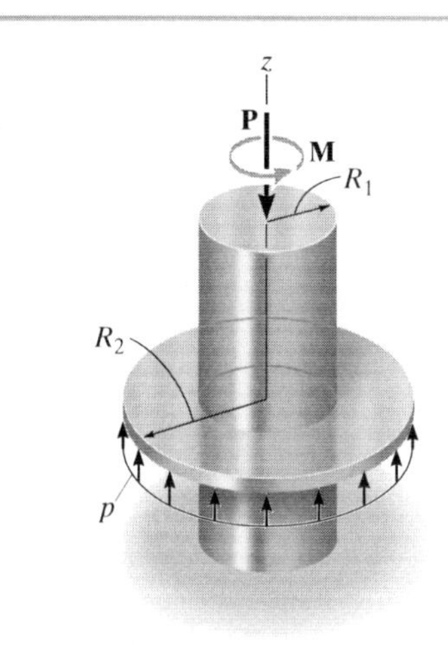
Journal Bearings When a moment is applied to a shaft in a nonlubricated or partially lubricated journal bearing, the shaft will tend to roll up the side of the bearing until slipping occurs. This defines the radius of a friction circle, and from it the moment needed to turn the shaft can be determined.	$M = Rr \sin \phi_k$ 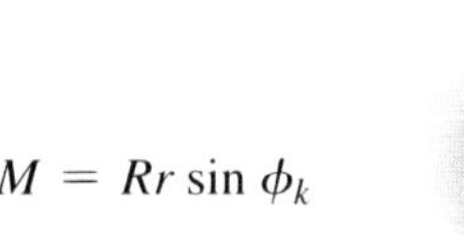 	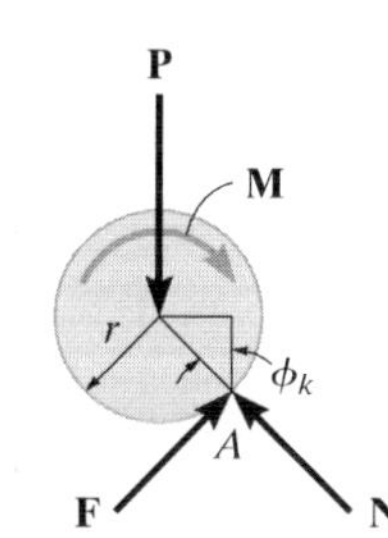
Rolling Resistance The resistance of a wheel to rolling over a surface is caused by localized *deformation* of the two materials in contact. This causes the resultant normal force acting on the rolling body to be inclined so that it provides a component that acts in the opposite direction of the applied force **P** causing the motion. This effect is characterized using the *coefficient of rolling resistance*, *a*, which is determined from experiment.	$P \approx \frac{Wa}{r}$	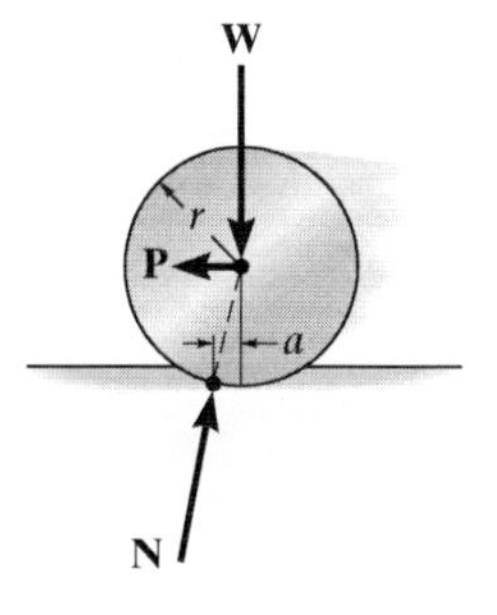

REVIEW PROBLEMS

8–133. Each of the cylinders has a mass of 50 kg. If the coefficients of static friction at the points of contact are $\mu_A = 0.5$, $\mu_B = 0.5$, $\mu_C = 0.5$, and $\mu_D = 0.6$, determine the smallest couple moment M needed to rotate cylinder E.

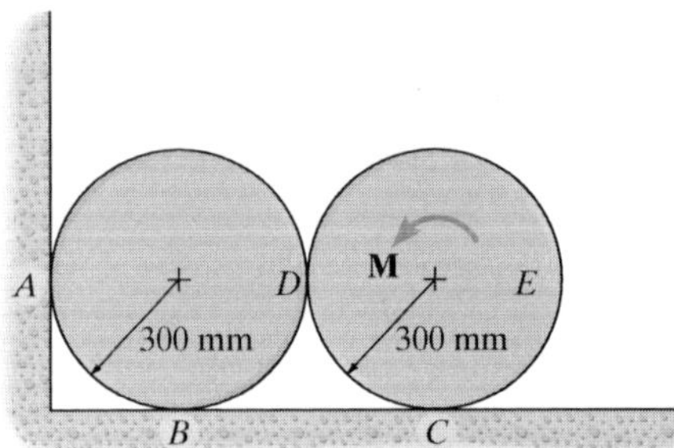

Prob. 8–133

8–134. The clamp is used to tighten the connection between two concrete drain pipes. Determine the least coefficient of static friction at A or B so that the clamp does not slip regardless of the force in the shaft CD.

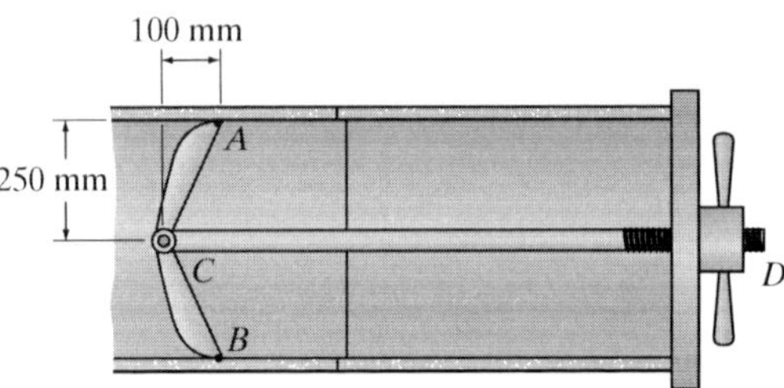

Prob. 8–134

8–135. If $P = 900$ N is applied to the handle of the bell crank, determine the maximum torque M the cone clutch can transmit. The coefficient of static friction at the contacting surface is $\mu_s = 0.3$.

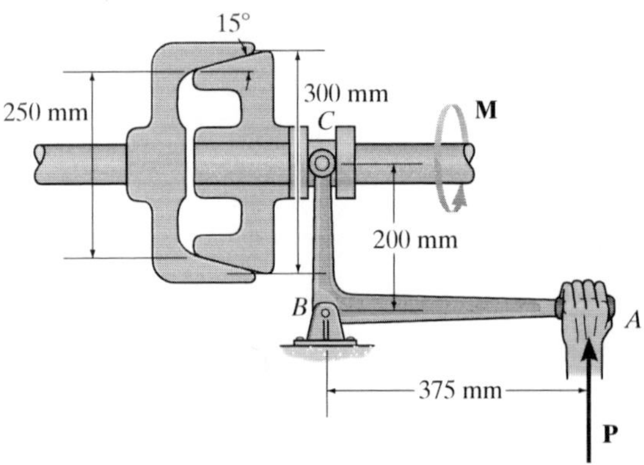

Prob. 8–135

***8–136.** The lawn roller has a mass of 80 kg. If the arm BA is held at an angle of 30° from the horizontal and the coefficient of rolling resistance for the roller is 25 mm, determine the force P needed to push the roller at constant speed. Neglect friction developed at the axle, A, and assume that the resultant force **P** acting on the handle is applied along arm BA.

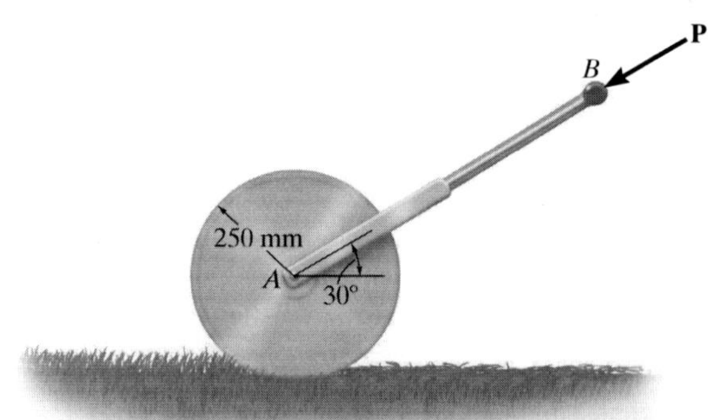

Prob. 8–136

8–137. Two blocks A and B, each having a mass of 6 kg, are connected by the linkage shown. If the coefficients of static friction at the contacting surfaces are $\mu_A = 0.2$ and $\mu_B = 0.8$, determine the largest vertical force P that may be applied to pin C without causing the blocks to slip. Neglect the weight of the links.

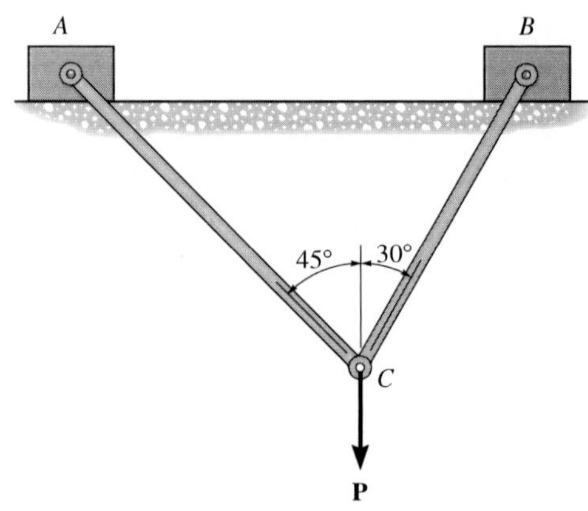

Prob. 8–137

8–138. The uniform 60-kg crate C rests uniformly on a 10-kg dolly D. If the front casters of the dolly at A are locked to prevent rolling while the casters at B are free to roll, determine the maximum force **P** that may be applied without causing motion of the crate. The coefficient of static friction between the casters and the floor is $\mu_f = 0.35$ and between the dolly and the crate, $\mu_d = 0.5$.

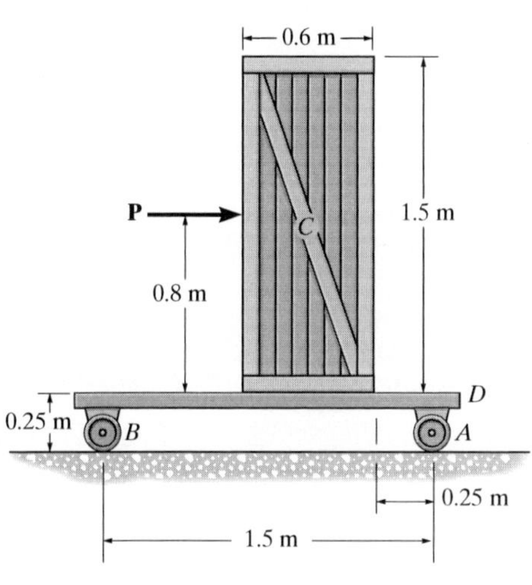

Prob. 8–138

8–139. The 3-Mg four-wheel-drive truck (SUV) has a center of mass at G. Determine the maximum mass of the log that can be towed by the truck. The coefficient of static friction between the log and the ground is $\mu_s = 0.8$, and the coefficient of static friction between the wheels of the truck and the ground is $\mu_s' = 0.4$. Assume that the engine of the truck is powerful enough to generate a torque that will cause all the wheels to slip.

***8–140.** A 3-Mg front-wheel-drive truck (SUV) has a center of mass at G. Determine the maximum mass of the log that can be towed by the truck. The coefficient of static friction between the log and the ground is $\mu_s = 0.8$, and the coefficient of static friction between the front wheels of the truck and the ground is $\mu_s' = 0.4$. The rear wheels are free to roll. Assume that the engine of the truck is powerful enough to generate a torque that will cause the front wheels to slip.

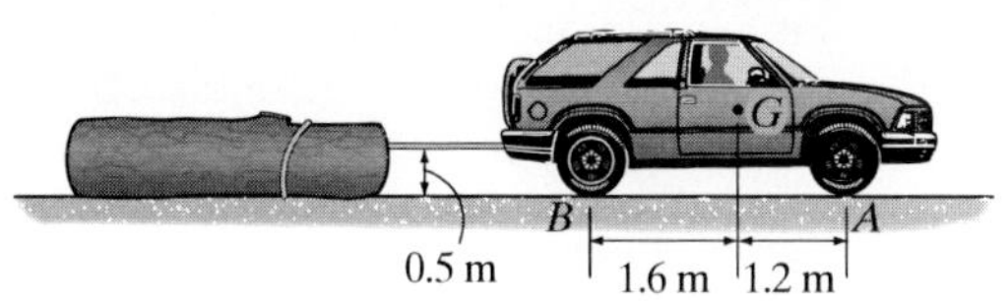

Probs. 8–139/140

8–141. A roofer, having a mass of 70 kg, walks slowly in an upright position down along the surface of a dome that has a radius of curvature of $r = 20$ m. If the coefficient of static friction between his shoes and the dome is $\mu_s = 0.7$, determine the angle θ at which he first begins to slip.

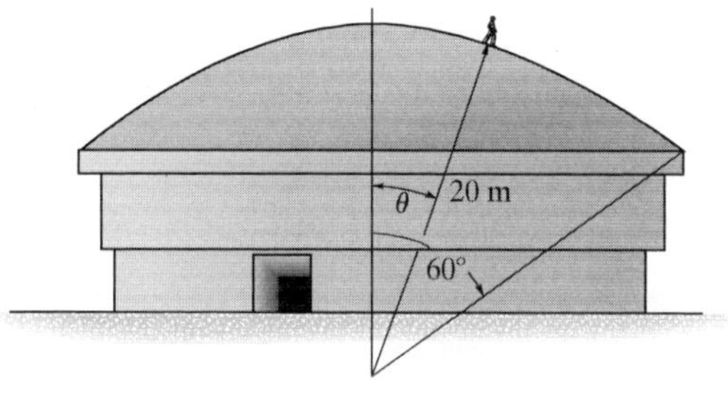

Prob. 8–141

8–142. Determine the minimum horizontal force P required to hold the crate from sliding down the plane. The crate has a mass of 50 kg and the coefficient of static friction between the crate and the plane is $\mu_s = 0.25$.

8–143. Determine the minimum force P required to push the crate up the plane. The crate has a mass of 50 kg and the coefficient of static friction between the crate and the plane is $\mu_s = 0.25$.

***8–144.** A horizontal force of $P = 100$ N is just sufficient to hold the crate from sliding down the plane, and a horizontal force of $P = 350$ N is required to just push the crate up the plane. Determine the coefficient of static friction between the plane and the crate, and find the mass of the crate.

8

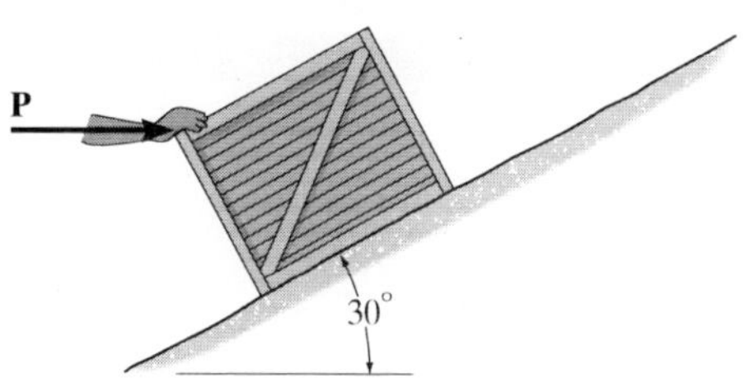

Probs. 8–142/143/144

Chapter 9

When a pressure tank of any shape is designed, it is important to be able to determine its center of gravity, calculate its volume and surface area, and determine the forces of the liquids they contain. All of these topics will be covered in this chapter.

Center of Gravity and Centroid

CHAPTER OBJECTIVES

- To discuss the concept of the center of gravity, center of mass, and the centroid.
- To show how to determine the location of the center of gravity and centroid for a system of discrete particles and a body of arbitrary shape.
- To use the theorems of Pappus and Guldinus for finding the surface area and volume for a body having axial symmetry.
- To present a method for finding the resultant of a general distributed loading and show how it applies to finding the resultant force of a pressure loading caused by a fluid.

Video Solutions are available for selected questions in this chapter.

9.1 Center of Gravity, Center of Mass, and the Centroid of a Body

In this section we will first show how to locate the center of gravity for a body, and then we will show that the center of mass and the centroid of a body can be developed using this same method.

Center of Gravity. A body is composed of an infinite number of particles of differential size, and so if the body is located within a gravitational field, then each of these particles will have a weight dW, Fig. 9–1*a*. These weights will form an approximately parallel force system, and the resultant of this system is the total weight of the body, which passes through a single point called the *center of gravity*, G, Fig. 9–1*b*.*

*This is true as long as the gravity field is assumed to have the same magnitude and direction everywhere. That assumption is appropriate for most engineering applications, since gravity does not vary appreciably between, for instance, the bottom and the top of a building.

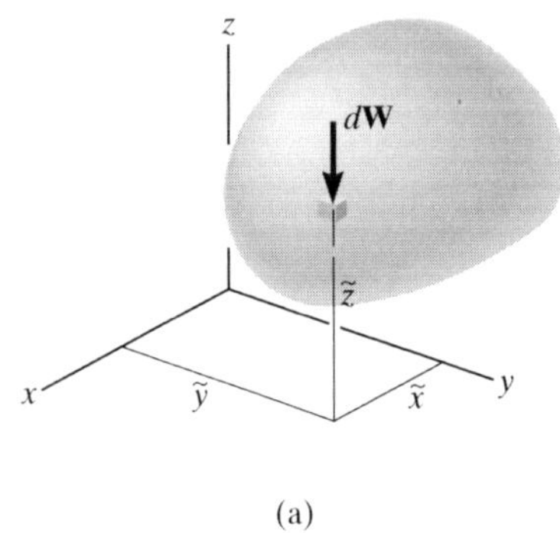

(a)

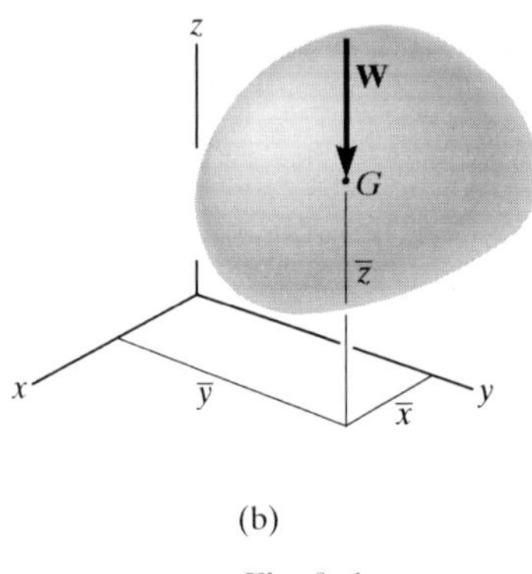

(b)

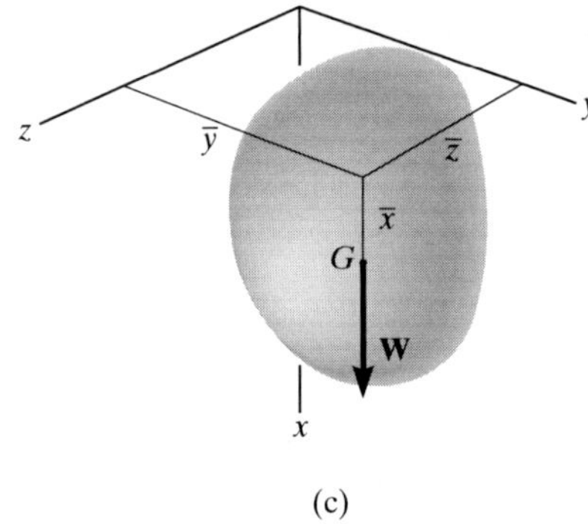

(c)

Fig. 9–1

Using the methods outlined in Sec. 4.8, the weight of the body is the sum of the weights of all of its particles, that is

$$+\downarrow F_R = \Sigma F_z; \qquad W = \int dW$$

The location of the center of gravity, measured from the y axis, is determined by equating the moment of W about the y axis, Fig. 9–1*b*, to the sum of the moments of the weights of the particles about this same axis. If dW is located at point $(\tilde{x}, \tilde{y}, \tilde{z})$, Fig. 9–1*a*, then

$$(M_R)_y = \Sigma M_y; \qquad \bar{x}W = \int \tilde{x}\,dW$$

Similarly, if moments are summed about the x axis,

$$(M_R)_x = \Sigma M_x; \qquad \bar{y}W = \int \tilde{y}\,dW$$

Finally, imagine that the body is fixed within the coordinate system and this system is rotated 90° about the y axis, Fig. 9–1*c*. Then the sum of the moments about the y axis gives

$$(M_R)_y = \Sigma M_y; \qquad \bar{z}W = \int \tilde{z}\,dW$$

Therefore, the location of the center of gravity G with respect to the x, y, z axes becomes

$$\bar{x} = \frac{\int \tilde{x}\,dW}{\int dW} \qquad \bar{y} = \frac{\int \tilde{y}\,dW}{\int dW} \qquad \bar{z} = \frac{\int \tilde{z}\,dW}{\int dW} \tag{9–1}$$

Here

$\bar{x}, \bar{y}, \bar{z}$ are the coordinates of the center of gravity G, Fig. 9–1*b*.
$\tilde{x}, \tilde{y}, \tilde{z}$ are the coordinates of each particle in the body, Fig. 9–1*a*.

9

Center of Mass of a Body. In order to study the *dynamic response* or accelerated motion of a body, it becomes important to locate the body's center of mass C_m, Fig. 9–2. This location can be determined by substituting $dW = g\,dm$ into Eqs. 9–1. Since g is constant, it cancels out, and so

$$\bar{x} = \frac{\int \tilde{x}\,dm}{\int dm} \qquad \bar{y} = \frac{\int \tilde{y}\,dm}{\int dm} \qquad \bar{z} = \frac{\int \tilde{z}\,dm}{\int dm} \tag{9–2}$$

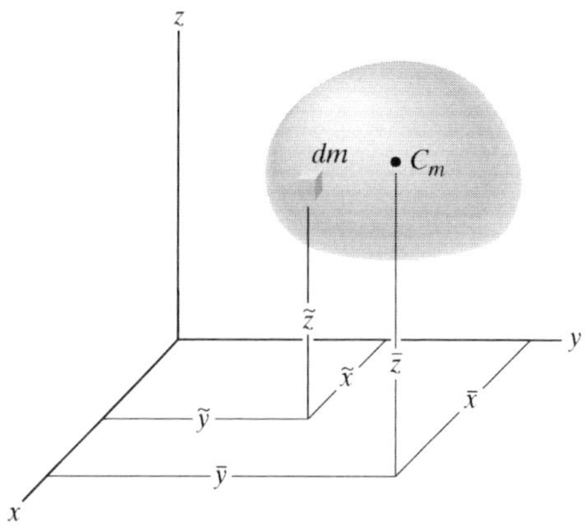

Fig. 9–2

Centroid of a Volume. If the body in Fig. 9–3 is made from a homogeneous material, then its density ρ (rho) will be constant. Therefore, a differential element of volume dV has a mass $dm = \rho\,dV$. Substituting this into Eqs. 9–2 and canceling out ρ, we obtain formulas that locate the *centroid C* or geometric center of the body; namely

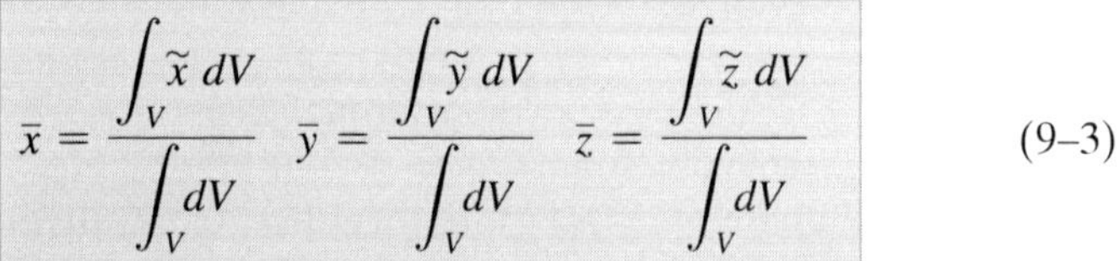

$$\bar{x} = \frac{\int_V \tilde{x}\,dV}{\int_V dV} \qquad \bar{y} = \frac{\int_V \tilde{y}\,dV}{\int_V dV} \qquad \bar{z} = \frac{\int_V \tilde{z}\,dV}{\int_V dV} \tag{9–3}$$

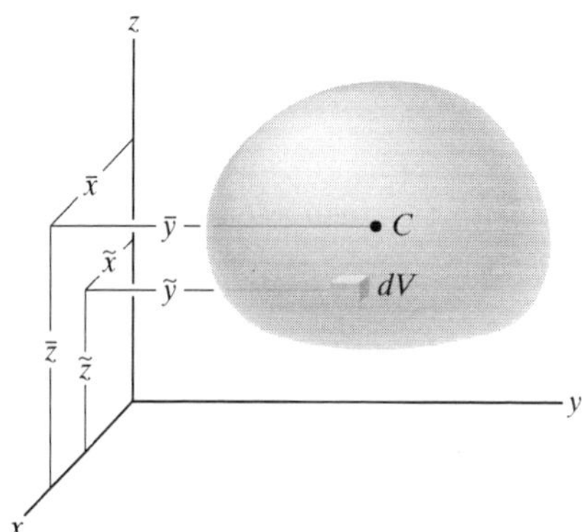

Fig. 9–3

These equations represent a balance of the moments of the volume of the body. Therefore, if the volume possesses two planes of symmetry, then its centroid must lie along the line of intersection of these two planes. For example, the cone in Fig. 9–4 has a centroid that lies on the y axis so that $\bar{x} = \bar{z} = 0$. The location $\bar{y}$ can be found using a single integration by choosing a differential element represented by a *thin disk* having a thickness dy and radius $r = z$. Its volume is $dV = \pi r^2\,dy = \pi z^2\,dy$ and its centroid is at $\tilde{x} = 0$, $\tilde{y} = y$, $\tilde{z} = 0$.

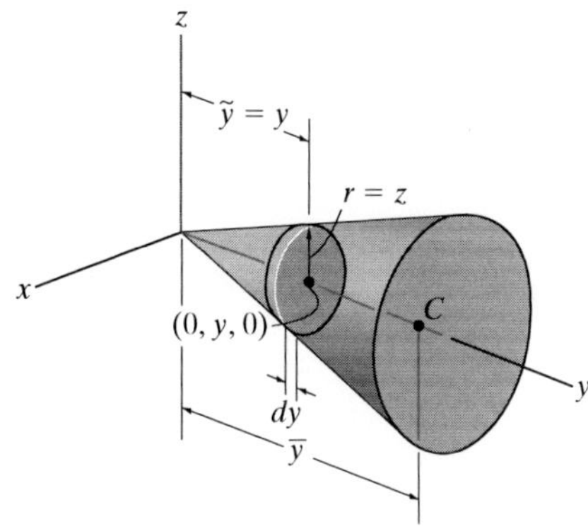

Fig. 9–4

9

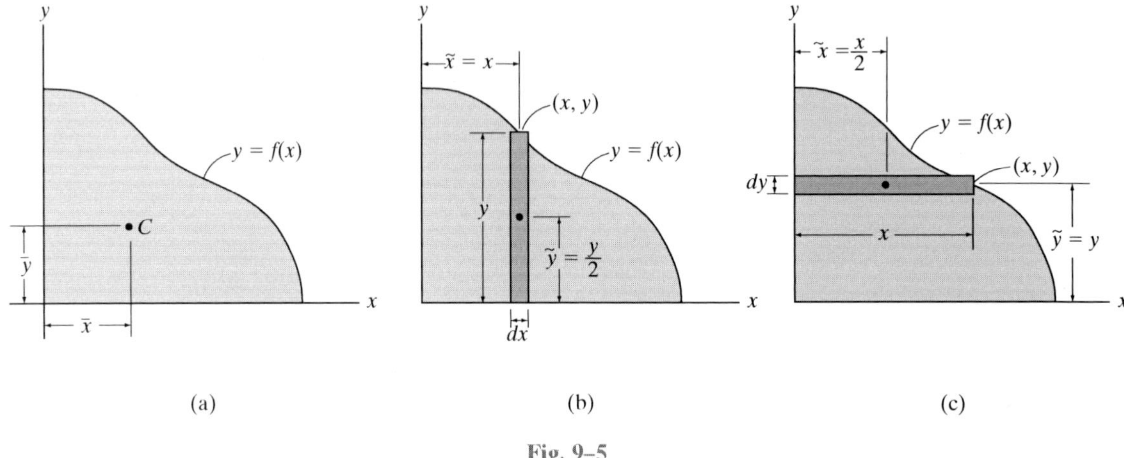

Fig. 9–5

Integration must be used to determine the location of the center of gravity of this goal post due to the curvature of the supporting member.

Centroid of an Area. If an area lies in the x–y plane and is bounded by the curve $y = f(x)$, as shown in Fig. 9–5*a*, then its centroid will be in this plane and can be determined from integrals similar to Eqs. 9–3, namely,

$$\bar{x} = \frac{\int_A \tilde{x}\, dA}{\int_A dA} \qquad \bar{y} = \frac{\int_A \tilde{y}\, dA}{\int_A dA} \tag{9–4}$$

These integrals can be evaluated by performing a *single integration* if we use a *rectangular strip* for the differential area element. For example, if a vertical strip is used, Fig. 9–5*b*, the area of the element is $dA = y\,dx$, and its centroid is located at $\tilde{x} = x$ and $\tilde{y} = y/2$. If we consider a horizontal strip, Fig. 9–5*c*, then $dA = x\,dy$, and its centroid is located at $\tilde{x} = x/2$ and $\tilde{y} = y$.

Centroid of a Line. If a line segment (or rod) lies within the x–y plane and it can be described by a thin curve $y = f(x)$, Fig. 9–6*a*, then its centroid is determined from

$$\bar{x} = \frac{\int_L \tilde{x}\, dL}{\int_L dL} \qquad \bar{y} = \frac{\int_L \tilde{y}\, dL}{\int_L dL} \tag{9–5}$$

9

Here, the length of the differential element is given by the Pythagorean theorem, $dL = \sqrt{(dx)^2 + (dy)^2}$, which can also be written in the form

$$dL = \sqrt{\left(\frac{dx}{dx}\right)^2 dx^2 + \left(\frac{dy}{dx}\right)^2 dx^2}$$

$$= \left(\sqrt{1 + \left(\frac{dy}{dx}\right)^2}\right) dx$$

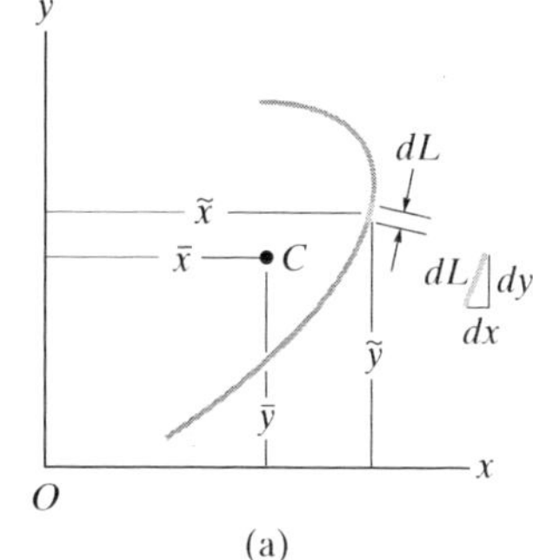

(a)

or

$$dL = \sqrt{\left(\frac{dx}{dy}\right)^2 dy^2 + \left(\frac{dy}{dy}\right)^2 dy^2}$$

$$= \left(\sqrt{\left(\frac{dx}{dy}\right)^2 + 1}\right) dy$$

Either one of these expressions can be used; however, for application, the one that will result in a simpler integration should be selected. For example, consider the rod in Fig. 9–6*b*, defined by $y = 2x^2$. The length of the element is $dL = \sqrt{1 + (dy/dx)^2}\, dx$, and since $dy/dx = 4x$, then $dL = \sqrt{1 + (4x)^2}\, dx$. The centroid for this element is located at $\tilde{x} = x$ and $\tilde{y} = y$.

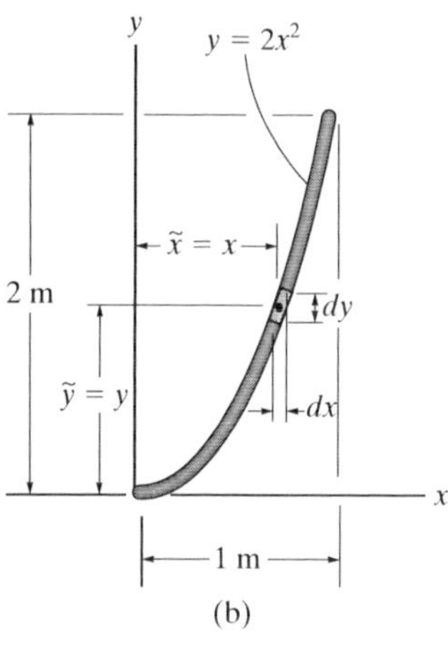

(b)

Fig. 9–6

Important Points

- The centroid represents the geometric center of a body. This point coincides with the center of mass or the center of gravity only if the material composing the body is uniform or homogeneous.
- Formulas used to locate the center of gravity or the centroid simply represent a balance between the sum of moments of all the parts of the system and the moment of the "resultant" for the system.
- In some cases the centroid is located at a point that is not on the object, as in the case of a ring, where the centroid is at its center. Also, this point will lie on any axis of symmetry for the body, Fig. 9–7.

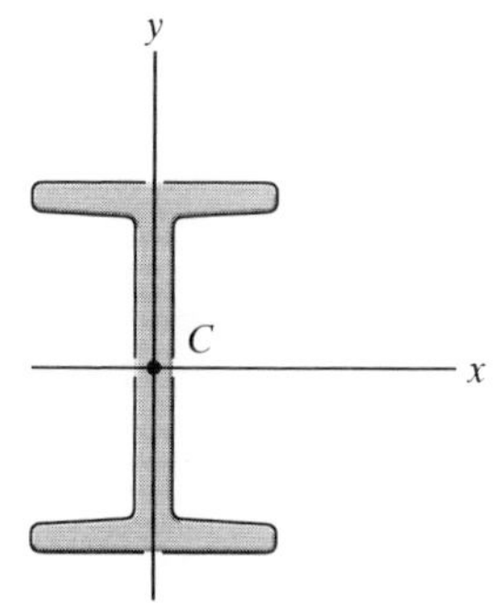

Fig. 9–7

9

Procedure for Analysis

The center of gravity or centroid of an object or shape can be determined by single integrations using the following procedure.

Differential Element.

- Select an appropriate coordinate system, specify the coordinate axes, and then choose a differential element for integration.
- For lines the element is represented by a differential line segment of length dL.
- For areas the element is generally a rectangle of area dA, having a finite length and differential width.
- For volumes the element can be a circular disk of volume dV, having a finite radius and differential thickness.
- Locate the element so that it touches the arbitrary point (x, y, z) on the curve that defines the boundary of the shape.

Size and Moment Arms.

- Express the length dL, area dA, or volume dV of the element in terms of the coordinates describing the curve.
- Express the moment arms $\tilde{x}$, $\tilde{y}$, $\tilde{z}$ for the centroid or center of gravity of the element in terms of the coordinates describing the curve.

Integrations.

- Substitute the formulations for $\tilde{x}$, $\tilde{y}$, $\tilde{z}$ and dL, dA, or dV into the appropriate equations (Eqs. 9–1 through 9–5).
- Express the function in the integrand in terms of the *same variable as the differential thickness of the element.*
- The limits of the integral are defined from the two extreme locations of the element's differential thickness, so that when the elements are "summed" or the integration performed, the entire region is covered.

EXAMPLE 9.1

Locate the centroid of the rod bent into the shape of a parabolic arc as shown in Fig. 9–8.

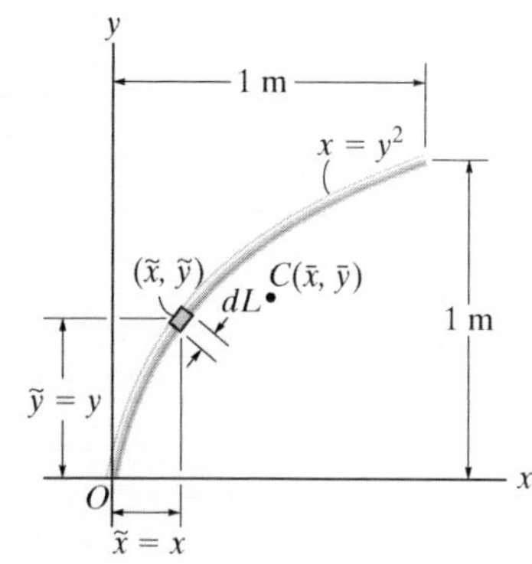

Fig. 9–8

SOLUTION

Differential Element. The differential element is shown in Fig. 9–8. It is located on the curve at the *arbitrary point* (x, y).

Area and Moment Arms. The differential element of length dL can be expressed in terms of the differentials dx and dy using the Pythagorean theorem.

$$dL = \sqrt{(dx)^2 + (dy)^2} = \sqrt{\left(\frac{dx}{dy}\right)^2 + 1}\, dy$$

Since $x = y^2$, then $dx/dy = 2y$. Therefore, expressing dL in terms of y and dy, we have

$$dL = \sqrt{(2y)^2 + 1}\, dy$$

As shown in Fig. 9–8, the centroid of the element is located at $\tilde{x} = x$, $\tilde{y} = y$.

Integrations. Applying Eq. 9–5 and using the integration formula to evaluate the integrals, we get

$$\bar{x} = \frac{\int_L \tilde{x}\, dL}{\int_L dL} = \frac{\int_0^{1\text{ m}} x\sqrt{4y^2 + 1}\, dy}{\int_0^{1\text{ m}} \sqrt{4y^2 + 1}\, dy} = \frac{\int_0^{1\text{ m}} y^2\sqrt{4y^2 + 1}\, dy}{\int_0^{1\text{ m}} \sqrt{4y^2 + 1}\, dy}$$

$$= \frac{0.6063}{1.479} = 0.410 \text{ m} \qquad \textit{Ans.}$$

$$\bar{y} = \frac{\int_L \tilde{y}\, dL}{\int_L dL} = \frac{\int_0^{1\text{ m}} y\sqrt{4y^2 + 1}\, dy}{\int_0^{1\text{ m}} \sqrt{4y^2 + 1}\, dy} = \frac{0.8484}{1.479} = 0.574 \text{ m} \qquad \textit{Ans.}$$

NOTE: These results for C seem reasonable when they are plotted on Fig. 9–8.

9

EXAMPLE 9.2

Locate the centroid of the circular wire segment shown in Fig. 9–9.

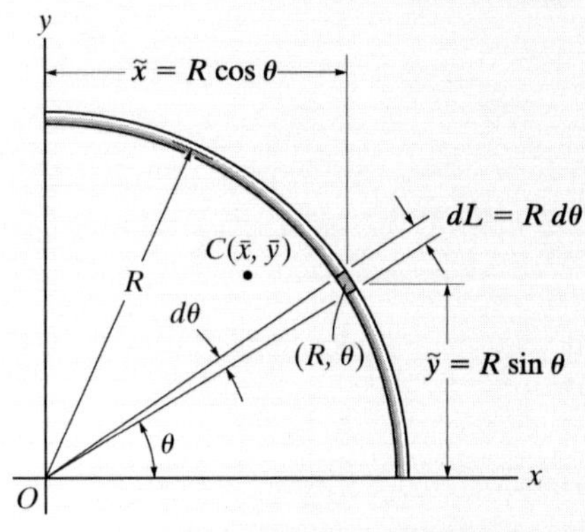

Fig. 9–9

SOLUTION

Polar coordinates will be used to solve this problem since the arc is circular.

Differential Element. A differential circular arc is selected as shown in the figure. This element lies on the curve at (R, θ).

Length and Moment Arm. The length of the differential element is $dL = R\,d\theta$, and its centroid is located at $\tilde{x} = R\cos\theta$ and $\tilde{y} = R\sin\theta$.

Integrations. Applying Eqs. 9–5 and integrating with respect to θ, we obtain

$$\bar{x} = \frac{\int_L \tilde{x}\,dL}{\int_L dL} = \frac{\int_0^{\pi/2} (R\cos\theta)R\,d\theta}{\int_0^{\pi/2} R\,d\theta} = \frac{R^2\int_0^{\pi/2} \cos\theta\,d\theta}{R\int_0^{\pi/2} d\theta} = \frac{2R}{\pi} \qquad \textit{Ans.}$$

$$\bar{y} = \frac{\int_L \tilde{y}\,dL}{\int_L dL} = \frac{\int_0^{\pi/2} (R\sin\theta)R\,d\theta}{\int_0^{\pi/2} R\,d\theta} = \frac{R^2\int_0^{\pi/2} \sin\theta\,d\theta}{R\int_0^{\pi/2} d\theta} = \frac{2R}{\pi} \qquad \textit{Ans.}$$

NOTE: As expected, the two coordinates are numerically the same due to the symmetry of the wire.

9

EXAMPLE 9.3

Determine the distance $\bar{y}$ measured from the x axis to the centroid of the area of the triangle shown in Fig. 9–10.

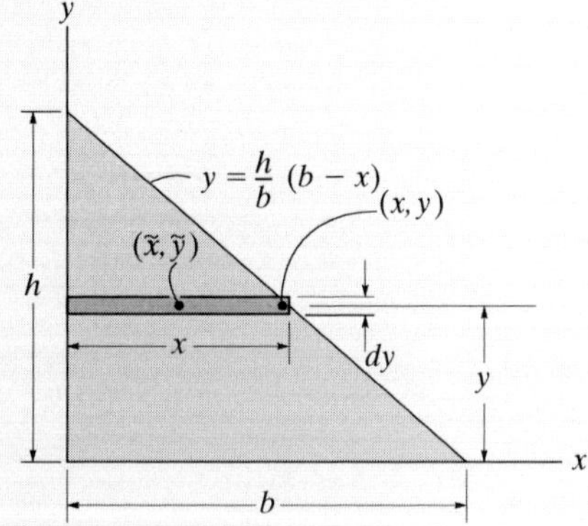

Fig. 9–10

SOLUTION

Differential Element. Consider a rectangular element having a thickness dy, and located in an arbitrary position so that it intersects the boundary at (x, y), Fig. 9–10.

Area and Moment Arms. The area of the element is $dA = x\,dy = \frac{b}{h}(h - y)\,dy$, and its centroid is located a distance $\tilde{y} = y$ from the x axis.

Integration. Applying the second of Eqs. 9–4 and integrating with respect to y yields

$$\bar{y} = \frac{\int_A \tilde{y}\,dA}{\int_A dA} = \frac{\int_0^h y\left[\frac{b}{h}(h - y)\,dy\right]}{\int_0^h \frac{b}{h}(h - y)\,dy} = \frac{\frac{1}{6}bh^2}{\frac{1}{2}bh}$$

$$= \frac{h}{3} \qquad \textit{Ans.}$$

NOTE: This result is valid for any shape of triangle. It states that the centroid is located at one-third the height, measured from the base of the triangle.

9

EXAMPLE 9.4

Locate the centroid for the area of a quarter circle shown in Fig. 9–11.

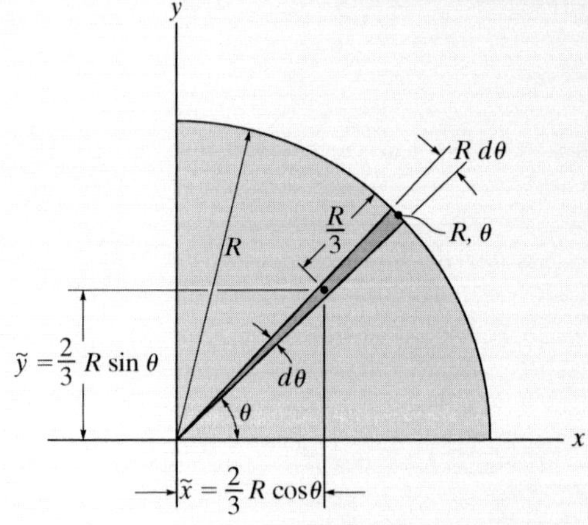

Fig. 9–11

SOLUTION

Differential Element. Polar coordinates will be used, since the boundary is circular. We choose the element in the shape of a *triangle*, Fig. 9–11. (Actually the shape is a circular sector; however, neglecting higher-order differentials, the element becomes triangular.) The element intersects the curve at point (R, θ).

Area and Moment Arms. The area of the element is

$$dA = \tfrac{1}{2}(R)(R\,d\theta) = \frac{R^2}{2}d\theta$$

and using the results of Example 9.3, the centroid of the (triangular) element is located at $\tilde{x} = \frac{2}{3}R\cos\theta$, $\tilde{y} = \frac{2}{3}R\sin\theta$.

Integrations. Applying Eqs. 9–4 and integrating with respect to θ, we obtain

$$\bar{x} = \frac{\int_A \tilde{x}\,dA}{\int_A dA} = \frac{\int_0^{\pi/2}\left(\frac{2}{3}R\cos\theta\right)\frac{R^2}{2}d\theta}{\int_0^{\pi/2}\frac{R^2}{2}d\theta} = \frac{\left(\frac{2}{3}R\right)\int_0^{\pi/2}\cos\theta\,d\theta}{\int_0^{\pi/2}d\theta} = \frac{4R}{3\pi} \qquad \textit{Ans.}$$

$$\bar{y} = \frac{\int_A \tilde{y}\,dA}{\int_A dA} = \frac{\int_0^{\pi/2}\left(\frac{2}{3}R\sin\theta\right)\frac{R^2}{2}d\theta}{\int_0^{\pi/2}\frac{R^2}{2}d\theta} = \frac{\left(\frac{2}{3}R\right)\int_0^{\pi/2}\sin\theta\,d\theta}{\int_0^{\pi/2}d\theta} = \frac{4R}{3\pi} \qquad \textit{Ans.}$$

9

EXAMPLE 9.5

Locate the centroid of the area shown in Fig. 9–12*a*.

SOLUTION I

Differential Element. A differential element of thickness dx is shown in Fig. 9–12*a*. The element intersects the curve at the *arbitrary point* (x, y), and so it has a height y.

Area and Moment Arms. The area of the element is $dA = y\,dx$, and its centroid is located at $\tilde{x} = x$, $\tilde{y} = y/2$.

Integrations. Applying Eqs. 9–4 and integrating with respect to x yields

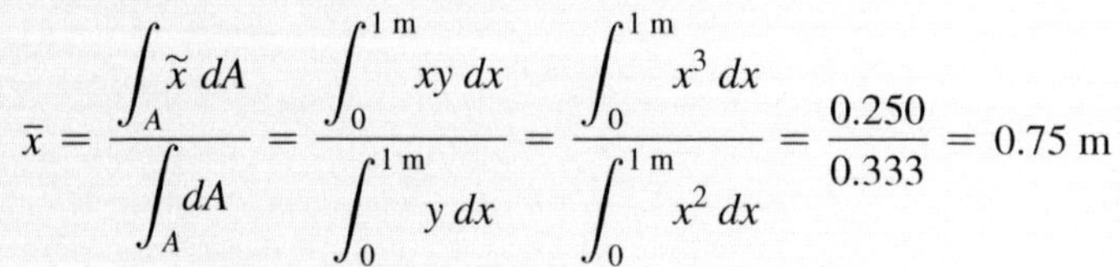

$$\bar{x} = \frac{\int_A \tilde{x}\,dA}{\int_A dA} = \frac{\int_0^{1\text{ m}} xy\,dx}{\int_0^{1\text{ m}} y\,dx} = \frac{\int_0^{1\text{ m}} x^3\,dx}{\int_0^{1\text{ m}} x^2\,dx} = \frac{0.250}{0.333} = 0.75\text{ m} \qquad \textit{Ans.}$$

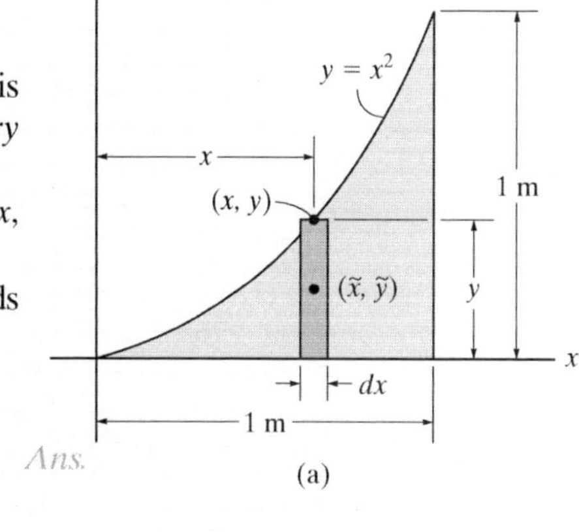

$$\bar{y} = \frac{\int_A \tilde{y}\,dA}{\int_A dA} = \frac{\int_0^{1\text{ m}} (y/2)y\,dx}{\int_0^{1\text{ m}} y\,dx} = \frac{\int_0^{1\text{ m}} (x^2/2)x^2\,dx}{\int_0^{1\text{ m}} x^2\,dx} = \frac{0.100}{0.333} = 0.3\text{ m} \qquad \textit{Ans.}$$

SOLUTION II

Differential Element. The differential element of thickness dy is shown in Fig. 9–12*b*. The element intersects the curve at the *arbitrary point* (x, y), and so it has a length $(1 - x)$.

Area and Moment Arms. The area of the element is $dA = (1 - x)\,dy$, and its centroid is located at

$$\tilde{x} = x + \left(\frac{1 - x}{2}\right) = \frac{1 + x}{2}, \quad \tilde{y} = y$$

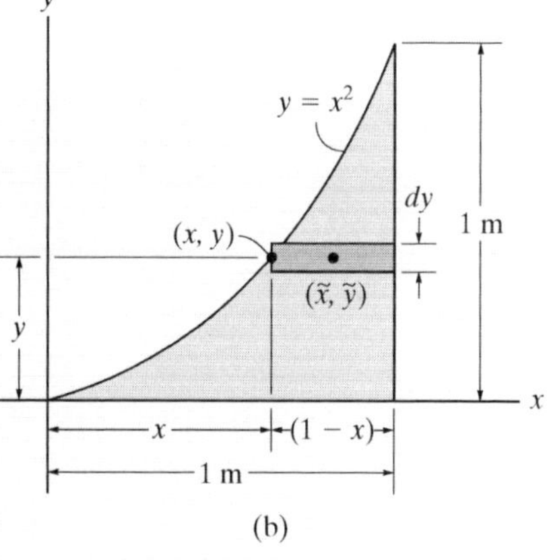

Fig. 9–12

Integrations. Applying Eqs. 9–4 and integrating with respect to y, we obtain

$$\bar{x} = \frac{\int_A \tilde{x}\,dA}{\int_A dA} = \frac{\int_0^{1\text{ m}} [(1 + x)/2](1 - x)\,dy}{\int_0^{1\text{ m}} (1 - x)\,dy} = \frac{\frac{1}{2}\int_0^{1\text{ m}} (1 - y)\,dy}{\int_0^{1\text{ m}} (1 - \sqrt{y})\,dy} = \frac{0.250}{0.333} = 0.75\text{ m} \qquad \textit{Ans.}$$

$$\bar{y} = \frac{\int_A \tilde{y}\,dA}{\int_A dA} = \frac{\int_0^{1\text{ m}} y(1 - x)\,dy}{\int_0^{1\text{ m}} (1 - x)\,dy} = \frac{\int_0^{1\text{ m}} (y - y^{3/2})\,dy}{\int_0^{1\text{ m}} (1 - \sqrt{y})\,dy} = \frac{0.100}{0.333} = 0.3\text{ m} \qquad \textit{Ans.}$$

NOTE: Plot these results and notice that they seem reasonable. Also, for this problem, elements of thickness dx offer a simpler solution.

9

EXAMPLE 9.6

Locate the centroid of the semi-elliptical area shown in Fig. 9–13*a*.

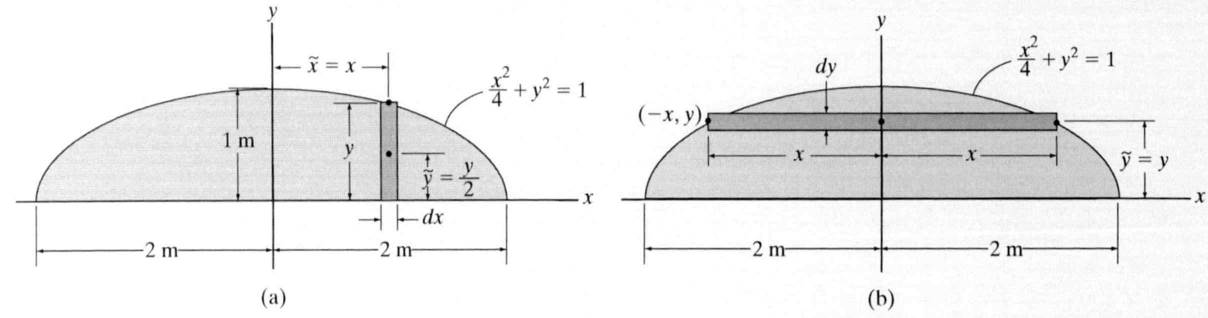

Fig. 9–13

SOLUTION I

Differential Element. The rectangular differential element parallel to the y axis shown shaded in Fig. 9–13*a* will be considered. This element has a thickness of dx and a height of y.

Area and Moment Arms. Thus, the area is $dA = y\,dx$, and its centroid is located at $\tilde{x} = x$ and $\tilde{y} = y/2$.

Integration. Since the area is symmetrical about the y axis,

$$\bar{x} = 0 \qquad \textit{Ans.}$$

Applying the second of Eqs. 9–4 with $y = \sqrt{1 - \frac{x^2}{4}}$, we have

$$\bar{y} = \frac{\int_A \tilde{y}\,dA}{\int_A dA} = \frac{\int_{-2\text{ m}}^{2\text{ m}} \frac{y}{2}(y\,dx)}{\int_{-2\text{ m}}^{2\text{ m}} y\,dx} = \frac{\frac{1}{2}\int_{-2\text{ m}}^{2\text{ m}}\left(1 - \frac{x^2}{4}\right)dx}{\int_{-2\text{ m}}^{2\text{ m}}\sqrt{1 - \frac{x^2}{4}}\,dx} = \frac{4/3}{\pi} = 0.424\text{ m} \qquad \textit{Ans.}$$

SOLUTION II

Differential Element. The shaded rectangular differential element of thickness dy and width $2x$, parallel to the x axis, will be considered, Fig. 9–13*b*.

Area and Moment Arms. The area is $dA = 2x\,dy$, and its centroid is at $\tilde{x} = 0$ and $\tilde{y} = y$.

Integration. Applying the second of Eqs. 9–4, with $x = 2\sqrt{1 - y^2}$, we have

$$\bar{y} = \frac{\int_A \tilde{y}\,dA}{\int_A dA} = \frac{\int_0^{1\text{ m}} y(2x\,dy)}{\int_0^{1\text{ m}} 2x\,dy} = \frac{\int_0^{1\text{ m}} 4y\sqrt{1 - y^2}\,dy}{\int_0^{1\text{ m}} 4\sqrt{1 - y^2}\,dy} = \frac{4/3}{\pi}\text{ m} = 0.424\text{ m} \qquad \textit{Ans.}$$

9

EXAMPLE 9.7

Locate the $\bar{y}$ centroid for the paraboloid of revolution, shown in Fig. 9–14.

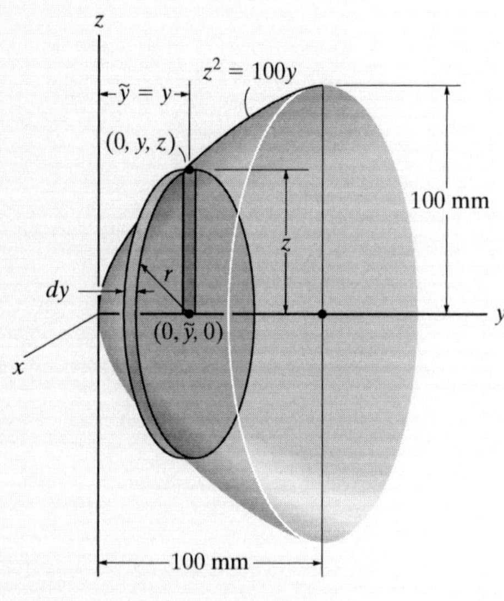

Fig. 9–14

SOLUTION

Differential Element. An element having the shape of a *thin disk* is chosen. This element has a thickness dy, it intersects the generating curve at the *arbitrary point* $(0, y, z)$, and so its radius is $r = z$.

Volume and Moment Arm. The volume of the element is $dV = (\pi z^2)\,dy$, and its centroid is located at $\tilde{y} = y$.

Integration. Applying the second of Eqs. 9–3 and integrating with respect to y yields.

$$\bar{y} = \frac{\int_V \tilde{y}\,dV}{\int_V dV} = \frac{\int_0^{100\text{ mm}} y(\pi z^2)\,dy}{\int_0^{100\text{ mm}} (\pi z^2)\,dy} = \frac{100\pi \int_0^{100\text{ mm}} y^2\,dy}{100\pi \int_0^{100\text{ mm}} y\,dy} = 66.7\text{ mm} \qquad \textit{Ans.}$$

9

EXAMPLE 9.8

Determine the location of the center of mass of the cylinder shown in Fig. 9–15 if its density varies directly with the distance from its base, i.e., $\rho = 200z\ \text{kg/m}^3$.

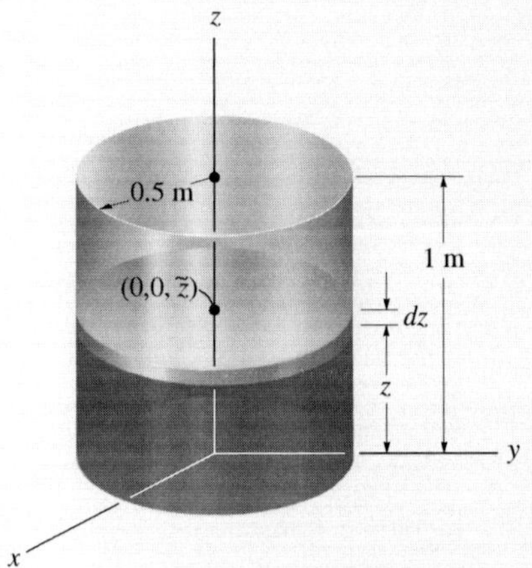

Fig. 9–15

SOLUTION

For reasons of material symmetry,

$$\bar{x} = \bar{y} = 0 \qquad \textit{Ans.}$$

Differential Element. A disk element of radius 0.5 m and thickness dz is chosen for integration, Fig. 9–15, since the *density of the entire element is constant* for a given value of z. The element is located along the z axis at the *arbitrary point* $(0, 0, z)$.

Volume and Moment Arm. The volume of the element is $dV = \pi(0.5)^2\,dz$, and its centroid is located at $\tilde{z} = z$.

Integrations. Using the third of Eqs. 9–2 with $dm = \rho\,dV$ and integrating with respect to z, noting that $\rho = 200z$, we have

$$\bar{z} = \frac{\int_V \tilde{z}\rho\,dV}{\int_V \rho\,dV} = \frac{\int_0^{1\text{ m}} z(200z)\left[\pi(0.5)^2\,dz\right]}{\int_0^{1\text{ m}} (200z)\pi(0.5)^2\,dz}$$

$$= \frac{\int_0^{1\text{ m}} z^2\,dz}{\int_0^{1\text{ m}} z\,dz} = 0.667\text{ m} \qquad \textit{Ans.}$$

9

FUNDAMENTAL PROBLEMS

F9–1. Determine the centroid $(\bar{x}, \bar{y})$ of the shaded area.

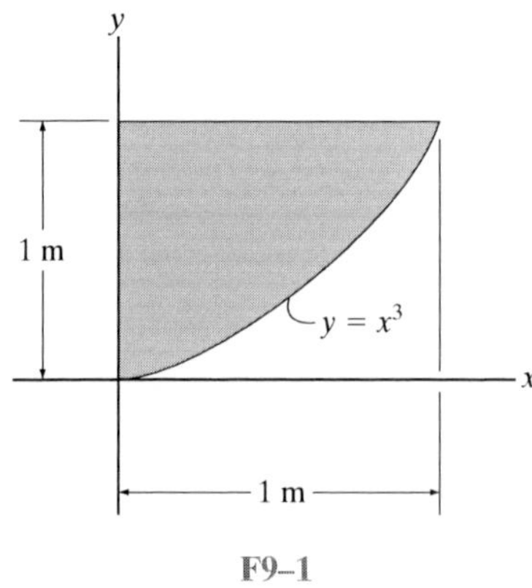

F9–1

F9–2. Determine the centroid $(\bar{x}, \bar{y})$ of the shaded area.

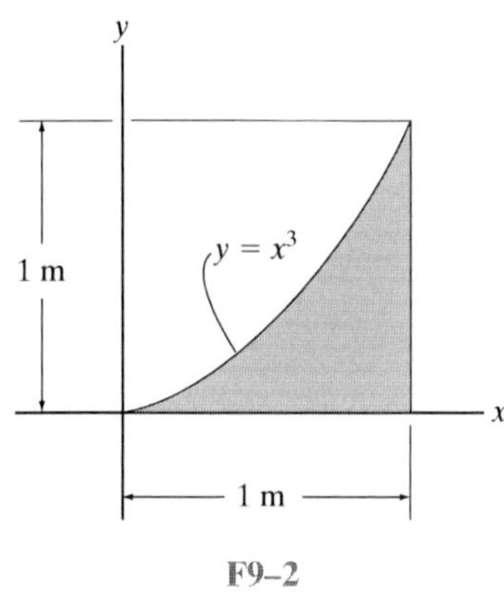

F9–2

F9–3. Determine the centroid $\bar{y}$ of the shaded area.

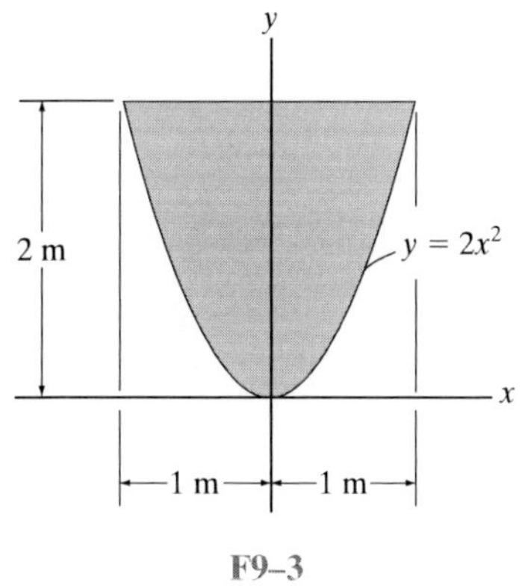

F9–3

F9–4. Locate the center mass $\bar{x}$ of the straight rod if its mass per unit length is given by $m = m_0(1 + x^2/L^2)$.

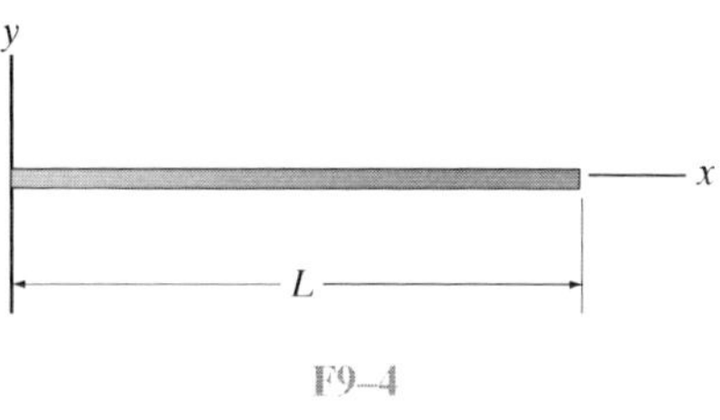

F9–4

F9–5. Locate the centroid $\bar{y}$ of the homogeneous solid formed by revolving the shaded area about the y axis.

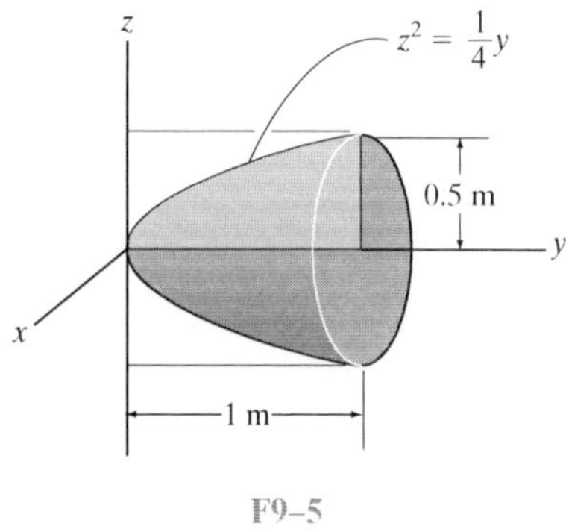

F9–5

F9–6. Locate the centroid $\bar{z}$ of the homogeneous solid formed by revolving the shaded area about the z axis.

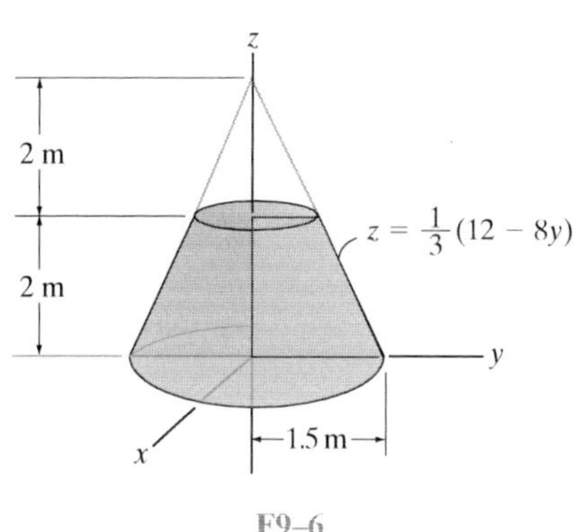

F9–6

PROBLEMS

9–1. Locate the center of mass of the homogeneous rod bent into the shape of a circular arc.

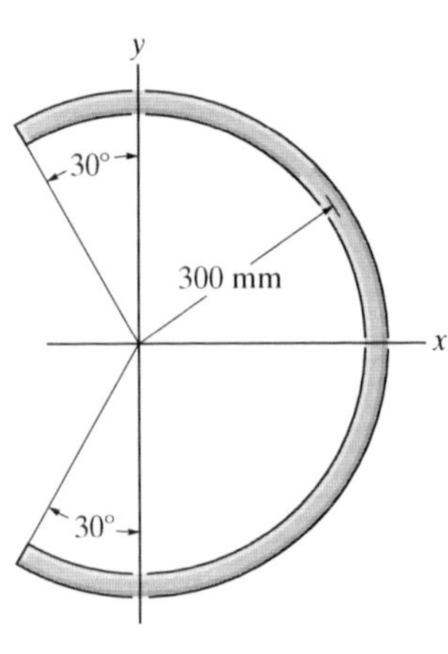

Prob. 9–1

9–2. Determine the distance $\bar{x}$ to the center of mass of the homogeneous rod bent into the shape shown. If the rod has a mass per unit length of 0.5 kg/m, determine the reactions at the fixed support O.

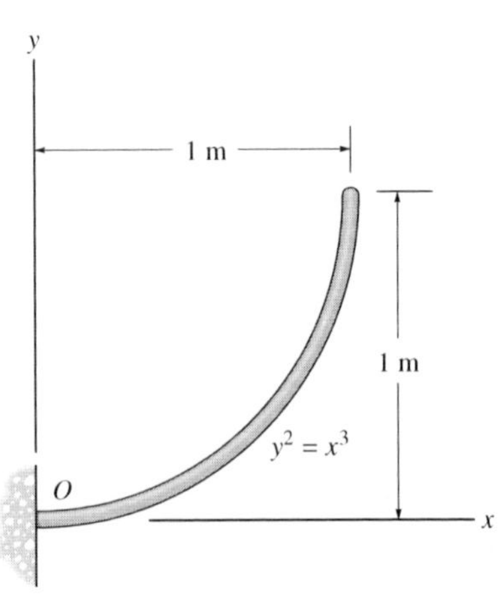

Prob. 9–2

9–3. Determine the mass and the location of the center of mass $\bar{x}$ of the rod if its mass per unit length is $m = m_0(1 + x/L)$.

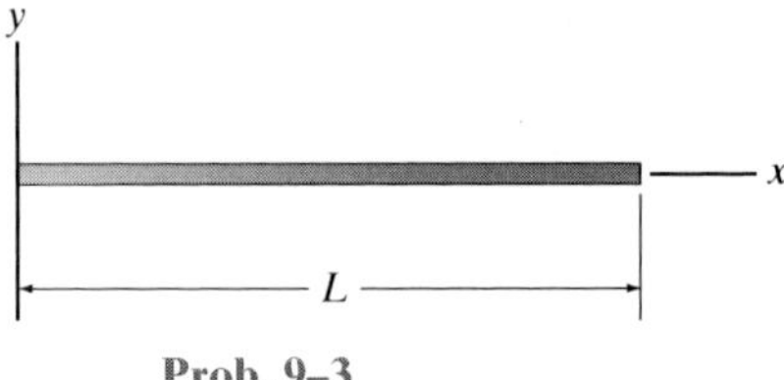

Prob. 9–3

***9–4.** Locate the centroid of the parabolic area.

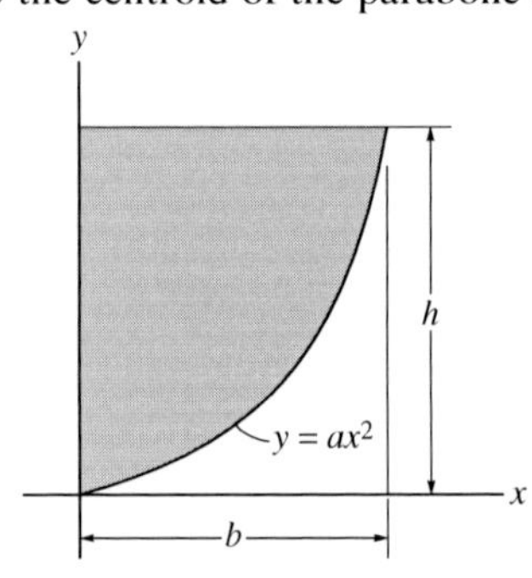

Prob. 9–4

9–5. Locate the centroid $(\bar{x}, \bar{y})$ of the uniform rod. Evaluate the integrals using a numerical method.

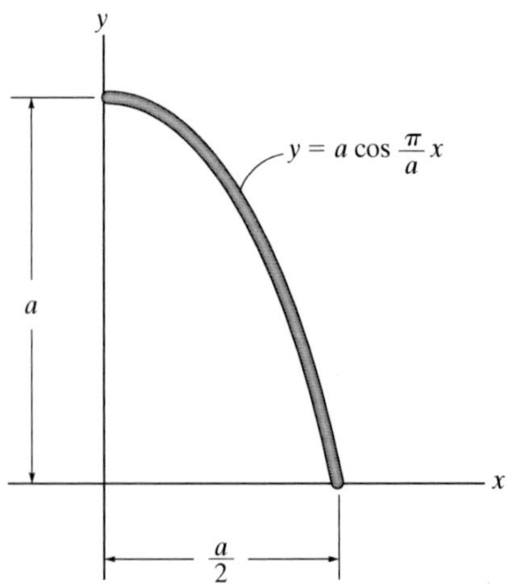

Prob. 9–5

9

9–6. Locate the centroid $\bar{y}$ of the area.

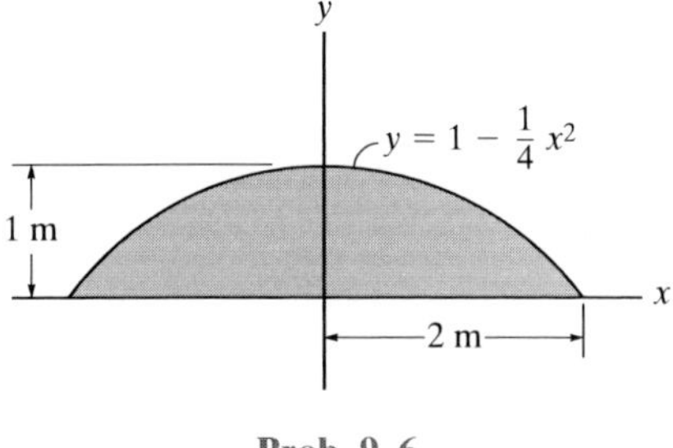

Prob. 9–6

9–7. Locate the centroid $\bar{x}$ of the parabolic area.

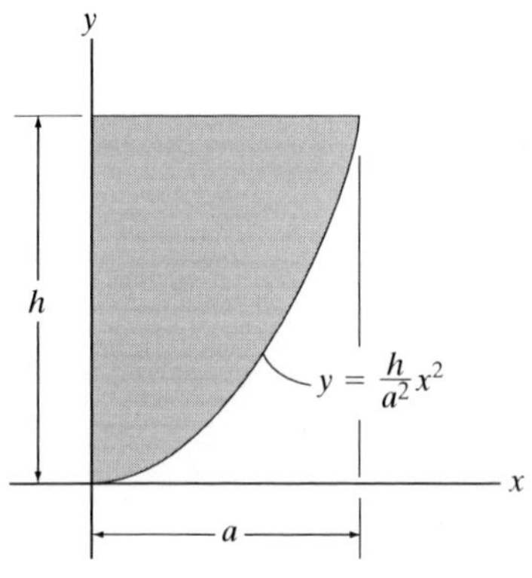

Prob. 9–7

***9–8.** Locate the centroid $\bar{y}$ of the parabolic area.

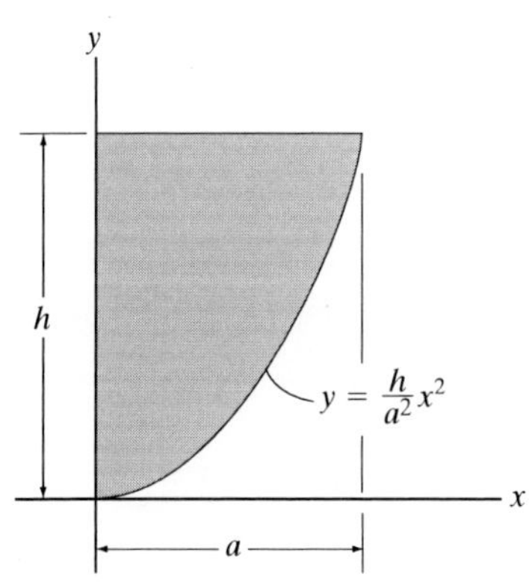

Prob. 9–8

9–9. Locate the centroid $\bar{x}$ of the area.

9–10. Locate the centroid $\bar{y}$ of the area.

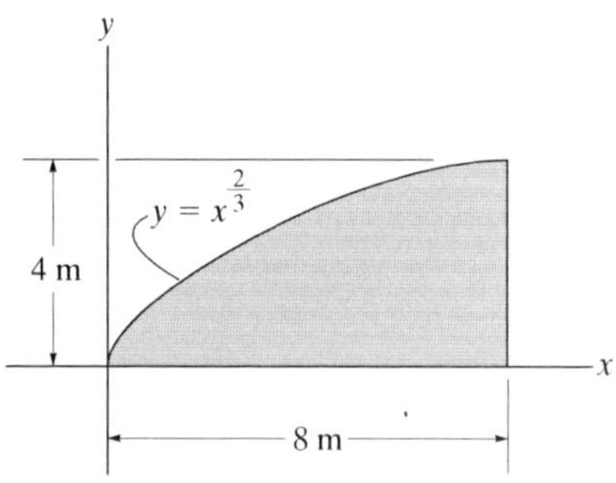

Probs. 9–9/10

9–11. Locate the centroid $\bar{x}$ of the area.

***9–12.** Locate the centroid $\bar{y}$ of the area.

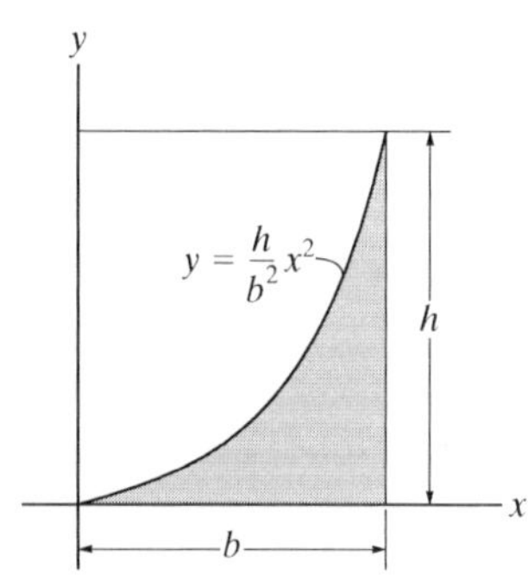

Probs. 9–11/12

9–13. Locate the centroid $\bar{x}$ of the area.

9–14. Locate the centroid $\bar{y}$ of the area.

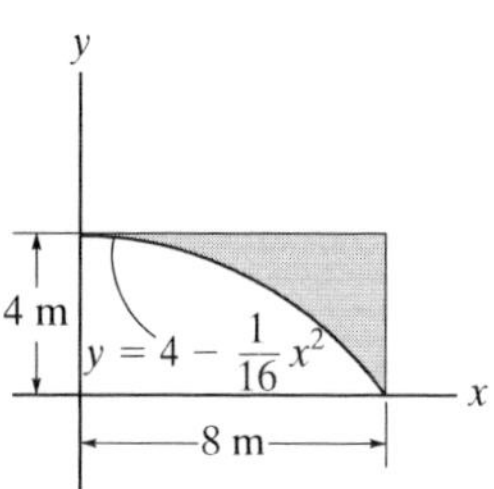

Probs. 9–13/14

9

9–15. Locate the centroid $\bar{x}$ of the area.

***9–16.** Locate the centroid $\bar{y}$ of the area.

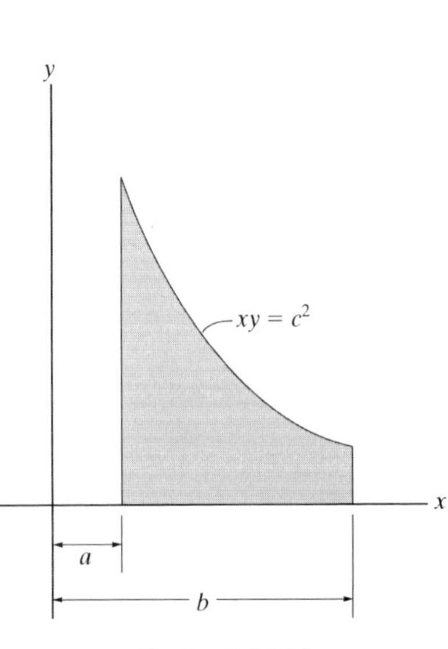

Probs. 9–15/16

9–17. Locate the centroid $\bar{x}$ of the shaded area.

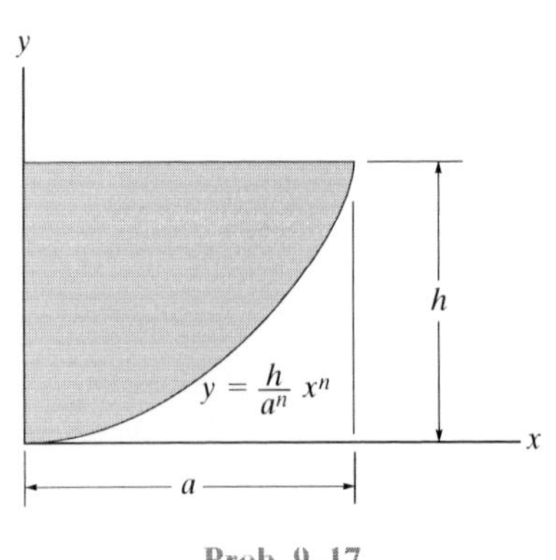

Prob. 9–17

9–18. Locate the centroid $\bar{x}$ of the area.

9–19. Locate the centroid $\bar{y}$ of the area.

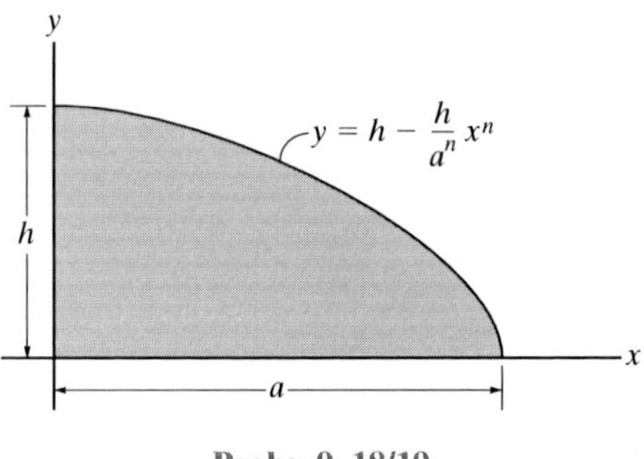

Probs. 9–18/19

***9–20.** Locate the centroid $\bar{y}$ of the shaded area.

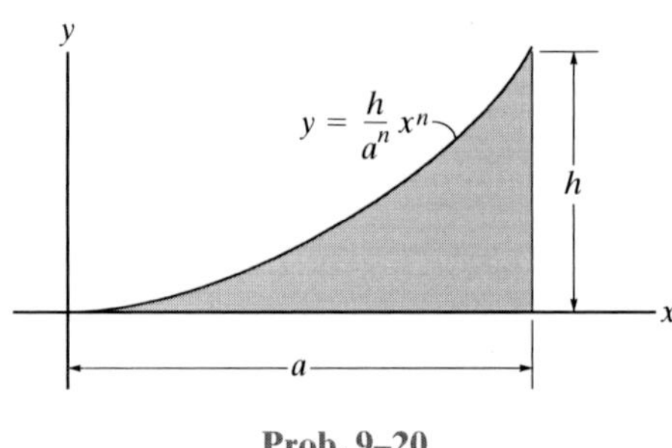

Prob. 9–20

9–21. Locate the centroid $\bar{x}$ of the area.

9–22. Locate the centroid $\bar{y}$ of the area.

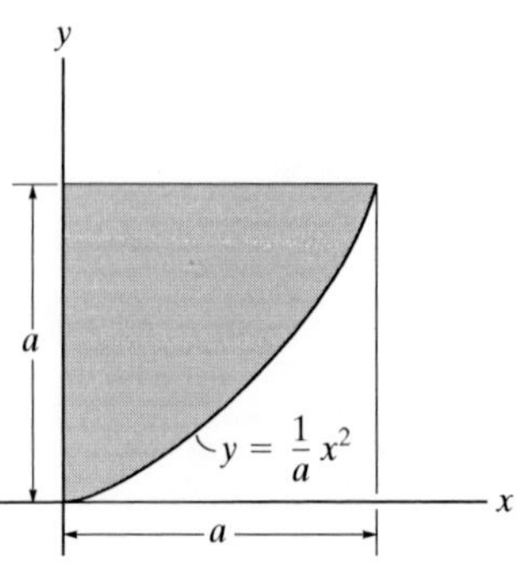

Probs. 9–21/22

9–23. Locate the centroid $\bar{x}$ of the quarter elliptical area.

***9–24.** Locate the centroid $\bar{y}$ of the quarter elliptical area.

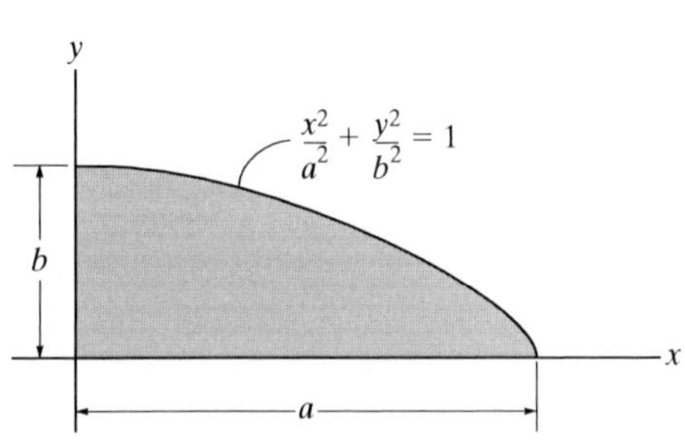

Probs. 9–23/24

9–25. Locate the centroid $\bar{y}$ of the shaded area. Solve the problem by evaluating the integrals using Simpson's rule.

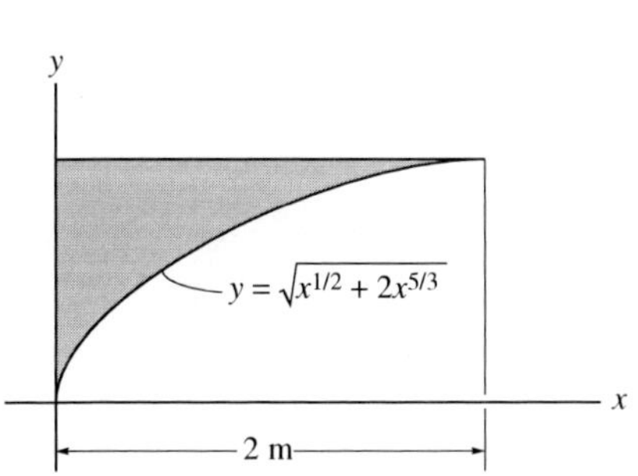

Prob. 9–25

9–26. Locate the centroid $\bar{x}$ of the area.

9–27. Locate the centroid $\bar{y}$ of the area.

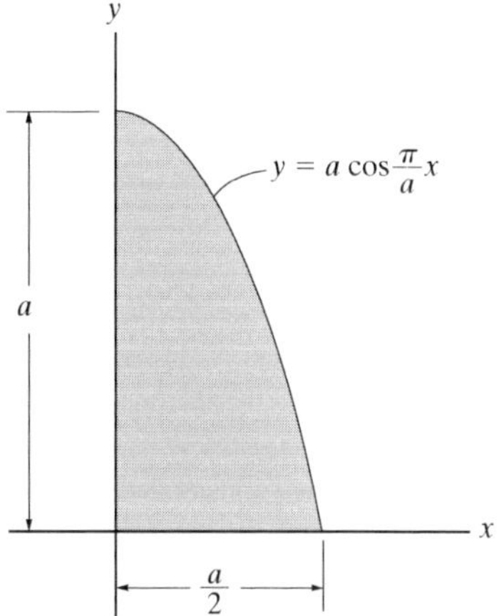

Probs. 9–26/27

***9–28.** Locate the centroid $\bar{x}$ of the area.

9–29. Locate the centroid $\bar{y}$ of the area.

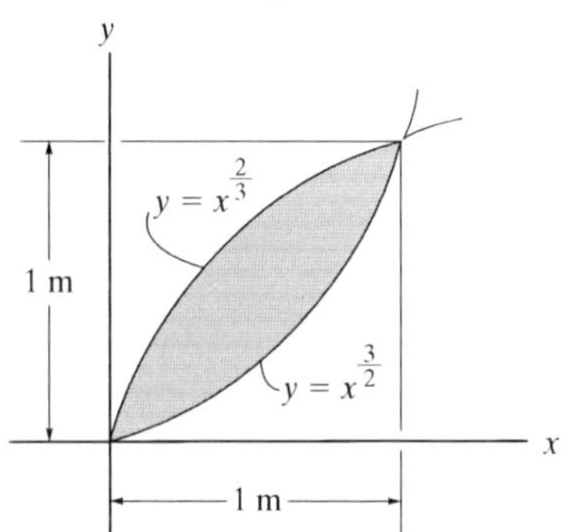

Probs. 9–28/29

9–30. Locate the centroid $\bar{x}$ of the area.

9–31. Locate the centroid $\bar{y}$ of the area.

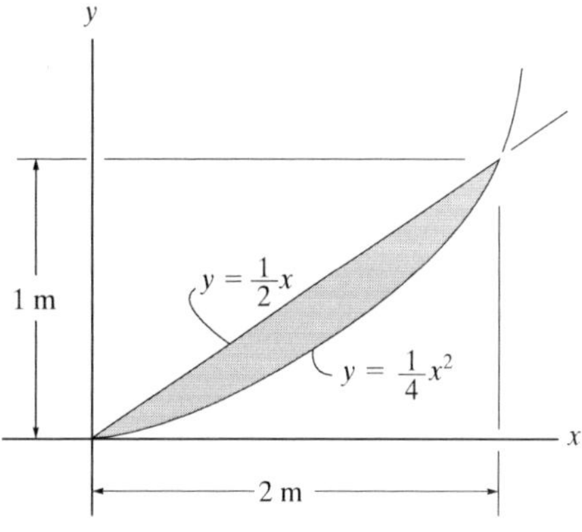

Probs. 9–30/31

9

***9–32.** Locate the centroid $\bar{x}$ of the area.

9–33. Locate the centroid $\bar{y}$ of the area.

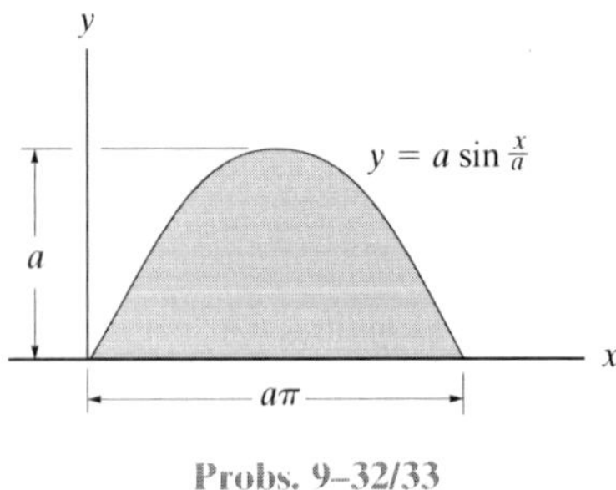

Probs. 9–32/33

9–34. The steel plate is 0.3 m thick and has a density of 7850 kg/m^3. Determine the location of its center of mass. Also find the reactions at the pin and roller support.

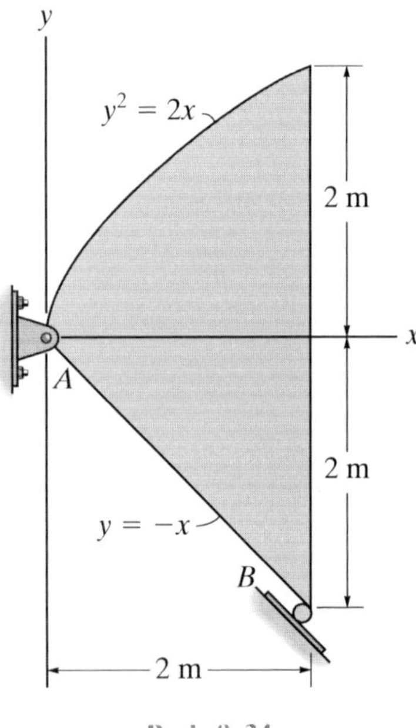

Prob. 9–34

9–35. Locate the centroid $\bar{x}$ of the shaded area.

***9–36.** Locate the centroid $\bar{y}$ of the shaded area.

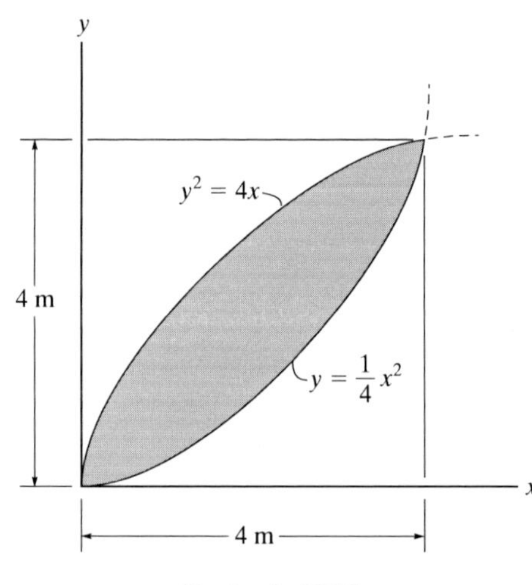

Probs. 9–35/36

9–37. If the density at any point in the quarter circular plate is defined by $\rho = \rho_0 xy$, where ρ_0 is a constant, determine the mass and locate the center of mass $(\bar{x}, \bar{y})$ of the plate. The plate has a thickness t.

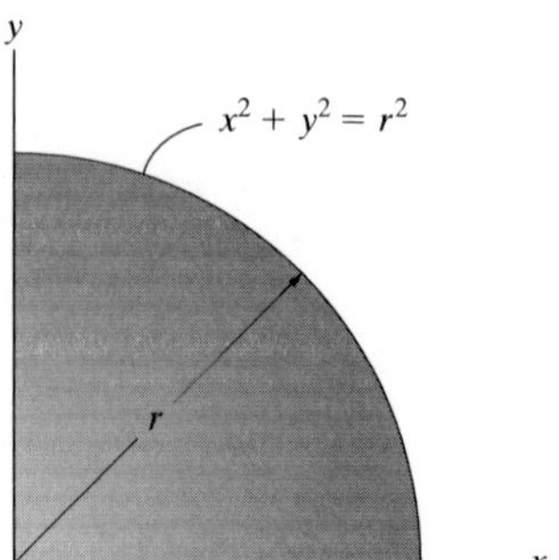

Prob. 9–37

9

9–38. Determine the location $\bar{r}$ of the centroid C of the cardioid, $r = a(1 - \cos\theta)$.

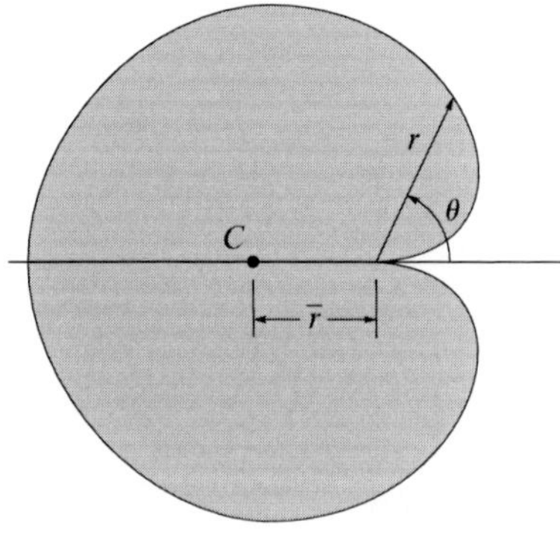

Prob. 9–38

9–39. Locate the centroid $\bar{y}$ of the paraboloid.

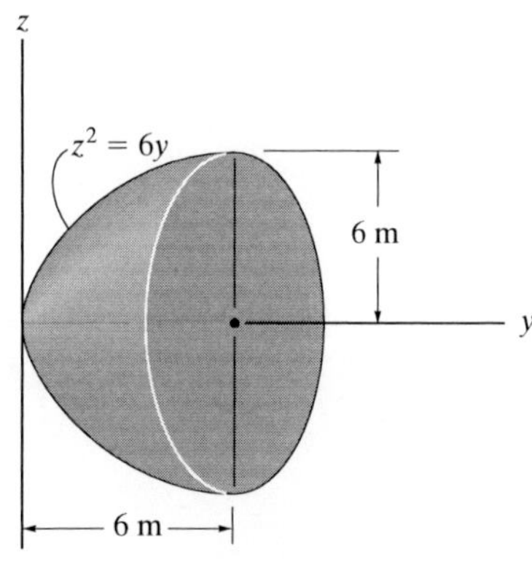

Prob. 9–39

***9–40.** Locate the center of gravity of the volume. The material is homogeneous.

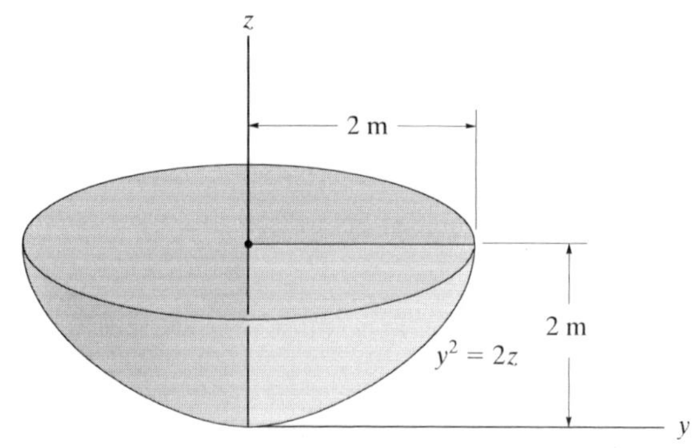

Prob. 9–40

9–41. Locate the centroid $\bar{z}$ of the hemisphere.

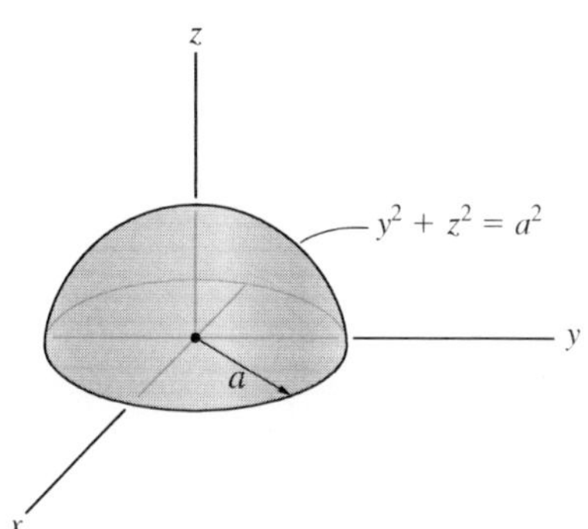

Prob. 9–41

9–42. Locate the centroid $\overline{y}$ of the paraboloid.

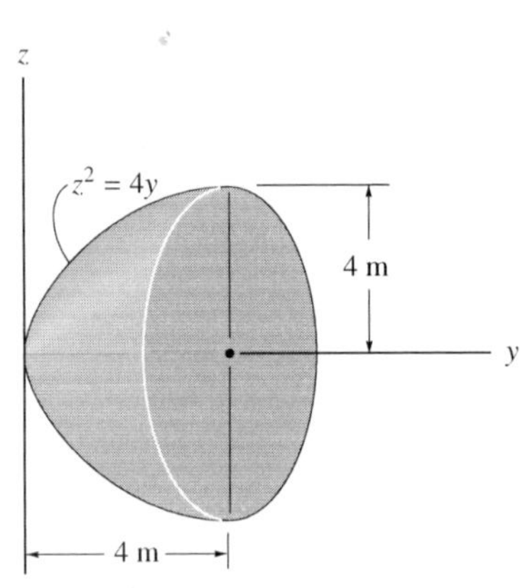

Prob. 9–42

9–43. Locate the centroid of the solid.

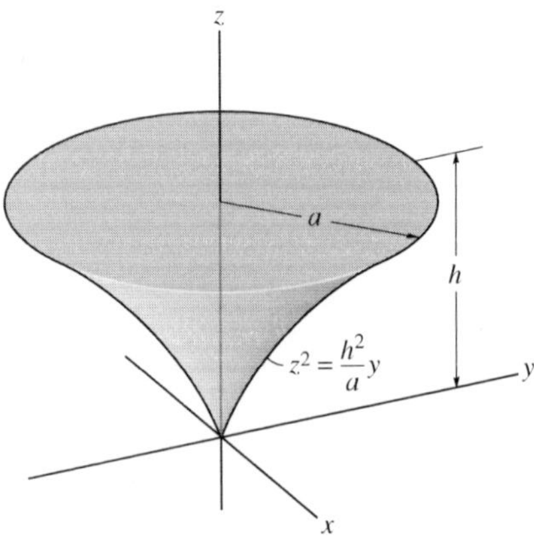

Prob. 9–43

***9–44.** Locate the centroid of the ellipsoid of revolution.

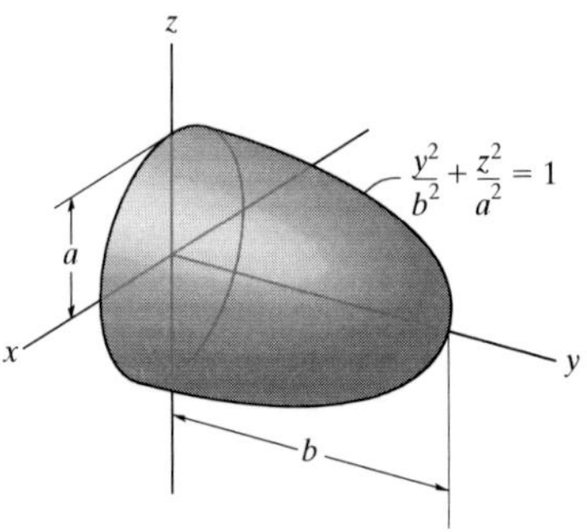

Prob. 9–44

9–45. Locate the center of gravity $\overline{z}$ of the frustum of the paraboloid. The material is homogeneous.

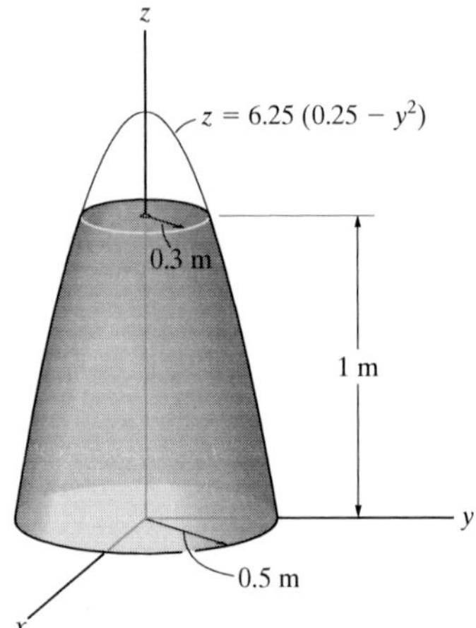

Prob. 9–45

9–46. The hemisphere of radius r is made from a stack of very thin plates such that the density varies with height, $\rho = kz$, where k is a constant. Determine its mass and the distance $\overline{z}$ to the center of mass G.

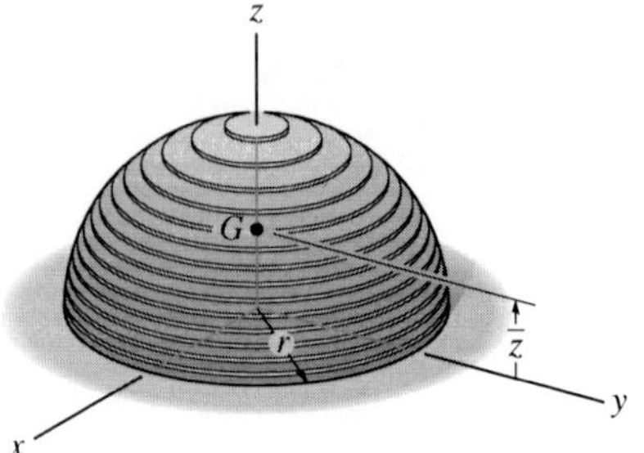

Prob. 9–46

9

9–47. Locate the centroid of the quarter-cone.

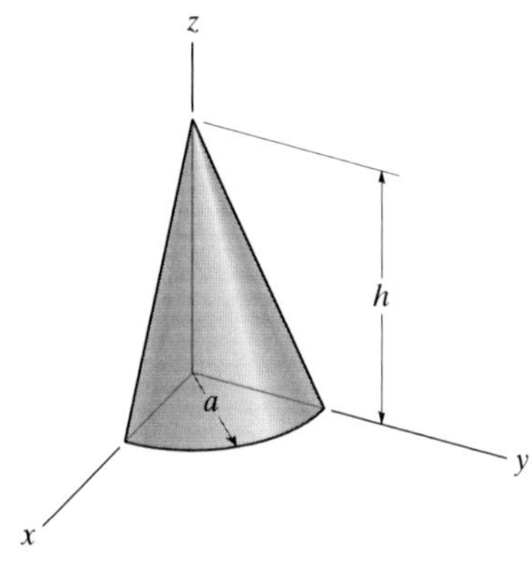

Prob. 9–47

***9–48.** Locate the centroid $\bar{z}$ of the frustum of the right-circular cone.

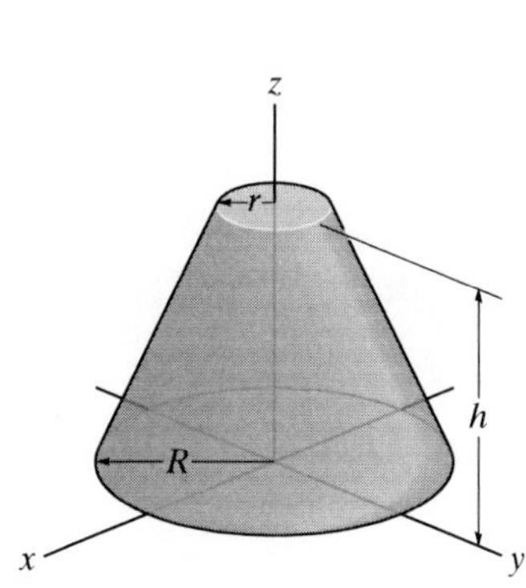

Prob. 9–48

9–49. The king's chamber of the Great Pyramid of Giza is located at its centroid. Assuming the pyramid to be a solid, prove that this point is at $\bar{z} = \frac{1}{4}h$. *Suggestion:* Use a rectangular differential plate element having a thickness dz and area $(2x)(2y)$.

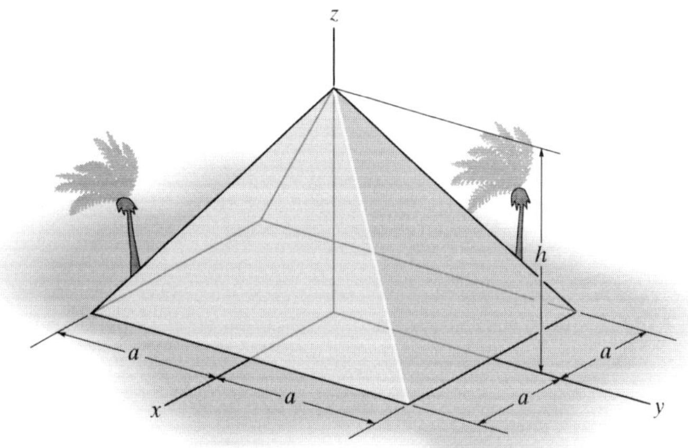

Prob. 9–49

9–50. Determine the location $\bar{z}$ of the centroid for the tetrahedron. *Suggestion:* Use a triangular "plate" element parallel to the x–y plane and of thickness dz.

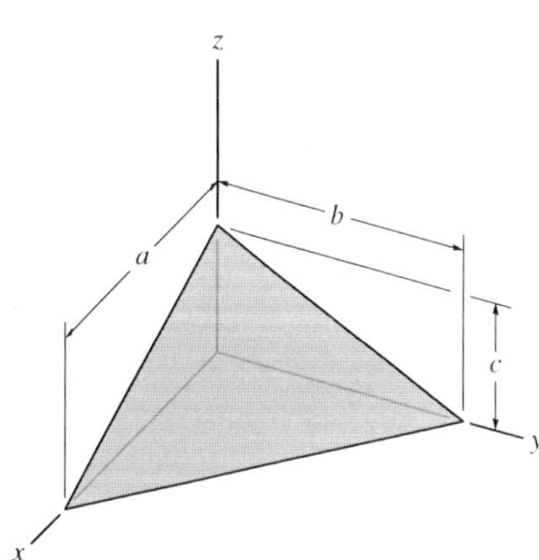

Prob. 9–50

9.2 Composite Bodies

A *composite body* consists of a series of connected "simpler" shaped bodies, which may be rectangular, triangular, semicircular, etc. Such a body can often be sectioned or divided into its composite parts and, provided the *weight* and location of the center of gravity of each of these parts are known, we can then eliminate the need for integration to determine the center of gravity for the entire body. The method for doing this follows the same procedure outlined in Sec. 9.1. Formulas analogous to Eqs. 9–1 result; however, rather than account for an infinite number of differential weights, we have instead a finite number of weights. Therefore,

$$\bar{x} = \frac{\Sigma \tilde{x} W}{\Sigma W} \quad \bar{y} = \frac{\Sigma \tilde{y} W}{\Sigma W} \quad \bar{z} = \frac{\Sigma \tilde{z} W}{\Sigma W} \tag{9–6}$$

Here

$\bar{x}, \bar{y}, \bar{z}$ represent the coordinates of the center of gravity G of the composite body.

$\tilde{x}, \tilde{y}, \tilde{z}$ represent the coordinates of the center of gravity of each composite part of the body.

ΣW is the sum of the weights of all the composite parts of the body, or simply the total weight of the body.

When the body has a *constant density or specific weight*, the center of gravity *coincides* with the centroid of the body. The centroid for composite lines, areas, and volumes can be found using relations analogous to Eqs. 9–6; however, the W's are replaced by L's, A's, and V's, respectively. Centroids for common shapes of lines, areas, shells, and volumes that often make up a composite body are given in the table on the inside back cover.

In order to determine the force required to tip over this concrete barrier it is first necessary to determine the location of its center of gravity G. Due to symmetry, G will lie on the vertical axis of symmetry.

9

Procedure for Analysis

The location of the center of gravity of a body or the centroid of a composite geometrical object represented by a line, area, or volume can be determined using the following procedure.

Composite Parts.

- Using a sketch, divide the body or object into a finite number of composite parts that have simpler shapes.
- If a composite body has a *hole*, or a geometric region having no material, then consider the composite body without the hole and consider the hole as an *additional* composite part having *negative* weight or size.

Moment Arms.

- Establish the coordinate axes on the sketch and determine the coordinates $\tilde{x}$, $\tilde{y}$, $\tilde{z}$ of the center of gravity or centroid of each part.

Summations.

- Determine $\bar{x}$, $\bar{y}$, $\bar{z}$ by applying the center of gravity equations, Eqs. 9–6, or the analogous centroid equations.
- If an object is *symmetrical* about an axis, the centroid of the object lies on this axis.

If desired, the calculations can be arranged in tabular form, as indicated in the following three examples.

The center of gravity of this water tank can be determined by dividing it into composite parts and applying Eqs. 9–6.

9

EXAMPLE 9.9

Locate the centroid of the wire shown in Fig. 9–16*a*.

SOLUTION

Composite Parts. The wire is divided into three segments as shown in Fig. 9–16*b*.

Moment Arms. The location of the centroid for each segment is determined and indicated in the figure. In particular, the centroid of segment ① is determined either by integration or by using the table on the inside back cover.

Summations. For convenience, the calculations can be tabulated as follows:

Segment	L (mm)	$\tilde{x}$ (mm)	$\tilde{y}$ (mm)	$\tilde{z}$ (mm)	$\tilde{x}L$ (mm^2)	$\tilde{y}L$ (mm^2)	$\tilde{z}L$ (mm^2)
1	$\pi(60) = 188.5$	60	−38.2	0	11 310	−7200	0
2	40	0	20	0	0	800	0
3	20	0	40	−10	0	800	−200
	$\Sigma L = 248.5$				$\Sigma\tilde{x}L = 11\,310$	$\Sigma\tilde{y}L = -5600$	$\Sigma\tilde{z}L = -200$

Thus,

$$\bar{x} = \frac{\Sigma\tilde{x}L}{\Sigma L} = \frac{11\,310}{248.5} = 45.5 \text{ mm} \qquad \textit{Ans.}$$

$$\bar{y} = \frac{\Sigma\tilde{y}L}{\Sigma L} = \frac{-5600}{248.5} = -22.5 \text{ mm} \qquad \textit{Ans.}$$

$$\bar{z} = \frac{\Sigma\tilde{z}L}{\Sigma L} = \frac{-200}{248.5} = -0.805 \text{ mm} \qquad \textit{Ans.}$$

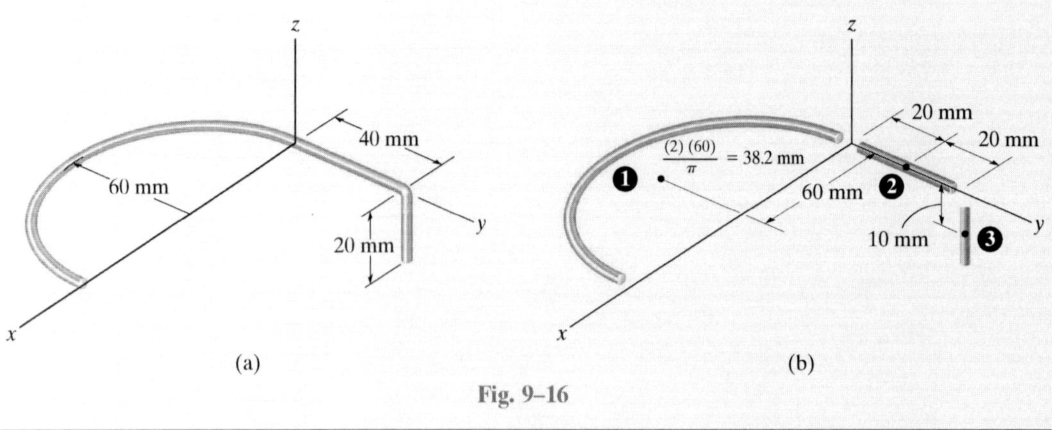

Fig. 9–16

EXAMPLE 9.10

Locate the centroid of the plate area shown in Fig. 9–17*a*.

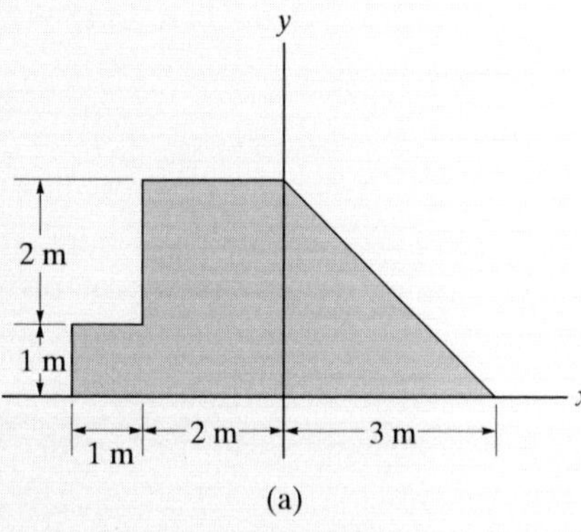

(a)

Fig. 9–17

SOLUTION

Composite Parts. The plate is divided into three segments as shown in Fig. 9–17*b*. Here the area of the small rectangle ③ is considered "negative" since it must be subtracted from the larger one ②.

Moment Arms. The centroid of each segment is located as indicated in the figure. Note that the $\tilde{x}$ coordinates of ② and ③ are *negative*.

Summations. Taking the data from Fig. 9–17*b*, the calculations are tabulated as follows:

Segment	A (m^2)	$\tilde{x}$ (m)	$\tilde{y}$ (m)	$\tilde{x}A$ (m^3)	$\tilde{y}A$ (m^3)
1	$\frac{1}{2}(3)(3) = 4.5$	1	1	4.5	4.5
2	$(3)(3) = 9$	-1.5	1.5	-13.5	13.5
3	$-(2)(1) = -2$	-2.5	2	5	-4
	$\Sigma A = 11.5$			$\Sigma \tilde{x}A = -4$	$\Sigma \tilde{y}A = 14$

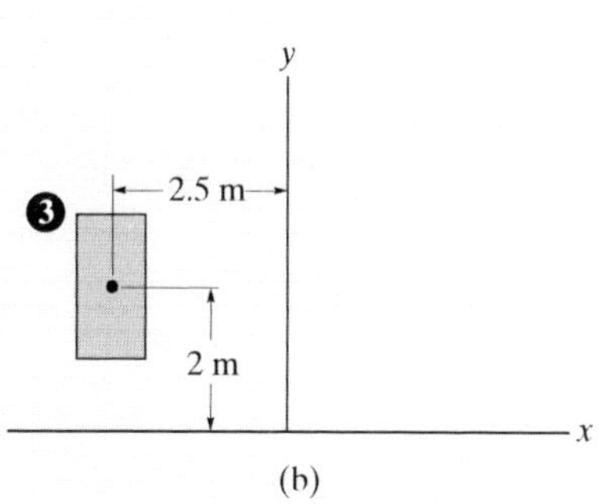

(b)

Thus,

$$\bar{x} = \frac{\Sigma \tilde{x}A}{\Sigma A} = \frac{-4}{11.5} = -0.348 \text{ m} \qquad \textit{Ans.}$$

$$\bar{y} = \frac{\Sigma \tilde{y}A}{\Sigma A} = \frac{14}{11.5} = 1.22 \text{ m} \qquad \textit{Ans.}$$

NOTE: If these results are plotted in Fig. 9–17*a*, the location of point *C* seems reasonable.

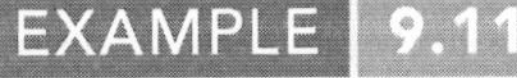

EXAMPLE 9.11

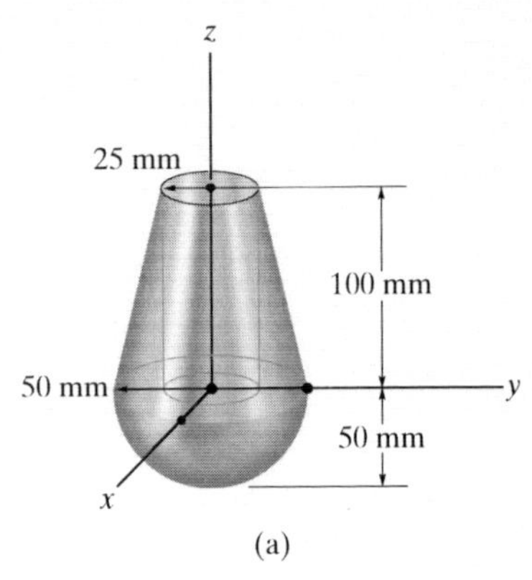

(a)

Fig. 9–18

Locate the center of mass of the assembly shown in Fig. 9–18*a*. The conical frustum has a density of $\rho_c = 8\ \text{Mg/m}^3$, and the hemisphere has a density of $\rho_h = 4\ \text{Mg/m}^3$. There is a 25-mm-radius cylindrical hole in the center of the frustum.

SOLUTION

Composite Parts. The assembly can be thought of as consisting of four segments as shown in Fig. 9–18*b*. For the calculations, ③ and ④ must be considered as "negative" segments in order that the four segments, when added together, yield the total composite shape shown in Fig. 9–18*a*.

Moment Arm. Using the table on the inside back cover, the computations for the centroid $\tilde{z}$ of each piece are shown in the figure.

Summations. Because of *symmetry*, note that

$$\bar{x} = \bar{y} = 0 \qquad \textit{Ans.}$$

Since $W = mg$, and g is constant, the third of Eqs. 9–6 becomes $\bar{z} = \Sigma\tilde{z}m/\Sigma m$. The mass of each piece can be computed from $m = \rho V$ and used for the calculations. Also, $1\ \text{Mg/m}^3 = 10^{-6}\ \text{kg/mm}^3$, so that

Segment	m (kg)	$\tilde{z}$ (mm)	$\tilde{z}m$ (kg · mm)
1	$8(10^{-6})\left(\frac{1}{3}\right)\pi(50)^2(200) = 4.189$	50	209.440
2	$4(10^{-6})\left(\frac{2}{3}\right)\pi(50)^3 = 1.047$	-18.75	-19.635
3	$-8(10^{-6})\left(\frac{1}{3}\right)\pi(25)^2(100) = -0.524$	$100 + 25 = 125$	-65.450
4	$-8(10^{-6})\pi(25)^2(100) = -1.571$	50	-78.540
	$\Sigma m = 3.142$		$\Sigma\tilde{z}m = 45.815$

$$\text{Thus,} \quad \bar{z} = \frac{\Sigma\tilde{z}m}{\Sigma m} = \frac{45.815}{3.142} = 14.6\ \text{mm} \qquad \textit{Ans.}$$

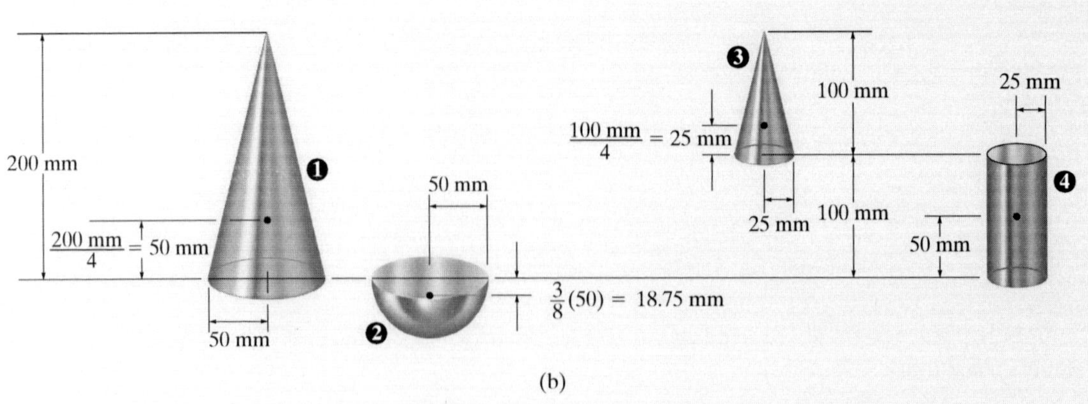

(b)

FUNDAMENTAL PROBLEMS

F9–7. Locate the centroid $(\bar{x}, \bar{y}, \bar{z})$ of the wire bent in the shape shown.

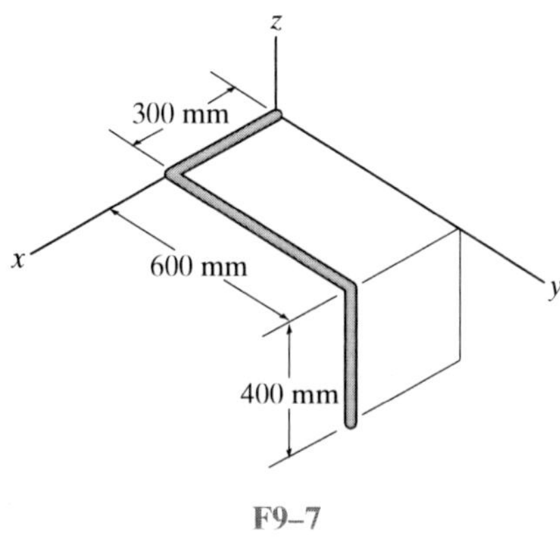

F9–7

F9–8. Locate the centroid $\bar{y}$ of the beam's cross-sectional area.

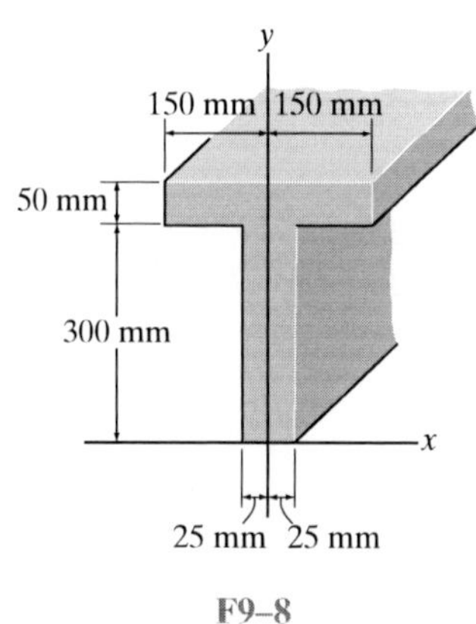

F9–8

F9–9. Locate the centroid $\bar{y}$ of the beam's cross-sectional area.

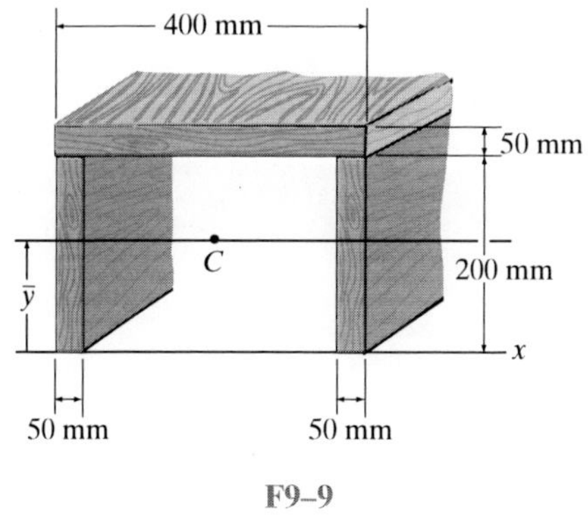

F9–9

F9–10. Locate the centroid $(\bar{x}, \bar{y})$ of the cross-sectional area.

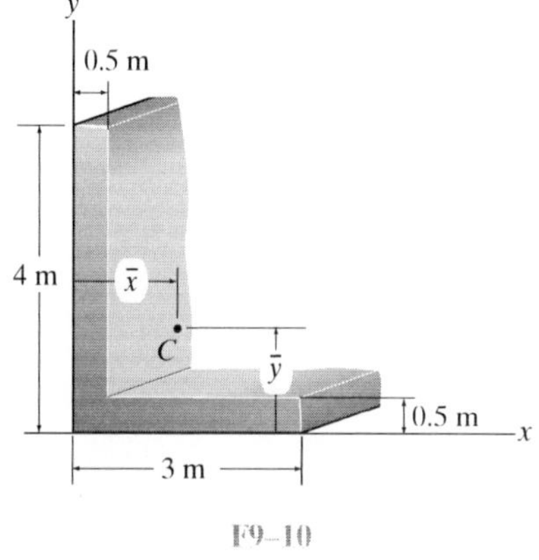

F9–10

F9–11. Locate the center of mass $(\bar{x}, \bar{y}, \bar{z})$ of the homogeneous solid block.

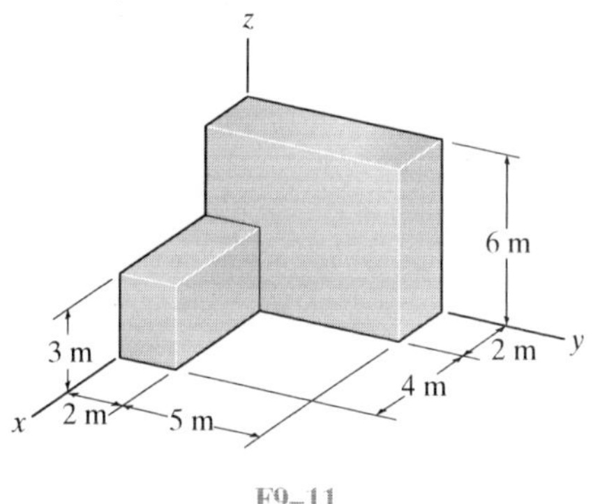

F9–11

F9–12. Determine the center of mass $(\bar{x}, \bar{y}, \bar{z})$ of the homogeneous solid block.

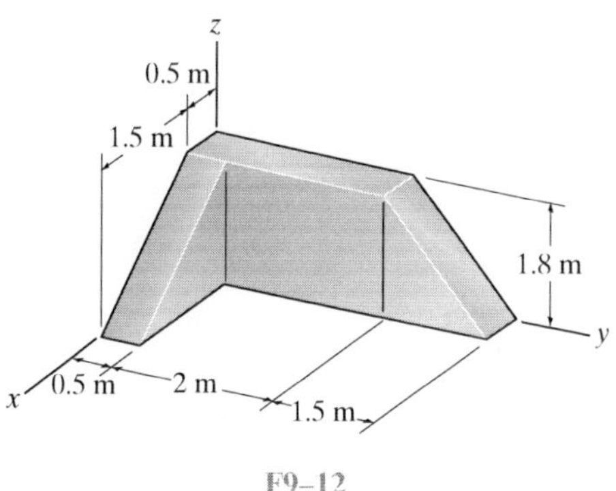

F9–12

PROBLEMS

9–51. The truss is made from five members, each having a length of 4 m and a mass of 7 kg/m. If the mass of the gusset plates at the joints and the thickness of the members can be neglected, determine the distance d to where the hoisting cable must be attached, so that the truss does not tip (rotate) when it is lifted.

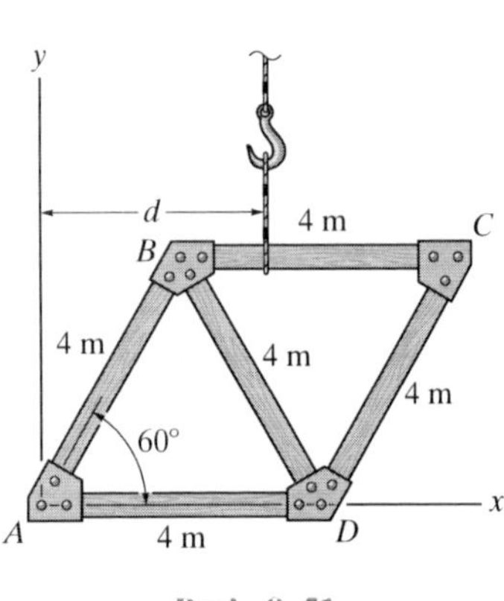

Prob. 9–51

***9–52.** Locate the centroid $(\bar{x}, \bar{y})$ of the uniform wire bent in the shape shown.

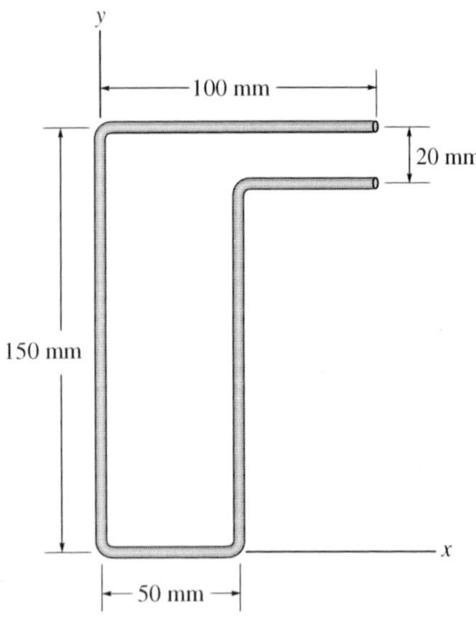

Prob. 9–52

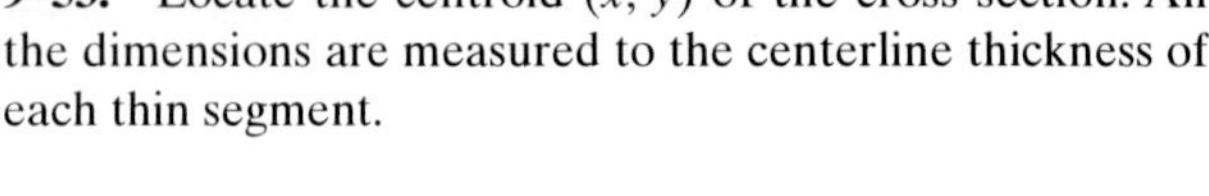

9–53. Locate the centroid $(\bar{x}, \bar{y})$ of the cross section. All the dimensions are measured to the centerline thickness of each thin segment.

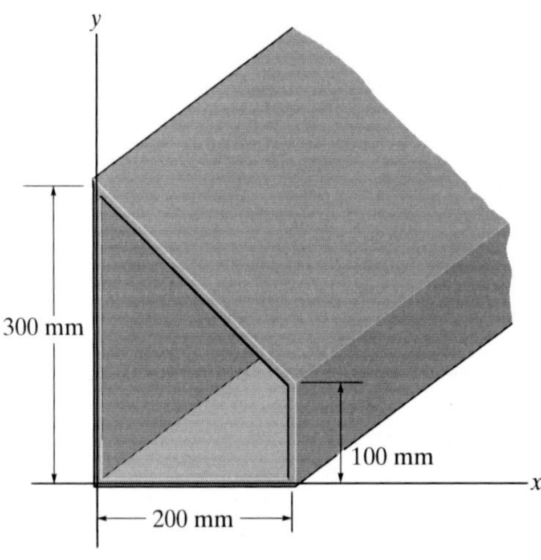

Prob. 9–53

9–54. Locate the centroid $(\bar{x}, \bar{y})$ of the metal cross section. Neglect the thickness of the material and slight bends at the corners.

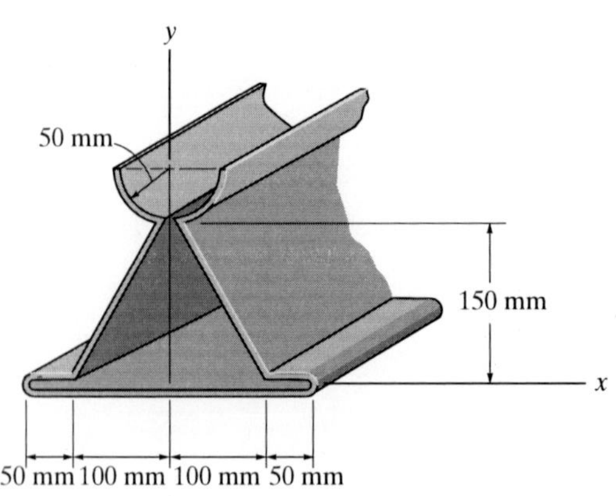

Prob. 9–54

9

9–55. Each of the three members of the frame has a mass per unit length of 6 kg/m. Locate the position $(\bar{x}, \bar{y})$ of the center of gravity. Neglect the size of the pins at the joints and the thickness of the members. Also, calculate the reactions at the pin A and roller E.

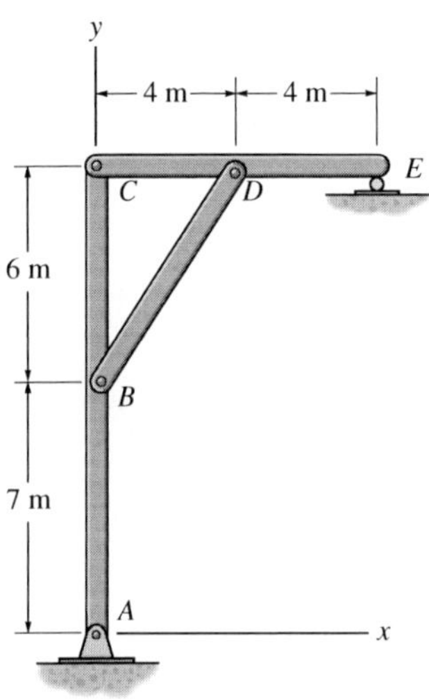

Prob. 9–55

***9–56.** The steel and aluminum plate assembly is bolted together and fastened to the wall. Each plate has a constant width in the z direction of 200 mm and thickness of 20 mm. If the density of A and B is $\rho_s = 7.85\ \text{Mg/m}^3$, and for C, $\rho_{al} = 2.71\ \text{Mg/m}^3$, determine the location $\bar{x}$ of the center of mass. Neglect the size of the bolts.

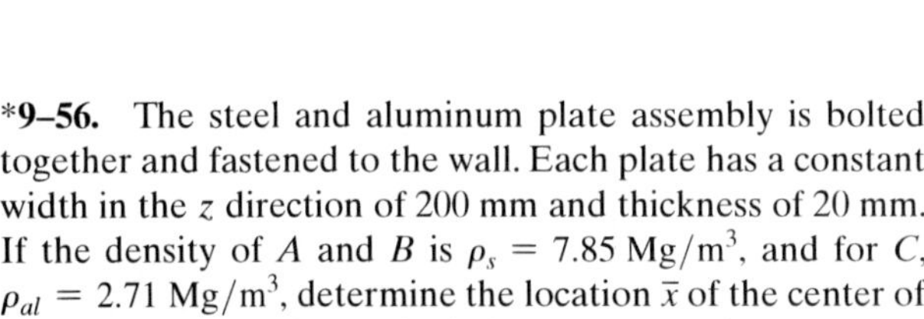

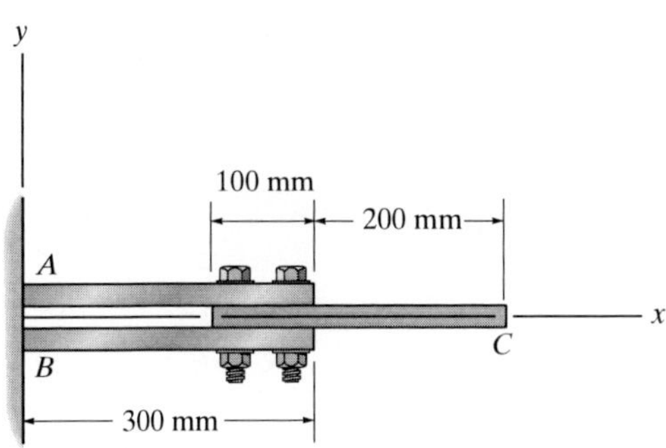

Prob. 9–56

9–57. To determine the location of the center of gravity of the automobile it is first placed in a *level position*, with the two wheels on one side resting on the scale platform P. In this position the scale records a reading of W_1. Then, one side is elevated to a convenient height c as shown. The new reading on the scale is W_2. If the automobile has a total weight of W, determine the location of its center of gravity $G(\bar{x}, \bar{y})$.

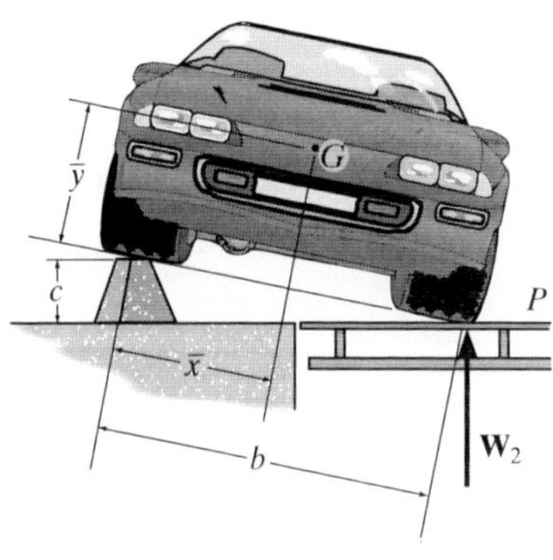

Prob. 9–57

9–58. Determine the location $\bar{y}$ of the centroidal axis $\bar{x}$–$\bar{x}$ of the beam's cross-sectional area. Neglect the size of the corner welds at A and B for the calculation.

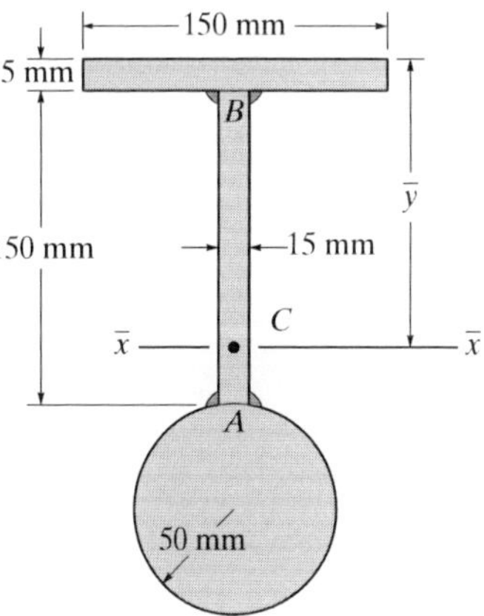

Prob. 9–58

9–59. Locate the centroid $\bar{y}$ of the cross-sectional area of the built-up beam.

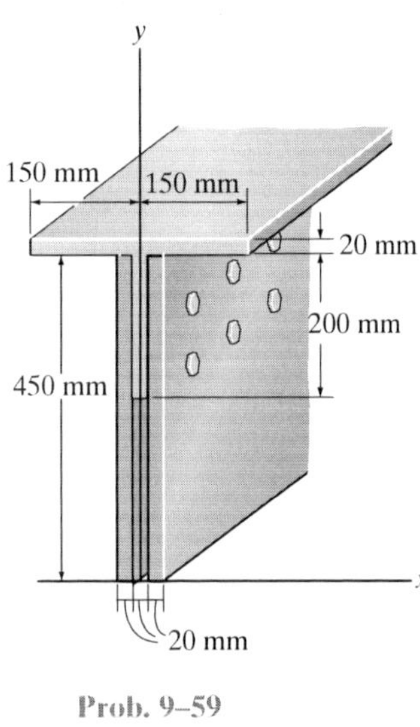

Prob. 9–59

*__9–60.__ Locate the centroid $\bar{y}$ of the cross-sectional area of the built-up beam.

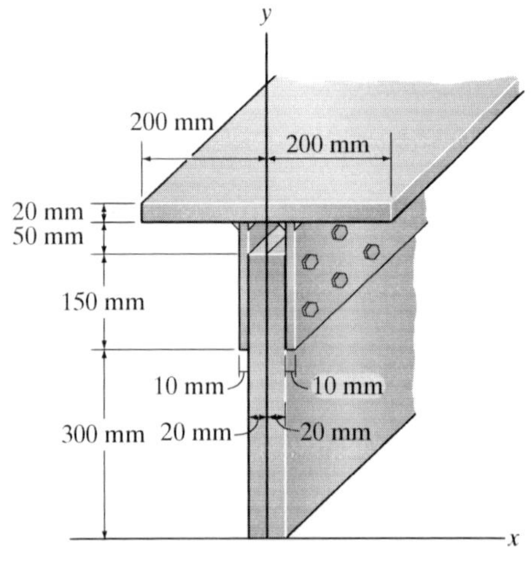

Prob. 9–60

9–61. Locate the centroid $(\bar{x}, \bar{y})$ of the member's cross-sectional area.

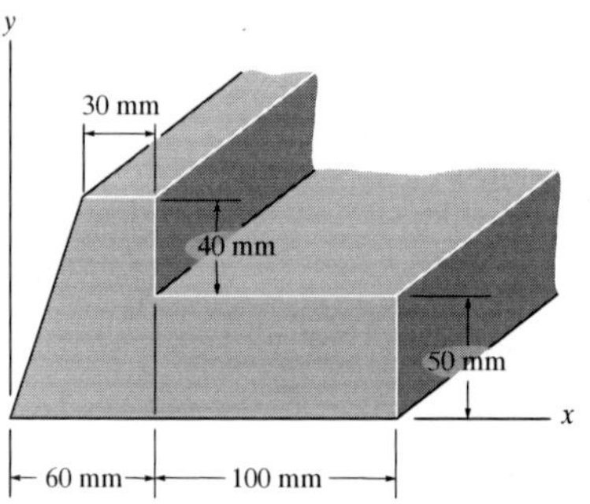

Prob. 9–61

9–62. Locate the centroid $\bar{y}$ of the bulb-tee cross section.

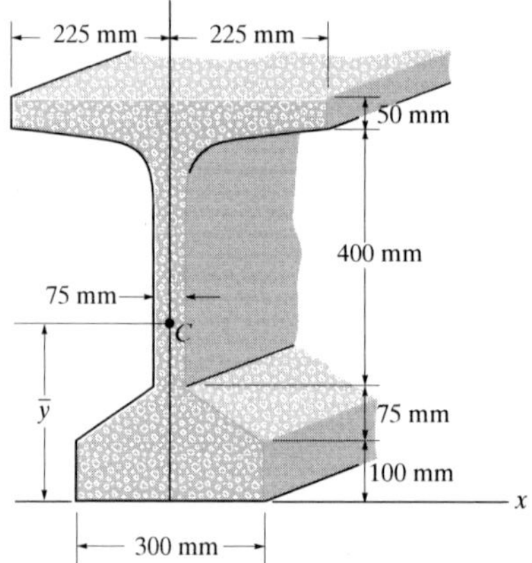

Prob. 9–62

9–63. The gravity wall is made of concrete. Determine the location $(\bar{x}, \bar{y})$ of the center of gravity G for the wall.

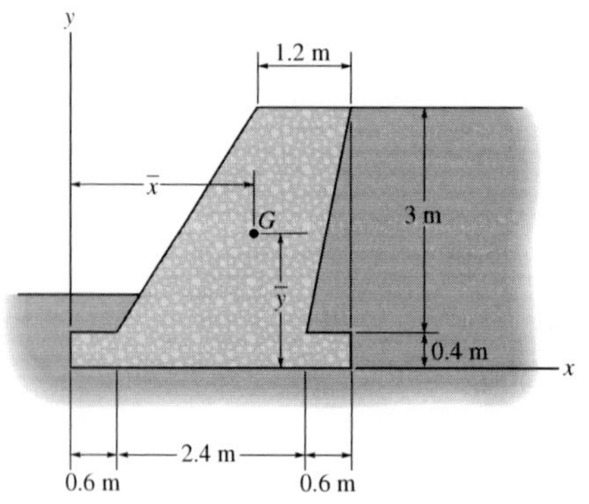

Prob. 9–63

9

*9–64. Locate the centroid $\bar{y}$ of the concrete beam having the tapered cross section shown.

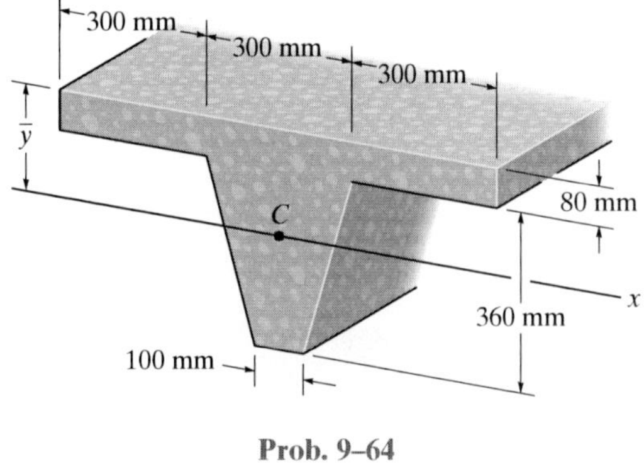

Prob. 9–64

9–65. Locate the center of gravity $G(\bar{x}, \bar{y})$ of the streetlight. Neglect the thickness of each segment. The mass per unit length of each segment is as follows: $\rho_{AB} = 12$ kg/m, $\rho_{BC} = 8$ kg/m, $\rho_{CD} = 5$ kg/m, and $\rho_{DE} = 2$ kg/m.

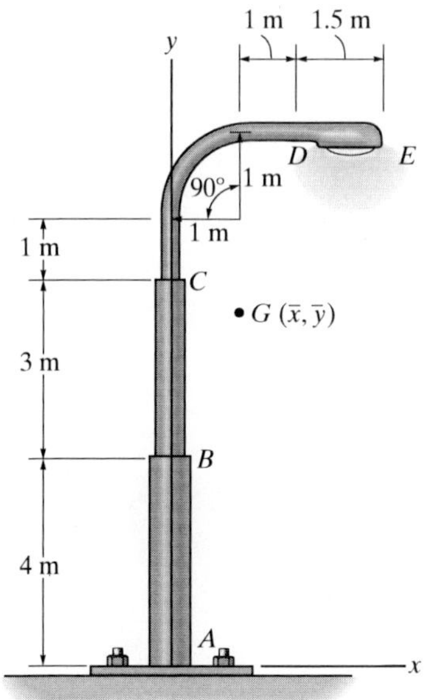

Prob. 9–65

9–66. Locate the centroid $\bar{y}$ of the cross-sectional area of the beam constructed from a channel and a plate. Assume all corners are square and neglect the size of the weld at A.

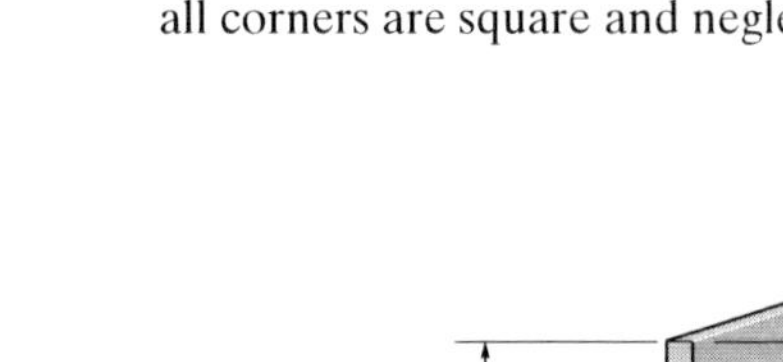

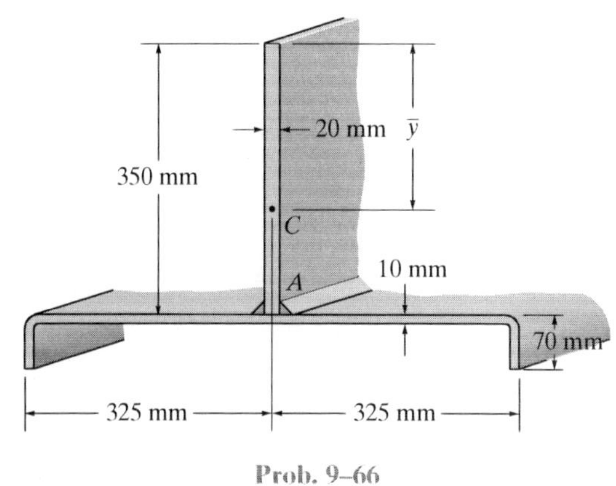

Prob. 9–66

9–67. An aluminum strut has a cross section referred to as a deep hat. Locate the centroid $\bar{y}$ of its area. Each segment has a thickness of 10 mm.

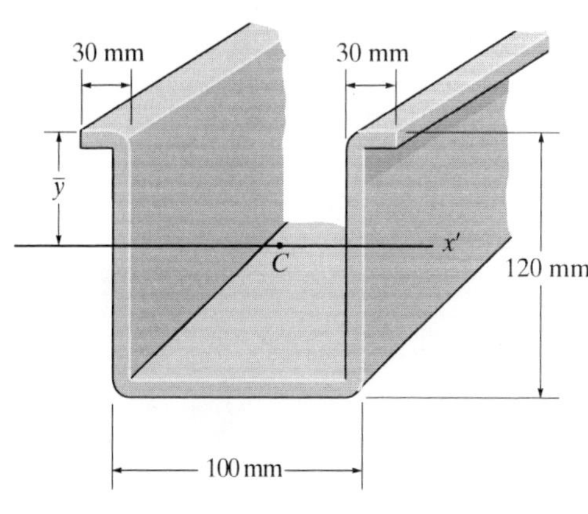

Prob. 9–67

9

***9–68.** Locate the centroid $\bar{y}$ of the beam's cross-sectional area.

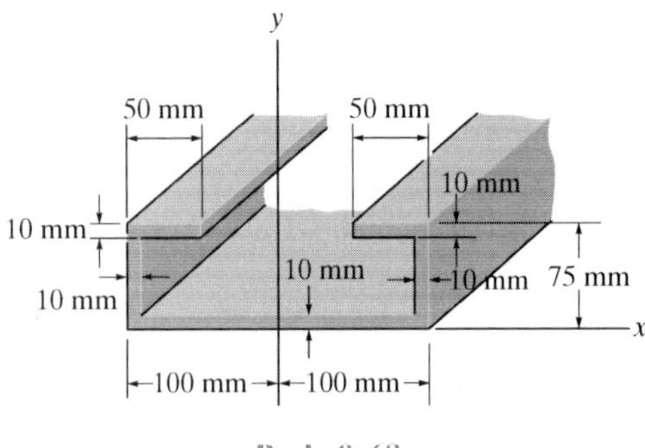

Prob. 9–68

9–69. Determine the location $\bar{x}$ of the centroid C of the shaded area that is part of a circle having a radius r.

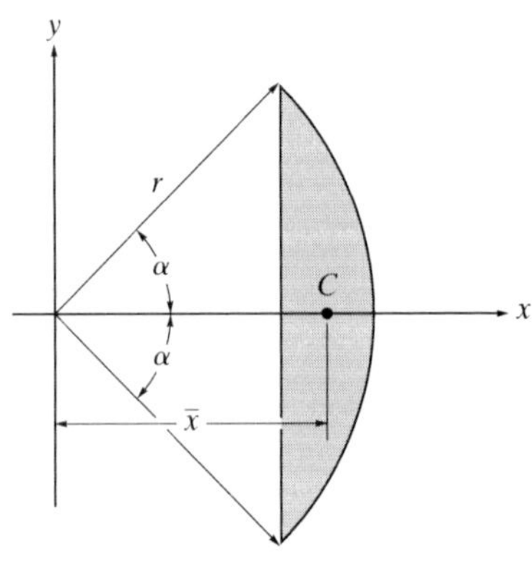

Prob. 9–69

9–70. Locate the centroid $\bar{y}$ for the cross-sectional area of the angle.

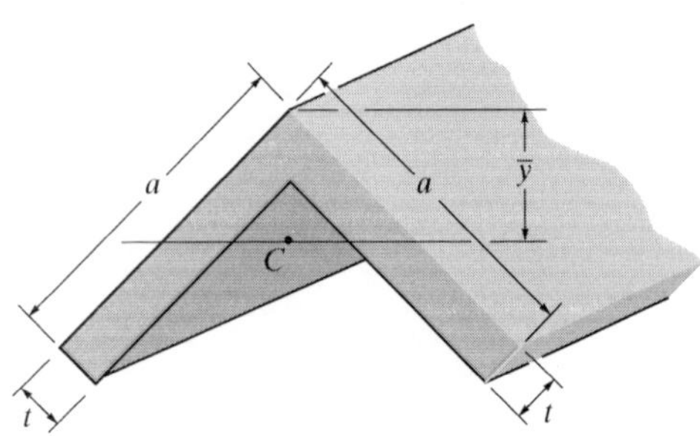

Prob. 9–70

9–71. Determine the location $\bar{y}$ of the centroid of the beam's cross-sectional area. Neglect the size of the corner welds at A and B for the calculation.

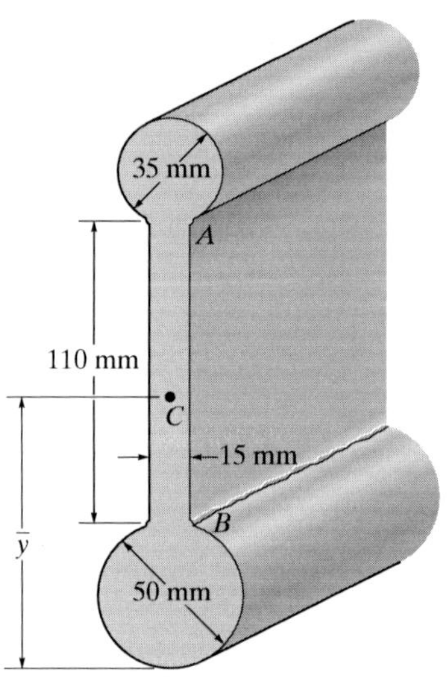

Prob. 9–71

***9–72.** Uniform blocks having a length L and mass m are stacked one on top of the other, with each block overhanging the other by a distance d, as shown. If the blocks are glued together, so that they will not topple over, determine the location $\bar{x}$ of the center of mass of a pile of n blocks.

9–73. Uniform blocks having a length L and mass m are stacked one on top of the other, with each block overhanging the other by a distance d, as shown. Show that the maximum number of blocks which can be stacked in this manner is $n < L/d$.

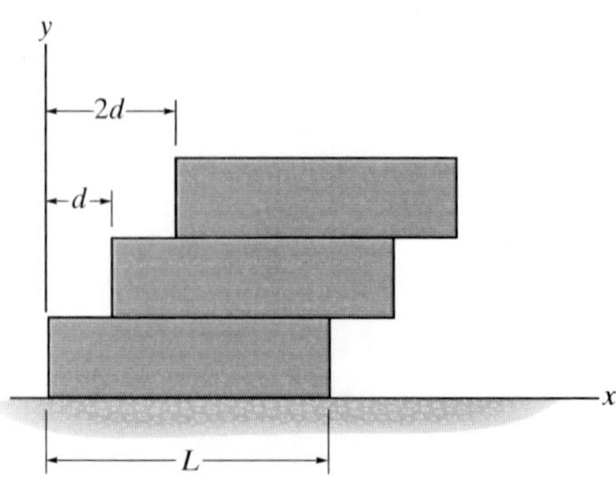

Probs. 9–72/73

9

9–74. The assembly is made from a steel hemisphere, $\rho_{st} = 7.80\ \text{Mg/m}^3$, and an aluminum cylinder, $\rho_{al} = 2.70\ \text{Mg/m}^3$. Determine the mass center of the assembly if the height of the cylinder is $h = 200$ mm.

9–75. The assembly is made from a steel hemisphere, $\rho_{st} = 7.80\ \text{Mg/m}^3$, and an aluminum cylinder, $\rho_{al} = 2.70\ \text{Mg/m}^3$. Determine the height h of the cylinder so that the mass center of the assembly is located at $\bar{z} = 160$ mm.

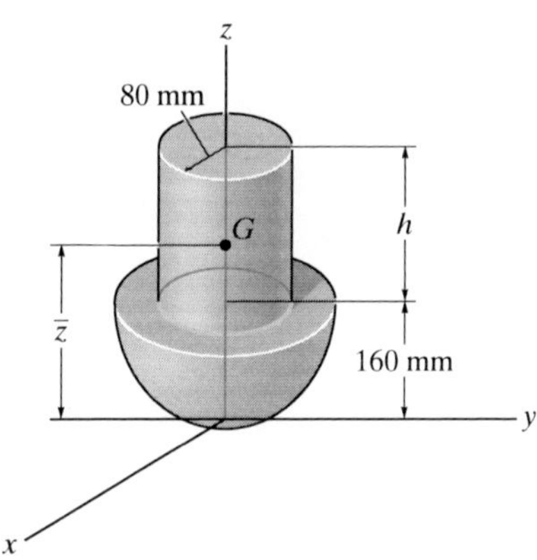

Probs. 9–74/75

9–77. Locate the centroid $\bar{y}$ for the strut's cross-sectional area.

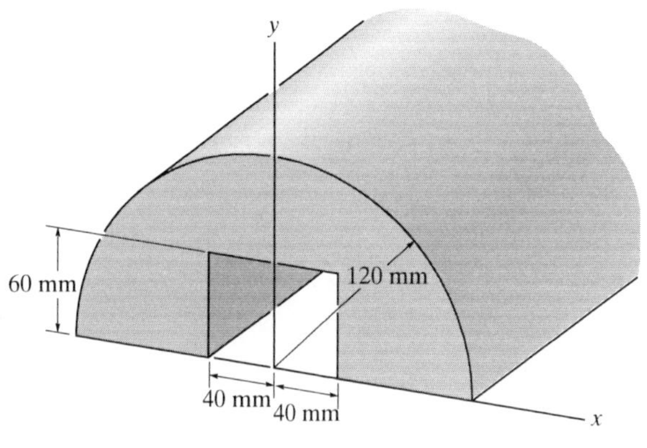

Prob. 9–77

***9–76.** Locate the centroid $\bar{y}$ for the beam's cross-sectional area.

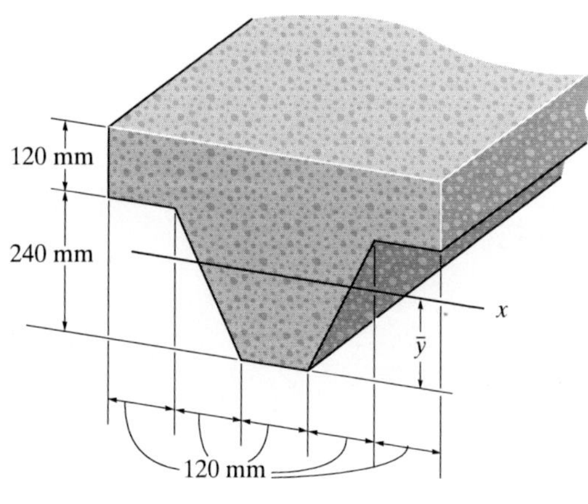

Prob. 9–76

9–78. An aluminum strut has a cross section referred to as a deep hat. Locate the centroid $\bar{y}$ of its area. Each segment has a thickness of 10 mm.

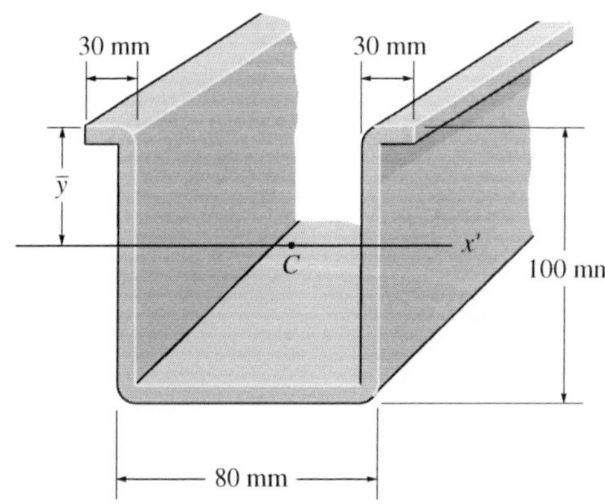

Prob. 9–78

9

9–79. Locate the center of mass of the block. Materials 1, 2, and 3 have densities of 2.70 Mg/m^3, 5.70 Mg/m^3, and 7.80 Mg/m^3, respectively.

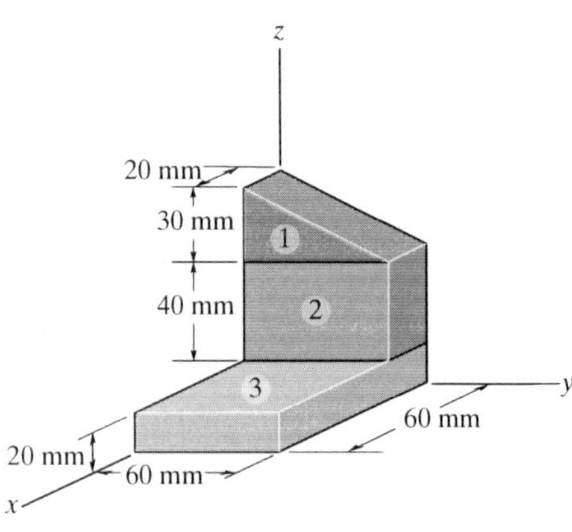

Prob. 9–79

***9–80.** Locate the centroid $\bar{z}$ of the homogenous solid formed by boring a hemispherical hole into the cylinder that is capped with a cone.

9–81. Locate the center of mass $\bar{z}$ of the solid formed by boring a hemispherical hole into a cylinder that is capped with a cone. The cone and cylinder are made of materials having densities of 7.80 Mg/m^3 and 2.70 Mg/m^3, respectively.

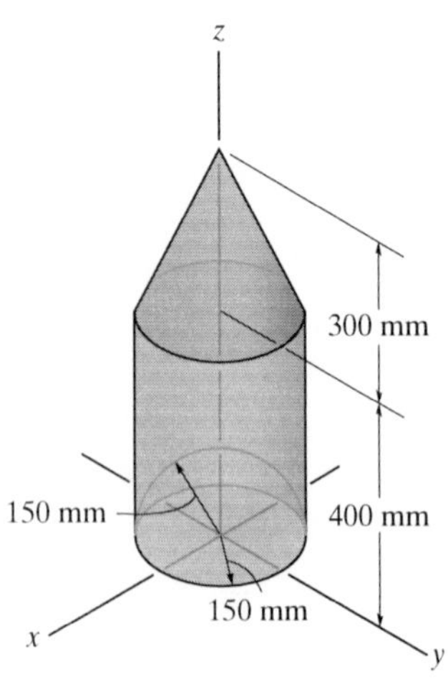

Probs. 9–80/81

9–82. Determine the distance h to which a 100-mm-diameter hole must be bored into the base of the cone so that the center of mass of the resulting shape is located at $\bar{z} = 115$ mm. The material has a density of 8 Mg/m^3.

9–83. Determine the distance $\bar{z}$ to the centroid of the shape that consists of a cone with a hole of height $h = 50$ mm bored into its base.

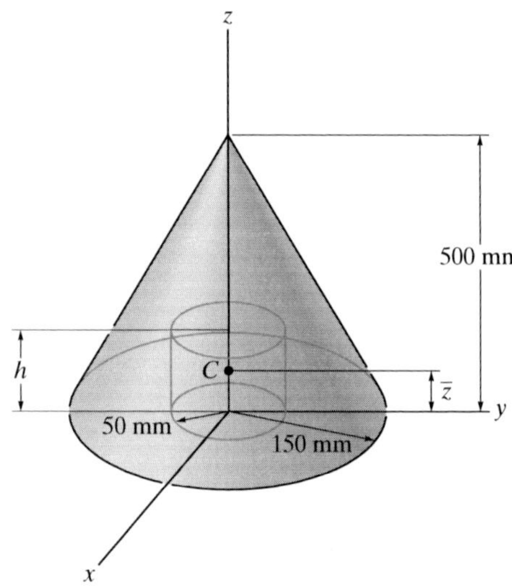

Probs. 9–82/83

***9–84.** Determine the distance $\bar{x}$ to the centroid of the solid which consists of a cylinder with a hole of length $h = 50$ mm bored into its base.

9–85. Determine the distance h to which a hole must be bored into the cylinder so that the center of mass of the assembly is located at $\bar{x} = 64$ mm. The material has a density of 8 Mg/m^3.

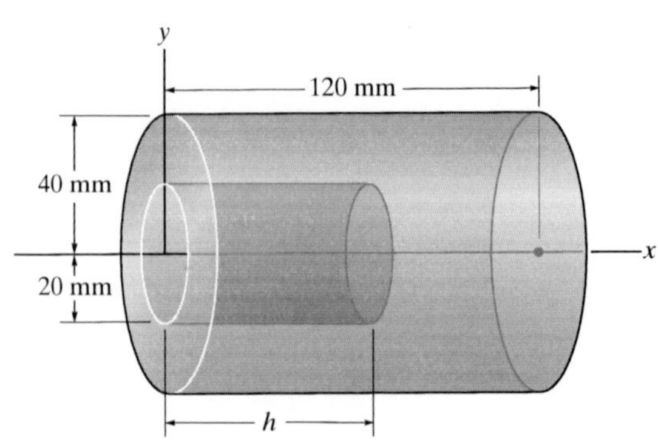

Probs. 9–84/85

9

9–86. Locate the center of mass $\bar{z}$ of the assembly. The assembly consists of a cylindrical center core, A, having a density of 7.90 Mg/m^3, and a cylindrical outer part, B, and a cone cap, C, each having a density of 2.70 Mg/m^3.

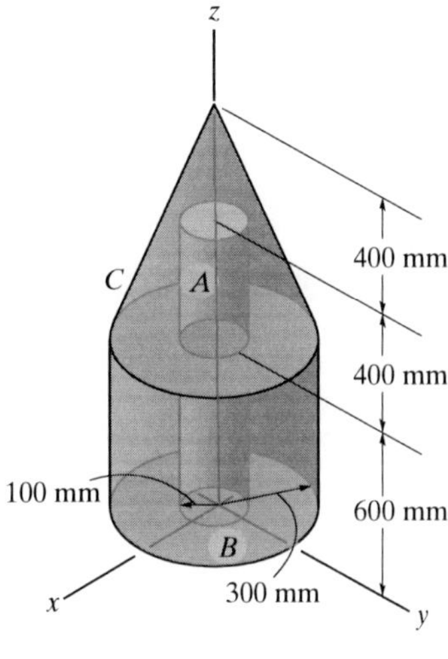

Prob. 9–86

9–87. Locate the center of mass $\bar{z}$ of the assembly. The material has a density of $\rho = 3$ Mg/m^3. There is a 30-mm diameter hole bored through the center.

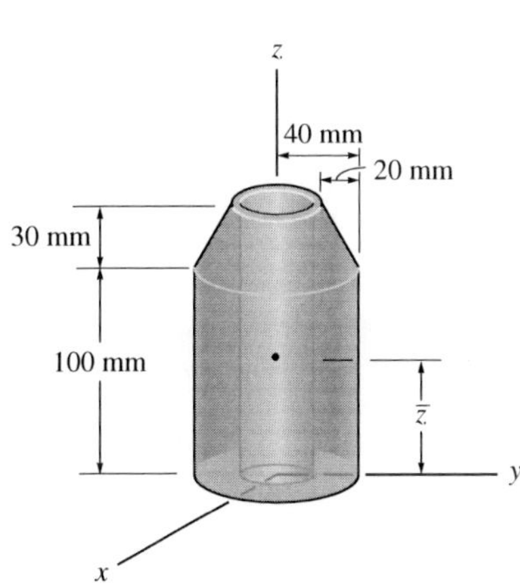

Prob. 9–87

***9–88.** A hole having a radius r is to be drilled in the center of the homogeneous block. Determine the depth h of the hole so that the center of gravity G is as low as possible.

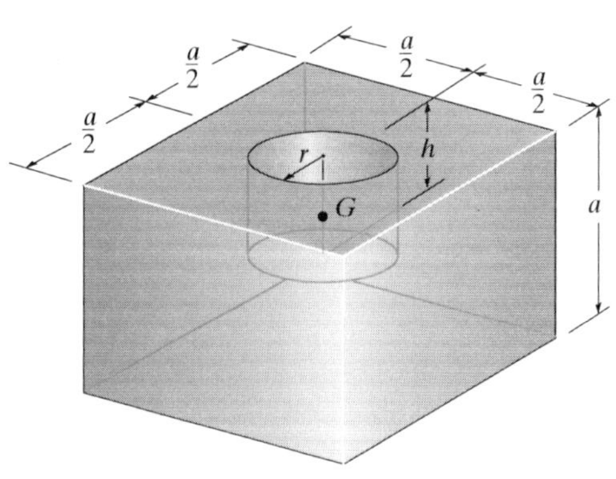

Prob. 9–88

9–89. Locate the center of mass $\bar{z}$ of the assembly. The cylinder and the cone are made from materials having densities of 5 Mg/m^3 and 9 Mg/m^3, respectively.

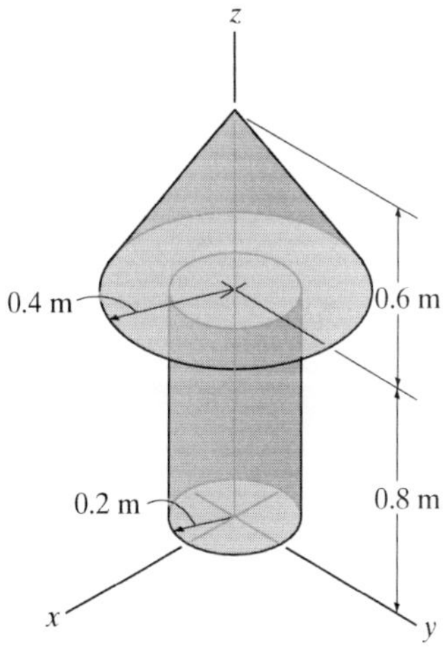

Prob. 9–89

9

*9.3 Theorems of Pappus and Guldinus

The two *theorems of Pappus and Guldinus* are used to find the surface area and volume of any body of revolution. They were first developed by Pappus of Alexandria during the fourth century A.D. and then restated at a later time by the Swiss mathematician Paul Guldin or Guldinus (1577–1643).

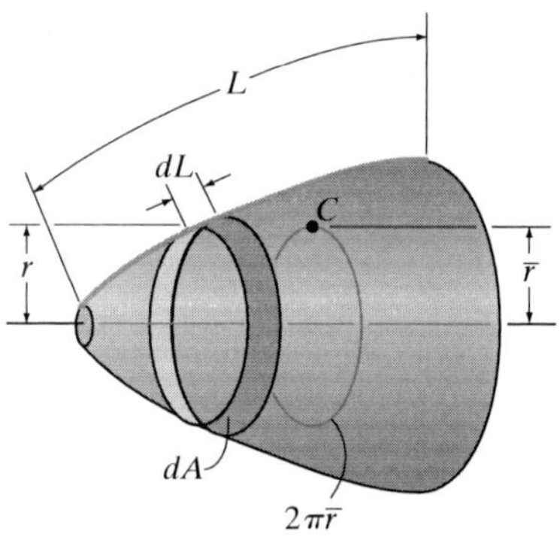

Fig. 9–19

The amount of roofing material used on this storage building can be estimated by using the first theorem of Pappus and Guldinus to determine its surface area.

Surface Area. If we revolve a *plane curve* about an axis that does not intersect the curve we will generate a *surface area of revolution*. For example, the surface area in Fig. 9–19 is formed by revolving the curve of length L about the horizontal axis. To determine this surface area, we will first consider the differential line element of length dL. If this element is revolved 2π radians about the axis, a ring having a surface area of $dA = 2\pi r\,dL$ will be generated. Thus, the surface area of the entire body is $A = 2\pi \int r\,dL$. Since $\int r\,dL = \bar{r}L$ (Eq. 9–5), then $A = 2\pi\bar{r}L$. If the curve is revolved only through an angle θ (radians), then

$$A = \theta\bar{r}L \qquad (9\text{–}7)$$

where

A = surface area of revolution

θ = angle of revolution measured in radians, $\theta \leq 2\pi$

$\bar{r}$ = perpendicular distance from the axis of revolution to the centroid of the generating curve

L = length of the generating curve

Therefore the first theorem of Pappus and Guldinus states that *the area of a surface of revolution equals the product of the length of the generating curve and the distance traveled by the centroid of the curve in generating the surface area.*

9

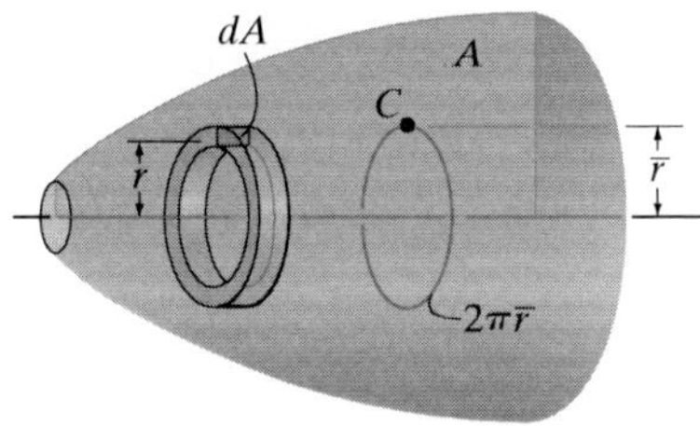

Fig. 9–20

Volume. A *volume* can be generated by revolving a *plane area* about an axis that does not intersect the area. For example, if we revolve the shaded area A in Fig. 9–20 about the horizontal axis, it generates the volume shown. This volume can be determined by first revolving the differential element of area dA 2π radians about the axis, so that a ring having the volume $dV = 2\pi r\, dA$ is generated. The entire volume is then $V = 2\pi \int r\, dA$. However, $\int r\, dA = \bar{r}\, A$, Eq. 9–4, so that $V = 2\pi\bar{r}A$. If the area is only revolved through an angle θ (radians), then

$$V = \theta\bar{r}A \tag{9–8}$$

where

V = volume of revolution

θ = angle of revolution measured in radians, $\theta \leq 2\pi$

$\bar{r}$ = perpendicular distance from the axis of revolution to the centroid of the generating area

A = generating area

Therefore the second theorem of Pappus and Guldinus states that *the volume of a body of revolution equals the product of the generating area and the distance traveled by the centroid of the area in generating the volume.*

Composite Shapes. We may also apply the above two theorems to lines or areas that are composed of a series of composite parts. In this case the total surface area or volume generated is the addition of the surface areas or volumes generated by each of the composite parts. If the perpendicular distance from the axis of revolution to the centroid of each composite part is $\tilde{r}$, then

$$A = \theta\Sigma(\tilde{r}L) \tag{9–9}$$

and

$$V = \theta\Sigma(\tilde{r}A) \tag{9–10}$$

Application of the above theorems is illustrated numerically in the following examples.

The volume of fertilizer contained within this silo can be determined using the second theorem of Pappus and Guldinus.

9

9

EXAMPLE 9.12

Show that the surface area of a sphere is $A = 4\pi R^2$ and its volume is $V = \frac{4}{3}\pi R^3$.

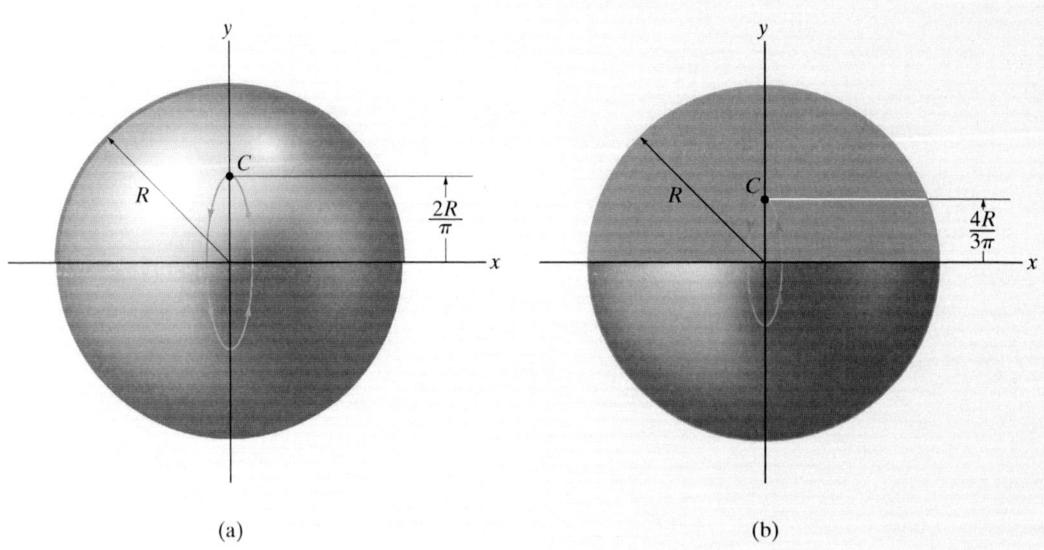

Fig. 9–21

SOLUTION

Surface Area. The surface area of the sphere in Fig. 9–21*a* is generated by revolving a semicircular *arc* about the *x* axis. Using the table on the inside back cover, it is seen that the centroid of this arc is located at a distance $\bar{r} = 2R/\pi$ from the axis of revolution (*x* axis). Since the centroid moves through an angle of $\theta = 2\pi$ rad to generate the sphere, then applying Eq. 9–7 we have

$$A = \theta\bar{r}L; \qquad A = 2\pi\left(\frac{2R}{\pi}\right)\pi R = 4\pi R^2 \qquad \textit{Ans.}$$

Volume. The volume of the sphere is generated by revolving the semicircular *area* in Fig. 9–21*b* about the *x* axis. Using the table on the inside back cover to locate the centroid of the area, i.e., $\bar{r} = 4R/3\pi$, and applying Eq. 9–8, we have

$$V = \theta\bar{r}A; \qquad V = 2\pi\left(\frac{4R}{3\pi}\right)\left(\frac{1}{2}\pi R^2\right) = \frac{4}{3}\pi R^3 \qquad \textit{Ans.}$$

EXAMPLE 9.13

Determine the surface area and volume of the full solid in Fig. 9–22*a*.

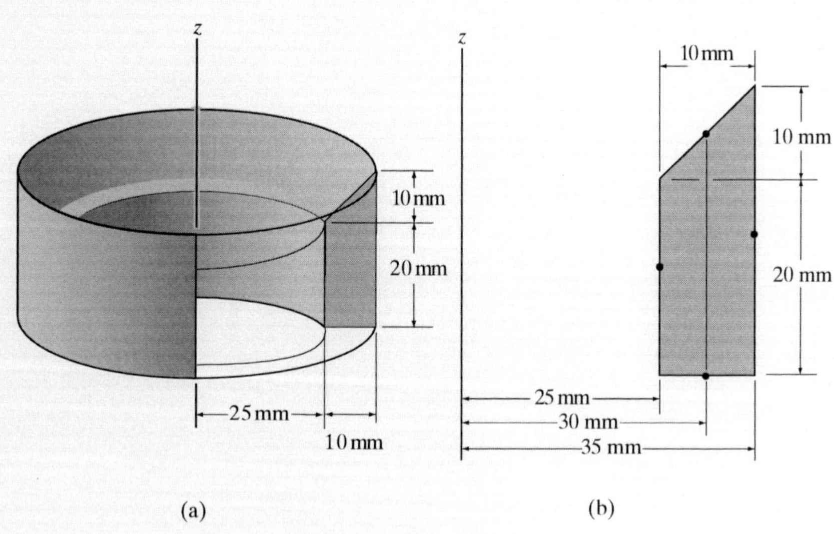

Fig. 9–22

SOLUTION

Surface Area. The surface area is generated by revolving the four line segments shown in Fig. 9–22*b* 2π radians about the z axis. The distances from the centroid of each segment to the z axis are also shown in the figure. Applying Eq. 9–7 yields

$$A = 2\pi\Sigma\tilde{r}L = 2\pi[(25\text{ mm})(20\text{ mm}) + (30\text{ mm})\left(\sqrt{(10\text{ mm})^2 + (10\text{ mm})^2}\right) + (35\text{ mm})(30\text{ mm}) + (30\text{ mm})(10\text{ mm})]$$
$$= 14\,290\text{ mm}^2 \qquad \textit{Ans.}$$

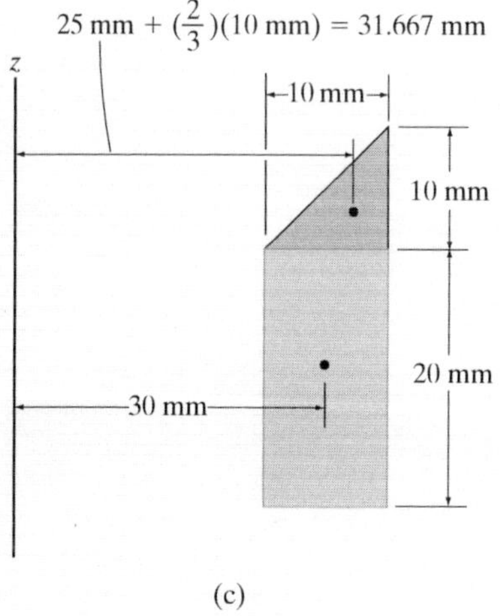

Volume. The volume of the solid is generated by revolving the two area segments shown in Fig. 9–22*c* 2π radians about the z axis. The distances from the centroid of each segment to the z axis are also shown in the figure. Applying Eq. 9–10, we have

$$V = 2\pi\Sigma\tilde{r}A = 2\pi\left\{(31.667\text{ mm})\left[\frac{1}{2}(10\text{ mm})(10\text{ mm})\right] + (30\text{ mm})[(20\text{ mm})(10\text{ mm})\right\}$$
$$= 47\,648\text{ mm}^3 \qquad \textit{Ans.}$$

FUNDAMENTAL PROBLEMS

F9–13. Determine the surface area and volume of the solid formed by revolving the shaded area 360° about the z axis.

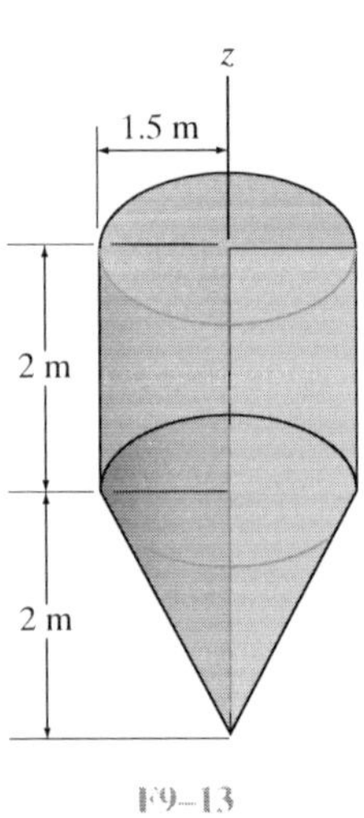

F9–13

F9–14. Determine the surface area and volume of the solid formed by revolving the shaded area 360° about the z axis.

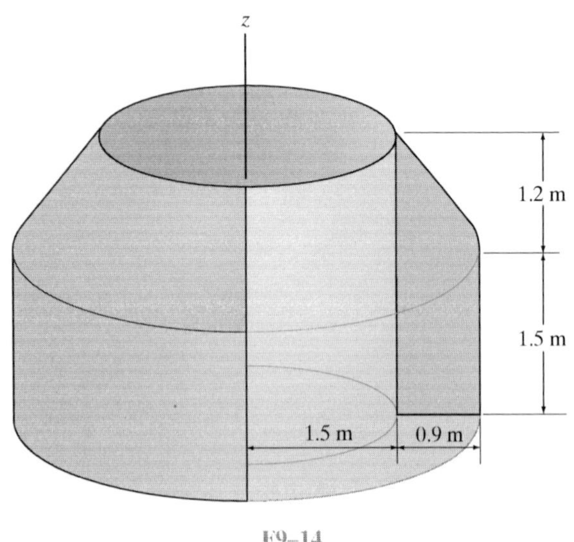

F9–14

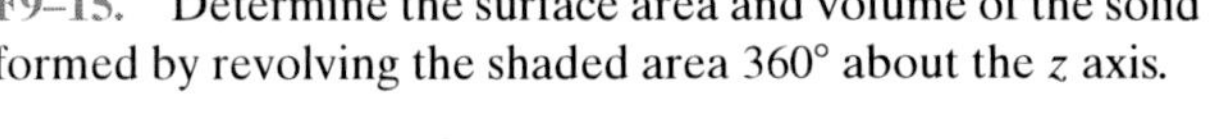

F9–15. Determine the surface area and volume of the solid formed by revolving the shaded area 360° about the z axis.

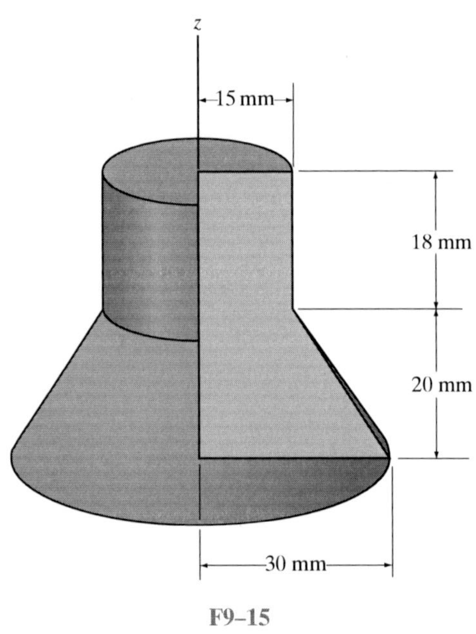

F9–15

F9–16. Determine the surface area and volume of the solid formed by revolving the shaded area 360° about the z axis.

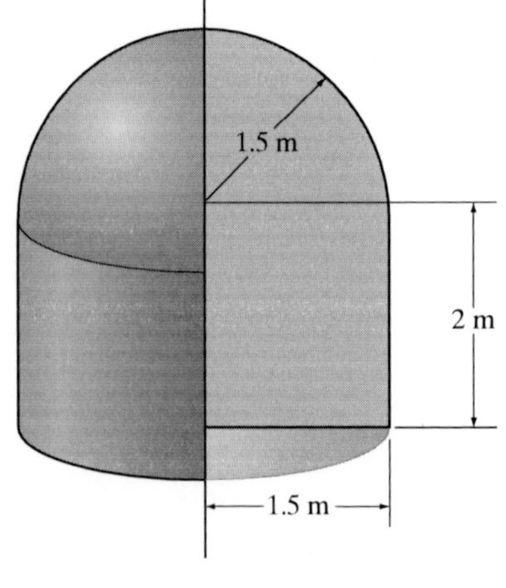

F9–16

PROBLEMS

9–90. The water tank AB has a hemispherical top and is fabricated from thin steel plate. Determine the volume within the tank.

9–91. The water tank AB has a hemispherical roof and is fabricated from thin steel plate. If a liter of paint can cover 3 m^2 of the tank's surface, determine how many liters are required to coat the surface of the tank from A to B.

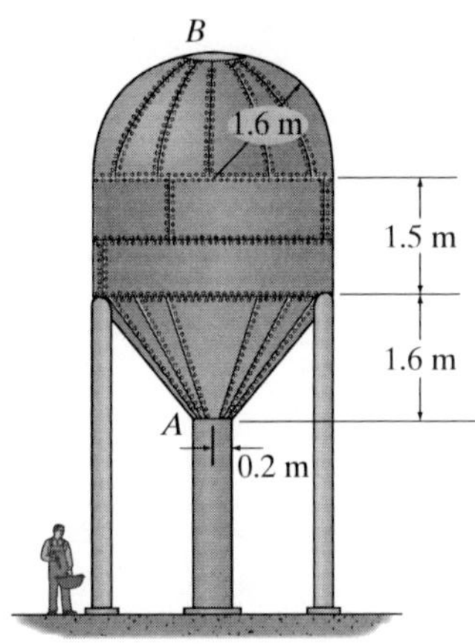

Probs. 9–90/91

*__9–92.__ Determine the outside surface area of the hopper.

9–93. The hopper is filled to its top with coal. Determine the volume of coal if the voids (air space) are 30 percent of the volume of the hopper.

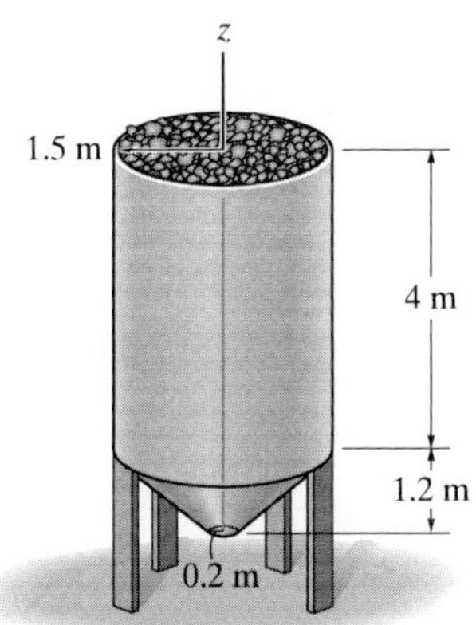

Probs. 9–92/93

9–94. The *rim* of a flywheel has the cross section A–A shown. Determine the volume of material needed for its construction.

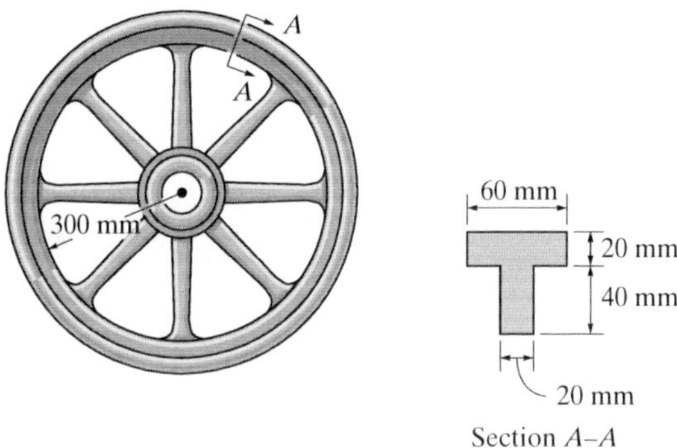

Prob. 9–94

9–95. Determine the surface area of the tank, which consists of a cylinder and hemispherical cap.

*__9–96.__ Determine the volume of the tank, which consists of a cylinder and hemispherical cap.

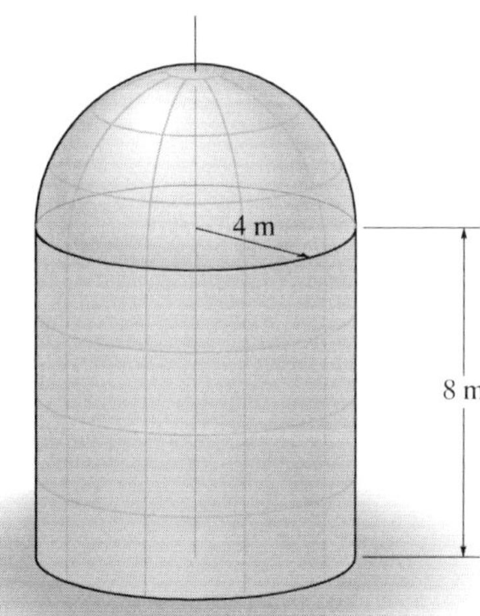

Probs. 9–95/96

9

9–97. The process tank is used to store liquids during manufacturing. Estimate the outside surface area of the tank. The tank has a flat top and the plates from which the tank is made have negligible thickness.

9–98. The process tank is used to store liquids during manufacturing. Estimate the volume of the tank. The tank has a flat top and the plates from which the tank is made have negligible thickness.

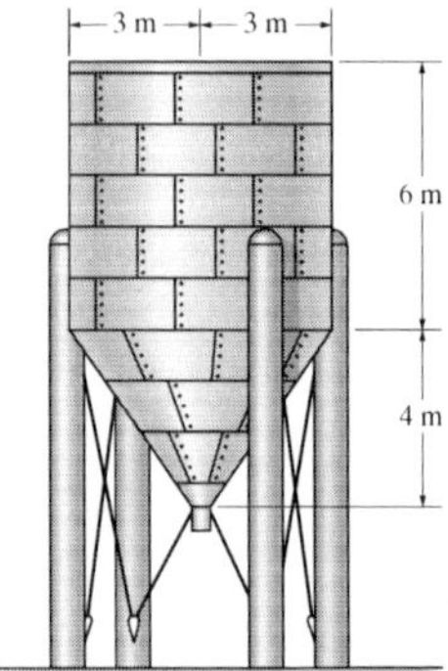

Probs. 9–97/98

9–99. The hopper is filled to its top with coal. Estimate the volume of coal if the voids (air space) are 35 percent of the volume of the hopper.

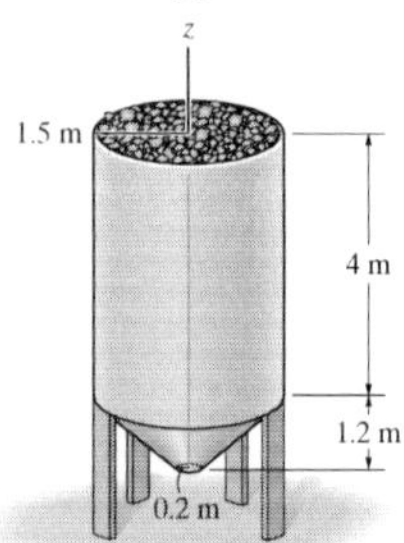

Prob. 9–99

***9–100.** Sand is piled between two walls as shown. Assume the pile to be a quarter section of a cone and that 26 percent of this volume is voids (air space). Use the second theorem of Pappus–Guldinus to determine the volume of sand.

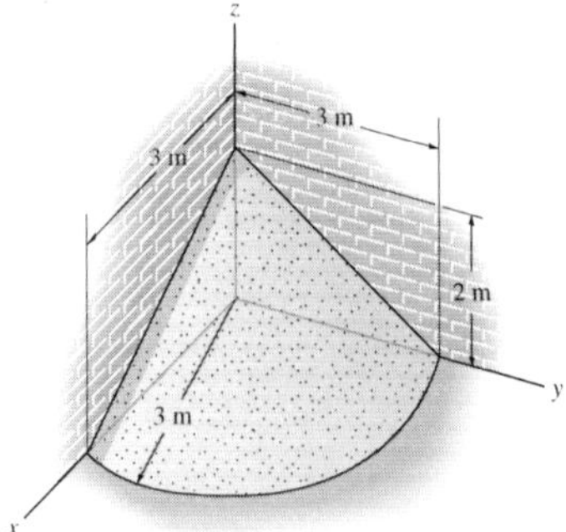

Prob. 9–100

9–101. Determine the surface area and the volume of the ring formed by rotating the square about the vertical axis.

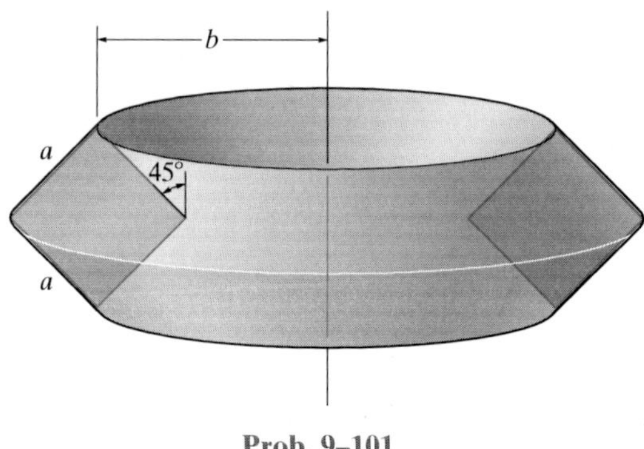

Prob. 9–101

9–102. Determine the volume of material needed to make the casting.

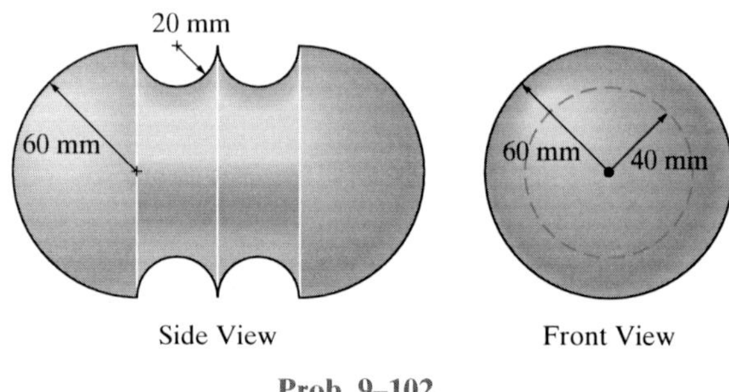

Prob. 9–102

9–103. Determine the surface area from A to B of the tank.

***9–104.** Determine the volume within the thin-walled tank from A to B.

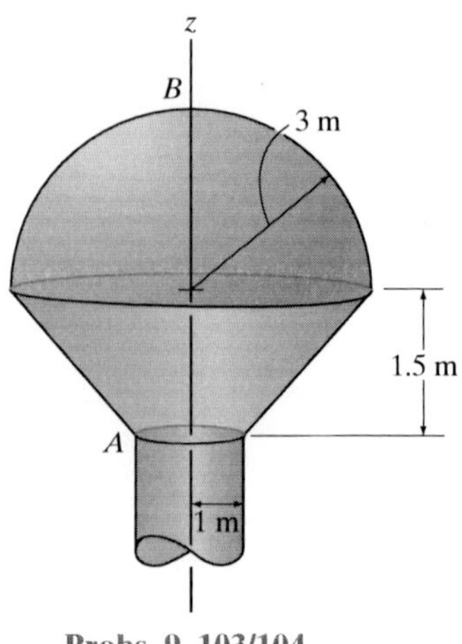

Probs. 9–103/104

9–105. A steel wheel has a diameter of 840 mm and a cross section as shown in the figure. Determine the total mass of the wheel if $\rho = 5\ \text{Mg/m}^3$.

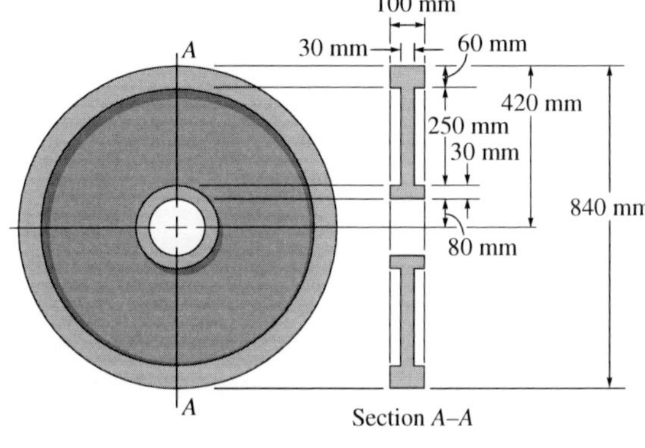

Prob. 9–105

9–107. Using integration, determine both the area and the centroidal distance $\bar{x}$ of the shaded area. Then, using the second theorem of Pappus–Guldinus, determine the volume of the solid generated by revolving the area about the y axis.

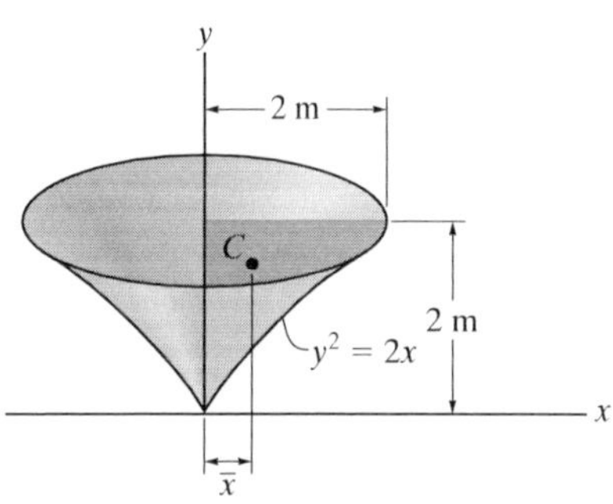

Prob. 9–107

9–106. Determine the volume of an ellipsoid formed by revolving the shaded area about the x axis using the second theorem of Pappus–Guldinus. The area and centroid y of the shaded area should first be obtained by using integration.

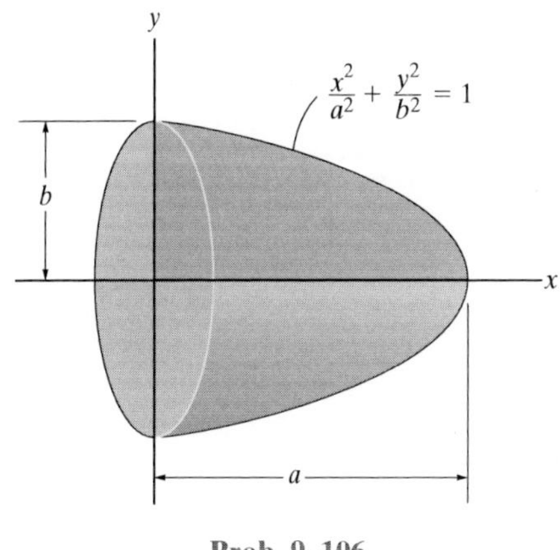

Prob. 9–106

***9–108.** Determine the height h to which liquid should be poured into the conical cup so that it contacts half the surface area on the inside of the cup.

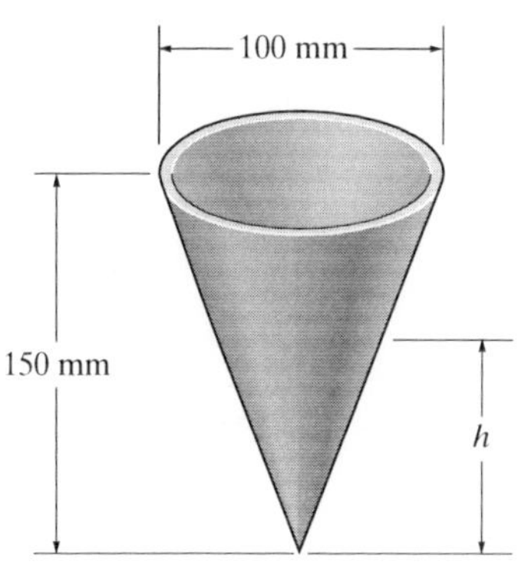

Prob. 9–108

9–109. Determine the interior surface area of the brake piston. It consists of a full circular part. Its cross section is shown in the figure.

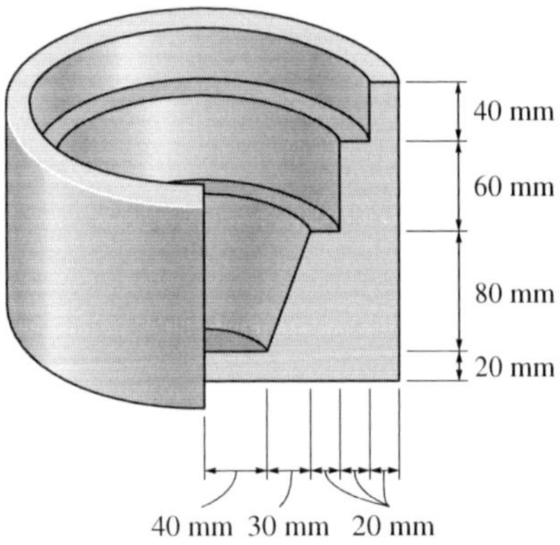

Prob. 9–109

9–110. Determine the surface area and the volume of the conical solid.

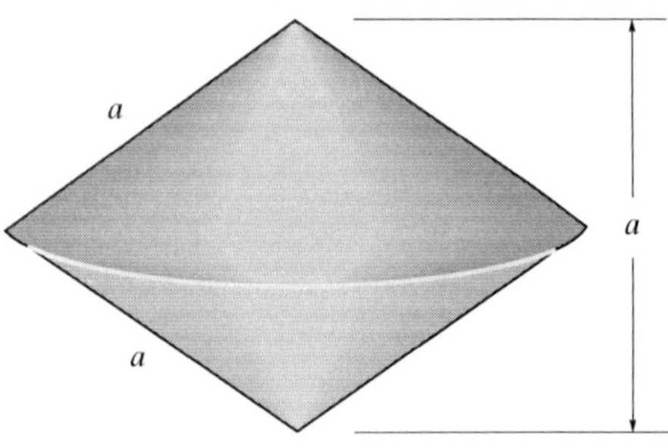

Prob. 9–110

9–111. Determine the height h to which liquid should be poured into the cup so that it contacts half the surface area on the inside of the cup. Neglect the cup's thickness for the calculation.

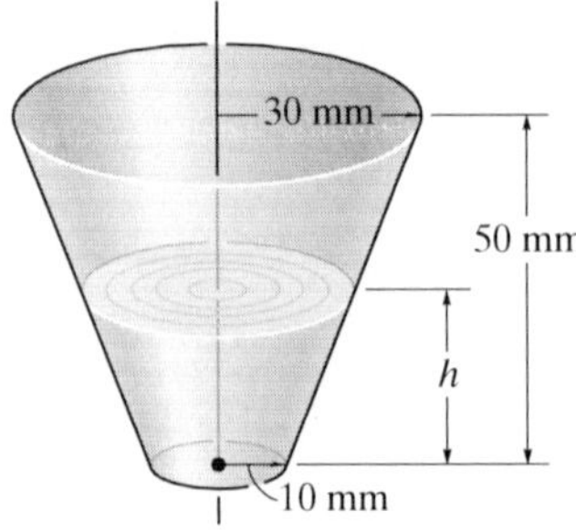

Prob. 9–111

***9–112.** The water tank has a paraboloid-shaped roof. If one liter of paint can cover 3 m² of the tank, determine the number of liters required to coat the roof.

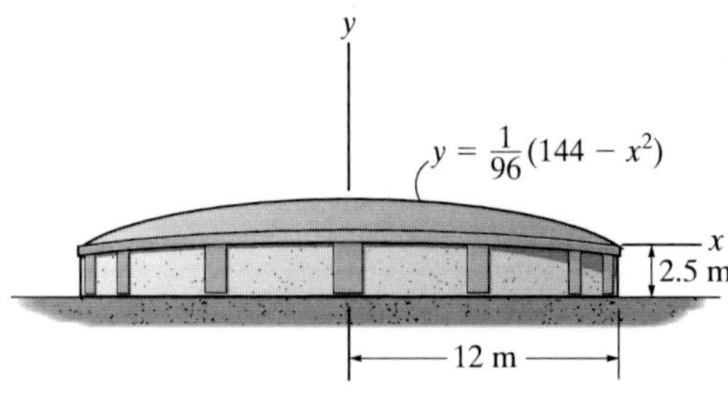

Prob. 9–112

9–113. A steel wheel has a diameter of 840 mm and a cross section as shown in the figure. Determine the total mass of the wheel if $\rho = 5 \text{ Mg/m}^3$.

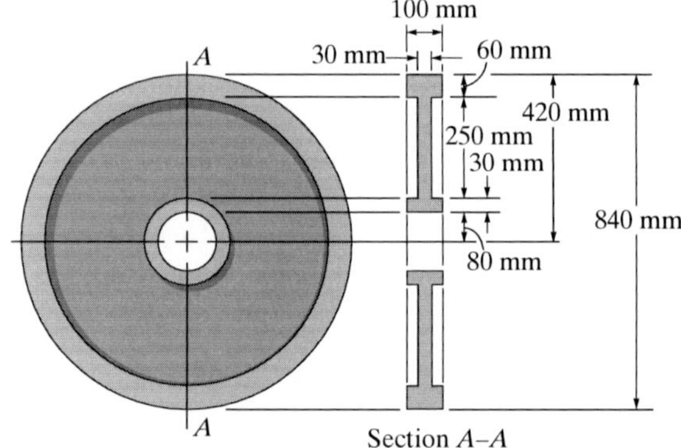

Prob. 9–113

9–114. Determine the surface area of the roof of the structure if it is formed by rotating the parabola about the y axis.

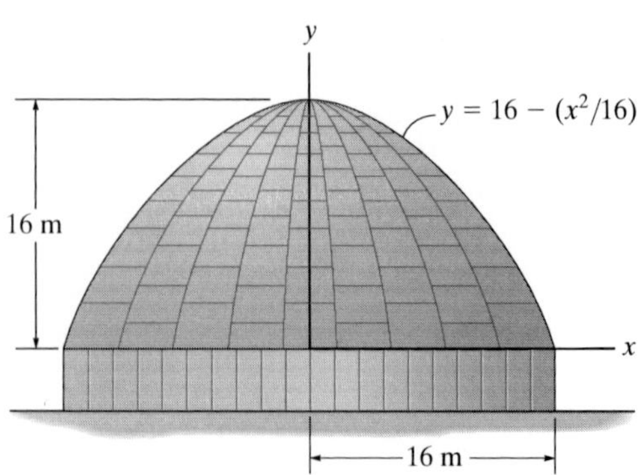

Prob. 9–114

*9.4 Resultant of a General Distributed Loading

In Sec. 4.9, we discussed the method used to simplify a two-dimensional distributed loading to a single resultant force acting at a specific point. In this section we will generalize this method to include flat surfaces that have an arbitrary shape and are subjected to a variable load distribution. Consider, for example, the flat plate shown in Fig. 9–23*a*, which is subjected to the loading defined by $p = p(x, y)$ Pa, where 1 Pa (pascal) $= 1\ \text{N/m}^2$. Knowing this function, we can determine the resultant force $\mathbf{F}_R$ acting on the plate and its location $(\bar{x}, \bar{y})$, Fig. 9–23*b*.

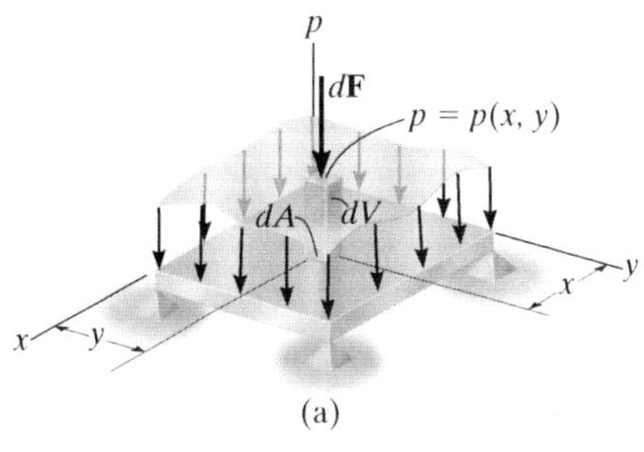

(a)

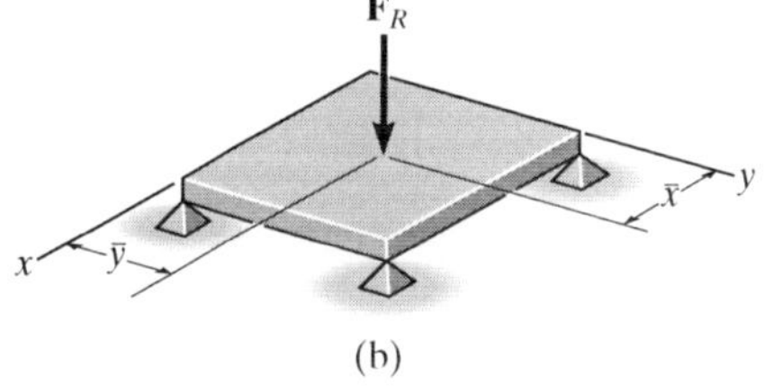

(b)

Fig. 9–23

Magnitude of Resultant Force. The force $d\mathbf{F}$ acting on the differential area dA m^2 of the plate, located at the arbitrary point (x, y), has a magnitude of $dF = [p(x, y)\ \text{N/m}^2](dA\ \text{m}^2) = [p(x, y)\ dA]$ N. Notice that $p(x, y)\ dA = dV$, the colored differential *volume element* shown in Fig. 9–23*a*. The *magnitude* of $\mathbf{F}_R$ is the sum of the differential forces acting over the plate's *entire surface area A*. Thus:

$$F_R = \Sigma F; \qquad F_R = \int_A p(x, y)\, dA = \int_V dV = V \qquad (9\text{–}11)$$

This result indicates that the *magnitude of the resultant force is equal to the total volume under the distributed-loading diagram.*

The resultant of a wind loading that is distributed on the front or side walls of this building must be calculated using integration in order to design the framework that holds the building together.

Location of Resultant Force. The location $(\bar{x}, \bar{y})$ of $\mathbf{F}_R$ is determined by setting the moments of $\mathbf{F}_R$ equal to the moments of all the differential forces $d\mathbf{F}$ about the respective y and x axes: From Figs. 9–23*a* and 9–23*b*, using Eq. 9–11, this results in

$$\bar{x} = \frac{\int_A x p(x, y)\, dA}{\int_A p(x, y)\, dA} = \frac{\int_V x\, dV}{\int_V dV} \qquad \bar{y} = \frac{\int_A y p(x, y)\, dA}{\int_A p(x, y)\, dA} = \frac{\int_V y\, dV}{\int_V dV} \qquad (9\text{–}12)$$

Hence, the *line of action of the resultant force passes through the geometric center or centroid of the volume under the distributed-loading diagram.*

9

*9.5 Fluid Pressure

According to Pascal's law, a fluid at rest creates a pressure p at a point that is the *same* in *all* directions. The magnitude of p, measured as a force per unit area, depends on the specific weight γ or mass density ρ of the fluid and the depth z of the point from the fluid surface.* The relationship can be expressed mathematically as

$$p = \gamma z = \rho g z \tag{9–13}$$

where g is the acceleration due to gravity. This equation is valid only for fluids that are assumed *incompressible*, as in the case of most liquids. Gases are compressible fluids, and since their density changes significantly with both pressure and temperature, Eq. 9–13 cannot be used.

To illustrate how Eq. 9–13 is applied, consider the submerged plate shown in Fig. 9–24. Three points on the plate have been specified. Since point B is at depth z_1 from the liquid surface, the *pressure* at this point has a magnitude $p_1 = \gamma z_1$. Likewise, points C and D are both at depth z_2; hence, $p_2 = \gamma z_2$. In all cases, the pressure acts *normal* to the surface area dA located at the specified point.

Using Eq. 9–13 and the results of Sec. 9.4, it is possible to determine the resultant force caused by a liquid and specify its location on the surface of a submerged plate. Three different shapes of plates will now be considered.

9

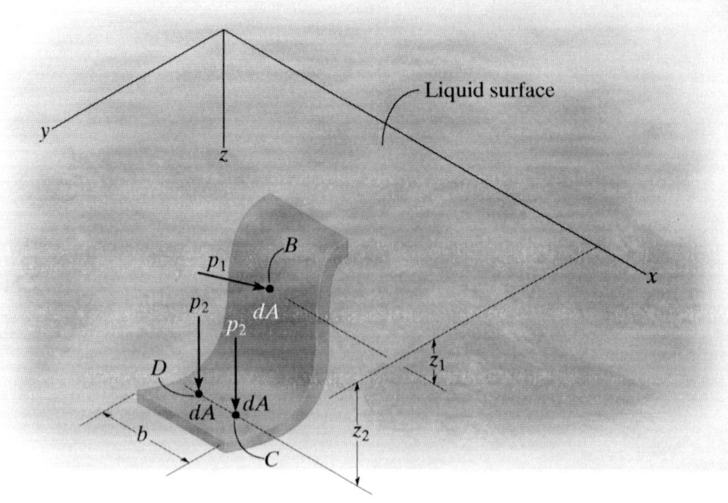

Fig. 9–24

*In particular, for water $\gamma = \rho g = 9810\ \text{N/m}^3$ since $\rho = 1000\ \text{kg/m}^3$ and $g = 9.81\ \text{m/s}^2$.

Flat Plate of Constant Width. A flat rectangular plate of constant width, which is submerged in a liquid having a specific weight γ, is shown in Fig. 9–25*a*. Since pressure varies linearly with depth, Eq. 9–13, the distribution of pressure over the plate's surface is represented by a trapezoidal volume having an intensity of $p_1 = \gamma z_1$ at depth z_1 and $p_2 = \gamma z_2$ at depth z_2. As noted in Sec. 9.4, the magnitude of the *resultant force* $\mathbf{F}_R$ is equal to the *volume* of this loading diagram and $\mathbf{F}_R$ has a *line of action* that passes through the volume's centroid C. Hence, $\mathbf{F}_R$ does *not* act at the centroid of the plate; rather, it acts at point P, called the *center of pressure*.

Since the plate has a *constant width*, the loading distribution may also be viewed in two dimensions, Fig. 9–25*b*. Here the loading intensity is measured as force/length and varies linearly from $w_1 = bp_1 = b\gamma z_1$ to $w_2 = bp_2 = b\gamma z_2$. The magnitude of $\mathbf{F}_R$ in this case equals the trapezoidal *area*, and $\mathbf{F}_R$ has a *line of action* that passes through the area's *centroid* C. For numerical applications, the area and location of the centroid for a trapezoid are tabulated on the inside back cover.

The walls of the tank must be designed to support the pressure loading of the liquid that is contained within it.

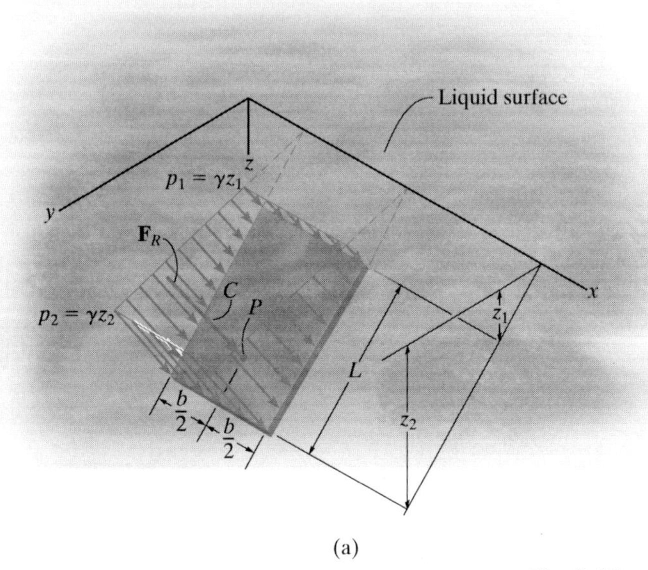

(a)

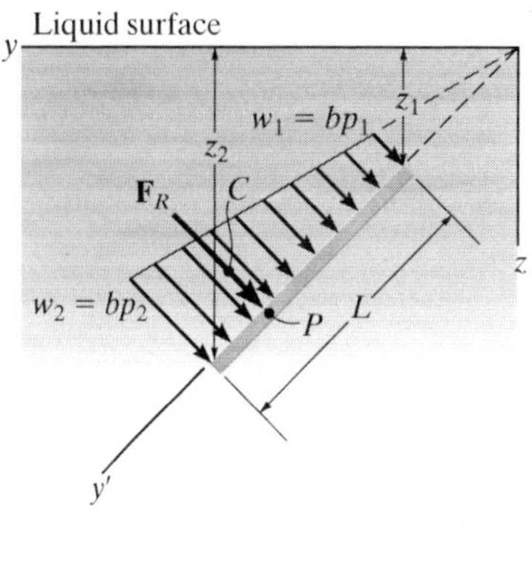

(b)

Fig. 9–25

9

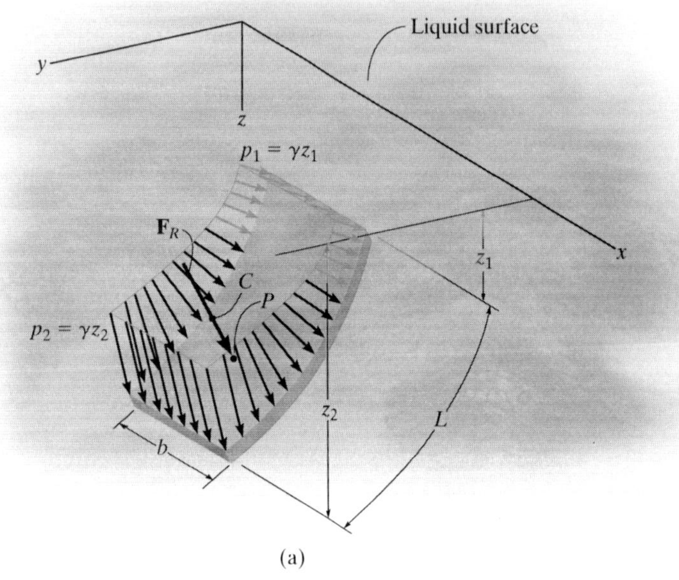

(a)

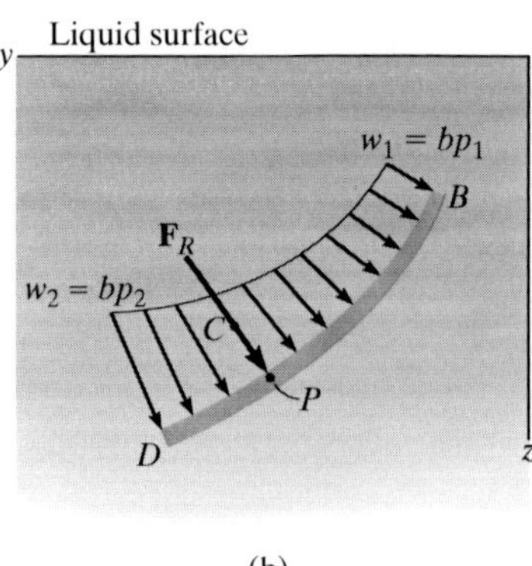

(b)

Fig. 9–26

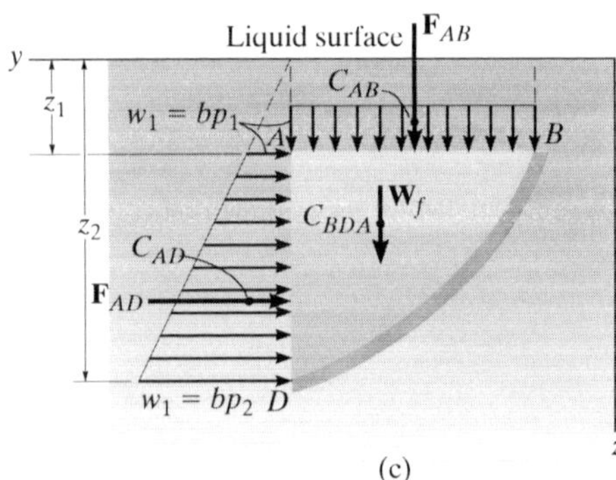

(c)

Curved Plate of Constant Width. When a submerged plate of constant width is curved, the pressure acting normal to the plate continually changes both its magnitude and direction, and therefore calculation of the magnitude of $\mathbf{F}_R$ and its location P is more difficult than for a flat plate. Three- and two-dimensional views of the loading distribution are shown in Figs. 9–26*a* and 9–26*b*, respectively. Although integration can be used to solve this problem, a simpler method exists. This method requires separate calculations for the horizontal and vertical *components* of $\mathbf{F}_R$.

For example, the distributed loading acting on the plate can be represented by the *equivalent loading* shown in Fig. 9–26*c*. Here the plate supports the weight of liquid W_f contained within the block BDA. This force has a magnitude $W_f = (\gamma b)(\text{area}_{BDA})$ and acts through the centroid of BDA. In addition, there are the pressure distributions caused by the liquid acting along the vertical and horizontal sides of the block. Along the vertical side AD, the force $\mathbf{F}_{AD}$ has a magnitude equal to the area of the trapezoid. It acts through the centroid C_{AD} of this area. The distributed loading along the horizontal side AB is *constant* since all points lying in this plane are at the same depth from the surface of the liquid. The magnitude of $\mathbf{F}_{AB}$ is simply the area of the rectangle. This force acts through the centroid C_{AB} or at the midpoint of AB. Summing these three forces yields $\mathbf{F}_R = \Sigma \mathbf{F} = \mathbf{F}_{AD} + \mathbf{F}_{AB} + \mathbf{W}_f$. Finally, the location of the center of pressure P on the plate is determined by applying $M_R = \Sigma M$, which states that the moment of the resultant force about a convenient reference point such as D or B, in Fig. 9–26*b*, is equal to the sum of the moments of the three forces in Fig. 9–26*c* about this same point.

9

Flat Plate of Variable Width. The pressure distribution acting on the surface of a submerged plate having a variable width is shown in Fig. 9–27. If we consider the force $d\mathbf{F}$ acting on the differential area strip dA, parallel to the x axis, then its magnitude is $dF = p\,dA$. Since the depth of dA is z, the pressure on the element is $p = \gamma z$. Therefore, $dF = (\gamma z)dA$ and so the resultant force becomes

$$F_R = \int dF = \gamma \int z\,dA$$

If the depth to the centroid C' of the area is $\bar{z}$, Fig. 9–27, then, $\int z\,dA = \bar{z}A$. Substituting, we have

$$F_R = \gamma\bar{z}A \tag{9–14}$$

In other words, *the magnitude of the resultant force acting on any flat plate is equal to the product of the area A of the plate and the pressure* $p = \gamma\bar{z}$ *at the depth of the area's centroid* C'. As discussed in Sec. 9.4, this force is also equivalent to the volume under the pressure distribution. Realize that its line of action passes through the centroid C of this *volume* and intersects the plate at the center of pressure P, Fig. 9–27. Notice that the location of C' does not coincide with the location of P.

The resultant force of the water pressure and its location on the elliptical back plate of this tank truck must be determined by integration.

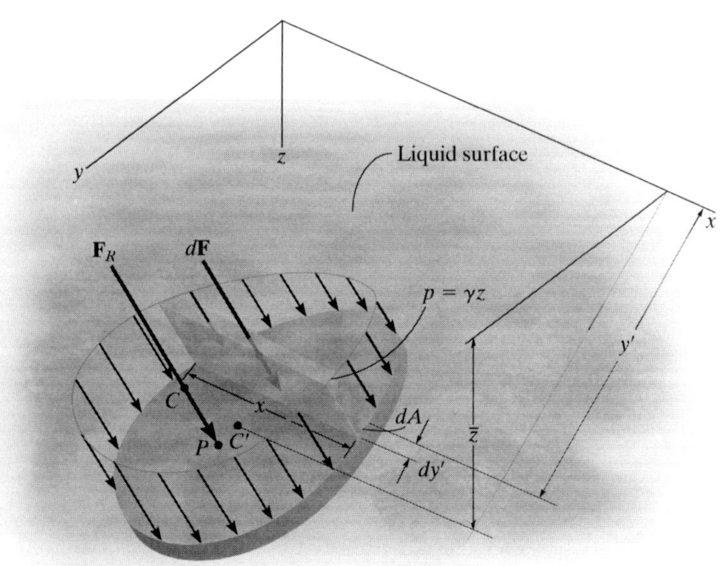

Fig. 9–27

9

EXAMPLE 9.14

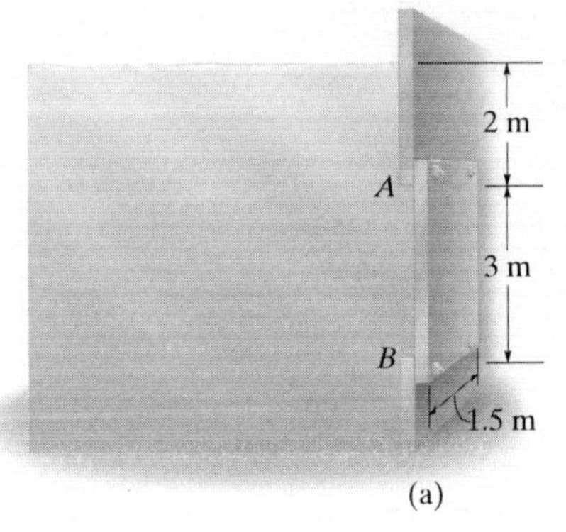

(a)

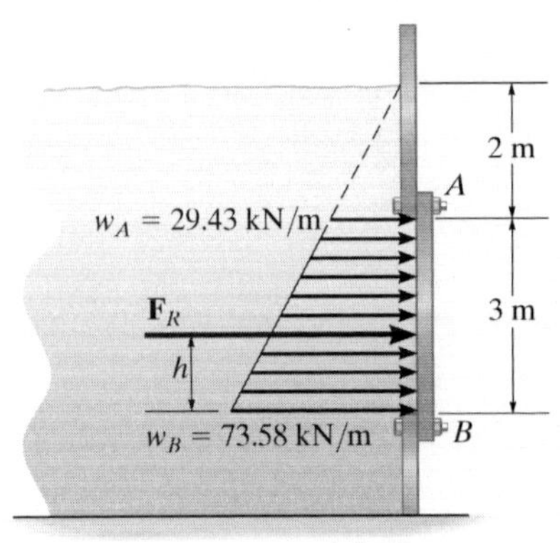

(b)

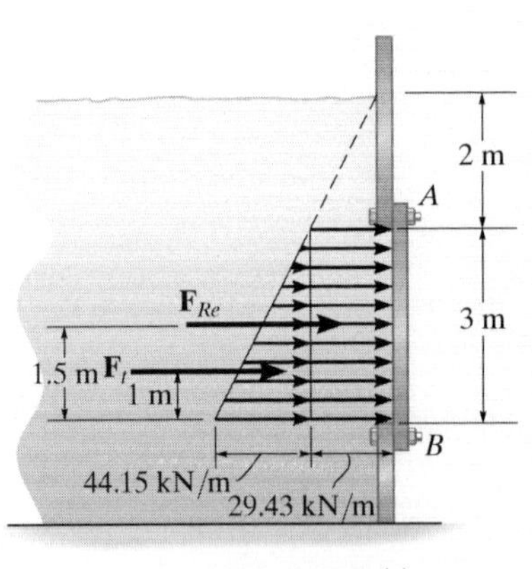

(c)

Fig. 9–28

Determine the magnitude and location of the resultant hydrostatic force acting on the submerged rectangular plate AB shown in Fig. 9–28*a*. The plate has a width of 1.5 m; $\rho_w = 1000\ \text{kg/m}^3$.

SOLUTION I

The water pressures at depths A and B are

$$p_A = \rho_w g z_A = (1000\ \text{kg/m}^3)(9.81\ \text{m/s}^2)(2\ \text{m}) = 19.62\ \text{kPa}$$

$$p_B = \rho_w g z_B = (1000\ \text{kg/m}^3)(9.81\ \text{m/s}^2)(5\ \text{m}) = 49.05\ \text{kPa}$$

Since the plate has a constant width, the pressure loading can be viewed in two dimensions as shown in Fig. 9–28*b*. The intensities of the load at A and B are

$$w_A = bp_A = (1.5\ \text{m})(19.62\ \text{kPa}) = 29.43\ \text{kN/m}$$

$$w_B = bp_B = (1.5\ \text{m})(49.05\ \text{kPa}) = 73.58\ \text{kN/m}$$

From the table on the inside back cover, the magnitude of the resultant force $\mathbf{F}_R$ created by this distributed load is

$$F_R = \text{area of a trapezoid} = \tfrac{1}{2}(3)(29.4 + 73.6) = 154.5\ \text{kN} \qquad \textit{Ans.}$$

This force acts through the centroid of this area,

$$h = \frac{1}{3}\left(\frac{2(29.43) + 73.58}{29.43 + 73.58}\right)(3) = 1.29\ \text{m} \qquad \textit{Ans.}$$

measured upward from B, Fig. 9–31*b*.

SOLUTION II

The same results can be obtained by considering two components of $\mathbf{F}_R$, defined by the triangle and rectangle shown in Fig. 9–28*c*. Each force acts through its associated centroid and has a magnitude of

$$F_{Re} = (29.43\ \text{kN/m})(3\ \text{m}) = 88.3\ \text{kN}$$

$$F_t = \tfrac{1}{2}(44.15\ \text{kN/m})(3\ \text{m}) = 66.2\ \text{kN}$$

Hence,

$$F_R = F_{Re} + F_t = 88.3 + 66.2 = 154.5\ \text{kN} \qquad \textit{Ans.}$$

The location of $\mathbf{F}_R$ is determined by summing moments about B, Figs. 9–28*b* and *c*, i.e.,

$$\circlearrowleft +(M_R)_B = \Sigma M_B;\ (154.5)h = 88.3(1.5) + 66.2(1)$$

$$h = 1.29\ \text{m} \qquad \textit{Ans.}$$

NOTE: Using Eq. 9–14, the resultant force can be calculated as $F_R = \gamma \bar{z} A = (9810\ \text{N/m}^3)(3.5\ \text{m})(3\ \text{m})(1.5\ \text{m}) = 154.5\ \text{kN}$.

9

EXAMPLE 9.15

Determine the magnitude of the resultant hydrostatic force acting on the surface of a seawall shaped in the form of a parabola as shown in Fig. 9–29*a*. The wall is 5 m long; $\rho_w = 1020\ \text{kg/m}^3$.

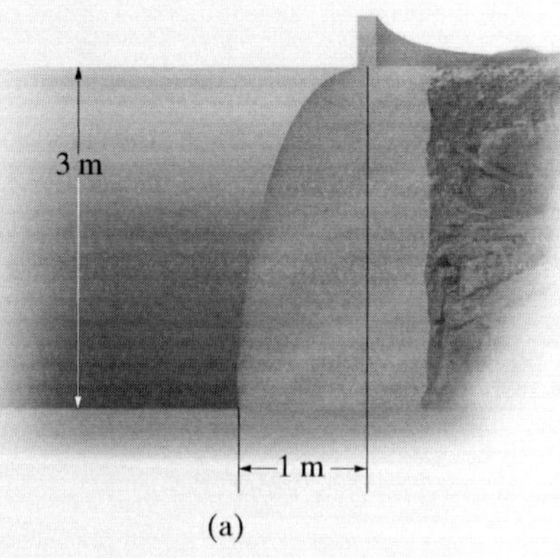

(a)

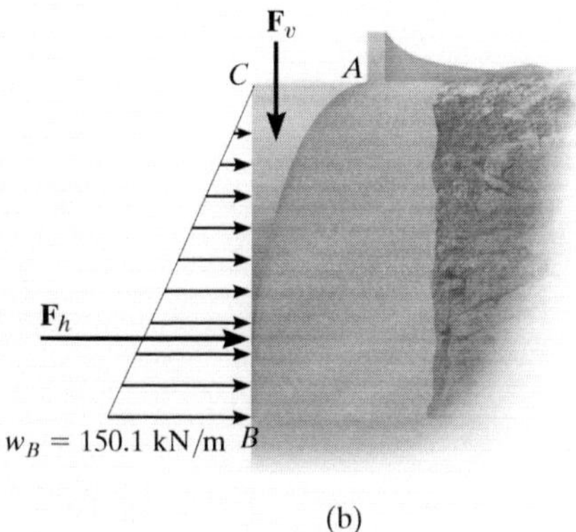

(b)

Fig. 9–29

SOLUTION

The horizontal and vertical components of the resultant force will be calculated, Fig. 9–29*b*. Since

$$p_B = \rho_w g z_B = (1020\ \text{kg/m}^3)(9.81\ \text{m/s}^2)(3\ \text{m}) = 30.02\ \text{kPa}$$

then

$$w_B = bp_B = 5\ \text{m}(30.02\ \text{kPa}) = 150.1\ \text{kN/m}$$

Thus,

$$F_h = \tfrac{1}{2}(3\ \text{m})(150.1\ \text{kN/m}) = 225.1\ \text{kN}$$

The area of the parabolic section ABC can be determined using the formula for a parabolic area $A = \frac{1}{3}ab$. Hence, the weight of water within this 5-m-long region is

$$\begin{aligned} F_v &= (\rho_w g b)(\text{area}_{ABC}) \\ &= (1020\ \text{kg/m}^3)(9.81\ \text{m/s}^2)(5\ \text{m})\left[\tfrac{1}{3}(1\ \text{m})(3\ \text{m})\right] = 50.0\ \text{kN} \end{aligned}$$

The resultant force is therefore

$$\begin{aligned} F_R &= \sqrt{F_h^2 + F_v^2} = \sqrt{(225.1\ \text{kN})^2 + (50.0\ \text{kN})^2} \\ &= 231\ \text{kN} \end{aligned}$$

Ans.

9

EXAMPLE 9.16

Determine the magnitude and location of the resultant force acting on the triangular end plates of the water trough shown in Fig. 9–30*a*; $\rho_w = 1000 \text{ kg/m}^3$.

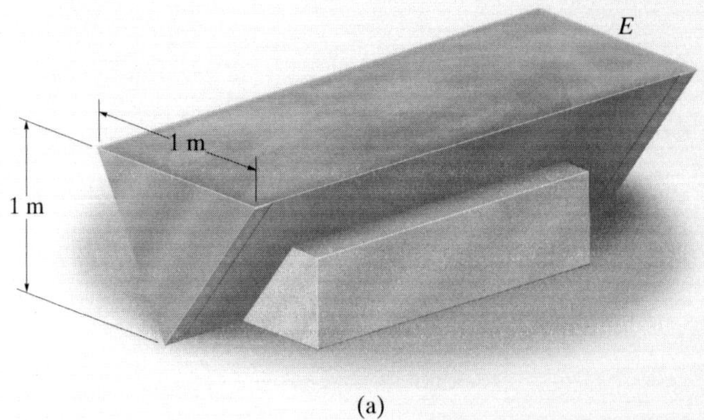

(a)

SOLUTION

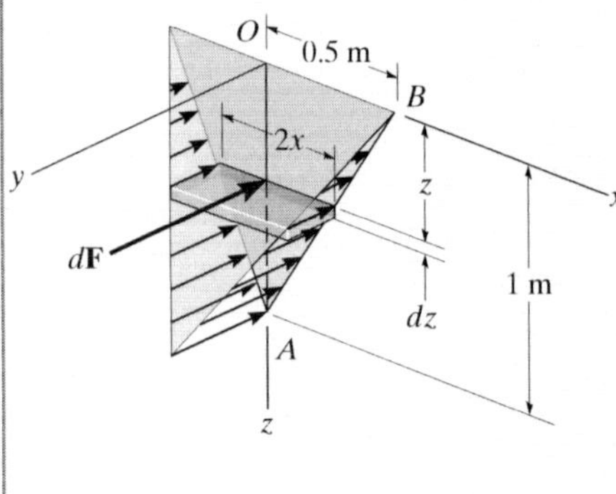

(b)

Fig. 9–30

The pressure distribution acting on the end plate E is shown in Fig. 9–30*b*. The magnitude of the resultant force is equal to the volume of this loading distribution. We will solve the problem by integration. Choosing the differential volume element shown in the figure, we have

$$dF = dV = p\,dA = \rho_w g z(2x\,dz) = 19\,620zx\,dz$$

The equation of line AB is

$$x = 0.5(1 - z)$$

Hence, substituting and integrating with respect to z from $z = 0$ to $z = 1$ m yields

$$F = V = \int_V dV = \int_0^{1\text{ m}} (19\,620)z[0.5(1 - z)]\,dz$$

$$= 9810\int_0^{1\text{ m}} (z - z^2)\,dz = 1635 \text{ N} = 1.64 \text{ kN} \qquad \textit{Ans.}$$

This resultant passes through the *centroid of the volume*. Because of symmetry,

$$\bar{x} = 0 \qquad \textit{Ans.}$$

Since $\tilde{z} = z$ for the volume element, then

$$\bar{z} = \frac{\int_V \tilde{z}\,dV}{\int_V dV} = \frac{\int_0^{1\text{ m}} z(19\,620)z[0.5(1 - z)]\,dz}{1635} = \frac{9810\int_0^{1\text{ m}} (z^2 - z^3)\,dz}{1635}$$

$$= 0.5 \text{ m} \qquad \textit{Ans.}$$

NOTE: We can also determine the resultant force by applying Eq. 9–14, $F_R = \gamma\bar{z}A = \left(9810 \text{ N/m}^3\right)\left(\frac{1}{3}\right)(1 \text{ m})\left[\frac{1}{2}(1 \text{ m})(1 \text{ m})\right] = 1.64 \text{ kN}$.

9

FUNDAMENTAL PROBLEMS

F9–17. Determine the magnitude of the hydrostatic force acting per meter length of the wall. Water has a density of $\rho = 1 \text{ Mg/m}^3$.

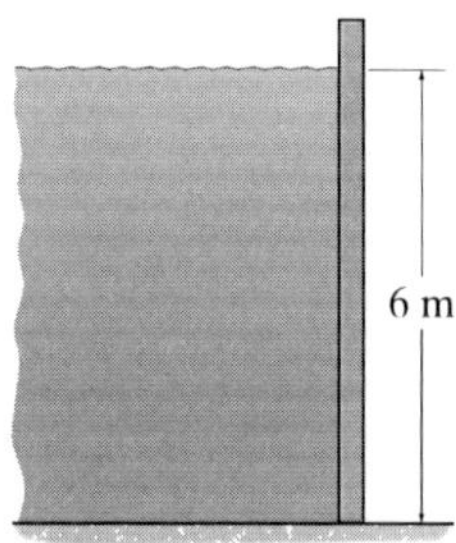

F9–17

F9–18. Determine the magnitude of the hydrostatic force acting on gate AB, which has a width of 1 m. The specific weight of water is $\gamma = 9.81 \text{ kN/m}^3$.

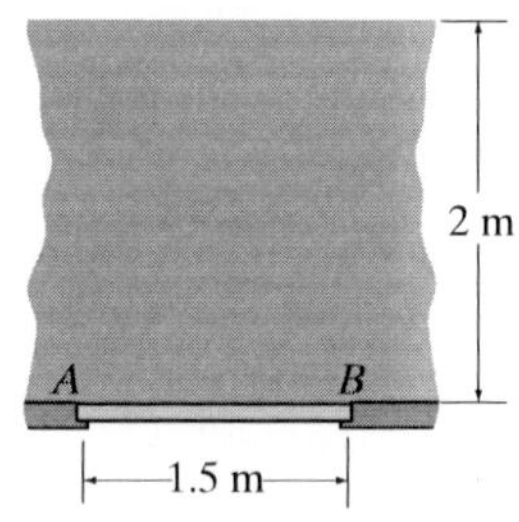

F9–18

F9–19. Determine the magnitude of the hydrostatic force acting on gate AB, which has a width of 1.5 m. Water has a density of $\rho = 1 \text{ Mg/m}^3$.

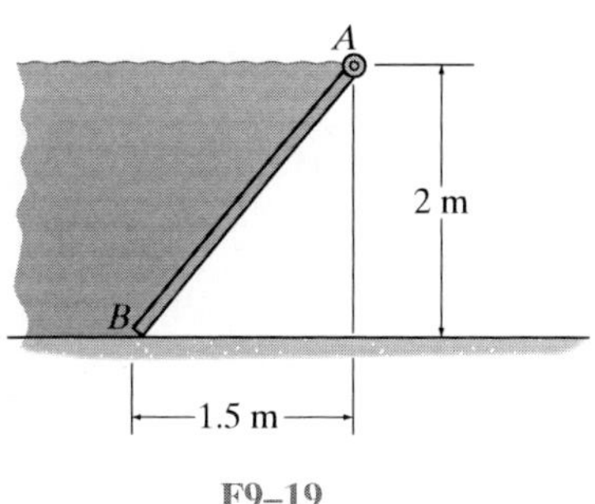

F9–19

F9–20. Determine the magnitude of the hydrostatic force acting on gate AB, which has a width of 2 m. Water has a density of $\rho = 1 \text{ Mg/m}^3$.

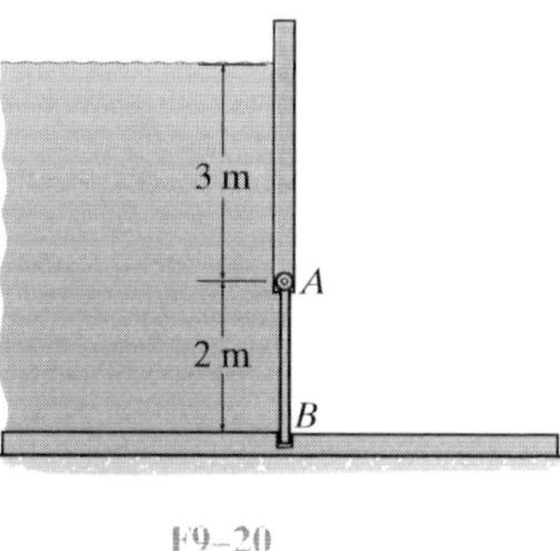

F9–20

F9–21. Determine the magnitude of the hydrostatic force acting on gate AB, which has a width of 0.6 m. The specific weight of water is $\gamma = 9.81 \text{ kN/m}^3$.

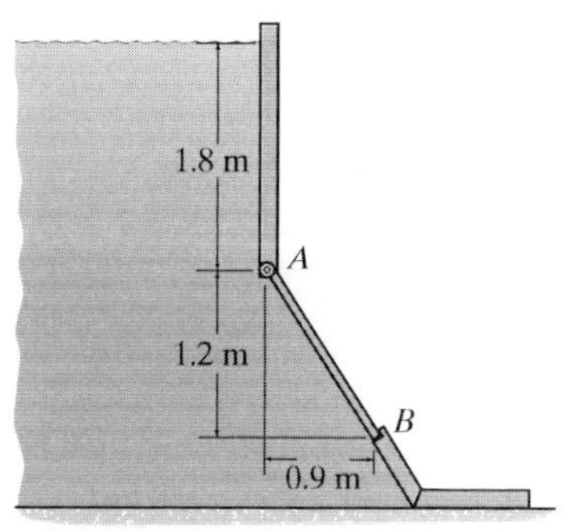

F9–21

9

PROBLEMS

9–115. The concrete "gravity" dam is held in place by its own weight. If the density of concrete is $\rho_c = 2.5\ \text{Mg/m}^3$, and water has a density of $\rho_w = 1.0\ \text{Mg/m}^3$, determine the smallest dimension d that will prevent the dam from overturning about its end A.

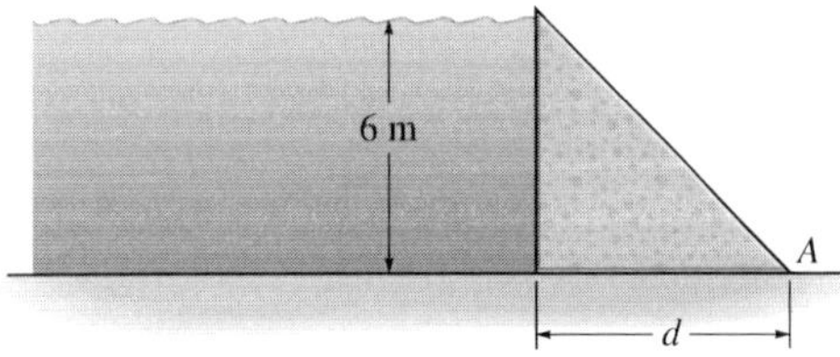

Prob. 9–115

*__9–116.__ The loading acting on a square plate is represented by a parabolic pressure distribution. Determine the magnitude of the resultant force and the coordinates $(\bar{x}, \bar{y})$ of the point where the line of action of the force intersects the plate. Also, what are the reactions at the rollers B and C and the ball-and-socket joint A? Neglect the weight of the plate.

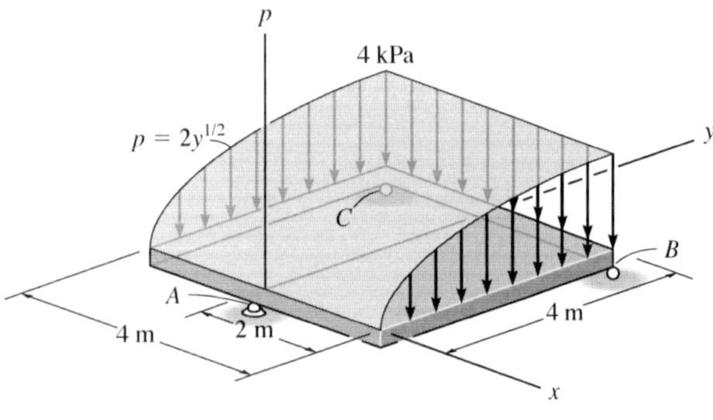

Prob. 9–116

9–117. The load over the plate varies linearly along the sides of the plate such that $p = \frac{2}{3}[x(4 - y)]$ kPa. Determine the resultant force and its position $(\bar{x}, \bar{y})$ on the plate.

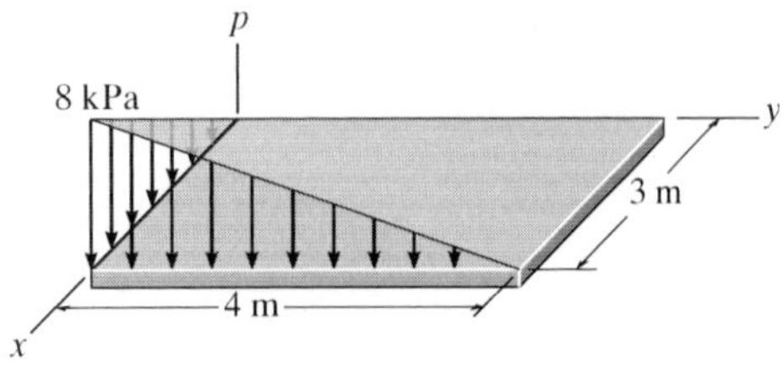

Prob. 9–117

9–118. The rectangular plate is subjected to a distributed load over its *entire surface*. The load is defined by the expression $p = p_0 \sin(\pi x/a)\sin(\pi y/b)$, where p_0 represents the pressure acting at the center of the plate. Determine the magnitude and location of the resultant force acting on the plate.

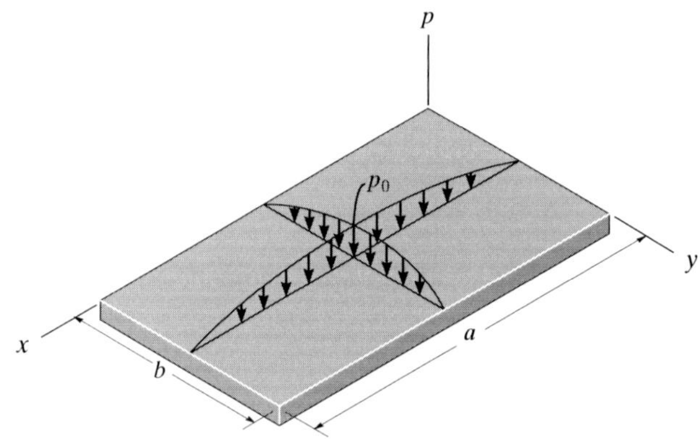

Prob. 9–118

9–119. The gate AB is 8 m wide. Determine the horizontal and vertical components of force acting on the pin at B and the vertical reaction at the smooth support A. $\rho_w = 1.0\ \text{Mg/m}^3$.

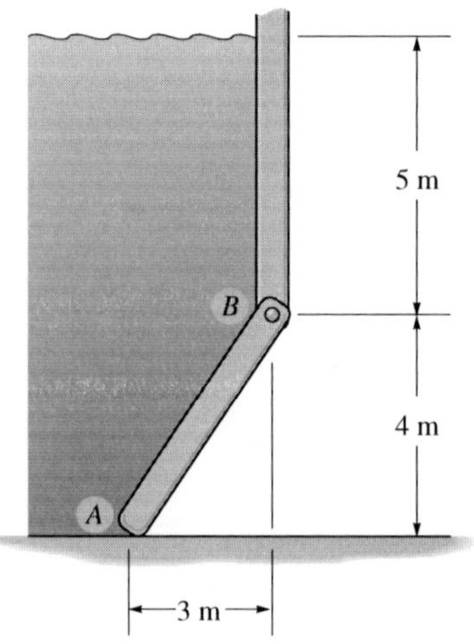

Prob. 9–119

9

***9–120.** The tank is filled with water to a depth of $d = 4$ m. Determine the resultant force the water exerts on side A and side B of the tank. If oil instead of water is placed in the tank, to what depth d should it reach so that it creates the same resultant forces? $\rho_o = 900\ \text{kg/m}^3$ and $\rho_w = 1000\ \text{kg/m}^3$.

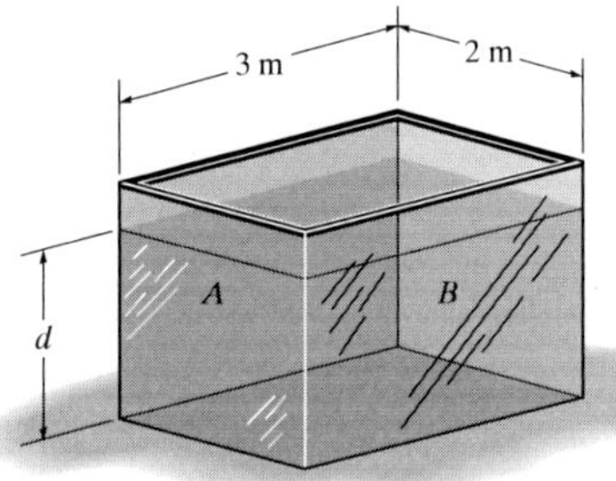

Prob. 9–120

9–121. When the tide water A subsides, the tide gate automatically swings open to drain the marsh B. For the condition of high tide shown, determine the horizontal reactions developed at the hinge C and stop block D. The length of the gate is 6 m and its height is 4 m. $\rho_w = 1.0\ \text{Mg/m}^3$.

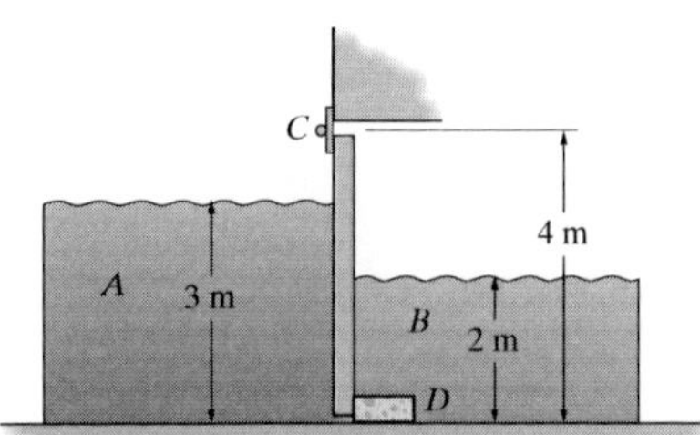

Prob. 9–121

9–122. The symmetric concrete "gravity" dam is held in place by its own weight. If the density of concrete is $\rho_c = 2.5\ \text{Mg/m}^3$, and water has a density of $\rho_w = 1.0\ \text{Mg/m}^3$, determine the smallest distance d at its base that will prevent the dam from overturning about its end A. The dam has a width of 8 m.

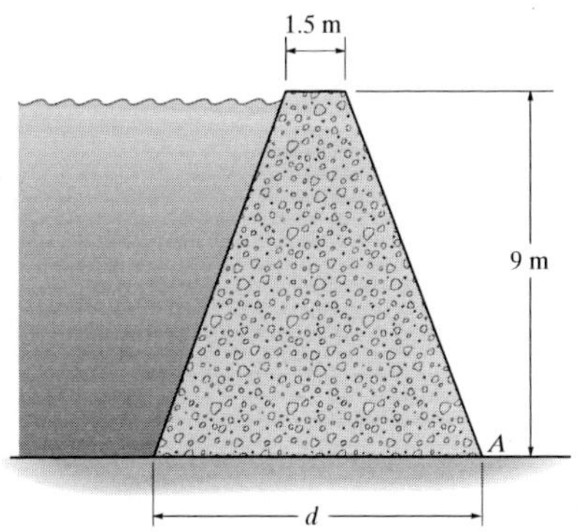

Prob. 9–122

9–123. Determine the magnitude of the resultant hydrostatic force acting on the dam and its location, measured from the top surface of the water. The width of the dam is 8 m; $\rho_w = 1.0\ \text{Mg/m}^3$.

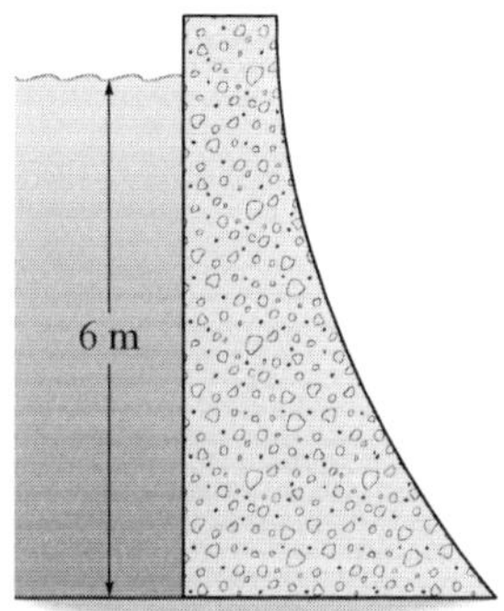

Prob. 9–123

***9–124.** The concrete dam in the shape of a quarter circle. Determine the magnitude of the resultant hydrostatic force that acts on the dam per meter of length. The density of water is $\rho_w = 1\ \text{Mg/m}^3$.

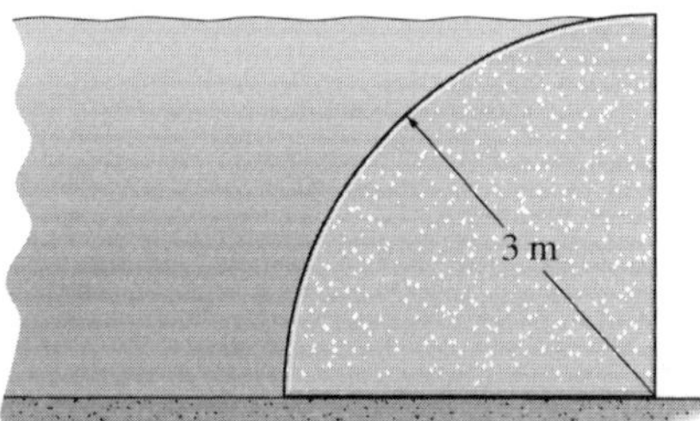

Prob. 9–124

9–125. The storage tank contains oil having a specific weight of $\gamma_o = 9\ \text{kN/m}^3$. If the tank is 6 m wide, calculate the resultant force acting on the inclined side BC of the tank, caused by the oil, and specify its location along BC, measured from B. Also compute the total resultant force acting on the bottom of the tank.

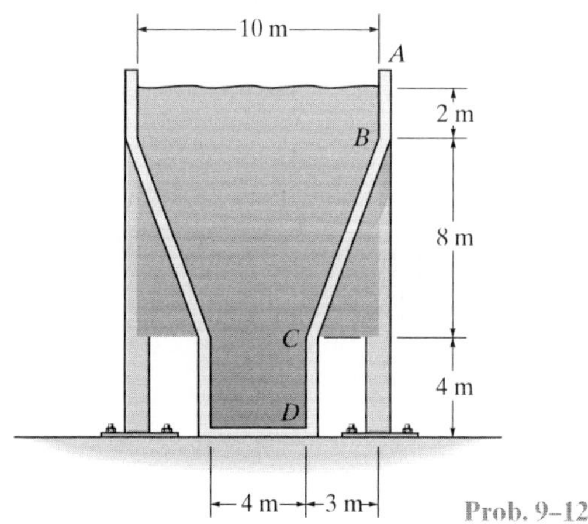

Prob. 9–125

9

9–126. The tank is filled to the top ($y = 0.5$ m) with water having a density of $\rho_w = 1.0\ \text{Mg/m}^3$. Determine the resultant force of the water pressure acting on the flat end plate C of the tank, and its location, measured from the top of the tank.

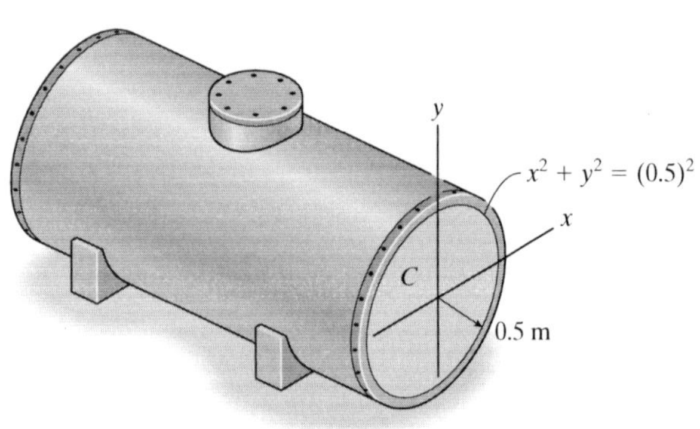

Prob. 9–126

9–127. The tank is filled with a liquid that has a density of $900\ \text{kg/m}^3$. Determine the resultant force that it exerts on the elliptical end plate, and the location of the center of pressure, measured from the x axis.

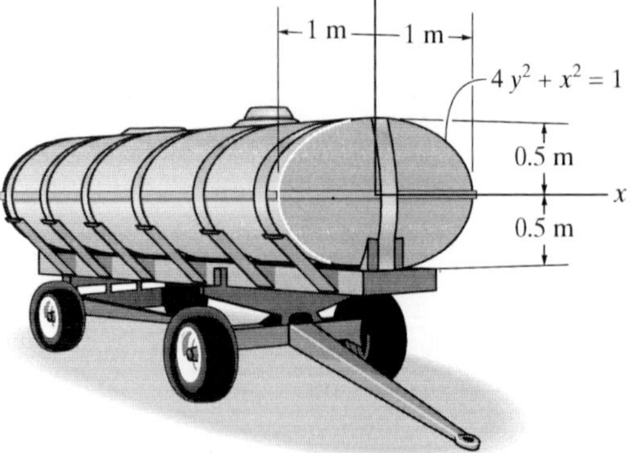

Prob. 9–127

***9–128.** The underwater tunnel in the aquatic center is fabricated from a transparent polycarbonate material formed in the shape of a parabola. Determine the magnitude of the hydrostatic force that acts per meter length along the surface AB of the tunnel. The density of the water is $\rho_w = 1000\ \text{kg/m}^3$.

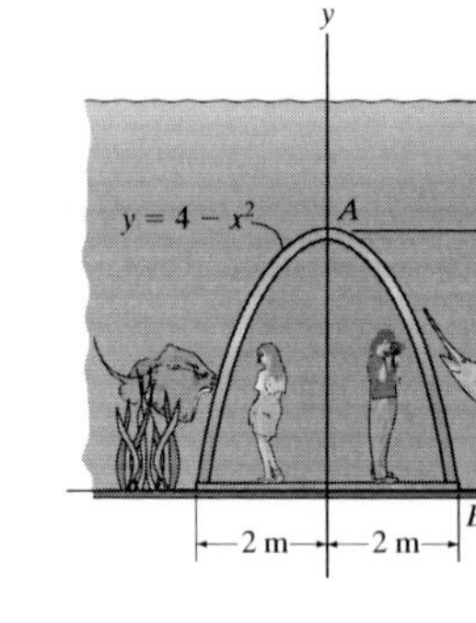

Prob. 9–128

9–129. The semicircular tunnel passes under a river that is 9 m deep. Determine the vertical resultant hydrostatic force acting per meter of length along the length of the tunnel. The tunnel is 6 m wide; $\rho_w = 1.0\ \text{Mg/m}^3$.

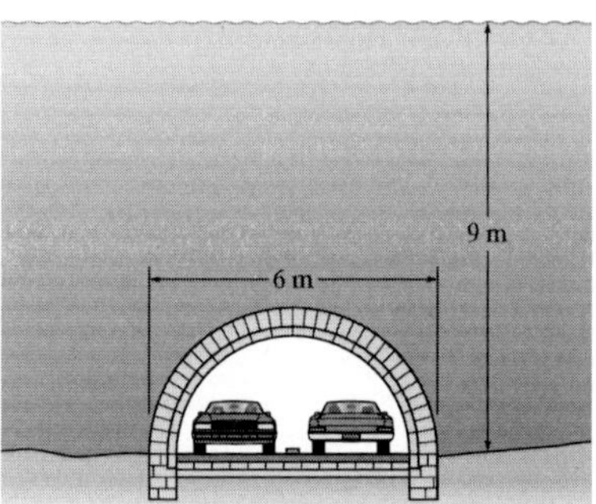

Prob. 9–129

9–130. The arched surface AB is shaped in the form of a quarter circle. If it is 8 m long, determine the horizontal and vertical components of the resultant force caused by the water acting on the surface; $\rho_w = 1.0\ \text{Mg/m}^3$.

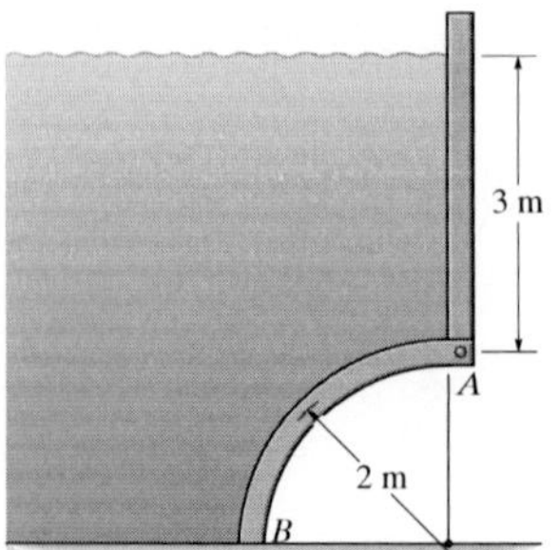

Prob. 9–130

9

CHAPTER REVIEW

Center of Gravity and Centroid

The *center of gravity G* represents a point where the weight of the body can be considered concentrated. The distance from an axis to this point can be determined from a balance of moments, which requires that the moment of the weight of all the particles of the body about this axis must equal the moment of the entire weight of the body about the axis.

$$\bar{x} = \frac{\int \tilde{x}\, dW}{\int dW}$$

$$\bar{y} = \frac{\int \tilde{y}\, dW}{\int dW}$$

$$\bar{z} = \frac{\int \tilde{z}\, dW}{\int dW}$$

The center of mass will coincide with the center of gravity provided the acceleration of gravity is constant.

$$\bar{x} = \frac{\int_L \tilde{x}\, dL}{\int_L dL} \qquad \bar{y} = \frac{\int_L \tilde{y}\, dL}{\int_L dL} \qquad \bar{z} = \frac{\int_L \tilde{z}\, dL}{\int_L dL}$$

$$\bar{x} = \frac{\int_A \tilde{x}\, dA}{\int_A dA} \qquad \bar{y} = \frac{\int_A \tilde{y}\, dA}{\int_A dA} \qquad \bar{z} = \frac{\int_A \tilde{z}\, dA}{\int_A dA}$$

The *centroid* is the location of the geometric center for the body. It is determined in a similar manner, using a moment balance of geometric elements such as line, area, or volume segments. For bodies having a continuous shape, moments are summed (integrated) using differential elements.

$$\bar{x} = \frac{\int_V \tilde{x}\, dV}{\int_V dV} \qquad \bar{y} = \frac{\int_V \tilde{y}\, dV}{\int_A dV} \qquad \bar{z} = \frac{\int_V \tilde{z}\, dV}{\int_V dV}$$

The center of mass will coincide with the centroid provided the material is homogeneous, i.e., the density of the material is the same throughout. The centroid will always lie on an axis of symmetry.

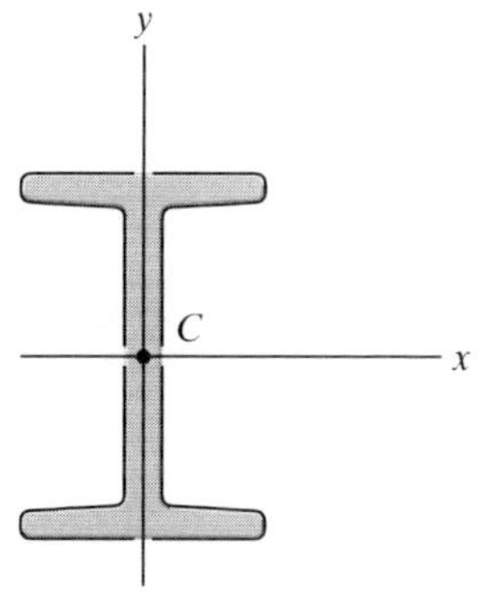

Composite Body If the body is a composite of several shapes, each having a known location for its center of gravity or centroid, then the location of the center of gravity or centroid of the body can be determined from a discrete summation using its composite parts.	$\bar{x} = \dfrac{\Sigma \tilde{x} W}{\Sigma W}$ $\bar{y} = \dfrac{\Sigma \tilde{y} W}{\Sigma W}$ $\bar{z} = \dfrac{\Sigma \tilde{z} W}{\Sigma W}$	z y x
Theorems of Pappus and Guldinus The theorems of Pappus and Guldinus can be used to determine the surface area and volume of a body of revolution. The *surface area* equals the product of the length of the generating curve and the distance traveled by the centroid of the curve needed to generate the area. The *volume* of the body equals the product of the generating area and the distance traveled by the centroid of this area needed to generate the volume.	$A = \theta \bar{r} L$ $V = \theta \bar{r} A$	

9

General Distributed Loading

The magnitude of the resultant force is equal to the total volume under the distributed-loading diagram. The line of action of the resultant force passes through the geometric center or centroid of this volume.

$$F_R = \int_A p(x, y)\, dA = \int_V dV$$

$$\bar{x} = \frac{\int_V x\, dV}{\int_V dV}$$

$$\bar{y} = \frac{\int_V y\, dV}{\int_V dV}$$

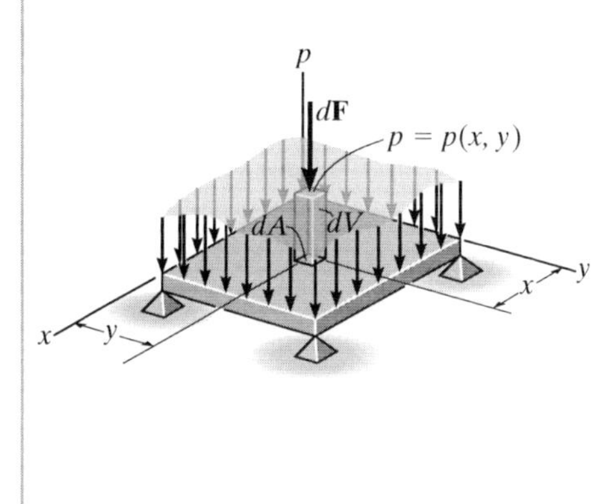

Fluid Pressure

The pressure developed by a liquid at a point on a submerged surface depends upon the depth of the point and the density of the liquid in accordance with Pascal's law, $p = \rho g h = \gamma h$. This pressure will create a *linear distribution* of loading on a flat vertical or inclined surface.

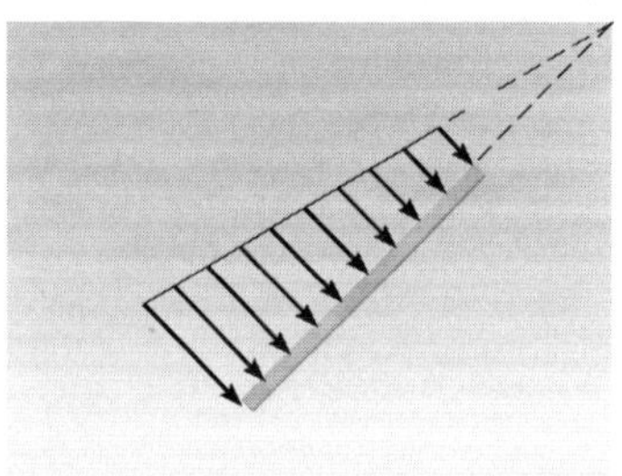

If the surface is horizontal, then the loading will be *uniform*.

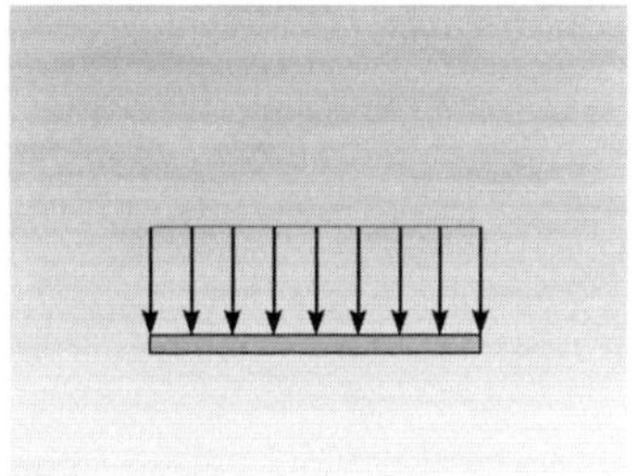

In any case, the resultants of these loadings can be determined by finding the volume under the loading curve or using $F_R = \gamma \bar{z} A$, where $\bar{z}$ is the depth to the centroid of the plate's area. The line of action of the resultant force passes through the centroid of the volume of the loading diagram and acts at a point P on the plate called the center of pressure.

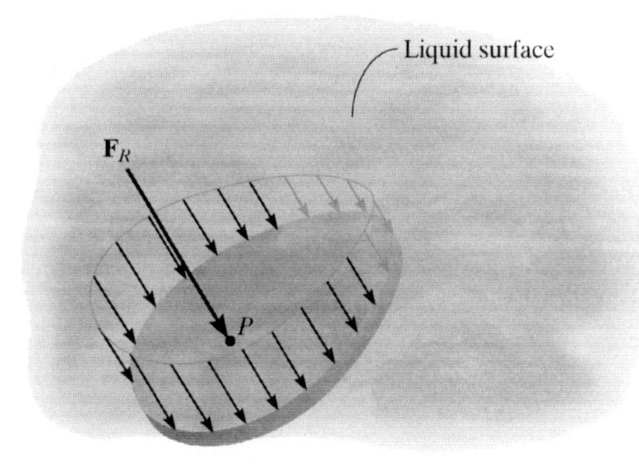

REVIEW PROBLEMS

9–131. A circular V-belt has an inner radius of 600 mm and a cross-sectional area as shown. Determine the volume of material required to make the belt.

***9–132.** A circular V-belt has an inner radius of 600 mm and a cross-sectional area as shown. Determine the surface area of the belt.

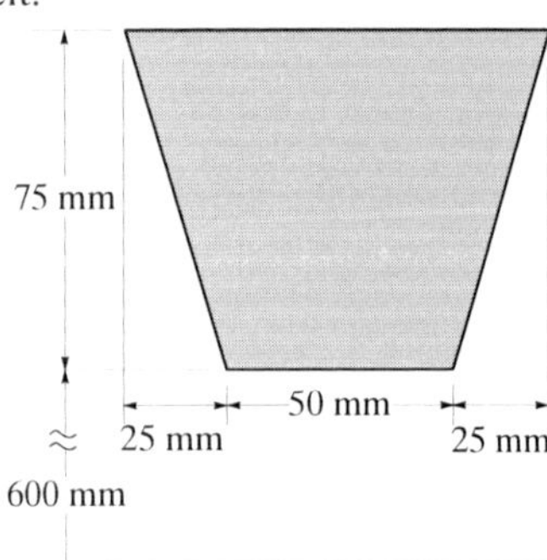

Probs. 9–131/132

9–133. Locate the centroid $\bar{y}$ of the beam's cross-sectional area.

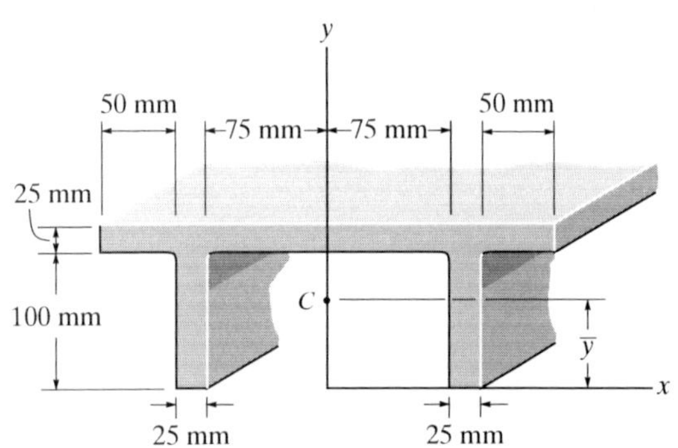

Prob. 9–133

9–134. Locate the centroid of the solid.

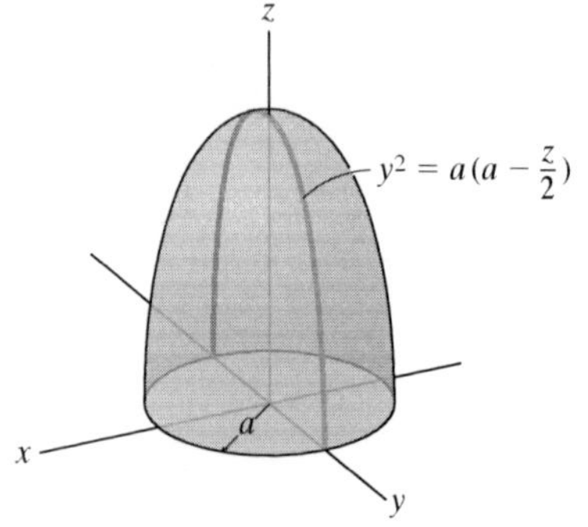

Prob. 9–134

9–135. Locate the centroid $\bar{x}$ of the triangular area.

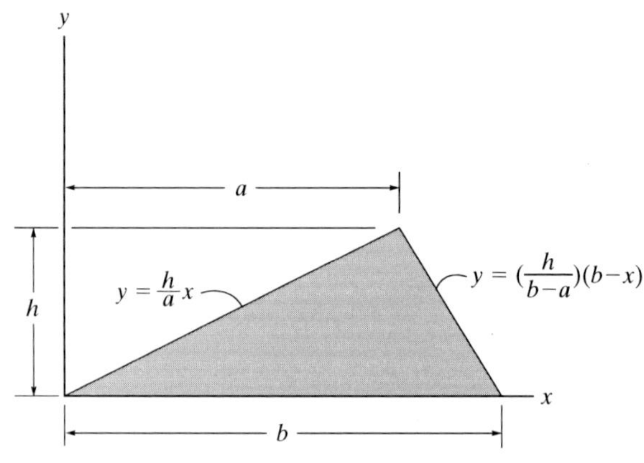

Prob. 9–135

***9–136.** Locate the centroid $\bar{y}$ of the bulb-tee cross section.

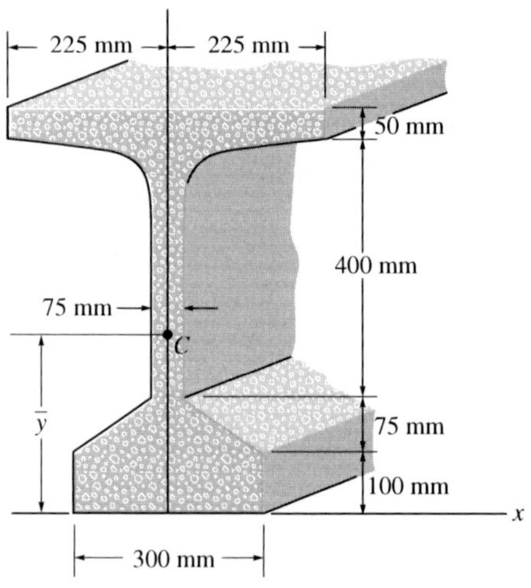

Prob. 9–136

9–137. Determine the location $(\bar{x}, \bar{y})$ of the center of mass of the turbine and compressor assembly. The mass and the center of mass of each of the various components are indicated below.

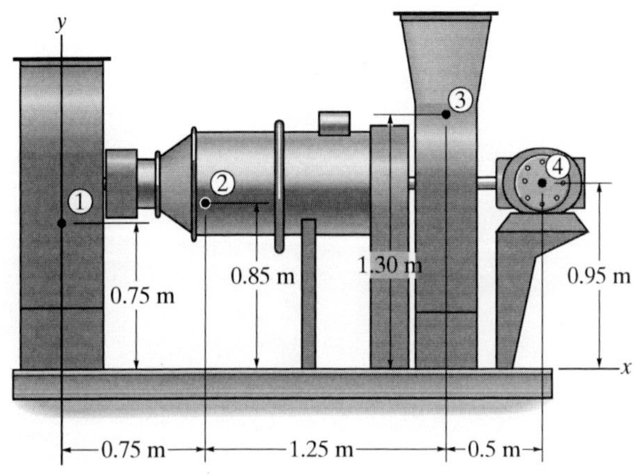

1. Intake housing 25 kg
2. Turbine 80 kg
3. Exhaust housing 30 kg
4. Compressor 105 kg

Prob. 9–137

9–138. Each of the three homogeneous plates welded to the rod has a density of $6\ Mg/m^3$ and a thickness of 10 mm. Determine the length l of plate C and the angle of placement, θ, so that the center of mass of the assembly lies on the y axis. Plates A and B lie in the x–y and z–y planes, respectively.

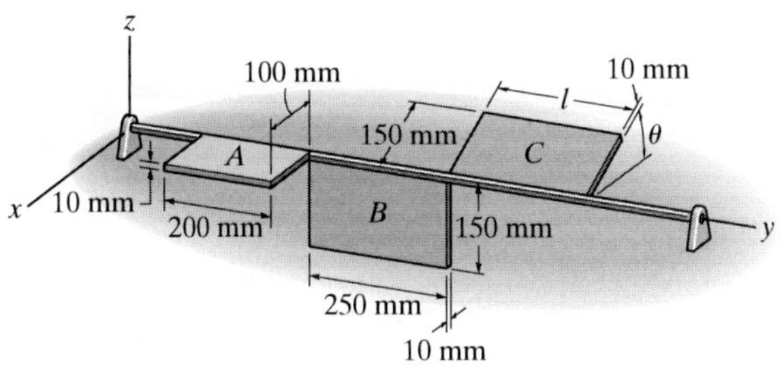

Prob. 9–138

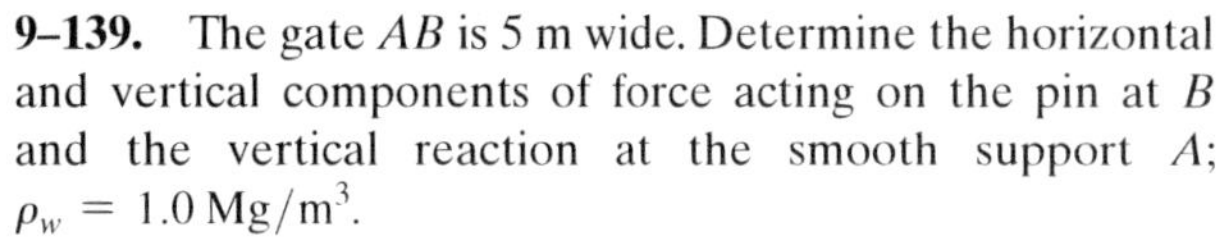

9–139. The gate AB is 5 m wide. Determine the horizontal and vertical components of force acting on the pin at B and the vertical reaction at the smooth support A; $\rho_w = 1.0\ Mg/m^3$.

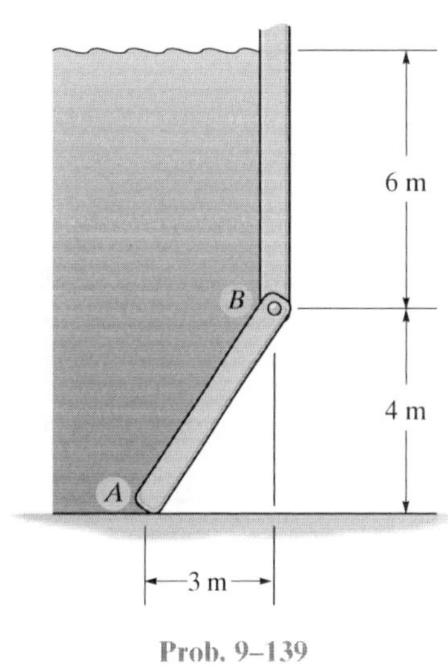

Prob. 9–139

***9–140.** The pressure loading on the plate is described by the function $p = \{-240/(x + 1) + 340\}$ Pa. Determine the magnitude of the resultant force and coordinates of the point where the line of action of the force intersects the plate.

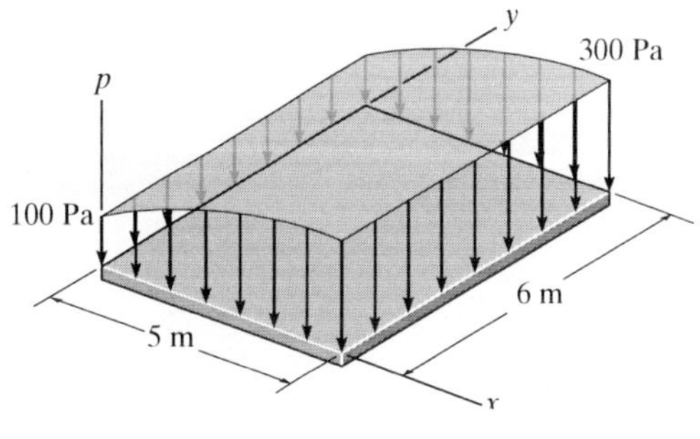

Prob. 9–140

Chapter 10

The design of these structural members requires calculation of their cross-sectional moment of inertia. In this chapter we will discuss how this is done.

Moments of Inertia

CHAPTER OBJECTIVES

- To develop a method for determining the moment of inertia for an area.
- To introduce the product of inertia and show how to determine the maximum and minimum moments of inertia for an area.
- To discuss the mass moment of inertia.

Video Solutions are available for selected questions in this chapter.

10.1 Definition of Moments of Inertia for Areas

Whenever a distributed loading acts perpendicular to an area and its intensity varies linearly, the computation of the moment of the loading distribution about an axis will involve a quantity called the *moment of inertia of the area*. For example, consider the plate in Fig. 10–1, which is subjected to a fluid pressure p. As discussed in Sec. 9.5, this pressure p varies linearly with depth, such that $p = \gamma y$, where γ is the specific weight of the fluid. Thus, the force acting on the differential area dA of the plate is $dF = p\,dA = (\gamma\,y)dA$. The moment of this force about the x axis is therefore $dM = y\,dF = \gamma y^2 dA$, and so integrating dM over the entire area of the plate yields $M = \gamma \int y^2 dA$. The integral $\int y^2 dA$ is called the *moment of inertia* I_x of the area about the x axis. Integrals of this form often arise in formulas used in fluid mechanics, mechanics of materials, structural mechanics, and mechanical design, and so the engineer needs to be familiar with the methods used for their computation.

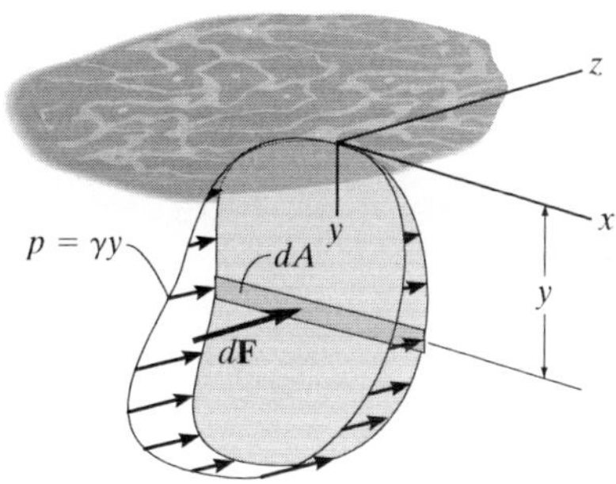

Fig. 10–1

Please refer to the Companion Website for the animation: *Moment of Inertia*

Moment of Inertia. By definition, the moments of inertia of a differential area dA about the x and y axes are $dI_x = y^2\,dA$ and $dI_y = x^2\,dA$, respectively, Fig. 10–2. For the entire area A the *moments of inertia* are determined by integration; i.e.,

$$I_x = \int_A y^2\,dA \qquad (10\text{–}1)$$
$$I_y = \int_A x^2\,dA$$

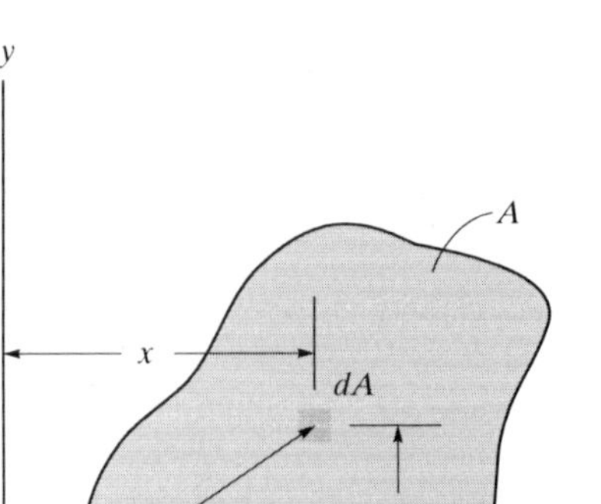

Fig. 10–2

We can also formulate this quantity for dA about the "pole" O or z axis, Fig. 10–2. This is referred to as the *polar moment of inertia*. It is defined as $dJ_O = r^2\,dA$, where r is the perpendicular distance from the pole (z axis) to the element dA. For the entire area the *polar moment of inertia* is

$$J_O = \int_A r^2\,dA = I_x + I_y \qquad (10\text{–}2)$$

This relation between J_O and I_x, I_y is possible since $r^2 = x^2 + y^2$, Fig. 10–2.

From the above formulations it is seen that I_x, I_y, and J_O will *always* be *positive* since they involve the product of distance squared and area. Furthermore, the units for moment of inertia involve length raised to the fourth power, e.g., m^4, mm^4.

10.2 Parallel-Axis Theorem for an Area

The *parallel-axis theorem* can be used to find the moment of inertia of an area about *any axis* that is parallel to an axis passing through the centroid and about which the moment of inertia is known. To develop this theorem, we will consider finding the moment of inertia of the shaded area shown in Fig. 10–3 about the x axis. To start, we choose a differential element dA located at an arbitrary distance y' from the *centroidal* x' axis. If the distance between the parallel x and x' axes is d_y, then the moment of inertia of dA about the x axis is $dI_x = (y' + d_y)^2\,dA$. For the entire area,

$$I_x = \int_A (y' + d_y)^2\,dA$$
$$= \int_A y'^2\,dA + 2d_y \int_A y'\,dA + d_y^2 \int_A dA$$

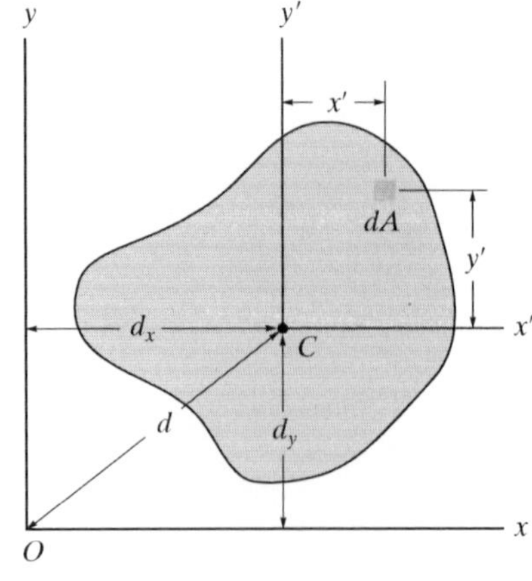

Fig. 10–3

The first integral represents the moment of inertia of the area about the centroidal axis, $\bar{I}_{x'}$. The second integral is zero since the x' axis passes through the area's centroid C; i.e., $\int y'\,dA = \bar{y}'\int dA = 0$ since $\bar{y}' = 0$. Since the third integral represents the total area A, the final result is therefore

$$I_x = \bar{I}_{x'} + Ad_y^2 \qquad (10\text{–}3)$$

A similar expression can be written for I_y; i.e.,

$$I_y = \bar{I}_{y'} + Ad_x^2 \qquad (10\text{–}4)$$

And finally, for the polar moment of inertia, since $\bar{J}_C = \bar{I}_{x'} + \bar{I}_{y'}$ and $d^2 = d_x^2 + d_y^2$, we have

$$J_O = \bar{J}_C + Ad^2 \qquad (10\text{–}5)$$

The form of each of these three equations states that *the moment of inertia for an area about an axis is equal to its moment of inertia about a parallel axis passing through the area's centroid plus the product of the area and the square of the perpendicular distance between the axes.*

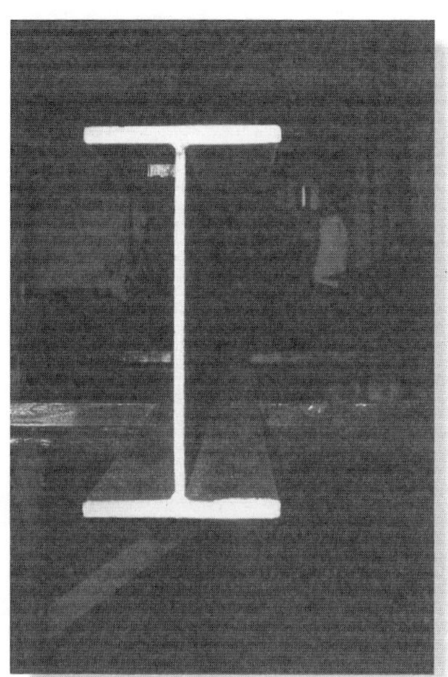

In order to predict the strength and deflection of this beam, it is necessary to calculate the moment of inertia of the beam's cross-sectional area.

10.3 Radius of Gyration of an Area

The *radius of gyration* of an area about an axis has units of length and is a quantity that is often used for the design of columns in structural mechanics. Provided the areas and moments of inertia are *known*, the radii of gyration are determined from the formulas

$$k_x = \sqrt{\frac{I_x}{A}}$$

$$k_y = \sqrt{\frac{I_y}{A}} \qquad (10\text{–}6)$$

$$k_O = \sqrt{\frac{J_O}{A}}$$

The form of these equations is easily remembered since it is similar to that for finding the moment of inertia for a differential area about an axis. For example, $I_x = k_x^2 A$; whereas for a differential area, $dI_x = y^2\,dA$.

10

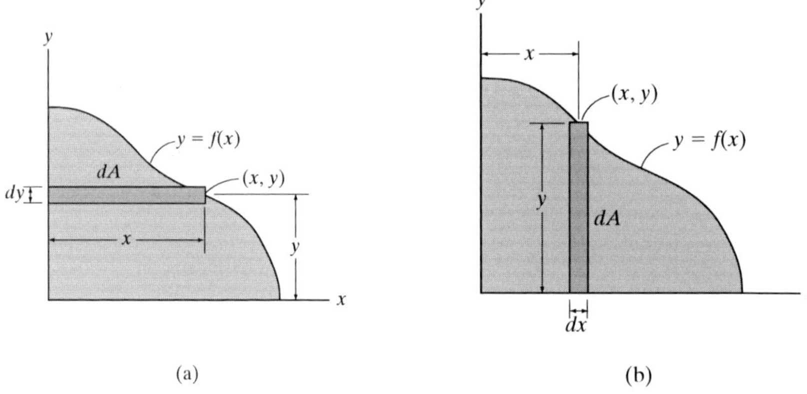

Fig. 10–4

Procedure for Analysis

In most cases the moment of inertia can be determined using a single integration. The following procedure shows two ways in which this can be done.

- If the curve defining the boundary of the area is expressed as $y = f(x)$, then select a rectangular differential element such that it has a finite length and differential width.
- The element should be located so that it intersects the curve at the *arbitrary point* (x, y).

Case 1.

- Orient the element so that its length is *parallel* to the axis about which the moment of inertia is computed. This situation occurs when the rectangular element shown in Fig. 10–4*a* is used to determine I_x for the area. Here the entire element is at a distance y from the x axis since it has a thickness dy. Thus $I_x = \int y^2 dA$. To find I_y, the element is oriented as shown in Fig. 10–4*b*. This element lies at the *same* distance x from the y axis so that $I_y = \int x^2 dA$.

Case 2.

- The length of the element can be oriented *perpendicular* to the axis about which the moment of inertia is computed; however, Eq. 10–1 *does not apply* since all points on the element will *not* lie at the same moment-arm distance from the axis. For example, if the rectangular element in Fig. 10–4*a* is used to determine I_y, it will first be necessary to calculate the moment of inertia of the *element* about an axis parallel to the y axis that passes through the element's centroid, and then determine the moment of inertia of the *element* about the y axis using the parallel-axis theorem. Integration of this result will yield I_y. See Examples 10.2 and 10.3.

10

EXAMPLE 10.1

Determine the moment of inertia for the rectangular area shown in Fig. 10–5 with respect to (a) the centroidal x' axis, (b) the axis x_b passing through the base of the rectangle, and (c) the pole or z' axis perpendicular to the x'–y' plane and passing through the centroid C.

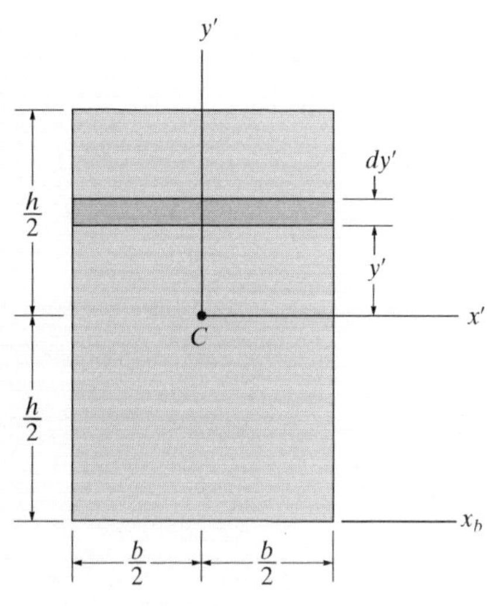

Fig. 10–5

SOLUTION (CASE 1)

Part (a). The differential element shown in Fig. 10–5 is chosen for integration. Because of its location and orientation, the *entire element* is at a distance y' from the x' axis. Here it is necessary to integrate from $y' = -h/2$ to $y' = h/2$. Since $dA = b\,dy'$, then

$$\bar{I}_{x'} = \int_A y'^2\,dA = \int_{-h/2}^{h/2} y'^2(b\,dy') = b\int_{-h/2}^{h/2} y'^2\,dy'$$

$$\bar{I}_{x'} = \frac{1}{12}bh^3 \qquad \textit{Ans.}$$

Part (b). The moment of inertia about an axis passing through the base of the rectangle can be obtained by using the above result of part (a) and applying the parallel-axis theorem, Eq. 10–3.

$$I_{x_b} = \bar{I}_{x'} + Ad_y^2$$

$$= \frac{1}{12}bh^3 + bh\left(\frac{h}{2}\right)^2 = \frac{1}{3}bh^3 \qquad \textit{Ans.}$$

Part (c). To obtain the polar moment of inertia about point C, we must first obtain $\bar{I}_{y'}$, which may be found by interchanging the dimensions b and h in the result of part (a), i.e.,

$$\bar{I}_{y'} = \frac{1}{12}hb^3$$

Using Eq. 10–2, the polar moment of inertia about C is therefore

$$\bar{J}_C = \bar{I}_{x'} + \bar{I}_{y'} = \frac{1}{12}bh(h^2 + b^2) \qquad \textit{Ans.}$$

10

EXAMPLE 10.2

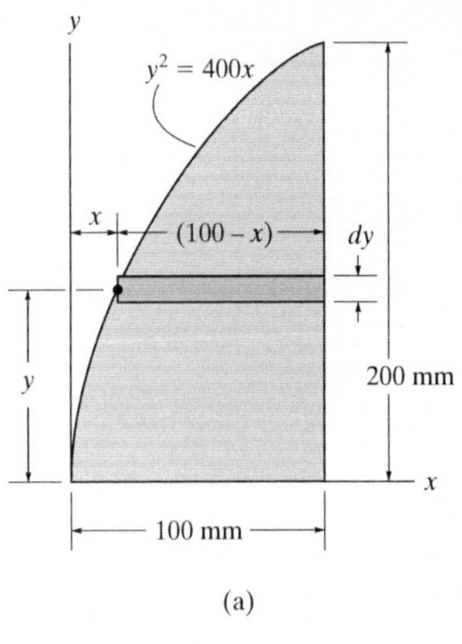

(a)

Determine the moment of inertia for the shaded area shown in Fig. 10–6*a* about the *x* axis.

SOLUTION I (CASE 1)

A differential element of area that is *parallel* to the *x* axis, as shown in Fig. 10–6*a*, is chosen for integration. Since this element has a thickness dy and intersects the curve at the *arbitrary point* (x, y), its area is $dA = (100 - x)\,dy$. Furthermore, the element lies at the same distance y from the x axis. Hence, integrating with respect to y, from $y = 0$ to $y = 200$ mm, yields

$$I_x = \int_A y^2\,dA = \int_0^{200\text{ mm}} y^2(100 - x)\,dy$$

$$= \int_0^{200\text{ mm}} y^2\left(100 - \frac{y^2}{400}\right)dy = \int_0^{200\text{ mm}}\left(100y^2 - \frac{y^4}{400}\right)dy$$

$$= 107(10^6)\text{ mm}^4 \qquad \textit{Ans.}$$

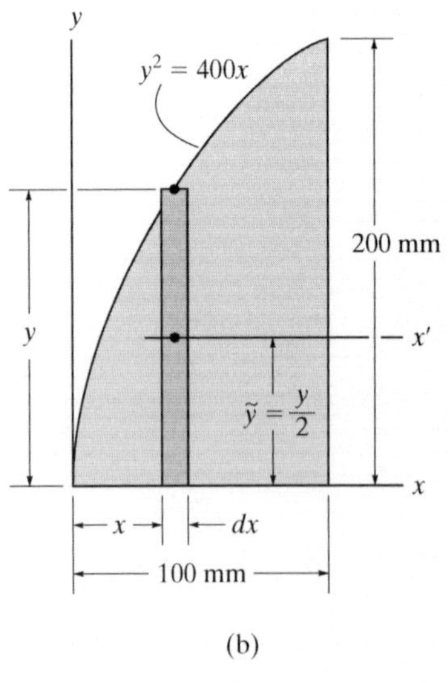

(b)

Fig. 10–6

SOLUTION II (CASE 2)

A differential element *parallel* to the *y* axis, as shown in Fig. 10–6*b*, is chosen for integration. It intersects the curve at the *arbitrary point* (x, y). In this case, all points of the element do *not* lie at the same distance from the x axis, and therefore the parallel-axis theorem must be used to determine the *moment of inertia of the element* with respect to this axis. For a rectangle having a base b and height h, the moment of inertia about its centroidal axis has been determined in part (a) of Example 10.1. There it was found that $\bar{I}_{x'} = \frac{1}{12}bh^3$. For the differential element shown in Fig. 10–6*b*, $b = dx$ and $h = y$, and thus $d\bar{I}_{x'} = \frac{1}{12}dx\,y^3$. Since the centroid of the element is $\tilde{y} = y/2$ from the x axis, the moment of inertia of the element about this axis is

$$dI_x = d\bar{I}_{x'} + dA\,\tilde{y}^2 = \frac{1}{12}dx\,y^3 + y\,dx\left(\frac{y}{2}\right)^2 = \frac{1}{3}y^3\,dx$$

(This result can also be concluded from part (b) of Example 10.1.) Integrating with respect to x, from $x = 0$ to $x = 100$ mm, yields

$$I_x = \int dI_x = \int_0^{100\text{ mm}} \frac{1}{3}y^3\,dx = \int_0^{100\text{ mm}} \frac{1}{3}(400x)^{3/2}\,dx$$

$$= 107(10^6)\text{ mm}^4 \qquad \textit{Ans.}$$

10

EXAMPLE 10.3

Determine the moment of inertia with respect to the x axis for the circular area shown in Fig. 10–7a.

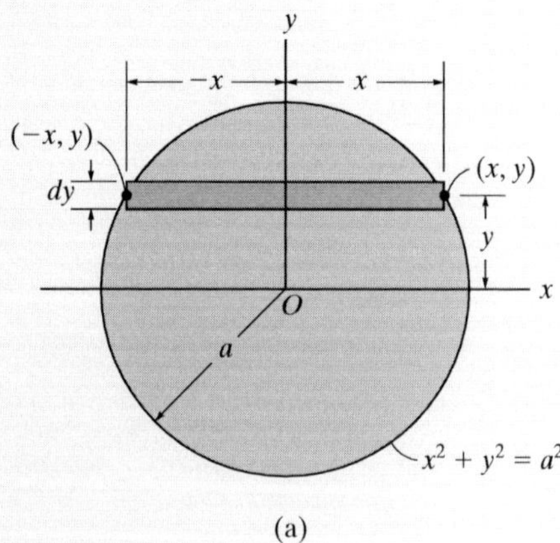

(a)

SOLUTION I (CASE 1)

Using the differential element shown in Fig. 10–7a, since $dA = 2x\,dy$, we have

$$I_x = \int_A y^2\,dA = \int_A y^2(2x)\,dy$$

$$= \int_{-a}^{a} y^2\left(2\sqrt{a^2 - y^2}\right)dy = \frac{\pi a^4}{4} \qquad \textit{Ans.}$$

SOLUTION II (CASE 2)

When the differential element shown in Fig. 10–7b is chosen, the centroid for the element happens to lie on the x axis, and since $\bar{I}_{x'} = \frac{1}{12}bh^3$ for a rectangle, we have

$$dI_x = \frac{1}{12}dx(2y)^3$$

$$= \frac{2}{3}y^3\,dx$$

Integrating with respect to x yields

$$I_x = \int_{-a}^{a} \frac{2}{3}(a^2 - x^2)^{3/2}\,dx = \frac{\pi a^4}{4} \qquad \textit{Ans.}$$

NOTE: By comparison, Solution I requires much less computation. Therefore, if an integral using a particular element appears difficult to evaluate, try solving the problem using an element oriented in the other direction.

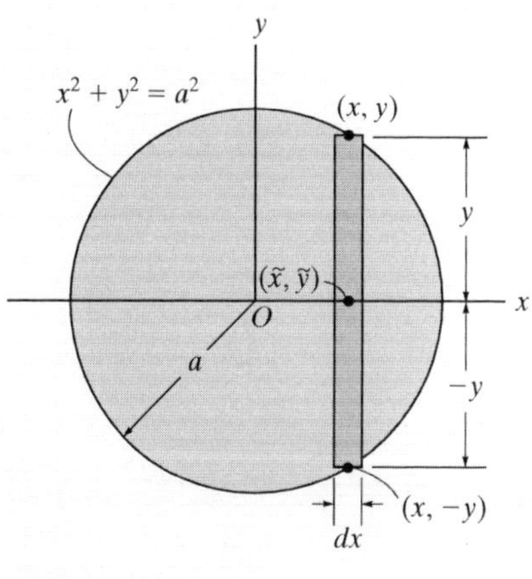

(b)

Fig. 10–7

10

FUNDAMENTAL PROBLEMS

F10–1. Determine the moment of inertia of the shaded area about the x axis.

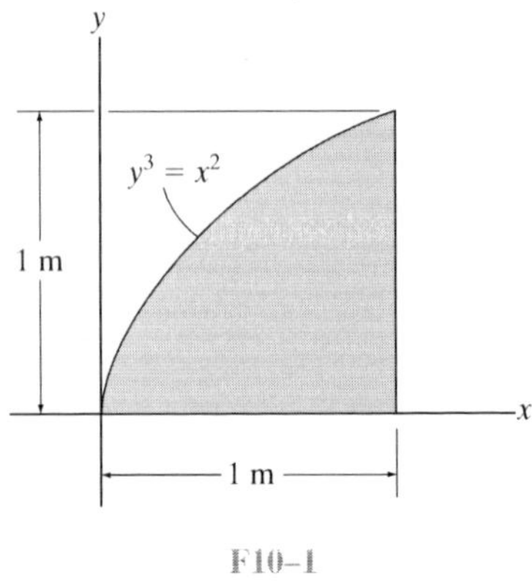

F10–1

F10–2. Determine the moment of inertia of the shaded area about the x axis.

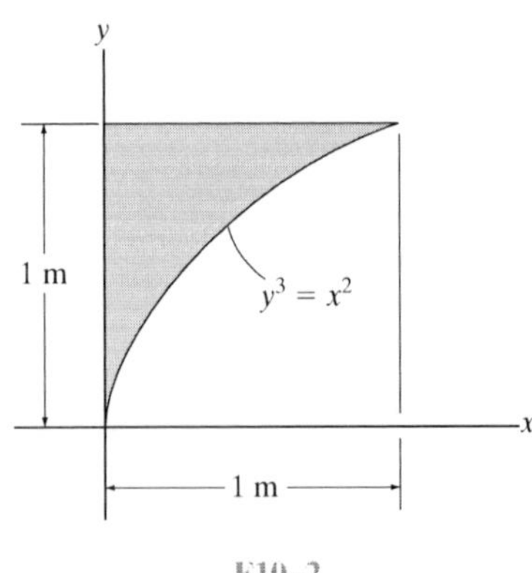

F10–2

F10–3. Determine the moment of inertia of the shaded area about the y axis.

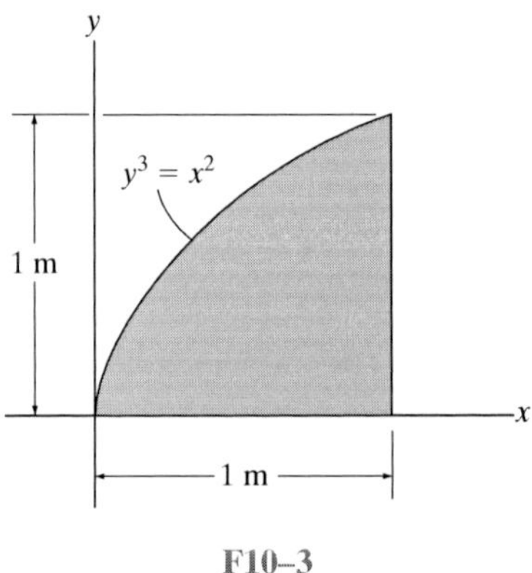

F10–3

F10–4. Determine the moment of inertia of the shaded area about the y axis.

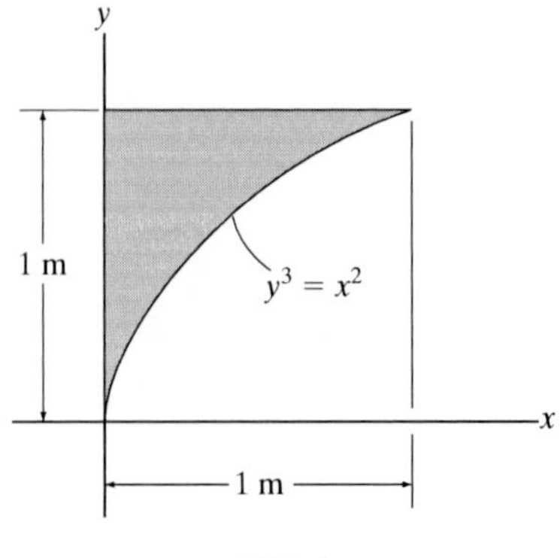

F10–4

10

PROBLEMS

10–1. Determine the moment of inertia of the shaded area about the x axis.

10–2. Determine the moment of inertia of the shaded area about the y axis.

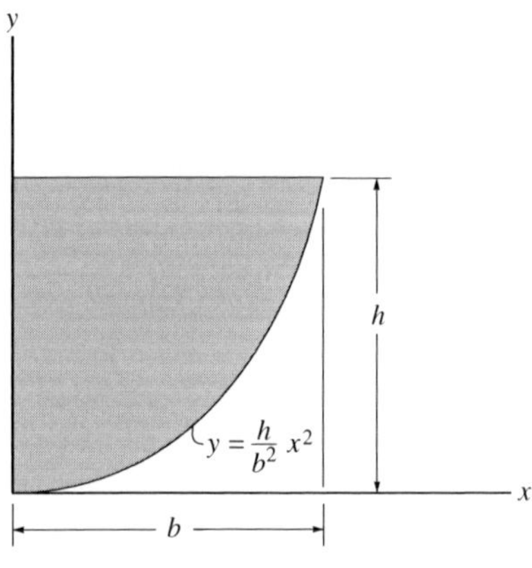

Probs. 10–1/2

10–5. Determine the moment of inertia of the area about the x axis.

10–6. Determine the moment of inertia of the area about the y axis.

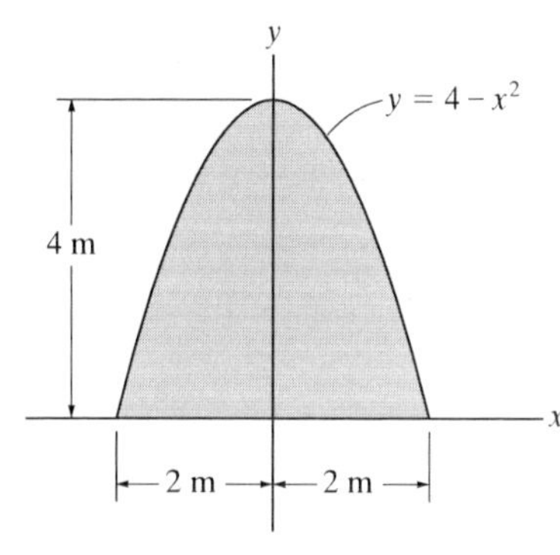

Probs. 10–5/6

10–3. Determine the moment of inertia of the area about the x axis.

***10–4.** Determine the moment of inertia of the area about the y axis.

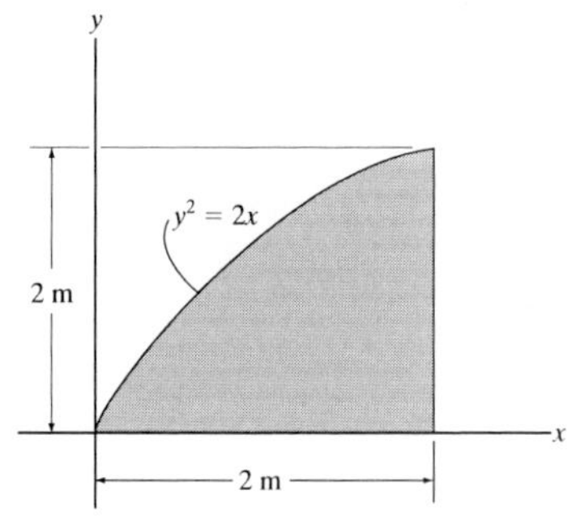

Probs. 10–3/4

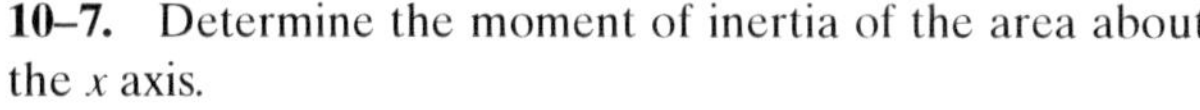

10–7. Determine the moment of inertia of the area about the x axis.

***10–8.** Determine the moment of inertia of the area about the y axis.

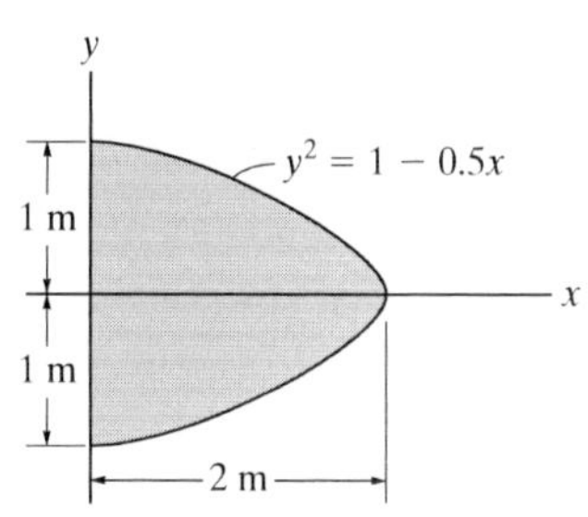

Probs. 10–7/8

10

10–9. Determine the moment of inertia of the area about the x axis.

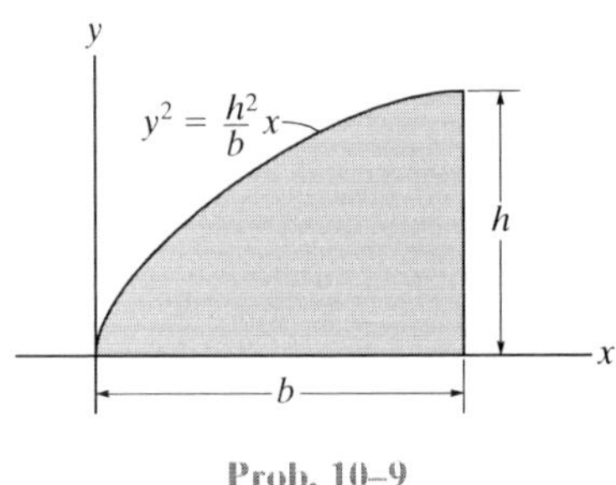

Prob. 10–9

10–10. Determine the moment of inertia for the thin strip of area about the x axis. The strip is oriented at an angle θ from the x axis. Assume that $t \ll l$.

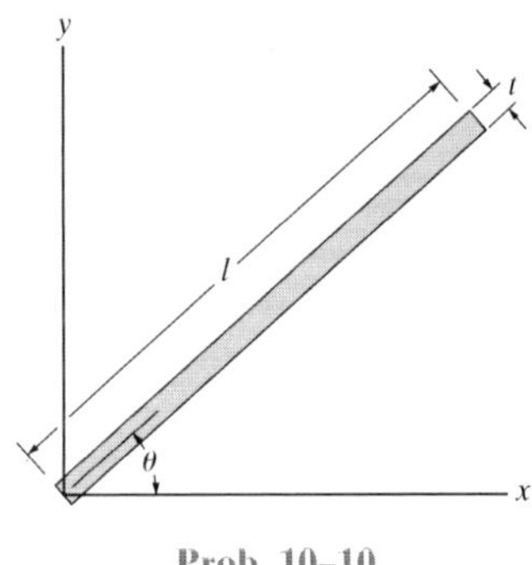

Prob. 10–10

10–11. Determine the moment of inertia of the area about the x axis.

***10–12.** Determine the moment of inertia of the area about the y axis.

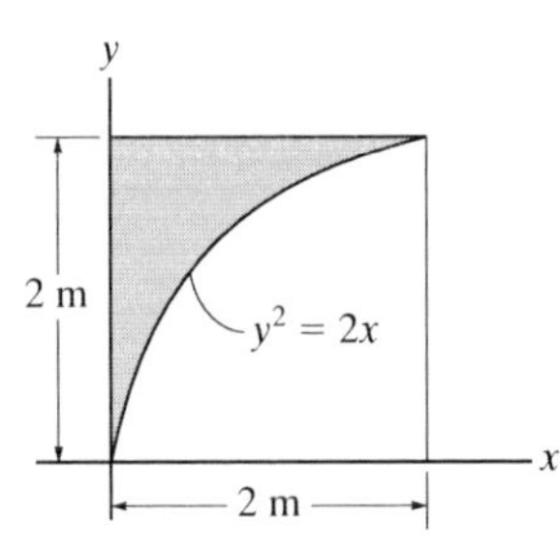

Probs. 10–11/12

10–13. Determine the moment of inertia for the shaded area about the x axis.

10–14. Determine the moment of inertia for the shaded area about the y axis.

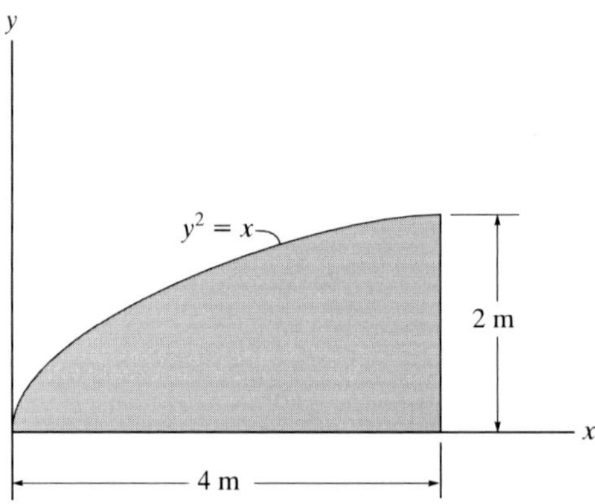

Probs. 10–13/14

10–15. Determine the moment of inertia of the shaded area about the y axis. Use Simpson's rule to evaluate the integral.

***10–16.** Determine the moment of inertia of the shaded area about the x axis. Use Simpson's rule to evaluate the integral.

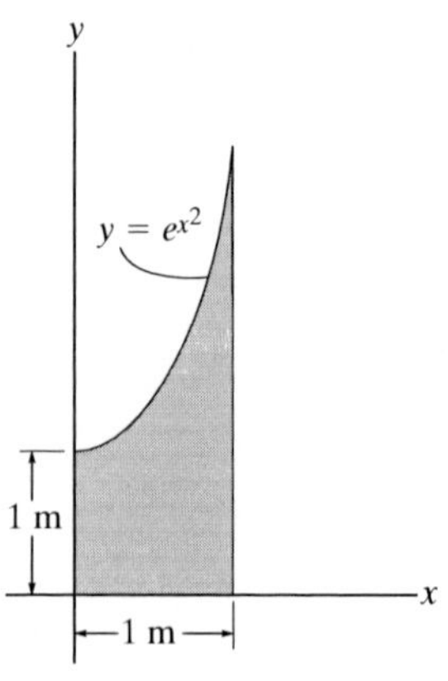

Probs. 10–15/16

10

10–17. Determine the moment of inertia of the shaded area about the x axis.

10–18. Determine the moment of inertia of the shaded area about the y axis.

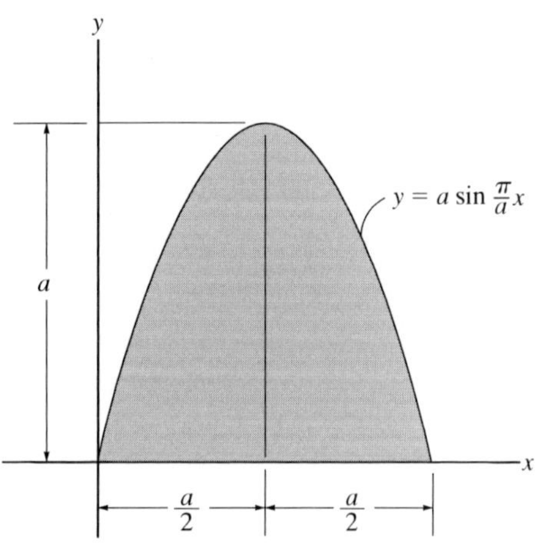

Probs. 10–17/18

10–19. Determine the moment of inertia of the shaded area about the x axis.

***10–20.** Determine the moment of inertia of the shaded area about the y axis.

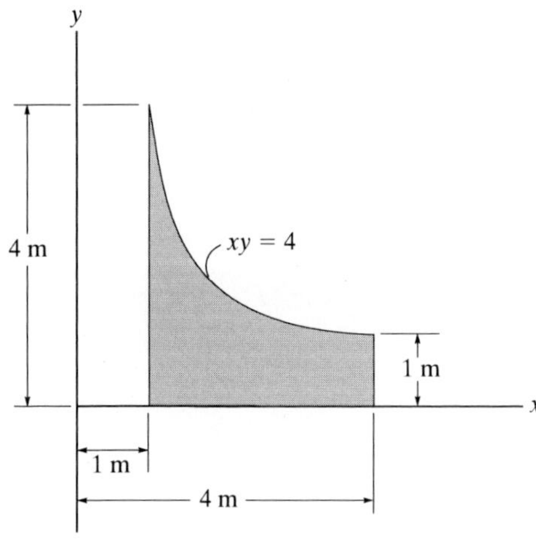

Probs. 10–19/20

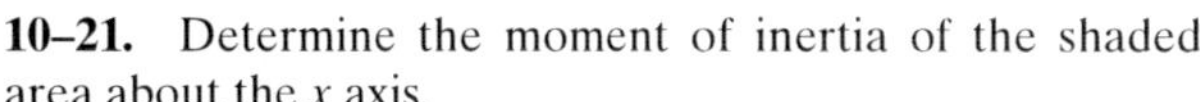

10–21. Determine the moment of inertia of the shaded area about the x axis.

10–22. Determine the moment of inertia of the shaded area about the y axis.

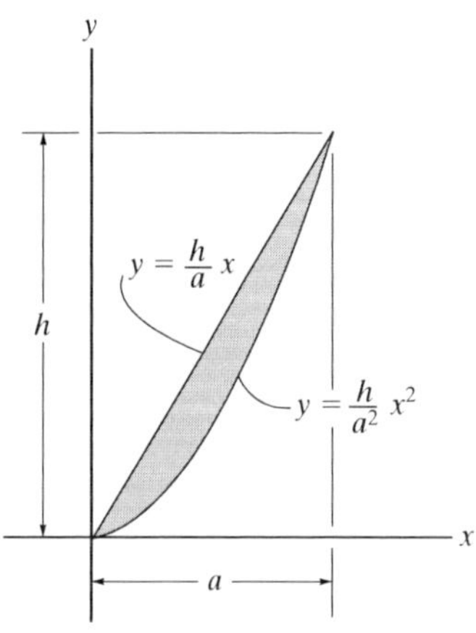

Probs. 10–21/22

10–23. Determine the moment of inertia of the shaded area about the x axis.

***10–24.** Determine the moment of inertia of the shaded area about the y axis.

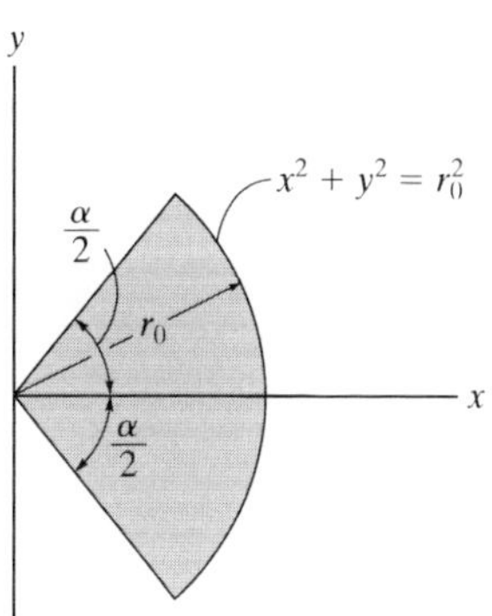

Probs. 10–23/24

10

10.4 Moments of Inertia for Composite Areas

A composite area consists of a series of connected "simpler" parts or shapes, such as rectangles, triangles, and circles. Provided the moment of inertia of each of these parts is known or can be determined about a common axis, then the moment of inertia for the composite area about this axis equals the *algebraic sum* of the moments of inertia of all its parts.

Procedure for Analysis

The moment of inertia for a composite area about a reference axis can be determined using the following procedure.

Composite Parts.

- Using a sketch, divide the area into its composite parts and indicate the perpendicular distance from the centroid of each part to the reference axis.

Parallel-Axis Theorem.

- If the centroidal axis for each part does not coincide with the reference axis, the parallel-axis theorem, $I = \bar{I} + Ad^2$, should be used to determine the moment of inertia of the part about the reference axis. For the calculation of $\bar{I}$, use the table on the inside back cover.

Summation.

- The moment of inertia of the entire area about the reference axis is determined by summing the results of its composite parts about this axis.
- If a composite part has a "hole", its moment of inertia is found by "subtracting" the moment of inertia of the hole from the moment of inertia of the entire part including the hole.

10

For design or analysis of this T-beam, engineers must be able to locate the centroid of its cross-sectional area, and then find the moment of inertia of this area about the centroidal axis.

EXAMPLE 10.4

Determine the moment of inertia of the area shown in Fig. 10–8*a* about the *x* axis.

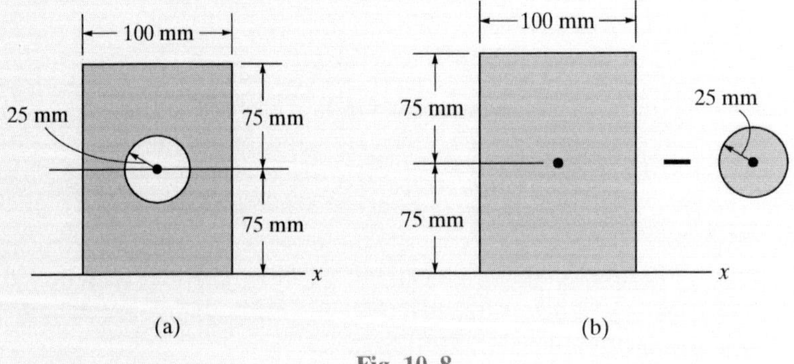

Fig. 10–8

SOLUTION

Composite Parts. The area can be obtained by *subtracting* the circle from the rectangle shown in Fig. 10–8*b*. The centroid of each area is located in the figure.

Parallel-Axis Theorem. The moments of inertia about the *x* axis are determined using the parallel-axis theorem and the geometric properties formulae for circular and rectangular areas $I_x = \frac{1}{4}\pi r^4$; $I_x = \frac{1}{12}bh^3$, found on the inside back cover.

Circle

$$I_x = \bar{I}_{x'} + Ad_y^2$$

$$= \frac{1}{4}\pi(25)^4 + \pi(25)^2(75)^2 = 11.4(10^6)\text{ mm}^4$$

Rectangle

$$I_x = \bar{I}_{x'} + Ad_y^2$$

$$= \frac{1}{12}(100)(150)^3 + (100)(150)(75)^2 = 112.5(10^6)\text{ mm}^4$$

Summation. The moment of inertia for the area is therefore

$$I_x = -11.4(10^6) + 112.5(10^6)$$

$$= 101(10^6)\text{ mm}^4 \qquad \textit{Ans.}$$

10

EXAMPLE 10.5

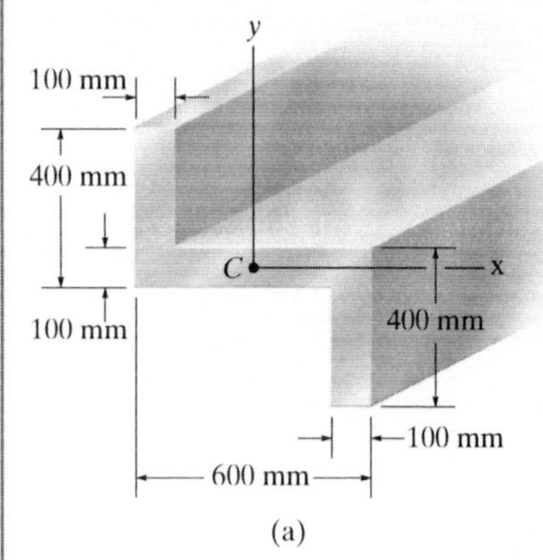

(a)

Determine the moments of inertia for the cross-sectional area of the member shown in Fig. 10–9*a* about the *x* and *y* centroidal axes.

SOLUTION

Composite Parts. The cross section can be subdivided into the three rectangular areas *A*, *B*, and *D* shown in Fig. 10–9*b*. For the calculation, the centroid of each of these rectangles is located in the figure.

Parallel-Axis Theorem. From the table on the inside back cover, or Example 10.1, the moment of inertia of a rectangle about its centroidal axis is $\bar{I} = \frac{1}{12}bh^3$. Hence, using the parallel-axis theorem for rectangles *A* and *D*, the calculations are as follows:

Rectangles A and D

$$I_x = \bar{I}_{x'} + Ad_y^2 = \frac{1}{12}(100)(300)^3 + (100)(300)(200)^2$$

$$= 1.425(10^9)\text{ mm}^4$$

$$I_y = \bar{I}_{y'} + Ad_x^2 = \frac{1}{12}(300)(100)^3 + (100)(300)(250)^2$$

$$= 1.90(10^9)\text{ mm}^4$$

(b)

Fig. 10–9

Rectangle B

$$I_x = \frac{1}{12}(600)(100)^3 = 0.05(10^9)\text{ mm}^4$$

$$I_y = \frac{1}{12}(100)(600)^3 = 1.80(10^9)\text{ mm}^4$$

Summation. The moments of inertia for the entire cross section are thus

$$I_x = 2[1.425(10^9)] + 0.05(10^9)$$

$$= 2.90(10^9)\text{ mm}^4 \qquad \textit{Ans.}$$

$$I_y = 2[1.90(10^9)] + 1.80(10^9)$$

$$= 5.60(10^9)\text{ mm}^4 \qquad \textit{Ans.}$$

FUNDAMENTAL PROBLEMS

F10–5. Determine the moment of inertia of the beam's cross-sectional area about the centroidal x and y axes.

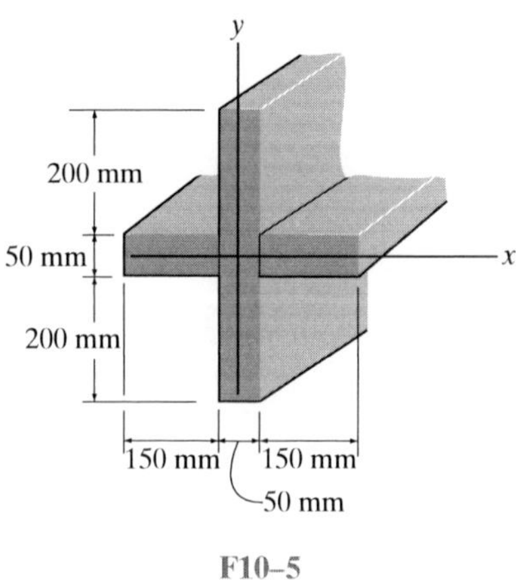

F10–5

F10–6. Determine the moment of inertia of the beam's cross-sectional area about the centroidal x and y axes.

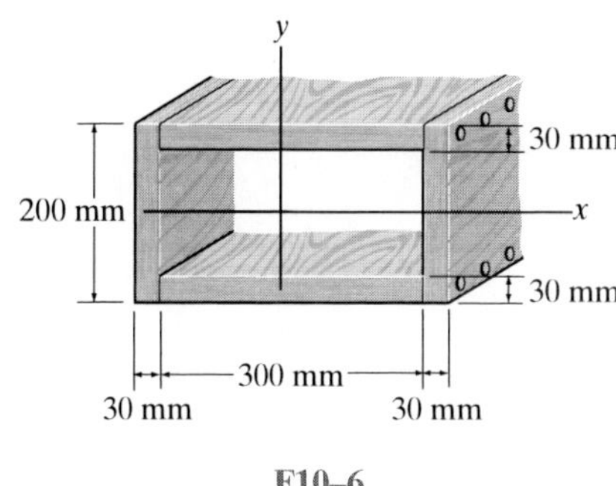

F10–6

F10–7. Determine the moment of inertia of the cross-sectional area of the channel with respect to the y axis.

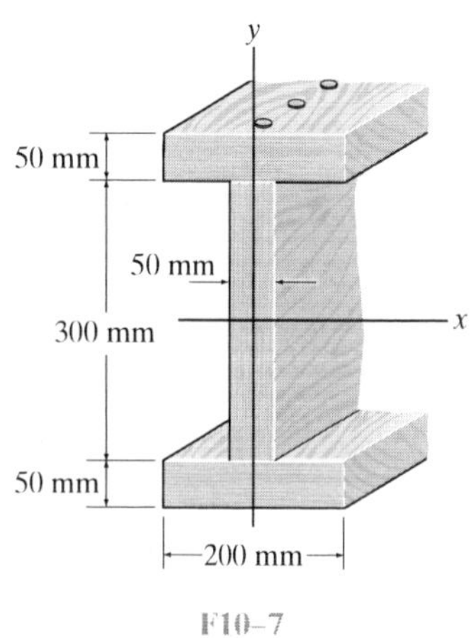

F10–7

F10–8. Determine the moment of inertia of the cross-sectional area of the T-beam with respect to the x' axis passing through the centroid of the cross section.

F10–8

PROBLEMS

10–25. Determine the moment of inertia of the composite area about the x axis.

10–26. Determine the moment of inertia of the composite area about the y axis.

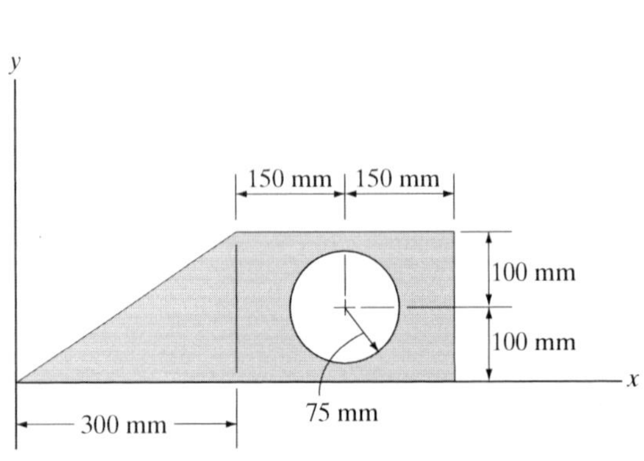

Probs. 10–25/26

***10–28.** Determine the moment of inertia of the beam's cross-sectional area about the x axis.

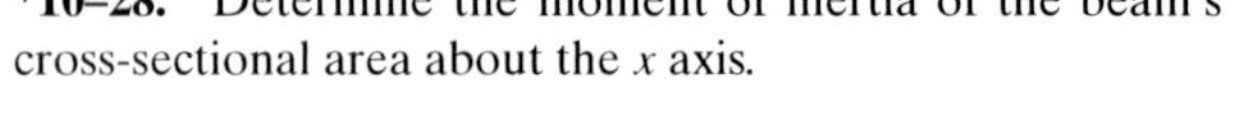
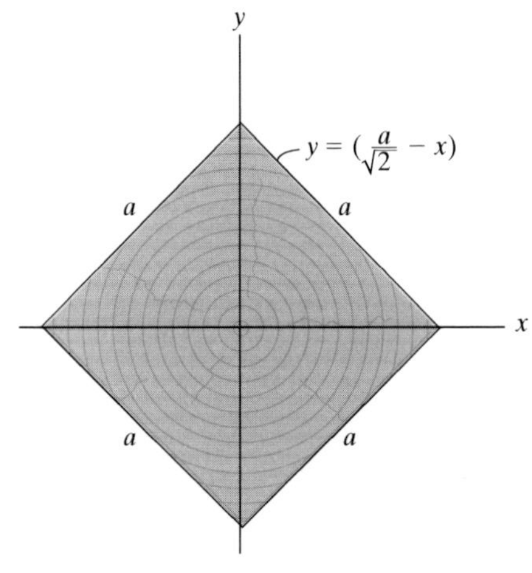

Prob. 10–28

10–27. Determine the radius of gyration k_x for the column's cross-sectional area.

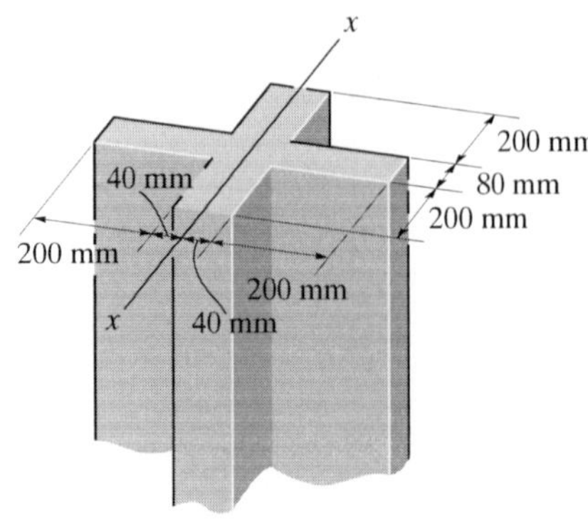

Prob. 10–27

10–29. Locate the centroid $\bar{y}$ of the cross section and determine the moment of inertia of the section about the x' axis.

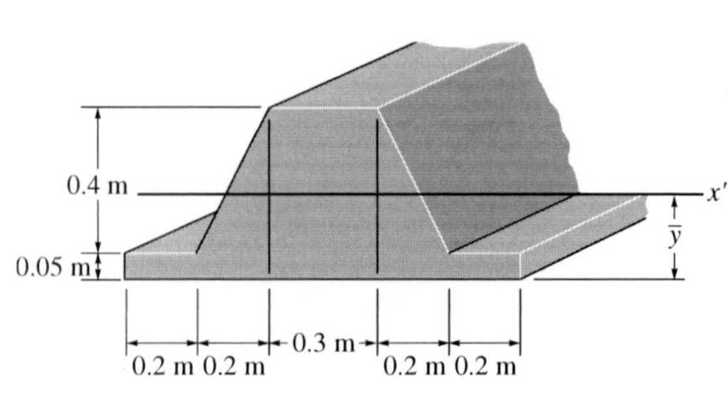

Prob. 10–29

10

10–30. Determine the distance $\bar{x}$ to the centroid of the beam's cross-sectional area, then find the moment of inertia about the y' axis.

10–31. Determine the moment of inertia of the beam's cross-sectional area about the x' axis.

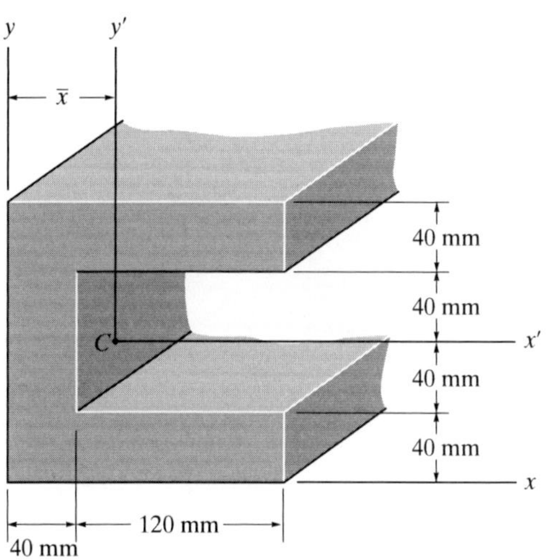

Probs. 10–30/31

*10–32. Determine the moment of inertia of the shaded area about the x axis.

10–33. Determine the moment of inertia of the shaded area about the y axis.

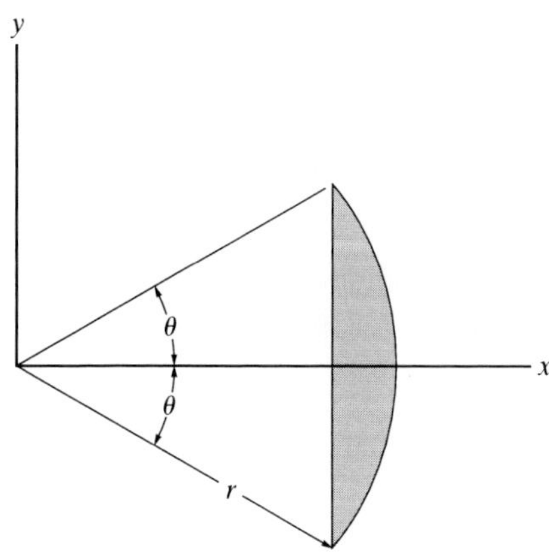

Probs. 10–32/33

10–34. Determine the moment of inertia of the beam's cross-sectional area about the y axis.

10–35. Determine $\bar{y}$, which locates the centroidal axis x' for the cross-sectional area of the T-beam, and then find the moment of inertia about the x' axis.

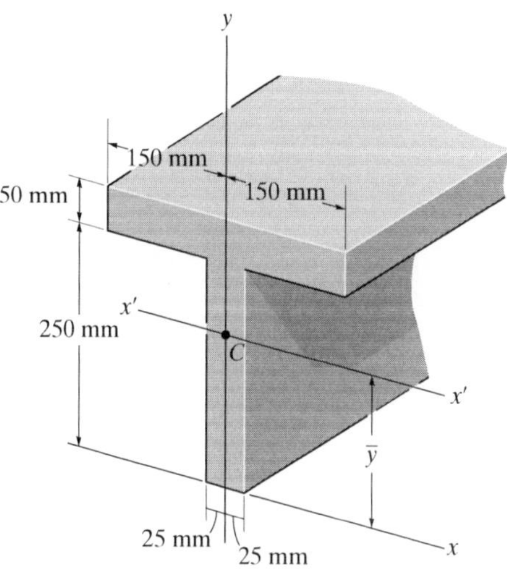

Probs. 10–34/35

*10–36. Determine the moment of inertia of the beam's cross-sectional area about the x axis.

10–37. Determine the moment of inertia of the beam's cross-sectional area about the y axis.

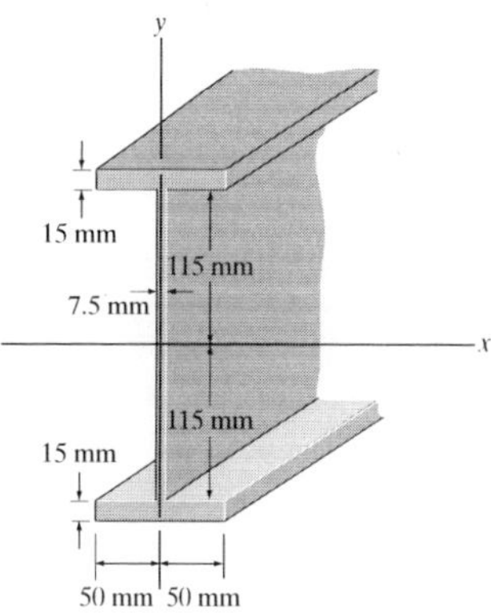

Probs. 10–36/37

10

10–38. Determine the moment of inertia of the beam's cross-sectional area about the x axis.

10–39. Determine the moment of inertia of the beam's cross-sectional area about the y axis.

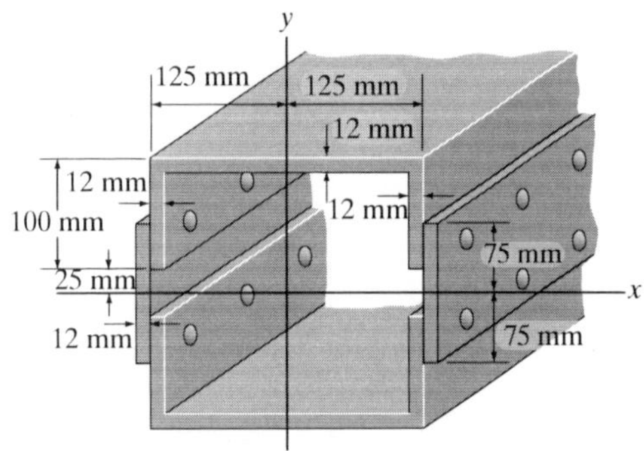

Probs. 10–38/39

***10–40.** Determine the moment of inertia of the cross-sectional area about the x axis.

10–41. Locate the centroid $\bar{x}$ of the beam's cross-sectional area, and then determine the moment of inertia of the area about the centroidal y' axis.

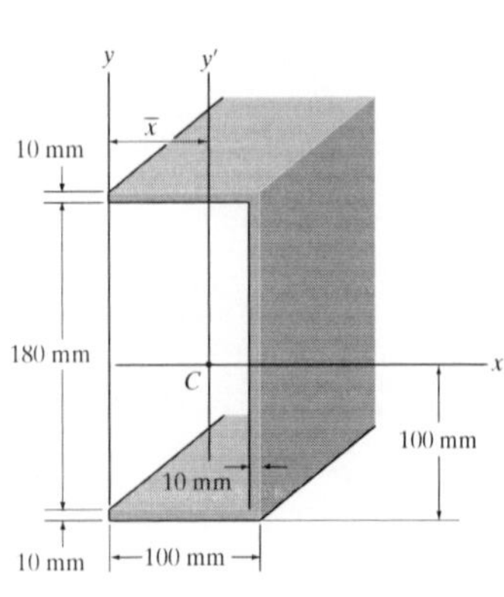

Probs. 10–40/41

10–42. Determine the moment of inertia of the beam's cross-sectional area about the x axis.

10–43. Determine the moment of inertia of the beam's cross-sectional area about the y axis.

***10–44.** Determine the distance $\bar{y}$ to the centroid C of the beam's cross-sectional area and then compute the moment of inertia $\bar{I}_{x'}$ about the x' axis.

10–45. Determine the distance $\bar{x}$ to the centroid C of the beam's cross-sectional area and then compute the moment of inertia $\bar{I}_{y'}$ about the y' axis.

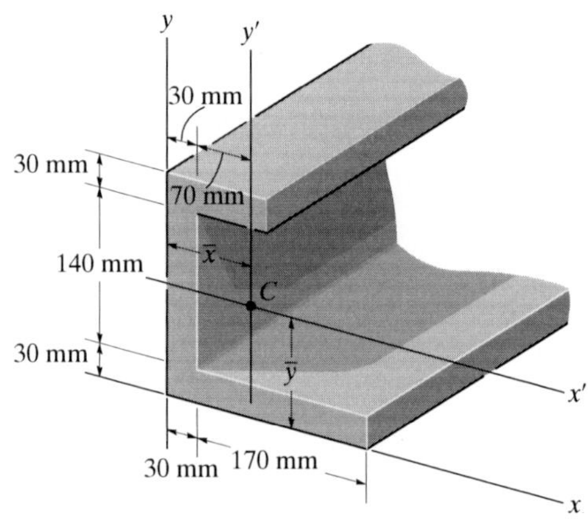

Probs. 10–42/43/44/45

10–46. Determine the distance $\bar{y}$ to the centroid for the beam's cross-sectional area; then determine the moment of inertia about the x' axis.

10–47. Determine the moment of inertia of the beam's cross-sectional area about the y axis.

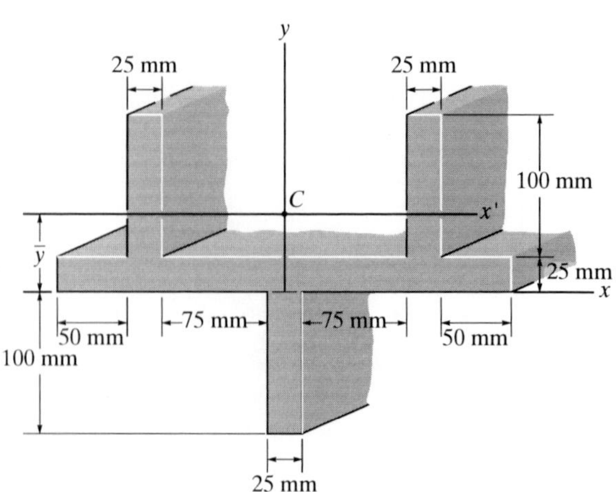

Probs. 10–46/47

10

***10–48.** Determine the beam's moment of inertia I_x about the centroidal x axis.

10–49. Determine the beam's moment of inertia I_y about the centroidal y axis.

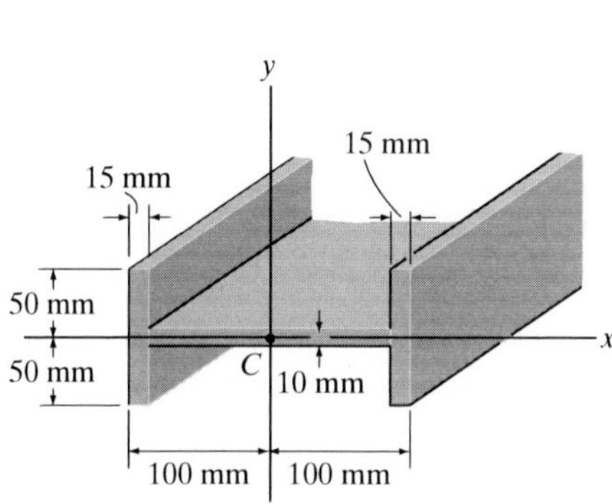

Probs. 10–48/49

10–51. Determine the moment of inertia of the beam's cross-sectional area with respect to the x' centroidal axis. Neglect the size of all the rivet heads, R, for the calculation. Handbook values for the area, moment of inertia, and location of the centroid C of one of the angles are listed in the figure.

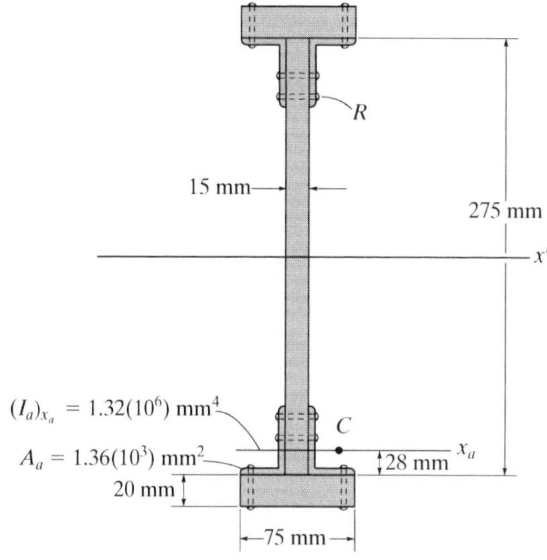

Prob. 10–51

10–50. Locate the centroid $\bar{y}$ of the cross section and determine the moment of inertia of the section about the x' axis.

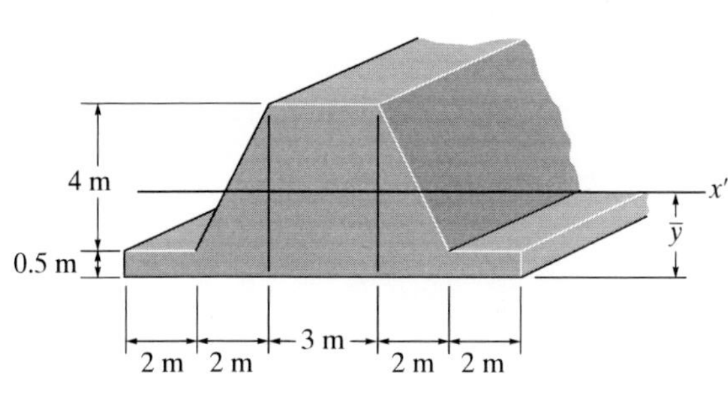

Prob. 10–50

***10–52.** Determine the moment of inertia of the parallelogram about the x' axis, which passes through the centroid C of the area.

10–53. Determine the moment of inertia of the parallelogram about the y' axis, which passes through the centroid C of the area.

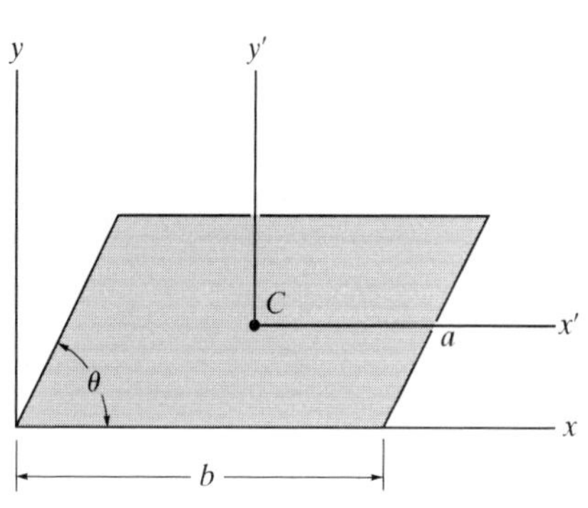

Probs. 10–52/53

*10.5 Product of Inertia for an Area

It will be shown in the next section that the property of an area, called the product of inertia, is required in order to determine the *maximum* and *minimum* moments of inertia for the area. These maximum and minimum values are important properties needed for designing structural and mechanical members such as beams, columns, and shafts.

The *product of inertia* of the area in Fig. 10–10 with respect to the x and y axes is defined as

$$I_{xy} = \int_A xy \, dA \qquad (10\text{–}7)$$

Fig. 10–10

If the element of area chosen has a differential size in two directions, as shown in Fig. 10–10, a double integration must be performed to evaluate I_{xy}. Most often, however, it is easier to choose an element having a differential size or thickness in only one direction in which case the evaluation requires only a single integration (see Example 10.6).

Like the moment of inertia, the product of inertia has units of length raised to the fourth power, e.g., m^4, mm^4. However, since x or y may be negative, the product of inertia may either be positive, negative, or zero, depending on the location and orientation of the coordinate axes. For example, the product of inertia I_{xy} for an area will be *zero* if either the x or y axis is an axis of *symmetry* for the area, as in Fig. 10–11. Here every element dA located at point (x, y) has a corresponding element dA located at $(x, -y)$. Since the products of inertia for these elements are, respectively, $xy\,dA$ and $-xy\,dA$, the algebraic sum or integration of all the elements that are chosen in this way will cancel each other. Consequently, the product of inertia for the total area becomes zero. It also follows from the definition of I_{xy} that the "sign" of this quantity depends on the quadrant where the area is located. As shown in Fig. 10–12, if the area is rotated from one quadrant to another, the sign of I_{xy} will change.

The effectiveness of this beam to resist bending can be determined once its moments of inertia and its product of inertia are known.

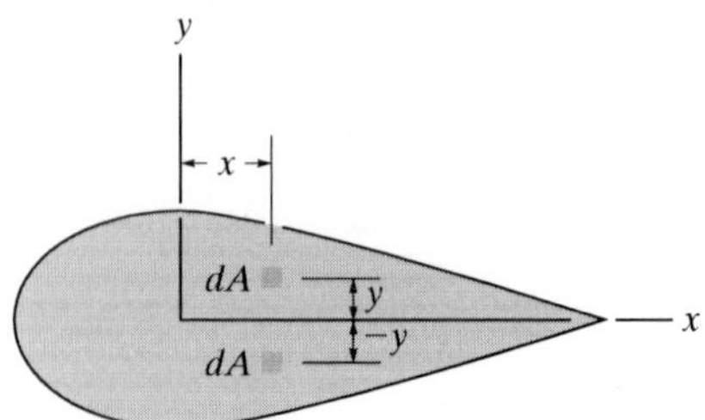

Fig. 10–11

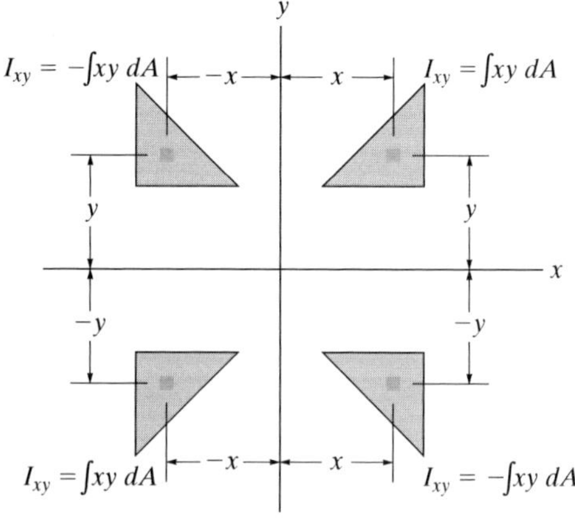

Fig. 10–12

Parallel-Axis Theorem. Consider the shaded area shown in Fig. 10–13, where x' and y' represent a set of axes passing through the *centroid* of the area, and x and y represent a corresponding set of parallel axes. Since the product of inertia of dA with respect to the x and y axes is $dI_{xy} = (x' + d_x)(y' + d_y)\,dA$, then for the entire area,

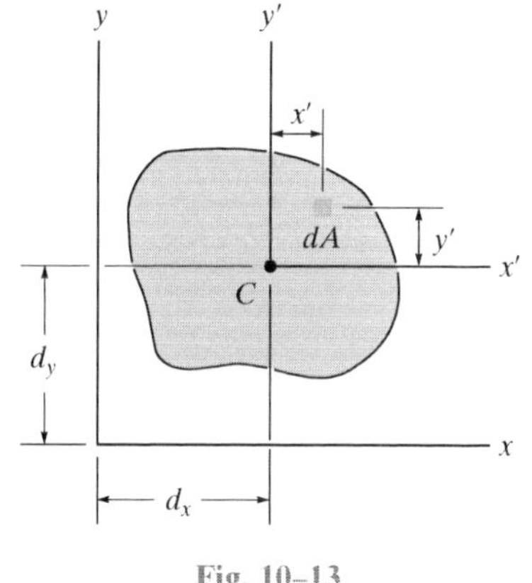

Fig. 10–13

$$\begin{aligned} I_{xy} &= \int_A (x' + d_x)(y' + d_y)\,dA \\ &= \int_A x'y'\,dA + d_x\int_A y'\,dA + d_y\int_A x'\,dA + d_xd_y\int_A dA \end{aligned}$$

The first term on the right represents the product of inertia for the area with respect to the centroidal axes, $\bar{I}_{x'y'}$. The integrals in the second and third terms are zero since the moments of the area are taken about the centroidal axis. Realizing that the fourth integral represents the entire area A, the parallel-axis theorem for the product of inertia becomes

$$I_{xy} = \bar{I}_{x'y'} + Ad_xd_y \qquad (10\text{–}8)$$

It is important that the *algebraic signs* for d_x and d_y be maintained when applying this equation.

10

EXAMPLE 10.6

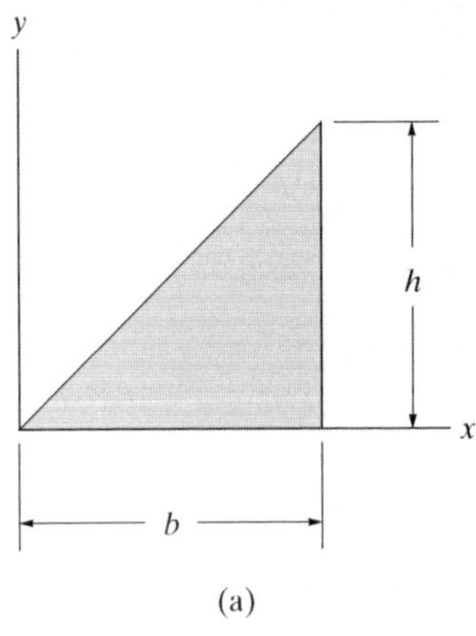

(a)

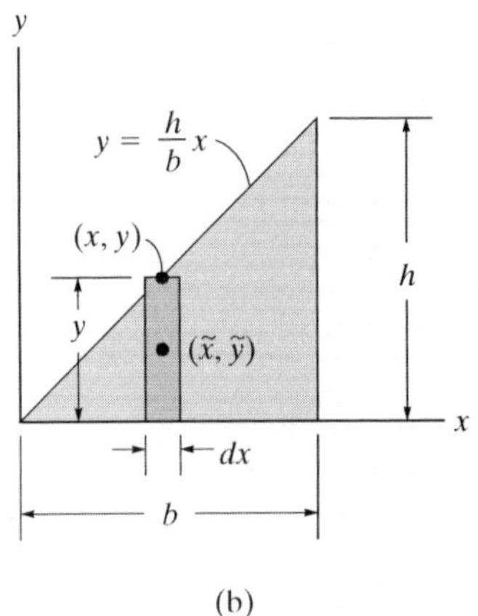

(b)

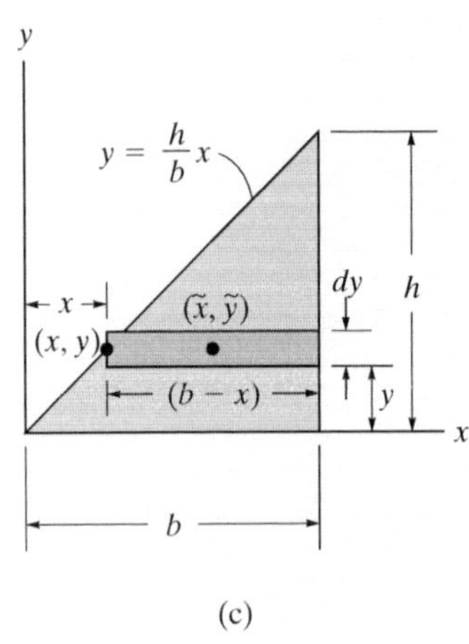

(c)

Fig. 10–14

Determine the product of inertia I_{xy} for the triangle shown in Fig. 10–14*a*.

SOLUTION I

A differential element that has a thickness dx, as shown in Fig. 10–14*b*, has an area $dA = y\,dx$. The product of inertia of this element with respect to the x and y axes is determined using the parallel-axis theorem.

$$dI_{xy} = d\bar{I}_{x'y'} + dA\,\tilde{x}\,\tilde{y}$$

where $\tilde{x}$ and $\tilde{y}$ locate the *centroid* of the element or the origin of the x', y' axes. (See Fig. 10–13.) Since $d\bar{I}_{x'y'} = 0$, due to symmetry, and $\tilde{x} = x$, $\tilde{y} = y/2$, then

$$dI_{xy} = 0 + (y\,dx)x\left(\frac{y}{2}\right) = \left(\frac{h}{b}x\,dx\right)x\left(\frac{h}{2b}x\right)$$
$$= \frac{h^2}{2b^2}x^3\,dx$$

Integrating with respect to x from $x = 0$ to $x = b$ yields

$$I_{xy} = \frac{h^2}{2b^2}\int_0^b x^3\,dx = \frac{b^2h^2}{8} \qquad \textit{Ans.}$$

SOLUTION II

The differential element that has a thickness dy, as shown in Fig. 10–14*c*, can also be used. Its area is $dA = (b - x)\,dy$. The *centroid* is located at point $\tilde{x} = x + (b - x)/2 = (b + x)/2$, $\tilde{y} = y$, so the product of inertia of the element becomes

$$dI_{xy} = d\bar{I}_{x'y'} + dA\,\tilde{x}\,\tilde{y}$$
$$= 0 + (b - x)\,dy\left(\frac{b + x}{2}\right)y$$
$$= \left(b - \frac{b}{h}y\right)dy\left[\frac{b + (b/h)y}{2}\right]y = \frac{1}{2}y\left(b^2 - \frac{b^2}{h^2}y^2\right)dy$$

Integrating with respect to y from $y = 0$ to $y = h$ yields

$$I_{xy} = \frac{1}{2}\int_0^h y\left(b^2 - \frac{b^2}{h^2}y^2\right)dy = \frac{b^2h^2}{8} \qquad \textit{Ans.}$$

10

EXAMPLE 10.7

Determine the product of inertia for the cross-sectional area of the member shown in Fig. 10–15*a*, about the *x* and *y* centroidal axes.

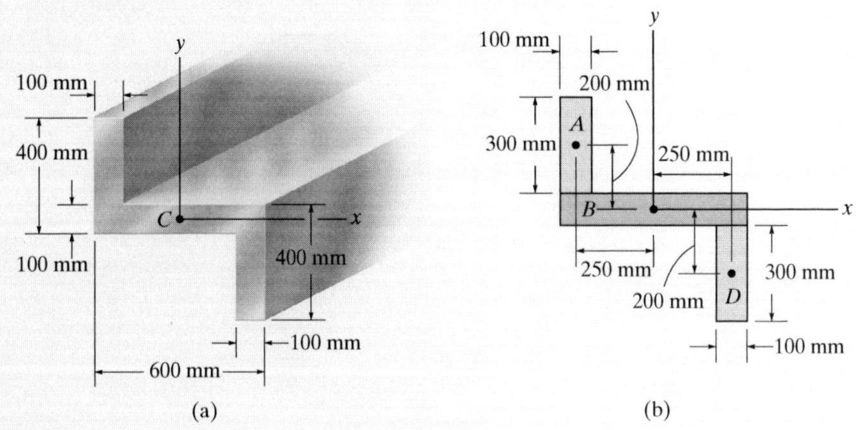

Fig. 10–15

SOLUTION

As in Example 10.5, the cross section can be subdivided into three composite rectangular areas *A*, *B*, and *D*, Fig. 10–15*b*. The coordinates for the centroid of each of these rectangles are shown in the figure. Due to symmetry, the product of inertia of *each rectangle* is *zero* about a set of x', y' axes that passes through the centroid of each rectangle. Using the parallel-axis theorem, we have

Rectangle A

$$I_{xy} = \bar{I}_{x'y'} + Ad_xd_y$$
$$= 0 + (300)(100)(-250)(200) = -1.50(10^9)\text{ mm}^4$$

Rectangle B

$$I_{xy} = \bar{I}_{x'y'} + Ad_xd_y$$
$$= 0 + 0 = 0$$

Rectangle D

$$I_{xy} = \bar{I}_{x'y'} + Ad_xd_y$$
$$= 0 + (300)(100)(250)(-200) = -1.50(10^9)\text{ mm}^4$$

The product of inertia for the entire cross section is therefore

$$I_{xy} = -1.50(10^9) + 0 - 1.50(10^9) = -3.00(10^9)\text{ mm}^4 \quad \textit{Ans.}$$

NOTE: This negative result is due to the fact that rectangles *A* and *D* have centroids located with negative *x* and negative *y* coordinates, respectively.

10

*10.6 Moments of Inertia for an Area about Inclined Axes

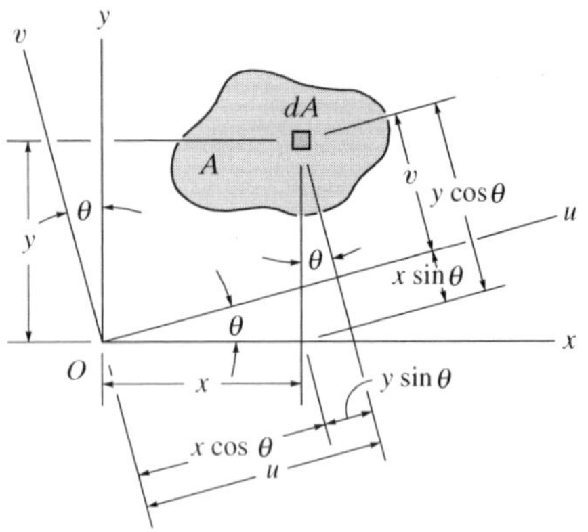

Fig. 10–16

In structural and mechanical design, it is sometimes necessary to calculate the moments and product of inertia I_u, I_v, and I_{uv} for an area with respect to a set of inclined u and v axes when the values for θ, I_x, I_y, and I_{xy} are *known*. To do this we will use *transformation equations* which relate the x, y and u, v coordinates. From Fig. 10–16, these equations are

$$u = x\cos\theta + y\sin\theta$$
$$v = y\cos\theta - x\sin\theta$$

With these equations, the moments and product of inertia of dA about the u and v axes become

$$dI_u = v^2\,dA = (y\cos\theta - x\sin\theta)^2\,dA$$
$$dI_v = u^2\,dA = (x\cos\theta + y\sin\theta)^2\,dA$$
$$dI_{uv} = uv\,dA = (x\cos\theta + y\sin\theta)(y\cos\theta - x\sin\theta)\,dA$$

Expanding each expression and integrating, realizing that $I_x = \int y^2\,dA$, $I_y = \int x^2\,dA$, and $I_{xy} = \int xy\,dA$, we obtain

$$I_u = I_x\cos^2\theta + I_y\sin^2\theta - 2I_{xy}\sin\theta\cos\theta$$
$$I_v = I_x\sin^2\theta + I_y\cos^2\theta + 2I_{xy}\sin\theta\cos\theta$$
$$I_{uv} = I_x\sin\theta\cos\theta - I_y\sin\theta\cos\theta + I_{xy}(\cos^2\theta - \sin^2\theta)$$

Using the trigonometric identities $\sin 2\theta = 2\sin\theta\cos\theta$ and $\cos 2\theta = \cos^2\theta - \sin^2\theta$ we can simplify the above expressions, in which case

$$\begin{aligned} I_u &= \frac{I_x + I_y}{2} + \frac{I_x - I_y}{2}\cos 2\theta - I_{xy}\sin 2\theta \\ I_v &= \frac{I_x + I_y}{2} - \frac{I_x - I_y}{2}\cos 2\theta + I_{xy}\sin 2\theta \\ I_{uv} &= \frac{I_x - I_y}{2}\sin 2\theta + I_{xy}\cos 2\theta \end{aligned} \qquad (10\text{–}9)$$

Notice that if the first and second equations are added together, we can show that the polar moment of inertia about the z axis passing through point O is, as expected, *independent* of the orientation of the u and v axes; i.e.,

$$J_O = I_u + I_v = I_x + I_y$$

10

Principal Moments of Inertia. Equations 10–9 show that I_u, I_v, and I_{uv} depend on the angle of inclination, θ, of the u, v axes. We will now determine the orientation of these axes about which the moments of inertia for the area are maximum and minimum. This particular set of axes is called the *principal axes* of the area, and the corresponding moments of inertia with respect to these axes are called the *principal moments of inertia*. In general, there is a set of principal axes for every chosen origin O. However, for structural and mechanical design, the origin O is located at the centroid of the area.

The angle which defines the orientation of the principal axes can be found by differentiating the first of Eqs. 10–9 with respect to θ and setting the result equal to zero. Thus,

$$\frac{dI_u}{d\theta} = -2\left(\frac{I_x - I_y}{2}\right)\sin 2\theta - 2I_{xy}\cos 2\theta = 0$$

Therefore, at $\theta = \theta_p$,

$$\tan 2\theta_p = \frac{-I_{xy}}{(I_x - I_y)/2} \qquad (10\text{–}10)$$

The two roots θ_{p_1} and θ_{p_2} of this equation are 90° apart, and so they each specify the inclination of one of the principal axes. In order to substitute them into Eq. 10–9, we must first find the sine and cosine of $2\theta_{p_1}$ and $2\theta_{p_2}$. This can be done using these ratios from the triangles shown in Fig. 10–17, which are based on Eq. 10–10.

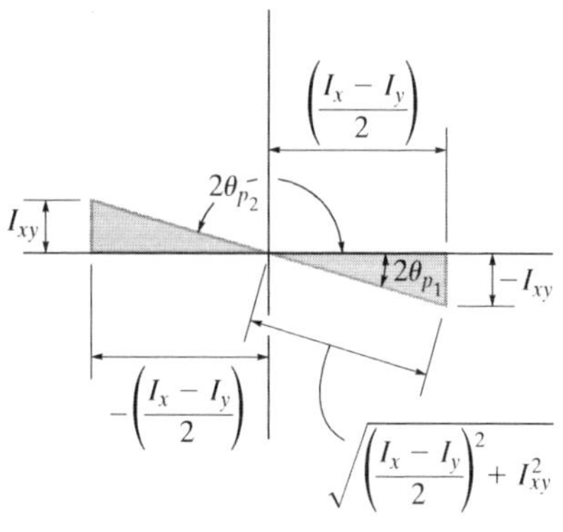

Fig. 10–17

Substituting each of the sine and cosine ratios into the first or second of Eqs. 10–9 and simplifying, we obtain

$$I_{\substack{max \\ min}} = \frac{I_x + I_y}{2} \pm \sqrt{\left(\frac{I_x - I_y}{2}\right)^2 + I_{xy}^2} \qquad (10\text{–}11)$$

Depending on the sign chosen, this result gives the maximum or minimum moment of inertia for the area. Furthermore, if the above trigonometric relations for θ_{p_1} and θ_{p_2} are substituted into the third of Eqs. 10–9, it can be shown that $I_{uv} = 0$; that is, the *product of inertia with respect to the principal axes is zero*. Since it was indicated in Sec. 10.6 that the product of inertia is zero with respect to any symmetrical axis, it therefore follows that *any symmetrical axis represents a principal axis of inertia for the area.*

10

EXAMPLE 10.8

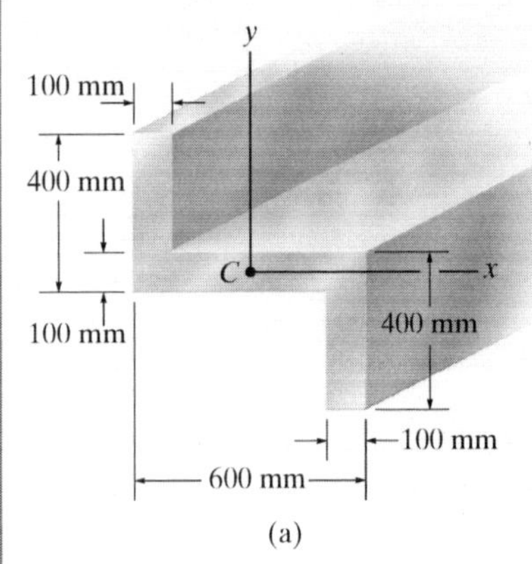

(a)

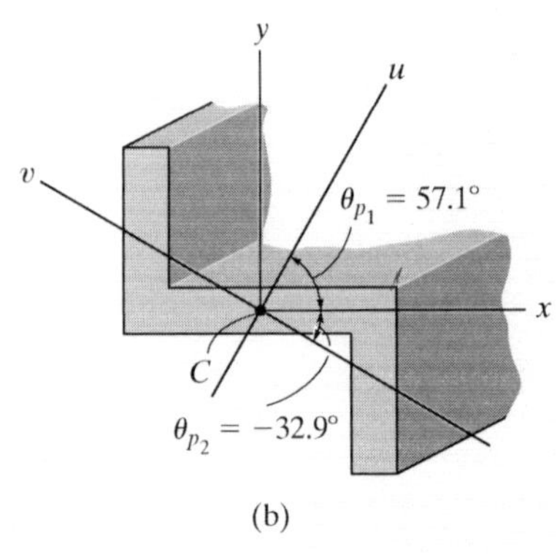

(b)

Fig. 10–18

Determine the principal moments of inertia and the orientation of the principal axes for the cross-sectional area of the member shown in Fig. 10–18*a* with respect to an axis passing through the centroid.

SOLUTION

The moments and product of inertia of the cross section with respect to the *x*, *y* axes have been determined in Examples 10.5 and 10.7. The results are

$$I_x = 2.90(10^9)\text{ mm}^4 \quad I_y = 5.60(10^9)\text{ mm}^4 \quad I_{xy} = -3.00(10^9)\text{ mm}^4$$

Using Eq. 10–10, the angles of inclination of the principal axes *u* and *v* are

$$\tan 2\theta_p = \frac{-I_{xy}}{(I_x - I_y)/2} = \frac{-[-3.00(10^9)]}{[2.90(10^9) - 5.60(10^9)]/2} = -2.22$$

$$2\theta_p = -65.8° \text{ and } 114.2°$$

Thus, by inspection of Fig. 10–18*b*,

$$\theta_{p_2} = -32.9° \quad \text{and} \quad \theta_{p_1} = 57.1° \qquad \textit{Ans.}$$

The principal moments of inertia with respect to these axes are determined from Eq. 10–11. Hence,

$$I_{\min}^{\max} = \frac{I_x + I_y}{2} \pm \sqrt{\left(\frac{I_x - I_y}{2}\right)^2 + I_{xy}^2}$$

$$= \frac{2.90(10^9) + 5.60(10^9)}{2}$$

$$\pm \sqrt{\left[\frac{2.90(10^9) - 5.60(10^9)}{2}\right]^2 + [-3.00(10^9)]^2}$$

$$I_{\min}^{\max} = 4.25(10^9) \pm 3.29(10^9)$$

or

$$I_{\max} = 7.54(10^9)\text{ mm}^4 \quad I_{\min} = 0.960(10^9)\text{ mm}^4 \qquad \textit{Ans.}$$

NOTE: The maximum moment of inertia, $I_{\max} = 7.54(10^9)\text{ mm}^4$, occurs with respect to the *u* axis since *by inspection* most of the cross-sectional area is farthest away from this axis. Or, stated in another manner, $I_{\max}$ occurs about the *u* axis since this axis is located within ±45° of the *y* axis, which has the larger value of I ($I_y > I_x$). Also, this can be concluded by substituting the data with $\theta = 57.1°$ into the first of Eqs. 10–9 and solving for I_u.

10

*10.7 Mohr's Circle for Moments of Inertia

Equations 10–9 to 10–11 have a graphical solution that is convenient to use and generally easy to remember. Squaring the first and third of Eqs. 10–9 and adding, it is found that

$$\left(I_u - \frac{I_x + I_y}{2}\right)^2 + I_{uv}^2 = \left(\frac{I_x - I_y}{2}\right)^2 + I_{xy}^2$$

Here I_x, I_y, and I_{xy} are *known constants*. Thus, the above equation may be written in compact form as

$$(I_u - a)^2 + I_{uv}^2 = R^2$$

When this equation is plotted on a set of axes that represent the respective moment of inertia and the product of inertia, as shown in Fig. 10–19, the resulting graph represents a *circle* of radius

$$R = \sqrt{\left(\frac{I_x - I_y}{2}\right)^2 + I_{xy}^2}$$

and having its center located at point $(a, 0)$, where $a = (I_x + I_y)/2$. The circle so constructed is called *Mohr's circle*, named after the German engineer Otto Mohr (1835–1918).

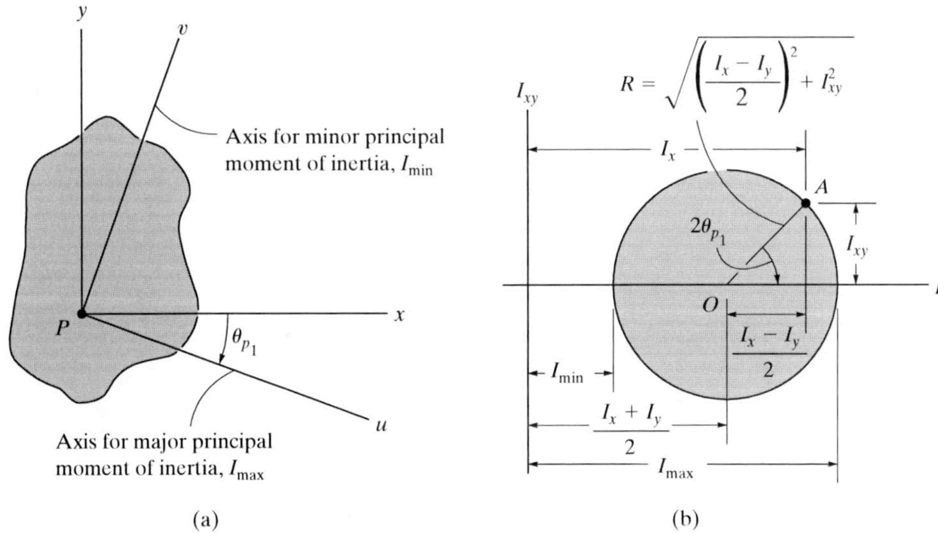

Fig. 10–19

10

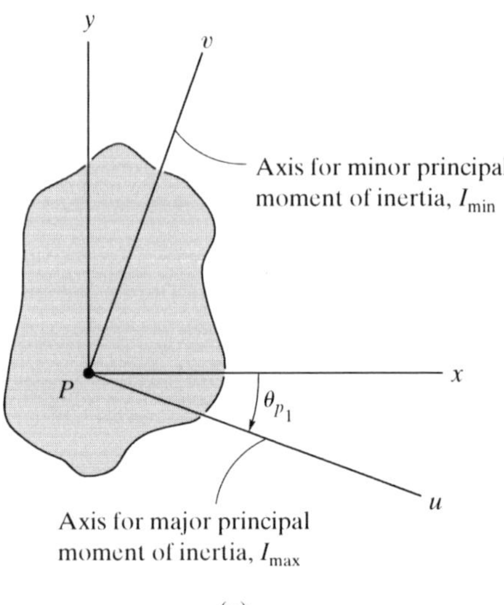

(a)

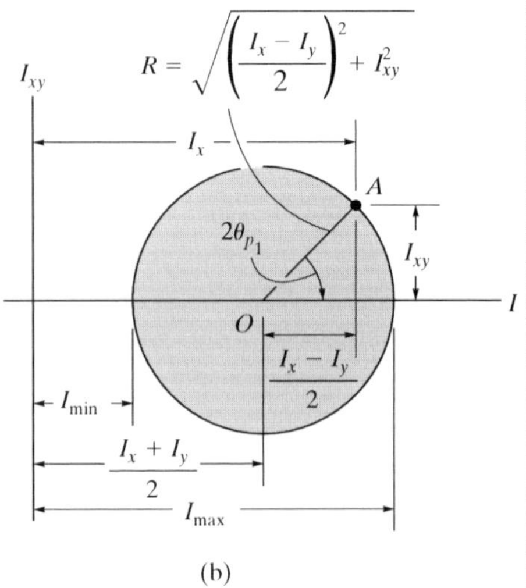

(b)

Fig. 10–19 (Repeated)

Procedure for Analysis

The main purpose in using Mohr's circle here is to have a convenient means for finding the principal moments of inertia for an area. The following procedure provides a method for doing this.

Determine I_x, I_y, and I_{xy}.

- Establish the x, y axes and determine I_x, I_y, and I_{xy}, Fig. 10–19*a*.

Construct the Circle.

- Construct a rectangular coordinate system such that the abscissa represents the moment of inertia I, and the ordinate represents the product of inertia I_{xy}, Fig. 10–19*b*.
- Determine the center of the circle, O, which is located at a distance $(I_x + I_y)/2$ from the origin, and plot the reference point A having coordinates (I_x, I_{xy}). Remember, I_x is always positive, whereas I_{xy} can be either positive or negative.
- Connect the reference point A with the center of the circle and determine the distance OA by trigonometry. This distance represents the radius of the circle, Fig. 10–19*b*. Finally, draw the circle.

Principal Moments of Inertia.

- The points where the circle intersects the I axis give the values of the principal moments of inertia I_{min} and I_{max}. Notice that, as expected, the *product of inertia will be zero at these points*, Fig. 10–19*b*.

Principal Axes.

- To find the orientation of the major principal axis, use trigonometry to find the angle $2\theta_{p_1}$, *measured from the radius OA to the positive I axis*, Fig. 10–19*b*. This angle represents *twice* the angle from the x axis to the axis of maximum moment of inertia I_{max}, Fig. 10–19*a*. Both the angle on the circle, $2\theta_{p_1}$, and the angle θ_{p_1} *must be measured in the same sense*, as shown in Fig. 10–19. The axis for minimum moment of inertia I_{min} is perpendicular to the axis for I_{max}.

10

Using trigonometry, the above procedure can be verified to be in accordance with the equations developed in Sec. 10.6.

EXAMPLE 10.9

Using Mohr's circle, determine the principal moments of inertia and the orientation of the major principal axes for the cross-sectional area of the member shown in Fig. 10–20*a*, with respect to an axis passing through the centroid.

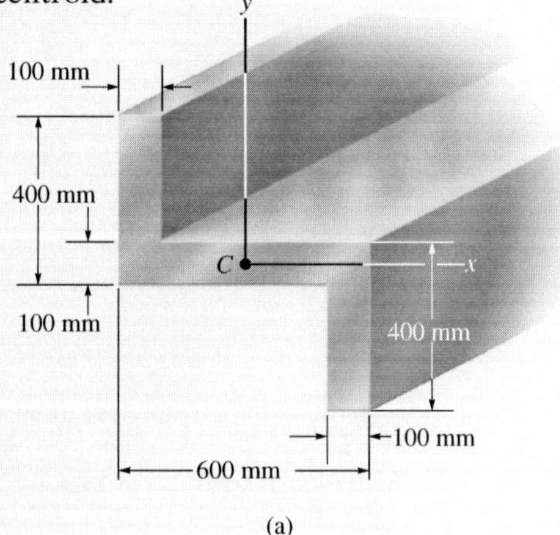

(a)

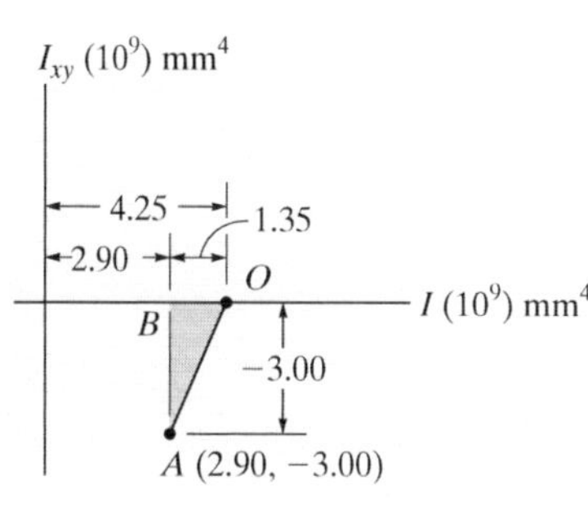

(b)

SOLUTION

Determine I_x, I_y, I_{xy}. The moments and product of inertia have been determined in Examples 10.5 and 10.7 with respect to the x, y axes shown in Fig. 10–20*a*. The results are $I_x = 2.90(10^9)\text{ mm}^4$, $I_y = 5.60(10^9)\text{ mm}^4$, and $I_{xy} = -3.00(10^9)\text{ mm}^4$.

Construct the Circle. The I and I_{xy} axes are shown in Fig. 10–20*b*. The center of the circle, O, lies at a distance $(I_x + I_y)/2 = (2.90 + 5.60)/2 = 4.25$ from the origin. When the reference point $A(I_x, I_{xy})$ or $A(2.90,-3.00)$ is connected to point O, the radius OA is determined from the triangle OBA using the Pythagorean theorem.

$$OA = \sqrt{(1.35)^2 + (-3.00)^2} = 3.29$$

The circle is constructed in Fig. 10–20*c*.

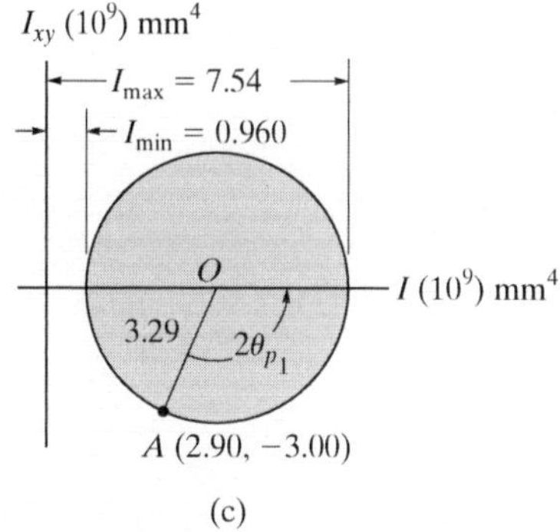

(c)

Principal Moments of Inertia. The circle intersects the I axis at points (7.54, 0) and (0.960, 0). Hence,

$$I_{max} = (4.25 + 3.29)10^9 = 7.54(10^9)\text{ mm}^4 \qquad \textit{Ans.}$$

$$I_{min} = (4.25 - 3.29)10^9 = 0.960(10^9)\text{ mm}^4 \qquad \textit{Ans.}$$

Principal Axes. As shown in Fig. 10–20*c*, the angle $2\theta_{p_1}$ is determined from the circle by measuring counterclockwise from OA to the direction of the *positive I* axis. Hence,

$$2\theta_{p_1} = 180^\circ - \sin^{-1}\left(\frac{|BA|}{|OA|}\right) = 180^\circ - \sin^{-1}\left(\frac{3.00}{3.29}\right) = 114.2^\circ$$

The principal axis for $I_{max} = 7.54(10^9)\text{ mm}^4$ is therefore oriented at an angle $\theta_{p_1} = 57.1^\circ$, measured *counterclockwise*, from the *positive x* axis to the *positive u* axis. The v axis is perpendicular to this axis. The results are shown in Fig. 10–20*d*.

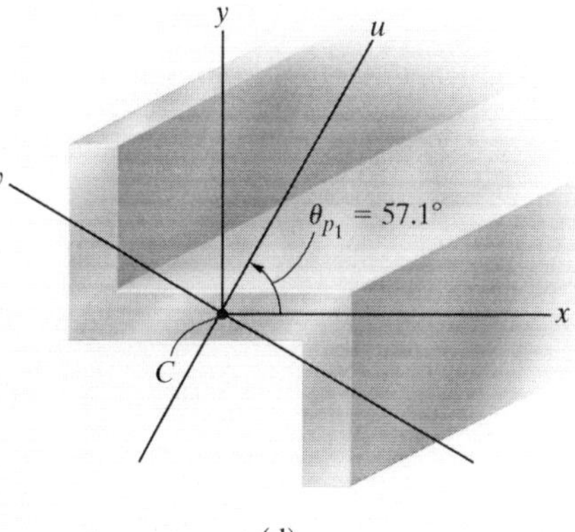

(d)

Fig. 10–20

10

PROBLEMS

10–54. Determine the product of inertia of the thin strip of area with respect to the x and y axes. The strip is oriented at an angle θ from the x axis. Assume that $t \ll l$.

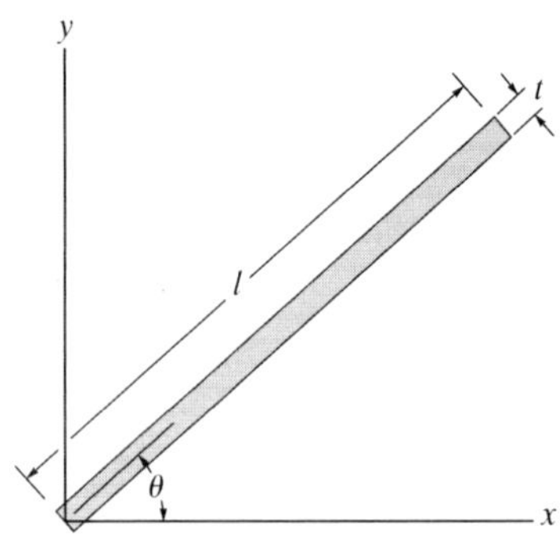

Prob. 10–54

10–55. Determine the product of inertia of the shaded area with respect to the x and y axes.

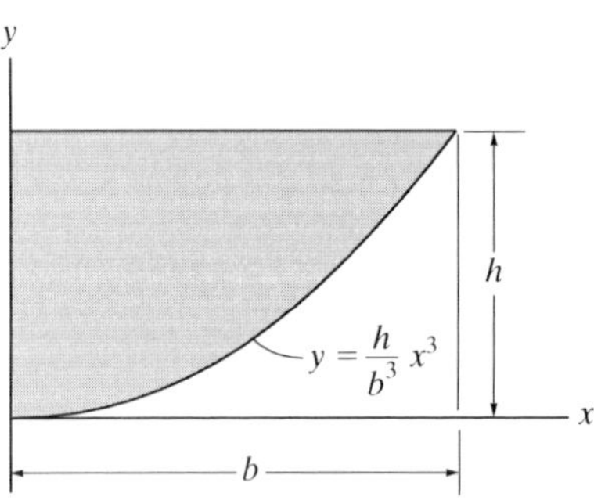

Prob. 10–55

***10–56.** Determine the product of inertia of the shaded portion of the parabola with respect to the x and y axes.

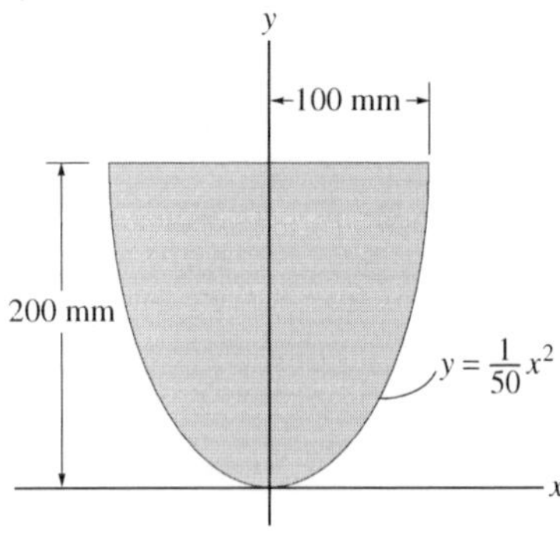

Prob. 10–56

10–57. Determine the product of inertia of the shaded area with respect to the x and y axes.

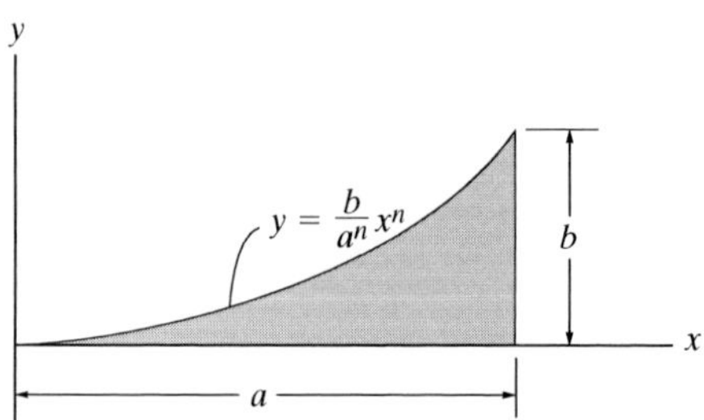

Prob. 10–57

10–58. Determine the product of inertia of the shaded area with respect to the x and y axes, and then use the parallel-axis theorem to find the product of inertia of the area with respect to the centroidal x' and y' axes.

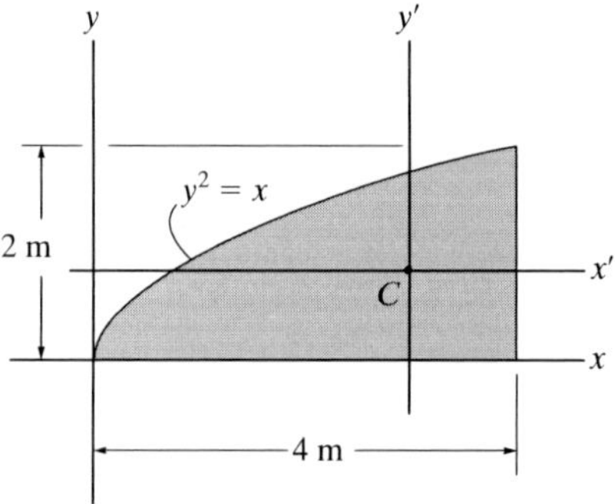

Prob. 10–58

10–59. Determine the product of inertia of the shaded area with respect to the x and y axes. Use Simpson's rule to evaluate the integral.

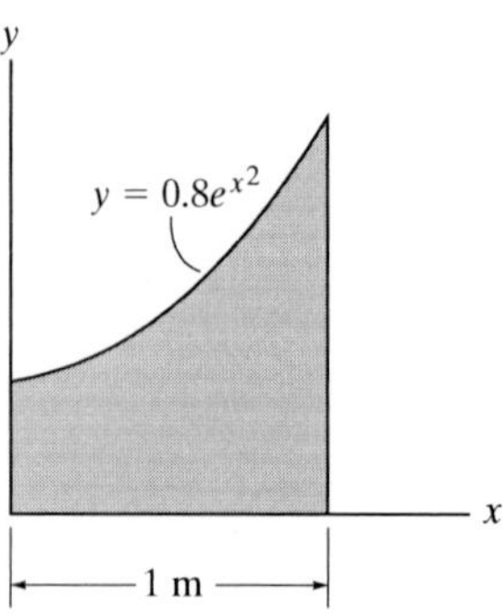

Prob. 10–59

*10–60. Determine the product of inertia of the shaded area with respect to the x and y axes.

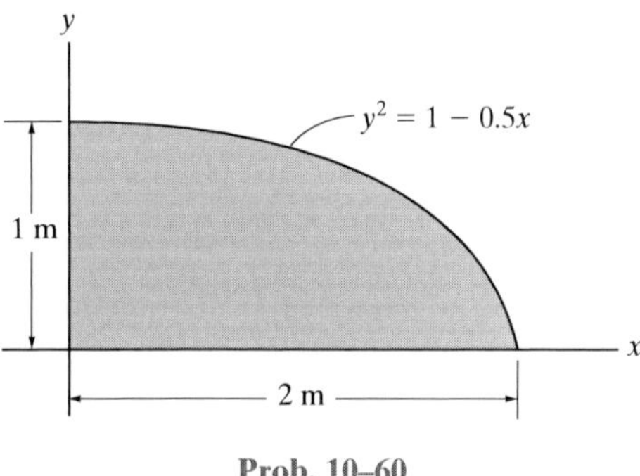

Prob. 10–60

10–61. Determine the product of inertia of the parallelogram with respect to the x and y axes.

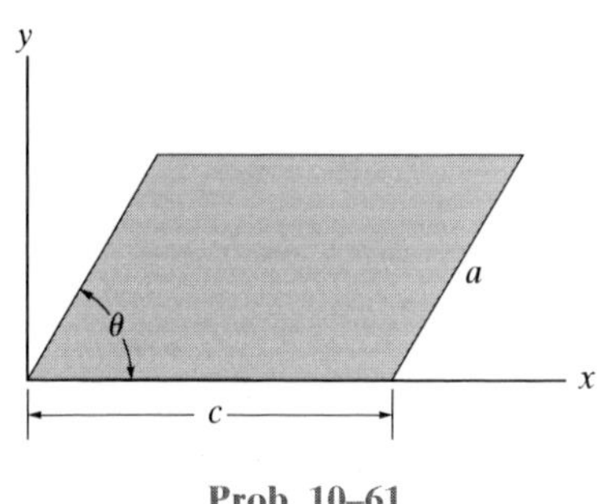

Prob. 10–61

10–62. Determine the product of inertia of the parabolic area with respect to the x and y axes.

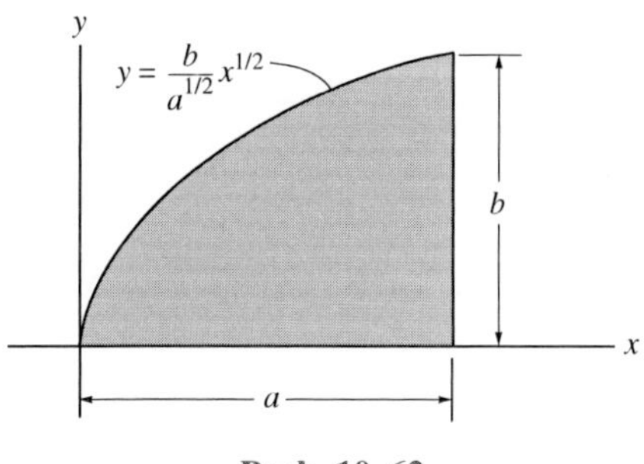

Prob. 10–62

10–63. Determine the product of inertia for the beam's cross-sectional area with respect to the u and v axes.

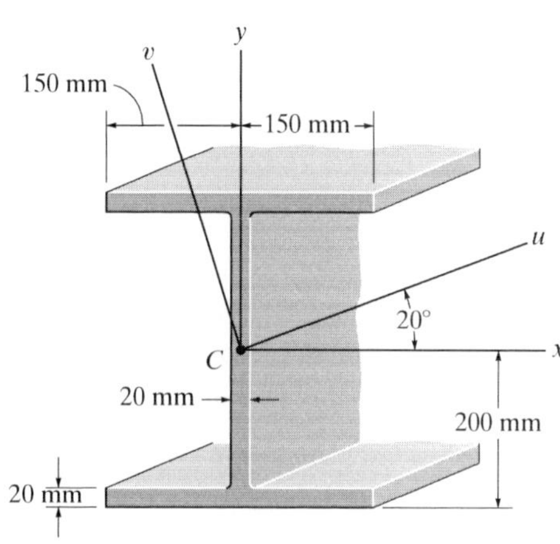

Prob. 10–63

*10–64. Determine the product of inertia of the beam's cross-sectional area with respect to the x and y axes that have their origin located at the centroid C.

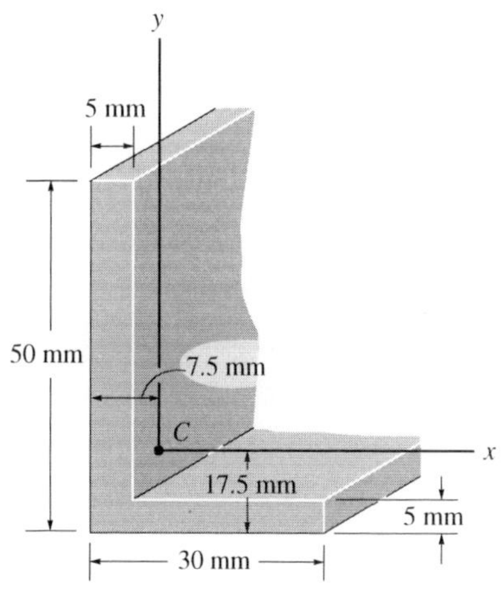

Prob. 10–64

10

10–65. Determine the product of inertia for the beam's cross-sectional area with respect to the x and y axes that have their origin located at the centroid C.

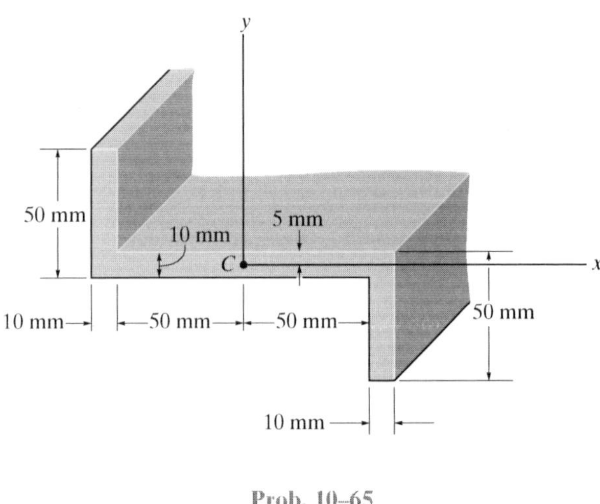

Prob. 10–65

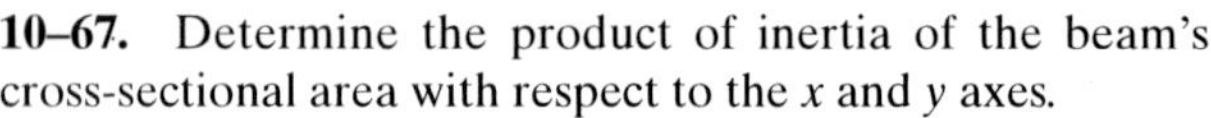
10–67. Determine the product of inertia of the beam's cross-sectional area with respect to the x and y axes.

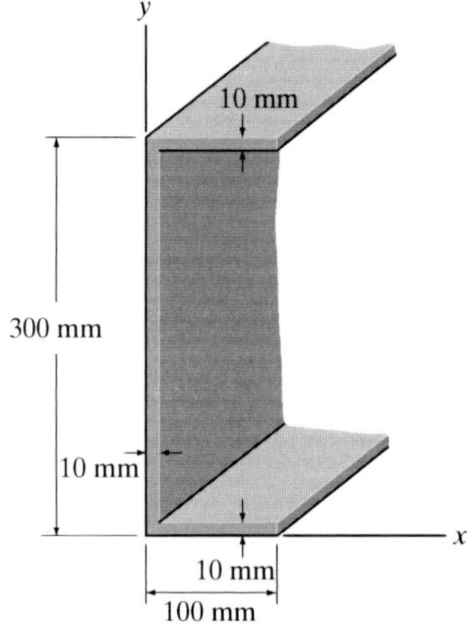

Prob. 10–67

10–66. Determine the product of inertia of the cross-sectional area with respect to the x and y axes.

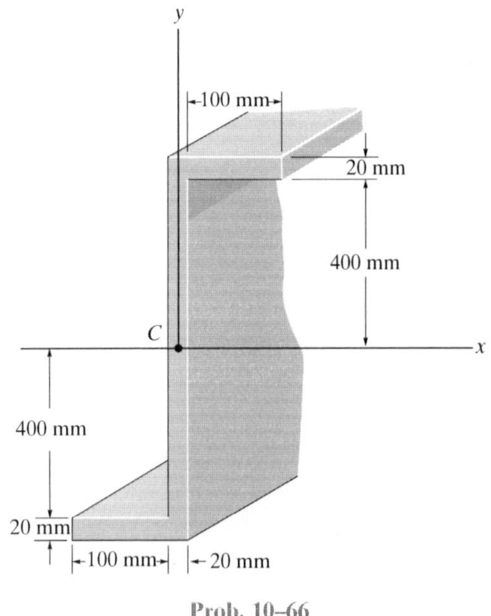

Prob. 10–66

***10–68.** Determine the distance $\bar{y}$ to the centroid of the area and then calculate the moments of inertia I_u and I_v of the channel's cross-sectional area. The u and v axes have their origin at the centroid C. For the calculation, assume all corners to be square.

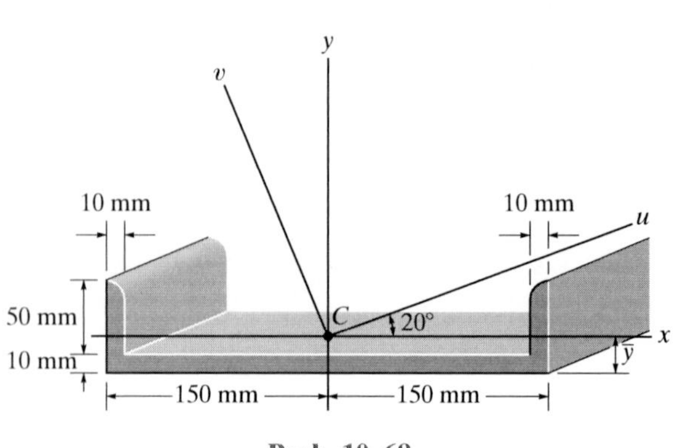

Prob. 10–68

10

10–69. Determine the moments of inertia I_u and I_v of the shaded area.

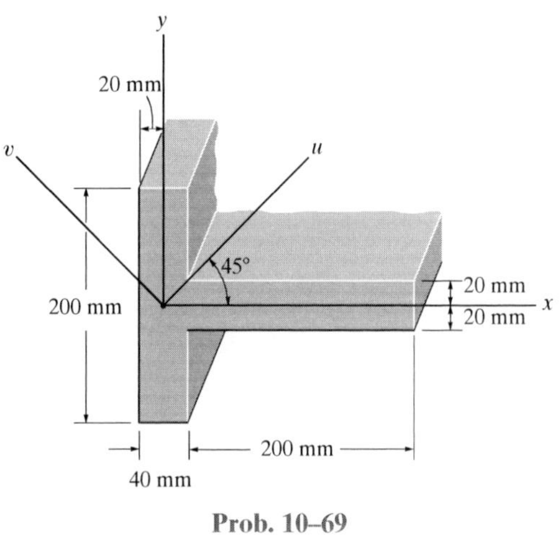

Prob. 10–69

10–70. Determine the moments of inertia and the product of inertia of the beam's cross sectional area with respect to the u and v axes.

10–71. Solve Prob. 10–70 using Mohr's circle. *Hint:* Once the circle is established, rotate $2\theta = 60°$ counterclockwise from the reference OA, then find the coordinates of the points that define the diameter of the circle.

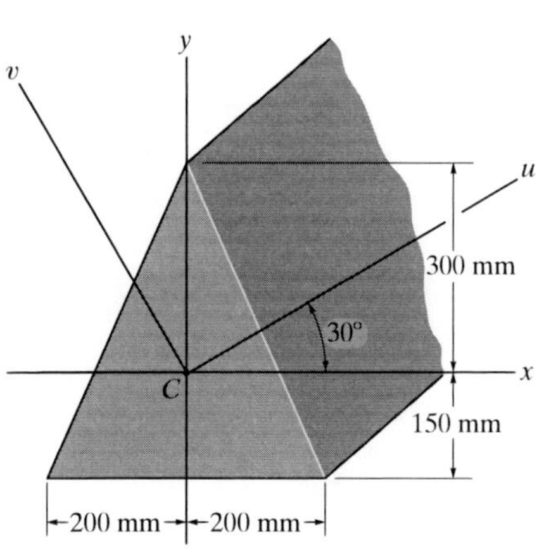

Probs. 10–70/71

*__10–72.__ Locate the centroid $\bar{y}$ of the beam's cross-sectional area and then determine the moments of inertia and the product of inertia of this area with respect to the u and v axes.

10–73. Solve Prob. 10–72 using Mohr's circle.

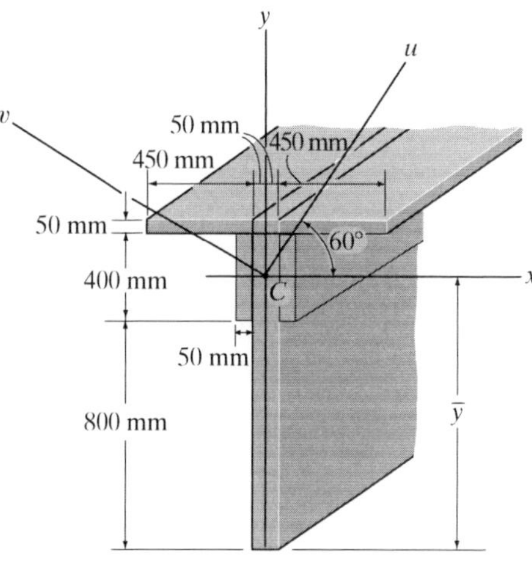

Probs. 10–72/73

10–74. Locate the centroid $\bar{y}$ of the beam's cross-sectional area and then determine the moments of inertia of this area and the product of inertia with respect to the u and v axes. The axes have their origin at the centroid C.

10–75. Solve Prob. 10–74 using Mohr's circle.

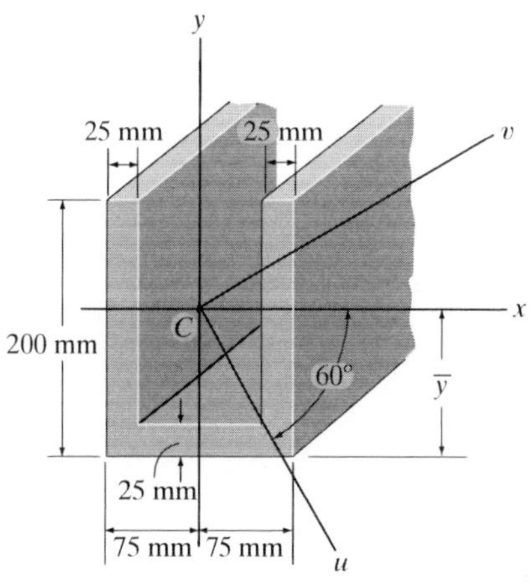

Probs. 10–74/75

10

***10–76.** Locate the centroid $\bar{x}$ of the beam's cross-sectional area and then determine the moments of inertia and the product of inertia of this area with respect to the u and v axes. The axes have their origin at the centroid C.

10–77. Solve Prob. 10–76 using Mohr's circle.

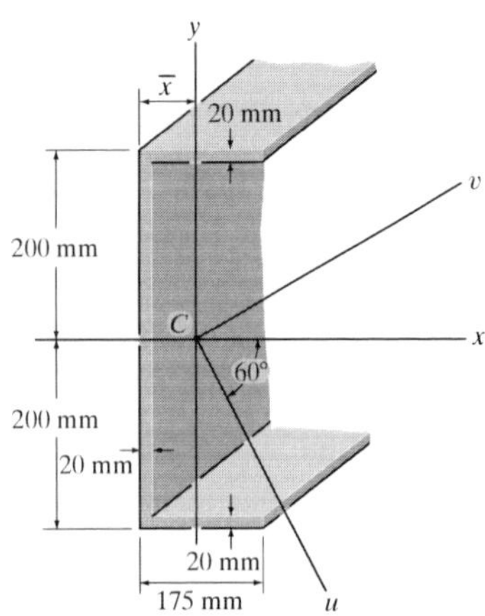

Probs. 10–76/77

10–78. Determine the principal moments of inertia for the angle's cross-sectional area with respect to a set of principal axes that have their origin located at the centroid C. Use the equation developed in Section 10.7. For the calculation, assume all corners to be square.

10–79. Solve Prob. 10–78 using Mohr's circle.

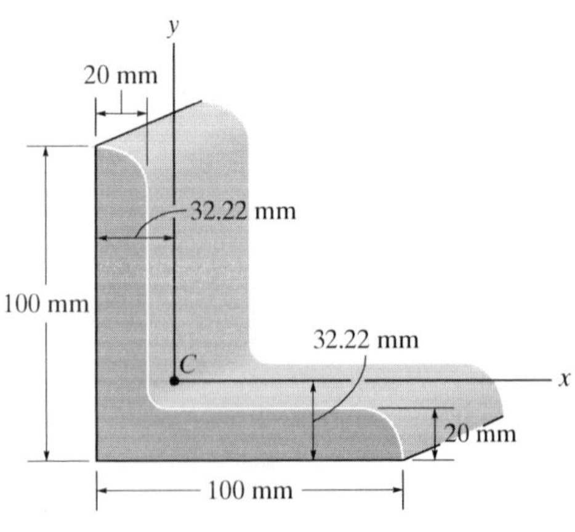

Probs. 10–78/79

***10–80.** Determine the orientation of the principal axes, which have their origin at centroid C of the beam's cross-sectional area. Also, find the principal moments of inertia.

10–81. Solve Prob. 10–80 using Mohr's circle.

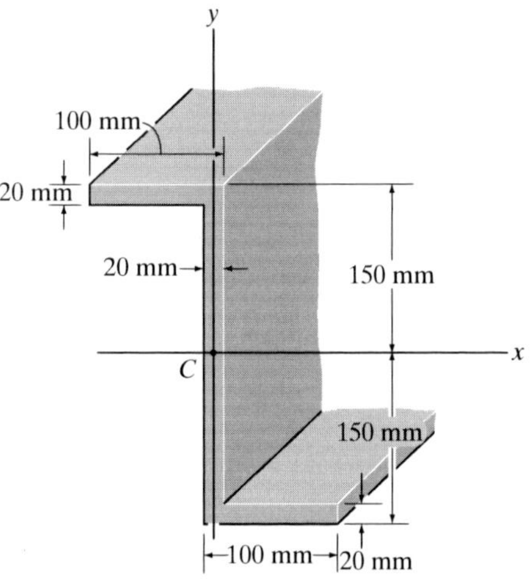

Probs. 10–80/81

10–82. Locate the centroid $\bar{y}$ of the beam's cross-sectional area and then determine the moments of inertia of this area and the product of inertia with respect to the u and v axes. The axes have their origin at the centroid C.

10–83. Solve Prob. 10–82 using Mohr's circle.

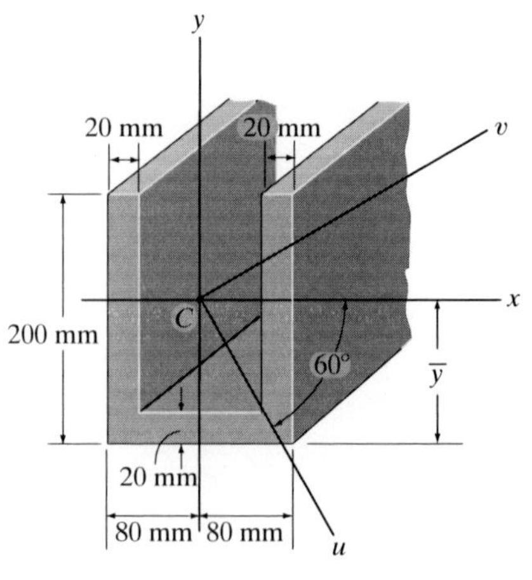

Probs. 10–82/83

10.8 Mass Moment of Inertia

The mass moment of inertia of a body is a measure of the body's resistance to angular acceleration. Since it is used in dynamics to study rotational motion, methods for its calculation will now be discussed.*

Consider the rigid body shown in Fig. 10–21. We define the *mass moment of inertia* of the body about the z axis as

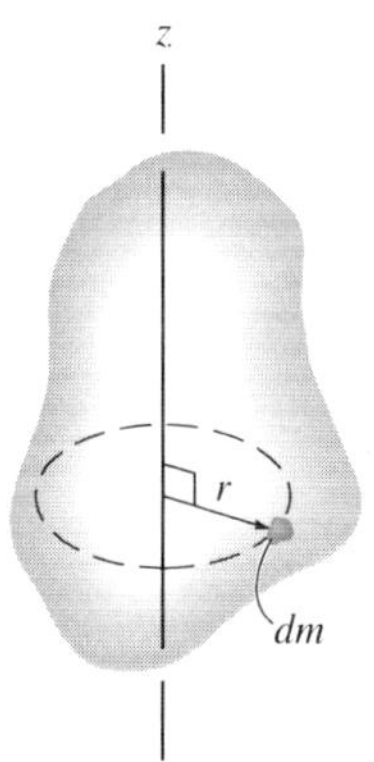

Fig. 10–21

$$I = \int_m r^2\, dm \tag{10–12}$$

Here r is the perpendicular distance from the axis to the arbitrary element dm. Since the formulation involves r, the value of I is *unique* for each axis about which it is computed. The axis which is generally chosen, however, passes through the body's mass center G. Common units used for its measurement are $\text{kg} \cdot \text{m}^2$.

If the body consists of material having a density ρ, then $dm = \rho\, dV$, Fig. 10–22*a*. Substituting this into Eq. 10–12, the body's moment of inertia is then computed using *volume elements* for integration; i.e.

$$I = \int_V r^2 \rho\, dV \tag{10–13}$$

For most applications, ρ will be a *constant*, and so this term may be factored out of the integral, and the integration is then purely a function of geometry.

$$I = \rho \int_V r^2\, dV \tag{10–14}$$

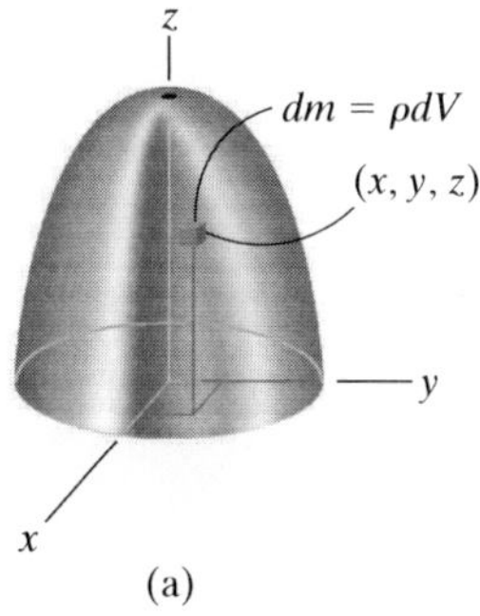

Fig. 10–22

*Another property of the body, which measures the symmetry of the body's mass with respect to a coordinate system, is the mass product of inertia. This property most often applies to the three-dimensional motion of a body and is discussed in *Engineering Mechanics: Dynamics* (Chapter 21).

10

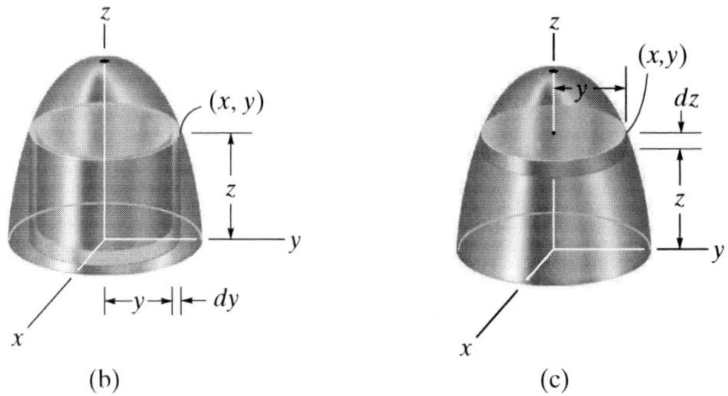

Fig. 10–22 (cont'd)

Procedure for Analysis

If a body is symmetrical with respect to an axis, as in Fig. 10–22, then its mass moment of inertia about the axis can be determined by using a single integration. Shell and disk elements are used for this purpose.

Shell Element.

- If a *shell element* having a height z, radius y, and thickness dy is chosen for integration, Fig. 10–22*b*, then its volume is $dV = (2\pi y)(z)\,dy$.
- This element can be used in Eq. 10–13 or 10–14 for determining the moment of inertia I_z of the body about the z axis since the *entire element*, due to its "thinness," lies at the *same* perpendicular distance $r = y$ from the z axis (see Example 10.10).

Disk Element.

- If a disk element having a radius y and a thickness dz is chosen for integration, Fig. 10–22*c*, then its volume is $dV = (\pi y^2)\,dz$.
- In this case the element is *finite* in the radial direction, and consequently its points *do not* all lie at the *same radial distance r* from the z axis. As a result, Eqs. 10–13 or 10–14 *cannot* be used to determine I_z. Instead, to perform the integration using this element, it is first necessary to determine the moment of inertia *of the element* about the z axis and then integrate this result (see Example 10.11).

10

EXAMPLE 10.10

Determine the mass moment of inertia of the cylinder shown in Fig. 10–23*a* about the z axis. The density of the material, ρ, is constant.

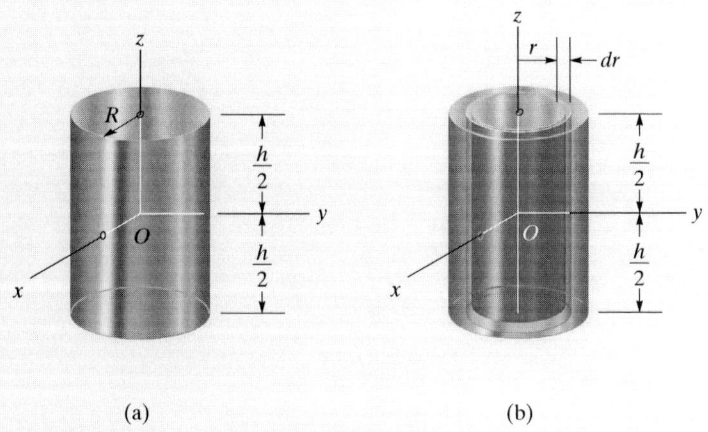

Fig. 10–23

SOLUTION

Shell Element. This problem will be solved using the *shell element* in Fig. 10–23*b* and thus only a single integration is required. The volume of the element is $dV = (2\pi r)(h)\,dr$, and so its mass is $dm = \rho\,dV = \rho(2\pi hr\,dr)$. Since the *entire element* lies at the same distance r from the z axis, the moment of inertia *of the element* is

$$dI_z = r^2\,dm = \rho 2\pi h r^3\,dr$$

Integrating over the entire cylinder yields

$$I_z = \int_m r^2\,dm = \rho 2\pi h \int_0^R r^3\,dr = \frac{\rho\pi}{2}R^4 h$$

Since the mass of the cylinder is

$$m = \int_m dm = \rho 2\pi h \int_0^R r\,dr = \rho\pi h R^2$$

then

$$I_z = \frac{1}{2}mR^2 \qquad \textit{Ans.}$$

10

EXAMPLE 10.11

A solid is formed by revolving the shaded area shown in Fig. 10–24*a* about the *y* axis. If the density of the material is 2 Mg/m^3, determine the mass moment of inertia about the *y* axis.

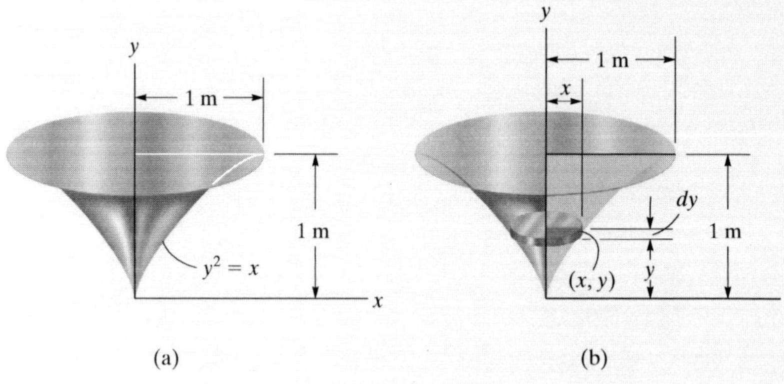

Fig. 10–24

SOLUTION

Disk Element. The moment of inertia will be determined using this *disk element*, as shown in Fig. 10–24*b*. Here the element intersects the curve at the arbitrary point (x, y) and has a mass

$$dm = \rho\, dV = \rho(\pi x^2)\, dy$$

Although all points on the element are *not* located at the same distance from the *y* axis, it is still possible to determine the moment of inertia dI_y *of the element* about the *y* axis. In the previous example it was shown that the moment of inertia of a homogeneous cylinder about its longitudinal axis is $I = \frac{1}{2}mR^2$, where m and R are the mass and radius of the cylinder. Since the height of the cylinder is not involved in this formula, we can also use this result for a disk. Thus, for the disk element in Fig. 10–24*b*, we have

$$dI_y = \frac{1}{2}(dm)x^2 = \frac{1}{2}[\rho(\pi x^2)\, dy]x^2$$

Substituting $x = y^2$, $\rho = 2\ \text{Mg/m}^3$, and integrating with respect to *y*, from $y = 0$ to $y = 1$ ft, yields the moment of inertia for the entire solid.

$$I_y = \frac{2\pi}{2}\int_0^{1\text{ m}} x^4\, dy = \frac{2\pi}{2}\int_0^{1\text{ m}} y^8\, dy = 0.349\ \text{Mg}\cdot\text{m}^2 = 349\ \text{kg}\cdot\text{m}^2 \quad \textit{Ans.}$$

10

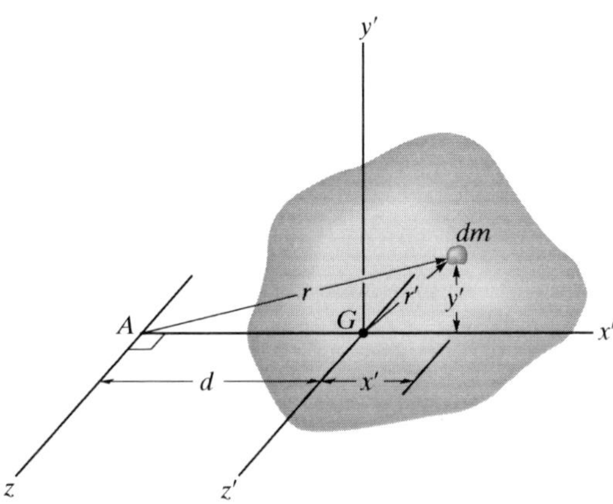

Fig. 10–25

Parallel-Axis Theorem. If the moment of inertia of the body about an axis passing through the body's mass center is known, then the moment of inertia about any other *parallel axis* can be determined by using the *parallel-axis theorem*. To derive this theorem, consider the body shown in Fig. 10–25. The z' axis passes through the mass center G, whereas the corresponding *parallel* z *axis* lies at a constant distance d away. Selecting the differential element of mass dm, which is located at point (x', y'), and using the Pythagorean theorem, $r^2 = (d + x')^2 + y'^2$, the moment of inertia of the body about the z axis is

$$I = \int_m r^2\, dm = \int_m [(d + x')^2 + y'^2]\, dm$$

$$= \int_m (x'^2 + y'^2)\, dm + 2d \int_m x'\, dm + d^2 \int_m dm$$

Since $r'^2 = x'^2 + y'^2$, the first integral represents I_G. The second integral is equal to *zero*, since the z' axis passes through the body's mass center, i.e., $\int x'\, dm = \bar{x} \int dm = 0$ since $\bar{x} = 0$. Finally, the third integral is the total mass m of the body. Hence, the moment of inertia about the z axis becomes

$$I = I_G + md^2 \tag{10–15}$$

where

I_G = moment of inertia about the z' axis passing through the mass center G

m = mass of the body

d = distance between the parallel axes

10

Radius of Gyration. Occasionally, the moment of inertia of a body about a specified axis is reported in handbooks using the *radius of gyration, k*. This value has units of length, and when it and the body's mass m are known, the moment of inertia can be determined from the equation

$$I = mk^2 \quad \text{or} \quad k = \sqrt{\frac{I}{m}} \tag{10–16}$$

Note the *similarity* between the definition of k in this formula and r in the equation $dI = r^2\,dm$, which defines the moment of inertia of a differential element of mass dm of the body about an axis.

Composite Bodies. If a body is constructed from a number of simple shapes such as disks, spheres, and rods, the moment of inertia of the body about any axis z can be determined by adding algebraically the moments of inertia of all the composite shapes calculated about the same axis. Algebraic addition is necessary since a composite part must be considered as a negative quantity if it has already been included within another part—as in the case of a "hole" subtracted from a solid plate. Also, the parallel-axis theorem is needed for the calculations if the center of mass of each composite part does not lie on the z axis. For calculations, a table of some simple shapes is given on the inside back cover.

This flywheel, which operates a metal cutter, has a large moment of inertia about its center. Once it begins rotating it is difficult to stop it and therefore a uniform motion can be effectively transferred to the cutting blade.

10

EXAMPLE 10.12

If the plate shown in Fig. 10–26*a* has a density of 8000 kg/m^3 and a thickness of 10 mm, determine its mass moment of inertia about an axis perpendicular to the page and passing through the pin at O.

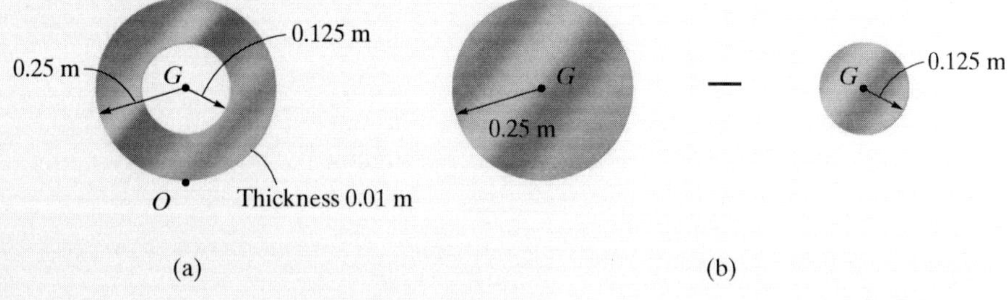

Fig. 10–26

SOLUTION

The plate consists of two composite parts, the 250-mm-radius disk *minus* a 125-mm-radius disk, Fig. 10–26*b*. The moment of inertia about O can be determined by finding the moment of inertia of each of these parts about O and then *algebraically* adding the results. The calculations are performed by using the parallel-axis theorem in conjunction with the mass moment of inertia formula for a circular disk, $I_G = \frac{1}{2}mr^2$, as found on the inside back cover.

Disk. The moment of inertia of a disk about an axis perpendicular to the plane of the disk and passing through G is $I_G = \frac{1}{2}mr^2$. The mass center of both disks is 0.25 m from point O. Thus,

$$m_d = \rho_d V_d = 8000\ \text{kg/m}^3\ [\pi(0.25\ \text{m})^2(0.01\ \text{m})] = 15.71\ \text{kg}$$

$$\begin{aligned}(I_O)_d &= \tfrac{1}{2}m_d r_d^2 + m_d d^2 \\ &= \tfrac{1}{2}(15.71\ \text{kg})(0.25\ \text{m})^2 + (15.71\ \text{kg})(0.25\ \text{m})^2 \\ &= 1.473\ \text{kg}\cdot\text{m}^2\end{aligned}$$

Hole. For the smaller disk (hole), we have

$$m_h = \rho_h V_h = 8000\ \text{kg/m}^3\ [\pi(0.125\ \text{m})^2(0.01\ \text{m})] = 3.93\ \text{kg}$$

$$\begin{aligned}(I_O)_h &= \tfrac{1}{2}m_h r_h^2 + m_h d^2 \\ &= \tfrac{1}{2}(3.93\ \text{kg})(0.125\ \text{m})^2 + (3.93\ \text{kg})(0.25\ \text{m})^2 \\ &= 0.276\ \text{kg}\cdot\text{m}^2\end{aligned}$$

The moment of inertia of the plate about the pin is therefore

$$\begin{aligned}I_O &= (I_O)_d - (I_O)_h \\ &= 1.473\ \text{kg}\cdot\text{m}^2 - 0.276\ \text{kg}\cdot\text{m}^2 \\ &= 1.20\ \text{kg}\cdot\text{m}^2\end{aligned}$$

Ans.

10

EXAMPLE 10.13

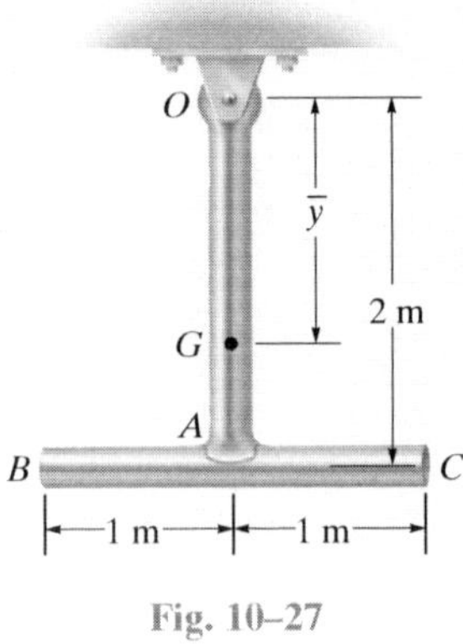

Fig. 10–27

The pendulum in Fig. 10–27 consists of two thin rods each having a mass of 9 kg. Determine the pendulum's mass moment of inertia about an axis passing through (a) the pin at O, and (b) the mass center G of the pendulum.

SOLUTION

Part (a). Using the table on the inside back cover, the moment of inertia of rod OA about an axis perpendicular to the page and passing through the end point O of the rod is $I_O = \frac{1}{3}ml^2$. Hence,

$$(I_{OA})_O = \frac{1}{3}ml^2 = \frac{1}{3}(9\text{ kg})(2\text{ m})^2 = 12\text{ kg}\cdot\text{m}^2$$

Realize that this same value may be determined using $I_G = \frac{1}{12}ml^2$ and the parallel-axis theorem; i.e.,

$$(I_{OA})_O = \frac{1}{12}ml^2 + md^2 = \frac{1}{12}(9\text{ kg})(2\text{ m})^2 + 9\text{ kg}(1\text{ m})^2$$
$$= 12\text{ kg}\cdot\text{m}^2$$

For rod BC we have

$$(I_{BC})_O = \frac{1}{12}ml^2 + md^2 = \frac{1}{12}(9\text{ kg})(2\text{ m})^2 + 9\text{ kg}(2\text{ m})^2$$
$$= 39\text{ kg}\cdot\text{m}^2$$

The moment of inertia of the pendulum about O is therefore

$$I_O = 12 + 39 = 51\text{ kg}\cdot\text{m}^2 \qquad \textit{Ans.}$$

Part (b). The mass center G will be located relative to the pin at O. Assuming this distance to be $\bar{y}$, Fig. 10–27, and using the formula for determining the mass center, we have

$$\bar{y} = \frac{\Sigma\tilde{y}m}{\Sigma m} = \frac{1(9) + 2(9)}{(9) + (9)} = 1.50\text{ m}$$

The moment of inertia I_G may be computed in the same manner as I_O, which requires successive applications of the parallel-axis theorem in order to transfer the moments of inertia of rods OA and BC to G. A more direct solution, however, involves applying the parallel-axis theorem using the result for I_O determined above; i.e.,

$$I_O = I_G + md^2; \qquad 51\text{ kg}\cdot\text{m}^2 = I_G + (18\text{ kg})(1.50\text{ m})^2$$

$$I_G = 10.5\text{ kg}\cdot\text{m}^2 \qquad \textit{Ans.}$$

10

PROBLEMS

***10–84.** Determine the moment of inertia of the thin ring about the z axis. The ring has a mass m.

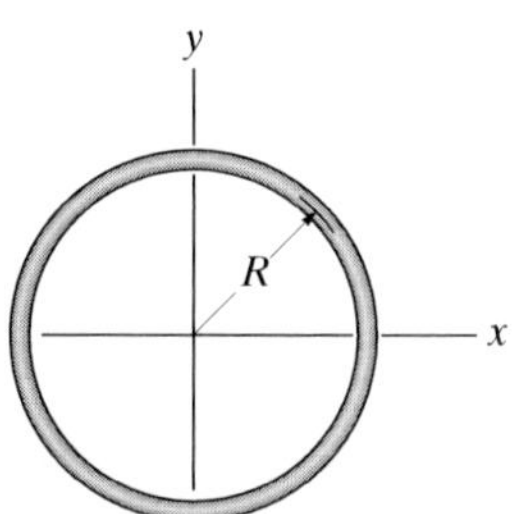

Prob. 10–84

10–85. Determine the moment of inertia of the semi-ellipsoid with respect to the x axis and express the result in terms of the mass m of the semiellipsoid. The material has a constant density ρ.

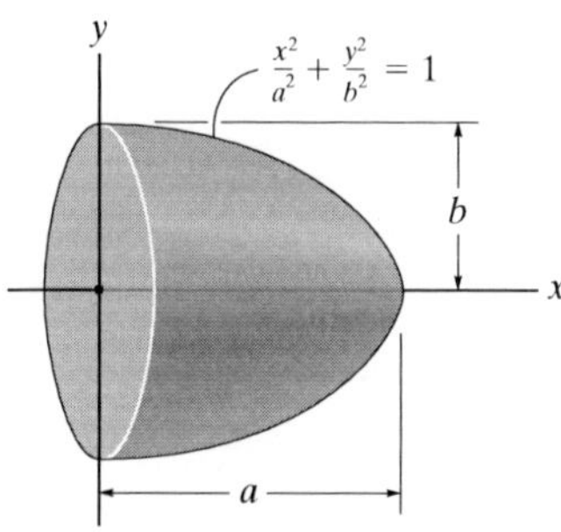

Prob. 10–85

10–86. Determine the moment of inertia I_x of the right circular cone and express the result in terms of the total mass m of the cone. The cone has a constant density ρ.

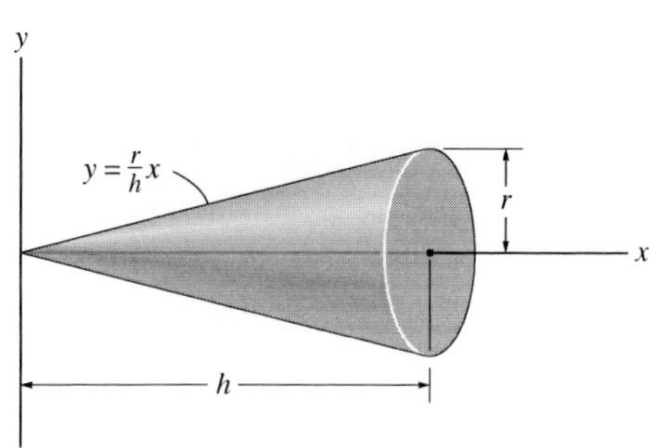

Prob. 10–86

10–87. Determine the radius of gyration k_x of the paraboloid. The density of the material is $\rho = 5\ \text{Mg/m}^3$.

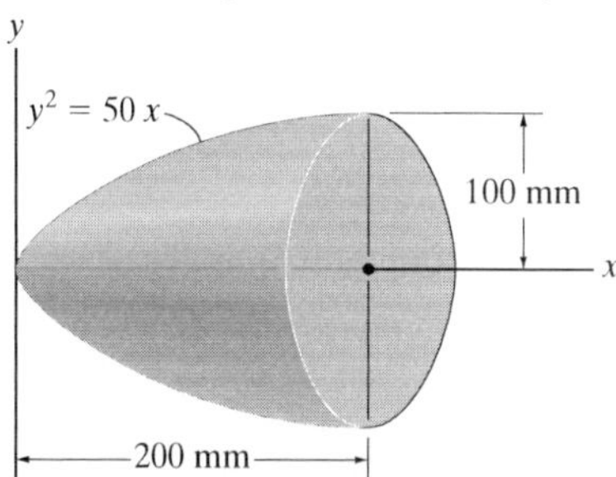

Prob. 10–87

***10–88.** Determine the moment of inertia of the ellipsoid with respect to the x axis and express the result in terms of the mass m of the ellipsoid. The material has a constant density ρ.

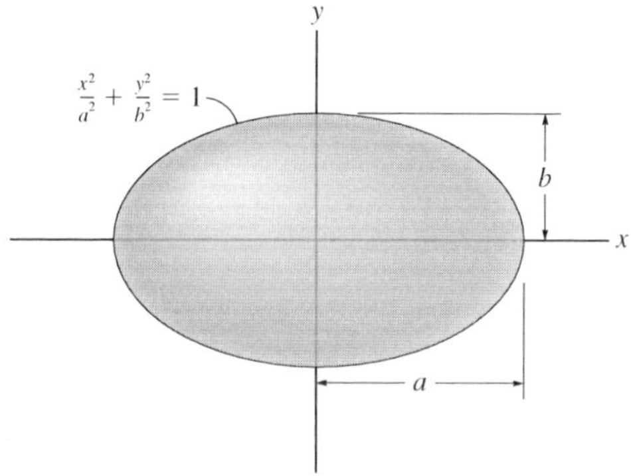

Prob. 10–88

10–89. Determine the moment of inertia of the homogenous triangular prism with respect to the y axis. Express the result in terms of the mass m of the prism. *Hint:* For integration, use thin plate elements parallel to the x–y plane having a thickness of dz.

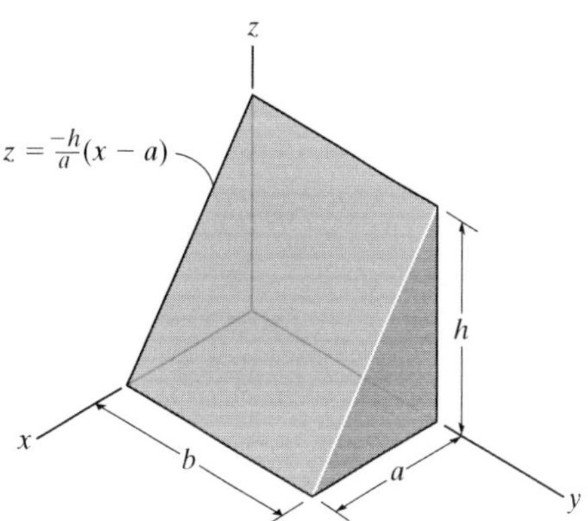

Prob. 10–89

10–90. Determine the mass moment of inertia I_z of the solid formed by revolving the shaded area around the z axis. The density of the materials is ρ. Express the result in terms of the mass m of the solid.

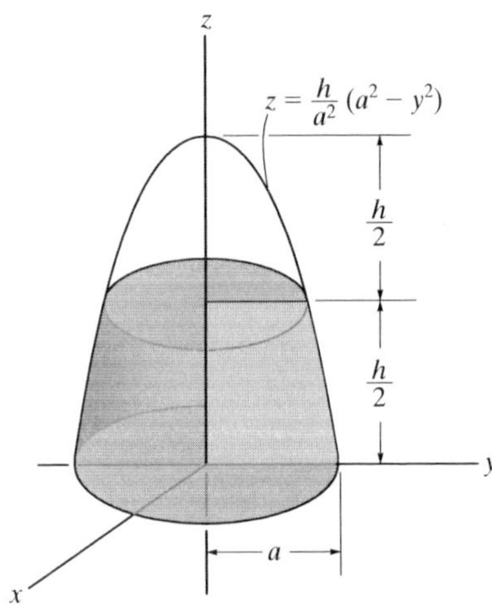

Prob. 10–90

10–91. Determine the moment of inertia I_x of the sphere and express the result in terms of the total mass m of the sphere. The sphere has a constant density ρ.

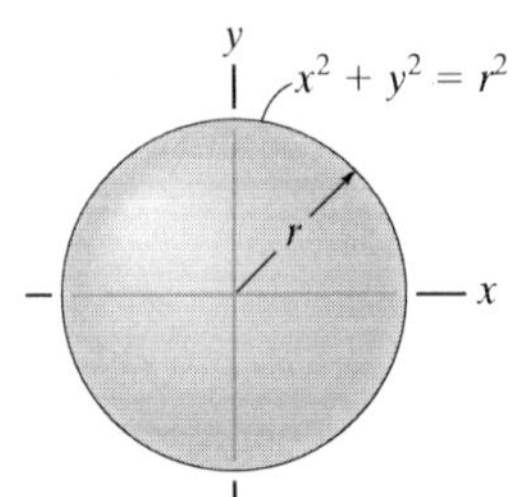

Prob. 10–91

***10–92.** Determine the mass moment of inertia of the 2-kg bent rod about the z axis.

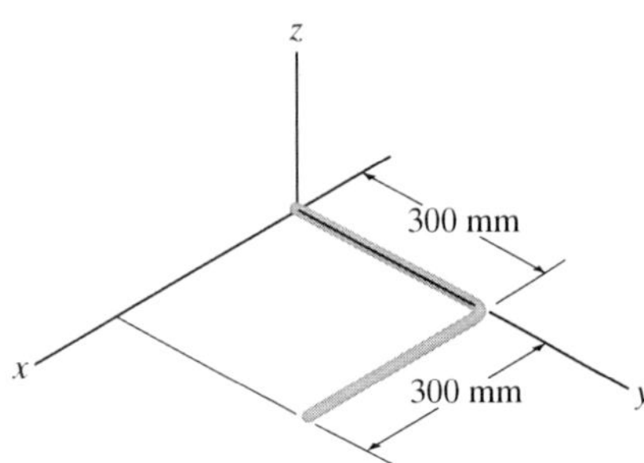

Prob. 10–92

10–93. Determine the mass moment of inertia I_y of the solid formed by revolving the shaded area around the y axis. The density of the material is ρ. Express the result in terms of the mass m of the solid.

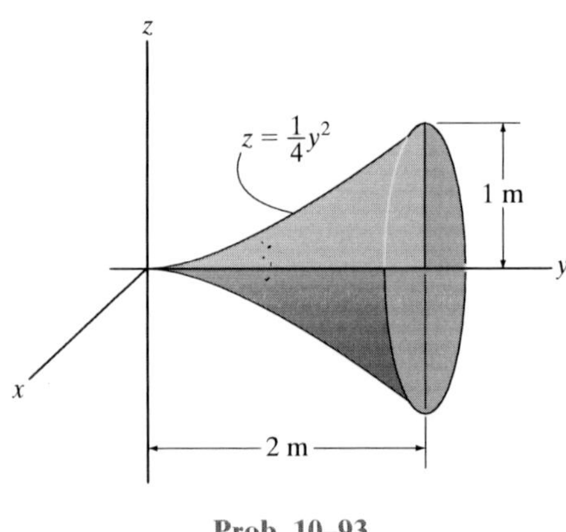

Prob. 10–93

10–94. Determine the mass moment of inertia I_y of the solid formed by revolving the shaded area around the y axis. The total mass of the solid is 1500 kg.

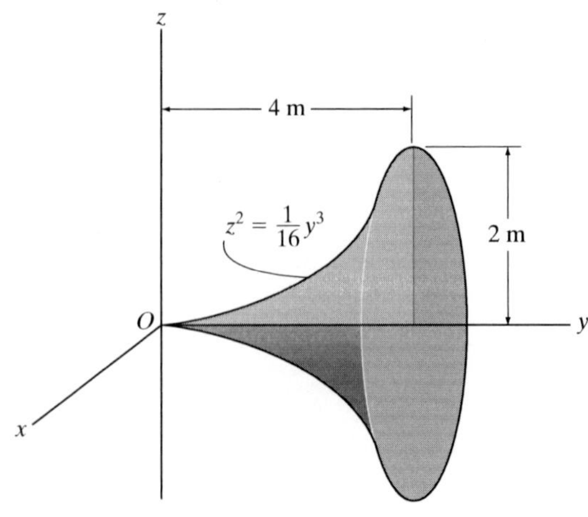

Prob. 10–94

10–95. Determine the moment of inertia I_z of the frustrum of the cone which has a conical depression. The material has a density of 200 kg/m^3.

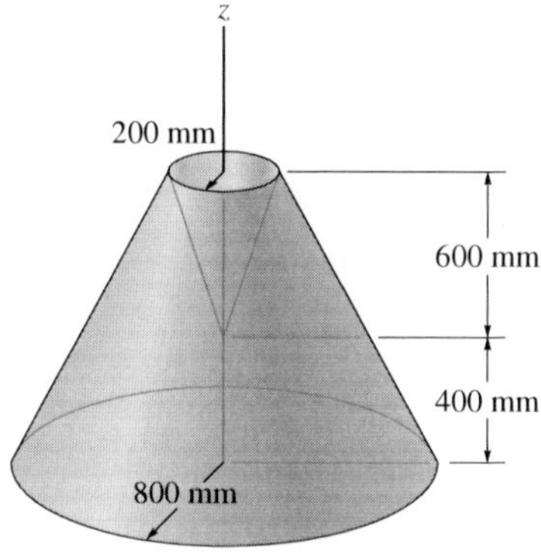

Prob. 10–95

***10–96.** The pendulum consists of a disk having a mass of 6 kg and slender rods AB and DC which have a mass of 2 kg/m. Determine the length L of DC so that the center of the mass is at the bearing O. What is the moment of inertia of the assembly about an axis perpendicular to the page and passing through point O?

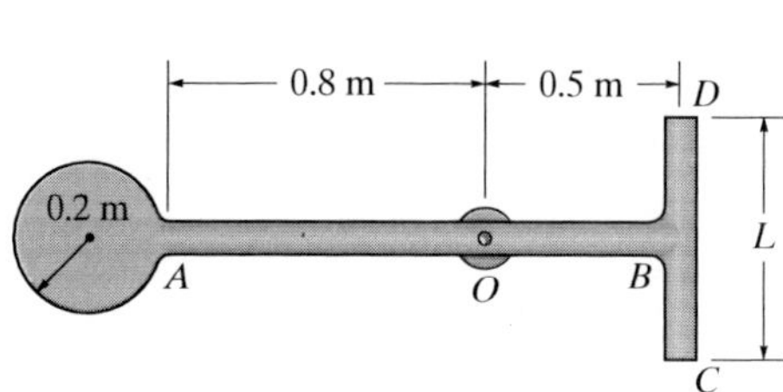

Prob. 10–96

10–97. The pendulum consists of the 3-kg slender rod and the 5-kg thin plate. Determine the location $\bar{y}$ of the center of mass G of the pendulum; then find the mass moment of inertia of the pendulum about an axis perpendicular to the page and passing through G.

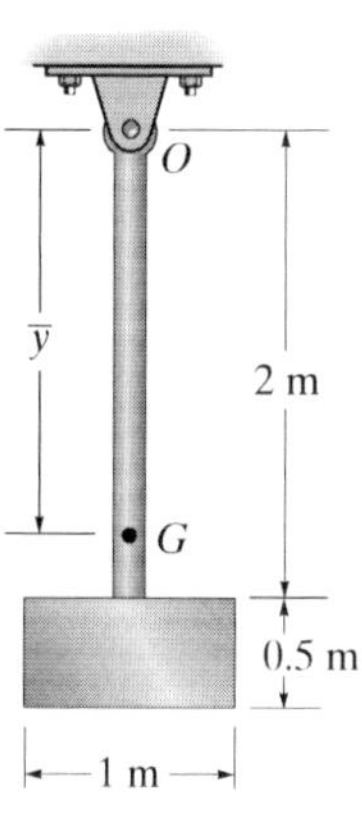

Prob. 10–97

10–98. Determine the location $\bar{y}$ of the center of mass G of the assembly and then calculate the moment of inertia about an axis perpendicular to the page and passing through G. The block has a mass of 3 kg and the mass of the semicylinder is 5 kg.

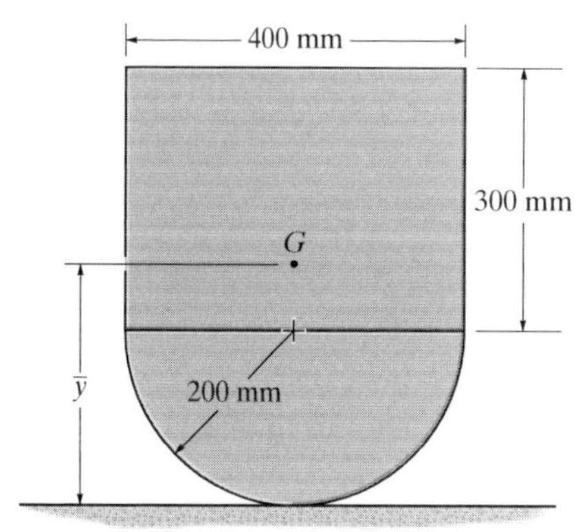

Prob. 10–98

10–99. If the large ring, small ring and each of the spokes weigh 500 N, 75 N, and 100 N, respectively, determine the mass moment of inertia of the wheel about an axis perpendicular to the page and passing through point A.

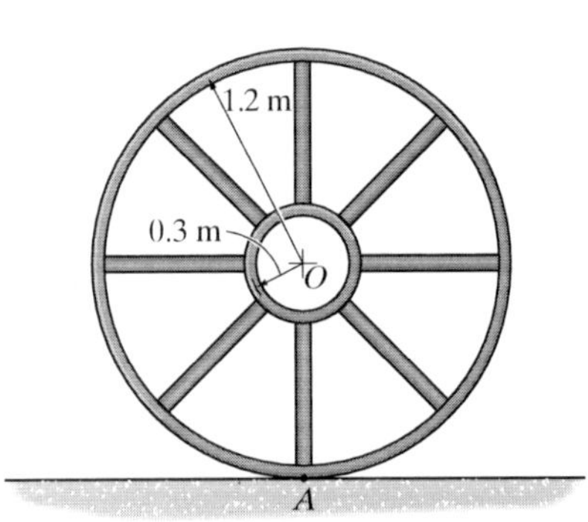

Prob. 10–99

*__10–100.__ Determine the mass moment of inertia of the assembly about the z axis. The density of the material is $7.85 \ \text{Mg/m}^3$.

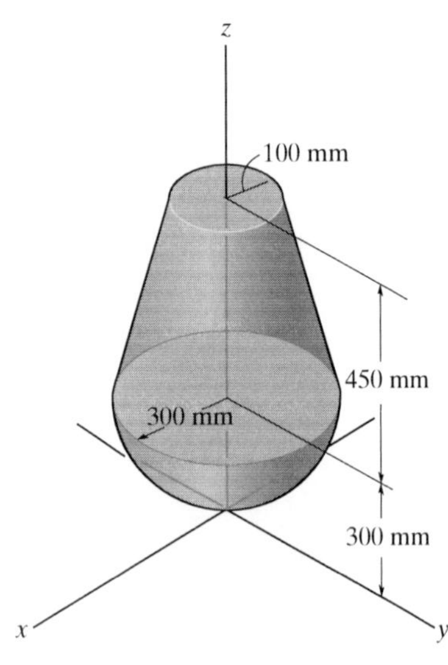

Prob. 10–100

10–101. Determine the moment of inertia I_z of the frustum of the cone which has a conical depression. The material has a density of $200 \ \text{kg/m}^3$.

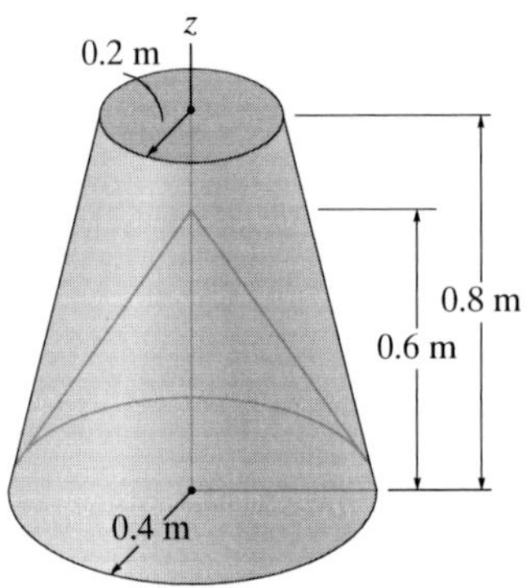

Prob. 10–101

10–102. The pendulum consists of a plate having a weight of 60 kg and a slender rod having a weight of 20 kg. Determine the radius of gyration of the pendulum about an axis perpendicular to the page and passing through point O.

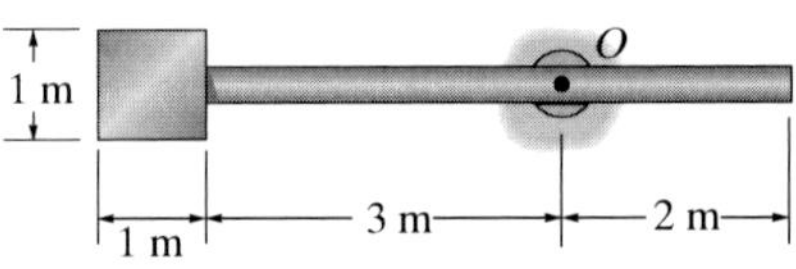

Prob. 10–102

10–103. The slender rods have a mass of $3 \ \text{kg/m}$. Determine the moment of inertia of the assembly about an axis perpendicular to the page and passing through point A.

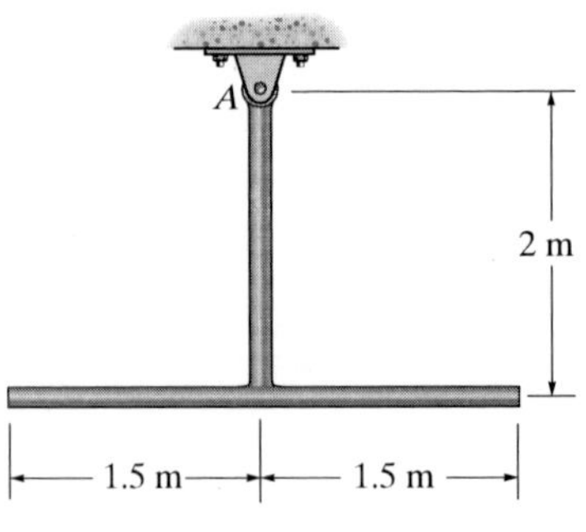

Prob. 10–103

10

***10–104.** Determine the moment of inertia I_z of the frustrum of the cone which has a conical depression. The material has a density of 2000 kg/m^3.

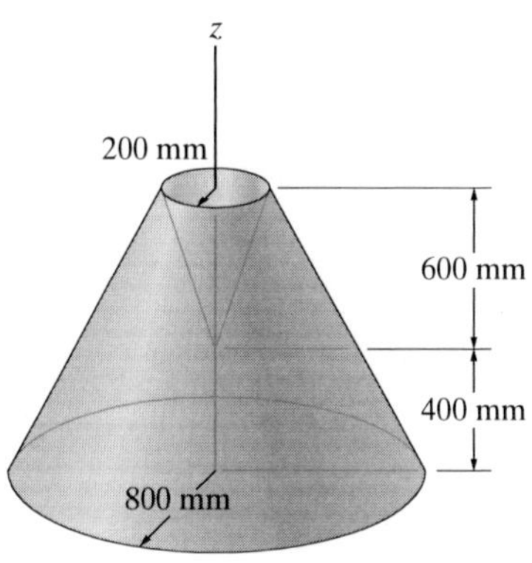

Prob. 10–104

10–105. Determine the moment of inertia of the wire triangle about an axis perpendicular to the page and passing through point O. Also, locate the mass center G and determine the moment of inertia about an axis perpendicular to the page and passing through point G. The wire has a mass of 0.3 kg/m. Neglect the size of the ring at O.

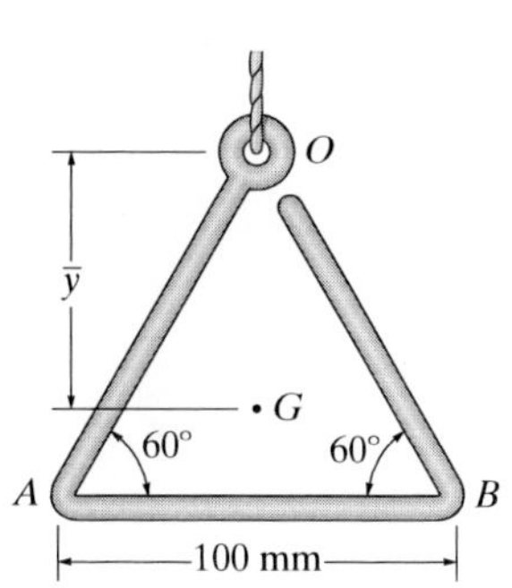

Prob. 10–105

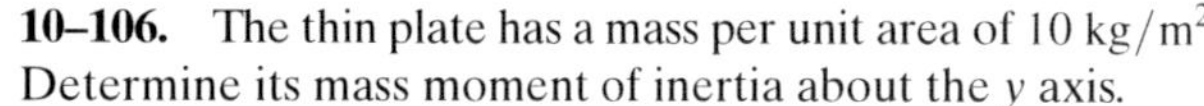

10–106. The thin plate has a mass per unit area of 10 kg/m^2. Determine its mass moment of inertia about the y axis.

10–107. The thin plate has a mass per unit area of 10 kg/m^2. Determine its mass moment of inertia about the z axis.

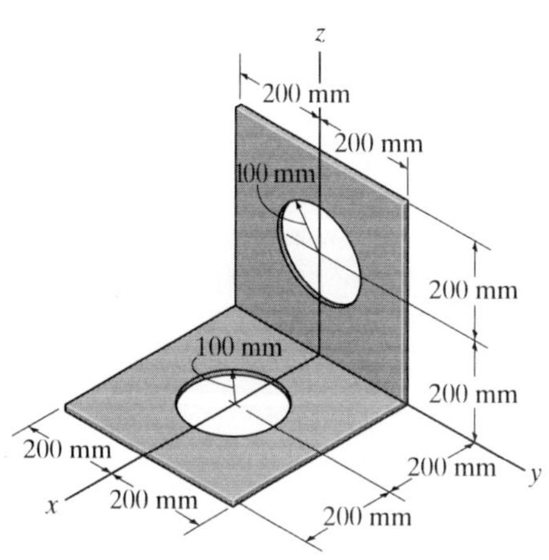

Probs. 10–106/107

***10–108.** Determine the mass moment of inertia of the overhung crank about the x axis. The material is steel having a density of $\rho = 7.85$ Mg/m^3.

10–109. Determine the mass moment of inertia of the overhung crank about the x' axis. The material is steel having a density of $\rho = 7.85$ Mg/m^3.

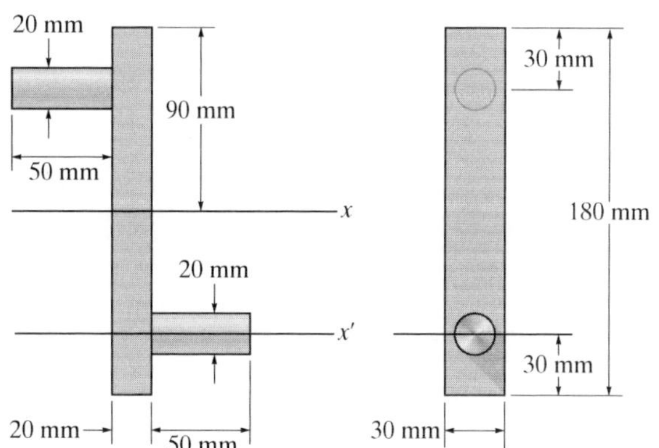

Probs. 10–108/109

CHAPTER REVIEW

Area Moment of Inertia

The *area moment of inertia* represents the second moment of the area about an axis. It is frequently used in formulas related to the strength and stability of structural members or mechanical elements.

$$I_x = \int_A y^2\, dA$$

If the area shape is irregular but can be described mathematically, then a differential element must be selected and integration over the entire area must be performed to determine the moment of inertia.

$$I_y = \int_A x^2\, dA$$

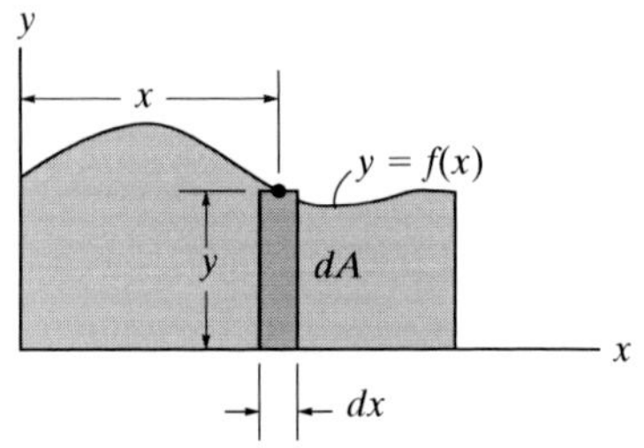

Parallel-Axis Theorem

If the moment of inertia for an area is known about a centroidal axis, then its moment of inertia about a parallel axis can be determined using the parallel-axis theorem.

$$I = \bar{I} + Ad^2$$

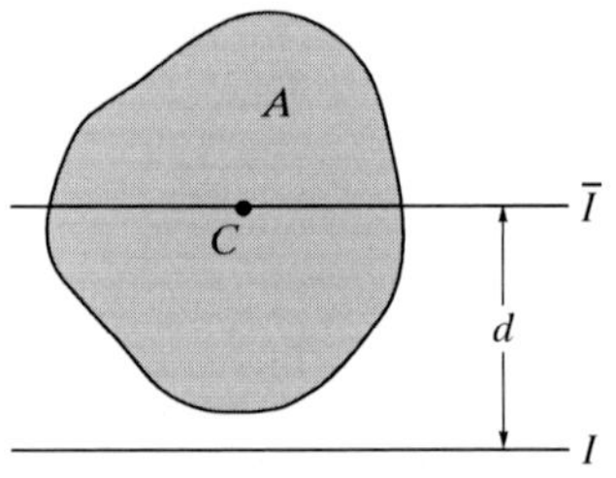

Composite Area

If an area is a composite of common shapes, as found on the inside back cover, then its moment of inertia is equal to the algebraic sum of the moments of inertia of each of its parts.

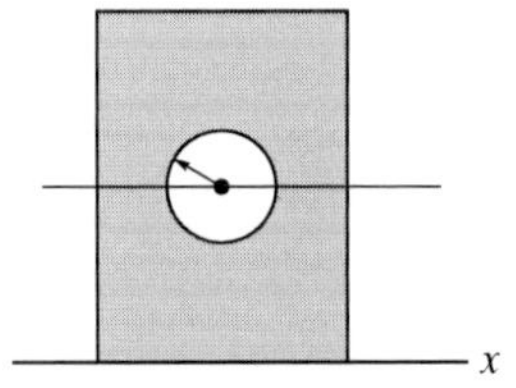

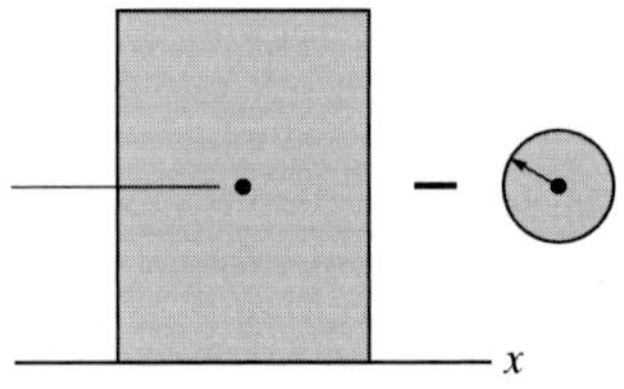

Product of Inertia

The *product of inertia* of an area is used in formulas to determine the orientation of an axis about which the moment of inertia for the area is a maximum or minimum.

$$I_{xy} = \int_A xy\, dA$$

If the product of inertia for an area is known with respect to its centroidal x', y' axes, then its value can be determined with respect to any x, y axes using the parallel-axis theorem for the product of inertia.

$$I_{xy} = \bar{I}_{x'y'} + Ad_xd_y$$

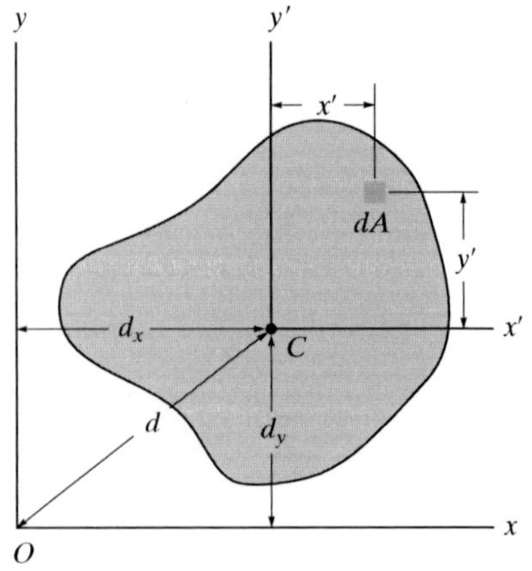

Principal Moments of Inertia

Provided the moments of inertia, I_x and I_y, and the product of inertia, I_{xy}, are known, then the transformation formulas, or Mohr's circle, can be used to determine the maximum and minimum or *principal moments of inertia* for the area, as well as finding the orientation of the principal axes of inertia.

$$I_{\substack{\max \\ \min}} = \frac{I_x + I_y}{2} \pm \sqrt{\left(\frac{I_x - I_y}{2}\right)^2 + I_{xy}^2}$$

$$\tan 2\theta_p = \frac{-I_{xy}}{(I_x - I_y)/2}$$

Mass Moment of Inertia

The *mass moment of inertia* is a property of a body that measures its resistance to a change in its rotation. It is defined as the "second moment" of the mass elements of the body about an axis.

$$I = \int_m r^2\, dm$$

For homogeneous bodies having axial symmetry, the mass moment of inertia can be determined by a single integration, using a disk or shell element.

$$I = \rho \int_V r^2\, dV$$

The mass moment of inertia of a composite body is determined by using tabular values of its composite shapes, found on the inside back cover, along with the parallel-axis theorem.

$$I = I_G + md^2$$

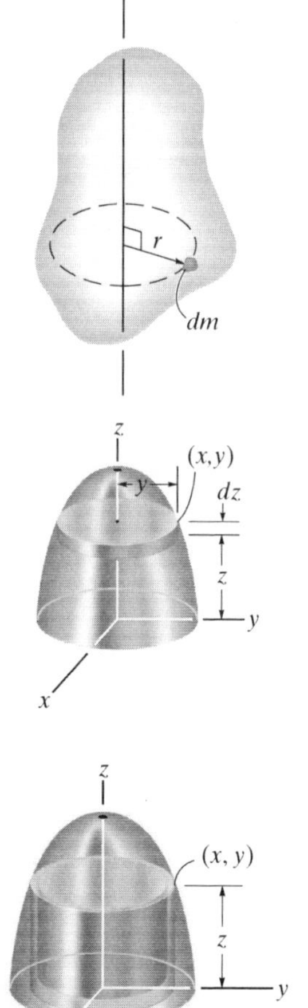

REVIEW PROBLEMS

10–110. Determine the radius of gyration k_x for the column's cross-sectional area.

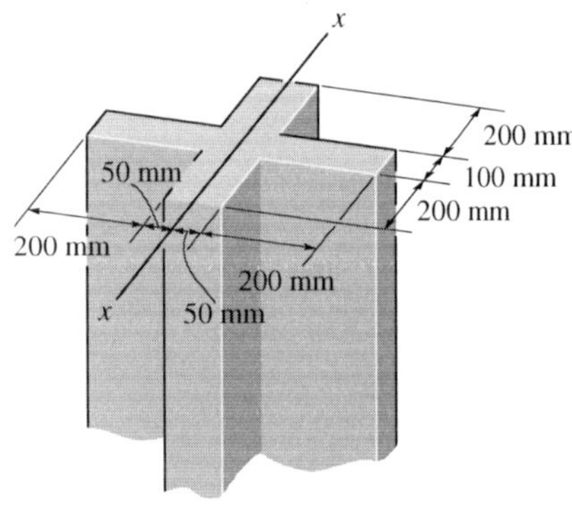

Prob. 10–110

10–111. Determine the area moment of inertia of the area about the x axis. Then, using the parallel-axis theorem, find the area moment of inertia about the x' axis that passes through the centroid C of the area. $\bar{y} = 120$ mm.

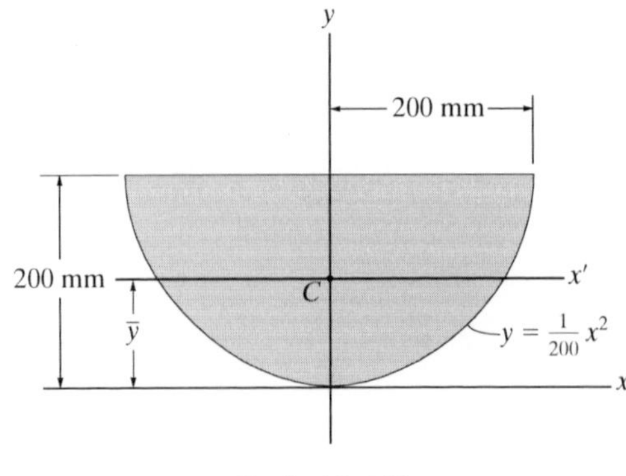

Prob. 10–111

***10–112.** Determine the product of inertia of the shaded area with respect to the x and y axes.

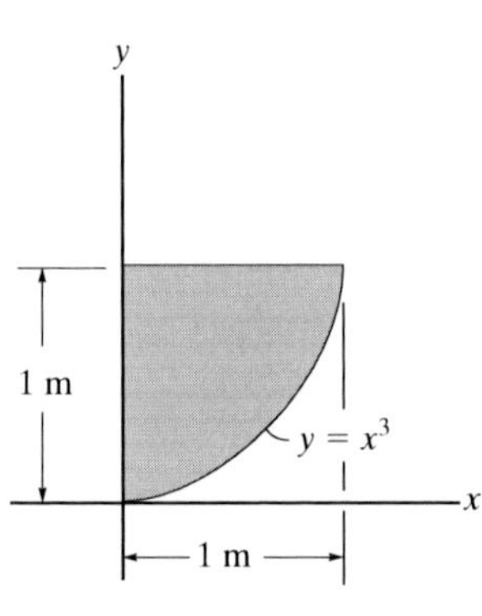

Prob. 10–112

10–113. Determine the area moment of inertia of the triangular area about (a) the x axis, and (b) the centroidal x' axis.

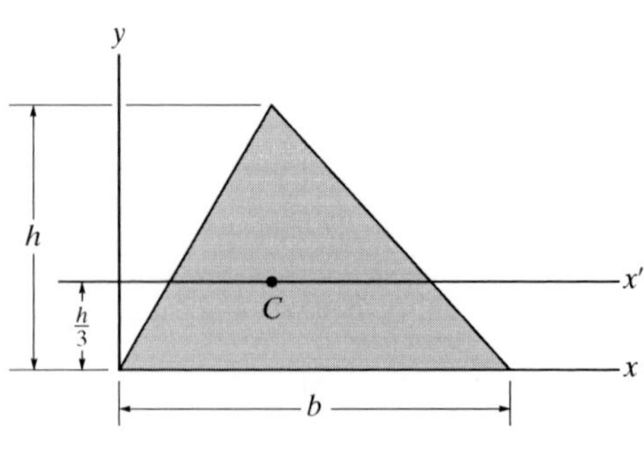

Prob. 10–113

10–114. Determine the mass moment of inertia I_x of the body and express the result in terms of the total mass m of the body. The density is constant.

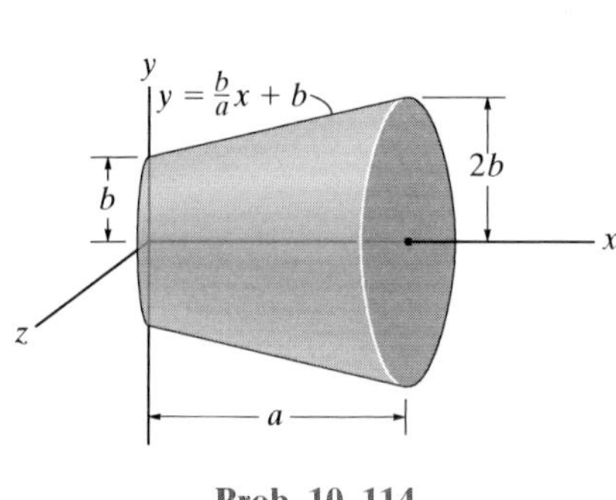

Prob. 10–114

10–115. Determine the moment of inertia for the shaded area about the x axis.

***10–116.** Determine the moment of inertia for the shaded area about the y axis.

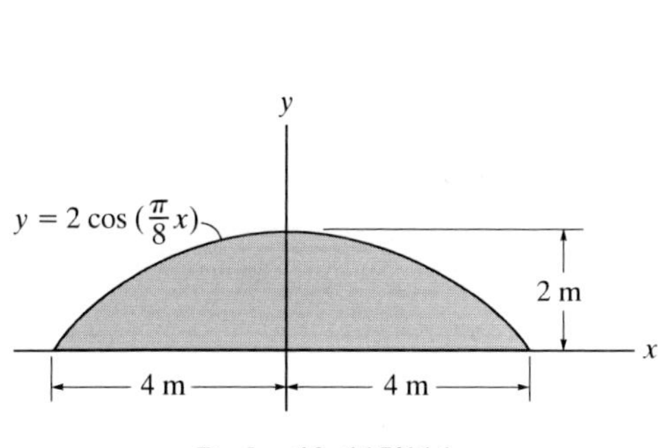

Probs. 10–115/116

10–117. Determine the area moments of inertia I_u and I_v and the product of inertia I_{uv} for the semicircular area.

Prob. 10–117

10–118. Determine the area moment of inertia of the beam's cross-sectional area about the x axis which passes through the centroid C.

10–119. Determine the area moment of inertia of the beam's cross-sectional area about the y axis which passes through the centroid C.

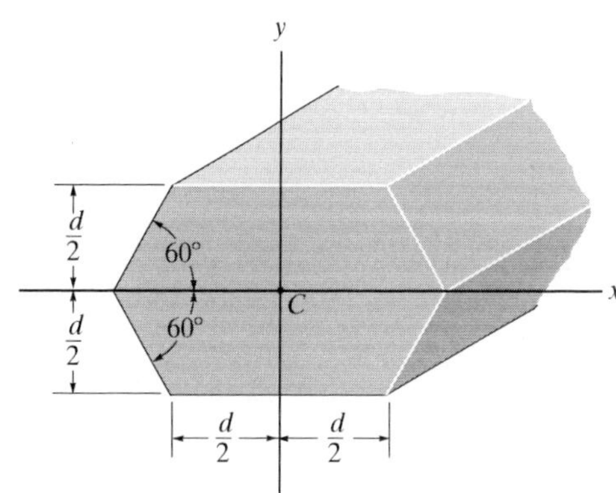

Probs. 10–118/119

10

Answers to Selected Problems

Chapter 5

5–10. $N_B = 3.46$ kN
$A_x = 1.73$ kN
$A_y = 1.00$ kN
5–11. $B_y = 586$ N
$F_A = 413$ N
5–13. $B_y = 736$ N
$A_x = 920$ N
$B_x = 920$ N
5–14. $F_{BC} = 3.92$ kN
$A_x = 3.13$ kN
$A_y = 950$ N
5–15. $F_B = 10.5$ N
$A_x = 42.0$ N
$A_y = 10.5$ N
5–17. $F_B = 105$ N
5–18. $N_A = 105$ kN
$B_x = 97.4$ kN
$B_y = 269$ kN
5–19. $B_x = 989$ N
$A_x = 989$ N
$B_y = 186$ N
5–21. $w_1 = 413$ kN/m
$w_2 = 407$ kN/m
5–22. $k = 1.33$ kN/m
$A_x = 398$ N
$A_y = 300$ N
5–23. $\theta = 23.1°$
$A_x = 353$ N
$A_y = 300$ N
5–25. $M_A = 10.6$ N · m
5–26. $N_B = 2.11$ N
$F_A = 2.81$ N
5–27. $F_B = 3.68$ kN, $x_A = 53.2$ mm, $x_B = 50.5$ mm

5–29. $\alpha = 1.02°$

5–30. $k_B = 2.50$ kN/m

5–31. $F_{AB} = 38.3$ kN
$C_x = 30.7$ kN
$C_y = 3.37$ kN

5–33. $F_{BD} = 628$ N
$C_x = 432$ N
$C_y = 68.2$ N

5–34. $\theta = a\cos\left(\dfrac{L+\sqrt{L^2+128r^2}}{16r}\right)$

5–35. $l = 0.67664$ m
$\theta = 2.1332$ N
$F_s = 2.1332$ N
$C_x = 432$ N
$N_B = 1.89$ N
$F_A = 2.52$ N

5–37. $F_{CB} = 782$ N
$A_x = 625$ N
$A_y = 681$ N

5–38. $F_2 = 724$ N
$F_1 = 1.45$ kN
$F_A = 1.75$ kN

5–39. $N_C = 213$ N
$A_x = 105$ N
$A_y = 118$ N

5–41. $T = \dfrac{W\cos\theta}{2\sin(\phi-\theta)}$

$A_x = \dfrac{W\cos\phi\cos\theta}{2\sin(\phi-\theta)}$

$A_x = \dfrac{W(\sin\phi\cos\theta - 2\cos\phi\sin\theta)}{2\sin(\phi-\theta)}$

5–42. $F = 10$ mN

5–43. $k = 250$ N/m

5–45. $F_B = 6.38$ N
$A_x = 3.19$ N
$A_y = 2.48$ N

5–46. $d = \dfrac{3a}{4}$

5–47. $\Delta_A = 0.1067$ m
$\Delta_B = 0.05333$ m

5–49. $P = 71.3$ N
$N = 446$ N

5–50. $\theta = \tan^{-1}\dfrac{b}{a}$

5–51. $F_B = 0.3\,P$
$F_C = 0.6\,P$
$x_C = 0.6\,P/k$

5–53. $T = \dfrac{W\cos\theta}{2\sin(\phi-\theta)}$

$A_x = \dfrac{W\cos\phi\cos\theta}{2\sin(\phi-\theta)}$

$A_y = \dfrac{W(\sin\phi\cos\theta - 2\cos\phi\sin\theta)}{2\sin(\phi-\theta)}$

5–54. $d = \dfrac{a}{\cos^3\theta}$

5–55. $N_A = 346$ N
$N_B = 693$ N
$a = 0.650$ m

5–57. $F = 5.20$ kN
$N_A = 17.3$ kN
$N_B = 24.9$ kN

5–58. $\theta = 63.4°$
$T = 29.2$ kN

5–59. $a = \sqrt{(4r^2l)^{\frac{2}{3}} - 4r^2}$

5–61. $h = \sqrt{\dfrac{s^2 - l^2}{3}}$

5–62. $T = 1.84$ kN
$F = 6.18$ kN

5–63. $R_D = 113.1$ kN
$R_E = 113.1$ kN
$R_F = 68.7$ kN

5–65. $N_C = 289$ N
$N_A = 213$ N
$N_B = 332$ N

5–66. $T_{BC} = T_{BD} = 17$ kN
$A_x = 0$
$A_y = 11.3$ kN
$A_z = -15.7$ kN

5–67. $T_B = 16.7$ kN
$A_x = 0$
$A_y = 5.00$ kN
$A_z = 16.7$ kN

5–69. $C_y = -318$ N
$C_z = 500$ N
$B_x = 239$ N
$B_z = -273$ N
$A_x = -239$ N
$A_z = 90.9$ N

5–70 $d = \dfrac{r}{2}\left(1+\dfrac{W}{P}\right)$

5–71. $N_B = 1000$ N
$(M_A)_z = 0$
$A_y = 0$
$(M_A)_y = -560$ N · m
$A_z = 400$ N

5–73. $T_{CD} = 3$ kN
$T_{EF} = 2.25$ kN
$T_{AB} = 0.75$ kN

5–74. $x = 0.667$ m
$y = 0.667$ m

5–75. $A_y = 0$
$T = 1.23$ kN
$B_x = -433$ N
$B_z = 1.42$ kN
$A_x = 867$ N
$A_z = 711$ N

5–77. $P = 0.5$ W

5–78. $C_y = 450$ N, $C_z = 250$ N
$B_x = 25$ N, $B_z = 1.125$ N
$A_x = 475$ N, $A_z = 125$ N

5–79. $F_{AC} = F_{BC} = 6.13$ kN
$F_{DE} = 19.62$ kN

5–81. $F_{BD} = 294$ N
$F_{BC} = 589$ N
$A_x = 0$
$A_y = 589$ N
$A_z = 490.5$ N

5–82. $T_{BD} = T_{CD} = 116.7$ N
$A_x = 66.7$ N
$A_y = 0$
$A_z = 100$ N

5–83. $T = 58.0$ N, $C_z = 87.0$ N, $C_y = 28.8$ N, $D_x = 0$
$D_y = 79.2$ N, $D_z = 58.0$ N

5–85. $F = 354$ N

5–86. $N_A = 8.00$ kN, $B_x = 5.20$ kN, $B_y = 5.00$ kN

5–87. $N_B = 400$ N, $F_A = 721$ N

5–89.
$$\theta = \tan^{-1}\left(\frac{1}{2}\cot\psi - \frac{1}{2}\cot\phi\right)$$

5–90. $A_x = 35.1$ N, $A_y = 343$ N, $N_B = 343$ N

5–91. $A_x = 0$
$A_y = -200$ N, $A_z = 150$ N, $(M_A)_x = 100$ N · m
$(M_A)_y = 0$, $(M_A)_z = 500$ N · m

Chapter 6

6–1. $F_{AB} = 60$ kN (T), $F_{BD} = 40$ kN (C)
$F_{AD} = 84.9$ kN (C), $F_{DC} = 141$ kN (T)
$F_{BC} = 60$ kN (T), $F_{DE} = 160$ kN (C)

6–2. $F_{AD} = 113$ kN (C)
$F_{AB} = 80$kN (T)
$F_{BD} = 0$
$F_{BC} = 80$ kN (T)
$F_{DC} = 113$ kN (T)
$F_{DE} = 160$ kN (C)

6–3. $F_{CD} = 5.21$ kN(C)
$F_{CB} = 4.17$ kN (T), $F_{AD} = 1.46$ kN (C)
$F_{AB} = 4.17$ kN (T), $F_{BD} = 4$ kN (T)

6–5. $F_{DC} = 400$ N (C)
$F_{DA} = 300$ N (C)
$F_{BA} = 250$ N (T)
$F_{BC} = 200$ N (T)
$F_{CA} = 283$ N (C)

6–6. $F_{DE} = 1.00$ kN (C)
$F_{DC} = 800$ N (T)
$F_{CE} = 900$ N(C)
$F_{CB} = 800$ N (T)
$F_{EB} = 750$ N (T)
$F_{EA} = 1.75$ kN (C)

6–7. $F_{JD} = 33.3$ kN (T)
$F_{AL} = F_{GH} = F_{LK} = F_{HI} = 28.3$ kN (C)
$F_{AB} = F_{GF} = F_{BC} = F_{FE} = F_{CD} = F_{ED} = 20$ kN (T)
$F_{BL} = F_{FH} = F_{LC} = F_{HE} = 0$
$F_{CK} = F_{EI} = 10$ kN (T)
$F_{KJ} = F_{IJ} = 23.6$ kN (C)
$F_{KD} = F_{ID} = 7.45$ kN (C)

6–9. $F_{AB} = 7.5$ kN (T), $F_{AE} = 4.5$ kN (C)
$F_{ED} = 4.5$ kN (C), $F_{EB} = 8$ kN (T)
$F_{BD} = 19.8$ kN (C), $F_{BC} = 18.5$ kN (T)

6–10. $F_{AB} = 196$ N (T), $F_{AE} = 118$ N (C)
$F_{ED} = 118$ N (C), $F_{EB} = 216$ N (T)
$F_{BD} = 1.04$ kN (C), $F_{BC} = 857$ N (T)

6–11. $F_{DE} = 16.3$ kN (C), $F_{DC} = 8.40$ kN (T)
$F_{EA} = 8.85$ kN (C), $F_{EC} = 6.20$ kN (C)
$F_{CF} = 8.77$ kN (T), $F_{CB} = 2.20$ kN (T)
$F_{BA} = 3.11$ kN (T), $F_{BF} = 6.20$ kN (C)
$F_{FA} = 6.20$ kN (T)

6–13. $F_{GB} = 30$ kN (T)
$F_{AF} = 20$ kN (C), $F_{AB} = 22.4$ kN (C)
$F_{BF} = 20$ kN (T), $F_{BC} = 20$ kN (T)
$F_{FC} = 28.3$ kN (C), $F_{FE} = 0$
$F_{ED} = 0$, $F_{EC} = 20.0$ kN(T), $F_{DC} = 0$

6–14. $F_{CB} = 3.00$ kN (T)
$F_{CD} = 2.60$ kN (C)
$F_{DE} = 2.60$ kN (C)
$F_{DB} = 2.00$ kN (T)
$F_{BE} = 2.00$ kN (C)
$F_{BA} = 5.00$ kN (T)

6–15. $F_{CB} = 8.00$ kN (T)
$F_{CD} = 6.93$ kN (C)
$F_{DE} = 6.93$ kN (C)
$F_{DB} = 4.00$ kN (T)
$F_{BE} = 4.00$ kN (C)
$F_{BA} = 12.0$ kN (T)

6–17. Maximum force:
$F_{DC} = F_{CB} = F_{CE} = F_{BE} = F_{BA} = 1.1547P$
$P = 5.20$ kN

6–18. $F_{CD} = 3.61$ kN (C), $F_{CB} = 3$ kN (T)
$F_{BA} = 3$ kN (T), $F_{BD} = 3$ kN (C)
$F_{DA} = 2.70$ kN (T), $F_{DE} = 6.31$ kN (C)

6–19. $F_{CD} = 467$ N (C), $F_{CB} = 389$ N (T)
$F_{BA} = 389$ N (T), $F_{BD} = 314$ N (C)
$F_{DE} = 1.20$ kN (C), $F_{DA} = 736$ N (T)

6–21. $P = 1.25$ kN

6–22. $F_{FE} = 0.667\,P$ (T), $F_{FD} = 1.67\,P$ (T)
$F_{AB} = 0.471\,P$ (C), $F_{AE} = 1.67\,P$ (T)
$F_{AC} = 1.49\,P$ (C), $F_{BF} = 1.41\,P$ (T),
$F_{BD} = 1.49\,P$ (C), $F_{EC} = 1.41\,P$ (T),
$F_{CD} = 0.471\,P$ (C)

6–23. $F_{CD} = 0.577\,P$ (C), $F_{DB} = 0.289\,P$ (T)
$F_{CE} = 0.577\,P$ (T), $F_{BC} = 0.577\,P$ (C)
$F_{BE} = 0.577\,P$ (T), $F_{AB} = 0.577\,P$ (C)
$F_{AE} = 0.577\,P$ (T)

6–25. $F_{BA} = P \csc 2\theta$(C), $F_{BC} = P \cot 2\theta$(C)
$F_{CA} = (\cot\theta \cos\theta - \sin\theta + 2\cos\theta)\,P$ (T)
$F_{CD} = (\cot 2\theta + 1)\,P$ (C)
$F_{DA} = (\cot 2\theta + 1)(\cos 2\theta)\,(P)$ (C)

6–26. Maximum force: $F_{CA} = 2.732\,P$ (T),
$F_{CD} = 1.577\,P$ (C), $P_{max} = 732$ N

6–27. $F_{GF} = 8.08$ kN (T)
$F_{BC} = 7.70$ kN (C)
$F_{CG} = 0.770$ kN (C)

6–29. $F_{LD} = 0$, $F_{LK} = -112.5$ kN (C)
$F_{CD} = 112.5$ kN(T), $F_{KD} = -50$ kN (C)

6–30. $F_{JI} = -75$ kN (C)
$F_{JE} = -50$ kN (C)
$F_{DE} = 75$ kN (T)

6–31. $F_{CB} = 5$ kN (T), $F_{BE} = 21.2$ kN (T)
$F_{EF} = -25$ kN (C)

6–33. $F_{BC} = -3.25$ kN (C)
$F_{CH} = 1.923$ kN (T)

6–34. $F_{CD} = -1.917$ kN (C), $F_{GF} = 1.533$ kN (T)

6–35. $F_{BC} = 10.4$ kN (C), $F_{HG} = 9.15$ kN (T),
$F_{HC} = 2.24$ kN (T)

6–37. $F_{KJ} = 3.07$ kN (C), $F_{CD} = 3.07$ kN (T)
$F_{ND} = 0.167$ kN (T), $F_{NJ} = 0.167$ kN (C)

6–38. $F_{JI} = 2.13$ kN (C)
$F_{DE} = 2.13$ kN (T)

6–39. $F_{IC} = 5.62$ kN (C), $F_{CG} = 9.00$ kN (T)

6–41. $F_{CB} = 3.60$ kN (T), $F_{GC} = 1.80$ kN (C)
$F_{FG} = 4.02$ kN (C)

6–42. $F_{ML} = 38.4$ kN (T)
$F_{DE} = -37.1$ kN (C)
$F_{DL} = -3.8$ kN (C)

6–43. $F_{EF} = -37.1$ kN (C)
$F_{EL} = 6$ kN (T)

6–45. $F_{AB} = P$ (T), $F_{EF} = P$ (C), $F_{BF} = 1.41\,P$ (C)

6–46. Method of Joints: By inspection, members *BN, NC, DO, OC, HJ, LE* and *G* are zero–force members.
$F_{CD} = 5.625$ kN (T), $F_{CM} = 2.00$ kN (T)

6–47. Method of Joints: By inspection, members *BN, NC, DO, OC HJ, LE* and *JG* are zero–force members.
$F_{EF} = 7.88$ kN (T), $F_{LK} = 9.25$ kN (C)
$F_{ED} = 1.94$ kN (T)

6–49. $F_{BG} = \{-600 \csc\theta\}$ N
$F_{BC} = -200L$ N
$F_{HG} = 400L$ N

6–50. $F_{AB} = 584$ N (T)
$F_{AC} = -1133$ N (C)
$F_{BC} = -142$ N (C)

6–51. $F_{AB} = 6.46$ kN (T), $F_{AC} = F_{AD} = 1.50$ kN (C)
$F_{BC} = F_{BD} = 3.70$ kN (C), $F_{BE} = 4.80$ kN (T)

6–53. $F_{BD} = 896$ N (C), $F_{DC} = 554$ N (T)
$F_{DA} = 146$ N (C), $F_{AB} = 52.1$ N (T)
$F_{AC} = 31.25$ N (T), $F_{CB} = 448$ N (C)

6–54. $F_{DB} = 474$N (C), $F_{DC} = 146$ N (T)
$F_{DA} = 1.08$ kN (T), $F_{AB} = 385$ N (C)
$F_{AC} = 231$ N (C), $F_{CB} = 281$ N (T)

6–55. $F_{BC} = F_{BD} = 1.34$ kN (C), $F_{AB} = 2.4$ kN (C)
$F_{AG} = F_{AE} = 1.01$ kN (T), $F_{BG} = 1.80$ kN (T)
$F_{BE} = 1.80$ kN (T)

6–57. $F_{CE} = 721$ N (T), $F_{BC} = 400$ N (C)
$F_{BE} = 0$, $F_{BF} = 2.10$ kN (T)

6–58. $F_{AB} = 1.50$ kN (C), $F_{AF} = 1.08$ kN (C)
$F_{AD} = 600$ N (T), $F_{FD} = 0$, $F_{ED} = 1.40$ kN (C)
$F_{BD} = 361$ N (C)

6–59. $F_{AE} = F_{AC} = 220$ N (T)
$F_{AB} = 583$ N (C)
$F_{BD} = 707$ N (C)
$F_{BE} = F_{BC} = 141.4$ N (T)

6–61. $P = 18.9$ N

6–62. $P = 368$ N

6–63. $P = 40.0$ N, $x = 240$ mm

6–65. $P = 2.24$ kn

6–66. $A_x = 4.20$ kN, $B_x = 4.20$ kN, $A_y = 4.00$ kN
$B_y = 3.20$ kN, $C_x = 3.40$ kN, $C_y = 4.00$ kN

6–67. $B_x = 4.00$ kN
$B_y = 2.00$ kN
$A_x = 6.00$ kN
$A_y = 0$

6–69. $C_x = 167$ N, $C_y = 111$ N
$A_x = 167$ N, $A_y = 389$ N

6–70. $3.64\,F$

6–71. $N_E = 5$ kN
$D_x = 0$
$N_C = 16.7$ kN
$A_x = 0$
$A_y = 2.67$ kN
$M_A = 21.5$ kN · m

6–73. $C_y = 0$
$B_y = 7.50$ kN
$M_A = 45.0$ kN · m
$A_y = -7.50$ kN
$A_x = 0$

6–74. $m = 1.71$ kg

6–75. $m = 106$ kg

6–77. $m = 366$ kg, $F_A = 2.93$ kN

6–78. $N_A = 3.67$ kN, $M_A = 5.55$ kN · m
$C_x = 2.89$ kN, $C_y = 1.32$ kN

6–79. $F_E = 4.55\ F$

6–81. $F_{AB} = 9.23$ kN, $C_x = 2.17$ kN, $C_y = 7.01$ kN
$D_x = 0$, $D_y = 1.96$ kN, $M_D = 2.66$ kN · m

6–82. $A_x = 2.98$ kN, $A_y = 235$ N
$B_x = 2.98$ kN, $B_y = 549$ N
$C_x = 2.98$ kN, $C_y = 1.33$ kN

6–83. $A_x = 500$ N, $A_y = 1000$ N
$C_x = 500$ N, $C_y = 500$ N

6–85. $F_A = 130$ N

6–86. $N_B = N_C = 49.5$ N

6–87. $R_E = 177$ N, $R_A = 128$ N

6–89. $m = 1.11$ Mg

6–90. $F_C = 28.8$ kN

6–91. $A_y = 3.09$ kN, $C_y = 1.52$ kN
$B_x = 3.5$ kN, $B_y = 23.5$ kN

6–93. $N_A = 490.5$ N, $N_B = 294.3$ N, $T = 353.70$ N,
$\theta = 33.7°$, $x = 177$ mm

6–94. **a.** $F = 875$ N, $N_C = 1750$ N
b. $F = 437.5$ N, $N_C = 437.5$ N

6–95. **a.** $F = 1025$ N, $N_C = 1900$ N
b. $F = 512.5$ N, $N_C = 362.5$ N

6–97. $T = 9.60$ N

6–98. $B_x = 4.28$ kN, $B_y = 3.15$ kN, $A_x = 4.28$ kN
$A_y = 4.05$ kN, $C_x = 3.38$ kN, $C_y = 4.05$ kN

6–99. $F = 370$ N

6–101. $N_A = 11.1$ kN, $N_B = 3.37$ kN, $F_{CD} = 6.5$ kN
$F_E = 5.88$ kN

6–102. $F_{AB} = 1.18$ kN, $F_E = 3.17$ kN, $F_{CD} = 19.6$ kN
$F_F = 16.8$ kN

6–103. $F_N = 26.25$ N

6–105. $F_{DB} = 2.601$ kN, $F_{FB} = 1.94$ kN

6–106. $F = 562.5$ N

6–107. $P = 80$ N

6–109. $E_x = 6.79$ kN, $E_y = 1.55$ kN
$D_x = 981$ N, $D_y = 981$ N

6–110. $M = 2.43$ kN · m

6–111. $F = 5.07$ kN

6–113. $A_x = 0.2629$ kN, $A_y = 0.6495$ kN
$B_x = 0.4871$ kN, $B_y = 0.6495$ kN

6–114. $E_x = 4.7243$ kN, $E_y = 2.5$ kN
$D_x = 4.7243$ kN, $D_y = 5$ kN

6–115. $F_s = 286$ N

6–117. $P = B_z = D_z = B_y = D_y = 283$ N
$B_x = D_x = 0$

6–118. $A_x = -172.3$ N, $A_y = -114.8$ N, $A_z = 0$ N
$C_x = 47.3$ N, $C_y = -61.9$ N, $C_z = -125$ N
$M_{Cy} = -429$ N · m, $M_{Cz} = 0$ N · m

6–119. $F_{AB} = 981$ N, $F_E = 2.64$ kN
$F_{CD} = 16.3$ kN, $F_F = 14.0$ kN

6–121. $A_y = 250$ N, $A_x = 1.40$ kN, $C_x = 500$ N,
$C_y = 1.70$ kN

6–122. $N_A = 4.60$ kN
$C_y = 7.05$ kN
$N_B = 7.05$ kN

6–123. $m = 3.86$ kg

6–125. $A_x = 117$ N, $A_y = 397$ N
$B_x = 97.4$ N, $B_y = 97.4$ N

6–126. $F_{AB} = 21.9$ kN (C), $F_{BG} = 17.5$ kN (T)
$F_{AG} = 13.1$ kN (T), $F_{BC} = 13.1$ kN (C)
$F_{GC} = 3.12$ kN (T), $F_{GF} = 11.2$ kN (T)
$F_{CF} = 3.12$ kN (C), $F_{CD} = 9.38$ kN (C)
$F_{DE} = 15.6$ kN (C), $F_{DF} = 12.5$ kN (T)
$F_{EF} = 9.38$ kN (T)

Chapter 7

7–1. $N_A = 5$ kN, $N_C = 4$ kN, $N_B = 3$ kN

7–2. $T_A = 100$ N · m, $T_B = 200$ N · m
$T_C = 200$ N · m, $T_D = 0$

7–3. $d = 0.200$ m

7–5. $V_A = 3$ kN, $N_A = 13.2$ kN, $M_A = 3.82$ kN · m
$V_B = 3$ kN, $N_B = 16.2$ kN, $M_B = 14.3$ kN · m

7–6. $a = \frac{L}{3}$

7–7. $N_E = 0$, $V_E = -50$ N, $M_E = -100$ N · m
$N_D = 0$, $V_D = 750$ N, $M_D = -1300$ N · m

7–9. $N_C = -30$ kN, $V_C = -8$ kN, $M_C = 6$ kN · m

7–10. $P = 0.533$ kN, $N_C = -2$ kN, $V_C = -0.533$ kN
$M_C = 0.400$ kN · m

7–11. $M_E = 7.5$ kN · m, $N_E = 0$, $V_E = 5$ kN
$M_D = -9.75$ kN · m, $N_D = 0$, $V_D = 8$ kN

7–13. $N_C = 0$, $V_C = -1$ kN, $M_C = 9$ kN · m

7–14. $N_D = -800$ N, $V_D = 0$, $M_D = 1.20$ kN · m

7–15. $w = 100$ N/m

7–17. $N_E = -1.92$ kN, $V_E = 800$ N, $M_E = 2.40$ kN · m

7–18. $N_C = 0$
$V_C = \frac{3w_0L}{8}$
$M_C = -\frac{5}{48}w_0L^2$

7–19. $N_E = 470$ N, $V_E = 215$ N
$M_E = 660$ N · m, $N_F = 0$
$V_F = -215$ N, $M_F = 660$ N · m

7–21. $N_D = 4$ kN, $V_D = -9$ kN, $M_D = -18$ kN · m
$N_E = 4$ kN, $V_E = 3.75$ kN, $M_E = -4.875$ kN · m

7–22. $N_E = 0$, $V_E = -1.17$ kN, $M_E = 4.97$ kN · m
$N_F = 0$, $V_F = 1.25$ kN, $M_F = 2.5$ kN · m

7–23. $V_D = 168$ N, $N_D = -110$ N, $M_D = 348$ N · m
$N_E = -168$ N, $V_E = -90.4$ N, $M_E = 190$ N · m

7–25. $N_C = 0$, $V_C = 0$, $M_C = 1.5$ kN · m

7–26. $A_y = \left(\frac{w}{6b}\right)(2a+b)(a-b), \frac{a}{b} = \frac{1}{4}$

7–27. $N_D = 2.40$ kN, $V_D = 50$ N, $M_D = 1.35$ kN · m

7–29. $N_D = 1.26$ kN, $V_D = 0$, $M_D = 500$ N · m

7–30. $N_E = -1.48$ kN
$V_E = 500$ N
$M_E = 1000$ N · m

7–31. $N_A = V_A = 2.83$ kN, $M_A = 212$ N · m

7–33. $a = \frac{2}{3}L$

7–34. $N_C = -\frac{wL}{2}\csc\theta,\ V_C = 0,\ M_C = \frac{wL^2}{8}\cos\theta$

7–35. $V_E = 0$, $N_E = 894$ N, $M_E = 0$, $V_F = 447$ N,
$N_F = -224$ N, $M_F = 224$ N · m

7–37. $N_C = -1.91$ kN, $V_C = 0$, $M_C = 382$ N · m

7–38. $N_D = 981$ N, $V_D = 0$, $M_D = 1.32$ kN · m
$N_E = 1.77$ kN, $V_E = 589$ N, $M_E = 1.68$ kN · m

7–39. $V = -0.293\, rw_0$, $N = -0.707\, rw_0$
$M = -0.0783\, r^2 w_0$

7–41. $C_x = -150$ kN, $C_y = -350$ kN, $C_z = 700$ kN
$M_{Cx} = 1.40$ MN · m, $M_{Cy} = -1.20$ MN · m,
$M_{Cz} = -750$ kN · m

7–42. $N_C = -170$ kN, $(V_C)_y = -50$ kN
$(V_C)_z = 500$ kN
$(M_C)_x = 1$ MN · m
$(M_C)_y = 900$ kN · m
$(M_C)_z = -260$ kN · m

7–43. $N_{Dy} = -65$ kN, $V_{Dx} = 116$ kN, $V_{Dz} = 0$
$M_{Dx} = 49.2$ kN · m
$M_{Dy} = 87.0$ kN · m
$M_{Dz} = 26.2$ kN · m

7–45. For $0 \le x < b$, $V = -\frac{Pa}{b}, M = -\frac{Pa}{b}x$.
For $b < x \le a + b$, $V = P$, $M = -P(a + b - x)$.

7–46. For $0 \le x < 2$ m, $V = 0.25$ kN, $M = (0.25x)$ kN · m.
For 2 m $< x \le 3$ m, $V = (3.25 - 1.5x)$ kN.
$M = (-0.75x^2 + 3.25x - 3.0)$ kN · m.

7–47. **a.** For $0 \le x < a$, $V = P$, $M = Px$.
For $a < x < L - a$, $V = 0$, $M = Pa$.
For $L - a < x \le L$, $V = -P$.
b. For $0 \le x < 1.5$ m, $V = 4$ kN, $M = (4x)$ kN · m.
For 1.5 m $< x < 2.1$ m, $V = 0$, $M = 6$ kN · m.
For 2.1 m $< x \le 3.6$ m, $V = -4$ kN,
$M = (14.4 - 4x)$ kN · m.

7–49. $M_{max} = 2$ kN · m

7–50. $V = (4 - 2x)$ kN
$M = (-x^2 + 4x - 10)$ kN · m

7–51. For $0 \le x < \frac{L}{2}$,
$V = \frac{wL}{8}$,
$M = \frac{wL}{8}x$.
For $\frac{L}{2} < x \le L$,
$V = \frac{w}{8}(5L - 8x)$,
$M = \frac{w}{8}(-L^2 + 5Lx - 4x^2)$.

7–53. For $0 \le x < 8$m, $V = 140 - 40x$ kN,
$M = (140x - 20x^2)$.
For 8m $< x \le 11$m, $V = 20$ kN,
$M = (20x - 320)$ kN · m.

7–54. $V = (525 - 300x)$ kN
$M = (-150x^2 + 525x + 300)$ kN · m

7–57. $V = 5(3 - x)$
$M = (15x - 2.5x^2 - 0.25)$ kN · m

7–58. For $0 \le x < L$, $V = \frac{w}{18}(7L - 18x)$,
$M = \frac{w}{18}\left(7Lx - 9x^2\right)$.
For $L \le x < 2L$, $V = \frac{w}{2}(3L - 2x)$,
$M = \frac{w}{18}(27Lx - 20L^2 - 9x^2)$.
For $2L < x \le 3L$, $V = \frac{w}{18}(47L - 18x)$,
$M = \frac{w}{18}(47Lx - 9x^2 - 60L^2)$.

7–59. $x = 1.732$ m
$M_{max} = 0.866$

7–61. $V = \left\{3.00 - \frac{x^2}{4}\right\}$ kN
$M = \left\{3.00x - \frac{x^3}{12}\right\}$ kN · m

7–62. $V = \frac{\gamma ht}{2d}x^2, M = -\frac{\gamma ht}{6d}x^3$

7–63. For $0 \le x < 5$ m, $V = 2.5 - 2x$, $M = 2.5x - x^2$.
For 5 m $< x \le 10$ m, $V = -7.5$, $M = -7.5x - 25$.

7–65. For $0 \le x < 3$ m, $V = \{-650 - 50.0x^2\}$ N,
$M = \{-650x - 16.7x^3\}$ N · m.
For 3 m $< x \le 7$ m, $V = \{2100 - 300x\}$ N,
$M = \{-150(7 - x)^2\}$ N · m.

7–66. $V = \frac{w}{12L}\left(4L^2 - 6Lx - 3x^2\right)$,
$M = \frac{w}{12L}\left(4L^2x - 3Lx^2 - x^3\right), M_{max} = 0.0940wL^2$

7–67. $N = \frac{P}{2}\cos\theta$
$V = \frac{P}{2}\sin\theta$
$M = \frac{Pr}{2}(1-\cos\theta)$

7–69. $V = |P|$
$M = |Pr\cos\theta|$
$T = |Pr(1-\sin\theta)|$

7–94. $T_{BD} = 390.9$ N, $T_{AC} = 378.4$ kN
$T_{CD} = 218.4$ N, $l = 4.674$ m

7–95. $P = 360$ N

7–97. $y_B = 2.60$ m, $y_D = 2.11$ m

7–98. $P_1 = 3288.1$ N, $y_D = 1.933$ m

7–99. $T_{AB} = 413$ N
$T_{BC} = 282$ N
$T_{CD} = 358$ N
$y_C = 3.08$ m

7–101. $T_{max} = 5.36$ kN, $L = 51.3$ m

7–102. $T_A = 46.4$ kN
$T_B = 54.5$ kN

7–103. $w_o = 4.40$ kN/m

7–105. $T_B = 48.7$ MN, $T_A = 51.4$ MN

7–106. $w_o = 77.8$ kN/m

7–107. $m_C = 478$ kg
$h = 0.827$ m
$L = 13.2$ m

7–109. $T_{max} = 594$ kN

7–110. $T_{min} = 552$ kN

7–111. $y = 4.5\left(1-\cos\frac{\pi}{24}x\right)$m
$T_{max} = 60.2$ kN

7–113. $w_o = 0.846$ kN/m

7–114. $h = 3.104$ m

7–115. $L = 10.39$ m

7–118. $h = 1.47$ m

7–119. $(F_V)_A = 165$ N, $(F_H)_A = 73.9$ N

7–121. $F_A = F_C = 11.1$ kN, $h = 23.5$ m

7–122. $\frac{h}{L} = 0.141$

7–123. $y = 45.512\{\cosh(0.0219722x) - 1\}$ m
$L = 52.55$ m

7–125. $N_D = -6.08$ kN
$V_D = -2.6$ kN, $M_D = -12.99$ kN · m

7–126. $V = 10 - 2x$
$M = 10x - x^2 - 30$

7–127. $a = 0.366$ L

7–129. $l_d = 5.67$ m, $d = 19.8$ m

7–130. For $0 \le x < 3$ m, $V = 1.50$ kN,
$M = \{1.50x\}$ kN · m.
For 3 m $< x \le 6$ m, $V = -4.50$ kN,
$M = \{27.0 - 4.50x\}$ kN · m.

7–131. $(V_C)_x = 450$ N, $N_C = 0$, $(V_C)_z = -550$ N
$(M_C)_x = -825$ N · m, $T_C = 30$ N · m,
$(M_C)_z = 675$ N · m

7–133. For $0 \le x < 5$ m, $V = (2.5 - 2x)$ kN,
$M = (2.5x - x^2)$ kN · m.
For 5 m $< x \le 10$ m, $V = -7.5$ kN,
$M = (25 - 7.5x)$ kN · m.

7–134. $N_C = 0$, $V_C = 9.0$ kN, $M_C = -62.5$ kN · m,
$N_B = 0$, $V_B = 27.5$ kN, $M_B = -184.5$ kN · m

7–135. $V_1(x) = A_y\frac{1}{\text{kN}}$
$M_1(x) = A_y x\frac{1}{\text{kN}\cdot\text{m}}$
$V_2(x) = \left[A_y - w(x-a)\right]\frac{1}{\text{kN}}$
$M_2(x) = \left[A_y x - w\frac{(x-a)^2}{2}\right]\frac{1}{\text{kN}\cdot\text{m}}$

7–137. For $0 \le x = 2$ m, $V = \{5.29 - 0.196x\}$ kN,
$M = \{5.29x - 0.0981x^2\}$ kN · m.
For 2 m $< x \le 5$ m, $V = \{-0.196x - 2.71\}$ kN,
$M = \{16.0 - 2.71x - 0.0981x^2\}$ kN · m.

7–139. For $0 \le x < 0.45$ m, $V = -1.47$ kN,
$M = (-1.47x)$ kN · m.
For 0.45 m $\le x \le 0.9$ m, $V = 1.47$ kN,
$M = (1.47x - 1.324)$ kN · m.

Chapter 8

8–1. $N_A = 16.5$ kN,
$N_B = 42.3$ kN, the **mine car does not move**.

8–2. $P = 12.8$ kN

8–3. $x = 0.5$ m

8–5. $O_x = 46.4$ N, $O_y = 400$ N

8–6. $O_x = 280$ N, $O_y = 945$ N

8–7. **a.** No
b. Yes

8–9. $\mu = 0.27$

8–10. It is not possible since
$\mu = 0.2 < 0.27$.

8–11. $n = 6$

8–13. $P = 147$ N

8–14. $P = 182$ N, $N_C = 606.6$ N

8–15. $\theta = 21.8°$

8–17. $\mu_s = 0.577$

8–18. $\mu = 0.354$

8–19. $d = 537$ mm

8–21. $\theta = 2\tan^{-1}\mu s$

8–22. Slip at clamp, n = 9.17, slip between end boards;
$n = 8.12$, *use* $n = 8$.

8–23. $P = 371.4$ N

8–25. $P = \frac{M_O(b - \mu_s c)}{\mu_s r a}$

8–26. $P = \dfrac{M_O(b+\mu_s c)}{\mu_s ra}$
$P < 0$ if $(b - \mu_s c) < 0$ i.e. if $\dfrac{b}{c} < \mu_s$

8–27. $P = \dfrac{M_O(b+\mu_s c)}{\mu_s ra}$

8–29. $\theta = 11.0°$

8–30. $\mu_s = 0.577$

8–31. $\theta = 31.0°$

8–33. $\mu_s = 0.268$

8–34. Thus, **he can move the crate**.

8–35. $\mu_{s'} = 0.376$

8–37. $P = 140$ N, $x_A = 523.5$ mm

8–38. $F_C = 30.5$ N, $N_C = 152.3$ N
$x_1 = 0.79$ m

8–39. $\theta = 7.50°$, $T = 452$ N

8–41. $F_A = 0.44$ kN, $N_A = 1.47$ kN, $N_B = 1.24$ kN

8–42. $\mu = 0.176$

8–43. $\mu_s = 0.509$

8–45. Can A will not move.

8–46. $P = 355$ N

8–47. $\mu_C = 0.0734$, $\mu_B = 0.0964$

8–49. $M = 77.3$ N · m

8–50. $F_A = 71.4$ N

8–53. $\phi = 42.6°$

8–54. $P = \dfrac{1}{2}\mu_s W$

8–55. $\theta = 33.4°$

8–57. $\mathbf{N} = (-a\mathbf{j} + h\mathbf{k}) \times (2a\mathbf{i} - a\mathbf{j})$, $n = \mathbf{N}/N$,
$h = \dfrac{2}{\sqrt{5}} a\mu$

8–58. $\theta = 33.4°$

8–59. $P = 5.53$ kN
Since a force P (> 0) is required to pull out the wedge, **the wedge will be self–locking when** $P = 0$.

8–61. $P = 34.5$ N

8–62. $x = 18.3$ mm

8–63. $P = 2.39$ kN

8–65. $P = 304$ N

8–66. $x = 32.9$ mm

8–67. $F = 3.84$ kN, $P = 73.3$ kN

8–69. $W = 7.19$kN

8–70. Since ϕs > θp, **the screw is self–locking.**

8–71. $F = 66.7$ N

8–73. $F_{AB} = 1.38$ kN (T), $F_{BD} = 828$ N (C)
$F_{BC} = 1.10$ kN (C), $F_{AC} = 828$ N (C)
$F_{AD} = 1.10$ kN (C), $F_{CD} = 1.38$ kN (T)

8–74. $M = 4.53$N · m

8–75. $M = 48.3$ N · m

8–77. $F_E = 72.7$ N
$F_D = 72.7$ N

8–78. $P = 880$ N
$M = 352$ N · m

8–79. $T = 4.02$ kN, $F = 11.6$ kN

8–81. $P = 1.98$ kN

8–82. $M = 0.202$ N · m

8–83. **a.** $F = 1.31$ kN
b. $F = 372$ N

8–85. 15.9 N < P < 217.4 N

8–86. $P = 1.54$ kN

8–87. $\mu_s = 0.0583$

8–89. Approx. 2 turns (695°)

8–90. $\theta = 99.2°$

8–91. $m_A = 2.22$ kg

8–93. $M = 177$ N · m

8–94. $m_D = 25.6$ kg

8–95. $m_A = 7.82$ kg

8–97. $P = 736$ N

8–99. $M = 187$ N · m

8–101. $m_C = 136$ kg, $M = 134$ N · m

8–102. $M = 3.93$ N · m

8–103. $F = 2.49$ kN

8–105. $M = 50.0$ N · m, $x = 286$ mm

8–106. $F_s = 85.4$ N

8–107. $M = 19$ N · m

8–109. $M = 54.4$ N · m

8–110. $M = 46.7$ N · m

8–111. $\mu_s = 0.321$

8–113. $P = 118$ N

8–114. $M = \dfrac{2\mu_s PR}{3\cos\theta}$

8–115. $M = 0.521$ PμR

8–117. $d = 0.140$ L

8–118. $(\gamma f)_A = 5$ mm, $(\gamma f)_B = 2$ mm

8–119. $(\gamma f)_A = 7.5$ mm, $(\gamma f)_B = 3$ mm

8–121. $P = 814$ N

8–122. $P = 68.97$ N

8–123. $P = 145.0$ N

8–125. $r = 20.6$ mm

8–126. $P = 299$ N

8–127. $P = 266$ N

8–129. $d = 38.5$ mm

8–130. $P = 245$ N

8–133. $M = 90.6$ N·m

8–134. $\mu_s = 0.4$

8–135. $M = 270$ N·m

8–137. $P = 40.2$ N

8–138. Check if crate slips on dolly, if crate tips, or if dolly tips. $P = 196$ N.

8–139. $m_l = 1500$ kg

8–141. $\theta = 35.0°$

8–142. $P = 140$ N

8–143. $P = 474$ N

Chapter 9

9–1. $\bar{x} = 124\text{ mm}$
$\bar{y} = 0$

9–2. $\bar{x} = 0.546\text{ m}, O_x = 0, O_y = 7.06\text{ N}$
$M_O = 3.85\text{ N}\cdot\text{m}$

9–3. $m = \frac{3}{2}m_o L,\ \bar{x} = \frac{5}{9}L$

9–5. $\bar{x} = 0.299a$
$\bar{y} = 0.537a$

9–6. $\bar{y} = \frac{2}{5}\text{ m}$

9–7. $\bar{x} = \frac{3}{8}a$

9–9. $\bar{x} = 5\text{ m}$

9–10. $\bar{y} = 1.43\text{ m}$

9–11. $\bar{x} = \frac{3b}{4}$

9–13. $\bar{x} = 6\text{ m}$

9–14. $\bar{y} = 2.8\text{ m}$

9–15. $\bar{x} = \frac{b-a}{\ln\frac{b}{a}}$

9–17. $\bar{x} = \frac{n+1}{2(n+2)}a$

9–18. $\bar{x} = \frac{(n+1)}{2(n+2)}a$

9–19. $\bar{y} = \frac{hn}{2n+1}$

9–21. $\bar{x} = \frac{3a}{8}$

9–22. $\bar{y} = \frac{3a}{5}$

9–23. $\bar{x} = \frac{4a}{3\pi}$

9–25. $\bar{x} = 0.649\text{ m}$

9–26. $\bar{x} = \left(\frac{\pi-2}{2\pi}\right)a$

9–27. $\bar{y} = \frac{\pi}{8}a$

9–29. $\bar{y} = \frac{25}{56}\text{ m}$

9–30. $\bar{x} = 1\text{ m}$

9–31. $\bar{y} = 0.4\text{ m}$

9–33. $\bar{y} = \frac{\pi a}{8}$

9–34. $\bar{x} = 1.26\text{ m}$
$\bar{y} = 0.143\text{ m}$
$N_B = 47.9\text{ kN}$
$A_x = 33.9\text{ kN}$
$A_y = 73.9\text{ kN}$

9–35. $x_c = 1.80\text{ m}$

9–37. $dm = \rho dV = \rho_0 xy\,(tydx),\ m = \frac{1}{8}\rho_0 r^4 t$
$\bar{x} = \frac{8}{15}r$
$\bar{y} = \frac{8}{15}r$

9–38. $\bar{r} = 0.833a$

9–39. $\bar{y} = 4\text{ m}$

9–41. $\bar{z} = \frac{3}{8}a$

9–42. $\bar{y} = 2.67\text{ m}$

9–43. $\bar{z} = \frac{5h}{6}$

9–45. $z_C = 0.422\text{ m}$

9–46. $\bar{z} = \frac{8}{15}r$

9–47. $\bar{z} = \frac{h}{4},\ \bar{x} = \bar{y} = \frac{a}{\pi}$

9–50. $\bar{z} = \frac{c}{4}$

9–51. $d = 3\text{ m}$

9–53. $\bar{x} = 77.3\text{ mm}$
$\bar{y} = 121\text{ mm}$

9–54. $\bar{x} = 0,\ \bar{y} = 58.3\text{ mm}$

9–55. $\bar{x} = 1.65\text{ m},\ \bar{y} = 9.25\text{ m}$
$A_x = 0,\ A_y = 1.32\text{ kN}$
$E_y = 342\text{ N}$

9–57. $\bar{x} = \frac{W_1}{W}b$
$\bar{y} = \frac{b(W_2 - W_1)\sqrt{b^2 - c^2}}{cW}$

9–58. $\bar{y} = 154\text{ mm}$

9–59. $\bar{y} = 293\text{ mm}$

9–61. $\bar{x} = 77.2\text{ mm},\ \bar{y} = 31.7\text{ mm}$

9–62. $\bar{y} = 291\text{ mm}$

9–63. $\bar{x} = 2.22\text{ m},\ \bar{y} = 1.41\text{ m}$

9–65. $x_c = 0.2\text{ m},\ y_c = 4.365\text{ m}$

9–66. $\bar{y} = 272\text{ mm}$

9–67. $\bar{y} = 66.1\text{ mm}$

9–69. $\bar{x} = \dfrac{\frac{2}{3} r \sin^3 \alpha}{\alpha - \frac{\sin 2\alpha}{2}}$

9–70. $\bar{y} = \dfrac{\sqrt{2}(a^2 + at - t^2)}{2(2a - t)}$

9–71. $\bar{y} = 85.9\,\text{mm}$

9–73. $n \le \dfrac{L}{d}$

9–74. $\bar{z} = 122\,\text{mm}$

9–75. $h = 385$ mm

9–77. $y_c = 56.6$ m

9–78. $\bar{y} = 53.0\,\text{mm}$

9–79. $\bar{x} = 22.7\,\text{mm}$
$\bar{y} = 29.5\,\text{mm}$
$\bar{z} = 22.6\,\text{mm}$

9–81. $\bar{z} = 359\,\text{mm}$

9–82. $h = 323$ mm

9–83. $\bar{z} = 128\,\text{mm}$

9–85. $h = 48$ mm

9–86. $\bar{z} = 463\,\text{mm}$

9–87. $\bar{z} = 58.1\,\text{mm}$

9–89. $\bar{z} = 754\,\text{mm}$

9–90. $V = 25.5\ \text{m}^3$

9–91. 14.4 liters

9–93. $V = 22.1\ \text{m}^3$

9–94. $V = 4.25\ (10^6)\ \text{mm}^3$

9–95. $A = 302\ \text{m}^2$

9–97. $A = 188\ \text{m}^2$

9–98. $V = 207\ \text{m}^3$

9–99. $V_c = 20.5\ \text{m}^3$

9–101. $A = 8\pi ba,\ V = 2\pi ba^2$

9–102. $A = 1.403\ (10^6)\ \text{mm}^3$

9–103. $A = 88.0\ \text{m}^2$

9–105. $m = 138$ kg

9–106. $V = \dfrac{2}{3}\pi ab^2$

9–107. $A = 1.33\ \text{m}^2$
$\bar{x} = 0.6\,\text{m}$
$V = 5.03\ \text{m}^3$

9–109. $A = 119\ (10^3)\ \text{mm}^2$

9–110. $A = \sqrt{3}\pi a^2$
$V = \dfrac{\pi}{4} a^3$

9–111. $h = 29.9$ mm

9–113. $m = 138$ kg

9–114. $A = 1365\ \text{m}^2$

9–115. $d = 2.68$ m

9–117. $F_R = \dfrac{2}{3}(xdx)[(4 - y)dy],\ F_R = 24.0\,\text{kN}$
$\bar{x} = 2.00\,\text{m},\ \bar{y} = 1.33\,\text{m}$

9–118. $F_R = \dfrac{4ab}{\pi^2} p_0$
$\bar{x} = \dfrac{a}{2}\quad \bar{y} = \dfrac{b}{2}$

9–119. $A_y = 2.51$ MN
$B_x = 2.20$ MN
$B_y = 859$ kN

9–121. $D_x = 101$ kN
$C_x = 46.6$ kN

9–122. $d = 3.65$ m

9–123. $F = 1.41$ MN, $h = 4$ m

9–125. $F_R = 2.77$ MN
$h = 5.22$ m, $F_{bot} = 3.02$ MN

9–126. $F = 3.85$ kN
$d' = 0.625$ m

9–127. $F_R = 6.93$ kN, $\bar{y} = -0.125\,\text{m}$

9–129. $F = 391$ kN/m

9–130. $F_x = 628$ kN
$F_y = 538$ kN

9–131. $V = 22.7\ (10)^{-3}\ \text{m}^3$

9–133. $\bar{y} = 87.5\,\text{mm}$

9–134. $\bar{x} = \bar{y} = 0,\ \bar{z} = \dfrac{2}{3}a$

9–135. $x_C = \dfrac{a + b}{3}$

9–137. $x_C = 1.594$ m, $y_C = 0.940$ m

9–138. $l = 265$ mm
$\theta = 70.4°$

9–139. $A_y = 1.77$ MN, $B_x = 1.57$ MN, $B_y = 594$ kN

Chapter 10

10–1. $I_x = \dfrac{2}{7} bh^3$

10–2. $I_y = \dfrac{2}{15} hb^3$

10–3. $I_x = 2.13\ \text{m}^4$

10–5. $I_x = 39.0\ \text{m}^4$

10–6. $I_y = 8.53\ \text{m}^4$

10–7. $I_x = 0.533\ \text{m}^4$

10–9. $I_x = \dfrac{2}{15} bh^3$

10–10. $I_x = \dfrac{1}{3} t l^3 \sin^2(\theta)$

10–11. $I_x = 3.20\ \text{m}^4$

10–13. $I_x = 4.27\ \text{m}^4$

10–14. $I_y = 36.7\ \text{m}^4$

10–15. $I_y = 0.628\ \text{m}^4$

10–17. $I_x = \dfrac{4a^4}{9\pi}$

10–18. $I_y = \left(\dfrac{\pi^2 - 4}{\pi^3}\right)a^4$

10–19. $I_x = 10\ \text{m}^4$

10–21. $I_x = \dfrac{ah^3}{28}$

10–22. $I_y = \dfrac{a^3h}{20}$

10–23. $dA = (rd\,\theta)dr,\ y = \text{r}\sin\theta,\ I_x = \dfrac{r_0^4}{8}(\alpha - \sin\alpha)$

10–25. $I_x = 798\ (10^6)\ \text{mm}^4$

10–26. $I_y = 10.3\ (10^9)\ \text{mm}^4$

10–27. $k_x = 103.5\ \text{mm}$

10–29. $\bar{y} = 0.181\text{m},\ I_{x'} = 4.23(10^{-3})\text{m}^4$

10–30. $\bar{x} = 68.0\text{mm},\ I_{y'} = 36.9(10^6)\text{mm}^4$

10–31. $I_{x'} = 49.5(10^6)\text{mm}^4$

10–33. $I_y = \dfrac{r^4}{4}(\theta + \dfrac{1}{2}\sin 2\theta - 2\sin\theta\cos^3\theta)$

10–34. $I_y = 115(10^6)\ \text{mm}^4$

10–35. $\bar{y} = 207\ \text{mm}$
$\bar{I}_{x'} = 222(10^6)\ \text{mm}^4$

10–37. $I_y = 2.51(10^6)\ \text{mm}^4$

10–38. $I_x = 115(10^6)\ \text{mm}^4$

10–39. $I_y = 153(10^6)\ \text{mm}^4$

10–41. $\bar{x} = 71.32\ \text{mm}$
$I_{y'} = 3.60(10^6)\ \text{mm}^4$

10–42. $I_x = 154(10^6)\ \text{mm}^4$

10–43. $I_y = 91.3(10^6)\ \text{mm}^4$

10–45. $\bar{x} = 61.6\ \text{mm},\ \bar{I}_{y'} = 41.2(10^6)\ \text{mm}^4$

10–46. $\bar{y} = 22.5\ \text{mm},\ I_{x'} = 34.4(10^6)\ \text{mm}^4$

10–47. $I_y = 122(10^6)\ \text{mm}^4$

10–49. $I_y = 29.8(10^6)\ \text{mm}^4$

10–50. $\bar{y} = 1.81\ \text{m},\ \bar{I}_x = 42.33\ \text{m}^4$

10–51. $\bar{I}_{x'} = 162(10^6)\ \text{mm}^4$

10–53. $\bar{I}_{y'} = \dfrac{ab\sin\theta}{12}(b^2 + a^2\cos^2\theta)$

10–54. $I_{xy} = \dfrac{1}{6}l^3 t\sin 2\theta$

10–55. $I_{xy} = \dfrac{3}{16}b^2h^2$

10–57. $I_{xy} = \dfrac{a^2b^2}{4(n+1)}$

10–58. $\bar{I}_{x'y'} = 1.07\text{m}^4$

10–59. $I_{xy} = 0.511\ \text{m}^4$

10–61. $I_{xy} = \dfrac{a^2c\sin^2\theta}{12}(4a\cos\theta + 3c)$

10–62. $I_{xy} = \dfrac{1}{6}a^2b^2$

10–63. $I_{uv} = 135(10)^6\ \text{mm}^4$

10–65. $I_{xy} = -1.10(10^6)\ \text{mm}^4$

10–66. $I_{xy} = 98.4(10^6)\ \text{mm}^4$

10–67. $I_{xy} = 17.1\ (10^6)\ \text{mm}^4$

10–69. $I_u = 85.3\ (10^6)\ \text{mm}^4$
$I_v = 85.3\ (10^6)\ \text{mm}^4$

10–70. $I_u = 909\ (10^6)\ \text{mm}^4$
$I_v = 703\ (10^6)\ \text{mm}^4$
$I_{uv} = 179\ (10^6)\ \text{mm}^4$

10–71. $I_u = 909\ (10^6)\ \text{mm}^4$
$I_v = 703\ (10^6)\ \text{mm}^4$
$I_{uv} = 179\ (10^6)\ \text{mm}^4$

10–73. $\bar{y} = 82.5\ \text{mm}$
$I_u = 109\ (10^8)\ \text{mm}^4$
$I_v = 238\ (10^8)\ \text{mm}^4$
$I_{uv} = 111\ (10^8)\ \text{mm}^4$

10–74. $\bar{y} = 82.5\ \text{mm}$
$\text{I}_u = 43.4\ (10^6)\ \text{mm}^4$
$\text{I}_v = 47.0\ (10^6)\ \text{mm}^4$
$\text{I}_{uv} = -3.08\ (10^6)\ \text{mm}^4$

10–75. $\bar{y} = 82.5\ \text{mm}$
$\text{I}_u = 43.4\ (10^6)\ \text{mm}^4$
$\text{I}_v = 47.0\ (10^6)\ \text{mm}^4$
$\text{I}_{uv} = -3.08\ (10^6)\ \text{mm}^4$

10–77. $\bar{x} = 48.2\text{mm}$
$\text{I}_u = 112\ (10^6)\ \text{mm}^4$
$\text{I}_v = 258\ (10^6)\ \text{mm}^4$
$\text{I}_{uv} = -126\ (10^6)\ \text{mm}^4$

10–78. $I_{max} = 4.92\ (10^6)\ \text{mm}^4,\ I_{min} = 1.36\ (10^6)\ \text{mm}^4$

10–79. $I_{max} = 4.92\ (10^6)\ \text{mm}^4,\ I_{min} = 1.36\ (10^6)\ \text{mm}^4$

10–81. $\theta = 12.3°$ (*Counterclockwise*)
$\theta = 77.7°$ (*Clockwise*)

10–82. $\bar{y} = 79.23\text{mm}$
$\text{I}_u = 42.2\ (10^6)\ \text{mm}^4$
$\text{I}_v = 41.9\ (10^6)\ \text{mm}^4$
$\text{I}_{uv} = 0.28\ (10^6)\ \text{mm}^4$

10–83. $\bar{y} = 79.23\text{mm}$
$\text{I}_u = 42.19\ (10^6)\ \text{mm}^4$
$\text{I}_v = 41.86\ (10^6)\ \text{mm}^4$
$\text{I}_{uv} = 0.28\ (10^6)\ \text{mm}^4$

10–85. $I_x = \dfrac{2}{5}\text{m}b^2$

10–86. $I_x = \dfrac{3}{10}\text{m}r^2$

10–87. $k_x = 57.7\ \text{mm}$

10–89. $I_y = \frac{m}{6}(a^2 + h^2)$

10–90. $I_z = \frac{7}{18}ma^2$

10–91. $I_x = \frac{2}{5}mr^2$

10–93. $I_y = \frac{5}{18}m$

10–94. $I_y = 1.71(10^3)\text{ kg}\cdot\text{m}^2$

10–95. $I_z = 34.2\text{ kg}\cdot\text{m}^2$

10–97. $\bar{y} = 1.78\text{ m}$, $I_G = 4.45\text{ kg}\cdot\text{m}^2$

10–98. $\bar{y} = 203\text{ mm}$, $I_G = 0.230\text{ kg}\cdot\text{m}^2$

10–99. $I_A = 327.3\text{ kg}\cdot\text{m}^2$

10–101. $I_z = 1.53\text{ kg}\cdot\text{m}^2$

10–102. $k_O = 3.15\text{ m}$

10–103. $I_A = 50.75\text{ kg}\cdot\text{m}^2$

10–105. $I_O = 0.450\,(10^{-3})\text{ kg}\cdot\text{m}^2$
$\bar{y} = 57.7\text{ mm}$
$I_G = 0.150\,(10^{-3})\text{ kg}\cdot\text{m}^2$

10–106. $I_y = 0.144\text{ kg}\cdot\text{m}^2$

10–107. $I_z = 0.113\text{ kg}\cdot\text{m}^2$

10–109. $I_{z'} = 7.19\text{ kg}\cdot\text{m}^2$

10–110. $k_x = 109\text{ mm}$

10–111. $I_x = 914\,(10^6)\text{ mm}^4$
$\bar{I}_{x'} = 146(10^6)\text{ mm}^4$

10–113. **a.** $I_x = \frac{bh^3}{12}$
b. $\bar{I}_{x'} = \frac{bh^3}{36}$

10–114. $I_x = \frac{93}{70}\text{m}b^2$

10–115. $I_x = 9.05\text{ m}^4$

10–117. $I_u = 5.09\,(10^6)\text{ mm}^4$, $I_v = 5.09\,(10^6)\text{ mm}^4$, $I_{uv} = 0$

10–118. $I_y = 0.0954\,d^4$

10–119. $I_y = 0.187\,d^4$

Section 3
Mechanics for Engineers – Dynamics

The content in this section is sourced from Hibbeler and Yap's Mechanics for Engineers – Dynamics, 13th SI edition

Hibbeler, R.C. & Yap, K.B. (2013). *Mechanics for Engineers SI – Mechanics* (13th ed.). Jurong, Singapore: Pearson Education South Asia Pte Ltd.

Chapter 12

Although each of these boats is rather large, from a distance their motion can be analyzed as if each were a particle.

Kinematics of a Particle

CHAPTER OBJECTIVES

- To introduce the concepts of position, displacement, velocity, and acceleration.
- To study particle motion along a straight line and represent this motion graphically.
- To investigate particle motion along a curved path using different coordinate systems.
- To present an analysis of dependent motion of two particles.
- To examine the principles of relative motion of two particles using translating axes.

Video Solutions are available for selected questions in this chapter.

12.1 Introduction

Mechanics is a branch of the physical sciences that is concerned with the state of rest or motion of bodies subjected to the action of forces. Engineering mechanics is divided into two areas of study, namely, statics and dynamics. *Statics* is concerned with the equilibrium of a body that is either at rest or moves with constant velocity. Here we will consider *dynamics*, which deals with the accelerated motion of a body. The subject of dynamics will be presented in two parts: *kinematics*, which treats only the geometric aspects of the motion, and *kinetics*, which is the analysis of the forces causing the motion. To develop these principles, the dynamics of a particle will be discussed first, followed by topics in rigid-body dynamics in two and then three dimensions.

Historically, the principles of dynamics developed when it was possible to make an accurate measurement of time. Galileo Galilei (1564–1642) was one of the first major contributors to this field. His work consisted of experiments using pendulums and falling bodies. The most significant contributions in dynamics, however, were made by Isaac Newton (1642–1727), who is noted for his formulation of the three fundamental laws of motion and the law of universal gravitational attraction. Shortly after these laws were postulated, important techniques for their application were developed by Euler, D'Alembert, Lagrange, and others.

There are many problems in engineering whose solutions require application of the principles of dynamics. Typically the structural design of any vehicle, such as an automobile or airplane, requires consideration of the motion to which it is subjected. This is also true for many mechanical devices, such as motors, pumps, movable tools, industrial manipulators, and machinery. Furthermore, predictions of the motions of artificial satellites, projectiles, and spacecraft are based on the theory of dynamics. With further advances in technology, there will be an even greater need for knowing how to apply the principles of this subject.

Problem Solving. Dynamics is considered to be more involved than statics since both the forces applied to a body and its motion must be taken into account. Also, many applications require using calculus, rather than just algebra and trigonometry. In any case, the most effective way of learning the principles of dynamics is *to solve problems*. To be successful at this, it is necessary to present the work in a logical and orderly manner as suggested by the following sequence of steps:

1. Read the problem carefully and try to correlate the actual physical situation with the theory you have studied.
2. Draw any necessary diagrams and tabulate the problem data.
3. Establish a coordinate system and apply the relevant principles, generally in mathematical form.
4. Solve the necessary equations algebraically as far as practical; then, use a consistent set of units and complete the solution numerically. Report the answer with no more significant figures than the accuracy of the given data.
5. Study the answer using technical judgment and common sense to determine whether or not it seems reasonable.
6. Once the solution has been completed, review the problem. Try to think of other ways of obtaining the same solution.

In applying this general procedure, do the work as neatly as possible. Being neat generally stimulates clear and orderly thinking, and vice versa.

12.2 Rectilinear Kinematics: Continuous Motion

We will begin our study of dynamics by discussing the kinematics of a particle that moves along a rectilinear or straight-line path. Recall that a *particle* has a mass but negligible size and shape. Therefore we must limit application to those objects that have dimensions that are of no consequence in the analysis of the motion. In most problems, we will be interested in bodies of finite size, such as rockets, projectiles, or vehicles. Each of these objects can be considered as a particle, as long as the motion is characterized by the motion of its mass center and any rotation of the body is neglected.

Rectilinear Kinematics. The kinematics of a particle is characterized by specifying, at any given instant, the particle's position, velocity, and acceleration.

Position. The straight-line path of a particle will be defined using a single coordinate axis s, Fig. 12–1*a*. The origin O on the path is a fixed point, and from this point the *position coordinate* s is used to specify the location of the particle at any given instant. The magnitude of s is the distance from O to the particle, usually measured in meters (m), and the sense of direction is defined by the algebraic sign on s. Although the choice is arbitrary, in this case s is positive since the coordinate axis is positive to the right of the origin. Likewise, it is negative if the particle is located to the left of O. Realize that position is a vector quantity since it has both magnitude and direction. Here, however, it is being represented by the algebraic scalar s since the direction always remains along the coordinate axis.

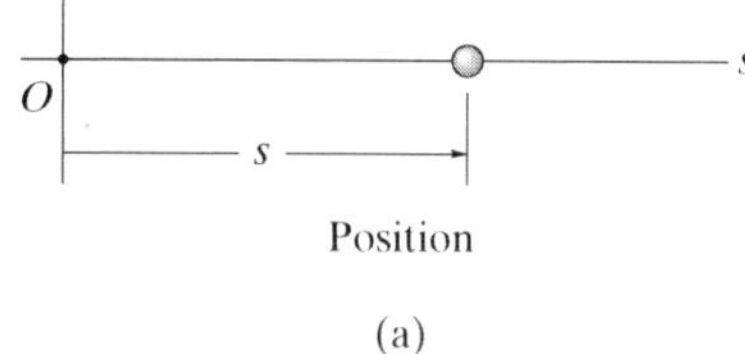

Position

(a)

Displacement. The *displacement* of the particle is defined as the *change* in its *position*. For example, if the particle moves from one point to another, Fig. 12–1*b*, the displacement is

$$\Delta s = s' - s$$

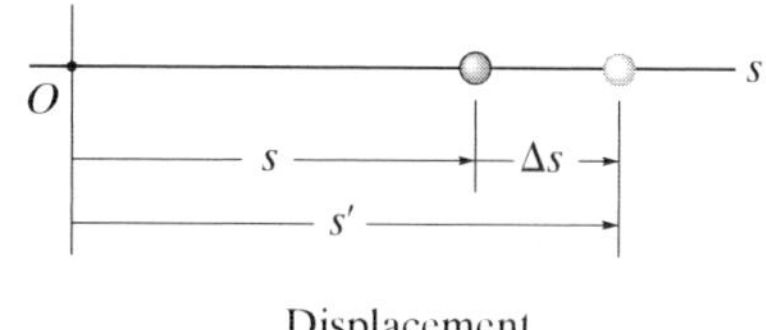

Displacement

(b)

Fig. 12–1

In this case Δs is *positive* since the particle's final position is to the *right* of its initial position, i.e., $s' > s$. Likewise, if the final position were to the *left* of its initial position, Δs would be *negative*.

The displacement of a particle is also a *vector quantity*, and it should be distinguished from the distance the particle travels. Specifically, the *distance traveled* is a *positive scalar* that represents the total length of path over which the particle travels.

Velocity. If the particle moves through a displacement Δs during the time interval Δt, the *average velocity* of the particle during this time interval is

$$v_{\text{avg}} = \frac{\Delta s}{\Delta t}$$

If we take smaller and smaller values of Δt, the magnitude of Δs becomes smaller and smaller. Consequently, the *instantaneous velocity* is a vector defined as $v = \lim_{\Delta t \to 0} (\Delta s/\Delta t)$, or

$$(\overset{+}{\rightarrow}) \qquad v = \frac{ds}{dt} \qquad (12\text{–}1)$$

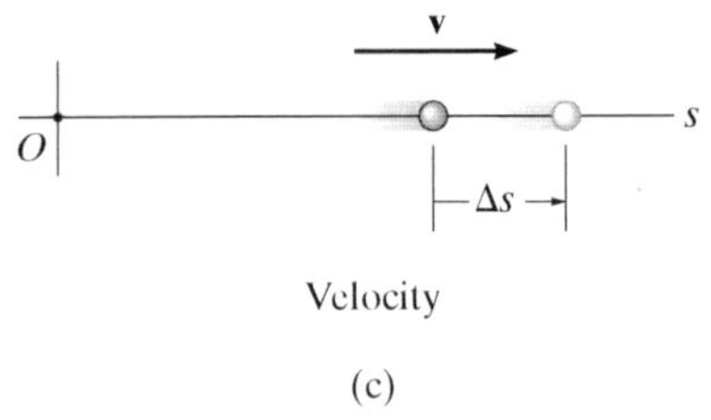

Velocity

(c)

Since Δt or dt is always positive, the sign used to define the *sense* of the velocity is the same as that of Δs or ds. For example, if the particle is moving to the *right*, Fig. 12–1*c*, the velocity is *positive;* whereas if it is moving to the *left*, the velocity is *negative*. (This is emphasized here by the arrow written at the left of Eq. 12–1.) The *magnitude* of the velocity is known as the *speed*, and it is generally expressed in units of m/s.

Occasionally, the term "average speed" is used. The *average speed* is always a positive scalar and is defined as the total distance traveled by a particle, s_T, divided by the elapsed time Δt; i.e.,

$$(v_{\text{sp}})_{\text{avg}} = \frac{s_T}{\Delta t}$$

For example, the particle in Fig. 12–1*d* travels along the path of length s_T in time Δt, so its average speed is $(v_{\text{sp}})_{\text{avg}} = s_T/\Delta t$, but its average velocity is $v_{\text{avg}} = -\Delta s/\Delta t$.

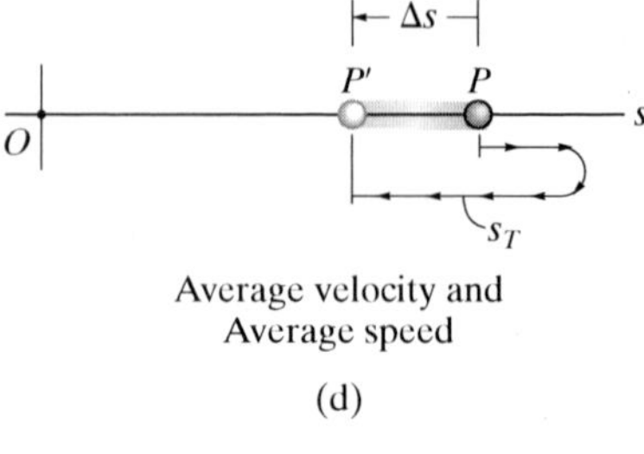

Average velocity and
Average speed

(d)

Fig. 12–1 (cont.)

Acceleration. Provided the velocity of the particle is known at two points, the *average acceleration* of the particle during the time interval Δt is defined as

$$a_{\text{avg}} = \frac{\Delta v}{\Delta t}$$

Here Δv represents the difference in the velocity during the time interval Δt, i.e., $\Delta v = v' - v$, Fig. 12–1*e*.

The *instantaneous acceleration* at time t is a vector that is found by taking smaller and smaller values of Δt and corresponding smaller and smaller values of Δv, so that $a = \lim_{\Delta t \to 0} (\Delta v / \Delta t)$, or

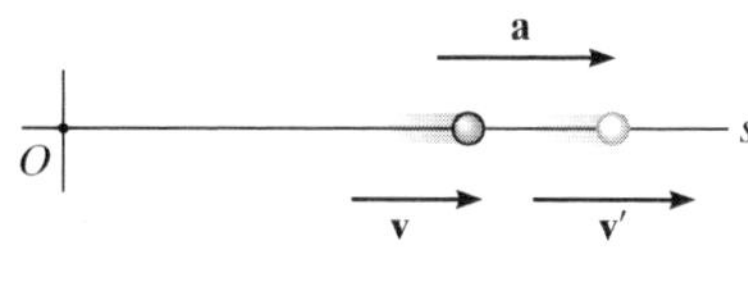

Acceleration

(e)

$$(\overset{+}{\rightarrow}) \qquad a = \frac{dv}{dt} \qquad (12\text{–}2)$$

Substituting Eq. 12–1 into this result, we can also write

$$(\overset{+}{\rightarrow}) \qquad a = \frac{d^2s}{dt^2}$$

Both the average and instantaneous acceleration can be either positive or negative. In particular, when the particle is *slowing down*, or its speed is decreasing, the particle is said to be *decelerating*. In this case, v' in Fig. 12–1*f* is *less* than v, and so $\Delta v = v' - v$ will be negative. Consequently, a will also be negative, and therefore it will act to the *left*, in the *opposite sense* to v. Also, notice that if the particle is originally at rest, then it can have an acceleration if a moment later it has a velocity v'; and, if the *velocity* is *constant*, then the *acceleration is zero* since $\Delta v = v - v = 0$. Units commonly used to express the magnitude of acceleration are m/s^2.

Deceleration

(f)

Fig. 12–1 (cont.)

Finally, an important differential relation involving the displacement, velocity, and acceleration along the path may be obtained by eliminating the time differential dt between Eqs. 12–1 and 12–2, which gives

$$(\overset{+}{\rightarrow}) \qquad a\,ds = v\,dv \qquad (12\text{–}3)$$

Although we have now produced three important kinematic equations, realize that the above equation is not independent of Eqs. 12–1 and 12–2.

Constant Acceleration, $a = a_c$. When the acceleration is constant, each of the three kinematic equations $a_c = dv/dt$, $v = ds/dt$, and $a_c\,ds = v\,dv$ can be integrated to obtain formulas that relate a_c, v, s, and t.

Velocity as a Function of Time. Integrate $a_c = dv/dt$, assuming that initially $v = v_0$ when $t = 0$.

$$\int_{v_0}^{v} dv = \int_0^t a_c\,dt$$

$$(\overset{+}{\rightarrow}) \qquad \boxed{v = v_0 + a_c t} \qquad (12\text{–}4)$$

Constant Acceleration

Position as a Function of Time. Integrate $v = ds/dt = v_0 + a_c t$, assuming that initially $s = s_0$ when $t = 0$.

$$\int_{s_0}^{s} ds = \int_0^t (v_0 + a_c t)\,dt$$

$$(\overset{+}{\rightarrow}) \qquad \boxed{s = s_0 + v_0 t + \tfrac{1}{2}a_c t^2} \qquad (12\text{–}5)$$

Constant Acceleration

Velocity as a Function of Position. Either solve for t in Eq. 12–4 and substitute into Eq. 12–5, or integrate $v\,dv = a_c\,ds$, assuming that initially $v = v_0$ at $s = s_0$.

$$\int_{v_0}^{v} v\,dv = \int_{s_0}^{s} a_c\,ds$$

$$(\overset{+}{\rightarrow}) \qquad \boxed{v^2 = v_0^2 + 2a_c(s - s_0)} \qquad (12\text{–}6)$$

Constant Acceleration

The algebraic signs of s_0, v_0, and a_c, used in the above three equations, are determined from the positive direction of the s axis as indicated by the arrow written at the left of each equation. Remember that these equations are useful *only when the acceleration is constant and when* $t = 0$, $s = s_0$, $v = v_0$. A typical example of constant accelerated motion occurs when a body falls freely toward the earth. If air resistance is neglected and the distance of fall is short, then the *downward* acceleration of the body when it is close to the earth is constant and approximately $9.81\ \text{m/s}^2$. The proof of this is given in Example 13.2.

Important Points

- Dynamics is concerned with bodies that have accelerated motion.
- Kinematics is a study of the geometry of the motion.
- Kinetics is a study of the forces that cause the motion.
- Rectilinear kinematics refers to straight-line motion.
- Speed refers to the magnitude of velocity.
- Average speed is the total distance traveled divided by the total time. This is different from the average velocity, which is the displacement divided by the time.
- A particle that is slowing down is decelerating.
- A particle can have an acceleration and yet have zero velocity.
- The relationship $a\,ds = v\,dv$ is derived from $a = dv/dt$ and $v = ds/dt$, by eliminating dt.

During the time this rocket undergoes rectilinear motion, its altitude as a function of time can be measured and expressed as $s = s(t)$. Its velocity can then be found using $v = ds/dt$, and its acceleration can be determined from $a = dv/dt$.

Procedure for Analysis

Coordinate System.

- Establish a position coordinate s along the path and specify its *fixed origin* and positive direction.
- Since motion is along a straight line, the vector quantities position, velocity, and acceleration can be represented as algebraic scalars. For analytical work the sense of s, v, and a is then defined by their *algebraic signs*.
- The positive sense for each of these scalars can be indicated by an arrow shown alongside each kinematic equation as it is applied.

Kinematic Equations.

- If a relation is known between any *two* of the four variables a, v, s, and t, then a third variable can be obtained by using one of the kinematic equations, $a = dv/dt$, $v = ds/dt$ or $a\,ds = v\,dv$, since each equation relates all three variables.*
- Whenever integration is performed, it is important that the position and velocity be known at a given instant in order to evaluate either the constant of integration if an indefinite integral is used, or the limits of integration if a definite integral is used.
- Remember that Eqs. 12–4 through 12–6 have only limited use. These equations apply *only* when the *acceleration is constant* and the initial conditions are $s = s_0$ and $v = v_0$ when $t = 0$.

*Some standard differentiation and integration formulas are given in Appendix A.

12

EXAMPLE 12.1

The car on the left in the photo and in Fig. 12–2 moves in a straight line such that for a short time its velocity is defined by $v = (3t^2 + 2t)$ m/s, where t is in seconds. Determine its position and acceleration when $t = 3$ s. When $t = 0$, $s = 0$.

© Csaba Vanyi/Shutterstock.com

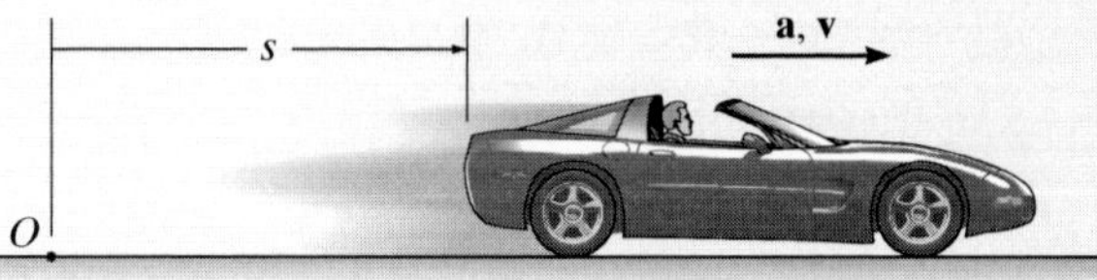

Fig. 12–2

SOLUTION

Coordinate System. The position coordinate extends from the fixed origin O to the car, positive to the right.

Position. Since $v = f(t)$, the car's position can be determined from $v = ds/dt$, since this equation relates v, s, and t. Noting that $s = 0$ when $t = 0$, we have*

$$(\overset{+}{\rightarrow}) \qquad v = \frac{ds}{dt} = (3t^2 + 2t)$$

$$\int_0^s ds = \int_0^t (3t^2 + 2t)dt$$

$$s\Big|_0^s = t^3 + t^2\Big|_0^t$$

$$s = t^3 + t^2$$

When $t = 3$ s,

$$s = (3)^3 + (3)^2 = 36 \text{ m} \qquad \textit{Ans.}$$

Acceleration. Since $v = f(t)$, the acceleration is determined from $a = dv/dt$, since this equation relates a, v, and t.

$$(\overset{+}{\rightarrow}) \qquad a = \frac{dv}{dt} = \frac{d}{dt}(3t^2 + 2t)$$

$$= 6t + 2$$

When $t = 3$ s,

$$a = 6(3) + 2 = 20 \text{ m/s}^2 \rightarrow \qquad \textit{Ans.}$$

NOTE: The formulas for constant acceleration *cannot* be used to solve this problem, because the acceleration is a function of time.

*The *same result* can be obtained by evaluating a constant of integration C rather than using definite limits on the integral. For example, integrating $ds = (3t^2 + 2t)dt$ yields $s = t^3 + t^2 + C$. Using the condition that at $t = 0$, $s = 0$, then $C = 0$.

EXAMPLE 12.2

A small projectile is fired vertically *downward* into a fluid medium with an initial velocity of 60 m/s. Due to the drag resistance of the fluid the projectile experiences a deceleration of $a = (-0.4v^3)\ \text{m/s}^2$, where v is in m/s. Determine the projectile's velocity and position 4 s after it is fired.

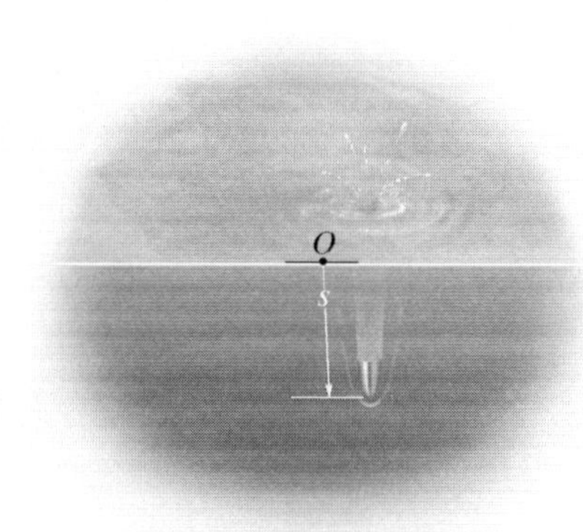

Fig. 12–3

SOLUTION

Coordinate System. Since the motion is downward, the position coordinate is positive downward, with origin located at O, Fig. 12–3.

Velocity. Here $a = f(v)$ and so we must determine the velocity as a function of time using $a = dv/dt$, since this equation relates v, a, and t. (Why not use $v = v_0 + a_c t$?) Separating the variables and integrating, with $v_0 = 60$ m/s when $t = 0$, yields

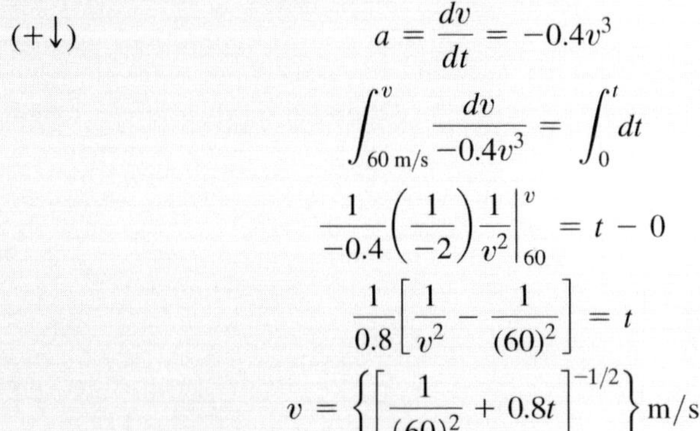

$$(+\downarrow) \qquad a = \frac{dv}{dt} = -0.4v^3$$

$$\int_{60\text{ m/s}}^{v} \frac{dv}{-0.4v^3} = \int_0^t dt$$

$$\frac{1}{-0.4}\left(\frac{1}{-2}\right)\frac{1}{v^2}\bigg|_{60}^{v} = t - 0$$

$$\frac{1}{0.8}\left[\frac{1}{v^2} - \frac{1}{(60)^2}\right] = t$$

$$v = \left\{\left[\frac{1}{(60)^2} + 0.8t\right]^{-1/2}\right\}\text{ m/s}$$

Here the positive root is taken, since the projectile will continue to move downward. When $t = 4$ s,

$$v = 0.559\text{ m/s}\downarrow \qquad \textit{Ans.}$$

Position. Knowing $v = f(t)$, we can obtain the projectile's position from $v = ds/dt$, since this equation relates s, v, and t. Using the initial condition $s = 0$, when $t = 0$, we have

$$(+\downarrow) \qquad v = \frac{ds}{dt} = \left[\frac{1}{(60)^2} + 0.8t\right]^{-1/2}$$

$$\int_0^s ds = \int_0^t \left[\frac{1}{(60)^2} + 0.8t\right]^{-1/2} dt$$

$$s = \frac{2}{0.8}\left[\frac{1}{(60)^2} + 0.8t\right]^{1/2}\bigg|_0^t$$

$$s = \frac{1}{0.4}\left\{\left[\frac{1}{(60)^2} + 0.8t\right]^{1/2} - \frac{1}{60}\right\}\text{ m}$$

When $t = 4$ s,

$$s = 4.43\text{ m} \qquad \textit{Ans.}$$

12

EXAMPLE 12.3

During a test a rocket travels upward at 75 m/s, and when it is 40 m from the ground its engine fails. Determine the maximum height s_B reached by the rocket and its speed just before it hits the ground. While in motion the rocket is subjected to a constant downward acceleration of 9.81 m/s^2 due to gravity. Neglect the effect of air resistance.

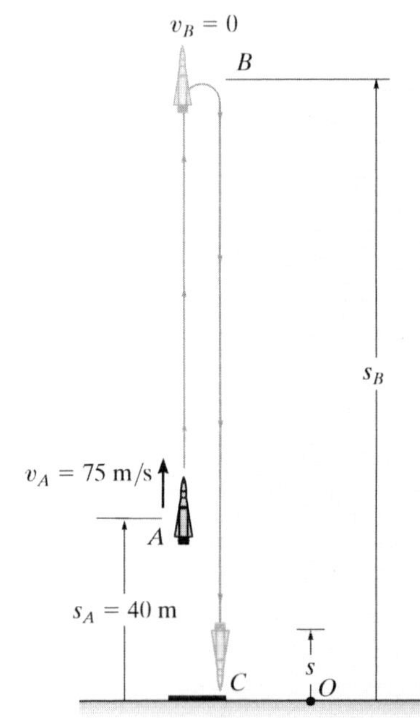

Fig. 12–4

SOLUTION

Coordinate System. The origin O for the position coordinate s is taken at ground level with positive upward, Fig. 12–4.

Maximum Height. Since the rocket is traveling *upward*, $v_A = +75$ m/s when $t = 0$. At the maximum height $s = s_B$ the velocity $v_B = 0$. For the entire motion, the acceleration is $a_c = -9.81$ m/s^2 (negative since it acts in the *opposite* sense to positive velocity or positive displacement). Since a_c is *constant* the rocket's position may be related to its velocity at the two points A and B on the path by using Eq. 12–6, namely,

$$(+\uparrow) \qquad v_B^2 = v_A^2 + 2a_c(s_B - s_A)$$

$$0 = (75\text{ m/s})^2 + 2(-9.81\text{ m/s}^2)(s_B - 40\text{ m})$$

$$s_B = 327\text{ m} \qquad \textit{Ans.}$$

Velocity. To obtain the velocity of the rocket just before it hits the ground, we can apply Eq. 12–6 between points B and C, Fig. 12–4.

$$(+\uparrow) \qquad v_C^2 = v_B^2 + 2a_c(s_C - s_B)$$

$$= 0 + 2(-9.81\text{ m/s}^2)(0 - 327\text{ m})$$

$$v_C = -80.1\text{ m/s} = 80.1\text{ m/s}\downarrow \qquad \textit{Ans.}$$

The negative root was chosen since the rocket is moving downward.

Similarly, Eq. 12–6 may also be applied between points A and C, i.e.,

$$(+\uparrow) \qquad v_C^2 = v_A^2 + 2a_c(s_C - s_A)$$

$$= (75\text{ m/s})^2 + 2(-9.81\text{ m/s}^2)(0 - 40\text{ m})$$

$$v_C = -80.1\text{ m/s} = 80.1\text{ m/s}\downarrow \qquad \textit{Ans.}$$

NOTE: It should be realized that the rocket is subjected to a *deceleration* from A to B of 9.81 m/s^2, and then from B to C it is *accelerated* at this rate. Furthermore, even though the rocket momentarily comes to *rest* at B ($v_B = 0$) the acceleration at B is still 9.81 m/s^2 downward!

12

EXAMPLE 12.4

A metallic particle is subjected to the influence of a magnetic field as it travels downward through a fluid that extends from plate A to plate B, Fig. 12–5. If the particle is released from rest at the midpoint C, $s = 100$ mm, and the acceleration is $a = (4s)\ \text{m/s}^2$, where s is in meters, determine the velocity of the particle when it reaches plate B, $s = 200$ mm, and the time it takes to travel from C to B.

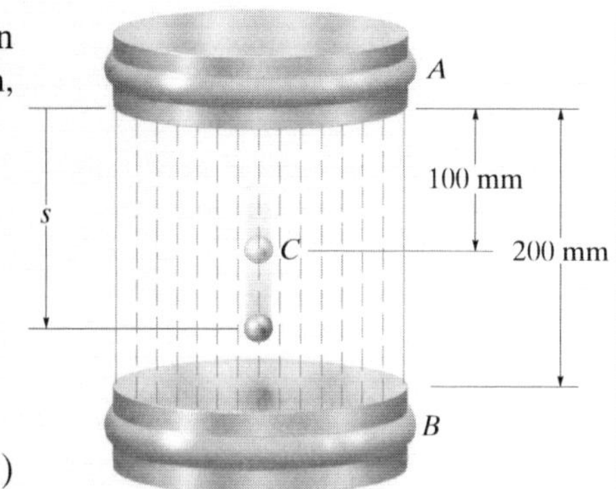

Fig. 12–5

SOLUTION

Coordinate System. As shown in Fig. 12–5, s is positive downward, measured from plate A.

Velocity. Since $a = f(s)$, the velocity as a function of position can be obtained by using $v\,dv = a\,ds$. Realizing that $v = 0$ at $s = 0.1$ m, we have

$(+\downarrow)$

$$v\,dv = a\,ds$$

$$\int_0^v v\,dv = \int_{0.1\text{ m}}^s 4s\,ds$$

$$\frac{1}{2}v^2\Big|_0^v = \frac{4}{2}s^2\Big|_{0.1\text{ m}}^s$$

$$v = 2(s^2 - 0.01)^{1/2}\ \text{m/s} \qquad (1)$$

At $s = 200$ mm $= 0.2$ m,

$$v_B = 0.346\ \text{m/s} = 346\ \text{mm/s}\downarrow \qquad \textit{Ans.}$$

The positive root is chosen since the particle is traveling downward, i.e., in the $+s$ direction.

Time. The time for the particle to travel from C to B can be obtained using $v = ds/dt$ and Eq. 1, where $s = 0.1$ m when $t = 0$. From Appendix A,

$(+\downarrow)$

$$ds = v\,dt$$

$$= 2(s^2 - 0.01)^{1/2}dt$$

$$\int_{0.1}^s \frac{ds}{(s^2 - 0.01)^{1/2}} = \int_0^t 2\,dt$$

$$\ln\left(\sqrt{s^2 - 0.01} + s\right)\Big|_{0.1}^s = 2t\Big|_0^t$$

$$\ln\left(\sqrt{s^2 - 0.01} + s\right) + 2.303 = 2t$$

At $s = 0.2$ m,

$$t = \frac{\ln\left(\sqrt{(0.2)^2 - 0.01} + 0.2\right) + 2.303}{2} = 0.658\ \text{s} \qquad \textit{Ans.}$$

NOTE: The formulas for constant acceleration cannot be used here because the acceleration changes with position, i.e., $a = 4s$.

EXAMPLE 12.5

A particle moves along a horizontal path with a velocity of $v = (3t^2 - 6t)$ m/s, where t is the time in seconds. If it is initially located at the origin O, determine the distance traveled in 3.5 s, and the particle's average velocity and average speed during the time interval.

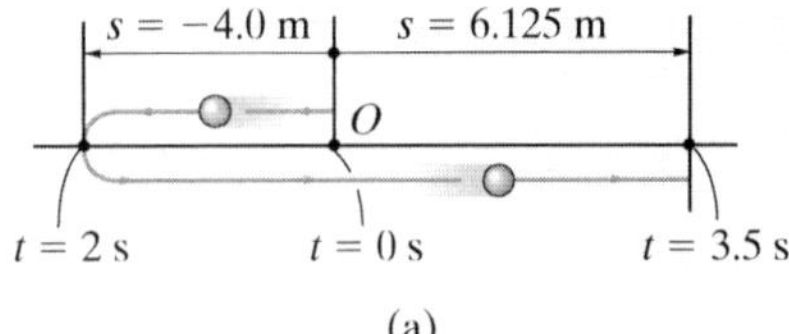

(a)

SOLUTION

Coordinate System. Here positive motion is to the right, measured from the origin O, Fig. 12–6a.

Distance Traveled. Since $v = f(t)$, the position as a function of time may be found by integrating $v = ds/dt$ with $t = 0$, $s = 0$.

$(\overset{+}{\rightarrow})$

$$ds = v\,dt$$

$$= (3t^2 - 6t)\,dt$$

$$\int_0^s ds = \int_0^t (3t^2 - 6t)\,dt$$

$$s = (t^3 - 3t^2)\text{ m} \qquad (1)$$

v (m/s)
v = 3t² − 6t
t (s)
(0, 0)
(2 s, 0)
(1 s, −3 m/s)

(b)

Fig. 12–6

In order to determine the distance traveled in 3.5 s, it is necessary to investigate the path of motion. If we consider a graph of the velocity function, Fig. 12–6b, then it reveals that for $0 < t < 2$ s the velocity is *negative*, which means the particle is traveling to the *left*, and for $t > 2$ s the velocity is *positive*, and hence the particle is traveling to the *right*. Also, note that $v = 0$ at $t = 2$ s. The particle's position when $t = 0$, $t = 2$ s, and $t = 3.5$ s can be determined from Eq. 1. This yields

$$s|_{t=0} = 0 \quad s|_{t=2\text{ s}} = -4.0\text{ m} \quad s|_{t=3.5\text{ s}} = 6.125\text{ m}$$

The path is shown in Fig. 12–6a. Hence, the distance traveled in 3.5 s is

$$s_T = 4.0 + 4.0 + 6.125 = 14.125\text{ m} = 14.1\text{ m} \qquad \textit{Ans.}$$

Velocity. The *displacement* from $t = 0$ to $t = 3.5$ s is

$$\Delta s = s|_{t=3.5\text{ s}} - s|_{t=0} = 6.125\text{ m} - 0 = 6.125\text{ m}$$

and so the average velocity is

$$v_{avg} = \frac{\Delta s}{\Delta t} = \frac{6.125\text{ m}}{3.5\text{ s} - 0} = 1.75\text{ m/s} \rightarrow \qquad \textit{Ans.}$$

The average speed is defined in terms of the *distance traveled* s_T. This positive scalar is

$$(v_{sp})_{avg} = \frac{s_T}{\Delta t} = \frac{14.125\text{ m}}{3.5\text{ s} - 0} = 4.04\text{ m/s} \qquad \textit{Ans.}$$

NOTE: In this problem, the acceleration is $a = dv/dt = (6t - 6)\text{ m/s}^2$, which is not constant.

It is highly suggested that you test yourself on the solutions to these examples, by covering them over and then trying to think about which equations of kinematics must be used and how they are applied in order to determine the unknowns. Then before solving any of the problems, try and solve some of the Fundamental Problems given below. The solutions and answers to all these problems are given in the back of the book. **Doing this throughout the book will help immensely in understanding how to apply the theory, and thereby develop your problem-solving skills.**

FUNDAMENTAL PROBLEMS

F12–1. Initially, the car travels along a straight road with a speed of 35 m/s. If the brakes are applied and the speed of the car is reduced to 10 m/s in 15 s, determine the constant deceleration of the car.

F12–1

F12–2. A ball is thrown vertically upward with a speed of 15 m/s. Determine the time of flight when it returns to its original position.

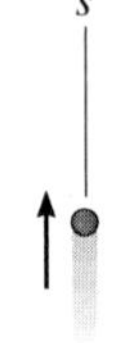

F12–2

F12–3. A particle travels along a straight line with a velocity of $v = (4t - 3t^2)$ m/s, where t is in seconds. Determine the position of the particle when $t = 4$ s. $s = 0$ when $t = 0$.

F12–4. A particle travels along a straight line with a speed $v = (0.5t^3 - 8t)$ m/s, where t is in seconds. Determine the acceleration of the particle when $t = 2$ s.

F12–5. The position of the particle is given by $s = (2t^2 - 8t + 6)$ m, where t is in seconds. Determine the time when the velocity of the particle is zero, and the total distance traveled by the particle when $t = 3$ s.

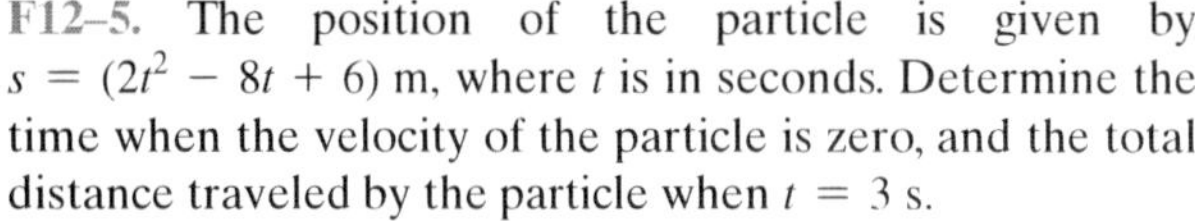

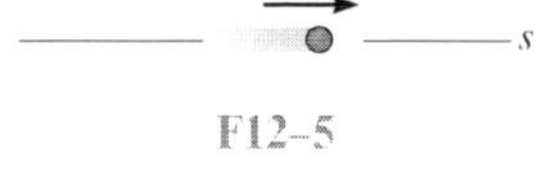

F12–5

F12–6. A particle travels along a straight line with an acceleration of $a = (10 - 0.2s)$ m/s^2, where s is measured in meters. Determine the velocity of the particle when $s = 10$ m if $v = 5$ m/s at $s = 0$.

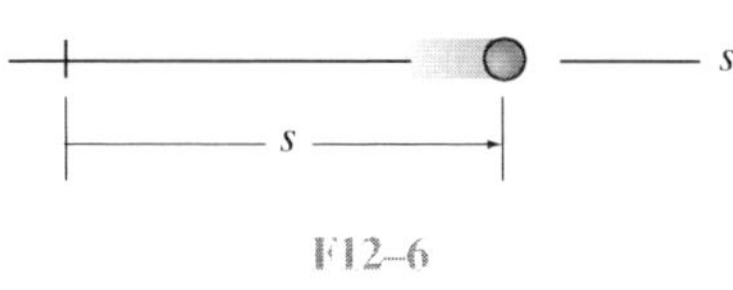

F12–6

F12–7. A particle moves along a straight line such that its acceleration is $a = (4t^2 - 2)$ m/s^2, where t is in seconds. When $t = 0$, the particle is located 2 m to the left of the origin, and when $t = 2$ s, it is 20 m to the left of the origin. Determine the position of the particle when $t = 4$ s.

F12–8. A particle travels along a straight line with a velocity of $v = (20 - 0.05s^2)$ m/s, where s is in meters. Determine the acceleration of the particle at $s = 15$ m.

PROBLEMS

12–1. Starting from rest, a particle moving in a straight line has an acceleration of $a = (2t - 6)\ \text{m/s}^2$, where t is in seconds. What is the particle's velocity when $t = 6$ s, and what is its position when $t = 11$ s?

12–2. When a train is traveling along a straight track at 2 m/s, it begins to accelerate at $a = (60\ v^{-4})\ \text{m/s}^2$, where v is in m/s. Determine its velocity v and the position 3 s after the acceleration.

Prob. 12–2

12–3. The acceleration of a particle as it moves along a straight line is given by $a = (2t - 1)\ \text{m/s}^2$, where t is in seconds. If $s = 1$ m and $v = 2$ m/s when $t = 0$, determine the particle's velocity and position when $t = 6$ s. Also, determine the total distance the particle travels during this time period.

***12–4.** Traveling with an initial speed of 70 km/h, a car accelerates at 6000 km/h^2 along a straight road. How long will it take to reach a speed of 120 km/h? Also, through what distance does the car travel during this time?

12–5. A bus starts from rest with a constant acceleration of 1 m/s^2. Determine the time required for it to attain a speed of 25 m/s and the distance traveled.

12–6. A stone A is dropped from rest down a well, and in 1 s another stone B is dropped from rest. Determine the distance between the stones another second later.

12–7. A bicyclist starts from rest and after traveling along a straight path a distance of 20 m reaches a speed of 30 km/h. Determine his acceleration if it is *constant*. Also, how long does it take to reach the speed of 30 km/h?

***12–8.** A particle moves along a straight line with an acceleration of $a = 5/(3s^{1/3} + s^{5/2})\ \text{m/s}^2$, where s is in meters. Determine the particle's velocity when $s = 2$ m, if it starts from rest when $s = 1$ m. Use a numerical method to evaluate the integral.

12–9. If it takes 3 s for a ball to strike the ground when it is released from rest, determine the height in meters of the building from which it was released. Also, what is the velocity of the ball when it strikes the ground?

12–10. A particle, initially at the origin, moves along a straight line through a fluid medium such that its velocity is defined as $v = 1.8(1 - e^{-0.3t})$ m/s, where t is in seconds. Determine the displacement of the particle during the first 3 s.

12–11. A truck, traveling along a straight road at 20 km/h, increases its speed to 120 km/h in 15 s. If its acceleration is constant, determine the distance traveled.

***12–12.** Determine the time required for a car to travel 1 km along a road if the car starts from rest, reaches a maximum speed at some intermediate point, and then stops at the end of the road. The car can accelerate at 1.5 m/s^2 and decelerate at 2 m/s^2.

12–13. The acceleration of a rocket traveling upward is given by $a = (6 + 0.02s)\ \text{m/s}^2$, where s is in meters. Determine the rocket's velocity when $s = 2$ km and the time needed to reach this altitude. Initially, $v = 0$ and $s = 0$ when $t = 0$.

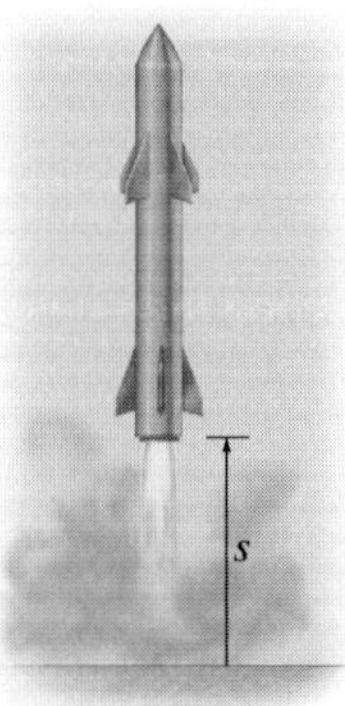

Prob. 12–13

12–14. The acceleration of a rocket traveling upward is given by $a = (6 + 0.02s)\ \text{m/s}^2$, where s is in meters. Determine the time needed for the rocket to reach an altitude of $s = 100$ m. Initially, $v = 0$ and $s = 0$ when $t = 0$.

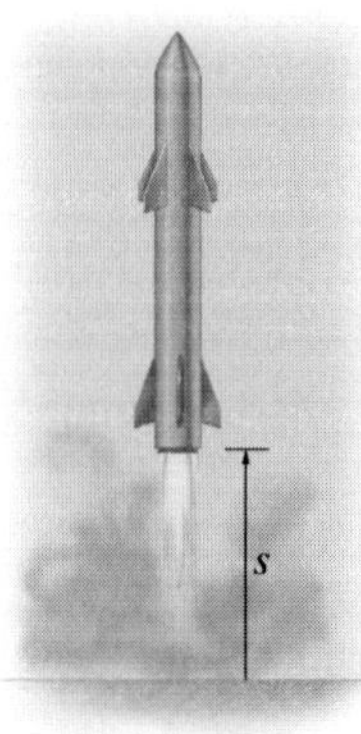

Prob. 12–14

12–15. A train starts from rest at station A and accelerates at $0.5\ \text{m/s}^2$ for 60 s. Afterwards it travels with a constant velocity for 15 min. It then decelerates at $1\ \text{m/s}^2$ until it is brought to rest at station B. Determine the distance between the stations.

***12–16.** A particle travels along a straight line such that in 2 s it moves from an initial position $s_A = +0.5$ m to a position $s_B = -1.5$ m. Then in another 4 s it moves from s_B to $s_C = +2.5$ m. Determine the particle's average velocity and average speed during the 6-s time interval.

12–17. The acceleration of a particle as it moves along a straight line is given by $a = (4t^3 - 1)\ \text{m/s}^2$, where t is in seconds. If $s = 2$ m and $v = 5$ m/s when $t = 0$, determine the particle's velocity and position when $t = 5$ s. Also, determine the total distance the particle travels during this time period.

12–18. A car can have an acceleration and a deceleration of $5\ \text{m/s}^2$. If it starts from rest, and can have a maximum speed of 60 m/s, determine the shortest time it can travel a distance of 1200 m when it stops.

12–19. A particle travels to the right along a straight line with a velocity $v = [5/(4 + s)]$ m/s, where s is in meters. Determine its position when $t = 6$ s if $s = 5$ m when $t = 0$.

***12–20.** A particle is moving along a straight line such that its acceleration is defined as $a = (4s^2)\ \text{m/s}^2$, where s is in meters. If $v = -100$ m/s when $s = 10$ m and $t = 0$, determine the particle's velocity as a function of position.

12–21. If the effects of atmospheric resistance are accounted for, a freely falling body has an acceleration defined by the equation $a = 9.81[1 - v^2(10^{-4})]\ \text{m/s}^2$, where v is in m/s and the positive direction is downward. If the body is released from rest at a *very high altitude*, determine (a) the velocity when $t = 5$ s, and (b) the body's terminal or maximum attainable velocity (as $t \rightarrow \infty$).

12–22. A particle moves along a straight line such that its position is defined by $s = (t^2 - 6t + 5)$ m. Determine the average velocity, the average speed, and the acceleration of the particle when $t = 6$ s.

12–23. When a particle falls through the air, its initial acceleration $a = g$ diminishes until it is zero, and thereafter it falls at a constant or terminal velocity v_f. If this variation of the acceleration can be expressed as $a = (g/v_f^2)(v_f^2 - v^2)$, determine the time needed for the velocity to become $v = v_f$. Initially the particle falls from rest.

***12–24.** A particle is moving along a straight line such that its velocity is defined as $v = (-4s^2)$ m/s, where s is in meters. If $s = 2$ m when $t = 0$, determine the velocity and acceleration as functions of time.

12–25. A sphere is fired downwards into a medium with an initial speed of 27 m/s. If it experiences a deceleration of $a = (-6t)\ \text{m/s}^2$, where t is in seconds, determine the distance traveled before it stops.

12–26. When two cars A and B are next to one another, they are traveling in the same direction with speeds v_A and v_B, respectively. If B maintains its constant speed, while A begins to decelerate at a_A, determine the distance d between the cars at the instant A stops.

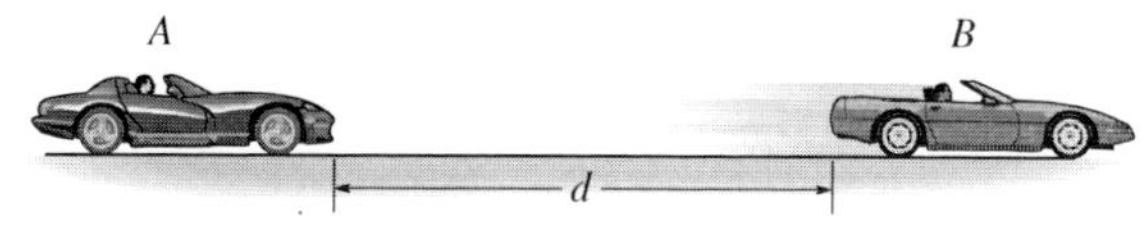

Prob. 12–26

12

12–27. A particle is moving along a straight line such that when it is at the origin it has a velocity of 4 m/s. If it begins to decelerate at the rate of $a = (-1.5v^{1/2})\ \text{m/s}^2$, where v is in m/s, determine the distance it travels before it stops.

***12–28.** A particle travels to the right along a straight line with a velocity $v = [5/(4 + s)]$ m/s, where s is in meters. Determine its deceleration when $s = 2$ m.

12–29. A particle moves along a straight line with an acceleration $a = 2\,v^{1/2}\ \text{m/s}^2$, where v is in m/s. If $s = 0$, $v = 4$ m/s when $t = 0$, determine the time for the particle to achieve a velocity of 20 m/s. Also, find the displacement of particle when $t = 2$ s.

12–30. As a train accelerates uniformly it passes successive kilometer marks while traveling at velocities of 2 m/s and then 10 m/s. Determine the train's velocity when it passes the next kilometer mark and the time it takes to travel the 2-km distance.

12–31. The acceleration of a particle along a straight line is defined by $a = (2t - 9)\ \text{m/s}^2$, where t is in seconds. At $t = 0$, $s = 1$ m and $v = 10$ m/s. When $t = 9$ s, determine (a) the particle's position, (b) the total distance traveled, and (c) the velocity.

***12–32.** The acceleration of a particle traveling along a straight line is $a = \frac{1}{4}s^{1/2}\ \text{m/s}^2$, where s is in meters. If $v = 0$, $s = 1$ m when $t = 0$, determine the particle's velocity at $s = 2$ m.

12–33. At $t = 0$ bullet A is fired vertically with an initial (muzzle) velocity of 450 m/s. When $t = 3$ s, bullet B is fired upward with a muzzle velocity of 600 m/s. Determine the time t, after A is fired, as to when bullet B passes bullet A. At what altitude does this occur?

12–34. A boy throws a ball straight up from the top of a 12-m high tower. If the ball falls past him 0.75 s later, determine the velocity at which it was thrown, the velocity of the ball when it strikes the ground, and the time of flight.

12–35. When a particle falls through the air, its initial acceleration $a = g$ diminishes until it is zero, and thereafter it falls at a constant or terminal velocity v_f. If this variation of the acceleration can be expressed as $a = (g/v_f^2)(v_f^2 - v^2)$, determine the time needed for the velocity to become $v = v_f/2$. Initially the particle falls from rest.

***12–36.** A particle is moving with a velocity of v_0 when $s = 0$ and $t = 0$. If it is subjected to a deceleration of $a = -kv^3$, where k is a constant, determine its velocity and position as functions of time.

12–37. As a body is projected to a high altitude above the earth's *surface*, the variation of the acceleration of gravity with respect to altitude y must be taken into account. Neglecting air resistance, this acceleration is determined from the formula $a = -g_0[R^2/(R + y)^2]$, where g_0 is the constant gravitational acceleration at sea level, R is the radius of the earth, and the positive direction is measured upward. If $g_0 = 9.81\ \text{m/s}^2$ and $R = 6356$ km, determine the minimum initial velocity (escape velocity) at which a projectile should be shot vertically from the earth's surface so that it does not fall back to the earth. *Hint:* This requires that $v = 0$ as $y \to \infty$.

12–38. Accounting for the variation of gravitational acceleration a with respect to altitude y (see Prob. 12–37), derive an equation that relates the velocity of a freely falling particle to its altitude. Assume that the particle is released from rest at an altitude y_0 from the earth's surface. With what velocity does the particle strike the earth if it is released from rest at an altitude $y_0 = 500$ km? Use the numerical data in Prob. 12–37.

12.3 Rectilinear Kinematics: Erratic Motion

When a particle has erratic or changing motion then its position, velocity, and acceleration *cannot* be described by a single continuous mathematical function along the entire path. Instead, a series of functions will be required to specify the motion at different intervals. For this reason, it is convenient to represent the motion as a graph. If a graph of the motion that relates any two of the variables s, v, a, t can be drawn, then this graph can be used to construct subsequent graphs relating two other variables since the variables are related by the differential relationships $v = ds/dt$, $a = dv/dt$, or $a\,ds = v\,dv$. Several situations occur frequently.

The s–t, v–t, and a–t Graphs. To construct the v–t graph given the s–t graph, Fig. 12–7*a*, the equation $v = ds/dt$ should be used, since it relates the variables s and t to v. This equation states that

$$\frac{ds}{dt} = v$$

slope of s–t graph = velocity

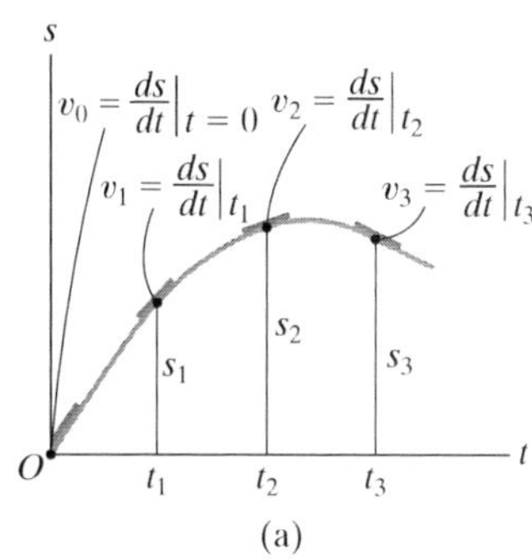

(a)

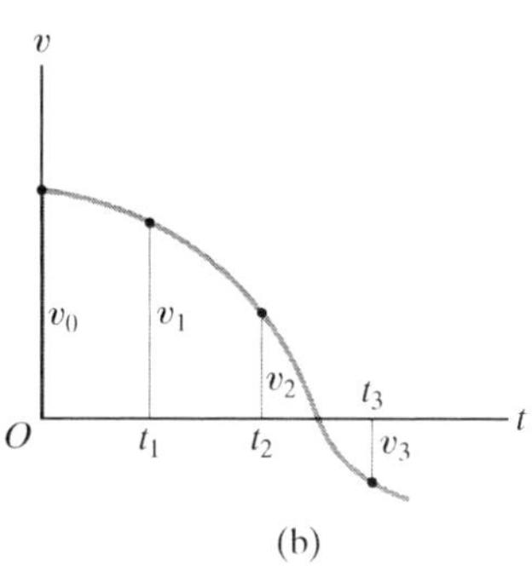

(b)

Fig. 12–7

For example, by measuring the slope on the s–t graph when $t = t_1$, the velocity is v_1, which is plotted in Fig. 12–7*b*. The v–t graph can be constructed by plotting this and other values at each instant.

The a–t graph can be constructed from the v–t graph in a similar manner, Fig. 12–8, since

$$\frac{dv}{dt} = a$$

slope of v–t graph = acceleration

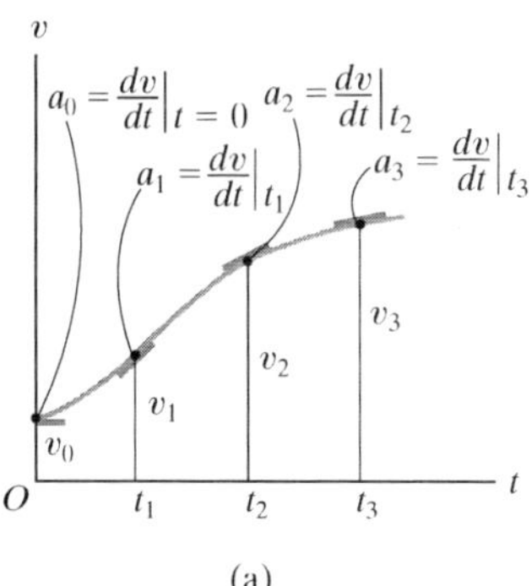

(a)

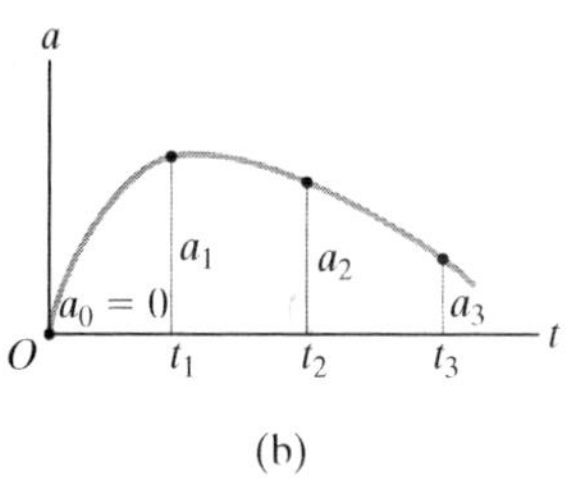

(b)

Fig. 12–8

Examples of various measurements are shown in Fig. 12–8*a* and plotted in Fig. 12–8*b*.

If the s–t curve for each interval of motion can be expressed by a mathematical function $s = s(t)$, then the equation of the v–t graph for the same interval can be obtained by differentiating this function with respect to time since $v = ds/dt$. Likewise, the equation of the a–t graph for the same interval can be determined by differentiating $v = v(t)$ since $a = dv/dt$. Since differentiation reduces a polynomial of degree n to that of degree $n - 1$, then if the s–t graph is parabolic (a second-degree curve), the v–t graph will be a sloping line (a first-degree curve), and the a–t graph will be a constant or a horizontal line (a zero-degree curve).

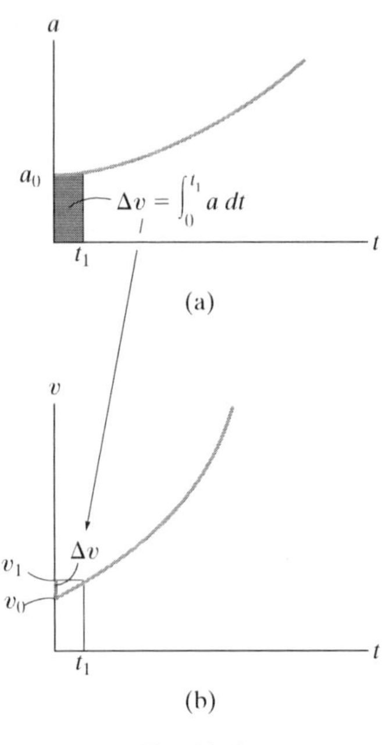

Fig. 12–9

If the a–t graph is given, Fig. 12–9a, the v–t graph may be constructed using $a = dv/dt$, written as

$$\Delta v = \int a\,dt$$

change in velocity = area under a–t graph

Hence, to construct the v–t graph, we begin with the particle's initial velocity v_0 and then add to this small increments of area (Δv) determined from the a–t graph. In this manner successive points, $v_1 = v_0 + \Delta v$, etc., for the v–t graph are determined, Fig. 12–9b. Notice that an algebraic addition of the area increments of the a–t graph is necessary, since areas lying above the t axis correspond to an increase in v ("positive" area), whereas those lying below the axis indicate a decrease in v ("negative" area).

Similarly, if the v–t graph is given, Fig. 12–10a, it is possible to determine the s–t graph using $v = ds/dt$, written as

$$\Delta s = \int v\,dt$$

displacement = area under v–t graph

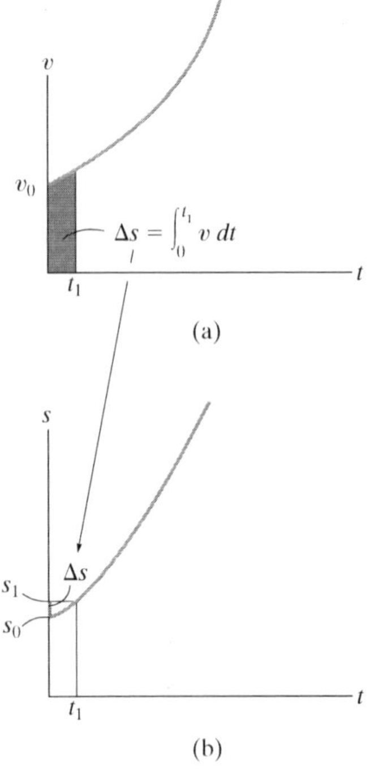

Fig. 12–10

In the same manner as stated above, we begin with the particle's initial position s_0 and add (algebraically) to this small area increments Δs determined from the v–t graph, Fig. 12–10b.

If segments of the a–t graph can be described by a series of equations, then each of these equations can be *integrated* to yield equations describing the corresponding segments of the v–t graph. In a similar manner, the s–t graph can be obtained by integrating the equations which describe the segments of the v–t graph. As a result, if the a–t graph is linear (a first-degree curve), integration will yield a v–t graph that is parabolic (a second-degree curve) and an s–t graph that is cubic (third-degree curve).

12

The v–s and a–s Graphs. If the a–s graph can be constructed, then points on the v–s graph can be determined by using $v\,dv = a\,ds$. Integrating this equation between the limits $v = v_0$ at $s = s_0$ and $v = v_1$ at $s = s_1$, we have,

$$\tfrac{1}{2}(v_1^2 - v_0^2) = \int_{s_0}^{s_1} a\,ds$$

area under a–s graph

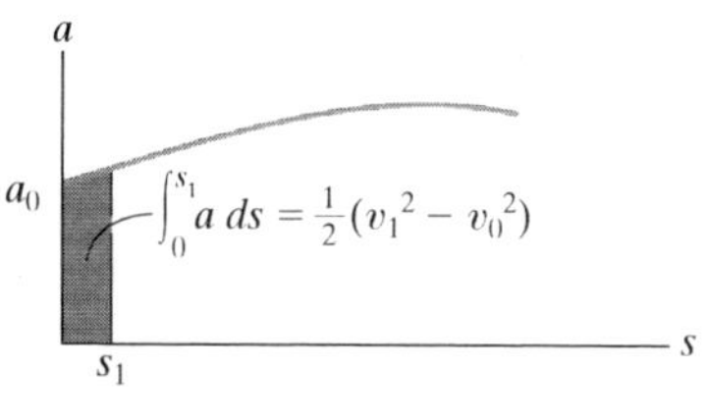

(a)

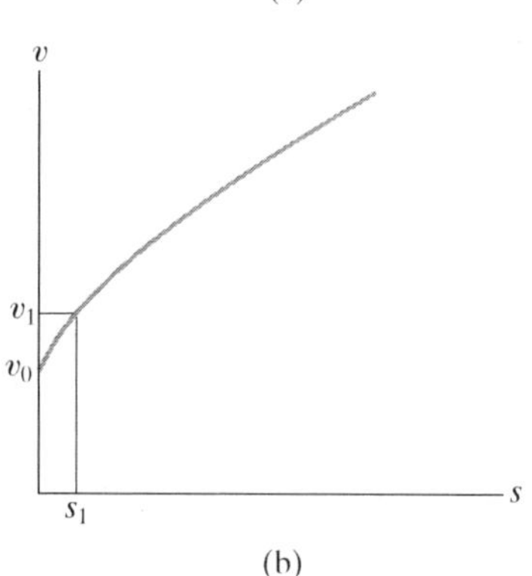

(b)

Fig. 12–11

Therefore, if the red area in Fig. 12–11a is determined, and the initial velocity v_0 at $s_0 = 0$ is known, then $v_1 = \left(2\int_0^{s_1} a\,ds + v_0^2\right)^{1/2}$, Fig. 12–11$b$. Successive points on the v–s graph can be constructed in this manner.

If the v–s graph is known, the acceleration a at any position s can be determined using $a\,ds = v\,dv$, written as

$$a = v\left(\frac{dv}{ds}\right)$$

acceleration = velocity times slope of v–s graph

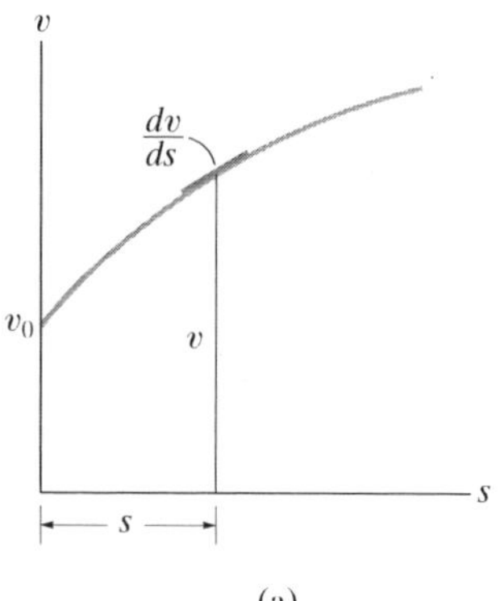

(a)

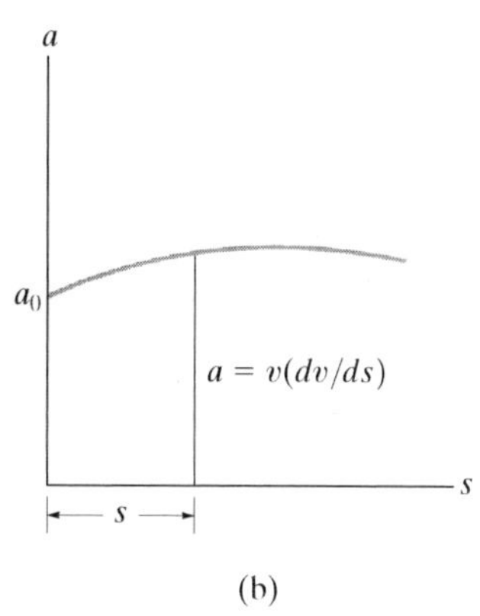

(b)

Fig. 12–12

Thus, at any point (s, v) in Fig. 12–12a, the slope dv/ds of the v–s graph is measured. Then with v and dv/ds known, the value of a can be calculated, Fig. 12–12b.

The v–s graph can also be constructed from the a–s graph, or vice versa, by approximating the known graph in various intervals with mathematical functions, $v = f(s)$ or $a = g(s)$, and then using $a\,ds = v\,dv$ to obtain the other graph.

12

EXAMPLE 12.6

A bicycle moves along a straight road such that its position is described by the graph shown in Fig. 12–13*a*. Construct the v–t and a–t graphs for $0 \le t \le 30$ s.

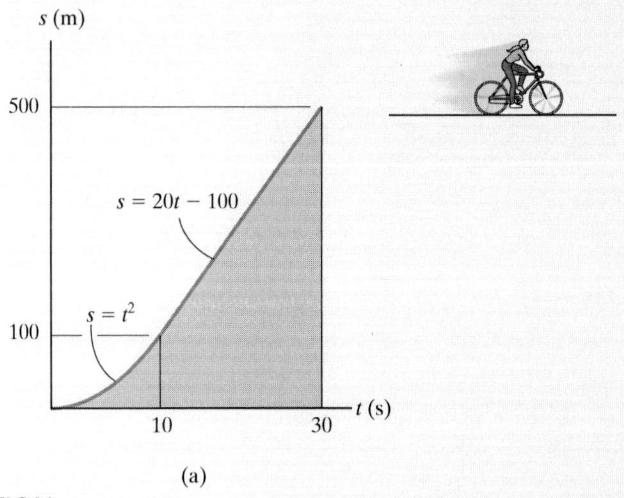

(a)

SOLUTION

v–t Graph. Since $v = ds/dt$, the v–t graph can be determined by differentiating the equations defining the s–t graph, Fig. 12–13*a*. We have

$$0 \le t < 10 \text{ s}; \qquad s = (t^2) \text{ m} \qquad v = \frac{ds}{dt} = (2t) \text{ m/s}$$

$$10 \text{ s} < t \le 30 \text{ s}; \qquad s = (20t - 100) \text{ m} \qquad v = \frac{ds}{dt} = 20 \text{ m/s}$$

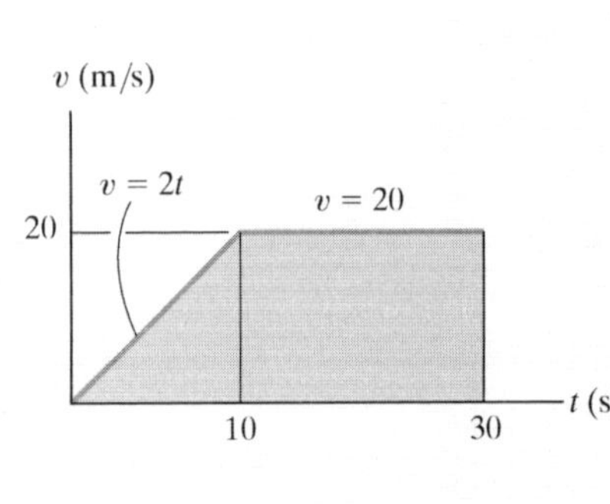

(b)

The results are plotted in Fig. 12–13*b*. We can also obtain specific values of v by measuring the *slope* of the s–t graph at a given instant. For example, at $t = 20$ s, the slope of the s–t graph is determined from the straight line from 10 s to 30 s, i.e.,

$$t = 20 \text{ s}; \qquad v = \frac{\Delta s}{\Delta t} = \frac{500 \text{ m} - 100 \text{ m}}{30 \text{ s} - 10 \text{ s}} = 20 \text{ m/s}$$

a–t Graph. Since $a = dv/dt$, the a–t graph can be determined by differentiating the equations defining the lines of the v–t graph. This yields

$$0 \le t < 10 \text{ s}; \qquad v = (2t) \text{ m/s} \qquad a = \frac{dv}{dt} = 2 \text{ m/s}^2$$

$$10 < t \le 30 \text{ s}; \qquad v = 20 \text{ m/s} \qquad a = \frac{dv}{dt} = 0$$

The results are plotted in Fig. 12–13*c*.

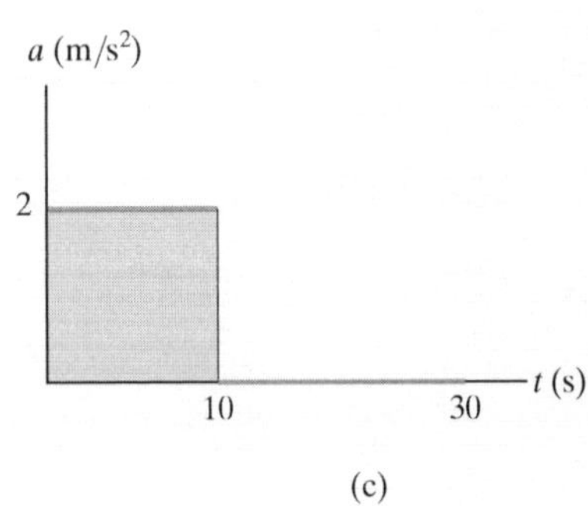

(c)

Fig. 12–13

NOTE: Show that $a = 2 \text{ m/s}^2$ when $t = 5$ s by measuring the slope of the v–t graph.

EXAMPLE 12.7

The car in Fig. 12–14*a* starts from rest and travels along a straight track such that it accelerates at $10\ \text{m/s}^2$ for 10 s, and then decelerates at $2\ \text{m/s}^2$. Draw the v–t and s–t graphs and determine the time t' needed to stop the car. How far has the car traveled?

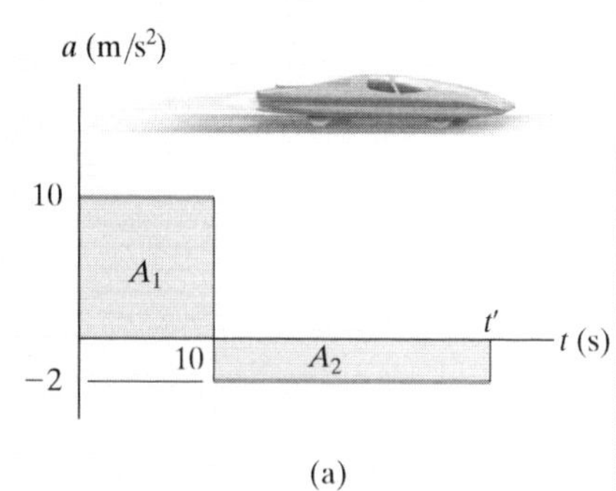

(a)

SOLUTION

v-t Graph. Since $dv = a\,dt$, the v–t graph is determined by integrating the straight-line segments of the a–t graph. Using the *initial condition* $v = 0$ when $t = 0$, we have

$$0 \le t < 10\ \text{s}; \qquad a = (10)\ \text{m/s}^2; \qquad \int_0^v dv = \int_0^t 10\,dt, \qquad v = 10t$$

When $t = 10$ s, $v = 10(10) = 100$ m/s. Using this as the *initial condition* for the next time period, we have

$$10\ \text{s} < t \le t';\ a = (-2)\ \text{m/s}^2;\ \int_{100\ \text{m/s}}^{v} dv = \int_{10\ \text{s}}^{t} -2\,dt,\ v = (-2t + 120)\ \text{m/s}$$

When $t = t'$ we require $v = 0$. This yields, Fig. 12–14*b*,

$$t' = 60\ \text{s} \qquad \textit{Ans.}$$

A more direct solution for t' is possible by realizing that the area under the a–t graph is equal to the change in the car's velocity. We require $\Delta v = 0 = A_1 + A_2$, Fig. 12–14*a*. Thus

$$0 = 10\ \text{m/s}^2(10\ \text{s}) + (-2\ \text{m/s}^2)(t' - 10\ \text{s})$$

$$t' = 60\ \text{s} \qquad \textit{Ans.}$$

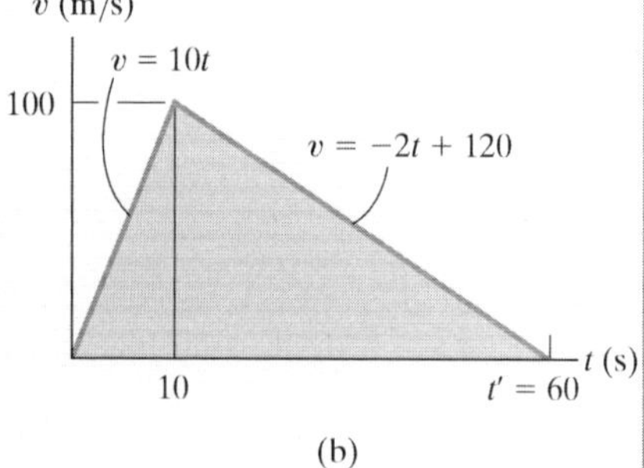

(b)

s-t Graph. Since $ds = v\,dt$, integrating the equations of the v–t graph yields the corresponding equations of the s–t graph. Using the *initial condition* $s = 0$ when $t = 0$, we have

$$0 \le t \le 10\ \text{s}; \qquad v = (10t)\ \text{m/s}; \qquad \int_0^s ds = \int_0^t 10t\,dt, \qquad s = (5t^2)\ \text{m}$$

When $t = 10$ s, $s = 5(10)^2 = 500$ m. Using this *initial condition*,

$$10\ \text{s} \le t \le 60\ \text{s};\ v = (-2t + 120)\ \text{m/s};\ \int_{500\ \text{m}}^{s} ds = \int_{10\ \text{s}}^{t} (-2t + 120)\,dt$$

$$s - 500 = -t^2 + 120t - [-(10)^2 + 120(10)]$$

$$s = (-t^2 + 120t - 600)\ \text{m}$$

When $t' = 60$ s, the position is

$$s = -(60)^2 + 120(60) - 600 = 3000\ \text{m} \qquad \textit{Ans.}$$

The s–t graph is shown in Fig. 12–14*c*.

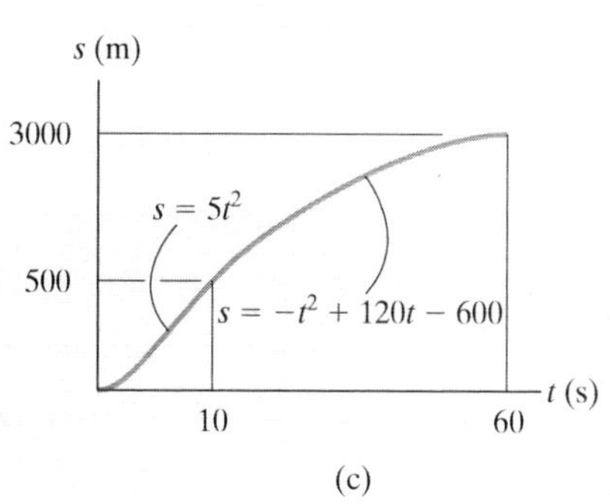

(c)

Fig. 12–14

NOTE: A direct solution for s is possible when $t' = 60$ s, since the *triangular area* under the v–t graph would yield the displacement $\Delta s = s - 0$ from $t = 0$ to $t' = 60$ s. Hence,

$$\Delta s = \tfrac{1}{2}(60\ \text{s})(100\ \text{m/s}) = 3000\ \text{m} \qquad \textit{Ans.}$$

EXAMPLE 12.8

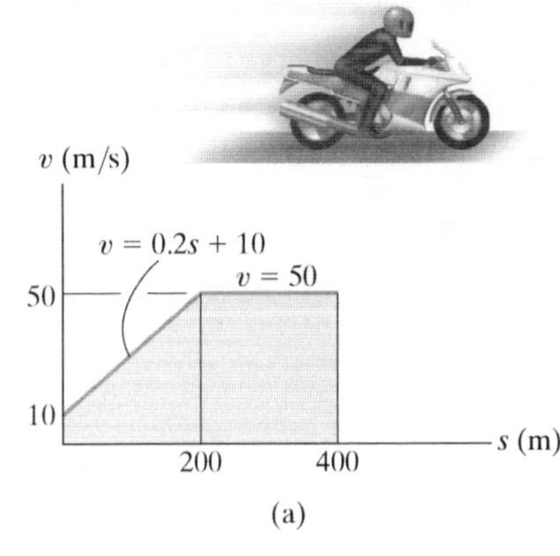

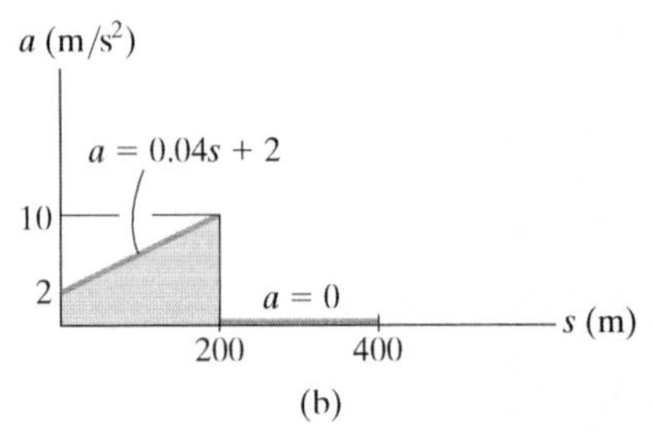

Fig. 12–15

The v–s graph describing the motion of a motorcycle is shown in Fig. 12–15*a*. Construct the a–s graph of the motion and determine the time needed for the motorcycle to reach the position $s = 400$ m.

SOLUTION

a–s Graph. Since the equations for segments of the v–s graph are given, the a–s graph can be determined using $a\,ds = v\,dv$.

$0 \le s < 200$ m; $\qquad v = (0.2s + 10)$ m/s

$$a = v\frac{dv}{ds} = (0.2s + 10)\frac{d}{ds}(0.2s + 10) = 0.04s + 2$$

$200 \text{ m} < s \le 400$ m; $\qquad v = 50$ m/s

$$a = v\frac{dv}{ds} = (50)\frac{d}{ds}(50) = 0$$

The results are plotted in Fig. 12–15*b*.

Time. The time can be obtained using the v–s graph and $v = ds/dt$, because this equation relates v, s, and t. For the first segment of motion, $s = 0$ when $t = 0$, so

$0 \le s < 200$ m; $\qquad v = (0.2s + 10)$ m/s; $\qquad dt = \dfrac{ds}{v} = \dfrac{ds}{0.2s + 10}$

$$\int_0^t dt = \int_0^s \frac{ds}{0.2s + 10}$$

$$t = (5\ln(0.2s + 10) - 5\ln 10) \text{ s}$$

At $s = 200$ m, $t = 5\ln[0.2(200) + 10] - 5\ln 10 = 8.05$ s. Therefore, using these initial conditions for the second segment of motion,

$200 \text{ m} < s \le 400$ m; $\qquad v = 50$ m/s; $\qquad dt = \dfrac{ds}{v} = \dfrac{ds}{50}$

$$\int_{8.05\text{ s}}^t dt = \int_{200\text{ m}}^s \frac{ds}{50};$$

$$t - 8.05 = \frac{s}{50} - 4; \quad t = \left(\frac{s}{50} + 4.05\right)\text{ s}$$

Therefore, at $s = 400$ m,

$$t = \frac{400}{50} + 4.05 = 12.1 \text{ s} \qquad \textit{Ans.}$$

NOTE: The graphical results can be checked in part by calculating slopes. For example, at $s = 0$, $a = v(dv/ds) = 10(50 - 10)/200 = 2\text{ m/s}^2$. Also, the results can be checked in part by inspection. The v–s graph indicates the initial increase in velocity (acceleration) followed by constant velocity ($a = 0$).

FUNDAMENTAL PROBLEMS

F12–9. The particle travels along a straight track such that its position is described by the s–t graph. Construct the v–t graph for the same time interval.

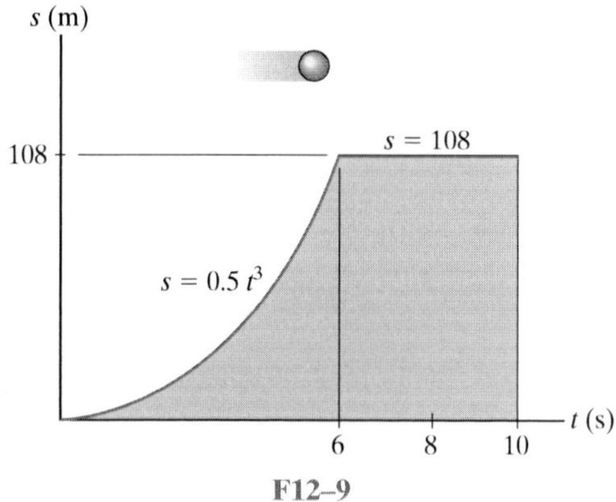

F12–9

F12–10. A van travels along a straight road with a velocity described by the graph. Construct the s–t and a–t graphs during the same period. Take $s = 0$ when $t = 0$.

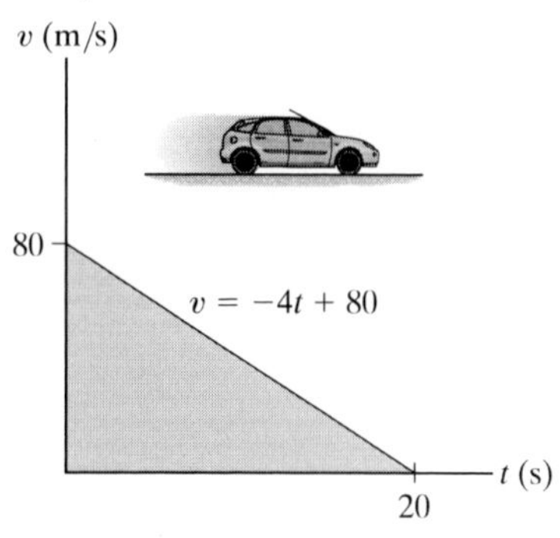

F12–10

F12–11. A bicycle travels along a straight road where its velocity is described by the v–s graph. Construct the a–s graph for the same time interval.

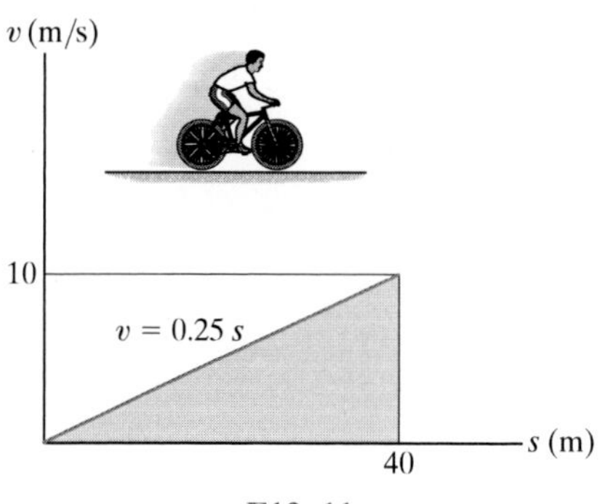

F12–11

F12–12. The sports car travels along a straight road such that its position is described by the graph. Construct the v–t and a–t graphs for the time interval $0 \leq t \leq 10$ s.

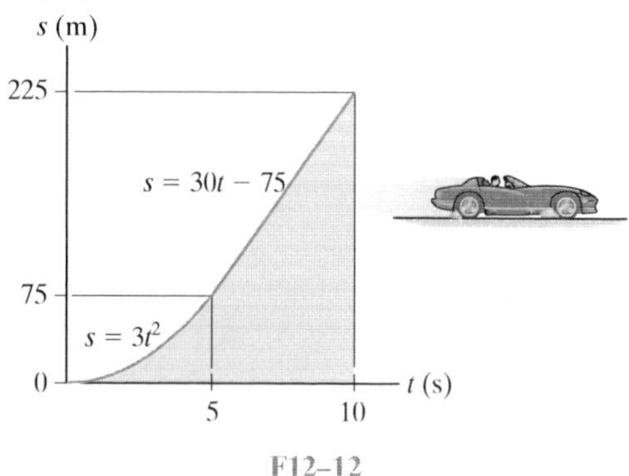

F12–12

F12–13. The dragster starts from rest and has an acceleration described by the graph. Construct the v–t graph for the time interval $0 \leq t \leq t'$, where t' is the time for the car to come to rest.

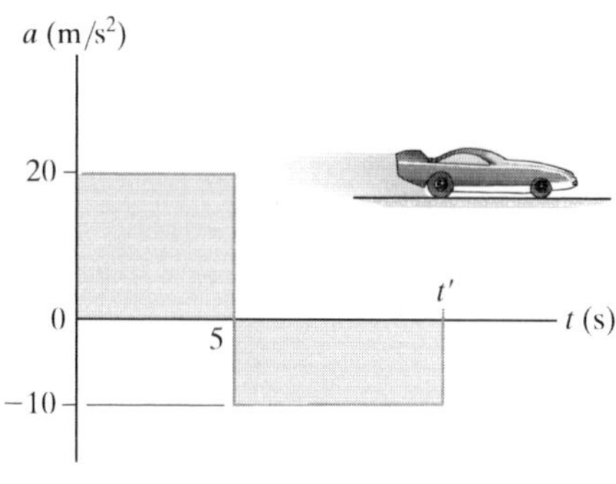

F12–13

F12–14. The dragster starts from rest and has a velocity described by the graph. Construct the s–t graph during the time interval $0 \leq t \leq 15$ s. Also, determine the total distance traveled during this time interval.

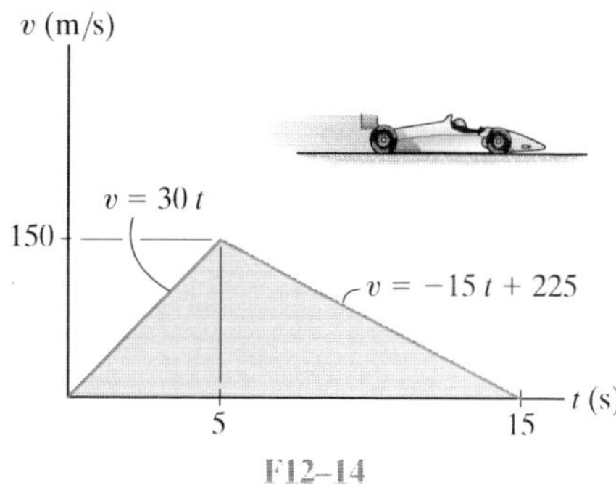

F12–14

12

PROBLEMS

12–39. A car starting from rest moves along a straight track with an acceleration as shown. Determine the time t for the car to reach a speed of 50 m/s and construct the v–t graph that describes the motion until the time t.

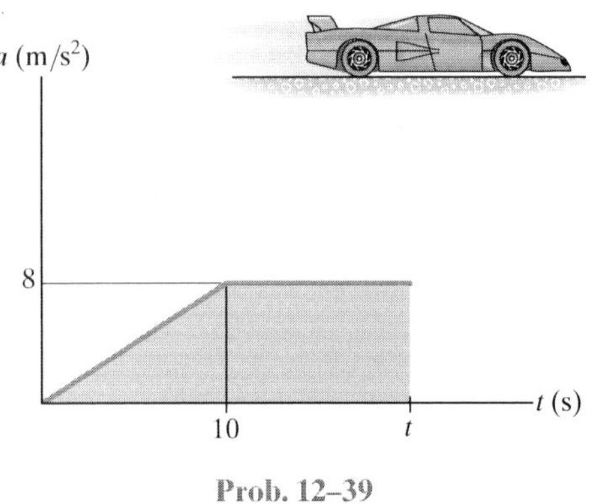

Prob. 12–39

***12–40.** If the position of a particle is defined by $s = [3\sin(\pi/4)t + 8]$ m, where t is in seconds, construct the s–t, v–t, and a–t graphs for $0 \le t \le 10$ s.

12–41. A train starts from station A and for the first kilometer, it travels with a uniform acceleration. Then, for the next two kilometers, it travels with a uniform speed. Finally, the train decelerates uniformly for another kilometer before coming to rest at station B. If the time for the whole journey is six minutes, draw the v–t graph and determine the maximum speed of the train.

12–42. A particle starts from $s = 0$ and travels along a straight line with a velocity $v = (t^2 - 4t + 3)$ m/s, where t is in seconds. Construct the v–t and a–t graphs for the time interval $0 \le t \le 4$ s.

12–43. If the position of a particle is defined by $s = [2\sin(\pi/5)t + 4]$ m, where t is in seconds, construct the s–t, v–t, and a–t graphs for $0 \le t \le 10$ s.

***12–44.** A particle travels along a curve defined by the equation $s = (t^3 - 3t^2 + 2t)$ m, where t is in seconds. Draw the $s - t$, $v - t$, and $a - t$ graphs for the particle for $0 \le t \le 3$ s.

12–45. The snowmobile moves along a straight course according to the v–t graph. Construct the s–t and a–t graphs for the same 50-s time interval. When $t = 0$, $s = 0$.

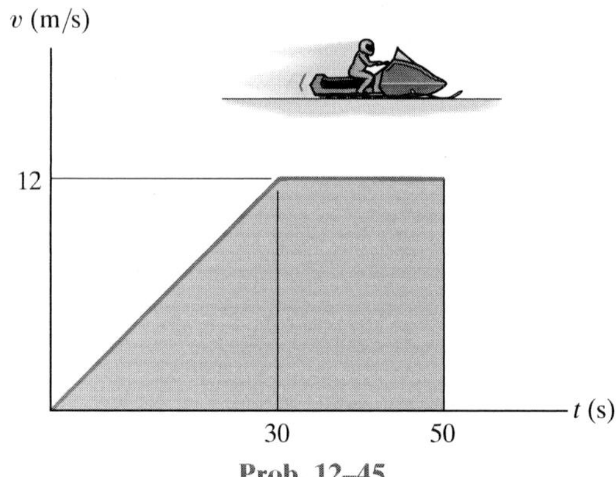

Prob. 12–45

12–46. The velocity of a car is plotted as shown. Determine the total distance the car moves until it stops ($t = 80$ s). Construct the a–t graph.

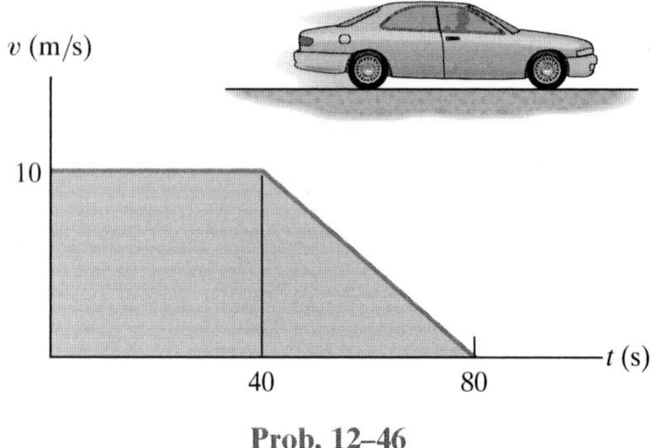

Prob. 12–46

12–47. The v–s graph for a go-cart traveling on a straight road is shown. Determine the acceleration of the go-cart at $s = 50$ m and $s = 150$ m. Draw the a–s graph.

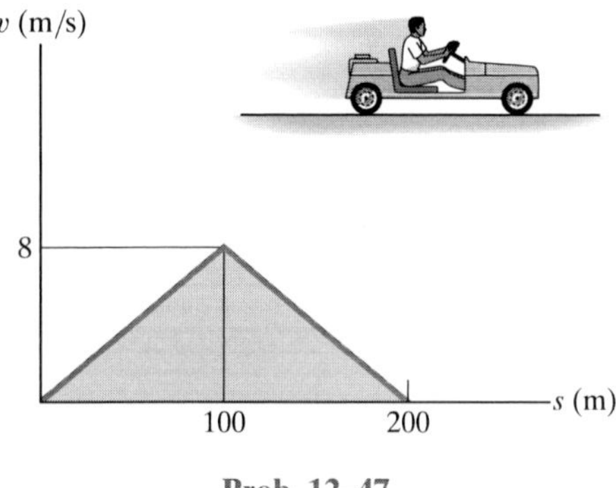

Prob. 12–47

***12–48.** The v–t graph for a particle moving through an electric field from one plate to another has the shape shown in the figure. The acceleration and deceleration that occur are constant and both have a magnitude of 4 m/s^2. If the plates are spaced 200 mm apart, determine the maximum velocity v_{max} and the time t' for the particle to travel from one plate to the other. Also draw the s–t graph. When $t = t'/2$ the particle is at $s = 100$ mm.

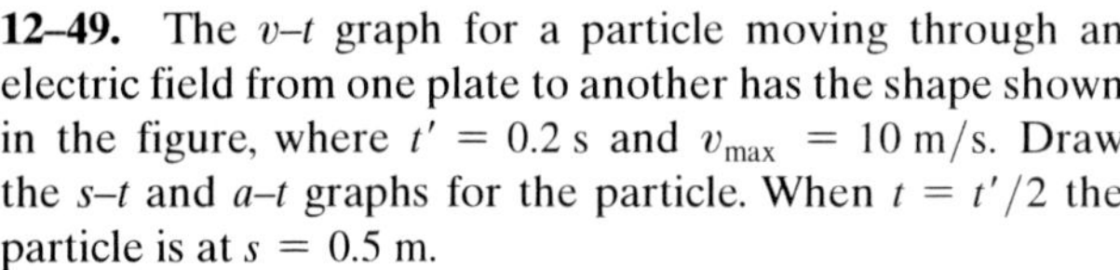

12–49. The v–t graph for a particle moving through an electric field from one plate to another has the shape shown in the figure, where $t' = 0.2$ s and $v_{max} = 10$ m/s. Draw the s–t and a–t graphs for the particle. When $t = t'/2$ the particle is at $s = 0.5$ m.

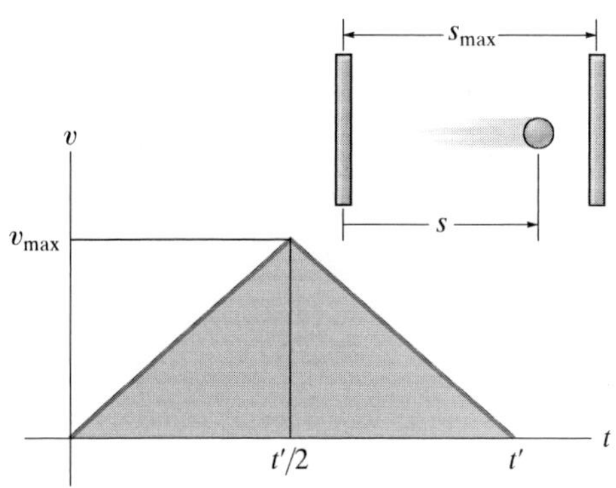

Probs. 12–48/49

12–50. The v–t graph of a car while traveling along a road is shown. Draw the s–t and a–t graphs for the motion.

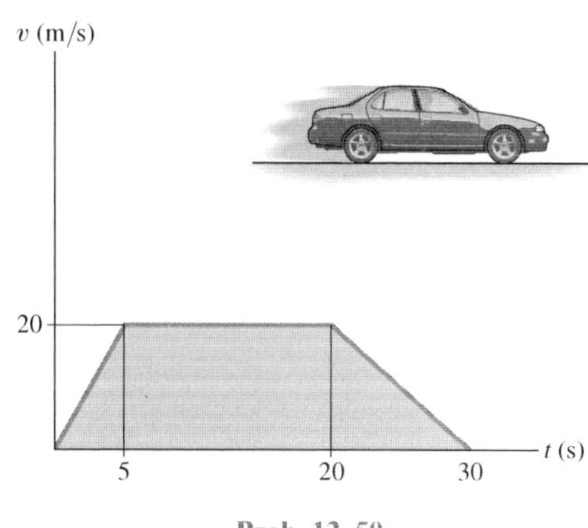

Prob. 12–50

12–51. The a–t graph of the bullet train is shown. If the train starts from rest, determine the elapsed time t' before it again comes to rest. What is the total distance traveled during this time interval? Construct the v–t and s–t graphs.

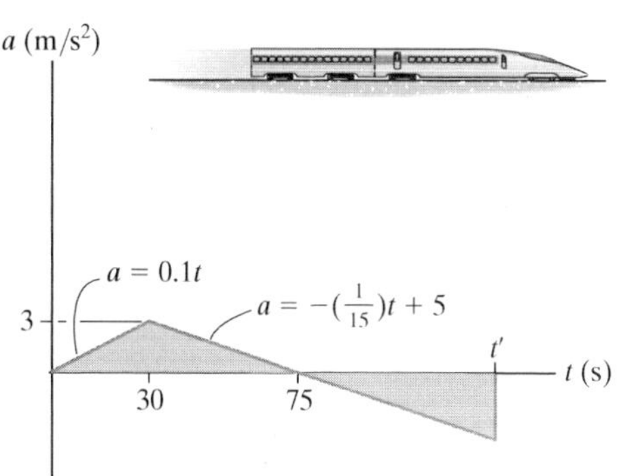

Prob. 12–51

***12–52.** The snowmobile moves along a straight course according to the v–t graph. Construct the s–t and a–t graphs for the same 100-s time interval. When $t = 0, s = 0$.

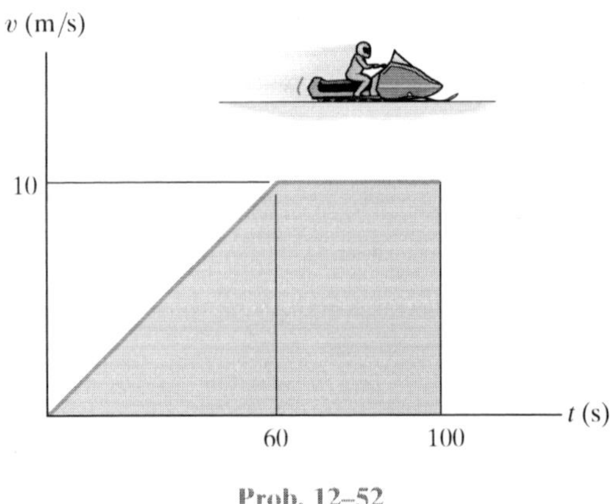

Prob. 12–52

12–53. A two-stage missile is fired vertically from rest with the acceleration shown. In 15 s the first stage A burns out and the second stage B ignites. Plot the v–t and s–t graphs which describe the two-stage motion of the missile for $0 \le t \le 20$ s.

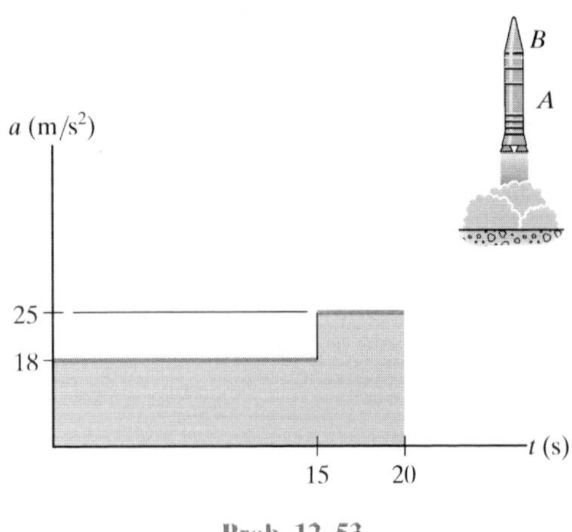

Prob. 12–53

12–54. A motorcycle starts from rest at $s = 0$ and travels along a straight road with the speed shown by the v–t graph. Determine the motorcycle's acceleration and position when $t = 8$ s and $t = 12$ s.

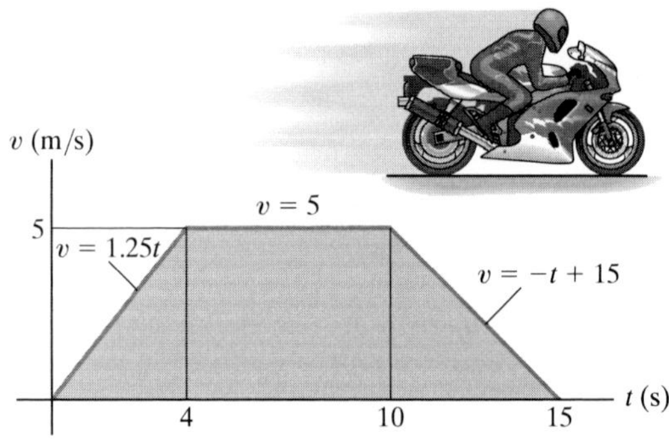

Prob. 12–54

12–55. A race car starting from rest travels along a straight road and for 10 s has the acceleration shown. Construct the v–t graph that describes the motion and find the distance traveled in 10 s.

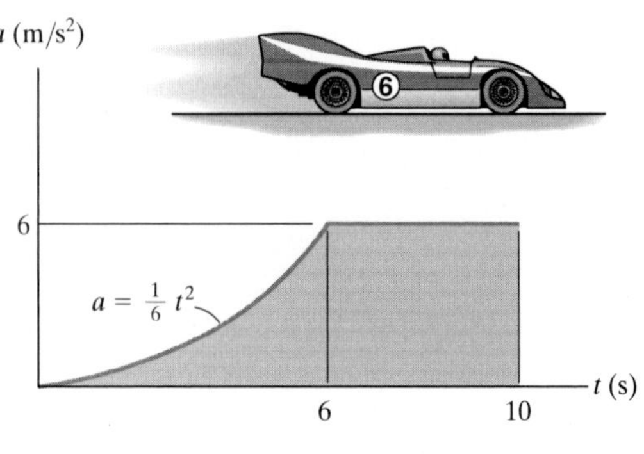

Prob. 12–55

*12–56. A car travels up a hill with the speed shown. Determine the total distance the car travels until it stops ($t = 60$ s). Plot the $a-t$ graph.

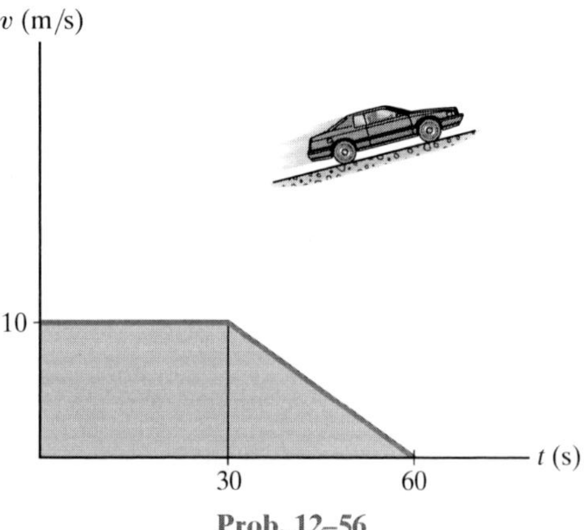

Prob. 12–56

12–57. A car starts from rest and travels along a straight road with a velocity described by the graph. Determine the total distance traveled until the car stops. Construct the s–t and a–t graphs.

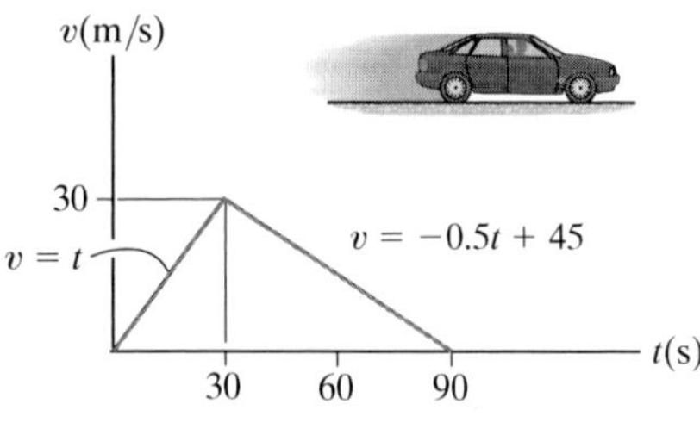

Prob. 12–57

12–58. The jet-powered boat starts from rest at $s = 0$ and travels along a straight line with the speed described by the graph. Construct the s–t and a–t graph for the time interval $0 \leq t \leq 50$ s.

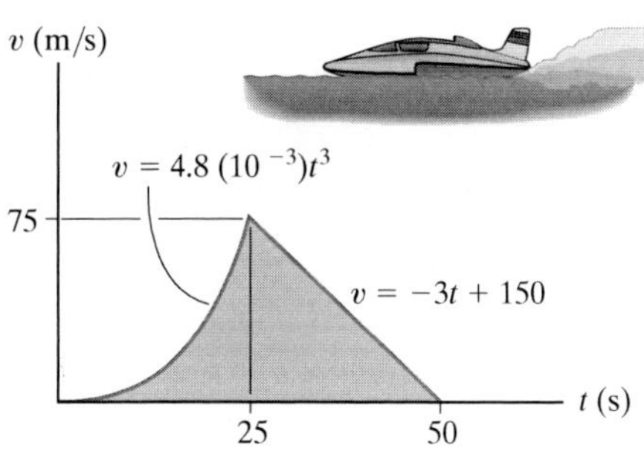

Prob. 12–58

12–59. From experimental data, the motion of a jet plane while traveling along a runway is defined by the v–t graph shown. Construct the s–t and a–t graphs for the motion.

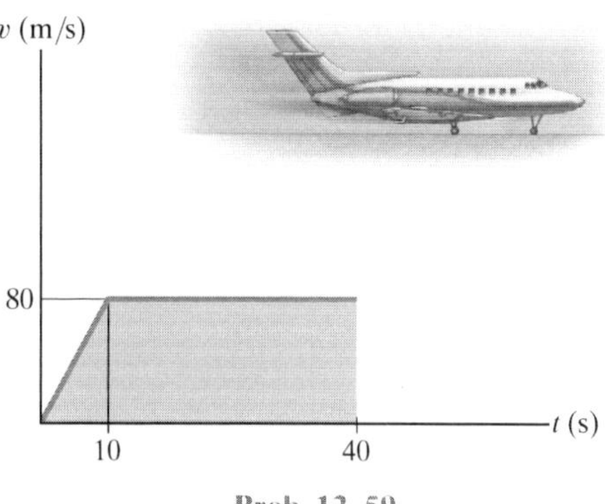

Prob. 12–59

*12–60. A car travels along a straight road with the speed shown by the v–t graph. Plot the a–t graph.

12–61. A car travels along a straight road with the speed shown by the v–t graph. Determine the total distance the car travels until it stops when $t = 48$ s. Also plot the s–t graph.

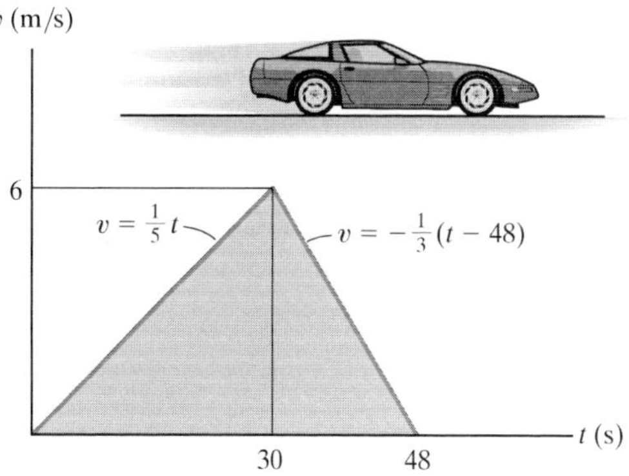

Probs. 12–60/61

12–62. The a–s graph for a train traveling along a straight track is given for the first 400 m of its motion. Plot the v–s graph. $v = 0$ at $s = 0$.

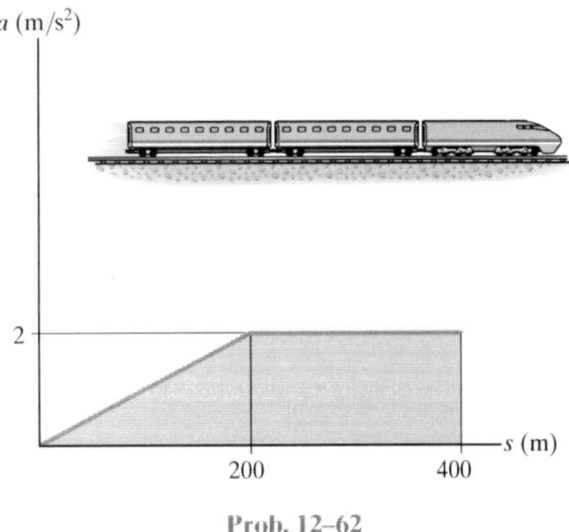

Prob. 12–62

12–63. The speed of a train during the first minute has been recorded as follows:

t (s)	0	20	40	60
v (m/s)	0	16	21	24

Plot the v–t graph, approximating the curve as straight-line segments between the given points. Determine the total distance traveled.

***12–64.** The v–$s$graph for a test vehicle is shown. Determine its acceleration when $s = 100$ m and when $s = 175$ m.

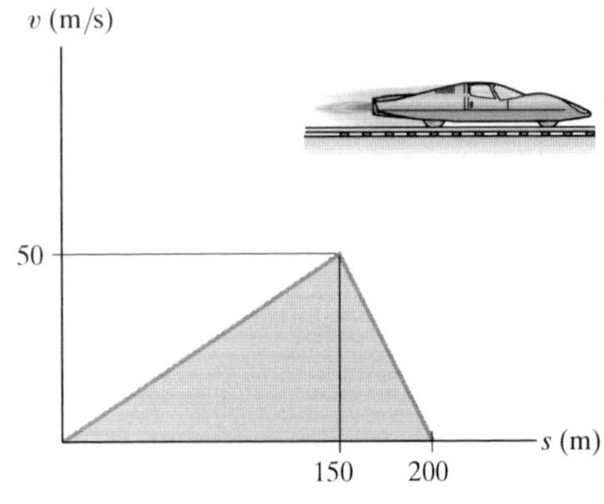

Prob. 12–64

12–65. Two cars start from rest side by side and travel along a straight road. Car A accelerates at 4 m/s^2 for 10 s and then maintains a constant speed. Car B accelerates at 5 m/s^2 until reaching a constant speed of 25 m/s and then maintains this speed. Construct the a–t, v–t, and s–t graphs for each car until $t = 15$ s. What is the distance between the two cars when $t = 15$ s?

12–66. A two-stage rocket is fired vertically from rest at $s = 0$ with an acceleration as shown. After 30 s the first stage A burns out and the second stage B ignites. Plot the v–t graph which describes the motion of the second stage for $0 \leq t \leq 60$ s.

12–67. A two-stage rocket is fired vertically from rest at $s = 0$ with an acceleration as shown. After 30 s the first stage A burns out and the second stage B ignites. Plot the s–t graph which describes the motion of the second stage for $0 \leq t \leq 60$ s.

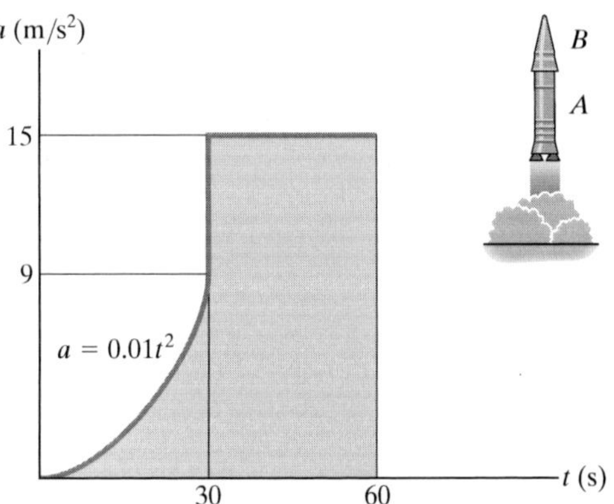

Probs. 12–66/67

***12–68.** The a–s graph for a jeep traveling along a straight road is given for the first 300 m of its motion. Construct the v–s graph. At $s = 0$, $v = 0$.

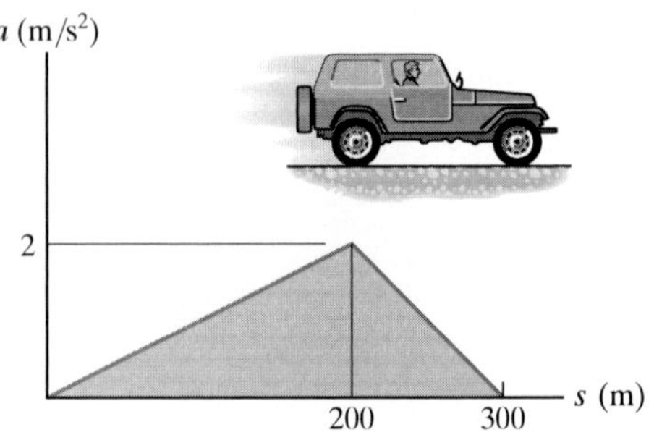

Prob. 12–68

12–69. The s–t graph for a train has been experimentally determined. From the data, construct the v–t and a–t graphs for the motion; $0 \le t \le 40$ s. For $0 \le t \le 30$ s, the curve is $s = (0.4t^2)$ m, and then it becomes straight for $t \ge 30$ s.

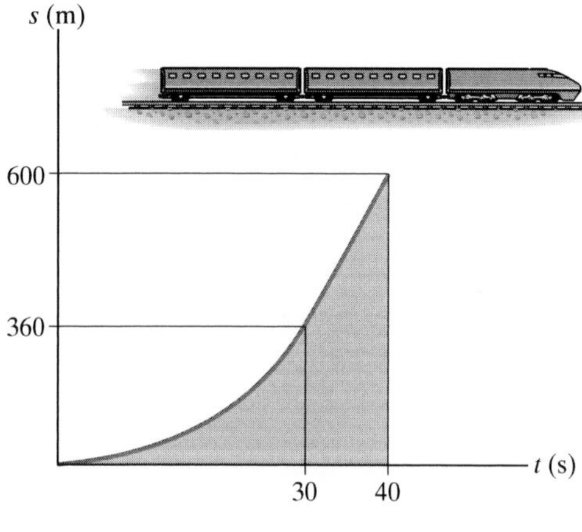

Prob. 12–69

12–71. The a–t graph for a car is shown. Construct the v–t and s–t graphs if the car starts from rest at $t = 0$. At what time t' does the car stop?

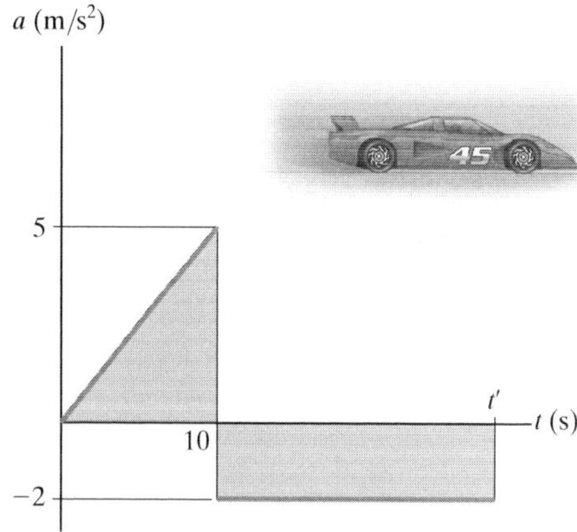

Prob. 12–71

12–70. The boat travels along a straight line with the speed described by the graph. Construct the s–t and a–s graphs. Also, determine the time required for the boat to travel a distance $s = 400$ m if $s = 0$ when $t = 0$.

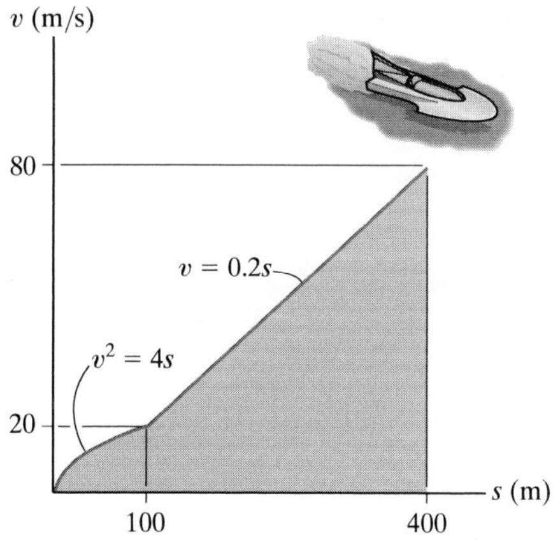

Prob. 12–70

***12–72.** The v–s graph for an airplane traveling on a straight runway is shown. Determine the acceleration of the plane at $s = 100$ m and $s = 150$ m. Draw the a–s graph.

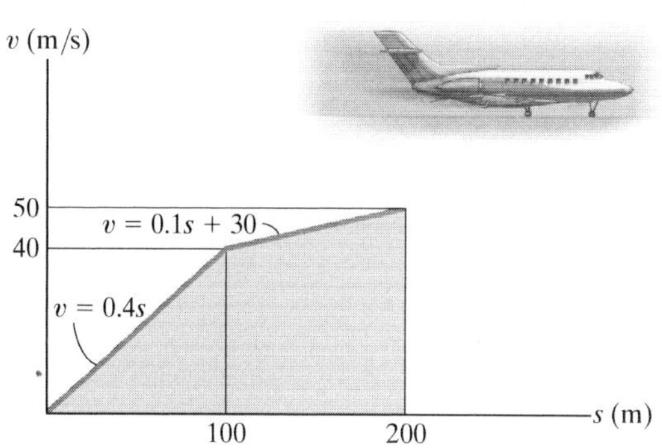

Prob. 12–72

12.4 General Curvilinear Motion

Curvilinear motion occurs when a particle moves along a curved path. Since this path is often described in three dimensions, vector analysis will be used to formulate the particle's position, velocity, and acceleration.* In this section the general aspects of curvilinear motion are discussed, and in subsequent sections we will consider three types of coordinate systems often used to analyze this motion.

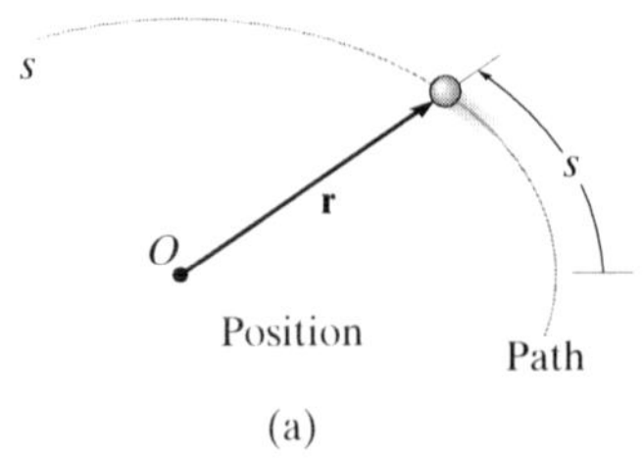

(a)

Position. Consider a particle located at a point on a space curve defined by the path function $s(t)$, Fig. 12–16*a*. The position of the particle, measured from a fixed point O, will be designated by the *position vector* $\mathbf{r} = \mathbf{r}(t)$. Notice that both the magnitude and direction of this vector will change as the particle moves along the curve.

Displacement. Suppose that during a small time interval Δt the particle moves a distance Δs along the curve to a new position, defined by $\mathbf{r}' = \mathbf{r} + \Delta\mathbf{r}$, Fig. 12–16*b*. The *displacement* $\Delta\mathbf{r}$ represents the change in the particle's position and is determined by vector subtraction; i.e., $\Delta\mathbf{r} = \mathbf{r}' - \mathbf{r}$.

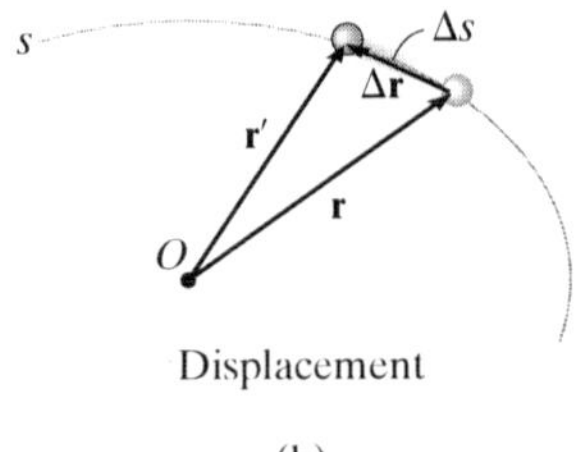

(b)

Velocity. During the time Δt, the *average velocity* of the particle is

$$\mathbf{v}_{avg} = \frac{\Delta\mathbf{r}}{\Delta t}$$

The *instantaneous velocity* is determined from this equation by letting $\Delta t \rightarrow 0$, and consequently the direction of $\Delta\mathbf{r}$ *approaches* the *tangent* to the curve. Hence, $\mathbf{v} = \lim_{\Delta t \to 0}(\Delta\mathbf{r}/\Delta t)$ or

$$\mathbf{v} = \frac{d\mathbf{r}}{dt} \qquad (12\text{–}7)$$

Since $d\mathbf{r}$ will be tangent to the curve, the *direction* of $\mathbf{v}$ is also *tangent to the curve*, Fig. 12–16*c*. The *magnitude* of $\mathbf{v}$, which is called the *speed*, is obtained by realizing that the length of the straight line segment $\Delta\mathbf{r}$ in Fig. 12–16*b* approaches the arc length Δs as $\Delta t \rightarrow 0$, we have $v = \lim_{\Delta t \to 0}(\Delta r/\Delta t) = \lim_{\Delta t \to 0}(\Delta s/\Delta t)$, or

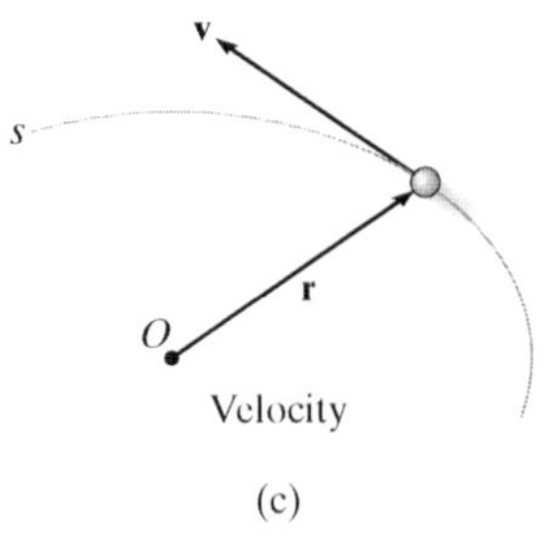

(c)

Fig. 12–16

$$v = \frac{ds}{dt} \qquad (12\text{–}8)$$

Thus, the *speed* can be obtained by differentiating the path function s with respect to time.

*A summary of some of the important concepts of vector analysis is given in Appendix B.

Acceleration. If the particle has a velocity $\mathbf{v}$ at time t and a velocity $\mathbf{v}' = \mathbf{v} + \Delta\mathbf{v}$ at $t + \Delta t$, Fig. 12–16*d*, then the *average acceleration* of the particle during the time interval Δt is

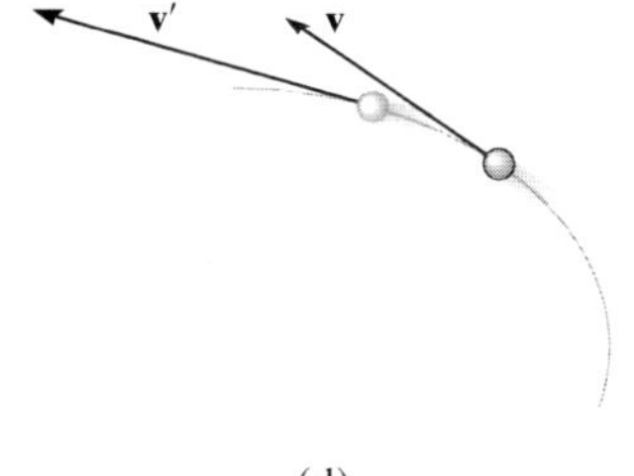

(d)

$$\mathbf{a}_{avg} = \frac{\Delta\mathbf{v}}{\Delta t}$$

where $\Delta\mathbf{v} = \mathbf{v}' - \mathbf{v}$. To study this time rate of change, the two velocity vectors in Fig. 12–16*d* are plotted in Fig. 12–16*e* such that their tails are located at the fixed point O' and their arrowheads touch points on a curve. This curve is called a *hodograph*, and when constructed, it describes the locus of points for the arrowhead of the velocity vector in the same manner as the *path s* describes the locus of points for the arrowhead of the position vector, Fig. 12–16*a*.

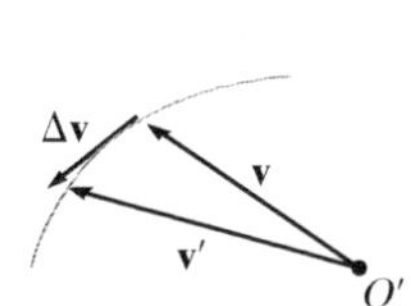

(e)

To obtain the *instantaneous acceleration*, let $\Delta t \rightarrow 0$ in the above equation. In the limit $\Delta\mathbf{v}$ will approach the *tangent to the hodograph*, and so $\mathbf{a} = \lim_{\Delta t \rightarrow 0}(\Delta\mathbf{v}/\Delta t)$, or

$$\mathbf{a} = \frac{d\mathbf{v}}{dt} \tag{12–9}$$

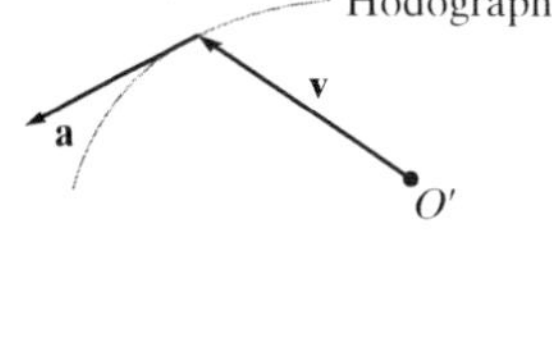

(f)

Substituting Eq. 12–7 into this result, we can also write

$$\mathbf{a} = \frac{d^2\mathbf{r}}{dt^2}$$

By definition of the derivative, $\mathbf{a}$ acts *tangent to the hodograph*, Fig. 12–16*f*, and, *in general it is not tangent to the path of motion*, Fig. 12–16*g*. To clarify this point, realize that $\Delta\mathbf{v}$ and consequently $\mathbf{a}$ must account for the change made in *both* the magnitude *and* direction of the velocity $\mathbf{v}$ as the particle moves from one point to the next along the path, Fig. 12–16*d*. However, in order for the particle to follow any curved path, the directional change always "swings" the velocity vector toward the "inside" or "concave side" of the path, and therefore $\mathbf{a}$ *cannot* remain tangent to the path. In summary, $\mathbf{v}$ is always tangent to the *path* and $\mathbf{a}$ is always tangent to the *hodograph*.

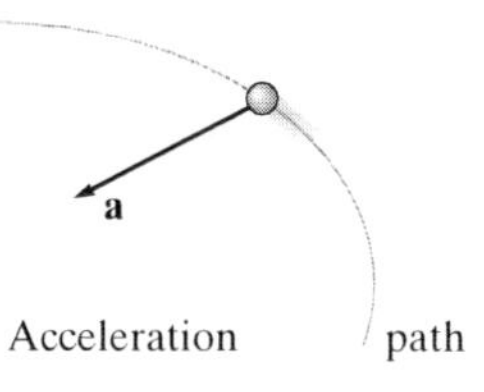

(g)

Fig. 12–16

12.5 Curvilinear Motion: Rectangular Components

Occasionally the motion of a particle can best be described along a path that can be expressed in terms of its x, y, z coordinates.

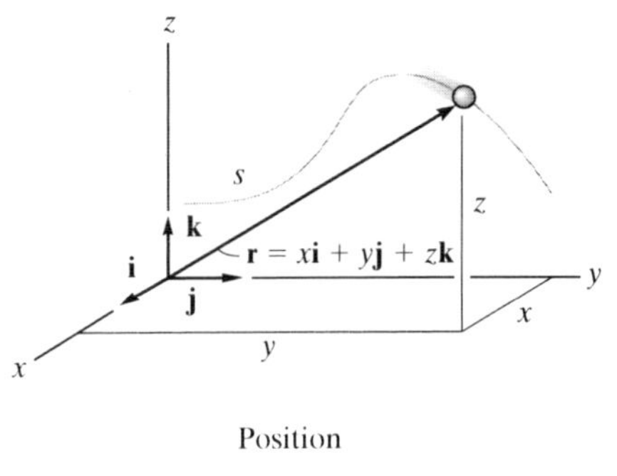

Position

(a)

Position. If the particle is at point (x, y, z) on the curved path s shown in Fig. 12–17a, then its location is defined by the *position vector*

$$\mathbf{r} = x\mathbf{i} + y\mathbf{j} + z\mathbf{k} \tag{12–10}$$

When the particle moves, the x, y, z components of $\mathbf{r}$ will be functions of time; i.e., $x = x(t)$, $y = y(t)$, $z = z(t)$, so that $\mathbf{r} = \mathbf{r}(t)$.

At any instant the *magnitude* of $\mathbf{r}$ is defined from Eq. B–3 in Appendix B as

$$r = \sqrt{x^2 + y^2 + z^2}$$

And the *direction* of $\mathbf{r}$ is specified by the unit vector $\mathbf{u}_r = \mathbf{r}/r$.

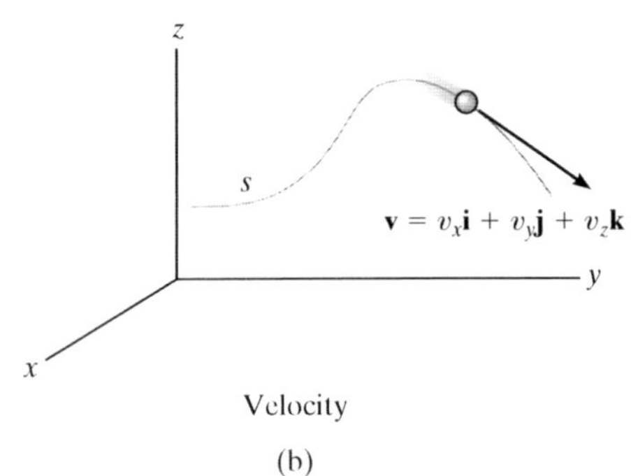

Velocity

(b)

Fig. 12–17

Velocity. The first time derivative of $\mathbf{r}$ yields the velocity of the particle. Hence,

$$\mathbf{v} = \frac{d\mathbf{r}}{dt} = \frac{d}{dt}(x\mathbf{i}) + \frac{d}{dt}(y\mathbf{j}) + \frac{d}{dt}(z\mathbf{k})$$

When taking this derivative, it is necessary to account for changes in *both* the magnitude and direction of each of the vector's components. For example, the derivative of the $\mathbf{i}$ component of $\mathbf{r}$ is

$$\frac{d}{dt}(x\mathbf{i}) = \frac{dx}{dt}\mathbf{i} + x\frac{d\mathbf{i}}{dt}$$

The second term on the right side is zero, provided the x, y, z reference frame is *fixed*, and therefore the *direction* (and the *magnitude*) of $\mathbf{i}$ does not change with time. Differentiation of the $\mathbf{j}$ and $\mathbf{k}$ components may be carried out in a similar manner, which yields the final result,

$$\mathbf{v} = \frac{d\mathbf{r}}{dt} = v_x\mathbf{i} + v_y\mathbf{j} + v_z\mathbf{k} \tag{12–11}$$

where

$$v_x = \dot{x} \quad v_y = \dot{y} \quad v_z = \dot{z} \tag{12–12}$$

The "dot" notation $\dot{x}, \dot{y}, \dot{z}$ represents the first time derivatives of $x = x(t)$, $y = y(t)$, $z = z(t)$, respectively.

The velocity has a *magnitude* that is found from

$$v = \sqrt{v_x^2 + v_y^2 + v_z^2}$$

and a *direction* that is specified by the unit vector $\mathbf{u}_v = \mathbf{v}/v$. As discussed in Sec. 12–4, this direction is *always tangent to the path*, as shown in Fig. 12–17*b*.

Acceleration. The acceleration of the particle is obtained by taking the first time derivative of Eq. 12–11 (or the second time derivative of Eq. 12–10). We have

$$\mathbf{a} = \frac{d\mathbf{v}}{dt} = a_x\mathbf{i} + a_y\mathbf{j} + a_z\mathbf{k} \qquad (12\text{–}13)$$

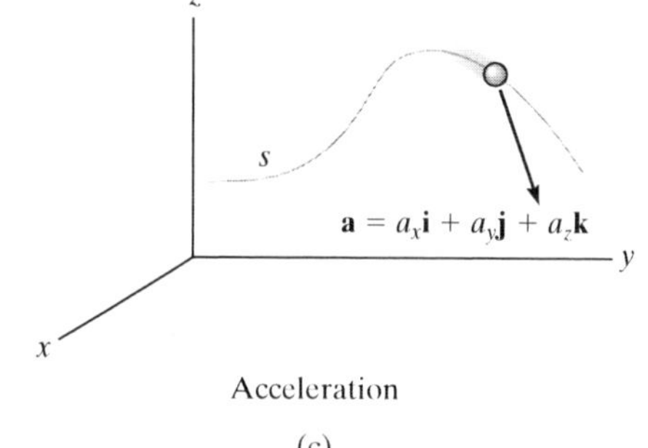

Acceleration

(c)

where

$$\begin{aligned} a_x &= \dot{v}_x = \ddot{x} \\ a_y &= \dot{v}_y = \ddot{y} \\ a_z &= \dot{v}_z = \ddot{z} \end{aligned} \qquad (12\text{–}14)$$

Here a_x, a_y, a_z represent, respectively, the first time derivatives of $v_x = v_x(t)$, $v_y = v_y(t)$, $v_z = v_z(t)$, or the second time derivatives of the functions $x = x(t)$, $y = y(t)$, $z = z(t)$.

The acceleration has a *magnitude*

$$a = \sqrt{a_x^2 + a_y^2 + a_z^2}$$

and a *direction* specified by the unit vector $\mathbf{u}_a = \mathbf{a}/a$. Since $\mathbf{a}$ represents the time rate of *change* in both the magnitude and direction of the velocity, in general $\mathbf{a}$ will *not* be tangent to the path, Fig. 12–17*c*.

Important Points

- Curvilinear motion can cause changes in *both* the magnitude and direction of the position, velocity, and acceleration vectors.
- The velocity vector is always directed *tangent* to the path.
- In general, the acceleration vector is *not* tangent to the path, but rather, it is tangent to the hodograph.
- If the motion is described using rectangular coordinates, then the components along each of the axes do not change direction, only their magnitude and sense (algebraic sign) will change.
- By considering the component motions, the change in magnitude and direction of the particle's position and velocity are automatically taken into account.

Procedure for Analysis

Coordinate System.

- A rectangular coordinate system can be used to solve problems for which the motion can conveniently be expressed in terms of its x, y, z components.

Kinematic Quantities.

- Since *rectilinear motion* occurs along *each coordinate axis*, the motion along each axis is found using $v = ds/dt$ and $a = dv/dt$; or in cases where the motion is not expressed as a function of time, the equation $a\,ds = v\,dv$ can be used.
- In two dimensions, the equation of the path $y = f(x)$ can be used to relate the x and y components of velocity and acceleration by applying the chain rule of calculus. A review of this concept is given in Appendix C.
- Once the x, y, z components of $\mathbf{v}$ and $\mathbf{a}$ have been determined, the magnitudes of these vectors are found from the Pythagorean theorem, Eq. B-3, and their coordinate direction angles from the components of their unit vectors, Eqs. B-4 and B-5.

EXAMPLE 12.9

At any instant the horizontal position of the weather balloon in Fig. 12–18*a* is defined by $x = (8t)$ m, where t is in seconds. If the equation of the path is $y = x^2/10$, determine the magnitude and direction of the velocity and the acceleration when $t = 2$ s.

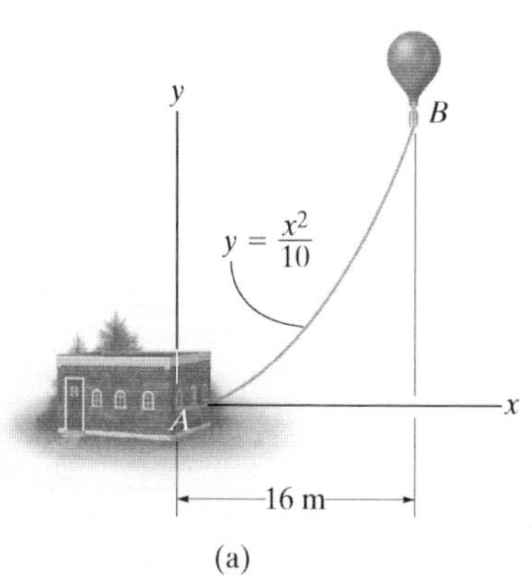

(a)

SOLUTION

Velocity. The velocity component in the x direction is

$$v_x = \dot{x} = \frac{d}{dt}(8t) = 8 \text{ m/s} \rightarrow$$

To find the relationship between the velocity components we will use the chain rule of calculus. When $t = 2$ s, $x = 8(2) = 16$ m, Fig. 12–18*a*, and so

$$v_y = \dot{y} = \frac{d}{dt}(x^2/10) = 2x\dot{x}/10 = 2(16)(8)/10 = 25.6 \text{ m/s} \uparrow$$

When $t = 2$ s, the magnitude of velocity is therefore

$$v = \sqrt{(8 \text{ m/s})^2 + (25.6 \text{ m/s})^2} = 26.8 \text{ m/s} \qquad \textit{Ans.}$$

The direction is tangent to the path, Fig. 12–18*b*, where

$$\theta_v = \tan^{-1}\frac{v_y}{v_x} = \tan^{-1}\frac{25.6}{8} = 72.6° \qquad \textit{Ans.}$$

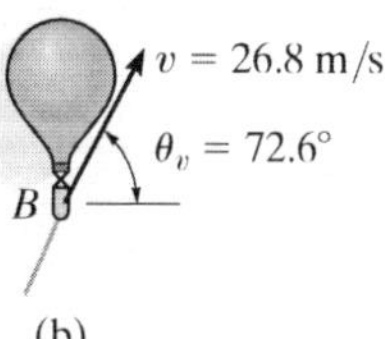

(b)

Acceleration. The relationship between the acceleration components is determined using the chain rule. (See Appendix C.) We have

$$a_x = \dot{v}_x = \frac{d}{dt}(8) = 0$$

$$a_y = \dot{v}_y = \frac{d}{dt}(2x\dot{x}/10) = 2(\dot{x})\dot{x}/10 + 2x(\ddot{x})/10$$

$$= 2(8)^2/10 + 2(16)(0)/10 = 12.8 \text{ m/s}^2 \uparrow$$

Thus,

$$a = \sqrt{(0)^2 + (12.8)^2} = 12.8 \text{ m/s}^2 \qquad \textit{Ans.}$$

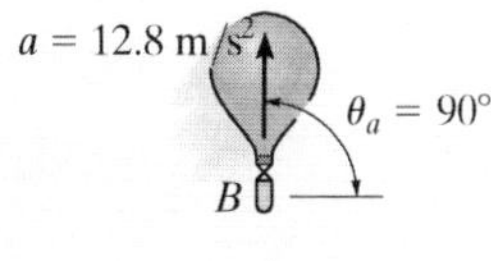

(c)

The direction of **a**, as shown in Fig. 12–18*c*, is

$$\theta_a = \tan^{-1}\frac{12.8}{0} = 90° \qquad \textit{Ans.}$$

Fig. 12–18

NOTE: It is also possible to obtain v_y and a_y by first expressing $y = f(t) = (8t)^2/10 = 6.4t^2$ and then taking successive time derivatives.

12

EXAMPLE 12.10

For a short time, the path of the plane in Fig. 12–19*a* is described by $y = (0.001x^2)$ m. If the plane is rising with a constant upward velocity of 10 m/s, determine the magnitudes of the velocity and acceleration of the plane when it reaches an altitude of $y = 100$ m.

SOLUTION

When $y = 100$ m, then $100 = 0.001x^2$ or $x = 316.2$ m. Also, due to constant velocity $v_y = 10$ m/s, so

$$y = v_y t; \qquad 100 \text{ m} = (10 \text{ m/s})\, t \qquad t = 10 \text{ s}$$

Velocity. Using the chain rule (see Appendix C) to find the relationship between the velocity components, we have

$$y = 0.001x^2$$

$$v_y = \dot{y} = \frac{d}{dt}(0.001x^2) = (0.002x)\dot{x} = 0.002xv_x \qquad (1)$$

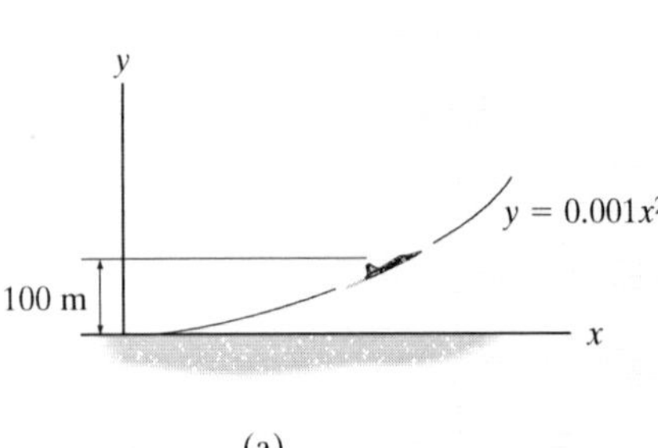

(a)

Thus

$$10 \text{ m/s} = 0.002(316.2 \text{ m})(v_x)$$
$$v_x = 15.81 \text{ m/s}$$

The magnitude of the velocity is therefore

$$v = \sqrt{v_x^2 + v_y^2} = \sqrt{(15.81 \text{ m/s})^2 + (10 \text{ m/s})^2} = 18.7 \text{ m/s} \qquad \textit{Ans.}$$

Acceleration. Using the chain rule, the time derivative of Eq. (1) gives the relation between the acceleration components.

$$a_y = \dot{v}_y = (0.002\dot{x})\dot{x} + 0.002x(\ddot{x}) = 0.002(v_x^2 + xa_x)$$

When $x = 316.2$ m, $v_x = 15.81$ m/s, $\dot{v}_y = a_y = 0$,

$$0 = 0.002\left[(15.81 \text{ m/s})^2 + 316.2 \text{ m}(a_x)\right]$$
$$a_x = -0.791 \text{ m/s}^2$$

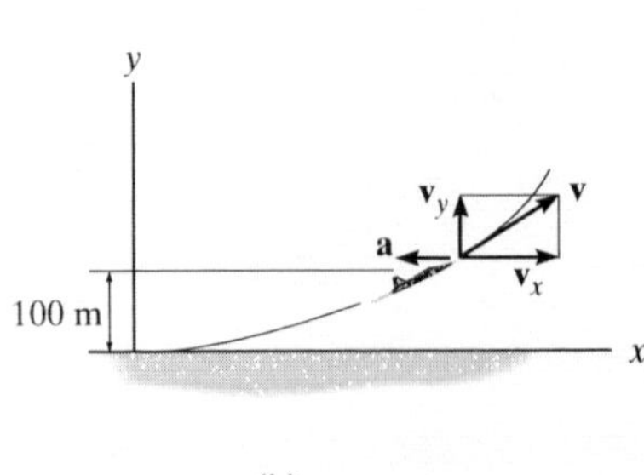

(b)

Fig. 12–19

The magnitude of the plane's acceleration is therefore

$$a = \sqrt{a_x^2 + a_y^2} = \sqrt{(-0.791 \text{ m/s}^2)^2 + (0 \text{ m/s}^2)^2}$$
$$= 0.791 \text{ m/s}^2 \qquad \textit{Ans.}$$

These results are shown in Fig. 12–19*b*.

12.6 Motion of a Projectile

The free-flight motion of a projectile is often studied in terms of its rectangular components. To illustrate the kinematic analysis, consider a projectile launched at point (x_0, y_0), with an initial velocity of $\mathbf{v}_0$, having components $(\mathbf{v}_0)_x$ and $(\mathbf{v}_0)_y$, Fig. 12–20. When air resistance is neglected, the only force acting on the projectile is its weight, which causes the projectile to have a *constant downward acceleration* of approximately $a_c = g = 9.81\ \text{m/s}^2$ or $g = 32.2\ \text{ft/s}^2$.*

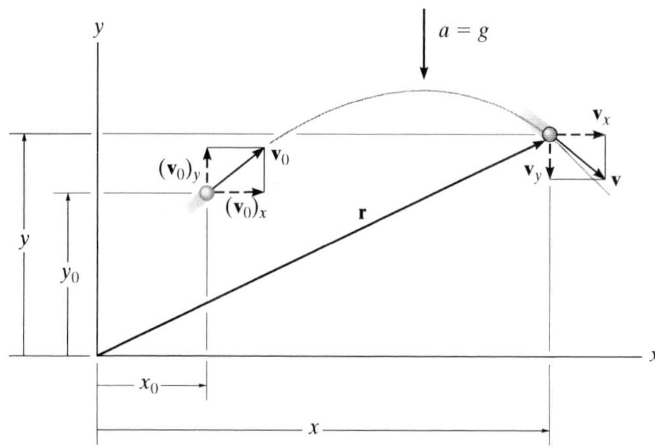

Fig. 12–20

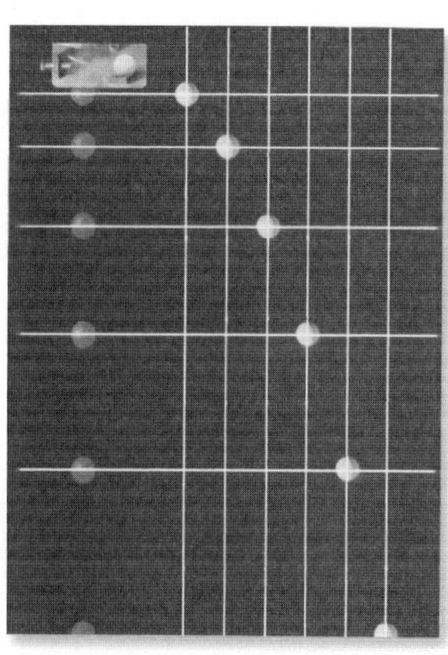

Each picture in this sequence is taken after the same time interval. The red ball falls from rest, whereas the yellow ball is given a horizontal velocity when released. Both balls accelerate downward at the same rate, and so they remain at the same elevation at any instant. This acceleration causes the difference in elevation between the balls to increase between successive photos. Also, note the horizontal distance between successive photos of the yellow ball is constant since the velocity in the horizontal direction remains constant.

Horizontal Motion. Since $a_x = 0$, application of the constant acceleration equations, 12–4 to 12–6, yields

$(\overset{+}{\rightarrow})$ $\quad v = v_0 + a_c t \qquad v_x = (v_0)_x$

$(\overset{+}{\rightarrow})$ $\quad x = x_0 + v_0 t + \frac{1}{2} a_c t^2; \qquad x = x_0 + (v_0)_x t$

$(\overset{+}{\rightarrow})$ $\quad v^2 = v_0^2 + 2a_c(x - x_0); \qquad v_x = (v_0)_x$

The first and last equations indicate that *the horizontal component of velocity always remains constant during the motion*.

Vertical Motion. Since the positive y axis is directed upward, then $a_y = -g$. Applying Eqs. 12–4 to 12–6, we get

$(+\uparrow)$ $\quad v = v_0 + a_c t; \qquad v_y = (v_0)_y - gt$

$(+\uparrow)$ $\quad y = y_0 + v_0 t + \frac{1}{2} a_c t^2; \qquad y = y_0 + (v_0)_y t - \frac{1}{2} g t^2$

$(+\uparrow)$ $\quad v^2 = v_0^2 + 2a_c(y - y_0); \qquad v_y^2 = (v_0)_y^2 - 2g(y - y_0)$

Recall that the last equation can be formulated on the basis of eliminating the time t from the first two equations, and therefore *only two of the above three equations are independent of one another*.

*This assumes that the earth's gravitational field does not vary with altitude.

To summarize, problems involving the motion of a projectile can have at most three unknowns since only three independent equations can be written; that is, *one* equation in the *horizontal direction* and *two* in the *vertical direction*. Once $\mathbf{v}_x$ and $\mathbf{v}_y$ are obtained, the resultant velocity $\mathbf{v}$, which is *always tangent* to the path, can be determined by the *vector sum* as shown in Fig. 12–20.

Procedure for Analysis

Coordinate System.

- Establish the fixed *x, y* coordinate axes and sketch the trajectory of the particle. Between any *two points* on the path specify the given problem data and identify the *three unknowns*. In all cases the acceleration of gravity acts downward and equals 9.81 m/s^2. The particle's initial and final velocities should be represented in terms of their *x* and *y* components.

- Remember that positive and negative position, velocity, and acceleration components always act in accordance with their associated coordinate directions.

Kinematic Equations.

- Depending upon the known data and what is to be determined, a choice should be made as to which three of the following four equations should be applied between the two points on the path to obtain the most direct solution to the problem.

Horizontal Motion.

- The *velocity* in the horizontal or *x* direction is *constant*, i.e., $v_x = (v_0)_x$, and

$$x = x_0 + (v_0)_x t$$

Vertical Motion.

- In the vertical or *y* direction *only two* of the following three equations can be used for solution.

$$v_y = (v_0)_y + a_c t$$

$$y = y_0 + (v_0)_y t + \tfrac{1}{2} a_c t^2$$

$$v_y^2 = (v_0)_y^2 + 2a_c(y - y_0)$$

For example, if the particle's final velocity v_y is not needed, then the first and third of these equations will not be useful.

© Dianne Maire/Shutterstock.com

Gravel falling off the end of this conveyor belt follows a path that can be predicted using the equations of constant acceleration. In this way the location of the accumulated pile can be determined. Rectangular coordinates are used for the analysis since the acceleration is only in the vertical direction.

EXAMPLE 12.11

A sack slides off the ramp, shown in Fig. 12–21, with a horizontal velocity of 12 m/s. If the height of the ramp is 6 m from the floor, determine the time needed for the sack to strike the floor and the range R where sacks begin to pile up.

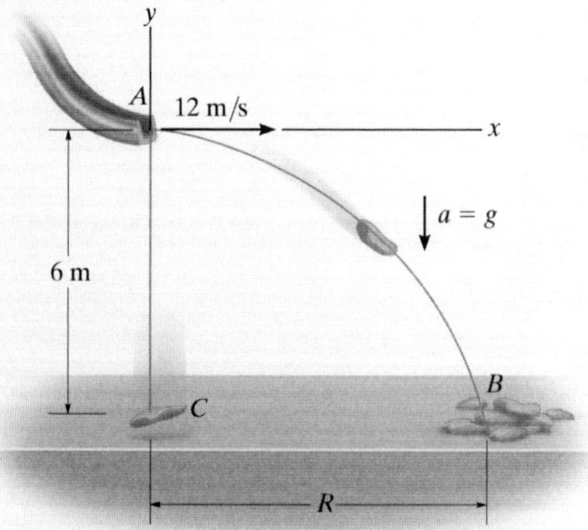

Fig. 12–21

SOLUTION

Coordinate System. The origin of coordinates is established at the beginning of the path, point A, Fig. 12–21. The initial velocity of a sack has components $(v_A)_x = 12\text{ m/s}$ and $(v_A)_y = 0$. Also, between points A and B the acceleration is $a_y = -9.81\text{ m/s}^2$. Since $(v_B)_x = (v_A)_x = 12\text{ m/s}$, the three unknowns are $(v_B)_y$, R, and the time of flight t_{AB}. Here we do not need to determine $(v_B)_y$.

Vertical Motion. The vertical distance from A to B is known, and therefore we can obtain a direct solution for t_{AB} by using the equation

$(+\uparrow)$
$$y_B = y_A + (v_A)_y t_{AB} + \tfrac{1}{2}a_c t_{AB}^2$$
$$-6\text{ m} = 0 + 0 + \tfrac{1}{2}(-9.81\text{ m/s}^2)t_{AB}^2$$
$$t_{AB} = 1.11\text{ s} \qquad \textit{Ans.}$$

Horizontal Motion. Since t_{AB} has been calculated, R is determined as follows:

$(\overset{+}{\rightarrow})$
$$x_B = x_A + (v_A)_x t_{AB}$$
$$R = 0 + 12\text{ m/s}\,(1.11\text{ s})$$
$$R = 13.3\text{ m} \qquad \textit{Ans.}$$

NOTE: The calculation for t_{AB} also indicates that if a sack were released *from rest* at A, it would take the same amount of time to strike the floor at C, Fig. 12–21.

12

EXAMPLE 12.12

The chipping machine is designed to eject wood chips at $v_O = 7.5$ m/s as shown in Fig. 12–22. If the tube is oriented at 30° from the horizontal, determine how high, h, the chips strike the pile if at this instant they land on the pile 6 m from the tube.

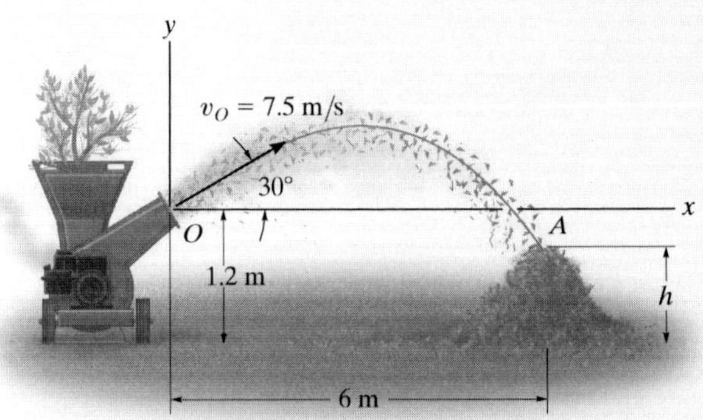

Fig. 12–22

SOLUTION

Coordinate System. When the motion is analyzed between points O and A, the three unknowns are the height h, time of flight t_{OA}, and vertical component of velocity $(v_A)_y$. [Note that $(v_A)_x = (v_O)_x$.] With the origin of coordinates at O, Fig. 12–22, the initial velocity of a chip has components of

$$(v_O)_x = (7.5 \cos 30°) \text{ m/s} = 6.50 \text{ m/s} \rightarrow$$

$$(v_O)_y = (7.5 \sin 30°) \text{ m/s} = 3.75 \text{ m/s}\uparrow$$

Also, $(v_A)_x = (v_O)_x = 6.50$ m/s and $a_y = -9.81$ m/s^2. Since we do not need to determine $(v_A)_y$, we have

Horizontal Motion.

$(\overset{+}{\rightarrow})$

$$x_A = x_O + (v_O)_x t_{OA}$$

$$6 \text{ m} = 0 + (6.50 \text{ m/s})t_{OA}$$

$$t_{OA} = 0.923 \text{ s}$$

Vertical Motion. Relating t_{OA} to the initial and final elevations of a chip, we have

$(+\uparrow)$ $y_A = y_O + (v_O)_y t_{OA} + \frac{1}{2}a_c t_{OA}^2$

$$(h - 1.2 \text{ m}) = 0 + (3.75 \text{ m/s})(0.923 \text{ s}) + \tfrac{1}{2}(-9.81 \text{ m/s}^2)(0.923 \text{ s})^2$$

$$h = 0.483 \text{ m}$$ *Ans.*

NOTE: We can determine $(v_A)_y$ by using $(v_A)_y = (v_O)_y + a_c t_{OA}$.

EXAMPLE 12.13

The track for this racing event was designed so that riders jump off the slope at 30°, from a height of 1 m. During a race it was observed that the rider shown in Fig. 12–23*a* remained in mid air for 1.5 s. Determine the speed at which he was traveling off the ramp, the horizontal distance he travels before striking the ground, and the maximum height he attains. Neglect the size of the bike and rider.

 (a)

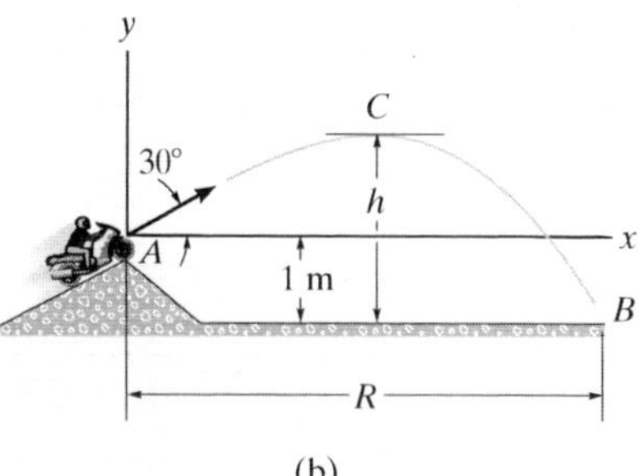

(b)

Fig. 12–23

SOLUTION

Coordinate System. As shown in Fig. 12–23*b*, the origin of the coordinates is established at *A*. Between the end points of the path *AB* the three unknowns are the initial speed v_A, range *R*, and the vertical component of velocity $(v_B)_y$.

Vertical Motion. Since the time of flight and the vertical distance between the ends of the path are known, we can determine v_A.

$(+\uparrow)$

$$y_B = y_A + (v_A)_y t_{AB} + \tfrac{1}{2}a_c t_{AB}^2$$

$$-1\text{ m} = 0 + v_A \sin 30°(1.5\text{ s}) + \tfrac{1}{2}(-9.81\text{ m/s}^2)(1.5\text{ s})^2$$

$$v_A = 13.38\text{ m/s} = 13.4\text{ m/s} \qquad \textit{Ans.}$$

Horizontal Motion. The range *R* can now be determined.

$(\overset{+}{\rightarrow})$

$$x_B = x_A + (v_A)_x t_{AB}$$

$$R = 0 + 13.38 \cos 30°\text{ m/s}(1.5\text{ s})$$

$$= 17.4\text{ m} \qquad \textit{Ans.}$$

In order to find the maximum height *h* we will consider the path *AC*, Fig. 12–23*b*. Here the three unknowns are the time of flight t_{AC}, the horizontal distance from *A* to *C*, and the height *h*. At the maximum height $(v_C)_y = 0$, and since v_A is known, we can determine *h directly* without considering t_{AC} using the following equation.

$$(v_C)_y^2 = (v_A)_y^2 + 2a_c[y_C - y_A]$$

$$0^2 = (13.38 \sin 30°\text{ m/s})^2 + 2(-9.81\text{ m/s}^2)[(h - 1\text{ m}) - 0]$$

$$h = 3.28\text{ m} \qquad \textit{Ans.}$$

NOTE: Show that the bike will strike the ground at *B* with a velocity having components of

$$(v_B)_x = 11.6\text{ m/s}\rightarrow, \quad (v_B)_y = 8.02\text{ m/s}\downarrow$$

FUNDAMENTAL PROBLEMS

F12–15. If the x and y components of a particle's velocity are $v_x = (32t)$ m/s and $v_y = 8$ m/s, determine the equation of the path $y = f(x)$. $x = 0$ and $y = 0$ when $t = 0$.

F12–16. A particle is traveling along the straight path. If its position along the x axis is $x = (8t)$ m, where t is in seconds, determine its speed when $t = 2$ s.

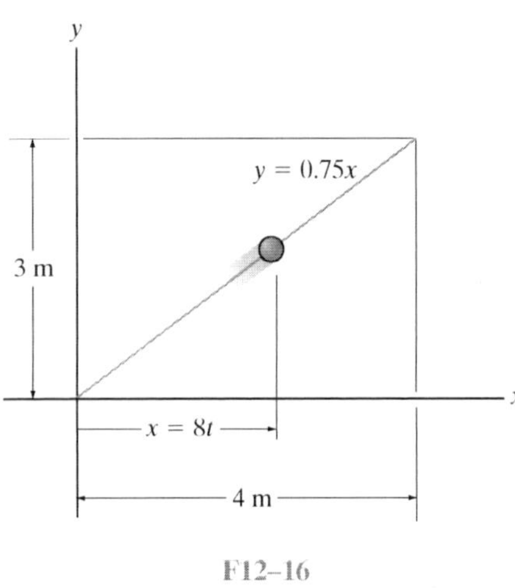

F12–16

F12–17. A particle is constrained to travel along the path. If $x = (4t^4)$ m, where t is in seconds, determine the magnitude of the particle's velocity and acceleration when $t = 0.5$ s.

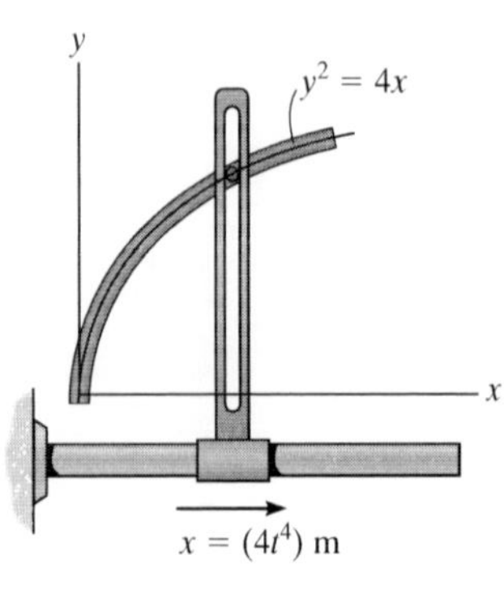

F12–17

F12–18. A particle travels along a straight-line path $y = 0.5x$. If the x component of the particle's velocity is $v_x = (2t^2)$ m/s, where t is in seconds, determine the magnitude of the particle's velocity and acceleration when $t = 4$ s.

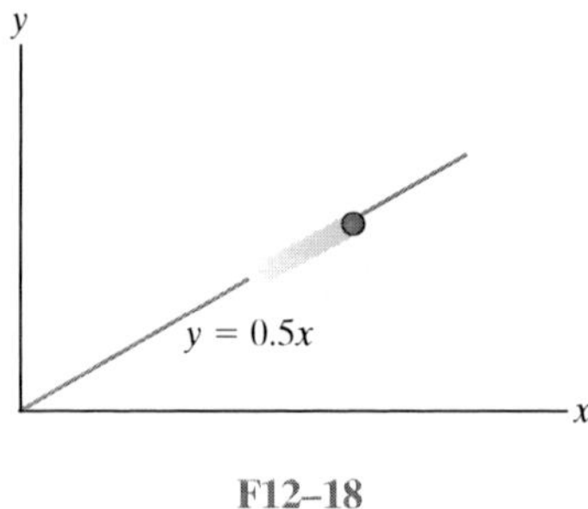

F12–18

F12–19. A particle is traveling along the parabolic path $y = 0.25x^2$. If $x = (2t^2)$ m, where t is in seconds, determine the magnitude of the particle's velocity and acceleration when $t = 2$ s.

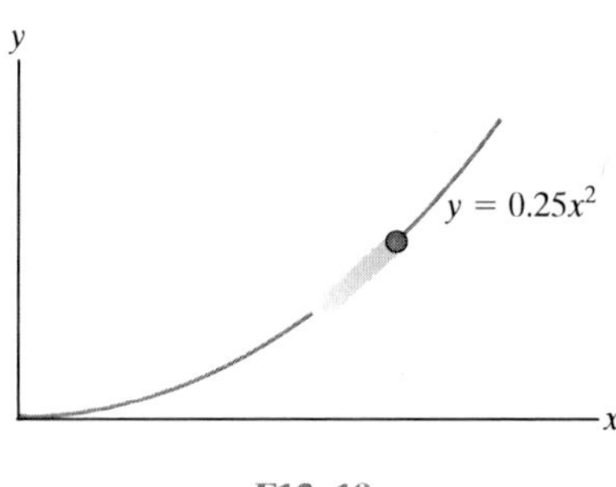

F12–19

F12–20. The box slides down the slope described by the equation $y = (0.05x^2)$ m, where x is in meters. If the box has x components of velocity and acceleration of $v_x = -3$ m/s and $a_x = -1.5$ m/s^2 at $x = 5$ m, determine the y components of the velocity and the acceleration of the box at this instant.

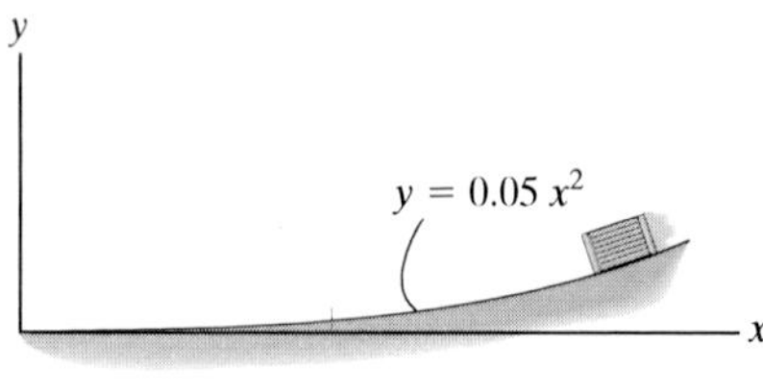

F12–20

F12–21. The ball is kicked from point A with the initial velocity $v_A = 10$ m/s. Determine the maximum height h it reaches.

F12–22. The ball is kicked from point A with the initial velocity $v_A = 10$ m/s. Determine the range R, and the speed when the ball strikes the ground.

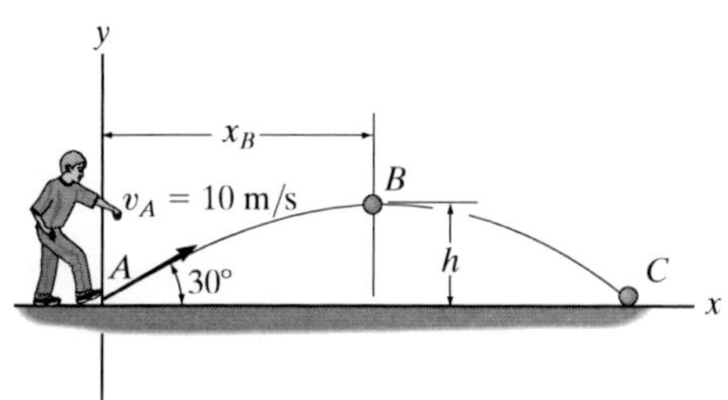

F12–21/22

F12–23. Determine the speed at which the basketball at A must be thrown at the angle of 30° so that it makes it to the basket at B.

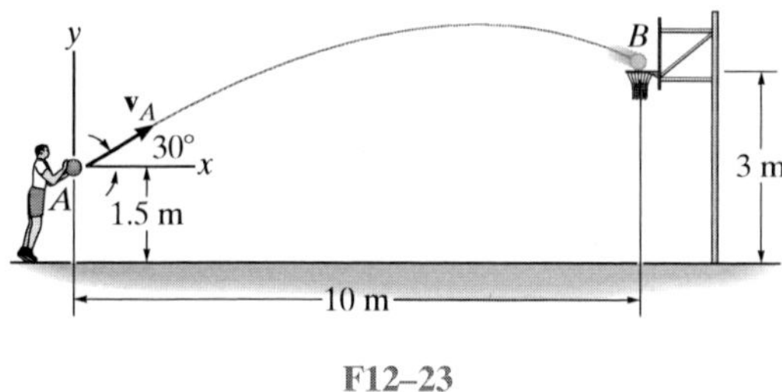

F12–23

F12–24. Water is sprayed at an angle of 90° from the slope at 20 m/s. Determine the range R.

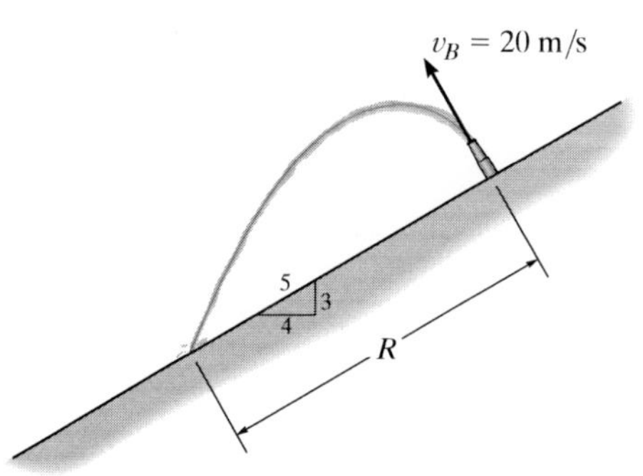

F12–24

F12–25. A ball is thrown from A. If it is required to clear the wall at B, determine the minimum magnitude of its initial velocity $\mathbf{v}_A$.

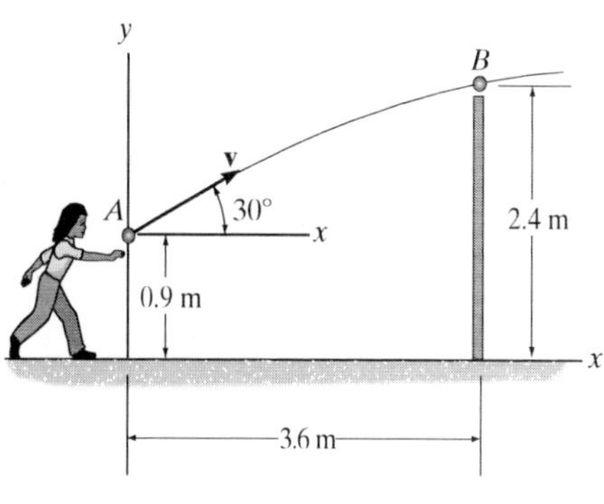

F12–25

F12–26. A projectile is fired with an initial velocity of $v_A = 150$ m/s off the roof of the building. Determine the range R where it strikes the ground at B.

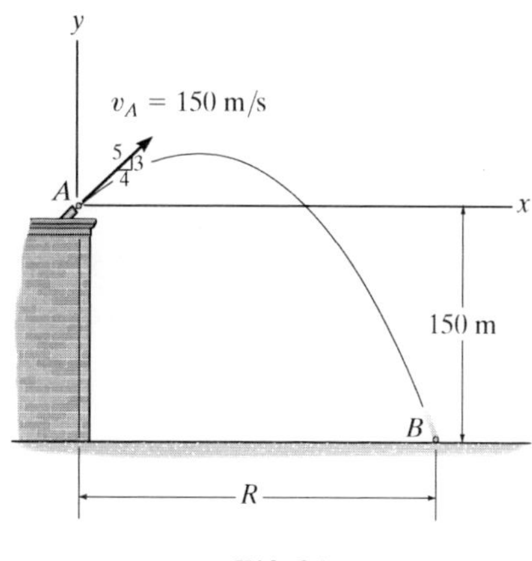

F12–26

PROBLEMS

12–73. The position of a particle is defined by $r = \{5(\cos 2t)\mathbf{i} + 4(\sin 2t)\mathbf{j}\}$ m, where t is in seconds and the arguments for the sine and cosine are given in radians. Determine the magnitudes of the velocity and acceleration of the particle when $t = 1$ s. Also, prove that the path of the particle is elliptical.

12–74. The velocity of a particle is $\mathbf{v} = \{3\mathbf{i} + (6 - 2t)\mathbf{j}\}$ m/s, where t is in seconds. If $\mathbf{r} = \mathbf{0}$ when $t = 0$, determine the displacement of the particle during the time interval $t = 1$ s to $t = 3$ s.

12–75. A particle is traveling with a velocity of $\mathbf{v} = \{3\sqrt{t}e^{-0.2t}\mathbf{i} + 4e^{-0.8t^2}\mathbf{j}\}$ m/s, where t is in seconds. Determine the magnitude of the particle's displacement from $t = 0$ to $t = 3$ s. Use Simpson's rule with $n = 100$ to evaluate the integrals. What is the magnitude of the particle's acceleration when $t = 2$ s?

***12–76.** The velocity of a particle is given by $v = \{16t^2\mathbf{i} + 4t^3\mathbf{j} + (5t + 2)\mathbf{k}\}$ m/s, where t is in seconds. If the particle is at the origin when $t = 0$, determine the magnitude of the particle's acceleration when $t = 2$ s. Also, what is the x, y, z coordinate position of the particle at this instant?

12–77. The car travels from A to B, and then from B to C, as shown in the figure. Determine the magnitude of the displacement of the car and the distance traveled.

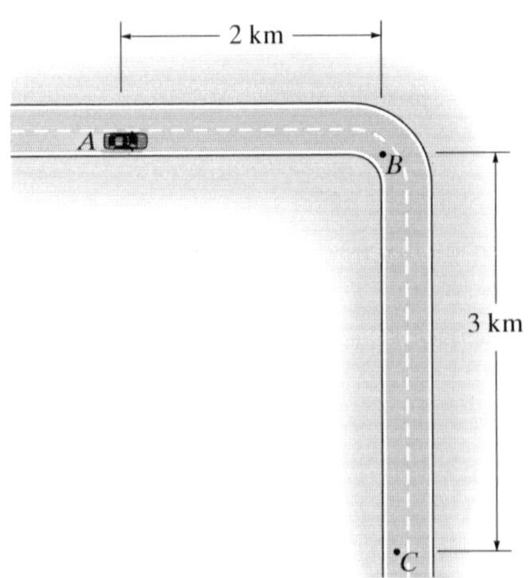

Prob. 12–77

12–78. A car travels east 2 km for 5 minutes, then north 3 km for 8 minutes, and then west 4 km for 10 minutes. Determine the total distance traveled and the magnitude of displacement of the car. Also, what is the magnitude of the average velocity and the average speed?

12–79. A car traveling along the straight portions of the road has the velocities indicated in the figure when it arrives at points A, B, and C. If it takes 3 s to go from A to B, and then 5 s to go from B to C, determine the average acceleration between points A and B and between points A and C.

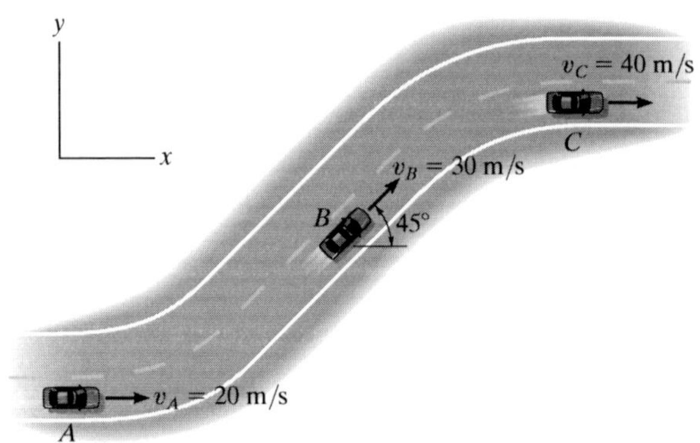

Prob. 12–79

***12–80.** A particle travels along the curve from A to B in 2 s. It takes 4 s for it to go from B to C and then 3 s to go from C to D. Determine its average speed when it goes from A to D.

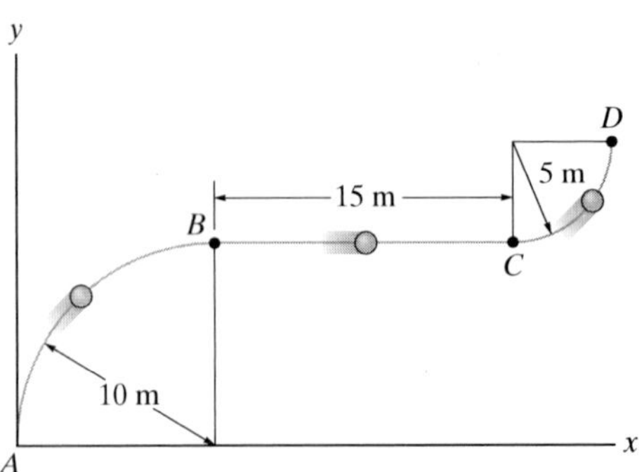

Prob. 12–80

12–81. The position of a crate sliding down a ramp is given by $x = (0.25t^3)$ m, $y = (1.5t^2)$ m, $z = (6 - 0.75t^{5/2})$ m, where t is in seconds. Determine the magnitude of the crate's velocity and acceleration when $t = 2$ s.

12–82. A rocket is fired from rest at $x = 0$ and travels along a parabolic trajectory described by $y^2 = [120(10^3)x]$ m. If the x component of acceleration is $a_x = (\frac{1}{4}t^2)$ m/s², where t is in seconds, determine the magnitude of the rocket's velocity and acceleration when $t = 10$ s.

12–83. The flight path of the helicopter as it takes off from A is defined by the parametric equations $x = (2t^2)$ m and $y = (0.04t^3)$ m, where t is the time in seconds. Determine the distance the helicopter is from point A and the magnitudes of its velocity and acceleration when $t = 10$ s.

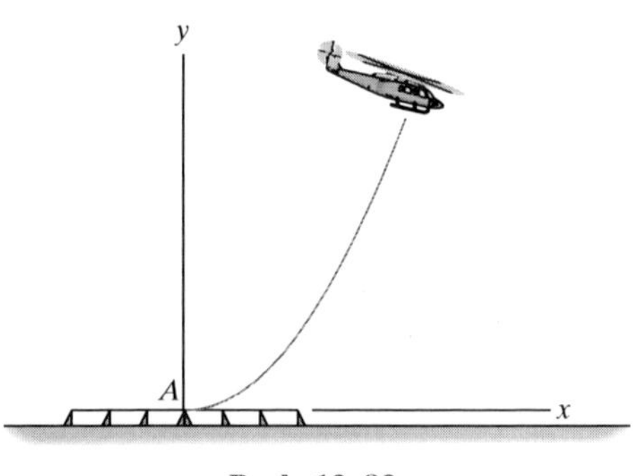

Prob. 12–83

***12–84.** The motorcycle travels with constant speed v_0 along the path that, for a short distance, takes the form of a sine curve. Determine the x and y components of its velocity at any instant on the curve.

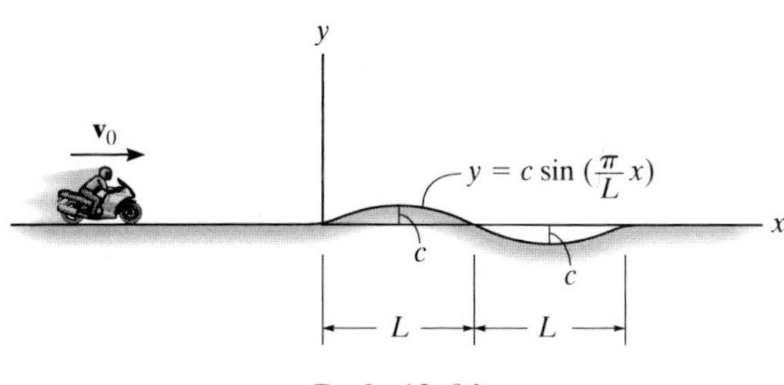

Prob. 12–84

12–85. A particle travels along the curve from A to B in 1 s. If it takes 3 s for it to go from A to C, determine its *average velocity* when it goes from B to C.

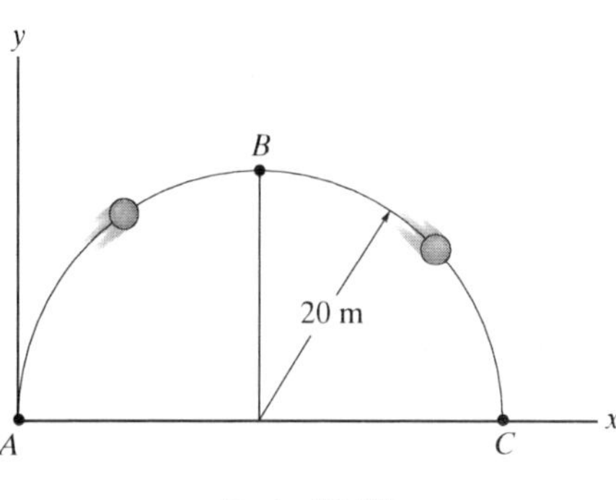

Prob. 12–85

12–86. When a rocket reaches an altitude of 40 m it begins to travel along the parabolic path $(y - 40)^2 = 160x$, where the coordinates are measured in meters. If the component of velocity in the vertical direction is constant at $v_y = 180$ m/s, determine the magnitudes of the rocket's velocity and acceleration when it reaches an altitude of 80 m.

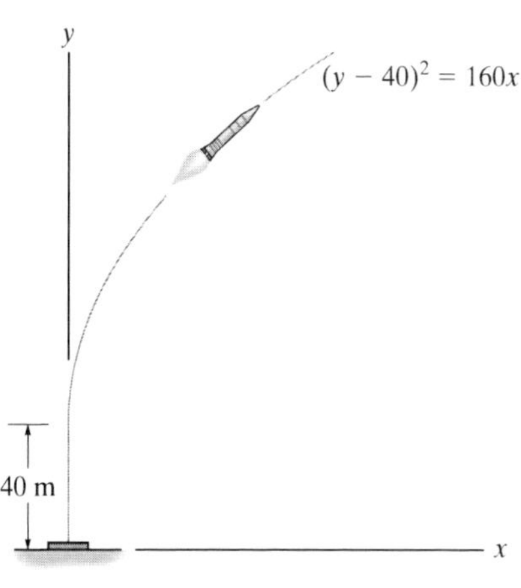

Prob. 12–86

12–87. Pegs A and B are restricted to move in the elliptical slots due to the motion of the slotted link. If the link moves with a constant speed of 10 m/s, determine the magnitude of the velocity and acceleration of peg A when $x = 1$ m.

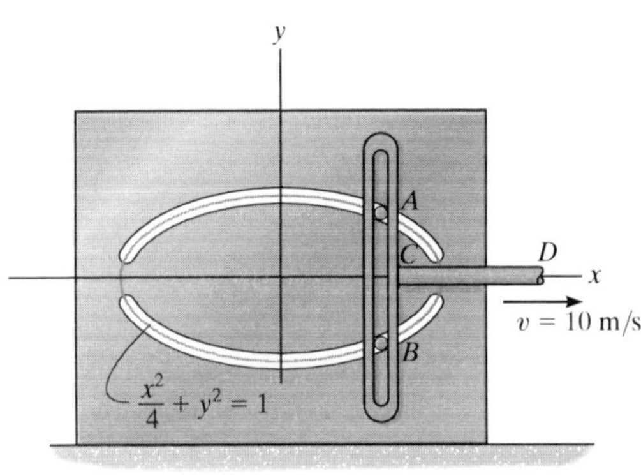

Prob. 12–87

*__12–88.__ Show that if a projectile is fired at an angle θ from the horizontal with an initial velocity $\mathbf{v}_0$, the *maximum* range the projectile can travel is given by $R_{max} = v_0^2/g$, where g is the acceleration of gravity. What is the angle θ for this condition?

12–89. It is observed that the time for the ball to strike the ground at B is 2.5 s. Determine the speed v_A and angle θ_A at which the ball was thrown.

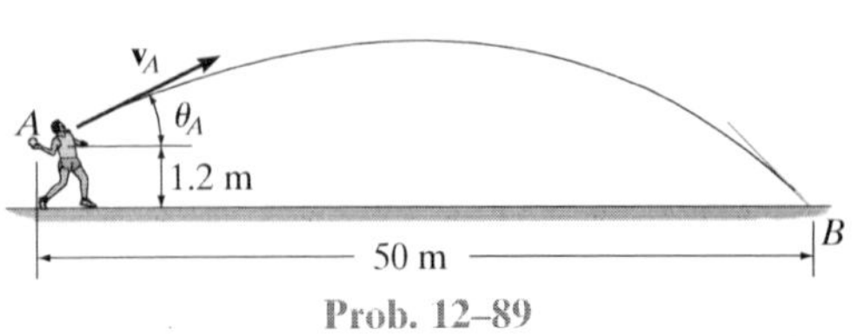

Prob. 12–89

12–90. Determine the minimum initial velocity v_0 and the corresponding angle θ_0 at which the ball must be kicked in order for it to just cross over the 3-m high fence.

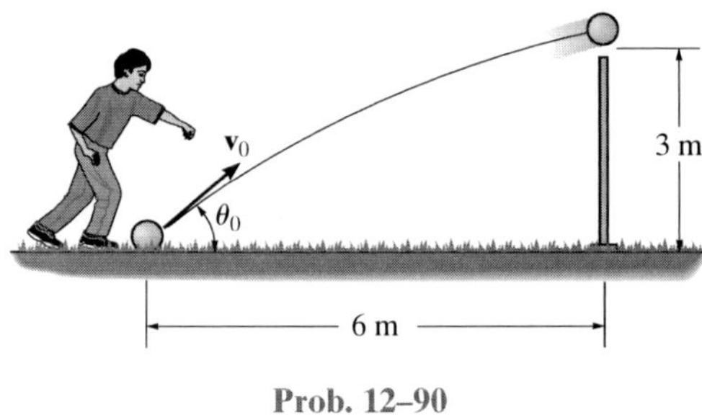

Prob. 12–90

12–91. The pitching machine is adjusted so that the baseball is launched with a speed of $v_A = 30$ m/s. If the ball strikes the ground at B, determine the two possible angles θ_A at which it was launched.

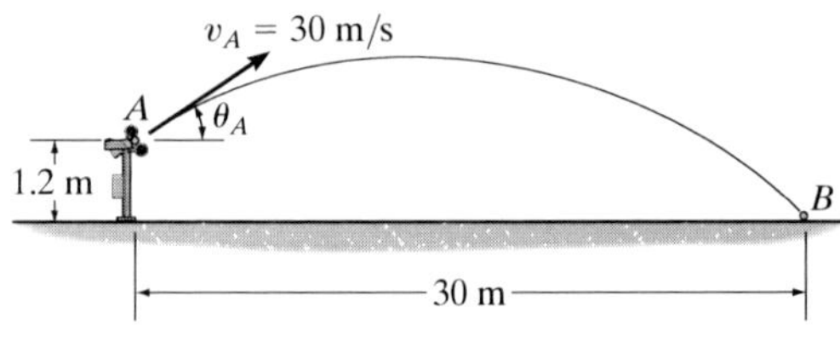

Prob. 12–91

*__12–92.__ The girl always throws the toys at an angle of 30° from point A as shown. Determine the time between throws so that both toys strike the edges of the pool B and C at the same instant. With what speed must she throw each toy?

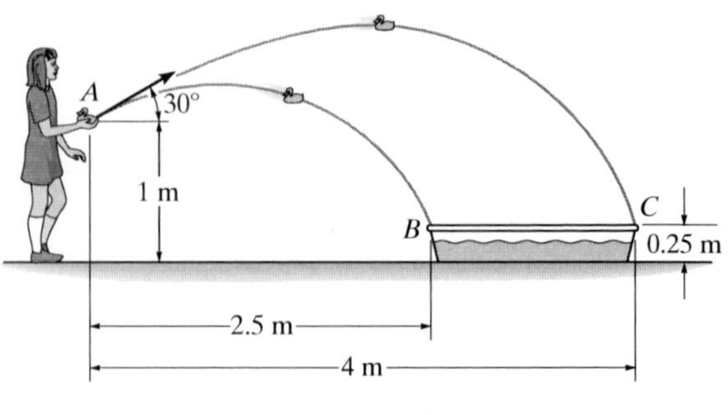

Prob. 12–92

12–93. The balloon A is ascending at the rate $v_A = 12$ km/h and is being carried horizontally by the wind at $v_w = 20$ km/h. If a ballast bag is dropped from the balloon at the instant $h = 50$ m, determine the time needed for it to strike the ground. Assume that the bag was released from the balloon with the same velocity as the balloon. Also, with what speed does the bag strike the ground?

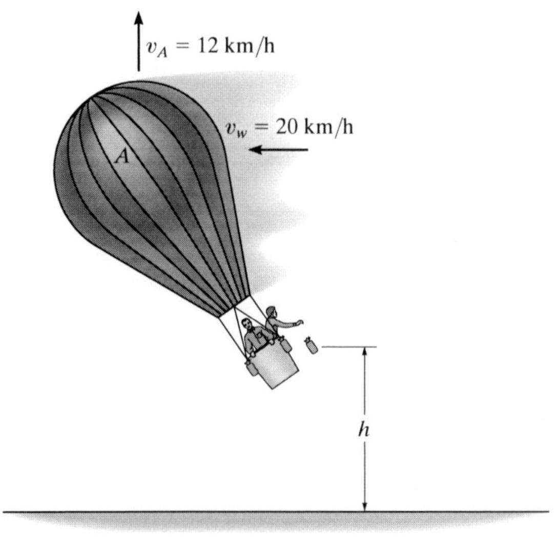

Prob. 12–93

12–94. The projectile is launched with a velocity $2\mathbf{v}_0$. Determine the range R, the maximum height h attained, and the time of flight. Express the results in terms of the angle θ and v_0. The acceleration due to gravity is g.

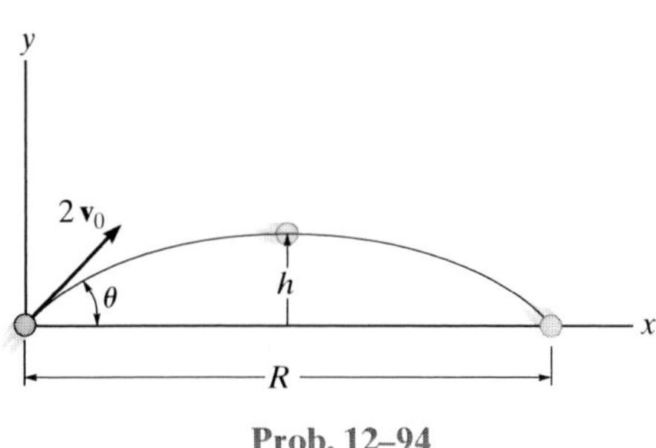

Prob. 12–94

12–95. A projectile is given a velocity $\mathbf{v}_0$ at an angle ϕ above the horizontal. Determine the distance d to where it strikes the sloped ground. The acceleration due to gravity is g.

***12–96.** A projectile is given a velocity $\mathbf{v}_0$. Determine the angle ϕ at which it should be launched so that d is a maximum. The acceleration due to gravity is g.

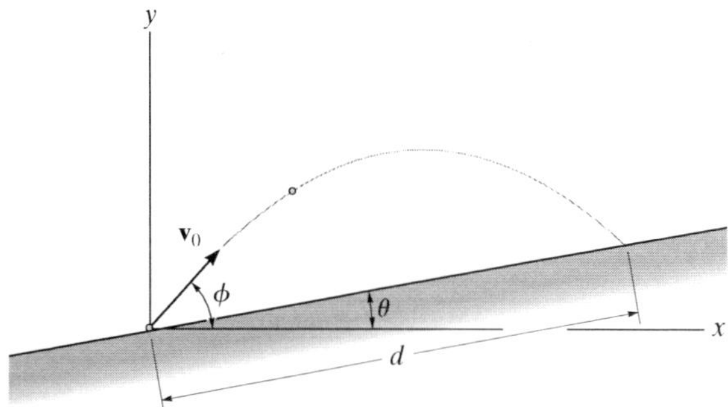

Probs. 12–95/96

12–97. Measurements of a shot recorded on a videotape during a basketball game are shown. The ball passed through the hoop even though it barely cleared the hands of the player B who attempted to block it. Neglecting the size of the ball, determine the magnitude v_A of its initial velocity and the height h of the ball when it passes over player B.

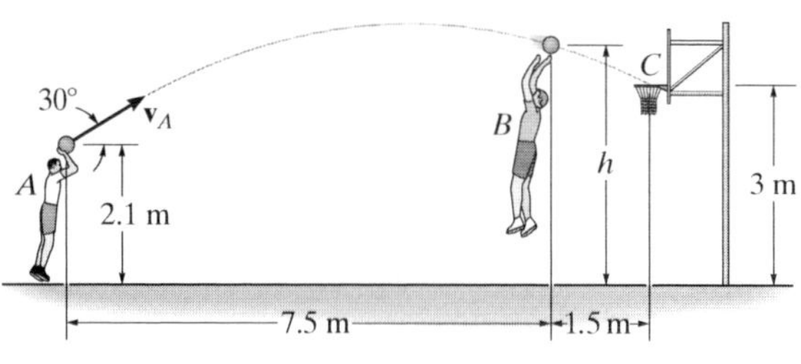

Prob. 12–97

12–98. A projectile is fired from the platform at B. The shooter fires his gun from point A at an angle of 30°. Determine the muzzle speed of the bullet if it hits the projectile at C.

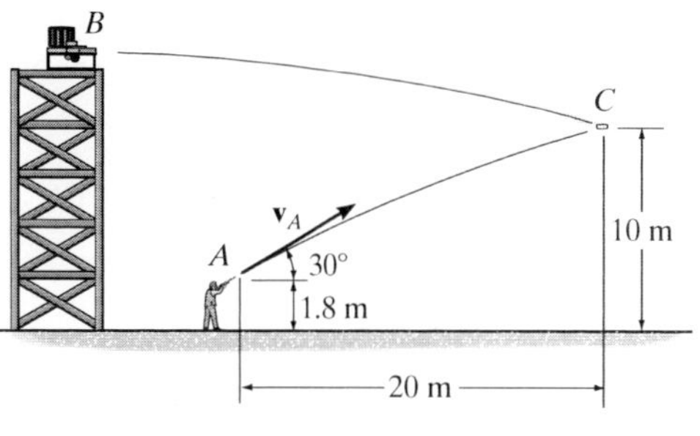

Prob. 12–98

12–99. The snowmobile is traveling at 10 m/s when it leaves the embankment at A. Determine the time of flight from A to B and the range R of the trajectory.

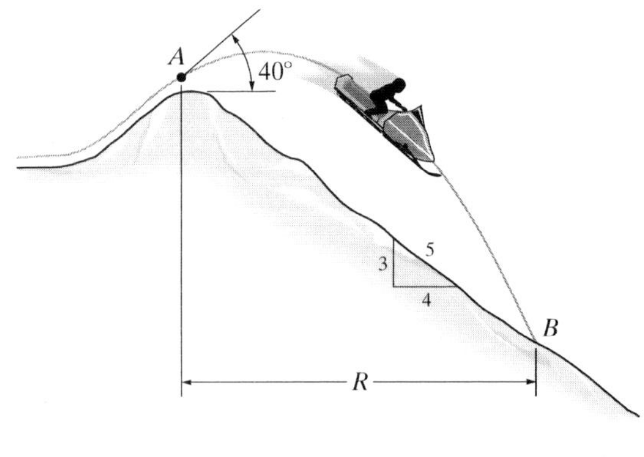

Prob. 12–99

*__12–100.__ It is observed that the skier leaves the ramp A at an angle $\theta_A = 25°$ with the horizontal. If he strikes the ground at B, determine his initial speed v_A and the time of flight t_{AB}.

12–101. It is observed that the skier leaves the ramp A at an angle $\theta_A = 25°$ with the horizontal. If he strikes the ground at B, determine his initial speed v_A and the speed at which he strikes the ground.

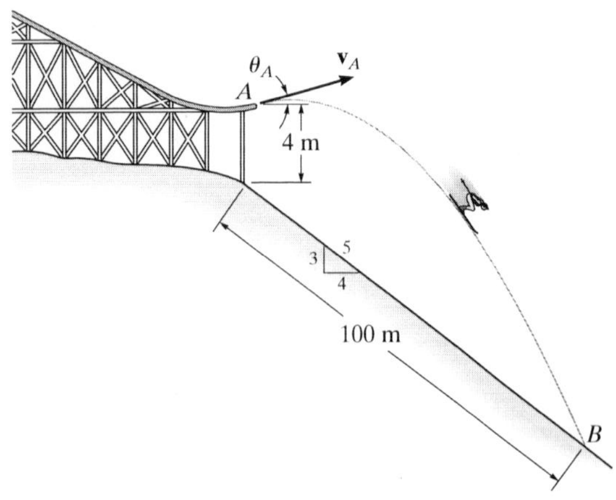

Probs. 12–100/101

12–102. The path of a particle is defined by $y^2 = 4kx$, and the component of velocity along the y axis is $v_y = ct$, where both k and c are constants. Determine the x and y components of acceleration when $y = y_0$.

12–103. The drinking fountain is designed such that the nozzle is located from the edge of the basin as shown. Determine the maximum and minimum speed at which water can be ejected from the nozzle so that it does not splash over the sides of the basin at B and C.

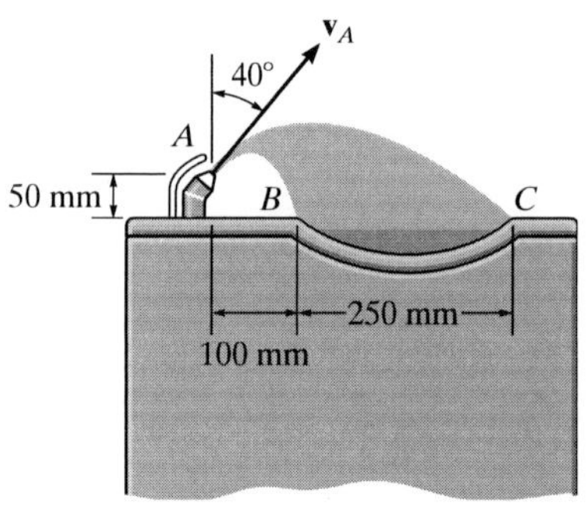

Prob. 12–103

*__12–104.__ The projectile is launched with a velocity $\mathbf{v}_0$. Determine the range R, the maximum height h attained, and the time of flight. Express the results in terms of the angle θ and v_0. The acceleration due to gravity is g.

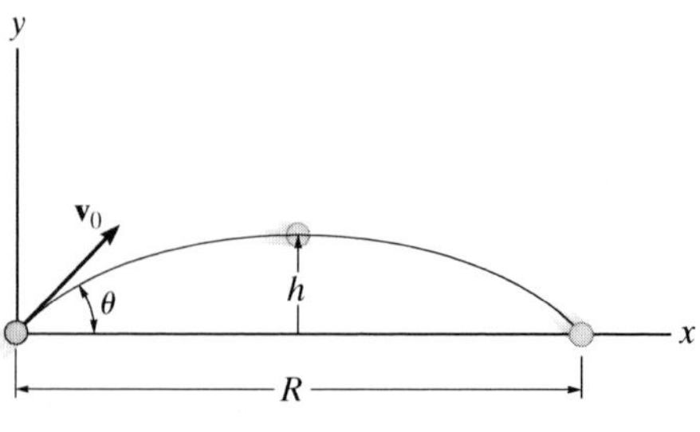

Prob. 12–104

12–105. The projectile is launched from a height h with a velocity $\mathbf{v}_0$. Determine the range R.

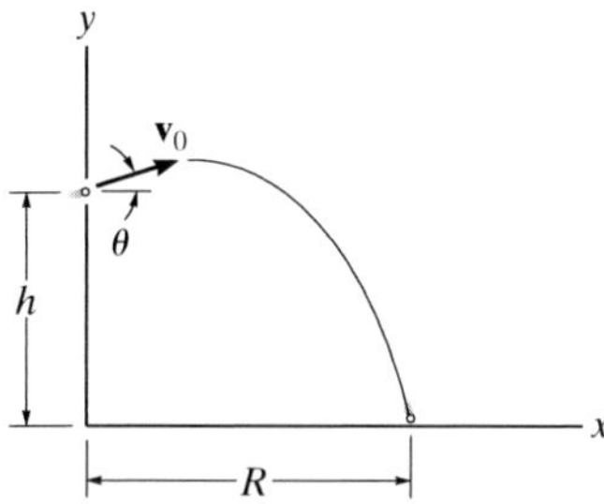

Prob. 12–105

12–106. The velocity of a particle is given by $v = \{12t^2\mathbf{i} + 5t^3\mathbf{j} + (6t + 3)\mathbf{k}\}$ m/s, where t is in seconds. If the particle is at the origin when $t = 0$, determine the magnitude of the particle's acceleration when $t = 4$ s. Also, what is the x, y, z coordinate position of the particle at this instant?

12–107. The golf ball is hit at A with a speed of $v_A = 40$ m/s and directed at an angle of 30° with the horizontal as shown. Determine the distance d where the ball strikes the slope at B.

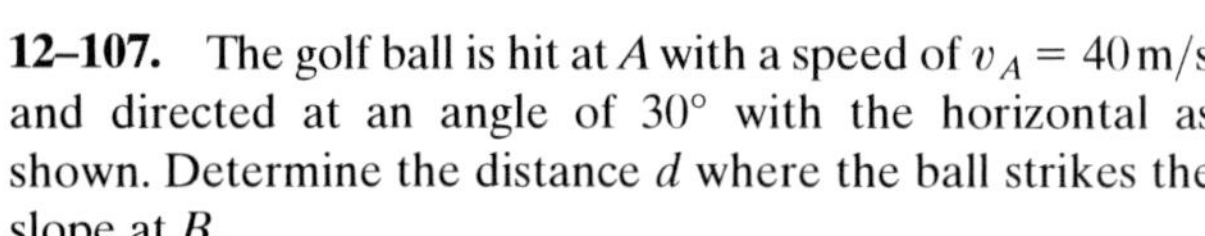

Prob. 12–107

***12–108.** The man at A wishes to throw two darts at the target at B so that they arrive at the *same time*. If each dart is thrown with a speed of 10 m/s, determine the angles θ_C and θ_D at which they should be thrown and the time between each throw. Note that the first dart must be thrown at θ_C $(>\theta_D)$, then the second dart is thrown at θ_D.

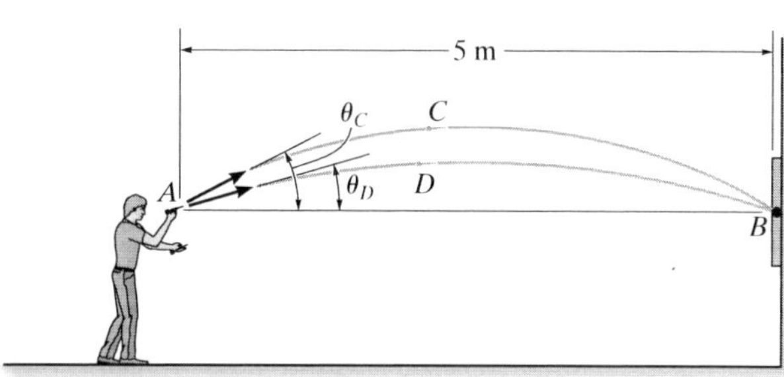

Prob. 12–108

12–109. A boy throws a ball at O in the air with a speed v_0 at an angle θ_1. If he then throws another ball with the same speed v_0 at an angle $\theta_2 < \theta_1$, determine the time between the throws so that the balls collide in midair at B.

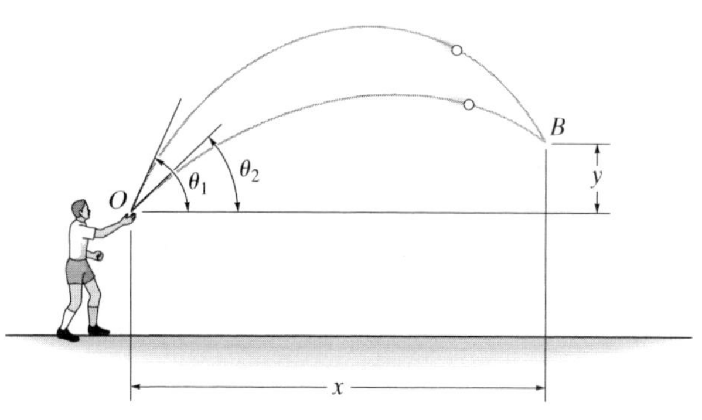

Prob. 12–109

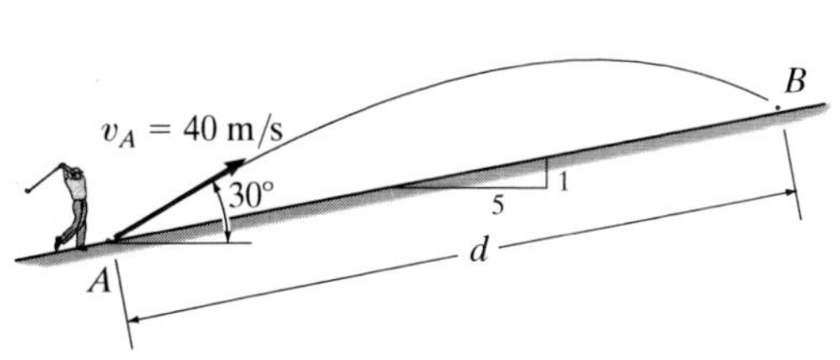

12

12–110. Small packages traveling on the conveyor belt fall off into a l-m-long loading car. If the conveyor is running at a constant speed of $v_C = 2$ m/s, determine the smallest and largest distance R at which the end A of the car may be placed from the conveyor so that the packages enter the car.

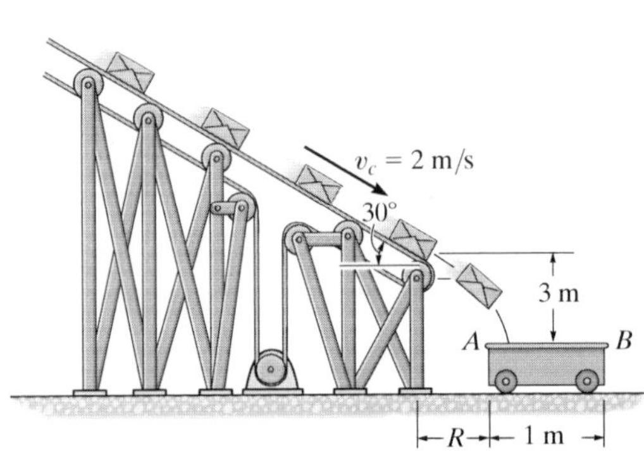

Prob. 12–110

12–111. A projectile is fired with a speed of $v = 60$ m/s at an angle of 60°. A second projectile is then fired with the same speed 0.5 s later. Determine the angle θ of the second projectile so that the two projectiles collide. At what position (x, y) will this happen?

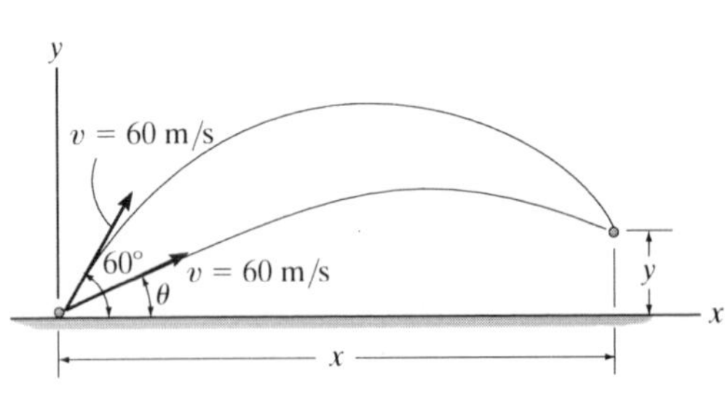

Prob. 12–111

***12–112.** The boy at A attempts to throw a ball over the roof of a barn with an initial speed of $v_A = 15$ m/s. Determine the angle θ_A at which the ball must be thrown so that it reaches its maximum height at C. Also, find the distance d where the boy should stand to make the throw.

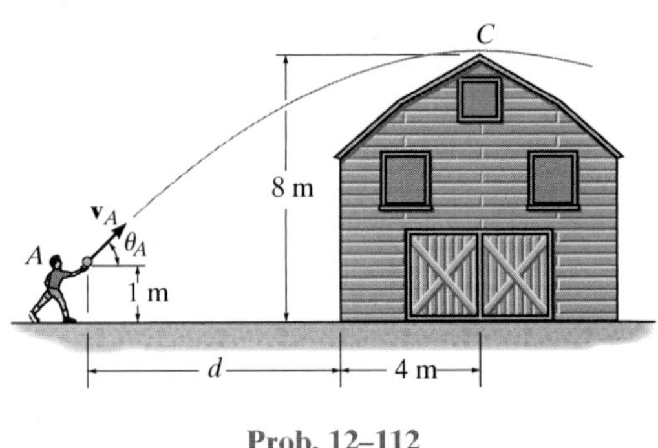

Prob. 12–112

12–113. The boy at A attempts to throw a ball over the roof of a barn such that it is launched at an angle $\theta_A = 40°$. Determine the minimum speed v_A at which he must throw the ball so that it reaches its maximum height at C. Also, find the distance d where the boy must stand so that he can make the throw.

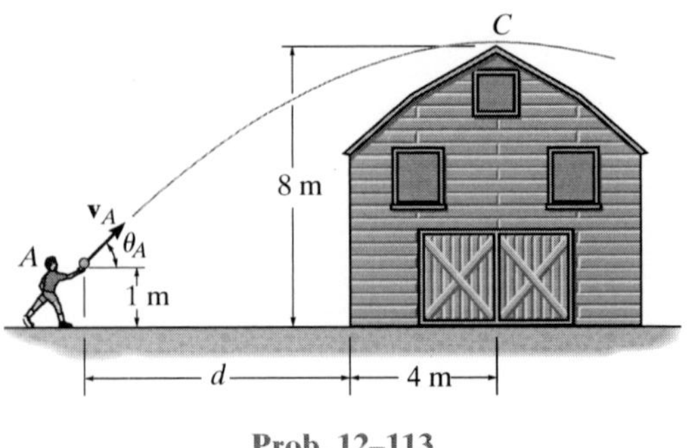

Prob. 12–113

12

12.7 Curvilinear Motion: Normal and Tangential Components

When the path along which a particle travels is *known*, then it is often convenient to describe the motion using n and t coordinate axes which act normal and tangent to the path, respectively, and at the instant considered have their *origin located at the particle*.

Planar Motion. Consider the particle shown in Fig. 12–24*a*, which moves in a plane along a fixed curve, such that at a given instant it is at position s, measured from point O. We will now consider a coordinate system that has its origin on the curve, and at the instant considered this origin happens to *coincide* with the location of the particle. The t axis is *tangent* to the curve at the point and is positive in the direction of *increasing* s. We will designate this positive direction with the unit vector $\mathbf{u}_t$. A unique choice for the *normal axis* can be made by noting that geometrically the curve is constructed from a series of differential arc segments ds, Fig. 12–24*b*. Each segment ds is formed from the arc of an associated circle having a *radius of curvature* ρ (rho) and *center of curvature* O'. The normal axis n is perpendicular to the t axis with its positive sense directed *toward* the center of curvature O', Fig. 12–24*a*. This positive direction, which is *always* on the concave side of the curve, will be designated by the unit vector $\mathbf{u}_n$. The plane which contains the n and t axes is referred to as the embracing or *osculating plane*, and in this case it is fixed in the plane of motion.*

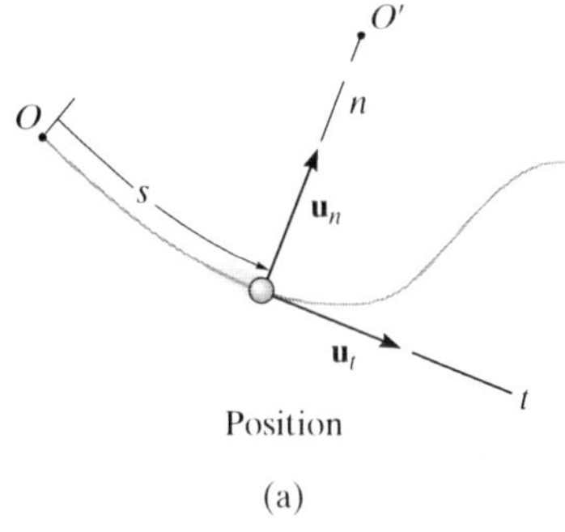

Position

(a)

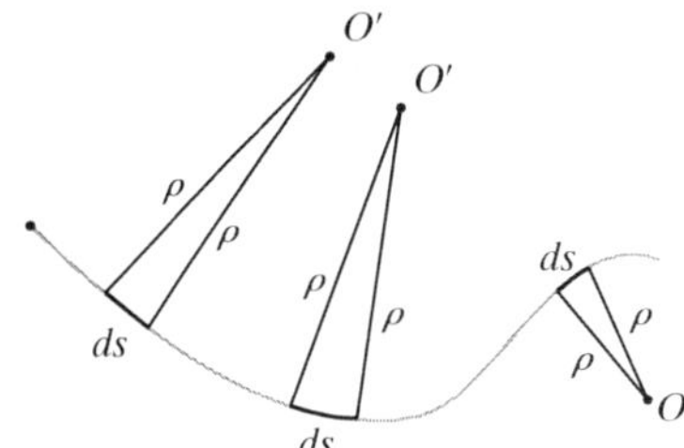

Radius of curvature

(b)

Velocity. Since the particle moves, s is a function of time. As indicated in Sec. 12.4, the particle's velocity $\mathbf{v}$ has a *direction* that is *always tangent to the path*, Fig. 12–24*c*, and a *magnitude* that is determined by taking the time derivative of the path function $s = s(t)$, i.e., $v = ds/dt$ (Eq. 12–8). Hence

$$\mathbf{v} = v\mathbf{u}_t \tag{12–15}$$

where

$$v = \dot{s} \tag{12–16}$$

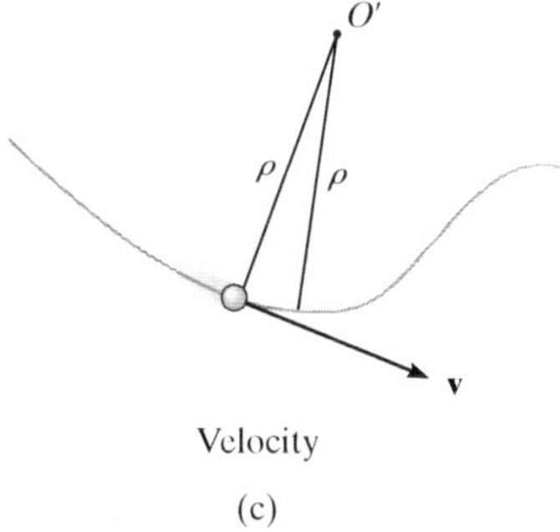

Velocity

(c)

Fig. 12–24

*The osculating plane may also be defined as the plane which has the greatest contact with the curve at a point. It is the limiting position of a plane contacting both the point and the arc segment ds. As noted above, the osculating plane is always coincident with a plane curve; however, each point on a three-dimensional curve has a unique osculating plane.

(d)

Acceleration. The acceleration of the particle is the time rate of change of the velocity. Thus,

$$\mathbf{a} = \dot{\mathbf{v}} = \dot{v}\mathbf{u}_t + v\dot{\mathbf{u}}_t \tag{12–17}$$

In order to determine the time derivative $\dot{\mathbf{u}}_t$, note that as the particle moves along the arc ds in time dt, $\mathbf{u}_t$ preserves its magnitude of unity; however, its *direction* changes, and becomes $\mathbf{u}_t'$, Fig. 12–24*d*. As shown in Fig. 12–24*e*, we require $\mathbf{u}_t' = \mathbf{u}_t + d\mathbf{u}_t$. Here $d\mathbf{u}_t$ stretches between the arrowheads of $\mathbf{u}_t$ and $\mathbf{u}_t'$, which lie on an infinitesimal arc of radius $u_t = 1$. Hence, $d\mathbf{u}_t$ has a *magnitude* of $du_t = (1)\,d\theta$, and its *direction* is defined by $\mathbf{u}_n$. Consequently, $d\mathbf{u}_t = d\theta\mathbf{u}_n$, and therefore the time derivative becomes $\dot{\mathbf{u}}_t = \dot{\theta}\mathbf{u}_n$. Since $ds = \rho d\theta$, Fig. 12–24*d*, then $\dot{\theta} = \dot{s}/\rho$, and therefore

$$\dot{\mathbf{u}}_t = \dot{\theta}\mathbf{u}_n = \frac{\dot{s}}{\rho}\mathbf{u}_n = \frac{v}{\rho}\mathbf{u}_n$$

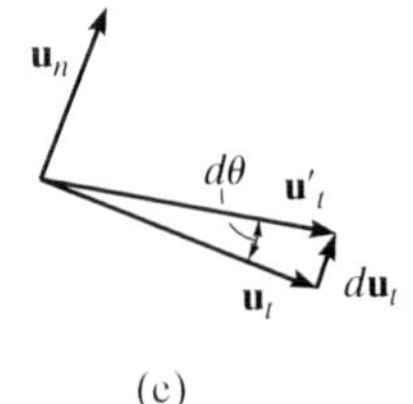

(e)

Substituting into Eq. 12–17, $\mathbf{a}$ can be written as the sum of its two components,

$$\mathbf{a} = a_t\mathbf{u}_t + a_n\mathbf{u}_n \tag{12–18}$$

where

$$a_t = \dot{v} \quad \text{or} \quad a_t ds = v\,dv \tag{12–19}$$

and

$$a_n = \frac{v^2}{\rho} \tag{12–20}$$

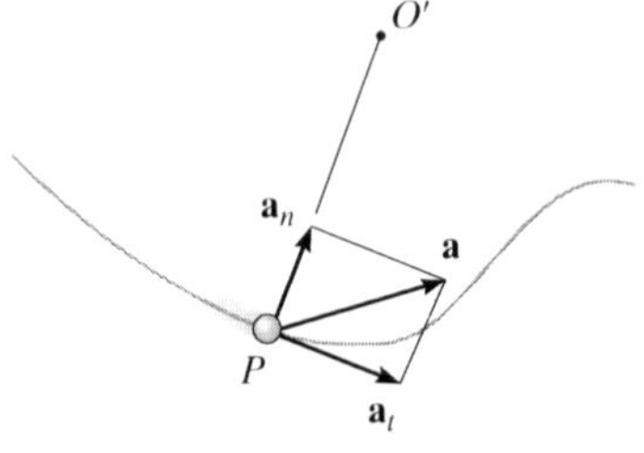

Acceleration

(f)

Fig. 12–24 (cont.)

These two mutually perpendicular components are shown in Fig. 12–24*f*. Therefore, the *magnitude* of acceleration is the positive value of

$$a = \sqrt{a_t^2 + a_n^2} \tag{12–21}$$

To better understand these results, consider the following two special cases of motion.

1. If the particle moves along a straight line, then $\rho \to \infty$ and from Eq. 12–20, $a_n = 0$. Thus $a = a_t = \dot{v}$, and we can conclude that the *tangential component of acceleration represents the time rate of change in the magnitude of the velocity.*
2. If the particle moves along a curve with a constant speed, then $a_t = \dot{v} = 0$ and $a = a_n = v^2/\rho$. Therefore, the *normal component of acceleration represents the time rate of change in the direction of the velocity*. Since $\mathbf{a}_n$ *always* acts towards the center of curvature, this component is sometimes referred to as the *centripetal* (or center seeking) *acceleration.*

As a result of these interpretations, a particle moving along the curved path in Fig. 12–25 will have accelerations directed as shown.

As the boy swings upward with a velocity $\mathbf{v}$, his motion can be analyzed using n–t coordinates. As he rises, the magnitude of his velocity (speed) is decreasing, and so a_t will be negative. The rate at which the direction of his velocity changes is a_n, which is always positive, that is, towards the center of rotation.

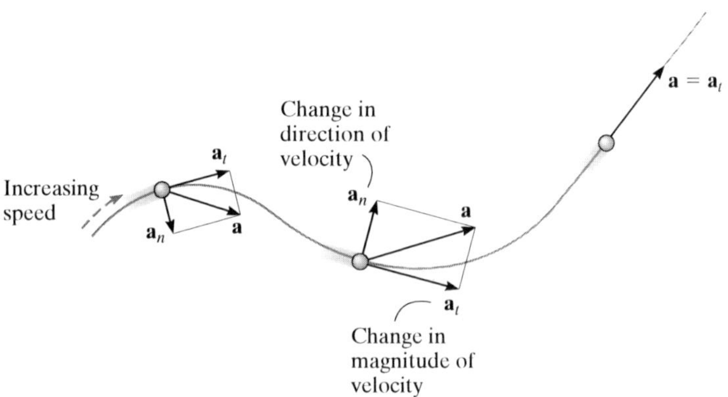

Fig. 12–25

Three-Dimensional Motion. If the particle moves along a space curve, Fig. 12–26, then at a given instant the t axis is uniquely specified; however, an infinite number of straight lines can be constructed normal to the tangent axis. As in the case of planar motion, we will choose the positive n axis directed toward the path's center of curvature O'. This axis is referred to as the *principal normal* to the curve. With the n and t axes so defined, Eqs. 12–15 through 12–21 can be used to determine $\mathbf{v}$ and $\mathbf{a}$. Since $\mathbf{u}_t$ and $\mathbf{u}_n$ are always perpendicular to one another and lie in the osculating plane, for spatial motion a third unit vector, $\mathbf{u}_b$, defines the *binormal axis b* which is perpendicular to $\mathbf{u}_t$ and $\mathbf{u}_n$, Fig. 12–26.

Since the three unit vectors are related to one another by the vector cross product, e.g., $\mathbf{u}_b = \mathbf{u}_t \times \mathbf{u}_n$, Fig. 12–26, it may be possible to use this relation to establish the direction of one of the axes, if the directions of the other two are known. For example, no motion occurs in the $\mathbf{u}_b$ direction, and if this direction and $\mathbf{u}_t$ are known, then $\mathbf{u}_n$ can be determined, where in this case $\mathbf{u}_n = \mathbf{u}_b \times \mathbf{u}_t$, Fig. 12–26. Remember, though, that $\mathbf{u}_n$ is always on the concave side of the curve.

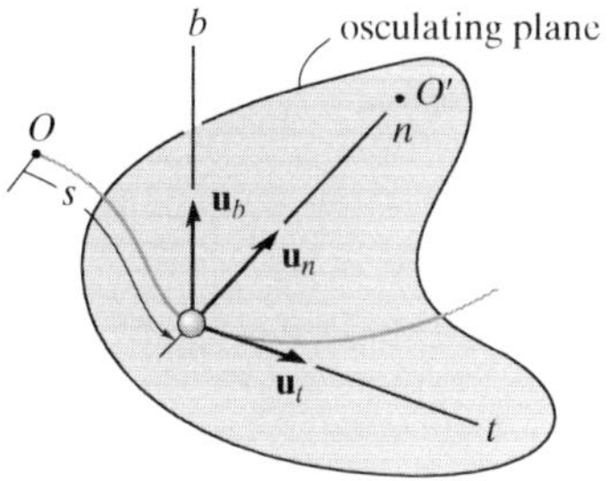

Fig. 12–26

12

Procedure for Analysis

Coordinate System.

- Provided the *path* of the particle is *known*, we can establish a set of n and t coordinates having a *fixed origin,* which is coincident with the particle at the instant considered.
- The positive tangent axis acts in the direction of motion and the positive normal axis is directed toward the path's center of curvature.

Velocity.

- The particle's *velocity* is always tangent to the path.
- The magnitude of velocity is found from the time derivative of the path function.

$$v = \dot{s}$$

Tangential Acceleration.

- The tangential component of acceleration is the result of the time rate of change in the *magnitude* of velocity. This component acts in the positive s direction if the particle's speed is increasing or in the opposite direction if the speed is decreasing.
- The relations between a_t, v, t and s are the same as for rectilinear motion, namely,

$$a_t = \dot{v} \quad a_t\,ds = v\,dv$$

- If a_t is constant, $a_t = (a_t)_c$, the above equations, when integrated, yield

$$s = s_0 + v_0 t + \tfrac{1}{2}(a_t)_c t^2$$
$$v = v_0 + (a_t)_c t$$
$$v^2 = v_0^2 + 2(a_t)_c(s - s_0)$$

Normal Acceleration.

- The normal component of acceleration is the result of the time rate of change in the *direction* of the velocity. This component is *always* directed toward the center of curvature of the path, i.e., along the positive n axis.
- The magnitude of this component is determined from

$$a_n = \frac{v^2}{\rho}$$

- If the path is expressed as $y = f(x)$, the radius of curvature ρ at any point on the path is determined from the equation

$$\rho = \frac{[1 + (dy/dx)^2]^{3/2}}{|d^2y/dx^2|}$$

The derivation of this result is given in any standard calculus text.

Please refer to the Companion Website for the animation: *The Dynamics of a Turning Vehicle Illustrating Force, f*

Motorists traveling along this cloverleaf interchange experience a normal acceleration due to the change in direction of their velocity. A tangential component of acceleration occurs when the cars' speed is increased or decreased.

EXAMPLE 12.14

When the skier reaches point A along the parabolic path in Fig. 12–27a, he has a speed of 6 m/s which is increasing at 2 m/s^2. Determine the direction of his velocity and the direction and magnitude of his acceleration at this instant. Neglect the size of the skier in the calculation.

Please refer to the Companion Website for the animation: *The Forward Velocity of Skateboard at Different Heights*

SOLUTION

Coordinate System. Although the path has been expressed in terms of its x and y coordinates, we can still establish the origin of the n, t axes at the fixed point A on the path and determine the components of $\mathbf{v}$ and $\mathbf{a}$ along these axes, Fig. 12–27a.

Velocity. By definition, the velocity is always directed tangent to the path. Since $y = \frac{1}{20}x^2$, $dy/dx = \frac{1}{10}x$, then at $x = 10$ m, $dy/dx = 1$. Hence, at A, $\mathbf{v}$ makes an angle of $\theta = \tan^{-1}1 = 45°$ with the x axis, Fig. 12–27b. Therefore,

$$v_A = 6 \text{ m/s} \quad 45° \swarrow \qquad \textit{Ans.}$$

The acceleration is determined from $\mathbf{a} = \dot{v}\mathbf{u}_t + (v^2/\rho)\mathbf{u}_n$. However, it is first necessary to determine the radius of curvature of the path at A (10 m, 5 m). Since $d^2y/dx^2 = \frac{1}{10}$, then

$$\rho = \frac{[1 + (dy/dx)^2]^{3/2}}{|d^2y/dx^2|} = \left.\frac{\left[1 + \left(\frac{1}{10}x\right)^2\right]^{3/2}}{\left|\frac{1}{10}\right|}\right|_{x=10\text{ m}} = 28.28 \text{ m}$$

The acceleration becomes

$$\mathbf{a}_A = \dot{v}\mathbf{u}_t + \frac{v^2}{\rho}\mathbf{u}_n$$
$$= 2\mathbf{u}_t + \frac{(6 \text{ m/s})^2}{28.28 \text{ m}}\mathbf{u}_n$$
$$= \{2\mathbf{u}_t + 1.273\mathbf{u}_n\} \text{ m/s}^2$$

As shown in Fig. 12–27b,

$$a = \sqrt{(2 \text{ m/s}^2)^2 + (1.273 \text{ m/s}^2)^2} = 2.37 \text{ m/s}^2$$

$$\phi = \tan^{-1}\frac{2}{1.273} = 57.5°$$

Thus, $45° + 90° + 57.5° - 180° = 12.5°$ so that,

$$a = 2.37 \text{ m/s}^2 \quad 12.5° \swarrow \qquad \textit{Ans.}$$

NOTE: By using n, t coordinates, we were able to readily solve this problem through the use of Eq. 12–18, since it accounts for the separate changes in the magnitude and direction of $\mathbf{v}$.

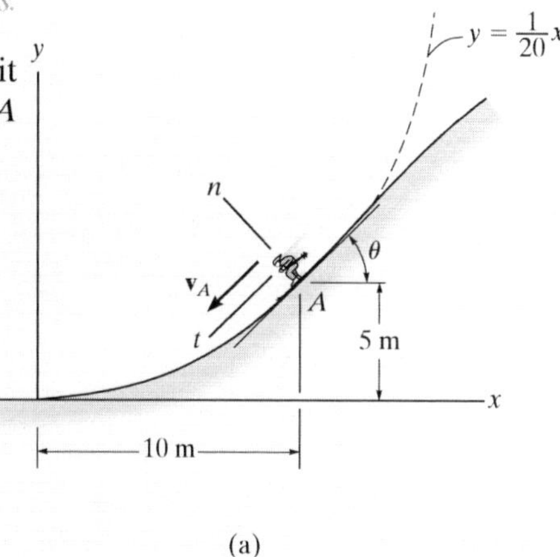

(a)

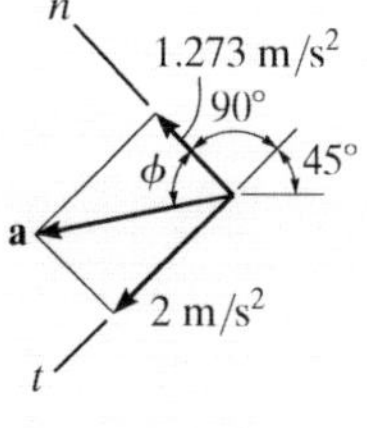

(b)

Fig. 12–27

EXAMPLE 12.15

A race car C travels around the horizontal circular track that has a radius of 300 m, Fig. 12–28. If the car increases its speed at a constant rate of 7 m/s^2, starting from rest, determine the time needed for it to reach an acceleration of 8 m/s^2. What is its speed at this instant?

Please refer to the Companion Website for the animation: *The Dynamics of a Turning Vehicle Illustrating Force, f*

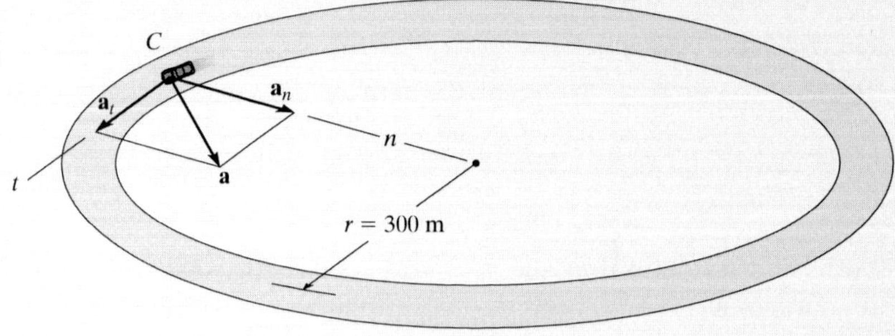

Fig. 12–28

SOLUTION

Coordinate System. The origin of the n and t axes is coincident with the car at the instant considered. The t axis is in the direction of motion, and the positive n axis is directed toward the center of the circle. This coordinate system is selected since the path is known.

Acceleration. The magnitude of acceleration can be related to its components using $a = \sqrt{a_t^2 + a_n^2}$. Here $a_t = 7 \text{ m/s}^2$. Since $a_n = v^2/\rho$, the velocity as a function of time must be determined first.

$$v = v_0 + (a_t)_c t$$
$$v = 0 + 7t$$

Thus

$$a_n = \frac{v^2}{\rho} = \frac{(7t)^2}{300} = 0.163t^2 \text{ m/s}^2$$

The time needed for the acceleration to reach 8 m/s^2 is therefore

$$a = \sqrt{a_t^2 + a_n^2}$$
$$8 \text{ m/s}^2 = \sqrt{(7 \text{ m/s}^2)^2 + (0.163t^2)^2}$$

Solving for the positive value of t yields

$$0.163t^2 = \sqrt{(8 \text{ m/s}^2)^2 - (7 \text{ m/s}^2)^2}$$
$$t = 4.87 \text{ s} \qquad \textit{Ans.}$$

Velocity. The speed at time $t = 4.87$ s is

$$v = 7t = 7(4.87) = 34.1 \text{ m/s} \qquad \textit{Ans.}$$

NOTE: Remember the velocity will always be tangent to the path, whereas the acceleration will be directed within the curvature of the path.

EXAMPLE 12.16

The boxes in Fig. 12–29*a* travel along the industrial conveyor. If a box as in Fig. 12–29*b* starts from rest at *A* and increases its speed such that $a_t = (0.2t)\ \text{m/s}^2$, where t is in seconds, determine the magnitude of its acceleration when it arrives at point *B*.

© Getty Images/The Image Bank

(a)

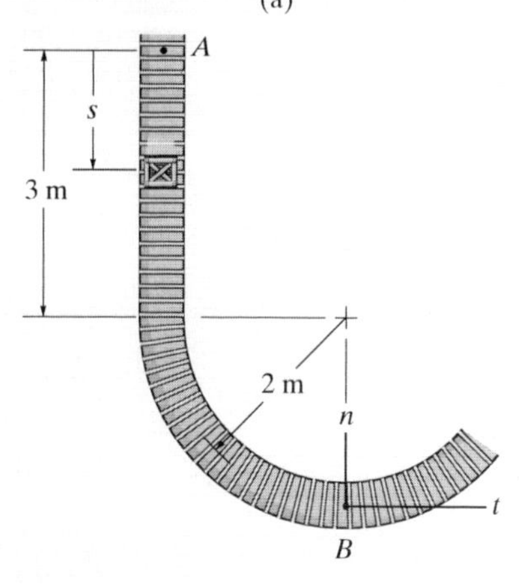

(b)

SOLUTION

Coordinate System. The position of the box at any instant is defined from the fixed point *A* using the position or path coordinate *s*, Fig. 12–29*b*. The acceleration is to be determined at *B*, so the origin of the *n, t* axes is at this point.

Acceleration. To determine the acceleration components $a_t = \dot{v}$ and $a_n = v^2/\rho$, it is first necessary to formulate v and $\dot{v}$ so that they may be evaluated at *B*. Since $v_A = 0$ when $t = 0$, then

$$a_t = \dot{v} = 0.2t \qquad (1)$$

$$\int_0^v dv = \int_0^t 0.2t\, dt$$

$$v = 0.1t^2 \qquad (2)$$

The time needed for the box to reach point *B* can be determined by realizing that the position of *B* is $s_B = 3 + 2\pi(2)/4 = 6.142$ m, Fig. 12–29*b*, and since $s_A = 0$ when $t = 0$ we have

$$v = \frac{ds}{dt} = 0.1t^2$$

$$\int_0^{6.142\text{ m}} ds = \int_0^{t_B} 0.1t^2 dt$$

$$6.142\text{ m} = 0.0333t_B^3$$

$$t_B = 5.690\text{s}$$

Substituting into Eqs. 1 and 2 yields

$$(a_B)_t = \dot{v}_B = 0.2(5.690) = 1.138\ \text{m/s}^2$$

$$v_B = 0.1(5.69)^2 = 3.238\ \text{m/s}$$

At B, $\rho_B = 2$ m, so that

$$(a_B)_n = \frac{v_B^2}{\rho_B} = \frac{(3.238\ \text{m/s})^2}{2\text{ m}} = 5.242\ \text{m/s}^2$$

The magnitude of $\mathbf{a}_B$, Fig. 12–29*c*, is therefore

$$a_B = \sqrt{(1.138\ \text{m/s}^2)^2 + (5.242\ \text{m/s}^2)^2} = 5.36\ \text{m/s}^2 \qquad \textit{Ans.}$$

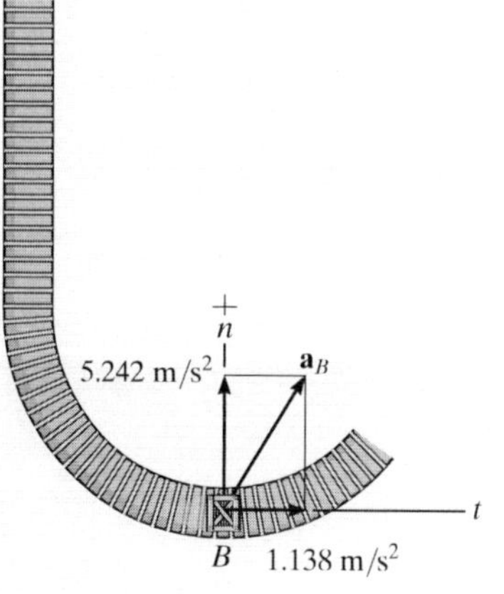

(c)

Fig. 12–29

FUNDAMENTAL PROBLEMS

F12–27. The boat is traveling along the circular path with a speed of $v = (0.0625t^2)$ m/s, where t is in seconds. Determine the magnitude of its acceleration when $t = 10$ s.

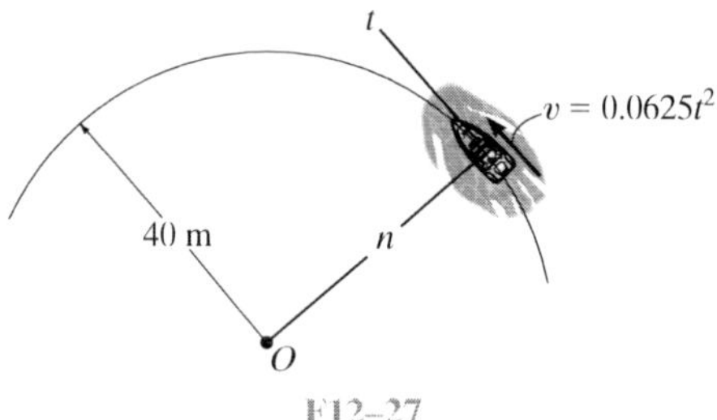

F12–27

F12–28. The car is traveling along the road with a speed of $v = (2\,s)$ m/s, where s is in meters. Determine the magnitude of its acceleration when $s = 10$ m.

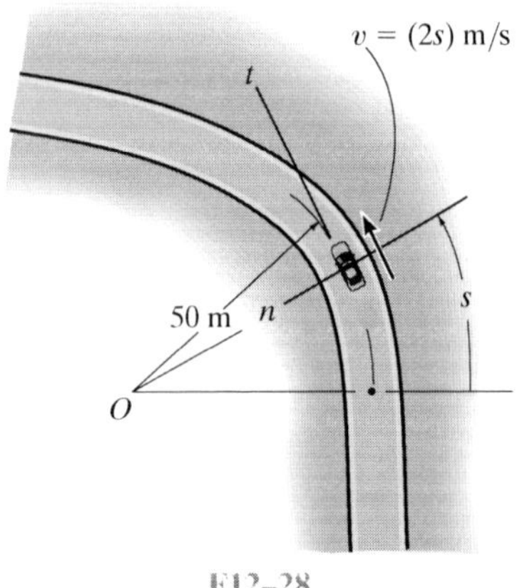

F12–28

F12–29. If the car decelerates uniformly along the curved road from 25 m/s at A to 15 m/s at C, determine the acceleration of the car at B.

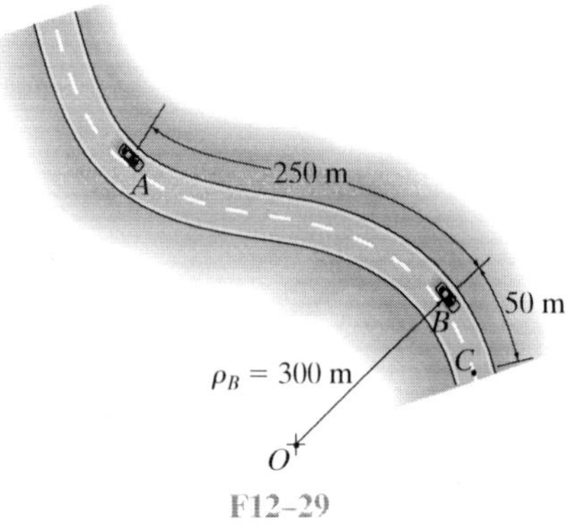

F12–29

F12–30. When $x = 3$ m, the crate has a speed of 6 m/s which is increasing at 2 m/s². Determine the direction of the crate's velocity and the magnitude of the crate's acceleration at this instant.

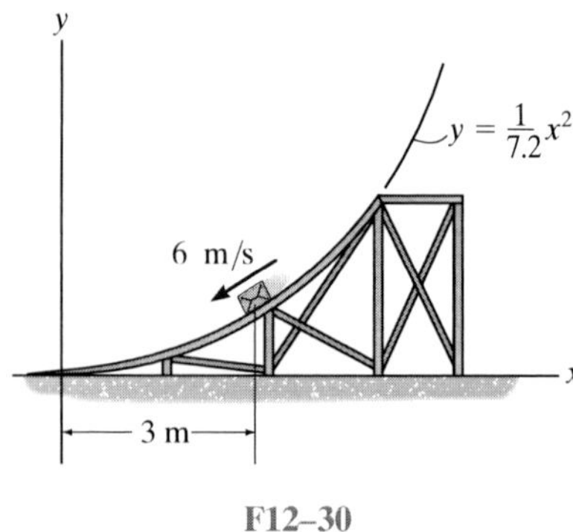

F12–30

F12–31. If the motorcycle has a deceleration of $a_t = -(0.001s)$ m/s² and its speed at position A is 25 m/s, determine the magnitude of its acceleration when it passes point B.

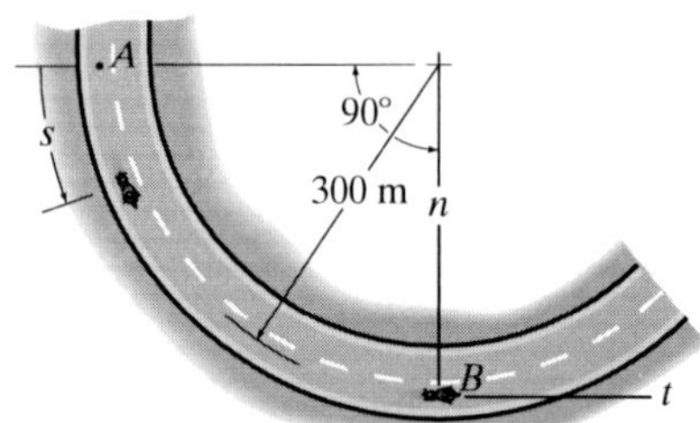

F12–31

F12–32. The car travels up the hill with a speed of $v = (0.2s)$ m/s, where s is in meters, measured from A. Determine the magnitude of its acceleration when it is at point $s = 50$ m, where $\rho = 500$ m.

F12–32

PROBLEMS

12–114. A car is traveling along a circular curve that has a radius of 50 m. If its speed is 16 m/s and is increasing uniformly at 8 m/s^2, determine the magnitude of its acceleration at this instant.

12–115. Determine the maximum constant speed a race car can have if the acceleration of the car cannot exceed 7.5 m/s^2 while rounding a track having a radius of curvature of 200 m.

***12–116.** The car travels around the circular track having a radius of $r = 300$ m such that when it is at point A it has a velocity of 5 m/s, which is increasing at the rate of $\dot{v} = (0.06t)$ m/s^2, where t is in seconds. Determine the magnitudes of its velocity and acceleration when it has traveled one-third the way around the track.

12–117. A car travels along a horizontal circular curved road that has a radius of 600 m. If the speed is uniformly increased at a rate of 2000 km/h^2, determine the magnitude of the acceleration at the instant the speed of the car is 60 km/h.

12–118. The truck travels in a circular path having a radius of 50 m at a speed of $v = 4$ m/s. For a short distance from $s = 0$, its speed is increased by $\dot{v} = (0.05s)$ m/s^2, where s is in meters. Determine its speed and the magnitude of its acceleration when it has moved $s = 10$ m.

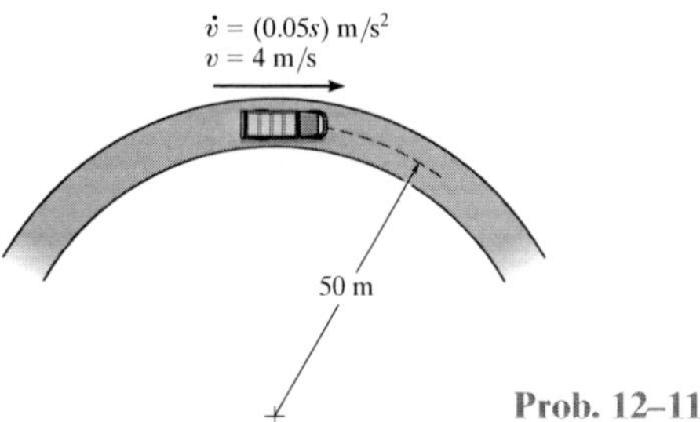

Prob. 12–118

12–119. Cars move around the "traffic circle" which is in the shape of an ellipse. If the speed limit is posted at 60 km/h, determine the maximum acceleration experienced by the passengers.

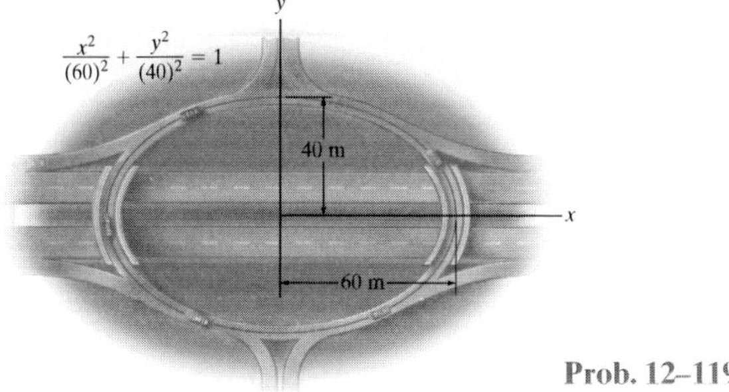

Prob. 12–119

***12–120.** Cars move around the "traffic circle" which is in the shape of an ellipse. If the speed limit is posted at 60 km/h, determine the minimum acceleration experienced by the passengers.

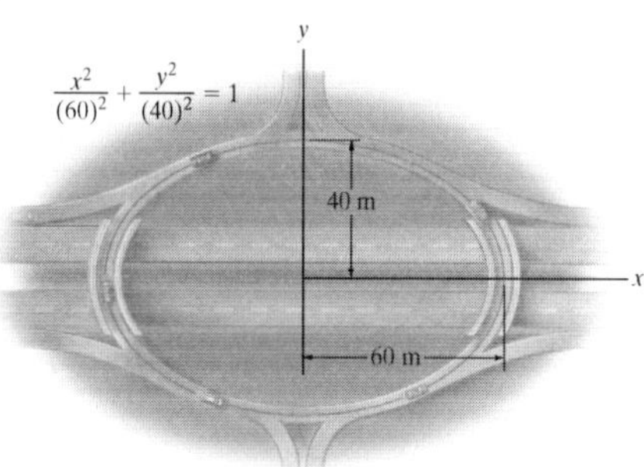

Prob. 12–120

12–121. When the roller coaster is at B, it has a speed of 25 m/s, which is increasing at $a_t = 3$ m/s^2. Determine the magnitude of the acceleration of the roller coaster at this instant and the direction angle it makes with the x axis.

12–122. If the roller coaster starts from rest at A and its speed increases at $a_t = (6 - 0.06s)$ m/s^2, determine the magnitude of its acceleration when it reaches B where $s_B = 40$ m.

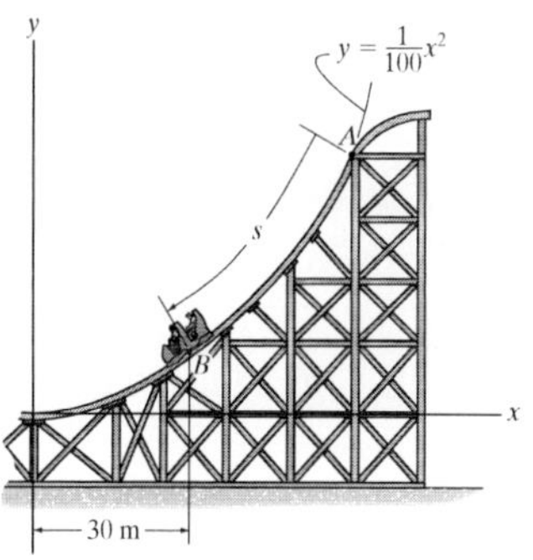

Probs. 12–121/122

12–123. The speedboat travels at a constant speed of 15 m/s while making a turn on a circular curve from A to B. If it takes 45 s to make the turn, determine the magnitude of the boat's acceleration during the turn.

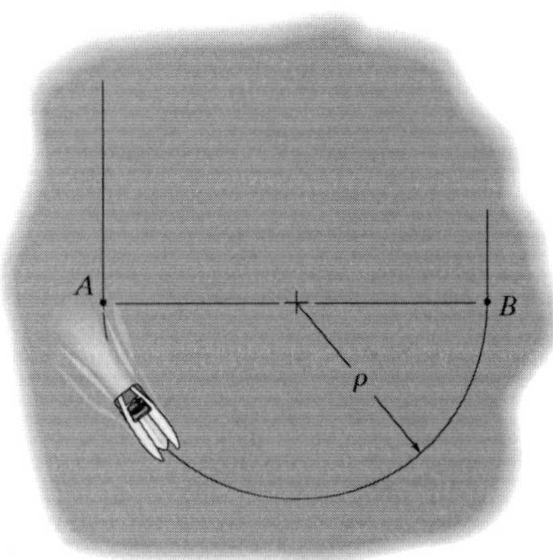
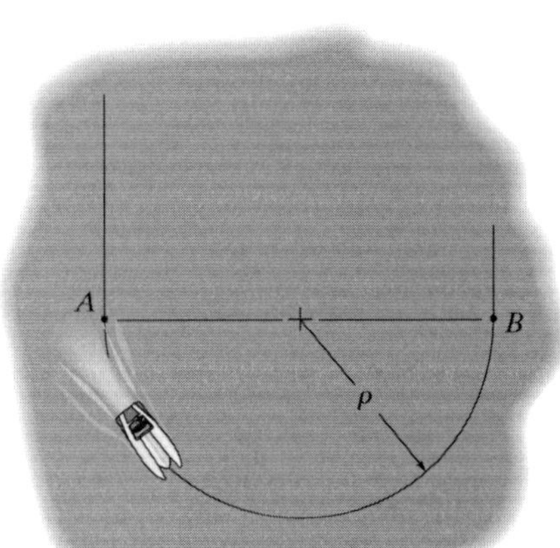

Prob. 12–123

12–125. The car passes point A with a speed of 25 m/s after which its speed is defined by $v = (25 - 0.15s)$ m/s. Determine the magnitude of the car's acceleration when it reaches point B, where $s = 51.5$ m and $x = 50$ m.

12–126. If the car passes point A with a speed of 20 m/s and begins to increase its speed at a constant rate of $a_t = 0.5\ \text{m/s}^2$, determine the magnitude of the car's acceleration when $s = 100$ m and $x = 0$.

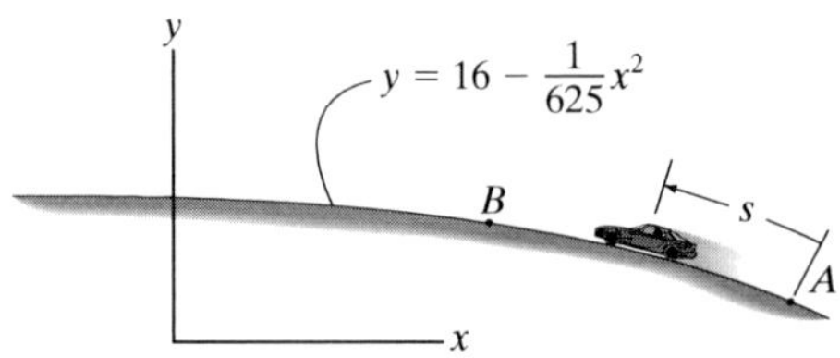

Probs. 12–125/126

*__12–124.__ The car travels along the circular path such that its speed is increased by $a_t = (0.5e^t)\ \text{m/s}^2$, where t is in seconds. Determine the magnitudes of its velocity and acceleration after the car has traveled $s = 18$ m starting from rest. Neglect the size of the car.

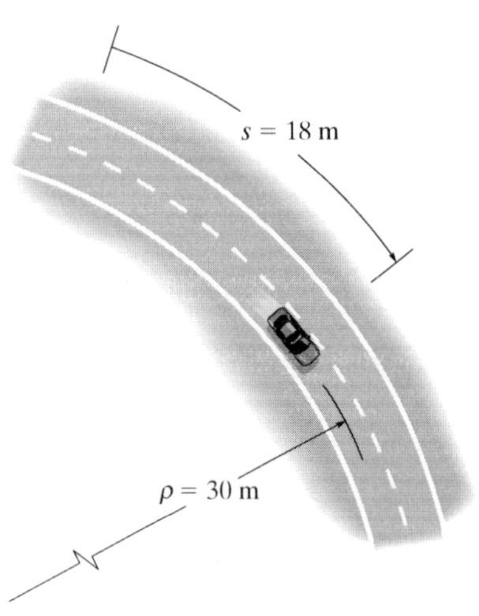

Prob. 12–124

12–127. A train is traveling with a constant speed of 14 m/s along the curved path. Determine the magnitude of the acceleration of the front of the train, B, at the instant it reaches point A $(y = 0)$.

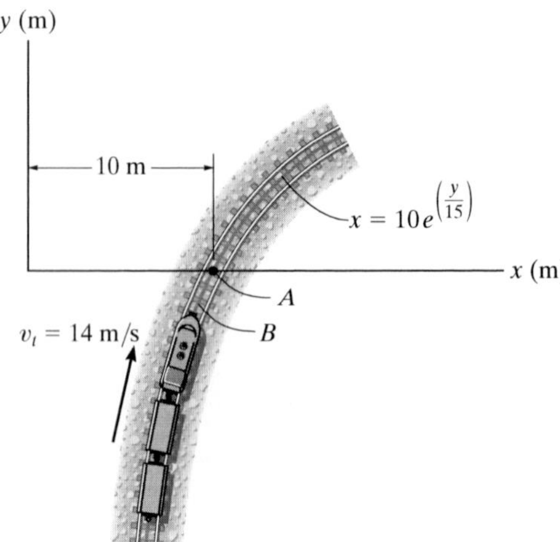

Prob. 12–127

***12–128.** The car travels along the curve having a radius of 300 m. If its speed is uniformly increased from 15 m/s to 27 m/s in 3 s, determine the magnitude of its acceleration at the instant its speed is 20 m/s.

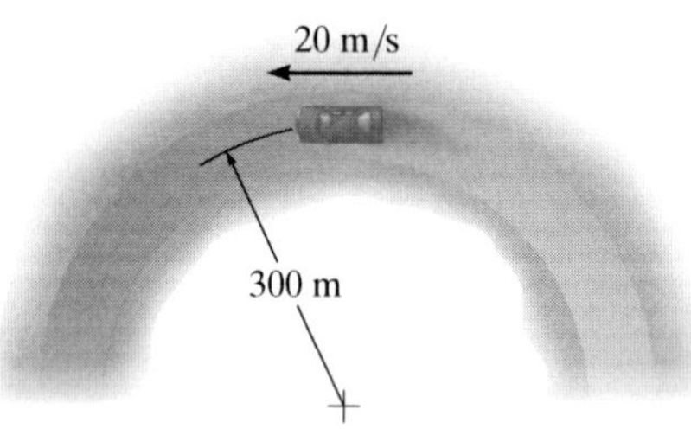

Prob. 12–128

12–129. When the car reaches point A it has a speed of 25 m/s. If the brakes are applied, its speed is reduced by $a_t = (-\frac{1}{4}t^{1/2})\ \text{m/s}^2$. Determine the magnitude of acceleration of the car just before it reaches point C.

12–130. When the car reaches point A, it has a speed of 25 m/s. If the brakes are applied, its speed is reduced by $a_t = (0.001s - 1)\ \text{m/s}^2$. Determine the magnitude of acceleration of the car just before it reaches point C.

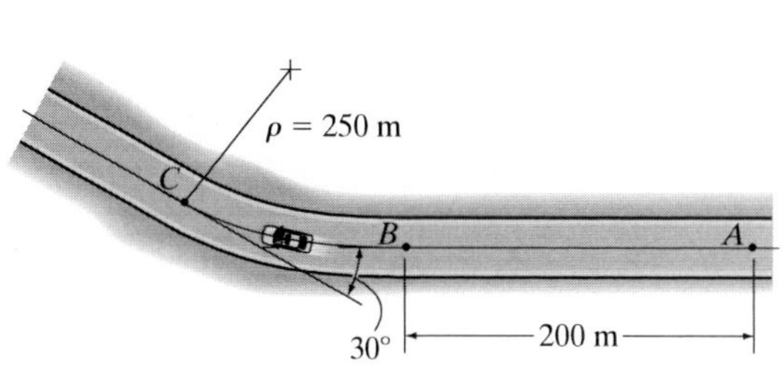

Probs. 12–129/130

12–131. At a given instant the train engine at E has a speed of 20 m/s and an acceleration of 14 m/s^2 acting in the direction shown. Determine the rate of increase in the train's speed and the radius of curvature ρ of the path.

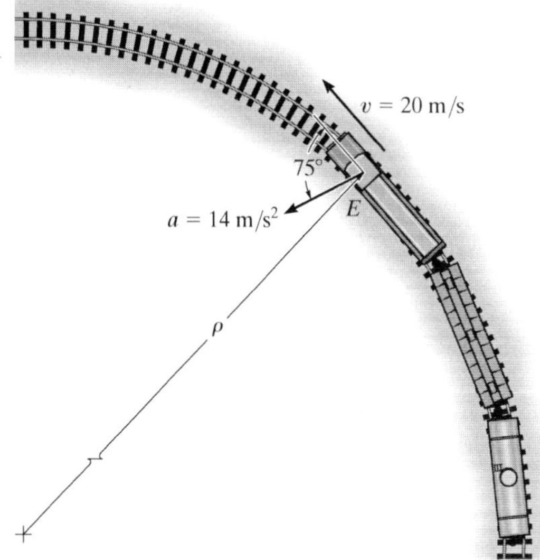

Prob. 12–131

***12–132.** Car B turns such that its speed is increased by $(a_t)_B = (0.5e^t)\ \text{m/s}^2$, where t is in seconds. If the car starts from rest when $\theta = 0°$, determine the magnitudes of its velocity and acceleration when the arm AB rotates $\theta = 30°$. Neglect the size of the car.

12–133. Car B turns such that its speed is increased by $(a_t)_B = (0.5e^t)\ \text{m/s}^2$, where t is in seconds. If the car starts from rest when $\theta = 0°$, determine the magnitudes of its velocity and acceleration when $t = 2$ s. Neglect the size of the car.

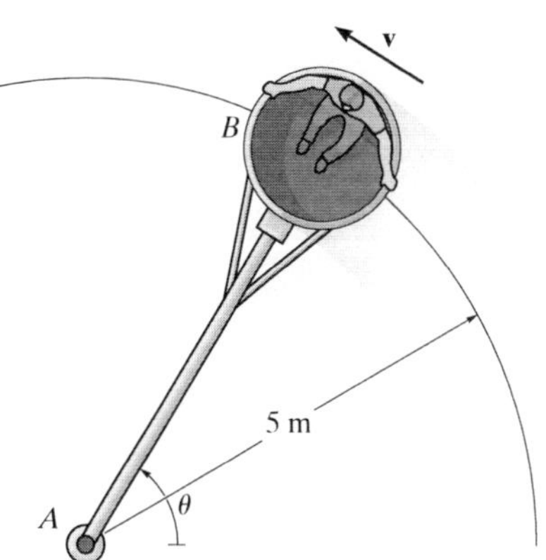

Probs. 12–132/133

12–134. At a given instant, a car travels along a circular curved road with a speed of 20 m/s while decreasing its speed at the rate of 3 m/s^2. If the magnitude of the car's acceleration is 5 m/s^2, determine the radius of curvature of the road.

12–135. A boat is traveling along a circular path having a radius of 20 m. Determine the magnitude of the boat's acceleration when the speed is $v = 5\ m/s$ and the rate of increase in the speed is $\dot{v} = 2\ m/s^2$.

***12–136.** Starting from rest, a bicyclist travels around a horizontal circular path, $\rho = 10\ m$, at a speed of $v = (0.09t^2 + 0.1t)\ m/s$, where t is in seconds. Determine the magnitudes of his velocity and acceleration when he has traveled $s = 3\ m$.

12–137. A particle travels around a circular path having a radius of 50 m. If it is initially traveling with a speed of 10 m/s and its speed then increases at a rate of $\dot{v} = (0.05\ v)\ m/s^2$, determine the magnitude of the particle's acceleraton four seconds later.

12–138. When the bicycle passes point A, it has a speed of 6 m/s, which is increasing at the rate of $\dot{v} = 0.5\ m/s^2$. Determine the magnitude of its acceleration when it is at point A.

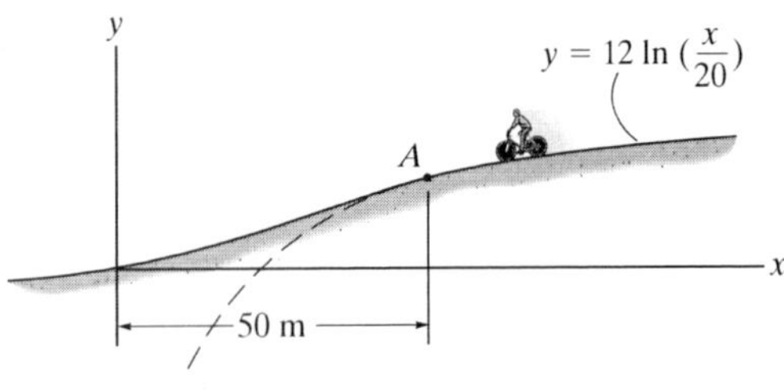

Prob. 12–138

12–139. The motorcycle is traveling at a constant speed of 60 km/h. Determine the magnitude of its acceleration when it is at point A.

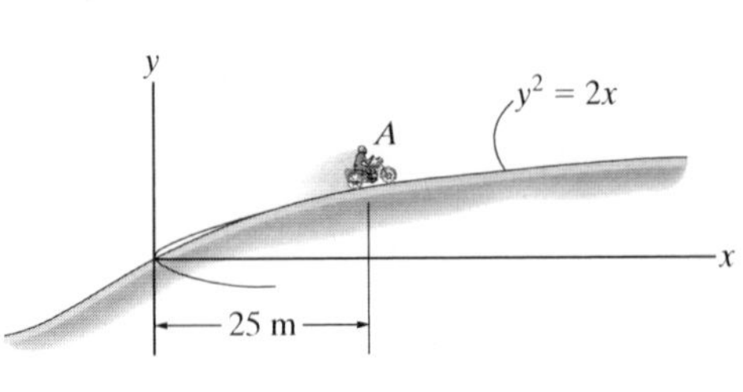

Prob. 12–139

***12–140.** The jet plane travels along the vertical parabolic path. When it is at point A it has a speed of 200 m/s, which is increasing at the rate of 0.8 m/s^2. Determine the magnitude of acceleration of the plane when it is at point A.

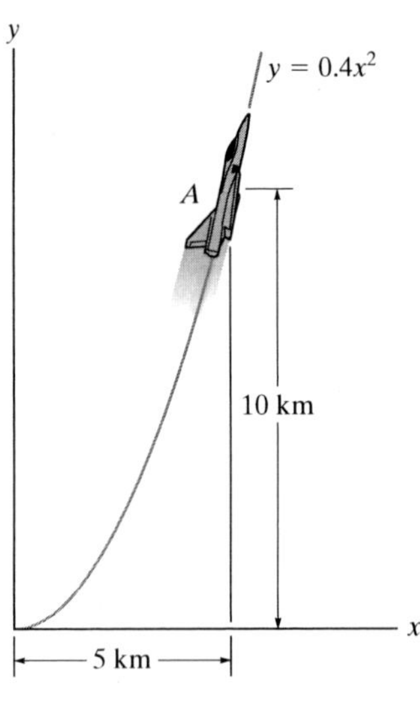

Prob. 12–140

12–141. The ball is ejected horizontally from the tube with a speed of 8 m/s. Find the equation of the path, $y = f(x)$, and then find the ball's velocity and the normal and tangential components of acceleration when $t = 0.25$ s.

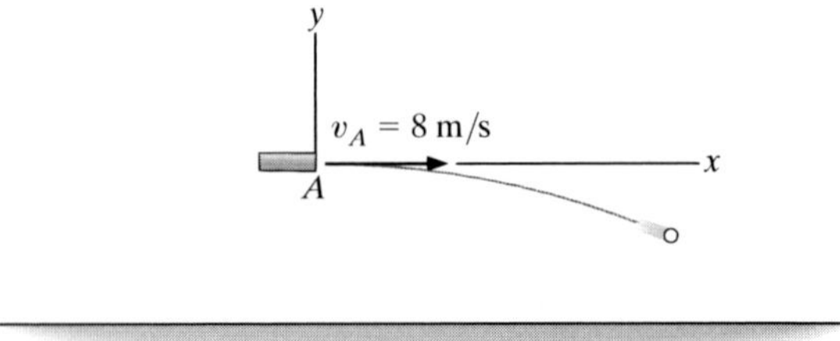

Prob. 12–141

12–142. A toboggan is traveling down along a curve which can be approximated by the parabola $y = 0.01x^2$. Determine the magnitude of its acceleration when it reaches point A, where its speed is $v_A = 10$ m/s, and it is increasing at the rate of $\dot{v}_A = 3$ m/s^2.

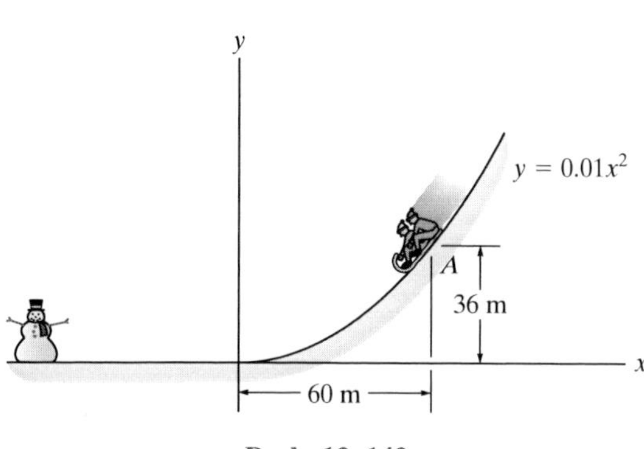

Prob. 12–142

12–143. A particle P moves along the curve $y = (x^2 - 4)$ m with a constant speed of 5 m/s. Determine the point on the curve where the maximum magnitude of acceleration occurs and compute its value.

***12–144.** The satellite S travels around the earth in a circular path with a constant speed of 20 Mm/h. If the acceleration is 2.5 m/s^2, determine the altitude h. Assume the earth's diameter to be 12 713 km.

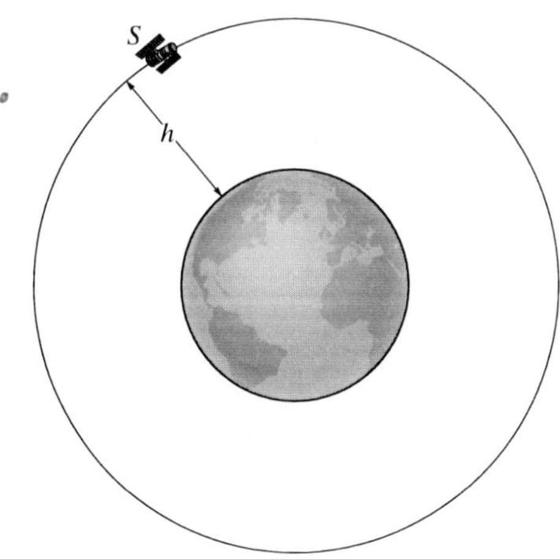

Prob. 12–144

12–145. The particle travels with a constant speed of 300 mm/s along the curve. Determine the particle's acceleration when it is located at point (200 mm, 100 mm) and sketch this vector on the curve.

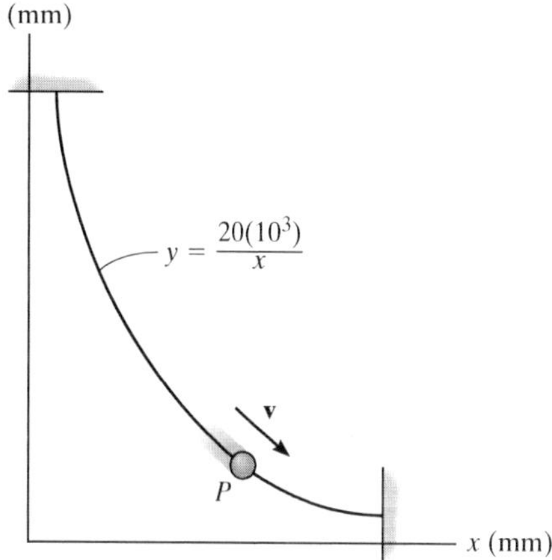

Prob. 12–145

12–146. The race car has an initial speed $v_A = 15$ m/s at A. If it increases its speed along the circular track at the rate $a_t = (0.4s)$ m/s^2, where s is in meters, determine the time needed for the car to travel 20 m. Take $\rho = 150$ m.

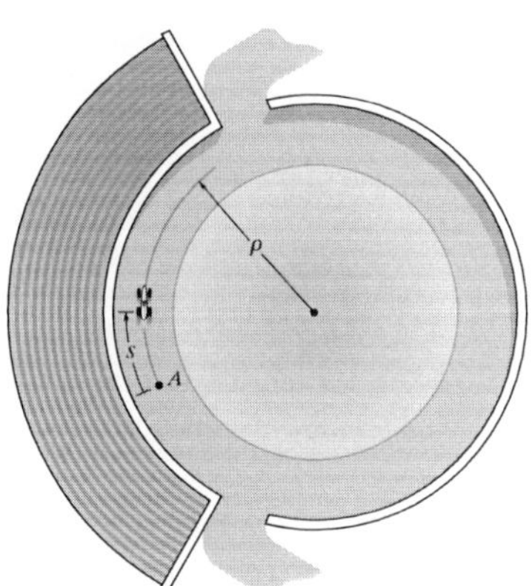

Prob. 12–146

12–147. A particle moves along the curve $y = \sin x$ with a constant speed $v = 2$ m/s. Determine the normal and tangential components of its velocity and acceleration at any instant.

***12–148.** A particle travels along the path $y = a + bx + cx^2$, where a, b, c are constants. If the speed of the particle is constant, $v = v_0$, determine the x and y components of velocity and the normal component of acceleration when $x = 0$.

12

12–149. The two particles A and B start at the origin O and travel in opposite directions along the circular path at constant speeds $v_A = 0.7\text{ m/s}$ and $v_B = 1.5\text{ m/s}$, respectively. Determine in $t = 2\text{ s}$, (a) the displacement along the path of each particle, (b) the position vector to each particle, and (c) the shortest distance between the particles.

12–150. The two particles A and B start at the origin O and travel in opposite directions along the circular path at constant speeds $v_A = 0.7\text{ m/s}$ and $v_B = 1.5\text{ m/s}$, respectively. Determine the time when they collide and the magnitude of the acceleration of B just before this happens.

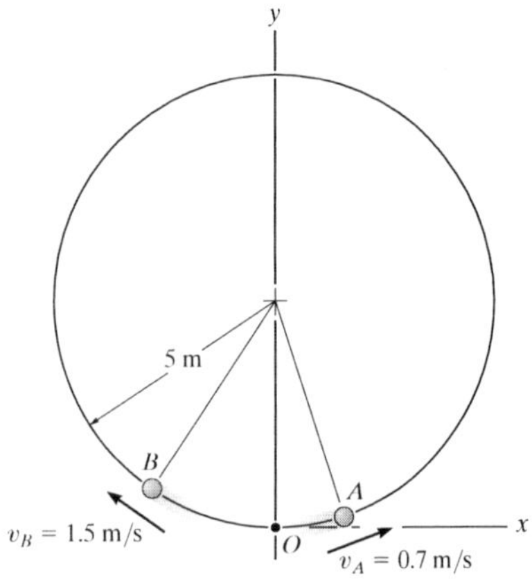

Probs. 12–149/150

12–151. The position of a particle traveling along a curved path is $s = (3t^3 - 4t^2 + 4)$ m, where t is in seconds. When $t = 2$ s, the particle is at a position on the path where the radius of curvature is 25 m. Determine the magnitude of the particle's acceleration at this instant.

***12–152.** Starting from rest the motorboat travels around the circular path, $\rho = 50$ m, at a speed $v = (0.8t)$ m/s, where t is in seconds. Determine the magnitudes of the boat's velocity and acceleration when it has traveled 20 m.

12–153. Starting from rest, the motorboat travels around the circular path, $\rho = 50$ m, at a speed $v = (0.2t^2)$ m/s, where t is in seconds. Determine the magnitudes of the boat's velocity and acceleration at the instant $t = 3$ s.

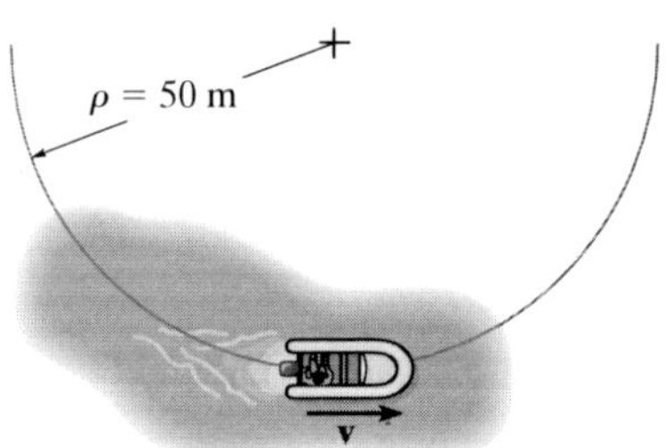

Probs. 12–152/153

12–154. The ball is kicked with an initial speed $v_A = 8\text{ m/s}$ at an angle $\theta_A = 40°$ with the horizontal. Find the equation of the path, $y = f(x)$, and then determine the ball's velocity and the normal and tangential components of its acceleration when $t = 0.25$ s.

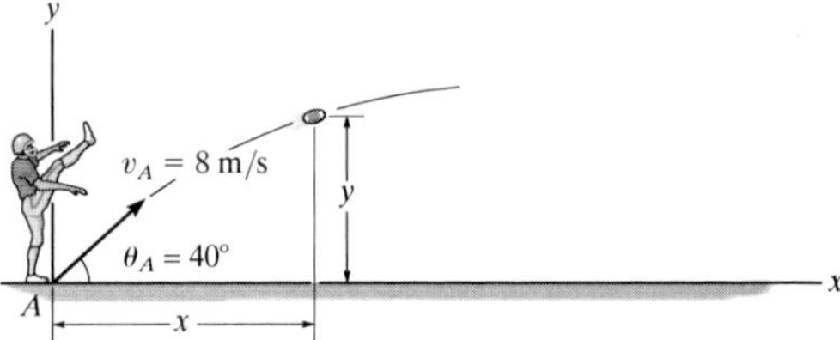

Prob. 12–154

12–155. The race car travels around the circular track with a speed of 16 m/s. When it reaches point A it increases its speed at $a_t = (\frac{4}{3}v^{1/4})\text{ m/s}^2$, where v is in m/s. Determine the magnitudes of the velocity and acceleration of the car when it reaches point B. Also, how much time is required for it to travel from A to B?

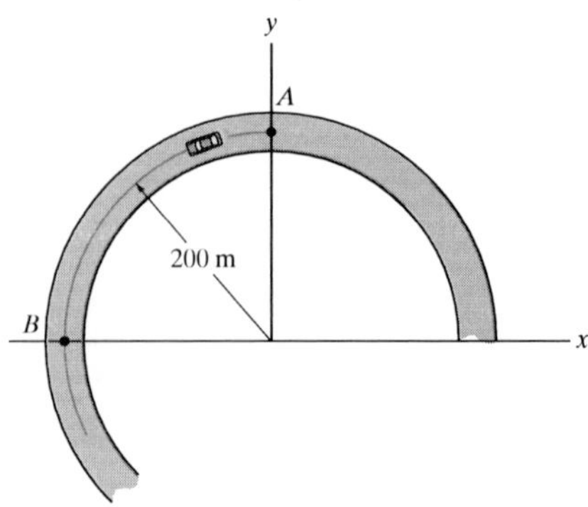

Prob. 12–155

*12–156. A particle P travels along an elliptical spiral path such that its position vector $\mathbf{r}$ is defined by $\mathbf{r} = \{2\cos(0.1t)\mathbf{i} + 1.5\sin(0.1t)\mathbf{j} + (2t)\mathbf{k}\}$ m, where t is in seconds and the arguments for the sine and cosine are given in radians. When $t = 8$ s, determine the coordinate direction angles α, β, and γ, which the binormal axis to the osculating plane makes with the x, y, and z axes. *Hint:* Solve for the velocity $\mathbf{v}_P$ and acceleration $\mathbf{a}_P$ of the particle in terms of their $\mathbf{i}$, $\mathbf{j}$, $\mathbf{k}$ components. The binormal is parallel to $\mathbf{v}_P \times \mathbf{a}_P$. Why?

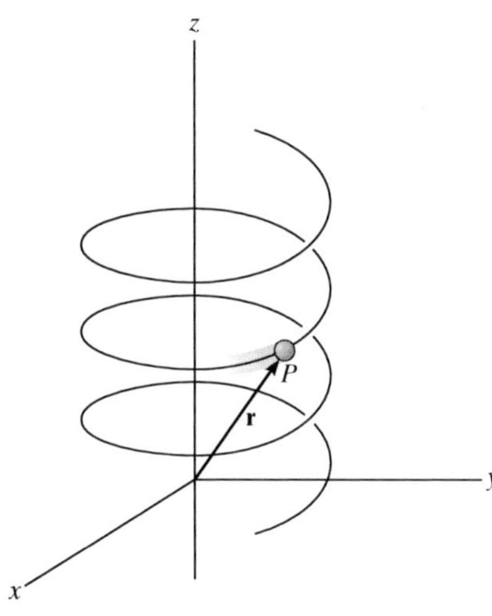

Prob. 12–156

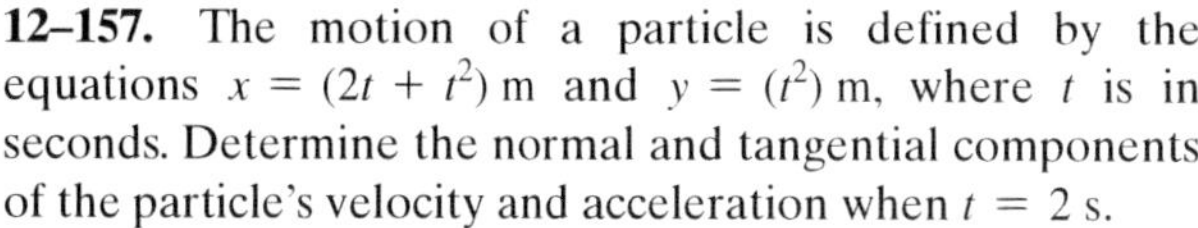

12–157. The motion of a particle is defined by the equations $x = (2t + t^2)$ m and $y = (t^2)$ m, where t is in seconds. Determine the normal and tangential components of the particle's velocity and acceleration when $t = 2$ s.

12–158. The motorcycle travels along the elliptical track at a constant speed v. Determine the greatest magnitude of the acceleration if $a > b$.

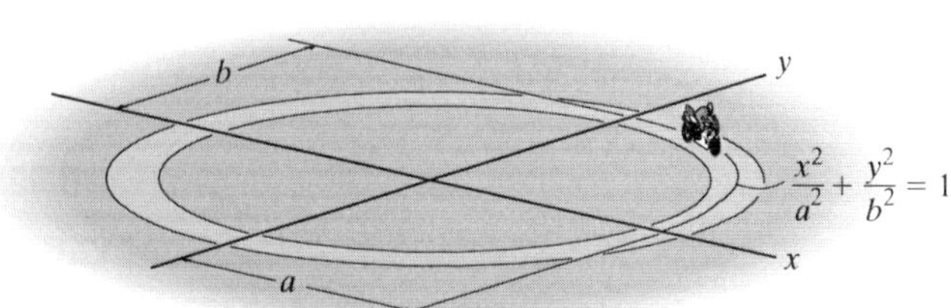

Prob. 12–158

12.8 Curvilinear Motion: Cylindrical Components

Sometimes the motion of the particle is constrained on a path that is best described using cylindrical coordinates. If motion is restricted to the plane, then polar coordinates are used.

Polar Coordinates. We can specify the location of the particle shown in Fig. 12–30*a* using a *radial coordinate* r, which extends outward from the fixed origin O to the particle, and a *transverse coordinate* θ, which is the counterclockwise angle between a fixed reference line and the r axis. The angle is generally measured in degrees or radians, where 1 rad $= 180°/\pi$. The positive directions of the r and θ coordinates are defined by the unit vectors $\mathbf{u}_r$ and $\mathbf{u}_\theta$, respectively. Here $\mathbf{u}_r$ is in the direction of increasing r when θ is held fixed, and $\mathbf{u}_\theta$ is in a direction of increasing θ when r is held fixed. Note that these directions are perpendicular to one another.

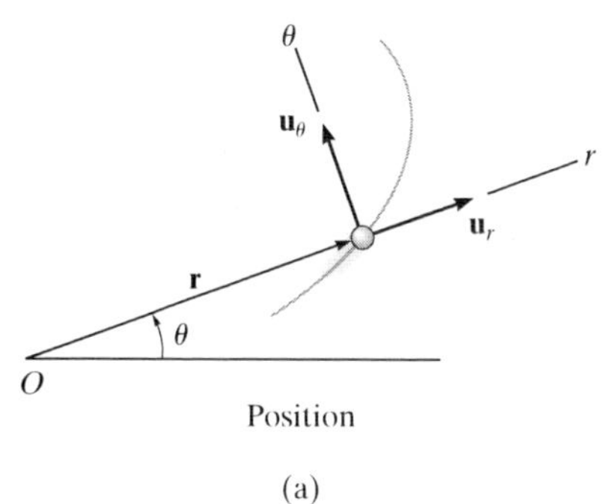

Position

(a)

Fig. 12–30

Position

(a)

Position. At any instant the position of the particle, Fig. 12–30*a*, is defined by the position vector

$$\mathbf{r} = r\mathbf{u}_r \tag{12–22}$$

Velocity. The instantaneous velocity $\mathbf{v}$ is obtained by taking the time derivative of $\mathbf{r}$. Using a dot to represent the time derivative, we have

$$\mathbf{v} = \dot{\mathbf{r}} = \dot{r}\mathbf{u}_r + r\dot{\mathbf{u}}_r$$

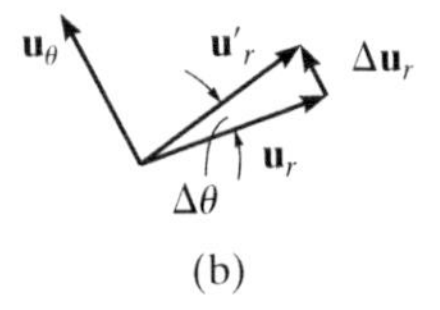

(b)

To evaluate $\dot{\mathbf{u}}_r$, notice that $\mathbf{u}_r$ only changes its direction with respect to time, since by definition the magnitude of this vector is always one unit. Hence, during the time Δt, a change Δr will not cause a change in the direction of $\mathbf{u}_r$; however, a change $\Delta\theta$ will cause $\mathbf{u}_r$ to become $\mathbf{u}'_r$, where $\mathbf{u}'_r = \mathbf{u}_r + \Delta\mathbf{u}_r$, Fig. 12–30*b*. The time change in $\mathbf{u}_r$ is then $\Delta\mathbf{u}_r$. For small angles $\Delta\theta$ this vector has a magnitude $\Delta u_r \approx 1(\Delta\theta)$ and acts in the $\mathbf{u}_\theta$ direction. Therefore, $\Delta\mathbf{u}_r = \Delta\theta\mathbf{u}_\theta$, and so

$$\dot{\mathbf{u}}_r = \lim_{\Delta t \to 0} \frac{\Delta\mathbf{u}_r}{\Delta t} = \left(\lim_{\Delta t \to 0} \frac{\Delta\theta}{\Delta t}\right)\mathbf{u}_\theta$$

$$\dot{\mathbf{u}}_r = \dot{\theta}\mathbf{u}_\theta \tag{12–23}$$

Substituting into the above equation, the velocity can be written in component form as

$$\mathbf{v} = v_r\mathbf{u}_r + v_\theta\mathbf{u}_\theta \tag{12–24}$$

where

$$\begin{aligned} v_r &= \dot{r} \\ v_\theta &= r\dot{\theta} \end{aligned} \tag{12–25}$$

These components are shown graphically in Fig. 12–30*c*. The *radial component* $\mathbf{v}_r$ is a measure of the rate of increase or decrease in the length of the radial coordinate, i.e., $\dot{r}$; whereas the *transverse component* $\mathbf{v}_\theta$ can be interpreted as the rate of motion along the circumference of a circle having a radius r. In particular, the term $\dot{\theta} = d\theta/dt$ is called the *angular velocity*, since it indicates the time rate of change of the angle θ. Common units used for this measurement are rad/s.

Since $\mathbf{v}_r$ and $\mathbf{v}_\theta$ are mutually perpendicular, the *magnitude* of velocity or speed is simply the positive value of

$$v = \sqrt{(\dot{r})^2 + (r\dot{\theta})^2} \tag{12–26}$$

and the *direction* of $\mathbf{v}$ is, of course, tangent to the path, Fig. 12–30*c*.

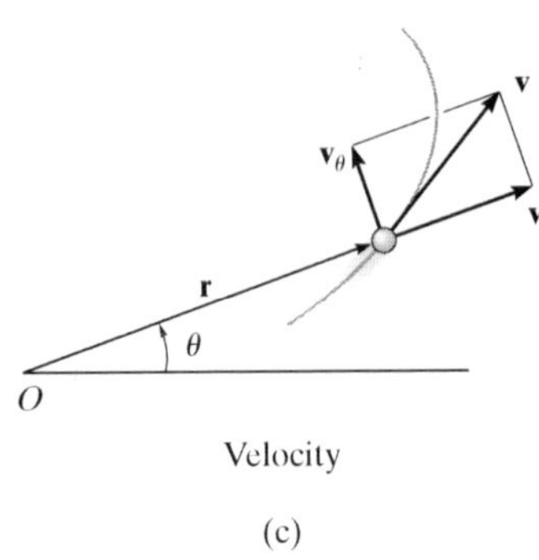

Velocity

(c)

Fig. 12–30 (cont.)

Acceleration. Taking the time derivatives of Eq. 12–24, using Eqs. 12–25, we obtain the particle's instantaneous acceleration,

$$\mathbf{a} = \dot{\mathbf{v}} = \ddot{r}\mathbf{u}_r + \dot{r}\dot{\mathbf{u}}_r + \dot{r}\dot{\theta}\mathbf{u}_\theta + r\ddot{\theta}\mathbf{u}_\theta + r\dot{\theta}\dot{\mathbf{u}}_\theta$$

To evaluate $\dot{\mathbf{u}}_\theta$, it is necessary only to find the change in the direction of $\mathbf{u}_\theta$ since its magnitude is always unity. During the time Δt, a change Δr will not change the direction of $\mathbf{u}_\theta$, however, a change $\Delta\theta$ will cause $\mathbf{u}_\theta$ to become $\mathbf{u}'_\theta$, where $\mathbf{u}'_\theta = \mathbf{u}_\theta + \Delta\mathbf{u}_\theta$, Fig. 12–30*d*. The time change in $\mathbf{u}_\theta$ is thus $\Delta\mathbf{u}_\theta$. For small angles this vector has a magnitude $\Delta u_\theta \approx 1(\Delta\theta)$ and acts in the $-\mathbf{u}_r$ direction; i.e., $\Delta\mathbf{u}_\theta = -\Delta\theta\mathbf{u}_r$. Thus,

(d)

$$\dot{\mathbf{u}}_\theta = \lim_{\Delta t \to 0} \frac{\Delta\mathbf{u}_\theta}{\Delta t} = -\left(\lim_{\Delta t \to 0} \frac{\Delta\theta}{\Delta t}\right)\mathbf{u}_r$$

$$\dot{\mathbf{u}}_\theta = -\dot{\theta}\mathbf{u}_r \qquad (12\text{–}27)$$

Substituting this result and Eq. 12–23 into the above equation for **a**, we can write the acceleration in component form as

$$\mathbf{a} = a_r\mathbf{u}_r + a_\theta\mathbf{u}_\theta \qquad (12\text{–}28)$$

where

$$a_r = \ddot{r} - r\dot{\theta}^2$$
$$a_\theta = r\ddot{\theta} + 2\dot{r}\dot{\theta} \qquad (12\text{–}29)$$

The term $\ddot{\theta} = d^2\theta/dt^2 = d/dt(d\theta/dt)$ is called the *angular acceleration* since it measures the change made in the angular velocity during an instant of time. Units for this measurement are rad/s^2.

Since $\mathbf{a}_r$ and $\mathbf{a}_\theta$ are always perpendicular, the *magnitude* of acceleration is simply the positive value of

$$a = \sqrt{(\ddot{r} - r\dot{\theta}^2)^2 + (r\ddot{\theta} + 2\dot{r}\dot{\theta})^2} \qquad (12\text{–}30)$$

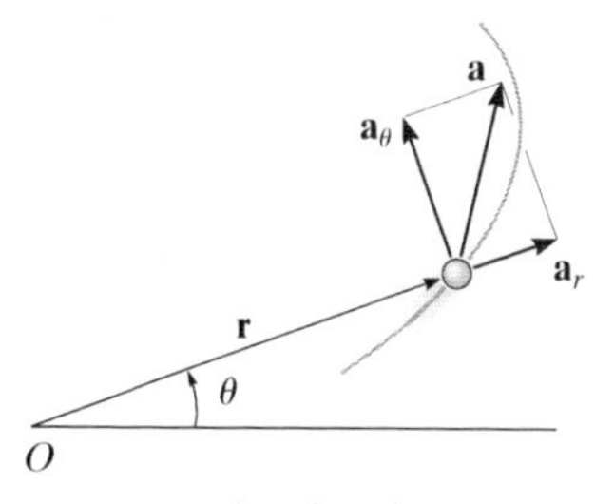

Acceleration

(e)

The *direction* is determined from the vector addition of its two components. In general, **a** will *not* be tangent to the path, Fig. 12–30*e*.

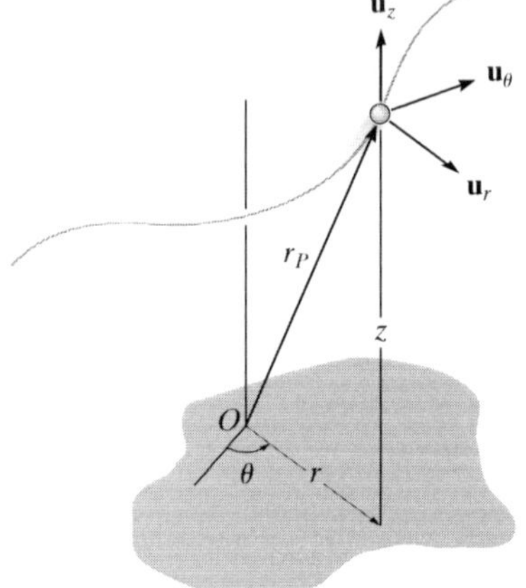

Fig. 12–31

© Orhan Cam/Shutterstock.com

The spiral motion of this girl can be followed by using cylindrical components. Here the radial coordinate r is constant, the transverse coordinate θ will increase with time as the girl rotates about the vertical, and her altitude z will decrease with time.

Cylindrical Coordinates. If the particle moves along a space curve as shown in Fig. 12–31, then its location may be specified by the three *cylindrical coordinates*, r, θ, z. The z coordinate is identical to that used for rectangular coordinates. Since the unit vector defining its direction, $\mathbf{u}_z$, is constant, the time derivatives of this vector are zero, and therefore the position, velocity, and acceleration of the particle can be written in terms of its cylindrical coordinates as follows:

$$\mathbf{r}_P = r\mathbf{u}_r + z\mathbf{u}_z$$

$$\mathbf{v} = \dot{r}\mathbf{u}_r + r\dot{\theta}\mathbf{u}_\theta + \dot{z}\mathbf{u}_z \qquad (12\text{–}31)$$

$$\mathbf{a} = (\ddot{r} - r\dot{\theta}^2)\mathbf{u}_r + (r\ddot{\theta} + 2\dot{r}\dot{\theta})\mathbf{u}_\theta + \ddot{z}\mathbf{u}_z \qquad (12\text{–}32)$$

Time Derivatives. The above equations require that we obtain the time derivatives $\dot{r}$, $\ddot{r}$, $\dot{\theta}$, and $\ddot{\theta}$ in order to evaluate the r and θ components of **v** and **a**. Two types of problems generally occur:

1. If the polar coordinates are specified as time parametric equations, $r = r(t)$ and $\theta = \theta(t)$, then the time derivatives can be found directly.
2. If the time-parametric equations are not given, then the path $r = f(\theta)$ must be known. Using the chain rule of calculus we can then find the relation between $\dot{r}$ and $\dot{\theta}$, and between $\ddot{r}$ and $\ddot{\theta}$. Application of the chain rule, along with some examples, is explained in Appendix C.

Procedure for Analysis

Coordinate System.

- Polar coordinates are a suitable choice for solving problems when data regarding the angular motion of the radial coordinate r is given to describe the particle's motion. Also, some paths of motion can conveniently be described in terms of these coordinates.
- To use polar coordinates, the origin is established at a fixed point, and the radial line r is directed to the particle.
- The transverse coordinate θ is measured from a fixed reference line to the radial line.

Velocity and Acceleration.

- Once r and the four time derivatives $\dot{r}$, $\ddot{r}$, $\dot{\theta}$, and $\ddot{\theta}$ have been evaluated at the instant considered, their values can be substituted into Eqs. 12–25 and 12–29 to obtain the radial and transverse components of **v** and **a**.
- If it is necessary to take the time derivatives of $r = f(\theta)$, then the chain rule of calculus must be used. See Appendix C.
- Motion in three dimensions requires a simple extension of the above procedure to include $\dot{z}$ and $\ddot{z}$.

EXAMPLE 12.17

The amusement park ride shown in Fig. 12–32*a* consists of a chair that is rotating in a horizontal circular path of radius *r* such that the arm *OB* has an angular velocity $\dot{\theta}$ and angular acceleration $\ddot{\theta}$. Determine the radial and transverse components of velocity and acceleration of the passenger. Neglect his size in the calculation.

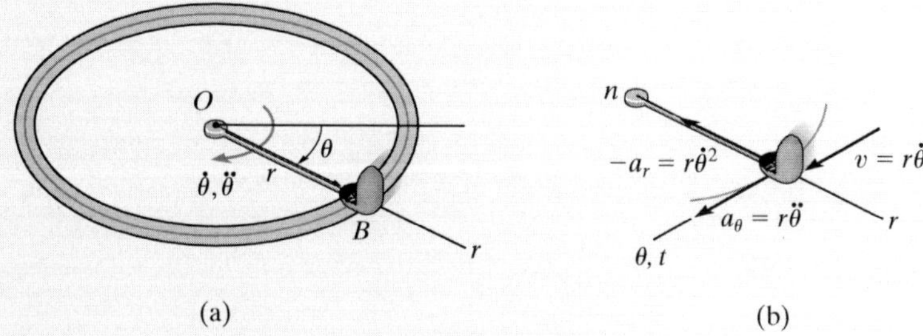

Fig. 12–32

SOLUTION

Coordinate System. Since the angular motion of the arm is reported, polar coordinates are chosen for the solution, Fig. 12–32*a*. Here θ is not related to *r*, since the radius is constant for all θ.

Velocity and Acceleration. It is first necessary to specify the first and second time derivatives of *r* and θ. Since *r* is *constant*, we have

$$r = r \qquad \dot{r} = 0 \qquad \ddot{r} = 0$$

Thus,

$$v_r = \dot{r} = 0 \qquad \textit{Ans.}$$

$$v_\theta = r\dot{\theta} \qquad \textit{Ans.}$$

$$a_r = \ddot{r} - r\dot{\theta}^2 = -r\dot{\theta}^2 \qquad \textit{Ans.}$$

$$a_\theta = r\ddot{\theta} + 2\dot{r}\dot{\theta} = r\ddot{\theta} \qquad \textit{Ans.}$$

These results are shown in Fig. 12–32*b*.

NOTE: The *n*, *t* axes are also shown in Fig. 12–32*b*, which in this special case of circular motion happen to be *collinear* with the *r* and θ axes, respectively. Since $v = v_\theta = v_t = r\dot{\theta}$, then by comparison,

$$-a_r = a_n = \frac{v^2}{\rho} = \frac{(r\dot{\theta})^2}{r} = r\dot{\theta}^2$$

$$a_\theta = a_t = \frac{dv}{dt} = \frac{d}{dt}(r\dot{\theta}) = \frac{dr}{dt}\dot{\theta} + r\frac{d\dot{\theta}}{dt} = 0 + r\ddot{\theta}$$

12

EXAMPLE 12.18

The rod OA in Fig. 12–33a rotates in the horizontal plane such that $\theta = (t^3)$ rad. At the same time, the collar B is sliding outward along OA so that $r = (100t^2)$ mm. If in both cases t is in seconds, determine the velocity and acceleration of the collar when $t = 1$ s.

(a)

SOLUTION

Coordinate System. Since time-parametric equations of the path are given, it is not necessary to relate r to θ.

Velocity and Acceleration. Determining the time derivatives and evaluating them when $t = 1$ s, we have

$$r = 100t^2\Big|_{t=1\text{ s}} = 100\,\text{mm} \quad \theta = t^3\Big|_{t=1\text{ s}} = 1\text{ rad} = 57.3^\circ$$

$$\dot{r} = 200t\Big|_{t=1\text{ s}} = 200\text{ mm/s} \quad \dot{\theta} = 3t^2\Big|_{t=1\text{ s}} = 3\text{ rad/s}$$

$$\ddot{r} = 200\Big|_{t=1\text{ s}} = 200\text{ mm/s}^2 \quad \ddot{\theta} = 6t\Big|_{t=1\text{ s}} = 6\text{ rad/s}^2.$$

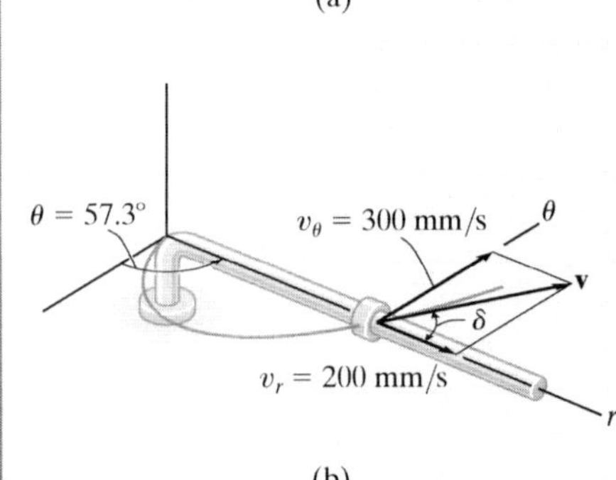

(b)

As shown in Fig. 12–33b,

$$\mathbf{v} = \dot{r}\mathbf{u}_r + r\dot{\theta}\mathbf{u}_\theta$$
$$= 200\mathbf{u}_r + 100(3)\mathbf{u}_\theta = \{200\mathbf{u}_r + 300\mathbf{u}_\theta\}\text{ mm/s}$$

The magnitude of $\mathbf{v}$ is

$$v = \sqrt{(200)^2 + (300)^2} = 361\text{ mm/s} \qquad \textit{Ans.}$$

$$\delta = \tan^{-1}\left(\frac{300}{200}\right) = 56.3^\circ \quad \delta + 57.3^\circ = 114^\circ \qquad \textit{Ans.}$$

As shown in Fig. 12–33c,

$$\mathbf{a} = (\ddot{r} - r\dot{\theta}^2)\mathbf{u}_r + (r\ddot{\theta} + 2\dot{r}\dot{\theta})\mathbf{u}_\theta$$
$$= [200 - 100(3)^2]\mathbf{u}_r + [100(6) + 2(200)3]\mathbf{u}_\theta$$
$$= \{-700\mathbf{u}_r + 1800\mathbf{u}_\theta\}\text{ mm/s}^2$$

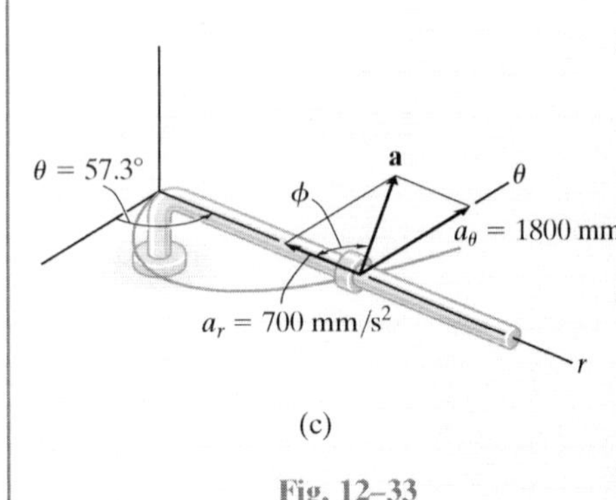

(c)

Fig. 12–33

The magnitude of $\mathbf{a}$ is

$$a = \sqrt{(-700)^2 + (1800)^2} = 1930\text{ mm/s}^2 \qquad \textit{Ans.}$$

$$\phi = \tan^{-1}\left(\frac{1800}{700}\right) = 68.7^\circ \quad (180^\circ - \phi) + 57.3^\circ = 169^\circ \qquad \textit{Ans.}$$

NOTE: The velocity is tangent to the path; however, the acceleration is directed within the curvature of the path, as expected.

12

EXAMPLE 12.19

The searchlight in Fig. 12–34*a* casts a spot of light along the face of a wall that is located 100 m from the searchlight. Determine the magnitudes of the velocity and acceleration at which the spot appears to travel across the wall at the instant $\theta = 45°$. The searchlight rotates at a constant rate of $\dot{\theta} = 4$ rad/s.

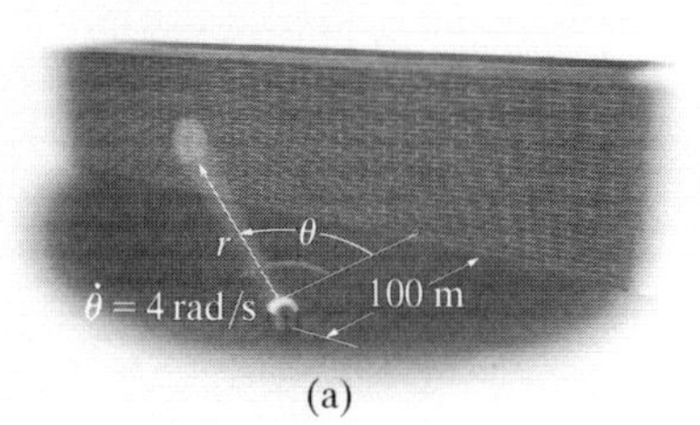

(a)

SOLUTION

Coordinate System. Polar coordinates will be used to solve this problem since the angular rate of the searchlight is given. To find the necessary time derivatives it is first necessary to relate r to θ. From Fig. 12–34*a*,

$$r = 100/\cos\theta = 100\sec\theta$$

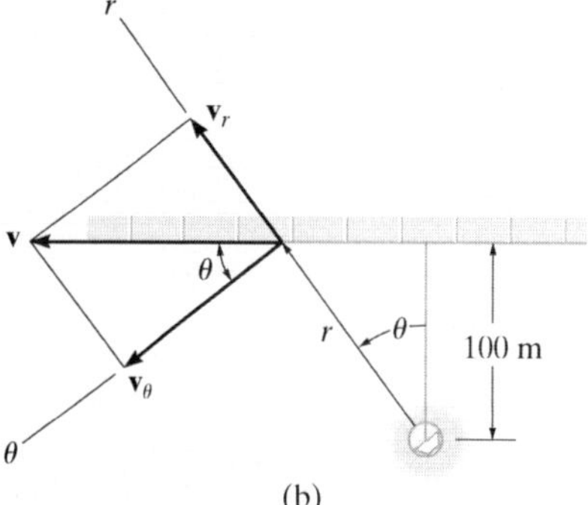

(b)

Velocity and Acceleration. Using the chain rule of calculus, noting that $d(\sec\theta) = \sec\theta\tan\theta\, d\theta$, and $d(\tan\theta) = \sec^2\theta\, d\theta$, we have

$$\begin{aligned}\dot{r} &= 100(\sec\theta\tan\theta)\dot{\theta}\\ \ddot{r} &= 100(\sec\theta\tan\theta)\dot{\theta}(\tan\theta)\dot{\theta} + 100\sec\theta(\sec^2\theta)\dot{\theta}(\dot{\theta})\\ &\quad + 100\sec\theta\tan\theta(\ddot{\theta})\\ &= 100\sec\theta\tan^2\theta(\dot{\theta})^2 + 100\sec^3\theta(\dot{\theta})^2 + 100(\sec\theta\tan\theta)\ddot{\theta}\end{aligned}$$

Since $\dot{\theta} = 4$ rad/s = constant, then $\ddot{\theta} = 0$, and the above equations, when $\theta = 45°$, become

$$\begin{aligned}r &= 100\sec 45° = 141.4\\ \dot{r} &= 400\sec 45°\tan 45° = 565.7\\ \ddot{r} &= 1600(\sec 45°\tan^2 45° + \sec^3 45°) = 6788.2\end{aligned}$$

As shown in Fig. 12–34*b*,

$$\begin{aligned}\mathbf{v} &= \dot{r}\mathbf{u}_r + r\dot{\theta}\mathbf{u}_\theta\\ &= 565.7\mathbf{u}_r + 141.4(4)\mathbf{u}_\theta\\ &= \{565.7\mathbf{u}_r + 565.7\mathbf{u}_\theta\}\text{ m/s}\\ v &= \sqrt{v_r^2 + v_\theta^2} = \sqrt{(565.7)^2 + (565.7)^2}\\ &= 800\text{ m/s}\end{aligned}$$ *Ans.*

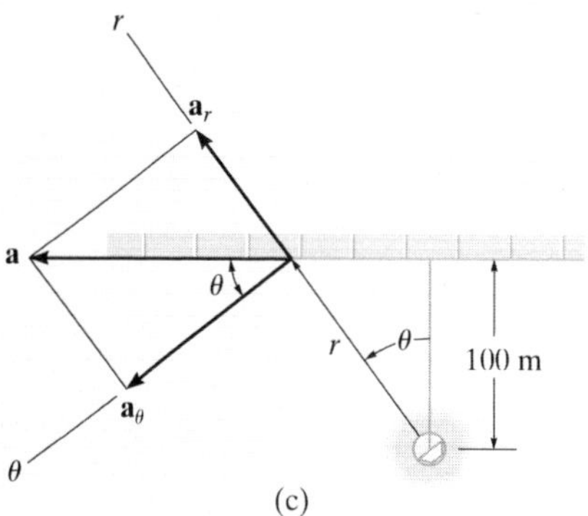

(c)

As shown in Fig. 12–34c,

$$\begin{aligned}\mathbf{a} &= (\ddot{r} - r\dot{\theta}^2)\mathbf{u}_r + (r\ddot{\theta} + 2\dot{r}\dot{\theta})\mathbf{u}_\theta\\ &= [6788.2 - 141.4(4)^2]\mathbf{u}_r + [141.4(0) + 2(565.7)4]\mathbf{u}_\theta\\ &= \{4525.5\mathbf{u}_r + 4525.5\mathbf{u}_\theta\}\text{ m/s}^2\\ a &= \sqrt{a_r^2 + a_\theta^2} = \sqrt{(4525.5)^2 + (4525.5)^2}\\ &= 6400\text{ m/s}^2\end{aligned}$$ *Ans.*

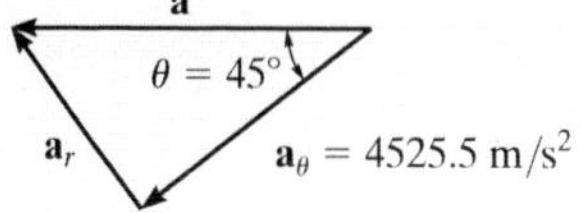

(d)

NOTE: It is also possible to find a without having to calculate $\ddot{r}$ (or a_r). As shown in Fig. 12–34*d*, since $a_\theta = 4525.5\text{ m/s}^2$, then by vector resolution, $a = 4525.5/\cos 45° = 6400\text{ m/s}^2$.

Fig. 12–34

12

EXAMPLE 12.20

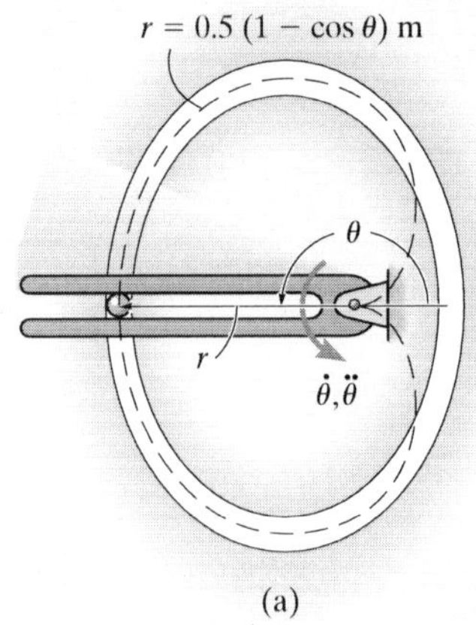

(a)

Due to the rotation of the forked rod, the ball in Fig. 12–35*a* travels around the slotted path, a portion of which is in the shape of a cardioid, $r = 0.15(1 - \cos\theta)$ m, where is in radians. If the ball's velocity is $v = 1.2$ m/s and its acceleration is $a = 9$ m/s^2 at the instant $\theta = 180°$, determine the angular velocity $\dot{\theta}$ and angular acceleration $\ddot{\theta}$ of the fork.

SOLUTION

Coordinate System. This path is most unusual, and mathematically it is best expressed using polar coordinates, as done here, rather than rectangular coordinates. Also, since $\dot{\theta}$ and $\ddot{\theta}$ must be determined, then r, θ coordinates are an obvious choice.

Velocity and Acceleration. The time derivatives of r and θ can be determined using the chain rule.

$$r = 0.15(1 - \cos\theta)$$

$$\dot{r} = 0.15(\sin\theta)\dot{\theta}$$

$$\ddot{r} = 0.15(\cos\theta)\dot{\theta}(\dot{\theta}) + 0.15(\sin\theta)\ddot{\theta}$$

Evaluating these results at $\theta = 180°$, we have

$$r = 0.3 \text{ m} \qquad \dot{r} = 0 \qquad \ddot{r} = -0.15\dot{\theta}^2$$

Since $v = 1.2$ m/s, using Eq. 12–26 to determine $\dot{\theta}$ yields

$$v = \sqrt{(\dot{r})^2 + (r\dot{\theta})^2}$$

$$1.2 = \sqrt{(0)^2 + (0.3\dot{\theta})^2}$$

$$\dot{\theta} = 4 \text{ rad/s} \qquad \textit{Ans.}$$

In a similar manner, $\ddot{\theta}$ can be found using Eq. 12–30.

$$a = \sqrt{(\ddot{r} - r\dot{\theta}^2)^2 + (r\ddot{\theta} + 2\dot{r}\dot{\theta})^2}$$

$$9 = \sqrt{[-0.15(4)^2 - 0.3(4)^2]^2 + [0.3\ddot{\theta} + 2(0)(4)]^2}$$

$$(9)^2 = (-7.2)^2 + 0.09\,\ddot{\theta}^2$$

$$\ddot{\theta} = 18 \text{ rad/s}^2 \qquad \textit{Ans.}$$

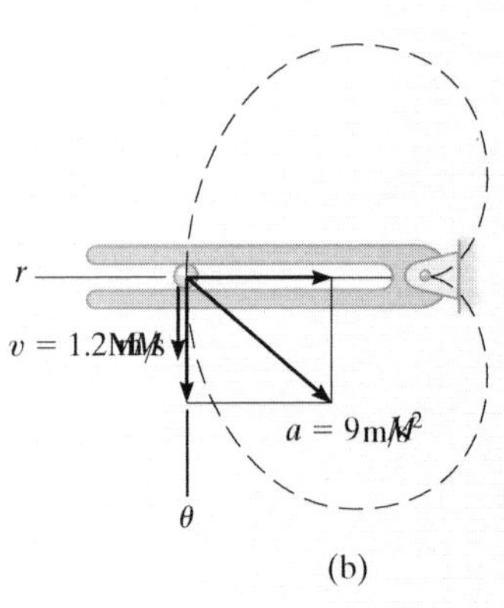

(b)

Fig. 12–35

Vectors **a** and **v** are shown in Fig. 12–35*b*.

NOTE: At this location, the θ and t (tangential) axes will coincide. The $+n$ (normal) axis is directed to the right, opposite to $+r$.

FUNDAMENTAL PROBLEMS

F12–33. The car has a speed of 55 m/s. Determine the angular velocity $\dot{\theta}$ of the radial line OA at this instant.

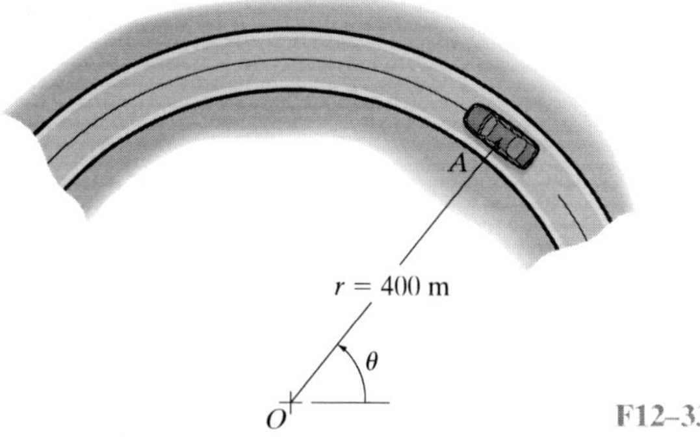

F12–33

F12–34. The platform is rotating about the vertical axis such that at any instant its angular position is $\theta = (4t^{3/2})$ rad, where t is in seconds. A ball rolls outward along the radial groove so that its position is $r = (0.1t^3)$ m, where t is in seconds. Determine the magnitudes of the velocity and acceleration of the ball when $t = 1.5$ s.

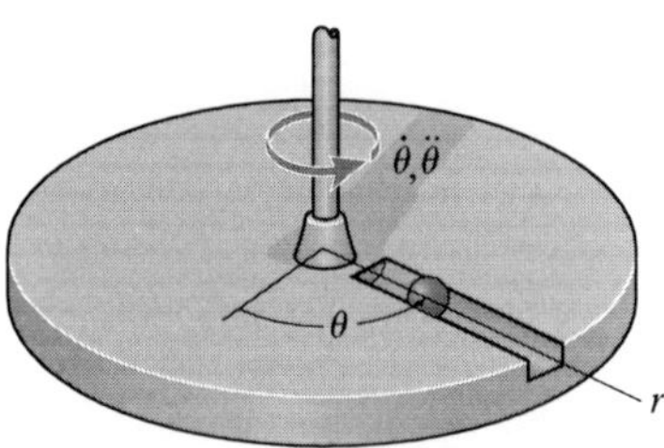

F12–34

F12–35. Peg P is driven by the fork link OA along the curved path described by $r = (2\theta)$ m. At the instant $\theta = \pi/4$ rad, the angular velocity and angular acceleration of the link are $\dot{\theta} = 3$ rad/s and $\ddot{\theta} = 1$ rad/s^2. Determine the magnitude of the peg's acceleration at this instant.

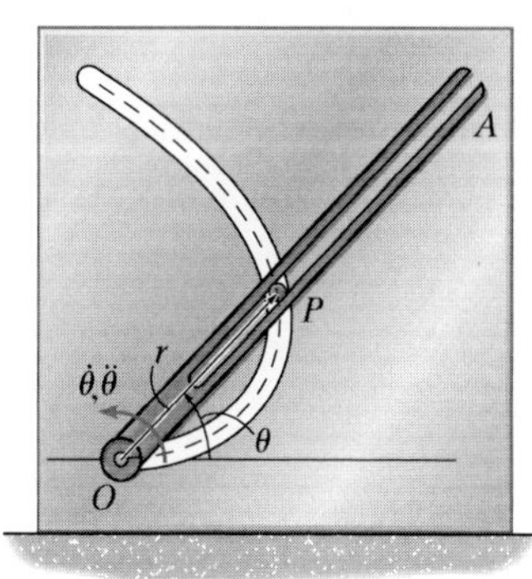

F12–35

F12–36. Peg P is driven by the forked link OA along the path described by $r = e^{\theta}$, where r is in meters. When $\theta = \frac{\pi}{4}$ rad, the link has an angular velocity and angular acceleration of $\dot{\theta} = 2$ rad/s and $\ddot{\theta} = 4$ rad/s^2. Determine the radial and transverse components of the peg's acceleration at this instant.

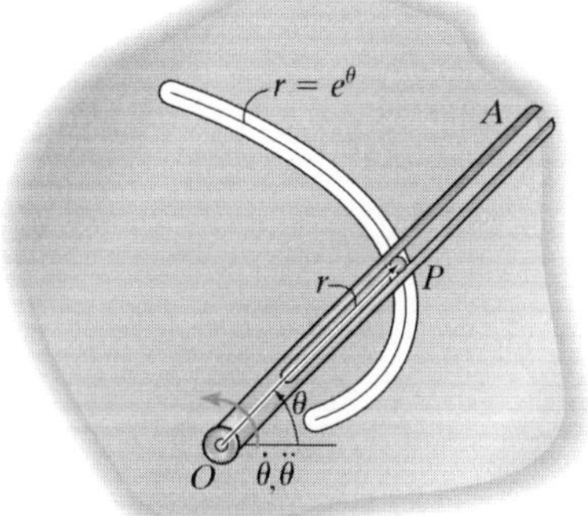

F12–36

F12–37. The collars are pin connected at B and are free to move along rod OA and the curved guide OC having the shape of a cardioid, $r = [0.2(1 + \cos\theta)]$ m. At $\theta = 30°$, the angular velocity of OA is $\dot{\theta} = 3$ rad/s. Determine the magnitude of the velocity of the collars at this point.

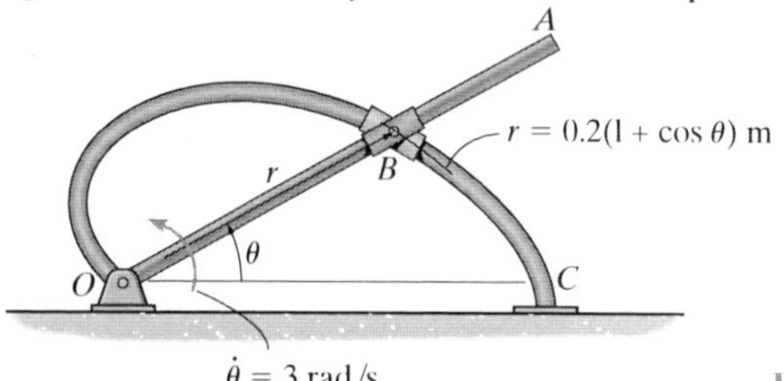

F12–37

F12–38. At the instant $\theta = 45°$, the athlete is running with a constant speed of 2 m/s. Determine the angular velocity at which the camera must turn in order to follow the motion.

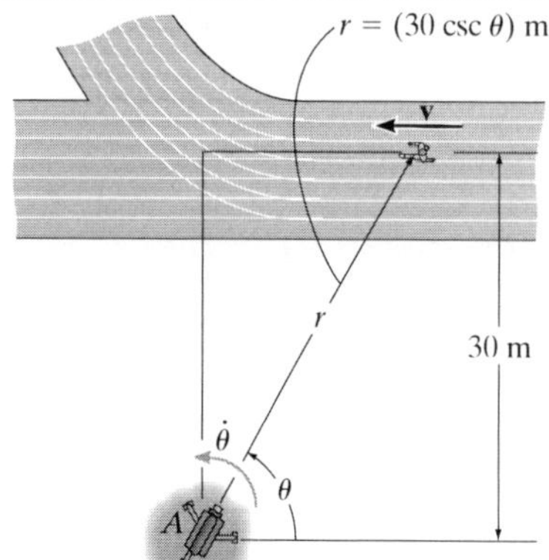

F12–38

PROBLEMS

12–159. The time rate of change of acceleration is referred to as the *jerk*, which is often used as a means of measuring passenger discomfort. Calculate this vector, $\frac{1}{2}\dot{\mathbf{a}}$, in terms of its cylindrical components, using Eq. 12–32.

***12–160.** A particle travels around a limaçon, defined by the equation $r = b - a \cos \theta$, where a and b are constants. Determine the particle's radial and transverse components of velocity and acceleration as a function of θ and its time derivatives.

12–161. If a particle's position is described by the polar coordinates $r = 4(1 + \sin t)$ m and $\theta = (2e^{-t})$ rad, where t is in seconds and the argument for the sine is in radians, determine the radial and tangential components of the particle's velocity and acceleration when $t = 2$ s.

12–162. A particle moves along a circular path of radius 300 mm. If its angular velocity is $\dot{\theta} = (2t^2)$ rad/s, where t is in seconds, determine the magnitude of the particle's acceleration when $t = 2$ s.

12–163. At the instant shown, the water sprinkler is rotating with an angular speed $\dot{\theta} = 2$ rad/s and an angular acceleration $\ddot{\theta} = 3$ rad/s^2. If the nozzle lies in the vertical plane and water is flowing through it at a constant rate of 3 m/s, determine the magnitudes of the velocity and acceleration of a water particle as it exits the open end, $r = 0.2$ m.

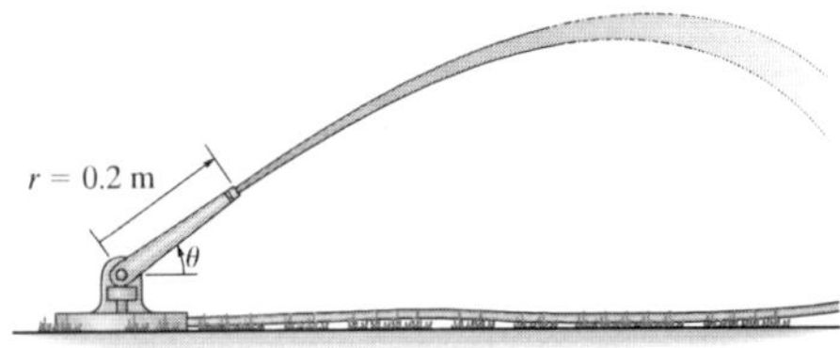

Prob. 12–163

***12–164.** A radar gun at O rotates with the angular velocity of $\dot{\theta} = 0.1$ rad/s and angular acceleration of $\ddot{\theta} = 0.025$ rad/s^2, at the instant $\theta = 45°$, as it follows the motion of the car traveling along the circular road having a radius of $r = 200$ m. Determine the magnitudes of velocity and acceleration of the car at this instant.

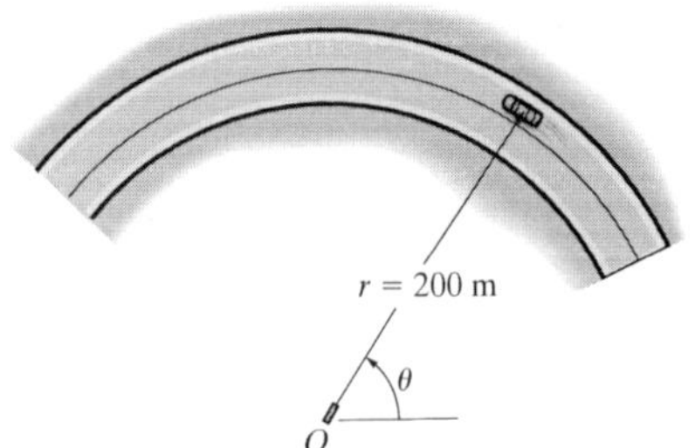

Prob. 12–164

12–165. A particle is moving along a circular path having a 400-mm radius. Its position as a function of time is given by $\theta = (2t^2)$ rad, where t is in seconds. Determine the magnitude of the particle's acceleration when $\theta = 30°$. The particle starts from rest when $\theta = 0°$.

12–166. If a particle's position is described by the polar coordinates $r = (2 \sin 2\theta)$ m and $\theta = (4t)$ rad, where t is in seconds, determine the radial and tangential components of its velocity and acceleration when $t = 1$ s.

12–167. If arm OA rotates counterclockwise with a constant angular velocity of $\dot{\theta} = 2$ rad/s, determine the magnitudes of the velocity and acceleration of peg P at $\theta = 30°$. The peg moves in the fixed groove defined by the lemniscate, and along the slot in the arm.

***12–168.** The peg moves in the curved slot defined by the lemniscate, and through the slot in the arm. At $\theta = 30°$, the angular velocity is $\dot{\theta} = 2$ rad/s, and the angular acceleration is $\ddot{\theta} = 1.5$ rad/s^2. Determine the magnitudes of the velocity and acceleration of peg P at this instant.

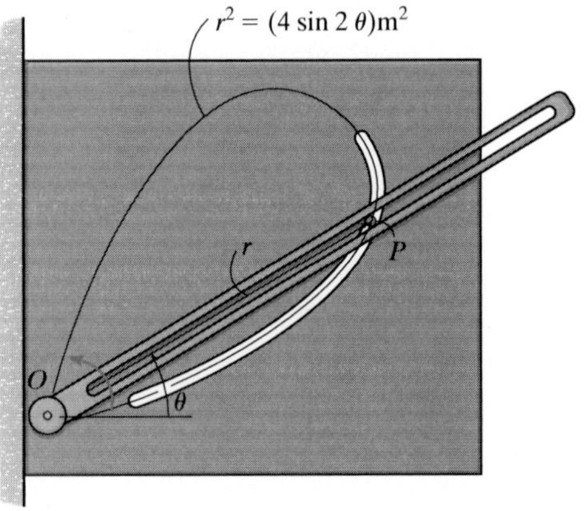

Probs. 12–167/168

12–169. The time rate of change of acceleration is referred to as the *jerk*, which is often used as a means of measuring passenger discomfort. Calculate this vector, $\dot{\mathbf{a}}$, in terms of its cylindrical components, using Eq. 12–32.

12–170. The pin follows the path described by the equation $r = (0.2 + 0.15 \cos \theta)$ m. At the instant $\theta = 30^\circ$, $\dot{\theta} = 0.7$ rad/s and $\ddot{\theta} = 0.5$ rad/s^2. Determine the magnitudes of the pin's velocity and acceleration at this instant. Neglect the size of the pin.

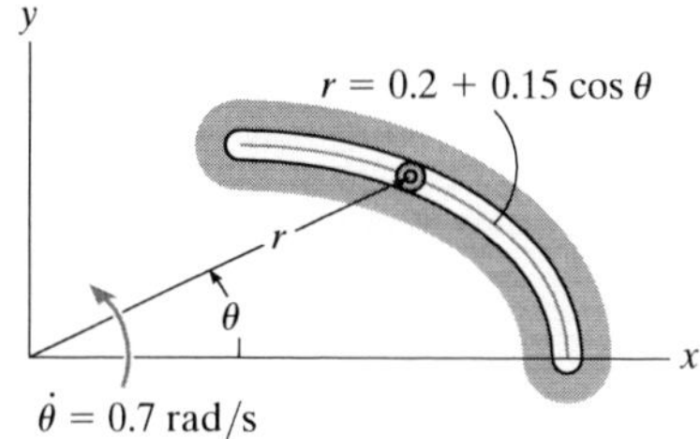

Prob. 12–170

12–171. The slotted link is pinned at O, and as a result of the constant angular velocity $\dot{\theta} = 3$ rad/s it drives the peg P for a short distance along the spiral guide $r = (0.4\,\theta)$ m, where θ is in radians. Determine the radial and transverse components of the velocity and acceleration of P at the instant $\theta = \pi/3$ rad.

***12–172.** Solve Prob. 12–171 if the slotted link has an angular acceleration $\ddot{\theta} = 8$ rad/s^2 when $\dot{\theta} = 3$ rad/s at $\theta = \pi/3$ rad.

12–173. The slotted link is pinned at O, and as a result of the constant angular velocity $\dot{\theta} = 3$ rad/s it drives the peg P for a short distance along the spiral guide $r = (0.4\,\theta)$ m, where θ is in radians. Determine the velocity and acceleration of the particle at the instant it leaves the slot in the link, i.e., when $r = 0.5$ m.

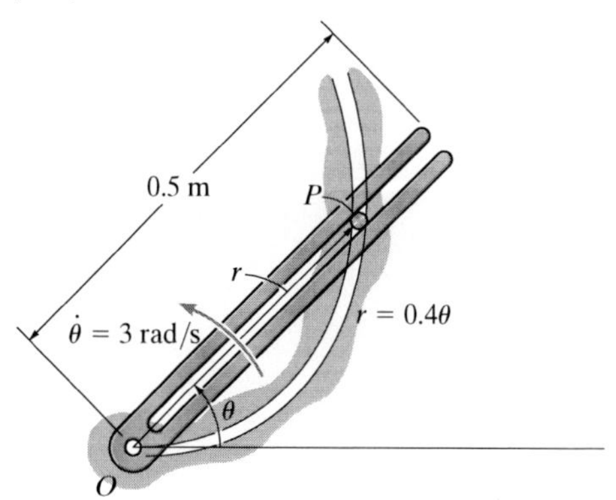

Probs. 12–171/172/173

12–174. A truck is traveling along the horizontal circular curve of radius $r = 60$ m with a constant speed $v = 20$ m/s. Determine the angular rate of rotation $\dot{\theta}$ of the radial line r and the magnitude of the truck's acceleration.

12–175. A truck is traveling along the horizontal circular curve of radius $r = 60$ m with a speed of 20 m/s which is increasing at 3 m/s^2. Determine the truck's radial and transverse components of acceleration.

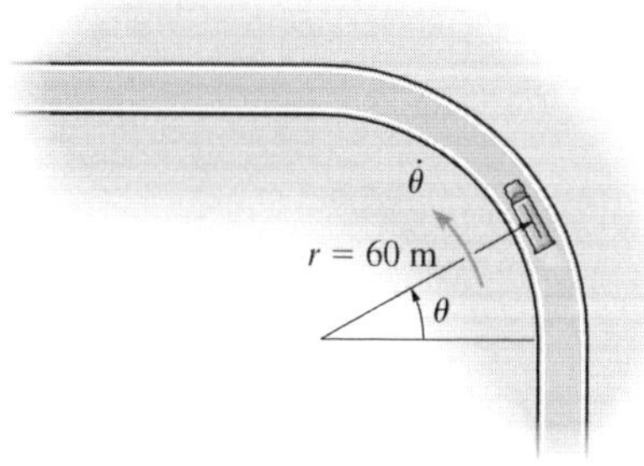

Probs. 12–174/175

***12–176.** The driver of the car maintains a constant speed of 40 m/s. Determine the angular velocity of the camera tracking the car when $\theta = 15^\circ$.

12–177. When $\theta = 15^\circ$, the car has a speed of 50 m/s which is increasing at 6 m/s^2. Determine the angular velocity of the camera tracking the car at this instant.

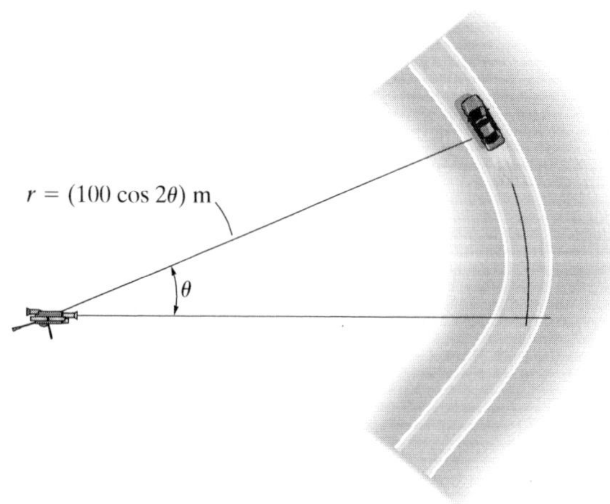

Probs. 12–176/177

12

12–178. The small washer slides down the cord OA. When it is at the midpoint, its speed is 200 mm/s and its acceleration is 10 mm/s^2. Express the velocity and acceleration of the washer at this point in terms of its cylindrical components.

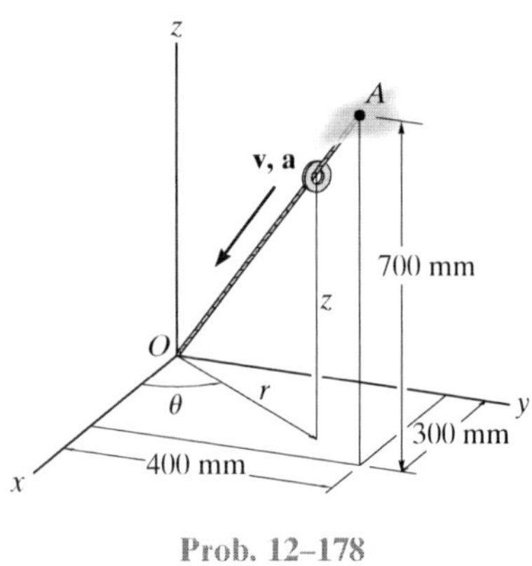

Prob. 12–178

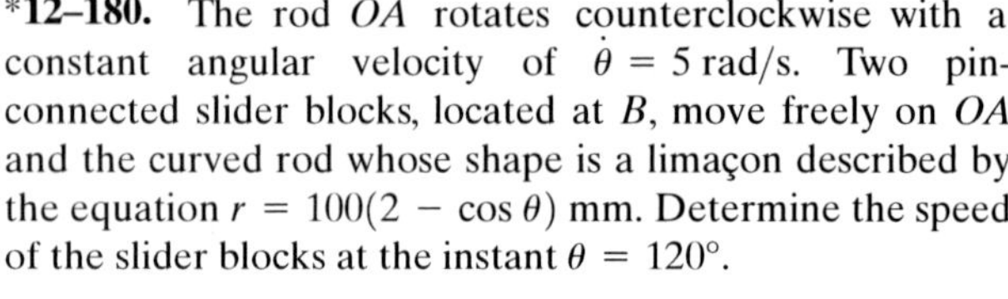

***12–180.** The rod OA rotates counterclockwise with a constant angular velocity of $\dot{\theta} = 5$ rad/s. Two pin-connected slider blocks, located at B, move freely on OA and the curved rod whose shape is a limaçon described by the equation $r = 100(2 - \cos\theta)$ mm. Determine the speed of the slider blocks at the instant $\theta = 120°$.

12–181. Determine the magnitude of the acceleration of the slider blocks in Prob. 12–180 when $\theta = 120°$.

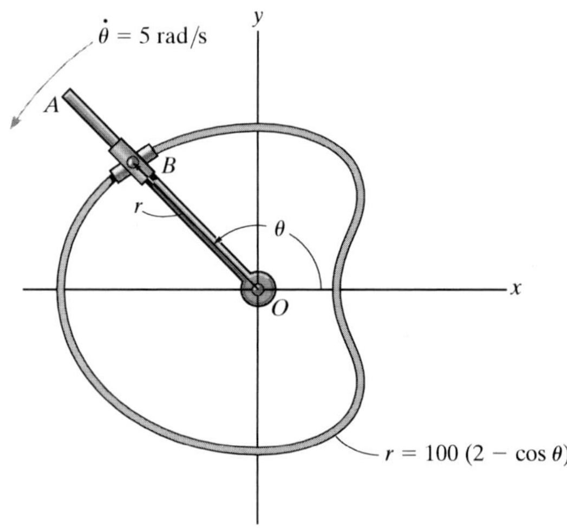

Probs. 12–180/181

12–179. A block moves outward along the slot in the platform with a speed of $\dot{r} = (4t)$ m/s, where t is in seconds. The platform rotates at a constant rate of 6 rad/s. If the block starts from rest at the center, determine the magnitudes of its velocity and acceleration when $t = 1$ s.

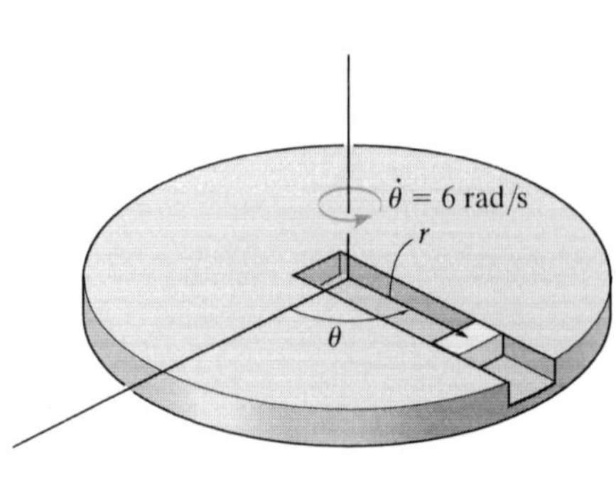

Prob. 12–179

12–182. A cameraman standing at A is following the movement of a race car, B, which is traveling around a curved track at a constant speed of 30 m/s. Determine the angular rate $\dot{\theta}$ at which the man must turn in order to keep the camera directed on the car at the instant $\theta = 30°$.

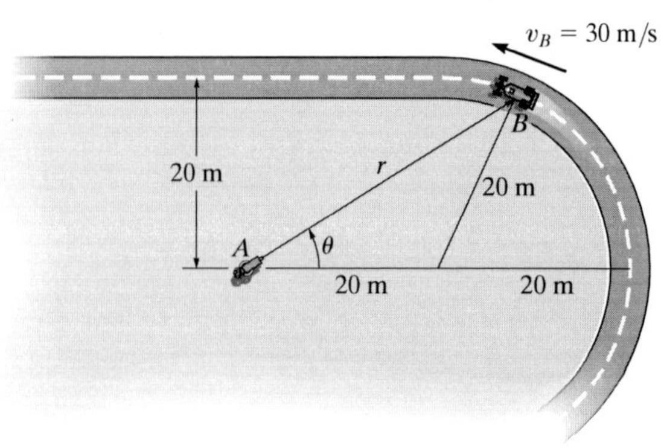

Prob. 12–182

12–183. The slotted arm AB drives pin C through the spiral groove described by the equation $r = a\theta$. If the angular velocity is constant at $\dot{\theta}$, determine the radial and transverse components of velocity and acceleration of the pin.

***12–184.** The slotted arm AB drives pin C through the spiral groove described by the equation $r = (1.5\,\theta)$ m, where θ is in radians. If the arm starts from rest when $\theta = 60°$ and is driven at an angular velocity of $\dot{\theta} = (4t)$ rad/s, where t is in seconds, determine the radial and transverse components of velocity and acceleration of the pin C when $t = 1$ s.

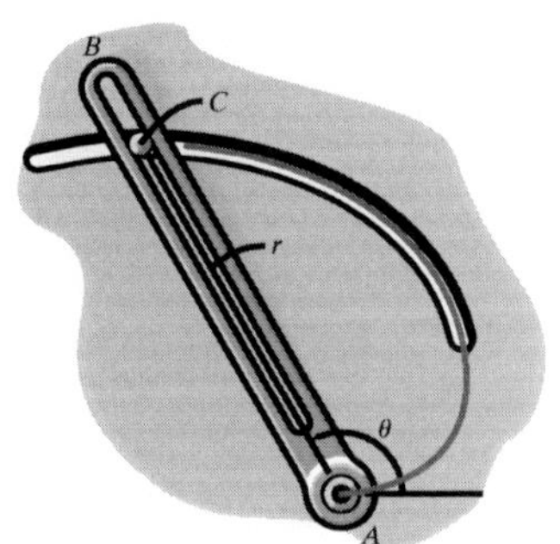

Probs. 12–183/184

12–185. The motion of peg P is constrained by the lemniscate curved slot in OB and by the slotted arm OA. If OA rotates counterclockwise with a constant angular velocity of $\dot{\theta} = 3$ rad/s, determine the magnitudes of the velocity and acceleration of peg P at $\theta = 30°$.

12–186. The motion of peg P is constrained by the lemniscate curved slot in OB and by the slotted arm OA. If OA rotates counterclockwise with an angular velocity of $\dot{\theta} = (3t^{3/2})$ rad/s, where t is in seconds, determine the magnitudes of the velocity and acceleration of peg P at $\theta = 30°$. When $t = 0$, $\theta = 0°$.

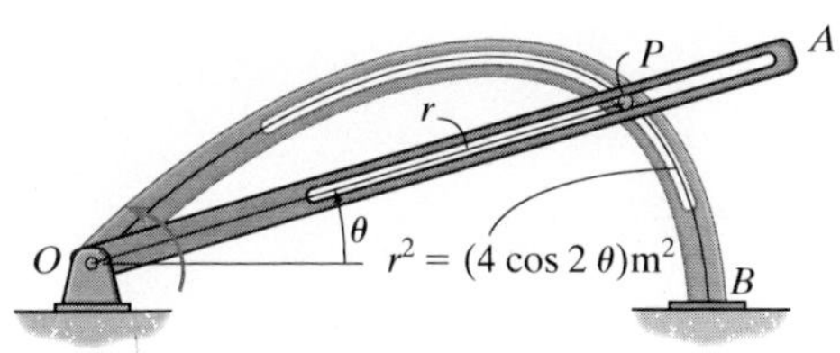

Probs. 12–185/186

12–187. If the circular plate rotates clockwise with a constant angular velocity of $\dot{\theta} = 1.5$ rad/s, determine the magnitudes of the velocity and acceleration of the follower rod AB when $\theta = 2/3\pi$ rad.

***12–188.** When $\theta = 2/3\pi$ rad, the angular velocity and angular acceleration of the circular plate are $\dot{\theta} = 1.5$ rad/s and $\ddot{\theta} = 3$ rad/s^2, respectively. Determine the magnitudes of the velocity and acceleration of the rod AB at this instant.

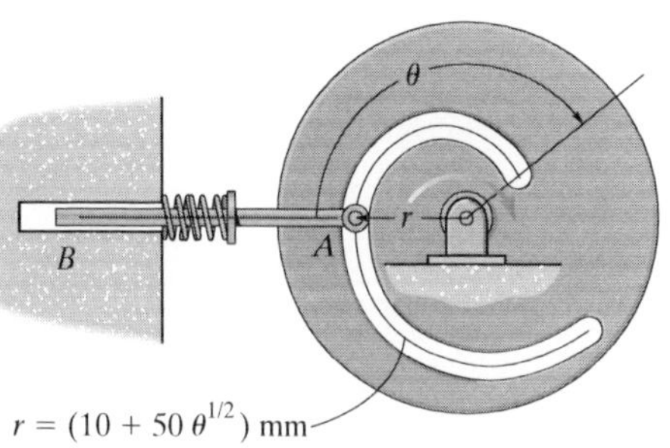

Probs. 12–187/188

12–189. The box slides down the helical ramp with a constant speed of $v = 2$ m/s. Determine the magnitude of its acceleration. The ramp descends a vertical distance of 1 m for every full revolution. The mean radius of the ramp is $r = 0.5$ m.

12–190. The box slides down the helical ramp such that $r = 0.5$ m, $\theta = (0.5t^3)$ rad, and $z = (2 - 0.2t^2)$ m, where t is in seconds. Determine the magnitudes of the velocity and acceleration of the box at the instant $\theta = 2\pi$ rad.

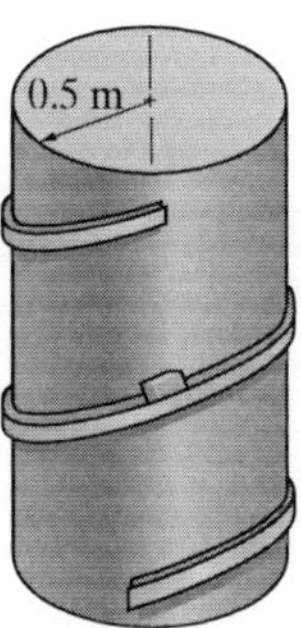

Probs. 12–189/190

12–191. For a *short distance* the train travels along a track having the shape of a spiral, $r = (1000/\theta)$ m, where θ is in radians. If it maintains a constant speed $v = 20$ m/s, determine the radial and transverse components of its velocity when $\theta = (9\pi/4)$ rad.

*__12–192.__ For a *short distance* the train travels along a track having the shape of a spiral, $r = (1000/\theta)$ m, where θ is in radians. If the angular rate is constant, $\dot{\theta} = 0.2$ rad/s, determine the radial and transverse components of its velocity and acceleration when $\theta = (9\pi/4)$ rad.

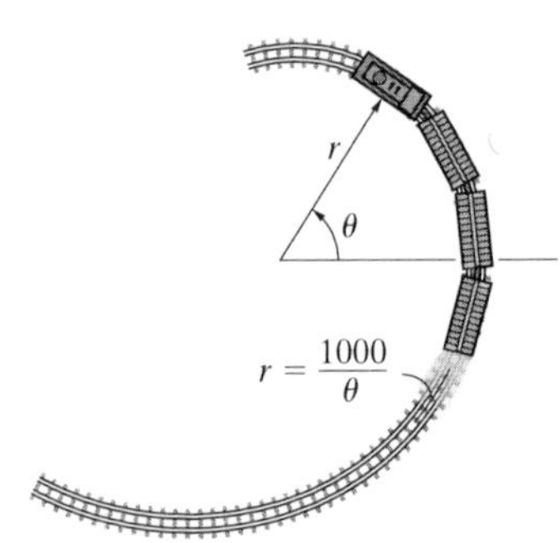

Probs. 12–191/192

12–193. If the cam rotates clockwise with a constant angular velocity of $\dot{\theta} = 5$ rad/s, determine the magnitudes of the velocity and acceleration of the follower rod AB at the instant $\theta = 30°$. The surface of the cam has a shape of limaçon defined by $r = (200 + 100\cos\theta)$ mm.

12–194. At the instant $\theta = 30°$, the cam rotates with a clockwise angular velocity of $\dot{\theta} = 5$ rad/s and angular acceleration of $\ddot{\theta} = 6$ rad/s^2. Determine the magnitudes of the velocity and acceleration of the follower rod AB at this instant. The surface of the cam has a shape of a limaçon defined by $r = (200 + 100\cos\theta)$ mm.

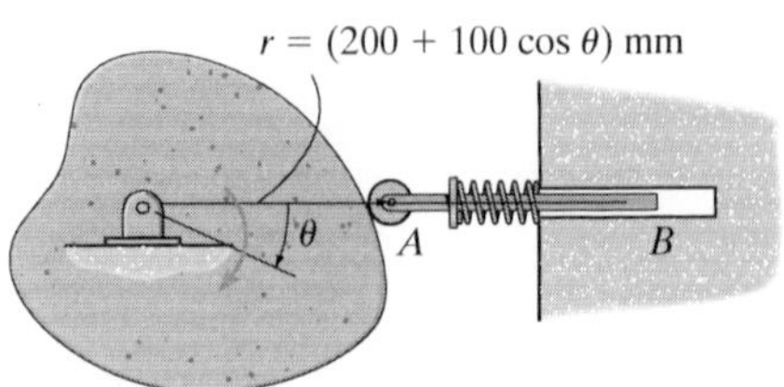

Probs. 12–193/194

12–195. A particle moves along an Archimedean spiral $r = (8\theta)$ m, where θ is given in radians. If $\dot{\theta} = 4$ rad/s (constant), determine the radial and transverse components of the particle's velocity and acceleration at the instant $\theta = \pi/2$ rad. Sketch the curve and show the components on the curve.

*__12–196.__ Solve Prob. 12–195 if the particle has an angular acceleration $\ddot{\theta} = 5$ rad/s^2 when $\dot{\theta} = 4$ rad/s at $\theta = \pi/2$ rad.

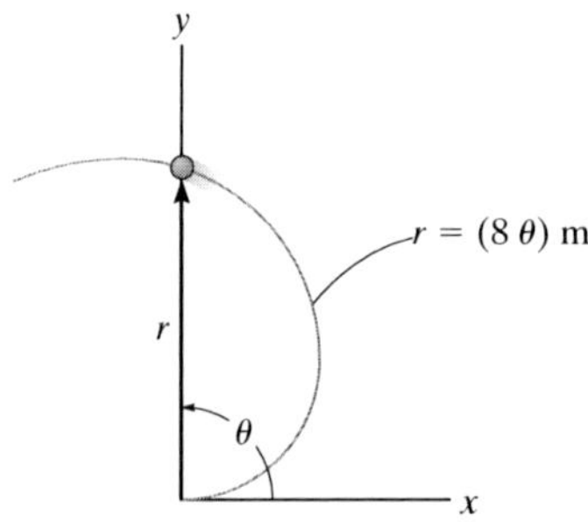

Probs. 12–195/196

12–197. The partial surface of the cam is that of a logarithmic spiral $r = (40e^{0.05\theta})$ mm, where θ is in radians. If the cam is rotating at a constant angular rate of $\dot{\theta} = 4$ rad/s, determine the magnitudes of the velocity and acceleration of the follower rod at the instant $\theta = 30°$.

12–198. Solve Prob. 12–197, if the cam has an angular acceleration of $\ddot{\theta} = 2$ rad/s^2 when its angular velocity is $\dot{\theta} = 4$ rad/s at $\theta = 30°$.

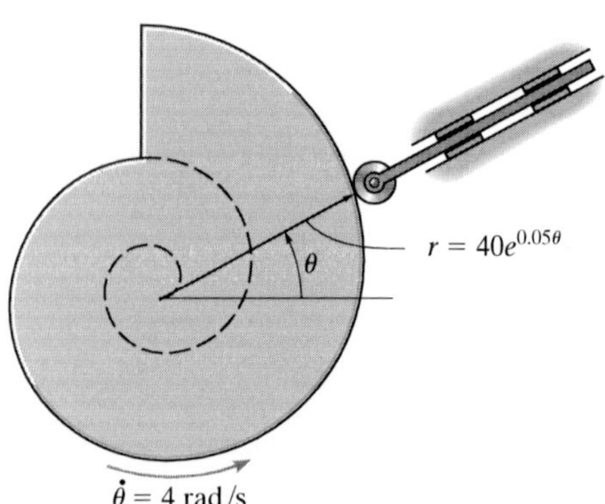

Probs. 12–197/198

12.9 Absolute Dependent Motion Analysis of Two Particles

In some types of problems the motion of one particle will *depend* on the corresponding motion of another particle. This dependency commonly occurs if the particles, here represented by blocks, are interconnected by inextensible cords which are wrapped around pulleys. For example, the movement of block A downward along the inclined plane in Fig. 12–36 will cause a corresponding movement of block B up the other incline. We can show this mathematically by first specifying the location of the blocks using *position coordinates* s_A and s_B. Note that each of the coordinate axes is (1) measured from a *fixed* point (O) or *fixed* datum line, (2) measured along each inclined plane *in the direction of motion* of each block, and (3) has a positive sense from the fixed datums to A and to B. If the total cord length is l_T, the two position coordinates are related by the equation

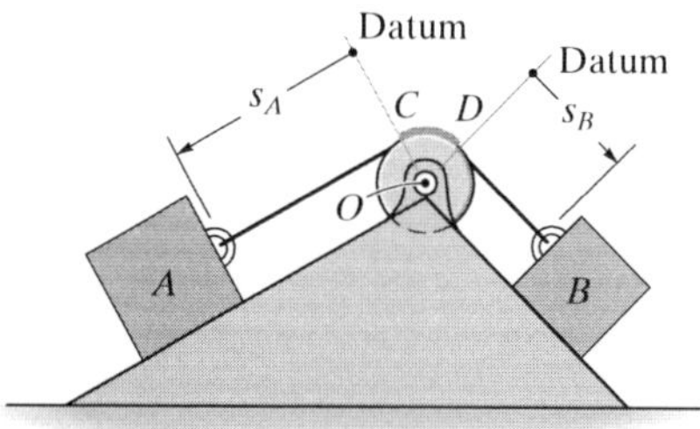

Fig. 12–36

$$s_A + l_{CD} + s_B = l_T$$

Here l_{CD} is the length of the cord passing over arc CD. Taking the time derivative of this expression, realizing that l_{CD} and l_T *remain constant*, while s_A and s_B measure the segments of the cord that change in length, we have

$$\frac{ds_A}{dt} + \frac{ds_B}{dt} = 0 \quad \text{or} \quad v_B = -v_A$$

The negative sign indicates that when block A has a velocity downward, i.e., in the direction of positive s_A, it causes a corresponding upward velocity of block B; i.e., B moves in the negative s_B direction.

In a similar manner, time differentiation of the velocities yields the relation between the accelerations, i.e.,

$$a_B = -a_A$$

A more complicated example is shown in Fig. 12–37*a*. In this case, the position of block A is specified by s_A, and the position of the *end* of the cord from which block B is suspended is defined by s_B. As above, we have chosen position coordinates which (1) have their origin at fixed points or datums, (2) are measured in the direction of motion of each block, and (3) from the fixed datums are positive to the right for s_A and positive downward for s_B. During the motion, the length of the red colored segments of the cord in Fig. 12–37*a remains constant*. If l represents the total length of cord minus these segments, then the position coordinates can be related by the equation

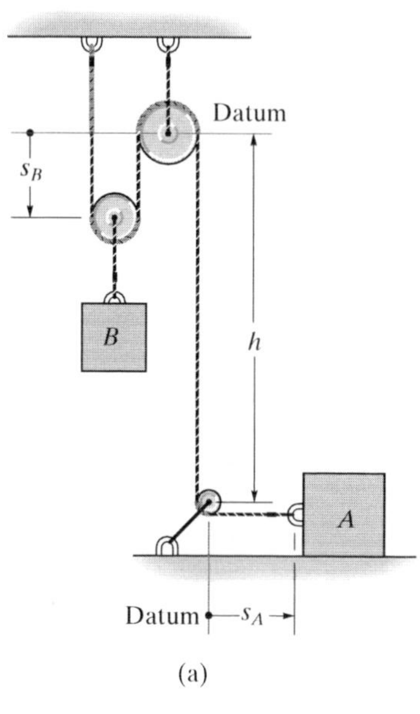

(a)

Fig. 12–37

$$2s_B + h + s_A = l$$

Since l and h are constant during the motion, the two time derivatives yield

$$2v_B = -v_A \qquad 2a_B = -a_A$$

Hence, when B moves downward ($+s_B$), A moves to the left ($-s_A$) with twice the motion.

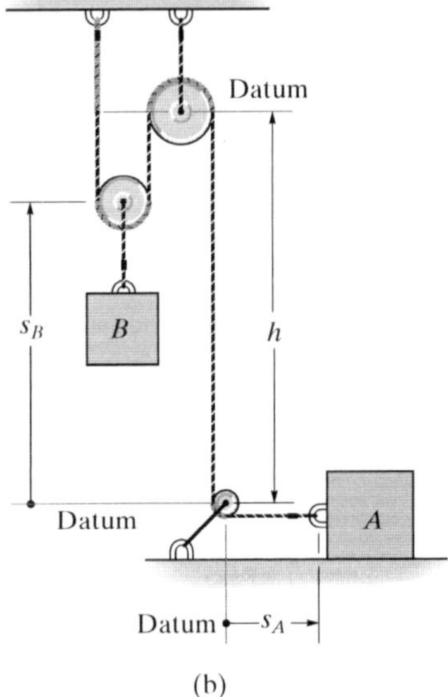

(b)

Fig. 12–37 (cont.)

This example can also be worked by defining the position of block B from the center of the bottom pulley (a fixed point), Fig. 12–37b. In this case

$$2(h - s_B) + h + s_A = l$$

Time differentiation yields

$$2v_B = v_A \qquad 2a_B = a_A$$

Here the signs are the same. Why?

© kosanperm/Fotolia.com

The motion of the lift on this crane depends upon the motion of the cable connected to the winch which operates it. It is important to be able to relate these motions in order to determine the power requirements of the winch and the force in the cable caused by any accelerated motion.

Procedure for Analysis

The above method of relating the dependent motion of one particle to that of another can be performed using algebraic scalars or position coordinates provided each particle moves along a rectilinear path. When this is the case, only the magnitudes of the velocity and acceleration of the particles will change, not their line of direction.

Position-Coordinate Equation.

- Establish each position coordinate with an origin located at a *fixed* point or datum.
- It is *not necessary* that the *origin* be the *same* for each of the coordinates; however, it is *important* that each coordinate axis selected be directed along the *path of motion* of the particle.
- Using geometry or trigonometry, relate the position coordinates to the total length of the cord, l_T, or to that portion of cord, l, which *excludes* the segments that do not change length as the particles move—such as arc segments wrapped over pulleys.
- If a problem involves a *system* of two or more cords wrapped around pulleys, then the position of a point on one cord must be related to the position of a point on another cord using the above procedure. Separate equations are written for a fixed length of each cord of the system and the positions of the two particles are then related by these equations (see Examples 12.22 and 12.23).

Time Derivatives.

- Two successive time derivatives of the position-coordinate equations yield the required velocity and acceleration equations which relate the motions of the particles.
- The signs of the terms in these equations will be consistent with those that specify the positive and negative sense of the position coordinates.

EXAMPLE 12.21

Determine the speed of block A in Fig. 12–38 if block B has an upward speed of 6 m/s.

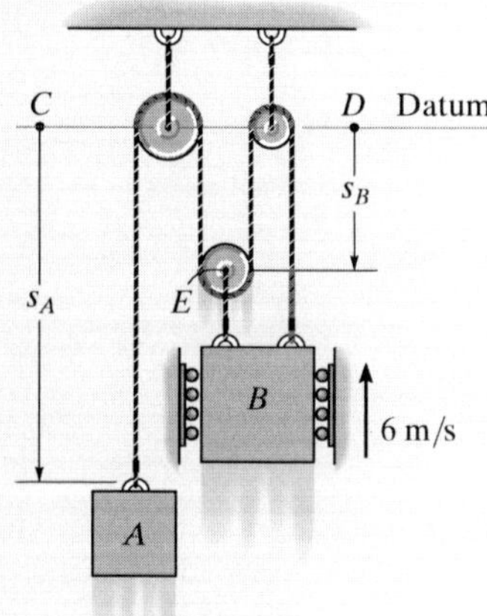

Fig. 12–38

SOLUTION

Position-Coordinate Equation. There is *one cord* in this system having segments which change length. Position coordinates s_A and s_B will be used since each is measured from a fixed point (C or D) and extends along each block's *path of motion*. In particular, s_B is directed to point E since motion of B and E is the *same*.

The red colored segments of the cord in Fig. 12–38 remain at a constant length and do not have to be considered as the blocks move. The remaining length of cord, l, is also constant and is related to the changing position coordinates s_A and s_B by the equation

$$s_A + 3s_B = l$$

Time Derivative. Taking the time derivative yields

$$v_A + 3v_B = 0$$

so that when $v_B = -6$ m/s (upward),

$$v_A = 18 \text{ m/s} \downarrow \qquad \textit{Ans.}$$

EXAMPLE 12.22

Determine the speed of A in Fig. 12–39 if B has an upward speed of 6 m/s.

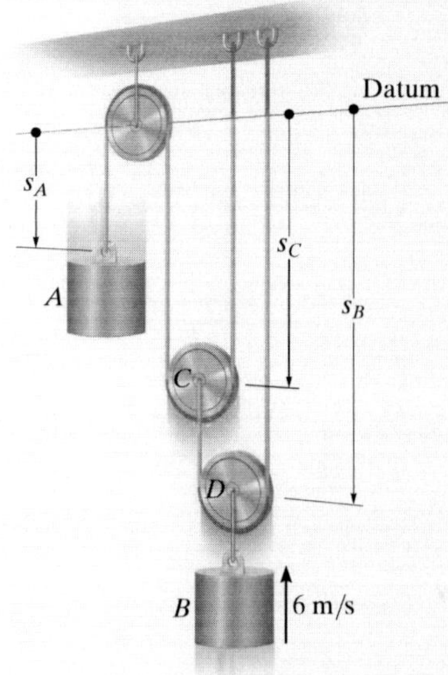

Fig. 12–39

SOLUTION

Position-Coordinate Equation. As shown, the positions of blocks A and B are defined using coordinates s_A and s_B. Since the system has *two cords* with segments that change length, it will be necessary to use a third coordinate, s_C, in order to relate s_A to s_B. In other words, the length of one of the cords can be expressed in terms of s_A and s_C, and the length of the other cord can be expressed in terms of s_B and s_C.

The red colored segments of the cords in Fig. 12–39 do not have to be considered in the analysis. Why? For the remaining cord lengths, say l_1 and l_2, we have

$$s_A + 2s_C = l_1 \qquad s_B + (s_B - s_C) = l_2$$

Time Derivative. Taking the time derivative of these equations yields

$$v_A + 2v_C = 0 \qquad 2v_B - v_C = 0$$

Eliminating v_C produces the relationship between the motions of each cylinder.

$$v_A + 4v_B = 0$$

so that when $v_B = -6$ m/s (upward),

$$v_A = +24 \text{ m/s} = 24 \text{ m/s} \downarrow \qquad \textit{Ans.}$$

EXAMPLE 12.23

Determine the speed of block B in Fig. 12–40 if the end of the cord at A is pulled down with a speed of 2 m/s.

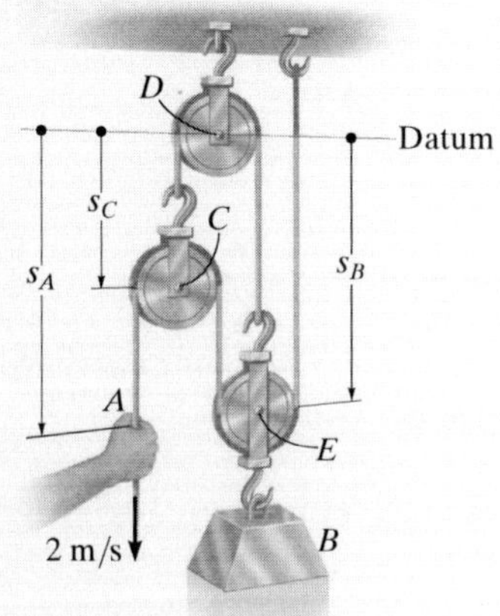

Fig. 12–40

SOLUTION

Position-Coordinate Equation. The position of point A is defined by s_A, and the position of block B is specified by s_B since point E on the pulley will have the *same motion* as the block. Both coordinates are measured from a horizontal datum passing through the *fixed* pin at pulley D. Since the system consists of *two* cords, the coordinates s_A and s_B cannot be related directly. Instead, by establishing a third position coordinate, s_C, we can now express the length of one of the cords in terms of s_B and s_C, and the length of the other cord in terms of s_A, s_B, and s_C.

Excluding the red colored segments of the cords in Fig. 12–40, the remaining constant cord lengths l_1 and l_2 (along with the hook and link dimensions) can be expressed as

$$s_C + s_B = l_1$$
$$(s_A - s_C) + (s_B - s_C) + s_B = l_2$$

Time Derivative. The time derivative of each equation gives

$$v_C + v_B = 0$$
$$v_A - 2v_C + 2v_B = 0$$

Eliminating v_C, we obtain

$$v_A + 4v_B = 0$$

so that when $v_A = 2$ m/s (downward),

$$v_B = -0.5 \text{ m/s} = 0.5 \text{ m/s} \uparrow \qquad \textit{Ans.}$$

EXAMPLE 12.24

A man at A is hoisting a safe S as shown in Fig. 12–41 by walking to the right with a constant velocity $v_A = 0.5\ \text{m/s}$. Determine the velocity and acceleration of the safe when it reaches the elevation of 10 m. The rope is 30 m long and passes over a small pulley at D.

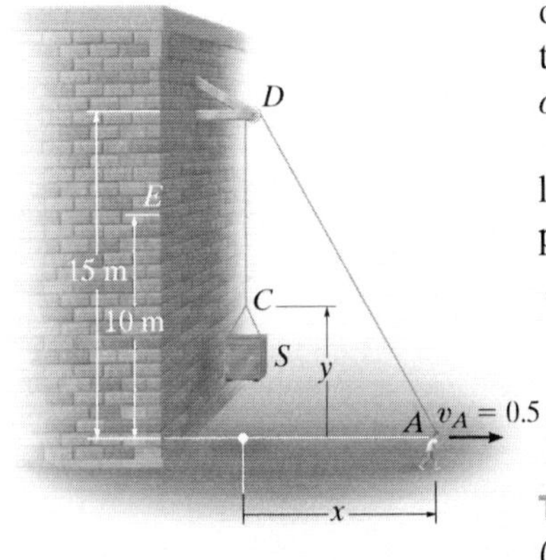

Fig. 12–41

SOLUTION

Position-Coordinate Equation. This problem is unlike the previous examples since rope segment DA changes *both direction and magnitude*. However, the ends of the rope, which define the positions of C and A, are specified by means of the x and y coordinates since they must be measured from a fixed point and *directed along the paths of motion* of the ends of the rope.

The x and y coordinates may be related since the rope has a fixed length $l = 30$ m, which at all times is equal to the length of segment DA plus CD. Using the Pythagorean theorem to determine l_{DA}, we have $l_{DA} = \sqrt{(15)^2 + x^2}$; also, $l_{CD} = 15 - y$. Hence,

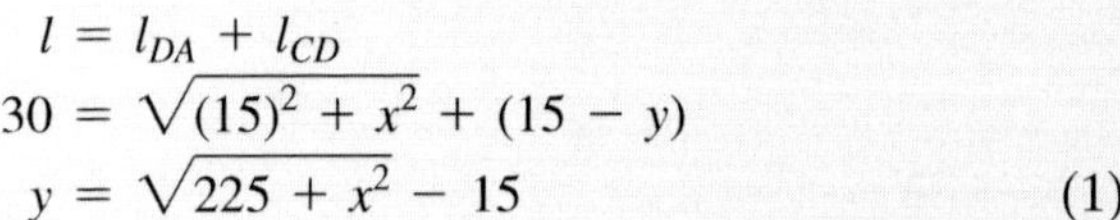

$$l = l_{DA} + l_{CD}$$

$$30 = \sqrt{(15)^2 + x^2} + (15 - y)$$

$$y = \sqrt{225 + x^2} - 15 \qquad (1)$$

Time Derivatives. Taking the time derivative, using the chain rule (see Appendix C), where $v_S = dy/dt$ and $v_A = dx/dt$, yields

$$v_S = \frac{dy}{dt} = \left[\frac{1}{2}\frac{2x}{\sqrt{225 + x^2}}\right]\frac{dx}{dt}$$

$$= \frac{x}{\sqrt{225 + x^2}}v_A \qquad (2)$$

At $y = 10$ m, x is determined from Eq. 1, i.e., $x = 20$ m. Hence, from Eq. 2 with $v_A = 0.5\ \text{m/s}$,

$$v_S = \frac{20}{\sqrt{225 + (20)^2}}(0.5) = 0.4\ \text{m/s} = 400\ \text{mm/s} \uparrow \qquad \textit{Ans.}$$

The acceleration is determined by taking the time derivative of Eq. 2. Since v_A is constant, then $a_A = dv_A/dt = 0$, and we have

$$a_S = \frac{d^2y}{dt^2} = \left[\frac{-x(dx/dt)}{(225 + x^2)^{3/2}}\right]xv_A + \left[\frac{1}{\sqrt{225 + x^2}}\right]\left(\frac{dx}{dt}\right)v_A + \left[\frac{1}{\sqrt{225 + x^2}}\right]x\frac{dv_A}{dt} = \frac{225v_A^2}{(225 + x^2)^{3/2}}$$

At $x = 20$ m, with $v_A = 0.5\ \text{m/s}$, the acceleration becomes

$$a_S = \frac{225(0.5\ \text{m/s})^2}{[225 + (20\ \text{m})^2]^{3/2}} = 0.00360\ \text{m/s}^2 = 3.60\ \text{mm/s}^2 \uparrow \qquad \textit{Ans.}$$

NOTE: The constant velocity at A causes the other end C of the rope to have an acceleration since $\mathbf{v}_A$ causes segment DA to change its direction as well as its length.

12.10 Relative-Motion of Two Particles Using Translating Axes

Throughout this chapter the absolute motion of a particle has been determined using a single fixed reference frame. There are many cases, however, where the path of motion for a particle is complicated, so that it may be easier to analyze the motion in parts by using two or more frames of reference. For example, the motion of a particle located at the tip of an airplane propeller, while the plane is in flight, is more easily described if one observes first the motion of the airplane from a fixed reference and then superimposes (vectorially) the circular motion of the particle measured from a reference attached to the airplane.

In this section *translating frames of reference* will be considered for the analysis.

Position. Consider particles A and B, which move along the arbitrary paths shown in Fig. 12–42. The *absolute position* of each particle, $\mathbf{r}_A$ and $\mathbf{r}_B$, is measured from the common origin O of the *fixed* x, y, z reference frame. The origin of a second frame of reference x', y', z' is attached to and moves with particle A. The axes of this frame are *only permitted to translate* relative to the fixed frame. The position of B measured relative to A is denoted by the *relative-position vector* $\mathbf{r}_{B/A}$. Using vector addition, the three vectors shown in Fig. 12–42 can be related by the equation

$$\mathbf{r}_B = \mathbf{r}_A + \mathbf{r}_{B/A} \tag{12–33}$$

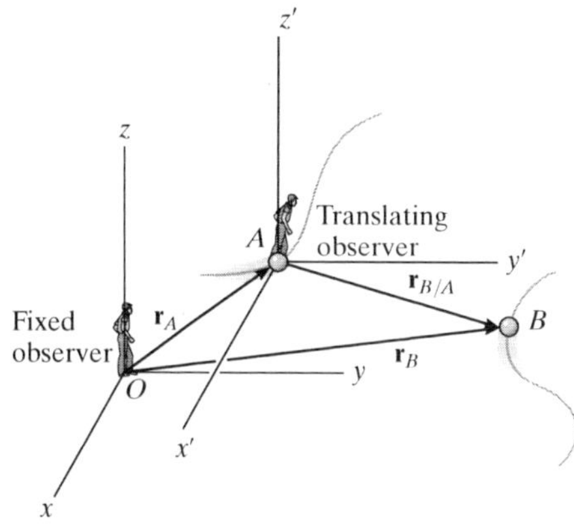

Fig. 12–42

Velocity. An equation that relates the velocities of the particles is determined by taking the time derivative of the above equation; i.e.,

$$\mathbf{v}_B = \mathbf{v}_A + \mathbf{v}_{B/A} \tag{12–34}$$

Here $\mathbf{v}_B = d\mathbf{r}_B/dt$ and $\mathbf{v}_A = d\mathbf{r}_A/dt$ refer to *absolute velocities*, since they are observed from the fixed frame; whereas the *relative velocity* $\mathbf{v}_{B/A} = d\mathbf{r}_{B/A}/dt$ is observed from the translating frame. It is important to note that since the x', y', z' axes translate, the *components* of $\mathbf{r}_{B/A}$ will *not* change direction and therefore the time derivative of these components will only have to account for the change in their magnitudes. Equation 12–34 therefore states that the velocity of B is equal to the velocity of A plus (vectorially) the velocity of "B with respect to A," as measured by the *translating observer* fixed in the x', y', z' reference frame.

12

Acceleration. The time derivative of Eq. 12–34 yields a similar vector relation between the *absolute* and *relative accelerations* of particles A and B.

$$\mathbf{a}_B = \mathbf{a}_A + \mathbf{a}_{B/A} \qquad (12\text{–}35)$$

Here $\mathbf{a}_{B/A}$ is the acceleration of B as seen by the observer located at A and translating with the x', y', z' reference frame.*

Procedure for Analysis

- When applying the relative velocity and acceleration equations, it is first necessary to specify the particle A that is the origin for the translating x', y', z' axes. Usually this point has a *known* velocity or acceleration.
- Since vector addition forms a triangle, there can be at most *two unknowns*, represented by the magnitudes and/or directions of the vector quantities.
- These unknowns can be solved for either graphically, using trigonometry (law of sines, law of cosines), or by resolving each of the three vectors into rectangular or Cartesian components, thereby generating a set of scalar equations.

© Nicholas Rjabow/Shutterstock.com

The pilots of these bi-planes flying close to one another must be aware of their relative positions and velocities at all times in order to avoid a collision.

* An easy way to remember the setup of these equations is to note the "cancellation" of the subscript A between the two terms, e.g., $\mathbf{a}_B = \mathbf{a}_{A} + \mathbf{a}_{B/A}$.

EXAMPLE 12.25

A train travels at a constant speed of 60 km/h and crosses over a road as shown in Fig. 12–43*a*. If the automobile *A* is traveling at 45 km/h along the road, determine the magnitude and direction of the velocity of the train relative to the automobile.

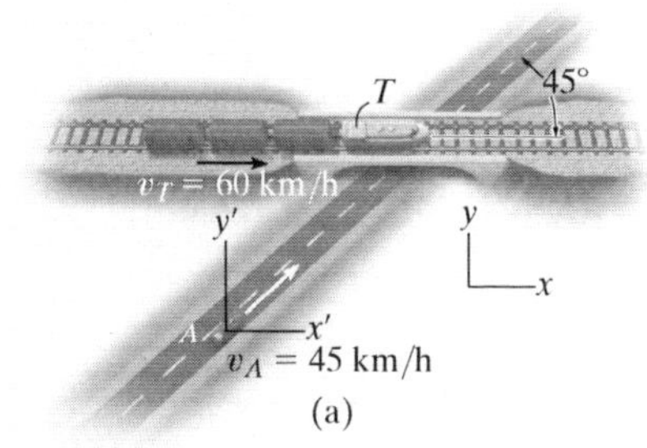

SOLUTION I

Vector Analysis. The relative velocity $\mathbf{v}_{T/A}$ is measured from the translating x', y' axes attached to the automobile, Fig. 12–43*a*. It is determined from $\mathbf{v}_T = \mathbf{v}_A + \mathbf{v}_{T/A}$. Since $\mathbf{v}_T$ and $\mathbf{v}_A$ are known in *both* magnitude and direction, the unknowns become the *x* and *y* components of $\mathbf{v}_{T/A}$. Using the *x*, *y* axes in Fig. 12–43*a*, we have

$$\mathbf{v}_T = \mathbf{v}_A + \mathbf{v}_{T/A}$$

$$60\mathbf{i} = (45 \cos 45°\mathbf{i} + 45 \sin 45°\mathbf{j}) + \mathbf{v}_{T/A}$$

$$\mathbf{v}_{T/A} = \{28.2\mathbf{i} - 31.8\mathbf{j}\} \text{ km/h}$$

The magnitude of $\mathbf{v}_{T/A}$ is thus

$$v_{T/A} = \sqrt{(28.2)^2 + (-31.8)^2} = 42.5 \text{ km/h} \qquad \textit{Ans.}$$

From the direction of each component, Fig. 12–43*b*, the direction of $\mathbf{v}_{T/A}$ is

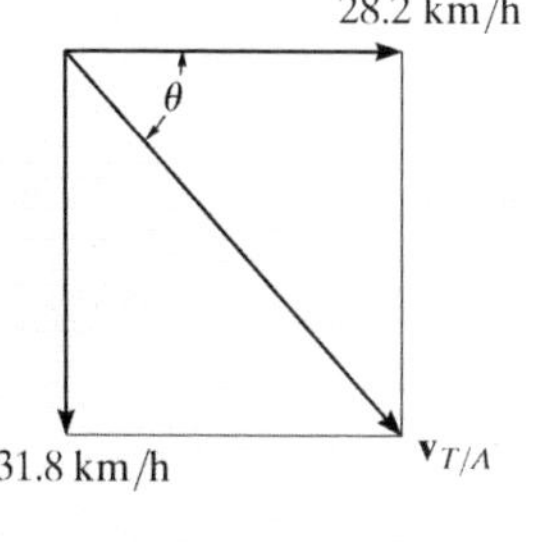

$$\tan\theta = \frac{(v_{T/A})_y}{(v_{T/A})_x} = \frac{31.8}{28.2}$$

$$\theta = 48.5° \;\circlearrowright \qquad \textit{Ans.}$$

Note that the vector addition shown in Fig. 12–43*b* indicates the correct sense for $\mathbf{v}_{T/A}$. This figure anticipates the answer and can be used to check it.

SOLUTION II

Scalar Analysis. The unknown components of $\mathbf{v}_{T/A}$ can also be determined by applying a scalar analysis. We will assume these components act in the *positive x* and *y* directions. Thus,

$$\mathbf{v}_T = \mathbf{v}_A + \mathbf{v}_{T/A}$$

$$\begin{bmatrix} 60 \text{ km/h} \\ \rightarrow \end{bmatrix} = \begin{bmatrix} 45 \text{ km/h} \\ \measuredangle^{45°} \end{bmatrix} + \begin{bmatrix} (v_{T/A})_x \\ \rightarrow \end{bmatrix} + \begin{bmatrix} (v_{T/A})_y \\ \uparrow \end{bmatrix}$$

Resolving each vector into its *x* and *y* components yields

$$(\overset{+}{\rightarrow}) \qquad 60 = 45 \cos 45° + (v_{T/A})_x + 0$$

$$(+\uparrow) \qquad 0 = 45 \sin 45° + 0 + (v_{T/A})_y$$

Solving, we obtain the previous results,

$$(v_{T/A})_x = 28.2 \text{ km/h} = 28.2 \text{ km/h} \rightarrow$$

$$(v_{T/A})_y = -31.8 \text{ km/h} = 31.8 \text{ km/h} \downarrow$$

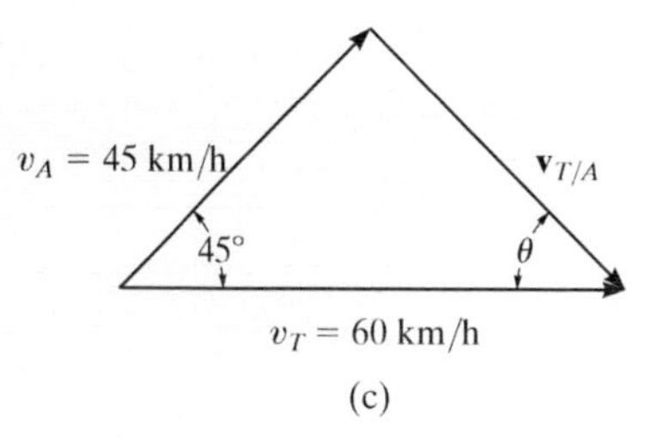

Fig. 12–43

EXAMPLE 12.26

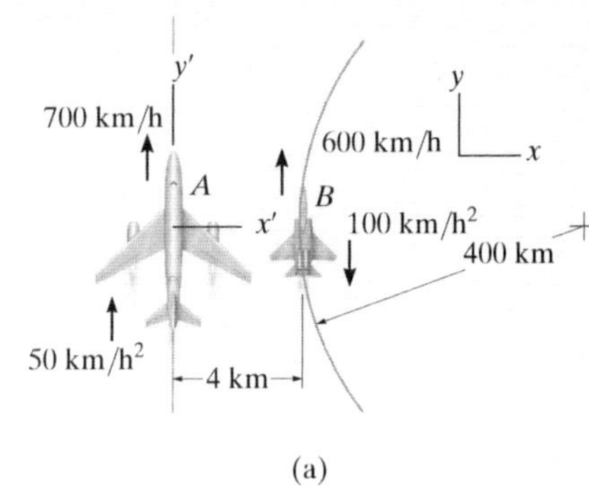

(a)

Plane A in Fig. 12–44*a* is flying along a straight-line path, whereas plane B is flying along a circular path having a radius of curvature of $\rho_B = 400$ km. Determine the velocity and acceleration of B as measured by the pilot of A.

SOLUTION

Velocity. The origin of the x *and* y axes are located at an arbitrary fixed point. Since the motion relative to plane A is to be determined, the *translating frame of reference* x', y' is attached to it, Fig. 12–44*a*. Applying the relative-velocity equation in scalar form since the velocity vectors of both planes are parallel at the instant shown, we have

$\mathbf{v}_{B/A}$

$v_A = 700$ km/h $\quad v_B = 600$ km/h

(b)

$$(+\uparrow) \qquad v_B = v_A + v_{B/A}$$

$$600 \text{ km/h} = 700 \text{ km/h} + v_{B/A}$$

$$v_{B/A} = -100 \text{ km/h} = 100 \text{ km/h} \downarrow \qquad \textit{Ans.}$$

The vector addition is shown in Fig. 12–44*b*.

Acceleration. Plane B has both tangential and normal components of acceleration since it is flying along a *curved path*. From Eq. 12–20, the magnitude of the normal component is

$$(a_B)_n = \frac{v_B^2}{\rho} = \frac{(600 \text{ km/h})^2}{400 \text{ km}} = 900 \text{ km/h}^2$$

Applying the relative-acceleration equation gives

$$\mathbf{a}_B = \mathbf{a}_A + \mathbf{a}_{B/A}$$

$$900\mathbf{i} - 100\mathbf{j} = 50\mathbf{j} + \mathbf{a}_{B/A}$$

Thus,

$$\mathbf{a}_{B/A} = \{900\mathbf{i} - 150\mathbf{j}\} \text{ km/h}^2$$

From Fig. 12–44*c*, the magnitude and direction of $\mathbf{a}_{B/A}$ are therefore

$$a_{B/A} = 912 \text{ km/h}^2 \quad \theta = \tan^{-1}\frac{150}{900} = 9.46^\circ \; \measuredangle \qquad \textit{Ans.}$$

NOTE: The solution to this problem was possible using a translating frame of reference, since the pilot in plane A is "translating." Observation of the motion of plane A with respect to the pilot of plane B, however, must be obtained using a *rotating* set of axes attached to plane B. (This assumes, of course, that the pilot of B is fixed in the rotating frame, so he does not turn his eyes to follow the motion of A.) The analysis for this case is given in Example 16.21.

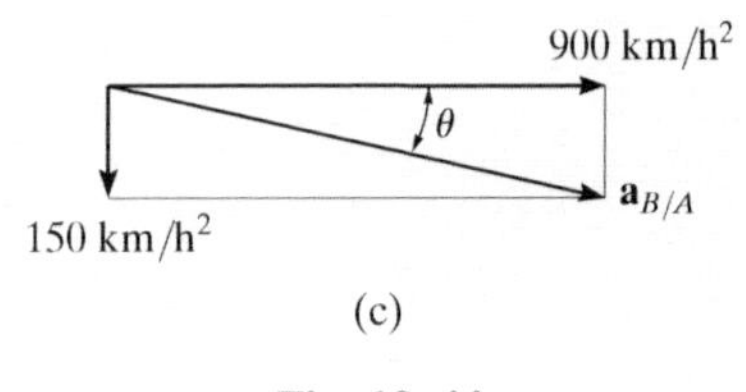

(c)

Fig. 12–44

12

EXAMPLE 12.27

At the instant shown in Fig. 12–45*a*, cars *A* and *B* are traveling with speeds of 18 m/s and 12 m/s, respectively. Also at this instant, *A* has a decrease in speed of 2 m/s², and *B* has an increase in speed of 3 m/s². Determine the velocity and acceleration of *B* with respect to *A*.

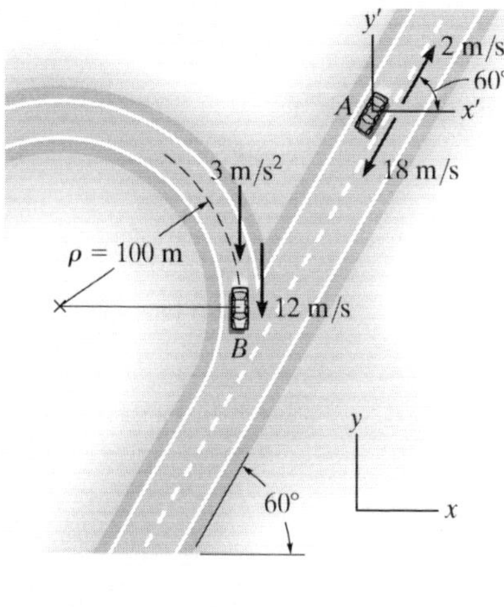

(a)

SOLUTION

Velocity. The fixed *x*, *y* axes are established at an arbitrary point on the ground and the translating *x'*, *y'* axes are attached to car *A*, Fig. 12–45*a*. Why? The relative velocity is determined from $\mathbf{v}_B = \mathbf{v}_A + \mathbf{v}_{B/A}$. What are the two unknowns? Using a Cartesian vector analysis, we have

$$\mathbf{v}_B = \mathbf{v}_A + \mathbf{v}_{B/A}$$

$$-12\mathbf{j} = (-18\cos 60°\mathbf{i} - 18\sin 60°\mathbf{j}) + \mathbf{v}_{B/A}$$

$$\mathbf{v}_{B/A} = \{9\mathbf{i} + 3.588\mathbf{j}\}\ \text{m/s}$$

Thus,

$$v_{B/A} = \sqrt{(9)^2 + (3.588)^2} = 9.69\ \text{m/s} \qquad \textit{Ans.}$$

Noting that $\mathbf{v}_{B/A}$ has $+\mathbf{i}$ and $+\mathbf{j}$ components, Fig. 12–45*b*, its direction is

$$\tan\theta = \frac{(v_{B/A})_y}{(v_{B/A})_x} = \frac{3.588}{9}$$

$$\theta = 21.7° \ \measuredangle \qquad \textit{Ans.}$$

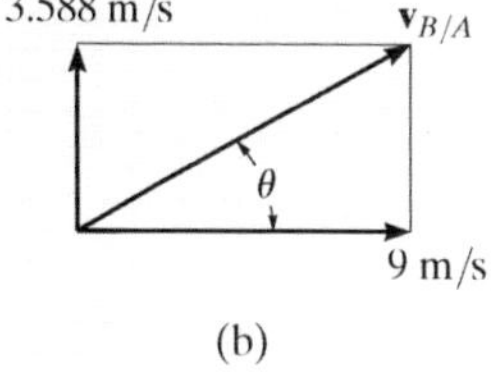

(b)

Acceleration. Car *B* has both tangential and normal components of acceleration. Why? The magnitude of the normal component is

$$(a_B)_n = \frac{v_B^2}{\rho} = \frac{(12\ \text{m/s})^2}{100\ \text{m}} = 1.440\ \text{m/s}^2$$

Applying the equation for relative acceleration yields

$$\mathbf{a}_B = \mathbf{a}_A + \mathbf{a}_{B/A}$$

$$(-1.440\mathbf{i} - 3\mathbf{j}) = (2\cos 60°\mathbf{i} + 2\sin 60°\mathbf{j}) + \mathbf{a}_{B/A}$$

$$\mathbf{a}_{B/A} = \{-2.440\mathbf{i} - 4.732\mathbf{j}\}\ \text{m/s}^2$$

Here $\mathbf{a}_{B/A}$ has $-\mathbf{i}$ and $-\mathbf{j}$ components. Thus, from Fig. 12–45*c*,

$$a_{B/A} = \sqrt{(2.440)^2 + (4.732)^2} = 5.32\ \text{m/s}^2 \qquad \textit{Ans.}$$

$$\tan\phi = \frac{(a_{B/A})_y}{(a_{B/A})_x} = \frac{4.732}{2.440}$$

$$\phi = 62.7° \ \measuredangle \qquad \textit{Ans.}$$

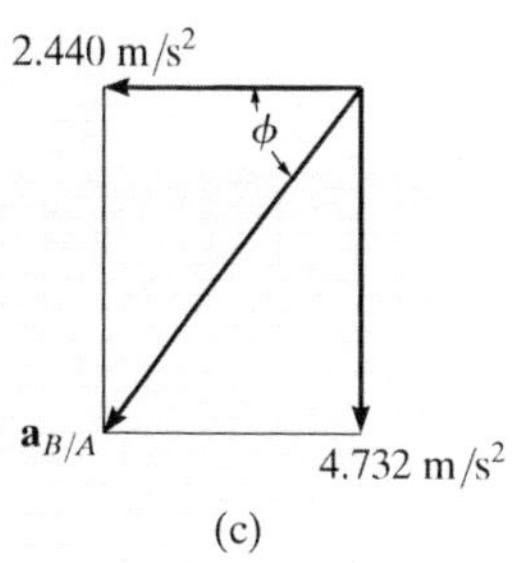

(c)

Fig. 12–45

NOTE: Is it possible to obtain the relative acceleration of $\mathbf{a}_{A/B}$ using this method? Refer to the comment made at the end of Example 12.26.

FUNDAMENTAL PROBLEMS

F12–39. Determine the velocity of block D if end A of the rope is pulled down with a speed of $v_A = 3$ m/s.

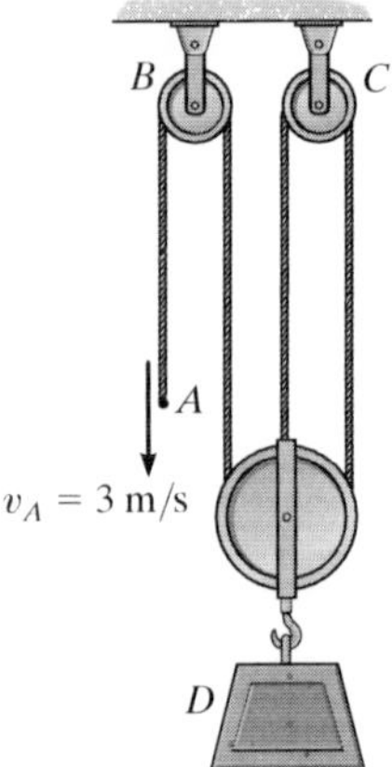

F12–39

F12–40. Determine the velocity of block A if end B of the rope is pulled down with a speed of 6 m/s.

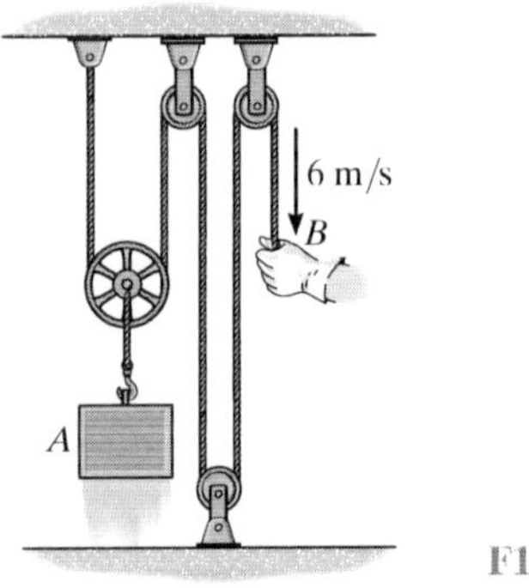

F12–40

F12–41. Determine the velocity of block A if end B of the rope is pulled down with a speed of 1.5 m/s.

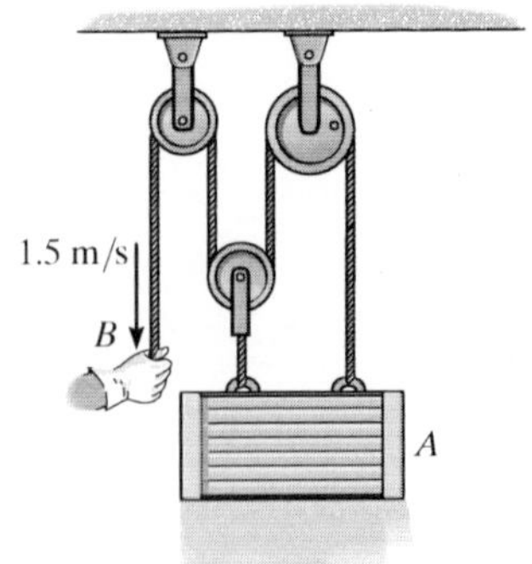

F12–41

F12–42. Determine the velocity of block A if end F of the rope is pulled down with a speed of $v_F = 3$ m/s.

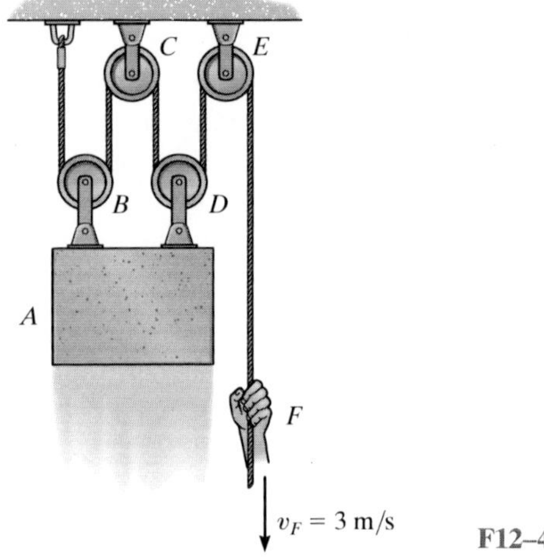

F12–42

F12–43. Determine the velocity of car A if point P on the cable has a speed of 4 m/s when the motor M winds the cable in.

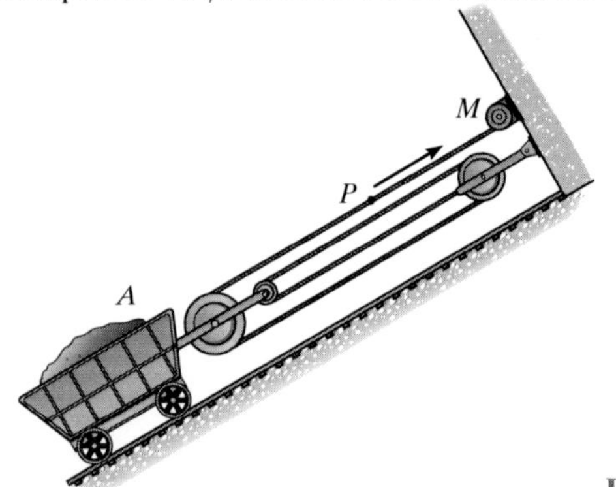

F12–43

F12–44. Determine the velocity of cylinder B if cylinder A moves downward with a speed of $v_A = 4$ m/s.

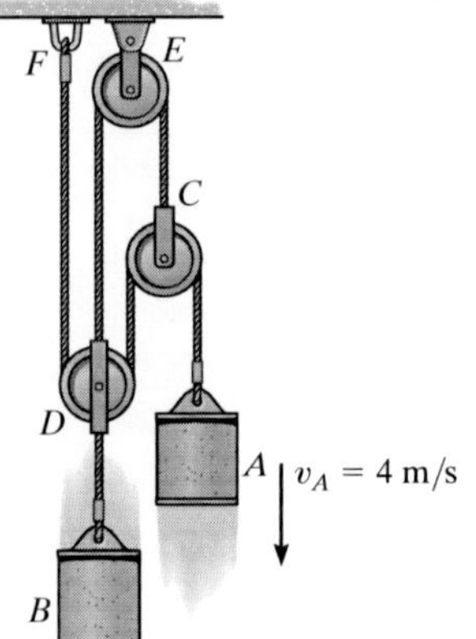

F12–44

F12–45. Car A is traveling with a constant speed of 80 km/h due north, while car B is traveling with a constant speed of 100 km/h due east. Determine the velocity of car B relative to car A.

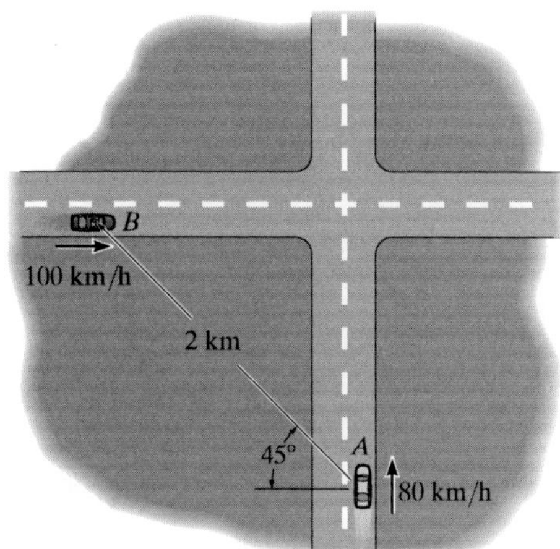

F12–45

F12–46. Two planes A and B are traveling with the constant velocities shown. Determine the magnitude and direction of the velocity of plane B relative to plane A.

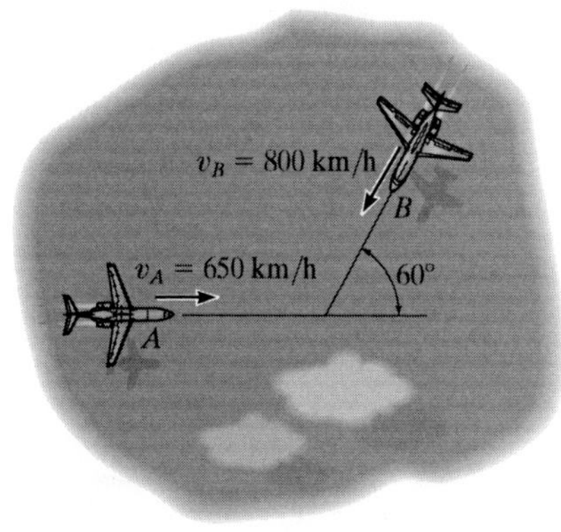

F12–46

F12–47. The boats A and B travel with constant speeds of $v_A = 15$ m/s and $v_B = 10$ m/s when they leave the pier at O at the same time. Determine the distance between them when $t = 4$ s.

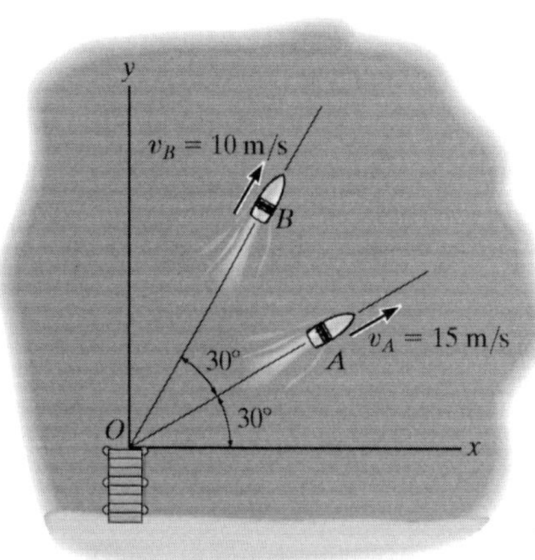

F12–47

F12–48. At the instant shown, cars A and B are traveling at the speeds shown. If B is accelerating at 1200 km/h^2 while A maintains a constant speed, determine the velocity and acceleration of A with respect to B.

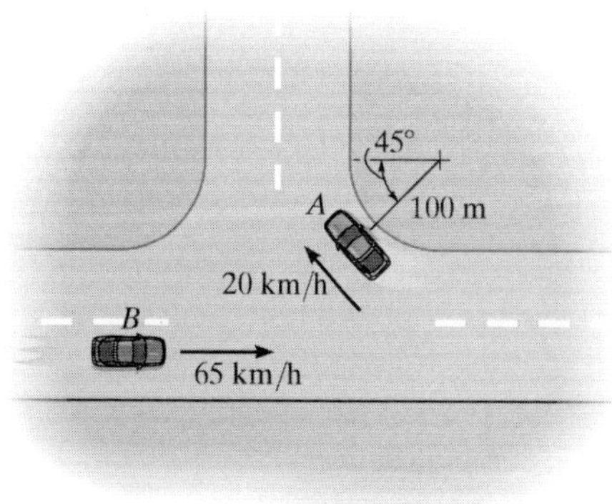

F12–48

PROBLEMS

12–199. If the end of the cable at A is pulled down with a speed of 2 m/s, determine the speed at which block B rises.

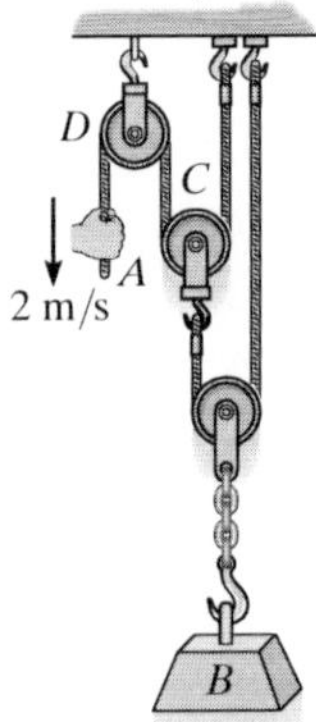

Prob. 12–199

*__12–200.__ The motor at C pulls in the cable with an acceleration $a_C = (3t^2)\ \text{m/s}^2$, where t is in seconds. The motor at D draws in its cable at $a_D = 5\ \text{m/s}^2$. If both motors start at the same instant from rest when $d = 3$ m, determine (a) the time needed from $d = 0$, and (b) the relative velocity of block A with respect to block B when this occurs.

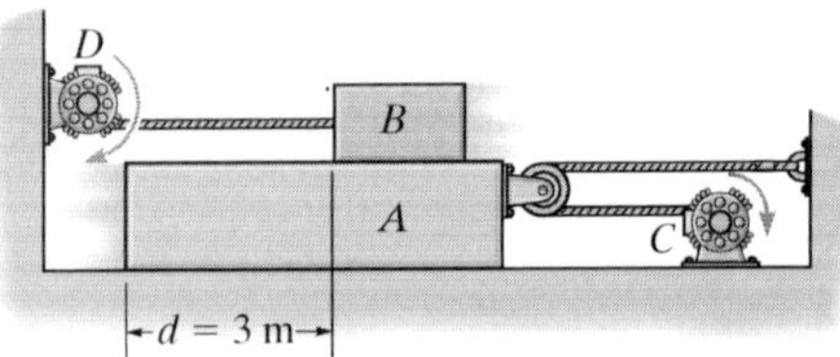

Prob. 12–200

12–201. If the end of the cable at A is pulled down with a speed of 2 m/s, determine the speed at which block B rises.

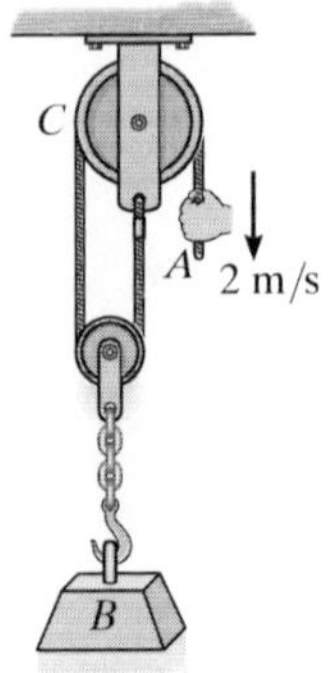

Prob. 12–201

12–202. Determine the time needed for the load at B to attain a speed of 8 m/s, starting from rest, if the cable is drawn into the motor with an acceleration of 0.2 m/s^2.

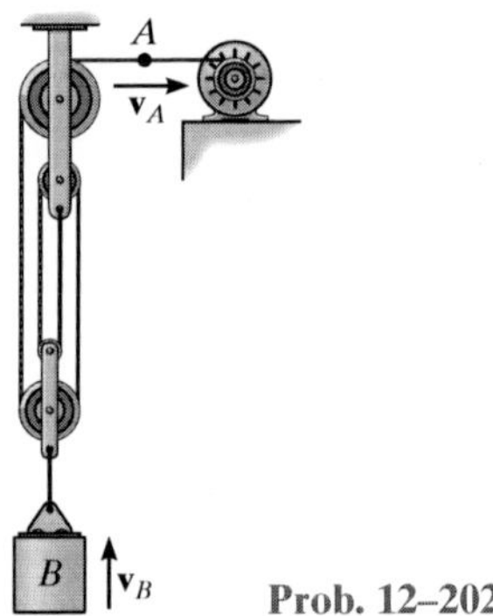

Prob. 12–202

12–203. Determine the displacement of the log if the truck at C pulls the cable 1.2 m to the right.

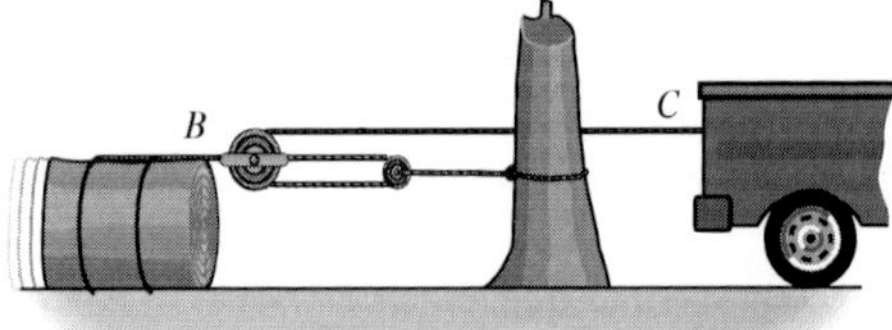

Prob. 12–203

*__12–204.__ Determine the speed of cylinder A, if the rope is drawn toward the motor M at a constant rate of 10 m/s.

12–205. If the rope is drawn toward the motor M at a speed of $v_M = (5t^{3/2})$ m/s, where t is in seconds, determine the speed of cylinder A when $t = 1$ s.

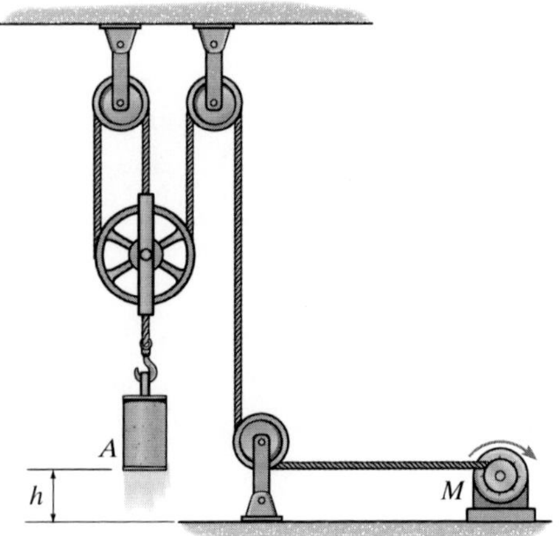

Probs. 12–204/205

12–206. If the hydraulic cylinder H draws in rod BC at 1 m/s, determine the speed of slider A.

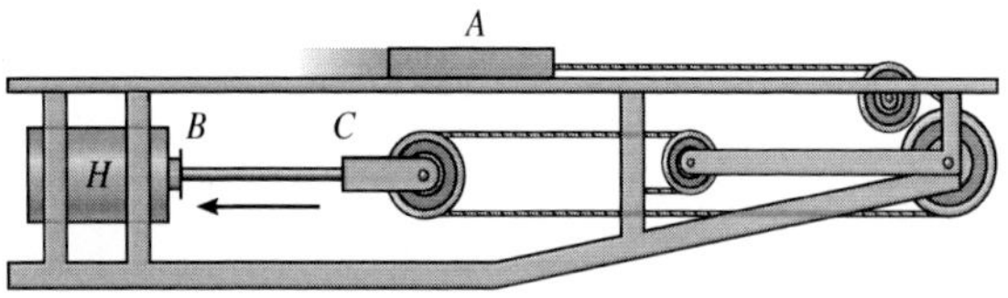

Prob. 12–206

12–207. If block A is moving downward with a speed of 2 m/s while C is moving up at 1 m/s, determine the speed of block B.

***12–208.** If block A is moving downward at 2 m/s while block C is moving down at 6 m/s, determine the speed of block B.

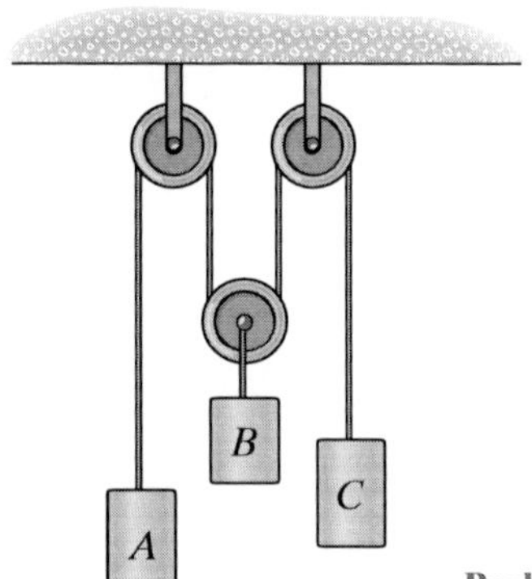

Probs. 12–207/208

12–209. If the end A of the cable is moving upwards at $v_A = 14$ m/s, determine the speed of block B.

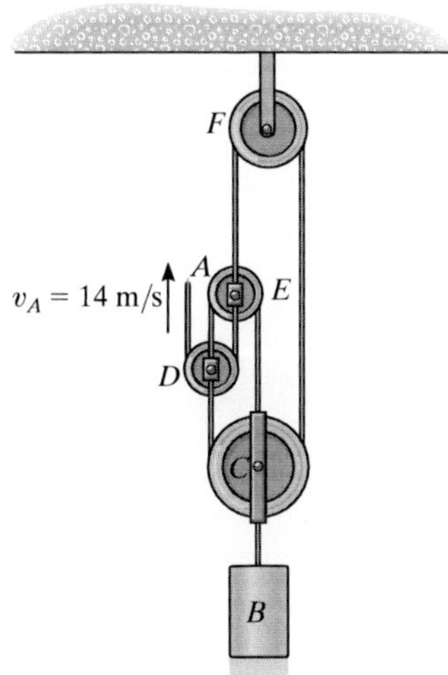

Prob. 12–209

12–210. The motor draws in the cable at C with a constant velocity of $v_C = 4$ m/s. The motor draws in the cable at D with a constant acceleration of $a_D = 8$ m/s^2. If $v_D = 0$ when $t = 0$, determine (a) the time needed for block A to rise 3 m, and (b) the relative velocity of block A with respect to block B when this occurs.

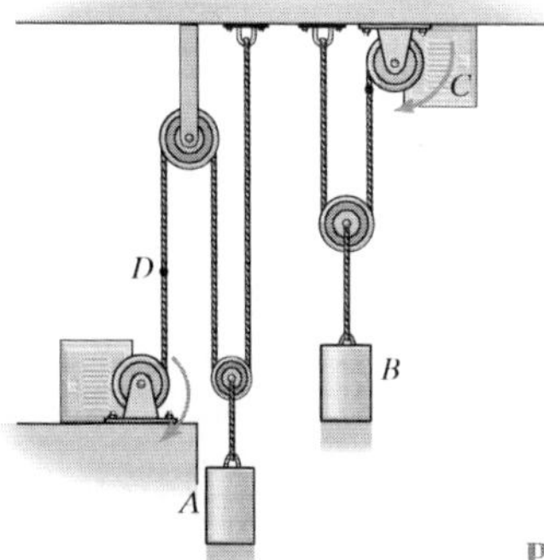

Prob. 12–210

12–211. Determine the speed of block A if the end of the rope is pulled down with a speed of 4 m/s.

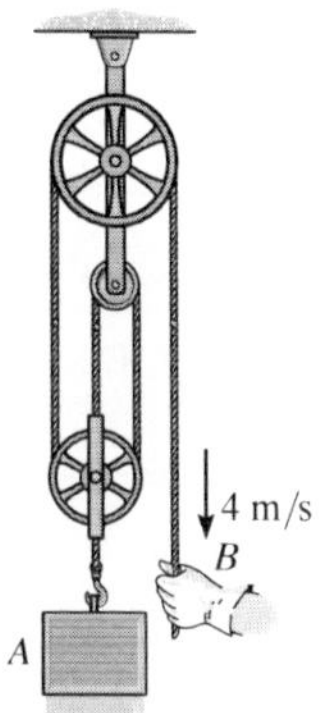

Prob. 12–211

***12–212.** The cylinder C is being lifted using the cable and pulley system shown. If point A on the cable is being drawn toward the drum with a speed of 2 m/s, determine the velocity of the cylinder.

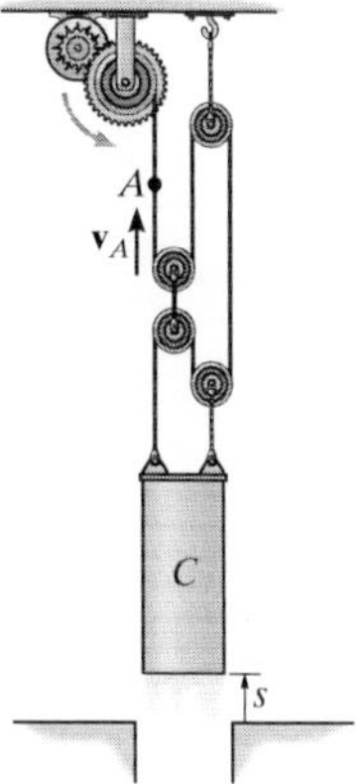

Prob. 12–212

12–213. The man pulls the boy up to the tree limb C by walking backward at a constant speed of 1.5 m/s. Determine the speed at which the boy is being lifted at the instant $x_A = 4$ m. Neglect the size of the limb. When $x_A = 0$, $y_B = 8$ m, so that A and B are coincident, i.e., the rope is 16 m long.

12–214. The man pulls the boy up to the tree limb C by walking backward. If he starts from rest when $x_A = 0$ and moves backward with a constant acceleration $a_A = 0.2$ m/s^2, determine the speed of the boy at the instant $y_B = 4$ m. Neglect the size of the limb. When $x_A = 0$, $y_B = 8$ m, so that A and B are coincident, i.e., the rope is 16 m long.

Probs. 12–213/214

12–215. A man can row a boat at 5 m/s in still water. He wishes to cross a 50-m-wide river to point B, 50 m downstream. If the river flows with a velocity of 2 m/s, determine the speed of the boat and the time needed to make the crossing.

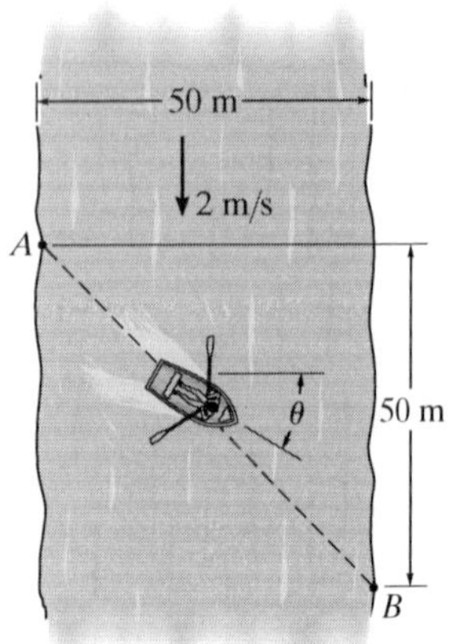

Prob. 12–215

***12–216.** The car is traveling at a constant speed of 100 km/h. If the rain is falling at 6 m/s in the direction shown, determine the velocity of the rain as seen by the driver.

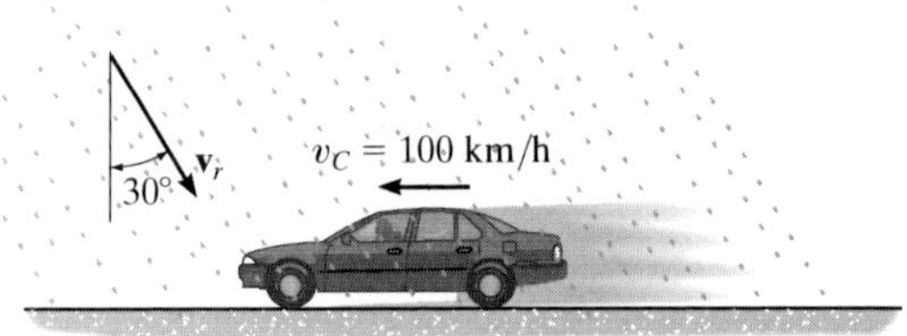

Prob. 12–216

12–217. The crate C is being lifted by moving the roller at A downward with a constant speed of $v_A = 2$ m/s along the guide. Determine the velocity and acceleration of the crate at the instant $s = 1$ m. When the roller is at B, the crate rests on the ground. Neglect the size of the pulley in the calculation. *Hint:* Relate the coordinates x_C and x_A using the problem geometry, then take the first and second time derivatives.

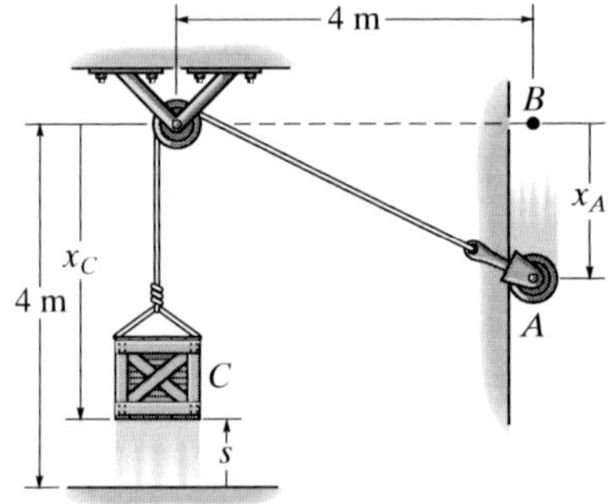

Prob. 12–217

12–218. The man can row the boat in still water with a speed of 5 m/s. If the river is flowing at 2 m/s, determine the speed of the boat and the angle θ he must direct the boat so that it travels from A to B.

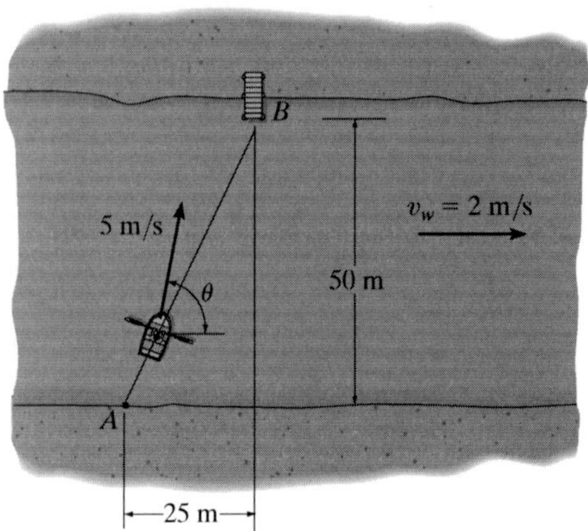

Prob. 12–218

12–219. A man walks at 5 km/h in the direction of a 20-km/h wind. If raindrops fall vertically at 7 km/h in *still air*, determine the direction in which the drops appear to fall with respect to the man. Assume the horizontal speed of the raindrops is equal to that of the wind.

Prob. 12–219

***12–220.** If block B is moving down with a velocity v_B and has an acceleration a_B, determine the velocity and acceleration of block A in terms of the parameters shown.

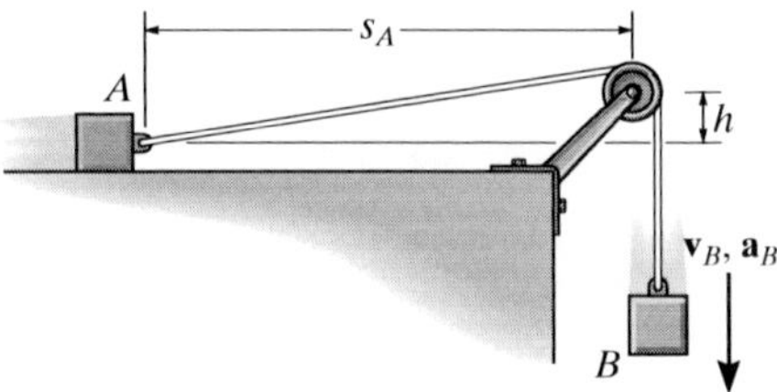

Prob. 12–220

12–221. At a given instant, two particles A and B are moving with a speed of 8 m/s along the paths shown. If B is decelerating at 6 m/s^2 and the speed of A is increasing at 5 m/s^2, determine the acceleration of A with respect to B at this instant.

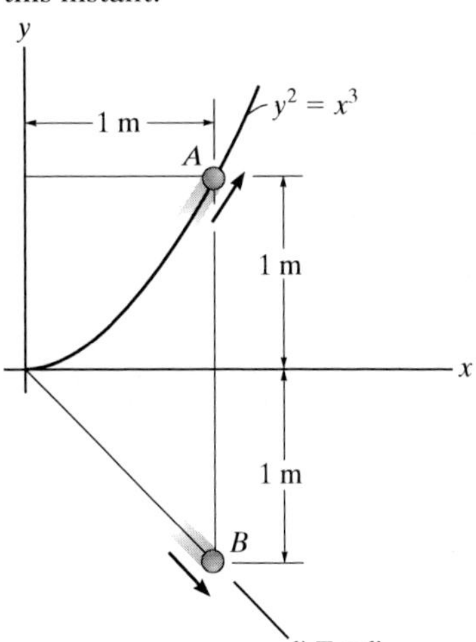

Prob. 12–221

12–222. Two planes, A and B, are flying at the same altitude. If their velocities are $v_A = 600$ km/h and $v_B = 500$ km/h such that the angle between their straight-line courses is $\theta = 75°$, determine the velocity of plane B with respect to plane A.

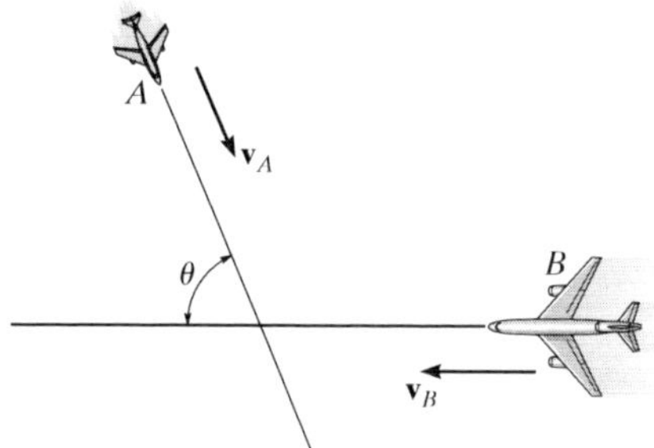

Prob. 12–222

12–223. If the truck travels at a constant speed of $v_T = 1.8$ m/s, determine the speed of the crate for any angle θ of the rope. The rope has a length of 30 m and passes over a pulley of negligible size at A. *Hint:* Relate the coordinates x_T and x_C to the length of the rope and take the time derivative. Then substitute the trigonometric relation between x_C and θ.

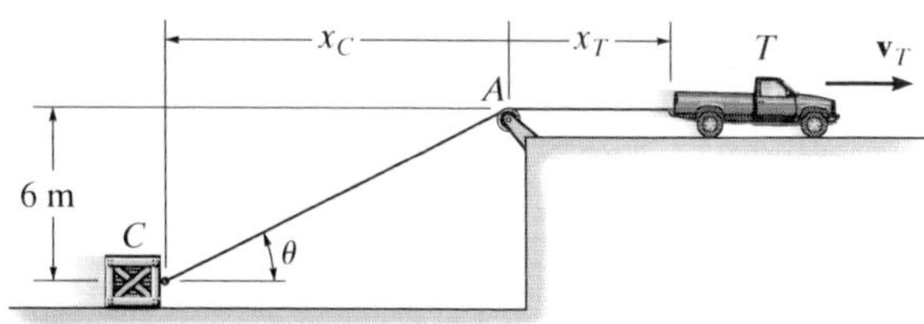

Prob. 12–223

***12–224.** At the instant shown, car A travels along the straight portion of the road with a speed of 25 m/s. At this same instant car B travels along the circular portion of the road with a speed of 15 m/s. Determine the velocity of car B relative to car A.

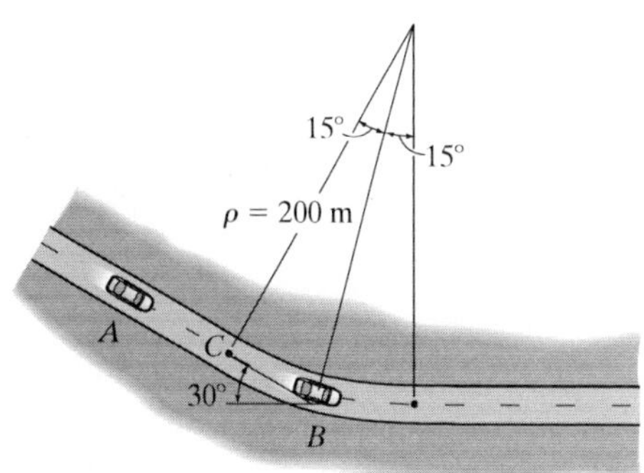

Prob. 12–224

12

12–225. An aircraft carrier is traveling forward with a velocity of 50 km/h. At the instant shown, the plane at A has just taken off and has attained a forward horizontal air speed of 200 km/h, measured from still water. If the plane at B is traveling along the runway of the carrier at 175 km/h in the direction shown, determine the velocity of A with respect to B.

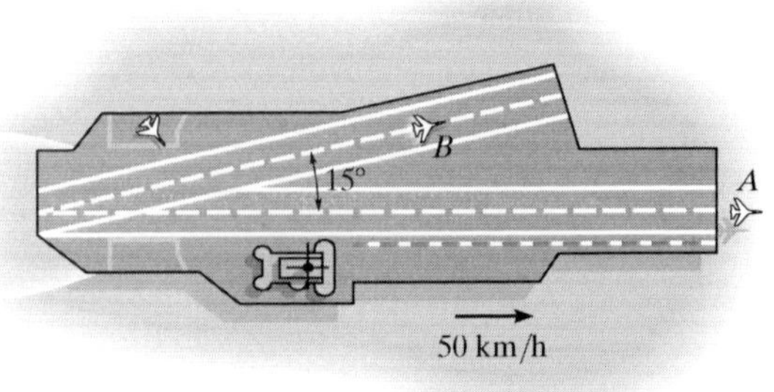

Prob. 12–225

12–226. A car is traveling north along a straight road at 50 km/h. An instrument in the car indicates that the wind is coming from the east. If the car's speed is 80 km/h, the instrument indicates that the wind is coming from the north-east. Determine the speed and direction of the wind.

12–227. The motor at C pulls in the cable with an acceleration $a_C = (3t^2)\ \text{m/s}^2$, where t is in seconds. The motor at D draws in its cable at $a_D = 5\ \text{m/s}^2$. If both motors start at the same instant from rest when $d = 3$ m, determine (a) the time needed for $d = 0$, and (b) the velocities of blocks A and B when this occurs.

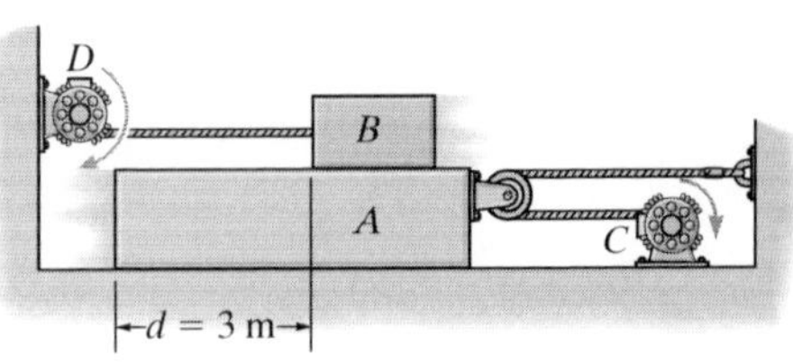

Prob. 12–227

***12–228.** At the instant shown, the bicyclist at A is traveling at 7 m/s around the curve on the race track while increasing his speed at $0.5\ \text{m/s}^2$. The bicyclist at B is traveling at 8.5 m/s along the straight-a-way and increasing his speed at $0.7\ \text{m/s}^2$. Determine the relative velocity and relative acceleration of A with respect to B at this instant.

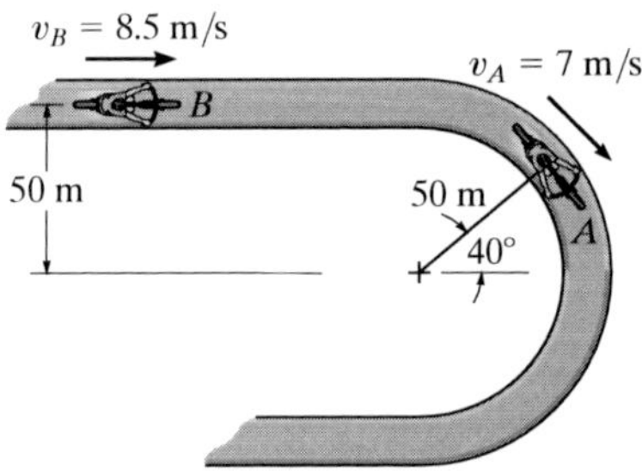

Prob. 12–228

12–229. At the instant shown car A is traveling with a velocity of 30 m/s and has an acceleration of $2\ \text{m/s}^2$ along the highway. At the same instant B is traveling on the trumpet interchange curve with a speed of 15 m/s, which is decreasing at $0.8\ \text{m/s}^2$. Determine the relative velocity and relative acceleration of B with respect to A at this instant.

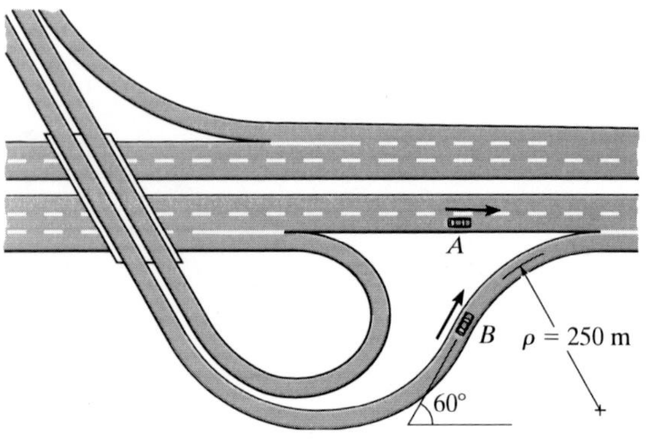

Prob. 12–229

12–230. The two cyclists A and B travel at the same constant speed v. Determine the speed of A with respect to B if A travels along the circular track, while B travels along the diameter of the circle.

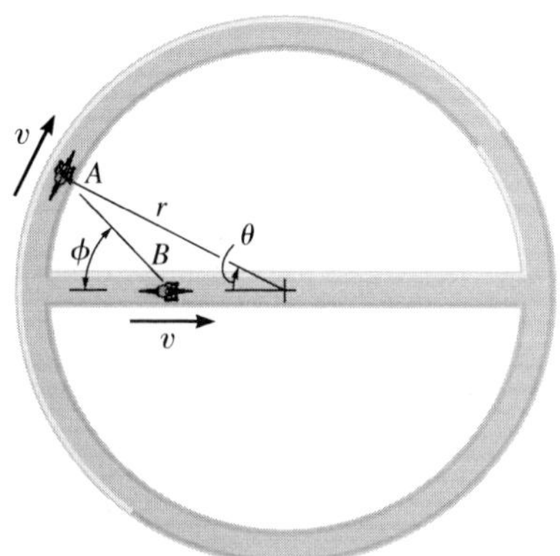

Prob. 12–230

12–231. At the instant shown, cars A and B travel at speeds of 70 km/h and 50 km/h, respectively. If B is increasing its speed by 1100 km/h^2, while A maintains a constant speed, determine the velocity and acceleration of B with respect to A. Car B moves along a curve having a radius of curvature of 0.7 km.

***12–232.** At the instant shown, cars A and B travel at speeds of 70 km/h and 50 km/h, respectively. If B is decreasing its speed at 1400 km/h^2 while A is increasing its speed at 800 km/h^2, determine the acceleration of B with respect to A. Car B moves along a curve having a radius of curvature of 0.7 km.

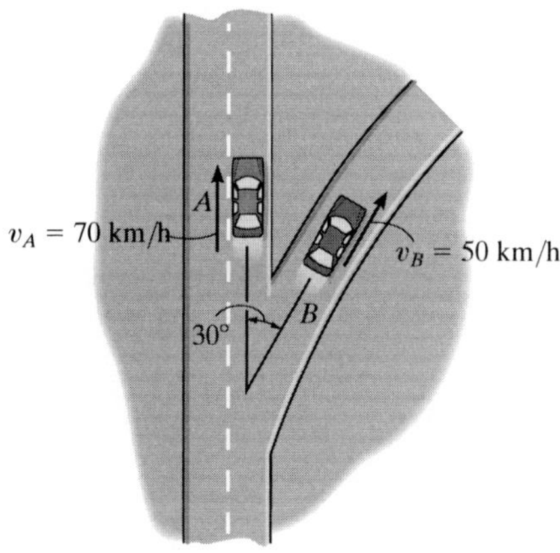

Probs. 12–231/232

12–233. A passenger in an automobile observes that raindrops make an angle of 30° with the horizontal as the auto travels forward with a speed of 60 km/h. Compute the terminal (constant) velocity $\mathbf{v}_r$ of the rain if it is assumed to fall vertically.

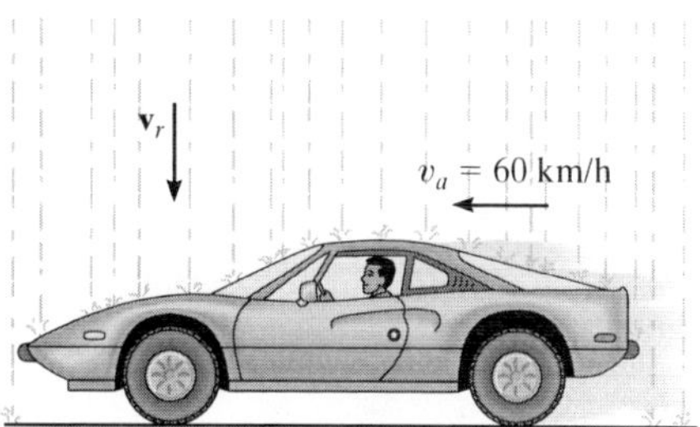

Prob. 12–233

12–234. At the instant shown, the car at A is traveling at 10 m/s around the curve while increasing its speed at 5 m/s^2. The car at B is traveling at 18.5 m/s along the straightaway and increasing its speed at 2 m/s^2. Determine the relative velocity and relative acceleration of A with respect to B at this instant.

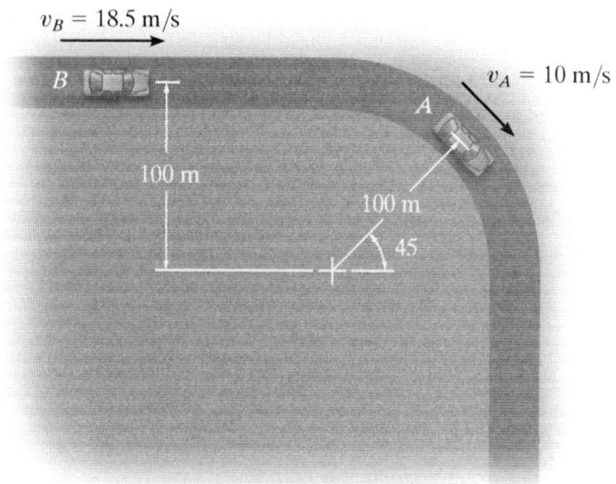

Prob. 12–234

12–235. The ship travels at a constant speed of $v_s = 20$ m/s and the wind is blowing at a speed of $v_w = 10$ m/s, as shown. Determine the magnitude and direction of the horizontal component of velocity of the smoke coming from the smoke stack as it appears to a passenger on the ship.

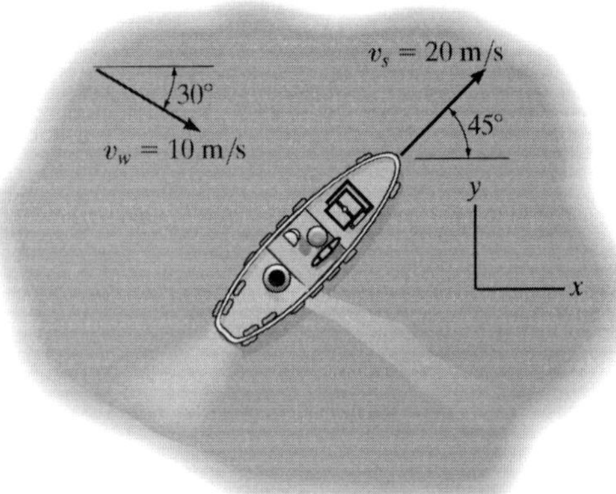

Prob. 12–235

***12–236.** Car A travels along a straight road at a speed of 25 m/s while accelerating at 1.5 m/s^2. At this same instant car C is traveling along the straight road with a speed of 30 m/s while decelerating at 3 m/s^2. Determine the velocity and acceleration of car A relative to car C.

12–237. Car B is traveling along the curved road with a speed of 15 m/s while decreasing its speed at 2 m/s^2. At this same instant car C is traveling along the straight road with a speed of 30 m/s while decelerating at 3 m/s^2. Determine the velocity and acceleration of car B relative to car C.

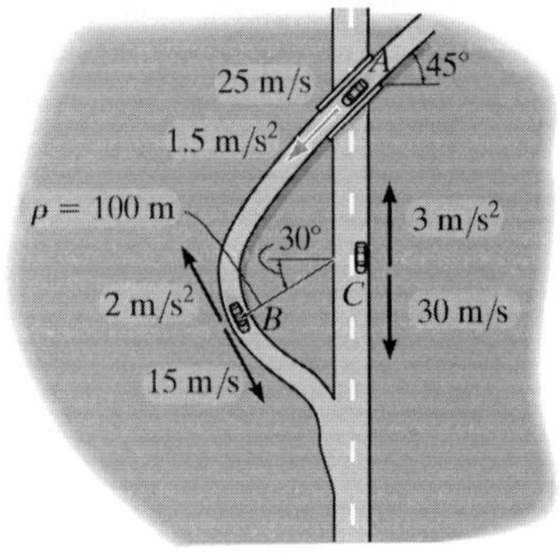

Probs. 12–236/237

12–238. At a given instant the football player at A throws a football C with a velocity of 20 m/s in the direction shown. Determine the constant speed at which the player at B must run so that he can catch the football at the same elevation at which it was thrown. Also calculate the relative velocity and relative acceleration of the football with respect to B at the instant the catch is made. Player B is 15 m away from A when A starts to throw the football.

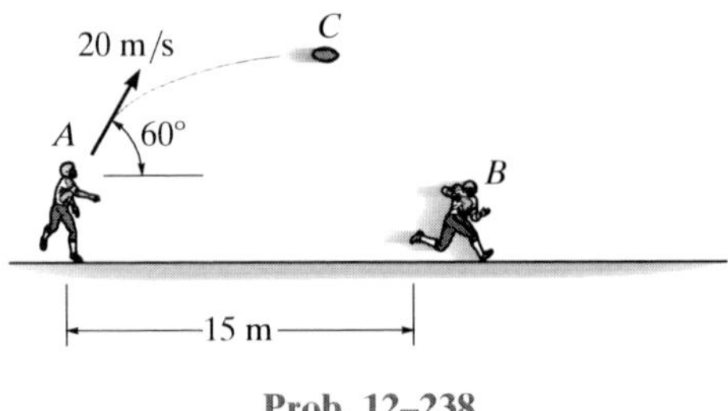

Prob. 12–238

12–239. Both boats A and B leave the shore at O at the same time. If A travels at v_A and B travels at v_B, write a general expression to determine the velocity of A with respect to B.

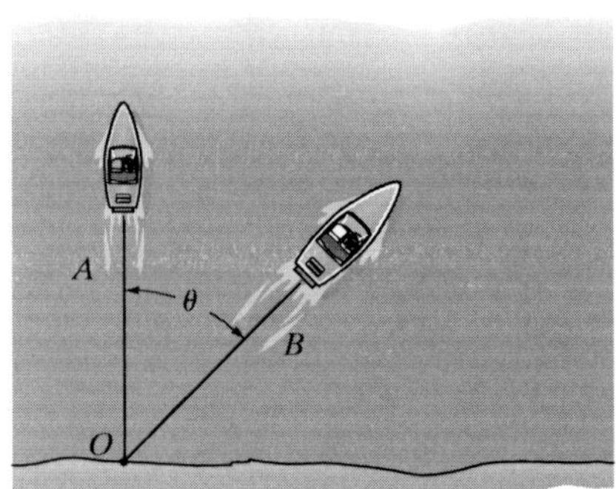

Prob. 12–239

CONCEPTUAL PROBLEMS

P12–1. If you measured the time it takes for the construction elevator to go from A to B, then B to C, and then C to D, and you also know the distance between each of the points, how could you determine the average velocity and average acceleration of the elevator as it ascends from A to D? Use numerical values to explain how this can be done.

P12–1

P12–2. If the sprinkler at A is 1 m from the ground, then scale the necessary measurements from the photo to determine the approximate velocity of the water jet as it flows from the nozzle of the sprinkler.

P12–2

P12–3. The basketball was thrown at an angle measured from the horizontal to the man's outstretched arm. If the basket is 3 m from the ground, make appropriate measurements in the photo and determine if the ball located as shown will pass through the basket.

P12–3

P12–4. The pilot tells you the wingspan of her plane and her constant airspeed. How would you determine the acceleration of the plane at the moment shown? Use numerical values and take any necessary measurements from the photo.

P12–4

CHAPTER REVIEW

Rectilinear Kinematics

Rectilinear kinematics refers to motion along a straight line. A position coordinate s specifies the location of the particle on the line, and the displacement Δs is the change in this position.

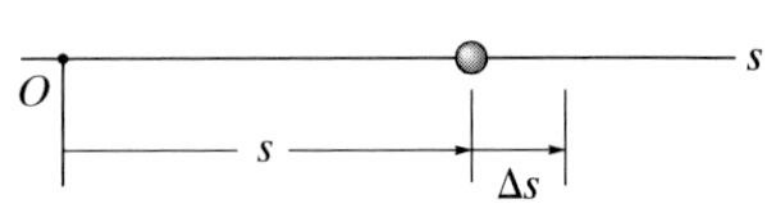

The average velocity is a vector quantity, defined as the displacement divided by the time interval.

$$v_{avg} = \frac{\Delta s}{\Delta t}$$

The average speed is a scalar, and is the total distance traveled divided by the time of travel.

$$(v_{sp})_{avg} = \frac{s_T}{\Delta t}$$

The time, position, velocity, and acceleration are related by three differential equations.

$$a = \frac{dv}{dt}, \quad v = \frac{ds}{dt}, \quad a\,ds = v\,dv$$

If the acceleration is known to be constant, then the differential equations relating time, position, velocity, and acceleration can be integrated.

$$v = v_0 + a_c t$$

$$s = s_0 + v_0 t + \tfrac{1}{2} a_c t^2$$

$$v^2 = v_0^2 + 2a_c(s - s_0)$$

Graphical Solutions

If the motion is erratic, then it can be described by a graph. If one of these graphs is given, then the others can be established using the differential relations between a, v, s, and t.

$$a = \frac{dv}{dt},$$

$$v = \frac{ds}{dt},$$

$$a\,ds = v\,dv$$

Curvilinear Motion, x, y, z

Curvilinear motion along the path can be resolved into rectilinear motion along the x, y, z axes. The equation of the path is used to relate the motion along each axis.

$$v_x = \dot{x} \qquad a_x = \dot{v}_x$$

$$v_y = \dot{y} \qquad a_y = \dot{v}_y$$

$$v_z = \dot{z} \qquad a_z = \dot{v}_z$$

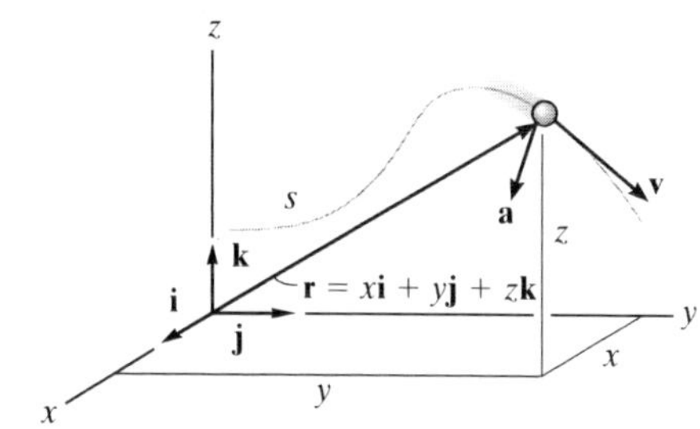

Projectile Motion

Free-flight motion of a projectile follows a parabolic path. It has a constant velocity in the horizontal direction, and a constant downward acceleration of $g = 9.81\ \text{m/s}^2$ in the vertical direction. Any two of the three equations for constant acceleration apply in the vertical direction, and in the horizontal direction only one equation applies.

$$(+\uparrow) \qquad v_y = (v_0)_y + a_c t$$

$$(+\uparrow) \qquad y = y_0 + (v_0)_y t + \tfrac{1}{2} a_c t^2$$

$$(+\uparrow) \qquad v_y^2 = (v_0)_y^2 + 2a_c(y - y_0)$$

$$(\overset{+}{\rightarrow}) \qquad x = x_0 + (v_0)_x t$$

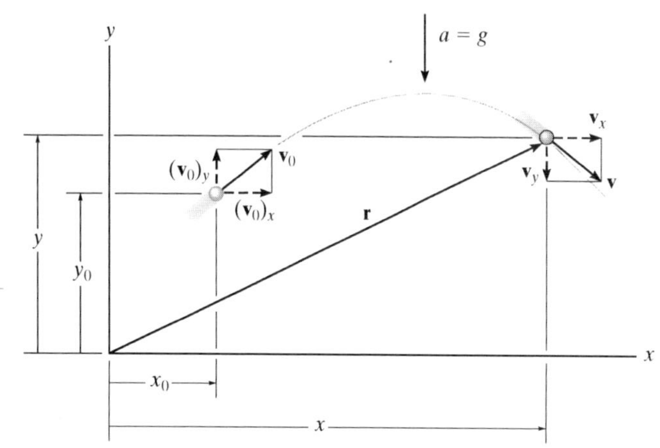

Curvilinear Motion *n*, *t*

If normal and tangential axes are used for the analysis, then **v** is always in the positive *t* direction.

The acceleration has two components. The tangential component, $\mathbf{a}_t$, accounts for the change in the magnitude of the velocity; a slowing down is in the negative *t* direction, and a speeding up is in the positive *t* direction. The normal component $\mathbf{a}_n$ accounts for the change in the direction of the velocity. This component is always in the positive *n* direction.

$$a_t = \dot{v} \quad \text{or} \quad a_t\,ds = v\,dv$$

$$a_n = \frac{v^2}{\rho}$$

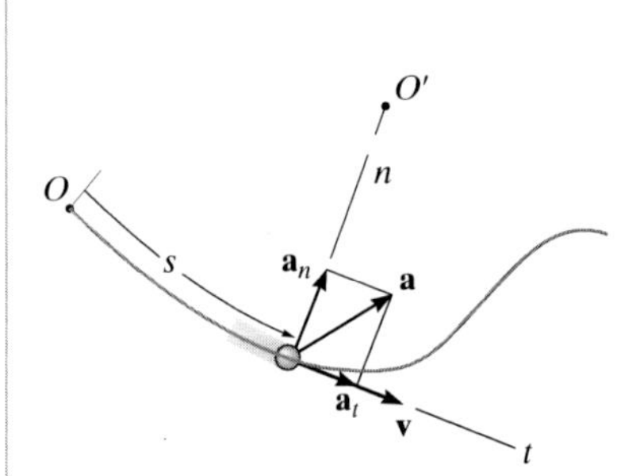

Curvilinear Motion *r*, *θ*

If the path of motion is expressed in polar coordinates, then the velocity and acceleration components can be related to the time derivatives of *r* and *θ*.

To apply the time-derivative equations, it is necessary to determine $r, \dot{r}, \ddot{r}, \dot{\theta}, \ddot{\theta}$ at the instant considered. If the path $r = f(\theta)$ is given, then the chain rule of calculus must be used to obtain time derivatives. (See Appendix C.)

Once the data are substituted into the equations, then the algebraic sign of the results will indicate the direction of the components of **v** or **a** along each axis.

$$v_r = \dot{r}$$

$$v_\theta = r\dot{\theta}$$

$$a_r = \ddot{r} - r\dot{\theta}^2$$

$$a_\theta = r\ddot{\theta} + 2\dot{r}\dot{\theta}$$

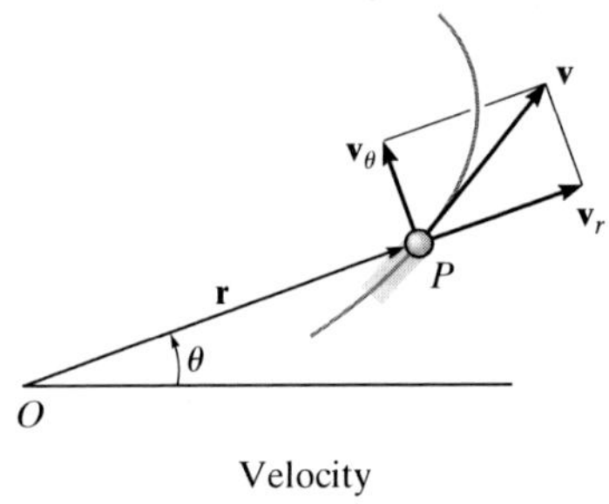

Velocity

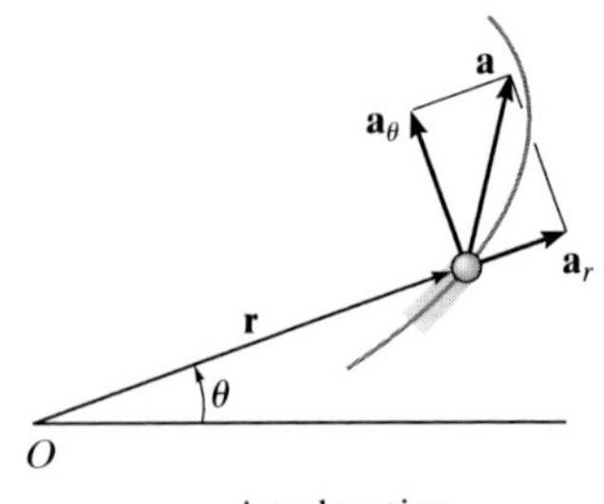

Acceleration

Absolute Dependent Motion of Two Particles

The dependent motion of blocks that are suspended from pulleys and cables can be related by the geometry of the system. This is done by first establishing position coordinates, measured from a fixed origin to each block. Each coordinate must be directed along the line of motion of a block.

Using geometry and/or trigonometry, the coordinates are then related to the cable length in order to formulate a position coordinate equation.

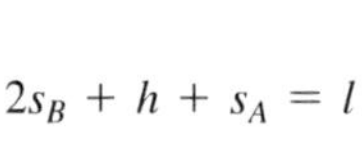

$$2s_B + h + s_A = l$$

The first time derivative of this equation gives a relationship between the velocities of the blocks, and a second time derivative gives the relation between their accelerations.

$$2v_B = -v_A$$

$$2a_B = -a_A$$

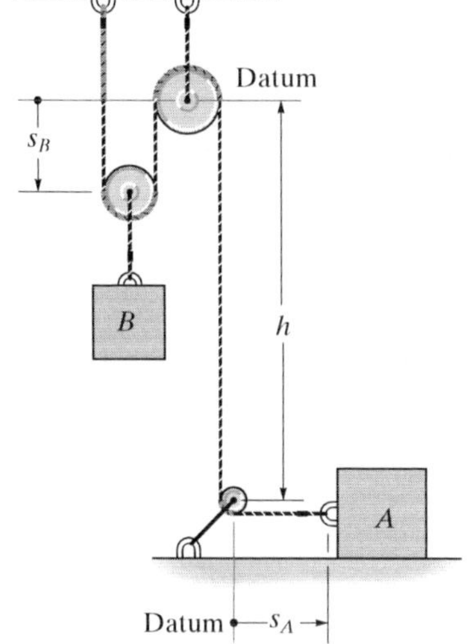

Relative-Motion Analysis Using Translating Axes

If two particles A and B undergo independent motions, then these motions can be related to their relative motion using a *translating set of axes* attached to one of the particles (A).

$$\mathbf{r}_B = \mathbf{r}_A + \mathbf{r}_{B/A}$$

For planar motion, each vector equation produces two scalar equations, one in the x, and the other in the y direction. For solution, the vectors can be expressed in Cartesian form, or the x and y scalar components can be written directly.

$$\mathbf{v}_B = \mathbf{v}_A + \mathbf{v}_{B/A}$$

$$\mathbf{a}_B = \mathbf{a}_A + \mathbf{a}_{B/A}$$

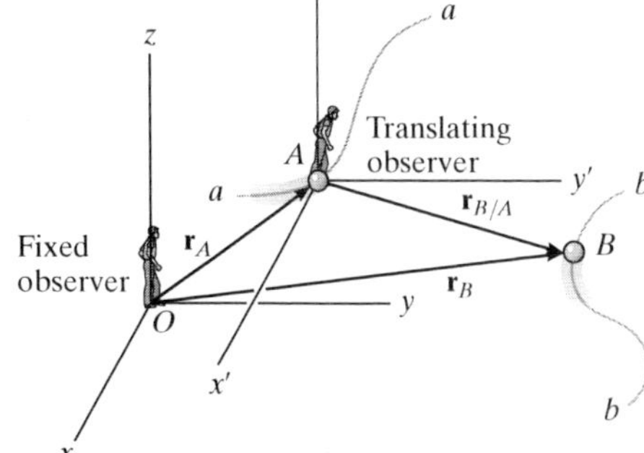

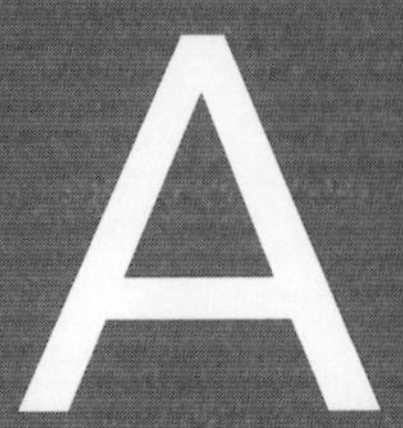

Mathematical Expressions

Quadratic Formula

If $ax^2 + bx + c = 0$, then $x = \dfrac{-b \pm \sqrt{b^2 - 4ac}}{2a}$

Hyperbolic Functions

$\sinh x = \dfrac{e^x - e^{-x}}{2}$, $\cosh x = \dfrac{e^x + e^{-x}}{2}$, $\tanh x = \dfrac{\sinh x}{\cosh x}$

Trigonometric Identities

$\sin\theta = \dfrac{A}{C}$, $\csc\theta = \dfrac{C}{A}$

$\cos\theta = \dfrac{B}{C}$, $\sec\theta = \dfrac{C}{B}$

$\tan\theta = \dfrac{A}{B}$, $\cot\theta = \dfrac{B}{A}$

$\sin^2\theta + \cos^2\theta = 1$

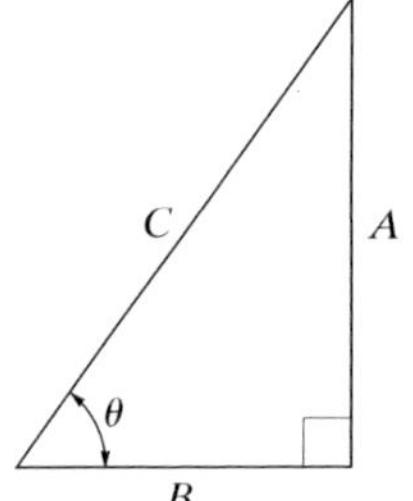

$\sin(\theta \pm \phi) = \sin\theta\cos\phi \pm \cos\theta\sin\phi$

$\sin 2\theta = 2\sin\theta\cos\theta$

$\cos(\theta \pm \phi) = \cos\theta\cos\phi \mp \sin\theta\sin\phi$

$\cos 2\theta = \cos^2\theta - \sin^2\theta$

$\cos\theta = \pm\sqrt{\dfrac{1 + \cos 2\theta}{2}}$, $\sin\theta = \pm\sqrt{\dfrac{1 - \cos 2\theta}{2}}$

$\tan\theta = \dfrac{\sin\theta}{\cos\theta}$

$1 + \tan^2\theta = \sec^2\theta \quad 1 + \cot^2\theta = \csc^2\theta$

Power-Series Expansions

$\sin x = x - \dfrac{x^3}{3!} + \cdots \qquad \sinh x = x + \dfrac{x^3}{3!} + \cdots$

$\cos x = 1 - \dfrac{x^2}{2!} + \cdots \qquad \cosh x = 1 + \dfrac{x^2}{2!} + \cdots$

Derivatives

$$\frac{d}{dx}(u^n) = nu^{n-1}\frac{du}{dx}$$

$$\frac{d}{dx}(uv) = u\frac{dv}{dx} + v\frac{du}{dx}$$

$$\frac{d}{dx}\left(\frac{u}{v}\right) = \frac{v\dfrac{du}{dx} - u\dfrac{dv}{dx}}{v^2}$$

$$\frac{d}{dx}(\cot u) = -\csc^2 u\frac{du}{dx}$$

$$\frac{d}{dx}(\sec u) = \tan u \sec u\frac{du}{dx}$$

$$\frac{d}{dx}(\csc u) = -\csc u \cot u\frac{du}{dx}$$

$$\frac{d}{dx}(\sin u) = \cos u\frac{du}{dx}$$

$$\frac{d}{dx}(\cos u) = -\sin u\frac{du}{dx}$$

$$\frac{d}{dx}(\tan u) = \sec^2 u\frac{du}{dx}$$

$$\frac{d}{dx}(\sinh u) = \cosh u\frac{du}{dx}$$

$$\frac{d}{dx}(\cosh u) = \sinh u\frac{du}{dx}$$

Integrals

$$\int x^n\,dx = \frac{x^{n+1}}{n+1} + C,\ n \neq -1$$

$$\int \frac{dx}{a+bx} = \frac{1}{b}\ln(a+bx) + C$$

$$\int \frac{dx}{a+bx^2} = \frac{1}{2\sqrt{-ba}}\ln\left[\frac{a+x\sqrt{-ab}}{a-x\sqrt{-ab}}\right] + C,\ ab < 0$$

$$\int \frac{x\,dx}{a+bx^2} = \frac{1}{2b}\ln(bx^2+a) + C$$

$$\int \frac{x^2\,dx}{a+bx^2} = \frac{x}{b} - \frac{a}{b\sqrt{ab}}\tan^{-1}\frac{x\sqrt{ab}}{a} + C,\ ab > 0$$

$$\int \frac{dx}{a^2-x^2} = \frac{1}{2a}\ln\left[\frac{a+x}{a-x}\right] + C,\ a^2 > x^2$$

$$\int \sqrt{a+bx}\,dx = \frac{2}{3b}\sqrt{(a+bx)^3} + C$$

$$\int x\sqrt{a+bx}\,dx = \frac{-2(2a-3bx)\sqrt{(a+bx)^3}}{15b^2} + C$$

$$\int x^2\sqrt{a+bx}\,dx = \frac{2(8a^2-12abx+15b^2x^2)\sqrt{(a+bx)^3}}{105b^3} + C$$

$$\int \sqrt{a^2-x^2}\,dx = \frac{1}{2}\left[x\sqrt{a^2-x^2} + a^2\sin^{-1}\frac{x}{a}\right] + C,\ a > 0$$

$$\int x\sqrt{x^2 \pm a^2}\,dx = \frac{1}{3}\sqrt{(x^2 \pm a^2)^3} + C$$

$$\int x^2\sqrt{a^2-x^2}\,dx = -\frac{x}{4}\sqrt{(a^2-x^2)^3} + \frac{a^2}{8}\left(x\sqrt{a^2-x^2} + a^2\sin^{-1}\frac{x}{a}\right) + C,\ a > 0$$

$$\int \sqrt{x^2 \pm a^2}\,dx = \frac{1}{2}\left[x\sqrt{x^2 \pm a^2} \pm a^2\ln\left(x + \sqrt{x^2 \pm a^2}\right)\right] + C$$

$$\int x\sqrt{a^2-x^2}\,dx = -\frac{1}{3}\sqrt{(a^2-x^2)^3} + C$$

$$\int x^2\sqrt{x^2 \pm a^2}\,dx = \frac{x}{4}\sqrt{(x^2 \pm a^2)^3} \mp \frac{a^2}{8}x\sqrt{x^2 \pm a^2} - \frac{a^4}{8}\ln\left(x + \sqrt{x^2 \pm a^2}\right) + C$$

$$\int \frac{dx}{\sqrt{a+bx}} = \frac{2\sqrt{a+bx}}{b} + C$$

$$\int \frac{x\,dx}{\sqrt{x^2 \pm a^2}} = \sqrt{x^2 \pm a^2} + C$$

$$\int \frac{dx}{\sqrt{a+bx+cx^2}} = \frac{1}{\sqrt{c}}\ln\left[\sqrt{a+bx+cx^2} + x\sqrt{c} + \frac{b}{2\sqrt{c}}\right] + C,\ c > 0$$

$$= \frac{1}{\sqrt{-c}}\sin^{-1}\left(\frac{-2cx-b}{\sqrt{b^2-4ac}}\right) + C,\ c < 0$$

$$\int \sin x\,dx = -\cos x + C$$

$$\int \cos x\,dx = \sin x + C$$

$$\int x\cos(ax)\,dx = \frac{1}{a^2}\cos(ax) + \frac{x}{a}\sin(ax) + C$$

$$\int x^2\cos(ax)\,dx = \frac{2x}{a^2}\cos(ax) + \frac{a^2x^2-2}{a^3}\sin(ax) + C$$

$$\int e^{ax}\,dx = \frac{1}{a}e^{ax} + C$$

$$\int xe^{ax}\,dx = \frac{e^{ax}}{a^2}(ax-1) + C$$

$$\int \sinh x\,dx = \cosh x + C$$

$$\int \cosh x\,dx = \sinh x + C$$

A

APPENDIX

B Vector Analysis

The following discussion provides a brief review of vector analysis. A more detailed treatment of these topics is given in *Mechanics for Engineers: Statics*.

Vector. A vector, **A**, is a quantity which has magnitude and direction, and adds according to the parallelogram law. As shown in Fig. B–1, **A** = **B** + **C**, where **A** is the *resultant vector* and **B** and **C** are *component vectors*.

Unit Vector. A unit vector, $\mathbf{u}_A$, has a magnitude of one "dimensionless" unit and acts in the same direction as **A**. It is determined by dividing **A** by its magnitude A, i.e,

$$\mathbf{u}_A = \frac{\mathbf{A}}{A} \qquad \text{(B–1)}$$

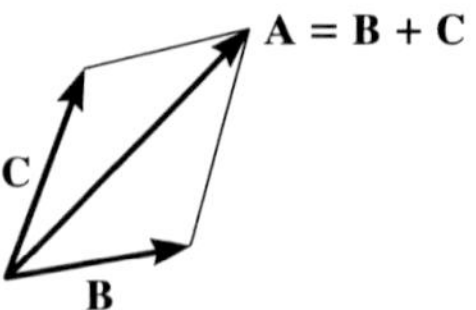

Fig. B–1

Cartesian Vector Notation. The directions of the positive x, y, z axes are defined by the Cartesian unit vectors **i**, **j**, **k**, respectively.

As shown in Fig. B–2, vector **A** is formulated by the addition of its x, y, z components as

$$\mathbf{A} = A_x\mathbf{i} + A_y\mathbf{j} + A_z\mathbf{k} \tag{B–2}$$

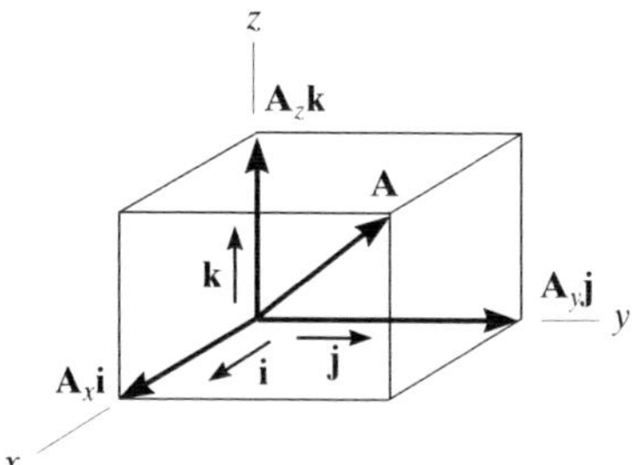

Fig. B–2

The *magnitude* of **A** is determined from

$$A = \sqrt{A_x^2 + A_y^2 + A_z^2} \tag{B–3}$$

The *direction* of **A** is defined in terms of its *coordinate direction angles*, α, β, γ, measured from the *tail* of **A** to the *positive* x, y, z axes, Fig. B–3. These angles are determined from the *direction cosines* which represent the **i**, **j**, **k** components of the unit vector $\mathbf{u}_A$; i.e., from Eqs. B–1 and B–2

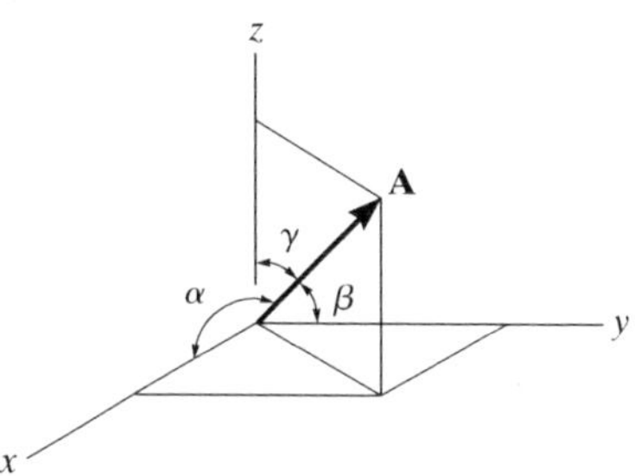

Fig. B–3

$$\mathbf{u}_A = \frac{A_x}{A}\mathbf{i} + \frac{A_y}{A}\mathbf{j} + \frac{A_z}{A}\mathbf{k} \tag{B–4}$$

so that the direction cosines are

$$\cos\alpha = \frac{A_x}{A} \quad \cos\beta = \frac{A_y}{A} \quad \cos\gamma = \frac{A_z}{A} \tag{B–5}$$

Hence, $\mathbf{u}_A = \cos\alpha\mathbf{i} + \cos\beta\mathbf{j} + \cos\gamma\mathbf{k}$, and using Eq. B–3, it is seen that

$$\cos^2\alpha + \cos^2\beta + \cos^2\gamma = 1 \tag{B–6}$$

The Cross Product. The cross product of two vectors **A** and **B**, which yields the resultant vector **C**, is written as

$$\mathbf{C} = \mathbf{A} \times \mathbf{B} \tag{B–7}$$

and reads **C** equals **A** "cross" **B**. The *magnitude* of **C** is

$$C = AB\sin\theta \tag{B–8}$$

where θ is the angle made between the *tails* of **A** and **B** ($0° \le \theta \le 180°$). The *direction* of **C** is determined by the right-hand rule, whereby the fingers of the right hand are curled *from* **A** *to* **B** and the thumb points in the direction of **C**, Fig. B–4. This vector is perpendicular to the plane containing vectors **A** and **B**.

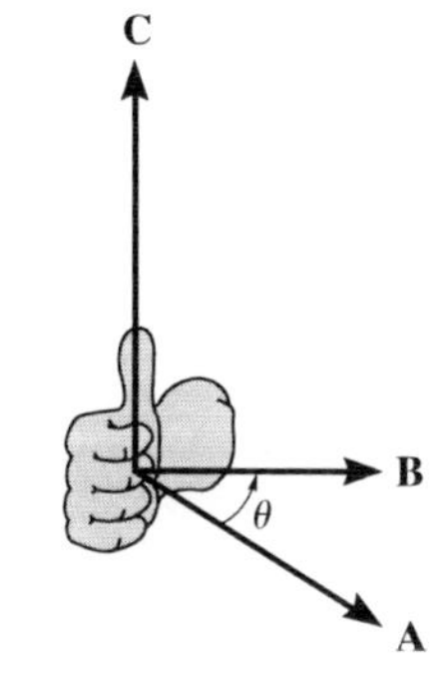

Fig. B–4

The vector cross product is *not* commutative, i.e., $\mathbf{A} \times \mathbf{B} \neq \mathbf{B} \times \mathbf{A}$. Rather,

$$\mathbf{A} \times \mathbf{B} = -\mathbf{B} \times \mathbf{A} \qquad \text{(B–9)}$$

The distributive law is valid; i.e.,

$$\mathbf{A} \times (\mathbf{B} + \mathbf{D}) = \mathbf{A} \times \mathbf{B} + \mathbf{A} \times \mathbf{D} \qquad \text{(B–10)}$$

And the cross product may be multiplied by a scalar m in any manner; i.e.,

$$m(\mathbf{A} \times \mathbf{B}) = (m\mathbf{A}) \times \mathbf{B} = \mathbf{A} \times (m\mathbf{B}) = (\mathbf{A} \times \mathbf{B})m \qquad \text{(B–11)}$$

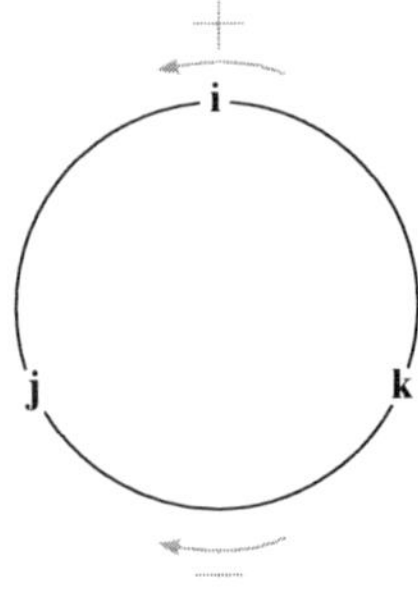

Fig. B–5

Equation B–7 can be used to find the cross product of any pair of Cartesian unit vectors. For example, to find $\mathbf{i} \times \mathbf{j}$, the magnitude is $(i)(j) \sin 90° = (1)(1)(1) = 1$, and its direction $+\mathbf{k}$ is determined from the right-hand rule, applied to $\mathbf{i} \times \mathbf{j}$, Fig. B–2. A simple scheme shown in Fig. B–5 may be helpful in obtaining this and other results when the need arises. If the circle is constructed as shown, then "crossing" two of the unit vectors in a *counterclockwise* fashion around the circle yields a *positive* third unit vector, e.g., $\mathbf{k} \times \mathbf{i} = \mathbf{j}$. Moving *clockwise*, a *negative* unit vector is obtained, e.g., $\mathbf{i} \times \mathbf{k} = -\mathbf{j}$.

If $\mathbf{A}$ and $\mathbf{B}$ are expressed in Cartesian component form, then the cross product, Eq. B–7, may be evaluated by expanding the determinant

$$\mathbf{C} = \mathbf{A} \times \mathbf{B} = \begin{vmatrix} \mathbf{i} & \mathbf{j} & \mathbf{k} \\ A_x & A_y & A_z \\ B_x & B_y & B_z \end{vmatrix} \qquad \text{(B–12)}$$

which yields

$$\mathbf{C} = (A_yB_z - A_zB_y)\mathbf{i} - (A_xB_z - A_zB_x)\mathbf{j} + (A_xB_y - A_yB_x)\mathbf{k}$$

Recall that the cross product is used in statics to define the moment of a force $\mathbf{F}$ about point O, in which case

$$\mathbf{M}_O = \mathbf{r} \times \mathbf{F} \qquad \text{(B–13)}$$

where $\mathbf{r}$ is a position vector directed from point O to *any point* on the line of action of $\mathbf{F}$.

B

The Dot Product. The dot product of two vectors **A** and **B**, which yields a scalar, is defined as

$$\mathbf{A} \cdot \mathbf{B} = AB \cos \theta \qquad \text{(B–14)}$$

and reads **A** "dot" **B**. The angle θ is formed between the *tails* of **A** and **B** ($0° \leq \theta \leq 180°$).

The dot product is commutative; i.e.,

$$\mathbf{A} \cdot \mathbf{B} = \mathbf{B} \cdot \mathbf{A} \qquad \text{(B–15)}$$

The distributive law is valid; i.e.,

$$\mathbf{A} \cdot (\mathbf{B} + \mathbf{D}) = \mathbf{A} \cdot \mathbf{B} + \mathbf{A} \cdot \mathbf{D} \qquad \text{(B–16)}$$

And scalar multiplication can be performed in any manner, i.e.,

$$m(\mathbf{A} \cdot \mathbf{B}) = (m\mathbf{A}) \cdot \mathbf{B} = \mathbf{A} \cdot (m\mathbf{B}) = (\mathbf{A} \cdot \mathbf{B})m \qquad \text{(B–17)}$$

Using Eq. B–14, the dot product between any two Cartesian vectors can be determined. For example, $\mathbf{i} \cdot \mathbf{i} = (1)(1) \cos 0° = 1$ and $\mathbf{i} \cdot \mathbf{j} = (1)(1) \cos 90° = 0$.

If **A** and **B** are expressed in Cartesian component form, then the dot product, Eq. C–14, can be determined from

$$\boxed{\mathbf{A} \cdot \mathbf{B} = A_x B_x + A_y B_y + A_z B_z} \qquad \text{(B–18)}$$

The dot product may be used to determine the *angle θ formed between two vectors*. From Eq. B–14,

$$\theta = \cos^{-1}\left(\frac{\mathbf{A} \cdot \mathbf{B}}{AB}\right) \qquad \text{(B–19)}$$

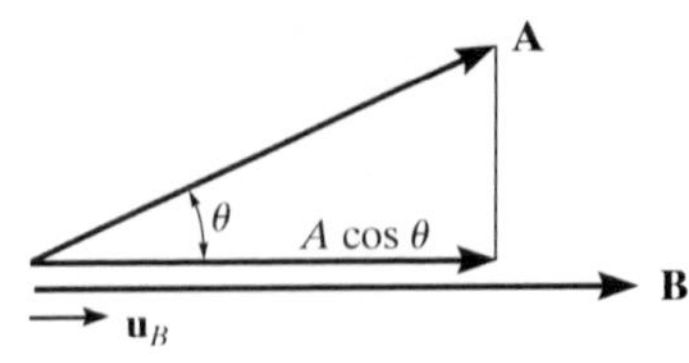

Fig. B–6

It is also possible to find the *component of a vector in a given direction* using the dot product. For example, the magnitude of the component (or projection) of vector **A** in the direction of **B**, Fig. B–6, is defined by $A\cos\theta$. From Eq. B–14, this magnitude is

$$A\cos\theta = \mathbf{A}\cdot\frac{\mathbf{B}}{B} = \mathbf{A}\cdot\mathbf{u}_B \tag{B–20}$$

where $\mathbf{u}_B$ represents a unit vector acting in the direction of **B**, Fig. B–6.

Differentiation and Integration of Vector Functions. The rules for differentiation and integration of the sums and products of scalar functions also apply to vector functions. Consider, for example, the two vector functions $\mathbf{A}(s)$ and $\mathbf{B}(s)$. Provided these functions are smooth and continuous for all s, then

$$\frac{d}{ds}(\mathbf{A} + \mathbf{B}) = \frac{d\mathbf{A}}{ds} + \frac{d\mathbf{B}}{ds} \tag{B–21}$$

$$\int (\mathbf{A} + \mathbf{B})\,ds = \int \mathbf{A}\,ds + \int \mathbf{B}\,ds \tag{B–22}$$

For the cross product,

$$\frac{d}{ds}(\mathbf{A} \times \mathbf{B}) = \left(\frac{d\mathbf{A}}{ds} \times \mathbf{B}\right) + \left(\mathbf{A} \times \frac{d\mathbf{B}}{ds}\right) \tag{B–23}$$

Similarly, for the dot product,

$$\frac{d}{ds}(\mathbf{A}\cdot\mathbf{B}) = \frac{d\mathbf{A}}{ds}\cdot\mathbf{B} + \mathbf{A}\cdot\frac{d\mathbf{B}}{ds} \tag{B–24}$$

Answers to Selected Problems

Chapter 12

12–1. $v_{(t_1)} = 0$ m/s
$d_{(t_2)} = 80.7$ m

12–2. $v = 3.93$ m/s
$s = 9.98$ m

12–3. $v_1 = 32.00$ m/s
$s_1 = 67.00$ m
$d = 66.00$ m

12–5. $t = 25$ s
$s = 312.5$ m

12–6. $\Delta s = 14.71$ m

12–7. $a_c = 1.74$ m/s^2
$t = 4.80$ s

12–9. $v = 29.4$ m/s
$h = 44.1$ m

12–10. $S_p(t_1) = 1.839$ m

12–11. $d = 291.7$ m

12–13. $v_{pl} = 322.49$ m/s
$t_1 = 19.27$ s

12–14. $t_1 = 5.62$ s

12–15. $s_3 = 28.4$ km

12–17. $v = 625$ m/s
$s = 639.5$ m
$d = 637.5$ m

12–18. $t_3 = 32$ s

12–19. $s = 7.87$ m

12–21. (a) When $t = 5$ s,
$v = 45.5$ m/s.
(b) $v_{max} = 100$ m/s

12–22. $V_{avevel} = 0$ m/s
$V_{avespeed} = 3$ m/s
$a_1 = 2$ m/s^2

12–23. $$t = \frac{v_f}{2g}\ln\left(\frac{v_f + v}{v_f - v}\right)$$

12–25. $s = 54.0$ m

12–26. $$s_{BA} = \left|\frac{2v_A v_B - v_A^2}{2a_A}\right|$$

12–27. $s|_{t=2.667} = 3.56$ m

12–29. $t = 2.47$ s
$s|_{t=2s} = 18.7$ m

12–30. $v = 14$ m/s
$t = 250$ s

12–31. (a) $s = -30.5$ m
(b) $s_{Tot} = 56.0$ m
(c) $v = 10$ m/s

12–33. $t = 10.3$ s
$h = 4.11$ km

12–34. $v_1 = 3.68$ m/s ↑
$t_2 = 1.98$ s
$v_2 = 15.8$ m/s ↓

12–35. $$t = 0.549\left(\frac{v_f}{g}\right)$$

12–37. $v = 11.2$ km/s

12–38. $$v = -R\sqrt{\frac{2g_0(y_0 - y)}{(R + y)(R + y_0)}}$$
$v = 3.02$ km/s ↓

12–39. $t = 11.25$ s

12–41. $v_{max} = 16.7$ m/s
$v = v_{max}$ for 2 min $< t <$ 4 min

12–42. $a|_{t=0} = -4$ m/s^2
$a|_{t=2\,s} = 0$
$a|_{t=4} = 4$ m/s^2
$v|_{t=0\,s} = 3$ m/s
$v|_{t=2\,s} = -1$ m/s
$v|_{t=4\,s} = 3$ m/s

12–43. $$s = 2\sin\left(\frac{\pi}{5}t\right) + 4$$
$$v = \frac{2\pi}{5}\cos\left(\frac{\pi}{5}t\right)$$
$$a = \frac{2\pi^2}{25}\sin\left(\frac{\pi}{5}t\right)$$

12–45. For $0 \le t < 30$ s, $s = \left(\frac{1}{5}t^2\right)$ m, $a = 0.4$ m/s^2.
For $30 < t \le 50$ s, $s = (12t - 180)$ m, $a = 0$.

12–46. $s = 600$ m
For $0 \le t < 40$ s, $a = 0$.
For 40 s $< t \le$ 80 s, $a = -0.250$ m/s^2.

12–47. At $s = 50$ m, $a = 0.32$ m/s^2.
At $s = 150$ m, $a = -0.32$ m/s^2.
At $s = 100$ m, a changes from $a_{max} = 0.64$ m/s^2 to $a_{min} = -0.64$ m/s^2.

12–49. When $t = 0.1$ s, $s = 0.5$ m and a changes from 100 m/s^2 to -100 m/s^2.
When $t = 0.2$ s, $s = 1$ m.

12–50. For $0 \le t < 5$ s, $a = 4$ m/s^2.
For 20 s $< t \le$ 30 s, $a = -2$ m/s^2.
At $t = 5$ s, $s = 50$ m. At $t = 20$ s, $s = 350$ m.
At $t = 30$ s, $s = 450$ m.

12–51. $t' = 133$ s
When $t = 30$ s, $v = 45$ m/s and s = 450 m.
When $t = 75$ s, $v = v_{max} = 112.5$ m/s and $s = 4500$ m.
When $t = 133$ s, $v = 0$ and $s = 8857$ m.

12–53. When $t = 15$ s,
$v = 270$ m/s,
$s = 2.025$ km.
When $t = 20$ s,
$v = 395$ m/s,
$s = 3.69$ km.

12–54. When $t = t_4$, $a_4 = 0$, $s_4 = 30.00$ m.
When $t = t_5$, $a = -1$ m/s^2, $s_5 = 48$ m.

12–55. When $t = 6$s, $v = 12.0$ m/s and $s = 18$ m.
When $t = 10$ s, $v = 36.0$ m/s and $s = 114$ m.

12–57. For $0 \le t < 30$ s, $s = \left(\dfrac{t^2}{2}\right)$ m, $a = 1$ m/s^2.
For $30\text{ s} < t \le 90$ s, $s = \left(-\dfrac{1}{4}t^2 + 45t - 675\right)$ m, $a = -0.5$ m/s^2.
$s|_{t=90\text{ s}} = 1350$ s

12–58. When $t = 25$ s, $a = a_{max} = 9$ m/s^2 and
$s = 469$ m.
When $t = 50$ s, $s = 1406$ m.

12–59. For $0 \le t < 10$ s, $s = (4t^2)$ m, $a = 8$ m/s^2.
For $10 < t \le 40$ s, $s = (80t - 400)$ m, $a = 0$.

12–61. When $t = 30$ s, $s = 90$ m.
When $t = 48$ s, $s = 144$ m.

12–62. When $s = 200$ m, $v = 20$ m/s.
When $s = 400$ m, $v = 34.64$ m/s.

12–63. $s_T = 980$ m

12–65. When $t = 5$ s, $s_B = 62.5$ m.
When $t = 10$ s, $v_A = (v_A)_{max} = 40$ m/s and
$s_A = 200$ m.
When $t = 15$ s, $s_A = 400$ m and $s_B = 312.5$ m.
$\Delta s = s_A - s_B = 87.5$ m

12–66. When $t = 30$ s, $v = 90$ m/s.
When $t = 60$ s, $v = 540$ m/s.

12–67. When $t = 30$ s, $s = 675$ m.
When $t = 60$ s $s = 10\,125$ m.

12–69. For $0 \le t < 30$ s,
$s = (0.4t^2)$ m, $v = (0.8t)$ m/s, $a = 0.8$ m/s^2.
For $30 < t \le 40$ s,
$s = (24t - 360)$ m, $v = 24$ m/s, $a = 0$.

12–70. When $s = 100$ m, $t = 10$ s.
When $s = 400$ m, $t = 16.9$ s.
$a|_{s=100\text{ m}} = 4$ m/s^2
$a|_{s=400\text{ m}} = 16$ m/s^2

12–71. For $0 \le t < 10$ s, a = $(0.5\,t)$ m/s^2,
$v = (0.25t^2)$ m/s, $s = \left(\dfrac{t^3}{12}\right)$ m.
For $10 < t \le t'$ (= 22.5 s), $a = -2$ m/s^2,
$v = (45 - 2t)$ m/s,
$s = (-(t-10)^2 + 25\,(t-10) + 83.33)$ m.

12–73. $v = 9.68$ m/s
$a = 16.8$ m/s^2

12–74. $\Delta\mathbf{r} = \{6\mathbf{i} + 4\mathbf{j}\}$ m

12–75. $d_1 = 8.34$ m, $a_2 = 0.541$ m/s^2

12–77. $\Delta r = 3.61$ km
$d = 5$ km

12–78. $s = 9$ km
$\Delta\mathbf{r} = 3.61$ km
$v_{avg} = 2.61$ m/s
$(v_{sp})_{avg} = 6.52$ m/s

12–79. $\mathbf{a}_{AB} = \{0.404\mathbf{i} + 7.07\mathbf{j}\}$ m/s^2
$\mathbf{a}_{AC} = \{2.50\mathbf{i}\}$ m/s^2

12–81. $v = 8.55$ m
$a = 5.82$ m/s

12–82. $v = 1003$ m/s
$a = 103$ m/s^2

12–83. $|\mathbf{r}_1| = 204$ m
$|\mathbf{v}_1| = 41.8$ m/s
$|\mathbf{a}_1| = 4.66$ m/s^2

12–85. $\mathbf{v}_{avg} = \{10\mathbf{i} - 10\mathbf{j}\}$ m/s

12–86. $v = 201$ m/s
$a = 405$ m/s^2

12–87. $v = 10.4$ m/s
$a = 38.5$ m/s^2

12–89. $\theta_A = 30.5°$
$v_A = 23.2$ m/s

12–90. $\theta = 58.3°$
$(v_0)_{min} = 9.76$ m/s

12–91. $\theta_A = 7.19°$ and $80.5°$

12–93. $t = 3.55$ s, $v = 32.0$ m/s

12–94. $t_R = \dfrac{4v_0 \sin(\theta)}{g}$

$$R = \frac{8v_0^{\ 2}}{g}\sin(\theta)\cos(\theta)$$

$$h = \frac{2v_0^{\ 2}\sin(\theta)^2}{g}$$

12–95. $d = \dfrac{v_0^2}{g\cos\theta}(\sin 2\phi - 2\tan\theta\cos^2\phi)$

12–97. $v_A = 11.1$ m/s, $h = 3.45$ m

12–98. $v_A = 28.0$ m/s

12–99. $t = 2.48$ s, $R = 19.01$ m

12–101. $v_A = 19.4$ m/s
$v_B = 40.4$ m/s

12–102. $a_x = \dfrac{c}{2k}(y + ct^2)$, $a_y = c$

12–103. $v_{min} = 0.838$ m/s, $v_{max} = 1.764$ m/s

12–105. $R = \dfrac{v_0^2\cos(\theta)^2}{g}\left[\tan(\theta) + \sqrt{\tan(\theta)^2 + \dfrac{2gh}{v_0^2\cos(\theta)^2}}\right]$

12–106. $a_{mag} = 258.6$ m/s^2
$x = 256.00$ m
$y = 320.00$ m
$z = 60.00$ m

12–107. $d = 94.1$ m

12–109. $\Delta t = \dfrac{2v_0 \sin(\theta_1 - \theta_2)}{g(\cos\theta_2 + \cos\theta_1)}$

12–110. $R_{\min} = 0.189$ m
$R_{\max} = 1.19$ m

12–111. $\theta = 57.6°$
$x = 222$ m
$y = 116$ m

12–113. $t = 1.195$ s, $d = 12.7$ m

12–114. $a = 9.50$ m/s^2

12–115. $v = 38.7$ m/s

12–117. $a = 0.488$ m/s^2

12–118. $v = 4.58$ m/s
$a = 0.653$ m/s^2

12–119. $a_{max} = 10.42$ m/s^2

12–121. $a = 8.43$ m/s^2
$\theta = 38.2°$ ⦣

12–122. $a = 6.03$ m/s^2

12–123. $a = 1.05$ m/s^2

12–125. $a = 2.75$ m/s^2

12–126. $a = 1.68$ m/s^2

12–127. $a_n = 5.02$ m/s^2

12–129. $a = 1.30$ m/s^2

12–130. $a = 0.730$ m/s^2

12–131. $a_t = 3.62$ m/s^2
$\rho = 29.6$ m

12–133. $v = 3.19$ m/s
$a = 4.22$ m/s^2

12–134. $p = 100$ m

12–135. $a = 2.36$ m/s^2

12–137. $a = 3.05$ m/s^2

12–138. $a = 0.525$ m/s^2

12–139. $a = 0.763$ m/s^2

12–141. $y = -0.0766x^2$
$v = 8.37$ m/s
$a_n = 9.38$ m/s^2
$a_t = 2.88$ m/s^2

12–142. $a = 3.05$ m/s^2

12–143. $x = 0$, $y = -4$ m
$(a)_{\max} = 50$ m/s^2

12–145. $|a| = 322$ mm/s^2, $a_x = 144$ mm/s^2,
$a_y = 288$ mm/s^2

12–146. $t = 1.21$ s

12–147. $a_n = \dfrac{v^2 bc\sin(cx)}{[1 + bc(\cos(cx))^2]^{3/2}}$
$a_t = v_t = v_n = 0$

12–149. $s_A = 1.40$ m
$s_B = 3$ m
$\mathbf{r}_A = \{1.38\mathbf{i} + 0.195\mathbf{j}\}$ m
$\mathbf{r}_B = \{-2.82\mathbf{i} + 0.873\mathbf{j}\}$ m
$\Delta r = 4.26$ m

12–150. $t = 14.3$ s
$a_B = 0.45$ m/s^2

12–151. $a = 32.2$ m/s^2

12–153. $v = 1.80$ m/s
$a = 1.20$ m/s^2

12–154. $y = \{0.839x - 0.131x^2\}$ m
$a_t = -3.94$ m/s^2
$a_n = 8.98$ m/s^2

12–155. $t = 10.1$ s
$v = 47.6$ m/s
$a = 11.8$ m/s^2

12–157. $v_n = 0$
$v_t = 7.21$ m/s
$a_n = 0.555$ m/s^2
$a_t = 2.77$ m/s^2

12–158. $a_{\max} = \dfrac{a}{b^2}v^2$

12–159. $\frac{1}{2}\mathbf{j} = \frac{1}{2}(\dddot{r} - 3\dot{r}\dot{\theta}^2 - 3r\dot{\theta}\ddot{\theta})\mathbf{u_r}$
$+\frac{1}{2}(r\dddot{\theta} + 3\dot{r}\ddot{\theta} + 3\ddot{r}\dot{\theta} - r\dot{\theta}^3)\mathbf{u_\theta}$
$+\frac{1}{2}(z)\mathbf{u}_z$

12–161. $v_r = -1.66$ m/s
$v_\theta = -2.07$ m/s
$a_r = -4.20$ m/s^2
$a_\theta = 2.97$ m/s^2

12–162. $v = 2.4$ m/s
$a = 19.3$ m/s^2

12–163. $v = 3.03$ m/s
$a = 12.63$ m/s^2

12–165. $a = 2.32$ m/s^2

12–166. $v_r = -2.33$ m/s
$v_\theta = 7.91$ m/s
$a_r = -158$ m/s^2
$a_\theta = -18.6$ m/s^2

12–167. $v = 4.30$ m/s
$a = 26.3$ m/s^2

12–169. $\dot{\mathbf{a}} = \left(\dddot{r} - 3\dot{r}\dot{\theta}^2 - 3r\dot{\theta}\ddot{\theta}\right)\mathbf{u}_r$
$+\left(3\ddot{r}\dot{\theta} + r\dddot{\theta} + 3\dot{r}\ddot{\theta} - r\dot{\theta}^3\right)\mathbf{u}_\theta$
$+\left(\dddot{z}\right)\mathbf{u}_z$

12–170. $v = 0.237$ m/s
$a = 0.278$ m/s^2

12–171. $v_r = 1.20$ m/s
$v_\theta = 1.26$ m/s
$a_r = -3.77$ m/s^2
$a_\theta = 7.20$ m/s^2

12–173. $v_r = 1.20$ m/s
$v_\theta = 1.50$ m/s
$a_r = -4.50$ m/s^2
$a_\theta = 7.20$ m/s^2

12–174. $\theta = 0.33$ rad/s
$a = 6.67$ m/s^2

12–175. $a_r = -6.67$ m/s^2
$a_\theta = 3.00$ m/s^2

12–177. $\dot{\theta} = 0.378$ rad/s
12–178. $\mathbf{v} = \{-116\mathbf{u}_r - 163\mathbf{u}_z\}$ mm/s
$\mathbf{a} = \{-5.81\mathbf{u}_r - 8.14\mathbf{u}_z\}$ mm/s^2
12–179. $v = 12.6$ m/s
$a = 83.2$ m/s^2
12–181. $a = 8.66$ m/s^2
12–182. $\dot{\theta} = 0.75$ rad/s
12–183. $v_r = a\dot{\theta}$
$v_\theta = a\theta\dot{\theta}$ m/s
$a_r = a\theta\dot{\theta}^2$
$a_\theta = 2a\dot{\theta}^2$
12–185. $v = 8.49$ m/s
$a = 88.2$ m/s
12–186. $v = 5.16$ m/s
$a = 39.1$ m/s
12–187. $v_r = 25.9$ mm/s
$a_r = -195$ mm/s^2
12–189. $a = 7.26$ m/s^2
12–190. $v = 4.16$ m/s
$a = 33.1$ m/s^2
12–191. $v_r = -2.80$ m/s
$v_\theta = 19.8$ m/s
12–193. $v_r = -250$ mm/s
$a_r = -9330$ mm/s^2
12–194. $v_r = -250$ mm/s
$a_r = -9630$ mm/s^2
12–195. $v_r = 32.0$ m/s
$v_\theta = 50.3$ m/s
$a_r = -201$ m/s^2
$a_\theta = 256$ m/s^2
12–197. $v = 8.21$ mm/s
$a = -665$ mm/s^2
12–198. $v = 8.21$ mm/s
$a = -659$ mm/s^2
12–199. $v_B = 0.5$ m/s
12–201. $v_B = -1$ m/s
12–202. $t = 160$ s
12–203. $\Delta s_B = 0.4$ m
12–205. $v_A = 1.67$ m/s
12–206. $v_A = 2$ m/s$\leftarrow$
12–207. $v_B = 0.5$ m/s$\uparrow$
12–209. $v_B = -2$ m/s$\uparrow$
12–210. (a)$t = 1.225$ s, (b) $v_{AB} = -2.90$ m/s
12–211. $v_A = 1.33$ m/s$\uparrow$
12–213. $v_B = 0.671$ m/s$\uparrow$
12–214. $v_B = 1.41$ m/s$\uparrow$
12–215. $v_b = 6.21$ m/s
$t = 11.4$ s
12–217. $v_C = 1.2$ m/s$\uparrow$
$a_C = 0.512$ m/s$^2\uparrow$
12–218. $v_b = 5.56$ m/s
$\theta = 84.4°$
12–219. $v_{r/m} = 16.6$ km/h
$\theta = 25.0°$ ◺
12–221. $|a_{AB}| = 4.47$ m/s^2
$a_{AB} = [0.2\ 4.46]$ m/s^2
12–222. $v_{B/A} = 875$ km/h
$\theta = 41.5°$ ◺
12–223. $v_c = [1.8 \sec\theta]$ m/s$\rightarrow$
12–225. $v_{A/B} = 49.1$ km/h
$\theta = 67.2°$ ◸
12–226. $v_w = 58.3$ km/h
$\theta = 59.0°$ ◺
12–227. $v_A = 0.605$ m/s
$v_B = 5.33$ m/s
$t = 1.07$ s
12–229. $\mathbf{v}_{BA} = [-22.5\ 12.99]$ m/s
$|\mathbf{v}_{BA}| = 26.0$ m/s
$\mathbf{a}_{BA} = [-1.621\ -1.143]$m/s^2
$|\mathbf{a}_{BA}| = 1.983$ m/s^2
12–230. $v_{A/B} = v\sqrt{2(1-\sin\theta)}$
12–231. $v_{B/A} = 36.6$ km/h
$\theta = 46.9°$ ◺
$a_{B/A} = 3737$ km/h^2
$\phi = 12.9°$ ◺
12–233. $v_r = 34.6$ km/h
12–234. $\mathbf{v}_{AB} = [-11.43 - 7.07]$ m/s
$\mathbf{a}_{AB} = [0.828 - 4.243]$ m/s^2
12–235. $v_{w/s} = 19.9$ m/s
$\theta = 74.0°$ ◸
12–237. $v_{B/C} = 18.6$ m/s
$\theta_v = 66.2°$ ◿
$a_{B/C} = 0.959$ m/s^2
$\theta_a = 8.57°$ ◺
12–238. $v_B = 5.75$ m/s
$v_{C/B} = 17.8$ m/s
$\theta = 76.2°$ ◺
$a_{C/B} = 9.81$ m/s$^2\downarrow$
12–239. $v_{A/B} = \sqrt{v_A^2 + v_B^2 - 2v_A v_B\cos\theta}$
$\theta = \tan^{-1}\left(\dfrac{v_A - v_B\cos\theta}{v_B\sin\theta}\right)$ ◺